Cannabis Use, Neurobiology, Psychology, and Treatment

Cannabis Use, Neurobiology, Psychology, and Treatment

Edited by

Colin R. Martin
Clinical Psychobiology and Applied Psychoneuroimmunology; Institute for Health and Wellbeing, University of Suffolk, Ipswich, United Kingdom

Vinood B. Patel
University of Westminster, London, United Kingdom

Victor R. Preedy
King's College London; King's College Hospital, London, United Kingdom

ELSEVIER

ACADEMIC PRESS
An imprint of Elsevier

Academic Press is an imprint of Elsevier
125 London Wall, London EC2Y 5AS, United Kingdom
525 B Street, Suite 1650, San Diego, CA 92101, United States
50 Hampshire Street, 5th Floor, Cambridge, MA 02139, United States
The Boulevard, Langford Lane, Kidlington, Oxford OX5 1GB, United Kingdom

ISBN: 978-0-323-89862-1

For information on all Academic Press publications
visit our website at https://www.elsevier.com/books-and-journals

Publisher: Nikki P. Levy
Acquisitions Editor: Anna Valutkevich
Editorial Project Manager: Timothy Bennett
Production Project Manager: Swapna Srinivasan
Cover Designer: Mark Rogers

Typeset by STRAIVE, India

Dedication

I dedicate this book to my daughter Dr. Caragh Brien, an incredible junior doctor of whom I am so proud.

Colin R. Martin

Contents

Part VIII
Impact on non-neurological systems

39. The impact of cannabis and cannabinoids on the skin

Kimberly Shao, Jane M. Grant-Kels, and Campbell Stewart

40. Cannabis consumption in reproductive function and teratogenicity

Susana Sá and Bruno M. Fonseca

41. Phytocannabinoids have cytotoxic, antiproliferative, and antimigratory activities on cancer cells and cancer stem cells

Hadar Peeri and Hinanit Koltai

Part IX
Treatments and strategies

42. Motivational interviewing for cannabis use disorders: A narrative review from research to practice

Maria Teresa Pons Cabrera, Clara Oliveras Salvà, Pablo Rodrigo Guzmán Cortez, and Hugo López-Pelayo

Contributors

Numbers in parentheses indicate the pages on which the authors' contributions begin.

Jean Addington (511), Department of Psychiatry, Hotchkiss Brain Institute, University of Calgary, Calgary, AB, Canada

Hanana AlTfaili (355), Department of Psychiatry and Behavioral Sciences, The University of Texas Health Science Center at San Antonio, San Antonio, TX, United States

Tarcísio Cícero de Lima Araújo (21), Central for Analysis of Drugs, Medicines and Food (CAFMA)—Federal University of Vale do São Francisco, Pernambuco, Brazil

Abagail Barazi (595), Mercy Health St. Vincent Medical Center, Toledo, OH, United States

Malia Bautista (475), Department of Neurobiology and Behavior, University of California Irvine, Irvine, CA, United States

Chris Bladen (383), Macquarie Medical School, Macquarie University, Sydney, NSW, Australia

Emilie Bouquet (127), Addictovigilance Center, Department of Clinical Pharmacology, Poitiers University Hospital, Poitiers, France

Zoe Bourgault (313), Translational Addiction Research Laboratory, Campbell Family Mental Health Research Institute, Centre for Addiction and Mental Health; Department of Pharmacology and Toxicology, University of Toronto, Toronto, ON, Canada

Christopher S. Breivogel (329,487), Department of Pharmaceutical & Clinical Sciences, Campbell University College of Pharmacy & Health Sciences, Buies Creek, NC, United States

Nicholas Butowski (279), Neurosurgery Department; Department of Molecular Science, Director of Translational Research, Neurosurgery Department, Division of Neuro-Oncology; Department of Molecular Science, University of California, San Francisco, CA, United States

Maria Teresa Pons Cabrera (567), Addictions Unit, Psychiatry and Psychology Service, Clinic Neurosciences Institute, Hospital Clínic de Barcelona, Grup de Recerca en Addiccions Clínic (GRAC), Institut d'Investigacions Biomèdiques August Pi i Sunyer (IDIBAPS), Barcelona, Spain

M. Mallar Chakravarty (457), Computational Brain Anatomy Laboratory, Cerebral Imaging Centre, Douglas Mental Health University Institute; Integrated Program in Neuroscience, McGill University; Cerebral Imaging Centre, Douglas Mental Health University Institute; Department of Psychiatry, McGill University; Department of Biomedical Engineering, McGill University, Montreal, QC, Canada

Pablo Rodrigo Guzmán Cortez (567), Addictions Unit, Psychiatry and Psychology Service, Clinic Neurosciences Institute, Hospital Clínic de Barcelona, Grup de Recerca en Addiccions Clínic (GRAC), Institut d'Investigacions Biomèdiques August Pi i Sunyer (IDIBAPS), Barcelona, Spain

Lani Cupo (457), Computational Brain Anatomy Laboratory, Cerebral Imaging Centre, Douglas Mental Health University Institute; Integrated Program in Neuroscience, McGill University; Cerebral Imaging Centre, Douglas Mental Health University Institute, Montreal, QC, Canada

Stephen R. Dager (55), Department of Radiology, Center on Human Development and Disability, School of Medicine, University of Washington, Seattle, WA, United States

Valéria de Almeida (407), Laboratory of Neuroproteomics, Department of Biochemistry and Tissue Biology, Institute of Biology, University of Campinas (UNICAMP), Campinas, Brazil; Department of Psychiatry and Psychotherapy, University Hospital, LMU Munich, Munich, Germany

Natacha De Genna (343), Department of Psychiatry, University of Pittsburgh School of Medicine, Pittsburgh, PA, United States

Iker Bengoetxea de Tena (417), Department of Pharmacology, Faculty of Medicine and Nursing, University of the Basque Country (UPV/EHU), Vizcaya, Spain

Roya Derwish (209), Krembil Research Institute, University Health Network; Centre for Cannabinoid Therapeutics, Toronto, ON, Canada

Angeline J. Dukes (475), Department of Neurobiology and Behavior, University of California Irvine, Irvine, CA, United States

Khalil Eldeeb (329), Department of Pharmacology, Campbell University School of Osteopathic Medicine, Lillington, NC, United States; Department of Pharmacology, Al Azhar Faculty of Medicine, Al Azhar University, New Damietta, Egypt

Norman E. Fenn, III (595), Manchester University College of Pharmacy & Health Sciences; Parkview Women's & Children's Hospital, Fort Wayne, IN, United States

Bruno M. Fonseca (541), UCIBIO – Applied Molecular Biosciences Unit, Laboratory of Biochemistry; Associate Laboratory i4HB – Institute for Health and Bioeconomy, Faculty of Pharmacy, University of Porto, Porto, Portugal

Christie D. Fowler (475), Department of Neurobiology and Behavior, University of California Irvine, Irvine, CA, United States

Brett C. Ginsburg (355), Department of Psychiatry and Behavioral Sciences, The University of Texas Health Science Center at San Antonio, San Antonio, TX, United States

Rohit Goyal (253), Department of Neuropharmacology, School of Pharmaceutical Sciences, Shoolini University, Solan, HP, India

Jane M. Grant-Kels (525), Department of Dermatology, University of Connecticut Health Center, Farmington, CT; Department of Dermatology, University of Florida, Gainesville, FL, United States

Raquel Guerrero-Alba (395), Departamento de Fisiología y Farmacología, Centro de Ciencias Básicas, Universidad Autónoma de Aguascalientes, Ciudad Universitaria, Aguascalientes, Mexico

Elisa Guma (457), Computational Brain Anatomy Laboratory, Cerebral Imaging Centre, Douglas Mental Health University Institute; Integrated Program in Neuroscience, McGill University; Cerebral Imaging Centre, Douglas Mental Health University Institute, Montreal, QC, Canada

Daniel Gyamfi (629), The Doctors Laboratory Ltd, London, United Kingdom

Meaghan E. Hazelet (343), Department of Psychology, Slippery Rock University, Slippery Rock, PA, United States

Mickel Hiebert-Giesbrecht (3), Department of Biological Sciences, University of Alberta, Edmonton, AB, Canada

E. Hoch (113), IFT Institut für Therapieforschung, München, Germany

Allyn C. Howlett (329), Department of Physiology and Pharmacology, Wake Forest University School of Medicine, Winston-Salem, NC, United States

Gary Kenneth Hulse (71,93), Department of Psychiatry, University of Western Australia, Perth; School of Health and Medical Sciences, Edith Cowan University, Joondalup, WA, Australia

Malliga R. Iyer (237), National Institute on Alcohol Abuse and Alcoholism, National Institutes of Health, Rockville, MD, United States

Sara Jalali (511), Department of Psychiatry, Hotchkiss Brain Institute, University of Calgary, Calgary, AB, Canada

Allegra J. Johnson (55), Department of Radiology, Integrated Brain Imaging Center, School of Medicine, University of Washington, Seattle, WA, United States

Emilie Jouanjus (127), Addictovigilance Center, Department of Medical and Clinical Pharmacology, Toulouse University Hospital, Toulouse, France

Mohamad N. Khalaifa (343), Department of Psychology, Slippery Rock University, Slippery Rock, PA, United States

Bareera Khan (253), Faculty of Applied Sciences and Biotechnology, Shoolini University of Biotechnology and Management Sciences, Solan, HP, India

Natalia M. Kleinhans (55), Department of Radiology, Integrated Brain Imaging Center, Center on Human Development and Disability, School of Medicine, University of Washington, Seattle, WA, United States

Tatiana Klimenko (621), V. Serbsky Federal Medical Research Center of Psychiatry and Addiction, Ministry of Health, Russian Federation—National Scientific Center of Narcology (Branch); The All-Russian State University of Justice, Ministry of Justice of the Russian Federation, Moscow, Russia

Hinanit Koltai (555), Department of Ornamental Horticulture and Biotechnology, Institute of Plant Sciences, ARO, Volcani Institute, Rishon LeZion, Israel

Lakshmi P. Kotra (209), Krembil Research Institute, University Health Network, Toronto; Department of Pharmaceutical Sciences, Leslie Dan Faculty of Pharmacy, University of Toronto; Centre for Cannabinoid Therapeutics, Toronto, ON, Canada

Deepak Kumar (253), Department of Neuropharmacology, School of Pharmaceutical Sciences, Shoolini University, Solan, HP, India

Maryse Lapeyre-Mestre (127), Addictovigilance Center, Department of Medical and Clinical Pharmacology, Toulouse University Hospital, Toulouse, France

Sarah F. Larsen (55), Department of Radiology, Integrated Brain Imaging Center, School of Medicine, University of Washington, Seattle, WA, United States

Bernard Le Foll (139,313), Translational Addiction Research Laboratory, Campbell Family Mental Health Research Institute, Centre for Addiction and Mental Health; Department of Psychiatry, Faculty of Medicine; Department of Family and Community Medicine; Institute of Medical Science; Department of Pharmacology and Toxicology, University of Toronto, Toronto; Waypoint Research Institute, Waypoint Centre for Mental Health Care, Penetanguishene, ON, Canada

Rachel Leis (595), Mercy Health St. Vincent Medical Center, Toledo, OH, United States

Naama Levy-Cooperman (155), Altreos Research Partners, Inc., Toronto, ON, Canada

Joshua A. Lile (605), Department of Behavioral Science, University of Kentucky College of Medicine, Lexington, KY, United States

Alberto Llorente-Ovejero (417), Department of Pharmacology, Faculty of Medicine and Nursing, University of the Basque Country (UPV/EHU), Vizcaya, Spain

Hugo López-Pelayo (567), Addictions Unit, Psychiatry and Psychology Service, Clinic Neurosciences Institute, Hospital Clínic de Barcelona, Grup de Recerca en Addiccions Clínic (GRAC), Institut d'Investigacions Biomèdiques August Pi i Sunyer (IDIBAPS), Barcelona, Spain

Dietrich E. Lorke (445), Department of Anatomy and Cellular Biology, College of Medicine and Health Sciences, Khalifa University, Abu Dhabi, United Arab Emirates

Mohamed Mahgoub (445), Department of Health and Medical Sciences, Khawarizmi International College, Abu Dhabi, United Arab Emirates

Iván Manuel (417), Department of Pharmacology, Faculty of Medicine and Nursing, University of the Basque Country (UPV/EHU), Vizcaya, Spain

Bruno A. Marichal-Cancino (395), Departamento de Fisiología y Farmacología, Centro de Ciencias Básicas, Universidad Autónoma de Aguascalientes, Ciudad Universitaria, Aguascalientes, Mexico

Justin Matheson (139,313), Translational Addiction Research Laboratory, Campbell Family Mental Health Research Institute, Centre for Addiction and Mental Health, Toronto, ON, Canada

Sara Mejía-Chávez (377), Institute of Neurobiology, Autonomous National University of Mexico, Juriquilla, Queretaro, Mexico

Santiago Monleón (365), Department of Psychobiology, Faculty of Psychology, University of Valencia, Valencia, Spain

Paula Morales (395), Instituto de Química Médica, CSIC, Madrid, Spain

Nuria Morales-Puerto (265), Departamento de Farmacología y Toxicología, Facultad de Medicina, Universidad Complutense; Instituto de Investigación Sanitaria, Hospital 12 de Octubre; Red de Trastornos Adictivos del Instituto de Salud Carlos III; Instituto Universitario de Investigación Neuroquímica (IUIN), Facultad de Medicina, Universidad Complutense, Madrid, Spain

Marta Moreno-Rodríguez (417), Department of Pharmacology, Faculty of Medicine and Nursing, University of the Basque Country (UPV/EHU), Vizcaya, Spain

Sharlene D. Newman (293), Alabama Life Research Institute, The University of Alabama, Tuscaloosa, AL, United States

Esther O'Shea (265), Departamento de Farmacología y Toxicología, Facultad de Medicina, Universidad Complutense; Instituto de Investigación Sanitaria, Hospital 12 de Octubre; Red de Trastornos Adictivos del Instituto de Salud Carlos III; Instituto Universitario de Investigación Neuroquímica (IUIN), Facultad de Medicina, Universidad Complutense, Madrid, Spain

Alexa O'Regan (193), Royal College of Surgeons Ireland, Dublin 2, Ireland

Murat Oz (445), Department of Pharmacology and Therapeutics, Faculty of Pharmacy, Kuwait University, Safat, Kuwait

Alison Paplaskas (595), Mercy Health St. Vincent Medical Center, Toledo, OH, United States

Vinood B. Patel (629), School of Life Sciences, University of Westminster, London, United Kingdom

Hadar Peeri (555), Department of Ornamental Horticulture and Biotechnology, Institute of Plant Sciences, ARO, Volcani Institute, Rishon LeZion, Israel

Georg Petroianu (445), Department of Pharmacology, College of Medicine and Health Sciences, Khalifa University, Abu Dhabi, United Arab Emirates

Ilaria Di Marco Pisciottano (33), Department of Chemistry, Istituto Zooprofilattico Sperimentale del Mezzogiorno, Naples, Italy

Victor R. Preedy (629), School of Life Course and Population Sciences, Faculty of Life Science and Medicine, King's College London, London, United Kingdom

Ulrich W. Preuss (113), Departments of Psychiatry, Psychotherapy and Psychosomatics, RKH Hospital Ludwigsburg, Martin-Luther-Universität Halle-Wittenberg, Ludwigsburg, Germany

Richard L. Price (193), Swift Institute; University of Nevada-Reno College of Medicine, Reno, NV, United States

Victor M. Pulgar (487), Department of Pharmaceutical & Clinical Sciences, Campbell University College of Pharmacy & Health Sciences, Buies Creek; Department of Obstetrics and Gynecology, Wake Forest School of Medicine, Winston-Salem, NC, United States

Rajkumar Rajendram (629), College of Medicine, King Saud bin Abdulaziz University for Health Sciences; Department of Medicine, King Abdulaziz Medical City, King Abdullah International Medical Research Center, Ministry of National Guard Health Affairs, Riyadh, Saudi Arabia

Albert Stuart Reece (71,93), Department of Psychiatry, University of Western Australia, Perth; School of Health and Medical Sciences, Edith Cowan University, Joondalup, WA, Australia

Matt Reid (171), Sociology Department, Widener University, Chester, PA, United States

Justin P. Reinert (595), The University of Toledo College of Pharmacy and Pharmaceutical Sciences; The University of Toledo Medical Center, Toledo, OH, United States

Gale A. Richardson (343), Department of Psychiatry, University of Pittsburgh School of Medicine, Pittsburgh, PA, United States

J. Eduardo Rodriguez Almaraz (279), Neurosurgery Department, Division of Neuro-Oncology; Department of Epidemiology & Biostatistics, University of California, San Francisco, CA, United States

Rafael Rodríguez-Puertas (417), Department of Pharmacology, Faculty of Medicine and Nursing, University of the Basque Country (UPV/EHU), Vizcaya, Spain

Larissa Araújo Rolim (21), Central for Analysis of Drugs, Medicines and Food (CAFMA)—Federal University of Vale do São Francisco; Department of Pharmacy, Federal University of Vale do São Francisco, José de Sá Maniçoba Avenue, Center, Petrolina, Pernambuco, Brazil

Thorsten Rudroff (301), Department of Health and Human Physiology, University of Iowa, Iowa City, IA, United States

Pedro Guilherme Sousa de Sá (21), Central for Analysis of Drugs, Medicines and Food (CAFMA)—Federal University of Vale do São Francisco, Pernambuco, Brazil

Susana Sá (541), Department of Biomedicine, Unit of Anatomy; Center for Health Technology and Services Research (CINTESIS), Faculty of Medicine, University of Porto, Porto, Portugal

Shivani Sachdev (383), Macquarie Medical School, Macquarie University, Sydney, NSW, Australia; Laboratory of Bioorganic Chemistry, National Institute of Diabetes, Digestive, and Kidney Diseases, National Institutes of Health, Bethesda, MD, United States

Clara Oliveras Salvà (567), Addictions Unit, Psychiatry and Psychology Service, Clinic Neurosciences Institute, Hospital Clínic de Barcelona, Grup de Recerca en Addiccions Clínic (GRAC), Institut d'Investigacions Biomèdiques August Pi i Sunyer (IDIBAPS), Barcelona, Spain

Olga Santesteban-Echarri (511), Department of Psychiatry, Hotchkiss Brain Institute, University of Calgary, Calgary, AB, Canada

Marina Santiago (383), Macquarie Medical School, Macquarie University, Sydney, NSW, Australia

Siddharth Sarkar (499), Department of Psychiatry and NDDTC, Room no 4096, Teaching Block, All India Institute of Medical Sciences, New Delhi, India

Kerri A. Schoedel (155), Altreos Research Partners, Inc., Toronto, ON, Canada

Nadja Schröder (431), Departamento de Fisiologia, Instituto de Ciências Básicas da Saúde, Universidade Federal do Rio Grande do Sul (UFRGS), Porto Alegre, Brazil

Mahadev Singh Sen (499), Department of Psychiatry, Institute of Human Behaviour and Allied Sciences, New Delhi, India,

Beatrice Setnik (225), Cabin Creek Ln, Raleigh, NC, United States

Svetlana Shakhova (621), Department of Drug Addiction Clinical Research, V. Serbsky Federal Medical Research Center of Psychiatry and Addiction, Ministry of Health, Russian Federation—National Scientific Center of Narcology (Branch), Moscow, Russia

Kimberly Shao (525), Department of Dermatology, University of Connecticut Health Center, Farmington, CT, United States

Aditi Sharma (253), Department of Neuropharmacology, School of Pharmaceutical Sciences, Shoolini University, Solan, HP, India

Camila Ribeiro Silva (21), Central for Analysis of Drugs, Medicines and Food (CAFMA)—Federal University of Vale do São Francisco, Pernambuco, Brazil

Yesh Chandra Singh (499), Department of Psychiatry and NDDTC, Room no 4096, Teaching Block, All India Institute of Medical Sciences, New Delhi, India

Kenneth William Sommerville (225), Sommerville LLC, Hampstead, NC, United States

Campbell Stewart (525), Department of Dermatology, University of Connecticut Health Center, Farmington, CT, United States

William W. Stoops (605), Department of Behavioral Science, University of Kentucky College of Medicine, Lexington, KY, United States

Claudia Torres-Calzada (3), Department of Biological Sciences, University of Alberta, Edmonton, AB, Canada

Ryan Vandrey (581), Department of Psychiatry and Behavioral Sciences, Johns Hopkins Bayview Medical Center, Behavioral Biology Research Center, Baltimore, MD, United States

Arturo Venebra-Muñoz (377), Neurobiology of Addiction and Brain Plasticity Laboratory, Faculty of Sciences, Autonomous University of the State of Mexico, Toluca, State of Mexico, Mexico

Concepción Vinader-Caerols (365), Department of Psychobiology, Faculty of Psychology, University of Valencia, Valencia, Spain

Michael J. Wesley (605), Department of Behavioral Science, University of Kentucky College of Medicine, Lexington, KY, United States

Jennifer A. Willford (343), Department of Psychology, Slippery Rock University, Slippery Rock, PA, United States

David S. Wishart (3), Department of Biological Sciences, University of Alberta, Edmonton, AB, Canada

David Wolinsky (581), Department of Psychiatry and Behavioral Sciences, Johns Hopkins Bayview Medical Center, Behavioral Biology Research Center, Baltimore, MD, United States

J.W.M. Wong (113), Psychosomatik Fachklinik Schoemberg GmbH, Schoemberg, Germany

Craig D. Workman (301), Department of Health and Human Physiology, University of Iowa, Iowa City, IA, United States

Keun-Hang Susan Yang (445), Department of Biological Sciences, Schmid College of Science and Technology, Chapman University, Orange, CA, United States

Yi Yang (209), Krembil Research Institute, University Health Network, Toronto; Department of Pharmaceutical Sciences, Leslie Dan Faculty of Pharmacy, University of Toronto; Centre for Cannabinoid Therapeutics, Toronto, ON, Canada

Erica Zamberletti (181), Department of Biotechnology and Life Sciences, Neuroscience Center, University of Insubria, Busto Arsizio, VA, Italy

Preface

Extracts from the plant *Cannabis sativa* contain scores of psychoactive compounds. The principal agent is tetrahydrocannabinol, but there are also other compounds that are neuroactive. Crude preparations of extracts from the cannabis plant is one of the most widely used drugs on a global basis, and the World Health Organization has indicated that 2.5% of the world's population consume cannabis. Both adverse and nonadverse effects of cannabis have been reported that are dependent on the different forms in which it is used. For example, a small proportion of cannabis users will develop psychoses or exhibit extreme behavioral problems such as paranoia. There are also neurological changes that may be transient or long term. Refined extracts of compounds that are related to the active ingredient tetrahydrocannabinol are called cannabinoids and include cannabidiol, cannabinol, cannabigerol, and cannabichromene, and approximately 80 other active compounds many of which have been reported to have neuroactive effects. Unfortunately, some synthetic cannabinoids have been used recreationally with drastic consequences. However, it is important to point out that the usage of any component or regimen requires scientifically vigorous trials and investigations. Treatments and pathways seen in modeling systems or in vitro need to be verified in vivo. Adverse effects also need to be investigated. The gold standard is randomized controlled trials and with due consideration of toxic and adverse effects; these chapters provide a framework only.

Effectively, there is a seamless biology between the science behind the neuroactive compounds extracted from *C. sativa* and the potential usage of synthetic cannabinoid mimics to procure beneficial effects and the detrimental use of synthetic compounds. However, most of this information is scattered throughout the literature or focused on those with specific scientific expertise.

The editors have sought to overcome these limitations and advance the understanding of neurosciences and cannabis with a series of three books collectively called **The Neuroscience of Cannabis and Cannabinoids**.

It brings together information relating to cannabis and cannabinoids in relation to neurological processes and systems.

The three books are

Book 1: ***Cannabis Use, Neurobiology, Psychology, and Treatment***

Book 2: *Medicinal Usage of Cannabis and Cannabinoids*

Book 3: *Neurobiology and Physiology of the Endocannabinoid System*

The Neuroscience of Cannabis and Cannabinoids transcends both the disciplinary and intellectual divides as each chapter has:

- **Abstract (published online)**
- **Applications to other areas**
- **Key facts**
- **Mini-dictionary of terms**
- **Summary points**

The section **Applications to other areas** is particularly important as it highlights the translational aspect of cannabis and cannabinoid-related research. The section **Key facts** gives important information about individual components in each chapter. The **Mini-dictionary of terms** is suited for both the novice and those working in other scientific fields or areas. The section **Summary points** encapsulates the entire chapter in brief sets of simple sentences.

This book ***Cannabis Use, Neurobiology, Psychology, and Treatment*** is divided into ten parts as follows:

1. **Setting the scene and introductory chapters**
2. **Cannabis usage**
3. **Pharmacological and physiological aspects**
4. **Neurobiology of cannabis**
5. **Neurobiology of cannabinoids**
6. **New and synthetic cannabinoids**

7. **Biomarkers, analysis and screening**
8. **Impact on non-neurological systems**
9. **Treatments and strategies**
10. **Resources**

It has been difficult to ascribe particular chapters to different parts of the book as many chapters can be placed into two or more sections. Nevertheless, the navigation of the chapters, areas, and key aspects related to cannabis and cannabinoids is aided by the excellent index at the end of the book.

This book is designed for research and teaching purposes. It is suitable for neurologists, psychologists, psychiatrists, behavioral scientists, health scientists, healthcare workers, pharmacologists, and research scientists. The audience also includes federal and state program directors. It is valuable as a personal reference book and also for academic libraries that covers the domains of neurology, health sciences, or substance misuse. Contributions are from leading national and international experts including those from world-renowned institutions. The book is suitable for undergraduates, postgraduate, lecturers, and academic professors.

Editors

Colin R. Martin
Clinical Psychobiology and Applied Psychoneuroimmunology; Institute for Health and Wellbeing, University of Suffolk, Ipswich, United Kingdom

Vinood B. Patel
University of Westminster, London, United Kingdom

Victor R. Preedy
King's College London; King's College Hospital, London, United Kingdom

Part I

Setting the scene and introductory chapters

Chapter 1

Metabolomics of the cannabis plant

Mickel Hiebert-Giesbrecht, Claudia Torres-Calzada, and David S. Wishart
Department of Biological Sciences, University of Alberta, Edmonton, AB, Canada

Abbreviations

ANOVA	analysis of variance
API	atmospheric pressure ionization
CBC	cannabichromene
CBCA	cannabichromenic acid
CBD	cannabidiol
CBDA	cannabidiolic acid
CBDV	cannabidivarin
CBDVA	cannabidivarinic acid
CBG	cannabigerol
CBGA	cannabigerolic acid
CBN	cannabinol
CBNA	cannabinolic acid
DAD	diode-array detection
Δ8-THC	delta-8-tetrahydrocannabinol
Δ9-THC	delta-9-tetrahydrocannabinol
Δ9-THCA	delta-9-tetrahydrocannabinolic acid
Δ9-THCAA	delta-9-tetrahydrocannabinolic acid-A
eCBS	endocannabinoid system
EI-MS	electron ionization mass spectrometry
ESI	electrospray ionization
FD	fluorescence detector
FID	flame ionization detection
GC	gas chromatography
GCxGC-MS	two-dimensional gas chromatography coupled to mass spectrometry
HCA	hierarchical clustering analysis
HPLC	high-performance liquid chromatography
HRMS	high-resolution mass spectrometry
LC	liquid chromatography
LC-ESI-MS	liquid chromatography-electrospray ionization-mass spectrometry
LC-MS	liquid chromatography coupled to mass spectrometry
LC-MS/MS	liquid chromatography coupled to tandem mass spectrometry
NIST	National Institute of Standards and Technology
NMR	nuclear magnetic resonance
OPLS-DA	orthogonal partial least square discriminant analysis
PCA	principal component analysis
PLS-DA	partial least square discriminant analysis
THCV	tetrahydrocannabivarin
THCVA	tetrahydrocannabivarinic acid
TMS	trimethyl silyl
HPLC-UV	ultrahigh-performance chromatography coupled to an ultraviolet detector

Cannabis Use, Neurobiology, Psychology, and Treatment. https://doi.org/10.1016/B978-0-323-89862-1.00002-7

Introduction

Cannabis (*Cannabis sativa* L.), like most plants, displays extraordinary chemical diversity. Most plants use their chemical arsenal to defend or protect themselves against insect herbivory, heat, cold, or drought. However, in the case of cannabis, this chemical arsenal has been found to be more useful to humans than to the plant itself. Indeed cannabis is particularly rich in medicinal and euphoriant chemicals (Pertwee, 2014). With many countries legalizing or relaxing the rules for medicinal or recreational cannabis use, there has been an exponential growth in the cannabis market. This rapid market growth has forced the medical community to seek a better understanding of what kinds of potentially beneficial or harmful compounds can be found in cannabis and how these chemicals affect human physiology and behavior. As a result, scientific research around cannabis has experienced significant progress over the last few years. High-throughput technologies, such as genomics, transcriptomics, proteomics, and more recently metabolomics, have been particularly useful in advancing our knowledge about cannabis and its effects. Because of the compelling need to better understand the chemical content of cannabis, metabolomics is particularly well suited for chemically characterizing this plant.

Metabolomics is a field of omics science that uses analytical chemistry techniques and advanced computational methods to characterize the metabolome (Wishart et al., 2016). The metabolome is defined as the complete collection of metabolites, or small molecules found in a given organelle, cell, tissue, organ, biofluid, or organism (Wishart, 2019). A key feature of metabolomics is the use of high-throughput technologies such as robotic mass spectrometry (MS) or automated nuclear magnetic resonance (NMR) spectroscopy to measure hundreds or even thousands of chemicals simultaneously.

Metabolomics studies can be grouped into two categories: qualitative or untargeted methods and quantitative or targeted methods. In untargeted metabolomics, the analyzed chemical compounds are unknown and only their chromatographic or spectral profiles are recorded. These profiles are then analyzed by statistical methods to identify relevant features that could distinguish samples or groups of samples. Untargeted metabolomics does not require prior knowledge of the samples' chemical composition, theoretically enabling every present chemical compound to be analyzed. With untargeted MS-based metabolomics, it is not uncommon to detect $>$10,000 chemical features. However, in practice, the number of compounds that can be unambiguously identified from these features is often less than a few hundred. Indeed identification of the analytes through untargeted metabolomics continues to be the main bottleneck for this type of analysis (De Jong & Beecher, 2012). While data collection for untargeted metabolomics is very fast, data processing is often very complex and slow. Furthermore, because of the unknown nature of the analytes in untargeted metabolomics, most untargeted metabolomics measurements are only qualitative.

Targeted metabolomics, on the other hand, quantitatively measures a defined set of known analytes using reference standards or spectral libraries derived from reference standards. While targeted methods do not detect 10,000+ features, they can analyze close to 1000 individual analytes and yield absolutely quantitative results. The development and validation of a targeted metabolomic assay can be quite time consuming and expensive, with a requirement for isotopically labeled internal standards as well as analytical external standards. However, once developed, targeted methods are easily automated, fast, cheap, and reliable. Because targeted studies generate tables of absolute analyte concentrations, it is much easier to share, compare, and analyze targeted metabolomic data than untargeted metabolomic data.

Both untargeted and targeted metabolomics methods use a wide diversity of analytical platforms. These include NMR, liquid chromatography mass spectrometry (LC-MS), gas chromatography mass spectrometry (GC-MS), and other instrumental variations. Each platform has its strengths and weaknesses, with some being more sensitive or others being more quantitative or still other being more automatable. The purpose of this chapter is to briefly describe the exciting potential of cannabis metabolomics, the analytical tools, and methods used in cannabis metabolomics, to summarize the findings of key cannabis metabolomic studies and to provide an overview of different metabolomics applications to cannabis research. A thorough compilation of the current literature is provided and contextualized to help readers understand this rapidly growing field.

Analytical tools and methods for metabolomics studies in cannabis

Traditional analytical chemistry approaches have been applied to cannabis for more than 60 years. These methods, which typically identify 5–10 specific compounds at a time, have catalogued more than 550 chemicals from a broad spectrum of chemical classes (Aizpurua-Olaizola et al., 2016). However, comprehensive metabolomic and genetic studies of cannabis and related plants suggest that cannabis likely contains $>$10,000 chemicals (Fang et al., 2019). This chemical complexity is obviously far too great for traditional analytical chemistry methods. Metabolomics, on the other hand, is well suited to such large-scale chemical analyses.

In a recently published review, Aliferis and Bernard-Perron (2020) highlighted the significant increase in the number of cannabis metabolomics studies published from 2000 to 2019. Most early metabolomic studies only reported a few dozen cannabis metabolites. However, more recent cannabis metabolomic studies have substantially increased their level of compound coverage (see Table 1). The data summarized in Table 1 also highlight the range of advanced analytical technologies

TABLE 1 Summary of published research applying metabolomics principles for the study of *Cannabis sativa*.

Platform	Type of assay	Number of analytes	Types of analytes	Brief description	References
GC-MS	Untargeted	236	Cannabinoids, fatty acids, sugars, carboxylic acid, amino acids, organic acids, polyamines, carbohydrates, and polyphenols	Untargeted GC-MS analysis of cannabis seeds of two different environments. High-altitude sample has more antioxidant power and nutraceutical benefits	Rashid et al. (2021)
HPLC-UV	Targeted	10	Cannabinoids	Cannabinoid content and gene expression is monitored during cultivation	Muntendam et al. (2012)
^{1}H-NMR	Untargeted	Undefined	Chemometric profiles		
HPLC-UV	Targeted	11	Cannabinoids	Chemometric analysis of 33 strains of cannabis and clustered into five groups. Eleven known cannabinoids and 21 unknown putative cannabinoids were used for this analysis	Mudge et al. (2018)
HPLC-UV	Untargeted	21	Unknown cannabinoids		
HPLC-UV	Targeted	10	Cannabinoids	Establish a connection between the chemical composition and the morphology of 11 phenotypes from three different varieties of cannabis based on their cannabinoid and essential oil profiles	Eržen et al. (2021)
GC-FID	Targeted	29	Terpenes (essential oils)		
H NMR	Targeted	5	Cannabinoids	^{1}H NMR-based metabolomics and RT-PCR were applied to monitor metabolite profiles of *C. sativa* trichomes during the last 4 weeks of the flowering period	Happyana and Kayser (2016)
	Untargeted	20	Sugars, amino acids, carboxylic acids, choline		
HPLC-UV	Targeted	32	Cannabinoids	A classification system based on cannabinoid content was used to identify relationships of different terpenes from five distinct chemovars	Mudge et al. (2019)
GC-MS	Untargeted	29	Monoterpenes		
GC-MS	Untargeted	38	Sesquiterpenes		
LC-MS	Targeted	14	Cannabinoids	Differential analysis of cannabis floral tissue	Welling et al. (2021)
	Untargeted	2599 features	Undefined		
HPLC-UV	Targeted	3	d9-THC, CBN, CBD	Evaluation of the optimal postharvest storage conditions based on the metabolomic profiling of Cannabis secondary metabolites	Milay et al. (2020)
LC-MS	Targeted	16	Cannabinoids		
GC-MS	Targeted	39	Terpenes		
LC-MS	Targeted	5	Cannabinoids	Evaluated the chemical composition of medicinal cannabis extracts. Quantified five terpenoids during plant extraction	Citti et al. (2017)
	Untargeted	Undefined	Undefined		
LC-MS	Untargeted	Undefined	Undefined	UV exposure of cannabis plants induced changes in some phenolic compounds	Marti et al. (2014)

Continued

TABLE 1 Summary of published research applying metabolomics principles for the study of *Cannabis sativa*—cont'd

Platform	Type of assay	Number of analytes	Types of analytes	Brief description	References
LC-MS	Untargeted	135	Phytocannabinoids	Eight cannabinoids were identified and 127 were putatively identified. Multivariate analysis was used to distinguish 50 different cannabis samples	Cerrato et al. (2021)
H-NMR	Untargeted	Undefined	Cannabinoids, sugars, amino acids	Chemometric analysis of cannabis varieties using NMR and PCA to explain variance	Choi et al. (2004)
LC-MS	Untargeted	Undefined	Cannabinoids, Flavonoids	Chemometric profiling for cultivar characterization	Li et al. (2022)
LC-MS GC-MS	Untargeted	46 134	22 Cannabinoids, 70 terpenoids, 31 lipids, 16 flavonoids, 6 amino acids, 5 organic acids, 4 benzenoids, 4 organic oxygen compounds, 3 hydro-carbons, 3 carbohydrates, 2 organoheterocyclic compounds, 2 organo-nitrogen compounds, and 1 alkaloid	Development of LC and GC-MS methods for untargeted analysis of polar and nonpolar *C. sativa* extracts	Delgado-Povedano et al. (2020)
LC-MS	Untargeted	94	Cannabinoids	Cannabinoids were used to profile 36 cannabis samples and correlate with anticonvulsant activity	Berman et al. (2018)
GC-MS/FID	Untargeted	71	Terpenoids	Analysis of essential oil composition and antibacterial activity	Iseppi et al. (2019)
GC-FID HPLC-DAD	Targeted Targeted	17 8	Terpenoids Cannabinoids	Analysis of the phenotypic and biological properties of different *C. sativa* strains to determine the chemotype of *C. sativa* used in preparations destined for human consumption	Giese et al. (2015)
LC-MS	Untargeted	62	Cannabinoids, flavonoids, fatty acids	Untargeted LC-MS study of cannabis samples before and after decarboxylation	Lewis et al. (2017)
GC-FID	Targeted	25	Cannabinoids and terpenoids	Identification of sources of variability in floral metabolites due to flower maturity and position on the plant	Richins et al. (2018)
HPLC-DAD GC-FID	Targeted Targeted	8 28	Cannabinoids Terpenoids	Evaluation of the cannabinoid and terpene content during growth	Aizpurua-Olaizola et al. (2016)
GC-FID	Targeted	44	Terpenoids, cannabinoids	Chemometric distinction of cannabis varieties	Hazekamp et al. (2016)
HPLC-FD GC-FID	Targeted Targeted	5 19	Tocopherols Fatty acids	Analysis of the composition of oil from cannabis seeds	Kriese et al. (2004)
GC-FID	Targeted	17	Terpenoids	Analysis of commercial samples and grouping by multivariate similarities of terpene content	Fischedick (2017)

TABLE 1 Summary of published research applying metabolomics principles for the study of *Cannabis sativa*—cont'd

Platform	Type of assay	Number of analytes	Types of analytes	Brief description	References
H NMR	Untargeted	31	Sugars, amino acids, organic acids	The chemical composition of monoecious varieties of cannabis changes over seasons and depending on the strain	Ingallina et al., (2020)
GC-MS, HPLC-PDA/FD	Targeted	49	Cannabinoids, terpenes, phenolics, biogenic amines		
GC-MS	Untargeted	536	Smoke compounds	Analysis of the smoke condensate of tabaco and cannabis	Graves et al. (2020)
LC-MS	Untargeted	189	Polar lipids (sulfolipids and phospholipids)	Characterization of the *C. sativa* lipidome	Antonelli et al. (2020)
H-NMR	Untargeted	Undefined	Undefined	Phytochemical characterization of leaves, male and female inflorescences of wild *C. sativa*	Nagy et al. (2019)
GC-MS	Targeted	84	Essential oil components, terpenoids, cannabinoids, alkanes		
LC-MS	Targeted	15	Phenolics, cannabinoids		
LC-MS	Targeted	206	Amino acids, organic acids, biogenic amines, acylcarnitines, phospholipids, and sugars	Development of a fully quantitative analytical method for the analysis of cannabis and other plant samples	Zheng et al. (2021)
GC-MS	Targeted	93	Terpenoids	Fully quantitative analytical MSMS method for terpenes in cannabis	Shapira et al. (2019)

used in cannabis metabolomic studies including NMR, GC-MS, gas chromatography coupled to flame ionization detection (GC-FID), LC-MS, and LC coupled to tandem mass spectrometry (LC-MS/MS). The complexity of the cannabis matrix, combined with the biases or limitations that different methods have for certain classes of chemicals (volatiles, nonvolatiles, charged or neutral) means that several technologies must often be used to achieve broad coverage of the cannabis metabolome. These instrumental limitations exist regardless of whether one performs targeted or untargeted metabolomics and must be duly considered when designing metabolomics experiments. Table 2 summarizes the features and drawbacks of the analytical platforms most commonly used in metabolomics (Wishart, 2016).

Independent of instrumental quality or resolution, metabolomics analyses are heavily influenced by preanalytical factors such as experimental design, sample preparation, handling, and storage. In other words, high-end instruments and many hours of human labor can be wasted if the proper steps are not taken to ensure sample quality and reproducibility (Kim & Verpoorte, 2009). A good experimental design is fundamental to a good cannabis metabolomics experiment and, many factors including plant cultivars, growth conditions, plant treatments, and plant age need to be considered. Because cannabis is a legally controlled substance in most countries, its growth is tightly regulated and usually restricted to commercial greenhouses holding a cannabis licence. Cannabis usually is grown under artificial environmental conditions, with optimized watering, continuous lighting, temperature control, and carefully controlled fertilization. If these confounding factors are not considered during the experimental design stage, the resulting variability in the analytical results may be too large to reach significant conclusions. However, even with an appropriate experimental design, sample preparation, handling, and storage can introduce another level of variability in a metabolomics data set (Martins et al., 2018). It is important to remember that metabolism continues long after a plant or a plant bud has been picked. This is why for metabolomics studies, it is important to quickly quench all metabolic activity by either freezing or applying organic solvents to plant materials. Factors such as temperature, pH, light, and solvent purity (Salem et al., 2020) can also affect metabolite extraction and stability.

In cannabis research, GC-MS and GC-FID methods have traditionally been used to identify and quantify relatively volatile analytes such as terpenes and cannabinoids. Moreover, GC-MS instruments are much more affordable and more robust than LC-MS instruments making GC data generally more reliable and comparable across studies. Additionally,

TABLE 2 A detailed comparison of metabolomics technologies (Wishart, 2016).

Technology	Advantages	Disadvantages
NMR spectroscopy	• Quantitative • Nondestructive • Fast (2–3 min per sample) • Requires no derivatization • Requires no separation • Detects most organic classes • Allows identification of novel chemicals • Most spectral features are identifiable • Robust, mature technology • Can be used for metabolite imaging (fMRI or MRS) • Can be fully automated • Compatible with liquids and solids • Long instrument lifetime (over 20 years)	• Low sensitivity (LOD = 5 μM) • High start-up cost (>US$1 million) • Large instrument footprint • Cannot detect or identify salts and inorganic ions • Cannot detect nonprotonated compounds • Requires larger sample volumes (0.1–0.5 mL)
GC-MS spectrometry	• Robust, mature technology • Modest start-up cost • Quantitative (with calibration) • Modest sample volume • Good sensitivity • Large body of software and databases for metabolite identification • Detects most organic and some inorganic molecules • Excellent separation reproducibility • Many spectral features are identifiable • Can be mostly automated • Compatible with gases and liquids	• Destructive (sample not recoverable) • May require sample derivatization • Requires separation • Slow (20–40 min per sample) • Cannot be used in imaging • Not compatible with solids • Novel compound identification is difficult
LC-MS spectrometry	• Superb sensitivity (LOD = 0.5 nM) • Very flexible technology • Detects most organic and some inorganic molecules • Small sample volumes (10–100 μL) • Can be used in metabolite imaging (MALDI) • Has the potential to detect the largest portion of metabolome • Can be mostly automated • Compatible with solids and liquids	• Destructive (sample not recoverable) • Not very quantitative • Higher start-up cost (>US$300,000) • Slow (15–50 min per sample) • Usually requires separation • Poor separation and lower reproducibility versus GC-MS • Less-robust instrumentation than NMR or GC-MS • Not compatible with gases • Most spectral features are not yet identifiable • Novel compound identification is difficult • Short instrument lifetime (<9 years)

GC-MS is a particularly rich source of structural information due to the fragmentation patterns obtained through electron ionization mass spectrometry (EI-MS). The reproducibility of the molecular fragmentation at a given energy level (commonly 70 eV) has allowed the creation of extensive spectral libraries, which are very useful for fast and reliable chemical identification. These features, together with its excellent chromatographic separation reproducibility, have made GC-MS ideal for performing metabolomics studies of high value cannabis volatiles. However, GC-MS is not so ideal when analyzing nonvolatile analytes, such as very polar compounds, triterpenoids, flavonoids, or larger organic molecules. Although this limitation can be partially overcome by chemical derivatization (reacting the sample with silicon-containing compounds to yield volatile derivatives), this approach is not suitable for all analytes. Derivatization also introduces an additional level of uncertainty to the measurements as many possible derivatives can be generated for each starting analyte. Another drawback for GC metabolomics also lies with the longer analytical processing time, typically ranging from 30 to 120 minutes. This can be a significant downside in metabolomics studies, particularly if a high number of samples need to be processed to draw statistically significant conclusions. To date GC-based analysis on cannabis account for approximately 50% of reported studies (Table 1).

Over the past decade, LC-MS has largely become the metabolomics technology of choice. The advantages of LC-MS over other analytical platforms include its superb sensitivity, its small sample requirements (10–100 μL), and its ability to detect a wide range of molecules (from hydrophobic to very hydrophilic and from small molecules to very large macromolecules). Like GC-MS, LC-MS is a destructive analytical method, meaning the injected sample cannot be recovered. LC-MS systems can be coupled (tandem MS or MS/MS or MS^n) to perform more complex analysis. This significantly increases the method's selectivity where, for example, a particular parent to daughter ion transition can be recorded, which is very specific to a given analyte. Like GC-MS, LC-MS also uses fragmentation patterns and spectral library comparisons to identify unknown molecules. However, the higher mass resolution of LC-MS instruments over GC-MS instruments means that accurate measurement of the parent ion mass is often sufficient to determine the chemical formula of an unknown molecule. The combination of these attributes, together with the flexibility of different LC separation systems, gives LC-MS-based metabolomics the potential for detecting the largest portion of the cannabis metabolome. This is why LC-MS has become the preferred platform in most cannabis metabolomics studies (Table 1).

Despite its many advantages, LC-MS still has its drawbacks. For untargeted analysis, LC-MS can introduce artifacts or adducts which are formed during the ionization process (e.g., M+Na, M+K). These artifacts are often difficult to identify during data analysis. Additionally, LC-MS databases used for metabolite identification are quite small compared to those used in GC-MS, although they are growing rapidly (Blaženović et al., 2018; Da Silva et al., 2015; Tsugawa, 2018; Wishart et al., 2016).

NMR is another platform that has been widely used in metabolic studies. While GC-MS and LC-MS are generally much more sensitive, NMR still offers a number of unique advantages for metabolomics studies. NMR is nondestructive, requires little sample work-up, and is well suited to handle highly polar metabolites (Medeiros Selegato et al., 2019). Each analytical run is fast, normally lasting five to six minutes. NMR instruments are much more robust than GC-MS or LC-MS instruments and the data are highly reproducible. Additionally, several NMR software and databases exist for metabolite identification and quantification (Wishart, 2008). The main disadvantages of NMR include its cost, sensitivity, and sample size requirements. NMR instruments are very expensive to purchase and maintain and occupy a significant footprint. NMR requires relatively large sample volumes (~200–500 μL) and its sensitivity (1–5 μM) is several orders of magnitude above those of most LC-MS instruments. However, with the appearance of higher field magnets, cryogenically cooled probes, and small-volume microprobes (60 μL), the issue of sensitivity is becoming less of a concern. To date, only five NMR metabolomic studies have been done on cannabis (Table 1).

Understanding cannabis through metabolomics

Major chemical classes in cannabis

Even though the psychoactive properties of cannabis had been known for millennia, it was not until 1964 that the main psychoactive component in cannabis, delta-9-tetrahydrocannabinol (Δ9-THC) was isolated and characterized by Gaoni and Mechoulam (1964). Since then, the chemical complexity of cannabis has been slowly revealed through dozens of small-scale analytical chemistry studies and a growing number of metabolomics and chemo-genomics studies. Thanks to these studies, more than 6100 chemicals have been directly measured or genetically inferred to be in cannabis (https://cannabisdatabase.ca/). This list includes 113 phytocannabinoids and 120 terpenes (Aizpurua-Olaizola et al., 2016) as well as 50 hydrocarbons, 46 phenolic or polyphenolic compounds, 34 sugars, 25 ketones and aldehydes along with fatty acids, organic acids, amino acids, alkaloids, esters, lactones, biogenic amines, vitamins, and phytosterols (ElSohly & Slade, 2005).

Phytocannabinoids (Fig. 1) are the group of cannabis metabolites that have attracted the most interest by cannabis researchers, mainly due to their psychoactive activity. Interestingly, only a handful of the phytocannabinoids are psychoactive, while many others are either inactive or appear to modulate the psychoactivity of other phytocannabinoids. A good example of this is Δ9-THC and cannabidiol (CBD). While CBD is not psychoactive on its own, it modulates the activity of Δ9-THC by affecting the binding of it to the cannabinoid receptors in the nervous system (Freeman et al., 2019). This relationship has led to a cannabis chemotyping system proposed by Small and Beckstead in 1973 where chemotypes are defined by the relative concentrations of Δ9-THC and CBD. Chemotype I includes plants with a high Δ9-THC/CBD ratio (<2.0) and with Δ9-THC levels above 0.3%; chemotype II consists of plants with a balanced ratio of Δ9-THC and CBD (0.5–2.0); chemotype III has mainly CBD and levels of Δ9-THC below 0.3% (Small & Beckstead, 1973). This system has further evolved to include two more chemotypes: chemotype IV that is dominated by a different cannabinoid, cannabigerol (CBG) (>0.30%) and, more recently, chemotype V that has been characterized by a virtual absence of all cannabinoids (Pacifico et al., 2008). The chemotyping classification of cannabis samples, based on reporting ratios of Δ9-THC and CBD, was devised for medical applications. However, it was quickly realized that metabolites other than Δ9-THC and CBD had useful medical benefits. This has led to an explosion of studies to identify, quantify, and characterize other cannabis metabolites and a reconsideration of the utility of Small and Beckstead's simplified system.

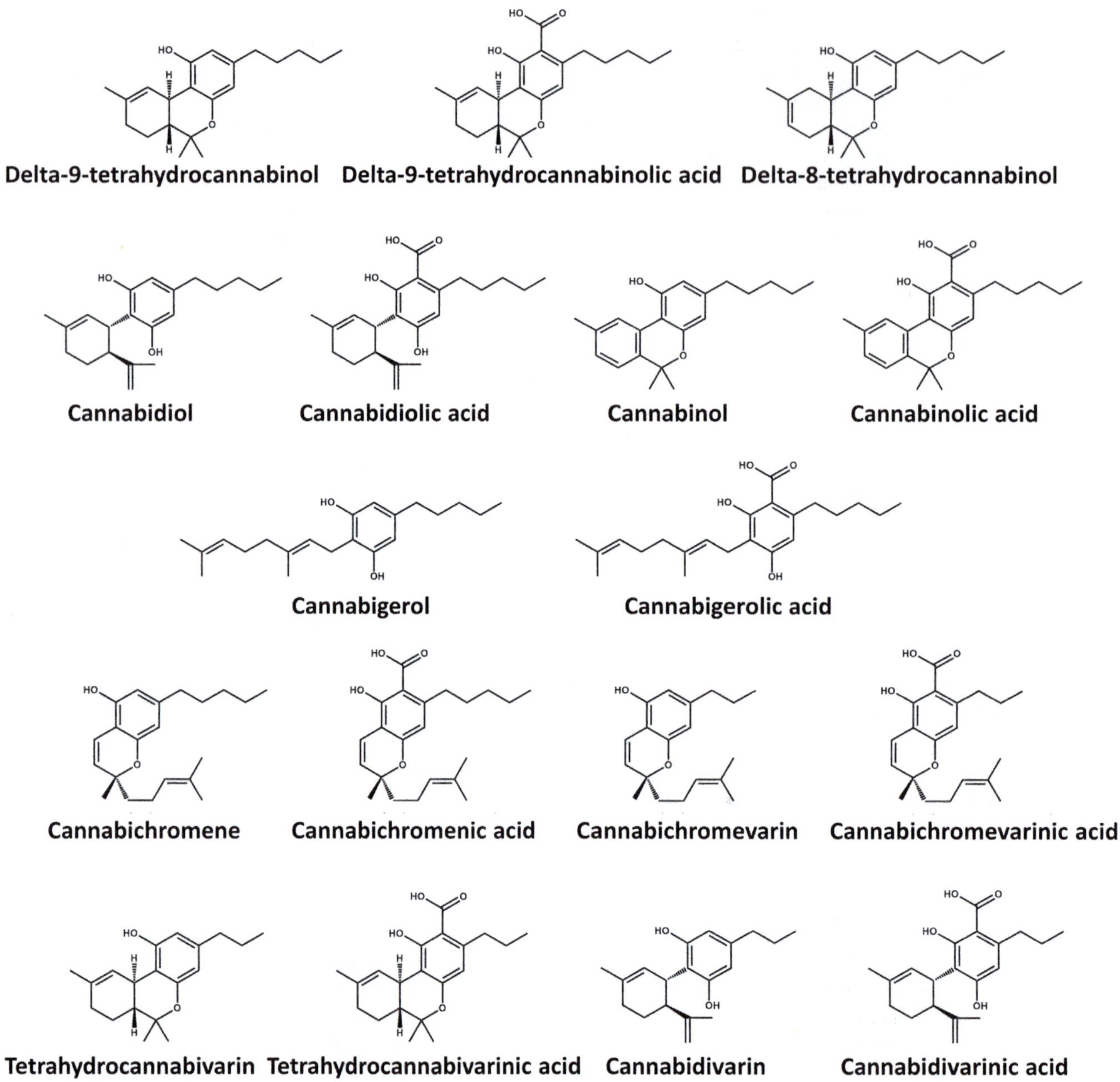

FIG. 1 Chemical structures of a select group of cannabinoids.

One group of metabolites not included in Small and Beckstead's chemotyping system, but of great interest to cannabis users and producers are the terpenes. Terpenes are the most diverse family of secondary metabolites in nature (Fig. 2). Many terpenes are volatile or scent-producing compounds characterized by the presence of five carbon segments known as isoprene units. The modular biosynthesis of terpenes gives rise to subgroups with increases of five carbon atoms every time: C5 hemiterpenes, C10 monoterpenes, C15 sesquiterpenes, C20 diterpenes, C25 sesterterpenes, C30 triterpenes, and C40 tetraterpenes. Although larger terpenes exist, the more commonly described members are monoterpenes to triterpenes. Mono- and sesquiterpenes are the main components in essential oils and other volatile fractions from cannabis and are the main contributors to cannabis aromas and fragrances. The composition of these terpenes, however, can be quite variable in different cannabis "strains," but this variability is not captured in the Δ9-THC/CBD defined chemotypes. Because fragrance and aroma are so important to cannabis users, and if commercially reproducible cannabis products and "strains" are to be made, then the terpenoid composition and Δ9-THC/CBD must be considered.

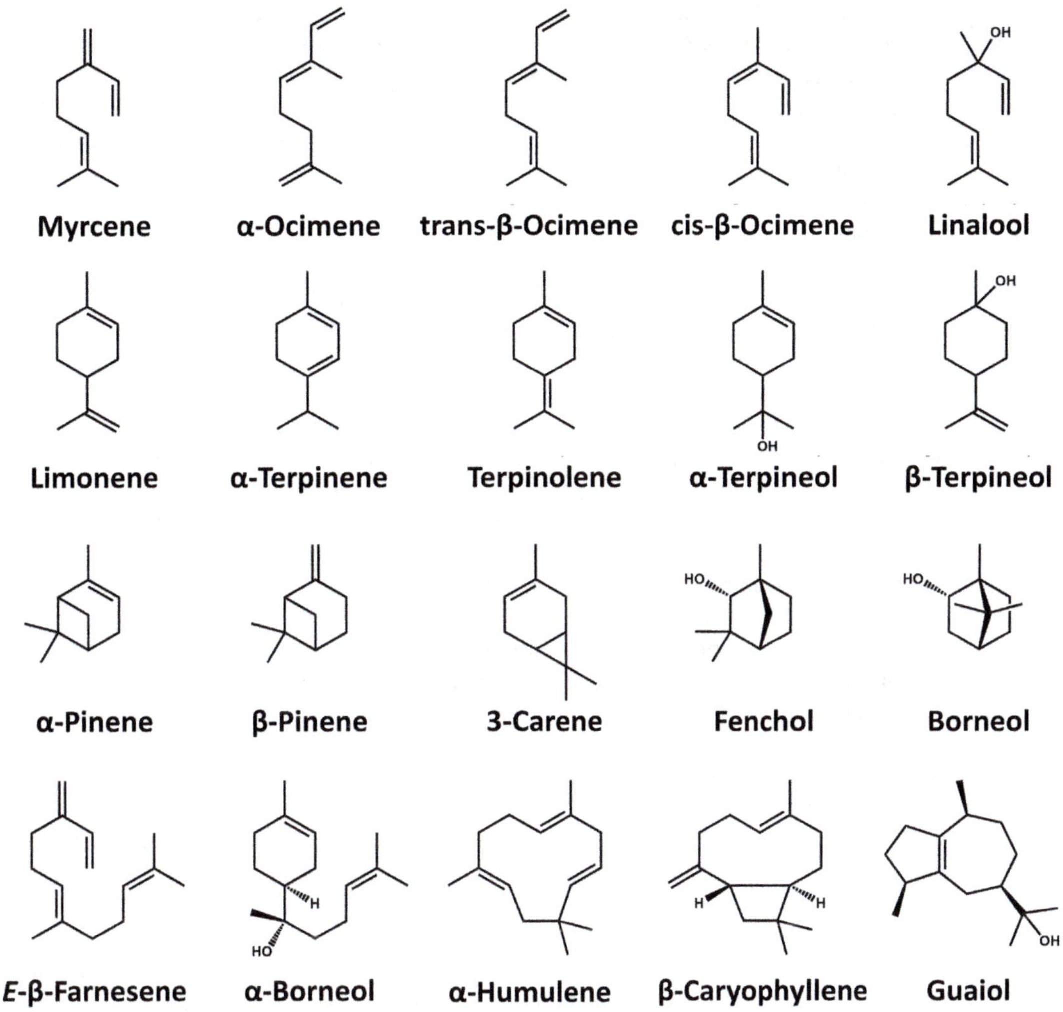

FIG. 2 Chemical structures of terpenes commonly reported in *Cannabis sativa*.

The most common monoterpenes in cannabis are myrcene, α-pinene, limonene, and linalool. The more abundant sesquiterpenes include β-caryophyllene, α-humulene, bisabolol, and (*E*)-β-farnesene. Although the exact terpenoid composition that gives rise to the characteristic aromas of cannabis is still to be determined, it is generally known that myrcene is responsible for the sweet and "earthy" notes (Surendran et al., 2021). Also, it has recently been reported that the characteristic "skunk" smell of some cannabis "strains" could be attributed to a newly discovered group of sulfur-containing hemiterpenes (Oswald et al., 2021). This is by no means a complete list of all known cannabis terpenes, as >120 of these molecules have been described in cannabis so far. However, it is difficult to identify and quantify them due to the lack of commercial standards. Terpene detection is also challenging due to their low concentrations in cannabis (Booth & Bohlmann, 2019). This makes terpene identification via metabolomics studies more difficult than the identification of other chemical classes.

Phytocannabinoids and terpenes are obviously not the only chemicals present in cannabis. Thousands of other endogenous metabolites as well as many exogenous molecules (pesticides and herbicides) also exist in or on cannabis plants or arise from cannabis combustion. These compounds also play a role in the physiological effects (positive and negative) of cannabis. Metabolomic studies offer a route to identify these compounds, some of which are described in more detail below.

Summary of cannabis metabolomic studies

Several research groups have taken advantage of multiple metabolomic platforms to more fully explore the chemical diversity of cannabis. Using a cutoff of 15 metabolites being identified or annotated to be classified as a cannabis metabolomic study, we identified a total of 29 metabolomic studies over the last 20 years. Most of these are summarized in Table 1. We found that 16 applied GC-MS or GC-FID methods (Aizpurua-Olaizola et al., 2016; Delgado-Povedano et al., 2020; Eržen et al., 2021; Fischedick, 2017; Giese et al., 2015; Graves et al., 2020; Hazekamp et al., 2016; Ingallina et al., 2020; Iseppi et al., 2019; Kriese et al., 2004; Milay et al., 2020; Mudge et al., 2019; Nagy et al., 2019; Rashid et al., 2021; Richins et al., 2018; Shapira et al., 2019), 12 used LC-MS methods (Antonelli et al., 2020; Berman et al., 2018; Cerrato et al., 2021; Citti et al., 2017; Delgado-Povedano et al., 2020; Lewis et al., 2017; Li et al., 2022; Marti et al., 2014; Milay et al., 2020; Nagy et al., 2019; Welling et al., 2021; Zheng et al., 2021), 10 reported the use of high performance liquid chromatography (HPLC) coupled to ultraviolet (UV), diode array detection (DAD) or fluorescence detection (FD) (Aizpurua-Olaizola et al., 2016; Eržen et al., 2021; Giese et al., 2015; Ingallina et al., 2020; Kriese et al., 2004; Milay et al., 2020; Mudge et al., 2018; Mudge et al., 2019; Muntendam et al., 2012), and 5 studies used NMR (Choi et al., 2004; Happyana & Kayser, 2016; Ingallina et al., 2020; Muntendam et al., 2012; Nagy et al., 2019). A total of 10 of these studies applied a targeted approach, 11 were untargeted analyses and seven used a combination of both. A wide range of metabolites is covered in these studies but most of them focused on cannabinoids, terpenoids, and to a lesser extent, primary metabolites, lipids, and flavonoids. Most of their findings are summarized in Table 1.

It is not possible to critically review or summarize every cannabis metabolomics study in Table 1, but we will try to highlight some of the more interesting methods or important findings according to the specific platform technology used. We will begin with studies that used GC-FID as their primary analytical platform. One GC-FID study of particular note was conducted by Giese et al. (2015). These researchers employed a GC-FID method to analyze 17 of the most common terpenes and 8 of the most common cannabinoids found in cannabis. The method used high-throughput homogenization to prepare sample extracts which were then profiled for cannabinoids via an HPLC-DAD assay and terpenes by GC-FID. This blended assay was then used to examine the phenotypic and biological properties of different cannabis cultivars destined for human consumption. A more recent GC-FID method developed and validated by Ibrahim et al., (2018) analyzed cannabinoids through trimethyl silyl (TMS) derivatization. This method reproducibly quantified CBD, Δ9-THC, CBG, tetrahydrocannabivarin (THCV), cannabichromene (CBC), trans-Δ8-tetrahydrocannabinol (Δ8-THC), cannabinol (CBN), cannabidiolic acid (CBDA), cannabigerolic acid (CBGA), and Δ9-tetrahydrocannabinolic acid-A (Δ9-THCAA) with a precision of $\leq 15\%$ and an accuracy of $\pm 5\%$. The method, applicable to leaves, stems, roots, and buds of cannabis plants, enabled differentiation of cannabis varieties and evaluated the genetic stability of micropropagated plants. More recently, another GC-FID study found notable differences in the cannabinoid and essential oil content in three different cannabis varieties (Eržen et al., 2021). Using eleven hemp phenotypes from three different varieties containing 0.09–338 mL of essential oil per 100 g of air-dried flower, they established a connection between the chemical composition and the morphological characteristics of the different cannabis plants. Twenty-nine essential oils, along with 10 cannabinoids, were detected by GC-FID and HPLC, respectively. A principal component analysis (PCA) and an analysis of variance (ANOVA) showed striking differences between the three varieties when the average of the essential oil components was compared among the phenotypes.

Besides GC-FID, GC-MS is another analytical method widely used for cannabis studies. Iseppi et al. (2019) used GC-MS and GC-FID to obtain a qualitative and a semi-quantitative profile of the essential oils of 17 different varieties of hemp. They also developed a GC-MS method to quantify CBD in the essential oils of all 17 varieties. This was done to establish the role of CBD in hemp's antibacterial activity against pathogenic and spoilage microorganisms. Another GC-MS study by Mudge et al. (2019) used headspace GC-MS analysis to characterize volatile terpenes in cannabis chemovars. This method led to the detection of 67 metabolites comprising 29 monoterpenes and 38 sesquiterpenes. These were identified using the National Institute of Standards and Technology (NIST) spectral database and reference standards. This metabolomics technique is notable for its identification of a high number of low-abundance terpenes, which has been quite challenging in the past. Another GC-MS cannabis study of note is one described by Graves et al. (2020). These researchers employed two-dimensional GC-MS (GCxGC-MS) metabolomics methods to tentatively identify 536 compounds found in mainstream cannabis smoke. This work is one of just a tiny handful of cannabis studies to look closely at the compounds inhaled by cannabis users.

NMR has also been used in cannabis characterization and classification. One of the first cannabis metabolomics studies used NMR to analyze 12 *C. sativa* cultivars (Choi et al., 2004). This study showed that Δ9-tetrahydrocannabinolic acid (Δ9-THCA) and CBDA, along with sucrose, glucose, asparagine, and glutamic acid can be used as major metabolites to distinguish cultivars. A more recent NMR-based metabolomics study monitored the production of cannabinoids in the trichomes during the flowering period (Happyana & Kayser, 2016). The samples were prepared by extracting the plant material with either chloroform or water to yield the organic and aqueous extracts, respectively. Samples harvested at different times during the flowering stage could be discerned by profiling asparagine, glutamine, glucose, and fructose from the aqueous extracts and through the contents of Δ9-THCA and CBDA from the organic preparations.

LC-MS is currently the most popular platform used in plant metabolomics (Zheng et al., 2021) and is the method of choice in most cannabis testing laboratories (Leghissa et al., 2018). Zheng et al. (2021) developed a targeted, fully quantitative, LC-MS-based method for analyzing over 200 plant metabolites, including amino acids, biogenic amines, organic acids (including phytohormones), acylcarnitines, phospholipids, and hexoses. This LC-MS assay was recently used to analyze commercial cannabis buds and could routinely detect 78 cannabis metabolites. A different untargeted LC-MS approach was used by (Li et al., 2022) to profile a diverse range of metabolites present in cannabis leaves collected in three different regions of China. Flavonoid glycosides and flavonoid glucuronides were identified as most differentiating chemical markers from the different provinces. Moreover, variations in the content of the flavonoids were used to differentiate *C. sativa* varieties and to elucidate their geographical location. LC-MS has also been used to improve cannabis breeding and selection. For instance, Welling et al. (2021) studied the floral tissues of male and female industrial hemp accessions to establish sampling requirements for germplasm comparison of cannabis chemotypes. Using UHPLC-ESI-HRMS, a targeted approach was used to identify the phytocannabinoids cannabichromenic acid (CBCA), cannabinolic acid (CBNA), cannabidivarinic acid (CBDVA), cannabidivarin (CBDV), tetrahydrocannabivarinic acid (THCVA), CBDA, CBD, Δ9-THCAA, Δ9-THC, CBC, CBGA, CBG, CBN, and THCV. This was coupled with an untargeted LC-MS approach to better understand the metabolomic variation among cannabis flowers.

In many cases, blended LC-MS and GC-MS approaches have been used to extend metabolite coverage in cannabis studies. For instance, Delgado-Povedano et al. (2020) used a combination of LC-MS and GC-MS methods to identify 169 metabolites including terpenoids, cannabinoids, flavonoids, benzenoids, amino acids, and hydrocarbons in 17 cannabis cultivars. Alternately, developing class-specific assays can greatly extend compound coverage. For instance, Antonelli et al. (2020) used high-resolution LC-MS lipidomics methods to putatively identify 189 different lipids, including 80 sulfolipids and 51 phospholipids from a single strain of *C. sativa*.

Metabolomics as a vehicle for cannabis classification

Cannabis strains, cultivars, and varietals have been under active development for decades, but with little-to-no quality control or almost no provenance tracking. As a result, cannabis classification is very challenging. The current scientific consensus states that the plant consists of the single variable species *C. sativa* L., as described by Leonard Fuchs. In 1976, two botanists (Small and Cronquist) established a formal botanical nomenclature for *C. sativa* that is still valid today (McPartland, 2018). This classification relies on a two-step process. The first step identifies two subspecies (either weakly or highly intoxicant) based on their Δ9-THC content in dried female flowering tops. These are *C. sativa* subsp. *sativa* (low Δ9-THC content), and *C. sativa* subsp. *indica* (high Δ9-THC content). The second step classifies the varieties within each subspecies, based on their adaptation to grow in the wild or in a cultivated environment, and therefore based on the domestication traits found in them.

Another approach that classifies cannabis plants based on their plant or bud morphology, their cannabinoid composition as well as their fragrance or aroma has led to the appearance of a very large number of so-called cannabis "cultivars" or "strains." While widely used and popular, the emergence of this vernacular classification system, which is independent of Small and Cronquist's taxonomic classification system, has been problematic. Indeed, the main concern with this vernacular classification is the lack of standards, rules, and regulations needed for strain naming (Small, 2015).

Recently, another cannabis plant classification based on their chemical diversity has been proposed. These "chemotaxonomic" approaches use metabolomics and the measured concentrations of dozens of cannabis metabolites to objectively characterize and classify cannabis strains. For example, Mudge et al. (2018) used a combination of targeted and untargeted metabolomics employing ultrahigh-performance liquid chromatography coupled to ultraviolet detection (UHPLC-UV) to generate a chemometric profile for 33 strains of commercially available cannabis. She then used multivariate statistics (PCA) to cluster these 33 strains and found that they fell into just five chemotaxonomic groups. Another study conducted by Fischedick (2017) used GC-FID and GC-MS metabolomic methods to measure terpenoid content for 233 cannabis samples belonging to 30 different strains or cultivars. Using multivariate statistics, they found that these strains separated into five major groups on their terpenoid content. These data suggest that most commercial cannabis strains can be objectively grouped into a relatively small number (i.e., five) chemically similar classes. No doubt further chemical differentiation of cannabis varietals via metabolomics methods is certainly possible. Indeed, it would be most helpful to see a standardized system that quantitatively measures both terpenoids and cannabinoids to classify cannabis cultivars. Overall, these early attempts at chemotaxonomic classification for cannabis certainly demonstrate the potential for metabolomics to create a reliable, chemically based, fully objective classification system for cannabis.

Using metabolomics for quality control and safety in cannabis production

The manufacture of standardized cannabis products that meet quality standards and safety set by regulatory agencies is another challenge for the cannabis industry. Cannabis products and cannabis plants need to be chemically consistent yielding the same Δ9-THC, CBN, and CBD content and the same aromas or fragrances (i.e., terpenoids) year-in and year-out. Mix-ups with strains, cultivars, seeds, and germplasm must be avoided. Similarly, storage and shipping must be done in a manner to ensure minimal breakdown or decay of the chemicals in cannabis products. Likewise, cannabis products need to be safe with minimal levels of toxic pesticides, toxic metals, mycotoxins, or other contaminants. Hence, good agricultural practices, good product management practices, and good quality control practices play a key role in ensuring consistency in product quality. In this regard, metabolomics can and is beginning to play an important role in monitoring cannabis quality and safety. As has been noted, a number of robust, quantitative metabolomic assays have been already developed that can measure most of the key cannabinoids and most of the essential terpenoids in cannabis. These include Δ9-THC, CBN, and CBD and myrcene, α-pinene, limonene, β-caryophyllene, and α-humulene (Milay et al., 2020). These can be used to monitor product quality and product consistency (Shapira et al., 2019). They can also be used to monitor product resilience to different storage or shipping conditions. Indeed, a number of metabolomics studies showed that the chemical composition of cannabis plants varies with storage conditions such as temperature and light. For instance, a study conducted by Trofin et al. (2011) found that the stability of cannabinoids in herbal cannabis changed significantly during long-term storage. When Δ9-THC, CBN, and CBD were measured over a 4-year period, Δ9-THC in samples exposed to light and stored at 22°C was found to be quickly degraded. Similarly, Grafström et al. (2019) analyzed a collection of cannabis resins by GC-FID and GC-MS and observed a loss or degradation of a number of compounds, including THCV, Δ1-tetrahydrocannabiorcol, β-caryophyllene, and α- and β-humulene, in the resins exposed to air and light under different storage conditions.

In addition to assays targeting endogenous compounds in cannabis, a number of metabolomics assays (LC-MS and GC-MS) have been developed for measuring herbicide and pesticide residues in plants, including cannabis (Atapattu & Johnson, 2020). Indeed, most commercially produced cannabis products must go through rigorous (metabolomics based) pesticide/herbicide testing at certified labs. Pesticide/herbicide residue limits in cannabis must conform to those found in most seed crops, fruits, and vegetables (Health Canada, 2019). Furthermore, testing for toxic metals (absorbed by cannabis plants from the soil) can also be done using MS-based metabolomics methods such as inductively coupled plasma (ICP) mass spectrometry (Nelson et al., 2021). Many of these assays are just starting to be offered in certified cannabis testing labs, but as the commercialization of the cannabis industry continues to expand, the need for these metabolomic tests and testing facilities will, no doubt, increase.

The future of cannabis metabolomics

While technological and methodological improvements in cannabis metabolomics are progressing well (see Table 1), there are still many areas where future improvements are possible. Three areas where we believe more work needs to be done or where more effort is being directed include (1) the development of standardized "kit-based" quantitative cannabis metabolomic assays; (2) the implementation of a standardized chemotaxonomy for cannabis; and (3) the establishment of centralized databases for the cannabis metabolome.

Cannabis metabolomics kits

Cannabis metabolomics lacks consistency and standardization. Every lab uses different technologies and measures different cannabis compounds—some quantitative but most qualitatively. This lack of consistency makes it very difficult for cannabis producers or cannabis researchers to compare or interpret their data. This has substantially slowed the adoption of metabolomics methods by the cannabis industry. Certainly, the development of standardized "kit-based" quantitative cannabis metabolomic assays would help matters considerably. Metabolomics kits for different platforms (LC-MS and NMR) for human metabolomics studies are already commercially available (https://biocrates.com/, https://www.bruker.com/it/applications/academia-life-science/metabolomics/metabolomics-solution.html). Likewise commercial systems for pesticide residue testing and drug testing are also available (https://www.thermofisher.com/ca/en/home/industrial/food-beverage/food-analytical-testing/pesticide-residues-analysis.html). It seems reasonable to think that similar metabolomics kits (GC-MS, LC-MS, and possibly NMR) would certainly help in standardize cannabis testing. It would also make cannabis metabolomics more accessible to many other researchers and organizations.

Cannabis chemotaxonomy

The establishment of quantitative, fully standardized cannabis testing "kits" would also open the door to standardized chemotaxonomic classification of cannabis cultivars and varietals. Objective chemotaxonomic analysis requires objective, quantitatively consistent metabolite profiling (Hazekamp et al., 2016). A standardized, targeted metabolomics kit assay would provide objective and quantitatively consistent metabolite profiling. The implementation of a standardized chemotaxonomy for cannabis would also help to establish more consistent reporting methods and create better quality control systems for cannabis breeding and cannabis labeling.

Cannabis metabolome databases

As Table 1 highlights, the growth over the past decade in cannabis testing and cannabis chemical characterization have been quite impressive. However, most of this measured chemical data are scattered in various papers, journals, and textbooks. Likewise, the spectral (MS, NMR) information needed to readily identify and quantify cannabis metabolites via metabolomics methods is also widely scattered in various commercial databases, online libraries, and journal articles. These issues of scattered, decentralized knowledge could be addressed with the development of centralized cannabis-specific metabolomic databases. Metabolomic databases have been a central pillar for the evolution of metabolomics (Tsugawa, 2018; Vinaixa et al., 2016). Modern metabolomic databases often contain detailed information about compounds, pathways, reactions, and MS or NMR spectral data to aid in the identification and characterization of metabolites (Wishart, 2007). A number of metabolomics databases, such as the HMDB (Wishart et al., 2022), Metlin (Guijas et al., 2018), and The Metabolomics Workbench (Sud et al., 2016) have already been established. However, these are largely focused on human or mammalian metabolomes and have little data on plants, much less on cannabis (Guijas et al., 2018).

The development of a cannabis-specific metabolome database similar in design and content to the HMDB would help centralize much of the existing cannabis metabolome data. In an effort to address this issue, we have created the Cannabis Compound Database or CCD (https://cannabisdatabase.ca/). This online database contains detailed information about the small molecules found in both the cannabis plant and in cannabis smoke condensate. The CCD is designed to provide comprehensive information about a vast collection of known chemicals found in cannabis, their abundance within different cultivars, their physiological effects, and their known protein targets in humans. Currently, the CCD contains 6172 metabolite entries and 2336 chemical concentrations for 115 cannabis cultivars. In addition to the metabolites, a list of 5207 human protein targets are linked to various cannabis chemical entries. Additionally, the CCD contains 50,817 MS/MS spectra (both, predicted and experimental), 5929 GC-MS spectra (predicted and experimental), and 1529 NMR spectra. The CCD contains detailed compound descriptions and supports extensive text, sequence, chemical structure, MS, and NMR spectral query searches. Many data fields in the CCD are hyperlinked to other databases such as the HMDB, FooDB (https://foodb.ca/), PubChem (S. Kim et al., 2021), KEGG (Kanehisa et al., 2021), and ChEBI (Hastings et al., 2016). The CCD represents a first step to consolidate the known information about cannabis. No doubt as our knowledge grows about the chemical complexity of cannabis, this database should continue to expand. In doing so, it will also help make metabolomic analysis and the interpretation of cannabis metabolomic data far easier.

Concluding remarks

Recent trends toward legalizing cannabis around the world have led to an explosion of cannabis research. Because the physiological effects of cannabis (both good and bad) are driven by the chemicals found in cannabis, it is clear that a better understanding of what chemicals are in cannabis must be gained. Furthermore, improvements in the ways that cannabis is classified, monitored, produced, and managed (especially with regard to quality control and safety) must also be chemically driven. Currently, the best route to obtaining that chemical information is through metabolomics. While advances in cannabis metabolomics are moving at an impressive pace, the lack of standardization is hindering its widespread application in general cannabis research, in the cannabis industry and within cannabis testing labs. The development of standardized "kit-based" cannabis assays, the implementation of a standardized cannabis chemotaxonomy, and the establishment of centralized databases for the cannabis metabolome could go a long way to firmly establishing metabolomics as a key tool for advancing both cannabis science and the cannabis industry.

Applications to other areas

In this chapter, we have reviewed how metabolomics can be used to analyze the chemical diversity in the cannabis plant. Metabolomics has proved itself as a valuable tool in cannabis due to its capacity to identify, quantify, and characterize a wide range of analytes. As a result, metabolomics can not only be used to explore and enumerate the chemical diversity of cannabis plants and cannabis products but it can also help to associate this information with important organoleptic and plant physiological properties. For instance, metabolomics can be used to relate specific volatile profiles with desired (or undesirable) aromas or tastes, it can be used to identify contaminants or toxic residues arising from production and processing, it can be used to monitor product quality and consistency, it can be used to detect the compounds in cannabis smoke, and it can be used to classify cannabis varieties or chemotypes. In addition to the application of metabolomics toward understanding the cannabis plant, metabolomics can also be used to understand how cannabis affects human physiology. By using metabolomics to monitor cannabis-induced changes in general human metabolism, human brain metabolism, and human gut metabolism, it is possible to gain a more complete understanding of the complex interactions of cannabinoids (and other alkaloids) on human physiology. As metabolomics technologies continue to mature and the needed computational infrastructure becomes more widespread, metabolomics will surely become more routine in cannabis research, not only in academia but also in regulatory organizations and in the cannabis industry.

Key facts

- Metabolomics is a field of the omics sciences that uses analytical chemistry techniques and advanced computational methods to characterize complex biochemical mixtures.
- Metabolomics studies can be grouped into two general categories: untargeted metabolomics and targeted metabolomics. Untargeted methods are ideal for hypothesis generation or discovery, while targeted methods are best for hypothesis testing and reference-based testing.
- Depending on the technology platform used, metabolomics studies measure and identify between 15 to more than 500 analytes in cannabis or cannabis products.
- Metabolomics relies on the use of advanced automated or robotic analytical chemistry techniques and cutting-edge computational methods to achieve optimal results.
- No single analytical method can accomplish full coverage of given metabolome; hence a combination of methods is often needed to achieve maximal coverage.

Mini-dictionary of terms

- Metabolome: The complete collection of metabolites, or small molecules found in a given organelle, cell, tissue, organ, biofluid, or organism.
- Metabolomics: The comprehensive characterization of metabolites and other chemical species (both exogenous and endogenous) in biological specimens in response to different perturbations or interventions.
- Multivariate statistical analysis: A branch of statistics that is concerned with the analysis of several simultaneous dependent variables and their relations with other variables. Some examples of multivariate statistical methods are principal component analysis (PCA) and orthogonal projections to latent structures discriminant analysis (OPLS-DA).
- Database: An organized collection of data that is stored and accessed electronically. Metabolomic databases often contain thousands of compounds, descriptions, chemical reactions, and metabolic pathways for many organisms and are designed to facilitate the exploration of metabolites. Metabolomic databases must support text, name, structure, and spectral searches

Summary points

- Cannabis is known to have a complex chemical profile with more than 6100 different compounds reported so far.
- The chemical diversity of cannabis represents a difficult challenge for traditional analytical chemistry methods.
- Metabolomics can provide a comprehensive overview of the chemical space in cannabis.

- Metabolomics is being successfully applied to answer an increasing number of biological questions about cannabis, including questions regarding its identity, its chemical composition, its quality, its consistency, and its effect on human physiology.
- The development of fully quantitative standardized methods and comprehensive metabolome databases will allow the field of cannabis metabolomics to move forward.

References

Aizpurua-Olaizola, O., Soydaner, U., Öztürk, E., Schibano, D., Simsir, Y., Navarro, P., Etxebarria, N., & Usobiaga, A. (2016). Evolution of the cannabinoid and terpene content during the growth of Cannabis sativa plants from different chemotypes. *Journal of Natural Products*, *79*(2), 324–331. https://doi.org/10.1021/acs.jnatprod.5b00949.

Aliferis, K. A., & Bernard-Perron, D. (2020). Cannabinomics: Application of metabolomics in Cannabis (Cannabis sativa L.) research and development. *Frontiers Plant Science*, *11*(May), 1–20. https://doi.org/10.3389/fpls.2020.00554.

Antonelli, M., Benedetti, B., Cannazza, G., Cerrato, A., Citti, C., Montone, C. M., Piovesana, S., & Laganà, A. (2020). New insights in hemp chemical composition: a comprehensive polar lipidome characterization by combining solid phase enrichment, high-resolution mass spectrometry, and cheminformatics. *Analytical and Bioanalytical Chemistry*, *412*(2), 413–423. https://doi.org/10.1007/s00216-019-02247-6.

Atapattu, S. N., & Johnson, K. R. D. (2020). Pesticide analysis in cannabis products. *Journal of Chromatography A*, *1612*, 460656. https://doi.org/10.1016/j.chroma.2019.460656.

Berman, P., Futoran, K., Lewitus, G. M., Mukha, D., Benami, M., Shlomi, T., & Meiri, D. (2018). A new ESI-LC/MS approach for comprehensive metabolic profiling of phytocannabinoids in Cannabis. *Scientific Reports*, *8*(1), 1–15. https://doi.org/10.1038/s41598-018-32651-4.

Blaženović, I., Kind, T., Ji, J., & Fiehn, O. (2018). Software tools and approaches for compound identification of LC-MS/MS data in metabolomics. *Metabolites*, *8*(2). https://doi.org/10.3390/metabo8020031.

Booth, J. K., & Bohlmann, J. (2019). Terpenes in Cannabis sativa—From plant genome to humans. *Plant Science*, *284*(January), 67–72. https://doi.org/10.1016/j.plantsci.2019.03.022.

Canada, C. H. (2019). *Mandatory Cannabis Testing for Pesticide Active Ingredients: Requirements. Government of Canada = Gouvernement du Canada.*

Cerrato, A., Citti, C., Cannazza, G., Capriotti, A. L., Cavaliere, C., Grassi, G., Marini, F., Montone, C. M., Paris, R., Piovesana, S., & Laganà, A. (2021). Phytocannabinomics: Untargeted metabolomics as a tool for cannabis chemovar differentiation. *Talanta*, *230*(March), 1–9. https://doi.org/10.1016/j.talanta.2021.122313.

Choi, Y. H., Kim, H. K., Hazekamp, A., Erkelens, C., Lefeber, A. W. M., & Verpoorte, R. (2004). Metabolomic differentiation of Cannabis sativa cultivars using 1H NMR spectroscopy and principal component analysis. *Journal of Natural Products*, *67*(6), 953–957. https://doi.org/10.1021/np049919c.

Citti, C., Battisti, U. M., Braghiroli, D., Ciccarella, G., Schmid, M., Vandelli, M. A., & Cannazza, G. (2017). A metabolomic approach applied to a liquid chromatography coupled to high-resolution tandem mass spectrometry method (HPLC-ESI-HRMS/MS): Towards the comprehensive evaluation of the chemical composition of cannabis medicinal extracts. *Phytochemical Analysis*, *29*(2), 144–155. https://doi.org/10.1002/pca.2722.

Da Silva, R. R., Dorrestein, P. C., & Quinn, R. A. (2015). Illuminating the dark matter in metabolomics. *Proceedings of the National Academy of Sciences of the United States of America*, *112*(41), 12549–12550. https://doi.org/10.1073/pnas.1516878112.

De Jong, F. A., & Beecher, C. (2012). Addressing the current bottlenecks of metabolomics: Isotopic Ratio Outlier Analysis^{TM}, an isotopic-labeling technique for accurate biochemical profiling. *Bioanalysis*, *4*(18), 2303–2314. https://doi.org/10.4155/bio.12.202.

Delgado-Povedano, M. M., Sánchez-Carnerero Callado, C., Priego-Capote, F., & Ferreiro-Vera, C. (2020). Untargeted characterization of extracts from Cannabis sativa L. cultivars by gas and liquid chromatography coupled to mass spectrometry in high resolution mode. *Talanta*, *208*. https://doi.org/10.1016/j.talanta.2019.120384.

ElSohly, M. A., & Slade, D. (2005). Chemical constituents of marijuana: The complex mixture of natural cannabinoids. *Life Sciences*, *78*(5), 539–548. https://doi.org/10.1016/j.lfs.2005.09.011.

Eržen, M., Košir, I. J., Ocvirk, M., Kreft, S., & Cerenak, A. (2021). Metabolomic analysis of cannabinoid and essential oil. *Plants*, *100*, 966.

Fang, C., Fernie, A. R., & Luo, J. (2019). Exploring the diversity of plant metabolism. *Trends in Plant Science*, *24*(1), 83–98. https://doi.org/10.1016/j.tplants.2018.09.006.

Fischedick, J. T. (2017). Identification of terpenoid chemotypes among high (−)-trans-Δ 9-tetrahydrocannabinol-producing Cannabis sativa L. Cultivars. *Cannabis and Cannabinoid Research*, *2*(1), 34–47. https://doi.org/10.1089/can.2016.0040.

Freeman, A. M., Petrilli, K., Lees, R., Hindocha, C., Mokrysz, C., Curran, H. V., Saunders, R., & Freeman, T. P. (2019). How does cannabidiol (CBD) influence the acute effects of delta-9-tetrahydrocannabinol (THC) in humans? A systematic review. *Neuroscience and Biobehavioral Reviews*, *107* (September), 696–712. https://doi.org/10.1016/j.neubiorev.2019.09.036.

Gaoni, Y., & Mechoulam, R. (1964). Isolation, structure, and partial synthesis of an active constituent of hashish. *Journal of the American Chemical Society*, *86*(8), 1646–1647. https://doi.org/10.1021/ja01062a046.

Giese, M. W., Lewis, M. A., Giese, L., & Smith, K. M. (2015). Development and validation of a reliable and robust method for the analysis of cannabinoids and terpenes in cannabis. *Journal of AOAC International*, *98*(6), 1503–1522. https://doi.org/10.5740/jaoacint.15-116.

Grafström, K., Andersson, K., Pettersson, N., Dalgaard, J., & Dunne, S. J. (2019). Effects of long term storage on secondary metabolite profiles of cannabis resin. *Forensic Science International*, *301*, 331–340. https://doi.org/10.1016/j.forsciint.2019.05.035.

Graves, B. M., Johnson, T. J., Nishida, R. T., Dias, R. P., Savareear, B., Harynuk, J. J., Kazemimanesh, M., Olfert, J. S., & Boies, A. M. (2020). Comprehensive characterization of mainstream marijuana and tobacco smoke. *Scientific Reports*, *10*(1). https://doi.org/10.1038/s41598-020-63120-6.

Guijas, C., Montenegro-Burke, R., Domingo-Almenara, X., Palermo, A., Warth, B., Hemann, G., Koellensperger, G., Huan, T., Uritboonthai, W., Aisporna, A., Wolan, D., Spilker, M., Benton, P., & Siuzdak, G. (2018). METLIN: A technology platform for identifying knowns and unknowns. *Analytical Chemistry*, *90*(5), 3156–3164. https://doi.org/10.1021/acs.analchem.7b04424.METLIN.

Happyana, N., & Kayser, O. (2016). Monitoring metabolite profiles of Cannabis sativa L trichomes during flowering period using 1H NMR-based metabolomics and real-time PCR. *Planta Medica*, *82*(13), 1217–1223. https://doi.org/10.1055/s-0042-108058.

Hastings, J., Owen, G., Dekker, A., Ennis, M., Kale, N., Muthukrishnan, V., Turner, S., Swainston, N., Mendes, P., & Steinbeck, C. (2016). ChEBI in 2016: Improved services and an expanding collection of metabolites. *Nucleic Acids Research*, *44*(D1), D1214–D1219. https://doi.org/10.1093/nar/gkv1031.

Hazekamp, A., Tejkalová, K., & Papadimitriou, S. (2016). Cannabis: From cultivar to chemovar II-A metabolomics approach to cannabis classification. *Cannabis and Cannabinoid Research*, *1*(1), 202–215. https://doi.org/10.1089/can.2016.0017.

Ibrahim, E., Gul, W., Gul, S., Stamper, B., Hadad, G., Salam, R., … ElSohly, M. (2018). Determination of acid and neutral cannabinoids in extracts of different strains of Cannabis sativa using GC-FID. *Planta Medica*, *84*(4), 250–259.

Ingallina, C., Sobolev, A. P., Circi, S., Spano, M., Fraschetti, C., Filippi, A., Di Sotto, A., Di Giacomo, S., Mazzoccanti, G., Gasparrini, F., Quaglio, D., Campiglia, E., Carradori, S., Locatelli, M., Vinci, G., Rapa, M., Ciano, S., Giusti, A. M., Botta, B., & Mannina, L. (2020). Cannabis sativa L. inflorescences from monoecious cultivars grown in central Italy: An untargeted chemical characterization from early flowering to ripening. *Molecules*, *25*(8). https://doi.org/10.3390/molecules25081908.

Iseppi, R., Brighenti, V., Licata, M., Lambertini, A., Sabia, C., Messi, P., Pellati, F., & Benvenuti, S. (2019). Chemical characterization and evaluation of the antibacterial activity of essential oils from fibre-type Cannabis sativa L. (Hemp). *Molecules*, *1*, 7–12.

Kanehisa, M., Furumichi, M., Sato, Y., Ishiguro-Watanabe, M., & Tanabe, M. (2021). KEGG: integrating viruses and cellular organisms. *Nucleic Acids Research*, *49*(D1), D545–D551. https://doi.org/10.1093/nar/gkaa970.

Kim, H. K., & Verpoorte, R. (2009). Sample preparation for plant metabolomics. *Phytochemical Analysis*, *21*(1), 4–13. https://doi.org/10.1002/pca.1188.

Kim, S., Chen, J., Cheng, T., Gindulyte, A., He, J., He, S., Li, Q., Shoemaker, B. A., Thiessen, P. A., Yu, B., Zaslavsky, L., Zhang, J., & Bolton, E. E. (2021). PubChem in 2021: New data content and improved web interfaces. *Nucleic Acids Research*, *49*(D1), D1388–D1395. https://doi.org/10.1093/nar/gkaa971.

Kriese, U., Schumann, E., Weber, W. E., Beyer, M., Brühl, L., & Matthäus, B. (2004). Oil content, tocopherol composition and fatty acid patterns of the seeds of 51 Cannabis sativa L. genotypes. *Euphytica*, *137*(3), 339–351. https://doi.org/10.1023/B:EUPH.0000040473.23941.76.

Leghissa, A., Hildenbrand, Z. L., & Schug, K. A. (2018). A review of methods for the chemical characterization of cannabis natural products. *Journal of Separation Science*, *41*(1), 398–415. https://doi.org/10.1002/jssc.201701003.

Lewis, M. M., Yang, Y., Wasilewski, E., Clarke, H. A., & Kotra, L. P. (2017). Chemical profiling of medical cannabis extracts. *ACS Omega*, *2*(9), 6091–6103. https://doi.org/10.1021/acsomega.7b00996.

Li, C.-R., Yang, L.-X., Guo, Z.-F., Yang, H., Zhang, Y., Wang, Y.-M., Zhang, G.-Z., Li, P., & Gao, W. (2022). LC-MS-based untargeted metabolomics reveals chemical differences of Cannabis leaves from different regions of China. *Industrial Crops and Products*, *176*(July), 114411. https://doi.org/10.1016/j.indcrop.2021.114411.

Marti, G., Schnee, S., Andrey, Y., Simoes-Pires, C., Carrupt, P. A., Wolfender, J. L., & Gindro, K. (2014). Study of leaf metabolome modifications induced by UV-C radiations in representative vitis, cissus and cannabis species by LC-MS based metabolomics and antioxidant assays. *Molecules*, *19*(9), 14004–14021. https://doi.org/10.3390/molecules190914004.

Martins, M., Caldana, C., Wolf, L. D., & Furlan de Abreu, L. G. (2018). The importance of experimental design, quality assurance, and control in plant metabolomics experiments. *Plant Metabolomics*, *1908*(January), 2–6. https://doi.org/10.1007/978-1-4939-7819-9_1.

McPartland, J. M. (2018). Cannabis systematics at the levels of family, genus, and species. *Cannabis and Cannabinoid Research*, *3*(1), 203–212. https://doi.org/10.1089/can.2018.0039.

Medeiros Selegato, D., Pilon, A. C., & Carnevale Neto, F. (2019). Plant metabolomics using NMR spectroscopy. *NMR-Based Metabolomics*, 345–362.

Milay, L., Berman, P., Shapira, A., Guberman, O., & Meiri, D. (2020). Metabolic profiling of cannabis secondary metabolites for evaluation of optimal postharvest storage conditions. *Frontiers in Plant Science*, *11*(October), 1–15. https://doi.org/10.3389/fpls.2020.583605.

Mudge, E. M., Murch, S. J., & Brown, P. N. (2018). Chemometric analysis of cannabinoids: Chemotaxonomy and domestication syndrome. *Scientific Reports*, *8*(1), 1–9. https://doi.org/10.1038/s41598-018-31120-2.

Mudge, E. M., Brown, P. N., & Murch, S. J. (2019). The terroir of cannabis: Terpene metabolomics as a tool to understand Cannabis sativa selections. *Planta Medica*, *85*(9–10), 781–796. https://doi.org/10.1055/a-0915-2550.

Muntendam, R., Happyana, N., Erkelens, T., Bruining, F., & Kayser, O. (2012). Time dependent metabolomics and transcriptional analysis of cannabinoid biosynthesis in Cannabis sativa var. bedrobinol and bediol grown under standardized condition and with genetic homogeneity. *Online International Journal of Medicinal Plant Research*, *1*(2), 31–40. https://doi.org/10.13140/RG.2.1.2263.5360.

Nagy, D. U., Cianfaglione, K., Maggi, F., Sut, S., & Dall'Acqua, S. (2019). Chemical characterization of leaves, male and female flowers from spontaneous cannabis (Cannabis sativa L.) growing in hungary. *Chemistry and Biodiversity*, *16*(3). https://doi.org/10.1002/cbdv.201800562.

Nelson, J., Jones, C., Heckle, S., & Anderson, L. (2021). Determination of heavy metals in a variety of cannabis and cannabis-derived products, First Action 2021.03. *The Journal of AOAC International*, *qsab173*. https://doi.org/10.1093/jaoacint/qsab173.

Oswald, I. W. H., Ojeda, M. A., Pobanz, R. J., Koby, K. A., Buchanan, A. J., Del Rosso, J., Guzman, M. A., & Martin, T. J. (2021). Identification of a new family of prenylated volatile sulfur compounds in cannabis revealed by comprehensive two-dimensional gas chromatography. *ACS Omega*, *6*(47), 31667–31676. https://doi.org/10.1021/acsomega.1c04196.

Pacifico, D., Miselli, F., Carboni, A., Moschella, A., & Mandolino, G. (2008). Time course of cannabinoid accumulation and chemotype development during the growth of Cannabis sativa L. *Euphytica*, *160*(2), 231–240. https://doi.org/10.1007/s10681-007-9543-y.

Pertwee, R. (2014). *Handbook of Cannabis*. (Issue July).

Rashid, A., Ali, V., Khajuria, M., Faiz, S., Gairola, S., & Vyas, D. (2021). GC–MS based metabolomic approach to understand nutraceutical potential of Cannabis seeds from two different environments. *Food Chemistry*, *339*(February), 128076. https://doi.org/10.1016/j.foodchem.2020.128076.

Richins, R. D., Rodriguez-Uribe, L., Lowe, K., Ferral, R., & O'Connell, M. A. (2018). Accumulation of bioactive metabolites in cultivated medical Cannabis. *PLoS One*, *13*(7), 1–20. https://doi.org/10.1371/journal.pone.0201119.

Salem, M. A., De Souza, L. P., Serag, A., Fernie, A. R., Farag, M. A., Ezzat, S. M., & Alseekh, S. (2020). Metabolomics in the context of plant natural products research: From sample preparation to metabolite analysis. *Metabolites*, *10*(1), 1–30. https://doi.org/10.3390/metabo10010037.

Shapira, A., Berman, P., Futoran, K., Guberman, O., & Meiri, D. (2019). Tandem mass spectrometric quantification of 93 terpenoids in cannabis using static headspace injections. *Analytical Chemistry*, *91*(17), 11425–11432. https://doi.org/10.1021/acs.analchem.9b02844.

Small, E. (2015). Evolution and classification of Cannabis sativa (Marijuana, Hemp) in relation to human utilization. *Botanical Review*, *81*(3), 189–294. https://doi.org/10.1007/s12229-015-9157-3.

Small, E., & Beckstead, H. D. (1973). Cannabinoid phenotypes in Cannabis sativa. *Nature*, *245*, 147–148.

Sud, M., Fahy, E., Cotter, D., Azam, K., Vadivelu, I., Burant, C., Edison, A., Fiehn, O., Higashi, R., Nair, K. S., Sumner, S., & Subramaniam, S. (2016). Metabolomics workbench: An international repository for metabolomics data and metadata, metabolite standards, protocols, tutorials and training, and analysis tools. *Nucleic Acids Research*, *44*(D1), D463–D470. https://doi.org/10.1093/nar/gkv1042.

Surendran, S., Qassadi, F., Surendran, G., Lilley, D., & Heinrich, M. (2021). Myrcene—What are the potential health benefits of this flavouring and aroma agent? *Frontiers in Nutrition*, *8*(July), 1–14. https://doi.org/10.3389/fnut.2021.699666.

Trofin, I. G., Vlad, C. C., Dabija, G., & Filipescu, L. (2011). Influence of storage conditions on the chemical potency of herbal cannabis. *Revista de Chimie*, *62*(6), 639–645.

Tsugawa, H. (2018). Advances in computational metabolomics and databases deepen the understanding of metabolisms. *Current Opinion in Biotechnology*, *54*, 10–17. https://doi.org/10.1016/j.copbio.2018.01.008.

Vinaixa, M., Schymanski, E. L., Neumann, S., Navarro, M., Salek, R. M., & Yanes, O. (2016). Mass spectral databases for LC/MS- and GC/MS-based metabolomics: State of the field and future prospects. *TrAC, Trends in Analytical Chemistry*, *78*, 23–35. https://doi.org/10.1016/j.trac.2015.09.005.

Welling, M. T., Deseo, M. A., Bacic, A., & Doblin, M. S. (2021). Untargeted metabolomic analyses reveal chemical complexity of dioecious cannabis flowers. *Australian Journal of Chemistry*, *74*(6), 463–479. https://doi.org/10.1071/CH21033.

Wishart, D. S. (2016). Emerging applications of metabolomics in drug discovery and precision medicine. *Nature Reviews Drug Discovery*, *15*(7), 473–484.

Wishart, D. S. (2007). Current progress in computational metabolomics. *Briefings in Bioinformatics*, *8*(5), 279–293. https://doi.org/10.1093/bib/bbm030.

Wishart, D. S. (2008). Quantitative metabolomics using NMR. *Trends in Analytical Chemistry*, *27*(3), 228–237. https://doi.org/10.1016/j.trac.2007.12.001.

Wishart, D. S. (2019). Metabolomics for investigating physiological and pathophysiological processes. *Physiological Reviews*, *99*(4), 1819–1875. https://doi.org/10.1152/physrev.00035.2018.

Wishart, D. S., Guo, A., Oler, E., Wang, F., Anjum, A., Peters, H., Dizon, R., Sayeeda, Z., Tian, S., Lee, B. L., Berjanskii, M., Mah, R., Yamamoto, M., Jovel, J., Torres-Calzada, C., Hiebert-Giesbrecht, M., Lui, V. W., Varshavi, D., Varshavi, D., & Gautam, V. (2022). HMDB 5.0: the Human Metabolome Database for 2022. *Nucleic Acids Research*, *50*, 622–631.

Wishart, D. S., Mandal, R., Stanislaus, A., & Ramirez-Gaona, M. (2016). Cancer metabolomics and the human metabolome database. *Metabolites*, *6* (1). https://doi.org/10.3390/metabo6010010.

Zheng, J., Johnson, M., Mandal, R., & Wishart, D. S. (2021). A comprehensive targeted metabolomics assay for crop plant sample analysis. *Metabolites*, *11* (5). https://doi.org/10.3390/metabo11050303.

Chapter 2

Cannabis sativa roots: Chemical and pharmacological profile

Tarcísio Cícero de Lima Araújo[a], Camila Ribeiro Silva[a], Pedro Guilherme Sousa de Sá[a], and Larissa Araújo Rolim[a,b]

[a]*Central for Analysis of Drugs, Medicines and Food (CAFMA)—Federal University of Vale do São Francisco, Pernambuco, Brazil,* [b]*Department of Pharmacy, Federal University of Vale do São Francisco, José de Sá Maniçoba Avenue, Center, Petrolina, Pernambuco, Brazil*

Abbreviations

ANVISA	Brazilian National Health Surveillance Agency
C. sativa	*Cannabis sativa*
CNS	central nervous system
Cs-CEE	*Cannabis sativa* crude ethanolic extract
Cs-CME	*Cannabis sativa* crude methanolic extract
EAqCs	aqueous Extract of *C. sativa* roots
ESI-IT-MS	electrospray ionization and ion-trap mass spectrometry
GC-MS	gas chromatography coupled mass spectrometry
HPLC-MS	high performance liquid chromatography coupled mass spectrometry
HPLC-PDA	high performance liquid chromatography with photodiode array detector
IR	infrared
LC	classic liquid chromatography
NMR	nuclear magnetic resonance
OECD	organization for economic co-operation and development
TLC	thin layer chromatography

Introduction

Cannabis sativa was part of the therapeutic arsenal of many ancient societies. The Chinese emperor Chen Nung elaborated the first Chinese Pharmacopeia approximately 5000 years ago. *C. sativa* was one of many medicinal plants described there. It was prescribed for the treatment of rheumatism, malaria, and fatigue. It also appeared in the Assyrian clay tablet and in the Ebers papyrus. The reports follow the history of a diverse array of civilizations, showing that this plant have been present for all this time in the society history (Bonini et al., 2018).

C. sativa is a dioecious herbaceous species, female, and male flowers develop on different plants, with morphological similarities between them, but considerable phenotypic variability (Elsohly et al., 2017). Female plants have a greater number of leaves, take longer to flower, and produce greater amounts of chemicals and fibers, whereas male plants are thinner, less leafy, and the flowering period is faster (Petit et al., 2020).

The species has an odd number of leaves with lanceolate leaflets and serrated margins. The stems are usually erect and branched. The flowers of the species are small, imperfect, and accumulate in inflorescences. The seeds are small and each plant can produce from 15 to 1000 seeds. The roots are branched and adapt to the soil and the amount of water available (Barrales-Cureño et al., 2020; Stewart-Wade et al., 2002).

Regarding the roots, the central part present itself as ligneous, with a variable and perpendicular length, with a thickness ranging from 3 to 6 cm, rigid, and with various formats. Commonly, it is easy to find lateral branches leaving this central portion of the root, with thickness ranging from 6 to 20 mm (Silva, 2018). The roots of *C. sativa* are very little explored, with the first mentions of its medicinal usage dated from the 17th century, when the historian Pliny described that the root decoct usage was able to reduce the muscle contractions and relieve joint problems, in addition to be presented as a possible cure for gout and other diseases (Ryz et al., 2017).

Cannabis Use, Neurobiology, Psychology, and Treatment. https://doi.org/10.1016/B978-0-323-89862-1.00001-5

FIG. 1 Brazilian *Cannabis sativa* L. Stem (A); Leaf (B); Inflorescence (C); Fruit (D), and Root (E). *(From Lima, K. S. B. (2018).*

In Brazil, the *Cannabis sativa* was macroscopically described by Rizzo (1972) and Souza et al. (2006) after an analysis of the plant material from a police drug seizing as a characteristic axial root plant in which a sole root gives origin to various secondary roots until they turn themselves into a fibrous, perpendicular, and white-brownish mass (Fig. 1).

The Middle Vale do São Francisco region, comprising the states of Pernambuco and Bahia, is considered a great area of *cannabis* cultivation, and it is very common to find traditional communities (including indigenous populations) that make the empirical usage of *cannabis* root preparations. Previous works carried out with the plant cultivated in this region demonstrated the presence of active chemical compounds in the roots, such as cannabisativine, anhydrocanabissativine, alkaloids, and terpenes. In addition, a first study with the roots of these plants showed potential biological activities such as anti-inflammatory and antiasthmatic effects with minimum toxic effects (Lima et al., 2021; Silva, 2018).

Although uncommon, recent studies with *C. sativa* roots were performed seeking to prove the ethnopharmacologic reports on this plant, once the usage of *cannabis* root-based preparations by societies dates from centuries. Ryz et al. (2017) list in their study diverse pathologic conditions reported as treatable by *cannabis* root extracts, highlighting illnesses such as rheumatic diseases, fever, inflammation, skin burn, tumors, bacterial infections, gastrointestinal disorders, and others.

Chemical aspects of *Cannabis sativa* roots

Few studies are available related to the pharmacological effects of the roots. However, ethnopharmacological studies evidenced medicinal properties of this part of the plant (Lima et al., 2021). Among the active compounds identified in the roots are mainly present: monoterpenes, triterpenoids, alkaloids, steroids, and amides (Table 1) (Elhendawy et al., 2019).

The carvone and dihydrocarvone monoterpenes, responsible for the characteristic mint aroma, were identified by gas-liquid chromatography in the hexane fraction of the root extract of *C. sativa* (Sethi et al., 1977). The main triterpenoids identified and isolated are friedelin-3-one and its correspondent alcohol epifriedelanol, and they were obtained from the root ethanolic extract (Slatkin et al., 1971). Cannabisativine and anhydrocanabisativine stand out among the identified alkaloids reported for the first time in *C. sativa* (Elsohly et al., 1978; Latter et al., 1975; Turner et al., 1976).

The *p*-coumaroiltyramine amide was also reported as a compound isolated from the root ethanolic extract of *C. sativa* (Slatkin et al., 1971; Yamamoto et al., 1991). Steroids such as sitosterol, campsterol, and stigmasterol are also commonly reported (Elhendawy et al., 2019; Latter et al., 1975; Ryz et al., 2017).

TABLE 1 Class of the active compounds identified in the *Cannabis sativa* L.

Molecular formula	Class	Compound	Structural formula
$C_{10}H_{14}O$	Monoterpene	Carvona	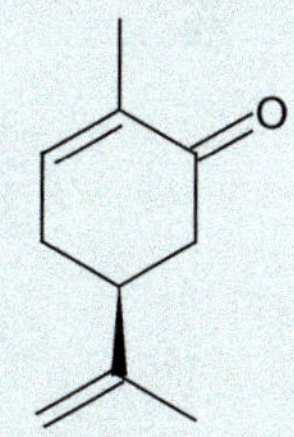
$C_{10}H_{14}O$	Monoterpene	Dihydrocarvone	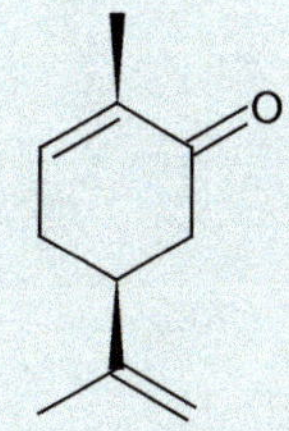
$C_{30}H_{50}O$	Triterpenoids	Friedelan-3-one	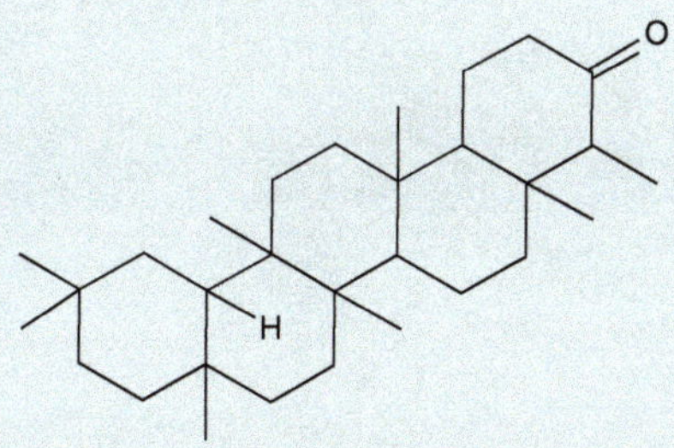
$C_{30}H_{50}O$	Triterpenoids	Friedelin	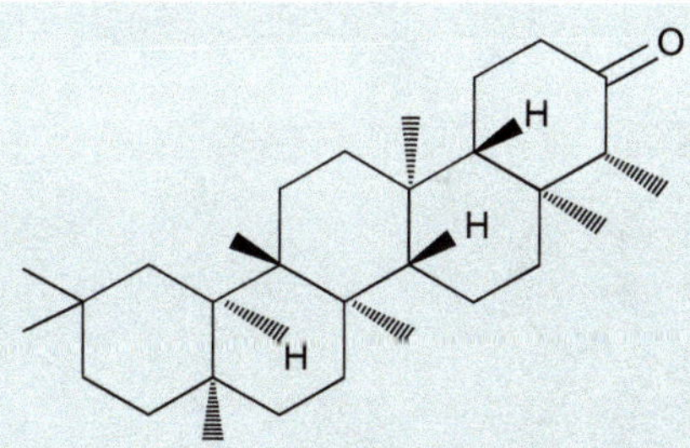
$C_{30}H_{52}O$	Triterpenoids	Epifriedelanol	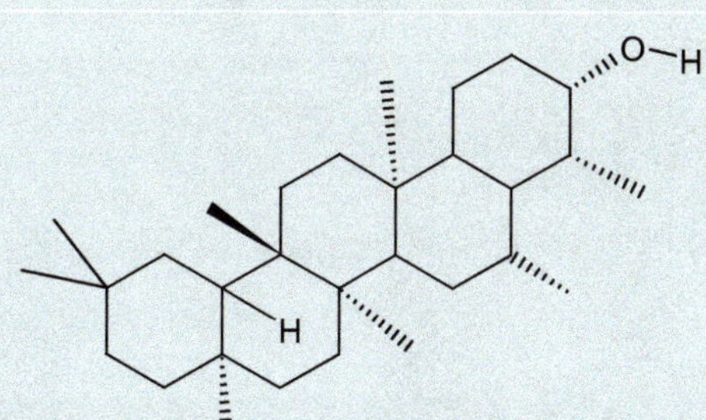
$C_{21}H_{39}N_3O_3$	Alkaloids	Cannabissativine	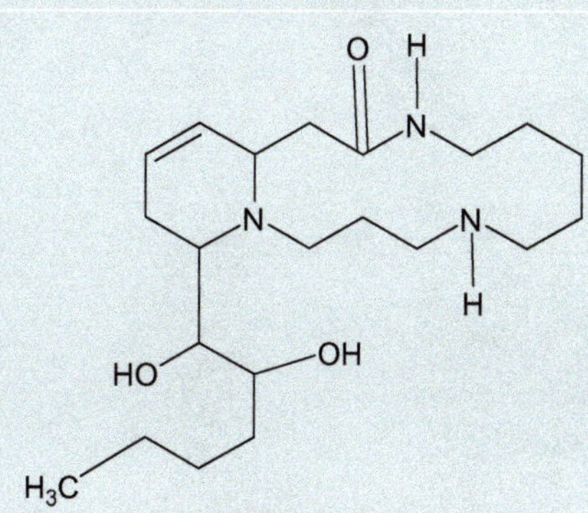

Continued

TABLE 1 Class of the active compounds identified in the *Cannabis sativa* L—cont'd

Molecular formula	Class	Compound	Structural formula
$C_{21}H_{38}N_3O_3$	Alkaloids	Anhydrocannabissativine	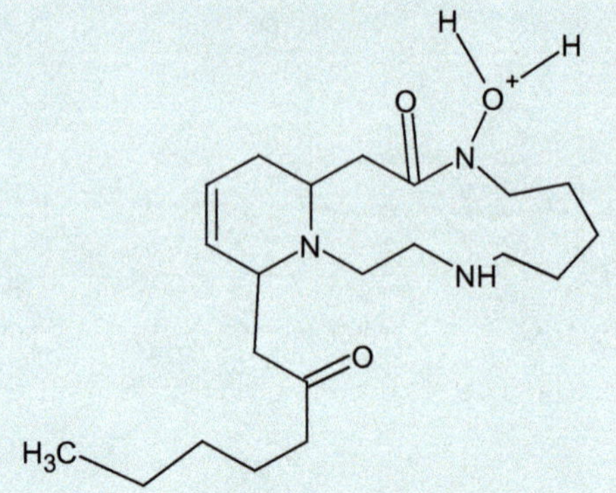
$C_{16}H_{15}NO_3$	Alkaloids	*p*-coumaroyltyramine	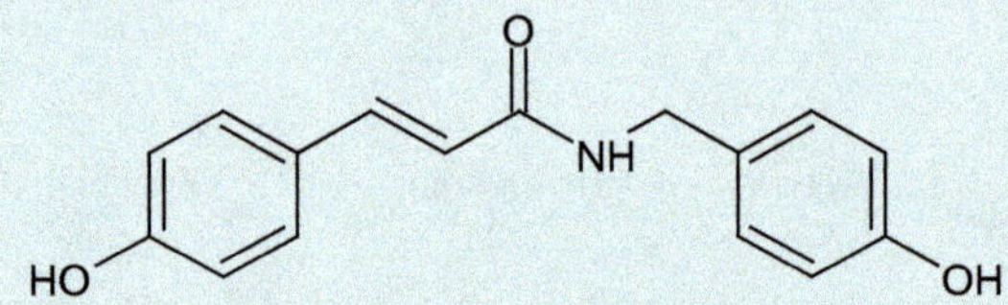
$C_{18}H_{19}NO_4$	Alkaloids	Feruloyltyramine	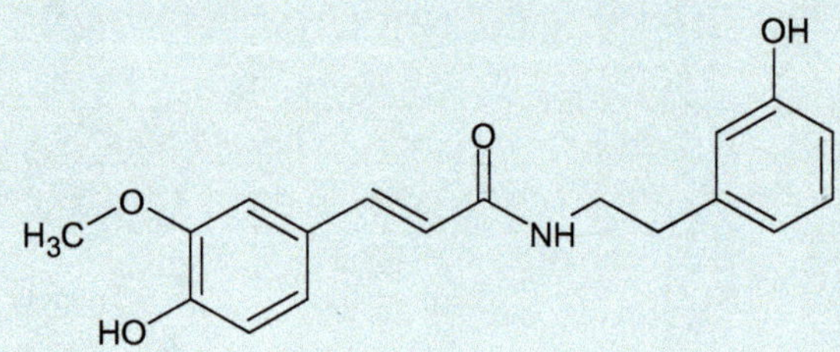
$C_{29}H_{50}O$	Steroids	Sitosterol	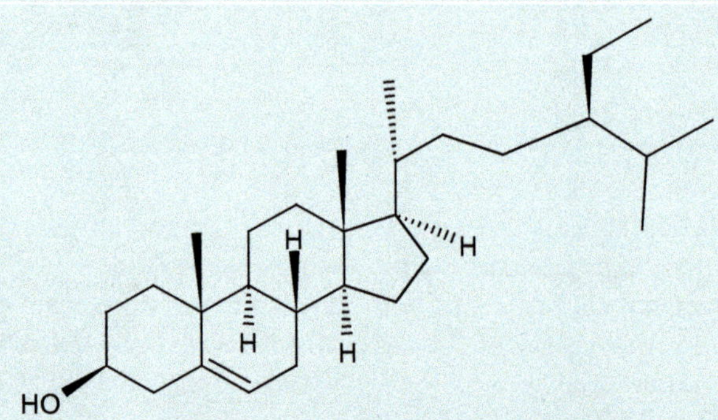
$C_{28}H_{48}O$	Steroids	Campsterol	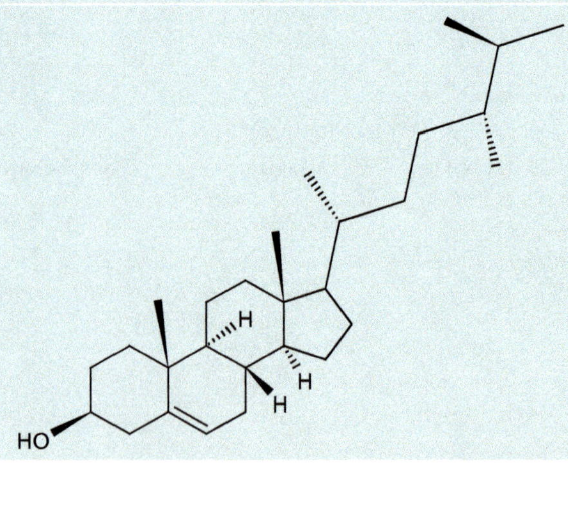

TABLE 1 Class of the active compounds identified in the *Cannabis sativa* L—cont'd

Molecular formula	Class	Compound	Structural formula
$C_{29}H_{48}O$	Steroids	Stigmasterol	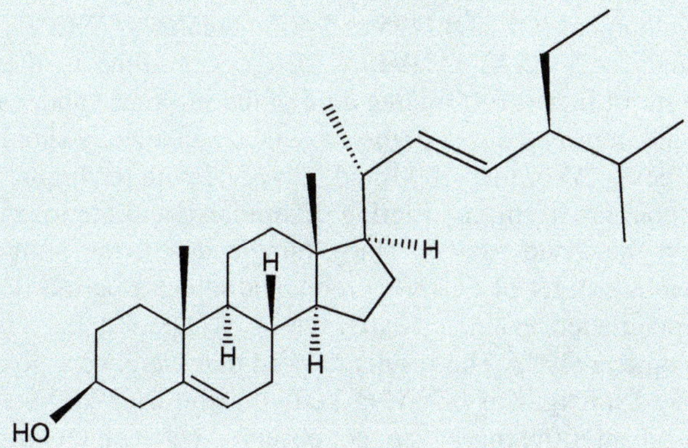

According to Silva (2018), when performing essays described in the Brazilian Pharmacopoeia with four samples of *C. sativa* root through the data obtained by HPLC-MS, the presence of cannabisativine was confirmed in the four samples (Table 1). Through the development of the analysis method by HPLC-PDA, it was verified that to obtain the best hierarchical factor of the chromatographic response, the flow rate should be 0.6 mL/min, the proportion of trifluoroacetic acid of 15%, and temperature of 25°C.

C. sativa root powder is classified as moderately thick, and the phytochemical screening showed that the main secondary metabolites present in it are phenolic compounds, coumarins, anthracenic derivatives, mono-, di, and sesquiterpenes, triterpenes, and steroids. In addition, the presence of Cannabisativine in the root may be confirmed by HPLC-MS (Lima et al., 2021).

Fresh roots can be washed, dried, grinded, and submitted to exhaustive maceration with ethanol, as well methanol, obtaining the *C. sativa* crude ethanolic extract (Cs-CEE) and methanolic extract (Cs-CME) at the end of the process. These extracts, when submitted to a phytochemical screening, may show the presence of alkaloids, phenolic compounds, coumarins, mono, di, and sesquiterpenes, anthracenic derivatives, as well triterpenes and steroids (Neto, 2021).

When the extracts described above were submitted to HPLC-ESI-IT-MS, gas chromatography coupled with mass spectrometry (GC-MS), classic liquid chromatography (LC), thin-layer chromatography (TLC), nuclear magnetic resonance (NMR), and infrared (IR) techniques, it was possible to identify three compounds: *p*-coumaroyltyramine, pheruloyltyramine, and cannabisativine. While GC-MS allowed the identification of the majority of the compounds (friedelan-3-one and β-sitosterol), LC and TLC techniques yielded the isolation of friedelan-3-one and the identification of the mixtures of epifridelanol with friedelan-3-one and β-sitosterol with stigmasterol (Table 1) (Neto, 2021).

Araújo (2021) analyzed an aqueous extract of *C. sativa* roots (EAqCs) by HPLC-MS and evaluated its pharmacological effect against uterine diseases *in vivo* and *in vitro* over rodent uterus. HPLC-MS data suggest that four chemical compounds may be present in the extract: canabissativine, anhydrocanabissativine, pheruloyltyramine, and p-coumaroyltyramine (Table 1). It is worth to mention that the equipment was being operated in the positive mode.

According to Ferrini et al. (2021), by submitting the extracts to chemical characterization and quantification using of GC-MS, it was possible to identify the following compounds: phytosterols, campsterol, stigmasterol, and β-sitosterol and the triterpenes epifriedelanol and friedelin (Table 1).

Toxicological aspects of *Cannabis sativa* roots

In vivo toxicity studies

C. sativa roots have chemical compounds from a diverse array of secondary metabolites, such as terpenes, phenols, and alkaloids (Gul et al., 2018). Therefore, it is essential to assess the ability of the crude extracts of *C. sativa* root to cause toxic effects through the administration to live organisms. These kinds of studies can guarantee the safety of the obtained extract, allowing the continuation of future works.

Toxicity studies should be carried out taking into consideration the standardized protocols well established in the literature, with these directives being created and approved by the regulatory agencies. Also, the wellness and rational use of the animals in *in vivo*, *in vitro*, and *ex vivo* experiments should be observed (ANVISA, 2013).

A study conducted by Lima et al. (2021) used *C. sativa* roots of plants cultivated in the northeastern region of Brazil. The toxicity study was based on the procedures and guidelines approved by the Organization for Economic Co-operation and Development (OECD) (425 and 407 guidelines) (OECD, 2008a, 2008b) and the Brazilian National Health Surveillance Agency (ANVISA) (ANVISA, 2013). According to these guidelines, the oral toxicity study should be performed in two steps: first, a 1000 mg/kg dose of the substance under study should be administered orally in the animals of both sexes; second, repeated doses of the substance under study should be orally administered in animals of both sexes with, at least, two doses lower than 1000 mg/kg/day and with the higher dose being selected under the expectative to cause toxic effects (without the death or suffering of animals), and the lower dose two or four times lower (ANVISA, 2013).

In the acute toxicity study (single dose), the animals were observed right after the dose administration of an aqueous extract of *C. sativa* roots and along a period of 14 days. Body weight and water and food consumption were the parameters evaluated. After the 14 days period, the animals were euthanized, and their organs were collected for macroscopic analysis. The results showed that there were no animal deaths or clinical signs of toxicity and also that animals showed no signs of behavior, body weight, food and water consumption, and size or appearance changes. Statistically significant differences were not observed between the organs' weight of the animals treated and that of the control group (Lima et al., 2021).

In the repeated doses of the toxicity study, the authors used two doses of the aqueous extract of *C. sativa*, which were selected from the results of the acute toxicity study, with the higher dose being 100 mg/kg and the lower dose being 25 mg/kg. The doses were administered orally once a day for 28 days (ANVISA, 2013; OECD, 2008a). After the administration, the animals were periodically observed for toxicity signs such as behavior changes, piloerection, abdominal contractions, aggressivity, and others. At the end of the study, the animals were subjected to fasting, anesthetized for blood, and organs harvesting. The blood collected was analyzed for biochemical and hematological parameters, while the organs were microscopically evaluated in search for tissue alterations.

The obtained results showed that there were no deaths, clinical signs of toxicity, or behavior changes of the animals treated with *C. sativa* root extracts at the doses of 25 and 100 mg/kg, as well there were no significant differences in the relative body weight of the treated animals when compared to the control group. In addition, no alteration in the structure of the tissues was observed, with exception of edema points that were found in the kidneys of the male animals treated with 100 mg/kg of the extract (Fig. 2) (Lima et al., 2021).

It can be concluded that the *C. sativa* roots have low oral toxicity in a single dose or repeated doses (Fig. 2), what stimulates pharmacological studies with this part of the plant once the potential to cause toxic effects *in vivo* at doses lower than 100 mg/kg is very low.

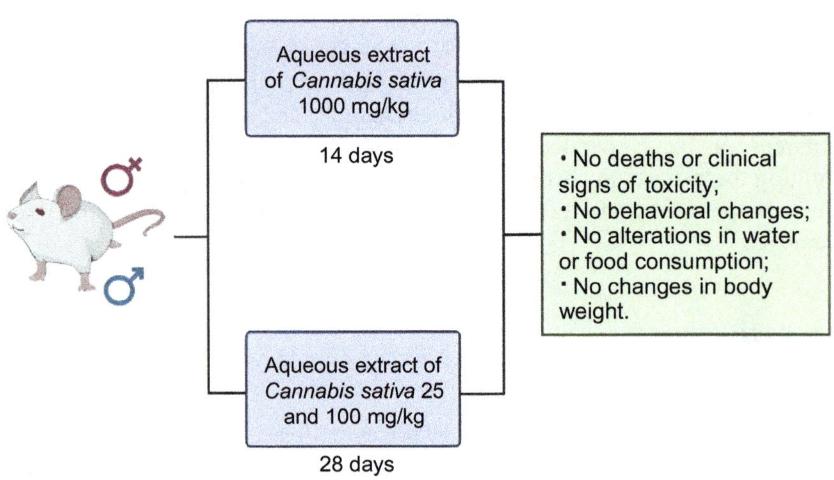

FIG. 2 Acute and repeated dose oral toxicity tests. This figure presents an illustrative scheme of the performance of oral toxicity tests. *(From ANVISA (2013). OECD (2008a). OECD (2008b). Created with BioRender.com).*

Pharmacological aspects of *Cannabis sativa* roots

Until the moment, a wide array of pharmacological effects was attributed to *C. sativa* roots by *in vivo* preclinical and ethnopharmacological studies carried out with the aim to provide a base and justify the medicinal interest in this part of the plant. However, the medicinal properties of the roots were ignored for many years, and for this reason the studies involving the chemical, toxicological, and pharmacological aspects are rare (Lima et al., 2021).

Along history, many empirical uses of *C. sativa* roots were reported in books and manuscripts around the world, mainly based on the oral administration and on the observation of the effects caused. In a systematic review published by Ryz et al. (2017), many biological effects, such as gout treatment, inflammation treatment, skin burn treatment, tumor treatment, emesis, and infections treatment, were reported, with these effects being described by many authors since the 16th century (Ryz et al., 2017).

Recent studies conducted with the *C. sativa* roots from the Brazilian northeastern region represent an advance in the discovery of biological and chemical aspects of the extracts of this part of the plant. In addition, it also provides an analysis of the best administration routes and its safety level, seeking to validate the ethnopharmacological indications and contribute with the diffusion of the knowledge about the roots of this plant, thus stimulating the conduction of new works with the roots of *C. sativa* in the whole world.

Effects on the central nervous system

The *C. sativa* roots have effects over the central nervous system (CNS) of mice without psychotropic activity. Lima et al. (2021) showed promising results regarding the usage of the roots cultivated in the Brazilian northeastern region after evaluating the motor coordination of animals in the rotarod test (described by Dunham & Miya, 1957), following the administration of 100 mg/kg of *C. sativa* roots extract. The number of fallings from the rotatory rod as well the permanence time were recorded (Fig. 3).

The authors conducted the open filed test (described by Prut & Belzung, 2003) for the evaluation of the exploratory and locomotor functions of the animals. At 100 mg/kg (oral route), crossed quadrants, self-cleaning behavior, standing on hind

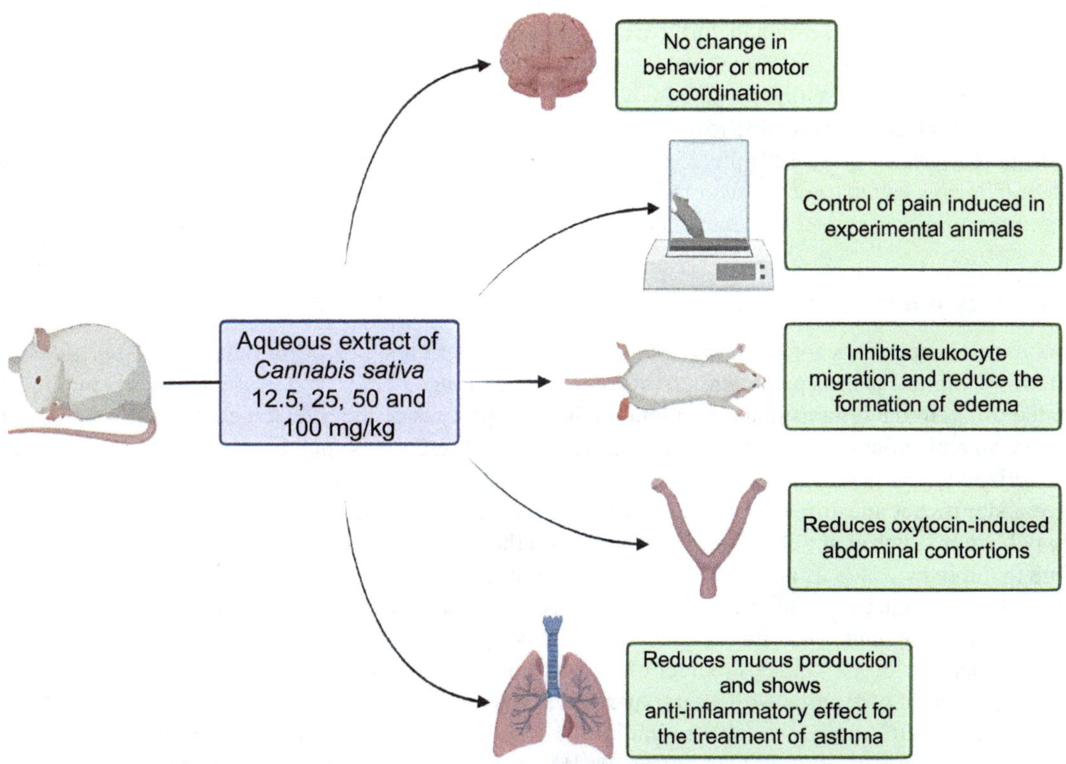

FIG. 3 Pharmacological effects presented by the roots of *Cannabis sativa*. This figure presents a schematic illustrating the performance of pharmacological tests in murine models. *(From Araújo, T. C. L. (2021). Lima et al. (2021), Menezes et al., (2021). Created with BioRender.com).*

legs behavior, and time of immobility were the evaluated parameters. For both protocols, the results showed that no animal presented any alteration of the behavior or motor coordination when compared to a negative control group (Lima et al., 2017), corroborating the fact that the concentrations of the classic phytocannabinoids in the roots are very low or even nonexistent, mainly THC, which is responsible for the psychoactive effects of the plant (Potter, 2009).

Antinociceptive effect

The nociception consists in the activity produced by the nervous system in response to a determined harmful or painful stimulus, different from pain, that involve a series of subjective factors inherent to the individual (Raja et al., 2020). Therefore, the nociception is presented as an initial and primordial parameter in the process of pain evaluation.

Menezes et al. (2021) conducted a study with the roots of *C. sativa* cultivated in Brazilian northeastern region. The authors provided data related to the capacity of this part of the plant to promote effect in the reduction of the mensurable painful stimuli in mice. For this, the abdominal contortions induced by acetic acid protocol, which is a test for evaluating central and/or peripheral analgesic activities, were employed using three doses of an aqueous extract of *C. sativa* roots (oral route). The contortions were observed for a period of 20 min after an intraperitoneal injection of acetic acid, observing signs as core rotations and extensions of the hind limbs. The three doses tested demonstrated significant activity; however, the dose of 12.5 mg/kg presented a great potential of reduction of the abdominal contortions, acting as a minimizing agent of the behavior related to acetic acid-induced pain (Fig. 3) (Menezes et al., 2021; Whittle, 1964; Yin et al., 2016).

Additionally, the formalin-induced nociception test was employed as a complementary protocol of antinociceptive activity. In this test, the nociception is presented in two phases, with two different mechanisms: the first phase is called neurogenic phase because it consists in the direct stimulation of the afferent nociceptive fibers (especially the C kind fibers); the second phase is called inflammatory phase, and it is initiated by the release of the chemical inflammatory mediators in the tissue related to the peripheric nervous system function (Gonçalves et al., 2008). The authors tested three doses of the aqueous extract of the *C. sativa* roots in mice measuring the time that the animals spent licking, shaking, and/or biting the injured paw in both phases (Liu et al., 2021). The 12.5 and 50 mg/kg doses showed a statistically significant effect in the nociceptive behavior in the first phase, while in the second phase they were able to reduce the reactivity time, showing a significant effect against the inflammatory pain (Menezes et al., 2021).

The hot plate test, widely employed as an acute neuropathic pain model stimulated by temperature (Deng et al., 2021), was used to establish the *C. sativa* root aqueous extract effect on the thermal hyperalgesia. Initially, the authors evaluated the 12.5, 25, and 50 mg/kg doses (oral route), subjecting the animals to the thermal stress after 30, 60, 90, and 120 min of extract administration. The latency time, which is the time it takes for a painful reaction by the animal, was the assessed parameter. As a result, all doses were able to significantly increase the reactivity time of the animals at 60 min (12.5 and 50 mg/kg), 90 min (12.5 mg/kg), and 120 min (25 mg/kg) when compared to the negative control group. This study suggests that the roots of *C. sativa* can control the pain induced in animal models based on the results observed in the protocols of nociception and thermal hyperalgesia (Menezes et al., 2021).

Anti-inflammatory effect

Inflammation, a vital mechanism for the body homeostasis, consists in a series of immunological reactions generated in response to harmful stimuli, such as pathogens, toxic compounds, strange bodies, and tissue injuries, actuating in such form to neutralize the threat to the organism and promote the healing process of the affected areas (Chen et al., 2018). The *C. sativa* roots have an anti-inflammatory effect well described in the literature, which is related to the active chemical compounds characterized in this part of the plant, once alkaloids, phytosteroids, friedelin, *p*-coumaroyltyramine, and feruloyltyramine have a significant anti-inflammatory effect (Ryz et al., 2017; Yamamoto et al., 1991).

In a study publish by Lima et al. (2021) and conducted with the roots of C. sativa from the Brazilian northeastern region, it was possible to put in evidence its anti-inflammatory effect, observed through preclinical methods of anti-inflammatory activity. The protocol of leucocytes migration induced by carrageenan evaluates the ability of the substance under study inhibiting the process of immune cells migration, such as leucocytes, to the inflammation site. The authors evaluated the anti-inflammatory activity of the aqueous extract of *C. sativa* at the doses of 25, 50, and 100 mg/kg after subjecting the animals to the induction of peritonitis with an intraperitoneal injection of carrageenan (Andrade et al., 2011). The results showed that the extract of *C. sativa* was able to inhibit the leucocytes migration at all doses (Fig. 3).

In addition, the carrageenan-induced paw edema model, employed as an initial method of evaluation, was conducted with the administration of four doses (12.5, 25, 50, and 100 mg/kg; one dose per group) of the *C. sativa* aqueous extract and an intraplantar injection of carrageenan for inducting the local inflammatory process and formation of edemas. The results, obtained after measuring the volume of the paws in a specific apparatus, demonstrated that all doses presented an effect of paw edema reduction, even in the lowest dose of 12.5 mg/kg, and this effect lasted for 4 h (Lima et al., 2021).

Anti-dysmenorrheic effect

The primary dysmenorrhea is characterized by a painful sensation that is established in women in childbearing age with diverse causes and a mechanism not totally elucidated (Sanctis et al., 2015). In a study conducted by Araújo (2021), it was possible to determine the effect of *C. sativa* roots aqueous extract on the abdominal contortions induced by oxytocin (0.4 UI) 30 min after the oral administration of the treatment (12.5, 25, and 50 mg/kg of extract).

Then, the observation continued with the registering of the contortions, which are characterized by the contraction of the abdominal wall accompanied by pelvic rotation, for a period of 30 min. The obtained results demonstrated that the roots of *C. sativa* have a reduction effect on the abdominal contortions induced by oxytocin at the doses of 25 and 50 mg/kg, inhibiting the contortions in 85% and 76%, respectively (Fig. 3) (Araújo, 2021).

Anti-asthmatic effect

The asthmatic condition consists in the appearing of allergic symptoms that affect the respiratory system producing oscillating respiration, caught, shortness of breath, and chest tightness, affecting mainly children (Reed, 2006; Vasconcelos et al., 2008). The preclinical model of asthma induction, with intraperitoneal injection of ovalbumin, is capable of providing data related to the behavior of the experimental animals in addition to the possibility of analyzing hematological parameters and cells differential counting through the blood and bronchoalveolar fluid harvesting at the end of the experiment, as well the microscopic evaluation of the lung tissue. The authors orally administered three doses of *C. sativa* root extract from the Brazilian northeastern region (12.5, 25, and 50 mg/kg) to the animals, and the experiment lasted for 26 days (Menezes et al., 2021).

The obtained results presented data related to a reduction in the total counting of leucocytes of the treated animals with the 25 mg/kg dose of the extract, decreasing the cells migration. In addition, all doses were able to reduce the differential counting of eosinophils in the bronchoalveolar fluid. In the lungs histopathological analyses, the 50 mg/kg dose was able to minimize the presence of inflammatory cells infiltrating the lung tissue, a dexamethasone-like effect (Fig. 3). In addition, all three doses were able to significatively reduce the mucus production in the animals immunized with ovalbumin, suggesting the roots of *C. sativa* as a potential candidate with anti-inflammatory effect for the treatment of asthma (Menezes et al., 2021).

Final considerations

Obtaining the chemical profile of the *C. sativa* roots revealed the presence of monoterpenes, triterpenoids, alkaloids, and steroids after subjecting the aqueous extract to the HPLC-MS analyses. In addition, the pentacyclic triterpenes friedelan-3-one and β-sitosterol were identified as the major compounds of the ethanolic and methanolic extracts by GC-MS and were reported in the roots of *C. sativa* cultivated in the Vale do São Francisco region for the first time.

Considering the rareness of studies carried out with *C. sativa* roots, the presented works represent an originality related to the medicinal potential of this part of the plant. Data referring to the oral toxicity, acute as well subchronic, demonstrated that, in the tested doses, *C. sativa* produced no clinically observed toxic effect in animal models, encouraging the continuity of studies with the plant roots.

In relation to the pharmacological aspects related to the roots of *C. sativa*, promising effects of the aqueous extract, when orally administered in mice, were presented, demonstrating an antinociceptive, anti-inflammatory, anti-dysmenorrheic, and anti-asthmatic effects, causing no toxic effect to the central nervous system in the methods employed. This study represents the first therapeutic evidences of the use of the *C. sativa* roots cultivated in the Vale do São Francisco region, in the Brazilian northeastern region, assuring its safety, efficacy, and confirming the ethnopharmacological indications described in the literature.

Other areas of application

The usage of *C. sativa* roots with medical purposes has fundamental empirical importance, taking into consideration the therapeutic indications passed down through generations. In addition, reported recommendations by physicians and herbalists in old writings (dated from the 17th century) demonstrated that the roots were clinically employed in the treatment of many ills such as gout, rheumatic diseases, and burns (Ryz et al., 2017).

Despite the empirical reports presenting data related to the applications of the extracts and cataplasms for the treatment of many diseases, until this moment no clinical study was conducted with the roots of *C. sativa* aiming to prove the ethnopharmacological indications. Studies on the chemical composition and pharmacological effects of the plant roots have been carried out all over the world, especially in the last years, demonstrating the importance of the preliminary studies, always considering the report on the ethnopharmacological usage (Elhendawy et al., 2019; Gul et al., 2018; Ryz et al., 2017).

The data presented in this chapter may support future research involving the roots of *C. sativa* cultivated in the Brazilian northeastern region, as well as enable, through the dissemination of the knowledge, the reproducibility of the protocols for a wide array of *C. sativa* cultivars worldwide (Araújo, 2021; Lima et al., 2021; Menezes et al., 2021).

The clinical applicability of the experimental data presented in this chapter is, surely, the next step of the studies, considering the ethnopharmacological reports and therapeutical indications of the roots along the history. Obviously, theses future steps should be preceded by safety evaluation, quality control of the raw materials, in addition to preformulation studies with the root extracts, compatibility and stability studies, aiming to obtain a robust, effective, and safe product from the medical and pharmaceutical viewpoint.

Mini-dictionary of terms

- Ethnopharmacology: The science dedicated to the study of the folk knowledge about the usage of medicinal plants and their correlation with the technical-scientific knowledge.
- Medium Vale do São Francisco: Region at the São Francisco River edge, in the semi-arid region of the Brazilian northeastern region, comprising the states of Bahia and Pernambuco.
- Clinical signs of animal toxicity: Set of signs and symptoms presented by animals after the administrations of substances with toxic potential, including piloerection, depression, shivering, loss of motor coordination, and others.
- Abdominal contortions: Painful responses induced by the administration of substances with central or peripheral actions characterized by trunk rotation and extension of the hind limbs of the animals.
- Nociception: Process characterized by the physiological steps of pain, including transmission and modulation of painful signals from external stimuli.
- Dysmenorrhea: Normal painful response characterized by cramps in the abdominal region of women in childbearing age, appearing in the beginning of the menstrual period.

Key facts of marijuana roots

- The root structure of *Cannabis sativa* has many branches, connected to a thicker stem.
- Empirically, the roots are used in the form of infusions.
- In Vale do São Francisco, the roots are popularly used to treat inflammation, arthritis, and other diseases.
- In the chemical composition of the roots, there is an absence of classical cannabinoids and the presence of several classes of secondary metabolites.
- Pharmacological studies with the roots are scarce, with first reports in the literature dating back to 1971.

Summary points

- The *C. sativa* roots from the Brazilian northeastern region present chemical compounds from pentacyclic triterpenoids, alkaloids, steroids, and class amides.
- Single and repeated doses oral toxicity studies have demonstrated that the aqueous extract of the roots does not produce toxic effects in mice.
- Toxic effects on the central nervous system were not observed in the animals.
- The roots presented a significant anti-inflammatory effect in animal models of acute and chronic inflammation, as well as reduced abdominal contortions in the animals treated with oxytocin in the primary dysmenorrhea animal model.
- The antinociceptive and antiasthmatic effects were observed from the inflammatory pain and ovalbumin-induced asthma protocols, respectively.
- This chapter will support new studies with *C. sativa* roots, exploring all its medicinal potential.

References

Andrade, G. S., Guimarães, A. G., Santana, M. T., Siqueira, R. S., Passos, L. O., Machado, S. M. F., Ribeiro, A. S., Sobral, M., Almeida, J. R. G. S., & Quintans-Júnior, L. J. (2011). Phytochemical screening, antinociceptive and anti-inflammatory effects of the essential oil of *Myrcia pubiflora* in mice. *Revista Brasileira de Farmacognosia*, *22*(1), 181–188. https://doi.org/10.1590/S0102-695X2011005000205.

ANVISA. (2013). Guia para a condução de estudos não clínicos de toxicologia e segurança farmacológica necessários ao desenvolvimento de medicamentos. *Agência de Vigilância Sanitária (ANVISA).*, 1–48. https://www.gov.br/anvisa/pt-br/centraisdeconteudo/publicacoes/medicamentos/publicacoes-sobre-medicamentos/guia-para-a-conducao-de-estudos-nao-clinicos-de-toxicologia-e-seguranca-farmacologica-necessarios-ao-desenvolvimento-de-medicamentos.pdf/view.

Araújo, T. C. L. (2021). *Estudo químico e farmacológico, em útero de roedores, do extrato aquoso das raízes de Cannabis sativa L* (Unpublished master thesis). Petrolina, Brazil: Federal University of San Francisco Valley.

Barrales-Cureño, H. J., Lópes-Valdez, L. G., Reyes, C., Cetina-Alcalá, V. M., Vasquez-García, I., Diaz-Lira, O. F., & Herrera-Cabrera, B. E. (2020). Chemical characteristics, therapeutic uses, and legal aspects of the cannabinoids of *Cannabis sativa*: A review. *Brazilian Archives of Biology and Technology*, *63*(e20190222), 1–14. https://doi.org/10.1590/1678-4324-2020190222.

Bonini, S. A., Premoli, M., Tambaro, S., Kumar, A., Maccarinelli, G., Memo, M., & Mastinu, A. (2018). *Cannabis sativa*: A comprehensive ethnopharmacological review of a medicinal plant with a long history. *Journal of Ethnopharmacology*, *227*, 300–315. https://doi.org/10.1016/j.jep.2018.09.004.

Chen, L., Deng, H., Cui, H., Fang, J., Zuo, Z., Deng, J., Li, Y., Wang, X., & Zhao, L. (2018). Oncotarget 7204 www.impactjournals.com/oncotarget inflammatory responses and inflammation-associated diseases in organs. *Oncotarget*, *9*(6), 7204–7218. www.impactjournals.com/oncotarget/.

Deng, J., Han, J., Chen, J., Zhang, Y., Huang, Q., Wang, Y., Qi, X., Liu, Z., Leung, E. L. H., Wang, D., Feng, Q., & Lu, L. (2021). Comparison of analgesic activities of aconitine in different mice pain models. *PLoS One*, *16*, 1–10. https://doi.org/10.1371/journal.pone.0249276.

Dunham, N. W., & Miya, T. S. (1957). A note on a simple apparatus for detecting neurological deficit in rats and mice. college of pharmacy, University of Nebraska, Lincoln 8. *Journal of the American Pharmaceutical Association (Scientific Ed.)*, *46*(3), 208–209. https://doi.org/10.1002/jps.3030460322.

Elhendawy, M. A., Wanas, A. S., Radwan, M. M., Azzaz, N. A., Toson, E. S., & ElSohly, M. A. (2019). Chemical and biological studies of *Cannabis sativa* roots. *Medical Cannabis and Cannabinoids*, *1*(2), 104–111. https://doi.org/10.1159/000495582.

Elsohly, M. A., Radwan, M. M., Gul, W., Chandra, S., & Galal, A. (2017). Phytochemistry of *Cannabis sativa* L. *Progress in the Chemistry of Organic Natural Products*, *103*, 1–36. https://doi.org/10.1007/978-3-319-45541-9_1.

Elsohly, M. A., Turner, C. E., Phoebe-Jr, C. H., Knapp, J. E., Schiff-Jr, P. L., & Slatkin, D. J. (1978). Anhydrocannabisativine, a new alkaloid from *Cannabis sativa* L. *Journal of Pharmaceutical Sciences*, *67*(1), 124. https://doi.org/10.1002/jps.2600670135.

Ferrini, F., Fraternale, D., Zeppa, S. D., Verardo, G., Gorassini, A., Carrabs, V., Albertini, M. C., & Sestili, P. (2021). Yield, characterization and possible exploitation of *Cannabis Sativa* L. roots grown under aeroponics cultivation. *Preprints*, 1–24. https://doi.org/10.20944/preprints202106.0532.v1.

Gonçalves, J. C. R., Oliveira, F. D. S., Benedito, R. B., de Sousa, D. P., Almeida, R. N., & Araújo, D. A. M. (2008). Antinociceptive activity of (-)-carvone: Evidence of association with decreased peripheral nerve excitability. *Biological and Pharmaceutical Bulletin*, *31*(5), 1017–1020. https://doi.org/10.1248/bpb.31.1017.

Gul, W., Gul, S. W., Chandra, S., Lata, H., Ibrahim, E. A., & Elsohly, M. A. (2018). Detection and quantification of cannabinoids in extracts of *Cannabis sativa* roots using LC-MS/MS. *Planta Medica*, *84*(4), 267–271. https://doi.org/10.1055/s-0044-100798.

Latter, H. L., Abraham, D. J., Turner, C. E., Knapp, J. E., Schiff-Jr, P. L., & Slatkin, D. J. (1975). Cannabisativine, a new alkaloid from *Cannabis sativa* L. root. *Tetrahedron Letters*, *33*(16), 2815–2818. https://doi.org/10.1016/S0040-4039(00)75003-9.

Lima, K. S. B., Silva, M. E. G. C., Araújo, T. C. L., Silva, C. P. F., Santos, B. L., Ribeiro, L. A. A., Menezes, P. M. N., Silva, M. G., Lavor, É. M., Silva, F. S., Nunes, X. P., & Rolim, L. A. (2021). *Cannabis* roots: Pharmacological and toxicological studies in mice. *Journal of Ethnopharmacology*, *271*. https://doi.org/10.1016/j.jep.2021.113868.

Liu, X., Zhang, J. T., Hu, Y., Shan, W. Q., Wang, Z. H., Fu, Q. Y., Fu, D. N., Ji, J., & Liu, T. (2021). Formalin itch test: Low-dose formalin induces histamine-independent, TRPA1-mediated itch in mice. *Frontiers in Medicine*, *8*, 1–13. https://doi.org/10.3389/fmed.2021.627725.

Menezes, P. M. N., Araújo, T. C. L., Pereira, E. C. V., Neto, J. A., Silva, D. S., Brito, M. C., Lima, K. S. B., Monte, A. P. O., Matos, M. H. T., Lima, R. S., Ribeiro, L. A. A., Silva, F. S., & Rolim, L. A. (2021). Investigation of antinociceptive, antipyretic, antiasthmatic, and spasmolytic activities of Brazilian *Cannabis sativa* L. roots in rodents. *Journal of Ethnopharmacology*, *278*. https://doi.org/10.1016/j.jep.2021.114259.

Neto, J. A. (2021). *Isolamento, purificação e caracterização dos constituintes químicos da Cannabis sativa L. do semiárido brasileiro* (Unpublished master thesis). Petrolina, Brazil: Federal University of San Francisco Valley.

OECD. (2008a). Guidelines for the testing of chemicals. Guideline 407. Repeated Dose 28-day Oral Toxicity Study in Rodents. *Drug and Chemical Toxicology*, *34*(1), 13. http://www.oecd-ilibrary.org/environment/test-no-407-repeated-dose-28-day-oral-toxicity-study-in-rodents_9789264070684-en.

OECD. (2008b). *Guidelines for the testing of chemicals. Guideline 425. Acute Oral Toxicity–Up-and-Down-Procedure. Test No. 404: Acute Dermal Irritation/Corrosion, October*. https://doi.org/10.1787/9789264070622-en.

Petit, L., Salentijn, E. M. J., Paulo, M., Thouminot, C., Van-Dinter, B. J., Magagnini, G., Gusovius, H., Tang, K., Amaducci, S., Wang, S., Uhrlaub, B., Müssig, J., & Trindade, L. M. (2020). Genetic variability of morphological, flowering, and biomass quality traits in Hemp (*Cannabis sativa* L.). *Frontiers in Plant Science*, *11*(102), 1–17. https://doi.org/10.3389/fpls.2020.00102.

Potter, J. P. D. (2009). *The propagation, characterisation and optimisation of Cannabis Sativa L as phytopharmaceutical. (Doktorska disertacija)*. London, Department of Pharmaceutical Science Research, King's College London: 224 str. March, 224.

Prut, L., & Belzung, C. (2003). The open field as a paradigm to measure the effects of drugs on anxiety-like behaviors: A review. *European Journal of Pharmacology*, *463*(1–3), 3–33. https://doi.org/10.1016/S0014-2999(03)01272-X.

Raja, S. N., Carr, D. B., Cohen, M., Finnerup, N. B., Flor, H., Gibson, S., Keefe, F. J., Mogil, J. S., Ringkamp, M., Sluka, K. A., Song, X. J., Stevens, B., Sullivan, M. D., Tutelman, P. R., Ushida, T., & Vader, K. (2020). The revised International Association for the Study of Pain definition of pain: Concepts, challenges, and compromises. *Pain, 161*(9), 1976–1982. https://doi.org/10.1097/j.pain.0000000000001939.

Reed, C. E. (2006). The natural history of asthma. *Journal of Allergy and Clinical Immunology, 118*(3), 543–548. https://doi.org/10.1016/j.jaci.2006.06.020.

Rizzo, J. A. (1972). *Cannabis sativa* L. (Maconha). *Journal of Tropical Pathology, 1*(3). https://doi.org/10.5216/rpt.v1i3.23221.

Ryz, N. R., Remillard, D. J., & Russo, E. B. (2017). *Cannabis* roots: A traditional therapy with future potential for treating inflammation and pain. *Cannabis and Cannabinoid Research, 2*(1), 210–216. https://doi.org/10.1089/can.2017.0028.

Sanctis, V., Soliman, A., Bernasconi, S., Bianchin, L., Bona, G., Bozzola, M., Buzi, F., de Sanctis, C., Tonini, G., Rigon, F., & Perissinotto, E. (2015). Primary dysmenorrhea in adolescents: Prevalence, impact and recent knowledge. *Pediatric Endocrinology Reviews, 13*(2), 512–520.

Sethi, V. K., Jain, M. P., & Thakur, R. S. (1977). Chemical investigation of wild *Cannabis sativa* L. roots. *Planta medica, 32*(4), 378–379. https://doi.org/10.1055/s-0028-1097617.

Silva, M. E. G. C. (2018). *Caracterização físico-química e perfil químico dos compostos presentes nas raízes da Cannabis sativa L* (Unpublished master thesis). Petrolina, Brazil: Federal University of San Francisco Valley.

Slatkin, D. J., Doorenbos, N. J., Harris, L. S., Masoud, A. N., Quimby, M. W., & Schiff-Jr, P. L. (1971). Chemical constituents of *Cannabis sativa* L. root. *Journal of Pharmaceutical Sciences, 60*(12), 1887–1892. https://doi.org/10.1002/jps.2600601232.

Souza, D. Z., Michellin, K., Holler, M. G., Soares, G. L. G., Ritter, M. R., & Bianchi, N. R. (2006). Roteiro ilustrado para identificação morfológica da *Cannabis sativa* L. *Revista Perícia Federal, 24*, 16–22. https://apcf.org.br/wp-content/uploads/2020/06/24.pdf.

Stewart-Wade, S. M., Neumann, S., Collins, L. L., & Boland, G. J. (2002). The biology of Canadian weeds. 117. *Taraxacum officinale* G. H. Weber ex Wiggers. *Canadian Journal of Plant Science, 82*, 825–853. https://doi.org/10.4141/P01-010.

Turner, C. E., Hsu, M.-F. H., Knapp, J. E., Schiff Jr., P. L., & Slatkin, D. J. (1976). Isolation of cannabisativine, an alkaloid, from *Cannabis sativa* L. root. *Journal of Pharmaceutical Sciences, 65*(7), 1084–1085. https://doi.org/10.1002/jps.2600650736.

Vasconcelos, J. F., Teixeira, M. M., Barbosa-Filho, J. M., Lúcio, A. S. S. C., Almeida, J. R. G. S., de Queiroz, L. P., Ribeiro-dos-Santos, R., & Soares, M. B. P. (2008). Corrigendum to "The triterpenoid lupeol attenuates allergic airway inflammation in a murine model". *International Immunopharmacology, 8*(12), 1714. https://doi.org/10.1016/j.intimp.2008.07.008.

Whittle, B. A. (1964). Release of a kinin by intraperitoneal injection of chemical agents in mice. *International Journal of Neuropharmacology, 3*, 369–378.

Yamamoto, I., Matsunaga, T., Kobayashi, H., Watanabe, K., & Yoshimura, H. (1991). Analysis and pharmacotoxicity of feruloyltyramine as a new constituent and p-coumaroyltyramine in *Cannabis sativa* L. *Pharmacology, Biochemistry and Behavior, 40*(3), 465–469. https://doi.org/10.1016/0091-3057(91)90348-6.

Yin, Z. Y., Li, L., Chu, S. S., Sun, Q., Ma, Z. L., & Gu, X. P. (2016). Antinociceptive effects of dehydrocorydaline in mouse models of inflammatory pain involve the opioid receptor and inflammatory cytokines. *Scientific Reports, 6*(321), 1–9. https://doi.org/10.1038/srep27129.

Chapter 3

Cannabis-related compounds in beverages and food

Ilaria Di Marco Pisciottano
Department of Chemistry, Istituto Zooprofilattico Sperimentale del Mezzogiorno, Naples, Italy

Abbreviations

ARfD	acute reference dose
CBC	cannabichromene
CBCA	cannabichromenic acid
CBD	cannabidiol
CBDA	cannabidiolic acid
CBDV	cannabidivarin
CBDVA	cannabidivarinic acid
CBG	cannabigerol
CBGA	cannabigerolic acid
CBN	cannabinol
DAD	diode array detector
EFSA	European Food Safety Authority
FAO	Food and Agriculture Organization of the United Nations
GC	gas chromatography
GC-FID	gas chromatography coupled to flame ionization detector
GC-MS	gas chromatography coupled to mass spectrometry
HPLC-DAD	high-performance liquid chromatography coupled to diode array detector
HPLC-FLD	high-performance liquid chromatography coupled to fluorescence detector
HPLC-MS/MS	high-performance liquid chromatography coupled to tandem mass spectrometry
HPLC-UV	high-performance liquid Chromatography coupled to ultraviolet detector
HPLC-UV-DAD	high-performance liquid chromatography coupled to ultraviolet and diode array detector
LC-MS	liquid chromatography coupled to mass spectrometry
LC-MS/MS	liquid chromatography coupled to tandem mass spectrometry
LOD	limit of detection
LOQ	limit of quantification
nd	not detected
NIR	near-infrared
THC	tetrahydrocannabinol
THCA	tetrahydrocannabinolic acid
THCA-A	tetrahydrocannabinolic acid A
THCA-B	tetrahydrocannabinolic acid B
THC-COOH	11-nor-9-carboxy-Δ^9-THC
THC-OH	11-hidroxy-Δ^9-THC
THCV	tetrahydrocannabivarin
THCVA	tetrahydrocannabivarin acid
TLC	thin-layer chromatography
UHPLC-HRMS	ultra high-performance chromatography coupled to high resolution mass spectrometry
UV	ultraviolet detector
Δ^8-THC	Δ^8-tetrahydrocannabinol
Δ^9-THC	Δ^9-tetrahydrocannabinol

Cannabis Use, Neurobiology, Psychology, and Treatment. https://doi.org/10.1016/B978-0-323-89862-1.00020-9

Introduction

Nowadays, *Cannabis sativa* L., also known as industrial hemp, has a wide panel of applications, such as in textile and paper industry, as well as in nutrition, medicine, energy production, and cosmetic industry (Crini et al., 2020; Das et al., 2017; Ranalli & Venturi, 2004). The interest in hemp seeds as food and supplements ingredient has grown through the years because of their acknowledged nutritional properties. In fact, they have a high content in essential fatty acids, minerals, vitamins, and fibers, as well as essential amino acids embedded in highly digestible proteins (Callaway, 2004; Chen et al., 2010). Moreover, among the more than 100 phytocannabinoids produced by hemp (Hanuš et al., 2016), CBD has been suggested as beneficial for human health because of its antiinflammatory, analgesic, anxiolytic, antipsychotic, spasmolytic, as well as anticonvulsant effects (Ellis et al., 2018; Hobbs et al., 2020; Khan et al., 2020; Lazarini-Lopes et al., 2020; Mechoulam et al., 2002; Wong & Cairns, 2019). Following that, hemp seeds and its by-products, such as hempseed flour and oil, have increasingly been used as food and ingredients for food, beverages, and supplements because of the nutritional properties of hemp but also because of the claimed health benefits of CBD. In fact, a lot of hemp-containing products claim a high CBD content or to even be enriched with CBD to boost its effects, thus being more tempting for the consumers. However, the consumption of hemp-derived products could represent a risk for public health because of the undesired content of Δ^9-THC and other psychoactive cannabinoids naturally present in hemp. The potential risk for the consumers' health has led worldwide to investigate the cannabinoid content in hemp seeds, hempseed oil, hempseed flour, and dietary products containing hemp as ingredient. This review aims to report the results of all the monitoring studies performed to determine Δ^9-THC and the other relevant cannabinoids in food and beverages since today.

Analytical methods for determination of cannabinoids in dietary products

For a long time, the gas chromatography coupled to mass spectrometry detection (GC-MS) has been the analytical technique of choice for the determination of mainly Δ^9-THC and its acidic precursor THCA-A. However, the major issue of this technique, when it comes to cannabinoid detection, is that it cannot effectively distinguish Δ^9-THC from THCA-A. In the Cannabis plants, in fact, most of the cannabinoids are initially synthesized as carboxylic acids, such as THCA-A, CBDA, and CBGA. Then, these nonpsychoactive precursors are decarboxylated and converted into their neutral forms, that is, Δ^9-THC, CBD, and CBG, respectively, by drying, heating, combustion, or aging (Thomas & ElSohly, 2016). Therefore, when analyzing by GC-MS, the high temperature in the GC injector causes the thermal conversion of the acidic cannabinoids in their neutral forms, thus making difficult to quantify them separately (Lehmann et al., 1997). The only way to overcome this issue is to derivatise the sample; this way the acidic forms of the cannabinoids are unavailable for decarboxylation, and they can be determined individually.

Another analytical technique that has been and still is used for cannabinoid determination in dietary products is the high-performance liquid chromatography (HPLC) coupled to diode array detector (HPLC-DAD), UV-visible detector (HPLC-UV), fluorescence detector (HPLC-FLD), or their combination. These instruments are widely employed for a large number of analyses, since they are quite simple to use, affordable, and suitable for the determination of the majority of compounds.

However, since the liquid chromatography coupled to mass spectrometry (LC-MS) has started to become popular, the difference with the older techniques has been evident. The LC-MS, in fact, allows to unambiguously identify and to precisely quantify compounds at very low concentrations and in very complex mixtures. Moreover, it allows to simultaneously analyze hundreds of compounds with the minimum error and with high selectivity, specificity, and sensitivity. That said, it is not difficult to understand why this technique has been increasingly employed in the determination of cannabinoids in food and beverages, gradually becoming one of the most preferred techniques.

Table 1 reported all the analytical techniques and purification procedures used for cannabinoid detection in hemp seeds, hempseed oil, hempseed flour, and food and beverages containing hemp as ingredient since today.

Cannabinoid determination in hemp-derived ingredients

Hemp seeds, hempseed oil, and hempseed flour, as well as protein powder, have been the first dietary products to be studied for their cannabinoid content; the occurrence data of cannabinoids in these products are reported in Table 2. Generally, hemp seeds for human consumption are produced from industrial hemp, that is, a cannabis plant with a low content in psychoactive cannabinoids. Moreover, the correct harvesting procedure should be made by minimizing all the contacts between the seeds and the apical part of the plant. The hemp inflorescences, in fact, are richer in Δ^9-THC and THCA-A than the seeds and an incorrect manufacturing could result in a cross-contamination between the two parts,

TABLE 1 Analytical techniques and purification procedures used for determination of cannabinoids in hemp-derived ingredients and hemp-containing dietary products.

Work	Country	Samples (*n*)	Cannabinoids	Purification	Analytical technique
Matsunaga et al. (1990)	Japan	Hemp seeds sold as feed for birds (6; 2 without bracts and 4 surrounded by bracts)	THC, CBN	Extraction with benzene, evaporation	TLC, GC, GC-MS
Lehmann et al. (1997)	Switzerland	Cannabis seed oil (10 samples from different commercial source in Switzerland +15 samples produced in their laboratory using cannabis seeds of different types and origins)	THC, THC-COOH in the urine of volunteers that ingested the tested Cannabis oil	No purification	GC-MS
Bosy and Cole (2000)	United States, Canada	Commercially available hempseed oil (8)	Δ^9-THC	Extraction with acetonitrile, evaporation, dissolution with n-hexane and purification by SPE	GC-MS
Ross et al. (2000)	United States, The Netherlands	Drug-type (4) and fiber-type (4) hemp seeds	Total Δ^9-THC	Sequence of extractions with different solvents, each followed by mixing ang centrifugation, then the final extract was purified by silica column	GC-MS
Zoller et al. (2000)	Switzerland	Hemp seeds Hempseed oil Biscuits Herbal hemp Hemp tea	Δ^9-THC, THCA-A	Hemp seeds, biscuits, herb: extraction with methanol/ dichloromethane Hempseed oil: dissolution with methanol Hemp tea: infusion in boiling water and dilution with methanol	HPLC-UV (Δ^9-THC, THCA) HPLC-FLD (Δ^9-THC)
Lachenmeier et al. (2004)	Germany	Hemp seeds (4) Hempseed oil (3) Hemp tea (5) Chocolate (4) Flour (1) Bar and snack (7) Soft drink (1) Beer (1) Pastilles (1)	CBD, CBN, Δ^9-THC	Headspace solid-phase microextraction after homogenization/ shaking/infusion for solid/liquid/herbal samples. Liquid-liquid extraction	GC-MS
Pellegrini et al. (2005)	Italy	Hemp seed Hempseed oil Pastilles Beer Liqueur Scented grass (the values are the mean of 5 samplings of the same product batch)	Δ^9-THC, CBD, CBN	Liquid-liquid extraction with hexane/isopropanol	GC-MS

Continued

TABLE 1 Analytical techniques and purification procedures used for determination of cannabinoids in hemp-derived ingredients and hemp-containing dietary products—cont'd

Work	Country	Samples (*n*)	Cannabinoids	Purification	Analytical technique
Holler et al. (2008)	United States	Hemp seeds (4) Hempseed oil (44) Bars (6) Lollipop (6) Protein (2) Gummy treats (2) Butter (2) Tablets (2) Tea (2) Beer (1) Chips (1) Cereals (1) Flour (1) Pretzel (1) Vodka (1)	Δ^9-THC	Liquid-liquid extraction for liquid samples Solid-liquid extraction for solid samples	GC-MS
Lalge et al. (2016)	Czech Republic	Hempseed powder Hempseed oil Hemp protein	Δ^9-THC, CBD, CBN	Ultrasound-assisted solid-liquid extraction	GC-FID
Chang et al. (2017)	Taiwan, China, Canada	Hemp nuts products (18)	Δ^9-THC, CBD, CBN THC-COOH (11-nor-9-carboxy-Δ^9-THC) in the urine	Ultrasound-assisted solid-liquid extraction	HPLC-MS/MS
Escrivá et al. (2017)	Spain	Milk (13) Junior formula (5) Liver (10) Hemp seeds (10)	Δ^9-THC, THC-COOH, THC-OH	Extraction with organic solvent and purification by SPE	LC-MS/MS
Ciolino et al. (2018a)	United States	Hempseed oil (2), 60 commercial cannabis products (oral supplements, foods, candies, beverages, vapes/eliquids, and topical)	CBD, CBDA, Δ^9-THC, THCA, CBN, Δ^8-THC, CBG, CBGA, CBDV, THCV, and CBC	Ultrasound-assisted solid-liquid extraction	Qualitative GC-MS
Ciolino et al. (2018b)	United States	Hempseed oil (2), 60 commercial cannabis products (oral supplements, foods, candies, beverages, vapes/eliquids, and topical)	CBD, CBDA, Δ^9-THC, THCA, CBN, Δ^8-THC, CBG, CBGA, CBDV, THCV, and CBC	Ultrasound-assisted solid-liquid extraction	Quantitative HPLC-DAD
Citti et al. (2018)	Italy	Hempseed oil (13)	CBD, CBDA, Δ^9-THC, THCA, CBN, CBG, CBDV	Extraction of the oil from cold pressing of hemp seeds and dilution with organic solvent	HPLC-UV
Meng et al. (2018)	Canada	40 Commercial products (oil, plant material, cream)	CBD, CBDA, Δ^9-THC, THCA-A	Ultrasound-assisted extraction	LC-MS/MS

TABLE 1 **Analytical techniques and purification procedures used for determination of cannabinoids in hemp-derived ingredients and hemp-containing dietary products—cont'd**

Work	Country	Samples (*n*)	Cannabinoids	Purification	Analytical technique
Brighenti et al. (2019)	Italy	1 Hemp honey 4 Honey	CBD, CBDA, Δ^9-THC, THCA, CBG, CBGA	Quechers	LC-MS/MS
Christinat et al. (2020)	Switzerland	Hemp seeds (3) Hempseed oil (4) Hemp protein (3) Chocolate (3) Mayonnaise (1) CBD oil (1) Tea pod (2) Roasted coffee (1) Coffee pod (1) Raw milk (1)	CBD, CBDA, Δ^9-THC, THCA-A, CBN, CBC, CBCA, CBDV, CBDVA, CBG, CBGA, THCV, THCVA, Δ^8-THC, THC-COOH	Quechers	LC-MS/MS
Izzo et al. (2020)	Italy	Hempseed oil (13)	CBDA	Formato et al. (2020)	HPLC-UV-DAD
Jang et al. (2020)	Several Countries	Hemp seeds Hempseed oil	THC, CBD, CBN	Ultrasound-assisted extraction	GC-MS
Piccolella et al. (2020)	Italy	Hemp coffee (3)	CBD, CBDA, THC, THCA	Infusion by moka	UHPLC-HRMS
Risoluti et al. (2020)	Italy	Hemp flours (10)	CBD, THC, CBG	Solid-liquid extraction	Screening by MicroNIR/ Chemometrics (plus GC-MS for comparison)
Di Marco Pisciottano et al. (2021)	Italy	Flour (7) Pasta (30) Bakery product (22) Coffee (6) Herbs for infusion (3) Honey (2) Egg (2) Chocolate (1) Others (5) Alcoholic beverage (11) Beer (3) Milk (2) Hemp seeds for birds (6)	CBD, CBDA, Δ^9-THC, THCA, CBN, Δ^8-THC, CBG, CBGA, THCV	Solid-liquid extraction or dilution with organic solvent	LC-MS/MS
Dubrow et al. (2021)	United States	Nonalcoholic beverages (8) Edibles (17) Gummies (17) Pet products (22) Powder (1) Tinctures/oils (82)	CBDA, CBD, THCA, Δ^9-THC, Δ^8-THC, CBN, CBC, CBG, CBGA, CBDV, THCV	Several matrix-related extraction procedures	HPLC-DAD
Kladar et al. (2021)	European Countries	Hemp seeds (5 peeled +5 unpeeled) Hempseed oil (17) Hemp protein (6) Hemp tea (6) Confectioneries (5)	Total Δ^9-THC, CBD, CBN	Extraction with acetonitrile or hexane	GC-MS

Continued

TABLE 1 Analytical techniques and purification procedures used for determination of cannabinoids in hemp-derived ingredients and hemp-containing dietary products—cont'd

Work	Country	Samples (*n*)	Cannabinoids	Purification	Analytical technique
		Alcoholic beverages (2)			
Madej et al. (2021)	Poland	Hemp oil supplements (7)	CBD, CBDA, THC, CBN	Liquid-liquid extraction	HPLC-DAD
Dräger et al. (2022)	German	Hemp seeds (93) Hempseed oil (137) Hemp tea (89) Food supplements (131) Nonalcoholic beverages (61)	THC	Not reported	GC-MS and LC-MS/MS

In the table are also reported information about the provenience of the samples (country), the type, and number of samples and the cannabinoids analyzed.

TABLE 2 Occurrence data of cannabinoids in hemp-derived ingredients, that is, hemp seeds, hempseed oil, flour, and protein powder.

Work	Samples (*n*)	LOQ/LOD	Concentration levels
Matsunaga et al. (1990)	**Hemp seeds** sold as feed for birds (6; 2 without bracts and 4 surrounded by bracts)	Not reported	**Seeds without bracts** THC = 0.3 mg/100 g CBN = 0.4 mg/100 g **Seeds with bracts** THC = (47.2 ± 3.5) mg/100 g CBN = (34.9 ± 3.3) mg/100 g
Lehmann et al. (1997)	**Cannabis seed oil** (10 samples from different commercial source in Switzerland +15 samples produced in their laboratory using cannabis seeds of different types and origins)	LOQ = 1 μg/g seed oil	In the range 3–1500 μg/g. Only one oil contained an extraordinarily high THC content (1500). Eighty eight percent of the samples were below the tolerated maximum concentration set by Swiss Federal Office of Public Health, that is 50 μg/g. The 15 samples prepared in laboratory were obtained by both fiber and drug type of hemp and had a THC content of 30 μg/g or lower, whereas 30% of the commercial oils were higher than the maximum tolerated value
Bosy and Cole (2000)	Commercially available **hempseed oil** (8)	Not reported	Δ^9-THC = 11.5–117.5 mg/g
Ross et al. (2000)	Drug-type (4) and fiber-type (4) **hemp seeds**	Not reported	**Fiber type** Total Δ^9-THC = 0–12 μg/g **Drug type** Total Δ^9-THC = 41–124 μg/g
Zoller et al. (2000)	**Hemp seeds** **Hempseed oil**	LOD (FLD) = 0.1 ng LOD (UV) = 0.01 (THC) and 0.05 (THCA) ng	**Hemp seeds** (UV) THC = 3.9 and 5.2 mg/kg **Hempseed oil** (UV) THC = 4.1–880 mg/kg
Lachenmeier et al. (2004)	**Hemp seeds** (4) **Hempseed oil** (3) **Flour** (1)	**Oil** THC = 0.05 mg/kg (LOD) THC = 0.30 mg/kg (LOQ)	**Hemp seeds** THC = 0.49–3.36 mg/kg CBD = 0.41–33.81 mg/kg

TABLE 2 Occurrence data of cannabinoids in hemp-derived ingredients, that is, hemp seeds, hempseed oil, flour, and protein powder—cont'd

Work	Samples (*n*)	LOQ/LOD	Concentration levels
		CBD = 0.09 mg/kg (LOD) CBD = 0.34 mg/kg (LOQ) CBN = 0.15 mg/kg (LOD) CBN = 0.43 mg/kg (LOQ)	CBN = 0.11–0.32 mg/kg **Hempseed oil** THC = 1.22–14.01 mg/kg CBD = 2.30–23.10 mg/kg CBN = 2.24–4.19 mg/kg **Flour** THC = 0.26–0.29 mg/kg CBD = 10.06–11.06 mg/kg CBN = 0.23–0.25 mg/kg
Pellegrini et al. (2005)	**Hemp seed** **Hempseed oil**	**All matrices** THC = 0.3 ng/g or ng/mL (LOD) THC = 1 ng/g (LOQ) CBD = 0.6 ng/g (LOD) CBD = 2 ng/g (LOQ) CBN = 0.3 ng/g (LOD) CBN = 1 ng/g (LOQ)	**Hemp seed** THC = 328.3 ± 36.3 ng/g CBD = nd CBN = 146.0 ± 4.5 ng/g **Hempseed oil** THC = 25.0 ± 0.5 ng/mL CBD = 3.7 ± 4.2 ng/mL CBN = 2.4 ± 0.3 ng/mL
Holler et al. (2008)	**Hemp seeds** (4) **Hempseed oil** (44) **Protein** (2) **Flour** (1)	LOD (Δ^9-THC) = 1.0–2.5 μg/g; 2.5 ng/mL (aqueous products) LOQ (Δ^9-THC) = 1 μg/g (seeds); 2.5 μg/g (oil)	**Hemp seed** Δ^9-THC = nd **Hempseed oil** Δ^9-THC = nd–117.5 μg/g **Protein** Δ^9-THC = nd **Flour** Δ^9-THC = <LOQ
Ciolino et al. (2018b)	**Hempseed oil** (2)	LOQ = 10.0 μg/g LOD = 0.20 μg/mL	**Hempseed oil** CBD = 12–26 μg/g Δ^9-THC = trace THCA = 1.5–3.1 μg/g CBDA = 14–56 μg/g CBN = nd
Lalge et al. (2016)	**Hempseed powder** **Hempseed oil** **Hemp protein**	Not reported	**Hemp grits** Δ^9-THC = nd CBD = nd CBN = nd **Hemp protein** Δ^9-THC = 0.9312 mg/g CBD = 0.6250 mg/g CBN = nd **Hemp flour** Δ^9-THC = 1.5270 mg/g CBD = 0.7620 mg/g CBN = nd
Chang et al. (2017)	**Hemp nuts products** (18)	THC = 0.3 ng/mL (LOD) THC = 1 ng/mL (LOQ) CBD = 0.4 ng/mL (LOD) CBD = 1 ng/mL (LOQ) CBN = 0.2 ng/mL (LOD) CBN = 0.5 ng/mL (LOQ)	THC = <LOQ–8.18 μg/g CBD = 0.03–25.17 μg/g CBN = 0.05–4.12 μg/g
Escrivá et al. (2017)	**Hemp seeds** (10)	**Hemp seed** Δ^9-THC = 3.10 ng/g (CCα); 3.52 ng/g (CCβ) THC-OH = 3.91 ng/g (CCα); 4.59 ng/g (CCβ) THC-COOH = 6.78 ng/g (CCα); 7.22 ng/g (CCβ)	**Hemp seeds** Δ^9-THC = 0.82 μg/g (mean value)

Continued

TABLE 2 Occurrence data of cannabinoids in hemp-derived ingredients, that is, hemp seeds, hempseed oil, flour, and protein powder—cont'd

Work	Samples (*n*)	LOQ/LOD	Concentration levels
Citti et al. (2018)	**Hempseed oil** (13)	For all analytes LOQ = 1 μg/mL LOD = 0.2 μg/mL	CBDV = <LOQ–75.39 mg/kg CBDA = <LOQ–821.1 mg/kg CBG = <LOD–1.381 mg/kg CBD = <LOD–1056 mg/kg CBN = <LOQ–12.41 mg/kg THC = <LOD–1.804 mg/kg THCA = <LOQ–9.462 mg/kg
Meng et al. (2018)	40 Commercial products (**oil, plant material, cream**)	LLOQ = 0.195 ng/mL LOD (CBD and THC) = 0.048 ng/mL LOD (CBDA and THCA-A) = 0.024 ng/mL	Δ^9-THC = <LOQ–337 μg/g CBD = <LOQ–8410 μg/g CBDA = <LOQ–51.8 μg/g THCA-A = <LOQ–4.97 μg/g
Christinat et al. (2020)	**Hemp seeds** (3) **Hempseed oil** (4) **Hemp protein** (3) **CBD oil** (1)	LOQ = 0.15 mg/kg (hemp seeds and hemp proteins) LOQ = 0.6 mg/kg (hempseed oil)	**Hemp seeds** CBD = 0.49–1.06 mg/kg CBDA = 1.22–6.36 mg/kg Δ^9-THC = <LOQ–0.21 mg/kg THCA-A = <LOQ–0.68 mg/kg CBN = <LOQ Δ^8-THC = <LOQ CBC = <LOQ CBCA = <LOQ–0.36 mg/kg CBG = <LOQ CBGA = <LOQ–0.18 mg/kg CBDV = <LOQ CBDVA = <LOQ–0.41 mg/kg THCV = <LOQ THCVA = <LOQ THC-COOH = <LOQ **Hempseed oil** CBD = <LOQ–15.5 mg/kg CBDA = <LOQ–50.6 mg/kg Δ^9-THC = <LOQ–9.96 mg/kg THCA-A = <LOQ–6.05 mg/kg CBN = <LOQ–1.88 mg/kg Δ^8-THC = <LOQ CBC = <LOQ–0.97 mg/kg CBCA = <LOQ–1.59 mg/kg CBG = <LOQ–0.60 mg/kg CBGA = <LOQ–1.08 mg/kg CBDV = <LOQ–0.61 mg/kg CBDVA = <LOQ–2.01 mg/kg THCV = <LOQ–6.46 mg/kg THCVA = <LOQ THC-COOH = <LOQ **Hemp protein** CBD = <LOQ–4.96 mg/kg CBDA = 0.17–12.2 mg/kg Δ^9-THC = 0.20–0.38 mg/kg THCA-A = 0.32–0.70 mg/kg CBN = <LOQ Δ^8-THC = <LOQ CBC = <LOQ–0.31 mg/kg CBCA = <LOQ–0.75 mg/kg CBG = <LOQ CBGA = <LOQ–0.17 mg/kg

TABLE 2 Occurrence data of cannabinoids in hemp-derived ingredients, that is, hemp seeds, hempseed oil, flour, and protein powder—cont'd

Work	Samples (*n*)	LOQ/LOD	Concentration levels
			CBDV = <LOQ–0.19 mg/kg CBDVA = <LOQ–0.98 mg/kg THCV = <LOQ–0.40 mg/kg THCVA = <LOQ–0.80 mg/kg THC-COOH = <LOQ **CBD oil** CBD = 45,029 mg/kg CBDA = 502 mg/kg Δ^9-THC = 683 mg/kg THCA-A = 2.22 mg/kg CBN = 17.4 mg/kg Δ^8-THC = 4.96 mg/kg CBC = 831 mg/kg CBCA = 12.7 mg/kg CBG = 272 mg/kg CBGA = 8.37 mg/kg CBDV = 129 mg/kg CBDVA = 2.97 mg/kg THCV = 4.46 mg/kg THCVA = 0.85 mg/kg THC-COOH = <LOQ
Izzo et al. (2020)	**Hempseed oil** (13)	Not reported	CBDA = 0.28–3.8 mg/g
Jang et al. (2020)	**Hemp seeds** **Hempseed oil**	**Hemp seeds** LOQ = 0.05 (THC), 0.01 (CBD), 0.01 (CBN) μg/g **Hempseed oil** LOQ = 0.05 (THC), 0.001 (CBD), 0.001 (CBN) μg/mL	**Hemp seeds** THC = 0.06–5.91 μg/g CBD = 0.32–25.55 μg/g CBN = 0.01–1.50 μg/g **Hempseed oil** THC = 0.3–19.73 μg/mL CBD = 6.66–63.40 μg/mL CBN = 0.11–2.31 μg/mL
Risoluti et al. (2020)	**Hemp flours** (10)	Sensitivity: CBD = 0.10% THC and CBG = 0.2%	Only one sample containing 0.03% of CBD
Di Marco Pisciottano et al. (2021)	**Flour** (5) **Hemp seeds for birds** (6)	**Flour** LOQ = 0.020 mg/kg **Hemp seeds for birds** LOQ = 0.100 mg/kg	**Flour** Δ^9-THC = 0.021–0.428 mg/kg THCA-A = 0.499–2.04 mg/kg Δ^8-THC = <LOQ–0.880 mg/kg CBD = 1.70–14.7 mg/kg CBDA = 8.31–37.0 mg/kg CBG = <LOQ–1.68 mg/kg CBGA = <LOQ–2.48 mg/kg CBN = <LOQ–0.155 mg/kg THCV = <LOQ–0.094 mg/kg **Hemp seeds for birds** Δ^9-THC = <LOQ–1.57 mg/kg THCA-A = <LOQ–1.64 mg/kg Δ^8-THC = <LOQ CBD = <LOQ–8.39 mg/kg CBDA = <LOQ–39.2 mg/kg CBG = <LOQ–0.126 mg/kg CBGA = <LOQ–0.576 mg/kg CBN = <LOQ–0.200 mg/kg THCV = <LOQ

Continued

TABLE 2 Occurrence data of cannabinoids in hemp-derived ingredients, that is, hemp seeds, hempseed oil, flour, and protein powder—cont'd

Work	Samples (*n*)	LOQ/LOD	Concentration levels
Dubrow et al. (2021)	**Powder** (1) **Tinctures/oils** (82)	LOQ (matrix-related) = 0.01–0.1 mg/g LOD (matrix-related) = 0.004–0.04 mg/g	CBN, CBC, CBGA, CBDV, THCA, Δ^8-THC, and THCV were present only infrequently and at low concentrations **Powder** CBDA = 4.08 mg/g CBD = 0.21 mg/g CBN = <LOD Δ^9-THC = 0.01 mg/g THCA-A = 0.12 mg/g **Tinctures/oils** CBDA = <LOD–10.05 mg/g CBD = <LOD–171.85 mg/g CBN = <LOD–1.42 mg/g Δ^9-THC = <LOD–3.34 mg/g THCA-A = <LOD–0.07 mg/g
Kladar et al. (2021)	**Hemp seeds** (5 peeled +5 unpeeled) **Hempseed oil** (17) **Hemp protein** (6)	**Hemp seeds** CBD = 1 mg/kg (LOD), 2 mg/kg (LOQ) Total Δ^9-THC = 0.5 mg/kg (LOD), 1 mg/kg (LOQ) CBN = 0.5 mg/kg (LOD), 1 mg/kg (LOQ) **Hempseed oil** CBD = 0.5 mg/kg (LOD), 1 mg/kg (LOQ) Total Δ^9-THC = 0.5 mg/kg (LOD), 1 mg/kg (LOQ) CBN = 0.1 mg/kg (LOD), 0.5 mg/kg (LOQ) **Hemp protein** CBD = 1 mg/kg (LOD), 1.5 mg/kg (LOQ) Total Δ^9-THC = 0.2 mg/kg (LOD), 0.50 mg/kg (LOQ) CBN = 0.5 mg/kg (LOD), 1 mg/kg (LOQ)	**Hemp seeds** CBD = <LOQ–71.97 mg/kg Total Δ^9-THC = <LOQ–3.36 mg/kg CBN = <LOQ–1.99 mg/kg **Hempseed oil** CBD = <LOQ–116.84 mg/kg Total Δ^9-THC = <LOQ–8.10 mg/kg CBN = <LOQ–2.82 mg/kg **Hemp protein** CBD = <LOQ–11.14 mg/kg Total Δ^9-THC = 1.33–2.17 mg/kg CBN = 1.36–1.82 mg/kg
Madej et al. (2021)	**Hemp oil supplements** (7)	CBDA = 0.17 μg/mL (LOD), 0.78 μg/mL (LOQ) CBD = 1.94 μg/mL (LOD), 3.12 μg/mL (LOQ) CBN = 1.68 μg/mL (LOD), 5.03 μg/mL (LOQ) Δ^9-THC = 0.65 μg/mL (LOD), 1.95 μg/mL (LOQ)	CBDA = 0.035–25.75% *w/w* CBD = 1.45–18.25% w/w CBN = 0.003–0.35% w/w Δ^9-THC = <LOD–0.84% w/w
Dräger et al. (2022)	**Hemp seeds** (93) **Hempseed oil** (137)	Not reported	**Hemp seeds** THC = 0.0035–196 mg/kg **Hempseed oil** THC = 0.025–232.8 mg/kg

The levels of cannabinoids are reported as concentration ranges and information about the limits of detection (LOD) and the limits of quantification (LOQ) of the methods are also given.

causing a higher presence of psychoactive cannabinoids in the edible hemp seeds. Hempseed oil is made by cold pressing the seeds into an omega-rich culinary oil, whereas CBD oil is made by extracting the compound CBD from the leaves, flowers, and stalks of a hemp plant. The different production process justifies the higher CBD content that was generally found in CBD oil in respect to that of hemp oil (Christinat et al., 2020). Hemp flour and protein powder are both made by grinding the whole hemp seeds, with the difference that in hemp flour the seeds have been previously stripped of the oil. These raw materials are 100% hemp; therefore, it is just reasonable to suppose that their cannabinoid content would be higher than in their derived products. The work of Matsunaga et al. (1990) has been the first to investigate the content of some cannabinoids, that is, Δ^9-THC and CBN, in hemp seeds sold as bird feed. They observed that hemp seeds with bracts had a higher content of Δ^9-THC than those without bracts, thus suggesting that they should be dehulled to be safer for the consumers. Since then, most of the scientific works about cannabinoid occurrence in food and beverages have been performed on these matrices, being them the ingredients for all the other dietary products. As expected, due to lipophilic nature of these compounds, the cannabinoid content is overall higher in hempseed oil than in hemp seeds, flour, and protein (Christinat et al., 2020; Dräger et al., 2022; Dubrow et al., 2021; Holler et al., 2008; Jang et al., 2020; Kladar et al., 2021; Lachenmeier et al., 2004). Also, it is noteworthy to underline that CBD and its acidic precursor CBDA, when analyzed, are generally found at concentrations higher than those of the Δ^9-THC (Ciolino et al., 2018b; Citti et al., 2018; Di Marco Pisciottano et al., 2021; Dubrow et al., 2021; Jang et al., 2020; Kladar et al., 2021; Lachenmeier et al., 2004; Madej et al., 2021; Meng et al., 2018). That is coherent with the (THC+CBN)/CBD *ratio* indicated to identify the phenotype of the cannabis plants: a fiber-type hemp, in fact, should have a *ratio* lower than 1 (Avico et al., 1985; de Meijer et al., 1992; Fetterman et al., 1971).

Cannabinoids monitoring in food and beverages containing hemp-derived ingredients

Since hemp seeds and hempseed-derived products started to be used as ingredients in food and beverage production, the need to monitor the content of psychoactive cannabinoids also in the final product has increased. Those products, in fact, often claim for healthy nutritional properties due to the presence of hemp and even to the high content of CBD, the cannabinoid considered as the most beneficial for human's wellness. That leads to an increased commercialization of hemp-derived food because of the appeal of the promised beneficial effects derived from the hemp. However, the consumption of these products could represent a risk for the consumers' health, for example, if the cannabis cultivar employed has a too much high content of psychoactive cannabinoids or if a cross contamination from the inflorescences has occurred during the seed harvesting. The concern about the risk for consumers' health has led many countries to intensify the monitoring of cannabinoids in hemp-derived food and beverages. In 2016, for example, the European Commission recommended the Member States to issue monitoring plan about the levels of Δ^9-THC, THCA-A, and THCA-B in food, to evaluate possible exposition of consumers to these substances (European Commission, 2016). All the works from the scientific literature reporting cannabinoid levels in food and beverages containing hemp-derived food are hereafter reported grouped by food and beverage category.

Cannabinoids content in cereals, cereal-derived, and bakery products

Cereals and cereal-derived products can be produced with hemp seeds or using hempseed flour; for example, pasta can be made using a small amount of hempseed flour or cereal bars and snacks can also contain hemp seeds. In the same way, bakery products can be produced with a percentage of hempseed flour and there are even flours intended for bakery that are a mix of different flours, including the hemp-derived one. Table 3 reports all the results from the scientific literature about the cannabinoids content in this kind of products. As can be seen, there are not so much works, but all of their results are coherent with the data about cannabinoid contamination in hemp-derived ingredients. In fact, also in this case CBD and CBDA are the predominant cannabinoids, whereas Δ^9-THC and THCA-A are determined at low concentration, with the only exception of the bakery products analyzed by Di Marco Pisciottano et al. (2021). Actually, those values are indicated as a range, but in the 22 bakery products they analyzed, only one samples had Δ^9-THC at 8.87 mg/kg and THCA-A at 47.7 mg/kg, whereas all the other results were below 3.13 and 2.89 mg/kg, respectively. For what concerns the work by Dubrow et al. (2021), they determined CBDA, CBD, THCA, Δ^9-THC, Δ^8-THC, CBN, CBC, CBG, CBGA, CBD, and THCV also in 17 edibles, including honey, chocolate bars, candies, cookies, cake, brownies, and popped corn. However, since the details of each sample are not expressly given, it was not possible to include the results in the correct category, and the data about all the 17 edibles are herein reported.

TABLE 3 **Occurrence data of cannabinoids in cereals, cereal-derived products, and bakery products.**

Work	Samples (*n*)	LOQ/LOD	Concentration levels
Zoller et al. (2000)	**Biscuits**	LOD (FLD) = 0.1 ng LOD (UV) = 0.01 (THC) and 0.05 (THCA) ng	**Biscuits** THC (UV) = 3.38 mg/kg
Lachenmeier et al. (2004)	**Bar and snack** (7)	Not reported	**Bar and snack** THC = 0.04–0.20 mg/kg CBD = 0.29–10.02 mg/kg CBN = 0.03–0.18 mg/kg
Holler et al. (2008)	**Bars** (6) **Chips** (1) **Cereals** (1) **Pretzel** (1)	LOD (Δ^9-THC) = 1.0–2.5 μg/g LOQ (Δ^9-THC) = 1 μg/g (bars)	**Bars** Δ^9-THC = nd – <LOQ **Chips** Δ^9-THC = nd **Cereals** Δ^9-THC = nd **Pretzel** Δ^9-THC = nd
Di Marco Pisciottano et al. (2021)	**Mixed flour** (7) **Pasta** (30) **Bakery product** (22)	LOQ = 0.020 mg/kg	**Mixed flour** Δ^9-THC = 0.076–0.093 mg/kg THCA-A = 0.050–0.090 mg/kg Δ^8-THC = <LOQ CBD = 1.48–1.83 mg/kg CBDA = 2.45–2.67 mg/kg CBG = 0.049–0.058 mg/kg CBGA = 0.026–0.045 mg/kg CBN = <LOQ THCV = <LOQ **Pasta** Δ^9-THC = <LOQ–0.140 mg/kg THCA-A = <LOQ–0.994 mg/kg Δ^8-THC = <LOQ CBD = <LOQ–2.70 mg/kg CBDA = <LOQ–11.9 mg/kg CBG = <LOQ–0.110 mg/kg CBGA = <LOQ–0.277 mg/kg CBN = <LOQ–0.024 mg/kg THCV = <LOQ **Bakery products** Δ^9-THC = <LOQ–8.87 mg/kg THCA-A = <LOQ–47.7 mg/kg Δ^8-THC = <LOQ–0.320 mg/kg CBD = <LOQ–47.1 mg/kg CBDA = <LOQ–159.6 mg/kg CBG = <LOQ–2.51 mg/kg CBGA = <LOQ–2.04 mg/kg CBN = <LOQ–0.446 mg/kg THCV = <LOQ–0.182 mg/kg
Dubrow et al. (2021)	**Edibles** (17) including honey, chocolate bars, candies, cookies, cake, brownies and popped corn	LOQ (matrix-related) = 0.01–0.1 mg/g LOD (matrix-related) = 0.004–0.04 mg/g	CBN, CBC, CBGA, CBDV, THCA, Δ^8-THC, and THCV were present only infrequently and at low concentrations **Edibles** CBDA = <LOD–0.23 mg/g CBD = LOD–7.53 mg/g CBN = <LOD–0.04 mg/g Δ^9-THC = <LOD–0.31 mg/g THCA-A = <LOD

The levels of cannabinoids are reported as concentration ranges and information about the limits of detection (LOD) and the limits of quantification (LOQ) of the methods are also given.

Cannabinoids content in products for infusion and animal-derived products

In Table 4, all occurrence data about cannabinoids both in products for infusion (coffee, tea, and herbs) and in animal-derived products (egg, milk, honey, and their by-products) are reported. In the first subcategory, we can find ground coffee or coffee pod, tea, and herbal mixtures all containing hemp, that is, both hemp seeds and inflorescences. As can be clearly seen, a high concentration of the cannabinoids and, in particular, of the psychoactive ones was observed in all the samples, up to 8831.04 mg/kg of total Δ^9-THC (Kladar et al., 2021). These products are, in fact, mainly made by hemp and even hemp inflorescences, that is, the richest part in psychoactive cannabinoid content; therefore, it is evident that the high levels of cannabinoids are a direct consequence of the sample composition. It is fitting to specify that the results reported for the group of Lachenmeier et al. (2004), Christinat et al. (2020), and Kladar et al. (2021) refer to the samples analyzed using the purification procedure reported in their works. However, to estimate the real consumer's intake from these products, the above-mentioned authors have also analyzed the sample infusions prepared according to the manufacturer's instructions reported on the packaging. All the results showed an undeniable lower level of the cannabinoids found in the infusions in respect with those determined after the extraction with organic solvents, which being coherent with the lipophilic nature of these compounds. Moving on the second subcategory of this food group, we have the animal-derived products, that is, egg, milk, honey, and some of their by-products, such as butter and junior formula. These foods are overall from animals that have been fed with hemp or hemp seeds, such as cows and chicken, but there are also three honey samples that were added with hemp and that show the highest level of cannabinoids from this subcategory (Brighenti et al., 2019; Di Marco Pisciottano et al., 2021). Another not only interesting but also alarming information is that about the junior formula reported by Escrivá et al. (2017) containing Δ^9-THC. All the other samples were mainly not contaminated, suggesting that no significant cannabinoid transfer from feed to food is observable from the collected results.

TABLE 4 Occurrence data of cannabinoids in products for infusion and animal-derived products.

Work	Samples (*n*)	LOQ/LOD	Concentration levels
Coffee, tea and herbs for infusion			
Zoller et al. (2000)	**Herbal hemp** **Hemp tea**	LOD (FLD) =0.1 ng LOD (UV)=0.01 (THC) and 0.05 (THCA) ng	**Herbal hemp** (UV) THC=1480–1020 mg/kg **Hemp tea** (UV) THC=1.0 mg/kg
Lachenmeier et al. (2004)	**Hemp tea** (5)	**Tea leaves** THC=0.01 mg/kg (LOD) THC=0.08 mg/kg (LOQ) CBD=0.12 mg/kg (LOD) CBD=0.32 mg/kg (LOQ) CBN=0.01 mg/kg (LOD) CBN=0.09 mg/kg (LOQ)	**Hemp tea** THC=0.03–19.72 mg/kg CBD=0.51–79.35 mg/kg CBN=0.01–1.36 mg/kg
Pellegrini et al. (2005)	**Scented grass**	**All matrices** THC=0.3 ng/g or ng/mL (LOD) THC=1 ng/g (LOQ) CBD=0.6 ng/g (LOD) CBD=2 ng/g (LOQ) CBN=0.3 ng/g (LOD) CBN=1 ng/g (LOQ)	**Scented grass** THC=350.0±45.4 ng/g CBD=21.9±1.3 ng/g CBN=159.7±2.3 ng/g
Ciolino et al. (2018b)	60 Commercial cannabis products (**oral supplements, foods, candies, beverages, vapes/ eliquids, and topical**)	LOQ=10.0 μg/g LOD=0.20 μg/mL	**Coffee** CBD=0.042–0.82 mg/g Δ^9-THC=0.034 mg/g CBDA=nd CBN=trace

Continued

TABLE 4 Occurrence data of cannabinoids in products for infusion and animal-derived products—cont'd

Work	Samples (*n*)	LOQ/LOD	Concentration levels
Christinat et al. (2020)	**Tea pod** (2) **Roasted coffee** (1) **Coffee pod** (1)	LOQ = 0.15 mg/kg	**Tea pod** CBD = 1268–1348 mg/kg CBDA = 136–149 mg/kg Δ^9-THC = 11.2–12.9 mg/kg THCA-A = 0.70–0.74 mg/kg CBN = 11.6–13.1 mg/kg Δ^8-THC = 1.04–1.42 mg/kg CBC = 73.5–80.2 mg/kg CBCA = 6.45–7.69 mg/kg CBG = 29.7–30 mg/kg CBGA = 4.18–4.59 mg/kg CBDV = 21.1–22.4 mg/kg CBDVA = 4.28–4.59 mg/kg THCV = 0.36–0.49 mg/kg THCVA = 0.17 mg/kg THC-COOH = <LOQ **Coffees** CBD = 63.9–223 mg/kg CBDA = 229–482 mg/kg Δ^9-THC = 4.3–34 mg/kg THCA-A = 7.37–49.2 mg/kg CBN = <LOQ–2.42 mg/kg Δ^8-THC = <LOQ–0.34 mg/kg CBC = 3.25–14.1 mg/kg CBCA = 8.92–20.3 mg/kg CBG = 2.44–15.3 mg/kg CBGA = 10.7–37.8 mg/kg CBDV = 0.61–2.4 mg/kg CBDVA = 2.59–6.26 mg/kg THCV = <LOQ–0.59 mg/kg THCVA = 0.17–0.76 mg/kg THC-COOH = <LOQ
Piccolella et al. (2020)	**Hemp coffee** (3)	Not reported	CBDA = 0.282–0.333 wt% CBD = 0.004–0.005 wt% THCA = 0.0294–0.0440 wt% THC = nd
Di Marco Pisciottano et al. (2021)	**Coffee** (6) **Herbs for infusion** (3)	LOQ = 0.020 mg/kg	**Coffee** Δ^9-THC = <LOQ–30.9 mg/kg THCA-A = <LOQ–91.4 mg/kg Δ^8-THC = <LOQ–1.72 mg/kg CBD = <LOQ–189.6 mg/kg CBDA = 0.191–1360.8 mg/kg CBG = <LOQ–8.03 mg/kg CBGA = <LOQ–18.9 mg/kg CBN = <LOQ–0.560 mg/kg THCV = <LOQ–0.470 mg/kg **Herbs for infusion** Δ^9-THC = <LOQ–131.4 mg/kg THCA-A = <LOQ–411.9 mg/kg Δ^8-THC = <LOQ–10.2 mg/kg CBD = <LOQ–1172.3 mg/kg CBDA = <LOQ–4406.0 mg/kg CBG = <LOQ–94.3 mg/kg CBGA = <LOQ–493.9 mg/kg CBN = <LOQ–9.61 mg/kg THCV = <LOQ–2.58 mg/kg

TABLE 4 Occurrence data of cannabinoids in products for infusion and animal-derived products—cont'd

Work	Samples (*n*)	LOQ/LOD	Concentration levels
Kladar et al. (2021)	**Hemp tea** (6)	**Tea** CBD = 10 mg/kg (LOD), 20 mg/kg (LOQ) Total Δ^9-THC = 3 mg/kg (LOD), 10 mg/kg (LOQ) CBN = 1.5 mg/kg (LOD), 10 mg/kg (LOQ)	**Hemp tea** CBD = 443.89–25,003.51 mg/kg Total Δ^9-THC = 13.41–8831.04 mg/kg CBN = 13.86–265.56 mg/kg
Dräger et al. (2022)	**Hemp tea** (89)	Not reported	**Hemp tea** THC = 0.09 μg/kg–588 mg/kg
Honey, egg, milk and derived products			
Holler et al. (2008)	**Butter** (2)	LOD (Δ^9-THC) = 1.0–2.5 μg/g LOQ (Δ^9-THC) = 1 μg/g	**Butter** Δ^9-THC = nd – <LOQ
Escrivá et al. (2017)	**Milk** (13) **Junior formula** (5)	**Milk** Δ^9-THC = 4.28 ng/g (CCα); 4.67 ng/g (CCβ) THC-OH = 4.13 ng/g (CCα); 4.44 ng/g (CCβ) THC-COOH = 8.73 ng/g (CCα); 8.93 ng/g (CCβ)	**Milk** Δ^9-THC = nd THC-OH = nd THC-COOH = nd **Junior formula** Δ^9-THC = nd–56.11 ng/g THC-OH = nd THC-COOH = nd
Brighenti et al. (2019)	1 **Hemp honey** 4 **honey**	**For all analytes** LOD = 0.3 ng/g LOQ = 0.5 ng/g	All cannabinoids < LOD in the 4 samples of **honey** **Hemp honey** Δ^9-THC and THCA = <LOD CBD = 4.2 ± 0.3 ng/g CBDA = 40.8 ± 6.9 ng/g CBG = 0.5 ± 0.1 ng/g CBGA = 4.8 ± 0.9 ng/g
Christinat et al. (2020)	**Raw milk** (1)	LOQ = 0.005 mg/kg (raw milk and milk powder)	**Raw milk** CBD = 0.0095 mg/kg All the other cannabinoids were < LOQ
Di Marco Pisciottano et al. (2021)	**Honey** (2) **Egg** (2) **Milk** (2)	**Honey** and **egg** LOQ = 0.020 mg/kg **Milk** LOQ = 0.002 mg/L	In egg and milk samples were not contaminated by the cannabinoids **Honey** Δ^9-THC = 0.550–74.9 mg/kg THCA-A = <LOQ–1.47 mg/kg Δ^8-THC = 0.098–11.8 mg/kg CBD = 6.83–2733.7 mg/kg CBDA = 5.70–66.4 mg/kg CBG = 0.359–38.0 mg/kg CBGA = <LOQ–1.61 mg/kg CBN = <LOQ–17.1 mg/kg THCV = <LOQ–0.109 mg/kg

The levels of cannabinoids are reported as concentration ranges and information about the limits of detection (LOD) and the limits of quantification (LOQ) of the methods are also given.

Cannabinoids content in alcoholic and nonalcoholic beverages

Beverages can be divided into two major groups, that is, nonalcoholic and alcoholic beverages. Nonalcoholic beverages can be, for example, iced tea, soft, sport or energy, and drinks added with hemp, whereas alcoholic beverages are often liqueur made by alcoholic infusion of hemp seeds or even hemp inflorescences. As can be expected because of the lipophilicity of these compounds, alcoholic infusions should contain a higher level of cannabinoids, probably even of the psychoactive ones when the inflorescences are used. Table 5 reports the cannabinoid content of the alcoholic and nonalcoholic beverages analyzed since today. As can be observed, nonalcoholic beverages have indeed a lower content of all cannabinoids because

TABLE 5 Occurrence data of cannabinoids in alcoholic and nonalcoholic beverages.

Work	Samples (*n*)	LOQ/LOD	Concentration levels
Lachenmeier et al. (2004)	**Soft drink** (1) **Beer** (1)	Not reported	**Beverages** THC = nd CBD = 0.02–0.03 mg/kg CBN = 0.02–0.05 mg/kg
Pellegrini et al. (2005)	**Beer** **Liqueur**	**All matrices** THC = 0.3 ng/g or ng/mL (LOD) THC = 1 ng/g (LOQ) CBD = 0.6 ng/g (LOD) CBD = 2 ng/g (LOQ) CBN = 0.3 ng/g (LOD) CBN = 1 ng/g (LOQ)	**Beer** THC = 5.7 ± 1.1 ng/mL CBD = 4.8 ± 0.8 ng/mL CBN = nd **Liqueur** THC = 8.0 ± 0.9 ng/mL CBD = 91.9 ± 3.5 ng/mL CBN = nd
Holler et al. (2008)	**Tea** (2) **Beer** (1) **Vodka** (1)	LOD (Δ^9-THC) = 2.5 ng/mL LOQ (Δ^9-THC) = 5 ng/mL	**Tea** Δ^9-THC = nd–<LOQ **Beer** Δ^9-THC = nd **Vodka** Δ^9-THC = nd
Di Marco Pisciottano et al. (2021)	**Alcoholic beverage** (11) **Beer** (3)	LOQ = 0.002 mg/L	**Alcoholic beverages** Δ^9-THC = <LOQ–1.48 mg/L THCA-A = <LOQ–10.8 mg/L Δ^8-THC = <LOQ–1.32 mg/L CBD = <LOQ–48.3 mg/L CBDA = <LOQ–155.7 mg/L CBG = <LOQ–1.16 mg/L CBGA = <LOQ–5.36 mg/L CBN = <LOQ–0.231 mg/L THCV = <LOQ–0.289 mg/L **Beers** Δ^9-THC = <LOQ–0.020 mg/L THCA-A = <LOQ–0.066 mg/L Δ^8-THC = <LOQ CBD = <LOQ–0.236 mg/L CBDA = <LOQ–0.064 mg/L CBG = <LOQ–0.008 mg/L CBGA = <LOQ–0.009 mg/L CBN = <LOQ THCV = <LOQ
Dubrow et al. (2021)	**Nonalcoholic beverages** (8)	LOQ (matrix-related) = 0.01–0.1 mg/g LOD (matrix-related) = 0.004–0.04 mg/g	CBN, CBC, CBGA, CBDV, THCA, Δ^8-THC, and THCV were present only infrequently and at low concentrations **Beverages** CBDA = <LOD CBD = 0.02–2.79 mg/g CBN = <LOD–0.02 mg/g Δ^9-THC = <LOD THCA-A = <LOD

TABLE 5 Occurrence data of cannabinoids in alcoholic and nonalcoholic beverages—cont'd

Work	Samples (*n*)	LOQ/LOD	Concentration levels
Kladar et al. (2021)	**Alcoholic beverages** (2)	Not reported	**Alcoholic beverages** CBD=<LOD Total Δ^9-THC=<LOD CBN=<LOD
Dräger et al. (2022)	**Nonalcoholic beverages** (61)	Not reported	**Beverages** THC=0.009–529.3125 µg/kg

The levels of cannabinoids are reported as concentration ranges and information about the limits of detection (LOD) and the limits of quantification (LOQ) of the methods are also given.

of their mainly aqueous composition (Dubrow et al., 2021; Holler et al., 2008; Lachenmeier et al., 2004), whereas the concentration levels in alcoholic beverages are way higher (Di Marco Pisciottano et al., 2021; Pellegrini et al., 2005). However, that is also related to the type of hemp-derived ingredient used to produce the beverage: when hempseed or CBD oil is added to the drink, a higher concentration of cannabinoids could be observed even in nonalcoholic beverages. It is also interesting to observe that beer samples show cannabinoid levels between alcoholic and nonalcoholic beverages, probably due to the low but still present alcoholic content (Di Marco Pisciottano et al., 2021; Pellegrini et al., 2005). Even in this case, the CBD is the predominant cannabinoid in all works, this being once more coherent with the cannabinoid profile found in hemp-derived ingredients.

Cannabinoids content in other dietary products

Lastly, Table 6 reports the cannabinoid occurrence in all the dietary products not belonging to any of the other groups and being present in a small number of works, such as various sweets, pastilles, jam, vape/eliquids, mayonnaise, creams, pet products, and food supplements. Since the scarce number of samples of the same type, it is difficult to make an evaluation of the contamination profile of these products, but it is fitting to report the data anyway. What is immediately evident is that the levels of cannabinoids are generally lower than those of the other categories and that there is no type of product that stands out against the others, if not for the food supplements analyzed by the group of Dräger et al. (2022). Actually, those results are not so surprising when considering that the authors included in the group also CBD oil and hemp-containing supplements; therefore, the high Δ^9-THC content was quite predictable. Also, the CBD and CBDA content was significantly higher in the pet products and gummies analyzed by Dubrow et al. (2021), but, as the authors described in their work, those products were added with low-THC hemp or CBD, which being also reported on the packaging label. Overall, all the other products had from low to not detectable levels of cannabinoids.

TABLE 6 Occurrence data of cannabinoids in other dietary products, such as various sweets, pastilles, jam, vape/eliquids, mayonnaise, creams, pet products, and food supplements.

Work	Samples (*n*)	LOQ/LOD	Concentration levels
Lachenmeier et al. (2004)	**Chocolate** (4) **Pastilles** (1)	**Chocolate** THC=0.03 mg/kg (LOD) THC=0.06 mg/kg (LOQ) CBD=0.17 mg/kg (LOD) CBD=0.35 mg/kg (LOQ) CBN=0.09 mg/kg (LOD) CBN=0.34 mg/kg (LOQ)	**Chocolate** THC=0.03–0.11 mg/kg CBD=0.03–1.34 mg/kg CBN=0.03–0.06 mg/kg **Pastille** THC=0.16 mg/kg CBD=0.13–0.34 mg/kg CBN=0.10–0.11 mg/kg

Continued

TABLE 6 Occurrence data of cannabinoids in other dietary products, such as various sweets, pastilles, jam, vape/eliquids, mayonnaise, creams, pet products, and food supplements—cont'd

Work	Samples (*n*)	LOQ/LOD	Concentration levels
Pellegrini et al. (2005)	**Pastilles**	**All matrices** THC = 0.3 ng/g or ng/mL (LOD) THC = 1 ng/g (LOQ) CBD = 0.6 ng/g (LOD) CBD = 2 ng/g (LOQ) CBN = 0.3 ng/g (LOD) CBN = 1 ng/g (LOQ)	**Pastilles** THC = 23.2 ± 2.8 ng/g CBD = 8.2 ± 1.2 ng/g CBN = 5.0 ± 0.7 ng/g
Holler et al. (2008)	**Lollipop** (6) **Gummy treats** (2) **Tablets** (2)	LOD (Δ^9-THC) = 1.0–2.5 μg/g LOQ (Δ^9-THC) = 1 μg/g (lollipop)	**Lollipop** Δ^9-THC = nd–1.04 μg/g **Gummy treats** Δ^9-THC = nd **Tablets** Δ^9-THC = nd
Ciolino et al. (2018b)	60 COMMERCIAL cannabis products (**oral supplements, foods, candies, beverages, vapes/eliquids, and topical**)	LOQ = 10.0 μg/g LOD = 0.20 μg/mL	**Jam** CBD = 0.62 mg/g Δ^9-THC = 0.023 mg/g CBDA = nd CBN = nd **Candies** CBD = 0.31 mg/g Δ^9-THC = 0.018 mg/g CBDA = nd CBN = nd **Oral supplements and vapes/eliquids** CBD = 0.011–1 45 mg/g Δ^9-THC = nd–5.3 mg/g CBDA = nd–0.54 mg/g CBN = nd–0.21 mg/g
Christinat et al. (2020)	**Chocolate** (3) **Mayonnaise** (1)	LOQ = 0.15 mg/kg	**Chocolate** CBD = <LOQ–0.43 mg/kg CBDA = <LOQ–1.81 mg/kg Δ^9-THC = <LOQ THCA-A = <LOQ CBN = <LOQ Δ^8-THC = <LOQ CBC = <LOQ CBCA = <LOQ–0.16 mg/kg CBG = <LOQ CBGA = <LOQ CBDV = <LOQ CBDVA = <LOQ THCV = <LOQ THCVA = <LOQ THC-COOH = <LOQ **Mayonnaise** All cannabinoids were <LOQ
Di Marco Pisciottano et al. (2021)	**Chocolate** (1) **Others** (5) including creams, food preparation and dehulled hemp seed	LOQ = 0.020 mg/kg	**Chocolate** THCA-A = 0.194 mg/kg THCV = 0.023 mg/kg All the other cannabinoids were not detected **Others** Δ^9-THC = 0.036–0.311 mg/kg

TABLE 6 Occurrence data of cannabinoids in other dietary products, such as various sweets, pastilles, jam, vape/eliquids, mayonnaise, creams, pet products, and food supplements—cont'd

Work	Samples (n)	LOQ/LOD	Concentration levels
			THCA-A = 0.359–1.90 mg/kg Δ^8-THC = <LOQ–0.020 mg/kg CBD = <LOQ–1.64 mg/kg CBDA = 0.066–11.4 mg/kg CBG = <LOQ–0.044 mg/kg CBGA = <LOQ–0.073 mg/kg CBN = <LOQ THCV = <LOQ–0.238 mg/kg
Dubrow et al. (2021)	**Gummies** (17) **Pet products** (22)	LOQ (matrix-related) = 0.01–0.1 mg/g LOD (matrix-related) = 0.004–0.04 mg/g	CBN, CBC, CBGA, CBDV, THCA, Δ^8-THC, and THCV were present only infrequently and at low concentrations **Gummies** CBDA = <LOD–0.11 mg/g CBD = 0.07–14.05 mg/g CBN = <LOD–0.01 mg/g Δ^9-THC = <LOD–0.15 mg/g THCA-A = <LOD **Pet products** CBDA = <LOD–15.36 mg/g CBD = <LOD–51.23 mg/g CBN = <LOD–0.63 mg/g Δ^9-THC = <LOD–1.85 mg/g THCA-A = <LOD–0.49 mg/g
Kladar et al. (2021)	**Confectioneries** (5)	**Chocolate** CBD = 1 mg/kg (LOD), 2 mg/kg (LOQ) Total Δ^9-THC = 0.2 mg/kg (LOD), 0.5 mg/kg (LOQ) CBN = 0.5 mg/kg (LOD), 1 mg/kg (LOQ)	**Chocolate** CBD = <LOQ–2.01 mg/kg Total Δ^9-THC = <LOQ–0.76 mg/kg CBN = <LOQ
Dräger et al. (2022)	**Food supplements** (131)	Not reported	**Food supplements** THC = 0.0200–3400 mg/kg

The levels of cannabinoids are reported as concentration ranges and information about the limits of detection (LOD) and the limits of quantification (LOQ) of the methods are also given.

Conclusions

The use of hemp-derived ingredients in dietary products has brought the need to investigate the cannabinoid content, in particular the psychoactive ones, in food and beverages to estimate consumers' exposure. The occurrence data collected from all the works from the scientific literature suggest that there is indeed an intake of Δ^9-THC and THCA-A due to the ingestion of hemp-containing dietary products. Furthermore, in 2020, EFSA published a scientific opinion about acute human exposure assessment to Δ^9-THC (European Food Safety Authority (EFSA) et al., 2020) based on the occurrence data from the monitoring activities of the Member States. The conclusion was that most of the hemp and hemp-containing products exceeded the ARfD of 1 μg/kg established by EFSA but also that the limited amount of data for some food categories and some analytical limitations represented an important source of uncertainty. Therefore, the EFSA stated that the exposure evaluation is expected to be an overestimation of acute exposure to Δ^9-THC in the EU and that more occurrence data are needed to refine the exposure assessment. Other than the necessity of implementing the monitoring activity about cannabinoid content in hemp-containing dietary products, the problem should be addressed also in its origin. In fact, a more attentive hemp manufacturing should be encouraged, focused not only on avoiding cross contamination between seeds and inflorescences during the harvesting but also on selecting the correct hemp cultivar, that is, the one with a low content in psychoactive cannabinoids.

Applications to other areas

In this chapter, we have reported all the works from the scientific literature about the determination of the cannabinoid content in hemp-derived ingredients, food, and beverages. We have summarized all the analytical technique and purification procedures used since today to determine the major cannabinoids in the above-mentioned matrices. Also, we have reported and evaluated all the analytical results found out about cannabinoid levels in food and beverages containing hemp as ingredient. The contamination data collected since today suggest that the ingestion of those dietary products could represent a risk to the consumers' health because of the possible presence of the psychoactive cannabinoids. Also, only a few countries have regulations about cannabinoid content in food and beverages containing hemp and the monitoring activities are still scarce. Those data could be useful to implement the monitoring activities worldwide and maybe to focus the attention on some categories of food and beverages that have already been found to contain high levels of psychoactive cannabinoids. That should lead to a greater attention in both the selection of a hemp cultivar with a low content of Δ^9-THC, Δ^8-THC, and THCA-A and a proper manufacturing procedure in order to avoid cross contamination between the seeds and the apical part of the plant. Moreover, these data could be used to estimate the consumers' intake of cannabinoids from the consumption of the analyzed food and beverages, also considering the different population groups. The risk related to the cannabinoid intake, in fact, should be evaluated also considering population age and susceptibility, as in the case of the junior formula containing Δ^9-THC reported by Escrivá et al. (2017).

Mini-dictionary of terms

- **ARfD**. The ARfD of a chemical is defined from FAO as "an estimate of a substance in food or drinking water, expressed on body weight basis, that can be ingested over a short period of time, usually during one meal or one day, without appreciable health risk to the consumer on the basis of all known facts at the time of evaluation".
- **CBN**. The main product of the oxidation of Δ^9-THC with mild psychoactivity; its concentration rises with the plant aging.
- **Chromatography**. A separation technique always coupled to a detector, being it UV, FLD, DAD or Mass Spectrometry detector. It allows to separate the molecules basing on their affinity with a stationary and a mobile phase.
- **Δ^8-THC**. A structural isomer of Δ^9-THC with lower psychoactivity.
- **THCA-B**. A minor structural isomer of THCA-A.
- **Mass spectrometry**. An instrumental technique allowing to identify molecules basing on their mass-to-charge *ratio*.
- **Tandem mass spectrometry**. A particularly selective experimental mode used in mass spectrometry detection based on the isolation of a precursor ion, its fragmentation, and the isolation of its product ions.

Summary points

- The hemp has been used as food and beverage ingredient for a long time because of its proved nutritional properties and, more recently, because of the CBD supposed beneficial effects.
- The use of the wrong hemp cultivar or the incorrect manufacturing could cause the presence of psychoactive cannabinoids in food and beverages.
- The concern about the risk for consumers' health has led to monitor the cannabinoid content in some categories of dietary products.
- The most analyzed dietary products have been the hemp-derived ingredients, such as hemp seeds, hempseed oil, flour, and protein powder, which are composed by 100% hemp.
- Most of the scientific works analyzed only the major cannabinoids, such as Δ^9-THC, THCA-A, CBD, and CBDA, not considering all the other cannabinoids still significantly present in hemp.
- The spread of more performing analytical techniques caused the implement of the monitoring activity, both in terms of matrix variety and of number of tested cannabinoids.
- The occurrence data show that the presence of psychoactive cannabinoids in dietary products containing hemp is a real problem that should not be underestimated.
- The results suggest that the monitoring of the hemp manufacturing procedures and of the cannabinoid content in dietary products should be implemented.

References

Avico, U., Pacifici, R., & Zuccaro, P. (1985). Variations of tetrahydrocannabinol content in cannabis plants to distinguish the fibre-type from drug-type plants. *Bulletin on Narcotics*, *37*(4), 61–65.

Bosy, T. Z., & Cole, K. A. (2000). Consumption and quantitation of Δ9-tetrahydrocannabinol in commercially available hemp seed oil products. *Journal of Analytical Toxicology*, *24*(7), 562–566.

Brighenti, V., Licata, M., Pedrazzi, T., Maran, D., Bertelli, D., Pellati, F., & Benvenuti, S. (2019). Development of a new method for the analysis of cannabinoids in honey by means of high-performance liquid chromatography coupled with electrospray ionisation-tandem mass spectrometry detection. *Journal of Chromatography A*, *1597*, 179–186.

Callaway, J. C. (2004). Hempseed as a nutritional resource: An overview. *Euphytica*, *140*, 65–72.

Chang, C. W., Tung, C. W., Tsai, C. C., Wu, Y. T., & Hsu, M. C. (2017). Determination of cannabinoids in hemp nut products in Taiwan by HPLC-MS/MS coupled with chemometric analysis: Quality evaluation and a pilot human study. *Drug Testing and Analysis*, *9*(6), 888–897.

Chen, T., He, J., Zhang, J., Zhang, H., Qian, P., Hao, J., & Li, L. (2010). Analytical characterization of hempseed (seed of Cannabis sativa L.) oil from eight regions in China. *Journal of Dietary Supplements*, *7*(2), 117–129.

Christinat, N., Savoy, M. C., & Mottier, P. (2020). Development, validation and application of a LC-MS/MS method for quantification of 15 cannabinoids in food. *Food Chemistry*, *318*, 126469.

Ciolino, L. A., Ranieri, T. L., & Taylor, A. M. (2018a). Commercial cannabis consumer products part 1: GC–MS qualitative analysis of cannabis cannabinoids. *Forensic Science International*, *289*, 429–437.

Ciolino, L. A., Ranieri, T. L., & Taylor, A. M. (2018b). Commercial cannabis consumer products part 2: HPLC-DAD quantitative analysis of cannabis cannabinoids. *Forensic Science International*, *289*, 438–447.

Citti, C., Pacchetti, B., Vandelli, M. A., Forni, F., & Cannazza, G. (2018). Analysis of cannabinoids in commercial hemp seed oil and decarboxylation kinetics studies of cannabidiolic acid (CBDA). *Journal of Pharmaceutical and Biomedical Analysis*, *149*, 532–540.

Crini, G., Lichtfouse, E., Chanet, G., & Morin-Crini, N. (2020). Applications of hemp in textiles, paper industry, insulation and building materials, horticulture, animal nutrition, food and beverages, nutraceuticals, cosmetics and hygiene, medicine, agrochemistry, energy production and environment: A review. *Environmental Chemistry Letters*, *18*(5), 1451–1476.

Das, L., Liu, E., Saeed, A., Williams, D. W., Hu, H., Li, C., … Shi, J. (2017). Industrial hemp as a potential bioenergy crop in comparison with kenaf, switchgrass and biomass sorghum. *Bioresource Technology*, *244*, 641–649.

de Meijer, E. P. M., van der Kamp, H. J., & van Eeuwijk, F. A. (1992). Characterisation of *Cannabis* accessions with regard to cannabinoid content in relation to other plant characters. *Euphytica*, *62*, 187–200.

Di Marco Pisciottano, I., Guadagnuolo, G., Soprano, V., Esposito, M., & Gallo, P. (2021). A survey of Δ9-THC and relevant cannabinoids in products from the Italian market: A study by LC–MS/MS of food, beverages and feed. *Food Chemistry*, *346*, 128898.

Dräger, H., Barthlott, I., Golombek, P., Walch, S. G., & Lachenmeier, D. W. (2022). Time trends of tetrahydrocannabinol (THC) in a 2008–2021 German National Survey of hemp food products. *Food*, *11*(3), 486.

Dubrow, G. A., Pawar, R. S., Srigley, C., Sam, J. F., Talavera, C., Parker, C. H., & Noonan, G. O. (2021). A survey of cannabinoids and toxic elements in hemp-derived products from the United States marketplace. *Journal of Food Composition and Analysis*, *97*, 103800.

Ellis, L. D., Berrue, F., Morash, M., Achenbach, J. C., Hill, J., & McDougall, J. J. (2018). Comparison of cannabinoids with known analgesics using a novel high throughput zebrafish larval model of nociception. *Behavioural Brain Research*, *337*, 151–159.

Escrivá, Ú., Andrés-Costa, M. J., Andreu, V., & Picó, Y. (2017). Analysis of cannabinoids by liquid chromatography–mass spectrometry in milk, liver and hemp seed to ensure food safety. *Food Chemistry*, *228*, 177–185.

European Commission. (2016). Commission recommendation (EU) 2016/2115 of 1 December 2016 on the monitoring of the presence of Δ^9-- tetrahydrocannabinol, its precursors and other cannabinoids in food. *Official Journal*, *L327*, 103–104.

European Food Safety Authority (EFSA), Arcella, D., Cascio, C., & Mackay, K. (2020). Acute human exposure assessment to tetrahydrocannabinol (Δ9-THC). *EFSA Journal*, *18*(1), e05953.

Fetterman, P. S., Keith, E. S., Waller, C. W., Guerrero, O., Doorenbos, N. J., & Quimby, M. W. (1971). Mississippi-grown Cannabis sativa L.: Preliminary observation on chemical definition of phenotype and variations in tetrahydrocannabinol content versus age, sex, and plant part. *Journal of Pharmaceutical Sciences*, *60*(8), 1246–1249.

Formato, M., Crescente, G., Scognamiglio, M., Fiorentino, A., Pecoraro, M. T., Piccolella, S., et al.…. (2020). (–)-Cannabidiolic acid, a still overlooked bioactive compound: An introductory review and preliminary research. *Molecules*, *25*(11), 2638. https://doi.org/10.3390/molecules25112638.

Hanuš, L. O., Meyer, S. M., Muñoz, E., Taglialatela-Scafati, O., & Appendino, G. (2016). Phytocannabinoids: A unified critical inventory. *Natural Product Reports*, *33*(12), 1357–1392.

Hobbs, J. M., Vazquez, A. R., Remijan, N. D., Trotter, R. E., McMillan, T. V., Freedman, K. E., … Weir, T. L. (2020). Evaluation of pharmacokinetics and acute anti-inflammatory potential of two oral cannabidiol preparations in healthy adults. *Phytotherapy Research*, *34*(7), 1696–1703.

Holler, J. M., Bosy, T. Z., Dunkley, C. S., Levine, B., Past, M. R., & Jacobs, A. (2008). Δ9-Tetrahydrocannabinol content of commercially available hemp products. *Journal of Analytical Toxicology*, *32*(6), 428–432.

Izzo, L., Pacifico, S., Piccolella, S., Castaldo, L., Narváez, A., Grosso, M., & Ritieni, A. (2020). Chemical analysis of minor bioactive components and cannabidiolic acid in commercial hemp seed oil. *Molecules*, *25*(16), 3710.

Jang, E., Kim, H., Jang, S., Lee, J., Baeck, S., In, S., … Han, E. (2020). Concentrations of THC, CBD, and CBN in commercial hemp seeds and hempseed oil sold in Korea. *Forensic Science International*, *306*, 110064.

Khan, R., Naveed, S., Mian, N., Fida, A., Raafey, M. A., & Aedma, K. K. (2020). The therapeutic role of cannabidiol in mental health: A systematic review. *Journal of Cannabis Research*, *2*(1), 1–21.

Kladar, N., Čonić, B. S., Božin, B., & Torović, L. (2021). European hemp-based food products–health concerning cannabinoids exposure assessment. *Food Control*, *129*, 108233.

Lachenmeier, D. W., Kroener, L., Musshoff, F., & Madea, B. (2004). Determination of cannabinoids in hemp food products by use of headspace solid-phase microextraction and gas chromatography–mass spectrometry. *Analytical and Bioanalytical Chemistry*, *378*(1), 183–189.

Lalge, A. B., Mendel, P., Vyhnanek, T., Trojan, V., Fiserova, H., Hřivna, L., … Havel, L. (2016). GC-FID analysis of food samples made of hemp. In *MendelNet 2016: Proceedings of international PhD students conference* (pp. 600–604).

Lazarini-Lopes, W., Do Val-da Silva, R. A., da Silva-Júnior, R. M., Leite, J. P., & Garcia-Cairasco, N. (2020). The anticonvulsant effects of cannabidiol in experimental models of epileptic seizures: From behavior and mechanisms to clinical insights. *Neuroscience & Biobehavioral Reviews*, *111*, 166–182.

Lehmann, T., Sager, F., & Brenneisen, R. (1997). Excretion of cannabinoids in urine after ingestion of cannabis seed oil. *Journal of Analytical Toxicology*, *21*(5), 373–375.

Madej, K., Chmiołek, A., Szlachta, K., & Piekoszewski, W. (2021). HPLC-DAD analysis of hemp oil supplements for determination of four cannabinoids: Cannabidiol, cannabidiolic acid, cannabinol and delta 9-tetrahydrocannabinol. *Separations*, *8*(12), 227.

Matsunaga, T., Nagatomo, H., Yamamoto, I., & Yoshimura, H. (1990). Identification and determination of cannabinoids in commercially available cannabis seeds. *Eisei kagaku*, *36*(6), 545–547.

Mechoulam, R., Parker, L. A., & Gallily, R. (2002). Cannabidiol: An overview of some pharmacological aspects. *The Journal of Clinical Pharmacology*, *42*(S1), 11S–19S.

Meng, Q., Buchanan, B., Zuccolo, J., Poulin, M. M., Gabriele, J., & Baranowski, D. C. (2018). A reliable and validated LC-MS/MS method for the simultaneous quantification of 4 cannabinoids in 40 consumer products. *PLoS One*, *13*(5), e0196396.

Pellegrini, M., Marchei, E., Pacifici, R., & Pichini, S. (2005). A rapid and simple procedure for the determination of cannabinoids in hemp food products by gas chromatography-mass spectrometry. *Journal of Pharmaceutical and Biomedical Analysis*, *36*(5), 939–946.

Piccolella, S., Crescente, G., Formato, M., & Pacifico, S. (2020). A cup of hemp coffee by Moka pot from southern Italy: An UHPLC-HRMS investigation. *Food*, *9*(8), 1123.

Ranalli, P., & Venturi, G. (2004). Hemp as a raw material for industrial applications. *Euphytica*, *140*(1), 1–6.

Risoluti, R., Gullifa, G., Battistini, A., & Materazzi, S. (2020). Monitoring of cannabinoids in hemp flours by MicroNIR/chemometrics. *Talanta*, *211*, 120672.

Ross, S. A., Mehmedic, Z., Murphy, T. P., & ElSohly, M. A. (2000). GC-MS analysis of the total δ9-thc content of both drug-and fiber-type cannabis seeds. *Journal of Analytical Toxicology*, *24*(8), 715–717.

Thomas, B. F., & ElSohly, M. A. (2016). Biosynthesis and pharmacology of phytocannabinoids and related chemical constituents. In *The analytical chemistry of cannabis* (pp. 27–41). Amsterdam, The Netherlands: Elsevier.

Wong, H., & Cairns, B. E. (2019). Cannabidiol, cannabinol and their combinations act as peripheral analgesics in a rat model of myofascial pain. *Archives of Oral Biology*, *104*, 33–39.

Zoller, O., Rhyn, P., & Zimmerli, B. (2000). High-performance liquid chromatographic determination of Δ9-tetrahydrocannabinol and the corresponding acid in hemp containing foods with special regard to the fluorescence properties of Δ9-tetrahydrocannabinol. *Journal of Chromatography A*, *872*(1–2), 101–110.

Chapter 4

The impact of prenatal cannabis exposure: An overview

Sarah F. Larsen[a], Allegra J. Johnson[a], Stephen R. Dager[b], and Natalia M. Kleinhans[c]

[a]*Department of Radiology, Integrated Brain Imaging Center, School of Medicine, University of Washington, Seattle, WA, United States,* [b]*Department of Radiology, Center on Human Development and Disability, School of Medicine, University of Washington, Seattle, WA, United States,* [c]*Department of Radiology, Integrated Brain Imaging Center, Center on Human Development and Disability, School of Medicine, University of Washington, Seattle, WA, United States*

Abbreviations

ABCD	Adolescent Brain Cognitive Development Study
ADHD	attention deficit hyperactivity disorder
CBD	cannabidiol
CB1	cannabinoid receptor type 1
D1	dopamine receptor D1, also known as DRD1
D2	dopamine receptor D2, also known as D2R
fMRI	functional magnetic resonance imaging
GD	gestational day
MHPCD	Maternal Health Practices and Child Development Study
mRNA	messenger ribonucleic acid
NSAID	nonsteroidal antiinflammatory drug
OPPS	Ottawa Prenatal Prospective Study
OTC	over the counter
PCE	prenatal cannabis exposure
POMC	pro-opiomelanocortin
SEM	standard error of the mean
SES	socioeconomic status
SSRI	selective serotonin reuptake inhibitor
THC	tetrahydrocannabinol
5-HT	serotonin (5-hydroxytryptamine)

Introduction

Current trends toward legalization and greater availability of cannabis products have led to increased consumption in general (Hasin, 2018), including during pregnancy. Cannabis use during pregnancy has increased 62% from 2002 through 2014 (Brown et al., 2017) and more than doubled when comparing 2002–03 to 2016–17 rates (Volkow et al., 2019). In 2019, cannabis use occurred in at least 1 in every 15 pregnancies (Young-Wolff et al., 2021). Factors contributing to the escalation of cannabis use while pregnant include exposure to internet resources that tout the medicinal benefits for morning sickness, anecdotal experiences, advice from friends and co-workers, and a perceived lack of evidence, or contradictory evidence, about the harms of prenatal cannabis exposure (PCE) in humans (Jarlenski et al., 2016).

Public perspectives are shifting toward regarding cannabis as a benign treatment alternative for a variety of pregnancy symptoms (Bayrampour et al., 2019). These shifting perspectives about potential cannabis harms to their fetus are balanced with challenging decisions about their own self-care. Pregnant people are looking to cannabis as an alternative to medications that are potentially more dangerous to the baby such as opioids, NSAIDS, and antinausea medications that they might be prescribed for their own health conditions (Barbosa-Leiker et al., 2020). For example, pregnant people are increasingly self-medicating with cannabis for its antiemetic properties to treat "morning sickness," which is a hallmark of the first trimester of pregnancy and also a critical time window for fetal development (Volkow et al., 2017; Westfall et al., 2006).

Cannabis Use, Neurobiology, Psychology, and Treatment. https://doi.org/10.1016/B978-0-323-89862-1.00017-9

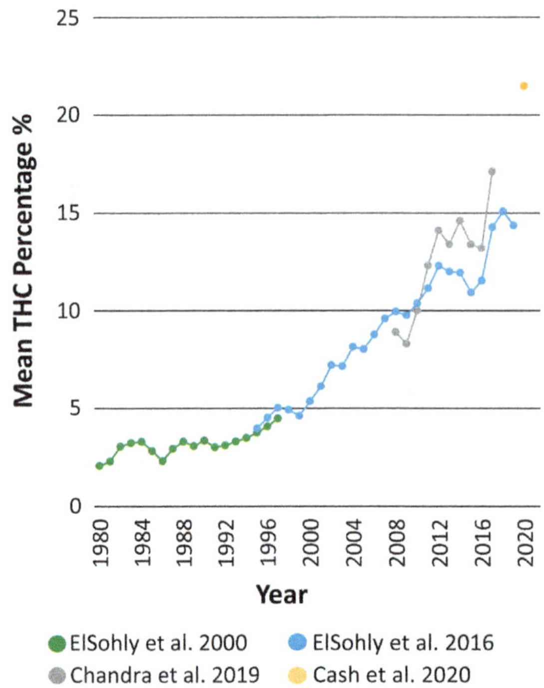

FIG. 1 US nationwide THC potency trends for cannabis flower. Between 1980 and 2020, the percentage of THC in cannabis flower has increased from 2% to over 20%. *(Data from Cash, M. C., Cunnane, K., Fan, C., & Romero-Sandoval, E. A. (2020). Mapping cannabis potency in medical and recreational programs in the United States.* PLoS One, *15(3), e0230167. https://doi.org/10.1371/journal.pone.0230167; Chandra, S., Radwan, M. M., Majumdar, C. G., Church, J. C., Freeman, T. P., & Elsohly, M. A. (2019). New trends in cannabis potency in USA and Europe during the last decade (2008–2017).* European Archives of Psychiatry and Clinical Neuroscience, *269(1), 5–15. https://doi.org/10.1007/s00406-019-00983-5; ElSohly, M. A., Mehmedic, Z., Foster, S., Gon, C., Chandra, S., & Church, J. C. (2016). Changes in cannabis potency over the last 2 decades (1995–2014): Analysis of current data in the United States.* Biological Psychiatry, *79(7), 613–619. https://doi.org/10.1016/j.biopsych.2016.01.004; ElSohly, M. A., Ross, S. A., Mehmedic, Z., Arafat, R., Yi, B., & Banahan III, B. F. (2000). Potency trends of delta9-THC and other cannabinoids in confiscated marijuana from 1980-1997.* Journal of Forensic Sciences, *45(1), 24-30. https://www.ncbi.nlm.nih.gov/pubmed/10641915.)*

From a public health perspective, understanding the impacts of PCE is becoming urgent due to increasing cannabis potencies. The content of THC in the flower bud (Fig. 1) has increased tenfold in the last 40 years in the United States, from 2% to over 20% (Cash et al., 2020; Chandra et al., 2019; ElSohly et al., 2016, 2000). For an average 0.5 g cannabis joint in 1980, with 2% THC potency, 10 mg of THC was consumed. Today that same joint delivers a dose of 100 mg. In addition to increased potency of flower bud, concentrated extracts made from the cannabis resin, termed "hash," "wax," and "shatter" are gaining in market share. In Washington State, the mean THC content for these extracts was 68.7% (Smart et al., 2017). This trend toward utilization of higher potency concentrates by the public at large is also reflected in the use patterns during pregnancy. Pregnant women in Washington State consume cannabis flower 44% of the time, concentrates 42%, edibles 8%, tinctures 2% with the remaining 4% consisting of other forms (N. Kleinhans unpublished data). Thus, concentrates are used by pregnant people almost as often as flower, and combined, these two categories comprise 86% of products used.

Given that all medications have a dose-response curve, and higher doses typically come with more adverse side effects, this dramatic increase in potency along with increased utilization of more potent extracts suggest adverse outcomes may be more evident among PCE infants born today as compared with those studied in years past.

Ethical challenges to studies during pregnancy

The title of this chapter highlights the experience of the unborn child; however, with a pregnancy, the health of the mother is inseparable from that of the child. The safety and efficacy of medications used during pregnancy are often poorly studied, and this is particularly true for cannabis. One of the biggest challenges to studying PCE is the same challenge that researchers face when studying pregnant people for any prescription drug. Pregnancy historically has precluded individuals from being included in studies for new medications. Very few drugs have undergone clinical trials on pregnant people. Ethical questions juxtapose the idea of beneficence, what is in the best interest of a fetus who is not able to provide informed

TABLE 1 Therapeutics used for common pregnancy ailments.

Drug category	Most prevalent form	Ailment	Women taking medication (%)	Women taking medication during pregnancy (%)	1st trimester (%)
Any drug category			99.3	97.1	95.7
Gastrointestinal/ antiemetic	5-HT_3 antagonists	Nausea, morning sickness	35.6	34.3	19.5
Antibiotics	Penicillin derivatives	Bacterial infection	28.3	25.5	12.6
Analgesics total	–	Pain/ inflammation	25.9	23.7	15.6
	Acetaminophen		NR	19.9	11.7
	NSAIDs		NR	6.6	NR
	Opioids		NR	3.6	1.4
Antidepressants	SSRI	Depression	NR	6.1	5.3

Percent of prescription and over the counter (OTC) drugs taken by women in general, at any time during pregnancy, and during the developmentally sensitive first trimester of pregnancy. *NR*, Not reported.
Data from Haas, D. M., Marsh, D. J., Dang, D. T., Parker, C. B., Wing, D. A., Simhan, H. N., Grobman, W. A., Mercer, B. M., Silver, R. M., Hoffman, M. K., Parry, S., Iams, J. D., Caritis, S. N., Wapner, R. J., Esplin, M. S., Elovitz, M. A., Peaceman, A. M., Chung, J., Saade, G. R., & Reddy, U. M. (2018). Prescription and other medication use in pregnancy. *Obstetrics & Gynecology, 131*(5), 789–798. https://doi.org/10.1097/aog.0000000000002579.

consent, with the principle of justice, that pregnant people deserve effective and safe treatment during pregnancy (Matsui, 2015). Acknowledging that as previously observed (Baylis, 2010), "Pregnant women get sick, and sick women get pregnant," Table 1 shows an abbreviated list of common prescription drugs that women take, and how these numbers change with pregnancy. Women are treated for a variety of ailments which do not disappear during pregnancy.

Pregnant people have problems that start during pregnancy, such as morning sickness, infection, back pain, depression, or have preexisting conditions that continue during pregnancy. To treat these symptoms, commonly used prescription and OTC treatment options are used; however, these therapeutics also carry some risk to the unborn fetus (Table 2). Each woman, with the help or guidance of medical professionals, therefore must carefully weigh the risks to the fetus as best as can be determined against the benefits of her own treatment for a given medical condition. Within this dose-dependent risk benefit context of all therapeutics, cannabis is now entering the mainstream. Cannabis is a potential therapeutic for nausea, neuropathic pain, sleep issues, and depression (Russo et al., 2007; Westfall et al., 2006). Given that other prescription or OTC treatment options have many concerning side effects, some opt to self-medicate with cannabis. This situation, in which there is no risk-free option, increases the need to understand at what doses therapeutic effect is obtained by THC, and at what doses negative side effects occur so that informed choices for medical treatment may be made during pregnancy.

Current state of PCE research

Animal studies

Basic science and preclinical studies using animal models provide evidence suggesting possible long-term effects that need to be studied further. Animal studies have demonstrated that PCE is associated with long-term effects on motor skills, cognition, socioemotional behaviors, and reward processing. For example, perinatal exposure to delta-9-tetrahydrocannabinol (THC) alters dopamine-mediated motor behaviors (Moreno et al., 2003) including immobility, inhibited locomotion and a reduction in the stereotyped behaviors typically observed after injection of dopamine receptor D2 agonists. THC administration during gestation and lactation caused reduced memory retention and social discrimination in those offspring, which persisted into adulthood (Campolongo et al., 2007). THC during gestation also produced offspring with increased

TABLE 2 Prenatal exposure risks of commonly utilized prescription and OTC drugs.

Drug category	Prenatal exposure risk
Antiemetic	Cleft palate Cardiovascular defects[a]
Antibiotics	Asthma[b] Childhood obesity[c]
Acetaminophen	ADHD[d] Asthma[e]
NSAID	Miscarriage[f] Impairs central nervous system development, compromises neural activity, impairs sciatic nerve morphology[g]
Opioid	Smaller brain volumes[h] Reduced cognitive and motor scores[i]
SSRI	Brain structure and connectivity issues[j] Miscarriage[k]

Adverse outcomes attributed to fetal exposure to drug categories commonly used during pregnancy.
[a]*Danielsson et al. (2014).*
[b]*Metsälä et al. (2015).*
[c]*Wang et al. (2018).*
[d]*Masarwa et al. (2018).*
[e]*Liew et al. (2021).*
[f]*Nakhai-Pour et al. (2011).*
[g]*Canan et al. (2008).*
[h]*Merhar et al. (2021).*
[i]*Yeoh et al. (2019).*
[j]*Lugo-Candelas et al. (2018).*
[k]*Rahimi et al. (2006).*

emotionality, including higher rates of isolation-induced ultrasonic vocalizations, reduced play-related behavior, and increased anxiety and emotional reactivity that persisted across life stages (Trezza et al., 2008). Animal tissue studies have provided strong evidence that PCE alters neural development through its interaction with the endogenous cannabinoid signaling system. THC passes through the placenta (Grotenhermen, 2003) and binds to cannabinoid 1 (CB1) receptors, which are found in all layers of the placental membrane (Park et al., 2003). Stimulation of CB1 receptors impairs fetal growth by inhibiting cytotrophoblast cell proliferation (Khare, 2006). Animal studies have also demonstrated altered protein and chemical expression in the brains of THC-exposed offspring. Adult male offspring of mothers exposed to THC during gestation, and via lactation, show reduced prefrontal cortex concentrations of glutamate and norepinephrine (Campolongo et al., 2007), increased susceptibility to seizures (De Salas-Quiroga et al., 2015), and upregulation of genes involved in apoptosis and downregulation of genes involved in myelination and glutamate production. Critically, repeated THC exposure in utero can disrupt endocannabinoid signaling, rewire cortical circuitry (Tortoriello et al., 2014), and interfere with the development of cortical projection neurons, resulting in fewer subcerebral projection neurons in THC-exposed rats compared to controls (De Salas-Quiroga et al., 2015).

White matter microstructure, and corresponding functional networks, may be particularly vulnerable to cannabis exposure during the prenatal period. Comparative research using a rodent model has provided evidence that during the pre/perinatal stage, cannabinoid receptors are distributed throughout white matter tracts (Berrendero et al., 1999). Notably, major commissural pathways (i.e., corpus callosum, anterior commissure, and stria terminalis) have higher levels of CB1 receptors than the nucleus accumbens, caudate-putamen, and the hippocampus in the pre/perinatal stage (Berrendero et al., 1999). This pattern of higher receptor binding in white matter is transient: during childhood, cannabinoid receptor density becomes depleted, acquiring the classic adult pattern. The opposite trend occurs in neural tissue, with CB1 receptor density increasing over time and peaking in adulthood (Fig. 2). Consistent with this developmental pattern, Gómez et al. (2003) found that transcripts for neural adhesion molecule L1, which plays an important role in brain development, were selectively increased in several major white matter tracts including the corpus callosum, fimbria, and the stria terminalis, but not grey matter structures such as the amygdala, hippocampus, or the striatum in animals exposed to THC in utero. L1 is found in neuronal and nonneuronal cells, including immature oligodendrocytes and Schwann cells, which are involved in the formation of myelin.

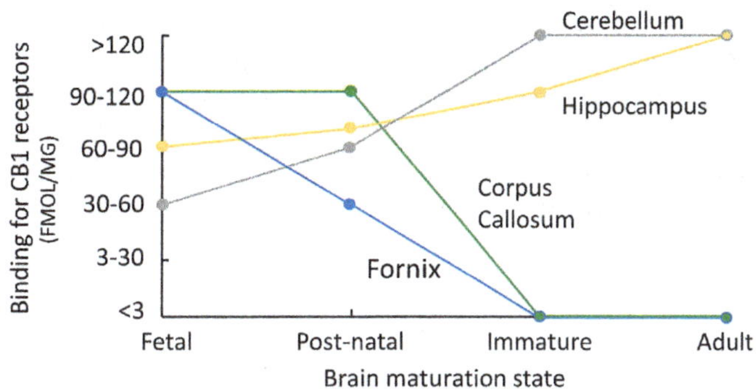

FIG. 2 Developmental changes in CB1 receptor binding. Cannabinoid receptor type 1 (CB1) binding rates for the cerebellum and hippocampus brain regions and the corpus callosum and fornix white matter tracts. *(From Zalesky, A., Solowij, N., Yucel, M., Lubman, D. I., Takagi, M., Harding, I. H., ... Seal, M. (2012). Effect of long-term cannabis use on axonal fibre connectivity.* Brain, *135(7), 2245–2255. https://doi.org/10.1093/brain/aws136, used with permission.)*

Postmortem human studies corroborate animal results

Although behavioral studies suggest altered neurodevelopment in PCE infants, very little information is available related to the neurobiological consequences of PCE in humans. A series of postmortem studies on a sample of human fetal tissue have corroborated some of the animal literature that reported alterations in neural development. Hurd and colleagues (Wang et al., 2004) measured mRNA expression of CB1, D1, and D2 in the putamen, amygdala, and the hippocampus in 21 controls and 21 cannabis-exposed mid-gestation fetuses. Pregnant mothers were categorized according to their amount of regular cannabis use, which included none, low (average <0.4 joints/day), moderate (>0.4 and <0.89 joints/day), and high (>0.89 joints/day). Significantly reduced D2 mRNA expression levels were observed in the amygdala across all cannabis exposed fetuses, exposed to an average of 1.2 joints/day (Wang et al., 2004). Reductions in D2 expression were apparent in even the lightest users. Further, reduced D2 expression in the fetal tissue was correlated to higher levels of maternal cannabis intake during pregnancy (Wang et al., 2004). No differences were found in CB1 or D1. Related studies by the same group found reduced D2 receptors (D2R) and DA expression in the nucleus accumbens (Spano et al., 2007). *Reductions in D2 expression and DA release are characteristics of addiction in adults* (Volkow et al., 2007), and this work indicates that D2R impairments predate adult drug use in individuals with PCE, suggesting *neurobiological changes associated with addiction may be present at birth.*

Sex differences in PCE outcome?

There is mounting evidence suggesting that PCE affects male and female fetuses differently. Many animal studies have shown that perinatal administration of THC alters the expression of genes involved in brain development differently in males and females. The opioid peptide precursors proenkephalin (PENK) mRNA levels increase in females but decrease in males following administration of cannabinoids at gestational day (GD) 21 in the caudate-putamen, hypothalamic, paraventricular and ventromedial nuclei, and cerebral cortex (Pérez-Rosado et al., 2000) (sexual dimorphic effects in PENK were not seen at earlier periods). A follow-up study showed similar effects in proopiomelanocortin (POMC): prenatal THC exposure altered POMC mRNA levels in the arcuate nucleus and the cerebral cortex, with sex differences appearing at GD21 (Perez-Rosado et al., 2002) (females higher, males lower). Notably, animals that were perinatally exposed to THC show sex differences in signs of opioid vulnerability. THC-exposed males, but not females, show withdrawal symptoms following a single dose of naloxone, tolerance to analgesic effects (Vela et al., 1995), and higher morphine preference in a conditioned place preference test (Rubio, 1998). The dopaminergic system also appears to show sex differences, with the impact of PCE having a more deleterious effect in males than in females (De Fonseca et al., 1991). Perinatal exposure to THC has been shown to alter D2-mediated motor behaviors in rats (Moreno et al., 2003). Sex differences in locomotor behavior have been observed in animals with perinatal exposure to doses of THC (2 or 5 mg/kg): adult male offspring showed reduced locomotor behavior relative to controls, while females showed increased locomotor activity relative to controls, suggesting that prenatally THC-exposed females habituate faster than expected to new environments (Moreno et al., 2005; Navarro et al., 1994). Emotional behavior also appears to differ between male and female prenatally THC-exposed animals (Rubio, 1998). In a study measuring exploratory behaviors in a novel environment, THC produced disinhibition in males, who emerge 2–3 min sooner from the safe chamber than male rats without THC exposure.

In contrast, no differences were observed between female rats with and without THC exposure. Finally, a human fetal tissue study (described above) (Wang et al., 2004) reported that D2 receptor expression in the amygdala during embryonic development may be lower in males than females with PCE. This finding was specific to D2 receptors; no differences were observed in D1 or CB1 receptors (Wang et al., 2004). Fig. 3, published in (Wang et al., 2004), shows D2 mRNA expression levels in male and female fetuses at 18–22 gestational weeks of age (A). Panel (B) shows the distribution pattern of the D2 mRNA expression in the amygdala of one control and one matched cannabis-exposed male subject at 20-gestational weeks of age.

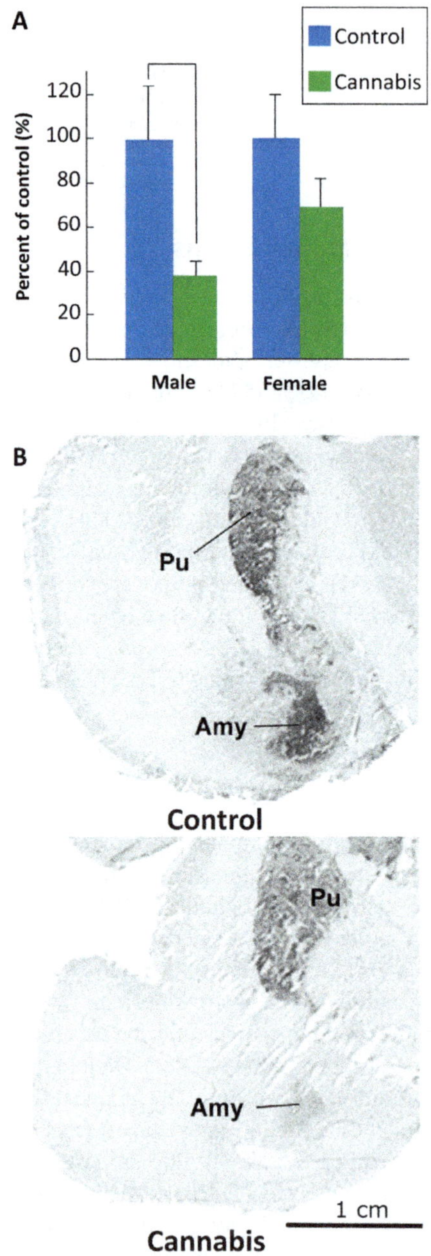

FIG. 3 Dopamine D2 mRNA expression levels in the basal amygdala nucleus. The dopamine D2 mRNA expression levels (mean ± SEM) in the basal amygdala nucleus of male and female fetuses (18–22 gestational weeks) (A) and the distribution pattern of the D2 mRNA expression in the amygdala of a control and cannabis-exposed (B) 20-week male subject. The graphs show the cannabis-exposed group (*green* bar) expressed as a percentage of control (*blue* bar). *P < .05. Key: *SEM*, standard error of the mean; *Pu*, putamen; *amy*, amygdala. *(From Wang, X., Dow-Edwards, D., Anderson, V., Minkoff, H., & Hurd, Y. L. (2004). In utero marijuana exposure associated with abnormal amygdala dopamine D2 gene expression in the human fetus. Biological Psychiatry, 56(12), 909-915. https://doi.org/10.1016/j.biopsych.2004.10.015, with permission.)*

Human studies

Cannabis use during pregnancy can alter animal brain development and lay the neurological foundation for developing behavioral problems, executive dysfunction, and addiction later in life (Spano et al., 2007; Szutorisz et al., 2014). However, neural correlates of PCE in humans have been reported in only a few studies: one found increased cortical thickness in 6- to 8-year-old children with PCE after controlling for prenatal tobacco exposure (El Marroun et al., 2016) and the other used fMRI to investigate executive dysfunction in young adults with PCE (Smith et al., 2016). Grewen and colleagues found altered connectivity (hypo-connectivity) of subcortical seed regions having high fetal CB1R expression in association with PCE for areas involved in visual spatial and motor learning, attention, and fine-tuning of motor outputs (Grewen et al., 2015). In that study, statistical approaches were employed in an attempt to isolate cannabis effects from effects of concurrent use of opioids, nicotine, alcohol, and/or selective serotonin reuptake inhibitors (Grewen et al., 2015). A second fetal imaging study reported increased hippocampal connectivity in association with cannabis exposure in utero (Thomason et al., 2021). While these studies support the view that PCE alters brain development, the paucity of data in this area is striking, and further challenged by many postpartum factors also associated with altered brain development and behavioral challenges, see for example (Krugers et al., 2017).

Large, elegant studies from the 1980s and 1990s (Day & Richardson, 1991; Fried, 1980) likely underestimate the sequelae associated with PCE today, given that the potency of THC and the ratio of THC to CBD have increased substantially since these foundational studies were conducted. Within the United States, cannabis potency has increased 10-fold between 1980 and 2020, Fig. 1, from 2% to over 20% for cannabis flower (Cash et al., 2020; Chandra et al., 2019; ElSohly et al., 2016, 2000). Further, the ratio of THC to CBD has increased from 14:1 to 104:1 between 1995 and 2017 (Chandra et al., 2019; Elsohly et al., 2016), which also may result in a greater potential for harm given that CBD has been found to have opposing effects to THC (Pertwee, 2008). To gain an appreciation of the comparative dose difference, in the 1980s MHPCD study (Day & Richardson, 1991) and the OPPS study (Fried, 1980), one joint was approx. 10 mg THC. Heavy users were considered to have been using one or more joints per day or a weekly dose of >70 mg. In 2005, the time frame of the cannabis consumption for the Generation R and ABCD studies (El Marroun et al., 2009; Winiger & Hewitt, 2020), a joint would be approximately 8% THC, 40 mg of THC per joint, and a weekly dose of >280 mg for a heavy user. One joint a day in 2020 has a potency of over 20% yielding >700 mg/week for a heavy user. Using number of joints as a metric does not yield comparable results. These older studies, however, even at the relatively low cannabis dosages, have shown some concerning trends as children age. See Table 3 for summary of four large studies OPPS, MHPCD, and Generation R, and a current longitudinal study, ABCD, which is looking specifically at older children >9 year old and retrospectively categorizing for PCE.

TABLE 3 Developmental timeline of findings for major PCE studies.

	OPPS[1,a,b,c]	MHPCD[2,d,e]	Generation R[3,f,g,h,i,j,k,l,m,n]	ABCD[4,o,p]
PCE year	1978	1982	2001–06	2005–09
Fetus (in utero)	↓ Gestational age	↓ Birth length—1st tri exposure ↑ Birth weight—3rd tri exposure	↓ Birth weight ↑ Growth—2nd tri to parturition	
Neonate (0–28 days)	↓ Response to light ↑ Startle response, tremors	↓ Body length subpopulation: slight changes in EEG traces		
9 month		↓ Mental development ↓ BSID scores		
18 month			↓ Attention ↑ Aggression (for girls only)	

Continued

TABLE 3 Developmental timeline of findings for major PCE studies—cont'd

	OPPS[1,a,b,c]	MHPCD[2,d,e]	Generation R[3,f,g,h,i,j,k,l,m,n]		ABCD[4,o,p]
3 year	↑ Motor skills	↓ Short-term memory ↓ Verbal reasoning (African Americans only) ↓ Sleep efficiency ↑ Nocturnal arousals ↑ Wake-time after sleep onset			
4 year	↓ Memory ↓ Verbal scores				
6 year	↓ Attention ↑ Impulsivity ↑ Hyperactivity	↓ Concentration ↓ Overall IQ score ↓ Verbal reasoning ↓ Quantitative reasoning ↓ Short-term memory ↑ Impulsivity ↑ Hyperactivity ↑ Delinquency	↑ Cortisol	↑ Frontal cortical thickness	
7 year			↑ Externalization of problems		
8 year					
9 year 10 year	↓ Visual perception ↑ Impulsivity	↓ Abstract reasoning ↓ Visual reasoning ↓ Concentration ↓ Internalization (Implosion) ↓ Learning and memory ↓ IQ score ↑ Externalization (Explosion) ↑ Depression ↑ Impulsivity ↑ Hyperactivity ↑ Delinquency		↑ Psychotic-like experiences	↑ Psychopathology characteristics (i.e., psychotic-like experiences and internalizing, externalizing, attention, thought and, social problems) ↑ Disorders of sleep: initiating and maintaining sleep, arousal, sleep wake, excessive somnolence, summed sleep disorder score
11 year					
12 year					
13 year	↓ Concentration ↓ Visual memory ↓ Verbal reasoning				
14 year		↑ Delinquency			
15 year					
16 year		↓ In fine motor coordination ↑ In visual-motor coordination			

TABLE 3 Developmental timeline of findings for major PCE studies—cont'd

	OPPS[1,a,b,c]	MHPCD[2,d,e]	Generation R[3,f,g,h,i,j,k,l,m,n]	ABCD[4,o,p]
18 year	Response inhibition (fMRI): ↓ Cerebellum activity ↑ Bilateral PFC activity ↑ Premotor cortex activity Working memory (fMRI): ↓ Medial PFC activity ↓ Dorsolateral PFC activity ↓ Ventrolateral PFC activity ↑ Left medial PFC activity ↑ Inferior frontal gyrus activity ↑ Left cerebellum activity			
19 year				
20 year				
21 year				
22 year				

Four large community-based observational studies and the long-term impacts of PCE on child development. Key: ↑, increase; ↓, decrease; EEG, electroencephalogram; fMRI, functional magnetic resonance imaging; BSID, Bayley scales of infant development; PFC, prefrontal cortex.
[1]***Ottawa Prenatal Prospective Study.***
[a]*Fried (1980).*
[b]*Fried and Watkinson (1988).*
[c]*Fried et al. (1998).*
[2]***Maternal Health Practices and Child Development Study.***
[d]*Day and Richardson (1991).*
[e]*Day et al. (1991).*
[3]***Generation R study.***
[f]*Bolhuis et al. (2018).*
[g]*Cajachagua-Torres et al. (2021).*
[h]*El Marroun et al. (2019).*
[i]*El Marroun et al. (2011).*
[j]*El Marroun et al. (2016).*
[k]*El Marroun et al. (2009).*
[l]*Hofman et al. (2004).*
[m]*Jaddoe et al. (2012).*
[n]*Jaddoe et al. (2010).*
[4]***Adolescent brain cognitive development.***
[o]*Paul et al. (2021).*
[p]*Winiger and Hewitt (2020).*
Modified from McLemore, G. L., & Richardson, K. A. (2016). Data from three prospective longitudinal human cohorts of prenatal marijuana exposure and offspring outcomes from the fetal period through young adulthood. *Data in Brief, 9*, 753–757. https://doi.org/10.1016/j.dib.2016.10.005.

PCE and adverse neonatal outcomes and long-term neurodevelopment

Well-controlled studies have linked PCE to neonatal outcome measures such as low birth weight, small for gestational age, and admission to the neonatal intensive unit (Koto et al., 2022; Nguyen & Harley, 2022). There is also evidence for long-term impacts on behavior and cognition in both children and adults. For example, by early childhood, PCE children acquire visual-perceptual tasks and language skills more slowly and show increased levels of aggression and poorer attention skills (El Marroun et al., 2011). PCE in the first and second trimesters has been associated with poorer verbal and quantitative reasoning skills, as well as deficits in short-term memory (Fried et al., 2003; Goldschmidt et al., 2008). Executive dysfunction is also apparent in PCE school-age children, particularly on tasks that require impulse control (Fried et al., 1998). Finally, there is evidence that symptoms of depression are higher in children and adolescents with PCE compared to nonexposed peers (Day et al., 2011), although residual environmental and genetic confounds likely have played a role in these findings (El Marroun et al., 2019).

Confounds

Clarifying the true risk of PCE involves both not attributing risk factors to cannabis that are instead due to other sources and identifying true risks that increase with THC dose, or ratio with CBD. In prior studies, effects of maternal cannabis use have been difficult to separate from other potential contributing factors such as concurrent alcohol and/or tobacco use, adverse parenting styles, and lifestyle issues such as poor nutrition, low SES, and stress (Conner et al., 2016; Jaques et al., 2014).

Impact of CBD

Although the role of prenatal CBD exposure is unknown, there is some evidence that CBD may have neuroprotective effects (Beale et al., 2018; Ceprián et al., 2017; Mohammed et al., 2017; Wall et al., 2019; Yücel et al., 2016), suggesting that modern strains, which have been bred to have extremely high levels of THC and only trace levels of CBD, may be more harmful than what one would predict from the increase in THC potency alone. For these reasons, effects that appeared mild and/or unreliable in previous work, such as birth outcomes, may become more evident and detrimental in this modern generation of cannabis users.

Summary

Cannabis has the potential to provide therapeutic benefits, for treating ailments such as nausea, sleep disturbances, and depression, which are common ailments of pregnancy (efficacy of cannabinoids as therapeutics will be covered in other chapters). More pregnant people are turning to cannabis as a therapeutic or consuming recreationally in states in which it is legal. This, combined with the dramatic 10-fold increase of potency for commercially available cannabis raises concern that the PCE dose-dependent negative effects that previously were equivocal will become more prevalent. THC has been found to induce changes in multiple brain regions of animal offspring. Specific regions impacted in the animal studies include the putamen, with changes in dopamine mediated motor activity, the hippocampus with memory impacts, and the amygdala with emotional changes, in particular anxiety. Further, changes in the wiring of cortical circuitry were found with changes in myelination and neural projections for animal models. Postmortem human infant data have replicated some of these results showing a reduction in dopamine receptor D2 expression in the amygdala, which is a concern for predisposition to addiction.

PCE is a risk factor for adverse human neonatal outcomes such as birth weight and head circumference. Differences in later development have also been noted that may stem from changes in the putamen, hippocampus, and amygdala or their associated white matter signaling tracts. The putamen is involved in learning and motor control, including language functions and speech articulation—studies have found that PCE children acquire visual-perceptual tasks and language skills more slowly and have poorer verbal skills. The hippocampus is involved with memory formation, and PCE children have been found to exhibit short-term memory deficits. The amygdala is involved in processing emotions, and PCE children have been shown to exhibit increased levels of aggression. Together both the hippocampus and amygdala play roles in depression, which is also found to be more prevalent in PCE children and teens. All these associations have been identified in offspring of pregnant mothers who took much lower amounts of THC than are consumed today.

Applications to other areas

Gaining an understanding of the dose-related risks of cannabis consumption as well as the therapeutic effect relative to other treatment options will aid both prenatal care providers and public health policy makers in providing accurate information to expectant mothers. The key take-away relating to PCE is that more cannabis research is required to inform decisions about THC use during pregnancy and the potential risks to exposed fetuses.

Future directions:

1. What is the minimum effective THC dose required for therapeutic benefit? Do effective dose ranges differ while pregnant?
2. What THC dose confers measurable risks to the child?
3. How do the risks of cannabis use compare with those of other treatment options?
4. Does CBD modify the risk profile for THC, or independently provide therapeutic benefit without psychoactive effects or impacts on fetal development?
5. Can PCE lead to increased risk of drug use and dependency later in life?

Future study designs can take advantage of new regulatory requirements, such as product potency labeling in states where cannabis is regulated, to more accurately determine dosage consumed in observational studies, also allowing comparison between studies.

Key facts of PCE

- Cannabis is currently consumed in 6%–7% of pregnancies and usage is trending higher.
- The potency of cannabis flower has increased tenfold in potency during the last 40 years, from 2% to well over 20%.
- High potency extracts (68.7% THC in Washington State) are used during pregnancy 42% of the time (Unpublished data N. Kleinhans).
- Cannabis has demonstrated therapeutic efficacy for common pregnancy symptoms, including nausea "morning sickness," sleep disturbances, and back pain.
- PCE has been found to impact birth outcomes (low birth weight, small for gestational age, and reduced head circumference).
- PCE has been found to affect multiple brain regions including: the putamen, hippocampus, and the amygdala, as well as changes to white matter tracts interconnecting those regions.
- PCE reduces D2 expression in the amygdala. Reductions in D2 expression and DA release are characteristics of addiction in adults, suggesting that neurobiological changes associated with addiction may be present at birth in PCE infants.
- PCE neurobiological effects may be differentially impacted by fetal sex; for example, there is greater impact of PCE on male D2 expression.

Mini-dictionary of terms

Adverse birth outcomes: These are metrics that have been linked to poor neonatal outcomes and include low birth weight, small for gestational age, decreased head circumference, and length.
Amygdala: An almond-shaped deep brain region involved in emotion processing, in particular fear, anxiety, and aggression.
Hippocampus: A brain region in the temporal lobes involved with learning and memory formation.
Putamen: A brain region involved in learning and motor control, including language functions and speech articulation.
White matter tracts: Bundles of myelinated axons that connect neural cells bodies in one brain region to another. Myelination is a fatty substance that coats axons that increases signal transmission rates and appears white, as opposed to the unmyelinated neuron cell bodies which appear grey and are termed grey matter.

Summary points

- Frequency of cannabis use and potency of cannabis used are increasing among pregnant people.
- Cannabis has the potential as a therapeutic for treating common pregnancy ailments such as morning sickness, back pain, and sleep disruption.
- Previous outcome studies, which reported near equivocal results, do not reflect current self-dosing levels. Modern strains have the potential to induce significant fetal health impacts.
- Even at lower doses, past studies indicate that THC consumption during pregnancy (PCE) is a risk factor for adverse birth outcomes such as being born with low birth weight or small for gestational age.
- PCE has been found to affect numerous brain regions including: the putamen (motor control), hippocampus (short-term memory), and the amygdala (emotional regulation), as well as changes to white matter tracts interconnecting those brain regions.
- Updated studies needed
  - Studies that reflect current dose levels to identify dose-related harms.
  - Double-blind studies that identify the minimum effective cannabinoid dose for targeted therapeutic efficacy during pregnancy.

References

Barbosa-Leiker, C., Burduli, E., Smith, C. L., Brooks, O., Orr, M., & Gartstein, M. (2020). Daily cannabis use during pregnancy and postpartum in a state with legalized recreational cannabis. *Journal of Addiction Medicine*, *14*(6), 467–474. https://doi.org/10.1097/ADM.0000000000000625.

Baylis, F. (2010). Pregnant women deserve better. *Nature*, *465*(7299), 689–690. https://doi.org/10.1038/465689a.

Bayrampour, H., Zahradnik, M., Lisonkova, S., & Janssen, P. (2019). Women's perspectives about cannabis use during pregnancy and the postpartum period: An integrative review. *Preventive Medicine*, *119*, 17–23. https://doi.org/10.1016/j.ypmed.2018.12.002.

Beale, C., Broyd, S. J., Chye, Y., Suo, C., Schira, M., Galettis, P., Martin, J. H., Yücel, M., & Solowij, N. (2018). Prolonged cannabidiol treatment effects on hippocampal subfield volumes in current cannabis users. *Cannabis and Cannabinoid Research*, *3*(1), 94–107. https://doi.org/10.1089/can.2017.0047.

Berrendero, F., Sepe, N., Ramos, J. A., Di Marzo, V., & Fernandez-Ruiz, J. J. (1999). Analysis of cannabinoid receptor binding and mRNA expression and endogenous cannabinoid contents in the developing rat brain during late gestation and early postnatal period. *Synapse*, *33*(3), 181–191. https://doi.org/10.1002/(SICI)1098-2396(19990901)33:3<181::AID-SYN3>3.0.CO;2-R.

Bolhuis, K., Kushner, S. A., Yalniz, S., Hillegers, M. H. J., Jaddoe, V. W. V., Tiemeier, H., & El Marroun, H. (2018). Maternal and paternal cannabis use during pregnancy and the risk of psychotic-like experiences in the offspring. *Schizophrenia Research*, *202*, 322–327. https://doi.org/10.1016/j.schres.2018.06.067.

Brown, Q. L., Sarvet, A. L., Shmulewitz, D., Martins, S. S., Wall, M. M., & Hasin, D. S. (2017). Trends in marijuana use among pregnant and nonpregnant reproductive-aged women, 2002-2014. *JAMA*, *317*(2), 207. https://doi.org/10.1001/jama.2016.17383.

Cajachagua-Torres, K. N., Jaddoe, V. W. V., de Rijke, Y. B., van den Akker, E. L. T., Reiss, I. K. M., van Rossum, E. F. C., & El Marroun, H. (2021). Parental cannabis and tobacco use during pregnancy and childhood hair cortisol concentrations. *Drug and Alcohol Dependence*, *225*, 108751. https://doi.org/10.1016/j.drugalcdep.2021.108751.

Campolongo, P., Trezza, V., Cassano, T., Gaetani, S., Morgese, M. G., Ubaldi, M., Soverchia, L., Antonelli, T., Ferraro, L., Massi, M., Ciccocioppo, R., & Cuomo, V. (2007). Perinatal exposure to delta-9-tetrahydrocannabinol causes enduring cognitive deficits associated with alteration of cortical gene expression and neurotransmission in rats. *Addiction Biology*, *12*(3-4), 485–495. https://doi.org/10.1111/j.1369-1600.2007.00074.x.

Canan, S., Aktaş, A., Ulkay, M. B., Colakoglu, S., Ragbetli, M. C., Ayyildiz, M., Geuna, S., & Kaplan, S. (2008). Prenatal exposure to a non-steroidal anti-inflammatory drug or saline solution impairs sciatic nerve morphology: A stereological and histological study. *International Journal of Developmental Neuroscience*, *26*(7), 733–738. https://doi.org/10.1016/j.ijdevneu.2008.07.005.

Cash, M. C., Cunnane, K., Fan, C., & Romero-Sandoval, E. A. (2020). Mapping cannabis potency in medical and recreational programs in the United States. *PLoS One*, *15*(3), e0230167. https://doi.org/10.1371/journal.pone.0230167.

Ceprián, M., Jiménez-Sánchez, L., Vargas, C., Barata, L., Hind, W., & Martínez-Orgado, J. (2017). Cannabidiol reduces brain damage and improves functional recovery in a neonatal rat model of arterial ischemic stroke. *Neuropharmacology*, *116*, 151–159. https://doi.org/10.1016/j.neuropharm.2016.12.017.

Chandra, S., Radwan, M. M., Majumdar, C. G., Church, J. C., Freeman, T. P., & Elsohly, M. A. (2019). New trends in cannabis potency in USA and Europe during the last decade (2008–2017). *European Archives of Psychiatry and Clinical Neuroscience*, *269*(1), 5–15. https://doi.org/10.1007/s00406-019-00983-5.

Conner, S. N., Bedell, V., Lipsey, K., Macones, G. A., Cahill, A. G., & Tuuli, M. G. (2016). Maternal marijuana use and adverse neonatal outcomes: a systematic review and meta-analysis. *Obstetrics and Gynecology*, *128*(4), 713–723. https://doi.org/10.1097/AOG.0000000000001649.

Danielsson, B., Wikner, B. N., & Källén, B. (2014). Use of ondansetron during pregnancy and congenital malformations in the infant. *Reproductive Toxicology*, *50*, 134–137. https://doi.org/10.1016/j.reprotox.2014.10.017.

Day, N. L., Leech, S. L., & Goldschmidt, L. (2011). The effects of prenatal marijuana exposure on delinquent behaviors are mediated by measures of neurocognitive functioning. *Neurotoxicology and Teratology*, *33*(1), 129–136. https://doi.org/10.1016/j.ntt.2010.07.006.

Day, N. L., & Richardson, G. A. (1991). Prenatal marijuana use: Epidemiology, methodologic issues, and infant outcome. *Clinics in Perinatology*, *18*(1), 77–91. https://www.ncbi.nlm.nih.gov/pubmed/2040119.

Day, N., Sambamoorthi, U., Taylor, P., Richardson, G., Robles, N., Jhon, Y., Scher, M., Stoffer, D., Cornelius, M., & Jasperse, D. (1991). Prenatal marijuana use and neonatal outcome. *Neurotoxicology and Teratology*, *13*(3), 329–334. https://doi.org/10.1016/0892-0362(91)90079-c.

De Fonseca, F. R., Cebeira, M., Fernández-Ruiz, J. J., Navarro, M., & Ramos, J. A. (1991). Effects of pre- and perinatal exposure to hashish extracts on the ontogeny of brain dopaminergic neurons. *Neuroscience*, *43*(2-3), 713–723. https://doi.org/10.1016/0306-4522(91)90329-m.

De Salas-Quiroga, A., Díaz-Alonso, J., García-Rincón, D., Remmers, F., Vega, D., Gómez-Cañas, M., Lutz, B., Guzmán, M., & Galve-Roperh, I. (2015). Prenatal exposure to cannabinoids evokes long-lasting functional alterations by targeting CB1 receptors on developing cortical neurons. *Proceedings of the National Academy of Sciences*, *112*(44), 13693–13698. https://doi.org/10.1073/pnas.1514962112.

El Marroun, H., Bolhuis, K., Franken, I. H. A., Jaddoe, V. W. V., Hillegers, M. H., Lahey, B. B., & Tiemeier, H. (2019). Preconception and prenatal cannabis use and the risk of behavioural and emotional problems in the offspring; a multi-informant prospective longitudinal study. *International Journal of Epidemiology*, *48*(1), 287–296. https://doi.org/10.1093/ije/dyy186.

El Marroun, H., Hudziak, J. J., Tiemeier, H., Creemers, H., Steegers, E. A. P., Jaddoe, V. W. V., Hofman, A., Verhulst, F. C., Van Den Brink, W., & Huizink, A. C. (2011). Intrauterine cannabis exposure leads to more aggressive behavior and attention problems in 18-month-old girls. *Drug and Alcohol Dependence*, *118*(2-3), 470–474. https://doi.org/10.1016/j.drugalcdep.2011.03.004.

El Marroun, H., Tiemeier, H., Franken, I. H. A., Jaddoe, V. W. V., Van Der Lugt, A., Verhulst, F. C., Lahey, B. B., & White, T. (2016). Prenatal cannabis and tobacco exposure in relation to brain morphology: A prospective neuroimaging study in young children. *Biological Psychiatry*, *79*(12), 971–979. https://doi.org/10.1016/j.biopsych.2015.08.024.

El Marroun, H., Tiemeier, H., Steegers, E. A., Jaddoe, V. W., Hofman, A., Verhulst, F. C., van den Brink, W., & Huizink, A. C. (2009). Intrauterine cannabis exposure affects fetal growth trajectories: The Generation R Study. *Journal of the American Academy of Child and Adolescent Psychiatry*, *48*(12), 1173–1181. https://doi.org/10.1097/CHI.0b013e3181bfa8ee.

ElSohly, M. A., Mehmedic, Z., Foster, S., Gon, C., Chandra, S., & Church, J. C. (2016). Changes in cannabis potency over the last 2 decades (1995–2014): Analysis of current data in the United States. *Biological Psychiatry*, *79*(7), 613–619. https://doi.org/10.1016/j.biopsych.2016.01.004.

ElSohly, M. A., Ross, S. A., Mehmedic, Z., Arafat, R., Yi, B., & Banahan, B. F., III. (2000). Potency trends of delta9-THC and other cannabinoids in confiscated marijuana from 1980-1997. *Journal of Forensic Sciences*, *45*(1), 24–30. https://pubmed.ncbi.nlm.nih.gov/10641915/.

Fried, P. A. (1980). Marihuana use by pregnant women: Neurobehavioral effects in neonates. *Drug and Alcohol Dependence*, *6*(6), 415–424. https://doi.org/10.1016/0376-8716(80)90023-x.

Fried, P. A., & Watkinson, B. (1988). 12- and 24-month neurobehavioural follow-up of children prenatally exposed to marihuana, cigarettes and alcohol. *Neurotoxicology and Teratology*, *10*(4), 305–313. https://doi.org/10.1016/0892-0362(88)90032-3.

Fried, P. A., Watkinson, B., & Gray, R. (1998). Differential effects on cognitive functioning in 9- to 12-Year olds prenatally exposed to cigarettes and marihuana. *Neurotoxicology and Teratology*, *20*(3), 293–306. https://doi.org/10.1016/s0892-0362(97)00091-3.

Fried, P. A., Watkinson, B., & Gray, R. (2003). Differential effects on cognitive functioning in 13- to 16-year-olds prenatally exposed to cigarettes and marihuana. *Neurotoxicology and Teratology*, *25*(4), 427–436. https://doi.org/10.1016/s0892-0362(03)00029-1.

Goldschmidt, L., Richardson, G. A., Willford, J., & Day, N. L. (2008). Prenatal marijuana exposure and intelligence test performance at age 6. *Journal of the American Academy of Child and Adolescent Psychiatry*, *47*(3), 254–263. https://doi.org/10.1097/CHI.0b013e318160b3f0.

Gómez, M. A., Hernández, M., Johansson, B., De Miguel, R., Ramos, J. A., & Fernández-Ruiz, J. (2003). Prenatal cannabinoid exposure and gene expression for neural adhesion molecule L1 in the fetal rat brain. *Developmental Brain Research*, *147*(1-2), 201–207. https://doi.org/10.1016/j.devbrainres.2003.10.016.

Grewen, K., Salzwedel, A. P., & Gao, W. (2015). Functional connectivity disruption in neonates with prenatal marijuana exposure. *Frontiers in Human Neuroscience*, *9*, 601. https://doi.org/10.3389/fnhum.2015.00601.

Grotenhermen, F. (2003). Pharmacokinetics and pharmacodynamics of cannabinoids. *Clinical Pharmacokinetics*, *42*(4), 327–360. https://doi.org/10.2165/00003088-200342040-00003.

Hasin, D. S. (2018). US epidemiology of cannabis use and associated problems. *Neuropsychopharmacology*, *43*(1), 195–212. https://doi.org/10.1038/npp.2017.198.

Hofman, A., Jaddoe, V. W. V., Mackenbach, J. P., Moll, H. A., Snijders, R. F. M., Steegers, E. A. P., Verhulst, F. C., Witteman, J. C. M., & Buller, H. A. (2004). Growth, development and health from early fetal life until young adulthood: the Generation R Study. *Paediatric and Perinatal Epidemiology*, *18*(1), 61–72. https://doi.org/10.1111/j.1365-3016.2003.00521.x.

Jaddoe, V. W., van Duijn, C. M., Franco, O. H., van der Heijden, A. J., van Iizendoorn, M. H., de Jongste, J. C., van der Lugt, A., Mackenbach, J. P., Moll, H. A., Raat, H., Rivadeneira, F., Steegers, E. A., Tiemeier, H., Uitterlinden, A. G., Verhulst, F. C., & Hofman, A. (2012). The generation R Study: Design and cohort update 2012. *European Journal of Epidemiology*, *27*(9), 739–756. https://doi.org/10.1007/s10654-012-9735-1.

Jaddoe, V. W. V., Van Duijn, C. M., Van Der Heijden, A. J., Mackenbach, J. P., Moll, H. A., Steegers, E. A. P., Tiemeier, H., Uitterlinden, A. G., Verhulst, F. C., & Hofman, A. (2010). The Generation R Study: Design and cohort update 2010. *European Journal of Epidemiology*, *25*(11), 823–841. https://doi.org/10.1007/s10654-010-9516-7.

Jaques, S. C., Kingsbury, A., Henshcke, P., Chomchai, C., Clews, S., Falconer, J., Abdel-Latif, M. E., Feller, J. M., & Oei, J. L. (2014). Cannabis, the pregnant woman and her child: Weeding out the myths. *Journal of Perinatology*, *34*(6), 417–424. https://doi.org/10.1038/jp.2013.180.

Jarlenski, M., Tarr, J. A., Holland, C. L., Farrell, D., & Chang, J. C. (2016). Pregnant women's access to information about perinatal marijuana use: A qualitative study. *Women's Health Issues*, *26*(4), 452–459. https://doi.org/10.1016/j.whi.2016.03.010.

Khare, M. (2006). 9-Tetrahydrocannabinol inhibits cytotrophoblast cell proliferation and modulates gene transcription. *Molecular Human Reproduction*, *12*(5), 321–333. https://doi.org/10.1093/molehr/gal036.

Koto, P., Allen, V. M., Fahey, J., & Kuhle, S. (2022). Maternal cannabis use during pregnancy and maternal and neonatal outcomes: A retrospective cohort study. *BJOG: An International Journal of Obstetrics and Gynaecology*. https://doi.org/10.1111/1471-0528.17114.

Krugers, H. J., Arp, J. M., Xiong, H., Kanatsou, S., Lesuis, S. L., Korosi, A., Joels, M., & Lucassen, P. J. (2017). Early life adversity: Lasting consequences for emotional learning. *Neurobiology of Stress*, *6*, 14–21. https://doi.org/10.1016/j.ynstr.2016.11.005.

Liew, Z., Yuan, Y., Meng, Q., Von Ehrenstein, O. S., Cui, X., Flores, M. E. S., & Ritz, B. (2021). Prenatal exposure to acetaminophen and childhood asthmatic symptoms in a population-based cohort in Los Angeles, California. *International Journal of Environmental Research and Public Health*, *18*(19), 10107. https://doi.org/10.3390/ijerph181910107.

Lugo-Candelas, C., Cha, J., Hong, S., Bastidas, V., Weissman, M., Fifer, W. P., Myers, M., Talati, A., Bansal, R., Peterson, B. S., Monk, C., Gingrich, J. A., & Posner, J. (2018). Associations between brain structure and connectivity in infants and exposure to selective serotonin reuptake inhibitors during pregnancy. *JAMA Pediatrics*, *172*(6), 525. https://doi.org/10.1001/jamapediatrics.2017.5227.

Masarwa, R., Levine, H., Gorelik, E., Reif, S., Perlman, A., & Matok, I. (2018). Prenatal exposure to acetaminophen and risk for attention deficit hyperactivity disorder and autistic spectrum disorder: A systematic review, meta-analysis, and meta-regression Analysis of Cohort Studies. *American Journal of Epidemiology*, *187*(8), 1817–1827. https://doi.org/10.1093/aje/kwy086.

Matsui, D. (2015). Ethics of studies of drugs in pregnancy. *Pediatric Drugs*, *17*(1), 31–35. https://doi.org/10.1007/s40272-014-0104-2.

Merhar, S. L., Kline, J. E., Braimah, A., Kline-Fath, B. M., Tkach, J. A., Altaye, M., He, L., & Parikh, N. A. (2021). Prenatal opioid exposure is associated with smaller brain volumes in multiple regions. *Pediatric Research*, *90*(2), 397–402. https://doi.org/10.1038/s41390-020-01265-w.

Metsälä, J., Lundqvist, A., Virta, L. J., Kaila, M., Gissler, M., & Virtanen, S. M. (2015). Prenatal and post-natal exposure to antibiotics and risk of asthma in childhood. *Clinical and Experimental Allergy*, *45*(1), 137–145. https://doi.org/10.1111/cea.12356.

Mohammed, N., Ceprian, M., Jimenez, L., Pazos, M. R., & Martinez-Orgado, J. (2017). Neuroprotective effects of cannabidiol in hypoxic ischemic insult. The therapeutic window in newborn mice. *CNS & Neurological Disorders Drug Targets*, *16*(1), 102–108. https://doi.org/10.2174/1871527315666160927110305.

Moreno, M., Escuredo, L., Munoz, R., Rodriguez de Fonseca, F., & Navarro, M. (2005). Long-term behavioural and neuroendocrine effects of perinatal activation or blockade of CB1 cannabinoid receptors. *Behavioural Pharmacology*, *16*(5-6), 423–430. https://doi.org/10.1097/00008877-200509000-00015.

Moreno, M., Trigo, J. M., Escuredo, L., Rodríguez de Fonseca, F., & Navarro, M. (2003). Perinatal exposure to Δ9-tetrahydrocannabinol increases presynaptic dopamine D2 receptor sensitivity: A behavioral study in rats. *Pharmacology Biochemistry and Behavior*, *75*(3), 565–575. https://doi.org/10.1016/s0091-3057(03)00117-5.

Nakhai-Pour, H. R., Broy, P., Sheehy, O., & Bérard, A. (2011). Use of nonaspirin nonsteroidal anti-inflammatory drugs during pregnancy and the risk of spontaneous abortion. *Canadian Medical Association Journal*, *183*(15), 1713–1720. https://doi.org/10.1503/cmaj.110454.

Navarro, M., Rubio, P., & Rodriguez de Fonseca, F. (1994). Sex-dimorphic psychomotor activation after perinatal exposure to (-)-delta 9-tetrahydrocannabinol. An ontogenic study in wistar rats. *Psychopharmacology*, *116*(4), 414–422. https://doi.org/10.1007/BF02247471.

Nguyen, V. H., & Harley, K. G. (2022). Prenatal cannabis use and infant birth outcomes in the pregnancy risk assessment monitoring system. *The Journal of Pediatrics*, *240*, 87–93. https://doi.org/10.1016/j.jpeds.2021.08.088.

Park, B., Gibbons, H. M., Mitchell, M. D., & Glassa, M. (2003). Identification of the CB1 cannabinoid receptor and fatty acid amide hydrolase (FAAH) in the human placenta. *Placenta*, *24*(5), 473–478. https://doi.org/10.1053/plac.2002.0926.

Paul, S. E., Hatoum, A. S., Fine, J. D., Johnson, E. C., Hansen, I., Karcher, N. R., Moreau, A. L., Bondy, E., Qu, Y., Carter, E. B., Rogers, C. E., Agrawal, A., Barch, D. M., & Bogdan, R. (2021). Associations between prenatal cannabis exposure and childhood outcomes. *JAMA Psychiatry*, *78*(1), 64. https://doi.org/10.1001/jamapsychiatry.2020.2902.

Perez-Rosado, A., Gomez, M., Manzanares, J., Ramos, J. A., & Fernandez-Ruiz, J. (2002). Changes in prodynorphin and POMC gene expression in several brain regions of rat fetuses prenatally exposed to delta(9)-tetrahydrocannabinol. *Neurotoxicity Research*, *4*(3), 211–218. https://doi.org/10.1080/10298420290023936.

Pérez-Rosado, A., Manzanares, J., Fernández-Ruiz, J., & Ramos, J. A. (2000). Prenatal Δ9-tetrahydrocannabinol exposure modifies proenkephalin gene expression in the fetal rat brain: Sex-dependent differences. *Developmental Brain Research*, *120*(1), 77–81. https://doi.org/10.1016/s0165-3806(99)00170-4.

Pertwee, R. G. (2008). The diverse CB1and CB2receptor pharmacology of three plant cannabinoids: Δ9-tetrahydrocannabinol, cannabidiol and Δ9-tetrahydrocannabivarin. *British Journal of Pharmacology*, *153*(2), 199–215. https://doi.org/10.1038/sj.bjp.0707442.

Rahimi, R., Nikfar, S., & Abdollahi, M. (2006). Pregnancy outcomes following exposure to serotonin reuptake inhibitors: A meta-analysis of clinical trials. *Reproductive Toxicology*, *22*(4), 571–575. https://doi.org/10.1016/j.reprotox.2006.03.019.

Rubio, P. (1998). Maternal exposure to low doses of Δ9-tetrahydrocannabinol facilitates morphine-induced place conditioning in adult male offspring. *Pharmacology Biochemistry and Behavior*, *61*(3), 229–238. https://doi.org/10.1016/s0091-3057(98)00099-9.

Russo, E. B., Guy, G. W., & Robson, P. J. (2007). Cannabis, pain, and sleep: Lessons from therapeutic clinical trials of Sativex®, a cannabis-based medicine. *Chemistry & Biodiversity*, *4*(8), 1729–1743. https://doi.org/10.1002/cbdv.200790150.

Smart, R., Caulkins, J. P., Kilmer, B., Davenport, S., & Midgette, G. (2017). Variation in cannabis potency and prices in a newly legal market: Evidence from 30 million cannabis sales in Washington state. *Addiction*, *112*(12), 2167–2177. https://doi.org/10.1111/add.13886.

Smith, A. M., Mioduszewski, O., Hatchard, T., Byron-Alhassan, A., Fall, C., & Fried, P. A. (2016). Prenatal marijuana exposure impacts executive functioning into young adulthood: An fMRI study. *Neurotoxicology and Teratology*, *58*, 53–59. https://doi.org/10.1016/j.ntt.2016.05.010.

Spano, M. S., Ellgren, M., Wang, X., & Hurd, Y. L. (2007). Prenatal cannabis exposure increases heroin seeking with allostatic changes in limbic enkephalin systems in adulthood. *Biological Psychiatry*, *61*(4), 554–563. https://doi.org/10.1016/j.biopsych.2006.03.073.

Szutorisz, H., Dinieri, J. A., Sweet, E., Egervari, G., Michaelides, M., Carter, J. M., Ren, Y., Miller, M. L., Blitzer, R. D., & Hurd, Y. L. (2014). Parental THC exposure leads to compulsive heroin-seeking and altered striatal synaptic plasticity in the subsequent generation. *Neuropsychopharmacology*, *39*(6), 1315–1323. https://doi.org/10.1038/npp.2013.352.

Thomason, M. E., Palopoli, A. C., Jariwala, N. N., Werchan, D. M., Chen, A., Adhikari, S., Espinoza-Heredia, C., Brito, N. H., & Trentacosta, C. J. (2021). Miswiring the brain: Human prenatal Δ9-tetrahydrocannabinol use associated with altered fetal hippocampal brain network connectivity. *Developmental Cognitive Neuroscience*, *51*, 101000. https://doi.org/10.1016/j.dcn.2021.101000.

Tortoriello, G., Morris, C. V., Alpar, A., Fuzik, J., Shirran, S. L., Calvigioni, D., Keimpema, E., Botting, C. H., Reinecke, K., Herdegen, T., Courtney, M., Hurd, Y. L., & Harkany, T. (2014). Miswiring the brain: 9-tetrahydrocannabinol disrupts cortical development by inducing an SCG10/stathmin-2 degradation pathway. *The EMBO Journal*, *33*(7), 668–685. https://doi.org/10.1002/embj.201386035.

Trezza, V., Campolongo, P., Cassano, T., Macheda, T., Dipasquale, P., Carratù, M. R., Gaetani, S., & Cuomo, V. (2008). Effects of perinatal exposure to delta-9-tetrahydrocannabinol on the emotional reactivity of the offspring: a longitudinal behavioral study in wistar rats. *Psychopharmacology*, *198*(4), 529–537. https://doi.org/10.1007/s00213-008-1162-3.

Vela, G., Fuentes, J. A., Bonnin, A., Fernández-Ruiz, J., & Ruiz-Gayo, M. (1995). Perinatal exposure to Δ9-tetrahydrocannabinol (Δ9-THC) leads to changes in opioid-related behavioral patterns in rats. *Brain Research*, *680*(1-2), 142–147. https://doi.org/10.1016/0006-8993(95)00255-o.

Volkow, N. D., Compton, W. M., & Wargo, E. M. (2017). The risks of marijuana use during pregnancy. *JAMA*, *317*(2), 129. https://doi.org/10.1001/jama.2016.18612.

Volkow, N. D., Fowler, J. S., Wang, G.-J., Swanson, J. M., & Telang, F. (2007). Dopamine in drug abuse and addiction. *Archives of Neurology*, *64*(11), 1575. https://doi.org/10.1001/archneur.64.11.1575.

Volkow, N. D., Han, B., Compton, W. M., & McCance-Katz, E. F. (2019). Self-reported medical and nonmedical cannabis use among pregnant women in the United States. *JAMA*, *322*(2), 167. https://doi.org/10.1001/jama.2019.7982.

Wall, M. B., Pope, R., Freeman, T. P., Kowalczyk, O. S., Demetriou, L., Mokrysz, C., Hindocha, C., Lawn, W., Bloomfield, M. A., Freeman, A. M., Feilding, A., Nutt, D., & Curran, H. V. (2019). Dissociable effects of cannabis with and without cannabidiol on the human brain's resting-state functional connectivity. *Journal of Psychopharmacology*, *33*(7), 822–830. https://doi.org/10.1177/0269881119841568.

Wang, X., Dow-Edwards, D., Anderson, V., Minkoff, H., & Hurd, Y. L. (2004). In utero marijuana exposure associated with abnormal amygdala dopamine D2 gene expression in the human fetus. *Biological Psychiatry*, *56*(12), 909–915. https://doi.org/10.1016/j.biopsych.2004.10.015.

Wang, B., Liu, J., Zhang, Y., Yan, C., Wang, H., Jiang, F., Li, F., & Zhang, J. (2018). Prenatal exposure to antibiotics and risk of childhood obesity in a Multicenter Cohort Study. *American Journal of Epidemiology*, *187*(10), 2159–2167. https://doi.org/10.1093/aje/kwy122.

Westfall, R. E., Janssen, P. A., Lucas, P., & Capler, R. (2006). Survey of medicinal cannabis use among childbearing women: Patterns of its use in pregnancy and retroactive self-assessment of its efficacy against 'morning sickness'. *Complementary Therapies in Clinical Practice*, *12*(1), 27–33. https://doi.org/10.1016/j.ctcp.2005.09.006.

Winiger, E. A., & Hewitt, J. K. (2020). Prenatal cannabis exposure and sleep outcomes in children 9–10 years of age in the adolescent brain cognitive development SM study. *Sleep Health*, *6*(6), 787–789. https://doi.org/10.1016/j.sleh.2020.05.006.

Yeoh, S. L., Eastwood, J., Wright, I. M., Morton, R., Melhuish, E., Ward, M., & Oei, J. L. (2019). Cognitive and motor outcomes of children with prenatal opioid exposure. *JAMA Network Open*, *2*(7), e197025. https://doi.org/10.1001/jamanetworkopen.2019.7025.

Young-Wolff, K. C., Ray, G. T., Alexeeff, S. E., Adams, S. R., Does, M. B., Ansley, D., & Avalos, L. A. (2021). Rates of prenatal cannabis use among pregnant women before and during the COVID-19 pandemic. *Journal of the American Medical Association*. https://doi.org/10.1001/jama.2021.16328.

Yücel, M., Lorenzetti, V., Suo, C., Zalesky, A., Fornito, A., Takagi, M. J., Lubman, D. I., & Solowij, N. (2016). Hippocampal harms, protection and recovery following regular cannabis use. *Translational Psychiatry*, *6*(1), e710. https://doi.org/10.1038/tp.2015.201.

Chapter 5

Cannabinoid genotoxicity and congenital anomalies: A convergent synthesis of European and USA data sets

Albert Stuart Reece[a,b] and Gary Kenneth Hulse[a,b]

[a]*Department of Psychiatry, University of Western Australia, Perth, WA, Australia,* [b]*School of Health and Medical Sciences, Edith Cowan University, Joondalup, WA, Australia*

Abbreviations

CA	congenital anomaly
CAR	congenital anomaly rate
CB1R	cannabinoid type 1 receptor
***E*-values**	expected values
FGF1R	fibroblast growth factor type 1
MB	megabases
mEV	minimum 95% lower bound of the *E*-value
PCE	prenatal cannabis exposure
***P*-values**	probability values
THC	Δ9-tetrahydrocannabinol
VACTERL	vertebral-anorectal-cardiovascular-tracheo esophageal-renal-limb anomaly syndrome
WDR	World Drug Report

Introduction

Ever since an initial report in 2007 from Hawaii (Forrester & Merz, 2007), there has been increasing evidence from various jurisdictions around the world including Colorado, Hawaii, Canada, and Australia of cannabis-related teratogenesis occurring following prenatal cannabis exposure (PCE) (Reece & Hulse, 2019a, 2019b, 2020, 2020a, 2020b, 2021a, 2021c). Most recently, several large nationwide or continent-wide studies have issued from USA (Reece & Hulse, 2021a, 2022b) and Europe (Reece & Hulse, 2022a), which strongly confirm these recent reports. In terms of the richness and depth of the USA (sample size = 13,520) and European (sample size = 76,904) data sets, these two massive data sets arguably define the majority of the world's available data on teratogenesis upon which the impact of population cannabis exposure can be assessed.

Accordingly, it would provide a great boost to evidence-based policy formation if harmony and synthesis existed between the USA and European data sets and their principle findings. While each report is obviously very valuable in its own right, it would be a great assistance to policy makers to have a unified evidentiary basis from which to formulate policy. Moreover, both data sets could potentially act as a check on the other to ascertain whether their findings truly reflect underlying causal pathways, or whether they can be merely dismissed as observational studies whose findings may be different in another time and circumstance. Clearly, were the two main global data sets to arrive at substantially similar findings then, in the correct background and context and as outlined by A.B. Hill long ago (Hill, 1965), the causal argument becomes robustly established.

The subject of cannabinoid teratogenesis is important and timely for several parallel and convergent reasons that are all occurring at a similar time. First, cannabinoid teratogenesis is part of a larger subject of cannabinoid genotoxicity generally. Cannabinoid genotoxicity can be expected to be reflected on cancer induction and cellular and organismal aging processes. Although the literature on cannabis and cancer is often controversial with many positive and negative studies in existence,

Cannabis Use, Neurobiology, Psychology, and Treatment. https://doi.org/10.1016/B978-0-323-89862-1.00040-4

recent reports have drawn increasing attention to cannabinoid carcinogenesis including its implication in several adult (liver, pancreatic, thyroid and breast) cancers (Reece & Hulse, 2021a). Cannabis has also been shown be driving the 100% rise in testicular cancer in young men (Reece & Hulse, 2021c), to be causally related to acute lymphoid leukaemia (Reece & Hulse, 2021b), the commonest cancer in children, acute myeloid leukaemia (Reece & Hulse, 2021b; Robison et al., 1989; Trivers et al., 2006; Wen et al., 2000), and to be driving the 50% rise in US pediatric cancer rates from 1975 to 2018 (Reece & Hulse, 2021d). These elevations in pediatric cancer reflect transgenerational impacts of adult exposure to known carcinogens and demonstrate the important of multigenerational inheritance of mutagenicity and malignant teratogenicity.

Second, a recent explosion of mechanistic understandings of the pathways leads to cannabinoid teratogenesis. Cannabis exposure has been related to dysfunction of the male and female reproductive tracts (Chioccarelli et al., 2010; Rossato et al., 2008), sperm motility (Chioccarelli et al., 2010; Rossato et al., 2008), germ cell division (Chioccarelli et al., 2010; Morishima, 1984; Rossato et al., 2008), major chromosomal dysfunction (Huang et al., 1999; Leuchtenberger & Leuchtenberger, 1971; Leuchtenberger et al., 1973; Stenchever et al., 1974), oxidation of the bases of DNA and single- and double-stranded DNA breaks (Huang et al., 1999; Nahas, 1990a, 1990b; Russo et al., 2019), and major epigenomic disruptions of both histone and protamine physiology (Chioccarelli et al., 2010; Mon et al., 1981b) and DNA methylation (DiNieri et al., 2011; Ellis et al., 2021; Murphy et al., 2018; Schrott et al., 2021; Szutorisz et al., 2014; Szutorisz & Hurd, 2016; Watson et al., 2015). Major epigenomic studies have recently been published which clearly demonstrate in both rats and humans how alteration at 127–163 DNA methylation signals can control and perturb several hundred genes affecting all the body systems described in the present report including cardiovascular, central nervous, gastrointestinal, chromosomal, embryonic growth, limb formation, sensory, and uronephrological development (Schrott et al., 2021).

Third, it has become very clear that moves to increase cannabis availability and access across the globe are associated with four convergent trends in both USA and Europe, namely increased prevalence of use, increased intensity of daily use, increased THC potency of cannabis herb and resin, and increased mental health pathology resulting in increasing needs for mental health treatment (Manthey et al., 2021; Reece & Hulse, 2021c, 2021e; Substance Abuse and Mental Health Services Administration, 2019; United National Office of Drugs and Crime, 2019, 2021). These clearly documented trends show that what is actually involved is a confluence or convergence of rising cannabinoid trends all of which tend to increase cannabinoid exposure in individuals and populations. This is an extremely grave and concerning matter given the abundantly documented exponential dose-response effects of cannabis genotoxicity demonstrated in both laboratory studies (Busch et al., 1979; Fisar et al., 2014; Fish et al., 2019; Koller et al., 2015; Russo et al., 2019; Singh et al., 2015) and human populations (Reece & Hulse, 2019b, 2020, 2020c, 2021c, 2021e, 2022a, 2022b). That is that people will not just be exposed to higher levels of cannabinoids under a liberalization paradigm, but they move into a qualitatively different higher dose genotoxic dose range where outcomes will be qualitatively very different to lower dose zones. This tendency will of course be exacerbated by the fat solubility of cannabinoids and their tendency to accumulate in fat stores with very protracted half-lives of excretion of many months (Nahas, 1990a). To imagine therefore that cannabis liberalization is simply associated with more people using cannabis is therefore a gross and rather dangerous oversimplification.

While the recent European and American data sets share significant size they also share many differences. For example, the USA through the Centres of Disease Control (CDC) in Atlanta, Georgia, currently tracks 42 congenital anomalies. In recent years, they have tracked 62 anomalies as the anomalies they track has changed across time (NBDPN, 2018). The presence of national levels of various cannabinoids makes it possible to estimate state-level exposure to individual cannabinoids. At present, Europe tracks 95 congenital anomalies many of which have no parallel in the USA data set (Commission, E, 2021). On the other hand, we know of no readily available national index which tracks the levels of individual cannabinoids across European nations. Importantly, Europe tracks all birth anomalies not just those born alive. The total anomaly rate from the European Eurocat database includes stillborn and early terminations of pregnancy for anomaly in addition to children born alive. This makes it possible to study the actual rates of abnormal pregnancies with accuracy. Similar data are not available from USA where projections to total anomaly rates have to be estimated from published reports.

While it is important to acknowledge that many smaller studies exist, which have usually returned a negative result in cannabis-related birth defects, this is relatively unsurprising given that most of the disorders involved are rare and that large populations are required to detect a meaningful signal. Moreover, if the study is not carefully performed the signal-to-noise ratio easily drops away and sensitivity can easily be lost with a lack of methodological rigour. All of which implies that the findings of larger population studies are inherently much more concerning than smaller underpowered studies.

Methodologies

Data. The data in this report have recently been published (Reece & Hulse, 2022a, 2022b). Data are drawn from these publications. Data are essentially re-packaged and processed and presented comparatively and differently to the form in which it was first published. The report was compiled in January 2022.

Supplementary Tables are available in the Mendeley Data Repository at this URL: doi: 10.17632/s38pmskz73.1.

Ethics. Ethical approval for this study was provided from the Human Research Ethics Committee of the University of Western Australia number RA/4/20/4724 on 24th September 2021.

Analyses

Continuous comparison

Supplementary Table 1 sets out 86 significant regression coefficients from bivariate regression of the congenital anomaly rate (CAR) against various metrics of cannabis exposure in USA. The table is organized by the regression covariate and by descending minimum *E*-value.

The *E*-value or expected value provides the degree of correlation required of some extraneous covariate with *both* the exposure of interest and the outcome of concern in order to explain away an apparently causal effect. It is therefore a key and central metric widely employed in formal quantitative causal inference. It has a confidence interval and the 97.5% lower bound provides 95% confidence that the true value of the *E*-value will lie above that point. For guidance it should be noted that nine is said to be quite high and is the association of tobacco with lung cancer (Pearl & Mackaenzie, 2019), while the cutoff for causality is generally said to be 1.25 (VanderWeele et al., 2019). It follows therefore that some of the *E*-value estimates and minimum *E*-values (mEV) are reported in this table, which range up to 9.00×10^{30} and 7.53×10^{19}, respectively, are very elevated indeed.

Supplementary Table 2 sets out the European data in a similar way and lists 225 regression results and their covariates.

Supplementary Table 3 extracts from the 86 USA regression equations listed in Supplementary Table 1 only the most significant single anomaly. These 41 anomalies are listed in this table.

Similarly Supplementary Table 4 extracts only unique anomalies from all of those presented in Supplementary Table 2 for Europe and lists 70 as being the most significant of these regression equations.

Table 1 lists the 41 congenital anomalies (CA's) of Supplementary Table 3 along with annotations as to whether they are confirmed or not by continuous regression on the European data set. As indicated in the Table the European data confirms the USA data for 24/41 (58.5%) of the CA's.

Table 2 lists the 70 anomalies of Supplementary Table 4 and indicates whether they are confirmed in the United States continuous regression results or not, or have no applicable data. As noted in that table in 35/70 (50.0%) of cases the European data is confirmed, in 14/70 (20.0%) it is not confirmed an in 21/70 (30.0%) there is no relevant data which makes sense as the European database has many more CA's than the USA data set.

Categorical comparison

Supplementary Table 5 sets out a categorical analysis comparing the highest cannabis or cannabidiol quintile with the lowest quintile. The applicable regressor in each case is listed in the table. 35 CA's are listed in descending order of minimum *E*-value (mEV). The table also provides key metrics including the relative risk, attributable fraction in the exposed (AFE), and population attributable risk (PAR, also known as the attributable fraction in the population, AFP) together with their confidence intervals. The values for P and the *E*-value estimates and lower bounds are also provided.

Supplementary Table 6 performs a similar function for the European data set and lists 621 CA's and their regressors. These are also listed by their covariate for cannabis exposure and ordered in descending order of mEV. The highest mEV in this table is 26.43.

Supplementary Table 7 extracts the most significant regression equation from the list of Supplementary Table 5 for each congenital anomaly remaining with 25 CA's remaining.

Supplementary Table 8 extracts the 89 most significant terms for each CA from Supplementary Table 6 and lists the CA's alphabetically.

TABLE 1 Continuous analysis of USA cannabis-related congenital anomalies and comparison with European data.

USA

	Congenital anomaly	*P*_value	*E*-value - Point	*E*-value - Lower	Positive	Negative
1	Anophthalmia/microphthalmia	0.0045	3.17	1.74	Yes	
2	Anotia/microtia	7.57E-04	7.51E+03	65.76		No
3	Aortic valve stenosis	0.0019	4.60E+03	36.41		No
4	Atrial septal defect	0.0378	313.06	2.08	Yes	
5	Biliary atresia	2.43E-04	2.22E+11	3.48E+05		No
6	Bladder exstrophy	0.0170	860.98	5.62		No
7	Choanal atresia	0.0033	3.15	1.78	Yes	
8	Cleft palate alone	0.0014	6.42E+03	48.45	Yes	
9	Clubfoot	3.16E-05	1.39E+07	1.10E+04	Yes	
10	Coarctation of the aorta	0.0372	300.37	2.12		No
11	Congenital cataract	0.0479	357.58	1.39		No
12	Congenital hip dislocation	6.32E-07	9.00E+30	7.53E+19	Yes	
13	Congenital posterior urethral valves	1.18E-04	1.64E+06	1.96E+03	Yes	
14	Deletion 22q11.2	0.0024	2.49E+05	155.04	Yes	
15	Diaphragmatic hernia	0.0065	4.96E+07	263.36	Yes	
16	Encephalocele	0.0013	5.29E+03	45.63		No
17	Epispadias	0.0111	1.02E+07	90.57		No
18	Esophageal atresia/tracheoesophageal fistula	4.83E-06	1.49E+05	1.34E+03		No
19	Hirschsprung disease (congenital megacolon)	0.0084	2.22E+15	2.67E+04	Yes	
20	Holoprosencephaly	0.0104	3.54	1.68	Yes	
21	Hypoplastic left heart syndrome	0.0047	1.65E+03	15.88		No
22	Hypospadias	1.16E-05	9.34E+04	842.36	Yes	
23	Interrupted aortic arch	3.40E-06	4.28E+07	4.68E+04	Yes	
24	**Limb deficiencies (reduction defects)**	0.0134	3.45E+03	9.53		No
25	Obstructive genitourinary defect	0.0072	9.69E+14	3.51E+04	Yes	
26	Omphalocele	0.0025	3.55E+03	29.18	Yes	
27	Pulmonary valve atresia	0.0047	1.02E+04	29.43		No
28	Rectal and large intestinal atresia/stenosis	0.0040	1.54E+08	751.61	Yes	
29	**Reduction deformity, Lower limbs**	0.0420	3.39E+04	2.57		No
30	Renal agenesis/hypoplasia	7.34E-04	7.45E+03	66.37	Yes	
31	Single ventricle	0.0394	2.99	1.22		No
32	Small intestinal atresia/stenosis	3.39E-06	8.48E+20	3.86E+12		No
33	Spina bifida without anencephalus	0.0008	3.23	1.96	Yes	
34	Tetralogy of Fallot	0.0168	553.33	5.16		No
35	Total anomalous pulmonary venous connection	0.0299	2.61E+03	3.71		No
36	Transposition of great arteries	0.0479	3.21E+05	1.71	Yes	
37	Trisomy 13	2.08E-07	1.11E+06	8.58E+03	Yes	
38	Trisomy 18	2.34E-06	1.99E+05	1.85E+03	Yes	
39	Trisomy 21 (Down syndrome)	2.03E-16	6.97E+08	8.30E+06	Yes	
40	Turner syndrome	0.0021	1.54E+05	137.32	Yes	
41	Ventricular septal defect	0.0021	3.92E+03	32.64	Yes	
	Count				24	17
	Total				41	41
	Percentage				58.5%	41.5%

Table 3 lists the 25 CA's from Supplementary Table 7 and notes that all 25 (100%) are identifiable in the European data set.

Table 4 lists the 89 CA's from Supplementary Table 8 and finds that 26/89 (29.2%) are identifiable in the USA data set, 29/89 (32.6%) are not identifiable and for 34/89 (38.2%) there is no data from which to comment. When one considers only the cases where data does exist then 26/55 (47.3%) are identified in the USA data set.

TABLE 2 **Continuous analysis of European cannabis-related congenital anomalies and comparison with USA data.**

Europe

	Congenital anomaly	*P*_value	*E*-value - Point	*E*-value - Lower	Positive	Negative	N/A	
1	Abdominal wall defects	1.51E-07	386.98	61.05	Yes			Abdominal wall defects
2	All Anomalies	0.0006	55.76	8.41			N/A	All Anomalies
3	Anencephalus and similar	1.95E-06	223.60	35.09		No		Anencephalus and similar
4	Annular pancreas	0.0521	12.15	1.08			N/A	Annular pancreas
5	Anophthalmos/micropthalmos	0.0013	7.87E+13	6.18E+05	Yes			Anophthalmos/micropthalmos
6	Ano-rectal atresia and stenosis	0.0135	20.65	2.75		No		Ano-rectal atresia and stenosis
7	Aortic atresia/interrupted aortic arch	0.0241	9.36E+09	48.63			N/A	Aortic atresia/interrupted aortic arch
8	Arhinencephaly/holoprosencephaly	0.0052	41.11	4.61	Yes			Arhinencephaly/holoprosencephaly
9	Atresia of bile ducts	6.26E-06	6.50E+19	5.10E+11	Yes			Atresia of bile ducts
10	Atresia or stenosis of other parts of small intestin	0.0445	230.98	1.66	Yes			Atrial septal defect (ASD)
11	Atrial septal defect (ASD)	0.0003	67.07	10.21	Yes			Atrioventricular septal defect (AVSD)
12	Atrioventricular septal defect (AVSD)	0.0181	545.31	5.05		No		Bilateral renal agenesis including Potter syndrome
13	Bilateral renal agenesis including Potter syndrom	0.0014	4.13E+03	41.97	Yes			Choanal atresia
14	Choanal atresia	4.09E-06	20.55	7.63	Yes			Chromosomal
15	Chromosomal	0.0008	51.45	7.72	Yes			Cleft lip with or without palate
16	Cleft lip with or without palate	0.0235	1.11E+03	4.58	Yes			Cleft palate
17	Cleft palate	0.0027	2.64E+03	26.58	Yes			Club foot - talipes equinovarus
18	Club foot - talipes equinovarus	1.33E-04	13.14	4.74	Yes			Congenital cataract
19	Congenital cataract	0.0201	5.73	1.73	Yes			Congenital constriction bands/amniotic band
20	Congenital constriction bands/amniotic band	1.60E-07	773.45	95.71			N/A	Congenital glaucoma
21	Congenital glaucoma	6.48E-09	6.95E+07	3.22E+05			N/A	Congenital heart defects
22	Congenital heart defects	0.0023	2.97E+03	30.00	Yes			Congenital hydronephrosis
23	Congenital hydronephrosis	2.14E-05	3.75E+18	2.94E+10		No		Craniosynostosis
24	Craniosynostosis	4.49E-07	26.57	9.97		No		Cystic adenomatous malf of lung
25	Cystic adenomatous malf of lung	0.0146	6.11	1.90			N/A	Diaphragmatic hernia
26	Diaphragmatic hernia	5.04E-05	2.22E+05	1.03E+03	Yes			Digestive system
27	Digestive system	3.18E-05	3.09E+05	1.43E+03	Yes			Double outlet right ventricle
28	Double outlet right ventricle	3.25E-06	369.89	45.54			N/A	Down Syndrome
29	Down Syndrome	4.84E-06	20.13	7.46	Yes			Edward syndrome/trisomy 18
30	Edward syndrome/trisomy 18	0.0012	9.51	3.30	Yes			Eye
31	Eye	0.0039	7.84	2.63	Yes			Fetal alcohol syndrome
32	Fetal alcohol syndrome	0.0037	31.88	4.58			N/A	Gastroschisis
33	Gastroschisis	1.11E-06	2.98E+21	2.34E+13		No		Genetic syndromes + microdeletions
34	Genetic syndromes + microdeletions	0.0003	6.15E+04	284.15	Yes			Genital
35	Genital	0.0020	4.43E+13	2.33E+05	Yes			Hip dislocation and/or dysplasia
36	Hip dislocation and/or dysplasia	2.17E-05	224.78	27.50		No		Hirschsprung's disease
37	Hirschsprung's disease	0.0023	2.91E+03	29.39	Yes			Hydrocephalus
38	Hydrocephalus	0.0004	61.61	9.34		No		Hypoplastic right heart
39	Hypoplastic right heart	0.0005	3.78E+04	174.71		No		Hypospadias
40	Hypospadias	0.0004	96.35	11.52	Yes			Klinefelter syndrome
41	Klinefelter syndrome	0.0039	3.10E+12	2.44E+04			N/A	Klinefelter syndrome
42	Lateral anomalies	1.31E-05	5.76E+05	2.67E+03			N/A	Lateral anomalies
43	Limb	1.22E-06	2.81E+06	1.30E+04	Yes			Limb
44	Maternal infections resulting in malformations	1.05E-05	9.51E+04	976.69			N/A	Maternal infections resulting in malformations
45	Mitral valve anomalies	2.11E-05	1.07E+19	5.61E+10			N/A	Mitral valve anomalies
46	Multicystic renal dysplasia	0.0025	8.46	2.88			N/A	Multicystic renal dysplasia
47	Nervous system	3.18E-05	3.89E+18	2.04E+10	Yes			Nervous system
48	Neural Tube Defects	7.61E-06	164.37	25.68	Yes			Neural Tube Defects
49	Omphalocele	0.0012	44.56	6.62	Yes			Omphalocele
50	Oro-facial clefts	1.29E-04	1.11E+05	512.89	Yes			Oro-facial clefts
51	Patau syndrome/trisomy 13	1.92E-04	121.51	14.65	Yes			Patau syndrome/trisomy 13
52	PDA as only CHD in term infants (>=37 weeks)	0.0071	1.21E+03	11.95		No		PDA as only CHD in term infants (>=37 weeks)
53	Polydactyly	0.0086	1.04E+03	10.17			N/A	Polydactyly
54	Posterior urethral valve and/or prune belly	0.0261	388.43	3.41	Yes			Posterior urethral valve and/or prune belly
55	Pulmonary valve stenosis	0.0080	35.08	3.85		No		Pulmonary valve stenosis
56	Respiratory	3.08E-07	1.68E+23	8.81E+14		No		Respiratory
57	Severe CHD	2.34E-05	126.20	19.61			N/A	Severe CHD
58	Severe microcephaly	9.13E-07	4.59E+21	3.60E+13		No		Severe microcephaly
59	Single ventricle	0.0378	13.91	1.56	Yes			Single ventricle
60	Situs inversus	5.60E-05	3.53E+04	362.44			N/A	Situs inversus
61	Skeletal dysplasias	0.0034	32.78	4.73			N/A	Skeletal dysplasias
62	Spina Bifida	1.52E-05	139.95	21.80	Yes			Spina Bifida
63	Syndactyly	0.0101	22.95	3.13			N/A	Syndactyly
64	Teratogenic syndromes with malformations	5.65E-06	8.21E+19	6.44E+11			N/A	Teratogenic syndromes with malformations
65	Transposition of great vessels	0.0012	4.79E+03	48.76		No		Transposition of great vessels
66	Turner syndrome	0.0015	63.88	7.47	Yes			Turner syndrome
67	Urinary	0.0032	2.26E+03	22.70	Yes			Urinary
68	Vascular disruption anomalies	0.0024	41.52	5.63			N/A	Vascular disruption anomalies
69	VATER/VACTERL	6.62E-06	1.70E+20	8.94E+11			N/A	VATER/VACTERL
70	Ventricular septal defect (VSD)	0.0051	28.78	4.08	Yes			Ventricular septal defect (VSD)
	Count				35	14	21	
	Total				70	70	70	
	Percentage				50.0%	20.0%	30.0%	
	Percentage with data				71.4%	28.6%		

TABLE 3 Categorical analysis of USA cannabis-related congenital anomalies and comparison with European data.

		USA—Categorical				
		Significance	*E*-values		Confirmation	
No.	Congenital anomaly	*P*-value	*E*-value—Point	*E*-value—Lower	Yes	No
1	Atrioventricular septal defect	0.0470	1.29	1.03	Yes	
2	Biliary atresia	0.0233	1.69	1.18	Yes	
3	Cleft lip with and without cleft palate	0.0159	2.41	1.37	Yes	
4	Cloacal exstrophy	2.13E-86	9.17	7.61	Yes	
5	Clubfoot	0.0048	1.43	1.20	Yes	
6	Coarctation of the aorta	9.74E-45	2.11	1.97	Yes	
7	Congenital hip dislocation	7.27E-70	3.99	3.57	Yes	
8	Congenital posterior urethral valves	1.35E-04	1.99	1.56	Yes	
9	Deletion 22q11.2	0.0051	2.05	1.42	Yes	
10	Diaphragmatic hernia	2.11E-08	1.80	1.57	Yes	
11	Double outlet right ventricle	7.31E-04	1.66	1.36	Yes	
12	Epispadias	0.0287	1.95	1.20	Yes	
13	Esophageal atresia/tracheoesophageal fistula	0.0195	1.45	1.15	Yes	
14	Hirschsprung disease (congenital megacolon)	6.69E-06	2.27	1.77	Yes	
15	Holoprosencephaly	2.90E-12	1.78	1.60	Yes	
16	Hypoplastic left heart syndrome	0.0048	1.43	1.20	Yes	
17	Obstructive genitourinary defect	2.22E-15	3.25	2.64	Yes	
18	Pulmonary valve atresia	1.02E-05	2.04	1.64	Yes	
19	Single ventricle	0.0060	1.72	1.30	Yes	
20	Small intestinal atresia/stenosis	5.93E-06	1.83	1.54	Yes	
21	Spina bifida without anencephalus	0.0181	1.28	1.10	Yes	
22	Trisomy 13	3.50E-06	1.89	1.75	Yes	
23	Trisomy 18	1.06E-61	1.94	1.85	Yes	
24	Trisomy 21 (Down syndrome)	4.02E-26	1.55	1.49	Yes	
25	Turner syndrome	7.69E-49	2.50	2.31	Yes	
	Count				25	0
	Total				25	25
	Percentage				100.0%	0.0%

TABLE 4 Categorical analysis of European cannabis-related congenital anomalies and comparison with USA data.

		Europe					
			E-values		Confirmation		
	Congenital anomaly	*P*-value	*E*-value—Point	*E*-value—Lower	Yes	No	N/A
1	Abdominal wall Defx	3.01E-120	3.48	3.23	Yes		
2	All anomalies	0.00E+00	2.63	2.60			N/A
3	Amniotic band	1.09E-47	10.69	8.08		No	
4	An/Microphthalmos	1.25E-55	5.77	4.87		No	
5	Anencephalus & ~	1.53E-212	5.47	5.04		No	
6	Annular Pancreas	1.52E-13	8.00	5.02			N/A
7	Anophthalmos	1.06E-06	3.99	2.59		No	
8	Ano-rectal S/A	4.03E-39	2.71	2.45			N/A
9	Anotia	4.63E-13	4.36	3.22		No	
10	Aortic atresia ~	5.71E-25	4.58	3.68			N/A
11	Aortic valve S/A	7.14E-13	2.20	1.91			N/A
12	Arterial truncus	9.92E-07	2.41	1.87			N/A
13	ASD	0.00E+00	4.02	3.89		No	
14	AVSD	1.65E-101	3.68	3.38	Yes		
15	Bilat renal agenesis	1.08E-47	3.86	3.38		No	
16	Bile duct A	1.00E-40	24.75	15.08		No	
17	Blader Extr/Epispad	1.56E-18	3.74	3.01	Yes		
18	Choanal atresia	7.34E-94	7.58	6.48		No	
19	Chromosomal	0.00E+00	5.17	5.03	Yes		
20	Cleft lip + palate	1.80E-101	2.65	2.50	Yes		
21	Cleft palate	1.79E-34	2.33	2.13	Yes		
22	Club foot	4.23E-292	3.53	3.37	Yes		
23	Coarctation aorta	5.78E-33	2.48	2.25	Yes		
24	Cong. cataract	4.88E-66	5.36	4.63		No	
25	Cong. glaucoma	1.52E-43	15.26	10.61			N/A
26	Congenital heart	0.00E+00	3.60	3.52	Yes		
27	Conjoined twins	8.62E-14	12.70	6.86			N/A
28	Craniosynostosis	5.72E-155	5.27	4.81		No	
29	Cystic Lung	4.05E-80	13.77	10.74			N/A
30	Diaphragmatic hernia	8.77E-57	3.19	2.88	Yes		
31	Digestive system	1.61E-264	3.37	3.22	Yes		
32	Double outlet RV	1.28E-46	7.05	5.69			N/A
33	Down Syndrome	0.00E+00	4.24	4.10		No	
34	Duodenal S/A	1.50E-10	2.27	1.91			N/A
35	Ear, face and neck	3.38E-44	2.88	2.60	Yes		

Continued

TABLE 4 Categorical analysis of European cannabis-related congenital anomalies and comparison with USA data—cont'd

		Europe					
			E-values		Confirmation		
	Congenital anomaly	*P*-value	*E*-value—Point	*E*-value—Lower	Yes	No	N/A
36	Ebstein's anomaly	3.23E-17	5.61	4.09		No	
37	Edward syndrome	0.00E+00	8.97	8.28	Yes		
38	Encephalocele	4.76E-21	2.96	2.53		No	
39	Eye	2.27E-175	4.65	4.30			N/A
40	Fetal alcohol	5.88E-57	104.82	43.19			N/A
41	Gastroschisis	6.55E-39	3.13	2.77		No	
42	Genetic syndromes	1.38E-228	5.58	5.15			N/A
43	Genital	2.55E-243	2.94	2.82	Yes		
44	Hip dysplasia	0.00E+00	23.58	21.11			N/A
45	Hirschsprungs	2.54E-88	9.07	7.56	Yes		
46	Holoprosencephaly ~	1.22E-72	5.29	4.61	Yes		
47	Hydrocephalus	1.76E-110	3.64	3.36		No	
48	Hydronephrosis	0.00E+00	4.16	3.99		No	
49	Hypoplastic left Ht	3.37E-61	3.54	3.18	Yes		
50	Hypoplastic right Ht	2.85E-59	7.52	6.17		No	
51	Hypospadias	2.92E-177	2.77	2.64			N/A
52	Klinefelter	1.75E-41	8.51	6.56			N/A
53	Lateral anomalies	2.36E-48	6.60	5.39			N/A
54	Limb	0.00E+00	4.00	3.88			N/A
55	Limb reductions	8.20E-65	2.46	2.30		No	
56	Matern infect malform	4.15E-87	20.09	14.94			N/A
57	Mitral valve anomalies	8.99E-58	7.50	6.13			N/A
58	Multicystic renal Dys	6.70E-251	6.60	6.05		No	
59	Nervous system	0.00E+00	5.28	5.08			N/A
60	Neural tube defects	9.97E-269	3.66	3.48			N/A
61	Oesophageal S/A	3.49E-44	2.97	2.67		No	
62	Omphalocele	4.94E-131	5.13	4.65		No	
63	Oro-facial clefts	3.99E-133	2.69	2.56	Yes		
64	Patau syndrome	1.08E-144	7.79	6.86	Yes		
65	PDA	2.14E-20	2.32	2.06		No	
66	Polydactyly	1.46E-292	3.51	3.35			N/A
67	Posterior Ureth Valv	1.28E-42	4.14	3.57	Yes		
68	PV atresia	1.42E-27	3.48	2.96	Yes		
69	PV stenosis	2.09E-95	3.43	3.16		No	

TABLE 4 Categorical analysis of European cannabis-related congenital anomalies and comparison with USA data—cont'd

		Europe					
			E-values		Confirmation		
	Congenital anomaly	*P*-value	*E*-value—Point	*E*-value—Lower	Yes	No	N/A
70	Respiratory	2.57E-203	11.48	10.00			N/A
71	Severe CHD	1.81E-317	2.83	2.73			N/A
72	Severe microcephaly	3.17E-148	6.69	5.97			N/A
73	Single ventricle	1.03E-25	4.08	3.36	Yes		
74	Situs inversus	1.42E-44	6.19	5.07			N/A
75	Skeletal dysplasias	5.12E-74	4.58	4.05			N/A
76	Small intestine S/A	8.23E-31	5.31	4.27	Yes		
77	Spina bifida	3.93E-84	3.01	2.79	Yes		
78	Syndactyly	3.47E-16	1.89	1.71			N/A
79	Teratogenic synds	1.42E-139	15.00	12.32			N/A
80	Tetralogy of fallot	3.12E-47	2.77	2.52		No	
81	Tot Anom Pul V Ret	4.07E-09	2.71	2.14		No	
82	Transpos Grt Vess	9.96E-33	2.50	2.26		No	
83	Turner syndrome	1.10E-146	7.95	6.99	Yes		
84	TV S/A	6.86E-24	4.55	3.65		No	
85	Urinary	0.00E+00	3.28	3.19	Yes		
86	Valproate syndrome	1.57E-07	30.11	7.00			N/A
87	Vascular disruptions	3.46E-101	3.22	2.99			N/A
88	VATER/VACTERL	2.43E-36	108.62	34.61			N/A
89	VSD	0.00E+00	2.82	2.74		No	
	Count				26	29	34
	Total				89	89	89
	Percentage				29.2%	32.6%	38.2%
	Percentage with data				47.3%	52.7%	

Continuous—Categorical comparison

When the 17 CA's which were not matched in Table 1 were matched against the categorical analysis of Europe all 17 (100%) were matched as shown in Supplementary Table 9.

When the 37 CA's in the European data set which did not match in the preceding analyses were studied along with the categorical analysis of the European data set, 17/37 (51.5%) were matched up, 11/37 (29.7%) were likely to matched based on similar terminology and external evidence, 4/37 (10.8%) did not match and 4/37 (10.8%) had no data (Supplementary Table 10). When one considers only those CA's for which data existed, these percentages went to 51.5%, 33.3%, and 12.1%, making a total of 84.8% matched or likely matched.

Supplementary Table 11 provides the list of 45 significantly cannabis-associated CA's for USA.

Table 5 lists the 93 European CA's noted to be cannabis related in the continuous or categorical analyses in alphabetical order. From the foot of this table, it is noted that 49/93 (52.7%) were positively matched, 15/93 (16.1%) were likely matched, 5/93 (5.4%) were not matched, and in 27/93 (29.0%) there was no comparable data available. These percentages

TABLE 5 Alphabetical summary of cannabis-related congenital anomalies in Europe.

Europe

No.	Congenital anomaly	*P*_Value	*E*-value - Point	*E*-value - Lower	Positive	Likely	Negative	N/A	Model type	
1	Abdominal wall defects	1.51E-07	386.98	61.05	Yes				Cont	Abdominal wall defects
2	All Anomalies	0.00E+00	2.63	2.60				N/A	Cont	All Anomalies
3	Amniotic band	1.09E-47	10.69	8.08			No		Catt	Amniotic band
4	Anencephalus and similar	1.25E-55	5.77	4.87		Yes			Cont-Catt	An/Microphthalmos
5	Annular Pancreas	1.53E-212	223.60	35.09				N/A	Catt	Anencephalus and similar
6	Anophthalmos	1.52E-13	8.00	5.02			No		Catt	Annular Pancreas
7	Anophthalmos/micropthalmos	1.06E-06	3.99	2.59	Yes				Cont	Anophthalmos
8	Ano-rectal atresia and stenosis	0.0013	7.87E+13	6.18E+05		Yes			Cont-Catt	Anophthalmos/micropthalmos
9	Anotia	0.0135	20.65	2.75		Yes			Catt	Ano-rectal atresia and stenosis
10	Aortic atresia/interrupted aortic arch	4.63E-13	4.36	3.22		Yes		N/A	Cont	Anotia
11	Aortic Valve S/A	5.71E-25	9.36E+09	48.63				N/A	Catt	Aortic atresia/interrupted aortic arch
12	Arterial Truncus	7.14E-13	2.20	1.91				N/A	Catt	Aortic Valve S/A
13	Atresia of bile ducts	0.0052	41.11	4.61	Yes				Cont	Arhinencephaly/holoprosencephaly
14	Atresia or stenosis of other parts of small intestine	9.92E-07	2.41	1.87	Yes				Cont	Arterial Truncus
15	Atrial Septal Defect (ASD)	0.00E+00	4.02	3.89	Yes				Cont-Catt	Atrial septal defect (ASD)
16	Atrioventricular septal defect (AVSD)	1.65E-101	67.07	10.21	Yes				Cont-Catt	Atrioventricular septal defect (AVSD)
17	Bilateral renal agenesis including Potter syndrome	1.08E-47	545.31	5.05	Yes				Cont	Bilat Renal Agenesis
18	Bile Duct Atresia	1.00E-40	6.50E+19	5.10E+11	Yes				Catt	Bile Duct Atresia
19	Bladder Extrophy / Epispadias	1.56E-18	3.74	3.01	Yes				Catt	Blader Extr / Epispad
20	Choanal atresia	7.34E-94	4.13E+03	41.97	Yes				Cont	Choanal Atresia
21	Chromosomal	0.00E+00	20.55	7.63	Yes				Cont	Chromosomal
22	Cleft lip +/- palate	0.0008	51.45	7.72	Yes				Catt	Cleft lip with or without palate
23	Cleft palate	1.79E-34	1.11E+03	4.58	Yes				Cont	Cleft palate
24	Club foot - talipes equinovarus	4.23E-292	2.64E+03	26.58	Yes				Cont	Club Foot
25	Coarctation Aorta	5.78E-33	2.48	2.25	Yes				Catt	Coarctation Aorta
26	Congenital cataract	4.88E-66	5.36	4.63	Yes				Cont	Cong. Cataract
27	Congenital constriction bands/amniotic band	0.0201	5.73	1.73				N/A	Cont	Congenital constriction bands/amniotic band
28	Congenital glaucoma	1.60E-07	773.45	95.71				N/A	Cont	Congenital glaucoma
29	Congenital heart defects	0.00E+00	6.95E+07	3.22E+05	Yes				Cont	Congenital Heart
30	Congenital hydronephrosis	0.0023	2.97E+03	30.00		Yes			Cont-Catt	Congenital hydronephrosis
31	Conjoined twins	8.62E-14	12.70	6.86				N/A	Catt	Conjoined twins
32	Craniosynostosis	5.72E-155	3.75E+18	2.94E+10			No		Cont	Craniosynostosis
33	Cystic adenomatous malf of lung	4.05E-80	13.77	10.74				N/A	Cont	Cystic adenomatous malf of lung
34	Diaphragmatic Hernia	8.77E-57	3.19	2.88	Yes				Catt	Diaphragmatic Hernia
35	Digestive system	1.61E-264	2.22E+05	1.03E+03	Yes				Cont	Digestive system
36	Double outlet right ventricle	1.28E-46	3.09E+05	1.43E+03				N/A	Cont	Double outlet right ventricle
37	Down Syndrome	0.00E+00	369.89	45.54	Yes				Cont	Down Syndrome
38	Duodenal S/A	1.50E-10	2.27	1.91				N/A	Catt	Duodenal S/A
39	Ear, face and neck	3.38E-44	2.88	2.60	Yes				Catt	Ear, face and neck
40	Ebstein's Anomaly	3.23E-17	5.61	4.09		Yes			Cont-Catt	Ebstein's Anomaly
41	Edward syndrome/trisomy 18	0.00E+00	8.97	8.28	Yes				Cont	Edward syndrome
42	Encephalocele	4.76E-21	2.96	2.53	Yes				Cont-Catt	Encephalocele
43	Eye	2.27E-175	4.65	4.30	Yes				Cont	Eye
44	Fetal alcohol syndrome	5.88E-57	104.82	43.19				N/A	Cont	Fetal Alcohol
45	Gastroschisis	6.55E-39	31.88	4.58		Yes			Cont-Catt	Gastroschisis
46	Genetic syndromes + microdeletions	1.38E-228	2.98E+21	2.34E+13	Yes				Cont	Genetic syndromes
47	Genital	2.55E-243	6.15E+04	284.15	Yes				Cont	Genital
48	Hip dislocation and/or dysplasia	0.0020	4.43E+13	2.33E+05			No		Cont	Hip dislocation and/or dysplasia
49	Hirschsprung's disease	2.54E-88	224.78	27.50	Yes				Cont	Hirschsprungs
50	Holoprosencephaly / Arhinencephaly	1.22E-72	5.29	4.61	Yes				Cont	Holoprosencephaly / Arhinencephaly
51	Hydrocephalus	1.76E-110	2.91E+03	29.39		Yes			Cont	Hydrocephalus
52	Hydronephrosis	0.00E+00	4.16	3.99		Yes			Cont-Catt	Hydronephrosis
53	Hypoplastic Left Ht	3.37E-61	3.54	3.18	Yes				Catt	Hypoplastic Left Ht
54	Hypoplastic right heart	2.85E-59	61.61	9.34			No		Cont	Hypoplastic right heart
55	Hypospadias	2.92E-177	3.78E+04	174.71	Yes				Cont	Hypospadias
56	Klinefelter syndrome	1.75E-41	3.10E+12	2.44E+04		Yes		N/A	Cont	Klinefelter syndrome
57	Lateral anomalies	2.36E-48	5.76E+05	2.67E+03				N/A	Cont	Lateral anomalies
58	Limb	0.00E+00	2.81E+06	1.30E+04	Yes				Cont	Limb anomalies
59	Limb reductions	8.20E-65	2.46	2.30	Yes				Cont-Catt	Limb reductions
60	Maternal infections resulting in malformations	4.15E-87	9.51E+04	976.69				N/A	Cont	Matern Infect Malformations
61	Mitral valve anomalies	8.99E-58	1.07E+19	5.61E+10				N/A	Cont	Mitral valve anomalies
62	Multicystic renal dysplasia	6.70E-251	6.60	6.05				N/A	Cont	Multicystic Renal Dys
63	Nervous system	0.00E+00	3.89E+18	2.04E+10	Yes				Cont	Nervous system
64	Neural Tube Defects	9.97E-269	164.37	25.68	Yes				Cont	Neural Tube Defects
65	Oesophageal S/A	3.49E-44	2.97	2.67	Yes				Cont-Catt	Oesophageal stenosis or atresia
66	Omphalocele	4.94E-131	44.56	6.62	Yes				Cont	Omphalocele
67	Oro-facial clefts	3.99E-133	1.11E+05	512.89	Yes				Cont	Oro-facial clefts
68	Patau syndrome/trisomy 13	1.08E-144	121.51	14.65	Yes				Cont	Patau syndrome
69	PDA as only CHD in term infants (>=37 weeks)	2.14E-20	1.21E+03	11.95		Yes			Cont-Catt	PDA as only CHD in term infants (>=37 weeks)
70	Polydactyly	1.46E-292	1.04E+03	10.17		Yes		N/A	Cont	Polydactyly
71	Posterior urethral valve and/or prune belly	1.28E-42	388.43	3.41	Yes				Catt	Posterior urethral valve and/or prune belly
72	Pulmonary valve stenosis	2.09E-95	35.08	3.85		Yes			Cont-Catt	Pulmonary valve stenosis
73	PV Atresia	1.42E-27	3.48	2.96	Yes				Catt	PV Atresia
74	Respiratory	2.57E-203	1.68E+23	8.81E+14		Yes			Cont	Respiratory
75	Severe CHD	1.81E-317	126.20	19.61				N/A	Cont	Severe CHD
76	Severe microcephaly	3.17E-148	4.59E+21	3.60E+13		Yes			Cont	Severe microcephaly
77	Single ventricle	1.03E-25	4.08	3.36	Yes				Catt	Single ventricle
78	Situs inversus	1.42E-44	3.53E+04	362.44				N/A	Cont	Situs inversus
79	Skeletal dysplasias	5.12E-74	32.78	4.73				N/A	Cont	Skeletal dysplasias
80	Small Intestine S/A	8.23E-31	5.31	4.27	Yes				Catt	Small Intestine stenosis or atresia
81	Spina Bifida	3.93E-84	139.95	21.80	Yes				Cont	Spina Bifida
82	Syndactyly	3.47E-16	22.95	3.13				N/A	Cont	Syndactyly
83	Teratogenic syndromes with malformations	1.42E-139	8.21E+19	6.44E+11				N/A	Cont	Teratogenic syndromes with malformations
84	Tetralogy of Fallot	3.12E-47	2.77	2.52	Yes				Catt	Tetralogy of Fallot
85	Tot Anom Pul V Ret	4.07E-09	2.71	2.14	Yes				Cont-Catt	Total Anomalous Pulm Venous Return
86	Transposition of great vessels	9.96E-33	4.79E+03	48.76	Yes				Cont-Catt	Transposition of great vessels
87	Turner syndrome	1.10E-146	7.95	6.99	Yes				Cont	Turner syndrome
88	TV S/A	6.86E-24	4.55	3.65				N/A	Cont-Catt	Tricuspid valve stenosis / atresia
89	Urinary	0.00E+00	2.26E+03	22.70	Yes				Cont	Urinary
90	Valproate syndrome	1.57E-07	30.11	7.00				N/A	Catt	Valproate syndrome
91	Vascular disruption anomalies	3.46E-101	41.52	5.63				N/A	Cont	Vascular disruption anomalies
92	VATER/VACTERL	2.43E-36	1.70E+20	8.94E+11				N/A	Cont	VATER/VACTERL
93	Ventricular septal defect (VSD)	0.00E+00	28.78	4.08	Yes				Cont	Ventricular septal defect (VSD)
	Count				49	15	5	27		
	Total				93	93	93	93		
	Percentage				52.7%	16.1%	5.4%	29.0%		
	Percentage with data				74.2%	22.7%	7.6%			
	Total with data					97.0%				

TABLE 6 Summary table for USA-European congenital anomaly list comparisons.

	Confirmed anomalies	
Parameter	**USA**	**Europe**
Continuous—Continuous	24	35
Categorical—Categorical	25	26
Continuous—Categorical	17	17
Total	45	49
Number likely	0	16
Total confirmed	45	65
Number assessed	45	93
Percentage confirmed	100.0%	52.7%
Inadequate data	0	24
Number assessed with adequate data	45	69
Percentage confirmed with adequate data	100.0%	94.2%
Exceptions		
		Amniotic band
		Anophthalmos
		Craniosynostosis
		Hip dislocation and/or dysplasia
		Hypoplastic right heart
Negative		
		Indeterminate sex

rose to 74.2%, 22.7%, and 7.6% when missing data were not considered. This made a total of 97.0% matched or likely matched from the European data set.

Supplementary Table 12 re-presents Table 5 grouped by the categorization of the match response. It is noteworthy that this table includes information on 23 cardiovascular anomalies, 14 gastrointestinal anomalies, 12 general anomalies, 11 central nervous system anomalies, 10 facial anomalies, 10 genitourinary anomalies, 6 chromosomal anomalies, 6 limb anomalies, four body wall anomalies and two respiratory anomalies. Thus, all major body systems are impacted by cannabis teratogenesis.

Table 6 summarizes all of this data concisely. As noted all 45 of the USA CA's were confirmed in the European data set. For the European data 49 CA's were directly confirmed, 16 were considered likely (these are discussed further below) and 24 had missing information. This indicates that 94.2% of the European CA's were (likely) matched in the USA data set where such data existed. The five anomalies listed were negative findings in the USA data set. Indeterminate sex did not test positively in any of the European analyses and was not found in the USA data set.

Consideration of analyses

Main results

The main results of this study are that the 45 congenital anomalies reported as being elevated in association with cannabis use on bivariate analysis in the United States were confirmed in the European data set. Moreover, 65/69 (94.2%) CA's for which data existed in the European data set were confirmed either by the USA data set or by similar results from elsewhere as described in further detail below.

These important results have a number of critical internal and external ramifications. Since these are arguably the two largest most comprehensive data sets in the world their individual findings naturally carry great weight internationally relating to cannabis teratogenesis. However, it is of paramount importance that cannabis teratogenesis was shown to be so consistent between the two data sets. This gives researchers and policy makers much greater confidence that these findings are not "merely" based on an observational study but actually carry external validity beyond the circumstances of each study individually.

Another implication of this internal consistency is the very striking breadth of the teratogenic impact of cannabis exposure across every major body system. While it is well known that cannabinoids have plethoric effects in many systems the observed breadth of effects implies that cannabinoids interfere with fundamental cellular processes including chromosomal disorders, energy generation and mitochondrial metabolism, signaling at the mitochondrial and plasma membranes, and multimodal dysregulation of epigenomic regulation.

Second, as enumerated in the Introduction section, cannabis teratogenesis is part of the broader picture of cannabis genotoxicity that can be expected to be expressed in birth defects, cancer incidence, inheritable cancer incidence, and cellular and organismal aging. It is for this reason that strong and robust proof of reproducible cannabis teratogenicity forms one of the major bases of the broader understanding of cannabinoid genotoxicity and its clinical implications. Therefore, in this regard, the present unifying synthesis has widespread ramifications far beyond the subject merely of cannabis teratogenesis itself.

Findings need to be interpreted in the light of the well-described exponential dose-response effect curve of cannabinoids to their genotoxic outcomes. The present findings therefore go beyond the merely descriptive and may be predictive of dramatically increased congenital teratogenesis among communities experiencing increased widespread cannabin exposure.

Although a human epidemiological study, it is imperative to recognize that this broad spectrum of severe congenital anomalies has been confirmed by earlier reports in hamsters, white rabbits, and rats (Geber & Schramm, 1969a, 1969b; Graham, 1976). White rabbits and hamsters are believed to form the most predictive preclinical models for human teratogenesis.

Specific congenital anomalies of interest

All anomalies

It is noteworthy and more than a little concerning that the category of all anomalies, which exists in the European data set but not in the USA data, was elevated with P-value of 0.0006, E-value estimate 55.76, and minimum E-value of 8.41 (Table 2). Naturally, this standalone finding by itself is very serious indeed.

Chromosomal anomalies

It is of interest that the chromosomal anomalies listed include Trisomies 13, 18, and 21, the monosomy turners syndrome, the disomy Klinefelters syndrome, genetic syndromes, and microdeletions. This confirms using European data strong evidence of direct damage to the human genome, which was recently demonstrated in the USA data relating to the totality of chromosomal disorders tracked (Reece & Hulse, 2021a, 2022a, 2022b). Indeed, human genome project data show that chromosomes 13, 18, 21s and X have lengths of 113 MB, 76 MB, 46 MB, and 153 MB so that they together account for 388 MB or 12.9% of the human genome (which is 3000 MB in length). Moreover, if one adds to this number the length of the five chromosomes commonly associated with acute lymphoid leukaemia in which cannabis has now been implicated (chromosomes 4, 9, 10, 11, and 22) totalling 645 MB and the chromosomes known to be implicated in testicular cancer which cannabis is also known to cause (chromosomes 1, 7, 8, 11, 12, 13, 18, 21, X and Y) totalling 1254 MB, one finds that, removing the duplicated chromosomes, cannabis is actually implicated in direct damage to 1765 MB of the 3000 MB of the human genome or 58.8%. This is a very large segment of genetic damage for direct cannabis-induced mutagenic damage and derangement.

Limb anomalies

It is of interest that the limb anomalies related to cannabis exposure in Europe include limb anomalies, limb reductions, polydactyly, and syndactyly. The maximum minimum E-value for limb anomalies in Europe was 1.30×10^4 and for limb reductions 2.30 (Table 2). The P-value for this relationship was below the limit of detectability ($P < 2.2 \times 10^{-320}$). Indeed, these minimum E-values were among the highest recorded in Table 2 of the 70 anomalies listed there. The equivalent data for USA were limb reduction deficiencies mEV = 9.53 and reduction deformities lower limbs mEV = 2.57 (Table 1). Moreover, polydactyly and syndactyly were first reported to be elevated in Hawaii providing further validation to the present findings (Forrester & Merz, 2007). Club foot/talipes equinovarus and hip dislocation and dysplasia were also on this list of limb anomalies. These findings are significant as limb anomalies are some of the most dramatic and obvious birth defects encountered.

It should also be noted that recent media reports mention epidemics of babies born with absent limbs occurring in hospitals in Germany (Robinson, 2019) and in parts of northeast France at rates up to 60 times historical background rates (Gant, 2019; Willsher, 2018). Moreover in these parts of France, cows are being born without forelimbs (Agence France-Presse in Paris, 2018). Large crops of cannabis are being cultivated in these rural parts of the country with cannabis used as animal fodder. However, similar reports are not occurring from Switzerland where cannabinoids are banned from entering the food chain. Based on the data presented in the present analysis, food chain contamination by cannabis products is an obvious strong suspect for this unforeseen wave of limb teratogenesis.

VACTERL syndrome

The presence of VACTERL syndrome on this list is of great interest. VACTERL, which is an acronym standing for vertebral, cardiac, tracheo-esophageal fistula/oesophageal atresia, renal abnormalities, and limb abnormalities, was recently described from Great Ormond St Hospital as a collection of anomalies that occurred together more than would have been expected by chance although initially its etiology was unknown (Hosptial, 2022). It has since been shown that it is likely due to derangement of the key human embryological morphogen sonic hedgehog an action, which has now been ascribed to several cannabinoids including Δ9-tetrahydrocannabinol (THC) and cannabidiol (Carlson, 2019; Fish et al., 2019). The presence of vertebral, limb, cardiac, and renal anomalies on this list is noted, just as is clearly seen with fetal cannabinoid exposure (Tables 1–5). VACTERL syndrome has not been reported in the USA data set.

Indeed the effects that were described herein with relation to VACTERL were strong and dramatic. Table 2 lists the effects of bivariate comparisons of anomalies with various metrics of cannabis exposure VACTERL syndrome has the lowest P-value (1.70×10^{20}), the highest E-value estimate (8.94×10^{11}), and one of the highest minimum E-values (4.93×10^{3}). In Supplementary Table 6, VACTERL has the second highest E-value estimate (49.76) and minimum E-value (26.43) of the 621 associations listed there placing it at the 99.68th percentile.

This becomes a serious issue. Data show therefore that cannabinoid exposure is linked with a multisystem generalized body dysmorphic syndrome including skeletal, cardiovascular renal, and limb anomalies in exactly the manner predicted by its cellular mechanism of action in inhibiting one of the embryo's major morphogenic gradients. It also means that the notion that cannabis is a soft and harmless drug is not only egregiously in error but also presents a very real danger to fetal-maternal health. It is worth underscoring again that limb anomalies are part of this body dysmorphia.

Cardiovascular anomalies

Cardiovascular anomalies are the commonest form of congenital anomalies so it was not unexpected that they would be well represented. Indeed in the initial Hawaiian report six cardiovascular anomalies were reported to be elevated with prenatal cannabis exposure. These were hypoplastic left heart syndrome, tetralogy of fallot, pulmonary valve atresia and or stenosis, ventricular septal defect, and atrial septal defect. It is therefore important that congenital heart disease was the most significant of all anomalies on bivariate continuous testing in Europe (Table 2) with P-value below the limits of detectability ($P < 22 \times 10^{-320}$) associated with the highest E-value estimate (6.95×10^{7}) and minimum E-value (3.22×10^{5}) of the 225 noted in Supplementary Table 2. Moreover, severe congenital heart disease was also very elevated in this bivariate analysis ($P = 0.0002$, E-value estimate $= 1.67 \times 10^{4}$, minimum E-value 171.06). Other cardiovascular anomalies are described further in other reports from our group (Reece & Hulse, 2019b, 2020, 2020a, 2020c, 2022a, 2022b).

Central nervous system anomalies

Anomalies of the central nervous system (CNS) were also well represented in the current analyses. This group included 11 anomalies including severe microcephaly. While slightly smaller heads are well recognized to be related to cannabis use with head circumference minimally smaller by several millimeters (Volkow et al., 2017), it is important to note that in Table 5 severe microcephalus recorded the fourth highest minimum E-value ($P = 4.34 \times 10^{-5}$, E-value $= 1.48 \times 10^{7}$, mEV $= 9.75 \times 10^{3}$). Clearly markedly smaller heads necessarily implies smaller brains with reduced computational power.

Mechanisms

As noted below, the demonstration of robust mechanisms and pathways by which cannabinoid-related mutagenesis may occur is central and critical to any consideration of potentially causative relationships. As mentioned in the Introduction section, there are a myriad of cellular and molecular pathways described by previous reports for cannabis mutagenesis and teratogenesis (Chioccarelli et al., 2010; DiNieri et al., 2011; Ellis et al., 2021; Huang et al., 1999; Leuchtenberger & Leuchtenberger, 1971; Leuchtenberger et al., 1973; Mon et al., 1981b; Morishima, 1984; Murphy et al., 2018; Nahas, 1990a, 1990b; Rossato et al., 2008; Russo et al., 2019; Schrott et al., 2021; Stenchever et al., 1974; Szutorisz et al., 2014; Szutorisz & Hurd, 2016; Watson et al., 2015). While space does not allow a detailed exploration of these features, it is worthwhile mentioning three areas of cannabinoid-related mutagenicity, namely inhibition of the gradients of key morphogens that control, regulate and direct body patterning in the embryo, chromosomal damage, and widespread epigenomic dysregulation.

Morphogenic gradients

Cannabis disrupts signaling of most of the major morphogen gradients involved in body pattern and organ formation including sonic hedgehog (Fish et al., 2019), bone morphogenetic proteins (Birerdinc et al., 2013; Richard & Picard, 2011; Xu et al., 2013), retinoic acid signaling (Fraher et al., 2015; Kučukalić et al., 2019; Lee & Jeong, 2012), Wnt signaling (McKenzie et al., 2019; Nallathambi et al., 2018; Nalli et al., 2019; Petko et al., 2018; Vallee et al., 2017; Xian et al., 2018), fibroblast growth factor (FGF) (Aguado et al., 2007; Williams et al., 2003), including transactivation of the FGF1R by CB1R (Asimaki et al., 2011), notch signaling (Frampton et al., 2010; Kim et al., 2014; Niu et al., 2015; Tanveer et al., 2012; Xapelli et al., 2013) (which is very involved in colorectal cancer), and the hippo pathway (Murphy et al., 2018). Body dysmorphogenesis then is very much to be expected in such a context of chaotic disruption of most of the key controlling morphogenic gradients.

Chromosomal breakage-fusion-bridge cycling

Direct damage to chromosomes by many cannabinoids including cannabidiol, THC, cannabinol, cannabigerol, cannabichromene, and many synthetic cannabinoids is well described. In particular, single- and double-stranded breaks were demonstrated from 1971 (Leuchtenberger & Leuchtenberger, 1971), end-to-end translocations (Huang et al., 1999), and chromosomal bridges between separating pronuclei at anaphase are all well described in the literature (Morishima, 1984) often after quite minimal cannabis exposures, such as four puffs of low strength cannabis smoke across a plate of cultured cells two weeks earlier (Leuchtenberger et al., 1973). This gross chromosomal damage results in cannabinoids being known to test positively in the micronucleus assay and the comet assay. Micronuclei results from chromosomes becoming dislodged from the mitotic spindle at anaphase and becoming isolated in the cell in a tiny nucleus, where lacking the normal complement of DNA maintenance machinery, the normal processes of DNA replication, and transcription causes it to become broken up and degraded and becomes a powerful engine for the generation of genetic aberrations resulting in severe outcomes including cell death, cancer, congenital anomalies, and mental retardation in surviving cells. These processes, however, have previously been described as being separate processes in the literature.

One important and instructive lesson in this regard comes from testicular cancer another disorder characterized by severe genetic damage to 11 chromosomes especially chromosome 12 (Shen et al., 2018). Testicular cancer has been described in metaanalysis as occurring 2.6 times more frequently in cannabis smokers (Gurney et al., 2015), and dose-response effects have been well demonstrated for metrics of duration, intensity, and total dose smoked (Callaghan et al., 2017; Daling et al., 2009; Lacson et al., 2012; Trabert et al., 2011). Moreover, since the median age of cannabis smoking in young men is about 20 years, and since the median age of testicular cancer and the usual oncogenic incubation period is about 33 years, these data imply that testicular cancer occurs in cannabis exposed males with an acceleration of the incidence—oncogenic incubation metric of ($2.6 \times 2.5 =$) 6.5-fold, which is clearly dramatic. Moreover, it is known that during oncogenesis of cannabis-related testicular cancer, which is mostly nonseminomatous germ cell tumor (NSGCT), one or two whole genome doubling events occur followed by loss of 50–70 chromosomal arms.

This observation is very dramatic because it implies major disruption to the meiotic and mitotic machinery of germ cell division to induce the whole genome doubling event/s and then the dramatic loss of up to 70 chromosomal arms at 6.5 times the rate normally seen in males who are not cannabis exposed. This is an extraordinary conclusion. This implies that the features of chromosomal breakage, fusion, and bridge formation are not isolated events but are in fact occurring in a cycle as the breakage-fusion-bridge cycle which was first described in 1938. This explains the necessity for the oncogenic incubation period while the malignant germ cells expand, gain clonal dominance and convert to overt clinical malignancy. This linkage between chromosomal damage formats into the breakage-fusion-bridge cycle has not been described formerly in the literature to our knowledge.

Epigenomic dysregulation

Several papers have reported profound dysregulation of the epigenome transgenerationally. Studies of DNA methylation (DiNieri et al., 2011; Ellis et al., 2021; Murphy et al., 2018; Schrott et al., 2021; Szutorisz et al., 2014; Szutorisz & Hurd, 2016; Watson et al., 2015) and histone deficiency and dysfunction (Mon et al., 1981b) are well documented. Together, they form a powerful and carefully documented literature of considerable public heath importance.

One of the most fascinating papers in this regard is a recent paper from North Carolina and collaborators documenting that human (and rat) cannabis dependence was linked with the induction of 163 sites of differential DNA methylation (using strict Bonferroni cutoff criteria) which together affected hundreds of genes (Schrott et al., 2021). Happily they found that after a 17-week period of cannabis abstinence that most of these changes resolved. However what was very concerning was the finding that after the period of abstinence, 127 new signals of differentially methylated DNA appeared that reflected the process of cannabis withdrawal after that washout period. This is clinically of great relevance as the level of use of many cannabis users will obviously fluctuate over time and periods of drug usage will naturally fluctuate with periods of withdrawal, so that this scenario closely mirrors in the experimental context real-world experience for many patients. It also documents in the epigenomic signature the cause and reflections of many of the illnesses and clinical symptomatology which patients describe.

Among the most prominent systems affected were organismal growth and agenesis, cardiogenesis, cerebral development, neurodevelopmental disorders, ganglionic neuronal growth, liver damage, gastrointestinal system, and growth and maturation of the lymphatic and hematological systems (Schrott et al., 2021).

Among pathway analysis conducted during cannabis withdrawal pathways involved in cerebral and cardiovascular organogenesis, hippocampal formation, area of blood vessel development, organismal death, quantity of pyramidal neurons, cognitive and learning impairments, encephalopathy, embryonal development, behavior, organismal survival, endocrine system damage, and damage to the seminiferous tubules that house the male germ stem cells were all annotated.

Importantly, these workers also provided a detailed data appendix in supplementary material along with their report documenting these changes. This compendious material ran to 360 pages. It is fascinating in the present context that genes controlling vertebral development (page 353), bone morphology (pages 319, 334, 339), bone mineralization (page 349), ossification (page 343), and bone marrow cell proliferation and bone mineral density (page 353) were mentioned in view of the above findings relating to skeletal anomalies, VACTERL syndrome, and limb anomalies. Disorders of myogenesis (page 320) may speak to limb anomalies and reduced foetal growth. Disorders of microtubule dynamics (page 300) were also reported and may in part explain the prominent findings of chromosomal disorders in the both European and USA data sets.

There were 84 hits for growth control; 101 hits for embryonic development, growth, embryogenesis embryonic morphogenesis and body axis development, and differentiation of embryonic stem cells. 256 gene hits were found for DNA regulation, transcription, promoter activation, repair replication, and recombination.

Page 308 lists abnormalities of the nervous system, synaptic transmission, and abnormalities of enteric ganglion formation. Such findings would well account for the presence of Hirschsprung disease (congenital megacolon due to absence of lower intestinal enteric ganglia) on both European and USA cannabis-related congenital anomaly lists.

Descriptions of involvement of neuronal growth pathways (pages 298, 299) and synapse development (page 308) along with many descriptions of altered neuronal and cerebral development account for clinical syndromes such as severe microcephaly found in the European data set.

Functional annotations for the cardiovascular (pages 288, 318, 321, and 324), gastrointestinal (pages 324 and 333), and uronephrological (pages 351 and 352) systems both in cannabis dependence and withdrawal were also very notable.

Together, this vast and rich body of evidence provides strong support for the notion that the pan-organismal pattern of foetotoxicity observed in teratological reviews such as this are strongly underpinned by multiple epigenomic mechanisms.

Exponentiation of genotoxic dose-response effects

A rich laboratory research literature demonstrates that exponential dose-responses are prototypical for many of the processes upon which genomic and epigenomic integrity depends including inhibition of cell growth (Tahir & Zimmerman, 1991), cell viability (Tahir et al., 1992), actin synthesis and mechanisms (Tahir & Zimmerman, 1991), mitochondrial ATP charge (Hebert-Chatelain et al., 2014; Mahoney & Harris, 1972), mitochondrial energy supply and supply of epigenomic substrates (Canto et al., 2015), histone synthesis (Mon et al., 1981a), mutagenesis (Busch et al., 1979), inhibition of tubulin and RNA synthesis (Tahir & Zimmerman, 1991), and micronucleus formation (Koller et al., 2015; Zimmerman & Raj, 1980).

These laboratory findings are supported by several human epidemiological studies documenting increased clinical evidence of cannabinoid genotoxicity at higher doses has also been compiled including Deletion 22q11.2, pulmonary valve atresia, congenital posterior urethral valves, trisomies 13,18 and 21, Turner syndrome, omphalocele, atrial septal defect, orofacial clefts, hydrocephalus, encephalocele, large intestinal stenosis or atresia, diaphragmatic hernia, Ebstein anomaly, microcephalus, microtia, hypoplastic left heart, and others (Reece & Hulse, 2020).

Qualitative causal inference

Professor Austin Bradford Hill was a foremost British epidemiologist who played a leading role in determining the relationship between tobacco smoking and lung cancer in the 1950s when the issue was unclear amongst statisticians generally. Hill described nine criteria required to demonstrate a causal relationship, namely strength of association, consistency among studies, specificity, temporal sequence, coherence with known data, biological plausibility, biological dose-response curve, analogy with similar situations elsewhere, and experimental confirmation (Hill, 1965). It is clear that the present report, along with the papers cited herein, amply fulfil these criteria.

Quantitative causal inference

As described above, *E*-values form one of the major pillars of the demonstration of causality on epidemiological criteria as they quantify the strength of association required of extraneous controlled covariates to obviate an apparently causal effect.

In addition to the bivariate analyses described herein multivariable regression techniques can also be used to control for other common classes of covariates particularly socioeconomic and ethnodemographic factors. Such regression modeling can also be inverse probability weighted, which makes the important conceptual advance from purely associations in an observational study into casual relationships from a pseud-randomized paradigm from which it is entirely proper to drawn causal inferences. Moreover space-time regression can also be performed to evaluate these changes in their native space-time context. Many recent reports perform just these kinds of advanced analyses albeit it is beyond our brief to examine these issues in the present context. The interested reader is referred to these reports (Reece & Hulse, 2019a, 2020a, 2020b, 2020c, 2022a, 2022b).

However, the general principal to emerge from this large body of work is that strong relationships such as those reported here are not removed by multivariable adjustment, inverse probability weighting, or space-time analysis but indeed are often enhanced.

We reiterate again that *E*-values exceeding nine are considered to be in the high range (Pearl & Mackaenzie, 2019). Thus, although the present work is not multivariable adjusted, the very large magnitude of the associations reported quantified with the very large *E*-values are stereotypical for inherently causal relationships. For these reasons, the present bivariate and unadjusted data sets and findings are weighty indeed notwithstanding their clearly acknowledged limitations.

Generalizability

In that this significant chapter combines the two largest and most comprehensive global databases on congenital teratogenesis and is able to demonstrate great internal and external concordance both within and external to this study with results reported in many other jurisdictions (Forrester & Merz, 2007; Reece & Hulse, 2019a, 2019b, 2020, 2020a, 2020b, 2021a, 2021b, 2022a, 2022b), we are confident that our data are widely generalizable across the international community wherever data are of sufficient quality to reliably demonstrate these relationships.

Strengths and limitations

The present work has several strengths including the use and comparison of two of the largest teratogenesis databases in the world together with strongly confirmatory findings between the two. Like many epidemiological studies, this work did not have available to it individual participant-level exposure data.

Future work can take these analyses further including multivariable studies and spatiotemporal studies. That is a very large project indeed which would be well suited to a dedicated public health institution with an interest in the field. Much of this work has already been done and has appeared in published reports or are presently making its way through the process to publication.

Conclusion

The present report confirms quantitative findings of cannabinoid-related teratogenesis of the USA data set using the European database, and near complete confirmation of the European data wherever adequate data exists. Since these two comprehensive databases comprise the majority of the global data in the area of teratogenesis the concordance of their findings provide strong evidence to support the veracity of the findings internally and are also externally concordant with other recent and similar reports. Very elevated *E*-values here strongly imply causal relationships, findings which have been repeatedly confirmed by inverse probability weighted multivariable and space-time modeling reports which have recently appeared. Overall, the concordance of these two major international databases strongly reinforces the importance and salience of cannabinoid teratogenicity-mutagenicity in the transgenerational context form a major pillar of cannabinoid-related genotoxicity syndrome overall and particularly in the context of well-demonstrated exponential effects at higher doses. All this strongly argues for strict control of community cannabinoid exposure in line with rational policies for all other major genotoxins to protect the integrity of both the food chain and the gene pool for several generations to come.

Applications to other areas

Cannabinoid-related teratogenicity, along with cannabinoid-related carcinogenicity, and cannabinoid-accelerated cellular aging form the major clinical manifestations of cannabinoid genotoxicity. This chapter on cannabinoid-related teratogenicity therefore needs to be read in the light of this larger literature to properly form a wholistic understanding of the whole corpus of the epidemiology of the generic topic of cannabinoid-related genotoxicity.

Many cannabinoids have been implicated as having teratogenic activities including Δ9THC, Δ8THC, cannabidiol, cannabinol, cannabichromene, and cannabigerol. This is consistent with the observation made long ago that the teratogenic chemical moiety is the olevitol cannabinoid nucleus. This in turn implies that the teratogenicity is a cannabinoid class effect and is relatively independent of which particular cannabinoid is under discussion, for all are now generically implicated.

Several studies have shown that the teratogenic effects of cannabis and cannabinoids are generally more powerful than those of tobacco and alcohol and are often more powerful than tobacco and alcohol combined. It follows therefore that cannabis should be considered with other major teratogenic and genotoxic agents and treated and classified as such and the community protected accordingly. This especially applies to vulnerable populations such as pregnant women, babies, and adolescents but equally applies to the whole population generally.

It especially applies to protection of the food chain from cannabinoid contamination.

Exponential genotoxic and teratogenic effects have been demonstrated in both laboratory and epidemiological studies that come into play at higher dose ranges. Rising prevalence of cannabis use, intensity of daily use, and THC potency have relatively abruptly launched the community into a much higher zone of cannabinoid exposure in which severe genotoxic outcomes will be commonplace as shown by the steep dose-response curve in the present studies for VACTERL syndrome a multisystem congenital anomaly, which includes major limb development anomalies.

Overall findings suggest that cannabinoid teratogenicity, carcinogenicity, pro-aging effects, and cannabinoid genotoxicity generally are highly important epidemiological and clinical phenomena capable of impacting population health profoundly and negatively for multiple generations to come. Given their obvious potency from extant population and registry based data sets on both sides of the Atlantic rational and evidence-based drug policy development would mandate that they be carefully regulated in line with all other potent genotoxins and teratogens in western communities.

Mini-dictionary of terms

Birth defect	Abnormal birth deformity or cellular structure
Cannabinoid	Chemical unique to *Cannabis sativa* plant
Congenital anomaly	Anomaly present at or soon after birth
Mutagenesis	Change in genetic sequence of DNA
Teratogenesis	Process of making a congenital deformity

Key facts of cannabinoid teratogenesis

- 45/62 congenital anomalies assessed in USA could be related on bivariate analysis to indices of cannabis or cannabinoid exposure.
- 89/95 congenital anomalies in Europe could be related to indices of simple or compound metrics of cannabis exposure.
- 65/67 European anomalies for which adequate data existed in the USA data set were confirmed in the transatlantic comparison.
- Congenital anomalies particularly affected were the cardiovascular, chromosomal, limb, gastrointestinal, uronephrological, and facial and central nervous systems.
- All anomalies were implicated as being cannabis related.
- The multisystem VACTERL syndrome (including limb defects) was also strongly implicated and was one of the strongest hits identified.
- The chromosomal anomalies identified accounted for 12.9% of the human genome. Together with cannabis-related cancer chromosomal aberrations this rose to 59% of the genome.
- Hence the findings in the United States and European data sets are strongly confirmatory and mutually externally validate each other.

Summary points

- Cannabis is an important human teratogen
- Cannabis is a more much more serious human teratogen than alcohol or tobacco
- 45/62 congenital anomalies in USA were shown to be cannabis related
- 89/95 congential anomalies in Europe could be related to metrics of cannabis exposure
- All USA cannabs-related anomalies were identified in the European data set
- 65/67 European anomalies that had comparators were confirmed in the USA data set
- Hence, findings from the two largest congential anomaly data sets in the world strongly support each other and are thus externally validated.
- The exponential genotoxic effects of cannabis teratogenesis are a major concern as the community moves into higher dose ranges of cannabis exposure.
- The importance of protecting the food chain is particularly salient, for the population in general and for vulnerable populations such as pregnant women, babies, and adolescents in particular.
- Cannabinoid teratogensis is a subset of cannabinoid genotoxicity and should be understood along with cannabinoid carcinogensis and cannabinoid acceleration of cellular aging to properly inform a global and wholistic view of this subject.

References

Agence France-Presse in Paris. (2018). France to investigate cause of upper limb defects in babies. *The Guardian*. Retrieved from: https://www.theguardian.com/world/2018/oct/21/france-to-investigate-cause-of-upper-limb-defects-in-babies.

Aguado, T., Romero, E., Monory, K., Palazuelos, J., Sendtner, M., Marsicano, G., ... Galve-Roperh, I. (2007). The CB1 cannabinoid receptor mediates excitotoxicity-induced neural progenitor proliferation and neurogenesis. *The Journal of Biological Chemistry*, *282*(33), 23892–23898. https://doi.org/10.1074/jbc.M700678200.

Asimaki, O., Leondaritis, G., Lois, G., Sakellaridis, N., & Mangoura, D. (2011). Cannabinoid 1 receptor-dependent transactivation of fibroblast growth factor receptor 1 emanates from lipid rafts and amplifies extracellular signal-regulated kinase 1/2 activation in embryonic cortical neurons. *Journal of Neurochemistry*, *116*(5), 866–873. https://doi.org/10.1111/j.1471-4159.2010.07030.x.

Birerdinc, A., Jarrar, M., Stotish, T., Randhawa, M., & Baranova, A. (2013). Manipulating molecular switches in brown adipocytes and their precursors: a therapeutic potential. *Progress in Lipid Research*, *52*(1), 51–61. https://doi.org/10.1016/j.plipres.2012.08.001.

Busch, F. W., Seid, D. A., & Wei, E. T. (1979). Mutagenic activity of marihuana smoke condensates. *Cancer Letters*, *6*(6), 319–324. Retrieved from: http://www.ncbi.nlm.nih.gov/pubmed/378366.

Callaghan, R. C., Allebeck, P., Akre, O., McGlynn, K. A., & Sidorchuk, A. (2017). Cannabis use and incidence of testicular cancer: A 42-year follow-up of Swedish men between 1970 and 2011. *Cancer Epidemiology, Biomarkers & Prevention*, *26*(11), 1644–1652. https://doi.org/10.1158/1055-9965.EPI-17-0428.

Canto, C., Menzies, K. J., & Auwerx, J. (2015). NAD(+) metabolism and the control of energy homeostasis: A balancing act between mitochondria and the nucleus. *Cell Metabolism*, *22*(1), 31–53. https://doi.org/10.1016/j.cmet.2015.05.023.

Carlson, B. M. (2019). *Human embryology and developmental biology. Vol. 1* (6th ed.). Philadelphia: Elsevier.

Chioccarelli, T., Cacciola, G., Altucci, L., Lewis, S. E., Simon, L., Ricci, G., … Cobellis, G. (2010). Cannabinoid receptor 1 influences chromatin remodeling in mouse spermatids by affecting content of transition protein 2 mRNA and histone displacement. *Endocrinology*, *151*(10), 5017–5029. https://doi.org/10.1210/en.2010-0133.

Commission, E. (2021). *Eurocat data: Prevalence charts and tables*. Eurocat. Retrieved from https://eu-rd-platform.jrc.ec.europa.eu/eurocat/eurocat-data/prevalence_en.

Daling, J. R., Doody, D. R., Sun, X., Trabert, B. L., Weiss, N. S., Chen, C., … Schwartz, S. M. (2009). Association of marijuana use and the incidence of testicular germ cell tumors. *Cancer*, *115*(6), 1215–1223. https://doi.org/10.1002/cncr.24159.

DiNieri, J. A., Wang, X., Szutorisz, H., Spano, S. M., Kaur, J., Casaccia, P., … Hurd, Y. L. (2011). Maternal cannabis use alters ventral striatal dopamine D2 gene regulation in the offspring. *Biological Psychiatry*, *70*(8), 763–769. https://doi.org/10.1016/j.biopsych.2011.06.027.

Ellis, R. J., Bara, A., Vargas, C. A., Frick, A. L., Loh, E., Landry, J., … Hurd, Y. L. (2021). Prenatal Δ(9)-tetrahydrocannabinol exposure in males leads to motivational disturbances related to striatal epigenetic dysregulation. *Biological Psychiatry*. https://doi.org/10.1016/j.biopsych.2021.09.017.

Fisar, Z., Singh, N., & Hroudova, J. (2014). Cannabinoid-induced changes in respiration of brain mitochondria. *Toxicology Letters*, *231*(1), 62–71. https://doi.org/10.1016/j.toxlet.2014.09.002.

Fish, E. W., Murdaugh, L. B., Zhang, C., Boschen, K. E., Boa-Amponsem, O., Mendoza-Romero, H. N., … Parnell, S. E. (2019). Cannabinoids exacerbate alcohol teratogenesis by a CB1-hedgehog interaction. *Scientific Reports*, *9*(1), 16057. https://doi.org/10.1038/s41598-019-52336-w.

Forrester, M. B., & Merz, R. D. (2007). Risk of selected birth defects with prenatal illicit drug use, Hawaii, 1986-2002. *Journal of Toxicology and Environmental Health. Part A*, *70*(1), 7–18. Retrieved from: http://www.ncbi.nlm.nih.gov/entrez/query.fcgi?cmd=Retrieve&db=PubMed&dopt=Citation&list_uids=17162495.

Fraher, D., Ellis, M. K., Morrison, S., McGee, S. L., Ward, A. C., Walder, K., & Gibert, Y. (2015). Lipid abundance in zebrafish embryos is regulated by complementary actions of the endocannabinoid system and retinoic acid pathway. *Endocrinology*, *156*(10), 3596–3609. https://doi.org/10.1210/en.2015-1315.

Frampton, G., Coufal, M., Li, H., Ramirez, J., & DeMorrow, S. (2010). Opposing actions of endocannabinoids on cholangiocarcinoma growth is via the differential activation of notch signaling. *Experimental Cell Research*, *316*(9), 1465–1478. https://doi.org/10.1016/j.yexcr.2010.03.017.

Gant, J. (2019). Scientists are baffled by spatter of babies born without hands or arms in France, as investigation fails to discover a cause. *Daily Mail*. Sunday 14th July. Retrieved from: https://www.dailymail.co.uk/news/article-7242059/Scientists-baffled-babies-born-without-hands-arms-France-probe-fails-discover-cause.html.

Geber, W. F., & Schramm, L. C. (1969a). Effect of marihuana extract on fetal hamsters and rabbits. *Toxicology and Applied Pharmacology*, *14*(2), 276–282. Retrieved from: http://www.ncbi.nlm.nih.gov/pubmed/5772851.

Geber, W. F., & Schramm, L. C. (1969b). Teratogenicity of marihuana extract as influenced by plant origin and seasonal variation. *Archives Internationales de Pharmacodynamie et de Thérapie*, *177*(1), 224–230. Retrieved from: http://www.ncbi.nlm.nih.gov/pubmed/5353879.

Graham, J. D. P. (1976). Cannabis and health. In J. D. P. Graham (Ed.), *Vol. 1. Cannabis and health* (1st ed., pp. 271–320). London, New York, San Francisco: Academic Press.

Gurney, J., Shaw, C., Stanley, J., Signal, V., & Sarfati, D. (2015). Cannabis exposure and risk of testicular cancer: A systematic review and meta-analysis. *BMC Cancer*, *15*, 897. https://doi.org/10.1186/s12885-015-1905-6.

Hebert-Chatelain, E., Reguero, L., Puente, N., Lutz, B., Chaouloff, F., Rossignol, R., … Marsicano, G. (2014). Cannabinoid control of brain bioenergetics: Exploring the subcellular localization of the CB1 receptor. *Molecular Metabolism*, *3*(4), 495–504. https://doi.org/10.1016/j.molmet.2014.03.007.

Hill, A. B. (1965). The environment and disease: Association or causation? *Proceedings of the Royal Society of Medicine*, *58*, 295–300. Retrieved from: http://www.ncbi.nlm.nih.gov/entrez/query.fcgi?cmd=Retrieve&db=PubMed&dopt=Citation&list_uids=14283879.

Hosptial, G. O. S. (2022). *VACTERL association*. Retrieved from: https://www.gosh.nhs.uk/conditions-and-treatments/conditions-we-treat/vacterl-association-0/.

Huang, H. F. S., Nahas, G. G., & Hembree, W. C. (1999). Effects of marijuana inhalation on spermatogenesis of the rat. In G. G. Nahas, K. M. Sutin, D. J. Harvey, & S. Agurell (Eds.), *1. Marijuana in medicine* (pp. 359–366). Totowa, New York: Human Press.

Kim, D., Lim, S., Park, M., Choi, J., Kim, J., Han, H., … Park, S. (2014). Ubiquitination-dependent CARM1 degradation facilitates notch1-mediated podocyte apoptosis in diabetic nephropathy. *Cellular Signalling*, *26*(9), 1774–1782. https://doi.org/10.1016/j.cellsig.2014.04.008.

Koller, V. J., Ferk, F., Al-Serori, H., Misik, M., Nersesyan, A., Auwarter, V., … Knasmuller, S. (2015). Genotoxic properties of representatives of alkylindazoles and aminoalkyl-indoles which are consumed as synthetic cannabinoids. *Food and Chemical Toxicology*, *80*, 130–136. https://doi.org/10.1016/j.fct.2015.03.004.

Kučukalić, S., Ferić Bojić, E., Babić, R., Avdibegović, E., Babić, D., Agani, F., … Džubur Kulenović, A. (2019). Genetic susceptibility to posttraumatic stress disorder: Analyses of the oxytocin receptor, retinoic acid receptor-related orphan receptor a and cannabinoid receptor 1 genes. *Psychiatria Danubina*, *31*(2), 219–226. https://doi.org/10.24869/psyd.2019.219.

Lacson, J. C., Carroll, J. D., Tuazon, E., Castelao, E. J., Bernstein, L., & Cortessis, V. K. (2012). Population-based case-control study of recreational drug use and testis cancer risk confirms an association between marijuana use and nonseminoma risk. *Cancer*, *118*(21), 5374–5383. https://doi.org/10.1002/cncr.27554.

Lee, Y. S., & Jeong, W. I. (2012). Retinoic acids and hepatic stellate cells in liver disease. *Journal of Gastroenterology and Hepatology*, *27*(Suppl 2), 75–79. https://doi.org/10.1111/j.1440-1746.2011.07007.x.

Leuchtenberger, C., & Leuchtenberger, R. (1971). Morphological and cytochemical effects of marijuana cigarette smoke on epithelioid cells of lung explants from mice. *Nature*, *234*(5326), 227–229. https://doi.org/10.1038/234227a0.

Leuchtenberger, C., Leuchtenberger, R., & Schneider, A. (1973). Effects of marijuana and tobacco smoke on human lung physiology. *Nature*, *241*(5385), 137–139. https://doi.org/10.1038/241137a0.

Mahoney, J. M., & Harris, R. A. (1972). Effect of 9-tetrahydrocannabinol on mitochondrial processes. *Biochemical Pharmacology, 21*(9), 1217–1226. Retrieved from https://www.ncbi.nlm.nih.gov/pubmed/5038669.

Manthey, J., Freeman, T. P., Kilian, C., Lopez-Pelayo, H., & Rehm, J. (2021). Public health monitoring of cannabis use in Europe: Prevalence of use, cannabis potency, and treatment rates. *The Lancet Regional Health—Europe, 10*, 100227. https://doi.org/10.1016//j.lanepe.2021.100227.

McKenzie, M. G., Cobbs, L. V., Dummer, P. D., Petros, T. J., Halford, M. M., Stacker, S. A., … Au, E. (2019). Non-canonical wnt signaling through ryk regulates the generation of somatostatin- and parvalbumin-expressing cortical interneurons. *Neuron, 103*(5), 853–864 e854. https://doi.org/10.1016/j.neuron.2019.06.003.

Mon, M. J., Haas, A. E., Stein, J. L., & Stein, G. S. (1981a). Influence of psychoactive and nonpsychoactive cannabinoids on cell proliferation and macromolecular biosynthesis in human cells. *Biochemical Pharmacology, 30*(1), 31–43. https://doi.org/10.1016/0006-2952(81)90281-1.

Mon, M. J., Haas, A. E., Stein, J. L., & Stein, G. S. (1981b). Influence of psychoactive and nonpsychoactive cannabinoids on chromatin structure and function in human cells. *Biochemical Pharmacology, 30*(1), 45–58. https://doi.org/10.1016/0006-2952(81)90282-3.

Morishima, A. (1984). Effects of cannabis and natural cannabinoids on chromosomes and ova. *NIDA Research Monograph, 44*, 25–45. Retrieved from: http://www.ncbi.nlm.nih.gov/entrez/query.fcgi?cmd=Retrieve&db=PubMed&dopt=Citation&list_uids=6090908.

Murphy, S. K., Itchon-Ramos, N., Visco, Z., Huang, Z., Grenier, C., Schrott, R., … Kollins, S. H. (2018). Cannabinoid exposure and altered DNA methylation in rat and human sperm. *Epigenetics, 13*(12), 1208–1221. https://doi.org/10.1080/15592294.2018.1554521.

Nahas, G. G. (1990a). *Cannabis physiopathology epidemiology detection. Vol. 1*. CRC Press Revivals.

Nahas, G. G. (1990b). *Keep off the grass. Vol. 1*. Middlebury: Paul S. Eriksson.

Nallathambi, R., Mazuz, M., Namdar, D., Shik, M., Namintzer, D., Vinayaka, A. C., … Koltai, H. (2018). Identification of synergistic interaction between cannabis-derived compounds for cytotoxic activity in colorectal cancer cell lines and colon polyps that induces apoptosis-related cell death and distinct gene expression. *Cannabis and Cannabinoid Research, 3*(1), 120–135. https://doi.org/10.1089/can.2018.0010.

Nalli, Y., Dar, M. S., Bano, N., Rasool, J. U., Sarkar, A. R., Banday, J., … Ali, A. (2019). Analyzing the role of cannabinoids as modulators of Wnt/beta-catenin signaling pathway for their use in the management of neuropathic pain. *Bioorganic & Medicinal Chemistry Letters, 29*(9), 1043–1046. https://doi.org/10.1016/j.bmcl.2019.03.013.

NBDPN. (2018). *National birth defects prevention network*. Houston, TX: National Birth Defects Prevention Network. Retrieved from https://www.nbdpn.org/ar.php.

Niu, F., Zhao, S., Xu, C. Y., Sha, H., Bi, G. B., Chen, L., … Nie, T. H. (2015). Potentiation of the antitumor activity of adriamycin against osteosarcoma by cannabinoid WIN-55,212-2. *Oncology Letters, 10*(4), 2415–2421. https://doi.org/10.3892/ol.2015.3525.

Pearl, J., & Mackaenzie, D. (2019). *The book of why. The new science of cause and effect. Vol. 1*. New York: Basic Books.

Petko, J., Tranchina, T., Patel, G., Levenson, R., & Justice-Bitner, S. (2018). Identifying novel members of the Wntless interactome through genetic and candidate gene approaches. *Brain Research Bulletin, 138*, 96–105. https://doi.org/10.1016/j.brainresbull.2017.07.004.

Reece, A. S., & Hulse, G. K. (2019a). Cannabis consumption patterns explain the east-west gradient in canadian neural tube defect incidence: An ecological study. *Globai Pediatric Health, 6*. https://doi.org/10.1177/2333794x19894798. 2333794x19894798.

Reece, A. S., & Hulse, G. K. (2019b). Cannabis teratology explains current patterns of coloradan congenital defects: The contribution of increased cannabinoid exposure to rising teratological trends. *Clinical Pediatrics (Phila), 58*(10), 1085–1123. https://doi.org/10.1177/0009922819861281.

Reece, A. S., & Hulse, G. K. (2020). Cannabis in pregnancy—Rejoinder, exposition and cautionary tales. *Psychiatric Times. October 10, 2020* https://www.bing.com/search?q=Cannabis+in+Pregnancy+%E2%80%93+Rejoinder%82C+Exposition+and+Cautionary+Tales&cvid=22538e20124c04711b92017489c92063214a&aqs=edge..92017469i92017457.92017439j92017480j92017481&pglt=92017443&FORM=ANSPA92017481&PC=U92017531. Retrieved from: https://www.bing.com/search?q=Cannabis+in+Pregnancy+%E2%80%93+Rejoinder%2C+Exposition+and+Cautionary+Tales&cvid=22538e0124c04711b92017489c63214a&aqs=edge..69i57.439j0j1&pglt=43&FORM=ANSPA1&PC=U531.

Reece, A. S., & Hulse, G. K. (2020a). Broad spectrum epidemiological contribution of cannabis and other substances to the teratological profile of northern New South Wales: Geospatial and causal inference analysis. *BMC Pharmacology and Toxicology, 21*(1), 75. https://doi.org/10.1186/s40360-020-00450-1.

Reece, A. S., & Hulse, G. K. (2020b). Canadian cannabis consumption and patterns of congenital anomalies: An ecological geospatial analysis. *Journal of Addiction Medicine, 14*(5), e195–e210. https://doi.org/10.1097/adm.0000000000000638.

Reece, A. S., & Hulse, G. K. (2020c). Contemporary epidemiology of rising atrial septal defect trends across USA 1991-2016: A combined ecological geospatiotemporal and causal inferential study. *BMC Pediatrics, 20*(1), 539. https://doi.org/10.1186/s12887-020-02431-z.

Reece, A. S., & Hulse, G. K. (2021a). Epidemiological overview of multidimensional chromosomal and genome toxicity of cannabis exposure in congenital anomalies and cancer development. *Scienitifc Reports, 11*(1), 13892. https://doi.org/10.10389/s41598-021-93411-5.

Reece, A. S., & Hulse, G. K. (2021b). Cannabinoid exposure as a major driver of pediatric acute lymphoid Leukaemia rates across the USA: Combined geospatial, multiple imputation and causal inference study. *BMC Cancer, 21*(1), 984. https://doi.org/10.1186/s12885-021-08598-7.

Reece, A. S., & Hulse, G. K. (2021c). Causal inference multiple imputation investigation of the impact of cannabinoids and other substances on ethnic differentials in US testicular cancer incidence. *BMC Pharmacology and Toxicology, 22*(1), 40. https://doi.org/10.1186/s40360-021-00505-x.

Reece, A. S., & Hulse, G. K. (2021d). A geospatiotemporal and causal inference epidemiological exploration of substance and cannabinoid exposure as drivers of rising US pediatric cancer rates. *BMC Cancer, 21*(1), 197. https://doi.org/10.1186/s12885-021-07924-3.

Reece, A. S., & Hulse, G. K. (2021e). Quadruple convergence—Rising cannabis prevalence, intensity, concentration and use disorder treatment. *The Lancet Regional Health—Europe*, 100245. https://doi.org/10.1016/j.lanepe.2021.100245.

Reece, A. S., & Hulse, G. K. (2022a). Cannabinoid- and substance-relationships of european congenital anomaly patterns: A space-time panel regression and causal inferential study. *Environmental Epigenetics, 8*(1). https://doi.org/10.1093/eep/dvab015. dvab015.

Reece, A. S., & Hulse, G. K. (2022b). Geotemporospatial and causal inference epidemiological analysis of US survey and overview of cannabis, cannabidiol and cannabinoid genotoxicity in relation to congenital anomalies 2001–2015. *BMC Pediatrics*, *22*(1), 47. https://doi.org/10.1186/s12887-021-02996-3.

Richard, D., & Picard, F. (2011). Brown fat biology and thermogenesis. *Frontiers in Bioscience (Landmark Edition)*, *16*, 1233–1260. https://doi.org/10.2741/3786.

Robinson, M. (2019). *Babies born with deformed hands spark investigation in Germany*. Retrieved from: https://edition.cnn.com/2019/09/16/health/hand-deformities-babies-gelsenkirchen-germany-intl-scli-grm/index.html.

Robison, L. L., Buckley, J. D., Daigle, A. E., Wells, R., Benjamin, D., Arthur, D. C., & Hammond, G. D. (1989). Maternal drug use and risk of childhood nonlymphoblastic leukemia among offspring. An epidemiologic investigation implicating marijuana (a report from the Childrens Cancer Study Group). *Cancer*, *63*(10), 1904–1911. Retrieved from: http://www.ncbi.nlm.nih.gov/pubmed/2649219.

Rossato, M., Pagano, C., & Vettor, R. (2008). The cannabinoid system and male reproductive functions. *Journal of Neuroendocrinology*, *20*(Suppl 1), 90–93. Retrieved from: http://www.ncbi.nlm.nih.gov/entrez/query.fcgi?cmd=Retrieve&db=PubMed&dopt=Citation&list_uids=18426506.

Russo, C., Ferk, F., Mišík, M., Ropek, N., Nersesyan, A., Mejri, D., … Knasmüller, S. (2019). Low doses of widely consumed cannabinoids (cannabidiol and cannabidivarin) cause DNA damage and chromosomal aberrations in human-derived cells. *Archives of Toxicology*, *93*(1), 179–188. https://doi.org/10.1007/s00204-018-2322-9.

Schrott, R., Murphy, S. K., Modliszewski, J. L., King, D. E., Hill, B., Itchon-Ramos, N., … Mitchell, J. T. (2021). Refraining from use diminishes cannabis-associated epigenetic changes in human sperm. *Environmental Epigenetics*, *7*(1), dvab009. https://doi.org/10.1093/eep/dvab009.

Shen, H., Shih, J., Hollern, D. P., Wang, L., Bowlby, R., Tickoo, S. K., … Hoadley, K. A. (2018). Integrated molecular characterization of testicular germ cell tumors. *Cell Reports*, *23*(11), 3392–3406. https://doi.org/10.1016/j.celrep.2018.05.039.

Singh, N., Hroudova, J., & Fisar, Z. (2015). Cannabinoid-induced changes in the activity of electron transport chain complexes of brain mitochondria. *Journal of Molecular Neuroscience*, *56*(4), 926–931. https://doi.org/10.1007/s12031-015-0545-2.

Stenchever, M. A., Kunysz, T. J., & Allen, M. A. (1974). Chromosome breakage in users of marihuana. *American Journal of Obstetrics and Gynecology*, *118*(1), 106–113. https://doi.org/10.1016/s0002-9378(16)33653-5.

Substance Abuse and Mental Health Services Administration. (2019). *National survey of drug use and health (NSDUH 2018)*. National Survey of Drug Use and Health, Substance Abuse and Mental Health Services Administration. Retrieved from: https://www.datafiles.samhsa.gov/study/national-survey-drug-use-and-health-nsduh-2018-nid18757.

Szutorisz, H., DiNieri, J. A., Sweet, E., Egervari, G., Michaelides, M., Carter, J. M., … Hurd, Y. L. (2014). Parental THC exposure leads to compulsive heroin-seeking and altered striatal synaptic plasticity in the subsequent generation. *Neuropsychopharmacology*, *39*(6), 1315–1323. https://doi.org/10.1038/npp.2013.352.

Szutorisz, H., & Hurd, Y. L. (2016). Epigenetic effects of cannabis exposure. *Biological Psychiatry*, *79*(7), 586–594. https://doi.org/10.1016/j.biopsych.2015.09.014.

Tahir, S. K., Trogadis, J. E., Stevens, J. K., & Zimmerman, A. M. (1992). Cytoskeletal organization following cannabinoid treatment in undifferentiated and differentiated PC12 cells. *Biochemistry and Cell Biology*, *70*(10-11), 1159–1173. Retrieved from http://www.ncbi.nlm.nih.gov/pubmed/1297339.

Tahir, S. K., & Zimmerman, A. M. (1991). Influence of marihuana on cellular structures and biochemical activities. *Pharmacology, Biochemistry, and Behavior*, *40*(3), 617–623. https://doi.org/10.1016/0091-3057(91)90372-9.

Tanveer, R., Gowran, A., Noonan, J., Keating, S. E., Bowie, A. G., & Campbell, V. A. (2012). The endocannabinoid, anandamide, augments notch-1 signaling in cultured cortical neurons exposed to amyloid-beta and in the cortex of aged rats. *The Journal of Biological Chemistry*, *287*(41), 34709–34721. https://doi.org/10.1074/jbc.M112.350678.

Trabert, B., Sigurdson, A. J., Sweeney, A. M., Strom, S. S., & McGlynn, K. A. (2011). Marijuana use and testicular germ cell tumors. *Cancer*, *117*(4), 848–853. https://doi.org/10.1002/cncr.25499.

Trivers, K. F., Mertens, A. C., Ross, J. A., Steinbuch, M., Olshan, A. F., Robison, L. L., & Children's Cancer, G. (2006). Parental marijuana use and risk of childhood acute myeloid leukaemia: a report from the children's cancer group (United States and Canada). *Paediatric and Perinatal Epidemiology*, *20*(2), 110–118. https://doi.org/10.1111/j.1365-3016.2006.00700.x.

United National Office of Drugs and Crime. (2019). *World drug report 2019*. Geneva, Switzerland: United National World Health Organization. Retrieved from: https://wdr.unodc.org/wdr2019/index.html.

United National Office of Drugs and Crime. (2021). *World Drug Report 2021*. Geneva, Switzerland: United National World Health Organization. Retrieved from: https://www.unodc.org/unodc/en/data-and-analysis/wdr2021.html.

Vallee, A., Lecarpentier, Y., Guillevin, R., & Vallee, J. N. (2017). Effects of cannabidiol interactions with Wnt/beta-catenin pathway and PPARgamma on oxidative stress and neuroinflammation in Alzheimer's disease. *Acta Biochimica et Biophysica Sinica Shanghai*, *49*(10), 853–866. https://doi.org/10.1093/abbs/gmx073.

VanderWeele, T. J., Ding, P., & Mathur, M. (2019). Technical considerations in the use of the E-value. *Journal of Causal Inference*, *7*(2), 1–11. https://doi.org/10.1515/jci-2018-0007.

Volkow, N. D., Compton, W. M., & Wargo, E. M. (2017). The risks of marijuana use during pregnancy. *JAMA*, *317*(2), 129–130. https://doi.org/10.1001/jama.2016.18612.

Watson, C. T., Szutorisz, H., Garg, P., Martin, Q., Landry, J. A., Sharp, A. J., & Hurd, Y. L. (2015). Genome-wide DNA methylation profiling reveals epigenetic changes in the rat nucleus accumbens associated with cross-generational effects of adolescent THC exposure. *Neuropsychopharmacology*, *40*(13), 2993–3005. https://doi.org/10.1038/npp.2015.155.

Wen, W. Q., Shu, X. O., Steinbuch, M., Severson, R. K., Reaman, G. H., Buckley, J. D., & Robison, L. L. (2000). Paternal military service and risk for childhood leukemia in offspring. *American Journal of Epidemiology, 151*(3), 231–240. Retrieved from: http://www.ncbi.nlm.nih.gov/pubmed/10670547.

Williams, E. J., Walsh, F. S., & Doherty, P. (2003). The FGF receptor uses the endocannabinoid signaling system to couple to an axonal growth response. *The Journal of Cell Biology, 160*(4), 481–486. https://doi.org/10.1083/jcb.200210164.

Willsher, K. (2018). Baby arm defects prompt nationwide investigation in France. *Guardian*. Retrieved from: https://www.theguardian.com/world/2018/oct/31/baby-arm-defects-prompt-nationwide-investigation-france.

Xapelli, S., Agasse, F., Sarda-Arroyo, L., Bernardino, L., Santos, T., Ribeiro, F. F., … Malva, J. O. (2013). Activation of type 1 cannabinoid receptor (CB1R) promotes neurogenesis in murine subventricular zone cell cultures. *PLoS One, 8*(5), e63529. https://doi.org/10.1371/journal.pone.0063529.

Xian, X., Tang, L., Wu, C., & Huang, L. (2018). miR-23b-3p and miR-130a-5p affect cell growth, migration and invasion by targeting CB1R via the Wnt/beta-catenin signaling pathway in gastric carcinoma. *Oncotargets and Therapy, 11*, 7503–7512. https://doi.org/10.2147/ott.S181706.

Xu, T. R., Yang, Y., Ward, R., Gao, L., & Liu, Y. (2013). Orexin receptors: Multi-functional therapeutic targets for sleeping disorders, eating disorders, drug addiction, cancers and other physiological disorders. *Cellular Signalling, 25*(12), 2413–2423. https://doi.org/10.1016/j.cellsig.2013.07.025.

Zimmerman, A. M., & Raj, A. Y. (1980). Influence of cannabinoids on somatic cells in vivo. *Pharmacology, 21*(4), 277–287. Retrieved from: http://www.ncbi.nlm.nih.gov/pubmed/6252564.

Chapter 6

Cannabis genotoxicity and cancer incidence: Highly concordant synthesis of European and USA datasets

Albert Stuart Reece[a,b] and Gary Kenneth Hulse[a,b]

[a]*Department of Psychiatry, University of Western Australia, Perth, WA, Australia,* [b]*School of Health and Medical Sciences, Edith Cowan University, Joondalup, WA, Australia*

Abbreviations

ALL	acute lymphoid leukemia
AML	acute myeloid leukemia
ATP	adenosine triphosphate
CLL	chronic lymphoid leukemia
CML	chronic myeloid leukemia
DNA	deoxyribonucleic acid
DSB	double-stranded breaks in DNA
EMCDDA	European Monitoring Centre for Drugs and Drug Addiction
***E*-value**	expected values
NH_Lymphoma	non-Hodgkin's Lymphoma
PTM	posttranslational modifications
***P*-value**	probability value
RNA	ribonucleic acid
V(D)J	variable, joining, diversity regions of genes

Introduction

The subject of cannabinoid-related carcinogenesis has long been controversial with both positive (Reece, 2009) and negative (Hashibe et al., 2006) studies being extant in the published literature. However, the area was recently given new impetus from sophisticated epidemiological studies, which demonstrated that cannabis is driving the 100% rise in testicular cancer, 1975–2018, in the USA (Reece & Hulse, 2021d) and is also involved in several common or rapidly growing cancer incidences including breast, pancreatic, thyroid, and liver cancer and many others in adults (Reece and Hulse, 2021a, 2022e, 2022h, 2022i; Reece and Hulse, 2021a). Of even greater concern is the recent demonstration that cannabis is driving acute lymphoid leukemia, the commonest cancer of childhood in the USA (Reece & Hulse, 2021a), and is also the driving force for the 50% rise in pediatric cancer, 1975–2018 (Reece & Hulse, 2021c). Cannabis has also been linked with acute myeloid leukemia in children (Reece & Hulse, 2021e; Robison et al., 1989; Trivers et al., 2006; Wen et al., 2000). This epidemiology of rising pediatric cancer is of concern as it is known to represent an evidence of heritable carcinogenic teratogenesis from one generation to another (Grobner et al., 2018; Ma et al., 2018).

Previous studies have linked cannabis exposure with adult tumors of the lung, head, and neck, larynx, prostate, testis, brain, and urothelium (Aldington et al., 2008; Berthiller et al., 2008; Chacko et al., 2006; Daling et al., 2009; Efird et al., 2004; Hashibe et al., 2002; Moiche Bokobo et al., 2001; Nieder et al., 2006; Sidney et al., 1997; Voirin et al., 2006; Zhang et al., 1999) and in several dose-response relationships have been demonstrated (Aldington et al., 2008; Berthiller et al., 2008; Daling et al., 2009; Zhang et al., 1999). Moreover several childhood cancers have also been described after parental exposure including neuroblastoma, rhabdomyosarcoma, and non-lymphoblastic leukemia (Bluhm et al., 2006; Hashibe et al., 2005; Robison et al., 1989) These tumors represent a presumptive evidence of heritable carcinogenesis (Grobner et al., 2018; Ma et al., 2018).

Cannabis Use, Neurobiology, Psychology, and Treatment. https://doi.org/10.1016/B978-0-323-89862-1.00028-3

Triple confluence of recent cannabis trends

Recent epidemiological studies have elegantly presented the cannabis epidemiology of Europe (Manthey et al., 2021) based on data from the European Monitoring Centre for Drugs and Drug Addiction (European Monitoring Centre for Drugs and Drug Addiction (EMCDDA), 2021). This has allowed studies to be conducted in Europe similar to those in the USA on the epidemiology of cannabis- and substance-related cancer incidence. Importantly, the collection of the European and USA datasets on these subjects comprises the bulk of the publicly available comprehensive data globally. For these reasons, the present study was undertaken to assess and compare the findings of these two globally important datasets to provide a unified evidence-based voice to properly inform drug policymakers and direct policy development.

Two other exciting developments make this investigation timely and important. It has been formally demonstrated that recent trends associated with cannabis liberalization in Europe and in the USA are each associated with first, rising prevalence rates of cannabis use; second, cannabis use intensity as daily use, and third, increased THC concentration of smoked cannabis herb and resins (European Monitoring Centre for Drugs and Drug Addiction (EMCDDA), 2021; Reece & Hulse, 2021d). This means that it is not one trend that is occurring but at least three, or indeed, four convergent and confluent trends if one includes treatment seeking for cannabis abuse and dependence (European Monitoring Centre for Drugs and Drug Addiction (EMCDDA), 2021; Reece & Hulse, 2021d). The net effect of this confluent convergence is a dramatically rising total exposure to cannabinoids acutely and because of their long half life in body fat stores and slow rate of excretion, in terms of cumulative cannabinoid dose sustained. This triple trend is exacerbated by cannabis apparently entering the food chain in some jurisdictions, for example, in the form of animal fodder, where such an accumulation is likely occurring clandestinely without even the consumer necessarily being aware of that exposure occurring.

Exponential cannabinoid genotoxicity dose-response curves

This feature achieves particular significance in view of the exponential dose-response relationship of cannabinoids to numerous measures of genotoxicity in cells, chromosomes, molecules, and organisms (Huang et al., 1999; Nahas et al., 1977; Reece & Hulse, 2016, 2021c; Russo et al., 2019). The confrontation of this triply rising convergence of cannabis prevalence, intensity, and potency with this exponential dose-response relationship implies that a steep "switch-like" effect will occur in areas where cannabis liberalization becomes entrenched with the relatively abrupt appearance of multiple severe genotoxic pathologies that could be expected to manifest. Indeed in parts of the USA, this may already be occurring in relation to gastroschisis, autism (Reece & Hulse, 2019a, 2019b, 2019c), and atrial septal defect (Reece & Hulse, 2020a, 2020b). It is also likely happening in parts of France and Germany in relation to limb reduction defects (Agence France-Presse in Paris, 2018; Robinson, 2019; Willsher, 2018).

Modern epigenomic insights

The other major development that allows greater insight into the relationship between cannabis and the development of malignant tumorigensis and inheritable carcinogenesis is studies in rats and humans on cannabinoid oncogenesis and epigenomic pathways. Two recent studies confirm that epigenomic pathways are widely perturbed by cannabis use, and pathways that are prominently implicated are the pathways in cancer and the hippo pathway, which itself is involved in cancer (Murphy et al., 2018; Schrott et al., 2021). Indeed, one recent paper from this group provided 359 pages of epigenomic and pathway data from their studies, which included no less than 487 hits to cancer-related pathways and mechanisms and 972 for carcinoma (Schrott et al., 2021).

It should also be mentioned that cannabis-related carcinogenicity is itself a part of the overall syndrome of cannabis-related genotoxicity. Together cannabis-related genotoxicity can be expected to be reflected clinically as increased rates of birth defects, cancer induction, and accelerated patterns of aging. For these reasons, we believe that it is important to read the cannabinoid oncogenesis literature in parallel with the cannabinoid teratogenesis and cannabinoid-related aging literature to properly form an understanding across these three large and interrelated knowledge domains.

All the above lines of evidence imply that the present investigation seeking to compare and assess the validity of findings in the USA and European datasets on cannabis- and substance-related carcinogenicity is timely not only in the general epidemiological sense but also in the sense of being a tipping point for cancer-related studies of population health at a time when epigenomic and mechanistic background knowledge of the biological pathways involved is reaching a very exciting cusp. For these reasons, we see the present epidemiological comparison as timely and important and constituting a significant contribution on the global stage to our understanding of cannabinoid genotoxicity more generally.

Methodological considerations

Data. Data informing this analysis appear as separate papers and manuscripts (Reece and Hulse, 2022e, 2022h, 2022i, 2023a). For comparative purposes, data have been re-presented differently to the form in which it was first published. It is noted that the sample size of the European dataset is 47,920, and the sample size of the USA dataset is 19,877. It should be observed that some cancer registries refer to the area of the head and neck generally as the oropharynx (International Classification of Disease (ICD) 10 codes C00-C14) and some more specifically the oropharynx proper as (ICD10 C10). Where this has been specified by the European cancer registries, this is taken into account in the analysis. This is the meaning of the term "Oropharynx" (narrowly defined, ICD10 C10) and Oropharynx_Broadly (-ly defined, ICD10 C00-C14) in the following tables.

Statistics. Published data are supplemented by some epidemiological calculations with two-by-two tables in R-Studio 4.1.1 using the epiR and dplyr packages from tidyverse (Stevenson et al., 2020; Wickham et al., 2019) to calculate relative risk, attributable fraction in the exposed and attributable fraction in the population for more covariates than what is presently available in the European dataset. The analysis was performed in January 2022.

The use of the *E*-value is an important technique in formal quantitative causal inference. The *E*-value or expected value is the degree of association that some extraneous unmeasured confounder must have with both an exposure of concern and an outcome of interest in order to explain away some apparently causal effect (VanderWeele & Ding, 2017). For example, the association of lung cancer with tobacco smoking is nine on this scale, which is considered to be a high value (Pearl & Mackaenzie, 2019). 1.25 is the usually quoted value required of causal effects (VanderWeele et al., 2019). The *E*-value has a confidence interval and the one with which we are mainly concerned is the lower 97.5% bound of this confidence interval, which establishes with 95% certainty that the true *E*-value is above this limit. It is a critically important metric to formal quantitative techniques of causal inference generally.

Data availability. Data that are already publicly available (see key references) and accompanying new R-code are available with the present report in the Mendeley data repositories at URL doi: https:/data.mendeley.com/datasets/10.17632/wszg9j6tjf, https:/data.mendeley.com/datasets/v4xfnxwv3z, https:/data.mendeley.com/datasets/dxc3br6863.

Comparative analyses

Continuous: Continuous comparison

Table 1 lists the 20 cancers that were identified as being cannabis-associated at continuous bivariate analysis in the USA along with their key statistics and notes whether they were identified in the European continuous dataset. It is seen that 18/20 (90%) are matched between the two lists. One tumor type, acute myeloid leukemia, cannot be matched from the present form of the European data at this time.

Table 2 performs a similar role for the European dataset. In this table, 20/27 (74.1%) cancer types associated with cannabis exposure in the European dataset could be matched from the US dataset. However, there is no comparable metric in the USA datasets for four of the cancer studies in Europe. When these missing data are removed, it is noted that 20/23 (87.0%) of the cannabis-associated cancers in Europe for which data are available can be matched.

Categorical: Categorical analysis

Table 3 compares the 23 cancers that were significantly elevated on categorical analysis in the USA with those found to be elevated on categorical analysis in Europe. It is noted that 17/21 (80.9%) tumor types for which data are available are confirmed in the European dataset.

Table 4 plays a similar role for the reverse scenario where the cannabis-associated tumors identified to be elevated in the European datasets are compared with those in the USA. Here 18/22 (81.8%) of those for which comparable data exist were confirmed.

Both analyses combined

Table 5 shows a list of 25 cancers associated with cannabinoids in the USA albeit it should be noted that the correlation of stomach cancer is with cannabidiol, potency and prevalence of which have been falling throughout the period of this study. Table 5 also lists 31 cancers in Europe identified as being linked with various metrics of cannabis exposure. This table shows that 23 of the 23 (100%) cancers identified in the USA dataset were identified in the European dataset where

TABLE 1 Cancers identified at continuous bivariate analysis in the USA confirmed in Europe.

No.	Cancer	*P*-Value	*E*-Value estimate	*E*-Value—Lower	Yes	No	N/A
1	All_Cancer	6.87E−20	1.02E+14	1.43E+11	Yes		
2	AML	1.34E−05	6.47E+08	1.03E+05			N/A
3	Bladder	2.15E−31	1.32E+18	1.84E+15	Yes		
4	Brain	2.94E−10	4.21E+09	5.88E+06	Yes		
5	Breast	3.64E−11	2.22E+13	3.53E+09	Yes		
6	CML	2.36E−04	552.10	27.73	Yes		
7	Colorectal	1.10E−19	8.46E+13	1.18E+11	Yes		
8	Esophagus	9.84E−08	1.84E+08	2.35E+05	Yes		
9	Hodgkin's	1.63E−13	1.85E+11	2.59E+08	Yes		
10	Kidney	0.0067	83.70	5.16	Yes		
11	Liver	2.42E−20	5.99E+18	9.53E+14	Yes		
12	Lung	6.07E−09	7.70E+08	1.08E+06	Yes		
13	NH_Lymphoma	7.05E−09	7.06E+08	9.87E+05	Yes		
14	Oropharynx	3.21E−06	1.27E+03	85.68	Yes		
15	Ovary	5.09E−31	9.91E+17	1.38E+15		No	
16	Pancreas	2.94E−06	2.80E+09	4.46E+05	Yes		
17	Prostate	2.28E−41	1.67E+21	2.34E+18	Yes		
18	Stomach	0.0058	2.19E+04	30.11	Yes		
19	Testis	1.47E−05	56.28	12.14	Yes		
20	Thyroid	2.70E−25	1.72E+21	2.74E+17	Yes		
	Count				18	1	1
	Total				20	20	20
	Percentage				90.0%	5.0%	5.0%

TABLE 2 Cancers identified at continuous bivariate analysis in Europe confirmed in the USA.

No.	Cancer	*P*-Value	*E*-Value estimate	*E*-Value 95% lower bound	Yes	No	N/A
1	All Cancers	6.70E−102	3.15E+10	5.46E+09	Yes		
2	All Cancers nNMSC	4.69E−180	4.88E+13	8.28E+12	Yes		
3	Anus	1.61E−30	2.77E+06	2.68E+05			N/A
4	Bladder	8.18E−04	7.87E+00	3.04	Yes		
5	Brain	3.09E−04	1.99E+01	5.26	Yes		
6	Breast	1.36E−65	3.51E+06	7.37E+05	Yes		
7	Colorectum	4.61E−45	2.03E+05	4.33E+04	Yes		
8	Corpus Uteri	5.00E−03	1.64E+01	3.24			N/A
9	Gallbladder and Biliary	5.57E−03	1.63E+11	3.93E+03			N/A

TABLE 2 Cancers identified at continuous bivariate analysis in Europe confirmed in the USA—cont'd

No.	Cancer	*P*-Value	*E*-Value estimate	*E*-Value 95% lower bound	Yes	No	N/A
10	Hodgkin's	2.21E−08	2.27E+33	1.57E+22	Yes		
11	Kidney	1.64E−47	3.32E+05	6.95E+04	Yes		
12	Larynx	8.82E−30	5.37E+20	3.88E+17		No	
13	Leukemia - Lymphoid	5.50E−06	2.82E+09	5.18E+05	Yes		
14	Leukemia - Myeloid	1.75E−11	4.42E+14	8.32E+10			N/A
15	Liver	6.44E−10	6.09E+42	1.01E+30	Yes		
16	Lung	5.35E−32	3.40E+04	7.01E+03	Yes		
17	Melanoma	8.89E−57	5.59E+05	1.28E+05		No	
18	Myeloma	3.63E−07	4.78E+08	3.41E+05		No	
19	Non-Hodgkin's lymphoma	7.36E−38	6.82E+04	1.46E+04	Yes		
20	Oesophagus	1.71E−19	2.58E+03	5.58E+02	Yes		
21	Oropharynx	3.70E−06	6.94E+01	15.59	Yes		
22	Oropharynx_Broad	2.82E−08	5.05E+02	74.07	Yes		
23	Pancreas	1.04E−43	1.67E+05	3.55E+04	Yes		
24	Prostate	1.55E−86	4.59E+07	9.56E+06	Yes		
25	Stomach	4.60E−02	3.43E+03	1.67	Yes		
26	Testis	2.84E−41	5.74E+07	5.26E+06	Yes		
27	Thyroid	3.26E−38	8.56E+07	6.63E+06	Yes		
	Count				20	3	4
	Total				27	27	27
	Percentage				74.1%	11.1%	14.8%

TABLE 3 Cancers identified at categorical bivariate analysis in USA confirmed in Europe.

			E-Values				
No.	Cancer	*P*-Value	*E*-Value—Estimate	*E*-Value—Lower	Yes	No	N/A
1	ALL	7.65E−25	1.51	1.44			N/A
2	All_Cancer	2.2E−320	1.36	1.35	Yes		
3	AML	3.11E−110	1.57	1.53			N/A
4	Bladder	2.2E−320	1.65	1.63	Yes		
5	Brain	5.67E−33	1.34	1.30		No	
6	Breast	8.06E−146	1.26	1.24	Yes		
7	CLL	2.98E−12	1.25	1.20	Yes		
8	CML	1.52E−12	1.35	1.28	Yes		
9	Colorectal	2.2E−320	1.63	1.62	Yes		

Continued

TABLE 3 Cancers identified at categorical bivariate analysis in USA confirmed in Europe—cont'd

			E-Values				
No.	Cancer	*P*-Value	*E*-Value—Estimate	*E*-Value—Lower	Yes	No	N/A
10	Esophagus	2.31E−43	1.40	1.37		No	
11	Hodgkin's	1.22E−30	1.46	1.41	Yes		
12	Kaposi	4.75E−29	2.03	1.87	Yes		
13	Liver	2.2E−320	2.23	2.20	Yes		
14	Lung	6.87E−194	1.29	1.28	Yes		
15	Melanoma	2.2E−320	3.75	3.73		No	
16	Myeloma	1.73E−30	1.32	1.29	Yes		
17	NH_Lymphoma	3.15E−145	1.38	1.36		No	
18	Oropharynx	9.76E−06	1.15	1.11	Yes		
19	Ovary	2.49E−312	1.77	1.73	Yes		
20	Pancreas	4.57E−166	1.45	1.43	Yes		
21	Prostate	2.2E−320	2.14	2.13	Yes		
22	Stomach	2.30E−192	1.60	1.57	Yes		
23	Thyroid	2.2E−320	2.34	2.31	Yes		
	Count				17	4	2
	Total				23	23	23
	Percentage				73.9%	17.4%	8.7%

TABLE 4 Cancers identified at categorical bivariate analysis in Europe confirmed in the USA.

No.	Cancer	*P*-Value	*E*-Value estimate	*E*-Value lower bound	Yes	No	N/A
1	Liver	<2.2E−320	2.92	2.91	Yes		
2	Thyroid	<2.2E−320	2.77	2.76	Yes		
3	Stomach	<2.2E−320	2.76	2.75	Yes		
4	Oropharynx_Broad	<2.2E−320	2.62	2.60	Yes		
5	Larynx	<2.2E−320	2.56	2.54		No	
6	Breast	<2.2E−320	2.34	2.34	Yes		
7	All Cancers	<2.2E−320	2.16	2.16	Yes		
8	Hodgkin's	<2.2E−320	1.92	1.91	Yes		
9	Bladder	<2.2E−320	1.90	1.89	Yes		
10	Kidney	<2.2E−320	1.89	1.89		No	
11	Pancreas	<2.2E−320	1.89	1.88	Yes		
12	Prostate	<2.2E−320	1.86	1.86	Yes		
13	Lung	<2.2E−320	1.86	1.85	Yes		

TABLE 4 Cancers identified at categorical bivariate analysis in Europe confirmed in the USA—cont'd

No.	Cancer	*P*-Value	*E*-Value estimate	*E*-Value lower bound	Yes	No	N/A
14	Leukemia	<2.2E−320	1.80	1.79			N/A
15	Colorectum	<2.2E−320	1.73	1.73	Yes		
16	All Cancers nNMSC	<2.2E−320	1.72	1.72	Yes		
17	Myeloma	<2.2E−320	1.45	1.44	Yes		
18	Corpus Uteri	<2.2E−320	1.33	1.32			N/A
19	Cervix	<2.2E−320	1.30	1.29		No	
20	Gallbladder and Biliary	1.91E−252	1.46	1.44			N/A
21	Kaposi	1.86E−170	3.58	3.36	Yes		
22	Testis	1.76E−119	1.24	1.23		No	
23	Leukemia-Lymphoid	9.73E−101	1.38	1.35	Yes		
24	Ovary	2.24E−92	1.18	1.17	Yes		
25	Leukemia-Myeloid	6.17E−64	1.41	1.38	Yes		
26	Anus	1.47E−08	1.19	1.15			N/A
	Count				18	4	4
	Total				26	26	26
	Percentage				69.2%	15.4%	15.4%

TABLE 5 Combined continuous and categorical analyses in Europe and the USA.

No. USA	USA	No. Europe	Europe	Yes	No	N/A
1	Acute Lymphoid Leukemia					N/A
2	Acute Myeloid Leukemia					N/A
3	All_Cancer	1	All Cancers	Yes		
		2	All Cancers nNMSC	Yes		
		3	Anus			N/A
4	Bladder	4	Bladder	Yes		
5	Brain	5	Brain	Yes		
6	Breast	6	Breast	Yes		
		7	Cervix		No	
7	Colorectum	8	Colorectum	Yes		
		9	Corpus Uteri			N/A
		10	Gallbladder and Biliary			N/A
8	Hodgkin's	11	Hodgkin's	Yes		
9	Kaposi	12	Kaposi	Yes		

Continued

TABLE 5 Combined continuous and categorical analyses in Europe and the USA—cont'd

No. USA	USA	No. Europe	Europe	Yes	No	N/A
10	Kidney	13	Kidney	Yes		
		14	Larynx		No	
		15	Leukemia			N/A
11	Chronic Lymphoid Leukemia	16	Leukemia-Lymphoid	Yes		
12	Chronic Myeloid Leukemia	17	Leukemia-Myeloid	Yes		
13	Liver	18	Liver	Yes		
14	Lung	19	Lung	Yes		
15	Melanoma	20	Melanoma	Yes		
16	Multiple myeloma	21	Multiple myeloma	Yes		
17	Non-Hodgkin's lymphoma	22	Non-Hodgkin's lymphoma	Yes		
18	Oesophagus	23	Oesophagus	Yes		
19	Oropharynx	24	Oropharynx	Yes		
		25	Oropharynx_Broad	Yes		
20	Ovary	26	Ovary	Yes		
21	Pancreas	27	Pancreas	Yes		
22	Prostate	28	Prostate	Yes		
23	Stomach	29	Stomach	Yes		
24	Testis	30	Testis	Yes		
25	Thyroid	31	Thyroid	Yes		
				25	2	6
				27	27	33
				92.6%	7.4%	18.2%

comparable data existed. 25/27 (95.6%) cancers in the European dataset were confirmed in the American dataset where comparable data existed.

Consideration of findings

Main findings

The main finding of this report is the very close association of cannabinoid-related carcinogenesis in the USA and Europe, with almost identical lists being produced in both continents.

This robust evidence of the concordance of the cannabis-cancer relationships in Europe and the USA is strongly supportive of views previously expressed in the literature (Aldington et al., 2008; Berthiller et al., 2008; Bluhm et al., 2006; Chacko et al., 2006; Daling et al., 2009; Efird et al., 2004; Hashibe et al., 2002, 2005; Moiche Bokobo et al., 2001; Nieder et al., 2006; Reece & Hulse, 2021d; Robison et al., 1989; Sidney et al., 1997; Voirin et al., 2006; Zhang et al., 1999) that cannabis is related to numerous adult and pediatric cancers and strongly refutes the more conservative frequently expressed view that only the evidence for testicular cancer can be considered reliable (Volkow et al., 2014).

The cancers involved read like the high points of clinical oncology and include all cancers that were linked with cannabidiol exposure in the USA, namely breast cancer, which is the commonest cancer; liver cancer, which is the cancer whose incidence is growing fastest; pancreatic cancer, the tumor with the fastest growing death rate; testicular cancer, the tumor responsible for the most years of life lost of any adult cancer; and prostate cancer, the commonest cancer in males.

Mechanistically, one of the most confronting aspects of the present study is the broad range of different tumors that appear to be induced or exacerbated by cannabis exposure. Still there are obvious precedents for environmental toxicants and carcinogens being associated with a host of tumors including both tobacco and alcohol, where each accounts for more than 12 different tumor types (British Lung Foundation, 2005; Centers of Disease Control, 2022; Jameson et al., 2018; World Health Organization, 2004). It may be possible that cannabis is either acting on different tissue types in different ways or alternatively is acting on fundamental processes related to biological processes that have downstream cancerogenic implications. For these reasons, consideration of the cellular mechanisms underlying these findings is of no little importance.

Comments on specific cancers

Testicular cancer

It is very important to observe that the list also includes ovarian and testicular cancers, which, as they involve the germ cells, have potential for intergenerational transmission of malignant genotoxicity. Testicular cancer is the tumor that is most strongly linked with cannabis exposure with a relative rate in meta-analysis of 2.59-fold (Gurney et al., 2015). Both prostate and breast tissue are on this list of tumors. It is known that cancers release extracellular DNA circles, RNAs, and oncoproteins into the tissue around them. It is therefore possible for these tumors of the reproductive tract to have intergenerational impacts. Importantly, detailed inverse probability-weighted multivariable regression, space-time, and causal inferential studies of prostate and ovarian cancer were recently conducted in the USA, which confirm the positive findings reported in the present paper (Reece and Hulse, 2022e).

Testicular carcinogenesis has now been deconvoluted and shown to involve one or two whole genome-doubling events, genomic demethylation, and then the pruning of 30–50 chromosomal arms in seminoma and 50–70 chromosomal arms in non-seminomatous germ cell tumor (Shen et al., 2018). If one assumes that the median age of cannabis use by young men is about 20 years and the median age of testicular carcinoma incidence is about 33 years, then the involvement of cannabis as a bone fide testicular carcinogen necessarily involves the rapid truncation and acceleration of the normal process of testicular oncogenesis, which begins prior to birth, into just 13 years, a 2.5-fold acceleration. This means that the whole process of testicular carcinogenesis must be accelerated (2.6×2.5=) 6.5-fold pursuant to cannabinoid exposure.

Cannabis has long been known to cause chromosomal breakages (Leuchtenberger & Leuchtenberger, 1971), end-to-end fusions (Huang et al., 1999), and bridge formations (Morishima, 1984). What has not been formerly explained in this context is that the breakage-fusion-bridge (BFB) process can cycle endlessly causing increasing amounts of chromosomal and genomic damage. With each round of cell division, the not (telomere) capped ends of the broken chromosomes are susceptible to erratic and random fusion events, which become disorganized by the normal process of cell division. This BFB cycle was described long ago and is known to be a major engine, which generates genetic damage and oncogenic change (McClintock, 1941). If operating in the testicular carcinogenesis context, it would explain the very rapid accrual of malignant transformation by premalignant testicular clones. This hypothetical mechanism merits further laboratory research. It is easy to see how such a cycle could "breed" or incubate oncogenic transformation over successive cell division cycles and thereby endow the nascent malignant clone with a growth advantage.

Interestingly, male genital neoplasia was identified on the recent epigenomic screen of cannabis-affected tissues (See Schrott Appendix pages 279 and 341 (Schrott et al., 2021)).

However, it is also clear that cannabinoids are not acting by just one mechanism alone to recapitulate the normal path to testicular oncogenesis 6.5 times faster than usual. Whatever the secrets of this process in the case of cannabis, it does not seem to us unlikely that an environmental agent, which is acting so powerfully in the testicular germ cell niche, may also act in a similar way in other stem cell tissue beds.

Breast cancer

Breast cancer is the commonest cancer in many Western nations so its inclusion on both the USA and European lists of cannabinoid-related cancers is of significant concern. Interestingly, it was identified in the recent detailed screen of cannabis-exposed tissues by Schrott and collaborators (See (Schrott et al., 2021) Appendix pages 281, 290, 304, 315, 346, 355). It was also identified in space-time and causal inferential studies in the USA (Reece & Hulse, 2022g).

Lung and oropharyngeal cancers

There has long been a controversy in relation to cannabis and lung and oropharyngeal cancer. Since they are both linked with tobacco, and since cannabis carries all the same tars and carcinogens as tobacco, it has long been expected that both cannabis and tobacco would show as a strong signal in the epidemiology of cancers of both the upper and lower aerodigestive tracts. Moreover, cannabis smoking has been shown to induce premalignant change in both the oropharynx and the bronchus. For that reason, the presence of both tumors in the data of both USA and Europe is noteworthy and important and worthy of serious investigation.

Liver cancer

Liver cancer was identified on both lists. Cannabinoids have been shown to induce their own feedback loop and promote both cirrhosis and liver cancer through an autocrine positive feedback loop including an angiogenic-inducing signal (Mukhopadhyay et al., 2015). Hepatocarcinogenesis was also recently defined in an epigenomic screen of cannabinoid effects in both humans and rodents (See (Schrott et al., 2021) Supplementary data page 284).

Pancreas cancer

Similarly, malignant transformation of the epigenome of the pancreas was identified on the epigenomic screen of cannabis exposure (See (Schrott et al., 2021) Appendix pages 279, 288, 298, 299, 300). Cannabis has recently been shown to be a major drive of pancreatic cancer incidence in patients younger than 50 years (Reece & Hulse, 2023b, 2023e).

Bladder cancer

The presence of bladder cancer, a well-recognized tobacco-related cancer, on the list of cannabis-related cancers is of interest. The pathological basis for tobacco and bladder cancer is that water-soluble carcinogens derived from tobacco sit in the bladder for extended periods saturating its transitional epithelium with multiple oncogenic products. It is therefore not at all surprising that cannabis should be acting in a similar manner. Moreover, cancer of the urinary tract was identified in a functional screen of a recent major epigenomic study on cannabinoid exposure ((Schrott et al., 2021) appendix page 320).

Thyroid cancer

Thyroid cancer was noted to be elevated in the USA recently in a bivariate relationship with cannabinoids (Reece and Hulse, 2021c, 2022e, 2022h, 2022i). The finding of a positive signal for this relationship in Europe in the present study represents the first time this signal has been confirmed outside of the USA to our knowledge. Interestingly, hundreds of hits for thyroid cancer were recently defined in the major epigenomic screen of cannabis exposure both in cannabis dependency and in cannabis withdrawal states (See (Schrott et al., 2021) appendix Pages 241, 242 and 345).

Acute myeloid leukemia

The list of tumors also includes acute myeloid leukemia (AML), which has been shown several times previously to be cannabis-related (Reece & Hulse, 2021d; Robison et al., 1989; Trivers et al., 2006; Wen et al., 2000). While this is known to be a tumor largely of later life, it has been demonstrated that its genomic seeds frequently first appear in prenatal life (Williams et al., 2022), which would make gestational teratogenic and carcinogenic exposures of primary importance in leukemogenesis in addition to demonstrating heritability of genotoxic mutagenicity (Grobner et al., 2018; Ma et al., 2018).

Acute lymphoid leukemia

Similarly, most cases of acute lymphoid leukemia (ALL) occur in childhood, and this tumor was also recently linked with cannabis exposure in an epidemiologically causal manner (Reece & Hulse, 2021c). Indeed, ALL is the commonest tumor of early childhood. Therefore, its presence on this list is a similarly proof of intergenerational transmission of teratogenic malignancy via genotoxic and epigenotoxic pathways (Grobner et al., 2018; Ma et al., 2018). The observation that ALL is also linked with cannabis exposure in Europe as well as USA data is original and extends earlier reports.

One of the surprising and robust features of the present study was the strong signals produced by the group of hematological cancers, acute and chronic lymphoid and myeloid leukemias (ALL, CLL, AML, CML), multiple myeloma, and

Non-Hodgkin's lymphoma, and Hodgkin's disease. While AML and ALL have previously been reported to be cannabis-related, the others to our knowledge have not been. Hematological malignancies are among the most common of the cancers, doubtless due to the fact that many immunocytes rearrange their receptors using V(D)J joining in a physiological process of DNA hypermutation to generate the enormous antigenic variability required for environmental surveillance. For this reason, lymphoid tissues are classically at risk of genomic instability and malignant transformation.

Biological mechanisms

Many mechanisms of genomic damage by cannabinoids have been described affecting the male and female reproductive tract (Chioccarelli et al., 2010; Rossato et al., 2008) including spermatocyte morphology (Huang et al., 1999), oocyte cell division including high rates of cell loss after single induced division (Morishima, 1984), single- and double- stranded DNA breaks (Russo et al., 2019), chromosomal fusions and translocations (Huang et al., 1999), oxidation of DNA bases (Russo et al., 2019), reduced DNA, RNA, nucleoprotein, and histone synthesis (McClean & Zimmerman, 1976), reduced formation of acetylated and phosphorylated histone proteins (Mon et al., 1981), interfering with histone-protamine transfer during sperm maturation (Chioccarelli et al., 2010; Rossato et al., 2008), and major disruption to the epigenomic machinery (Szutorisz et al., 2014; Watson et al., 2015). There is not sufficient space in the present paper to describe all of these in detail, but they have previously been considered and reviewed (Reece, 2022a, 2022b, 2022c; Reece & Hulse, 2019d, 2022f, 2022j, 2023c, 2023d, 2023g; Szutorisz et al., 2014; Watson et al., 2015).

Notwithstanding, three mechanisms of genomic damage to the oncogenic processes stand out as deserving of special attention, in part because they lead to positive feed-forward self-perpetuating and self-amplifying cycles and become powerful engines generating ongoing genotoxic damage and oncogenic transformation.

Cannabinoid-induced chromosomal pathophysiology

Given the above reports of a number of cancers, it is worthwhile to consider some arithmetic to appreciate the scale of the major genomic damage, which is entailed in some of the described pathologies. Testicular cancer is known to involved chromosomes 1, 7, 8, 11, 12, 13, 18, 21, *X*, and *Y* (Reece & Hulse, 2021d; Shen et al., 2018), which together have lengths of 246, 158, 146, 134, 1325, 113, 76, 46, 153, and 50 megabases totaling 1254 MB (41.8%) of the 3000 MB in the normal human genome. Moreover, testicular cancer is also known to involve one or two rounds of whole genome-doubling events indicating a major disruption to the machinery controlling meiosis (Shen et al., 2018).

ALL commonly involved translocations and breakages of chromosomes 4, 9, 10, 11, and 22, which have total lengths of 191, 136, 135, 134, and 49 megabases, together totaling some 645 megabases or 21.5% of the human genome.

Published reports have identified that chromosomal trisomies such as trisomies 13, 18, 21, and Turners syndrome are known complications of prenatal cannabis exposure (Reece & Hulse, 2021f, 2022c, 2022h, 2022i). These chromosomes are of lengths 113, 76, 46, and 153 megabases, which together total 388 megabases or 12.9% of the human genome.

If one adds together the three lists of chromosomes supplied above without duplication, it can be shown that 1,765 megabases (58.8%) of the human genome are directly implicated in known cannabis-related toxicity. By any measure this is a massive amount of direct genomic damage and makes chromosomal toxicity one of the major hallmarks of human genomic damage from cannabinoids. Moreover, it is important to appreciate that such major chromosomal toxicity has been described after many cannabinoids including Δ9THC, cannabinol, cannabidiol, and tetrahydrocannabivarin (Huang et al., 1999; Russo et al., 2019).

When the above clinical presentations are considered, it becomes apparent that there are essentially two groups of mechanisms disrupting normal chromosomal physiology in the development of congenital genetic disorders and cancers. One group is related to chromosome mis-segregation errors at the time of mitosis or meiosis where one or a few of the chromosomes become separated from the main mass and become isolated alone forming a micronucleus. The other group of disruptions involves chromosomal breakages and secondary end-to-end fusions to the uncapped and unstable broken chromosomal ends.

Given that cannabinoids have long been known to test positively in the micronucleus assay the first question is "Why might a chromosome come loose and disengage from the mitotic spindle at anaphase?" Four possible reasons suggest themselves that are not mutually exclusive. However, because the system is so complex, there are likely several more explanations. First, cannabinoids have been shown to reduce tubulin synthesis (Wang et al., 2011). Since the microtubules of the spindle are made of polymerized α- and β-tubulin polymers, one simplistic possibility is that they just slide off the free plus ends of the microtubule half spindles. Secondly, like many proteins, tubulin is known to undergo a complex code of posttranslational modifications (PTM) involving particularly glycosylation, acetylation, glutaminylation, and polyglutaminylation, which is thought to govern its intracellular address and function in much the same way as an address does on a letter. This PTM process has been referred to as a "tubulin code." In sperm polymerized tubulin forms, the microtubules of the elongated flagellum and interference with this tubulin code make the sperm quite incapable of making forward linear progress (Gadadhar et al., 2021). Cannabis was shown to epigenomically inhibit the main enzyme responsible for tubulin acetylation ATAT1 (Schrott et al., 2021), which controls the tensile strength of the microtubules it constitutes (Janke & Magiera, 2020; Moutin et al., 2021; Zheng et al., 2022). Cannabinoids may also perturb the tubulin code by other means.

Thirdly, it was recently shown in an epigenomic screen of cannabis-affected tissues that cannabis withdrawal affects microtubule function ((Schrott et al., 2021) Appendix page 300)). So, there may be an epigenomic explanation for this feature.

Fourthly, the process of chromosomal binding to the free-growing ends of the microtubule half spindles is not passive but active. Chromosomes are bound to about 25–30 microtubules by a complex multiprotein complex of 90 proteins called the kinetochore, which is actively regulated in many respects (Ryu & Hochstrasser, 2021). Moreover, most of the kinetochore proteins themselves carry complex PTMs, which are thought to control protein intracellular address and function. Key among these PTMs is sumoylation or the addition of a small ubiquitin-like modifier protein substituent molecule, which acts like a key and foundational PTM to control the subsequent addition of later PTMs (Ryu & Hochstrasser, 2021). THC blocks the action of the desumoylation machinery (Gowran et al., 2009), thereby interfering with the integrity of this key kinetochore PTM code and disrupting the kinetochore function. However, as mentioned above, there are doubtless many other possible ways to disrupt this elegant, sophisticated, and delicate machinery, and these pathways need not be operating in isolation.

The other key pathology to affect chromosomes comes from faulty repair of chromosomes after DSB breakages. As noted above, breakage, fusions, and bridges have all been observed following cannabinoid exposure, and it may be that the cycling of this process from one cell division to the next is also a potent engine for the generation of clinically significant genetic self-amplifying lesions (McClintock, 1941).

Cannabinoid-induced disruption of mitochondria: Epigenome interactions

It is well known that many cannabinoids inhibit mitochondrial function (Mahoney & Harris, 1972). This has been shown in the neurons, lymphocytes, cardiac cells, sperm, and many other tissues across many species. Indeed, mitochondria carry in and between their inner and outer membranes all of the signaling machinery required to transduced lipid-soluble cannabinoid signals in the same way that the plasmalemma does (Benard et al., 2012; Hebert-Chatelain et al., 2014). Since mitochondria supply not only energy (as ATP) but also epigenomic substrates for the epigenomic machinery (including methyl, acetyl, phosphoryl, adenosyl, ubiquitin, and many others), and since mitochondria and the nucleus have to maintain a fine mitonuclear balance and mitochondrial hormesis to keep the cell metabolically stable, it follows that any process that systematically disrupts mitochondrial function will necessarily perturb epigenomic regulation. Thus, since the mitochondrial inhibitory action of cannabinoids has been thoroughly demonstrated, it follows that that cannabinoids must necessarily disrupt epigenomic signaling also.

Cannabinoid-induced epigenomic disruption of cancer-related genes

A fascinating study of these issues recently appeared (Schrott et al., 2021). A group of USA investigators recently studied the whole epigenome of cannabis-exposed rats and humans before and after a 77-day period of documented abstinence. They studied patterns of DNA methylation by both whole-genome bisulfite sequencing and reduced representation bisulfite sequencing with virtually identical responses. The study was annotated by functional annotations from an Ingenuity Pathway Analysis (IPA). While their study produced numerous functional annotations for genes involved in the morphogenesis of most major body systems, much other useful data were also produced.

In particular, their 359-page supplementary material documenting their whole epigenome review identified the term "cancer" 487 times, "carcinoma" 84 times, "neoplasm" 28 times, "leukemia" 32 times, and "lymphoma" 20 times and other terms totaling some 810 mentions for neoplasia in all which thereby demonstrated that this strong emphasis on cancerogenic pathways was one of the major findings of their work. Since these findings were tumor-specific, they may be directly related to the 25 tumor types identified in this study.

These data may then be collated and reviewed. For example, it is apparent that the cancer with the lowest median P-value is thyroid cancer with three functional annotations, which have previously been shown to be strongly related to cannabis (Reece & Hulse, 2021c). Its overall P-value is 1.26×10^{-17}, and the cumulative number of genes implicated is 752. In cannabis dependence, its *P*-value falls to 1.24×10^{-17} compared with 0.0014 in cannabis withdrawal so that the ratio between these two *P*-values is 1.14×10^{14}. The median number of genes affected is 318.5 in cannabis dependence and 115.0 in cannabis withdrawal, and the ratio between these two numbers is 2.77.

The cancer with the next lowest overall median *P*-value is testicular cancer, which also has three annotations, an overall *P*-value of 1.60×10^{-8}, and a total of 364 genes affected. In cannabis dependence, its median *P*-value is 1.37×10^{-8} and in withdrawal, it rises to 6.75×10^{-4}. The median number of genes affected in cannabis dependence is 152 falling to 66 in withdrawal for a dependence-to-withdrawal ratio of 2.53.

However, the situation is very different if one considers acute myeloid leukemia, AML. Some cases of AML occur early in childhood. For AML, there were three gene annotations, a median overall *P*-value of 8.96×10^{-4}, and 47 genes implicated overall.

However, when one compares cannabis dependence to withdrawal, the pattern observed above and for all the other genes is actually reversed. Here the median *P*-value in cannabis dependence is 0.0017 and 6.26×10^{-4} in cannabis withdrawal. The median gene number affected in cannabis dependence is 5.5 and in cannabis withdrawal 36 so that the dependence withdrawal ratio is reversed for this tumor to 0.15.

The apparent clinical scenario from these data may be that babies born to cannabis-dependent parents experience cannabis withdrawal, which launches the gene cassettes mentioned especially that for AML. As a result, a large number of AML genes become activated to high levels of statistical significance, and the children experience inherited teratogenic leukemogenesis in their first years of life.

Cannabinoid-induced disruption of chromatin looping

One recent oncogenic mechanism that has come to light is the fact that it has been shown that cannabis epigenomically interferes with the basic machinery of gene expression in different ways. It was shown that cannabis exposure inhibits both the expression of CTCF, which forms the main boundary markers between topologically defined domains on chromatin and the cohesin motors, which extrude the chromatin loops (Schrott et al., 2021), which together compose the transcriptional factories from which the genes are transcribed (Ahn et al., 2021; Alexanian et al., 2021; Hua et al., 2021; Johnson et al., 2021; Kim et al., 2019; Prieler et al., 2021; Winick-Ng et al., 2021; Yang et al., 2021). Cohesin molecules form a ring or lasso around chromatin, and they actively extrude it forming loops (Gabriele et al., 2022; Kim et al., 2019; Shaltiel et al., 2022). Cohesin is a multimolecular complex, and its core components SMC1[a], SMC3, RAD51[b], and STAG[c] were all shown recently to be inhibited epigenetically by cannabis (Schrott et al., 2021).

Exponentiation of cannabinoid-related genotoxicity

The theme of exponential dose-response to multiple genomic and epigenomic processes has been amply demonstrated throughout the cannabinoid metabolic, epigenetic, and genotoxic literature. Thus, studies have shown exponential effects relating to the formation of micronuclei (Zimmerman & Raj, 1980), micronucleus formation (Koller et al., 2015); histone synthesis (Mon et al., 1981), mitochondrial energy supply and supply of epigenomic substrates (Canto et al., 2015), inhibition of tubulin and RNA synthesis (Tahir & Zimmerman, 1991), mutagenesis (Busch et al., 1979), actin synthesis and mechanisms (Tahir & Zimmerman, 1991), cell viability (Tahir et al., 1992), inhibition of cell growth (Tahir & Zimmerman, 1991), and mitochondrial ATP charge (Mahoney & Harris, 1972).

These laboratory findings are supported by several human epidemiological studies documenting that increased clinical evidence of cannabinoid genotoxicity at higher doses has also been compiled including Ebstein anomaly, microcephalus, microtia, intestinal stenosis or atresia, diaphragmatic hernia, omphalocele, atrial septal defect, orofacial clefts, hydrocephalus, encephalocele, large congenital posterior urethral valves, trisomies 13,18, and 21, Turner syndrome, Deletion 22q11.2, pulmonary valve atresia, hypoplastic left heart, and others (Reece & Hulse, 2020a, 2020b). Moreover, this effect has also been demonstrated in space-time regression studies for prostate and ovarian cancer development under the influence of cannabidiol (Reece and Hulse, 2022d).

---

a. SMC—Structural Maintenance of Chromosomes.

b. RAD51—Recombinase/Recombination Protein A/RAD51 (*S. cerevisiae*) Homolog (*E. coli* RecA Homolog).

c. STAG—Stromal Antigen. STAG1 is also called SCC1. STAG2 is also called SCC3B. STAG3 is also called SCC3H3.

Qualitative causal inference

It is worth considering how the present findings measure up against the qualitative criteria of causal inference outlined by Hill in 1965. Hill described the nine factors of strength of association, consistency among studies, specificity, temporality, coherence with known data, biological plausibility, a biological dose-response curve, analogy with similar situations elsewhere, and experimental confirmation. It is notable that the present results obviously fulfill the first eight of these criteria. In the light of the above findings, while much experimental work has been undertaken on cannabinoid genotoxicity, more needs to be done on tissue-specific interactions and the interplay of the epigenome and the genome in oncogenesis. Notwithstanding, it is well established that multiple cannabinoids (including cannabidiol) test positively in both the comet assay and the micronucleus assay (Hall & Degenhardt, 2009; Russo et al., 2019), which are long established tests of genotoxicity.

Quantitative causal inference

As alluded to above *E*-values are one of the pillars of quantitative causal inference. The exceedingly high *E*-values reported in many of the above tables place the strength of these associations beyond the bounds of mere association and well into the territory of causal relationships. Indeed as was noted above, an *E*-value of 9 is said to be high (VanderWeele & Ding, 2017) and above 1.25 is said to be required to demonstrate a putatively causal effect (VanderWeele et al., 2019). Many of the values noted herein far surpass that level.

Generalizability

The data reviewed constitute the largest most comprehensive publicly available datasets in the world to assess the involvement of cannabis in malignant tumorigensis and inheritable carcinogenesis. The results from the USA and European studies are convergent to a most remarkable degree, which naturally has the effect of building confidence in each set of results independently. Moreover, very high levels of both statistical significance and of *E*-values exist, which demonstrate not only a high degree of association but also extremely high *E*-values as is typical of truly causal relationships. Importantly, the *E*-values reported herein are much higher than that for the tobacco-lung cancer association, which is said to be high at 9. Finally, the cancers reported are underlain by multiple pathways, which are well established and support biologically plausible routes to oncogenic transformation by a known environmental carcinogen and mutagen.

For these reasons, we are confident that our results are widely generalizable wherever data of sufficient quality exist in order to investigate these matters.

Strengths and limitations

The present study has a number of strengths and limitations. Its strengths include the fact that it utilizes the two largest datasets in the world in this area, the use of data that have already been peer-reviewed and published, the relatively simple level of data processing involved, which essentially involves minimal data transformation, and the use of large sample sizes from multiple jurisdictions. Additionally, *E*-values have been widely employed, which is one of the key metrics for formal quantitative causal inferential analysis. One of the limitations of the present report is that only bivariate analysis has been used. However, these present results are supported by voluminous inverse probability-weighted multivariable adjustment regression studies and space-time regression so that confidence can be placed in them regardless (Reece, 2022c; Reece & Hulse, 2016, 2021b, 2021c, 2021d, 2021f, 2022a, 2022b, 2022d, 2022e, 2022f, 2022g, 2022h, 2022k, 2023a, 2023b, 2023c, 2023f). Moreover, the exceedingly high level of many of the *E*-values reported herein implies that further adjustment is unlikely to completely explain away all of the signals reported. Finally, like many epidemiological studies, we did not have access to individual participant-level data.

Conclusion

The present study demonstrates the remarkable concordance of USA and European lists of cannabis-associated cancers to 23 tumors including all cancers, common cancers such as breast and prostate cancers, and confirms the well-established finding in the literature relating to testicular carcinogenesis. The remarkably high degree of concordance in these very large datasets from both sides of the Atlantic and the very high *E*-value estimates imply a high degree of confidence in the results and are consistent with causal relationships. The analysis fulfills the criteria of both qualitative and quantitative epidemiological causality. This survey of cannabis-related carcinogenesis is underpinned by a host of genotoxic mechanisms including damage to gametes, chromosomes, DNA, genetic and epigenetic material, and energy and substrate supply to the epigenomic machinery through their well-established adverse mitochondrial effects.

Applications to other areas

In this chapter, we have reviewed the correspondence of cannabis-related cancer incidence on bivariate analysis between the European and USA datasets and have confirmed that they correspond very closely indeed. Data relate to similar datasets pertaining to the broader question of cannabis-related genotoxicity, which is also demonstrated clinically in cannabinoid-induced cancerogenesis (Chapter 5) and cannabinoid-induced cellular aging. Cannabinoid genotoxicity also includes cannabinoid-related epigenotoxicity at multiple levels. Data are particularly relevant and concerning given the well-substantiated exponential dose-response relationship between many cannabinoids and indices of cannabinoid genotoxicity both in the laboratory (Busch et al., 1979; Russo et al., 2019) and in epidemiological studies (Reece and Hulse, 2021c, 2022e, 2022h, 2022i). Prior work has demonstrated that these relationships are specific to cannabis and cannabinoids with effects far exceeding those that may be attributable to tobacco and alcohol (Reece and Hulse, 2021c, 2022e, 2022h, 2022i). The involvement of pediatric acute myeloid and lymphoid leukemia in this data demonstrates transgenerational transmission of mutagenicity and is clearly of particular concern. Food chain contamination would appear to be a particularly important hazard requiring careful vigilance to avoid accidental and casual exposure of vulnerable groups such as pregnant mothers, developing infants, and adolescents. Cannabinoid genotoxicity also relates to cannabinoid neurotoxicity as the former has been shown to be one cause of the latter both directly and via epigenomic pathways (Murphy et al., 2018; Schrott et al., 2021). Cannabinoid neurotoxicity has also been shown to be transgenerationally inheritable (Reece & Hulse, 2019a, 2019b; Schrott et al., 2019). Together the exponential risks posed by cannabinoid genotoxicity and neurotoxicity paint a portrait of serious multigenerational public health implications, which have yet to be properly factored into public policy and the protection of vulnerable populations.

Mini-dictionary of terms

- **Cancerogenesis**. The process of cancer induction and transformation.
- **Carcinogenesis**. The process of cancer induction and transformation.
- **Epigenotoxin**. A substance or insult that damages the epigenetic material.
- **Genotoxin**. A substance or insult that damages the genetic or epigenetic material.
- **Mutagenesis**. The process of alteration of the bases of DNA.
- **Oncogenesis**. The process of cancer induction and transformation.

Key facts of cannabinoid cancerogenicity

- Many cannabinoids are implicated in genotoxic activities including THC, cannabidiol, cannabinol, and synthetic cannabinoids. The genotoxic moiety is the cannabinoid olevitol nucleus.
- Cannabinoids are known to be toxic to the mitochondria, which support and subserve numerous reactions involving the genome and epigenome, which generally require substrates and energy.
- Cannabinoids are known to have many genotoxic actions including on oocytes, sperm, chromosomes, genes, DNA bases, and key epigenomic mechanisms including DNA methylation and histone synthesis.
- The European and USA datasets comprise the majority of the publicly available datasets in the world relating to these subjects.
- On both sides of the Atlantic, indices of cannabis consumption are epidemiologically connected to more than 20 different cancers.
- There is close correspondence between the cancers identified as being cannabis-related in the USA and Europe.
- This provides strong internal and external validation for the results from both continents.
- Cancers of the testis, liver, pancreas, and breast that have risen over recent decades in the USA and have been shown to be driven by cannabis.
- Cancers of the testis and pancreas have become more common in Europe.
- Both the USA and Europe have seen recent surges in the prevalence of cannabis use, the intensity of daily cannabis use, and the THC potency of widely available products.

- Exponential dose-response relationships are of particular concern in this triply convergent environment of sharply increasing community cannabinoid exposure.
- Populations need to be carefully protected from cannabinoid genotoxicity both for their own health and for that of subsequent generations.
- Food chain contamination is a special risk especially for vulnerable populations such as pregnant women, babies, and adolescents with growing brains.

Summary points

- 25/28 cancers monitored in the USA could be shown to be related to various cannabinoids.
- 27/41 cancers in Europe could be shown to be related to various metrics of cannabis exposure.
- 100% of the USA cannabis-related cancers and 96.5% of the European cannabis-related cancers were verified by comparison with the transatlantic dataset (where data were available).
- Data therefore provide good internal and external validation of findings on both continents.
- Data are consistent with multiple basic cellular and molecular genotoxic and epigenotoxic mechanisms.
- Data reflect exponential genotoxic dose-response relationships at higher doses.
- It is important to protect populations from cannabinoid genotoxicity and carefully limit cannabinoid exposure.
- Food chain contamination is a particular harzard and must be closely guarded against.
- Cannabinoid carcinogensis is more potent on many analyses than that from tobacco and alcohol.
- Findings justify careful protection of vulnerable populations from genotoxic and carcinogenic effects of cannabinoids.

References

Agence France-Presse in Paris. (2018). *France to investigate cause of upper limb defects in babies*. The Guardian. Retrieved from: https://www.theguardian.com/world/2018/oct/21/france-to-investigate-cause-of-upper-limb-defects-in-babies.

Ahn, J. H., Davis, E. S., Daugird, T. A., Zhao, S., Quiroga, I. Y., Uryu, H., & Wang, G. G. (2021). Phase separation drives aberrant chromatin looping and cancer development. *Nature*, *595*(7868), 591–595. https://doi.org/10.1038/s41586-021-03662-5.

Aldington, S., Harwood, M., Cox, B., Weatherall, M., Beckert, L., Hansell, A., & Beasley, R. (2008). Cannabis use and risk of lung cancer: A case-control study. *The European Respiratory Journal*, *31*(2), 280–286. Retrieved from: http://www.ncbi.nlm.nih.gov/entrez/query.fcgi?cmd=Retrieve&db=PubMed&dopt=Citation&list_uids=18238947.

Alexanian, M., Przytycki, P. F., Micheletti, R., Padmanabhan, A., Ye, L., Travers, J. G., & Srivastava, D. (2021). A transcriptional switch governs fibroblast activation in heart disease. *Nature*, *595*(7867), 438–443. https://doi.org/10.1038/s41586-021-03674-1.

Benard, G., Massa, F., Puente, N., Lourenco, J., Bellocchio, L., Soria-Gomez, E., & Marsicano, G. (2012). Mitochondrial CB(1) receptors regulate neuronal energy metabolism. *Nature Neuroscience*, *15*(4), 558–564. https://doi.org/10.1038/nn.3053.

Berthiller, J., Straif, K., Boniol, M., Voirin, N., Benhaim-Luzon, V., Ayoub, W. B., & Sasco, A. J. (2008). Cannabis smoking and risk of lung cancer in men: A pooled analysis of three studies in Maghreb. *Journal of Thoracic Oncology*, *3*(12), 1398–1403. Retrieved from: http://www.ncbi.nlm.nih.gov/entrez/query.fcgi?cmd=Retrieve&db=PubMed&dopt=Citation&list_uids=19057263.

Bluhm, E. C., Daniels, J., Pollock, B. H., & Olshan, A. F. (2006). Maternal use of recreational drugs and neuroblastoma in offspring: A report from the Children's Oncology Group (United States). *Cancer Causes & Control*, *17*(5), 663–669. Retrieved from: http://www.ncbi.nlm.nih.gov/entrez/query.fcgi?cmd=Retrieve&db=PubMed&dopt=Citation&list_uids=16633913.

British Lung Foundation. (2005). *Cannabis: A smoking gun*. Retrieved from: http://www.lunguk.org/downloads/A_Smoking_Gun.pdf.

Busch, F. W., Seid, D. A., & Wei, E. T. (1979). Mutagenic activity of marihuana smoke condensates. *Cancer Letters*, *6*(6), 319–324. Retrieved from: http://www.ncbi.nlm.nih.gov/pubmed/378366.

Canto, C., Menzies, K. J., & Auwerx, J. (2015). NAD(+) metabolism and the control of energy homeostasis: A balancing act between mitochondria and the nucleus. *Cell Metabolism*, *22*(1), 31–53. https://doi.org/10.1016/j.cmet.2015.05.023.

Centers of Disease Control. (2022). *Health effects of cigaratte smoking: Smoking and cancer*. Retrieved from: https://www.cdc.gov/tobacco/data_statistics/fact_sheets/health_effects/effects_cig_smoking/index.htm#cancer.

Chacko, J. A., Heiner, J. G., Siu, W., Macy, M., & Terris, M. K. (2006). Association between marijuana use and transitional cell carcinoma. *Urology*, *67*(1), 100–104. Retrieved from: http://www.ncbi.nlm.nih.gov/entrez/query.fcgi?cmd=Retrieve&db=PubMed&dopt=Citation&list_uids=16413342.

Chioccarelli, T., Cacciola, G., Altucci, L., Lewis, S. E., Simon, L., Ricci, G., & Cobellis, G. (2010). Cannabinoid receptor 1 influences chromatin remodeling in mouse spermatids by affecting content of transition protein 2 mRNA and histone displacement. *Endocrinology*, *151*(10), 5017–5029. https://doi.org/10.1210/en.2010-0133.

Daling, J. R., Doody, D. R., Sun, X., Trabert, B. L., Weiss, N. S., Chen, C., & S.M., S. (2009). Association of marijuana use and the incidence of testicular germ cell tumors. *Cancer*, *115*(6), 1215–1223. https://doi.org/10.1002/cncr.24159.

Efird, J. T., Friedman, G. D., Sidney, S., Klatsky, A., Habel, L. A., Udaltsova, N. V., & Nelson, L. M. (2004). The risk for malignant primary adult-onset glioma in a large, multiethnic, managed-care cohort: Cigarette smoking and other lifestyle behaviors. *Journal of Neuro-Oncology*, *68*(1), 57–69. Retrieved from: http://www.ncbi.nlm.nih.gov/entrez/query.fcgi?cmd=Retrieve&db=PubMed&dopt=Citation&list_uids=15174522.

European Monitoring Centre for Drugs and Drug Addiction (EMCDDA). (2021). *European Monitoring Centre for Drugs and Drug Addiction (EMCDDA): Statistical Bulletin 2021—Prevalence of drug use*. Statistical Bulletins. Retrieved from: https://www.emcdda.europa.eu/data/stats2021/gps_en.

Gabriele, M., Brandão, H. B., Grosse-Holz, S., Jha, A., Dailey, G. M., Cattoglio, C., & Hansen, A. S. (2022). Dynamics of CTCF- and cohesin-mediated chromatin looping revealed by live-cell imaging. *Science*, *376*(6592), 496–501. https://doi.org/10.1126/science.abn6583.

Gadadhar, S., Alvarez Viar, G., Hansen, J. N., Gong, A., Kostarev, A., Ialy-Radio, C., & Janke, C. (2021). Tubulin glycylation controls axonemal dynein activity, flagellar beat, and male fertility. *Science*, *371*(6525). https://doi.org/10.1126/science.abd4914.

Gowran, A., Murphy, C. E., & Campbell, V. A. (2009). Delta(9)-tetrahydrocannabinol regulates the p53 post-translational modifiers Murine double minute 2 and the Small Ubiquitin MOdifier protein in the rat brain. *FEBS Letters*, *583*(21), 3412–3418. https://doi.org/10.1016/j.febslet.2009.09.056.

Grobner, S. N., Worst, B. C., Weischenfeldt, J., Buchhalter, I., Kleinheinz, K., Rudneva, V. A., & Pfister, S. M. (2018). The landscape of genomic alterations across childhood cancers. *Nature*, *555*(7696), 321–327. https://doi.org/10.1038/nature25480.

Gurney, J., Shaw, C., Stanley, J., Signal, V., & Sarfati, D. (2015). Cannabis exposure and risk of testicular cancer: A systematic review and meta-analysis. *BMC Cancer*, *15*, 897–906. https://doi.org/10.1186/s12885-015-1905-6.

Hall, W., & Degenhardt, L. (2009). Adverse health effects of non-medical cannabis use. *Lancet*, *374*(9698), 1383–1391. https://doi.org/10.1016/S0140-6736(09)61037-0.

Hashibe, M., Ford, D. E., & Zhang, Z. F. (2002). Marijuana smoking and head and neck cancer. *Journal of Clinical Pharmacology*, *42*(11 Suppl), 103S–107S. Retrieved from: http://www.ncbi.nlm.nih.gov/entrez/query.fcgi?cmd=Retrieve&db=PubMed&dopt=Citation&list_uids=12412843.

Hashibe, M., Morgenstern, H., Cui, Y., Tashkin, D. P., Zhang, Z. F., Cozen, W., & Greenland, S. (2006). Marijuana use and the risk of lung and upper aerodigestive tract cancers: Results of a population-based case-control study. *Cancer Epidemiology, Biomarkers & Prevention*, *15*(10), 1829–1834. Retrieved from: http://www.ncbi.nlm.nih.gov/entrez/query.fcgi?cmd=Retrieve&db=PubMed&dopt=Citation&list_uids=17035389.

Hashibe, M., Straif, K., Tashkin, D. P., Morgenstern, H., Greenland, S., & Zhang, Z. F. (2005). Epidemiologic review of marijuana use and cancer risk. *Alcohol*, *35*(3), 265–275. https://doi.org/10.1016/j.alcohol.2005.04.008.

Hebert-Chatelain, E., Reguero, L., Puente, N., Lutz, B., Chaouloff, F., Rossignol, R., & Marsicano, G. (2014). Cannabinoid control of brain bioenergetics: Exploring the subcellular localization of the CB1 receptor. *Molecular Metabolism*, *3*(4), 495–504. https://doi.org/10.1016/j.molmet.2014.03.007.

Hua, P., Badat, M., Hanssen, L. L. P., Hentges, L. D., Crump, N., Downes, D. J., & Davies, J. O. J. (2021). Defining genome architecture at base-pair resolution. *Nature*, *595*(7865), 125–129. https://doi.org/10.1038/s41586-021-03639-4.

Huang, H. F. S., Nahas, G. G., & Hembree, W. C. (1999). Effects of Marijuana inhalation on spermatogenesis of the rat. In G. G. Nahas, K. M. Sutin, D. J. Harvey, & S. Agurell (Eds.), *Vol. 1. Marijuana in medicine* (pp. 359–366). Totowa, New York: Human Press.

Jameson, J. L., Fauci, A. S., Hauser, S. L., Longo, D. L., Jameson, J. L., & Loscalzo, J. (Eds.). (2018) (Vols. 1 & 2). *Harrison's Principles of Internal Medicine* (20th ed.). New York: McGraw Hill.

Janke, C., & Magiera, M. M. (2020). The tubulin code and its role in controlling microtubule properties and functions. *Nature Reviews. Molecular Cell Biology*, *21*(6), 307–326. https://doi.org/10.1038/s41580-020-0214-3.

Johnson, D., Crawford, M., Cooper, T., Claeys Bouuaert, C., Keeney, S., Llorente, B., & Neale, M. J. (2021). Concerted cutting by Spo11 illuminates meiotic DNA break mechanics. *Nature*, *594*(7864), 572–576. https://doi.org/10.1038/s41586-021-03389-3.

Kim, Y., Shi, Z., Zhang, H., Finkelstein, I. J., & Yu, H. (2019). Human cohesin compacts DNA by loop extrusion. *Science*, *366*(6471), 1345–1349. https://doi.org/10.1126/science.aaz4475.

Koller, V. J., Ferk, F., Al-Serori, H., Misik, M., Nersesyan, A., Auwarter, V., & Knasmuller, S. (2015). Genotoxic properties of representatives of alkylindazoles and aminoalkyl-indoles which are consumed as synthetic cannabinoids. *Food and Chemical Toxicology*, *80*, 130–136. https://doi.org/10.1016/j.fct.2015.03.004.

Leuchtenberger, C., & Leuchtenberger, R. (1971). Morphological and cytochemical effects of marijuana cigarette smoke on epithelioid cells of lung explants from mice. *Nature*, *234*(5326), 227–229. https://doi.org/10.1038/234227a0.

Ma, X., Liu, Y., Liu, Y., Alexandrov, L. B., Edmonson, M. N., Gawad, C., & Zhang, J. (2018). Pan-cancer genome and transcriptome analyses of 1,699 paediatric leukaemias and solid tumours. *Nature*, *555*(7696), 371–376. https://doi.org/10.1038/nature25795.

Mahoney, J. M., & Harris, R. A. (1972). Effect of 9 -tetrahydrocannabinol on mitochondrial processes. *Biochemical Pharmacology*, *21*(9), 1217–1226. Retrieved from: https://www.ncbi.nlm.nih.gov/pubmed/5038669.

Manthey, J., Freeman, T. P., Kilian, C., Lopez-Pelayo, H., & Rehm, J. (2021). Public health monitoring of cannabis use in Europe: Prevalence of use, cannabis potency, and treatment rates. *The Lancet Regional Health-Europe*, *10*, 100227. https://doi.org/10.1016/j.lanepe.2021.100227.

McClean, D. K., & Zimmerman, A. M. (1976). Action of delta 9-tetrahydrocannabinol on cell division and macromolecular synthesis in division-synchronized protozoa. *Pharmacology*, *14*(4), 307–321. Retrieved from: http://www.ncbi.nlm.nih.gov/entrez/query.fcgi?cmd=Retrieve&db=PubMed&dopt=Citation&list_uids=819944.

McClintock, B. (1941). The stability of broken ends of chromosomes in Zea mays. *Genetics*, *26*(2), 234–282. https://doi.org/10.1093/genetics/26.2.234.

Moiche Bokobo, P., Atxa de la Presa, M. A., & Cuesta Angulo, J. (2001). Transitional cell carcinoma in a young heavy marihuana smoker. *Archivos Españoles de Urología*, *54*(2), 165–167. Retrieved from: http://www.ncbi.nlm.nih.gov/entrez/query.fcgi?cmd=Retrieve&db=PubMed&dopt=Citation&list_uids=11341122.

Mon, M. J., Haas, A. E., Stein, J. L., & Stein, G. S. (1981). Influence of psychoactive and nonpsychoactive cannabinoids on cell proliferation and macromolecular biosynthesis in human cells. *Biochemical Pharmacology*, *30*(1), 31–43. https://doi.org/10.1016/0006-2952(81)90281-1.

Morishima, A. (1984). Effects of cannabis and natural cannabinoids on chromosomes and ova. *NIDA Research Monograph, 44*, 25–45. Retrieved from: http://www.ncbi.nlm.nih.gov/entrez/query.fcgi?cmd=Retrieve&db=PubMed&dopt=Citation&list_uids=6090908.

Moutin, M. J., Bosc, C., Peris, L., & Andrieux, A. (2021). Tubulin post-translational modifications control neuronal development and functions. *Developmental Neurobiology, 81*(3), 253–272. https://doi.org/10.1002/dneu.22774.

Mukhopadhyay, B., Schuebel, K., Mukhopadhyay, P., Cinar, R., Godlewski, G., Xiong, K., & Kunos, G. (2015). Cannabinoid receptor 1 promotes hepatocellular carcinoma initiation and progression through multiple mechanisms. *Hepatology, 61*(5), 1615–1626. https://doi.org/10.1002/hep.27686.

Murphy, S. K., Itchon-Ramos, N., Visco, Z., Huang, Z., Grenier, C., Schrott, R., & Kollins, S. H. (2018). Cannabinoid exposure and altered DNA methylation in rat and human sperm. *Epigenetics, 13*(12), 1208–1221. https://doi.org/10.1080/15592294.2018.1554521.

Nahas, G. G., Morishima, A., & Desoize, B. (1977). Effects of cannabinoids on macromolecular synthesis and replication of cultured lymphocytes. *Federation Proceedings, 36*(5), 1748–1752.

Nieder, A. M., Lipke, M. C., & Madjar, S. (2006). Transitional cell carcinoma associated with marijuana: Case report and review of the literature. *Urology, 67*(1), 200. Retrieved from: http://www.ncbi.nlm.nih.gov/entrez/query.fcgi?cmd=Retrieve&db=PubMed&dopt=Citation&list_uids=16413373.

Pearl, J., & Mackaenzie, D. (2019). *The book of why. The new science of cause and effect. Vol. 1*. New York: Basic Books.

Prieler, S., Chen, D., Huang, L., Mayrhofer, E., Zsótér, S., Vesely, M., & Klein, F. (2021). Spo11 generates gaps through concerted cuts at sites of topological stress. *Nature, 594*(7864), 577–582. https://doi.org/10.1038/s41586-021-03632-x.

Reece, A. S. (2009). Chronic toxicology of cannabis. *Clinical Toxicology (Philadelphia, Pa.), 47*(6), 517–524. https://doi.org/10.1080/15563650903074507.

Reece, A. S. (2022a). Limblessness: Cannabinoids inhibit key embryonic morphogens both directly and epigenomically. *British Medical Journal, 376*, n3114. https://doi.org/10.1136/bmj.n3114.

Reece, A. S. (2022b). Cannabinoid epigenotoxicity and genotoxicity. *JAMA, E3*. https://doi.org/10.1001/jama.2022.20620.

Reece, A. S. (2022c). Disruption of interlocking synchrony between metabolome and epigenome key to understanding widespread embyrotoxicity and carcinogenicity of diverse cannabinoids. *British Medical Journal*. Retrieved from: https://www.bmj.com/content/377/bmj.o1567/rr-0.

Reece, A. S., & Hulse, G. K. (2016). Chromothripsis and epigenomics complete causality criteria for cannabis- and addiction-connected carcinogenicity, congenital toxicity and heritable genotoxicity. *Mutation Research, 789*, 15–25. https://doi.org/10.1016/j.mrfmmm.2016.05.002.

Reece, A. S., & Hulse, G. K. (2019a). Epidemiological associations of various substances and multiple cannabinoids with autism in USA. *Clinical Pediatrics: Open Access, 4*(2), 1–20. https://doi.org/10.24105/2572-0775.4.155.

Reece, A. S., & Hulse, G. K. (2019b). Effect of cannabis legalization on US autism incidence and medium term projections. *Clinical Pediatrics: Open Access, 4*(2), 1–17. https://doi.org/10.4172/2572-0775.1000154.

Reece, A. S., & Hulse, G. K. (2019c). Gastroschisis and autism-dual canaries in the Californian coalmine. *JAMA Surgery, 154*(4), 366–367. https://doi.org/10.1001/jamasurg.2018.4694.

Reece, A. S., & Hulse, G. K. (2019d). Impacts of cannabinoid epigenetics on human development: reflections on Murphy et al. 'cannabinoid exposure and altered DNA methylation in rat and human sperm' epigenetics 2018;13:1208-1221. *Epigenetics, 14*(11), 1041–1056. https://doi.org/10.1080/15592294.2019.1633868.

Reece, A. S., & Hulse, G. K. (2020a). Cannabis in pregnancy—Rejoinder, exposition and cautionary tales. *Psychiatric Times, 37*. https://www.psychiatrictimes.com/view/cannabis-pregnancy-rejoinder-exposition-cautionary-tales. Retrieved from: https://www.bing.com/search?q=Cannabis+in+Pregnancy+%E2%80%93+Rejoinder%2C+Exposition+and+Cautionary+Tales&cvid=22538e0124c04711b92017489c63214a&aqs=edge..69i57.439j0j1&pglt=43&FORM=ANSPA1&PC=U531.

Reece, A. S., & Hulse, G. K. (2020b). Contemporary epidemiology of rising atrial septal defect trends across USA 1991-2016: A combined ecological geospatiotemporal and causal inferential study. *BMC Pediatrics, 20*(1), 539–550. https://doi.org/10.1186/s12887-020-02431-z.

Reece, A. S., & Hulse, G. K. (2021a). Cannabinoid exposure as a major driver of pediatric acute lymphoid Leukaemia rates across the USA: Combined geospatial, multiple imputation and causal inference study. *BMC Cancer, 21*(1), 984–1017.

Reece, A. S., & Hulse, G. K. (2021b). Geospatiotemporal and causal inferential epidemiological survey and exploration of cannabinoid- and substance-related carcinogenesis in USA 2003-2017. In *Vol. 1. Epidemiology of cannabis: Genotoxicity and neurotoxicity, epigenomics and aging* (p. 2500). New York, USA: Elsevier.

Reece, A. S., & Hulse, G. K. (2021c). Epidemiological overview of multidimensional chromosomal and genome toxicity of cannabis exposure in congenital anomalies and cancer development. *Scienitifc Reports, 11*(1), 13892–13912. https://doi.org/10.10389/s41598-021-93411-5.

Reece, A. S., & Hulse, G. K. (2021d). Causal inference multiple imputation investigation of the impact of cannabinoids and other substances on ethnic differentials in US testicular cancer incidence. *BMC Pharmacology and Toxicology, 22*(1), 40–71. https://doi.org/10.1186/s40360-021-00505-x.

Reece, A. S., & Hulse, G. K. (2021e). Quadruple convergence—Rising cannabis prevalence, intensity, concentration and use disorder treatment. *The Lancet Regional Health-Europe, 10*, 100245–100246. https://doi.org/10.1016/j.lanepe.2021.100245.

Reece, A. S., & Hulse, G. K. (2021f). A geospatiotemporal and causal inference epidemiological exploration of substance and cannabinoid exposure as drivers of rising US pediatric cancer rates. *BMC Cancer, 21*(1), 197–230. https://doi.org/10.1186/s12885-021-07924-3.

Reece, A. S., & Hulse, G. K. (2022a). Epidemiology of Δ8THC-related carcinogenesis in USA: A panel regression and causal inferential study. *International Journal of Environmental Research and Public Health, 19*(13), 7726–7752. Retrieved from: https://www.mdpi.com/1660-4601/19/13/7726.

Reece, A. S., & Hulse, G. K. (2022b). Cannabis- and substance-related epidemiological patterns of chromosomal congenital anomalies in Europe: Geospatiotemporal and causal inferential study. *International Journal of Environmental Research and Public Health, 19*, 11208–11258. Retrieved from: https://www.mdpi.com/1660-4601/19/18/11208.

Reece, A. S., & Hulse, G. K. (2022c). Geotemporospatial and causal inference epidemiological analysis of US survey and overview of cannabis, cannabidiol and cannabinoid genotoxicity in relation to congenital anomalies 2001–2015. *BMC Pediatrics, 22*(1), 47–124. https://doi.org/10.1186/s12887-021-02996-3.

Reece, A. S., & Hulse, G. K. (2022d). Cannabis genotoxicity and cancer incidence: A highly concordant synthesis of European and USA datasets. In V. Preedy, & V. Patel (Eds.), *Vol. 1. Cannabis, cannabinoids and endocannabinoids*. London, UK: Elsevier.

Reece, A. S., & Hulse, G. K. (2022e). Geotemporospatial and causal inferential epidemiological overview and survey of USA cannabis, cannabidiol and cannabinoid genotoxicity expressed in cancer incidence 2003–2017: Part 3—Spatiotemporal, multivariable and causal inferential pathfinding and exploratory analyses of prostate and ovarian cancers. *Archives of Public Health, 80*, 100–136. https://doi.org/10.1186/s13690-022-00813-6. Retrieved from: https://www.dropbox.com/sh/v45uyms8jzas0vb/AADMJqHo3l9ORCX1cKuyiAu7a?dl=0.

Reece, A. S., & Hulse, G. K. (2022f). Epigenomic and other evidence for cannabis-induced aging contextualized in a synthetic epidemiologic overview of cannabinoid-related teratogenesis and cannabinoid-related carcinogenesis. *International Journal of Environmental Research and Public Health, 19* (24), 16721–16776.

Reece, A. S., & Hulse, G. K. (2022g). Geospatiotemporal and causal inference study of cannabis and other drugs as risk factors for female breast cancer USA 2003-2017. *Environmental Epigenetics*, (1), 1–22. Retrieved from: https://academic.oup.com/eep/article/8/1/dvac006/6539905.

Reece, A. S., & Hulse, G. K. (2022h). Geotemporospatial and causal inferential epidemiological overview and survey of USA cannabis, cannabidiol and cannabinoid genotoxicity expressed in cancer incidence 2003–2017: Part 1—Continuous bivariate analysis. *Archives of Public Health, 80*, 99–133. https://doi.org/10.1186/s13690-022-00811-8. Retrieved from: https://www.dropbox.com/sh/v45uyms8jzas0vb/AADMJqHo3l9ORCX1cKuyiAu7a?dl=0.

Reece, A. S., & Hulse, G. K. (2022i). Geotemporospatial and causal inferential epidemiological overview and survey of USA cannabis, cannabidiol and cannabinoid genotoxicity expressed in cancer incidence 2003–2017: Part 2—Categorical bivariate analysis and attributable fractions. *Archives of Public Health, 80*, 100–135. https://doi.org/10.1186/s13690-022-00812-7. Retrieved from: https://www.dropbox.com/sh/v45uyms8jzas0vb/AADMJqHo3l9ORCX1cKuyiAu7a?dl=0.

Reece, A. S., & Hulse, G. K. (2022j). Novel insights into potential cannabis-related cancerogenesis from recent key whole epigenome screen of cannabis dependence and withdrawal: Epidemiological comment and explication of Schrott et al. *Genes (Basel), 14*(32), 1–14. Retrieved from: https://www.mdpi.com/2073-4425/14/1/32.

Reece, A. S., & Hulse, G. K. (2022k). State trends of cannabis liberalization as a causal driver of increasing testicular cancer rates across the USA. *International Journal of Environmental Research and Public Health, 19*, 12759–12796. Retrieved from: https://www.mdpi.com/1660-4601/19/19/12759/htm.

Reece, A. S., & Hulse, G. K. (2023a). *Cannabis- and substance-related carcinogenesis in Europe: A lagged causal inferential panel regression modelling study*. Manuscript submitted.

Reece, A. S., & Hulse, G. K. (2023b). *Cannabis could be the missing environmental pancreatic carcinogen hiding in plain view*. Manuscript submitted.

Reece, A. S., & Hulse, G. K. (2023c). Clinical epigenomic explanation of the epidemiology of cannabinoid genotoxicity manifesting as transgenerational teratogenesis, cancerogenesis and aging acceleration. *International Journal of Environmental Research and Public Health, 20*(4), 3360–3383. Retrieved from: https://www.mdpi.com/1660-4601/20/4/3360.

Reece, A. S., & Hulse, G. K. (2023d). *Epidemiology of cannabis: Genotoxicity and neurotoxicity, epigenomics and aging. Vol. 1*. New York, USA: Elsevier.

Reece, A. S., & Hulse, G. K. (2023e). *Sociodemographically stratified exploration of pancreatic cancer incidence in younger US patients: Implication of cannabis exposure as a risk factor*. Manuscript submitted.

Reece, A. S., & Hulse, G. K. (2023f). Chapter 4: Geospatiotemporal and causal inferential epidemiological survey and exploration of cannabinoid- and substance related carcinogenesis in USA 2003-2017. In *Vol. 1. Epidemiology of cannabis: Genotoxicity and neurotoxicity, epigenomics and aging* (p. 2500). New York, USA: Elsevier.

Reece, A. S., & Hulse, G. K. (2023g). Chapter 5: Multivalent cannabinoid epigenotoxicities and multigenerational aging. In *Vol. 1. Epidemiology of cannabis: Genotoxicity and neurotoxicity, epigenomics and aging* (p. 2500). New York, USA: Elsevier.

Robinson, M. (2019). *Babies born with deformed hands spark investigation in Germany*. Retrieved from: https://edition.cnn.com/2019/09/16/health/hand-deformities-babies-gelsenkirchen-germany-intl-scli-grm/index.html.

Robison, L. L., Buckley, J. D., Daigle, A. E., Wells, R., Benjamin, D., Arthur, D. C., & Hammond, G. D. (1989). Maternal drug use and risk of childhood nonlymphoblastic leukemia among offspring. An epidemiologic investigation implicating marijuana (a report from the Childrens Cancer Study Group). *Cancer, 63*(10), 1904–1911. Retrieved from: http://www.ncbi.nlm.nih.gov/pubmed/2649219.

Rossato, M., Pagano, C., & Vettor, R. (2008). The cannabinoid system and male reproductive functions. *Journal of Neuroendocrinology, 20*(Suppl 1), 90–93. Retrieved from: http://www.ncbi.nlm.nih.gov/entrez/query.fcgi?cmd=Retrieve&db=PubMed&dopt=Citation&list_uids=18426506.

Russo, C., Ferk, F., Mišík, M., Ropek, N., Nersesyan, A., Mejri, D., & Knasmüller, S. (2019). Low doses of widely consumed cannabinoids (cannabidiol and cannabidivarin) cause DNA damage and chromosomal aberrations in human-derived cells. *Archives of Toxicology, 93*(1), 179–188. https://doi.org/10.1007/s00204-018-2322-9.

Ryu, H. Y., & Hochstrasser, M. (2021). Histone sumoylation and chromatin dynamics. *Nucleic Acids Research, 49*(11), 6043–6052. https://doi.org/10.1093/nar/gkab280.

Schrott, R., Acharya, K., Itchon-Ramos, N., Hawkey, A. B., Pippen, E., Mitchell, J. T., & Murphy, S. K. (2019). Cannabis use is associated with potentially heritable widespread changes in autism candidate gene DLGAP2 DNA methylation in sperm. *Epigenetics, 15*(1-2), 161–173. https://doi.org/10.1080/15592294.2019.1656158.

Schrott, R., Murphy, S. K., Modliszewski, J. L., King, D. E., Hill, B., Itchon-Ramos, N., & Mitchell, J. T. (2021). Refraining from use diminishes cannabis-associated epigenetic changes in human sperm. *Environmental Epigenetics, 7*(1), 1–10. https://doi.org/10.1093/eep/dvab009.

Shaltiel, I. A., Datta, S., Lecomte, L., Hassler, M., Kschonsak, M., Bravo, S., & Haering, C. H. (2022). A hold-and-feed mechanism drives directional DNA loop extrusion by condensin. *Science, 376*(6597), 1087–1094. https://doi.org/10.1126/science.abm4012.

Shen, H., Shih, J., Hollern, D. P., Wang, L., Bowlby, R., Tickoo, S. K., & Hoadley, K. A. (2018). Integrated molecular characterization of testicular germ cell tumors. *Cell Reports*, *23*(11), 3392–3406. https://doi.org/10.1016/j.celrep.2018.05.039.

Sidney, S., Quesenberry, C. P., Jr., Friedman, G. D., & Tekawa, I. S. (1997). Marijuana use and cancer incidence (California, United States). *Cancer Causes & Control*, *8*(5), 722–728. Retrieved from: http://www.ncbi.nlm.nih.gov/entrez/query.fcgi?cmd=Retrieve&db=PubMed&dopt=Citation&list_uids=9328194.

Stevenson, M., Sergeant, E., Nunes, T., Heuer, C., Marshall, J., Sanchez, J., … Reynard, C. (2020). *epiR: Tools for the analysis of epidemiological data.* Retrieved from: https://fvas.unimelb.edu.au/research/groups/veterinary-epidemiology-melbourne. https://www.ausvet.com.au/.

Szutorisz, H., DiNieri, J. A., Sweet, E., Egervari, G., Michaelides, M., Carter, J. M., & Hurd, Y. L. (2014). Parental THC exposure leads to compulsive heroin-seeking and altered striatal synaptic plasticity in the subsequent generation. *Neuropsychopharmacology*, *39*(6), 1315–1323. https://doi.org/10.1038/npp.2013.352.

Tahir, S. K., Trogadis, J. E., Stevens, J. K., & Zimmerman, A. M. (1992). Cytoskeletal organization following cannabinoid treatment in undifferentiated and differentiated PC12 cells. *Biochemistry and Cell Biology*, *70*(10-11), 1159–1173. Retrieved from: http://www.ncbi.nlm.nih.gov/pubmed/1297339.

Tahir, S. K., & Zimmerman, A. M. (1991). Influence of marihuana on cellular structures and biochemical activities. *Pharmacology, Biochemistry, and Behavior*, *40*(3), 617–623. https://doi.org/10.1016/0091-3057(91)90372-9.

Trivers, K. F., Mertens, A. C., Ross, J. A., Steinbuch, M., Olshan, A. F., Robison, L. L., & Children's Cancer, G. (2006). Parental marijuana use and risk of childhood acute myeloid leukaemia: A report from the Children's Cancer Group (United States and Canada). *Paediatric and Perinatal Epidemiology*, *20*(2), 110–118. https://doi.org/10.1111/j.1365-3016.2006.00700.x.

VanderWeele, T. J., & Ding, P. (2017). Sensitivity analysis in observational research: Introducing the E-value. *Annals of Internal Medicine*, *167*(4), 268–274. https://doi.org/10.7326/M16-2607.

VanderWeele, T. J., Ding, P., & Mathur, M. (2019). Technical considerations in the use of the E-Value. *Journal of Causal Inference*, *7*(2), 1–11. https://doi.org/10.1515/jci-2018-0007.

Voirin, N., Berthiller, J., Benhaim-Luzon, V., Boniol, M., Straif, K., Ayoub, W. B., & Sasco, A. J. (2006). Risk of lung cancer and past use of cannabis in Tunisia. *Journal of Thoracic Oncology*, *1*(6), 577–579.

Volkow, N. D., Baler, R. D., Compton, W. M., & Weiss, S. R. (2014). Adverse health effects of marijuana use. *The New England Journal of Medicine*, *370* (23), 2219–2227. https://doi.org/10.1056/NEJMra1402309.

Wang, J., Yuan, W., & Li, M. D. (2011). Genes and pathways co-associated with the exposure to multiple drugs of abuse, including alcohol, amphetamine/methamphetamine, cocaine, marijuana, morphine, and/or nicotine: A review of proteomics analyses. *Molecular Neurobiology*, *44*(3), 269–286. https://doi.org/10.1007/s12035-011-8202-4.

Watson, C. T., Szutorisz, H., Garg, P., Martin, Q., Landry, J. A., Sharp, A. J., & Hurd, Y. L. (2015). Genome-wide DNA methylation profiling reveals epigenetic changes in the rat nucleus accumbens associated with cross-generational effects of adolescent THC exposure. *Neuropsychopharmacology*, *40*(13), 2993–3005. https://doi.org/10.1038/npp.2015.155.

Wen, W. Q., Shu, X. O., Steinbuch, M., Severson, R. K., Reaman, G. H., Buckley, J. D., & Robison, L. L. (2000). Paternal military service and risk for childhood leukemia in offspring. *American Journal of Epidemiology*, *151*(3), 231–240. Retrieved from: http://www.ncbi.nlm.nih.gov/pubmed/10670547.

Wickham, H., Averick, M., Bryan, J., Chang, W., McGowan, L. D., Francios, R., & Yutani, H. (2019). Welcome to the tidyverse. *Journal of Open Source Software*, *4*(43), 1686–1691. https://doi.org/10.21105/joss.01686.

Williams, N., Lee, J., Mitchell, E., Moore, L., Baxter, E. J., Hewinson, J., & Nangalia, J. (2022). Life histories of myeloproliferative neoplasms inferred from phylogenies. *Nature*, *602*(7895), 162–168. https://doi.org/10.1038/s41586-021-04312-6.

Willsher, K. (2018). *Baby arm defects prompt nationwide investigation in France.* Guardian. Retrieved from: https://www.theguardian.com/world/2018/oct/31/baby-arm-defects-prompt-nationwide-investigation-france.

Winick-Ng, W., Kukalev, A., Harabula, I., Zea-Redondo, L., Szabó, D., Meijer, M., & Pombo, A. (2021). Cell-type specialization is encoded by specific chromatin topologies. *Nature*, *599*(7886), 684–691. https://doi.org/10.1038/s41586-021-04081-2.

World Health Organization. (2004). *Tobacco smoke and involuntary smoking: IARC monographs on the evaluation of carcinogenic risks to humans. Vol. 83*. Geneva, Switzerland: World Health Organization. Retrieved from: https://publications.iarc.fr/101.

Yang, F., Tanasa, B., Micheletti, R., Ohgi, K. A., Aggarwal, A. K., & Rosenfeld, M. G. (2021). Shape of promoter antisense RNAs regulates ligand-induced transcription activation. *Nature*, *595*(7867), 444–449. https://doi.org/10.1038/s41586-021-03589-x.

Zhang, Z. F., Morgenstern, H., Spitz, M. R., Tashkin, D. P., Yu, G. P., Marshall, J. R., & Schantz, S. P. (1999). Marijuana use and increased risk of squamous cell carcinoma of the head and neck. *Cancer Epidemiology, Biomarkers & Prevention*, *8*(12), 1071–1078. Retrieved from: http://www.ncbi.nlm.nih.gov/entrez/query.fcgi?cmd=Retrieve&db=PubMed&dopt=Citation&list_uids=10613339.

Zheng, P., Obara, C. J., Szczesna, E., Nixon-Abell, J., Mahalingan, K. K., Roll-Mecak, A., & Blackstone, C. (2022). ER proteins decipher the tubulin code to regulate organelle distribution. *Nature*, *601*(7891), 132–138. https://doi.org/10.1038/s41586-021-04204-9.

Zimmerman, A. M., & Raj, A. Y. (1980). Influence of cannabinoids on somatic cells in vivo. *Pharmacology*, *21*(4), 277–287. Retrieved from: http://www.ncbi.nlm.nih.gov/pubmed/6252564.

Chapter 7

Cannabis, cognitive impairment and car crash risk

Ulrich W. Preuss[a], E. Hoch[b], and J.W.M. Wong[c]
[a]*Departments of Psychiatry, Psychotherapy and Psychosomatics, RKH Hospital Ludwigsburg, Martin-Luther-Universität Halle-Wittenberg, Ludwigsburg, Germany,* [b]*IFT Institut für Therapieforschung, München, Germany,* [c]*Psychosomatik Fachklinik Schoemberg GmbH, Schoemberg, Germany*

Abbreviations

BAC	blood alcohol concentration
CBD	cannabidiol
CI	confidence interval
DUIC	driving under the influence of cannabis
OR	odds ratio
ES	effect size
PK	pharmacokinetics
SDLP	standard deviation of lateral position
THC	delta-9 tetrahydrocannabinol
THCCOOH	11-OH-THC metabolites of THC

Introduction

Next to alcohol, cannabis is the second most frequently found substance in the bodies of drivers involved in fatal and non-fatal accidents (Pearlson et al., 2021). Rates of driving under the influence of cannabis rose in recent years (Asbridge et al., 2012). For instance, the DRUID (Driving Under the Influence of Drugs, Alcohol and Medicines) project reported on 50,000 drivers from 13 different countries of whom 1.32% used cannabis (EMCDDA, 2012). On weekends, the rates of positively screened cannabis users in traffic were 10%–12% and in subjects involved in a car crash 26%–27% (Compton, 2017), while 0.5%–7.6% of persons involved were severely injured.

Similarly, in Washington State, the average yearly percentage of DUIC (driving under the influence of cannabis) cases positive for delta-9-tetrahydrocannabinol (THC) and its principal metabolite THCCOOH increased from 19.1% and 27.9% in 2009–12 to 24.9% and 40.0%, respectively, after the legalization of cannabis (Couper & Peterson, 2014).

"Cannabis-impaired driving" describes the impairment caused by cognitive and psychomotor effects of THC, which negatively influences a driver of a motor vehicle after THC consumption. In contrast, a "cannabis-positive driver" is someone driving a motor vehicle with any detectable THC concentration in blood, oral fluid, or urine with/without showing impairments in his driving. DUIC is a judicial term that refers to a driver exhibiting a measured reduction in cognitive or psychomotor skills in conjunction with a defined THC concentration in blood, oral fluid, or urine (EMCDDA, 2018).

Recent studies and review reported that results on the relationship between driving performance and traffic crash risk under the influence of cannabis were inconsistent (Hartman et al., 2015; Elvik, 2013; Ramaekers et al., 2004).

Methodological heterogeneity may explain these mixed findings. As summarized in a review and meta-analysis, (Asbridge et al., 2012), several study types need to be considered for a comprehensive evaluation of the relationship between cannabis use, driving skills, and car crash risk (e.g., sample surveys, laboratory experiments, and epidemiological studies like case-control and their variant, "culpability" studies, Elvik, 2013).

Cannabis Use, Neurobiology, Psychology, and Treatment. https://doi.org/10.1016/B978-0-323-89862-1.00027-1

Each of the study approaches has strengths and weaknesses. On the one hand, experimental studies have a poor or a lack of validation of simulated driving scenario with on-the road- driving situations. Direct intraindividual comparisons between cannabis-induced effects in simulated and on-the-road driving are, indeed, sparse (Bondallaz et al., 2016).

On the other hand, it is challenging in epidemiological studies to obtain adequate proportion of cannabis users in their samples. Some studies rely on self-reports that may underestimate the actual fraction of cannabis users (Preuss et al., 2021). Only a minority of all studies assesses cannabis content in blood or other body tissues or fluids at the time of the crash. It is unclear whether the cannabis consumption is occasional or frequent, or when the last cannabis intake occurred prior to the crash. Among other factors that may contribute to crash risk under the influence of cannabis are the use of additional substances (e.g., alcohol), the frequency, and chronicity of cannabis use. Again, only a fraction of studies reports a combined assessment of various legal and illicit substances or controls for other confounding factors (Elvik, 2013).

The aim of this narrative review is first to present results from studies on impairment of driving-relevant cognitive abilities due to cannabis and subsequent reduction of driving skills from laboratory and simulator settings. These include several influencing factors like dose-response relationship, development of tolerance, the significance of THC biomarkers, the role of CBD (Cannabidiol), and medical cannabis. The relevance of cannabis-induced impairment on car crash risk is briefly reviewed by reporting on results from a series of meta-analyses. Results from case-control (culpability) studies are presented regarding the relationship between Cannabis use, Cannabis content in body tissues and fatal and nonfatal crashes.

Cannabis-induced impairments: Results from lab and simulation studies

Laboratory studies of the effects of cannabis on driving date back many decades. The findings of these early studies need to be considered within the context of the changing landscape of nonmedical cannabis use. Over time, it is known that the potency of recreational cannabis has steadily increased, and cannabis now has upward of 10% THC, with some potency estimates as high as 19%–22% (Cascini et al., 2012; Cash et al., 2020; Chandra et al., 2019). Thus, the utility of some of the early research using low potencies of THC (e.g., 1%–3%) is limited.

Results from these and other studies show that cannabis-induced cognitive impairments include increased reaction time, mild and short-lived reduced motor coordination, reduced short-term memory, temporal distortion, poor divided attention, and poor decision making in rapidly changing situations (Capler & Bilsker, 2017).

SDLP and cannabis dose

In lab and simulator studies, standard deviation of lateral position (SDLP) appeared to be one of the most sensitive road-tracking measures for revealing THC-induced impairment in participants under the influence of a 20mg THC dose (Cook et al., 2017). Subsequent studies have demonstrated similar results after a range of 13–38mg smoked THC doses (Watson & Erickson, 2019), or after ingestion of 16.5 or 45.7mg THC decoction or 20mg synthetic THC (dronabinol) (Liguori et al., 2002). Other experiments, however, found no significant increases in SDLP after a 13mg dose (Tank et al., 2019) or after a 22.9mg dose (Arkell et al., 2019). Such discrepancies could be explained by the variability in subjects' cannabis history as well as in the time-lapse between the end of cannabis consumption and the beginning of the task (Brands et al., 2021).

Alteration of subject's behavior under the influence of cannabis in an unexpected situation showed lesser risk taking (Smink et al., 2008). More recent studies confirmed a dose-related increase in headway mean and standard deviation after smoking a range of 13–38mg THC doses, as well as a decrease in mean speed (Arkell et al., 2019; Attwood et al., 1980; Cook et al., 2017; Tank et al., 2019; Watson & Erickson, 2019).

Further, previous lab and simulator studies reported that cannabis peak impairment on driving performance is evident 20–40min following inhalation (Sewell et al., 2009), even as THC blood levels are long past their peak and continue to diminish. By 1.0–2.5h postinhalation, behavioral impairment is still present but already beginning to decrease (Sewell et al., 2008). Because of this time-limited impairment, several authors suggest that following acute use, cannabis consumers should wait a minimum of 3–4h before attempting to drive (Fischer et al., 2011). A recent report (Arkell et al., 2020) used a double-blind, within-participant randomized clinical trial with an active THC dose of 13.75mg consumed by inhalation following vaporization and measurements of driving performance in an actual vehicle on a real road. The major findings were that weaving (assessed by standard deviation of lane position/SDLP) was significantly greater at 40–100min, but not at 240–300min postdose. Subjects' self-rated confidence to drive safely tracked poorly with actual measured SDLP, with participants significantly rating themselves as more impaired 4–5h following active THC compared to placebo, despite SDLP being unimpaired by that time. It is important to note that consumption of "edibles," with delayed onset and greater persistence of intoxication effects, and dosing via smoking or vaporization at a higher dose than used in this study (Arkell et al., 2020) would likely result in a greater period of impairment.

Dose-response relationship

Lab and simulator studies also report that effects of cannabis on cognition and motor coordination are mild at low doses and increase in magnitude at higher doses (Capler & Bilsker, 2017; Preuss et al., 2021; Weinstein et al., 2008). This conclusion is supported by many studies that show that the odds ratio for accidents increases with blood level of THC, and experimental studies showing that impairment as measured by SDLP gradually increases with increasing doses of THC (Capler & Bilsker, 2017) as also case-control studies.

Tolerance

One issue that complicates studying the relationship between dose and impairment is the differing effects of cannabis depending on the user's history of use. This phenomenon is known as *tolerance*. Current evidence indicates that in general, infrequent users are more likely to be negatively impacted by use than frequent users, who show a degree of resistance (or tolerance) to the negative cognitive effects of cannabis (Desrosiers et al., 2015; Ramaekers et al., 2009; Ramaekers et al., 2011).

This effect has been demonstrated in experimental studies comparing impairment as measured by neurocognitive tests in infrequent versus frequent cannabis users. Frequent users were defined as those individuals consuming cannabis four or more times a week and infrequent users as using cannabis three or less times a week (Hartman et al., 2015; Ramaekers et al., 2009). Accordingly, infrequent users showed impairment on a variety of neurocognitive tests after being administered cannabis, while frequent users showed no significant impairment effect at the same dose (a single cigarette between 6.8% and 13% THC content).

Tolerance to impairment effects was also noted in an on-road study of driving ability (Bosker et al., 2012). In this study, regular cannabis users (who mostly used cannabis daily ≥ 4 times a week) and occasional cannabis users (who used cannabis 5–36 times in a year) were administered 10 and 20 mg of oral THC on separate occasions prior to participating in an on-road driving session. They found that while impairment as measured by SDLP was seen in both groups, the magnitude of the impairment effect was stronger in occasional cannabis users.

Also, there was greater variability in effect amongst the regular users, suggesting differing degrees of tolerance. While tolerance is mostly a factor of dose, it may also be related to the strain of cannabis used. Different strains of cannabis have different ratios of cannabinoids and terpenoids, and tolerance to the psychoactivity of one strain that is used repeatedly may not be present with the use of another strain that has a different array of compounds (Russo and Taming, 2011). This may be an additional factor to consider in relation to driving impairment.

THC biomarkers and driving impairment

Previous studies investigating the relationships between THC-related biomarkers and impairment of driving and driving-related cognitive skills have generated somewhat inconsistent results (Arkell et al., 2021; Ramaekers et al., 2006; Schlienz et al., 2020; Vandrey et al., 2017). For example, Arkell et al. (2021) found no significant relationship between the change (from placebo) in SDLP and plasma or oral fluid THC concentrations (differences $= -0.011$ and -0.033) following vaporization of THC (13.75 mg; $n = 14$), whereas Schlienz et al. (2020) observed a number of significant correlations (ranging from weak to strong in strength) between the change (from placebo) in performance on two discrete cognitive tests and blood THC, 11-OH-THC, and 11-COOH-THC concentrations following oral ingestion of THC (10, 25 and 50 mg; $n = 17$, metabolism of THC in humas see Fig. 1). Other studies have reported a mixture of weak, significant and nonsignificant correlations between similar biomarkers and outcome measures (Ramaekers et al., 2006; Vandrey et al., 2017). This inconsistency could partly reflect the fact these studies have administered fixed doses of THC, via fixed routes, at fixed times relative to their performance test(s).

The authors of a recent meta-analysis (McCartney et al., 2021) concluded that their results on the relationships between THC-related biomarkers, subjective intoxication, and impairment of driving and driving-related cognitive skills indicate that blood THC, 11-OH-THC, and 11-COOH-THC concentrations, oral fluid THC concentrations, and subjective ratings of intoxication are relatively poor indicators of cannabis-induced impairment. The use of fixed limits as a means of identifying cannabis-impaired drivers should therefore be re-considered. Indeed, it seems there is a significant risk of unimpaired individuals being mistakenly identified as "cannabis impaired" (and vice versa) under this approach.

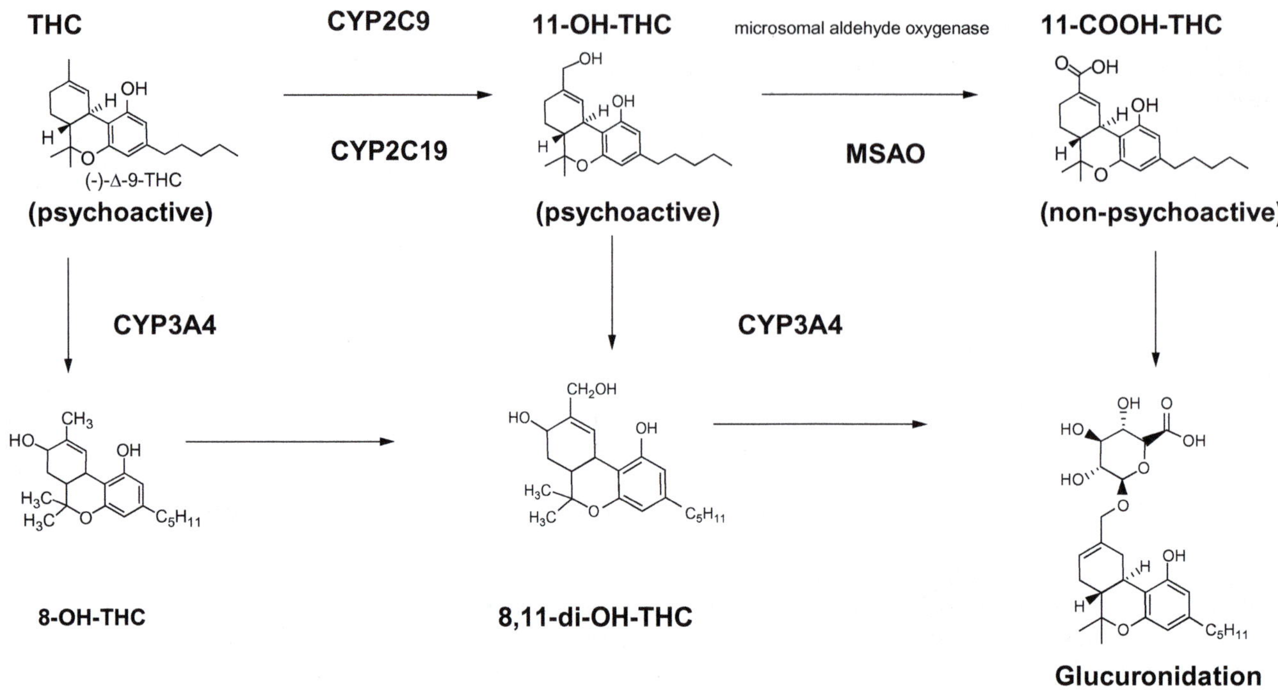

FIG. 1 Metabolism of THC in humans. *(Derived from Hoffmann, G., Studer, A., & Schürenkamp, J. (2020) Investigation of phase II metabolism of 11-hydroxy-Δ-9-tetrahydrocannabinol and metabolite verification by chemical synthesis of 11-hydroxy-Δ-9-tetrahydrocannabinol-glucuronide.* International Journal of Legal Medicine, 134, *2105–2119; Musshoff, F., & Madea, B. (2006) Review of biologic matrices (urine, blood, hair) as indicators of recent or ongoing cannabis use.* Therapeutic Drug Monitoring, 28(2), *155–163.)*

Role of CBD (Cannabidiol)

The second relevant ingredient of phytocannabinoids, beside THC, only recently came into focus regarding its effect on cognitive impairment and decreased driving abilities. Cannabidiol (CBD), administered simultaneously in vaporized cannabis, has been reported not to significantly diminish THC-induced driving impairment (Arkell et al., 2019). The assumption that CBD may alter either the pharmacokinetics (PK) of THC or modulate behavioral effects of the latter (Liu et al., 2020; Solowij et al., 2019) was unproven. A population PK model of THC and CBD (Liu et al., 2020) was proposed. When high-dose CBD was inhaled at the same time as THC, the systemic availability of the latter decreased significantly. Interestingly, in the same set of experiments, frequent users of cannabis appeared to have higher systemic availability of both THC and CBD when high-dose CBD was administered. It is worth noting that the majority of driving studies has been performed on inhaled cannabis in younger subjects, and there is a paucity of studies on driving performance following oral administration of the drug, where there is likely to be increased variability of both onset and duration of impairment. In addition, despite increasing use of cannabis in individuals aged over 60 (typically to help manage insomnia and chronically painful conditions), there are very few studies quantifying cannabis-related driving impairment in such older individuals following any route of drug administration, although age-related alterations in PK are likely.

THC and driving after the medical use of cannabis

Many jurisdictions internationally have adopted per se laws in an attempt to curb DUIC, regardless of whether cannabis is used recreationally or for medical reasons. Despite these laws, courts in some jurisdictions of the world have seen challenges, in which medical users assert that their driving is unaffected by cannabis use due to the development of a drug tolerance associated with more frequent use. However, to date, research specifically on the effects of medical cannabis use on driving is scarce and highly relevant. The prevalence of medical cannabis use in Canada, for example, is estimated 13% of all cannabis users of whom only half had medical authorization (Kruger & Kruger, 2019). It is further estimated that over half of those who use cannabis for medical purposes have driven within 2 h of using cannabis (Bonar et al., 2019), and most of these users indicated that there was "no risk" or "slight risk" of driving (Di Ciano et al., 2020).

In the same study, the effects of cannabis on simulated driving in participants who use cannabis for medical reasons were investigated (Di Ciano et al., 2020), and no difference in decreased speed between recreational and medical users was detected. The levels of THC in the blood did not show any evidence of tolerance when compared to THC levels after smoking cannabis for recreational purposes. Thus, medical users of cannabis should exercise caution in driving after the use of cannabis for medical purposes (Di Ciano et al., 2020). Eventually, despite some evidence of cognitive impairment due to medical cannabis, the question remains open, whether fatal or nonfatal car crash risk is significantly influenced by use of medical cannabis.

Meta-analyses: Cannabis-induced cognitive changes and crash risks

Based on the results of cannabis-induced driving ability cognitive impairments, the subsequent question is, whether these also influence the risk for fatal- and nonfatal car crash risk and whether this is further evidence of a dose-response relationship. Study results that may answer this question come from several meta-analyses during the last decade. Several of these meta-analyses have been published that conclude that the acute use of cannabis does increase collision risk (Asbridge et al., 2012; Li et al., 2012, Elvik, 2013, Hostiuc et al., 2018), although the extent of the increase differed across studies (Review by Preuss et al., 2021).

Evidence from case-control (culpability) studies: Relationship between different THC blood concentrations and car vehicle crashes

Another question is, how different doses of THC as the main psychoactive component of cannabis consumed and in body fluids and tissues influence the car crash risk. Further, it is of relevance, if DUIC also has a differential effect on fatal- and nonfatal car crash risk.

Five large case-control studies (Brubacher et al., 2019; Drummer et al., 2004, 2020; Laumon, Gadegbeku, Martin, Biecheler, and the SAM Group, 2005) assessed this question. The first of these studies (Drummer et al., 2004) investigated $n = 3398$ fatal crashes under the influence of alcohol and drugs from three Australian states, including alcohol- and substance-free controls. Using logistic regression analyses; influence of alcohol, cannabis, and other substances as well as other factors like age and gender on likelihood of culpability were determined. Drivers with blood THC concentrations of equal or more than 5 ng/ml had a greater and more statistically significant odds ratio (OR 6.6, 95% CI 1.5–28.0). In their study, the estimated odds ratio is greater than that for drivers with a blood alcohol concentration (BAC) of 0.10%–0.15% (OR 3.7, 95%CI 1.5–9.1).

The same research group recently conducted a culpability study that included 5000 injured motor vehicle drivers who underwent a comprehensive blood toxicology testing (Drummer et al., 2020). The sample included 1,000 drivers for each of 5 years from $\geq$5000 to 6000 drivers injured and taken to hospital in Victoria, Australia. In a logistic regression, drivers under the influence of THC showed a modest increase in culpability odds, when concentrations of all assessed substances were considered vs. controls (OR 1.9, 95% CI 1.2–3.1, $P = 0.007$). The increase in odds was most obvious at higher blood THC concentrations. At 5 ng/mL and above the OR was 3.2 ($P = 0.01$), and at THC concentrations of equal or more than 10 ng/mL, the OR was 10 ($P = 0.03$). At 1–5 ng/ml, the OR was 1.6 (95% CI 0.9–2.7), 5–9 ng/ml 1.9 (95% CI 0.7–5.0), 5 ng/ml and above the OR was 3.2 (95% CI 1.3–7.2, $P = 0.01$), and at THC concentrations of 10 ng/mL and above the OR was 10 (95% CI 1.3–82, $P = 0.03$). These results indicated increasing odds of culpability with rising concentration. THC was the third most prevalent drug detected in 11.1% of all drivers, following alcohol and stimulants. Only 1% of the 5000 drivers had THC only at 10 ng/mL or higher. The authors conclude that their results supported the hypothesis that elevated odds of culpability correlate with increasing THC levels.

Culpability studies from France

The first study (Laumon, Gadegbeku, Martin, Biecheler, and the SAM Group, 2005) conducted a case-control study on cannabis intoxication and fatal road crashes in France. They analyzed $n = 10{,}748$ drivers, with known drug and alcohol concentrations, involved in fatal crashes in France from October 2001 to September 2003. A total of 6,766 of these were considered culpable at the crash while $n = 3306$ drivers served as controls. In addition, $n = 681$ drivers were positive for cannabis (cases 8.8%, controls 2.8%), including 285 with an illegal blood alcohol-concentration ($\geq$0.5 g/l). Positive cannabis detection was associated with increased risk of responsibility (OR 3.32, 95% CI 2.63–4.18). A significant dose effect was identified; the odds ratio increased from 2.18 (1.22–3.89) for $0 < \text{THC} < 1$ ng/ml to THC$\geq$5 ng/ml OR 4.72 (3.04–7.33), uncorrected. The effect of cannabis remains significant after adjustment for different cofactors, including alcohol, with which no statistical interaction was observed. At least 2.5% (1.5%–3.5%) of fatal crashes were estimated as being attributable to cannabis, compared with 28.6% for alcohol (26.8%–30.5%). The authors concluded that driving under the influence of cannabis increases the risk of involvement in a fatal car crash, while the risk in their study was lower than that with positive blood alcohol concentrations.

More than a decade later, a similar study on another French cohort was published (Martin et al., 2017). This group of researchers included n=4,059 drivers. More than 300 of their characteristics, including DUIC and fatal crash risk, were obtained. The proportion of persons driving under the influence of alcohol was 2.1% (95% CI: 1.4 ± 2.8) and under the influence of cannabis 3.4% (2.9%–3.9%). Drivers under the influence of cannabis multiplied their risk for causing a fatal crash by 1.65 (1.16 ± 2.34). Strength of the study is certainly its presentation of separate crash ORs for various doses of cannabis detected in blood samples, after adjusting for age, gender, vehicle category, and time of crash, which replicated results from a previous publication of the same research group (Laumon, Gadegbeku, Martin, Biecheler, and the SAM Group, 2005) and the above-mentioned Australian study (Drummer et al., 2004). Surprisingly, another finding of the second French sample is that there was no significant alcohol × cannabis interaction on crash risk in their study.

Recent studies

A more recent study (Brubacher et al., 2019) conducted a responsibility analysis in a case-control sample to determine whether drivers injured in motor vehicle collisions who tested positive for THC or other drugs are more likely to have contributed to the crash than those who tested negative. Participants included $n = 2318$ injured drivers who required blood tests for clinical purposes following a motor vehicle collision, of whom 8.3% were screened positively for THC and 14.4% for alcohol. In addition, excess whole blood remaining after clinical use was obtained and broad-spectrum toxicology testing performed. To analyze data, unconditional logistic regression to determine ORs of crash risk for drivers with $0 < \text{THC} < 2$ ng/ml, $2\,\text{ng/ml} \leq \text{THC} < 5$ ng/ml and $\text{THC} \geq 5$ ng/ml (all vs. THC = 0 ng/ml). Risk estimates were adjusted for age, gender, and presence of other impairing substances. Odds ratios for all three ranges of blood tested THC were low and not statistically significant. Alcohol × cannabis interactions were detected for positive BAC × THC < 2 ng/ml OR = 1.75 (0.37, 17.1) and × THC ≥ 2 ng/ml OR = 1.62 (0.34, 15.7) which did not reach statistical significance. Unadjusted OR for car crash risk under any THC blood level was 1.13 (1.03–1.28). This study, like other culpability and case-control studies, relied on blood samples instead of self-reports. Additional strengths are the assessment of alcohol and recreational drug use. However, in comparison to the three previous case-control/culpability studies which included fatal crashes, nonfatal motor crashes with injuries only were analyzed in this study (Brubacher et al., 2019).

In general, if THC blood levels are compared regarding fatal and nonfatal crashes, the odds ratios for any THC blood levels are similar (fatal crashes: 2.7 ng/ml (Drummer et al., 2004), 1.78 ng/ml (Laumon, Gadegbeku, Martin, Biecheler, and the SAM Group, 2005), 1.65 ng/ml (Martin et al., 2017) vs. nonfatal crashes 1.13 ng/ml (Brubacher et al., 2019), 1.9 ng/ml (Drummer et al., 2020), while the levels in fatal crashes tended to be higher. A previous meta-analysis and systematic review, which excluded low-quality studies, reported cannabis-associated risk for nonfatal crashes (OR = 1.74; 95% CI = 0.88–3.46) and for fatal crashes (OR = 2.1; 95% CI = 1.31–3.36) (Asbridge et al., 2012) and support in general the findings for the five culpability studies.

If meta-analytical statistics across the 5 studies are employed, a general effect size for car crash risk of 0.178 for any cannabis detected is found (see Fig. 2). For THC values greater than 5ng/ml, the effect size increases to 0.640 indicating a stronger relationship for higher THC blood values with car crashes (see Fig. 3).

Summary of case-control study findings

Taken together, there is a potential relationship between high (but not low doses) THC blood levels and car crash risk. Results from the Australian Study (Drummer et al., 2004) suggested that cannabis showed a biological gradient with higher doses having a higher OR of culpability in fatal crashes. This significant dose effect was also reported from the first French (Laumon, Gadegbeku, Martin, Biecheler, and the SAM Group, 2005) and the Canadian Study (Brubacher et al., 2019), while in the second French study (Martin et al., 2017), the ORs show an inverse U-curve trend with a maximum at THC concentrations between >2 and <5 ng/ml, after controlling for alcohol intoxication. All authors across studies consistently argue that their studies yield a marked dose effect and a potential causal role of DUIC in particular in fatal crashes.

However, studies continue to appear suggesting that more research is needed to understand how cannabis might increase collision risk (Brubacher et al., 2019). According to several authors, research has yet to establish the characteristics of cannabis-involved collisions, or the impact of cannabis on injury severity or fatality resulting from collisions (Smink et al., 2005; Brands et al., 2021).

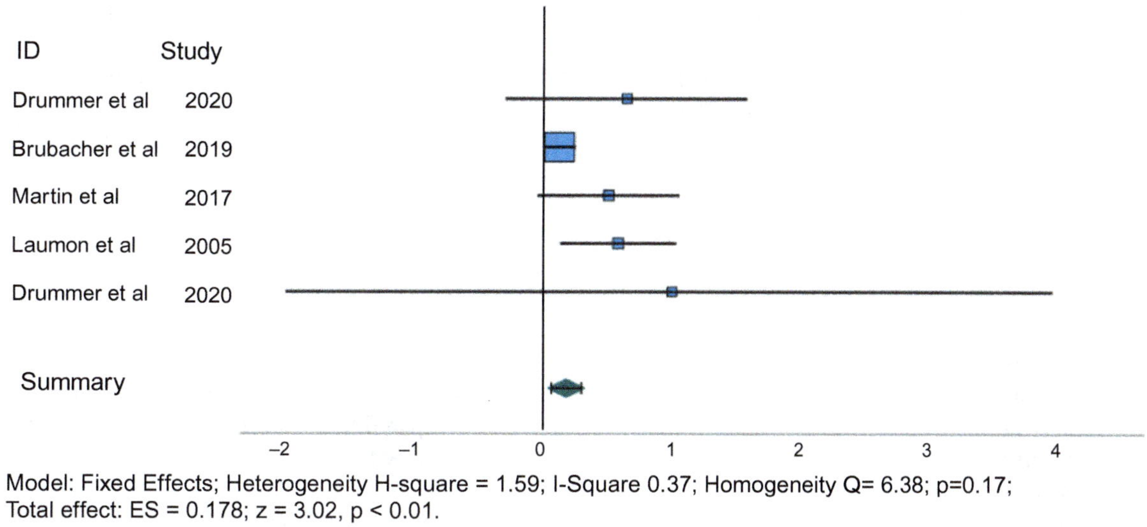

FIG. 2 Forest plot of culpability studies any THC and car crashes (log odds ratio and standard error of 95% CI). Log odds ratios and standard errors of 05% CI of each study were computed according to https://www.escal.site/ and https://handbook-5-1.cochrane.org/chapter_7/7_7_7_2_obtaining_standard_errors_from_confidence_intervals_and.htm. Meta-analytical statistics were computed using SPSS (Statistical Package for social sciences), Ver. 28.

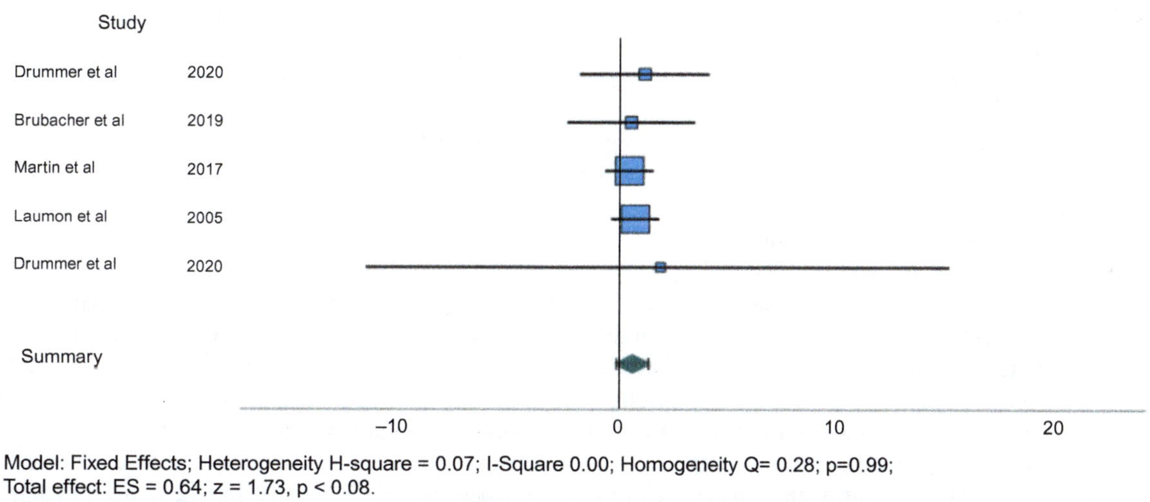

FIG. 3 Forest plot of culpability studies THC > 5 ng/ml and car crashes (log odds ratio and standard error of 95% CI). Log odds ratios and standard errors of 05% CI of each study were computed according to https://www.escal.site/ and https://handbook-5-1.cochrane.org/chapter_7/7_7_7_2_obtaining_standard_errors_from_confidence_intervals_and.htm. Meta-analytical statistics were computed using SPSS (Statistical Package for social sciences), Ver. 28.

Discussion

In this review, we first reported impairment of driving-related cognitive skills by cannabinoids and their potential relationship to (fatal and nonfatal) accidents. Second, we compared results from lab and simulator research, meta-analyses, and case-control (culpability) studies, to determine how cannabis use affects car crash risk in the real world.

To analyze cognitive impairment of cannabinoids, the first step involves the use of laboratory studies and driving simulators. The simulators range anywhere from videogame-like apparatuses linked to a typical desktop-sized computer display screen, steering wheel, and gas/brake pedals at one end of the spectrum, to an actual, repurposed, instrumented motor vehicle chassis on a motion base, situated in front of a wall-sized projection screen at the more sophisticated extreme. The advantages of such setups are obvious: subjects can be intoxicated with the study drug/placebo in the lab and subsequently asked to drive in a number of preprogrammed scenarios to quantify their degree of impairment.

Simulators in general provide a controlled, safe environment that theoretically translates into real-world driving performance.

Furthermore, because of the ease of manipulating the environment, driving scenarios can be easily constructed that would be unsafe or impossible to create on a real roadway.

In summary, these laboratory and simulator test studies reported that cannabis may acutely impair several driving-related skills in a dose-related fashion but that the effects between individuals vary more than they do with alcohol because of tolerance, differences in smoking technique, and different absorptions of THC (Sewell et al., 2009).

The majority of lab and simulator studies also report that cannabis smoking (THC-dominant and THC-CBD-equivalent) increases lane weaving after 40–100 min following vaporization and show impaired cognitive function consistently as well as critical-tracking tasks, reaction times, divided-attention tasks, and lane-position variability. All of these impaired functions may increase car crash risk. In general, however, there is no clear overall relationship with THC blood or serum levels and driving skills or crash risk from experimental studies, even if time of use and duration of consumption are considered.

Further, several studies indicate that chronic users of cannabis may have developed tolerance to the effects of cannabinoids and may have therefore less impairment of driving abilities compared to occasional users (Bosker et al., 2012). Evaluation of current driving ability impairment and potential increased risk for car crashes have to consider occasional vs. chronic use of cannabinoids too.

From the perspective of analysis of body tissues and fluids, the main THC metabolite, 11-nor-9-carboxy-THC (THC-COOH), is not psychoactive and persists in blood and urine long after impairment resolves (see Fig. 1: metabolism of Delta-9-THC). Thus, THC-COOH provides evidence of previous cannabis exposure but does not necessarily indicate impairment or recent use (Huestis et al., 2005). The active ingredient behind most of the effects of cannabis that impair driving ability is THC (metabolism of THC is illustrated in Fig. 1). Metabolites such as THC-COOH are present and detectable for a significant time after consumption but lack any proven psychoactive effects which impair driving ability.

Beside the finding that cannabis-induced cognitive effects vary in a dose-related fashion, there is also evidence that they are more pronounced with highly automatic driving functions, but more complex tasks that require conscious control are less affected. Because of both this and an increased awareness that they are impaired, cannabis smokers tend to compensate effectively for their impairment by utilizing a variety of behavioral strategies such as driving more slowly, passing less, and leaving more space between themselves and cars in front of them (Sewell et al., 2009).

As well, more investigation of the potential impairing effects of medical cannabis on therapeutic users is also warranted. Recent research indicates that, not surprisingly, effects of medical cannabis on driving abilities does not differ from the impairments found from use of phytocannabinioids (Di Ciano et al., 2020).

Further, it is important to recognize that an important compound of cannabis-products used is Cannabidiol (CBD). As for driving skills, it is a substance with sedative effects and therefore may play a role in impairing driving abilities and increase risks for accidents. However, the assumption that CBD may alter either the PK of THC or modulate behavioral effects of the latter is hitherto unproven. However, studies investigating the effects of CBD alone (without THC) on cognition, driving abilities, and car crash risks are still lacking (Liu et al., 2020), in particular in individuals aged 60 or over who may use cannabinoids more frequently to manage insomnia or chronic pain conditions.

Previous reviews also pointed out that the impairment of cognitive functioning and driving abilities influence car crash risk (Brands et al., 2021). Several meta-analyses have been published that conclude that the acute use of cannabis does increase collision risk (Asbridge et al., 2012; Li et al., 2012; Rogeberg & Elvik, 2016), although the extent of the increase differed across studies. Case-control (culpability) studies include data on potential dose-response relationships. Further, two culpability studies were conducted in nonfatal crashes (Brubacher et al., 2019; Drummer et al., 2020), while the majority included fatal car accidents (Drummer et al., 2004; Laumon, Gadegbeku, Martin, Biecheler, and the SAM Group, 2005; Martin et al., 2017). Therefore, there are no data for a direct comparison of the risk for fatal and nonfatal crashes. While THC blood levels tended to be higher in fatal vs. nonfatal crashes, this relationship has to be critically considered in context of the limitations of case-control and culpability studies. First, the assessment of the degree of intoxication for cannabis is a more difficult task than for alcohol (Brubacher et al., 2019; Drummer et al., 2004). As far as cannabis is concerned, after smoking and vaporization, there is a delay in maximal cannabis effects compared to the peak THC blood concentration after consumption, with peak effects after >30 min depending upon study design and time of testing. Effects on cognition and psychomotor function do not decline as quickly as THC concentration decreases. Thus, when smoking a "joint," whole-blood THC concentrations typically peak at >100 ng/ml during smoking and then drop so rapidly that THC is usually <2 ng/ml within 4 h of a single acute exposure. Psychotropic effects typically peak at 20–30 min and resolve by 4 h. Ingesting cannabis delays the onset and extends the duration of these effects. Therefore, it is difficult to determine, whether increased THC blood levels relate to cognitive and driving ability impairment and subsequent increased crash risks. Therefore, measures of THC and its nonpsychoactive metabolite THC-COOH within hours of an accident suffer from several limitations and may not reflect driving skill impairment.

Second, even higher THC concentrations of >5 ng/ml in recent studies yielded higher risks of injuries and fatal concentrations, but not all studies reported statistically significant relationships. This finding is supported by meta-analytical statistics (see Figs. 2 and 3), where an effect size of 0.17 was found for any THC and a higher effect size of 0.64 was detected for THC greater 5 ng/ml, see Figs. 2 and 3). Instead, the assessment of both biological measures and psychomotor characteristics may render evaluation of driving fitness more accurately (more valid to "real" impairment) and is probably applicable in practical traffic control situations.

Third, future experimental and laboratory research on cannabinoid's effects on fitness to drive should consider designs which include measures of psychomotor and cognitive functions like reaction times and decision making after long and monotonous drives and divided attention tasks. These cognitive characteristics may be most impaired by cannabis according to available research (Preuss et al., 2021).

Also, the results of case-control and culpability studies must be compared with findings from experimental and laboratory studies. As mentioned above, laboratory and simulator studies certainly suffer from the limitation that they are conducted in an artificial environment. Thus, experimental data show that drivers attempt to compensate for cannabis intoxication by slower driving after smoking cannabis, but their control deteriorates with increasing task complexity (Hartman et al., 2015). These behaviors limit the generalizability of experimental study results to authentic traffic situations (Asbridge et al., 2012).

Important questions also remain as to the duration of the effects of cannabis on driving and the time course of safe use of cannabis. Time course studies are especially important for different routes of administration that may have different PK.

Also, more research is needed to understand gender differences in the effects of cannabis. Other questions include the comparative pharmacology of different modes of administration of THC (smoking vs. edibles) and different doses and concentrations of THC in phytocannabinoids used (increased THC content in cannabis strains during the last decades) and the extent to which regular or frequent uses may develop and display tolerance to the impairing effects of cannabis on driving behavior.

In the meantime, patients who smoke marijuana should be counseled to have a designated driver if possible, to wait at least three hours after smoking before driving. They must know that marijuana is particularly likely to impair monotonous or prolonged driving and that mixing marijuana with alcohol will produce much more impairment than either drug used alone. According to the NHTSA, 72% of all alcohol-related fatalities are in unrestrained drivers (in comparison with only 45% in nonalcohol-related motor vehicle fatalities) (NHTSA, 2005), and it is reasonable to suspect that similar lack of attention to use of seat belts is true of cannabis-intoxicated drivers as well.

Summary

The review reports state-of-the art evidence on the relationship between cannabis use, cannabis-induced cognitive impairment, and traffic crash risks. Results on cognitive impairment from lab and simulator studies indicate that cognitive impairment includes increased reaction time, mild and short-lived reduced motor coordination, reduced short-term memory, temporal distortion, poor divided attention, and poor decision making in rapidly changing situations. Subsequent crash risk is certainly related to dosage used and tolerance while intoxicated individuals also use compensatory behavior due to their impairment. SDLP appeared to be one of the most sensitive road-tracking measures while THC biomarkers were found to be poor indicators of impaired driving skills. Further, there is not much information on the influence of CBD on driving abilities. Potential dose-response relationships on fatal and nonfatal car crash risks show that here may be a relationship for hither THC doses.

Applications to other areas

As indicated in the review, use of cannabis is widespread in societies and cultures over the world, so is the potential impairment of cognition and driving skills and the potential increased risk for car crashes. While there is some evidence that cannabis dose used and subsequent impairment are related to an increased risk for fatal and nonatal cases, certainly further research is needed, e.g., across several age groups including medical and, nonmedical cannabis formulations and different ways of applications. Further, various contents of THC, CBD and other ingredients, their combination with other substances (e.g., alcohol) may also influence cognition, driving abilities, and car crash risk. For practical use, more research on combined assessment of blood THC and cognitive abilities may be more appropriate to assess the ability to drive. Discussion on limits of THC blood levels and driving ability should certainly consider that blood levels alone do not reflect driving ability impairment and car crash risk.

Key facts

- Recreational and medical cannabis use lead to cognitive impairment in several domains
- These impairment results in impaired driving skills, dependent on cannabis dose, and duration and frequency of use
- Blood levels of THC and metabolites are no good indicators of driving ability and its impairment.
- Risk of fatal and nonfatal crashes are related to THC doses at least in high concentrations
- In culpability studies, THC levels were higher in fatal compared to nonfatal crashes. However, not direct comparison was conducted.

Summary points

- Cannabis results in cognitive impairment
- Cognitive impairment may in part influence and driving skills and lead to compensatory behavior
- Impaired driving skills may increase risk for fatal and nonfatal car crashes
- While medical cannabis influences the car crash risk in a similar fashion like recreational cannabis, the role of CBD is poorly studied.
- Results from culpability studies indicate that cannabis blood levels in fatal crashes may be higher compared to nonfatal crashes while no study conducted a direct comparison.

References

Arkell, T. R., Lintzeris, N., Kevin, R. C., Ramaekers, J. G., Vandrey, R., Irwin, C., Haber, P. S., & McGregor, I. S. (2019). Cannabidiol (CBD) content in vaporized cannabis does not prevent tetrahydrocannabinol (THC)-induced impairment of driving and cognition. *Psychopharmacology*, *236*, 2713–2724.

Arkell, T. R., Spindle, T. R., Kevin, R. C., Vandrey, R., & McGregor, I. S. (2021). The failings of per se limits to detect cannabis-induced driving impairment: Results from a simulated driving study. *Traffic Injury Prevention*, *22*(2), 102–107.

Arkell, T. R., Vinckenbosch, F., Kevin, R. C., Theunissen, E. L., McGregor, I. S., & Ramaekers, J. G. (2020). Effect of cannabidiol and Δ9-tetrahydrocannabinol on driving performance: A randomized clinical trial. *Journal of the American Medical Association*, *324*(21), 2177–2186.

Asbridge, M., Hayden, J. A., & Cartwright, J. L. (2012). Acute cannabis consumption and motor vehicle collision risk: Systematic review of observational studies and meta-analysis. *British Medical Journal*, *344*, e536.

Attwood, D. A., Williams, R. D., McBurney, L. J., & Frecker, R. C. (1980). Braking performance of drivers under the influence of alcohol and cannabis. In *Proceedings of the human factors society annual meeting, Los Angeles, CA*.

Bonar, E. E., Cranford, J. A., Arterberry, B. J., Walton, M. A., Bohnert, K. M., & Ilgen, M. A. (2019). Driving under the influence of cannabis among medical cannabis patients with chronic pain. *Drug and Alcohol Dependence*, *195*, 193–197.

Bondallaz, P., Favrat, B., Chtioui, H., Fornari, E., Maeder, P., & Giroud, C. (2016). Cannabis and its effects on driving skills. *Forensic Science International*, *268*, 92–102.

Bosker, W. M., Kuypers, K. P., Theunissen, E. L., Surinx, A., Blankespoor, R. J., Skopp, G., Jeffery, W. K., Walls, H., Leeuwen, C. J., & Ramaekers, J. G. (2012). Medicinal Δ9-tetrahydrocannabinol (dronabinol) impairs on-the-road driving performance of occasional and heavy cannabis users but is not detected in Standard Field Sobriety Tests. *Addiction*, *107*(10), 1837–1844.

Brands, B., Di Ciano, P., & Mann, R. E. (2021). Cannabis, impaired driving, and road safety: An overview of key questions and issues. *Frontiers in Psychiatry*, *12*, 641549.

Brubacher, J. R., Chan, H., Erdelyi, S., Macdonald, S., Asbridge, M., Mann, R. E., Eppler, J., Lund, A., MacPherson, A., Martz, W., Schreiber, W. E., Brant, R., & Purssell, R. A. (2019). Cannabis use as a risk factor for causing motor vehicle crashes: A prospective study. *Addiction*, *114*, 1616–1626.

Capler, R., & Bilsker, D. (2017). *Cannabis use and driving: Evidence review*. https://www.researchgate.net/publication/325128833_Cannabis_Use_and_Driving_Evidence_Review.

Cascini, F., Aiello, C., & Di Tanna, G. (2012). Increasing delta-9-tetrahydrocannabinol (Delta-9-THC) content in herbal cannabis over time: Systematic review and meta-analysis. *Current Drug Abuse Reviews*, *5*, 32–40.

Cash, M. C., Cunnane, K., Fan, C., & Romero-Sandoval, E. A. (2020). Mapping cannabis potency in medical and recreational programs in the United States. *PLoS One*, *15*, e0230167.

Chandra, S., Radwan, M. M., Majumdar, C. G., Church, J. C., Freeman, T. P., & ElSohly, M. A. (2019). New trends in cannabis potency in USA and Europe during the last decade (2008–2017). *European Archives of Psychiatry and Clinical Neuroscience*, *269*, 5–15.

Compton, R. (2017). Marijuana-impaired driving: A report to congress, National Highway Safety Transport Administration. Washington, DC. Available online at: https://www.nhtsa.gov/sites/nhtsa.dot.gov/files/documents/812440-marijuana-impaired-driving-report-to-congress.pdf.

Cook, S., Shank, D., Bruno, T., Turner, N. E., & Mann, R. E. (2017). Self-reported driving under the influence of alcohol and cannabis among Ontario students: Associations with graduated licensing, risk taking, and substance abuse. *Traffic Injury Prevention*, *18*, 449–455.

Couper, F. J., & Peterson, B. L. (2014). The prevalence of marijuana in suspected impaired driving cases in Washington state. *Journal of Analytical Toxicology*, *38*(8), 569–574.

Desrosiers, N. A., Ramaekers, J. G., Chauchard, E., Gorelick, D. A., & Huestis, M. A. (2015). Smoked cannabis' psychomotor and neurocognitive effects in occasional and frequent smokers. *Journal of Analytical Toxicology*, *39*(4), 251–261.

Di Ciano, P., Matamoros, A., Matheson, J., Fares, A., Hamilton, H. A., Wickens, C. M., Mann, R. E., Le Foll, B., Byrne, P. A., & Brands & B. (2020). Effects of therapeutic cannabis on simulated driving: A pilot study. *Journal of Concurrent Disorders*, *2*, 3–13.

Drummer, O. H., Gerostamoulos, J., Batziris, H., Chu, M., Caplehorn, J., Robertson, M. D., & Swann, P. (2004). The involvement of drugs in drivers of motor vehicles killed in Australian road traffic crashes. *Accident; Analysis and Prevention*, *36*, 239–248.

Drummer, O. H., Gerostamoulos, D., Di Rago, M., Woodford, N. W., Morris, C., Frederiksen, T., Jachno, K., & Wolfe, R. (2020). Odds of culpability associated with use of impairing drugs in injured drivers in Victoria, Australia. *Accident; Analysis and Prevention*, *135*, 105389.

Elvik, R. (2013). Risk of road accident associated with the use of drugs: A systematic review and meta-analysis of evidence from epidemiological studies. *Accident; Analysis and Prevention*, *60*, 254–267.

European Monitoring Centre for Drugs Drug Addiction EMCDDA. (2012). *Driving Under the Influence of Drugs, Alcohol and Medicines in Europe—Findings From the DRUID Project*. Luxembourg: Publications Office of the European Union.

European Monitoring Centre for Drugs Drug Addiction EMCDDA. (2018). *Cannabis and driving. Question and answers for policy making, Lisbon.*

Fischer, B., Jeffries, V., Hall, W., Room, R., Goldner, E., & Rehm, J. (2011). Lower Risk Cannabis use Guidelines for Canada (LRCUG): A narrative review of evidence and recommendations. *Canadian Journal of Public Health*, *102*(5), 324–327.

Hartman, R. L., Brown, T. L., Milavetz, G., Spurgin, A., Gorelick, D. A., Gaffney, G., & Huestis, M. A. (2015). Controlled cannabis vaporizer administration: Blood and plasma cannabinoids with and without alcohol. *Clinical Chemistry*, *61*, 850–869.

Hostiuc, S., Moldoveanu, A., Negoi, I., & Drima, E. (2018). The association of unfavorable traffic events and cannabis usage: A meta-analysis. *Frontiers in Pharmacology*, *9*, 99.

Huestis, M. A., Barnes, A., & Smith, M. L. (2005). Estimating the time of last cannabis use from plasma delta9- tetrahydrocannabinol and 11-nor-9-carboxydelta9- tetrahydrocannabinol concentrations. *Clinical Chemistry*, *51*, 2289–2295.

Kruger, D. J., & Kruger, J. S. (2019). Medical cannabis users' comparisons between medical cannabis and mainstream medicine. *Journal of Psychoactive Drugs*, *51*(1), 31–36.

Laumon, B., Gadegbeku, B., Martin, J. L., Biecheler, M. B., & the SAM Group. (2005). Cannabis intoxication and fatal road crashes in France: Population-based case-control study. *BMJ*, *331*, 1371.

Li, M. C., Brady, J. E., DiMaggio, C. J., Lusardi, A. R., Tzong, K. Y., & Li, G. (2012). Marijuana use and motor vehicle crashes. *Epidemiologic Reviews*, *34*, 65–72.

Liguori, A., Gatto, C. P., & Jarrett, D. B. (2002). Separate and combined effects of marijuana and alcohol on mood, equilibrium and simulated driving. *Psychopharmacology*, *163*, 399–405.

Liu, Z., Galettis, P., Broyd, S. J., Hell, H. V., Greenwood, L. M., Krey, P. D., Steigler, A., Zhu, X., Schneider, J., Solowij, N., & Martin, J. H. (2020). Model-based analysis on systemic availability of co-administered cannabinoids after controlled vaporised administration. *Internal Medicine Journal*, *50*, 8.

Martin, J. L., Gadegbeku, B., Wu, D., Viallon, V., & Laumon, B. (2017). Cannabis, alcohol and fatal road accidents. *PLoS One*, *12*, e0187320.

McCartney, D., Arkell, T. R., Irwin, C., & McGregor, I. S. (2021). Determining the magnitude and duration of acute Δ^9-tetrahydrocannabinol (Δ^9-THC)-induced driving and cognitive impairment: A systematic and meta-analytic review. *Neuroscience & Biobehavioral Reviews*, *126*, 175–193.

NHTSA. (2005). *2004 Annual Assessment of Fatal Motor Vehicle Traffic Crashes*. Washington, DC: National Center for Statistics and Analysis.

Pearlson, G. D., Stevens, M. C., & D'Souza, D. C. (2021). Cannabis and driving. *Frontiers in Psychiatry*, *12*, 689444.

Preuss, U. W., Huestis, M. A., Schneider, M., Hermann, D., Lutz, B., Hasan, A., Kambeitz, J., Wong, J. W. M., & Hoch, E. (2021). Cannabis use and car crashes: A review. *Frontiers in Psychiatry*, *12*, 643315.

Ramaekers, J. G., Berghaus, G., van Laar, M., & Drummer, O. H. (2004). Dose related risk of motor vehicle crashes after cannabis use. *Drug and Alcohol Dependence*, *73*, 109–119.

Ramaekers, J. G., Kauert, G., Theunissen, E. L., Toennes, S. W., & Moeller, M. R. (2009). Neurocognitive performance during acute THC intoxication in heavy and occasional cannabis users. *Journal of Psychopharmacology*, *23*(3), 266–277.

Ramaekers, J. G., Moeller, M. R., van Ruitenbeek, P., Theunissen, E. L., Schneider, E., & Kauert, G. (2006). Cognition and motor control as a function of Δ9-THC concentration in serum and oral fluid: Limits of impairment. *Drug and Alcohol Dependence*, *85*(2), 114–122.

Ramaekers, J. G., Theunissen, E. L., de Brouwer, M., Toennes, S. W., Moeller, M. R., & Kauert, G. (2011). Tolerance and cross-tolerance to neurocognitive effects of THC and alcohol in heavy cannabis users. *Psychopharmacology*, *214*(2), 391–401.

Rogeberg, O., & Elvik, R. (2016). The effects of cannabis intoxication on motor vehicle collision revisited and revised. *Addiction*, *111*, 1348–1359.

Russo, E. B., & Taming, T. H. C. (2011). Potential cannabis synergy and phytocannabinoid-terpenoid entourage effects. *British Journal of Pharmacology*, *163*(7), 1344–1364.

Schlienz, N. J., Spindle, T. R., Cone, E. J., Herrmann, E. S., Bigelow, G. E., Mitchell, J. M., Flegel, R., LoDico, C., & Vandrey, R. (2020). Pharmacodynamic dose effects of oral cannabis ingestion in healthy adults who infrequently use cannabis. *Drug and Alcohol Dependence*, *211*, 107969.

Sewell, R. A., Cohn, A. J., & Chawarski, M. C. (2008). Doubts about the role of cannabis in causing lung cancer. *European Respiratory Journal*, *32*(3), 815–816.

Sewell, R. A., Poling, J., & Sofuoglu, M. (2009). The effect of cannabis compared with alcohol on driving. *American Journal on Addictions*, *18*(3), 185–193.

Smink, B. E., Movig, K. L., Lusthof, K. J., De Gier, J. J., Uges, D. R., & Egberts, A. C. (2008). The relation between the use of psychoactive substances and the severity of the injury in a group of crash-involved drivers admitted to a regional trauma center. *Traffic Injury Prevention*, *9*, 105–108.

Smink, B. E., Ruiter, B., Lusthof, K. J., de Gier, J. J., Uges, D. R., & Egberts, A. C. (2005). Drug use and the severity of a traffic accident. *Accident; Analysis and Prevention*, *37*, 427–433.

Solowij, N., Broyd, S., Greenwood, L. M., van Hell, H., Martelozzo, D., Rueb, K., … Croft, R. (2019). A randomised controlled trial of vaporised Δ^9-tetrahydrocannabinol and cannabidiol alone and in combination in frequent and infrequent cannabis users: Acute intoxication effects. *European Archives of Psychiatry and Clinical Neuroscience*, *269*(1), 17–35.

Tank, A., Tietz, T., Daldrup, T., Schwender, H., Hellen, F., Ritz-Timme, S., & Hartung, B. (2019). On the impact of cannabis consumption on traffic safety: A driving simulator study with habitual cannabis consumers. *International Journal of Legal Medicine*, *133*, 1411–1420.

Vandrey, R., Herrmann, E. S., Mitchell, J. M., Bigelow, G. E., Flegel, R., LoDico, C., & Cone, E. J. (2017). Pharmacokinetic profile of oral cannabis in humans: Blood and oral fluid disposition and relation to pharmacodynamic outcomes. *Journal of Analytical Toxicology*, *41*(2), 83–99.

Watson, T. M., & Erickson, P. G. (2019). Cannabis legalization in Canada: How might 'strict' regulation impact youth? *Drugs Education, Prevent and Policy*, *26*, 1–5.

Weinstein, A., Brickner, O., Lerman, H., Greemland, M., Bloch, M., Lester, H., Chisin, R., Sarne, Y., Mechoulam, R., Bar-Hamburger, R., & Freedman, N. (2008). A study investigating the acute dose—response effects of 13 mg and 17 mg Δ 9-tetrahydrocannabinol on cognitive—motor skills, subjective and autonomic measures in regular users of marijuana. *Journal of Psychopharmacology*, *22*(4), 441–451.

Part II

Cannabis usage

Chapter 8

Recreational cannabis use: French perspective of adverse effects

Emilie Bouquet[a], Emilie Jouanjus[b], and Maryse Lapeyre-Mestre[b]
[a]*Addictovigilance Center, Department of Clinical Pharmacology, Poitiers University Hospital, Poitiers, France,* [b]*Addictovigilance Center, Department of Medical and Clinical Pharmacology, Toulouse University Hospital, Toulouse, France*

Abbreviations

AE	adverse event
ANSM	National Agency for Medicines and Health Products Safety
ARS	Regional Health Agencies
CB-receptors	cannabinoid receptors
CNS	central nervous system
COVID-19	coronavirus disease 2019
FAN	French Addictovigilance Network
MDMB-4en-PINACA	methyl 3,3-dimethyl-2-(1-(pent-4-en-1-yl)-1*H*-indazole-3-carboxamido)butanoate
THC	tetrahydrocannabinol
WHO	World Health Organization

Introduction

Cannabis is used since antiquity for both recreational and medicinal purposes. Actually, it is the most widely illicit substance used worldwide. In Europe, France holds a particular position due to the high prevalence of its use in the general population for a long time.

A large variety of cannabis products are actually available. Cannabis contains more than 120 phytocannabinoids, the main being tetrahydrocannabinol (THC) and cannabidiol (CBD). For recreational purpose, it can be consumed in three forms that differed by their THC content: marijuana (herb), resin, and oil.

As in other countries, THC content considerably increased these past 10 years. This may result from specific cultivation methods leading to high-potency products such as "shunk" and "sinsemilla." Changes in the characteristics of cannabis must be considered to understand the effects of cannabis.

The aim of this chapter is to present an overview of adverse health effects related to recreational cannabis use in France.

After a general presentation of the French monitoring system specifically dedicated to the potential of abuse and dependence of psychoactive substances including cannabis, the prevalence and the characteristics of its use are described and then its related adverse events (AEs) and their perspectives.

Addictovigilance in France

France is the only European country to have, as recommended by the World Health Organization (WHO), a surveillance system specifically dedicated to the potential of abuse and dependence of all psychoactive substances, except for alcohol and tobacco. Addictovigilance, which was created in 1990, is included in the French Public Health Code in the section related to poisoning substances (Baumevieille et al., 2019). Like any health surveillance system, addictovigilance is based on spontaneous reports of health professionals that are essential to detect signals and to generate alerts. Through a network of regional addictovigilance centers, specialized in pharmacology, addictology, toxicology, and/or epidemiology, it

Cannabis Use, Neurobiology, Psychology, and Treatment. https://doi.org/10.1016/B978-0-323-89862-1.00043-X

developed partnerships with physicians, toxicologists, addictology care centers, sentinelles networks of pharmacists, etc. (Jouanjus et al., 2019). This French Addictovigilance Network (FAN) also implemented several original pharmacoepidemiological tools repeated each year to complete the spontaneous reporting, such as the DRAMES survey (death in relation to the abuse of drugs and substances) in collaboration with toxicologists and the OPPIDUM survey (observation of illegal products and misuse of psychotropic medications) (Micallef et al., 2019).

The FAN is composed of 13 regional addictovigilance centers implemented in medical pharmacology departments of university hospitals throughout the country which provides a proactive surveillance and an expertise by considering also other indicators based on health insurance databases and health sentinel networks. Some of them are systematically analyzed for any addictovigilance national follow-up, as scientific literature data. They allow both a qualitative and quantitative assessment with the benefit of 20 years of hindsight for some of them. This multidimensional approach allows to detect early addictovigilance signals and changing use practices, leading to information and public health decisions via interactions at a national level with the French National Agency for Medicines and Health Products Safety (Agence nationale de sécurité du medicament et des produits de santé, ANSM) and/or at a regional level with Regional Health Agencies (Agences régionales de santé, ARS) (Micallef et al., 2019) (Fig. 1). This proactive approach highlighted, for example, an increased reporting of serious cardiovascular complications related to recreational cannabis between 2006 and 2010 and the possible favoring role of cannabis in young adults which was subsequently confirmed (Jouanjus et al., 2014). It also allowed to make an overview of reported AEs related to cannabis use which was recently published (Bouquet et al., 2021).

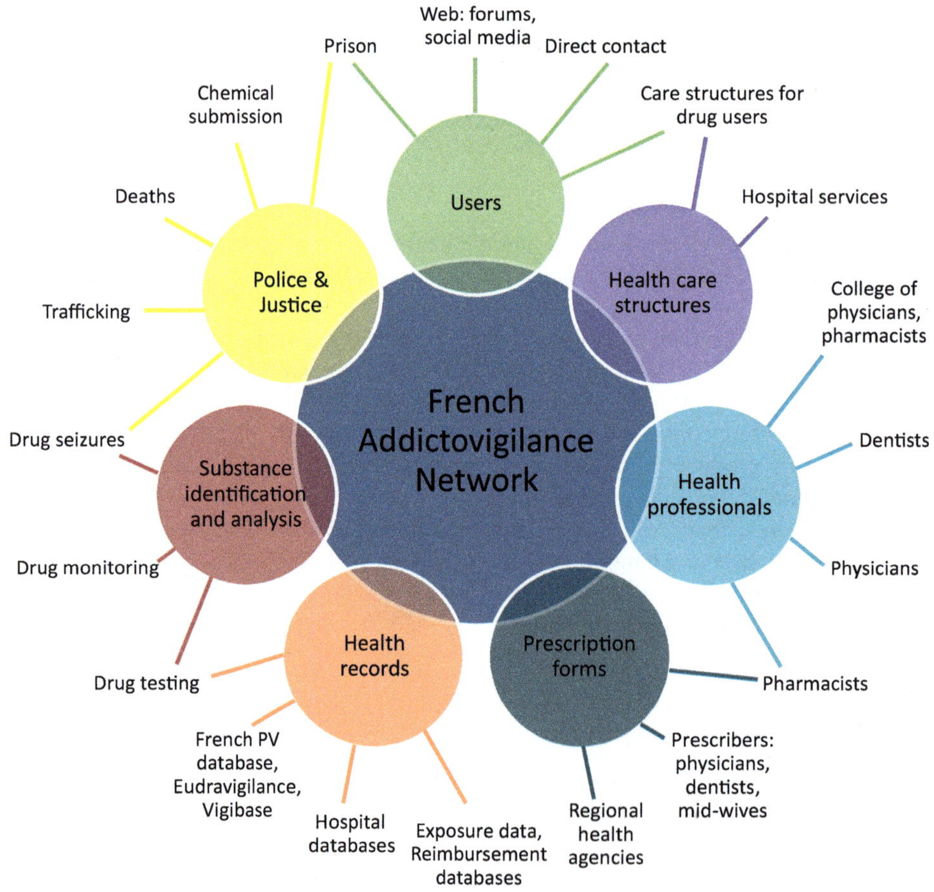

FIG. 1 The multidimensional approach of the French Addictovigilance Network. Partnership and datasources involved in the national addictovigilance surveillance system.

Rates of cannabis use in France

- **Prevalence of cannabis use**

France is one of the European countries with the highest prevalence of recreational cannabis use. The most recent data of the national Baromètre Santé health survey showed that cannabis is the first illicit substance used in France. Its consumption seems to be stabilizing after a constant increase until 2017, although it is still rising for men. In 2020, 46.1% of people 18–64 years (54.8% of men and 37.7% of women) have ever experienced it whereas they were nearly 44.8% (52.7% of men and 37.2% of women) in 2017 (Le Nezet et al., 2021). The proportion of regular consumers (more than 10 uses in the month) was 3.2% in 2020 versus 3.6% in 2017 (Le Nezet et al., 2021). Even if cannabis use is primarily male, it also concerns increasingly women (Le Nezet et al., 2021).

Cannabis use concerns mostly young people. According to the most recent data, France was in 2017 the European country with the highest prevalence of cannabis use in the last year among young adults (15–34) averaging 21.8% (European Monitoring Centre for Drugs and Drug Addiction EMCDDA, 2021; Spilka et al., 2018). Past-month use among middle-aged (15–24) markedly increased (Manthey et al., 2021). Moreover, its consumption among adolescents tends to decline. In 2017, 39% of adolescent aged 17 reported having experienced cannabis, 7.2% consuming it regularly, whereas problematic use was rising around 7.4% (Spilka et al., 2018).

In addition, the working population is also concerned. Among cohort CONSTANCES participants who have jobs, 8% of men and 4% of women used cannabis at least once a week (MILDECA, 2020).

- **THC content**

To understand the effects related to recreational cannabis use, the form and THC content must also be considered. Indeed, some events, particularly psychiatric events, are more serious than the contents are higher. As in other countries worldwide, THC contents have considerably increased in the past 10 years. In 2020, analyses of cannabis seizures have found average THC contents around 27% for resin and 13% for herbal, versus, respectively 10% and 7% in 2009 (Detrez, 2021; Dujourdy & Besacier, 2017). France is the European country in which the use of herbal is the least common and the use of both resin and herbal is the most common (European Monitoring Centre for Drugs and Drug Addiction (EMCDDA), 2019).

Adverse effects of recreational cannabis use in France

Cannabis use has multiple health effects. In the recreational context, it is most often consumed to experience feeling of euphoria, relaxation, and disinhibition. However, its use provides also undesired AEs. France is one of the European countries with the highest rates of treatment admissions for disorders related to this use, which increased by more than 50% between 2010 and 2019 (Manthey et al., 2021). A recent publication showed a threefold increase of cases of AEs notified to the FAN between 2012 and 2017 and a multitude of AEs, mainly affecting the central nervous system, the cardiovascular system, and the gastrointestinal tract (Bouquet et al., 2021). The main AEs of recreational cannabis use reported to the French Addictovigilance Network are summarized in Table 1.

- **Central nervous system (CNS)**

According to the FAN's experience, the main central nervous system AEs of cannabis reported are psychiatric (51.2% of all AEs), with in the first-place dependence which has doubled between 2012 and 2017, reaching 20.3% (Bouquet et al., 2021). Cannabis dependence is defined in the ICD-10 Classification of Mental and Behavioural Disorders and in the fifth revision of the Diagnostic and Statistical Manual of Mental Disorders (DSM-5), in particular by a craving, an unsuccessful attempt to cut down or control cannabis consumption as well as a tolerance leading to use cannabis in a larger amount than was intended or spending a lot of time to obtain, use or recover from cannabis effects (American Psychiatric Association, 2013; World Health Organization, 1993). The severity of dependence has been related to frequent use of high potency cannabis and to early cannabis initiation (Freeman & Winstock, 2015). The effects of cannabis on the central nervous system are essentially mediated by the interaction of THC on type 1 cannabinoid receptors. Using cannabis in adolescence may interfere with the endocannabinoid system and impair the brain development at this age. Regular cannabis use also promotes other psychiatric events such as psychotic disorders and in high cannabis and heavy users (Murray et al., 2016). Some cases of depressive and anxiety disorders are also reported, but the direction of the relationship is still unclear. Indeed, cannabis use appears to favor the occurrence of these disorders, but however some people report having initiated cannabis to self-treat these psychiatric symptoms (Kuhns et al., 2021; Murray et al., 2016). A French survey showed that cannabis

TABLE 1 Main adverse events of recreational cannabis use reported to the French Addictovigilance Network.

References	Objectives	Design	Main outcome measures	Measure of exposure	Participants: inclusion criteria and sample size	Results
Jouanjus et al. (2014)	- To describe all cases of CB-related CV complications - To estimate at the national level if there is a signal of increasing risk of these events	Case series	Cardiovascular disorders: acute coronary syndrome, stroke, peripheral arteriopathy	Self-report	**Inclusion criteria:** People being the subject for SRs to the FAN **Period**: 2006–2010 **Sample size:** $n = 35$, 30 men **Mean age:** 4.3 ± 8.8	During the inclusion period, the proportion of CV complications rose from 11 to 3.6% of all CB-related disorders reported to the FAN. 9 cases with CV history and 7 with CV familial history 24 reported associated substances use including 21 tobacco and 6 alcohol users 9 patients died during their hospitalizations CB consumption: 8 actual, 6 recent, and 16 daily use Duration of use varied between 2 and more than 25 years Toxicologic analyses were performed in 13 cases: each THC (+) and in 10 cases, only THC 3 of the 22 patients with no available toxicological data reported CB as being the only consumed product; and 13 reported using it with tobacco or alcohol Cardiac complications were 20 acute coronary syndrome and 2 heart rate disorders 10 peripheral and 3 cerebral complications cases were reported
Lapeyre-Mestre et al. (2017)	To investigate a potential increase of CB harms in France in young children	Case series	CB exposure in children	Self-report ± toxicological analysis	**Inclusion criteria:** children under 10 years old being subject for SRs to the FAN and included in the national hospitalization registry **Period:** 2010–2014 **Sample size:** $n = 140$, most of them were under 2 years old, 120 were hospitalized	Over the study period, 140 reports were collected. Hospitalizations related to CB ingestion in children increased from 66 cases in 2010 to 247 cases in 2014 (120 cases in total), most often it took place in the family context. Toxicological analysis were performed in 84 cases, including 16 THC (+).
Schreck et al. (2018)	To compare French cases of cannabinoid hyperemesis syndrome to those published in the international literature.	Case series Literature review	Cannabinoid hyperemesis syndromes	Self-report	**Inclusion criteria:** Patients with cyclical symptoms of nausea, vomiting, and abdominal pain in a context of chronic CB or synthetic cannabinoid use.	French cases had similar characteristics to those identified in the international literature. Cannabinoid hyperemesis syndrome concerned mainly young

					Period: up to June 2017 **Sample size:** - French cases: $n = 29$ - Literature cases: $n = 113$ **Mean age:** - French cases: 25.8 years - Literature cases: 30.7 years	men who inhaled CB daily for several years. Its physiopathology is not clearly understood. Symptoms were improved by hot baths/showers.
Revol et al. (2018)	To describe deaths related to CB use alone or in association	Case series	Deaths	Toxicological analyses	**Inclusion criteria**: cases of deaths with autopsy procedure included in the DRAMES survey **Period:** 2014–2016 **Sample size:** $n = 85$ deaths including 64 related to CB use alone.	All cases involved a cardiovascular disorder.
Bouquet et al. (2021)	To describe AEs of recreational CB in France	Cases series	All adverse events	Self-report ± toxicological analysis	**Inclusion criteria:** SRs involving CB alone or in association with tobacco ± alcohol to the FAN **Period:** 2012–2017 **Sample size:** $n = 2217$, 1694 men **Mean age:** 30 years	The number of SRs tripled between 2012 and 2017. A multitude of AEs were reported, mainly psychiatric (51.2%), neurological (15.6%), cardiac (7.8%), and gastrointestinal (7.7%). Some of them were unexpected (1.1%) or emerged such as cannabinoid hyperemesis syndrome (2.8%). Dependence that rose from 72 cases (10.1%) in 2012 to 145 cases (20.3%) in 2017. In 0.2% of cases, AEs resulted in a fatal condition.

Published data on adverse events notified to the French Addictovigilance Network these last years. *CB*, cannabis; *CV*, cardiovascular; *DRAMES*, death in relation to the abuse of drugs and substances; *FAN*, French Addictovigilance Network; *SRs*, spontaneous reports; *THC*, tetrahydrocannabinol.

was a risk factor for suicide attempts hospitalizations among adolescent girls (Du Roscoät et al., 2016). Furthermore, early cannabis consumption (before the age of 16) may cause amotivational syndromes, memory, and concentration disorders that may be responsible for low educational attainment as showed by the French TEMPO cohort study (Melchior et al., 2017).

Neurologic disorders were the second type of AEs reported to the FAN between 2012 and 2017, especially consciousness, memory and cognitive disorders, cerebrovascular accidents, and convulsions (Bouquet et al., 2021). Cognitive disorders are very frequent in cannabis users and are often accompanied by slowed reaction time and impaired executive function (Kuhns et al., 2021). Cerebrovascular accidents are dealt with in the following paragraph. Concerning convulsions, case series have been published in a context of cannabis intoxications (Bonkowsky et al., 2005; Spadari et al., 2009; Tilelli & Spack, 2006), whether clinical trials demonstrated efficacity of cannabis varieties with high level of CBD in treating epilepsy (Mechoulam & Parker, 2013) and preclinical studies showed that both THC and CBD have anticonvulsive effects, notably CBD (Detyniecki & Hirsch, 2015; Zaheer et al., 2018).

- **Cardiovascular disorders**

Cases of cardiovascular diseases in the context of cannabis exposure have been described sporadically from the 1970s. Later, they have been increasingly reported from the 2010s, particularly in patients with no known cardiovascular history nor risk factors and including among young adults or even adolescents.

These diseases both affect the cardiac system and the CNS, and thus include cardiovascular and cerebrovascular disorders, with the hypothesis of a common pathophysiological origin (Jouanjus et al., 2017). This latter hypothesis relies on the acute triggering of vascular contraction medicated by cannabinoids. The WHO considers cardiac syndromes and infarctions as well as strokes and cerebral ischemic events to be among the priority areas for research (World Health Organization, 2016).

Myocardial infarction remains the most frequently reported disease (Pacher et al., 2018). After the cornerstone study by Mittleman and colleagues in the beginning of the 2000s reporting an almost fivefold increase among subjects who had used cannabis in the hour preceding the onset of symptoms, cannabis has been included among the potential triggering factors of myocardial infarction, together with cocaine or caffeine for example (Mittleman et al., 2001; Nawrot et al., 2011). Case series or isolated cases of myocardial infarction published in the gray literature mainly refer to young men with an average age under 30 years old, in an apparent good health. Some of them refer to adolescents (Toce et al., 2019). Apart from the study by Mittleman et al. (2001), use of cannabis was associated to an increased risk for myocardial infarction independent from a list of confounders (including tobacco smoking) (Desai et al., 2017). Moreover, cardiovascular deaths were also higher among young cannabis users than among those unexposed (DeFilippis et al., 2018).

Cerebrovascular disease reported in the literature mainly corresponds to ischemic cerebral strokes or transient ischemic attacks in young adults, mostly males in their thirties, and with a strong temporal association between cannabis use and the onset of symptoms (Wolff et al., 2011). Besides, reversible cerebral vasoconstriction syndrome (RVCS) appears to be the best-documented type of cerebral strokes in young cannabis users. Case series or isolated cases of cannabis-related cerebral strokes mainly refer to young men with an average age under 35 years old. Use of cannabis was associated with an increased risk for cerebrovascular disorders in several studies (Barber et al., 2013; Hemachandra, McKetin, Cherbuin, & Anstey, 2016; Jouanjus et al., 2017). The association remained despite the identification of confounding factors such as cardiovascular risk factors, tobacco smoking, or cocaine intake.

The analysis of cannabis-related hospitalizations in a French hospital found several such cases of myocardial infarction and cerebral strokes in young patients with no known cardiovascular history nor risk factors (Jouanjus et al., 2011). Overall, the reports of serious adverse cases related with the use of cannabis collected by the FAN correspond to cardiovascular, cerebrovascular, and peripheral vascular disorders, with a total of 35 cases between 2006 and 2010 (Jouanjus et al., 2014). This first analysis emphasized a significant increase of the corresponding risk. The analysis of the data collected during the most recent period tend to confirm these initial findings and the increasing trend, with 182 cardiovascular disorders in the context of cannabis exposure between 2011 and 2017 (Jouanjus et al., 2021).

Use of cannabis should be systematically investigated before myocardial infarction or cerebral strokes, particularly in young patients and even in an unusual presentation and/or despite the absence of cardiovascular risk factor.

- **Gastrointestinal**

Cannabis may disrupt gastrointestinal homeostasis by modulating gastrointestinal mobility, inflammation, and immune tolerance. CB1 receptors have been found in the enteric nervous system and CB2 receptors in lymphatic plexus of the ileus that regulates inhibition of intestinal inflammation and motility as well as visceral pain. Thus, its use may result in various gastrointestinal AEs. Among them, the gastrointestinal AE that emerged is cannabinoid hyperemesis syndrome that is specific to cannabis. This new diagnostic entity, which was described for the first time by Allen et al. in 2004, is rising as demonstrated by the increased cases reported to the French Addictovigilance Network since 2015 (Bouquet et al., 2021; Schreck et al., 2018). It is distinguished from simple nausea and vomiting also described with cannabis, often after an acute exposure. What particularly characterizes this syndrome is that it occurs in chronic cannabis users and resolves with hot baths/showers and after stopping its use. Simonetto et al. (2012) defined some major and minor diagnostic criteria that were completed by Venkatesan et al. (2019). Major criteria were severe cyclic nausea and vomiting, resolution with cannabis stopping, symptom improvement with hot bath/shower, abdominal pain, and weekly cannabis consumption (Schreck et al., 2018; Simonetto et al., 2012; Venkatesan et al., 2019). Its physiopathology is not clearly understood and the cannabinoid hyperemesis syndrome is considered as a "paradox." Indeed, activation of CB1 receptors have been shown to decrease sensations of nausea (Galli et al., 2011) and THC has been shown to relieve severe nausea (Sallan et al., 1975). Several hypotheses have been proposed. Antiemetic effects of THC may be determined by a threshold and high THC content may favor the onset of symptoms (Darmani, 2010). High THC content may also promote desensitization and downregulation of cannabinoid receptors (Lundberg et al., 2005). Simonetto et al. also suggested that the central effect of chronic cannabis use may disturb the hypothalamic-pituitary-adrenal axis and involve stress-related autonomous nervous system, thus facilitating the occurrence of cannabinoid hyperemesis syndrome (Simonetto et al., 2012). Moreover, the transient receptor potential vanilloid 1 (TRPV1), located in the emesis center of the central nervous system, may also be involved (Dezieck et al., 2017).

In addition, cannabis use has also been implicated indirectly in steatosis in patient with chronic hepatitis C (Hézode et al., 2008).

Furthermore, several case series of acute pancreatitis have been published these last years (Azam et al., 2022; Belze, Legras, Ehrmann, Garot, & Perrotin, 2011; Bouquet et al., 2021). They occurred often in young people and especially in chronic and heavy cannabis consumers (daily and for more than 3 years), sometimes without other associated factor, in particular absence of chronic alcohol use (Culetto et al., 2017). Here again, reducing by 50% or stopping cannabis use can have a benefic impact to avoid recurrences (Culetto et al., 2017). Cannabis-related acute pancreatitis is rare but probably underdiagnosed and its physiopathology is to date unclear. Both CB1 and CB2 receptors are expressed in pancreatic cells. It has been showed that activation of CB1 receptors results in pancreatic fibrosis, whereas activation of CB2 receptor has antifibrogenic effect (Goyal & Singla, 2018). Another hypothesis is that cannabis could induce acute pancreatitis by hypersensitivity (Goyal & Singla, 2018). Mixed results were found for the effects of cannabis on the pancreas in the experimental models (Goyal & Singla, 2018).

- **Deaths**

In some cases, AEs related to recreational cannabis use can potentially lead to a fatal condition. Deaths represented 0.2% of the AEs reported to the FAN between 2012 and 2017: they concerned young adults averaging 35 years (Bouquet et al., 2021). When the causes of death were available, they were mainly cardiovascular (Bouquet et al., 2021). In addition, in France, the prospective DRAMES survey (death in relation to the abuse of drugs and substances) that collects annually deaths with autopsies and toxicological analysis showed an increase of cases related to cannabis use that reached 6.6% of cases in 2019. In all cases, a cardiovascular disease was involved (CEIP-A Grenoble, 2019; Revol et al., 2018).

- **Perinatal**

Very few data are available concerning perinatal events of cannabis use during pregnancy in France. According to the last French perinatal survey 2016, 2.1% of French women reported in the days following childbirth to have used cannabis during pregnancy, nearly half more than 3 times a month (INSERM—DREES, 2017). Furthermore, the French national OPPIDUM survey (Observation of illegal products and misuse of psychotropic medications) showed an increased use of cannabis among women taken care of specialized addiction care centers from 1.2% in 2005–2006 to 17.8% in 2017–2018 (Blayac et al., 2021). However, they do not describe the perinatal complications related to its use. The analysis of notifications reported to the FAN between 2012 and 2017 found 55 AEs (1.7% of all AEs) related to cannabis use during pregnancy, mainly cannabis dependence, intrauterine growth retardation, and prematurity (Bouquet et al., 2021). Some of them were unexpected, notably three cases of congenital malformations (duodenal atresia, horseshoe kidney, and cardiopathy) (Bouquet et al., 2021).

The international literature showed a normalization of cannabis use. In addition to its anxiolytic and sedative proprieties, cannabis may sometimes be consumed to manage nausea during pregnancy. Some studies have found an increased risk of adverse perinatal outcomes, in particular concerning prematurity and growth retardation. Cannabis use during pregnancy may affect the neurodevelopment of the child, but in many studies the role of confounding factors cannot be ruled out (Grant et al., 2020). No increased risk of congenital malformations related to prenatal cannabis use has been demonstrate to date (Grant et al., 2020). Recent geospatial studies reported a correlation between cannabis use and some congenital malformations; however, they do not allow to conclude either a causal relationship or an increase in individual risk (Grant et al., 2020).

The continued consumption of cannabis use during pregnancy and postpartum increased the risk of subsequent pediatric intoxications.

- **Pediatric intoxications**

Accidental exposure to cannabinoids in young children is a relatively recent phenomenon with first reports in the international scientific literature in the early 2010s. A review by Richards et al., which included 44 articles, emphasized that the majority of children were between 8 and 12 months of age, that almost half of the cases involved resin ingestion, and then cookies or joints (15%) (Chen & Klig, 2019; Richards et al., 2017).

Most US and Canadian publications emphasize the increased availability of cannabis-containing products in homes and the positive image conveyed by these products. As far as we know, only one case of death from accidental cannabis exposure was reported in the international literature. It concerned an 11-month-old boy (Colorado), who died from a cardiac arrest following an initial picture of respiratory depression, with a postmortem blood THC concentration of 7.8 g/L (Nappe & Hoyte, 2017).

In France, the first report concerning accidental exposure in children identified in the scientific literature from France described 93 cases reported to the Marseille Poison control center (South East France) between 1993 and 2007. Five serious cases were reported all over the 9 years, including two convulsive seizures (Spadari et al., 2009).

Following this first report, an increasing number of publications dealing with accidental cannabis exposure has been published from 2011 onward (Cesbron et al., 2015; Molly et al., 2012; Patissier et al., 2015; Pellissier et al., 2014).

The French addictovigilance network summarized data concerning children under 10 years old, from 2010 to 2014, through analysis of spontaneous reporting, and national hospitalization registry and/or toxicological database where available (Lapeyre-Mestre et al., 2017). For this period, a total of 140 reports were identified, most of these cases concerning toddlers (less than 2 years old): 120 were hospitalized, 9 were considered as "serious" (with a life-threatening condition needing an admission in intensive care unit). None of these children died. Most of cases were related to an unintentional ingestion in the family context. During this period, the number of cannabis-related hospitalizations in France hugely increased, from 66 cases in 2010 to 247 in 2014. The hospitalization rate per 100,000 children increased from 7.1 in 2010 to 23 in 2014 in French South areas, whereas it remains stable and low in other French areas. This geographical repartition may be superimposed on the regional prevalence of cannabis use in the French population. Although the number of cannabis pediatric exposures was low, this analysis highlights a signal of increasing rate of hospitalization of children following cannabis exposure, associated with an appearance of increasing seriousness of these cases (comas, seizures). This first report of the French addictovigilance network led the French health authorities to publish a warning about this phenomenon in 2015 and remind the population of the risk of these exposures in young children. Following updating of these data, showing a constant increase of cases between 2015 and 2017 led a second reminder in 2018. Publications based on the results of specific toxicological analyses complete this panorama (Feliu et al., 2017; Guidet et al., 2020; Kintz et al., 2017), suggesting the lack of a clear correlation between observed blood concentrations and symptom intensity. These trends were confirmed by recent investigations (Chartier et al., 2021), without any new information compared to the previous published data (most of outcomes were favorable and no death was reported).

Perspectives

Recently, the pandemic COVID-19 induced modification of consumption practices. During the first lockdown 2020, the frequencies of use increased. While a third of users mentioned smoking three or more joints a day before 2020, they were 40% to declare such consumption during this period (Brissot, Philippon, & Spilka, 2020). A high cannabis consumption persists and people using cannabis only in a festive context indicated having consumed it sometimes daily against anxiety and loneliness (Brissot et al., 2020). Varieties and forms of cannabis have varied (Gérome, 2021). Herbal consumption was privileged and chronic users indicated more frequently consuming in parallel resin (Brissot et al., 2020). The consumption of high-potency cannabis may result in an increase in AEs.

In addition, over the same period, cannabis adulterated with synthetic cannabinoid receptor agonists such as MDMB-4en-PINACA emerged in Europe, including France (Oomen et al., 2022). This emergence may potentially aggravate the risk of AEs. Many cases of events more serious than usual, sometimes after a single use of cannabis than nothing in appearance made it possible to distinguish from usual products, have been described by consumers. The considerable pharmacologic potency of the synthetic cannabinoid receptors agonists sprayed on cannabis may explain their occurrence, and specific toxicological analysis are required to identified them.

In 2022, cannabis has a dual status in France. It is both an illicit psychoactive substance listed as a nonmedical product in the 1961 international convention on narcotic drugs, and a medical product following the decree N°2020-1230 of 7 October 2020 establishing the experiment with medical cannabis. This dual status may impact the prevalence and practices of recreational cannabis use, promoting its consumption.

Conclusion

In France, a wide variety of AEs related to recreational cannabis use are reported, primarily concerning the central nervous system, the cardiovascular system and the gastrointestinal tract. Dependence continues to rise and diagnoses such as cannabinoid hyperemesis syndrome emerged. The high prevalence of cannabis use which THC contents increased, the modification of consumption practices since COVID-19 pandemic and the emergence of cannabis adulterated with synthetic cannabinoid receptor agonists are likely to aggravate AEs. Given the high prevalence of cannabis use in France, particularly among adolescents and young adults, and the description of unusual clinical presentations in this population, there is concern that serious AEs likely to be underdiagnosed occur.

Mini-dictionary of terms

Abuse: Abuse is a persistent or sporadic, intentional excessive use of substances.
Adverse events: According to the World Health Organization, adverse events are medical occurrence temporally associated with the use of a substances, which does not necessary have a causal relationship.
Cannabidiol: Cannabidiol is one of the most abundant phytocannabinoids found in *Cannabis sativa*, after tetrahydrocannabinol. It can modulate in particular the serotonin and dopamine receptors.
Cannabinoid receptors: Cannabinoid receptors are transmembrane receptors that modulate the central and peripheral actions of cannabis.
Cannabis: Cannabis designates products derived from the plant *C. sativa*. It contains more than 120 cannabinoids. It is available under different forms: marijuana (herb), resin, and oil. It is also called as weed, yobi, haschich, kif, etc.
Dependence: Dependence is defined in the ICD-10 Classification of the Mental and Behavioural Disorders and in the fifth revision of the Diagnostic and Statistical Manual of Mental Disorders (DSM-5) by a craving, an unsuccessful attempt to cut down or control cannabis consumption as well as tolerance leading to use cannabis in a larger amount than was intended or spending a lot of time to obtain, use, or recover from substance effects.
Tetrahydrocannabinol: Tetrahydrocannabinol is the most abundant phytocannabinoid found in cannabis with psychoactive effects.

Key facts of recreational cannabis use: French perspective of adverse effects

Key facts on the rates of cannabis use in France

- THC contents considerably increased these last 10 years worldwide, including France.
- France is the European country in which the use of herbal is the least common and the use of both herbal and resin is the most common.

Key facts on adverse events related to cannabis use in France

- Cases of AEs reported to the French Addictovigilance Network increased these last years.
- A multitude of AEs were reported, mainly affecting the central nervous system, the cardiovascular system, and the gastrointestinal tract.
- The recent pandemic COVID-19 induced modification of consumption practices that may impact AEs related to its use.

Summary points

- This chapter presents an overview of AEs related to recreational cannabis use in France.
- The recent modification of consumption practices since the pandemic COVID-19 was assessed.
- The data presented were based on the French Addictovigilance surveillance system completed by a literature review.
- The period considered was 2010–2021.
- Given the high prevalence of cannabis use in France, particularly among adolescents and young adults, and the description of unusual clinical presentations in this population, there is concern that serious adverse events likely to be underdiagnosed occur, such as cannabinoid hyperemesis syndromes for example.

References

American Psychiatric Association. (2013). *Diagnostic and statistical manual of mental disorders* (5th ed.). Washington, DC: APA.

Azam, C., Buscail, L., Culetto, A., & Lapeyre-Mestre, M. (2022). Cannabionoid-related acute pancreatitis: An update from international literature and individual case safety reports. *Drug Safety*. https://doi.org/10.1007/s40264-022-01146-7.

Barber, P. A., Pridmore, H. M., Krishnamurthy, V., Roberts, S., Spriggs, D. A., Carter, K. N., & Anderson, M. E. (2013). Cannabis, ischemic stroke, and transient ischemic attack: A case-control study. *Stroke*, *44*, 2327–2329.

Baumevieille, M., Perri-Plandé, J., Miremont-Salamé, G., Daveluy, A., & Haramburu, F. (2019). From psychoactive medicines to addictovigilance in French Public Health Code (1990-2017). *Therapie*, *74*, 375–382.

Belze, O. J., Legras, A., Ehrmann, S., Garot, D., & Perrotin, D. (2011). Cannabis-induced acute pancreatitis. *The American Journal of Emergency Medicine*, *29*, 131.

Blayac, L., Micallef, J., & Lapeyre-Mestre, M. (2021). Evolution des comportements de consommation de produits psychoactifs illicites ou détournés de leur usage médicamenteux chez les femmes enceintes en France: analyse des données OPPIDUM de 2005 à 2018. *Therapie*, *76*, 169.

Bonkowsky, J. L., Sarco, D., & Pomeroy, S. L. (2005). Ataxia and shaking in a 2-year-old girl. Acute marijuana intoxication presenting as seizure. *Pediatric Emergency Care*, *21*, 527–528.

Bouquet, E., Pain, S., Eiden, C., Jouanjus, E., Richard, N., Fauconneau, B., Pérault-Pochat, M. C., & French Addictovigilance Network. (2021). Adverse events of recreational cannabis use reported to the French addictovigilance network (2012-2017). *British Journal of Clinical Pharmacology*, *87*, 3925–3937.

Brissot, A., Philippon, A., & Spilka, S. (2020). *Résultats de l'enquête cannabis online. Observatoire Français des Drogues et Toxicomanies*. https://www.ofdt.fr/BDD/publications/docs/eisxab2ac.pdf.

CEIP-A Grenoble. (2019). *Décès en relation avec l'abus de médicaments et de substances. Résultats 2019*. Addictovigilance. https://addictovigilance.fr/programmes-dobservation/drames/.

Cesbron, A., Tokayeva, L., Loilier, M., Le Boisselier, R., Gaulier, J. M., & Coquerel, A. (2015). Intoxication cannabique sévère après ingestion accidentelle chez une enfant de 16 mois. *Toxicologie Analytique et Clinique*, *27*, S47.

Chartier, C., Penouil, F., Blanc-Brisset, I., Pion, C., Descatha, A., & Deguigne, M. (2021). Pediatric cannabis poisonings in France: More and more frequent and severe. *Clinical Toxicology*, *59*, 326–333.

Chen, Y. C., & Klig, J. E. (2019). Cannabis-related emergencies in children and teens. *Current Opinion in Pediatrics*, *31*, 291–296.

Culetto, A., Bournet, B., & Buscail, L. (2017). Clinical profile of cannabis-associated acute pancreatitis. *Digestive and Liver Disease*, *49*, 1284–1285.

Darmani, N. A. (2010). Cannabinoid-induced hyperemesis: A conundrum—From clinical recognition to basic science mechanisms. *Pharmaceuticals*, *3*, 2163–2177.

DeFilippis, E. M., Singh, A., Divakaran, S., Gupta, A., Collins, B. L., Biery, D., ... Blankstein, R. (2018). Cocaine and marijuana use among young adults with myocardial infarction. *Journal of the American College of Cardiology*, *5*, 2540–2551.

Desai, R., Patel, U., Sharma, S., Amin, P., Bhuva, R., Patel, M. S., Sharma, N., Shah, M., Patel, S., Savani, S., Batra, N., & Kumar, G. (2017). Recreational marijuana use and acute myocardial infarction: Insights from nationwide inpatient sample in the United States. *Cureus*, *9*, e1816.

Detrez, V. (2021). *Le Point SINTES N°7*. Note de synthèse N°2021-5 https://www.ofdt.fr/BDD/sintes/LePointSINTES07.pdf.

Detyniecki, K., & Hirsch, L. (2015). Marijuana use in epilepsy: The myth and the reality. *Current Neurology and Neuroscience Reports*, *15*, 65.

Dezieck, L., Hafez, Z., Conicella, A., Blohm, E., O'Connor, M. J., Schwarz, E. S., & Mullins, M. E. (2017). Resolution of cannabis hyperemesis syndrome with topical capsaicin in the emergency department: A case series. *Clinical Toxicology*, 908–913.

Du Roscoät, E., Legleye, S., Guignard, R., Husky, M., & Beck, F. (2016). Risk factors for suicide attempts and hospitalizations in a sample of 39,542 French adolescents. *Journal of Affective Disorders*, *190*, 517–521.

Dujourdy, L., & Besacier, F. (2017). A study of cannabis potency in France over 25 years period (1992–2016). *Forensic Science International*, *272*, 72–80.

European Monitoring Centre for Drugs and Drug Addiction (EMCDDA). (2021). *European drug report. Trends and developments*. https://www.emcdda.europa.eu/publications/edr/trends-developments/2021_en.

European Monitoring Centre for Drugs and Drug Addiction (EMCDDA). (2019). *European drug report. Trends and developments*. https://www.drugsandalcohol.ie/30619/1/EMCDDA_2019.pdf.

Feliu, C., Cazaubon, Y., Fouley, A., Guillemin, H., Millart, H., Gozalo, C., & Djerada, Z. (2017). Drowsiness and uncommon fever in a child after cannabis ingestion. *Annales de Biologie Clinique*, *75*, 462–465.

Freeman, T. P., & Winstock, A. R. (2015). Examining the profile of high-potency cannabis and its association with severity of cannabis dependence. *Psychological Medicine*, *45*, 3181–3189.

Galli, J. A., Sawaya, R. A., & Friedenberg, F. K. (2011). Cannabinoid hyperemesis syndrome. *Current Drug Abuse Reviews*, *4*, 241–249.

Gérome, C. (2021). *Usage et usagers de drogues en contexte de crise sanitaire*. Observatoire Français des Drogues et Toxicomanies. https://www.ofdt.fr/publications/collections/periodiques/lettre-tendances/usages-et-usagers-de-drogues-en-contexte-de-crise-sanitaire-tendances-147-decembre-2021/.

Goyal, H., & Singla, U. (2018). Cannabis and acute pancreatitis. *Pancreas*, *47*, e32–e33.

Grant, K. S., Conover, E., & Chambers, C. D. (2020). Update on the developmental consequences of cannabis use during pregnancy and lactation. *Birth Defects Research*, *112*, 1126–1138.

Guidet, C., Gregoire, M., Le Dreau, A., Vrignaud, B., Deslandes, G., & Monteil-Ganière, C. (2020). Cannabis intoxication after accidental ingestion in infants: Urine and plasma concentrations of Δ-9-tetrahydrocannabinol (THC), THC-COOH and 11-OH-THC in 10 patients. *Clinical Toxicology*, *58*, 421–423.

Hemachandra, D., McKetin, R., Cherbuin, N., & Anstey, K. J. (2016). Heavy cannabis users at elevated risk of stroke: Evidence from a general population survey. *Australian and New Zealand Journal of Public Health*, *40*, 226–230.

Hézode, C., Zafrani, E. S., Roudot-Thoraval, F., Costentin, C., Hessami, A., Bouvier-Alias, M., Medkour, E. S., Pawlostky, J. M., Lotersztajn, S., & Mallat, A. (2008). Daily cannabis use: A novel risk factor of steatosis severity in patients with chronic hepatitis C. *Gastroenterology*, *134*, 432–439.

INSERM—DREES. (Octobre 2017). *Enquête nationale périnatale. Rapport 2016. Les naissances et les établissements. Situation et évolution depuis 2010.* http://www.xn- -epop-inserm-ebb.fr/wp-content/uploads/2017/10/ENP2016_synthese_rapport.pdf.

Jouanjus, E., Lapeyre-Mestre, M., Micallef, J., & French Association of the Regional Abuse and Dependence Monitoring Centres (CEIP-A) Working Group on Cannabis Complications*. (2014). Cannabis use: Signal of increasing risk of serious cardiovascular disorders. *Journal of the American Heart Association*, *23*, e000638.

Jouanjus, E., Leymarie, F., Tubery, M., & Lapeyre-Mestre, M. (2011). Cannabis-related hospitalizations: Unexpected serious events identified through hospital databases. *British Journal of Clinical Pharmacology*, *71*, 758–765.

Jouanjus, E., Marqueyssat, G. S., Lapeyre-Mestre, M., & Micallef, J. (2021). Analysis of the cardiovascular complications related to use of cannabis reported to the French Addictovigilance Network in 2011-2017. Special Issue: Abstracts of the Annual Meeting of French Society of Pharmacology and Therapeutics *Fundamental & Clinical Pharmacology*, *35*, 1–225.

Jouanjus, E., Micallef, J., Mallaret, M., & Lapeyre-Mestre, M. (2019). Comment on: An insight into Z-Drug abuse and dependence: An examination of reports to the European Medicines Agency Database of suspected adverse drug reactions. *The International Journal of Neuropsychopharmacology*, *22*, 528–530.

Jouanjus, E., Raymond, V., Lapeyre-Mestre, M., & Wolff, V. (2017). What is the current knowledge about the cardiovascular risk for users of cannabis-based products? A systematic review. *Current Atherosclerosis Reports*, *19*, 26.

Kintz, P., Ameline, A., Eibel, A., Geraut, A., Farrugia, A., Berthelon, L., Farrugia, A., & Raul, J. S. (2017). Interpretation of cannabis findings in the hair of very young children: Mission impossible. *Current Pharmaceutical Biotechnology*, *18*, 791–795.

Kuhns, L., Kroon, E., Colyer-Patel, K., & Cousijn, J. (2021). Associations between cannabis use, cannabis use disorder, and mood disorders: Longitudinal, genetic, and neurocognitive evidence. *Psychopharmacology*. https://doi.org/10.1007/s00213-021-06001-8.

Lapeyre-Mestre, M., Chauvin, E., Jouanjus, E., Pauly, V., Lafaurie, M., & Roussin, A. (2017). Cannabis exposure in young children: Increasing rate of hospitalisation in France in the recent years. *Clinical Therapeutics*, *39*, e13.

Le Nezet, O., Philippon, A., Lahaie, E., & Andler, R. (2021). *Les niveaux d'usage de cannabis en France en 2020*. Note de synthèse N° 2021–06 https://www.ofdt.fr/publications/collections/notes/note-cannabis-barometre-sante/.

Lundberg, D. J., Daniel, A. R., & Thayer, S. A. (2005). Delta(9)-Tetrahydrocannabinol-induced desensitization of cannabinoid-mediated inhibition of synaptic transmission between hippocampal neurons in culture. *Neuropharmacology*, *49*, 1170–1177.

Manthey, J., Freeman, T. P., Kilian, C., López-Pelayo, H., & Rehm, J. (2021). Public health monitoring of cannabis use in Europe: Prevalence of use, cannabis potency, and treatment rates. *The Lancet Regional Health—Europe*, *10*, 100227.

Mechoulam, R., & Parker, L. A. (2013). The endocannabinoid system and the brain. *Annual Review of Psychology*, *64*, 21–47.

Melchior, M., Bolze, C., Fombonne, E., Surkan, P. J., Pryor, L., & Jauffret-Roustide, M. (2017). Early cannabis initiation and educational attainment: Is the association causal? Data from the French TEMPO study. *International Journal of Epidemiology*, *46*, 1641–1650.

Micallef, J., Jouanjus, É., Mallaret, M., & Lapeyre-Mestre, M. (2019). Safety signal detection by the French Addictovigilance Network: Innovative methods of investigation, examples and usefulness for public health. *Therapie*, *74*, 579–590.

MILDECA. (2020). *Les conduites addictives de la population active. Chiffres clés issus de la cohorte CONSTANCES*. https://www.drogues.gouv.fr/sites/drogues.gouv.fr/files/atoms/files/rapport_constances_def_2.pdf.

Mittleman, M. A., Lewis, R. A., Maclure, M., Schewood, J. B., & Muller, J. E. (2001). Triggering myocardial infarction by marijuana. *Circulation*, *103*, 2805–2809.

Molly, C., Mory, O., Basset, T., & Patural, H. (2012). Intoxication aiguë par cannabis chez un nourrisson de 10 mois. *Archives de Pédiatrie*, *19*, 729–732.

Murray, R. M., Quigley, H., Quattrone, D., Englund, A., & Di Forti, M. (2016). Traditional marijuana, high-potency cannabis and synthetic cannabinoids: Increasing risk for psychosis. *World Psychiatry*, *15*, 195–204.

Nappe, T. M., & Hoyte, C. O. (2017). Pediatric death due to myocarditis after exposure to cannabis. *Clinical Practice and Cases in Emergency Medicine*, *1*, 166–170.

Nawrot, T. S., Perez, L., Künzli, N., Munters, E., & Nemery, B. (2011). Public health importance of triggers of myocardial infarction: A comparative risk assessment. *Lancet*, *337*, 732–740.

Oomen, P. E., Schori, D., Tögel-Lins, K., Acreman, D., Chenorhokian, S., Luf, A., Karden, A., Paulos, C., Fornero, E., Gerace, E., Koning, R. P. J., Galindo, L., Smit-Rigter, L. A., Measham, F., & Ventura, M. (2022). Cannabis adulterated with the synthetic cannabinoid receptor agonist MDMN-4en-PINACA and the role of European drug checking services. *The International Journal on Drug Policy*, *100*, 103493. https://doi.org/10.1016/j.drugpo.2021.103493.

Pacher, P., Steffens, S., Hasko, G., Schindler, T. H., & Kunos, G. (2018). Cardiovascular effects of marijuana and synthetic cannabinoids: The good, the bad, and the ugly. *Nature Reviews Cardiology*, *15*, 151–166.

Patissier, C., Akhdar, M., Rosellini, D., Tambat, A., Tiprez, C., Wendremaire, P., & Renoux, M. C. (2015). A propos de 8 cas d'intoxication accidentelle au cannabis chez des nourrissons. *Archives de Pédiatrie*, *22*, 43–46.

Pellissier, F., Claudet, I., Pélissier-Alicot, A. L., & Franchitto, N. (2014). Parental cannabis abuse and accidental intoxications in children. *Pediatric Emergency Care*, *30*, 862–866.

Revol, B., Fouilhé Sam-Laï, N., Eysseric, H., Yahiaoui, N., & Mallaret, M. (2018). Décès en relation avec l'abus de médicaments et de substances (DRAMES) et cannabis: combien de cas depuis 3 ans? *Therapies*, *73*, 589–590.

Richards, J. R., Smith, N. E., & Moulin, A. K. (2017). Unintentional cannabis ingestion in children: A systematic review. *The Journal of Pediatrics*, *190*, 142–152.

Sallan, S. E., Zinberg, N. E., & Frei, E. (1975). Antiemetic effect of delta-9-tetrahydrocannabinol in patients receiving cancer chemotherapy. *The New England Journal of Medicine*, *293*, 795–797.

Schreck, B., Wagneur, N., Caillet, P., Gérardin, M., Cholet, J., Spadari, M., Authier, N., Tournebize, J., Gaillard, M., Serre, A., Carton, L., Pain, S., Jolliet, P., & Victorri-Vigneau, C. (2018). Cannabinoid hyperemesis syndrome: Review of the literature and of cases reported to the French addictovigilance network. *Drug and Alcohol Dependence*, *182*, 27–32.

Simonetto, D. A., Oxentenko, A. S., Herman, M. L., & Szostek, J. H. (2012). Cannabinoid hyperemesis: A case series of 98 patients. *Mayo Clinic Proceedings*, *87*, 114–119.

Spadari, M., Glaizal, M., Tichadou, L., Blanc, I., Drouet, G., Aymard, I., De Haro, L., Hayek-Langlois, M., & Arditti, J. (2009). Accidental cannabis poisoning in children: Experience of the Marseille poison center. *Presse Médicale*, *38*, 1563–1567.

Spilka, S., Le Nezet, O., Janssen, E., Brissot, A., Philippon, A., Shah, J., & Chyderiotis, S. (2018). *Les drogues à 17 ans: analyse de l'enquête ESCAPAD 2017. Tendances.* Observatoire français des drogues et toxicomanies. https://www.ofdt.fr/publications/collections/periodiques/lettre-tendances/les-drogues-17-ans-analyse-de-lenquete-escapad-2017-tendances-123-fevrier-2018/.

Tilelli, J. A., & Spack, L. D. (2006). Marijuana intoxication presenting as seizure-comment. *Pediatric Emergency Care*, *22*, 141.

Toce, M. S., Fariasn, M., Powelln, A. J., Dalyn, K. P., Vargasn, S. O., & Burnsn, M. M. (2019). Myocardial infarct after marijuana inhalation in a 16-year-old adolescent boy. *Pediatric and Developmental Pathology*, *22*, 80–86.

Venkatesan, T., Levinthal, D. J., Li, B. U. K., Tarbell, S. E., Adams, K. A., Issenman, R. M., Sarosiek, I., Jaradeh, S. S., Sharaf, R. N., Sultan, S., Stave, C. D., Monte, A. A., & Hasler, W. L. (2019). Role of chronic cannabis use: Cyclic vomiting syndrome vs cannabinoid hyperemesis syndrome. *Neurogastroenterology and Motility*, *31*, e13606.

Wolff, V., Lauer, V., Rouyer, O., Sellal, F., Meyer, N., Raul, J. S., Sabourdy, C., Boujan, F., Jahn, C., Beaujeux, R., & Marescaux, C. (2011). Cannabis use, ischemic stroke, and multifocal intracranial vasoconstriction: A prospective study in 48 consecutive young patients. *Stroke*, *42*, 1778–1780.

World Health Organization. (1993). *The ICD-10 classification of mental and behavioural disorders—Diagnostic criteria for research.* Geneva: World Health Organization.

World Health Organization. (2016). *The health and social effects of non medical cannabis use.* https://apps.who.int/iris/handle/10665/251056.

Zaheer, S., Kumar, D., Khan, M. T., Giyanwani, P. R., & Kiran, F. (2018). Epilepsy and cannabis: A literature review. *Cureus*, *10*, e3278.

Chapter 9

Acute and chronic impact of cannabis on human cognition

Justin Matheson[a] and Bernard Le Foll[a,b,c,d,e,f]

[a]*Translational Addiction Research Laboratory, Campbell Family Mental Health Research Institute, Centre for Addiction and Mental Health, Toronto, ON, Canada,* [b]*Department of Psychiatry, Faculty of Medicine, University of Toronto, Toronto, ON, Canada,* [c]*Department of Family and Community Medicine, University of Toronto, Toronto, ON, Canada,* [d]*Institute of Medical Science, University of Toronto, Toronto, ON, Canada,* [e]*Department of Pharmacology and Toxicology, University of Toronto, Toronto, ON, Canada,* [f]*Waypoint Research Institute, Waypoint Centre for Mental Health Care, Penetanguishene, ON, Canada*

Abbreviations

2-AG	2-arachidonoylglycerol
AEA	arachidonoylethanolamide (anandamide)
CBD	cannabidiol
ECS	endocannabinoid system
EEG	electroencephalogram
fMRI	functional magnetic resonance imaging
GPCR	G-protein coupled receptor
THC	delta-9-tetrahydrocannabinol
THCV	delta-9-tetrahydrocannabivarin

Introduction

Cannabis continues to be one of the most commonly used psychoactive drugs worldwide. Recent data from the United Nations Office on Drugs and Crime (UNODC) estimates that 4% of the global population used cannabis at least once in 2019, representing 200 million individuals (UNODC, 2021). Ongoing changes in the legal status of cannabis have dramatically shifted the landscape of cannabis products, increased the number of individuals who use cannabis, and changed attitudes toward use (Hall et al., 2019; Windle et al., 2019). Cannabis use is associated with numerous harms, including chronic bronchitis, cannabis use disorder, and increased risk of psychiatric illness, especially psychosis and mood and anxiety disorders (Hall et al., 2019). Given the estimated 200 million individuals using cannabis per year, even a small percent increase in the use of cannabis represents a substantial number of people impacted.

Impairment of cognition is one of the most consistent adverse effects associated with both acute and chronic use of cannabis (Broyd et al., 2016). Acute cannabis intoxication leads to impairment of cognitive skills that can increase the risk of motor vehicle collisions or other accidental injury (Hartman & Huestis, 2013) and exacerbate symptoms of psychiatric disorders such as psychosis (Curran et al., 2016). Human neuroimaging studies have found differences in brain structure and function when comparing individuals who use cannabis regularly to those who do not, though results have not always been consistent (Bloomfield et al., 2019; Sagar & Gruber, 2018). Importantly, an earlier age of onset and a heavier frequency and quantity of use have been associated with greater cannabis-related changes (Blest-Hopley et al., 2020; Sagar & Gruber, 2018). These findings support studies that have detected associations between cannabis use and poorer academic and vocational performance (Curran et al., 2016). This chapter will provide an overview of mechanisms of cannabis-induced cognitive impairment, discuss the current evidence for which cognitive domains are impacted by acannabis use, and explore some important moderators of cannabis-cognition relationships.

Cannabis Use, Neurobiology, Psychology, and Treatment. https://doi.org/10.1016/B978-0-323-89862-1.00004-0

Mechanisms of cannabis-induced cognitive impairment

The endocannabinoid system

The cognitive effects of cannabis are largely mediated by the interactions of its primary psychoactive compound, delta-9-tetrahydrocannabinol (THC), with the endocannabinoid system (ECS). The two primary cannabinoid receptors, cannabinoid receptors type-1 (CB_1) and type-2 (CB_2), are 7-transmembrane-domain G-protein coupled receptors (GPCRs) and are regarded as the most abundant GPCRs in the mammalian brain (Howlett et al., 2002). CB_1 is primarily localized in the central nervous system and plays a central role in mediating the cognitive effects of cannabinoids, while CB_2 is primarily localized in the periphery and is less involved in cognition (Howlett et al., 2002). The endogenous ECS ligands (the endocannabinoids) are phospholipid derivatives containing a poly-unsaturated fatty acid moiety (arachidonate) and a polar head group, either ethanolamine in the case of arachidonoylethanolamide (anandamide; AEA) or glycerol in the case of 2-arachidonoylglycerol (2-AG) (Pamplona & Takahashi, 2012). An important role of the endocannabinoids is that of retrograde messengers. For example, in response to a postsynaptic signal (e.g., an increase in intracellular calcium), 2-AG is produced by the postsynaptic neuron, released into the synaptic cleft, and activates the presynaptic CB_1 receptor, which in turn suppresses calcium influx into the neuron by inhibiting voltage-gated calcium channels (Lu & Mackie, 2016). The result is suppression of either inhibitory or excitatory neurotransmitter release (usually GABA or glutamate), dampening subsequent activity at the synapse for a brief duration.

Cellular mechanisms of THC effects

The maximal effects produced by THC stimulation of both CB_1 and CB_2 receptors are significantly lower than those of full receptor agonists, indicating that THC is a partial agonist at both receptors (Pertwee & Cascio, 2014). In rodent hippocampal, striatal, midbrain, and cerebellar slices and cell cultures, THC can inhibit glutamate transmission by direct activation of presynaptic CB_1 receptors (Hoffman & Lupica, 2013). However, THC has low efficacy at glutamatergic synapses and can antagonize the effects of higher efficacy agonists like WIN55,212-2 (Straiker & Mackie, 2005). In contrast, THC appears to behave as a full CB_1 receptor agonist at GABAergic synapses. The difference in the efficacy of THC is likely due to the density of CB_1 receptors, which is low at glutamatergic synapses but high at GABAergic synapses (Hoffman & Lupica, 2013). Puighermanal et al. (2009) demonstrated that targeted deletion of CB_1 receptors from GABAergic, but not glutamatergic, neurons prevented hippocampal-mediated amnesic effects in mice. Thus, it appears that in the hippocampus, THC inhibits GABA release from cholecystokinin-positive interneurons, which suppresses inhibitory control over pyramidal neurons, thus increasing glutamate release and disrupting synchronization of hippocampal network activity that is important for memory (Prini et al., 2020).

Neurocircuitry of THC effects

While the hippocampus has been a major focus of cannabis and cognition research, growing evidence suggests that THC exerts its cognitive effects through disruption of larger neural circuits including the mesocorticolimbic circuit and the cortical salience network (Ramaekers et al., 2021). THC administration can increase dopamine release in the nucleus accumbens by activating presynaptic CB_1 receptors at GABA and glutamate terminals in the ventral tegmental area (Bloomfield et al., 2016). In addition, THC administration leads to a loss of functional connectivity in the mesocorticolimbic circuit and likely also the cortical salience network (Ramaekers et al., 2021). Increased striatal dopamine release and decreased functional connectivity in these two important brain circuits are thought to interfere with several downstream cognitive processes, including distortion of attentional salience processing, reductions in attention, and impairment of psychomotor inhibition and working memory (Ramaekers et al., 2021). Thus, through its action at presynaptic CB_1 receptors in key brain regions including the hippocampus, striatum, and prefrontal cortex, THC can interfere with most major cognitive processes. Whether this impairment manifests in humans under typical conditions of cannabis use and to what extent will be considered in the following sections, after a brief discussion of methodological considerations.

Methods for studying cognitive effects of cannabis in humans

Cognitive domains

At its broadest level, cognition can be defined as the sum of mental processes involved in the acquisition of knowledge and understanding through thought, direct experience, and sensation. Cognition can be divided into different modalities (often referred to as "domains") that are thought to reflect relatively independent processes. As the reader will no doubt notice throughout this chapter, there is significant heterogeneity in the division and nomenclature of cognitive domains in the cannabis and cognition literature. Four primary cognitive domains will be highlighted here: memory (where working memory is distinguished from other memory types), attention, psychomotor function, and executive functions. In addition, tasks that probe speed of processing and general cognitive function, which do not fit into one of the previous cognitive domains, will be discussed separately.

Acute versus chronic effects

Cognitive impairment can be observed under conditions of acute cannabis intoxication and in individuals who use cannabis chronically who are not currently intoxicated. Thus, it is important to distinguish between cognitive effects that occur during acute intoxication that are a direct pharmacological effect of psychoactive cannabinoids ("acute effects") and cognitive effects that are the result of frequent, chronic use of cannabis that persist even when an individual has not recently used cannabis ("chronic effects"). Acute effects are typically measured in the human laboratory, where controlled doses of cannabis/cannabinoids and placebo are administered, and one or more validated cognitive tasks are used to probe-specific cognitive domains. Acute effects of cannabis vary by dose, route of administration, and other factors, but typically last for about 2–4 hours and return to baseline by 6 hours postadministration when smoked (Grotenhermen, 2003). Chronic effects of cannabis are typically assessed by comparing a sample of participants who use cannabis regularly to a sample of participants who do not use cannabis (or use infrequently). In addition, longitudinal studies have followed cohorts of individuals over time and assessed cognitive performance at multiple time points in a sample that included both cannabis-using and nonusing individuals.

Effects of cannabis and cannabinoids on human cognition

Several important metaanalyses have been published within the past 5 years to examine the relationship between acute cannabis intoxication (McCartney et al., 2021; Zhornitsky et al., 2021) or chronic cannabis use (Figueiredo et al., 2020; Krzyzanowski & Purdon, 2020; Lovell et al., 2020; Platt et al., 2019; Scott et al., 2018) and cognitive performance. These metaanalyses are summarized in Tables 1–6 to give the reader a broad picture of the acute and chronic cognitive effects of cannabis in humans.

TABLE 1 Metaanalyses of the acute and chronic effects of cannabis and cannabinoids on memory in humans (excluding working memory).

Reference	Target population and design	Cognitive domains	Effect size [95% confidence interval]
Acute effects			
Zhornitsky et al. (2021)	Healthy volunteers; placebo-controlled experimental design involving a CB_1 receptor partial agonist	Verbal learning	−0.688 [−0.888, −0.488]
		Verbal memory	−0.513 [−0.653, −0.374]
Chronic effects			
Figueiredo et al. (2020)	Case control, longitudinal or cross-sectional studies of adults aged 18 years or older (experiencing chronic cannabis use or DSM cannabis dependence, including a control group as appropriate)	Short-term memory	−0.48 [−0.65, −0.31]
		Long-term memory	−0.43 [−0.69, −0.16]
Krzyzanowski and Purdon (2020)	Cross-sectional or longitudinal studies (defined regular cannabis using group vs appropriate control group)	Total immediate recall	−0.366 [−0.483, −0.249]
		Short delay free recall	−0.273 [−0.440, −0.106]

Continued

TABLE 1 Metaanalyses of the acute and chronic effects of cannabis and cannabinoids on memory in humans (excluding working memory)—cont'd

Reference	Target population and design	Cognitive domains	Effect size [95% confidence interval]
		Long delay free recall	−0.338 [−0.516, −0.160]
		Recognition recall	−0.117 [−0.316, 0.082]
Lovell et al. (2020)	Cross-sectional studies of adults aged 18 years or older (mean 2 ≥ years and mean 4 ≥ days per week of cannabis use vs appropriate control group)	Learning and memory (overall)	−0.33 [−0.46, −0.19]
		Verbal learning	−0.37 [−0.52, −0.22]
		Retention	−0.36 [−0.51, −0.21]
Platt et al. (2019)	Case-control design	Event-based prospective memory	−0.49 [−0.90, −0.08]
		Time-based prospective memory	−0.70 [−0.80, −0.61]
Scott et al. (2018)	Cross-sectional studies of adolescent and/or young adults (mean age 26 years or younger) (heavy, frequent, and/or problematic use vs control group)	Learning	−0.33 [−0.42, −0.24]
		Delayed memory	−0.26 [−0.35, −0.16]

Effect size estimates are Cohen's *d* or Hedges' *g*; a negative effect size means that the active cannabis group performed worse than the placebo group (acute effects) or the cannabis use group performed worse than the control group (chronic effects); effect sizes are considered statistically significant with the 95% CI does not cross 0.

TABLE 2 Metaanalyses of the acute and chronic effects of cannabis and cannabinoids on working memory in humans.

Reference	Target population and design	Cognitive domains	Effect size [95% confidence interval]
Acute effects			
McCartney et al. (2021)	Placebo-controlled experimental design involving THC	Working memory	−0.36 [−0.52, −0.20]
Zhornitsky et al. (2021)	Healthy volunteers; placebo-controlled experimental design involving a CB_1 receptor partial agonist	Working memory	−0.514 [−0.655, −0.373]
Chronic effects			
Lovell et al. (2020)	Cross-sectional studies of adults aged 18 years or older (mean 2 ≥ years and mean 4 ≥ days per week of cannabis use vs appropriate control group)	Working memory	0.01 [−0.23, 0.25]
Scott et al. (2018)	Cross-sectional studies of adolescent and/or young adults (mean age 26 years or younger) (heavy, frequent, and/or problematic use vs control group)	Executive functioning—updating/working memory	−0.22 [−0.31, −0.12]

Effect size estimates are Cohen's *d* or Hedges' *g*; a negative effect size means that the active cannabis group performed worse than the placebo group (acute effects) or the cannabis use group performed worse than the control group (chronic effects); effect sizes are considered statistically significant with the 95% CI does not cross 0.

TABLE 3 Metaanalyses of the acute and chronic effects of cannabis and cannabinoids on attention in humans.

Reference	Target population and design	Cognitive domains	Effect size [95% confidence interval]
Acute effects			
McCartney et al. (2021)	Placebo-controlled experimental design involving THC	Divided attention	−0.28 [−0.36, −0.20]
		Sustained attention	−0.23 [−0.37, −0.10]
Zhornitsky et al. (2021)	Healthy volunteers; placebo-controlled experimental design involving a CB_1 receptor partial agonist	Attention	−0.223 [−0.348, −0.099]
Chronic effects			
Figueiredo et al. (2020)	Case control, longitudinal or cross-sectional studies of adults aged 18 years or older (experiencing chronic cannabis use or DSM cannabis dependence, including a control group as appropriate)	Attention	−0.16 [−0.30, −0.02]
Lovell et al. (2020)	Cross-sectional studies of adults aged 18 years or older (mean 2 ≥ years and mean 4 ≥ days per week of cannabis use vs appropriate control group)	Attention	0.05 [−0.21, 0.31]
Scott et al. (2018)	Cross-sectional studies of adolescent and/or young adults (mean age 26 years or younger) (heavy, frequent, and/or problematic use vs control group)	Attention	−0.21 [−0.31, −0.12]

Effect size estimates are Cohen's *d* or Hedges' *g*; a negative effect size means that the active cannabis group performed worse than the placebo group (acute effects) or the cannabis use group performed worse than the control group (chronic effects); effect sizes are considered statistically significant with the 95% CI does not cross 0.

TABLE 4 Metaanalyses of the acute and chronic effects of cannabis and cannabinoids on psychomotor function in humans.

Reference	Target population and design	Cognitive domains	Effect size [95% confidence interval]
Acute effects			
McCartney et al. (2021)	Placebo-controlled experimental design involving THC	Tracking performance	−0.42 [−0.58, −0.25]
		Fine motor function	−0.36 [−0.60, −0.12]
Chronic effects			
Figueiredo et al. (2020)	Case control, longitudinal or cross-sectional studies of adults aged 18 years or older (experiencing chronic cannabis use or DSM cannabis dependence, including a control group as appropriate)	Motor impulsivity	−0.05 [−0.34, 0.24]
Scott et al. (2018)	Cross-sectional studies of adolescent and/or young adults (mean age 26 years or younger) (heavy, frequent, and/or problematic use vs control group)	Motor functioning	−0.02 [−0.22, 0.18]

Effect size estimates are Cohen's *d* or Hedges' *g*; a negative effect size means that the active cannabis group performed worse than the placebo group (acute effects) or the cannabis use group performed worse than the control group (chronic effects); effect sizes are considered statistically significant with the 95% CI does not cross 0.

TABLE 5 Metaanalyses of the acute and chronic effects of cannabis and cannabinoids on executive functions in humans.

Reference	Target population and design	Cognitive domains	Effect size [95% confidence interval]
Acute effects			
McCartney et al. (2021)	Placebo-controlled experimental design involving THC	Conflict control	−0.34 [−0.42, −0.25]
		Fluid intelligence	−0.37 [−0.46, −0.27]
Zhornitsky et al. (2021)	Healthy volunteers; placebo-controlled experimental design involving a CB_1 receptor partial agonist	Executive functions	−0.370 [−0.485, −0.245]
		Impulsivity	−0.294 [−0.414, −0.174]
Chronic effects			
Figueiredo et al. (2020)	Case control, longitudinal or cross-sectional studies of adults aged 18 years or older (experiencing chronic cannabis use or DSM cannabis dependence, including a control group as appropriate)	Cognitive impulsivity	−0.30 [−0.58, −0.02]
		Cognitive flexibility	−0.33 [−0.54, −0.12]
Lovell et al. (2020)	Cross-sectional studies of adults aged 18 or older (mean 2 ≥ years and mean 4 ≥ days per week of cannabis use vs appropriate control group)	Executive function	−0.18 [−0.31, −0.05]
		Decision making	−0.52 [−0.93, −0.11]
Scott et al. (2018)	Cross-sectional studies of adolescent and/or young adults (mean age 26 years or younger) (heavy, frequent, and/or problematic use vs control group)	Executive functioning—abstraction/shifting	−0.30 [−0.40, −0.20]
		Executive functioning—inhibition	−0.25 [−0.38, −0.13]

Effect size estimates are Cohen's *d* or Hedges' *g*; a negative effect size means that the active cannabis group performed worse than the placebo group (acute effects) or the cannabis use group performed worse than the control group (chronic effects); effect sizes are considered statistically significant with the 95% CI does not cross 0.

TABLE 6 Metaanalyses of the acute and chronic effects of cannabis and cannabinoids on speed of processing and general cognition in humans.

Reference	Target population and design	Cognitive domains	Effect size [95% confidence interval]
Acute effects			
McCartney et al. (2021)	Placebo-controlled experimental design involving THC	Information processing	−0.38 [−0.55, −0.21]
		Reaction time	−0.28 [−0.43, −0.13]
Zhornitsky et al. (2021)	Healthy volunteers; placebo-controlled experimental design involving a CB_1 receptor partial agonist	Speed of processing	−0.384 [−0.492, −0.276]

TABLE 6 Metaanalyses of the acute and chronic effects of cannabis and cannabinoids on speed of processing and general cognition in humans—cont'd

Reference	Target population and design	Cognitive domains	Effect size [95% confidence interval]
Chronic effects			
Lovell et al. (2020)	Cross-sectional studies of adults aged 18 or older (mean 2 ≥ years and mean 4 ≥ days per week of cannabis use vs appropriate control group)	Global (cognition)	−0.25 [−0.35, −0.15]
		Information processing	−0.11 [−0.34, 0.12]
Scott et al. (2018)	Cross-sectional studies of adolescent and/or young adults (mean age 26 years or younger) (heavy, frequent, and/or problematic use vs control group)	Speed of information processing	−0.26 [−0.38, −0.15]
		Visuospatial function	−0.04 [−0.22, 0.18]

Effect size estimates are Cohen's *d* or Hedges' *g*; a negative effect size means that the active cannabis group performed worse than the placebo group (acute effects) or the cannabis use group performed worse than the control group (chronic effects); effect sizes are considered statistically significant with the 95% CI does not cross 0.

Memory

Acute effects

The most consistent impairment observed during acute cannabis intoxication is a decrease in performance on verbal recall tasks (Broyd et al., 2016; Dellazizzo et al., 2022), which are tasks that rely on episodic memory, yet have a working memory component. This acute effect of cannabis was first demonstrated experimentally in the early 1970s (Abel, 1970), and has been subsequently replicated across a range of doses and routes of administration (Ranganathan & D'Souza, 2006). As summarized in Table 1, a recent metaanalysis found moderate-to-large effects of acute cannabinoid exposure on both verbal learning and verbal memory (Zhornitsky et al., 2021). Immediate and delayed recall of words are impaired when lists are learned under the influence of intravenous THC, but not when the lists are learned prior to THC exposure, suggesting that THC specifically disrupts encoding of verbal memory (Ranganathan et al., 2017). Studies using prose recall (i.e., recall of a short story or scene) as a more ecologically valid task have provided additional support for a specific relationship between cannabis and impairment in verbal episodic memory (Hindocha et al., 2017).

Chronic effects

As observed with acute intoxication, chronic effects of cannabis on verbal episodic memory are some of the most consistent deficits in cross-sectional studies in both adolescent (Scott et al., 2018) and adult samples (Figueiredo et al., 2020). Notably, there was a significant impact of cannabis use on all 13 memory effect size estimates across the 5 metaanalyses summarized in Table 1 (Figueiredo et al., 2020; Krzyzanowski & Purdon, 2020; Lovell et al., 2020; Platt et al., 2019; Scott et al., 2018). While some studies have found improvement of verbal recall performance with sustained abstinence, the evidence for complete recovery of function is mixed (Broyd et al., 2016). In a metaanalysis, Krzyzanowski and Purdon (2020) found that duration of abstinence was an important predictor of verbal recall performance, where significant between-groups differences were observed only when cannabis-using groups were tested after fewer than 7 days of abstinence. This suggests that differences in verbal memory performance may be due to residual effects of cannabinoids rather than a chronic impact of cannabis use on brain structure and/or function, which warrants further clarification.

Working memory

Acute effects

Impairment of immediate verbal recall under the influence of cannabis is clear evidence that cannabis can adversely affect verbal working memory. Furthermore, cannabis can impair working memory across a range of other tasks, routes of administration, and doses (Broyd et al., 2016), which is emphasized by the significant impact of cannabinoids on working memory in the two metaanalyses summarized in Table 2 (McCartney et al., 2021; Zhornitsky et al., 2021). Cannabis seems to impair working memory at higher "load" (i.e., higher task difficulty) compared to lower load conditions (e.g., Hindocha et al., 2017; Ilan et al., 2004), which was supported by a pharmacological fMRI challenge that found vaporized THC interferes with the linear relationship between increasing working memory load and activity of the dorsolateral prefrontal cortex (Bossong et al., 2012). Inconsistency in working memory deficits during acute cannabis intoxication is likely related to the wide range of tasks available, e.g., digit span, *n*-back, delayed matching to sample, digit recall, or spatial working memory tasks (Broyd et al., 2016). Furthermore, it is difficult to isolate an effect of cannabis on working memory without involving other memory or attention processes. For example, an EEG study found that smoked cannabis interfered with neural activity related to both transient and sustained attention, which was paralleled by impaired working memory performance, suggesting that the effects of cannabis on working memory may be partly attributable to disrupted orienting of attention (Ilan et al., 2004).

Chronic effects

Of the two metaanalyses summarized in Table 2, one found a significant (small magnitude) effect of cannabis use on working memory impairment in adolescents and young adults (Scott et al., 2018), while the other found no significant impact of cannabis use on working memory in a sample of adults aged 18 years or older (Lovell et al., 2020). A recent study found that a positive urine THC test was associated with poorer working memory task performance, while other indicators of cannabis use (e.g., total lifetime use, age of first use) did not predict working memory performance, suggesting that the observed chronic effects of cannabis on working memory are due to subacute intoxication from residual cannabinoids (Owens et al., 2019), which is in line with the finding by Krzyzanowski and Purdon (2020) discussed above.

Attention

Acute effects

Acute cannabis intoxication leads to impairment in attention across a variety of tasks, dosing paradigms, and participant samples (Broyd et al., 2016; Crane et al., 2013). Results from the two metaanalyses summarized in Table 3 support this conclusion, as significant but small magnitude effect sizes were reported across three attention domains (McCartney et al., 2021; Zhornitsky et al., 2021). Cannabis appears to have a more consistent effect on divided attention (e.g., Bedi et al., 2013; Ramaekers et al., 2016; Spindle et al., 2018), which may represent an overall effect of cannabis on orienting attention. For example, one study found that intravenous THC impaired performance in a distractibility task and that study participants reported feeling more easily distracted by their environment (D'Souza et al., 2004).

Chronic effects

Chronic cannabis use has been associated with impairment of multiple modalities of attention (e.g., divided and sustained attention) in both adolescent and adult samples (Broyd et al., 2016). However, as summarized in Table 3, recent metaanalyses of cross-sectional studies in adolescents and young adults (Scott et al., 2018) and adults aged 18 years or older (Figueiredo et al., 2020) found small effect size associations between cannabis use and attention, while a third metaanalysis found no significant impact of cannabis use on attention (Lovell et al., 2020), suggesting that the chronic effects of cannabis on attention are less pronounced than effects on memory.

Psychomotor function

Acute effects

The existing literature suggests that cannabis impairs some components of psychomotor function across routes of administration (Broyd et al., 2016) and that this impairment is dose dependent (e.g., Boggs et al., 2018; Hunault et al., 2009). As summarized in Table 4, one recent metaanalysis by McCartney et al. (2021) found a significant effect of THC on tracking performance and fine motor function. In particular, the critical tracking task (which has participants control a cursor to follow a moving target on a screen) has been shown to be sensitive to cannabinoid-related impairment (e.g., Ramaekers et al., 2006, 2009).

Chronic effects

As summarized in Table 4, two recent metaanalyses failed to find a significant association between cannabis use and motor functioning in adolescents and young adults (Scott et al., 2018) or motor impulsivity in adults aged 18 or older (Figueiredo et al., 2020). Thus, it appears that either chronic cannabis use does not lead to meaningful impairment in psychomotor functions or that impairment manifests only in particularly tasks (e.g., the critical tracking task) or under certain conditions (e.g., more frequent use).

Executive functions

Acute effects

One systematic review found consistent evidence for an acute impairing effect of cannabis on response inhibition (Broyd et al., 2016). However, most of the evidence for an acute impairing effect of cannabis on response inhibition comes from the stop-signal task (e.g., McDonald et al., 2003; Ramaekers et al., 2006, 2009, 2016), a fairly simple task that may not actually be measuring executive function. For example, oral THC impaired stop reaction time yet had no effect on the more cognitively demanding go/no-go task, suggesting that THC may specifically impair motor inhibition without affecting higher-order cognitive inhibition (McDonald et al., 2003). In addition, all of the tasks that have been used to probe executive function performance depend on intact attentional processes, which are known to be impaired by cannabis, limiting the specificity of the observed effects (Oomen et al., 2018). More recently, as summarized in Table 5, two metaanalyses found a significant effect of THC on conflict control and fluid intelligence (McCartney et al., 2021) and of partial CB_1 receptor agonists on executive functions and impulsivity (Zhornitsky et al., 2021).

Chronic effects

Mixed evidence exists for a chronic effect of cannabis use on executive function. It has been noted that studies in older populations tend to find more evidence of cannabis-related executive dysfunction, whereas studies in younger samples often find no difference between cannabis-using and nonusing groups, suggesting a potential age-related effect of chronic cannabis use on executive functioning (Broyd et al., 2016). However, as summarized in Table 5, six effect size estimates across three recent metaanalyses have found a significant association between chronic cannabis use and impaired executive functioning, in adolescents and young adults (Scott et al., 2018) and in adults aged 18 years or older (Figueiredo et al., 2020; Lovell et al., 2020).

Speed of processing and general cognition

Acute effects

As summarized in Table 6, two recent metaanalyses found significant effects of THC on information processing and reaction time (McCartney et al., 2021) and of partial CB_1 receptor agonists on speed of processing (Zhornitsky et al., 2021). Of note, Zhornitsky et al. (2021) found that both sex and route of administration influenced effects on speed of processing, where male sex was associated with greater impairment, and oral route of administration was associated with less impairment.

Chronic effects

As summarized in Table 6, two recent metaanalyses have found mixed evidence for a chronic effect of cannabis on speed of processing and other measures of general cognitive function. While Scott et al. (2018) found a significant (small magnitude) association between chronic cannabis use and speed of information processing among adolescents and young adults, Lovell et al. (2020) did not find a significant association of chronic cannabis use with information processing in adults aged 18 years or older, though they did find a significant association with a global composite measure of cognition.

Factors influencing the cognitive effects of cannabis and cannabinoids

There are numerous factors that have an impact on the cognitive effects of cannabinoids, which makes it exceedingly difficult to make any absolute claims about the acute and chronic effects of cannabinoids. In general, we know that cannabinoids *can* impact virtually any cognitive domain, but the more relevant question is: under which conditions and in which individuals are these impairments likely to manifest? The following sections seek to shed some light on specific factors to consider that can help us to answer this question.

Route of administration

Smoking and vaping remain the most common methods of using cannabis, though oral formulations (e.g., edibles, oil capsules) are rising in popularity, and newer cannabis products such as topicals, sublingual preparations, and even rectal suppositories exist in rapidly diversifying cannabis retail markets (Spindle et al., 2019). Inhaled and oral formulations of cannabis products have been compared in multiple studies, which have confirmed significant differences in THC pharmacokinetics and time course of cognitive effects, e.g., the longer duration of acute effects observed with oral THC compared to inhaled THC (McCartney et al., 2021). Metaregression analysis of human laboratory studies of acute cannabis effects found that oral formulations of cannabis were associated with smaller speed of processing deficits than other routes such as smoking (Zhornitsky et al., 2021). Thus, the acute effects of cannabis on cognition differ in time course and magnitude depending on whether THC is ingested or inhaled, and this is likely true of other routes of administration, though this remains relatively untested.

Dose and potency

The acute cognitive effects of cannabinoids are often (but not always) dose-dependent in human laboratory experiments (Broyd et al., 2016; McCartney et al., 2021). In studies of smoked cannabis, instead of the exact dose of THC administered, the potency of the cannabis strain is often reported (along with an estimate of total THC exposure). Cannabis potency is typically quantified as the concentration of THC in a cannabis product. The mean THC potency of cannabis products has been increasing for decades; for example, potency of cannabis plant material in the United States increased from 8.9% in 2008 to 17.1% in 2017 (Chandra et al., 2019). This trend challenges the generalizability of previous studies of acute cognitive effects of cannabis, which have typically administered low-potency strains of cannabis, often as low as 1%–3% THC (e.g., Heishman et al., 1989).

An important source of variability in smoked cannabis studies is individual variation in smoking topography, which refers to the physical characteristics of smoking behavior such as puff duration, inhalational volume, and interpuff intervals. Since the effects of THC are felt very rapidly when cannabis is smoked, users have been described as "titrating to effect," i.e., regulating their intake to achieve a certain subjective high (Huestis, 2007). Thus, participants in a laboratory experiment administered smoked cannabis may choose to smoke different amounts and be exposed to different doses of THC as a result.

CBD and other phytocannabinoids

Converging evidence suggests that cannabidiol (CBD) may attenuate the cognitive impairing effects of THC—for example, consuming cannabis strains with a higher THC:CBD ratio has been associated with greater memory impairment and CBD has been shown to partially prevent cognitive impairment when co-administered with THC in human laboratory experiments (Colizzi & Bhattacharyya, 2017). However, this effect of CBD has not been demonstrated consistently, may be relevant only for memory and is of unclear relevance to chronic use of cannabis (Bourque & Potvin, 2021). In addition to CBD, there are hundreds of other phytocannabinoid compounds in the cannabis plant that could also interact with THC. For example, Englund et al. (2016) found protective effects of delta-9-tetrahydrocannabivarin (THCV), another phytocannabinoid which is a CB_1 receptor neutral antagonist, on THC-induced verbal recall impairment. Thus, further research is needed to clarify whether phytocannabinoids other than THC might have independent effects on cognition or may influence the acute effects of THC.

Tolerance

Studies comparing groups of participants with varying frequencies of cannabis use have found that acute cognitive effects of cannabis are less prominent in chronic, heavy users, suggesting tolerance develops to the cognitive effects of cannabis, likely via downregulation of CB_1 receptors (Ramaekers et al., 2020). A recent systematic review concluded that, though there was heterogeneity in research designs and outcomes, human studies have found that cognition was the outcome domain most likely to demonstrate tolerance when comparing heavy/frequent use of cannabis to no/limited use (Colizzi & Bhattacharyya, 2018). Tolerance seems to develop especially for attention, and possibly less so for memory, psychomotor function, and executive function (Broyd et al., 2016; Colizzi & Bhattacharyya, 2018). Prior exposure to cannabis is thus a crucial construct to evaluate in any study of cannabis and cognition.

Age

Adolescence is an important period for neurodevelopment, which has led researchers to examine whether the adolescent brain is more vulnerable to adverse effects of cannabis (Crane et al., 2013). In animal models, THC exposure during adolescence has a more pronounced cognitive impairing effect than adult exposure, and converging human evidence has similarly suggested a negative correlation between age of initiation of cannabis use and cognitive functioning (Crane et al., 2013). However, a recent metaanalysis of cross-sectional studies in adolescents and young adults failed to find an association between age of initiation of cannabis and cognitive effects (Scott et al., 2018), which was replicated in a separate metaanalysis that included adults aged 18 or older (Lovell et al., 2020). Similarly, Zhornitsky et al. (2021) found that age did not significantly influence the acute cognitive effects of partial CB_1 receptor agonists in any cognitive domain. Yet, a recent human laboratory study found that adolescents (aged 18–20 years) were more sensitive than adults (aged 30–40 years) to the effects of oral THC on reaction time, response accuracy, and time perception (Murray et al., 2022). Thus, more work is needed to characterize the relationship between age, cannabis, and cognition.

Sex

Animal studies have found sex differences in the ECS, in the behavioral effects of THC and in the pharmacokinetics of THC (Cooper & Craft, 2018). This is likely due to the interactions of gonadal hormones (especially estrogens) with the ECS; for example, estradiol can directly influence CB_1 receptor density in rats (Castelli et al., 2014). Sex differences in the acute effects of THC have been observed in humans, but the evidence is more equivocal. For example, our lab found no sex differences in the cognitive effects of smoked cannabis, despite significantly lower concentrations of THC and metabolites in female (vs male) participants (Matheson et al., 2020). One study found poorer psychomotor task performance in female compared to male participants after 10 mg of oral THC (Roser et al., 2009), though several other studies have not found sex differences (e.g., Anderson et al., 2010; Fogel et al., 2017). In their recent metaanalysis of placebo-controlled studies evaluating cognitive effects of partial CB_1 receptor agonists, Zhornitsky et al. (2021) found that sex impacted the domain of speed of processing, where a greater male to female ratio was associated with greater deficit. More human studies that are appropriately powered to examine sex differences in the cognitive effects of cannabinoids are needed to clarify this relationship.

Co-use of other drugs

Understanding the true cognitive effects of cannabis requires human participants to be tested in the absence of exposure to other psychoactive drugs, since different classes of drugs can have varying effects across cognitive domains (Fernandez-Serrano et al., 2011). However, it is challenging to find large enough samples of participants who use cannabis and no other psychoactive drugs (or use very infrequently), and there has been inconsistent reporting of drug use parameters in previous studies (Broyd et al., 2016). Furthermore, use of multiple classes of drugs seems to predict poorer cognitive functioning. For example, recent work from our group found that increasing number of substance use disorders was associated with poorer self-reported cognitive functioning, in both attention and executive function scores (Bourgault et al., 2022). Thus, prior exposure to other psychoactive drugs (e.g., alcohol, nicotine, and opioids) is likely a major contributor to heterogeneity in findings of acute and chronic cannabis effects on cognition.

Application to other areas

Studies of the acute and chronic impact of cannabis on cognition have applications in many different areas. One major area has been in understanding, evaluating, and legislating cannabis-involved driving. As can be expected from the broad array of cognitive decrements associated with acute cannabis intoxication, cannabis use prior to driving is associated with an increased risk of motor vehicle collisions (Hartman & Huestis, 2013). An important ongoing area of inquiry is in leveraging data from human laboratory experiments to relate blood concentrations of THC to driving impairment (McCartney et al., 2021), which would assist in developing evidence-based THC limits for defining cannabis-impaired driving at the roadside.

Another important application of cannabis and cognition research is in developing safe and effective cannabinoid drugs. There are many potential therapeutic applications of cannabinoid drugs, including in the treatment of mental health and substance use disorders (Kloiber et al., 2020), where there are significant concerns about adverse cognitive effects. Cognitive tasks that are sensitive to cannabis effects in healthy volunteers (e.g., verbal recall tasks) can be readily implemented in placebo-controlled trials of cannabinoids to monitor any unwanted cognitive adverse effects. Having a nuanced understanding of the effects of cannabis on cognition is important when undertaking studies in patient populations, as cognitive effects may diverge from effects seen in healthy populations. For example, patients with chronic pain, which has its own detrimental effects on cognitive functioning, may see improvements in cognition following treatment with cannabinoids, as the cannabinoids lead to alleviation of symptoms, which results in a net benefit to cognition (Ramaekers et al., 2021).

Mini-dictionary of terms

Attention: The cognitive ability to focus the information processing capability of the brain on task-relevant stimuli while ignoring irrelevant stimuli. The division of attention into distinct cognitive processes varies, but the most common task concepts include sustained attention, selective attention, divided attention, and attention switching.
Divided Attention: The ability to attend to or process more than one stimulus or piece of information at the same time. Typically assessed using a primary task (e.g., a tracking task) alone and with a secondary task (e.g., an arithmetic task) to determine how performance in the primary task is impacted by addition of a new task.
Episodic Memory: The recall of events that are personally meaningful. Experimentally, this can be assessed by having research participants learn and recall lists of words (verbal recall task) or short stories or vignettes (prose recall task).
Executive Functions: A family of "high-level" cognitive processes that select and monitor behaviors to work toward chosen goals. Often includes abstract reasoning, decision-making, risk-taking, inhibitory control, and impulsivity, among other constructs. Can be measured in a large variety of tasks.
Memory: The neural and psychological processes by which information is encoded, stored, and retrieved as needed. Many influential models of memory have been proposed, involving different divisions of memory into subtypes, which is reflected in the variety of memory tasks available.
Psychomotor Function: The sum of sensory processing, central processing and integration, overt motor behavior, and overall sensorimotor coordination to guide goal-directed movements. Often operationalized as gross motor behavior, which can be assessed using tasks that test ballistic activity (e.g., finger-tapping), gross body movement (e.g., body sway), fine motor control (e.g., trail-making task), or motor manipulation (e.g., grooved pegboard task).
Speed of Processing: The time it takes to performance a cognitive task. Speed of processing tasks is usually included in cognitive task batteries as a general measure of cognitive ability.
Sustained Attention: A process that allows for focus and continuous effort on an activity or stimulus over an extended period of time. Sustained attention tasks often involve monitoring performance on a simple, tedious task to examine if performance deteriorates over time.
Working Memory: The temporary storage and manipulation of a limited amount of information. In Baddeley's original multicomponent model of working memory, it is subserved by three systems (the phonological loop, which briefly stores verbal information; the visuospatial sketchpad, which briefly stores nonverbal information; and the episodic buffer, which briefly retrieves and stores episodic memories), which are coordinated by a process called the central executive.

Key facts of testing cognitive functions

There are a wide variety of cognitive tasks available to probe each cognitive domain, which often assess overlapping cognitive abilities, and thus the choice of appropriate tasks is vital to the correct interpretation of results.
Cognitive tasks are often administered in batteries, which involve multiple tasks that probe distinct and/or overlapping functions.
Choice of task should be based on whether the task is standardized and has adequate test-retest reliability, is appropriately difficult for the population being studied, can be administered during the time frame of the experimental session, and is not overly sensitive to practice effects.
Practice effects arise when a research participant improves in performance after completing the task multiple times, which is likely to happen in studies of acute cannabis effects or in longitudinal studies of chronic cannabis effects.
Practice effects can be mitigated by including a practice session to familiarize participants to the task ahead of data collection, or by administering alternate forms of a task (e.g., different word lists in a memory task).

Summary points

- This chapter focused on acute and chronic effects of cannabis on human cognition, as well as potential mechanisms of cannabis-related cognitive impairment, methodological considerations, and factors moderating cannabis-cognition relationships.
- THC (and other cannabinoids) can interfere with most major cognitive processes through action at presynaptic cannabinoid receptors in key brain regions including the hippocampus, striatum, and prefrontal cortex.
- Both acute cannabis intoxication and chronic use of cannabis are robustly and consistently associated with impairment of verbal recall performance, indicative of impaired verbal episodic and working memory.
- During acute cannabis intoxication, attention (especially divided attention), psychomotor function (especially critical tracking and motor inhibition), and executive functions are impaired and speed of processing is slowed, though these effects are not as consistently observed as verbal recall deficits.
- Chronic use of cannabis is associated with deficits in attention and executive function, while the evidence is mixed for speed of processing and does not clearly show an association with psychomotor function.
- There are many key factors that influence both acute and chronic cognitive effects of cannabis, including drug-related factors (e.g., dose, route of administration, presence of other phytocannabinoids such as CBD), prior exposure to cannabis (i.e., tolerance-related effects) and other psychoactive drugs, and individual difference measures such as age and sex.

References

Abel, E. L. (1970). Marijuana and memory. *Nature*, *227*(5263), 1151–1152. https://doi.org/10.1038/2271151b0.

Anderson, B. M., Rizzo, M., Block, R. I., Pearlson, G. D., & O'Leary, D. S. (2010). Sex, drugs, and cognition: Effects of marijuana. *Journal of Psychoactive Drugs*, *42*(4), 413–424. https://doi.org/10.1080/02791072.2010.10400704.

Bedi, G., Cooper, Z. D., & Haney, M. (2013). Subjective, cognitive and cardiovascular dose-effect profile of nabilone and dronabinol in marijuana smokers. *Addiction Biology*, *18*(5), 872–881. https://doi.org/10.1111/j.1369-1600.2011.00427.x.

Blest-Hopley, G., Giampietro, V., & Bhattacharyya, S. (2020). A systematic review of human neuroimaging evidence of memory-related functional alterations associated with cannabis use complemented with preclinical and human evidence of memory performance alterations. *Brain Sciences*, *10*(2). https://doi.org/10.3390/brainsci10020102.

Bloomfield, M. A. P., Ashok, A. H., Volkow, N. D., & Howes, O. D. (2016). The effects of Delta(9)-tetrahydrocannabinol on the dopamine system. *Nature*, *539*(7629), 369–377. https://doi.org/10.1038/nature20153.

Bloomfield, M. A. P., Hindocha, C., Green, S. F., Wall, M. B., Lees, R., Petrilli, K., … Freeman, T. P. (2019). The neuropsychopharmacology of cannabis: A review of human imaging studies. *Pharmacology & Therapeutics*, *195*, 132–161. https://doi.org/10.1016/j.pharmthera.2018.10.006.

Boggs, D. L., Cortes-Briones, J. A., Surti, T., Luddy, C., Ranganathan, M., Cahill, J. D., … Skosnik, P. D. (2018). The dose-dependent psychomotor effects of intravenous delta-9-tetrahydrocannabinol (Delta(9)-THC) in humans. *Journal of Psychopharmacology*, *32*(12), 1308–1318. https://doi.org/10.1177/0269881118799953.

Bossong, M. G., Jansma, J. M., van Hell, H. H., Jager, G., Oudman, E., Saliasi, E., … Ramsey, N. F. (2012). Effects of delta9-tetrahydrocannabinol on human working memory function. *Biological Psychiatry*, *71*(8), 693–699. https://doi.org/10.1016/j.biopsych.2012.01.008.

Bourgault, Z., Rubin-Kahana, D. S., Hassan, A. N., Sanches, M., & Le Foll, B. (2022). Multiple substance use disorders and self-reported cognitive function in U.S. adults: Associations and sex-differences in a nationally representative sample. *Frontiers in Psychiatry*, *12*. https://doi.org/10.3389/fpsyt.2021.797578.

Bourque, J., & Potvin, S. (2021). Cannabis and cognitive functioning: From acute to residual effects, from randomized controlled trials to prospective designs. *Frontiers in Psychiatry*, *12*, 596601. https://doi.org/10.3389/fpsyt.2021.596601.

Broyd, S. J., van Hell, H. H., Beale, C., Yucel, M., & Solowij, N. (2016). Acute and chronic effects of cannabinoids on human cognition—A systematic review. *Biological Psychiatry*, *79*(7), 557–567. https://doi.org/10.1016/j.biopsych.2015.12.002.

Castelli, M. P., Fadda, P., Casu, A., Spano, M. S., Casti, A., Fratta, W., & Fattore, L. (2014). Male and female rats differ in brain cannabinoid CB1 receptor density and function and in behavioural traits predisposing to drug addiction: Effect of ovarian hormones. *Current Pharmaceutical Design*, *20*(13), 2100–2113.

Chandra, S., Radwan, M. M., Majumdar, C. G., Church, J. C., Freeman, T. P., & ElSohly, M. A. (2019). New trends in cannabis potency in USA and Europe during the last decade (2008-2017). *European Archives of Psychiatry and Clinical Neuroscience*, *269*(1), 5–15. https://doi.org/10.1007/s00406-019-00983-5.

Colizzi, M., & Bhattacharyya, S. (2017). Does cannabis composition matter? Differential effects of delta-9-tetrahydrocannabinol and cannabidiol on human cognition. *Current Addiction Reports*, *4*(2), 62–74. https://doi.org/10.1007/s40429-017-0142-2.

Colizzi, M., & Bhattacharyya, S. (2018). Cannabis use and the development of tolerance: A systematic review of human evidence. *Neuroscience and Biobehavioral Reviews*, *93*, 1–25. https://doi.org/10.1016/j.neubiorev.2018.07.014.

Cooper, Z. D., & Craft, R. M. (2018). Sex-dependent effects of cannabis and cannabinoids: A translational perspective. *Neuropsychopharmacology*, *43*(1), 34–51. https://doi.org/10.1038/npp.2017.140.

Crane, N. A., Schuster, R. M., Fusar-Poli, P., & Gonzalez, R. (2013). Effects of cannabis on neurocognitive functioning: Recent advances, neurodevelopmental influences, and sex differences. *Neuropsychology Review*, *23*(2), 117–137. https://doi.org/10.1007/s11065-012-9222-1.

Curran, H. V., Freeman, T. P., Mokrysz, C., Lewis, D. A., Morgan, C. J., & Parsons, L. H. (2016). Keep off the grass? Cannabis, cognition and addiction. *Nature Reviews Neuroscience*, *17*(5), 293–306. https://doi.org/10.1038/nrn.2016.28.

Dellazizzo, L., Potvin, S., Giguère, S., & Dumais, A. (2022). Evidence on the acute and residual neurocognitive effects of cannabis use in adolescents and adults: A systematic meta-review of meta-analyses. *Addiction*. https://doi.org/10.1111/add.15764.

D'Souza, D. C., Perry, E., MacDougall, L., Ammerman, Y., Cooper, T., Wu, Y. T., … Krystal, J. H. (2004). The psychotomimetic effects of intravenous delta-9-tetrahydrocannabinol in healthy individuals: Implications for psychosis. *Neuropsychopharmacology*, *29*(8), 1558–1572. https://doi.org/10.1038/sj.npp.1300496.

Englund, A., Atakan, Z., Kralj, A., Tunstall, N., Murray, R., & Morrison, P. (2016). The effect of five day dosing with THCV on THC-induced cognitive, psychological and physiological effects in healthy male human volunteers: A placebo-controlled, double-blind, crossover pilot trial. *Journal of Psychopharmacology*, *30*(2), 140–151. https://doi.org/10.1177/0269881115615104.

Fernandez-Serrano, M. J., Perez-Garcia, M., & Verdejo-Garcia, A. (2011). What are the specific vs. generalized effects of drugs of abuse on neuropsychological performance? *Neuroscience and Biobehavioral Reviews*, *35*(3), 377–406. https://doi.org/10.1016/j.neubiorev.2010.04.008.

Figueiredo, P. R., Tolomeo, S., Steele, J. D., & Baldacchino, A. (2020). Neurocognitive consequences of chronic cannabis use: A systematic review and meta-analysis. *Neuroscience and Biobehavioral Reviews*, *108*, 358–369. https://doi.org/10.1016/j.neubiorev.2019.10.014.

Fogel, J. S., Kelly, T. H., Westgate, P. M., & Lile, J. A. (2017). Sex differences in the subjective effects of oral Delta(9)-THC in cannabis users. *Pharmacology, Biochemistry, and Behavior*, *152*, 44–51. https://doi.org/10.1016/j.pbb.2016.01.007.

Grotenhermen, F. (2003). Pharmacokinetics and pharmacodynamics of cannabinoids. *Clinical Pharmacokinetics*, *42*(4), 327–360. https://doi.org/10.2165/00003088-200342040-00003.

Hall, W., Stjepanovic, D., Caulkins, J., Lynskey, M., Leung, J., Campbell, G., & Degenhardt, L. (2019). Public health implications of legalising the production and sale of cannabis for medicinal and recreational use. *Lancet*, *394*(10208), 1580–1590. https://doi.org/10.1016/s0140-6736(19)31789-1.

Hartman, R. L., & Huestis, M. A. (2013). Cannabis effects on driving skills. *Clinical Chemistry*, *59*(3), 478–492. https://doi.org/10.1373/clinchem.2012.194381.

Heishman, S. J., Stitzer, M. L., & Yingling, J. E. (1989). Effects of tetrahydrocannabinol content on marijuana smoking behavior, subjective reports, and performance. *Pharmacology, Biochemistry, and Behavior*, *34*(1), 173–179.

Hindocha, C., Freeman, T. P., Xia, J. X., Shaban, N. D. C., & Curran, H. V. (2017). Acute memory and psychotomimetic effects of cannabis and tobacco both 'joint' and individually: A placebo-controlled trial. *Psychological Medicine*, *47*(15), 2708–2719. https://doi.org/10.1017/s0033291717001222.

Hoffman, A. F., & Lupica, C. R. (2013). Synaptic targets of Delta9-tetrahydrocannabinol in the central nervous system. *Cold Spring Harbor Perspectives in Medicine*, *3*(8). https://doi.org/10.1101/cshperspect.a012237.

Howlett, A. C., Barth, F., Bonner, T. I., Cabral, G., Casellas, P., Devane, W. A., … Pertwee, R. G. (2002). International Union of Pharmacology. XXVII. Classification of cannabinoid receptors. *Pharmacological Reviews*, *54*(2), 161–202.

Huestis, M. A. (2007). Human cannabinoid pharmacokinetics. *Chemistry & Biodiversity*, *4*(8), 1770–1804. https://doi.org/10.1002/cbdv.200790152.

Hunault, C. C., Mensinga, T. T., Bocker, K. B., Schipper, C. M., Kruidenier, M., Leenders, M. E., … Meulenbelt, J. (2009). Cognitive and psychomotor effects in males after smoking a combination of tobacco and cannabis containing up to 69 mg delta-9-tetrahydrocannabinol (THC). *Psychopharmacology*, *204*(1), 85–94. https://doi.org/10.1007/s00213-008-1440-0.

Ilan, A. B., Smith, M. E., & Gevins, A. (2004). Effects of marijuana on neurophysiological signals of working and episodic memory. *Psychopharmacology*, *176*(2), 214–222. https://doi.org/10.1007/s00213-004-1868-9.

Kloiber, S., Matheson, J., Kim, H. K., & Le Foll, B. (2020). Cannabinoid drugs in mental health disorders. In P. Riederer, G. Laux, T. Nagatsu, W. Le, & C. Riederer (Eds.), *NeuroPsychopharmacotherapy* (pp. 1–35). Cham: Springer International Publishing.

Krzyzanowski, D. J., & Purdon, S. E. (2020). Duration of abstinence from cannabis is positively associated with verbal learning performance: A systematic review and meta-analysis. *Neuropsychology*, *34*(3), 359–372. https://doi.org/10.1037/neu0000615.

Lovell, M. E., Akhurst, J., Padgett, C., Garry, M. I., & Matthews, A. (2020). Cognitive outcomes associated with long-term, regular, recreational cannabis use in adults: A meta-analysis. *Experimental and Clinical Psychopharmacology*, *28*(4), 471–494. https://doi.org/10.1037/pha0000326.

Lu, H. C., & Mackie, K. (2016). An introduction to the endogenous cannabinoid system. *Biological Psychiatry*, *79*(7), 516–525. https://doi.org/10.1016/j.biopsych.2015.07.028.

Matheson, J., Sproule, B., Di Ciano, P., Fares, A., Le Foll, B., Mann, R. E., & Brands, B. (2020). Sex differences in the acute effects of smoked cannabis: Evidence from a human laboratory study of young adults. *Psychopharmacology*, *237*(2), 305–316. https://doi.org/10.1007/s00213-019-05369-y.

McCartney, D., Arkell, T. R., Irwin, C., & McGregor, I. S. (2021). Determining the magnitude and duration of acute Δ(9)-tetrahydrocannabinol (Δ(9)-THC)-induced driving and cognitive impairment: A systematic and meta-analytic review. *Neuroscience and Biobehavioral Reviews*, *126*, 175–193. https://doi.org/10.1016/j.neubiorev.2021.01.003.

McDonald, J., Schleifer, L., Richards, J. B., & de Wit, H. (2003). Effects of THC on behavioral measures of impulsivity in humans. *Neuropsychopharmacology*, *28*(7), 1356–1365. https://doi.org/10.1038/sj.npp.1300176.

Murray, C. H., Huang, Z., Lee, R., & de Wit, H. (2022). Adolescents are more sensitive than adults to acute behavioral and cognitive effects of THC. *Neuropsychopharmacology*. https://doi.org/10.1038/s41386-022-01281-w.

Oomen, P. P., van Hell, H. H., & Bossong, M. G. (2018). The acute effects of cannabis on human executive function. *Behavioural Pharmacology*, *29*(7), 605–616. https://doi.org/10.1097/fbp.0000000000000426.

Owens, M. M., McNally, S., Petker, T., Amlung, M. T., Balodis, I. M., Sweet, L. H., & MacKillop, J. (2019). Urinary tetrahydrocannabinol is associated with poorer working memory performance and alterations in associated brain activity. *Neuropsychopharmacology*, *44*(3), 613–619. https://doi.org/10.1038/s41386-018-0240-4.

Pamplona, F. A., & Takahashi, R. N. (2012). Psychopharmacology of the endocannabinoids: Far beyond anandamide. *Journal of Psychopharmacology*, *26* (1), 7–22. https://doi.org/10.1177/0269881111405357.

Pertwee, R. G., & Cascio, M. G. (2014). Known pharmacological actions of delta-9-tetrahydrocannabinol and of four other chemical constituents of cannabis that activate cannabinoid receptors. In *Handbook of cannabis*. Oxford: Oxford University Press.

Platt, B., O'Driscoll, C., Curran, V. H., Rendell, P. G., & Kamboj, S. K. (2019). The effects of licit and illicit recreational drugs on prospective memory: A meta-analytic review. *Psychopharmacology*, *236*(4), 1131–1143. https://doi.org/10.1007/s00213-019-05245-9.

Prini, P., Zamberletti, E., Manenti, C., Gabaglio, M., Parolaro, D., & Rubino, T. (2020). Neurobiological mechanisms underlying cannabis-induced memory impairment. *European Neuropsychopharmacology*, *36*, 181–190. https://doi.org/10.1016/j.euroneuro.2020.02.002.

Puighermanal, E., Marsicano, G., Busquets-Garcia, A., Lutz, B., Maldonado, R., & Ozaita, A. (2009). Cannabinoid modulation of hippocampal long-term memory is mediated by mTOR signaling. *Nature Neuroscience*, *12*(9), 1152–1158. https://doi.org/10.1038/nn.2369.

Ramaekers, J. G., Kauert, G., Theunissen, E. L., Toennes, S. W., & Moeller, M. R. (2009). Neurocognitive performance during acute THC intoxication in heavy and occasional cannabis users. *Journal of Psychopharmacology*, *23*(3), 266–277. https://doi.org/10.1177/0269881108092393.

Ramaekers, J. G., Mason, N. L., Kloft, L., & Theunissen, E. L. (2021). The why behind the high: Determinants of neurocognition during acute cannabis exposure. *Nature Reviews Neuroscience*, *22*(7), 439–454. https://doi.org/10.1038/s41583-021-00466-4.

Ramaekers, J. G., Mason, N. L., & Theunissen, E. L. (2020). Blunted highs: Pharmacodynamic and behavioral models of cannabis tolerance. *European Neuropsychopharmacology*, *36*, 191–205. https://doi.org/10.1016/j.euroneuro.2020.01.006.

Ramaekers, J. G., Moeller, M. R., van Ruitenbeek, P., Theunissen, E. L., Schneider, E., & Kauert, G. (2006). Cognition and motor control as a function of Delta9-THC concentration in serum and oral fluid: Limits of impairment. *Drug and Alcohol Dependence*, *85*(2), 114–122. https://doi.org/10.1016/j.drugalcdep.2006.03.015.

Ramaekers, J. G., van Wel, J. H., Spronk, D. B., Toennes, S. W., Kuypers, K. P., Theunissen, E. L., & Verkes, R. J. (2016). Cannabis and tolerance: Acute drug impairment as a function of cannabis use history. *Scientific Reports*, *6*, 26843. https://doi.org/10.1038/srep26843.

Ranganathan, M., & D'Souza, D. C. (2006). The acute effects of cannabinoids on memory in humans: A review. *Psychopharmacology*, *188*(4), 425–444. https://doi.org/10.1007/s00213-006-0508-y.

Ranganathan, M., Radhakrishnan, R., Addy, P. H., Schnakenberg-Martin, A. M., Williams, A. H., Carbuto, M., … D'Souza, D. C. (2017). Tetrahydrocannabinol (THC) impairs encoding but not retrieval of verbal information. *Progress in Neuro-Psychopharmacology & Biological Psychiatry*, *79*(Pt B), 176–183. https://doi.org/10.1016/j.pnpbp.2017.06.019.

Roser, P., Gallinat, J., Weinberg, G., Juckel, G., Gorynia, I., & Stadelmann, A. M. (2009). Psychomotor performance in relation to acute oral administration of Delta9-tetrahydrocannabinol and standardized cannabis extract in healthy human subjects. *European Archives of Psychiatry and Clinical Neuroscience*, *259*(5), 284–292. https://doi.org/10.1007/s00406-009-0868-5.

Sagar, K. A., & Gruber, S. A. (2018). Interactions between recreational cannabis use and cognitive function: Lessons from functional magnetic resonance imaging. *Annals of the New York Academy of Sciences*. https://doi.org/10.1111/nyas.13990.

Scott, J. C., Slomiak, S. T., Jones, J. D., Rosen, A. F. G., Moore, T. M., & Gur, R. C. (2018). Association of cannabis with cognitive functioning in adolescents and young adults: A systematic review and meta-analysis. *JAMA Psychiatry*, *75*(6), 585–595. https://doi.org/10.1001/jamapsychiatry.2018.0335.

Spindle, T. R., Bonn-Miller, M. O., & Vandrey, R. (2019). Changing landscape of cannabis: Novel products, formulations, and methods of administration. *Current Opinion in Psychology*, *30*, 98–102. https://doi.org/10.1016/j.copsyc.2019.04.002.

Spindle, T. R., Cone, E. J., Schlienz, N. J., Mitchell, J. M., Bigelow, G. E., Flegel, R., … Vandrey, R. (2018). Acute effects of smoked and vaporized cannabis in healthy adults who infrequently use cannabis: A crossover trial. *JAMA Network Open*, *1*(7), e184841. https://doi.org/10.1001/jamanetworkopen.2018.4841.

Straiker, A., & Mackie, K. (2005). Depolarization-induced suppression of excitation in murine autaptic hippocampal neurones. *The Journal of Physiology*, *569*(Pt 2), 501–517. https://doi.org/10.1113/jphysiol.2005.091918.

UNODC. (2021). *World drug report 2021*. Retrieved from https://www.unodc.org/unodc/en/data-and-analysis/wdr2021.html.

Windle, S. B., Wade, K., Filion, K. B., Kimmelman, J., Thombs, B. D., & Eisenberg, M. J. (2019). Potential harms from legalization of recreational cannabis use in Canada. *Canadian Journal of Public Health*, *110*(2), 222–226. https://doi.org/10.17269/s41997-018-00173-1.

Zhornitsky, S., Pelletier, J., Assaf, R., Giroux, S., Li, C. R., & Potvin, S. (2021). Acute effects of partial CB(1) receptor agonists on cognition—A meta-analysis of human studies. *Progress in Neuro-Psychopharmacology & Biological Psychiatry*, *104*, 110063. https://doi.org/10.1016/j.pnpbp.2020.110063.

Chapter 10

Polydrug users, use of cannabinoids and abuse potential: A focus on cannabidiol (CBD)

Kerri A. Schoedel and Naama Levy-Cooperman
Altreos Research Partners, Inc., Toronto, ON, Canada

Abbreviations

AE	adverse event
CBD	cannabidiol
CB-XR	cannabinoid receptor (-1 or -2)
CUD	cocaine-use disorder
DSM-XV	Diagnostic and Statistical Manual of Mental Disorders (Fourth or Fifth Edition)
E_{max}	peak effect
FDA	Food and Drug Administration
OUD	opioid-use disorder
SUD	substance-use disorder
THC	9-delta-tetrahydrocannabinol
US	United States

Introduction

Cannabis species (*C. sativa*, *C. indica*) contain well over 500 known compounds (Rock & Parker, 2021), of which, 9-delta-tetrahydrocannabinol (THC) is well known for its psychoactive properties. This is the active compound considered responsible for most of the perceivable "abuse-related" effects of cannabis (i.e., the feelings of euphoria, "high," "stoned," etc.), which exerts its effects primarily through partial agonism of the cannabinoid-1 receptor (CB-1R). THC has shown significant abuse potential in animal and clinical laboratory studies and cannabis is one of the most widely abused substances in the world. In recent years, cannabidiol (CBD), present in *Cannabis* species in widely varying quantities, has garnered increasing interest for its potential therapeutic properties.

Although classified as a cannabinoid due to its presence in cannabis, CBD binds very weakly to CB-1 and CB-2 receptors (Mechoulam & Hanus, 2002; Parker et al., 2004). Instead, CBD affects CB-1R- and CB-2R-mediated processes indirectly by inhibiting the metabolism of naturally occurring endocannabinoids (*N*-arachidonoylethanolamine and 2-arachidonoylglycerol). Aside from its potential effects on endocannabinoid metabolism, CBD may also produce its effects via a number of other mechanisms, including positive allosteric modulation of gamma-aminobutyric acid-A and serotonin 1A receptors (Bakas et al., 2017), modulation of $\alpha 3$ and $\alpha 1$ glycine inhibition receptors (Devinsky, Cilio, & Cross, 2014), inhibition of muscarinic cholinergic and $\alpha 7$-nicotinic receptors (Ali et al., 1991; Mahgoub et al., 2013), modulation of intracellular calcium via G-protein coupled receptor 55 and transient receptor potential cation channel subfamily V member 1, and modulation of adenosine-mediated signaling (Iannotti et al., 2014; Sylantyev et al., 2013).

The purpose of this chapter is to review the abuse potential of CBD with a focus on polydrug users. Physical or psychological dependence, as important aspects of an "addiction" syndrome or substance-use disorder (SUD), will also be covered. A summary of the types of models/data examined in this chapter to determine abuse or dependence potential of CBD is shown in Table 1. In addition, CBD as a potential treatment for SUDs is briefly summarized.

Cannabis Use, Neurobiology, Psychology, and Treatment. https://doi.org/10.1016/B978-0-323-89862-1.00023-4

TABLE 1 Summary of types of models used to examine CBD abuse or dependence potential.

Abuse/dependence effect of interest	Species	Models
Rewarding effects	Rodents	Conditioned place preference studies
Reinforcing effects	Rodents Nonhuman primates	Self-administration studies
	Human	Human abuse potential studies (willingness to "take drug again" [indirect])
Perceivable/abuse-related subjective effects	Rodents Nonhuman primates Pigeons	Drug discrimination studies
	Human	Human abuse potential studies
		Side effects or mood measures from other clinical trials (indirect)
Physical dependence	Rodents	Physical dependence studies
	Human	Clinical physical dependence studies

Abuse and dependence potential of CBD in animals

Rewarding, reinforcing, and perceivable effects of CBD

Animal studies support the concept that CBD lacks reinforcing or rewarding effects. In a rodent model of "reward," i.e., conditioned place preference, whereby the animal is trained to associate the effects of a drug with a particular chamber in a specially designed cage, CBD administration to rats and mice over a wide range of doses was not associated with either a place preference (i.e., showing that the animals wanted to spend more time in the chamber where they could feel the effects of the drug) or place aversion (showing that they did not like the effects of the drug) (Genaro et al., 2017; Markos et al., 2017; Parker et al., 2004; Vann et al., 2008; Viudez-Martínez et al., 2019).

An animal model of "reinforcement" involves training an animal to self-administer a drug or other reward (such as food or saccharin) by pressing levers. The test drug is then substituted for the drug or food used for training. If the animals continue to press the lever associated with delivery of the test drug, it is considered to have reinforcing effects. In a modified version of this type of study, mice were trained to self-administer a saccharin/CBD (50 mg/kg oral) solution. Once the saccharin was removed from the solution, animals failed to continue self-administering CBD (Viudez-Martínez et al., 2019). In United States (US) Food and Drug Administration (FDA) review materials that are available to the public (Food and Drug Administration, 2018), a self-administration study with CBD oral solution is described, whereby rats were first trained to self-administer cocaine through lever pressing. When CBD was substituted for the cocaine, animals failed to self-administer the drug compared to saline. In contrast, the positive control (amphetamine) was self-administered more than saline when substituted for the cocaine. A second study is described where animals were similarly trained to deliver heroin via lever presses; these animals also did not self-administer CBD compared to vehicle control when it was substituted for heroin, although, additional benzodiazepine controls (diazepam and midazolam) were also not self-administered by these animals. Finally, a third study included monkeys trained to self-administer the benzodiazepine midazolam. When CBD was substituted for the midazolam, it was not self-administered more than vehicle control.

A third model of abuse potential in animals is the drug discrimination paradigm, whereby animals are trained to recognize the "subjective" effects of a drug of abuse (i.e., any effects that are perceivable to the animals, such as sedation, dizziness, etc.). Animals who have learned to distinguish a training drug of abuse from vehicle are then presented with the test drug; if the animals press more frequently on the training drug lever they are said to "generalize" to the training drug. That is, the test drug is thought to be producing perceivable effects that are similar to the drug of abuse. In published drug discrimination studies of monkeys, rats, and pigeons, CBD did not "generalize" to either THC or the CB-1R antagonist rimonabant, demonstrating that the animals did not consider its perceivable effects to be similar to these training drugs (Bueno et al., 1976; Henriksson et al., 1975; Hiltunen & Järbe, 1986; Järbe et al., 1977; Järbe & Mathis, 1991;

McMahon, 2016; Vann et al., 2008; Zuardi et al., 1981). A similar study is described in FDA review materials (Food and Drug Administration, 2018), where oral CBD did not generalize to a benzodiazepine in rats.

Physical dependence potential of CBD in animals

Physical dependence occurs with some drugs of abuse, including opioids and benzodiazepines, as well as drugs that are not associated with abuse, such as beta-blockers. When a drug also has abuse or "psychologic" dependence potential, such as opioids, the withdrawal syndrome experienced on discontinuation of the drug may result in drug-seeking or drug-taking, contributing to a dependence syndrome and loss of behavioral control. Animals are used to evaluate physical dependence with a test drug at doses and/or treatment durations that may not be ethical to study in humans. In these studies, the test drug is administered to animals daily for several days to weeks, and then abruptly discontinued. The animals are then closely observed for any signs of a withdrawal syndrome, such as changes in behavior, vital signs, food consumption, or body weight. Several relatively short-term studies have demonstrated that CBD administered daily for 3–6 days was not associated with rebound hyperexcitability, alterations in motor activity, or somatic or behavioral signs, indicating that it was not associated with a withdrawal syndrome after short-term use (Karler & Turkanis, 1980; Myers et al., 2018; Viudez-Martínez et al., 2019). A longer physical dependence study in adult and juvenile rats was performed with 20 days of twice daily oral CBD, using morphine and diazepam as positive controls. In this study, oral CBD did not show a clear withdrawal syndrome, although the validity of the findings is uncertain given that neither morphine nor diazepam produced a measurable withdrawal syndrome (Food and Drug Administration, 2018).

Abuse and dependence potential of CBD in humans

Subjective and mood effects of CBD in humans

Many studies of CBD have reported on side effects of CBD related to changes in mood or other neuropsychiatric symptoms. In healthy volunteers, oral CBD was associated with subjective side effects like somnolence (sleepiness) and dizziness at high doses (1500–6500 mg) (Taylor et al., 2018). These side effects have also been reported with CBD in various patient groups, such as those with Lennox-Gastaut and Dravet syndrome, as well as schizophrenia (Devinsky et al., 2017; Devinsky, Patel, Cross, et al., 2018; Devinsky, Patel, Thiele, et al., 2018; Sands, Rahdari, & Oldham, 2019). Other neuropsychiatric side effects have been reported occasionally, such as ataxia, anxiety, agitation, and psychomotor hyperactivity (Devinsky, Patel, Thiele, et al., 2018). Two metaanalyses of CBD oral solution have evaluated the incidence of common adverse events (AEs) across two and four trials, respectively. These studies found an overall incidence of somnolence of 22.1% for CBD vs 7.5% for placebo in the first metaanalysis and 24.5% vs 8.4% in the second metaanalysis (Lattanzi, Brigo, Cagnetti, et al., 2018; Lattanzi, Brigo, Trinka, et al., 2018). The side effect of somnolence could suggest abuse potential of the benzodiazepine type (i.e., use for relaxation).

Clinical studies of CBD have often included various measures of mood and subjective effects in healthy volunteers or other populations. With a few exceptions, the majority of studies in healthy volunteers, including those using oral CBD (100–6000 mg) or 32 mg of inhaled CBD, showed no effects on measures of mood and other effects, e.g., anxiety, addiction severity, euphoria, dissociation, psychotomimetic effects, sobriety/druggedness, alertness, inactivity, dizziness, tension/relaxation, dreaming, mental sedation, physical sedation, sleep, or sleepiness (e.g., Crippa et al., 2011; Fusar-Poli et al., 2009; Linares et al., 2018; Martin-Santos et al., 2012; Taylor et al., 2018; Zuardi et al., 2017). A few studies in healthy volunteers showed that oral CBD, at doses ranging from 300 to 900 mg, was associated with a decrease in subjective anxiety and energetic arousal, and increases in sedation (e.g., Crippa et al., 2011; Winton-Brown et al., 2011; Zuardi et al., 2017).

Human abuse potential studies of CBD in cannabis and polydrug users

Several human abuse potential studies with CBD have been published. In these studies, recreational drug users are asked to respond to scales that assess the test drug's perceivable effects, whether they like the effects of the test drug, and their willingness to take it again. These studies are usually conducted as single-dose, double-blind, randomized crossover studies where the test drug is compared to positive control(s) and placebo.

Studies with oral CBD doses up to 900 mg have shown minimal abuse-related effects relative to placebo, such as drug "liking," willingness to take the drug again, "good effects", feeling "high" or sedated in cannabis users (Arndt & de Wit, 2017; Babalonis et al., 2017; Haney et al., 2016). A study of recently detoxified habitual smoked marijuana users compared oral CBD doses up to 1200 mg with THC and diazepam and found no effects of CBD on a measure of "high" compared with

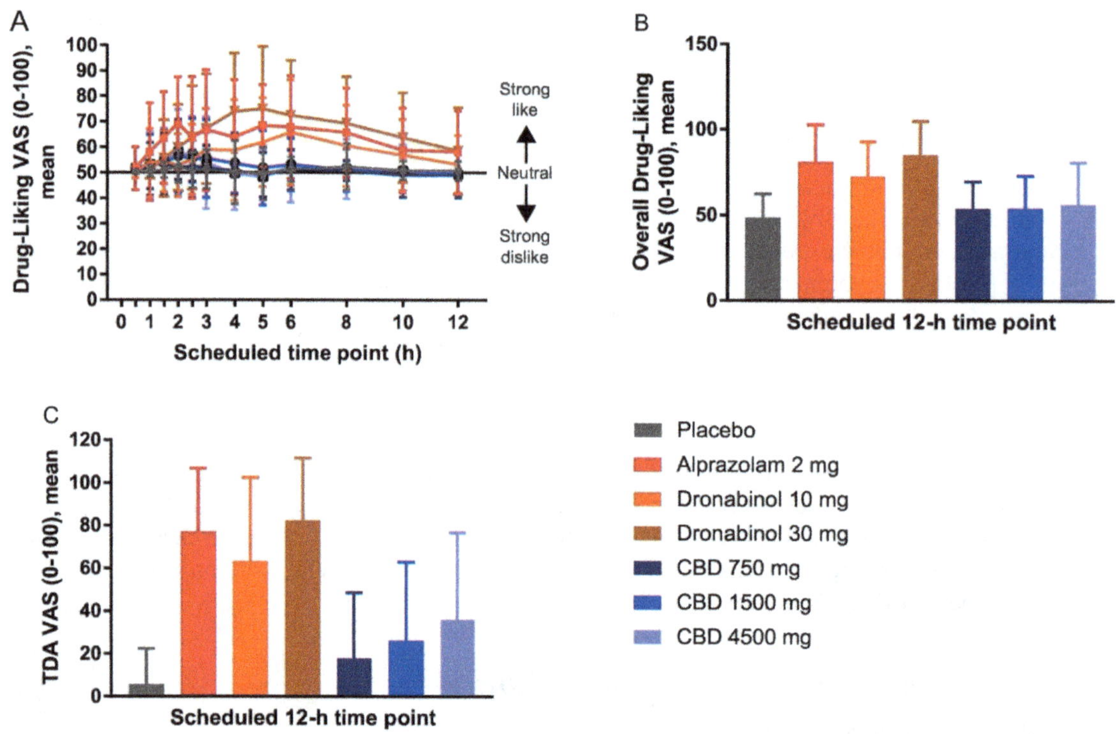

FIG. 1 Human abuse potential data for cannabidiol in recreational polydrug users. This figure shows "Drug Liking," "Overall Drug Liking," and willingness to "Take Drug Again" visual analog scale scores for oral cannabidiol, alprazolam, dronabinol, and placebo in recreational polydrug users. *(From Schoedel, K. A., Szeto, I., Setnik, B., et al. (2018). Abuse potential assessment of cannabidiol (CBD) in recreational polydrug users: A randomized, double-blind, controlled trial.* Epilepsy and Behavior, 88, *162–171, with permission.)*

placebo with either single doses or on Days 5, 12, and 19 of chronic dosing (Gong et al., 1984). A study of "smoked" CBD (150 μg/kg) in healthy male medical students found that CBD did not affect ratings of "high" (Dalton et al., 1976).

Another single-dose, randomized, double-blind, placebo- and active-controlled crossover study evaluated higher doses of CBD as an oral solution (750, 1500, and 4500 mg) compared with single oral doses of alprazolam (2 mg), synthetic THC (dronabinol 10 and 30 mg), and placebo in recreational polydrug users (Schoedel et al., 2018). In this study, peak "Drug Liking" was not significantly different for subjects taking 750 mg CBD compared with placebo (Fig. 1), which is consistent with findings of the Babalonis study (Babalonis et al., 2017). Although peak Drug Liking for 1500 and 4500 mg CBD was statistically different from placebo, the mean differences were <10 points on a 100-point VAS compared with >18-point differences between positive controls and placebo. In addition, alprazolam and dronabinol had significantly higher peak Drug Liking, Overall Liking, and Take Drug Again VAS compared with all doses of CBD. Overall, this study demonstrated minimal abuse-related effects of CBD at higher doses (1500 and 4500 mg).

Another study of volunteers with some cannabis experience found that 100 mg of vaporized CBD was associated with higher ratings of pleasant and liking compared to placebo and 100 mg oral CBD (Spindlea et al., 2020). A small study in noncannabis using volunteers found that only the 400 mg oral CBD dose increased ratings of good effects and willingness to take again compared to placebo, while the 800 mg tended to decrease positive ratings (Arout et al., 2021).

A number of other clinical studies conducted in cannabis users, typically in combination with THC, also included subjective effects or "intoxication" data with CBD alone. A randomized, placebo-controlled crossover study in current cannabis users and nonnaïve nonusers administered CBD (400 mg) by vaporization found no significant effect of CBD on a Clinician-Administered Dissociative States Scale, but CBD was associated with other subjective effects compared to placebo (e.g., intoxication, stoned, memory, perceptual distortion, etc.; Solowij et al., 2019). However, CBD scores were significantly lower than those of THC 8 mg.

Physical dependence potential of CBD in humans

As described above, physical dependence leading to withdrawal syndrome may not, on its own, be associated with "addiction" but in combination with other abuse-related effects or psychological dependence can lead to continued drug-taking/drug-seeking behavior. A few studies have evaluated aspects of CBD's physical dependence potential. In one study, oral CBD 200mg was administered for 10weeks (while subjects continued to use cannabis) and there was no evidence of withdrawal symptoms upon discontinuation (Solowij et al., 2018). However, this study is limited by the lack of placebo or other controls. Another study examined 750 or 1500mg CBD or placebo administered twice daily for 7days and found no clear evidence of any drug withdrawal syndrome upon discontinuation (Taylor et al., 2018). Finally, a dedicated physical dependence study found no evidence of withdrawal effects after discontinuation of 750mg oral CBD solution (administered twice daily for 4weeks) (Taylor et al., 2020). In this study, two withdrawal checklists did not show increases in the subjects randomized to placebo (i.e., those who abruptly discontinued CBD) (Fig. 2).

Evidence of CBD abuse in polydrug users

Despite the widespread availability of CBD products in many countries, at present no cases of CBD abuse, dependence, addiction, diversion, or withdrawal syndrome were identified in the medical literature and only a few studies have surveyed aspects associated with abuse or addiction potential of CBD. A study in Los Angeles surveyed young adult current cannabis users from March 2019 to 2020 (Fedorova et al., 2021). The sample was divided into users who primarily used CBD and those who primarily used THC. Within CBD-dominant users, 56.4% used only CBD, 17.8% used primarily CBD, and 25.7% used half CBD and half THC. While the study did not report on recreational (nonmedical) use of CBD, it did find that the most common reasons for use were to relieve physical pain (61.5% in CBD-dominant and 50.7% in THC-dominant users), as a sleep aid (42.3% and 29.0%, respectively), and to relieve anxiety (41.6% and 29.0%, respectively).

A national survey (2019) of US-based adults evaluated perceptions about medical use and abuse potential of CBD and THC and found that most respondents perceived CBD as having lower abuse potential compared to THC and marijuana (McFadden & Malone, 2021). About 8%–9% reported using CBD only or THC only, while 14% reported consuming both. Of CBD-only consumers, most reported using it for pain relief (58.8%), anxiety (52.6%), or sleep (36.1%), with only 3.1% using it for "fun or recreation". In users of both CBD/THC, primary reasons for CBD use were similar, although 24.2% reported using CBD for recreational reasons. In contrast, 59.8% of THC-only users reported using THC for recreational purposes.

The National Poison Data System in the US, which collects data on calls to poison control centers, added CBD to their list of monitored substances in 2019. In that year, there were 1131 case mentions of "cannabidiol or CBD", of which 912 were single exposures (Gummin et al., 2020). Most mentions of CBD were unintentional exposures (612), although there were 149 intentional exposures, a category that includes suspected suicide, misuse, abuse, or unknown. For context, in 2019, there were 1016 intentional exposures for marijuana (dried plant) and 4441 intentional exposures for the opioid, hydrocodone.

Misuse of CBD products

Although, at present, there does not appear to be significant recreational abuse or "addiction" to CBD, in many countries, misuse, which is the use of a substance for therapeutic reasons but in the absence of or contrary to medical advice, is widespread. Unregulated CBD products are widely available in many countries via retail stores, cannabis dispensaries, head shops, and Internet websites. A recent analysis found that a range of CBD products were accessible without a prescription in 7 of 9 countries examined; Australia and New Zealand being the exceptions where CBD could only be accessed with a prescription (McGregora et al., 2020). In most countries, including the US and Europe, the products were not legally available but were still widely accessible as unregulated/unapproved products. In Canada, where it is prohibited to promote cannabis or related products/services as having health or cosmetic benefits, an analysis of advertising claims of CBD products found that advertisements for the majority of CBD products made unfounded health claims in treating 171 medical conditions, most commonly, pain, anxiety, or inflammation but also more serious conditions, such as HIV/AIDS, cancer, and multiple sclerosis (Zenone et al., 2021). Despite a lack of evidence for efficacy in treating chronic pain (McDonagh et al., 2021), there is widespread belief that CBD products are effective, with a survey of patients from 7 pain management clinics in California showing that 62% reported trying a CBD product for treatment of pain (Schilling et al., 2021).

A further issue is that because many CBD products are either illegal or poorly regulated, they may not contain sufficient CBD to achieve any therapeutic benefits, even for conditions where there is some evidence of efficacy. For example, an analysis of CBD products showed that the daily recommended doses with orally administered nonprescription CBD products were typically well below 150mg, whereas doses used in published clinical trials were typically at least 2- to 10-fold higher (e.g., 300–1500mg) (McGregora et al., 2020). Further, many products do not contain the quantities of CBD described on the labels, with only 31% of products in the US having been accurately labeled with respect to

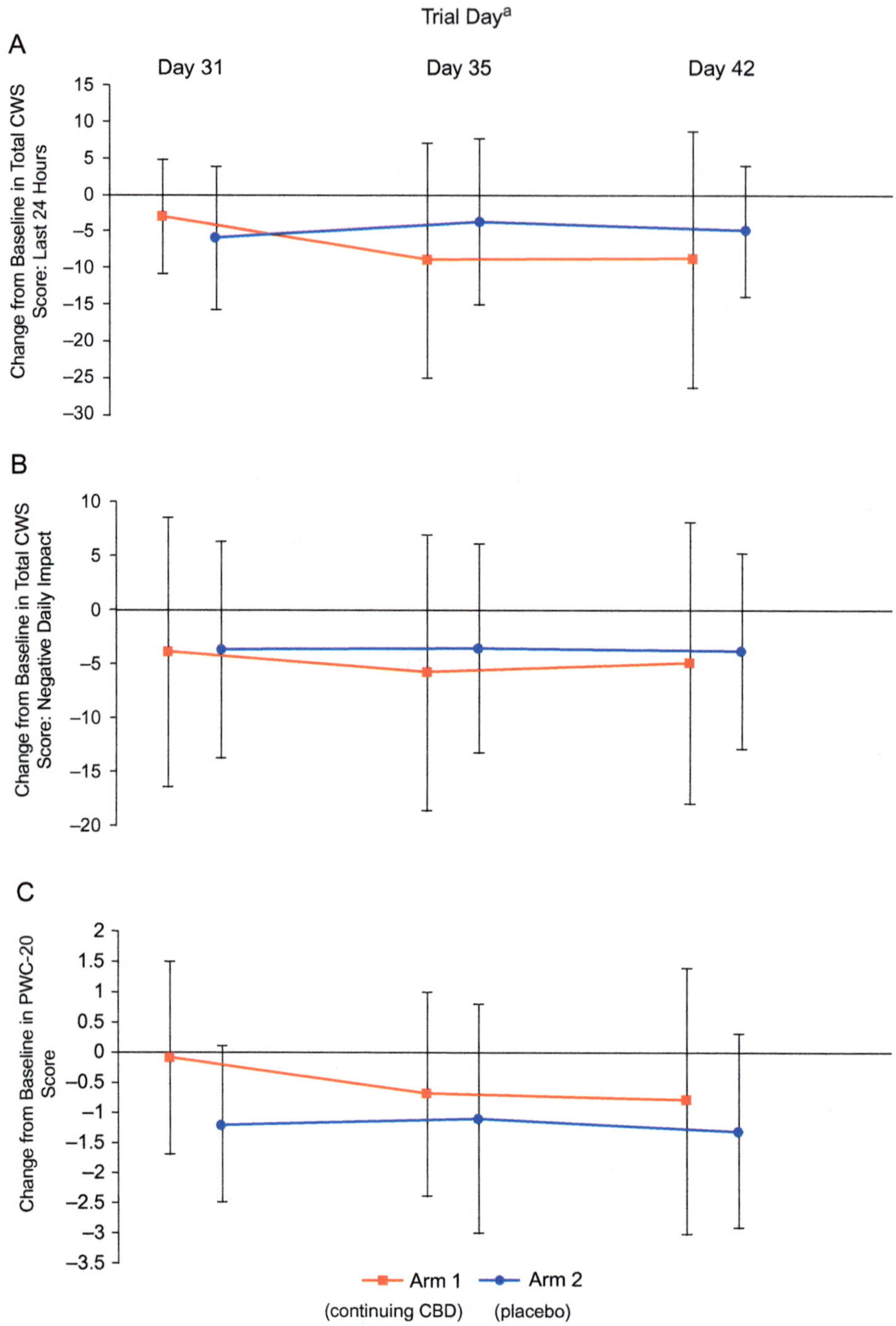

FIG. 2 Withdrawal checklist data from a clinical physical dependence study with cannabidiol. This figure shows change from baseline scores for Cannabis Withdrawal Scale (mean total ± standard deviation) scores in the past 24 hours (Panel A), negative impact on normal daily activities (Panel B) and Penn Physician Withdrawal Scale (Panel C) in healthy volunteers who received oral cannabidiol 750 mg twice daily for 4 weeks, followed by 14 days of cannabidiol or placebo. The figure shows that when participants were withdrawn from cannabidiol in a blinded manner, there was no evidence of physical dependence or withdrawal syndrome. *(From Taylor, L., Crockett, J., Tayo, B., Checketts, D., Sommerville, K. (2020). Abrupt withdrawal of cannabidiol (CBD): A randomized trial.* Epilepsy and Behavior, 104, *106938, with permission.)*

CBD content; most products contained far less CBD than claimed and 21% contained significant quantities of THC (Bonn-Miller et al., 2017).

There have been a few case reports of CBD poisoning with unregulated CBD products, including a pediatric patient who received CBD oil containing a substantial amount of THC (Herbst & Musgrave, 2020) and a further case where a man ingested several packages of unregulated CBD gummies (370 mg CBD in total) and presented to the emergency room with bizarre behavior, vomiting, slurred speech, and significant neurologic, cardiovascular, and respiratory depression (Bass & Linz, 2020). According to the report, the patient "felt the products were safe based on packaging and ate them as he would eat any other candy, not believing they would have any adverse effects." Given that 370 mg CBD would not be expected to result in the described effects, it is likely that the product was mislabeled or contained other substances or both.

Use of CBD in treatment of substance-use disorders

A small number of controlled clinical studies have evaluated CBD as a potential treatment for SUDs (Table 2). Study data are fairly limited as the doses, route, and specific SUD indication vary across the minimal number of studies completed. In addition, a few of the studies investigated the cannabis extract nabiximols (Sativex), a combination product containing both THC and CBD, and were therefore excluded from this summary.

Only one randomized controlled trial has evaluated the effect of CBD for the treatment of cannabis-use disorder. Freeman et al. (2020) evaluated 3 doses of oral CBD (200, 400, or 800 mg) administered daily over a 4-week duration in a group of patients ($N = 48$) who met Diagnostic and Statistical Manual of Mental Disorders-Version 5 (DSM-V) criteria for cannabis-use disorder. Findings showed a significant decrease in cannabis use in the 400 and 800 mg CBD treatment groups, as measured by urinary THC-COOH:creatinine levels and self-reported abstinence from cannabis. Two studies

TABLE 2 Summary of clinical studies of CBD effects on substance-use disorders.

Author, year	Design/study sample	CBD dose and route	Study methods and endpoints	Summary outcomes
Cannabis use disorder				
Freeman et al. (2020)	Phase 2a, randomized, double-blind, placebo-controlled, single site, parallel-group design $N = 48$ subjects meeting DSM-V criteria for cannabis-use disorder	CBD/placebo oral capsules (bid)—total daily dose: 200 mg 400 mg 800 mg Placebo	Screening Visit Baseline Visit (Week 0)—cannabis cessation attempt began Weeks 1 to 4—treatment phase (weekly visits) Follow up (Weeks 6, 8, 12, 16, 20, and 24) Motivational interviewing (6 × 30-min sessions) Primary outcomes/endpoints: Cannabis use—urine levels/reduction in urinary THC-COOH and self-reported days abstinent	**400 mg vs placebo:** ↓ cannabis use (THC-COOH: creatinine concentrations [−94.21 ng/mL] and increased abstinence [0.48 days/week]) **800 mg vs placebo:** ↓ cannabis use (THC-COOH: creatinine concentrations [−72.02 ng/L] and increased abstinence [0.27 days/week]) No differences observed at the 200 mg dose level CBD well tolerated with no severe AEs
Smoking cessation				
Morgan et al. (2013)	Randomized, double-blind, parallel-group, placebo-controlled study $N = 24$ smokers (>10 cigarettes per day) with an intention to quit smoking	Inhaler was provided and subjects instructed to use whenever they felt like smoking CBD 400 μg/puff Placebo	Subjects participated in 2 study visits separated by 1 week Measures: 4-item severity of dependence STAI BDI Behavior Impulsivity Scale	CBD vs placebo: ↓ in number of cigarettes smoked across the treatment week with CBD No significant reduction in craving No effects on sedation or depression

Continued

TABLE 2 Summary of clinical studies of CBD effects on substance-use disorders—cont'd

Author, year	Design/study sample	CBD dose and route	Study methods and endpoints	Summary outcomes
			Bond and Lader Mood Scale Craving VAS Primary endpoint: Number of cigarettes smoked in a 7-day period	
Hindocha et al. (2018)	Double-blind, placebo-controlled, crossover study *N* = 30 nicotine-dependent smokers (smoking ≥ 10 cigarettes a day for at least the last year)	CBD/placebo oral capsules: 800 mg Placebo	Subjects participated in 3 sessions separated by 3 days each Session 1: smoked normally before tasks (Satiated condition) Session 2 and 3: 12-h abstinence (minimum 1 -week washout between sessions) (abstinent condition) Measures: 12-h abstinence (minimum 1 -week washout between sessions) Measures: Visual probe task (attentional bias) Pleasantness Rating Task Mood and Physical symptom Scale QSU-Brief Drug effects VAS Fagerström Test STAI BDI	CBD vs placebo: ↓ attentional bias to cigarette cues with placebo compared to CBD (short stimulus [300 ms] only) ↓ ratings of pleasantness to cigarette cues with CBD compared to placebo No effect on measures of craving or withdrawal
Opioid-use disorder (OUD)				
Hurd et al. (2015) (Review & Pilot #2) Manini et al. (2016) Manini, Yiannoulos, and Bergamaschi (2015) (Pilot #1)	**Pilot study #1:** 2-session double-blind, placebo-controlled crossover Phase 1 study *N* = 17 healthy normal volunteers with prior opioid exposure (800 mg CBD *n* = 6; 400 mg CBD *n* = 6; Placebo *n* = 5) **Pilot study #2:** Double-blind, parallel-group study *N* = 3–6 opioid-dependent subjects (heroin-only based on DSM-IV) who had been abstinent for at least 7 days	**Pilot study #1:** CBD (oral) co-administered with intravenous fentanyl: 400 mg CBD 800 mg CBD Placebo and 0.5 (Session 1) and 1.0 mcg/kg (Session 2) fentanyl **Pilot study #2:** Oral CBD/placebo administered over 3 consecutive days: 400 mg 800 mg Placebo	**Pilot study #1:** 2 fentanyl sessions separated by 1 week PK blood samples collected Safety **Pilot study #2:** Cue-induced craving test session (opioid and neutral video cues) presented 1 h, 24 h, and 7 days postdose Heroin craving questionnaire at 24 h, Anxiety VAS	**Pilot study #1:** No difference in PK results Safety data similar across groups; CBD did not exacerbate fentanyl effects. **Pilot study #2:** *Craving:* Single dose of CBD reduced subjective cue-induced craving at 1 h postdose. This effect was maintained at 24 h and 7 days postdose based on heroin craving questionnaire. *Anxiety:* Effective reduction in Anxiety VAS
Hurd et al. (2019)	Randomized, double-blind, parallel-group, placebo-controlled study	Oral solution CBD/placebo administered over	Subjects completed 4 test sessions over 2 weeks First 3 sessions occurred on	400 and 800 mg CBD vs placebo: ↓ cue-induced craving and

TABLE 2 Summary of clinical studies of CBD effects on substance-use disorders—cont'd

Author, year	Design/study sample	CBD dose and route	Study methods and endpoints	Summary outcomes
	Opioid-dependent subjects ($N=48$; heroin-only based on DSM-IV) who had been abstinent for at least 7 days	3 consecutive days: 400 mg 800 mg Placebo	consecutive dosing days, last session occurred 1 week later Cue induced craving test session (opioid and neutral video cues) presented 1 h, 24 h, and 7 days postdose Heroin Craving Questionnaire Anxiety VAS Craving VAS PANAS Safety measures	Anxiety VAS scores 1 h after administration and continued effects at 7 days postdose ↓ measures of stress reactivity (e.g., increased heart rate and cortisol levels) Well tolerated with mild AEs
Cocaine-use disorder (CUD)				
Meneses-Gaya et al. (2021)	Randomized, double-blind, parallel-group, placebo-controlled study $N=31$ males with diagnosis of crack cocaine dependence (CBD $n=14$; placebo $n=17$)	CBD/placebo oral capsules: 300 mg (150 mg bid) Placebo	Patients underwent 5 days of inpatient adaptation and detoxification (Days 1 to 5) Patients underwent 10 days of treatment Measures: Cocaine Craving Questionnaire Minnesota Craving Questionnaire Drug Effect VAS BDI Beck Anxiety Questionnaire ASSIST UKU-Side Effect Rating Scale	CBD vs placebo: No significant effect of treatment observed on any of the endpoints
Mongeau-Pérusse et al. (2021)	Phase II double-blind, randomized, parallel-group, placebo-controlled superiority trial $N=78$ patients diagnosed with CUD (CBD $n=40$; placebo $n=38$)	CBD/placebo oral capsules: 300 mg Placebo	Phase 1: 10 days of inpatient detoxification Phase 2: 12-week outpatient follow-up (weekly visits) Primary Endpoints: Drug-cue-induced craving Time to cocaine relapse (over 12 weeks by counting the days between the detoxification discharge and the first day of cocaine use)	CBD vs placebo: No significant effect of treatment observed on any of the endpoints

AE = adverse event; ASSIST test = alcohol, smoking, and substance involvement screening test; BDI = Beck Depression Inventory; bid = twice daily; CBD = cannabidiol; CO = carbon monoxide; CUD = cocaine use disorder; DSM-V = Diagnostic and Statistical Manual of Mental Disorders-version 5; OUD = opioid-use disorder; PANAS = Positive and Negative Affect schedule; PK = pharmacokinetic; QSU-brief = questionnaire of smoking urges-brief; STAI = State-Trait Anxiety Index; Udvalg for Kliniske Undersøgelser; VAS = visual analog scale.

have evaluated the effect of CBD as a novel smoking cessation treatment. In a study by Morgan et al. (2013), 24 smokers were randomized to receive either a CBD (400 μg/puff) or placebo inhaler that they were to use every time they felt like smoking. A significant decrease in the number of cigarettes smoked over the treatment week was seen with CBD compared with placebo, but there were no differences in reported craving. In a more recent study, the effect of 800 mg oral CBD on nicotine withdrawal symptoms, as measured by attentional bias to cigarette cues, was investigated in 30 smokers. The study found that after abstaining from smoking, subjects were more biased to cigarette cues following placebo, whereas this bias was reversed after administration of CBD. In addition, CBD also reduced the reported pleasantness of cigarette-related images. Like the prior study, craving scores were not affected (Hindocha et al., 2018).

In a series of studies by Hurd and colleagues, oral doses of 400 and 800 mg CBD were evaluated as potential treatments for opioid-use disorder (OUD). Two pilot studies (Hurd et al., 2015 [$N = 6$] and Manini et al., 2015 [$N = 17$]) and one pivotal study ($N = 48$) found that administration of both CBD doses for 3 consecutive days resulted in lower scores on cue-induced measures of craving and anxiety, with no effect on cognition and minimal AEs. Two recent studies have evaluated the efficacy of 300 mg CBD as treatment for cocaine-use disorder (CUD). Meneses-Gaya et al. (2021) compared CBD and placebo administered over 10 days in 31 men diagnosed with crack-cocaine dependence and found no treatment differences between CBD and placebo on measures of cocaine craving, anxiety, or depression. In a second study (Mongeau-Pérusse, Brissette, Bruneau, et al., 2021), the effect of CBD ($N = 40$) on drug-cue induced craving and time to cocaine relapse was compared to placebo ($N = 38$) in patients diagnosed with CUD. No significant findings related to CBD treatment were reported.

Conclusions

Data from animal studies of reinforcing, rewarding, or "subjective" effects have demonstrated that, over a wide range of doses and routes of administration, CBD does not have reinforcing or rewarding effects. Further, CBD does not have perceivable effects that are similar to drugs of abuse like THC, ethanol, or benzodiazepines. Although data are limited, animal studies suggest that administration of CBD does not result in physical dependence or withdrawal signs upon abrupt discontinuation.

Overall, it is clear from human abuse potential studies in polydrug or cannabis users that CBD has minimal potential for abuse at supratherapeutic doses. Overall, these studies show that oral CBD at high doses ($\geq$1500 mg) or using routes associated with greater exposure (such as vaporization) produces some modest "abuse-related" subjective ratings, such as "liking" and feelings of sedation or relaxation (Fig. 3). However, across populations, these were markedly lower than those of the comparators, such as benzodiazepines or THC, indicating that the abuse potential of CBD is relatively low. Potentially abuse-related effects, including somnolence or relaxation, have been reported at high doses; however, effects are markedly lower than those of sedative drugs, and the abuse potential of CBD overall is much lower than that of THC.

CBD also appears to lack the potential for physical dependence and is not associated with a withdrawal syndrome upon abrupt discontinuation.

Consistent with research findings, limited postmarket data show that CBD does not appear to be recreationally abused or associated with development of dependence in polydrug users. However, misuse of unregulated or unlawfully marketed CBD products without a prescription is rampant, primarily for pain, anxiety, or sleep. In summary, the abuse and dependence profile of CBD is in marked contrast to that of THC (Table 3).

The results of studies with CBD as a potential treatment for SUDs vary depending both on indication treated and doses administered; however, promising results observed at doses between 400 and 800 mg, particularly in OUD, smoking cessation, and cannabis-use disorder, warrant additional research.

Until recently, CBD was regulated as a Schedule I controlled substances, along with THC and other "cannabinoids". At a 2018 meeting, the World Health Organization Expert Committee on Drug Dependence reviewed CBD and recommended that pure preparations of this substance be deregulated (i.e., should not be controlled) under International Drug Control Conventions (World Health Organization, 2018). The Committee recommended that a footnote be added to Schedule I of the 1961 Single Convention: "Preparations containing predominantly CBD and not more than 0.2 percent of THC are not under international control." In the US, FDA-approved products containing CBD (with THC no >0.1% by weight) were subsequently removed from Schedule V, with pending regulations in other countries. This is consistent with the conclusions that CBD lacks abuse and dependence potential.

Applications to other areas

In this chapter, we have reviewed the abuse and dependence potential of CBD. The effects determined in animal models translated well into human studies, including the absence of THC-like subjective effects, reinforcing or rewarding effects, and a lack of physical dependence potential. The clinical evidence of abuse potential, that is, subjective effects that may be of interest for individuals who abuse drugs, was gathered in various types of polydrug and/or cannabis users. The mood effects of CBD were generally found to be mild in polydrug and cannabis users, consisting mainly of mild sedation and/or dizziness. Because polydrug users have been shown to like the effects of drugs of abuse, they are generally considered the most sensitive model for determining if a drug or substance has abuse potential because nondrug using volunteers often

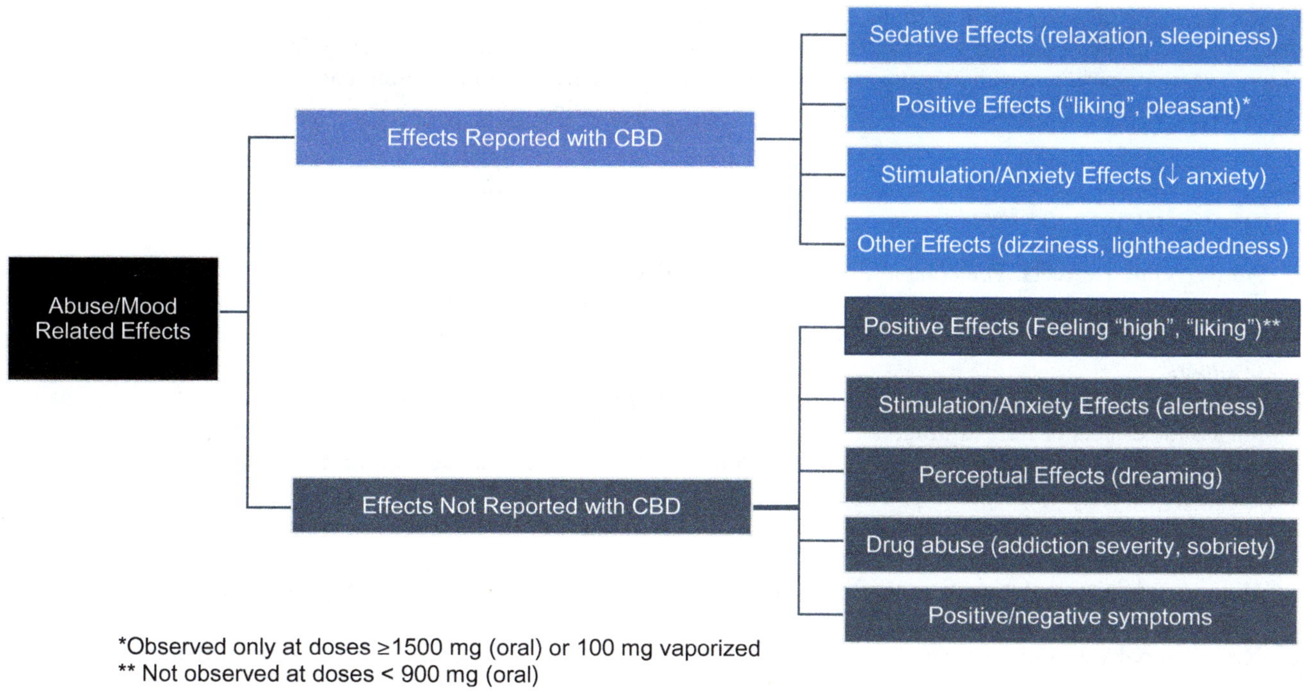

FIG. 3 Summary of neuropsychiatric/mood effects observed in clinical studies of cannabidiol.

TABLE 3 Comparison of key abuse and dependence potential data for 9-delta-tetrahydrocannabinol and cannabidiol.

Abuse/dependence measure	9-Delta-tetrahydrocannabinol	Cannabidiol
In vitro/animal data		
Binds to cannabinoid-1 receptor	+++	–
Reinforcing effects	++	–
Rewarding effects	++	–
Physical dependence	+	–
Human/clinical data		
Reports of drug "liking"	+++	+
Other abuse-related subjective effects (i.e., positive and sedative effects)	+++	+
Physical dependence potential	+	–
Postmarket/real-world data		
Evidence of recreational abuse	+++	(+/–)
Evidence of misuse	++	++
Evidence of psychological or physical dependence	+++	–

(–)=no effect; (+) (++) (+++)=small, medium, large effect; (+/–)=mixed or conflicting effects reported.

don't "like" the effects of drugs of abuse. Thus, while the mild sedation and dizziness effects observed in the human abuse potential studies may broadly apply to other groups, such as nondrug using volunteers and patient groups, these groups are less likely to report liking such effects and thus would be at even lower risk of abusing CBD than the polydrug users. The physical dependence study with CBD was performed in healthy volunteers. Again, this would be the most sensitive group to evaluate as any signs or symptoms of withdrawal could be considered drug related, in contrast to patient groups who may experience symptoms associated with the disease and who would be using concomitant medications, both of which could mask the emergence of a withdrawal syndrome. Thus, it is unlikely that patients, particularly those for seizure-related conditions, would experience withdrawal symptoms associated with CBD discontinuation, particularly as these patients are typically switched to other medications rather than abruptly discontinued.

Mini-dictionary of terms

- **Abuse-related effects**. Perceivable effects of a drug that are potentially of interest for abuse or misuse, such as euphoria/mood enhancement, sedation/relaxation, stimulation, or perceptual effects/hallucinations.
- **CBD**. Cannabidiol is one of many compounds found in *Cannabis* species and is used as a single ingredient for seizure disorders.
- **Drug "liking"**. Scale used in human abuse potential studies that asks research participants to rate whether they like or dislike the effects of the test drug.
- **[Drug/substance] schedule**. Level of control required by international convention or national law based on the substance's abuse and dependence potential.
- **Generalization**. When a compound is considered to feel similar to other another compound that animals are trained to recognize in the drug discrimination paradigm.
- **Human abuse potential study**. A type of clinical study where perceivable effects of single doses of a test drug are rated by recreational drug using volunteers.
- **Recreational use**. Episodic, nondependent use of a substance for its psychoactive effects rather than its therapeutic effects.
- **Reinforcing effects**. Effects associated with establishing or maintaining behavior, such as in the self-administration paradigm in animals.
- **Rewarding effects**. Hedonic properties, as evaluated in the conditioned place preference paradigm in animals.
- **Subjective effects**. Perceivable effects of a drug.
- **Substance-use disorders**. Disorders that affect a person's brain and their behavior, causing an inability to control their use of substances such as legal or illegal drugs, alcohol, or medications.
- **THC**. 9-delta-tetrahydrocannabinol, the psychoactive, abuseable ingredient found in *Cannabis* species.
- **Withdrawal syndrome**. Adverse clinical signs and symptoms associated with discontinuation of a drug after chronic use.

Key facts of CBD abuse potential

- The abuse potential of CBD was first evaluated in animals in the 1970s; however, the first CBD product was approved for therapeutic use in humans in 2018 in the US.
- Human abuse potential studies ask recreational drug users to rate drugs effects to see if they are "liked" and if these participants would seek to take the drug again. CBD shows only modest effects in these studies suggesting it has low "abuse potential." Animals don't like CBD either. These studies are considered important for determining whether a drug should be "scheduled" (i.e., controlled by agencies like the Drug Enforcement Administration in the US).
- There is currently one approved CBD product for use in Dravet's and Lennox-Gasteux Syndrome for use in the US and Europe; other CBD products are unregulated. Most unregulated CBD products do not contain the labelled quantity of CBD, may contain THC, and/or contain doses/concentrations of CBD that are lower than those that have shown therapeutic effects.
- In 2018, it was recommended to the World Health Organization that CBD be "descheduled," i.e., removed from international control. In 2020, the Drug Enforcement Administration published a rule proposing that FDA-approved preparations of CBD containing no more than 0.1% THC by weight be removed from the controlled substances list in the US.
- Only a handful of studies have evaluated CBD as a treatment for SUDs; however, there is promising data that suggest CBD may be able to treat some symptoms related to addiction. More controlled clinical studies are needed to confirm these findings.

Summary points

- Unlike THC, CBD does not bind to the CB-1R but rather works through a variety of other pharmacological mechanisms.
- CBD lacks reinforcing or rewarding effects in animal models; it is not self-administered by animals, and it is not associated with a conditioned place preference. CBD does not have effects that animals perceive to be similar to those of THC, benzodiazepines, or ethanol in animal drug discrimination studies.
- CBD does not appear to be associated with the development of physical dependence in animals.
- In clinical studies, CBD is associated with mild mood/neuropsychiatric effects at high doses, most commonly sleepiness (somnolence) and dizziness.
- In human abuse potential studies, oral CBD $\leq$900 mg is not associated with drug "liking" or abuse potential in recreational polydrug or cannabis users. At oral doses $\geq$1500 mg (or 100 mg vaporized), there is modest "liking" and other abuse-related effects in these subjects, demonstrating a minimal potential for abuse.
- A human physical dependence study showed no evidence of physical dependence with CBD or a withdrawal syndrome after abrupt discontinuation of oral CBD.
- There is limited evidence suggesting that CBD is not recreational abused in polydrug users or other populations, and there were no reports of psychological dependence or "addiction" identified in the medical literature. However, misuse of CBD (i.e., therapeutic use of unregulated products without medical advice) appears to be fairly widespread.
- Overall, studies with CBD as a potential treatment for SUDs vary depending both on indication treated and doses administered. CBD shows some promise in alleviating some clinical signs and symptoms associated with certain SUDs, particularly at doses between 400 and 800 mg; however, additional research is needed to confirm these findings.

References

Ali, S. F., Newport, G. D., Scallet, A. C., et al. (1991). Chronic marijuana smoke exposure in the rhesus monkey. IV: Neurochemical effects and comparison to acute and chronic exposure to delta-9-tetrahydrocannabinol (THC) in rats. *Pharmacology, Biochemistry, and Behavior*, *40*, 677–682.

Arndt, D. L., & de Wit, H. (2017). Cannabidiol does not dampen responses to emotional stimuli in healthy adults. *Cannabis and Cannabinoid Research*, *2*, 105–113.

Arout, C. A., Haney, M., Herrmann, E. S., Bedi, G., & Cooper, Z. D. (2021). A placebo-controlled investigation of the analgesic effects, abuse liability, safety and tolerability of a range of oral cannabidiol doses in healthy humans. *British Journal of Clinical Pharmacology*. https://doi.org/10.1111/bcp.14973. Online ahead of print.

Babalonis, S., Haney, M., Malcolm, R. J., et al. (2017). Oral cannabidiol does not produce a signal for abuse liability in frequent marijuana smokers. *Drug and Alcohol Dependence*, *172*, 9–13.

Bakas, T., van Nieuwenhuijzen, P. S., Devenish, S. O., et al. (2017). The direct actions of cannabidiol and 2- arachidonoyl glycerol at GABAA receptors. *Pharmacological Research*, *119*, 358–370.

Bass, J., & Linz, D. R. (2020). A case of toxicity from cannabidiol gummy ingestion. A case of toxicity from cannabidiol gummy ingestion. *Cureus*, *12*, e7688.

Bonn-Miller, M. O., Loflin, M. J. E., Thomas, B. F., et al. (2017). Labeling accuracy of cannabidiol extracts sold online. *JAMA*, *318*, 1708–1709.

Bueno, O. F., Carlini, E. A., Finkelfarb, E., et al. (1976). Delta 9-tetrahydrocannabinol, ethanol, and amphetamine as discriminative stimuli-generalization tests with other drugs. *Psychopharmacologia*, *46*, 235–243.

Crippa, J. A., Derenusson, G. N., Ferrari, T. B., et al. (2011). Neural basis of anxiolytic effects of cannabidiol (CBD) in generalized social anxiety disorder: A preliminary report. *Journal of Psychopharmacology*, *25*, 121–130.

Dalton, W. S., Martz, R., Lemberger, L., et al. (1976). Influence of cannabidiol on delta-9-tetrahydrocannabinol effects. *Clinical Pharmacology and Therapeutics*, *19*, 300–309.

Devinsky, O., Cilio, M. R., Cross, H., et al. (2014). Cannabidiol: Pharmacology and potential therapeutic role in epilepsy and other neuropsychiatric disorders. *Epilepsia*, *55*(6), 791–802.

Devinsky, O., Cross, J. H., Laux, L., et al. (2017). Cannabidiol in Dravet Syndrome Study Group. Trial of Cannabidiol for Drug-Resistant Seizures in the Dravet Syndrome. *The New England Journal of Medicine*, *376*, 2011–2020.

Devinsky, O., Patel, A. D., Cross, J. H., et al. (2018). GWPCARE3 Study Group. Effect of cannabidiol on drop seizures in the Lennox-Gastaut syndrome. *The New England Journal of Medicine*, *378*, 1888–1897.

Devinsky, O., Patel, A. D., Thiele, E. A., et al. (2018). GWPCARE1 Part A Study Group. Randomized, dose-ranging safety trial of cannabidiol in Dravet syndrome. *Neurology*, *90*, e1204–e1211.

Fedorova, E. V., Wong, C. F., Ataiants, J., Iverson, E., Conn, B. M., & Lankenau, S. E. (2021). Cannabidiol (CBD) and other drug use among young adults who use cannabis in Los Angeles. *Drug and Alcohol Dependence*, *221*, 108648.

Food and Drug Administration. (2018). *Other review(s)*. Application number: 210365Orig1s000; downloaded: https://www.accessdata.fda.gov/drugsatfda_docs/nda/2018/210365Orig1s000TOC.cfm. (Accessed 10 December 2021).

Freeman, T. P., Hindocha, C., Baio, G., Shaban, N. D. C., et al. (2020). Cannabidiol for the treatment of cannabis use disorder: A phase 2a, double-blind, placebo-controlled, randomised, adaptive Bayesian trial. *Lancet Psychiatry*, *7*, 865–874.

Fusar-Poli, P., Crippa, J. A., Bhattacharyya, S., et al. (2009). Distinct effects of {delta}9-tetrahydrocannabinol and cannabidiol on neural activation during emotional processing. *Archives of General Psychiatry*, *66*, 95–105.

Genaro, K., Fabris, D., Arantes, A. L. F., et al. (2017). Cannabidiol is a potential therapeutic for the affective-motivational dimension of incision pain in rats. *Frontiers in Pharmacology*, *8*, 391.

Gong, H., Tashkin, D. P., Simmons, M. S., et al. (1984). Acute and subacute bronchial effects of oral cannabinoids. *Clinical Pharmacology and Therapeutics*, *35*, 26–32.

Gummin, D. D., Mowry, J. B., Beuhler, M. C., Spyker, D. A., Brooks, D. E., Dibert, K. W., et al. (2020). 2019 annual report of the American association of poison control centers' national poison data system (NPDS): 37th annual report. *Clinical Toxicology*, *58*, 1360–1541.

Haney, M., Malcolm, R. J., Babalonis, S., et al. (2016). Oral cannabidiol does not alter the subjective, reinforcing or cardiovascular effects of smoked cannabis. *Neuropsychopharmacology*, *41*, 1974–1982.

Henriksson, B. G., Johansson, J. O., & Järbe, T. U. (1975). Delta 9-tetrahydrocannabinol produced discrimination in pigeons. *Pharmacology, Biochemistry, and Behavior*, *3*, 771–774.

Herbst, J., & Musgrave, G. (2020). Respiratory depression following an accidental overdose of a CBD-labeled product: A pediatric case report. *Journal of the American Pharmacists Association*, *60*(1), 248–252.

Hiltunen, A. J., & Järbe, T. U. (1986). Interactions between delta 9-tetrahydrocannabinol and cannabidiol as evaluated by drug discrimination procedures in rats and pigeons. *Neuropharmacology*, *25*, 133–142.

Hindocha, C., Freeman, T. P., Grabski, M., et al. (2018). Cannabidiol reverses attentional bias to cigarette cues in a human experimental model of tobacco withdrawal. *Addiction*, *113*, 1696–1705.

Hurd, Y. L., Spriggs, S., Alishayev, J., et al. (2019). Cannabidiol for the reduction of cue-induced craving and anxiety in drug-abstinent individuals with heroin use disorder: A double-blind randomized placebo-controlled trial. *The American Journal of Psychiatry*, *176*, 911–922.

Hurd, Y. L., Yoon, M., Manini, A. F., et al. (2015). Early phase in the development of cannabidiol as a treatment for addiction: Opioid relapse takes initial center stage. *Neurotherapeutics*, *12*, 807–815.

Iannotti, F. A., Hill, C. L., Leo, A., et al. (2014). Nonpsychotropic plant cannabinoids, cannabidivarin (CBDV) and cannabidiol (CBD), activate and desensitize transient receptor potential vanilloid 1 (TRPV1) channels in vitro: potential for the treatment of neuronal hyperexcitability. *ACS Chemical Neuroscience*, *5*, 1131–1141.

Järbe, T. U., Henriksson, B. G., & Ohlin, G. C. (1977). Delta9-THC as a discriminative cue in pigeons: effects of delta8-THC, CBD, and CBN. *Archives Internationales de Pharmacodynamie et de Thérapie*, *228*, 68–72.

Järbe, T. U., & Mathis, D. A. (1991). Discriminative stimulus functions of cannabinoids/cannabimimetics. *NIDA Research Monograph*, *116*, 75–99.

Karler, R., & Turkanis, S. A. (1980). Subacute cannabinoid treatment: Anticonvulsant activity and withdrawal excitability in mice. *British Journal of Pharmacology*, *68*, 479–484.

Lattanzi, S., Brigo, F., Cagnetti, C., et al. (2018). Efficacy and safety of adjunctive cannabidiol in patients with Lennox-Gastaut syndrome: A systematic review and meta-analysis. *CNS Drugs*, *32*, 905–916.

Lattanzi, S., Brigo, F., Trinka, E., et al. (2018). Efficacy and safety of cannabidiol in epilepsy: A systematic review and meta-analysis. *Drugs*, *78*, 1791–1804.

Linares, I. M. P., Guimaraes, F. S., Eckeli, A., et al. (2018). No acute effects of cannabidiol on the sleep-wake cycle of healthy subjects: A randomized, double-blind, placebo-controlled, crossover study. *Frontiers in Pharmacology*, *9*, 315.

Mahgoub, M., Keun-Hang, S. Y., Sydorenko, V., et al. (2013). Effects of cannabidiol on the function of α7-nicotinic acetylcholine receptors. *European Journal of Pharmacology*, *720*, 310–319.

Manini, A. F., Yiannoulos, G., Bergamaschi, M. M., et al. (2015). Safety and pharmacokinetics of oral cannabidiol when administered concomitantly with intravenous fentanyl in humans. *Journal of Addiction Medicine*, *9*(3), 204–210.

Markos, J. R., Harris, H. M., Gul, W., et al. (2017). Effects of cannabidiol on morphine conditioned place preference in mice. *Planta Medica*, *84*, 221–224.

Martin-Santos, R., Crippa, J. A., Batalla, A., et al. (2012). Acute effects of a single, oral dose of d9-tetrahydrocannabinol (THC) and cannabidiol (CBD) administration in healthy volunteers. *Current Pharmaceutical Design*, *18*, 4966–4979.

McDonagh, M. S., Wagner, J., Ahmed, A. Y., Fu, R., Morasco, B., Kansagara, D., et al. (2021). Living systematic review on cannabis and other plant-based treatments for chronic pain. *Comparative effectiveness review no. 250*. Rockville, MD: *AHRQ Publication*, 21(22)-EHC036.

McFadden, B. R., & Malone, T. (2021). Homegrown perceptions about the medical use and potential abuse of CBD and THC. *Addictive Behaviors*, *115*, 106799.

McGregora, I. S., Cairns, E. A., Abelev, S., Cohena, R., Hendersona, M., Couche, D., et al. (2020). Access to cannabidiol without a prescription: A cross-country comparison and analysis. *International Journal of Drug Policy*, *85*, 102935.

McMahon, L. R. (2016). Enhanced discriminative stimulus effects of Δ(9)-THC in the presence of cannabidiol and 8-OH-DPAT in rhesus monkeys. *Drug and Alcohol Dependence*, *165*, 87–93.

Mechoulam, R., & Hanus, L. (2002). Cannabidiol: An overview of some chemical and pharmacological aspects. Part I: Chemical aspects. *Chemistry and Physics of Lipids*, *121*, 35–43.

Meneses-Gaya, C., Crippa, J. A., Hallak, J. E., et al. (2021). Cannabidiol for the treatment of crack-cocaine craving: an exploratory double-blind study. *Brazilian Journal of Psychiatry*, *43*, 467–476.

Mongeau-Pérusse, V., Brissette, S., Bruneau, J., et al. (2021). Cannabidiol as a treatment for craving and relapse in individuals with cocaine use disorder: A randomized placebo-controlled trial. *Addiction*, *116*, 2431–2442.

Morgan, C. J., Das, R. K., Joye, A., et al. (2013). Cannabidiol reduces cigarette consumption in tobacco smokers: Preliminary findings. *Addictive Behaviors*, *38*, 2433–2436.

Myers, A. M., Siegele, P. B., Foss, J. D., et al. (2018). Single and combined effects of plant-derived and synthetic cannabinoids on cognition and cannabinoid-associated withdrawal signs in mice. *British Journal of Pharmacology*. https://doi.org/10.1111/bph.14147.

Parker, L. A., Burton, P., Sorge, R. E., et al. (2004). Effect of low doses of delta9-tetrahydrocannabinol and cannabidiol on the extinction of cocaine-induced and amphetamine-induced conditioned place preference learning in rats. *Psychopharmacology*, *175*, 360–366.

Rock, E. M., & Parker, L. A. (2021). Constituents of *Cannabis sativa*. *Advances in Experimental Medicine and Biology*, *1264*, 1–13.

Sands, T. T., Rahdari, S., Oldham, M. S., et al. (2019). Long-term safety, tolerability, and efficacy of cannabidiol in children with refractory epilepsy: Results from an expanded access program in the US. *CNS Drugs*, *33*(1), 47–60.

Schilling, J. M., Hughes, C. G., Wallace, M. S., Sexton, M., Backonja, M., & Moeller-Bertram, T. (2021). Cannabidiol as a treatment for chronic pain: A survey of patients' perspectives and attitudes. *Journal of Pain Research*, *14*, 1241–1250.

Schoedel, K. A., Szeto, I., Setnik, B., et al. (2018). Abuse potential assessment of cannabidiol (CBD) in recreational polydrug users: A randomized, double-blind, controlled trial. *Epilepsy & Behavior*, *88*, 162–171.

Solowij, N., Broyd, S. J., Beale, C., et al. (2018). Therapeutic effects of prolonged cannabidiol treatment on psychological symptoms and cognitive function in regular cannabis users: A pragmatic open label clinical trial. *Cannabis and Cannabinoid Research*, *3*, 21–34.

Solowij, N., Broyd, S., Greenwood, L. M., et al. (2019). A randomised controlled trial of vaporised Δ9-tetrahydrocannabinol and cannabidiol alone and in combination in frequent and infrequent cannabis users: Acute intoxication effects. *European Archives of Psychiatry and Clinical Neuroscience*, *269*, 17–35.

Spindlea, T. R., Conea, E. J., Goffia, E., Weertsa, E. M., Mitchell, J. M., Wineckerb, R. E., et al. (2020). Pharmacodynamic effects of vaporized and oral cannabidiol (CBD) and vaporized CBD-dominant cannabis in infrequent cannabis users. *Drug and Alcohol Dependence*, *211*, 107–137.

Sylantyev, S., Jensen, T. P., Ross, R. A., et al. (2013). Cannabinoid- and lysophosphatidylinositol-sensitive receptor GPR55 boosts neurotransmitter release at central synapses. *Proceedings of the National Academy of Sciences of the United States of America*, *110*, 5193–5198.

Taylor, L., Crockett, J., Tayo, B., Checketts, D., & Sommerville, K. (2020). Abrupt withdrawal of cannabidiol (CBD): A randomized trial. *Epilepsy and Behavior*, *104*, 106938.

Taylor, L., Gidal, B., Blakey, G., et al. (2018). A phase I, randomized, double-blind, placebo-controlled, single ascending dose, multiple dose, and food effect trial of the safety, tolerability and pharmacokinetics of highly purified cannabidiol in healthy subjects. *CNS Drugs*, *32*, 1053–1067.

Vann, R. E., Gamage, T. F., Warner, J. A., et al. (2008). Divergent effects of cannabidiol on the discriminative stimulus and place conditioning effects of delta(9)-tetrahydrocannabinol. *Drug and Alcohol Dependence*, *94*, 191–198.

Viudez-Martínez, A., García-Gutiérrez, M. S., Medrano-Relinque, J., et al. (2019). Cannabidiol does not display drug abuse potential in mice behavior. *Acta Pharmacologica Sinica*, *40*, 358–364.

Winton-Brown, T. T., Allen, P., Bhattacharyya, S., et al. (2011). Modulation of auditory and visual processing by delta-9-tetrahydrocannabinol and cannabidiol: an FMRI study. *Neuropsychopharmacology*, *36*, 1340–1348.

World Health Organization Expert Committee on Drug Dependence (WHO-ECDD) (2018). Cannabidiol (CBD).

Zenone, M. A., Snyder, J., & Crooks, V. (2021). Selling cannabidiol products in Canada: A framing analysis of advertising claims by online retailers. *BMC Public Health*, *21*, 1285.

Zuardi, A. W., Finkelfarb, E., Bueno, O. F., et al. (1981). Characteristics of the stimulus produced by the mixture of cannabidiol with delta 9-tetrahydrocannabinol. *Archives Internationales de Pharmacodynamie et de Thérapie*, *249*, 137–146.

Zuardi, A. W., Rodrigues, N. P., Silva, A. L., et al. (2017). Inverted u-shaped dose-response curve of the anxiolytic effect of cannabidiol during public speaking in real life. *Frontiers in Pharmacology*, *8*, 259.

Chapter 11

Cannabis stigmas: A narrative of features

Matt Reid
Sociology Department, Widener University, Chester, PA, United States

Abbreviations

AIDS	acquired immunodeficiency syndrome
FDA	U.S. Food and Drug Administration
HIV	human immunodeficiency virus

Introduction

This review details the nuanced social dynamics of cannabis stigmas, primarily as they manifest in Western societies. Yet even globally, cannabis has been a substance associated with outsiders in almost every society. Until the latter part of the 20th century, cannabis was chiefly "used by the poor, by the marginal, by the chronically ill, by the artistically and philosophically and spiritually inclined, by seekers after the meaning of life, and by social and religious nonconformists of various stripes" (Chasteen, 2016, p. 137). Succinctly stated, cannabis is a substance historically connected to marginalized groups.

As this review will demonstrate, the near-universal deviant status of cannabis stems from a variety of social factors, but one lesson is clear—all across the world, the social groups who used cannabis were rarely considered a part of respectable society (Chasteen, 2016). Despite the recent progress made in modern societies with liberal cannabis policies and growing social acceptance, people who associate with the plant still face powerful stigmas (Reid, 2020). This review begins with an overview of the concept of stigma and then proceeds to illuminate some of the social forces which fuel cannabis stigmas. It concludes with a brief summary of stigma management strategies and ways cannabis users have challenged these stigmas.

Main text

What are stigmas?... Stigmas are essentially devaluing labels applied to individuals who are associated with things deemed undesirable by mainstream society. The application of a stigma results in a disgraced identity because it is related to something larger society finds morally offensive. Modern research on stigmas was propelled by sociologist Erving Goffman and his canonical book, *Stigma: Notes of the Management of Spoiled Identity* (1963). It is important to note that stigmas are not objective, permanent, or universal fixtures in any given society. Rather, they are collectively created, situationally applied, and vary over time and between cultures. Stigmatizing features can include having physical or cognitive abnormalities, committing criminal or distasteful acts, and/or being a member of a repugnant social group. Moreover, stigmas are a form of shared cultural knowledge and exist beyond individual sentiments (Goffman, 1963; Herek, 2007). Common features of stigmas are outlined in Table 1.

Stigmas are also multidimensional and stem from various social forces, but they always serve to distance the stigmatized individual from "normal" individuals in a society (Goffman, 1963). Hence, stigmas may result from unequal power relations where privileged groups devalue the identity, lifestyle, and/or existence of groups with less social power (Herek, 2007). Since privileged groups wield more social power and cultural influence, they have a more considerable say in what becomes a widely accepted cultural attitude. In short, stigmas are a tool of social control that create in-groups and out-groups in any given society. Recent research on stigmas tends to be more attentive to such power inequalities, and those with social privileges are often able to resist applications of stigmatizing labels or prevent a stigma from becoming their master status (Livingston & Boyd, 2010).

Cannabis Use, Neurobiology, Psychology, and Treatment. https://doi.org/10.1016/B978-0-323-89862-1.00005-2

TABLE 1 Defining features of stigmas.

(1)	Stigmas tend to be stable once applied.
(2)	Stigmas are socially constructed in that they are created by society and their meanings vary across social contexts.
(3)	Stigmas result in negative evaluations by other people in a given society.
(4)	Stigmatized features often become a person's *master status*, or the primary identity through which others perceive them.
(5)	Stigmas commonly result in socially acceptable discrimination and restrictions on social rights

Modified from Herek, G. M. (2004). Beyond 'homophobia': Thinking about sexual prejudice and stigma in the twenty-first century. *Sexuality Research & Social Policy, 1*(2), 6–24.

Cannabis and stigmas… Like other drug-related stigmas, cannabis use is perceived to be a poor personal choice rather than an unavoidable situation, often leading others to view the application of cannabis stigmas as warranted or deserved (Lloyd, 2013; Satterlund et al., 2015). Cannabis use is a *discreditable stigma*, meaning the stigma is only applied to an individual once others learn about their use of cannabis. This differs from a *discrediting stigma* that is readily visible to others and not practically concealable (Goffman, 1963). While all stigmatized groups manage the impacts of stigmas applied to them in varying ways, those with discreditable stigmas commonly practice concealment strategies in attempts to pass as normal (Goffman, 1963; Herek, 2004). However, while keeping a stigmatized identity secret from others may protect the self from social rebuke, it does little to normalize the identity in the larger social environment (Reid, 2020). Suppressing the identities we possess may even be a detriment to one's sense of self and mental health (Satterlund et al., 2015). On the other hand, keeping cannabis use a secret protects the self from potential devaluation and discrimination. As such, the strategy of concealment is a double-edged sword with both personal benefits and compromises.

Why is cannabis stigmatized? Most simply stated, it is because society's power elite believe(d) cannabis to be against their group's interests. This was a stance adopted long ago and it is continually renewed. As such, perhaps the largest source of cannabis stigmas is its historic and ongoing criminalization (Bottorff et al., 2013; Hammer, 2015; McGettigan, 2019). While Harry Anslinger is often credited with single-handedly criminalizing cannabis in the United States in 1937, his active demonization of the plant was supported by the power structures of the era and still is today as evidenced in the Drug Enforcement Agency's classification of cannabis as a Schedule One prohibited substance (meaning no medical value and a high risk of abuse).

Timothy McGettigan (2019) uses the concept of a *truth regime* to describe this continued prohibition by the power structure. Regimes of truth are purely ideological and are largely immune from critiques based on scientific evidence and logical reasoning. The discourse that cannabis is bad and should be criminalized has crystallized in government bodies and public organizations, which in turn perpetuate this belief through policies, selective research, and other ideological tactics that prevent other truths from being legitimated in the public sphere (McGettigan, 2019). In fact, participants in several studies expressed that stigmas associated with cannabis result from societal ignorance and deliberate misinformation (Hathaway et al., 2011; Satterlund et al., 2015).

The legacy of the criminal definitions attached to the plant survive long after prohibition as well. Even in postprohibition environments, law enforcement personnel are commonly cited as experts in cannabis-related news articles, imparting the impression that the plant is still in the criminal arena (Boyd & Carter, 2012; Haines-Saah et al., 2014; Kim & Kim, 2018; Mortensen et al., 2019). Medical cannabis patients are also sometimes presumed to be engaging in illegal drug activity, particularly those who grow their own medicine (Bottorff et al., 2013). As such, legalizing cannabis for medical and/or adult-use does not entirely eliminate the criminal stigmas associated with the plant (Reid, 2021). Yet there is more nuance to unpack, as negative attitudes toward cannabis stem from a combination of overlapping social forces (Hammer, 2015). These are outlined in Fig. 1 and described throughout the chapter.

Capitalism and cannabis stigmas… At the most macro level, there seems to be a belief that cannabis detracts from the sacred mission of productivity that is a core value of capitalism. Sociologists have long argued that capitalism stems from the Protestant work ethic which holds life's pleasures are to be avoided as they get in the way of hard work (see Max Weber's *The Protestant Ethic and the Spirit of Capitalism*). Thus, capitalism as an extension of the protestant work ethic stigmatizes intoxication because it is thought to hinder productivity, despite there being limited evidence that cannabis reduces performance (Earleywine, 2010; Hathaway et al., 2011). This perspective also helps explain why other substances are stigmatized in capitalistic cultures as well as why caffeine holds a favored status (Pollan, 2021).

Capitalism also imparts the value that pleasure should only be achieved through diligent work, thus making pleasurable experiences through drug use largely immoral (Earleywine, 2010; Reinarman, 1994). Some researchers have even argued

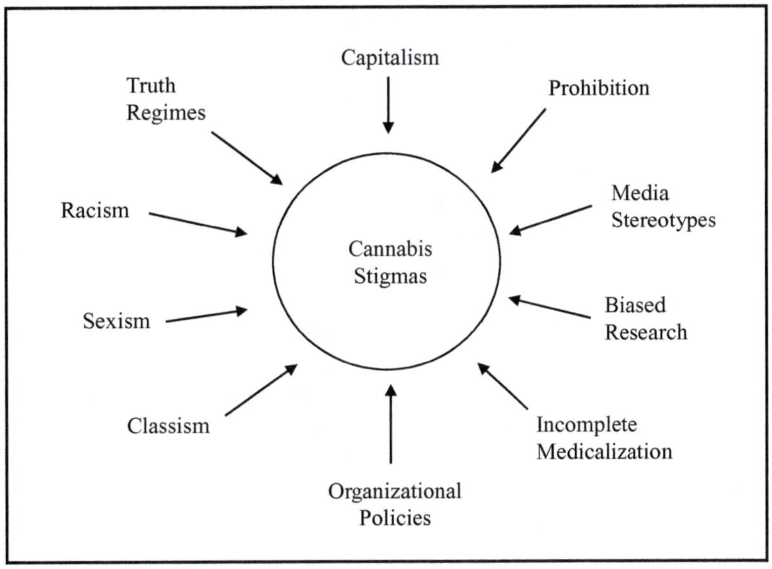

FIG. 1 Social forces contributing to cannabis stigmas.

the plant itself is not stigmatized so much as the intoxicating and pleasurable effects (Chapkis, 2007; Pedersen & Sandberg, 2013). Historian John Charles Chasteen (2016) says: "[M]arijuana's euphoriant qualities, the basis of its recreational appeal, worry many religious people, because the euphoria seems unearned and, therefore, immoral… It accompanies the worry that marijuana users are disrespecting traditional models" (pp. 136–137). All in all, capitalistic societies may stigmatize cannabis and other drugs because they are believed to conflict with productivity and the supposed pleasure intrinsic in hard work.

Cannabis stigmas can also be connected to capitalism's emphases on maturity, reliability, and diligence. Indeed, "stoner" stereotypes exaggerate qualities like immaturity, irresponsibility, and laziness (Mortensen et al., 2019; Newhart & Dolphin, 2019). The media still perpetuates these stereotypes, often portraying cannabis users as incompetent and immature (Reid, 2021). Such stereotypes are also partly based on cannabis use being culturally associated with the young adult phase of life as well as age-related biases in academic research (Newhart & Dolphin, 2019). Indeed, many academic studies on cannabis use are based on youth samples, reinforcing cultural assumptions related to age and cannabis use rather than challenging such stereotypes. In fact, cannabis is perceived to be incompatible with many social role expectations (Hathaway et al., 2011). Workers, parents, and even students who use cannabis are often assumed to be less proficient in these roles. For a specific example, recall how Tesla's stock price dropped over 9% and two executives quit after CEO Elon Musk smoked cannabis on a podcast in 2018.

Policies and cannabis stigmas… Even when cannabis is fully legalized, institutional and organizational policies may still problematize cannabis use. Testing positive for cannabis often makes a person ineligible for organ transplants in the United States (Newhart & Dolphin, 2019), and child welfare organizations have been known to target cannabis-using parents (Boyd, 2009; Reinarman et al., 2011). Similarly, because cannabis is still federally prohibited in the United States, cannabis users are technically forbidden from owning or operating a firearm. Federally regulated banks are also forbidden from serving businesses in the legal cannabis sector (at least as of the writing of this chapter). Even in states where it is legal, getting caught with cannabis on a college campus often carries greater penalties than those applied to alcohol. For prospective immigrants to the United States, admitting prior use or working in the legal cannabis industry can result in severe immigration penalties such as being denied naturalization or denial of eligibility for family immigration.

It has been theorized that as governments cede control over cannabis by ending prohibition, other institutions become forces of social control to deter cannabis use (O'Brien, 2013). The most evident example of this are the persistence of workplace drug testing programs that forbid employees from using cannabis, even in the employee's own leisure time (Reid, 2021). Notably, lower wage employees are more often subjected to this type of monitoring than those in higher salaried positions (Becker et al., 2013). My own research in this area has discovered that medical cannabis patients may experience a sense of *occupational immobility,* a feeling of being stuck in their current job because required drug testing prevents them from seeking work at more desirable employers (Reid, 2021).

Courtesy stigmas are also evident here. A *courtesy stigma* is when the stigma is applied to someone or something that associates with, but is not a member of, the stigmatized group (Goffman, 1963). In the context of cannabis, Lashley and Pollock (2019) found stigmas were sometimes applied to law offices, accounting firms, and public relations agencies who do business with those legally growing, distributing, or selling cannabis. As such, organizations may be hesitant to work with those in the cannabis industry due to cannabis-related courtesy stigmas.

Intersectional cannabis stigmas… Unsurprisingly, the application of a stigma tends to be easier and more severe when the individual possesses one or more marginalized identities, especially if said identities also carry a stigma as is the case with criminals and HIV patients. This speaks to the concept of *intersectionality*, or how our lives are shaped by multiple sources of privilege and oppression (Collins & Bilge, 2016). In other words, our social location in the larger power structure is determined by the intersections of our social class, gender, sexuality, race, ethnicity, ability, education level, and so much more. Intersectionality holds that people do not experience power relations based on one identity at a time and that the lived experience of any identity looks different when we factor in the other identities we possess. Applied here, intersectionality acknowledges that stigmas associated with cannabis are differentially experienced by our social location (Bottorff et al., 2013; Hammer, 2015; Hathaway, 2004; Reinarman et al., 2011; Satterlund et al., 2015). The convergence of multiple social identities and/or stigmatizing forces within the same individual is thus termed *intersectional stigma*.

Perhaps the most pressing examples of intersectional cannabis stigmas involve race. Some studies that suggest cannabis no longer carries a stigmatized master status tend to be based on the white experience (Hathaway et al., 2011; Jarvinen & Ravn, 2014; Sandberg, 2012), but due to pervasive racism, Western societies are more likely to problematize cannabis when it is associated with people of color. Content analysis studies of the media demonstrate this well. When stoner stereotypes or negative depictions of cannabis users are deployed in newspapers, they are more likely to feature racial minorities than white people (Haines-Saah et al., 2014; Mortensen et al., 2019). Similarly, sports reporters are more likely to problematize cannabis use by Black athletes, even going so far as to use the racist “thug” stereotype once a Black athlete is discovered to be using cannabis (Dickerson, 2018; Lewis & Proffitt, 2012). In short, claims that cannabis is normalized may come from a colorblind position of racial privileged (Reid, 2021).

One reason that racial identity strongly influences cannabis stigmas is that drug scares (at least in the United States) have historically been based on racialized fears. Craig Reinarman’s famous essay *On The Social Construction of Drug Scares* (1994) demonstrates how these recurring moral panics rely on ideologically constructing minority groups as being primarily responsible for any given drug problem. This long-standing theme of a racialized menace to society has influenced the prohibition of numerous substances as well as fueled public support for the war on drugs (Alexander, 2010). Yet demonizing minority groups for drug problems is not restricted to race. In terms of cannabis, the “dangerous class” (Reinarman, 1994) associated with cannabis has, at one or more points in history, included jazz musicians and patrons (a genre created by African Americans), Mexican immigrants, countercultural youth like beatniks and hippies, and gay HIV patients who fought to re-medicalized cannabis in the 1990s. Unfortunately, racial equity is rarely cited as a reason to legalize cannabis despite the criminalization of the plant being intensely racialized (Vitiello, 2019). Ironically, one of the more common mentions of race in American pro-legalization arguments is to vilify Latin American drug cartels (Bender, 2016).

Another common dimension of intersectional cannabis stigmas is that of gender. Research has demonstrated that cannabis is culturally coded as a masculine activity and thus may be more stigmatizing for women as they are thus perceived to be violating cultural gender prescriptions (Arnull & Ryder, 2019; Haines et al., 2009). Women also have the added stigma of being labeled negligent mothers if it becomes known they consume cannabis, a well-founded fear in light of child protective services intervening in the homes of families who use medical cannabis (Boyd, 2009; Reinarman et al., 2011).

Finally, my own research has uncovered an interesting manifestation of cannabis stigmas as they relate to masculinity. A participant in one of my former studies mentioned that men who use cannabis may be seen as failing to live up to the expectations of the traditional breadwinner/provider role. This is similar to a study by Dahl and Sandberg (2015) who found that men who use cannabis typically lack celebrated cultural characteristics of masculinity. Thus, even though men are not oppressed because of the gender, the way they experience cannabis stigmas is nonetheless influenced by their gender. Furthermore, since most American cultural images of stoners are males, this may make men more vulnerable to self-stigmatization related to their cannabis use (Hirst et al., 2018; Looby & Earleywine, 2010).

Stigmas and medical cannabis… Related to the above, cannabis patients (i.e., those who use cannabis for the explicit purpose of treating a medical issue) may be double stigmatized owing to their use of cannabis and underlying illness (Bottorff et al., 2013). They may have a disability or suffer from controversial illnesses like HIV/AIDS, fibromyalgia, or mental illness (Newhart & Dolphin, 2019; Werner, 2001). Moreover, people marginalized based on their race, gender, sexuality, and/or social class are already more likely to experience problematic interactions with medical professionals, potentially making cannabis use in the context of multiple marginalizations more likely to result in medical maltreatment (Bottorff et al., 2013; Chapkis & Webb, 2008; Werner, 2001).

Aside from the powerful way intersectional stigmas shape our lives, much of the stigma associated with medical cannabis comes from how it has yet to be embraced as an acceptable medicine in mainstream Western medicine. Cannabis is *incompletely medicalized* and best conceptualized in the arena of complementary and alternative medicines (Newhart & Dolphin, 2019). This incomplete medicalization is not necessarily due to a lack of evidence on cannabis' efficacy, but rather to how it is institutionally categorized and socially constructed. Because cannabis possesses a variety of meanings and can be used in a plethora of ways—as a recreational intoxicant, criminal offense, medicine, spiritual sacrament, and more—it is not given the full legitimacy or respect of other substances that are solely considered under medicine's purview (Reid, 2023).

Overall, the American medical establishment is still somewhat scornful of viewing cannabis as a medicine. The American Medical Association has taken a strong stance against medical cannabis, even going so far as to help overturn citizen-approved medical cannabis programs in states like Mississippi. One of their main arguments for adopting such a disapproving stance is that cannabis has not been subjected to the rigorous clinical trials required for other medications, even though the body of evidence we have to date is promising. Moreover, the FDA's approval process is not designed for multiple active compounds as are found in whole-plant cannabis (Bostwick, 2012). Thus, because cannabis (re)medicalized through various grassroots movements and has not gone through the formal process to become a medicine, the larger medical establishment is largely resistant to the idea of cannabis as a medicine (Chapkis & Webb, 2008; Jones & Hathaway, 2008; Kondrad & Reid, 2013; Penn, 2014; Robnett, 2019).

The active stigmatization of cannabis by the larger medical establishment is problematic for a number of reasons. The most obvious is that patients may experience unnecessary and/or prolonged pain if they forgo medical cannabis over concerns of stigma or pushback from medical professionals (Rudski, 2014; Ryan & Sharts-Hopko, 2017; Victorson et al., 2019). Perhaps because cannabis therapeutics are rarely included in medical education, there may be a mismatch between physicians' formal curriculum, which explicitly emphasizes stigma reduction, and the hidden curriculum, which implicitly reproduces anticannabis attitudes (Dubin et al., 2017). More disturbingly, my own research in this area has documented that some physicians may even withhold care and pain medicines should a patient test positive for cannabis (Reid, 2023). Because cannabis is lumped together with other prohibited substances, using it is seen as a violation of a treatment plan when it comes to narcotic pain killers.

Stigma management strategies… While decriminalization has lessened some stigmas attached to cannabis, legalization will not entirely eliminate such stigmas. As has been documented in this review, multiple social forces contribute to the stigmatization of cannabis, not just its historical or current illicit definition. This means that cannabis users will likely still practice *stigma management strategies*, or various ways to conceal or reduce the impacts of the stigma(s) applied to them. Some of these strategies are outlined in this section and Table 2.

Concealment can be an effective stigma management strategy for people facing discreditable stigmas, as is the case with cannabis (Bottorff et al., 2013; Herek, 2004). The idea here is to keep one's cannabis use secret from others, particularly those who are assumed to be disapproving or could impose consequences (e.g., employers). Research in this area has documented how cannabis users attempt to mask the smell of cannabis with perfumes, use eye drops to reduce red eyes, and even disguise cannabis smoking as cigarette smoking (Hathaway, 2004). Even cannabis growers may conceal the true nature of their work and simply describe themselves as small business owners when dealing with those outside of the cannabis sector (Adelman, 2013). Again, all of this is to avoid the application of the stigma in the first place. Research of medical cannabis patients finds similar concealment strategies in addition to patronizing dispensaries far outside of the patient's home area so as not to be spotted by members of one's community (Satterlund et al., 2015).

TABLE 2 Brief overview of cannabis stigma management and resistance strategies.

Concealing stigmas *Avoiding the application of the stigma*	Managing stigmas *Reducing the severity of an applied stigma*	Reducing stigmas *Addressing stigmatizing forces within society*
Masking smell	Strategies of inclusion	Legalization
Eye drops	Changing medical providers	Policy changes
Avoiding others	Relocation	Diversify media portrayals
Lying/fibbing	Cutting social ties	Align w/capitalistic values

When one's cannabis use is discovered by others, stigma management strategies often move from concealment to stigma reduction. The most extreme practice is perhaps cutting ties with others who disapprove of cannabis use, and medical cannabis patients have even been known to relocate to new or more accepting areas (Bottorff et al., 2013; Satterlund et al., 2015). They may also change medical providers if their current providers are disapproving of medical cannabis (Reid, 2023).

Yet perhaps more common when managing cannabis stigmas are *strategies of inclusion,* expressing one's conformity to social norms in hopes of reducing the severity of applied stigma (Roschelle & Kauffman, 2004). Cannabis users often attempt to portray their consumption as being safe, responsible, and controlled (Duff et al., 2012; Hathaway et al., 2011; Lau et al., 2015). The idea here is that risky, heavy, and/or impulsive cannabis use is problematic, so demonstrating self-control over one's use of cannabis makes it less likely one will be seen as abusing cannabis. This has also been documented with medical cannabis patients who balance symptom management with self-imposed limits on consumption to avoid becoming noticeably high (Newhart & Dolphin, 2019). Moreover, medical cannabis patients can avoid some stigma by emphasizing that their use of cannabis is principally medicinal with any resulting euphoria being a side effect (Bottorff et al., 2013; Ko et al., 2016). Relatedly, medical cannabis users may face less stigma if the method(s) they use to medicate do not resemble recreational styles of consumption, such as using boings (Rudski, 2014).

Resisting and challenging cannabis stigmas… For individual cannabis users, self-stigmatization can be reduced through mindful acceptance and/or cognitive restructuring (Mittal et al., 2012). The basic idea here is to resist the devaluing stereotypes associated with cannabis stigmas, coming to terms with the fact these are not objective truths but are rather social products. Yet a more powerful way to reduce cannabis stigmas is to work toward accommodating it within our social structure. Policy changes may stop institutions from treating cannabis users as criminals, mentally ill, and/or morally deficient (Hathaway et al., 2011). Legalization helps here, but it would be more effective to address the consequences of criminalization as well, particularly the racialized impacts of the war on drugs (Smith & Hattery, 2019).

Other ways to reduce prevailing cannabis stigmas include expanding media representations of cannabis users, moving away from stoner stereotypes and illuminating the diversity within the cannabis community (Loflin & Earleywine, 2014; Osborne & Fogel, 2008). Professionals and high achievers who use cannabis may also champion destigmatization by "coming out of the closet," so to speak (Garner, 2016). This helps counter the pervasive cannabis stigmas associated with incompetence, recklessness, and other juvenile behaviors (Newhart & Dolphin, 2019). Indeed, many cannabis users align with *tertiary deviance,* or deviants who embrace and seek to normalize their attitudes, behaviors, or conditions, often through advocacy and mobilization (Adler & Adler, 2016/1994).

Finally, and perhaps most powerfully, cannabis stigmas can be reduced by emphasizing how cannabis aligns with the values of capitalism. The burgeoning legal cannabis industry is just one example of this. Yet cannabis users often indicate the plant helps them focus (Chapkis & Webb, 2008; Garner, 2016), so perhaps more research into this purported effect may make cannabis more acceptable in productivity-oriented societies. Relatedly, if the effects of cannabis can be proven less debilitating than other common substances (e.g., the hangover associated with alcohol), employers may be more accommodating of off-the-clock cannabis use. Many athletes report using cannabis to aid in their body's recovery process, so perhaps future research can examine if this holds true in the context of long shifts and demanding work schedules.

Applications to other areas

The information in this review is applicable to anyone who associates with *Cannabis sativa*. The general lesson is that the plant still carries powerful social stigmas fueled by multiple social forces. As such, users, researchers, business owners, and others who associate with then plant can expect to experience the stigmas detailed in this review. Knowing where these stigmas come from as well as the various stigma management strategies can lessen the personal impact. High-profile individuals may also do well to be open about their use of or work with cannabis, as this helps destigmatized the plant. Yet everyone can resist perpetuating stigmatizing cannabis stereotypes by ceasing to assume all cannabis users must exhibit undesirable traits mentioned earlier.

More specifically, those in the health care profession can better educate themselves on cannabis' role in medicine and better support patients who use cannabis for therapeutic purposes. Medical providers may want to consider having explicit conversations about cannabis stigmas with patients who are interested in using cannabis. For those in the business world, it would be wise to drop cannabis from company drug-free workplace requirements, or at the very least, develop explicit policies that accommodate an employee's use of medical cannabis. Finally, for those working to legalize cannabis, it is important to confront the racialized legacy of prohibition and the intersectional cannabis stigmas detailed earlier.

Mini-dictionary of terms

- **Courtesy stigma—**A phenomenon where people who associate with stigmatized groups may experience stigmatization, even if they do not personally possess a stigmatized identity or engage in the stigmatized activity.
- **Incomplete medicalization**—A term describing how something may not be fully recognized as a medicine by others or the larger medical establishment, potentially because the thing may have widely accepted nonmedical applications.
- **Intersectionality—**The awareness that each person's social location is determined by the simultaneous interplay of multiple sources of privilege and oppression, not just by a single identity.
- **Master status—**A status, identity, or label that is always applied to us by others in every situation, other taking precedence over the other statuses we possess.
- **Stigma—**A devaluing social label applied to a person that often elicits moral outrage and disapproval from others in society.
- **Strategies of inclusion—**Emphasizing how one's deviant identity aligns with socially valued characteristics, thus allowing the individual to better fit in with respectable society.
- **Tertiary deviance—**Those who embrace their deviance and advocate for its social acceptance. Compare to primary deviance that is characterized by secrecy or denial, and secondary deviance, which is characterized by reluctant acceptance and social disapproval.
- **Truth regime—**A term from social philosopher Michel Foucault to describe how power structures shape discourse on what is considered to be true, often in spite of facts and often to serve the agenda of those in positions of power. In short, it is the politicization of truth.

Key facts of cannabis stigmas

- Cannabis has been a substance associated with outsiders and the marginalized in nearly all societies.
- Cannabis carries a discreditable (concealable) stigma, and it is a stigma that is seen as deserved because cannabis use is a choice.
- Cannabis stigmas are not exclusively related to its criminalization; they stem from capitalism, racism, organizational policy, and the media, amongst other social forces.
- Since cannabis is incompletely medicalized, medical cannabis patients face the same stigmas as nonmedical users and may also experience discrimination in mainstream medical settings.
- Stigma management strategies often practiced by cannabis users include concealment and risk reduction techniques.
- To help destigmatize cannabis, organizations can change their policies on cannabis use, the media can diversify portrayals of cannabis users, and research can illuminate how cannabis aligns with the values of capitalism.

Summary points

- Research on cannabis stigmas has demonstrated they are global and likely universal.
- Many different social forces inform the stigmatization of cannabis, including prohibition, capitalism, media portrayals, and organizational policies, to name a few.
- The incomplete medicalization of cannabis means it has not yet been fully embraced by (Western) medical institutions and cannabis patients may experiencing stigmatization.
- Considerations of social power dynamics and inequality are necessary when understanding how cannabis stigmas are experienced by various groups.
- Cannabis stigmas can be concealed or managed in various ways, but resisting and challenging such stigmas are important to destigmatize the plant on a larger social level.

References

Adelman, A. (2013). *Gender dynamics and social relations of production in Humboldt's marijuana industry*. Unpublished master's thesis at Humboldt State University. Accessed from: http://humboldt-dspace.calstate.edu/handle/2148/1457.

Adler, P., & Adler, P. (2016/1994). *Constructions of deviance: Social power, context, and interaction*. Boston: Cengage Learning.

Alexander, M. (2010). *The new Jim Crow: Mass incarceration in the age of colorblindness*. New York: The New Press.

Arnull, E., & Ryder, J. (2019). 'Because it's fun': English and American girls' counter-hegemonic stories of alcohol and marijuana use. *Journal of Youth Studies*. https://doi.org/10.1080/13676261.2019.1579898.

Becker, W. C., Meghani, S., Tetrault, J. M., & Fiellin, D. A. (2013). Racial/ethnic differences in report of drug testing practices at the workplace level in the U.S. *American Journal on Addictions*, *23*(4), 357–362.

Bender, S. W. (2016). The colors of cannabis: Race and marijuana. *U.C. Davis Law Review*, *50*(2), 689–706.

Bostwick, J. M. (2012). Blurred boundaries: The therapeutics and politics of medical marijuana. *Mayo Clinic Proceedings*, *87*(2), 172–186.

Bottorff, J. L., Bissell, L. J. L., Balneaves, L. G., Oliffe, J. L., Capler, N. R., & Buxton, J. (2013). Perceptions of cannabis as a stigmatized medicine: A qualitative descriptive study. *Harm Reduction Journal*, *10*(2), 1–10.

Boyd, S. C. (2009). High: Marijuana, women, and the law. *CJWL/RFD*, *21*, 35–54.

Boyd, S., & Carter, C. (2012). Using children: Marijuana grow-ops, media, and policy. *Critical Studies in Media Communication*, *29*(3), 238–257.

Chapkis, W. (2007). Cannabis, consciousness, and healing. *Contemporary Justice Review*, *10*(4), 443–460.

Chapkis, W., & Webb, R. J. (2008). *Dying to get high: Marijuana as medicine*. New York: New York University Press.

Chasteen, J. C. (2016). *Getting high: Marijuana through the ages*. Lanham: Rowman & Littlefield.

Collins, P. H., & Bilge, S. (2016). *Intersectionality*. Malden: Polity.

Dahl, S. L., & Sandberg, S. (2015). Female cannabis users and new masculinities: The gendering of cannabis use. *Sociology*, *49*(4), 696–711.

Dickerson, N. (2018). Ricky and sticky icky: Marijuana, sport, and the legibility/illegibility of black masculinity. *Sociology of Sport Journal*, *35*(4), 386–393.

Dubin, R. E., Kaplan, A., Graves, L., & Ng, V. K. (2017). Acknowledging stigma: Its presence in patient care and medical education. *Canadian Family Physician*, *63*, 906–908.

Duff, C., Asbridge, M., Brochu, S., Cousineau, M. M., Hathaway, A. D., Marsh, D., & Erickson, P. G. (2012). A Canadian perspective on cannabis normalization among adults. *Addiction Research and Theory*, *20*(4), 271–283.

Earleywine, M. (2010). Pot politics: Prohibition and morality. In D. Jacquette (Ed.), *Cannabis—Philosophy for everyone* (pp. 192–213). Malden: Wiley Blackwell.

Garner, E. (2016). A phenomenology of marijuana use among graduate students. *Indo-Pacific Journal of Phenomenology*, *16*, 111–127.

Goffman, E. (1963). *Stigma: Notes on the management of spoiled identity*. New York: Simon & Schuster.

Haines, R. J., Johnson, J. L., Carter, C. I., & Arora, K. (2009). 'I couldn't say I'm not a girl'—Adolescents talk about gender and marijuana use. *Social Science & Medicine*, *68*, 2029–2036.

Haines-Saah, R. J., Johnson, J. L., Repta, R., Ostry, A., Young, M. L., Shoveller, J., Sawatzky, R., Greaves, L., & Ratner, P. A. (2014). The privileged normalization of marijuana use- an analysis of Canadian newspaper reporting, 1997-2007. *Critical Public Health*, *24*(1), 47–61.

Hammer, M. (2015). Patients, police, and care providers: The cultural dimension of medical marijuana implementation. *Administration and Society*, *47*(3), 282–297.

Hathaway, A. D. (2004). Cannabis users' informal rules for managing stigma and risk. *Deviant Behavior*, *25*(6), 559–577.

Hathaway, A. D., Comeau, N. C., & Erickson, P. G. (2011). Cannabis normalization and stigma: Contemporary practices of moral regulation. *Criminology & Criminal Justice*, *11*(5), 451–469.

Herek, G. M. (2004). Beyond 'homophobia': Thinking about sexual prejudice and stigma in the twenty-first century. *Sexuality Research & Social Policy*, *1* (2), 6–24.

Herek, G. M. (2007). Confronting sexual stigma and prejudice: Theory and practice. *Journal of Social Issues*, *63*(4), 905–925.

Hirst, R. B., Conaboy, C., Vaughn, D., Enriquez, R. H., & Wickham, R. (2018). The 'Jay-Dar' phenomenon: Individuals discriminate cannabis users from nonusers based upon a photograph. *Substance Use & Misuse*, *53*(14), 2359–2367.

Jarvinen, M., & Ravn, S. (2014). Cannabis careers revisited: Applying Howard S. Becker's theory to present-day cannabis use. *Social Science & Medicine*, *100*, 133–140.

Jones, C., & Hathaway, A. D. (2008). Marijuana medicine and Canadian physicians: Challenges to meaningful drug policy reform. *Contemporary Justice Review*, *11*(2), 165–175.

Kim, H., & Kim, S. H. (2018). Framing marijuana: how U.S. newspapers frame marijuana legalization stories (1995-2014). *Preventive Medicine Reports*, *11*, 196–201.

Ko, G. D., Bober, S. L., Mindra, S., & Moreau, J. M. (2016). Medical cannabis—The Canadian perspective. *Journal of Pain Research*, *9*, 735–744.

Kondrad, E., & Reid, A. (2013). Colorado family physicians' attitudes toward medical marijuana. *Journal of the American Board of Family Medicine*, *26* (1), 52–60.

Lashley, K., & Pollock, T. G. (2019). Waiting to inhale: Reducing stigma in the medical cannabis industry. *Administrative Science Quarterly*, 1–49.

Lau, N., Sales, P., Averill, S., Murphy, F., Sato, S. O., & Murphy, S. (2015). Responsible and controlled use: Older cannabis users and harm reduction. *International Journal of Drug Policy*, *26*(8), 709–718.

Lewis, J., & Proffitt, J. M. (2012). Bong hits and water bottles: An analysis of news coverage of athletes and marijuana use. *Journal of Sports Media*, *7*(1), 1–21.

Livingston, J. D., & Boyd, J. E. (2010). Correlates and consequences of internalized stigma for people living with mental illness: a systematic review and meta-analysis. *Social Science & Medicine*, *71*, 2150–2161.

Lloyd, C. (2013). The stigmatization of problem drug users: A narrative literature review. *Drugs: Education, Prevention, and Policy*, *20*(2), 85–95.

Loflin, M., & Earleywine, M. (2014). A new method of cannabis ingestion: The dangers of dabs? *Addictive Behaviors*, *39*, 1430–1433.

Looby, A., & Earleywine, M. (2010). Gender moderates the impact of stereotype threat on cognitive function in cannabis users. *Addictive Behaviors*, *35*, 834–839.

McGettigan, T. (2019). *The politics of marijuana: A new paradigm*. New York: Peter Lang.

Mittal, D., Sullivan, G., Chekuri, L., Allee, E., & Corrigan, P. W. (2012). Empirical studies of self-stigma reduction strategies: A critical review of the literature. *Psychiatric Services*, *63*(10), 974–981.

Mortensen, T. M., Moscowitz, L., Wan, A., & Yang, A. (2019). The marijuana user in US news media: an examination of visual stereotypes of race, culture, criminality, and normification. *Visual Communication*, 1–25.

Newhart, M., & Dolphin, W. (2019). *The medicalization of marijuana: Legitimacy, stigma, and the patient experience*. New York: Routledge.

O'Brien, P. K. (2013). Medical marijuana and social control: Escaping criminalization and embracing medicalization. *Deviant Behavior*, *34*(6), 423–443.

Osborne, G. B., & Fogel, C. (2008). Understanding the motivations for recreational marijuana use among adult Canadians. *Substance Use & Misuse*, *43*, 539–572.

Pedersen, W., & Sandberg, S. (2013). The medicalization of revolt: A sociological analysis of medical cannabis users. *Sociology of Health & Illness*, *35*(1), 17–32.

Penn, R. A. (2014). Establishing expertise: Canadian community-based medical cannabis dispensaries as embodied health movement organisations. *International Journal of Drug Policy*, *25*, 372–377.

Pollan, M. (2021). *This is your mind on plants*. New York: Penguin Press.

Reid, M. (2020). A qualitative review of cannabis stigmas at the twilight of prohibition. *Journal of Cannabis Research*, *2*(46), 1–12. https://doi.org/10.1186/s42238-020-00056-8.

Reid, M. (2021). Troubling claims of normalization: Continuing stigmas within Michigan's medical cannabis community. *Deviant Behavior*, 1–20. https://doi.org/10.1080/01639625.2021.1953947.

Reid, M. (2023). How the incomplete medicalization of cannabis shapes patient experiences. *Sociological Focus*. https://doi.org/10.1080/00380237.2022.2163014.

Reinarman, C. (1994). The social construction of drug scares. In P. Adler, & P. Adler (Eds.), *Constructions of deviance: Social power, context, and interaction* (pp. 159–170). Boston: Cengage Learning.

Reinarman, C., Nunberg, H., Lanthier, F., & Heddleston, T. (2011). Who are medical marijuana patients? Population characteristics from nine California assessment clinics. *Journal of Psychoactive Drugs*, *43*(2), 128–135.

Robnett, M. T. (2019). Including PTSD in Colorado: The intersection between research, politics, and public opinion in medical cannabis policy. In T. McGettigan (Ed.), *The politics of marijuana: A new paradigm* (pp. 141–149). New York: Peter Lang.

Roschelle, A. R., & Kauffman, P. (2004). Fitting in and fighting back: Homeless kids' stigma management strategies. *Symbolic Interaction*, *27*(1), 23–36.

Rudski, J. M. (2014). Treatment acceptability, stigma, and legal consequences of medical marijuana are affected by method of administration. *Journal of Drug Issues*, *44*(3), 308–320.

Ryan, J., & Sharts-Hopko, N. (2017). The experiences of medical marijuana patients: A scoping review of the qualitative literature. *Journal of Neuroscience Nursing*, *49*(3), 185–190.

Sandberg, S. (2012). Is cannabis use normalized, celebrated or neutralized? Analyzing talk as action. *Addiction Research and Theory*, *20*(5), 372–381.

Satterlund, T. D., Lee, J. P., & Moore, R. S. (2015). Stigma among California's medical marijuana patients. *Journal of Psychoactive Drugs*, *47*(1), 10–17.

Smith, E., & Hattery, A. J. (2019). Policing the black body: Stop & frisk and the war on drugs. In T. McGettigan (Ed.), *The politics of marijuana: A new paradigm* (pp. 11–20). New York: Peter Lang.

Victorson, D., McMahon, M., Horowitz, B., Glickson, S., Parker, B., & Mendoza-Temple, L. (2019). Exploring cancer survivors' attitudes, perceptions, and concerns about using medical cannabis for symptom and side effect management: A qualitative focus group study. *Complementary Therapies in Medicine*, *47*, 1–8.

Vitiello, M. (2019). Marijuana legalization, racial disparity, and the hope for reform. *Lewis & Clark Law Review*, *23*(3), 789–822.

Werner, C. A. (2001). Medical marijuana and the AIDS crisis. *Journal of Cannabis Therapeutics*, *1*(3/4), 17–33.

Chapter 12

Alterations of THC and CBD ratios and impact on cognition

Erica Zamberletti

Department of Biotechnology and Life Sciences, Neuroscience Center, University of Insubria, Busto Arsizio, VA, Italy

Abbreviations

5-HT1A serotonin 1A receptor
AEA anandamide
AIDS acquired immunodeficiency syndrome
CB1 cannabinoid type 1 receptor
CB2 cannabinoid type 2 receptor
CBD cannabidiol
CNS central nervous system
FDA Food and Drug Administration
GABA gamma-aminobutyric acid
THC delta-9-tetrahydrocannabinol
TRPV1 transient receptor potential vanilloid type-1

Introduction

Cannabis is the most commonly used psychoactive substance worldwide, and it remains an illegal drug in the majority of countries. The recent widespread trend toward legalization and decriminalization of its recreational use in many states is accompanied by a general reduction of risk perception among both adolescents and adults. Alongside with this expanding legalization, also the increasing exploitation of cannabis for medical purposes is contributing to the decreased public perception of harm in the general population. The current scenario might influence the prevalence of cannabis use, and the global number of cannabis users could likely increase in the future. Because of the growing availability and use of cannabis as well as the approved medical uses of cannabinoid compounds, information regarding potential untoward effects and safety limits is needed to guide public policy. Decades of research have highlighted that cannabis can profoundly affect brain function and behavior, and some of its effects can be long lasting. Of primary concern are the potential consequences on cognitive functions (Kroon et al., 2021). Both acute and chronic cannabis use has been associated with changes in cognitive abilities (Schoeler & Bhattacharyya, 2013). While acute cannabis intoxication has been linked to transient mild-to-moderate effects on cognition, more detrimental and lasting effects have been reported as a consequence of repeated and heavy cannabis use, which represent a major social and healthcare issue.

Delta-9-tetrahydrocannabinol (THC) and cannabidiol (CBD)

Cannabis contains more than 100 different cannabinoids that possess proven psychoactive properties (ElSohly et al., 2017). Delta-9-tetrahydrocannabinol (THC) and cannabidiol (CBD) are thus far the two most widely researched cannabinoids.

THC is the principal psychotomimetic constituent of the cannabis plant and produces a wide range of effects in the body by acting as a partial agonist at both cannabinoid CB1 and CB2 receptors: in the central nervous system, THC effects are mainly mediated by the activation of CB1 receptors, while CB2 receptors are the primary mediators of its actions in peripheral districts (Pertwee, 2008). THC is the cannabinoid mostly responsible for the psychoactive effects in healthy individuals, including a feeling of euphoria, known as a "high," relaxation, mild anxiety and paranoia, increased heart rate, perceptual distortions, difficulties in concentration, and memory impairment (D'Souza et al., 2004). Activation of CB1 receptors by THC results in perturbation of GABAergic and glutamatergic neurotransmission, which can induce a transient

Cannabis Use, Neurobiology, Psychology, and Treatment. https://doi.org/10.1016/B978-0-323-89862-1.00007-6

excitatory-to-inhibitory imbalance in brain regions richly expressing CB1 receptors (Augustin & Lovinger, 2022). In addition, THC also promotes striatal dopamine release eventually leading to the development of dependence and addiction in some users (Bossong et al., 2015). Similar biological effects of the natural compounds are seen with the use of approved synthetic THC formulations for medical purposes. There are currently two FDA-approved synthetic THC compounds: dronabinol, approved for the management of chemotherapy-induced nausea and vomiting and for appetite stimulation in AIDS anorexia, and nabilone, approved for use in chemotherapy-induced nausea and vomiting.

Both human data and animal studies have highlighted that chronic intake of THC leads to detrimental effects on several cognitive domains, especially when regular and heavy use of THC is started during early developmental ages (Schoeler & Bhattacharyya, 2013; Zamberletti & Rubino, 2022). However, the great majority of findings has investigated the effect of THC alone without taking into account a possible modulatory role of other cannabinoids in the observed outcomes, making it difficult to translate those results to cannabis users.

CBD is the other major active cannabinoid present in cannabis. In contrast to THC, CBD has a low affinity for the cannabinoid CB1 and CB2 receptors, thus it is devoid of significant psychotomimetic effects. Many molecular targets of CBD have been identified so far, and some of its pharmacological actions have been associated with specific effects at the central nervous system (CNS) level (de Almeida & Devi, 2020). Namely, CBD agonism at serotonin 1A receptors (5-HT1A) was suggested to contribute to its anxiolytic and antidepressant effects in animal models (de Mello Schier et al., 2014). Its ability to activate and rapidly desensitize transient receptor potential vanilloid type-1 (TRPV1) cation channels has been implicated in its efficacy in ameliorating seizures in vitro and in vivo (Iannotti et al., 2014; Lazarini-Lopes et al., 2020). Notably, CBD recently received approval for the management and treatment of children who suffer from the seizure disorders Lennox-Gastaut syndrome and Dravet syndrome (Ryan, 2020). Molecular targets within the endocannabinoid system have also been identified. Indeed, CBD was shown to increase levels of the endogenous cannabinoid ligand anandamide (AEA) by inhibiting its reuptake and degradation (Bisogno et al., 2001), and this effect has been suggested to contribute to its antipsychotic properties (Leweke et al., 2012). Additionally, CBD was suggested to behave as a negative allosteric modulator of cannabinoid CB1 receptors thus promoting conformational changes toward an inactive or impaired signaling receptor state (Chung et al., 2019; Laprairie et al., 2015).

Overall, available findings suggest that many of the effects triggered by CBD are somehow opposite to those induced by THC, underlying the importance of investigating the effect of this cannabinoid in combination to THC in order to provide data that can be more effectively translated to human cannabis use.

THC:CBD ratio in modern cannabis

The composition of cannabis in terms of THC and CBD contents is largely unregulated, especially for preparations in the illicit market. Over the last decades, THC concentrations in cannabis rose dramatically, while CBD levels constantly declined (Fig. 1). The average concentration of THC in confiscated cannabis samples was approximately 4% in 1995 and rose up to 12% in 2014 (ElSohly et al., 2016), while CBD content went from approximately 0.5% in 2004 to less than 0.2% in 2014 (ElSohly et al., 2016), depicting the constant trend of increased potency of cannabis over time. Most worrisome, cannabis potency is still on the rise. Indeed, recent analysis of illegal cannabis preparations reported average THC

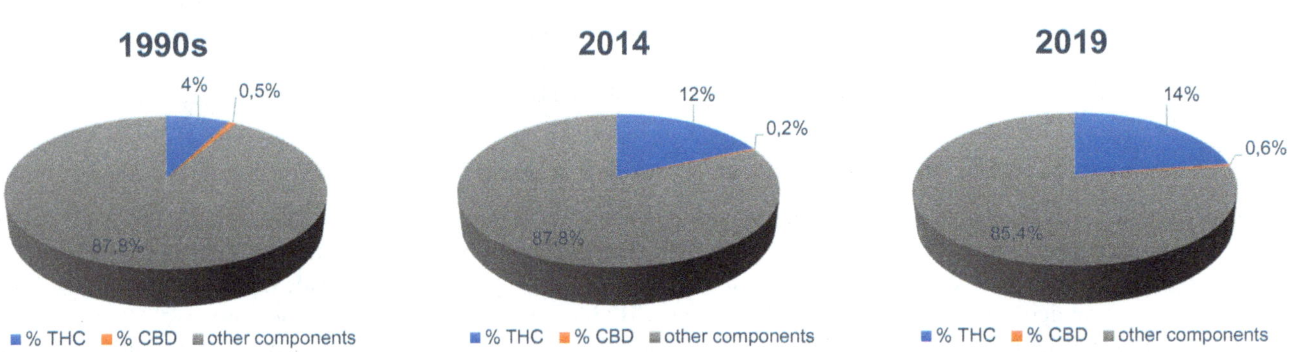

FIG. 1 Composition of cannabis in terms of THC and CBD contents over the last decades. The average concentration of THC in confiscated cannabis samples was approximately 4% in 1995 and rose up to 12% in 2014 (ElSohly et al., 2016), while CBD content went from approximately 0.5% in 2004 to less than 0.2% in 2014 (ElSohly et al., 2016). Analysis of illegal cannabis preparations in 2019 reported average THC contents of approximately 14%, while average contents of CBD were around 0.6% (Cash et al., 2020; ElSohly et al., 2021).

contents of approximately 14% in 2019, while average contents of CBD were around 0.6%, resulting in an overall mean THC:CBD ratio of 24.58 (Cash et al., 2020; ElSohly et al., 2021).

Therefore, while the doses of THC used in the clinical practice are relatively low, recreational cannabis can deliver significantly higher doses of THC and with THC:CBD ratios spanning from 1:1 to 80:1 (Madras, 2019).

It is likely that this ever-changing composition of cannabis in terms of THC and CBD contents might profoundly impact health outcomes especially among regular users. Particularly, the increased levels of THC found in present day cannabis, combined with reduced levels of CBD, raise concerns on the possible detrimental effects on measures of cognitive functions. However, the majority of data published so far compared the effects of THC and CBD alone whereas there is a paucity of research that specifically evaluated the effects of different THC-to-CBD ratio on cognitive domains when administered together.

Effect of different THC:CBD ratio on human cognition

Many studies have assessed the effect of cannabis or pure THC on cognitive functions and those studies have been extensively and comprehensively reviewed and discussed (Broyd et al., 2016; Colizzi & Bhattacharyya, 2017; Schoeler & Bhattacharyya, 2013). Overall, available data demonstrate a range of impaired cognitive functions, including attention, memory, executive and inhibitory processes, during both acute intoxication and following long-term use of cannabis. Cognitive dysfunction in long-term and/or heavy cannabis users has been shown to increase as a function of frequency, duration, dose and age of onset of cannabis use: adolescent cannabis users show similar deficits to those observed in adult users, but greater cognitive impairment is evident the earlier the cannabis use commences. Cognitive dysfunction in long-term users tends to persist for at least 1 month following the cessation of cannabis use, and may persist for longer if cannabis consumption occurs in adolescence (Kroon et al., 2021). Additionally, available data seem to suggest that, opposite to THC, CBD does not exert detrimental effects on cognitive functions when administered per se in healthy subjects (Bhattacharyya et al., 2010, 2012, 2015; Martin-Santos et al., 2012).

Conversely, literature on the effect of simultaneous administration of different THC:CBD ratio on cognitive functions is limited. The following paragraphs will describe available studies that have investigated possible consequences of THC and CBD when coadministered to (i) healthy participants; (ii) cannabis users, and (iii) individuals with underlying neurological disorders.

Studies in healthy subjects

Only few experimental laboratory-based studies have examined the effects of THC and CBD coadministration on cognition in healthy subjects (Table 1). In a prospective, double-blind, placebo-controlled crossover study in 20 healthy volunteers, administration of cannabis extract containing 2.5 mg THC and 1.35 mg CBD (THC:CBD ratio 1: 0.54) for four times over three consecutive weeks resulted in impaired attentional and working memory, similar to the effect triggered by THC alone, suggesting that CBD in the cannabis extract was not sufficient to reverse THC-induced deficits in attentional functioning and working memory, and thus, indicating that CBD may have no potential to acutely improve this aspect of cognitive

TABLE 1 Studies that evaluated the effects of THC and CBD combinations on cognitive performance in healthy individuals.

Study	Treatment protocol	Outcome
Roser et al. (2008)	Cannabis extract (2.5 mg THC and 1.35 mg CBD); four times over three consecutive weeks	Deficits in attentional functioning and working memory
Arkell et al. (2019)	Comparison between cannabis extract containing 11% THC (125 mg) or a 1:1 ratio to CBD (11% THC; 11% CBD)	Similar transient impairments of cognitive function
Pelletti et al. (2021)	Three light cannabis cigarettes containing a percentage 0.41% THC and 12.41% CBD	No impairment of attention, cognitive function, vigilance and decision-making

Administration of cannabis extract containing 2.5 mg THC and 1.35 mg CBD for four times over three consecutive weeks resulted in impaired attentional and working memory, similar to the effect triggered by THC alone (Roser et al., 2008). Vaporization of 125 mg THC-dominant (11% THC; <1% CBD) or a 1:1 THC: CBD ratio (11% THC, 11% CBD) showed similar transient impairments of cognitive function (Arkell et al., 2019). Administration of three light cannabis cigarettes containing a percentage of 0.41% of THC and of 12.41% of CBD did not produce impairment of attention, cognitive function, vigilance and decision-making (Pelletti et al., 2021). Data in healthy subjects seem to suggest that CBD might achieve protection only when combined with very low doses of THC.

performance (Roser et al., 2008). A recent randomized, double-blind, within-subjects crossover design, 14 healthy volunteers who received vaporization of 125 mg THC-dominant (11% THC; <1% CBD) or a 1:1 THC:CBD ratio (11% THC, 11% CBD) showed similar transient impairments of cognitive function, suggesting that CBD does not attenuate the detrimental effects of THC on cognition when administered in a 1:1 ratio (Arkell et al., 2019). In contrast, administration of three light cannabis cigarettes containing a percentage of 0.41% of THC and of 12.41% of CBD did not produce impairment of attention, cognitive function, vigilance, and decision-making, in a population of young healthy adults (Pelletti et al., 2021). However, it is difficult to establish whether this result could be interpreted as a protective effect of CBD toward THC, as THC content in light cannabis cigarettes used in this study was low and no comparison with the same THC dose alone was reported.

Studies in cannabis users

Naturalistic observations have provided evidence for a possible protective role of CBD toward THC-induced cognitive deficits in cannabis users, although defined and precise THC: CBD ratios cannot be extrapolated from these studies (Table 2). In addition, limited information with regard to the possible interactive effects of experimentally administered THC and CBD combinations can be extrapolated from these studies as clinical populations involved are very heterogenous in terms of baseline use of cannabis (THC and CBD contents, as well as frequency and duration of use).

Cross-sectional evaluations in recreational and heavy cannabis users in which verbal memory was examined either at baseline or after acute cannabis ingestion and blood THC and CBD levels were measured, showed that samples with higher levels of CBD were associated with better prose recall. When cannabis users were examined while not intoxicated, it was shown that higher hair THC concentrations were associated with worst performances on verbal recall (Morgan et al., 2010). In a subsequent study, the same research group also analyzed hair levels of CBD in relation to memory performance in cannabis users (Morgan et al., 2012). They found that prose recall and source memory were more impaired in daily cannabis users with higher THC in hair whereas CBD had no impact on performance on these tasks. However, individuals with higher CBD in hair showed better performance on a recognition memory task, irrespective of either THC level or degree of cannabis use (Morgan et al., 2012). These findings possibly suggest that CBD effects on cognition might be different depending on the cognitive domain considered.

Acute effects of different THC:CBD ratios on memory testing in a verbal recognition task were further investigated in a recent study (Curran et al., 2020). The authors demonstrated that memory performance was inversely correlated with THC blood levels, while no relationship to CBD blood levels was observed. Users consuming THC-based strains showed significantly worse memory performance in the post-use setting, whereas subjects using strains containing both THC and CBD showed no differences between pre- and post-use memory performance (Curran et al., 2020), suggesting that memory impairments triggered by acute THC intake are mitigated by concomitant presence of CBD in cannabis preparations.

In addition to naturalistic studies, clinical trials have tried to provide further information on the effect of defined THC: CBD ratios on cognitive performance.

A randomized, double-blind, placebo-controlled study in cannabis users demonstrated that administration of a combination of 8 mg THC and 16 mg CBD (1:2 ratio) did not affect the performance in an emotional facial recognition task compared to THC alone (Hindocha et al., 2015), suggesting that CBD can attenuate THC-induced impairment on emotion recognition. Additionally, acute CBD at the same dose improved emotional facial affect recognition with respect to placebo when administered alone (Hindocha et al., 2015). The same experimental design was used to assess possible effects of THC

TABLE 2 Clinical studies that evaluated the effects of THC and CBD combinations on cognitive performance in cannabis users.

Study	Treatment protocol	Outcome
Hindocha et al. (2015)	Combination of 8 mg THC and 16 mg CBD (1:2 ratio); single administration	No impairment of emotion recognition
Morgan et al. (2018)	Combination of 8 mg THC and 16 mg CBD (1:2 ratio); single administration	Impairment of episodic memory and working memory

Two randomized, double-blind, placebo-controlled studies in cannabis users demonstrated that administration of a combination of 8 mg THC and 16 mg CBD (1:2 ratio) did not affect the performance in an emotional facial recognition task (Hindocha et al., 2015), but resulted in detrimental effects on episodic and working memory (Morgan et al., 2018). This possibly suggest that CBD protection toward THC-induced cognitive impairments, when present, might be dependent upon the specific cognitive domain being assessed.

and CBD, either alone or in combination, on episodic memory and working memory in cannabis users (Morgan et al., 2018). Results from this study confirmed that THC had detrimental effects on episodic and working memory. However, unlike what previously reported for emotion recognition, coadministration of CBD failed to attenuate deficits in these cognitive domains (Morgan et al., 2018).

Overall, available evidence in cannabis users seems to suggest that the protective effect of CBD toward THC-induced cognitive impairments, when present, might be dependent upon the specific cognitive domain being assessed, rather than being general.

Studies in patients with underlying neurological disorders

Some studies have reported the effect of THC and CBD combinations on measures of cognition in clinical populations with underlying neurological disorders (Table 3).

A double-blind cross-over study in patients affected by a wide range of neurological symptoms mainly associated with multiple sclerosis and neuropathic pain conditions showed that CBD, administered in a combination 1:1 with THC (2.5 mg), was able to prevent THC-induced cognitive impairments in the short orientation-memory concentration test, without affecting cognitive abilities when administered per se at the same dose (Wade et al., 2003). When cognitive function was assessed in a longitudinal study in individuals using medical cannabis for relieving a variety of neurological conditions including anxiety, depression, chronic pain, and sleep, Gruber et al. (2016, 2018) reported improvement on measures of executive functioning and cognitive flexibility following 3 months of treatment compared to baseline performance. However, in this study, patients chose their own products and treatment regimen; therefore, no specific information was provided with respect to CBD and THC contents in the preparations resulting in a very heterogenous clinical sample.

Recently, improved cognitive functions have been reported when assessing the effect of CBD-enriched cannabis treatment (CBD: THC in a 20:1 ratio) in children diagnosed with autism spectrum disorder (Bilge & Ekici, 2021).

Available data in patients using cannabis for medical purposes, although very limited, seem to suggest that combinations of THC and CBD do not negatively impact cognition. In addition, in all these studies, cannabis treatment was also

TABLE 3 Data on the effects of medical use of THC and CBD combinations on cognitive performance.

Study	Treatment protocol	Population	Outcome
Wade et al. (2003)	Combination of THC: CBD 1:1 (2.5 mg each)	Patients affected by a wide range of neurological symptoms mainly associated with multiple sclerosis and neuropathic pain conditions	No detrimental effect in the short orientation-memory concentration test
Gruber et al. (2016, 2018)	Combination of THC: CBD 1:1 (2.5 mg each)	Patients affected by a wide range of neurological symptoms mainly associated with multiple sclerosis and neuropathic pain conditions	Improvement on measures of executive functioning and cognitive flexibility following 3 months of treatment
Bilge and Ekici (2021)	CBD-enriched cannabis treatment (CBD:THC in a 20:1 ratio)	Children diagnosed with autism spectrum disorder	Improved cognitive functions
Aragona et al. (2009)	Nabiximols	Cannabis-naïve patients with multiple sclerosis	No detrimental effect on cognitive functions
Schoedel et al. (2011)	Nabiximols; single-dose from 1 to 4 sprays	Healthy volunteers who were frequent recreational cannabis users	No clinically significant abnormalities on neurocognitive functions

No detrimental effect in the short orientation-memory concentration test were reported after use of a combination of THC:CBD 1:1 (2.5 mg each) in patients affected by a wide range of neurological symptoms mainly associated with multiple sclerosis and neuropathic pain conditions (Wade et al., 2003). Gruber et al. (2016, 2018) reported improvement on measures of executive functioning and cognitive flexibility following 3 months of treatment, with no specific information with respect to CBD and THC contents in the preparations. Improved cognitive functions have been reported when assessing the effect of CBD-enriched cannabis treatment (CBD: THC in a 20:1 ratio) in children diagnosed with autism spectrum disorder (Bilge & Ekici, 2021). However, improvement of cognitive performance is suggested to be secondary to the amelioration in the overall symptomatology (Sagar et al., 2021). The cannabinoid-based medication nabiximols administered to healthy volunteers who were frequent recreational cannabis did not induce any clinically significant abnormality on neurocognitive functions (Schoedel et al., 2011). The lack of detrimental effect of nabiximols on cognitive functions was also reported in a placebo-controlled crossover study in cannabis-naïve patients with multiple sclerosis (Aragona et al., 2009).

associated with an overall improvement of the neurological condition for which cannabis use was initiated, so it is not possible to rule out the possibility that the observed improvement of cognitive functions could be secondary to the amelioration of the overall symptomatology. This is also suggested by a recently published paper (Sagar et al., 2021). In this longitudinal and observational study in medical cannabis users, baseline cognitive and clinical assessments were performed prior to initiation of medical cannabis use and patients were reassessed at multiple time points following initiation of treatment (3, 6, and 12 months). Information regarding frequency of use and exposure to THC and CBD were also quantified to determine if medical cannabis use variables contributed to cognitive and clinical changes (Sagar et al., 2021). Over the course of 12 months of treatment, patients exhibited significant changes on several measures of executive functioning which occurred in the context of significant improvements on measures of mood, anxiety, and sleep. Interestingly, in the study, neither THC nor CBD use was directly correlated with observed improvements on measures of executive function but correlation analyses suggested that improved cognitive performance was associated with improvements in clinical symptoms.

Data from cannabinoid-based medication

Information on the possible effect of THC: CBD combinations on measures of cognitive functions could also be obtained from data in patients using cannabinoid-based medications. Unlike cannabis extracts, these products are characterized by known and constant THC and CBD contents, which allow a better correlation with the effect of the specific cannabinoids on clinical outcomes. Nabiximols (Sativex) is a cannabinoid-based medication consisting of a 1:1 THC: CBD combination approved for treatment of symptoms of spasticity in adults with multiple sclerosis in several countries and is available as an oromucosal spray. The impact of nabiximols has been evaluated in a single-dose, randomized, double-blind, balanced, placebo-controlled, crossover study in healthy volunteers who were frequent recreational cannabis users and showed no clinically significant abnormalities (Schoedel et al., 2011). In this work, neurocognitive functions were evaluated as secondary outcome measures in comparison with dronabinol and placebo. No significant cognitive and motor impairment effects were observed among treatment groups with the exception of a small effect of dronabinol on reaction time in the short-term memory test. The lack of detrimental effect of nabiximols on cognitive functions was also reported in a placebo-controlled crossover study in a small cohort of cannabis-naïve patients with multiple sclerosis (Aragona et al., 2009).

Effect of different THC:CBD ratio on cognition: Preclinical evidence

Studies in animal models have been useful for assessing the consequences of cannabinoid exposure on brain and behavior in the absence of other confounding factors which might influence the overall outcomes when trying to establish a causal association in humans. Using animal models, many research studies have highlighted the detrimental effects of THC on many cognitive domains which are strongly dependent on the dose, frequency, and age of exposure (Prini et al., 2020). Available evidence suggests that heavy and chronic THC exposure at early stages of brain development (i.e., from prenatal ages to adolescence) is associated with enduring cognitive defects in both nonhuman primates and rodent models (Zamberletti & Rubino, 2022). In contrast, very little information is available on the consequences of the administration of different THC-to-CBD combinations on measures of cognitive functioning (Table 4). Chronic coadministration of THC and CBD in a ratio 1:1 (3 mg each) to adolescent male mice did not result in enduring deficits in recognition memory and working memory (Murphy et al., 2017). In contrast, long-term deficits in short-term recognition memory were reported after coadministration of THC and CBD in ratios 3: 1 or 1: 33 in adolescent female rats (Gabaglio et al., 2021). Similarly, administration of a combination of THC and CBD in a ratio 1:1 to adolescent male squirrel monkeys impaired cognitive flexibility (Withey et al., 2021).

Interesting results have been obtained when investigating the effect of THC and CBD coadministration on cognitive performance in aged animals. Notably, in aged mice positive effects on measures of cognitive performance have been reported after administration of THC (Nidadavolu et al., 2021). In this case, administration of THC and CBD in a 1:1 ratio did not improve spatial learning memory deficits in aged mice (Nidadavolu et al., 2021), suggesting that CBD might prevent the procognitive effects of THC on age-related cognitive decline. A lack of protective effect on age-related cognitive impairments was also reported after administration of CBD-enriched botanical extract (62.7% CBD, 3.6% THC) to aged mice (Aso et al., 2016). Remarkably, while administration of THC and CBD combination did not protect against cognitive deficits associated with physiological aging, a beneficial effect was instead reported when cognitive deficits associated with pathological conditions were assessed in aged animals. Namely, chronic administration of CBD-enriched

TABLE 4 Consequences of the administration of different THC and CBD combinations on measures of cognitive functioning in animal models.

Study	THC:CBD ratio	Animal model	Outcome
Murphy et al. (2017)	Combination of THC:CBD 1:1	Adolescent male mice	Recognition memory and working memory unaffected
Gabaglio et al. (2021)	Combination of THC:CBD in ratios 3:1 or 1:33	Adolescent female rats	Impairments in short-term recognition memory
Withey et al. (2021)	Combination of THC:CBD 1:1	Adolescent male squirrel monkeys	Impaired cognitive flexibility
Nidadavolu et al. (2021)	Combination of THC:CBD 1:1	Aged male mice	No beneficial effect on spatial learning memory deficits associated with physiological aging
Aso et al. (2015)	CBD-enriched botanical extract (64.8% CBD, 2.3% THC)	Aged male AβPP/PS1 transgenic mice	Improved memory impairment in the mouse model of Alzheimer's disease
Aso et al. (2016)	CBD-enriched botanical extract (62.7% CBD, 3.6% THC)	Aged male wild-type mice or AβPP/PS1 transgenic mice	No effect on age-related cognitive impairments in wildtype mice but improved cognitive performance in the mouse model of Alzheimer's disease

Administration of THC and CBD in a ratio 1:1 to adolescent male mice did not result in enduring deficits in recognition memory and working memory (Murphy et al., 2017) but deficits in short-term recognition memory were reported after coadministration of THC and CBD in ratios 3: 1 or 1: 33 to adolescent female rats (Gabaglio et al., 2021). Similarly, administration of THC and CBD in a ratio 1:1 to adolescent male squirrel monkeys impaired cognitive flexibility (Withey et al., 2021). Administration of THC and CBD in a 1:1 ratio did not improve spatial learning memory deficits in aged mice (Nidadavolu et al., 2021). CBD-enriched botanical extract (62.7% CBD, 3.6% THC) did not ameliorate cognitive deficits associated with healthy brain aging but improved cognitive performance in mice models of Alzheimer-like phenotype (Aso et al., 2015, 2016).

botanical extract improved cognitive performance in mice models of Alzheimer-like phenotype even when administered at advanced stages of the disease (Aso et al., 2015, 2016).

As a whole, based on currently available preclinical evidence, the protective effect of CBD toward THC-induced cognitive defects does not seem to be present in young animals or during physiological brain aging. In contrast, combinations of THC and CBD are suggested to be effective in ameliorating memory impairment associated with pathological neurodegenerative conditions, such as Alzheimer's disease.

Conclusions

Experimental studies that specifically investigated whether the concurrent presence of CBD could modulate THC-induced cognitive deficits are limited. Findings in healthy subjects seem to suggest that CBD might not be able to dampen the detrimental effect of THC on cognition, while some potential to attenuate cognitive deterioration is seen when THC and CBD are coadministered to cannabis users. Administration of medical formulations containing THC and CBD in equal amounts to healthy volunteers did not affect cognitive performance. Evidence from animal studies in both rodents and nonhuman primates indicates that CBD might not be able to prevent the detrimental effects of THC on cognition especially in critical developmental windows. Overall, it is not possible to draw a definitive conclusion as available clinical and preclinical data are scarce and mixed. Further clinical studies with larger sample size and more homogenous populations are needed to clearly establish the potential modulatory effects of CBD toward THC-induced cognitive disturbances, especially following chronic administration, as both recreational and putative therapeutic human use of cannabis imply repeated exposure. Furthermore, future preclinical studies should be designed to better mimic human cannabis consumption, in terms of THC and CBD doses, duration of exposure, and delivery methods. Moreover, besides CBD, future studies should also take into account possible roles of other pharmacologically active components present in the cannabis plants that might possibly affect THC effects on cognitive functions.

Applications to other areas

In this chapter, we have reviewed available evidence on the possible modulatory effect of CBD on THC-induced cognitive impairments, when the two cannabinoids are administered simultaneously at specific ratios that could be reminiscent of those present in recreational and medical cannabis preparations. While experimental evidence has demonstrated the detrimental effects of THC on cognition, very few studies have dealt with the possible consequences of THC and CBD when administered in combination to both humans and animals. Understanding whether CBD could mitigate THC-induced detrimental effects on cognition would help to set personal and public decision making on the use of cannabis and to provide evidence on possible vulnerable populations as well as safer ratios to be used in the clinical setting.

Key facts of THC

- THC is the main psychotomimetic component of cannabis.
- THC can affect learning, memory, and reaction time.
- THC's effects on cognition are dose dependent.
- THC's effects on cognition are age dependent.

Key facts of CBD

- CBD is devoid of psychotomimetic activity.
- It is thought to modulate some of the adverse effects of THC on cognition.
- CBD does not dampen the detrimental effect of THC on cognition in healthy subjects.
- CBD partially attenuates cognitive deterioration in cannabis users.
- Medical formulations containing THC and CBD in equal amounts to healthy volunteers do not affect cognitive performance.
- Available animal studies do not demonstrate protective effects of CBD on cognitive deficits triggered by THC.

Mini-dictionary of terms

Cannabidiol. A major active cannabinoid present in cannabis; it is devoid of significant psychotomimetic effects.

Delta-9-tetrahydrocannabinol. The main psychotomimetic component present in cannabis; it is responsible for the psychoactive effects in healthy individuals, including a feeling of euphoria, known as a "high," relaxation, mild anxiety and paranoia, increased heart rate, perceptual distortions, difficulties in concentration, and memory impairment.

Dronabinol. Synthetic form of delta-9-tetrahydrocannabinol approved for the treatment of anorexia associated with weight loss in patients with AIDS and nausea and vomiting associated with cancer chemotherapy in patients who have failed to respond adequately to conventional antiemetic treatments.

Epidiolex. Prescription medicine containing purified cannabidiol that is used to treat seizures associated with Lennox-Gastaut syndrome, Dravet syndrome, or tuberous sclerosis complex in patients 1 year of age and older.

Nabilone. Synthetic analog of delta-9-tetrahydrocannabinol approved for the treatment of nausea and vomiting associated with cancer chemotherapy in patients who have failed to respond adequately to conventional antiemetic treatments.

Nabiximols. Whole-plant cannabis extract containing a 1:1 combination of THC:CBD approved for spasticity associated with multiple sclerosis and cancer-associated pain.

Summary points

- THC concentrations in cannabis rose dramatically in the last decades, while CBD levels declined.
- This change in cannabis composition could likely impact health outcomes, especially in terms of cognitive functioning.
- Despite the general interest in the possible modulatory effects of CBD toward THC-induced cognitive deterioration, few preclinical and clinical studies have specifically addressed this complex issue.
- Human evidence suggests that CBD might not be able to dampen the detrimental effect of THC on cognition in healthy volunteers, while some potential to attenuate cognitive deterioration is seen when THC and CBD are coadministered to cannabis users.
- The majority of evidence from animal studies suggests that CBD might not be able to prevent the detrimental effects of THC on cognition.

- Future clinical studies with larger sample size and more homogenous populations are needed to establish the potential modulatory effects of CBD toward THC-induced cognitive disturbances, especially following chronic exposure.
- Preclinical studies should be designed to better mimic human cannabis consumption, in terms of THC and CBD doses, duration of exposure, and delivery methods.

References

Aragona, M., Onesti, E., Tomassini, V., Conte, A., Gupta, S., Gilio, F., et al. (2009). Psychopathological and cognitive effects of therapeutic cannabinoids in multiple sclerosis: A double-blind, placebo controlled, crossover study. *Clinical Neuropharmacology*, *32*(1), 41–47. https://doi.org/10.1097/WNF.0B013E3181633497.

Arkell, T. R., Lintzeris, N., Kevin, R. C., Ramaekers, J. G., Vandrey, R., Irwin, C., et al. (2019). Cannabidiol (CBD) content in vaporized cannabis does not prevent tetrahydrocannabinol (THC)-induced impairment of driving and cognition. *Psychopharmacology*, *236*(9), 2713–2724. https://doi.org/10.1007/s00213-019-05246-8.

Aso, E., Andres-Benito, P., & Ferrer, I. (2016). Delineating the efficacy of a cannabis-based medicine at advanced stages of dementia in a murine model. *Journal of Alzheimer's Disease*, *54*(3), 903–912. https://doi.org/10.3233/JAD-160533.

Aso, E., Sanchez-Pla, A., Vegas-Lozano, E., Maldonado, R., & Ferrer, I. (2015). Cannabis-based medicine reduces multiple pathological processes in AbetaPP/PS1 mice. *Journal of Alzheimer's Disease*, *43*(3), 977–991. https://doi.org/10.3233/JAD-141014.

Augustin, S. M., & Lovinger, D. M. (2022). Synaptic changes induced by cannabinoid drugs and cannabis use disorder. *Neurobiology of Disease*, *167*, 105670. https://doi.org/10.1016/j.nbd.2022.105670.

Bhattacharyya, S., Crippa, J. A., Allen, P., Martin-Santos, R., Borgwardt, S., Fusar-Poli, P., et al. (2012). Induction of psychosis by Delta9-tetrahydrocannabinol reflects modulation of prefrontal and striatal function during attentional salience processing. *Archives of General Psychiatry*, *69*(1), 27–36. https://doi.org/10.1001/archgenpsychiatry.2011.161.

Bhattacharyya, S., Falkenberg, I., Martin-Santos, R., Atakan, Z., Crippa, J. A., Giampietro, V., et al. (2015). Cannabinoid modulation of functional connectivity within regions processing attentional salience. *Neuropsychopharmacology*, *40*(6), 1343–1352. https://doi.org/10.1038/npp.2014.258.

Bhattacharyya, S., Morrison, P. D., Fusar-Poli, P., Martin-Santos, R., Borgwardt, S., Winton-Brown, T., et al. (2010). Opposite effects of delta-9-tetrahydrocannabinol and cannabidiol on human brain function and psychopathology. *Neuropsychopharmacology*, *35*(3), 764–774. https://doi.org/10.1038/npp.2009.184.

Bilge, S., & Ekici, B. (2021). CBD-enriched cannabis for autism spectrum disorder: An experience of a single center in Turkey and reviews of the literature. *Journal of Cannabis Research*, *3*(1), 53. https://doi.org/10.1186/s42238-021-00108-7.

Bisogno, T., Hanus, L., De Petrocellis, L., Tchilibon, S., Ponde, D. E., Brandi, I., et al. (2001). Molecular targets for cannabidiol and its synthetic analogues: Effect on vanilloid VR1 receptors and on the cellular uptake and enzymatic hydrolysis of anandamide. *British Journal of Pharmacology*, *134*(4), 845–852. https://doi.org/10.1038/sj.bjp.0704327.

Bossong, M. G., Mehta, M. A., van Berckel, B. N., Howes, O. D., Kahn, R. S., & Stokes, P. R. (2015). Further human evidence for striatal dopamine release induced by administration of 9-tetrahydrocannabinol (THC): Selectivity to limbic striatum. *Psychopharmacology*, *232*(15), 2723–2729. https://doi.org/10.1007/s00213-015-3915-0.

Broyd, S. J., van Hell, H. H., Beale, C., Yucel, M., & Solowij, N. (2016). Acute and chronic effects of cannabinoids on human cognition—A systematic review. *Biological Psychiatry*, *79*(7), 557–567. https://doi.org/10.1016/j.biopsych.2015.12.002.

Cash, M. C., Cunnane, K., Fan, C., & Romero-Sandoval, E. A. (2020). Mapping cannabis potency in medical and recreational programs in the United States. *PLoS One*, *15*(3), e0230167. https://doi.org/10.1371/journal.pone.0230167.

Chung, H., Fierro, A., & Pessoa-Mahana, C. D. (2019). Cannabidiol binding and negative allosteric modulation at the cannabinoid type 1 receptor in the presence of delta-9-tetrahydrocannabinol: An in silico study. *PLoS One*, *14*(7), e0220025. https://doi.org/10.1371/journal.pone.0220025.

Colizzi, M., & Bhattacharyya, S. (2017). Does cannabis composition matter? Differential effects of Delta-9-tetrahydrocannabinol and cannabidiol on human cognition. *Current Addiction Reports*, *4*(2), 62–74. https://doi.org/10.1007/s40429-017-0142-2.

Curran, T., Devillez, H., YorkWilliams, S. L., & Bidwell, L. C. (2020). Acute effects of naturalistic THC vs. CBD use on recognition memory: A preliminary study. *Journal of Cannabis Research*, *2*(1), 28. https://doi.org/10.1186/s42238-020-00034-0.

de Almeida, D. L., & Devi, L. A. (2020). Diversity of molecular targets and signaling pathways for CBD. *Pharmacology Research & Perspectives*, *8*(6), e00682. https://doi.org/10.1002/prp2.682.

de Mello Schier, A. R., de Oliveira Ribeiro, N. P., Coutinho, D. S., Machado, S., Arias-Carrion, O., Crippa, J. A., et al. (2014). Antidepressant-like and anxiolytic-like effects of cannabidiol: A chemical compound of *Cannabis sativa*. *CNS & Neurological Disorders Drug Targets*, *13*(6), 953–960. https://doi.org/10.2174/1871527313666140612114838.

D'Souza, D. C., Perry, E., MacDougall, L., Ammerman, Y., Cooper, T., Wu, Y. T., et al. (2004). The psychotomimetic effects of intravenous delta-9-tetrahydrocannabinol in healthy individuals: Implications for psychosis. *Neuropsychopharmacology*, *29*(8), 1558–1572. https://doi.org/10.1038/sj.npp.1300496.

ElSohly, M. A., Chandra, S., Radwan, M., Majumdar, C. G., & Church, J. C. (2021). A comprehensive review of cannabis potency in the United States in the last decade. *Biological Psychiatry: Cognitive Neuroscience and Neuroimaging*, *6*(6), 603–606. https://doi.org/10.1016/j.bpsc.2020.12.016.

ElSohly, M. A., Mehmedic, Z., Foster, S., Gon, C., Chandra, S., & Church, J. C. (2016). Changes in cannabis potency over the last 2 decades (1995-2014): Analysis of current data in the United States. *Biological Psychiatry*, *79*(7), 613–619. https://doi.org/10.1016/j.biopsych.2016.01.004.

ElSohly, M. A., Radwan, M. M., Gul, W., Chandra, S., & Galal, A. (2017). Phytochemistry of *Cannabis sativa* L. *Progress in the Chemistry of Organic Natural Products*, *103*, 1–36. https://doi.org/10.1007/978-3-319-45541-9_1.

Gabaglio, M., Zamberletti, E., Manenti, C., Parolaro, D., & Rubino, T. (2021). Long-term consequences of adolescent exposure to THC-rich/CBD-poor and CBD-rich/THC-poor combinations: A comparison with pure THC treatment in female rats. *International Journal of Molecular Sciences*, *22* (16). https://doi.org/10.3390/ijms22168899.

Gruber, S. A., Sagar, K. A., Dahlgren, M. K., Gonenc, A., Smith, R. T., Lambros, A. M., … Lukas, S. E. (2018). The grass might be greener: Medical marijuana patients exhibit altered brain activity and improved executive function after 3 months of treatment. *Frontiers in Pharmacology*, *8* (983). https://doi.org/10.3389/fphar.2017.00983.

Gruber, S. A., Sagar, K. A., Dahlgren, M. K., Racine, M. T., Smith, R. T., & Lukas, S. E. (2016). Splendor in the grass? A pilot study assessing the impact of medical marijuana on executive function. *Frontiers in Pharmacology*, *7*, 355. https://doi.org/10.3389/fphar.2016.00355.

Hindocha, C., Freeman, T. P., Schafer, G., Gardener, C., Das, R. K., Morgan, C. J., et al. (2015). Acute effects of delta-9-tetrahydrocannabinol, cannabidiol and their combination on facial emotion recognition: A randomised, double-blind, placebo-controlled study in cannabis users. *European Neuropsychopharmacology*, *25*(3), 325–334. https://doi.org/10.1016/j.euroneuro.2014.11.014.

Iannotti, F. A., Hill, C. L., Leo, A., Alhusaini, A., Soubrane, C., Mazzarella, E., et al. (2014). Nonpsychotropic plant cannabinoids, cannabidivarin (CBDV) and cannabidiol (CBD), activate and desensitize transient receptor potential vanilloid 1 (TRPV1) channels in vitro: Potential for the treatment of neuronal hyperexcitability. *ACS Chemical Neuroscience*, *5*(11), 1131–1141. https://doi.org/10.1021/cn5000524.

Kroon, E., Kuhns, L., & Cousijn, J. (2021). The short-term and long-term effects of cannabis on cognition: Recent advances in the field. *Current Opinion in Psychology*, *38*, 49–55. https://doi.org/10.1016/j.copsyc.2020.07.005.

Laprairie, R. B., Bagher, A. M., Kelly, M. E., & Denovan-Wright, E. M. (2015). Cannabidiol is a negative allosteric modulator of the cannabinoid CB1 receptor. *British Journal of Pharmacology*, *172*(20), 4790–4805. https://doi.org/10.1111/bph.13250.

Lazarini-Lopes, W., Do Val-da Silva, R. A., da Silva-Junior, R. M. P., Leite, J. P., & Garcia-Cairasco, N. (2020). The anticonvulsant effects of cannabidiol in experimental models of epileptic seizures: From behavior and mechanisms to clinical insights. *Neuroscience and Biobehavioral Reviews*, *111*, 166–182. https://doi.org/10.1016/j.neubiorev.2020.01.014.

Leweke, F. M., Piomelli, D., Pahlisch, F., Muhl, D., Gerth, C. W., Hoyer, C., et al. (2012). Cannabidiol enhances anandamide signaling and alleviates psychotic symptoms of schizophrenia. *Translational Psychiatry*, *2*, e94. https://doi.org/10.1038/tp.2012.15.

Madras, B. K. (2019). Tinkering with THC-to-CBD ratios in marijuana. *Neuropsychopharmacology*, *44*(1), 215–216. https://doi.org/10.1038/s41386-018-0217-3.

Martin-Santos, R., Crippa, J. A., Batalla, A., Bhattacharyya, S., Atakan, Z., Borgwardt, S., et al. (2012). Acute effects of a single, oral dose of d9-tetrahydrocannabinol (THC) and cannabidiol (CBD) administration in healthy volunteers. *Current Pharmaceutical Design*, *18*(32), 4966–4979. https://doi.org/10.2174/138161212802884780.

Morgan, C. J. A., Freeman, T. P., Hindocha, C., Schafer, G., Gardner, C., & Curran, H. V. (2018). Individual and combined effects of acute delta-9-tetrahydrocannabinol and cannabidiol on psychotomimetic symptoms and memory function. *Translational Psychiatry*, *8*(1), 181. https://doi.org/10.1038/s41398-018-0191-x.

Morgan, C. J., Gardener, C., Schafer, G., Swan, S., Demarchi, C., Freeman, T. P., et al. (2012). Sub-chronic impact of cannabinoids in street cannabis on cognition, psychotic-like symptoms and psychological well-being. *Psychological Medicine*, *42*(2), 391–400. https://doi.org/10.1017/S0033291711001322.

Morgan, C. J., Schafer, G., Freeman, T. P., & Curran, H. V. (2010). Impact of cannabidiol on the acute memory and psychotomimetic effects of smoked cannabis: Naturalistic study: Naturalistic study [corrected]. *The British Journal of Psychiatry*, *197*(4), 285–290. https://doi.org/10.1192/bjp.bp.110.077503.

Murphy, M., Mills, S., Winstone, J., Leishman, E., Wager-Miller, J., Bradshaw, H., et al. (2017). Chronic adolescent delta(9)-tetrahydrocannabinol treatment of male mice leads to long-term cognitive and behavioral dysfunction, which are prevented by concurrent cannabidiol treatment. *Cannabis and Cannabinoid Research*, *2*(1), 235–246. https://doi.org/10.1089/can.2017.0034.

Nidadavolu, P., Bilkei-Gorzo, A., Kramer, M., Schurmann, B., Palmisano, M., Beins, E. C., et al. (2021). Efficacy of delta(9) -tetrahydrocannabinol (THC) alone or in combination with a 1:1 ratio of cannabidiol (CBD) in reversing the spatial learning deficits in old mice. *Frontiers in Aging Neuroscience*, *13*, 718850. https://doi.org/10.3389/fnagi.2021.718850.

Pelletti, G., Barone, R., Giorgetti, A., Garagnani, M., Rossi, F., Fais, P., et al. (2021). "light cannabis" consumption in a sample of young adults: Preliminary pharmacokinetic data and psychomotor impairment evaluation. *Forensic Science International*, *323*, 110822. https://doi.org/10.1016/j.forsciint.2021.110822.

Pertwee, R. G. (2008). The diverse CB1 and CB2 receptor pharmacology of three plant cannabinoids: Delta9-tetrahydrocannabinol, cannabidiol and delta9-tetrahydrocannabivarin. *British Journal of Pharmacology*, *153*(2), 199–215. https://doi.org/10.1038/sj.bjp.0707442.

Prini, P., Zamberletti, E., Manenti, C., Gabaglio, M., Parolaro, D., & Rubino, T. (2020). Neurobiological mechanisms underlying cannabis-induced memory impairment. *European Neuropsychopharmacology*, *36*, 181–190. https://doi.org/10.1016/j.euroneuro.2020.02.002.

Roser, P., Juckel, G., Rentzsch, J., Nadulski, T., Gallinat, J., & Stadelmann, A. M. (2008). Effects of acute oral delta9-tetrahydrocannabinol and standardized cannabis extract on the auditory P300 event-related potential in healthy volunteers. *European Neuropsychopharmacology*, *18*(8), 569–577. https://doi.org/10.1016/j.euroneuro.2008.04.008.

Ryan, M. (2020). Cannabidiol in epilepsy: The indications and beyond. *The Mental Health Clinician*, *10*(6), 317–325. https://doi.org/10.9740/mhc.2020.11.317.

Sagar, K. A., Dahlgren, M. K., Lambros, A. M., Smith, R. T., El-Abboud, C., & Gruber, S. A. (2021). An observational, longitudinal study of cognition in medical cannabis patients over the course of 12 months of treatment: Preliminary results. *Journal of the International Neuropsychological Society*, *27* (6), 648–660. https://doi.org/10.1017/S1355617721000114.

Schoedel, K. A., Chen, N., Hilliard, A., White, L., Stott, C., Russo, E., et al. (2011). A randomized, double-blind, placebo-controlled, crossover study to evaluate the subjective abuse potential and cognitive effects of nabiximols oromucosal spray in subjects with a history of recreational cannabis use. *Human Psychopharmacology*, *26*(3), 224–236. https://doi.org/10.1002/hup.1196.

Schoeler, T., & Bhattacharyya, S. (2013). The effect of cannabis use on memory function: An update. *Substance Abuse and Rehabilitation*, *4*, 11–27. https://doi.org/10.2147/SAR.S25869.

Wade, D. T., Robson, P., House, H., Makela, P., & Aram, J. (2003). A preliminary controlled study to determine whether whole-plant cannabis extracts can improve intractable neurogenic symptoms. *Clinical Rehabilitation*, *17*(1), 21–29. https://doi.org/10.1191/0269215503cr581oa.

Withey, S. L., Kangas, B. D., Charles, S., Gumbert, A. B., Eisold, J. E., George, S. R., et al. (2021). Effects of daily delta(9)-tetrahydrocannabinol (THC) alone or combined with cannabidiol (CBD) on cognition-based behavior and activity in adolescent nonhuman primates. *Drug and Alcohol Dependence*, *221*, 108629. https://doi.org/10.1016/j.drugalcdep.2021.108629.

Zamberletti, E., & Rubino, T. (2022). Dos(e)age: Role of dose and age in the long-term effect of cannabinoids on cognition. *Molecules*, *27*(4). https://doi.org/10.3390/molecules27041411.

Chapter 13

When cannabis is used for medicinal purposes: A focus on back pain

Alexa O'Regan[a] and Richard L. Price[b,c]
[a]*Royal College of Surgeons Ireland, Dublin 2, Ireland,* [b]*Swift Institute, Reno, NV, United States,* [c]*University of Nevada-Reno College of Medicine, Reno, NV, United States*

Abbreviations

A2A	adenosine A2A receptor
CBD	cannabidiol
CB1/CB2	cannabinoid 1/2
COX-2	cyclooxygenase-2
CSC	cannabis social club
IQ	intelligence quotient
MME	morphine milligram equivalents
MRI	magnetic resonance imaging
MSK	musculoskeletal
NSAID	nonsteroidal antiinflammatory drug
OIH	opioid-induced hyperanalgesia
PEG	pain intensity and interference
Ppar-λ	peroxisome proliferator-activated receptor gamma
PSQI	Pittsburgh sleep quality index
TENS	transcutaneous electrical nerve stimulator
THC	delta-9-tetrahydrocannabinol
5-HT1A	serotonin A1 receptor

Introduction

Chronic back pain is currently one of the most common complaints to primary care providers. In the United States alone, up to 80% of adults experience back pain in their lives, which contributes to disability, societal, and economic downfalls (First et al., 2020). The management of back pain is multimodal beginning with conservative therapies. This includes, but is not limited to, over the counter pain medicines, ice, heat, rest, exercise, physical therapy and activity modulation. Medicines used include topical analgesics, muscle relaxers, antidepressants, and anticonvulsants (i.e., gabapentin and opioids) (Fig. 1). If back pain persists, imaging is warranted with X-ray or advanced imaging such as MRI or CT scans. More invasive therapies may then be indicated such as corticosteroid injections or radiofrequency ablations. Surgical intervention may be indicated to address back pain depending on pathology and identification of pain generators based off clinical history and imaging. Finally, implanted pain management pumps or spinal cord stimulators can be utilized for patients suffering from refractory back pain (Casiano et al., 2022).

Types of back pain

The spine is a complex support structure for the human body. It not only protects the spinal cord but also keeps the body erect. It is composed of 33 vertebrae with associated intervertebral discs and articulating facet joints. This elaborate structure can fail and lead to back pain. Understanding types of back pain molds therapeutics to treat it. Types of back

Cannabis Use, Neurobiology, Psychology, and Treatment. https://doi.org/10.1016/B978-0-323-89862-1.00036-2

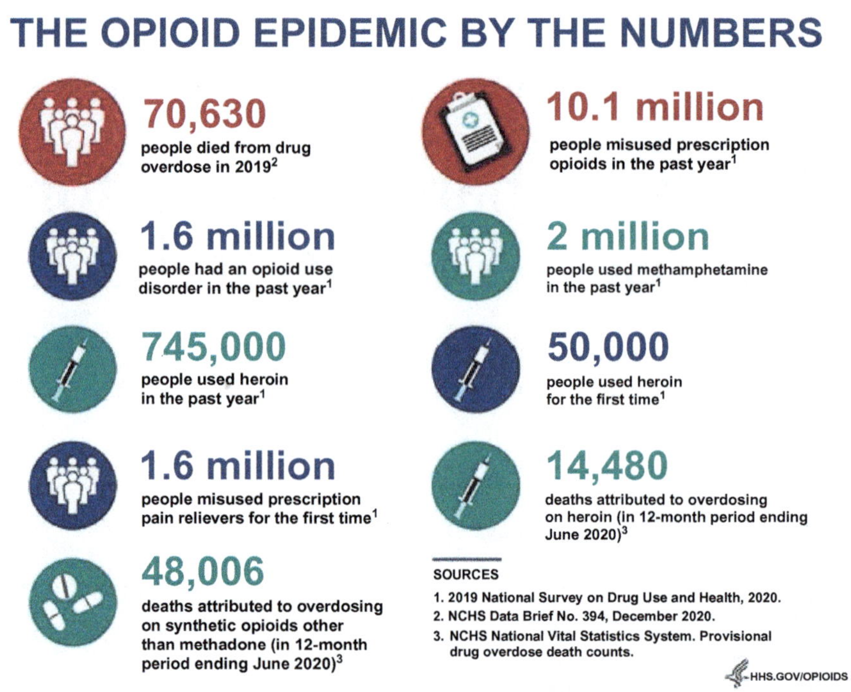

FIG. 1 The opioid epidemic by the numbers. *Adapted from the United States Department of Health and Human Services https://www.hhs.gov/opioids/about-the-epidemic/index.html (Assistant Secretary of Public Affairs (ASPA), n.d.)*

pain are distinguished based on patient history, clinical examination, and imaging. Specific questions in a patient history are important to differentiate key symptoms. These questions may include recent bowel or bladder incontinence, pain relief felt from leaning forward (positive shopping cart sign), unsteadiness in gait or footing, loss of sensation, loss of dexterity, and impacts on activities of daily living. Specific clinical examinations to narrow a differential diagnosis may include straight leg raise, Faber's test, Hoffmans sign, reflexes, and pinprick sensation. The first primary distinction that is key to make is the difference between mechanical and radicular back pain (Casiano et al., 2022). Different dermatomal descriptions of symptoms are very helpful in pointing toward different spinal pathologies. For example, C5 (cervical vertebrae 5) would be indicative of radiating pain to the deltoid. Being able to perform a competent and thorough history and clinical examination is pertinent to diagnosing the underlying type of back pain (Table 1).

Axial

Axial back pain can also be referred to as mechanical pain, and it is usually due to a muscle strain in the back. This can be caused or effected by being in certain positions or moving in certain ways. This pain can be described as dull aching or sharp and can be more or less severe depending on the time of day. This pain is normally relieved by rest. This tends to be a nonspecific pain. Mechanical back pain refers to pain that arises from within the spine, intervertebral disks, or the surrounding tissues (Murray et al., 2019).

Radicular

Radicular back pain is a pain that affects the nerve roots. Patient's classically present with a pain that radiates from their back into their extremities. Radicular pain commonly presents when a nerve root is being compressed or is inflamed. The pain can be associated with numbness, weakness, or tingling. Muscle strength is usually preserved in patients with true radiculopathy, due to the fact that the muscles are supplied by several nerve roots (Dydyk et al., 2021).

Degenerative

Degenerative back pain is caused by deterioration of the intervertebral disks in the spine. Degeneration is a natural process of aging. However, in some patients, degeneration can be accelerated and become symptomatic. Patients tend to present

TABLE 1 Modalities of back pain management.

Nondrug noninvasive therapies	**Physical therapy**
	Exercise
	Behavioral modification
	Rest
	Ice/heat
	Topical ointments/creams
Medicines	NSAIDs
	Gabapentin
	Muscle relaxers
	Anticonvulsants
	Antidepressants
	Narcotics
Procedures	Corticosteroid injections
	Radiofrequency ablations
Surgical	Spinal fusion surgeries
	Nerve decompression
Other	Spinal cord stimulator
	Implantable pain pump
	Electrical nerve stimulator
Alternative	Chiropractic care
	Acupuncture
	Transcutaneous electrical nerve stimulation (TENS)
	Massage
	Yoga

with a low-grade continuous pain around the area where the disk is degenerating. This area can also have severe flare-ups. These patients can also feel a spinal instability at the levels of degenerating disks (What Is Degenerative Disc Disease?, n.d.).

Post-surgical

Pain is common and normal following spinal surgery as tissues are healing, nerves are regaining their function, inflammation is decreasing, and patients are getting back to their activities of daily living. Post-surgical pain is often initially treated with narcotics and generally improves with time. Though sustained post-surgical pain can occur.

Musculoskeletal

Musculoskeletal (MSK) back pain is a diagnosis of exclusion that can be made once more specific causes of back pain are ruled out. This is usually sharp pain or stiffness of the back usually lateral to the spine. It is often generated by muscle or tendon attachment to the spine or ribs. This pain is not able to be specifically pinpointed and is caused by any combination of pathology involving the spinal vertebrae, disks, ligaments, facet joints, and surrounding supporting muscles (Degenerative cervical spine disease, 2022).

Multifactorial back pain

Back pain can be a multifaceted issue that can have many anatomical regions effected and several causes for these effects. Being able to localize patients' pain is helpful in targeting treatment. But as the spine is an interlinked structure with many parts that need to function harmoniously, an issue in one region can affect a differing region.

Opiate use for back pain

Despite multimodal therapies, patients with back pain often resort to taking potentially addicting opiates. Patients with chronic back pain are a main population that fall into long-term opioid use because they are a subset of pain patients that have nearly every aspect of day-to-day living affected. These patients continuously modify their physical activities due to pain anxiety, leading to disrupted ability to work, sleep, and function. Pain is with these patients everywhere they go. Due to the alarming global opiate crisis, novel pain control regimens are of critical importance to new clinical research (First et al., 2020).

Opiate-induced hyper analgesia (OIH) is a phenomenon demonstrated by chronic users of opiates in which their pain receptors are upregulated to noxious stimuli as a result of taking opiates. Just as the light receptors in the eye become more sensitive to light as they remain in the dark, pain receptors become more sensitive to painful stimuli after opiate use (Murray et al., 2019). OIH patients also have reduced sensitivity to the effects of opiates to relieve pain, therefore are caught in a vicious cycle between feeling more pain and not being able to take the pain away. This further supports the discussion that opioid-centered pain management is not optimal.

Pain management in the United States

In recent years, the field of pain management in the United States finds itself at a crossroads due to the current epidemic involving overuse, addiction, and mortality surrounding opioid use. A study conducted in 2017 in the United States noted 11.4 million Americans self-reported opiate misuse within the previous year. This resulted in over 47,000 drug overdose deaths and nearly 92,000 hospital admissions for opioid overdose in 2016 alone (Lavonas & Dezfulian, 2020). There are over 25 million adults who suffer with chronic pain in the United States (Coussens et al., 2019). In 2019, deaths associated with opiate overdoses soared to 70,630 with 1.6 million people misusing opiates for the first time (Fig. 2). The lack of effective and safe nonopioid options for pain control leaves these people with a life that lacks pain relief. A safe, effective, and nonaddictive substitute for pain management needs to be identified.

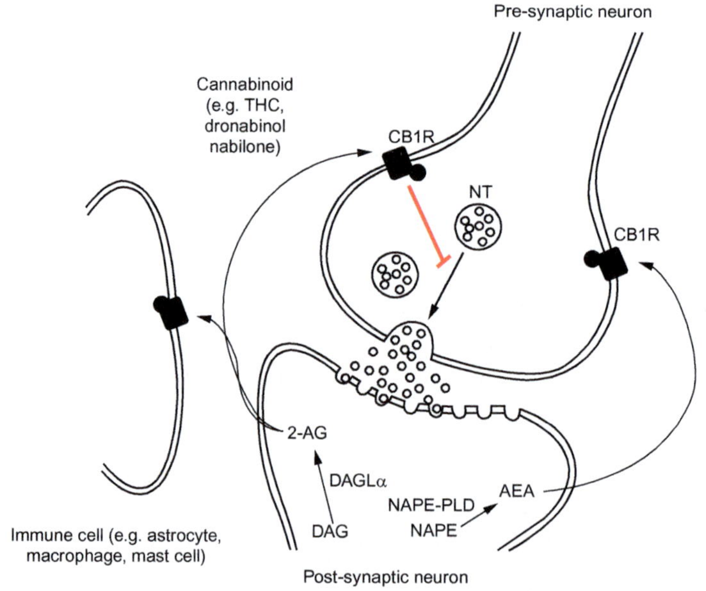

FIG. 2 Mechanism of action of the endocannabinoid system. *Cannabinoids bind to CB1R on the presynaptic neuron blocking neurotransmitter release (Price et al., 2022). Copyright permission granted.*

In the United States, Medicare expenses have skyrocketed in recent years trying to manage back pain. Spending on epidural corticosteroid injections has increased 629% and spending on opioid prescribing has increased 423% (Will et al., 2018). This is accompanied with increased spending in both magnetic resonance imaging (MRI) of the spine and spinal fusion surgeries. However, this increased spending is not correlated with better pain control or better patient outcomes (Upp & Waljee, 2020). This data obviously suggest that pain management strategy needs to change, and newer therapies need to start being considered more critically.

The opiate epidemic has led to a reform on US legislation. This has led to cracking down on physician prescribing of these medications, instead of focusing on patient care. This has left patients addicted to opiates looking elsewhere for pain management strategies. At the state and national level in the United States, policymakers continue to make reforms with the aim of reducing opiate-related morbidity and mortality. However, while the number of opiate-related deaths has seen a relative decrease, this is positively associated with a rise in heroin and synthetic opiate derivative-related deaths (Degenerative cervical spine disease, 2022; Dydyk et al., 2021; What Is Degenerative Disc Disease?, n.d.). Opiate misuse and addiction not only effects the user but nearly every reach of community and society as well as the economy.

Contrast with global pain management

Global pain management focuses on a biopsychosocial patient model. This begins with supportive management consisting of patient education, exercise, physiotherapy, and self-guided therapies (Foster et al., 2018). Once patients fail supportive therapies, they may be candidates for medicinal or surgical pain management strategies. Unfortunately, patients who either do not respond well or are not candidates for surgical intervention may end up relying on long-term opiate use for pain management. Unlike in the United States, global opiate therapy is typically reserved for patients in which back pain is refractory to other forms of therapy or are not surgical candidates.

Insufficiently managed back pain leads to disruption in regular medical treatment, self-medication, and increasing harm to patients and society. The goal of back pain management is keeping patients' pain at tolerable levels to continue carrying out activities of daily living and maintain quality of life. Adjunctive treatments such as nonopioid analgesics, antidepressants, regional analgesia, physical therapy, and psychological therapies should all be considered to enhance patients' pain control and bridge their opioid use. Given the challenge to effectively manage back pain, new therapies are desperately needed.

Legality of cannabis worldwide

The discussion surrounding the benefits of cannabis for pain along with the low side-effect profile has pushed for policy to occur ahead of science. Laws have been passed regarding its use long before the scientific knowledge following rigorous studies has been published (Hill & Palastro, 2017). Differing laws currently regulate cannabis usage medicinally and recreationally worldwide.

At least in the United States, cannabis has been regulated according to lawmakers and federal policy. Once the research of scientists potentially pointed in a different direction regarding the potential benefits of cannabis use, this was not congruent with the governments' views. For most of the U.S. policy history regarding cannabis, it has been led by lawmakers and law enforcement with nearly no input from scientists or medical professionals. This ever-growing gap in lack of pain control leaves a good place for discussion of a shift to different views of cannabis use and benefit.

In Europe, cannabis use tends to be more widely accepted for both medicinal and recreational purposes. In the Netherlands, Dutch "coffee shops" are available for people to buy and use cannabinoid products. These shops were made with the intent to decriminalize the use of marijuana and to separate the use of "hard drugs" such as heroin and amphetamines from the cannabis market (Monshouwer et al., 2011). The Dutch cannabis policy was designed with the aim of limiting risks of cannabis consumption for users, their environment, and society as a whole. This strategy of harm reduction has also been taken onboard internationally by places such as Uruguay and Spain.

Internationally, the legality of cannabis use remains highly controversial. The proponents for legalization highlight the potential to improve safety and quality control, to eradicate the cannabis black market, to improve availability of cannabis for medicinal purposes, to increase tax revenue, and to decrease drug violence. However, the opponents of legalization are concerned about the potentially addictive properties of the drug, highlighting the danger with worsening mental health outcomes and alterations in perception that could affect safety when carrying out daily tasks (Bahji & Stephenson, 2019).

Due to these differing views, the impact of cannabis legalization is unclear. Throughout the world, jurisdictions are legalizing marijuana use, leading policy to reform at a rapid rate. While the effects of this legalization are widely heterogeneous, it has led to increased research regarding the implications of cannabis use and legalization, leading to a more informed population (Bahji & Stephenson, 2019). The liberalization of marijuana around the globe will begin to lead to more clear answers regarding the benefits of its use for medicinal purposes.

Acceptance of cannabis for medicinal purposes

The use of medical cannabis is not a novel topic. Across the globe, including the United States, cannabis preparations have been widely available for use since the 1800s (Schmitz & Richert, 2003). However, acceptance of cannabis for medicinal purposes is a more recent phenomenon. The use of cannabis for pain control is on the rise, especially in jurisdictions of medicine that are trying to curb opioid use, but the current guideline recommendations of its use are inconsistent and lacking evidence (Busse et al., 2021). There is also a coexisting lack of knowledge from providers in medical education on safe use, distribution, and dosing of cannabis. This ultimately leads to physician uncertainty regarding its use. In many communities, cannabis use remains taboo. Cannabinoid acceptance for medicinal purposes, specifically pain management, is not currently a topic taught upon in most medical programs around the world. Thus, many physicians have a dearth of knowledge regarding this therapy.

Strong evidence supports the use of cannabinoids for pain management, but there is further research needed to look at which characteristics of pain specifically benefit from the effects of the cannabinoid system (Schmitz & Richert, 2003). Certainly, the use of cannabis for back pain is a burgeoning field. Furthermore, evidence is needed to justify optimal drug dosage and selection for each patient and risk-benefit ratios when combining cannabinoids with other prescription drugs.

Clinicians around the globe have noted improvements in patients with chronic pain after using medicinal cannabis. These improvements have also been associated with lessening of accompanying symptoms such as sleep difficulties, depression, and anxiety. These results have been documented in prospective cohort studies (Eisenberg et al., 2022). The European Pain Federation has recommended for clinicians to consider medicinal cannabis for chronic pain as a third line therapy and to trial individual patients based on needs and failure to improve with standardized pain management (Eisenberg et al., 2022). With recent experiences, medicinal cannabis may become more of a primary option for pain control with future research and better funding for cannabis research projects.

While many studies analyzing the effectiveness of cannabis for pain management reveal only modest effects, there is more here to consider. Chronic pain patients are multifactorial with extensive etiologies for their pain that are resistant to many traditional treatment methods along with several coexisting comorbidities. Unfortunately, there are no randomized control trials comparing cannabis to standard of care for back pain relief. The modest reduction in pain reports while using medicinal cannabis along with improvement in multiple related symptoms is clinically significant and should be taken into consideration.

Cannabis mechanism of action

Cannabis contains approximately 540 components, most predominantly studied and discussed are delta-9-tetrahydrocannabinol (THC) and cannabidiol (CBD). These are the two components of cannabis that are most critical to its analgesic effects. THC demonstrates agonist activity on the cannabinoid receptors, CB1 and CB2, resulting in both analgesic and psychotropic effects (Fig. 3). CB1 and CB2 are G-protein-coupled receptors that mediate both the psychoactive and medicinally therapeutic effects of cannabis (Kim et al., 2020). CBD demonstrates agonist activity on serotonin (5-HT1A), adenosine (A2A) and peroxisome proliferator-activated receptor gamma (PPAR-λ) receptors in order to produce anti-inflammatory responses and analgesia, without producing the psychotropic effects (Greis et al., 2022). CB1 is expressed at much higher levels in the brain when compared to the rest of the body, so it may have an integral role in the affective component of the experience of pain by modulating the response in the limbic system (Kim et al., 2020).

While cannabis does not directly play a role in the noradrenergic system which opiates utilize, it is postulated that these two systems overlap. This has started the discussion that lower opiate levels can be used to provide pain relief when concomitantly using cannabis (Takakuwa et al., 2020). This differs from opioid mechanism of action which is primarily inhibitory (Fig. 4).

Review of evidence

Systematic reviews

In a systemic review including four articles investigating the efficacy of cannabis to treat nonsurgical back pain, three of the four studies showed significant improvement in back pain for patients who used cannabis instead of placebo

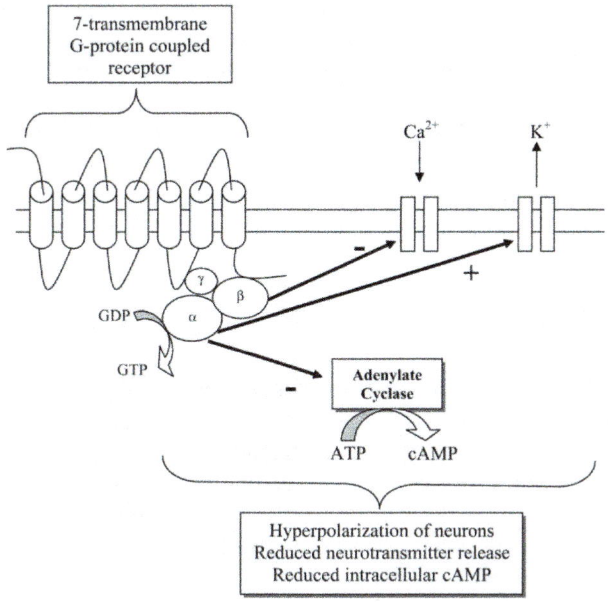

FIG. 3 Opioid mechanism of action. Drug binds to the opioid receptor, activating the G-protein. This results in the inhibition of voltage-gated calcium channels and adenylate cyclase. Potassium efflux channels are stimulated. All leading to hyperpolarization of neurons and reduced neurotransmitter release. *Copyright permission requested.*

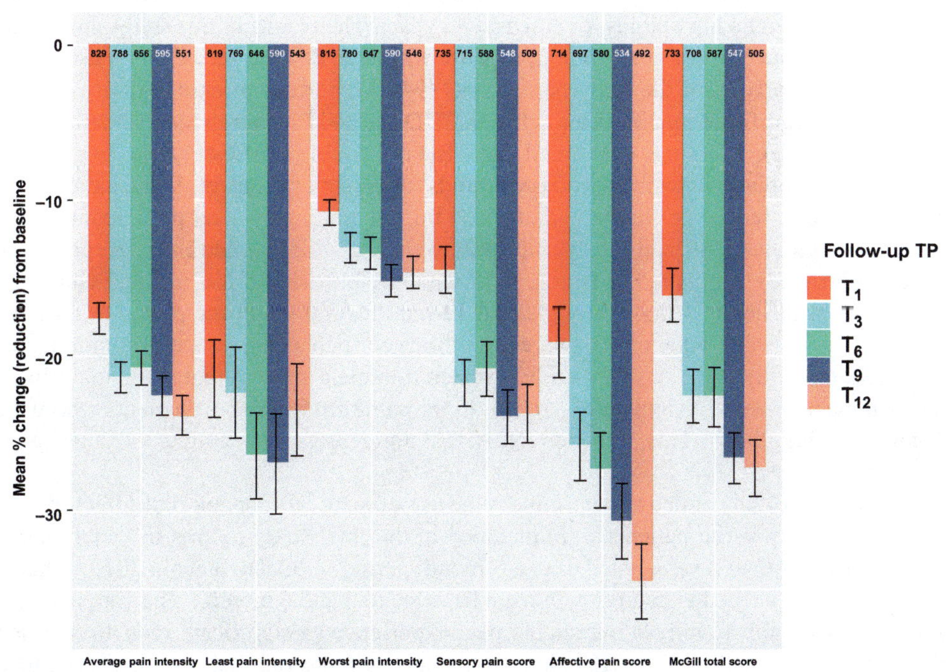

FIG. 4 Cannabis use improves chronic pain at different time points. Each color represents a different follow-up time point (TP) numbered by months. Numbers within the bars represent the number of patients who reported data at each time point. There is a sustained reduction of back pain at every time point for patients that use cannabis when compared to baseline (Aviram et al., 2021). *Copyright permission requested.*

(Price et al., 2022). In total, 110 participants met the criteria of this systematic review. In three of the four studies, there was a statistically significant reduction in back pain when comparing the cannabinoid group with the control group. The one study that did not have a statistically significant reduction in back pain had a very small sample size of seven patients included. There have not been any studies comparing oral versus inhaled cannabis use for pain management, so this is an area of research in need of attention before this is a widely accepted substance for pain control.

The National Academies of Science, Engineering, and Medicine conducted a thorough systematic review focusing on cannabinoid efficacy, which concluded that there is "extensive evidence" of cannabinoids' efficacy for chronic pain relief. The patients in this study also tolerated the cannabinoids to a great extent and reported no serious side effects, addiction potential, or adverse events (Meng et al., 2017).

A 2018 review included 16 studies with 1750 participants ranging from 2 to 26 weeks in length. These studies looked at the effects of oral mucosal spray containing plant-derived THC and CBD, nabilone, inhaled herbal cannabis, and dronabinol versus placebo and an analgesic (dihydrocodeine) (Mücke et al., 2018). The outcomes of this review concluded that cannabis-based medicines increased the number of people that achieved 50% or greater pain relief and 30% or greater pain relief when compared to placebo.

Clinical trials

Efficacy of cannabis use for spinal pain was included in a systematic review of randomized controlled trials in 2016. Here, it was reported that there were statistically significant improvements seen in pain, anxiety, and health-related overall quality of life. It was also reported that spine pain intensity was significantly reduced in the nabilone group compared to the control placebo group. THC and CBD were shown to be statistically significantly superior than placebo for managing pain on movement, at rest, and morning pain (Fitzcharles et al., 2016). The studies included in this review also reported statistically significant improvements in sleeping difficulties, contributing further to improved patient well-being and quality of life.

Cohort studies

A prospective cohort study conducted in 2019 sampled 97 patients from a private pain management center's population in an 8-week study (Capano et al., 2020). This study's aim was to investigate the impact of hemp extract cannabidiol (CBD) on opioid use and quality of life on chronic pain patients. The patients included in this study were between ages 30 and 65 years with chronic pain who have been taking opioids for at least two years. The results of this study showed that over half of the patients (53%) were able to reduce or eliminate their opioid use within an 8-week period when using CBD hemp extract to their pain control regimen. Nearly all of the study participants (94%) reported increased quality of life with CBD extract use. There were statistically significant relationships between CBD use and Pittsburgh Sleep Quality Index (PSQI) and Pain Intensity and Interference (PEG). This study concluded that CBD use could significantly improve chronic pain and quality of sleep while reducing use of opiates in patients who use opioids for pain management. Some patients included in the study reported they were also able to reduce or eliminate their need for polypharmacy—six participants were able to reduce or eliminate their antianxiety medications, and four participants were able to eliminate their sleeping medications.

In a prospective cohort study conducted in Israel, 1045 patients, licensed to use medical cannabis for chronic pain, completed questionnaires at different periods throughout a 12-month follow-up time period. Pain-level reports declined from baseline by 20% (Fig. 4), while other effects were also observed such as longer sleep duration, decreased body mass index, and lower depression ratings (Fig. 5). This study revealed treatment success that medicinal cannabis is a good therapeutic option for pain management (Aviram et al., 2021). In the same study, 43% of patients who were using analgesics before were no longer using them. This included NSAIDs, antidepressants, anticonvulsants, and over-the-counter analgesics.

A 2019 retrospectivecohort study aimed to determine if the use of medical cannabinoids affected the amount of opioids needed to control pain. Seventy-seven patients were included in the final study reports in which their needed daily milligram morphine equivalents (MMEs) were noted over a 6-month period (O'Connell et al., 2019). There was a statistically significant reduction in MMEs needed to control pain from baseline to 3 and 6 months. The conclusion of this study is that patient's using medicinal cannabis to relieve intractable pain experienced a significant reduction in necessary MME's to control their pain levels (O'Connell et al., 2019). This study supports the rational that medical cannabis shows strong opioid-sparing effects.

Cannabis reduces opiate use

Coupled with its ability to directly reduce pain, cannabinoids have also been shown through high-quality evidence to reduce opiate use. Recent reports have highlighted a lower mortality rate and morbidity rate from opiate overdoses in the United States since the more widespread acceptance and legalization of medicinal cannabis (Wiese & Wilson-Poe, 2018). Medicare prescription reports have also shown a lower amount of opioid prescriptions being written and filled since the medical cannabis legalization in the United States (Wiese & Wilson-Poe, 2018). For patients still taking opioids

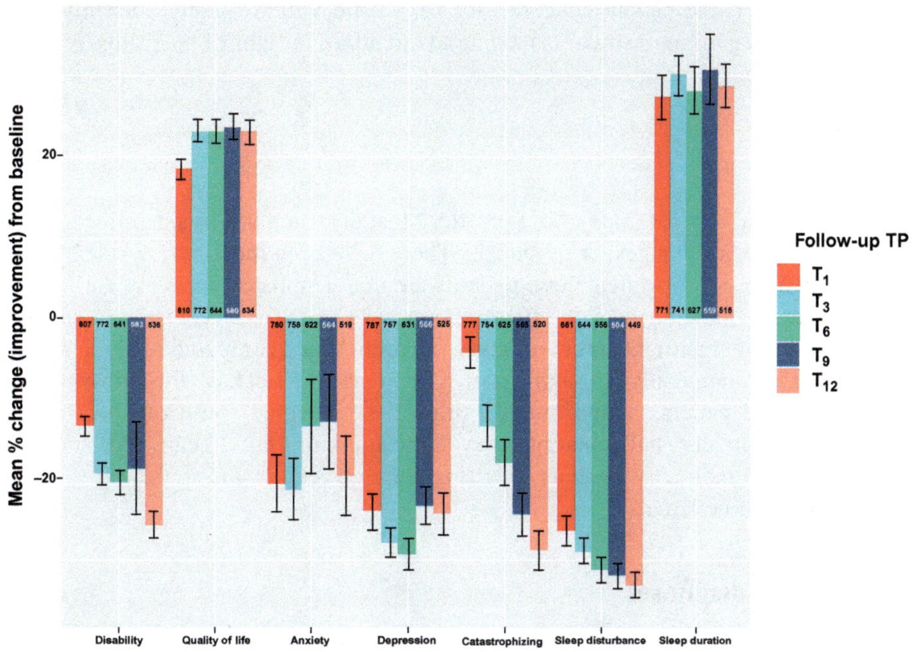

FIG. 5 Cannabis use improves quality of life and mental health indices for patients suffering from chronic pain. Negative values indicate improvement in disability, anxiety, depression, catastrophizing, and sleep disturbance. Positive values indicate improvement in quality of life and sleep duration. Each color represents a different follow-up time point (TP) numbered by months. There is improvement to baseline at everytime point (Aviram et al., 2021). *Copyright permission requested.*

and unable to eliminate their use completely, concomitant cannabis use has been shown to slow dose increase escalation over time and significantly reduce opiate intake (Bradford et al., 2018). CBD has been shown to have the ability to decrease the risk of opioid relapse by decreasing drug-seeking behaviors (Ren et al., 2009). CBD was able to decrease anxiety, cravings, and psychological stressors in patients with current or previous opiate dependency.

Adverse events

When considering prescribing a novel therapy, a clinician must thoroughly consider associated side effects and adverse events related to the therapeutic. The most common side effects and adverse events noted during these studies were drowsiness, dry mouth, dizziness, nausea, fatigue, and vertigo (Fitzcharles et al., 2016). Majority of the included studies reported no serious adverse events, and patient safety was never put at risk. One study reported one serious adverse event associated with nabilone use, in which a patient fell and sustained a fracture due to dizziness (Pinsger et al., 2006).

A potential shortcoming in the included studies is they mostly study the effects of synthetic cannabinoids rather than the flower form of cannabis. Due to the herbal form being more widely used and accepted for its medicinal properties, it is important for future research to include more studies on the plant itself rather than synthetic cannabinoids only. Another consideration may be to develop a standardized type of medicinal cannabis in order to better study it.

In the 2018 study (Mücke et al., 2018), psychiatric disorders occurred in 17% of participants using cannabis-based medicines and in 5% using placebo; however, they reported this as low-quality evidence based on the Cochrane "Risk of bias" tool. Additionally, cannabis use has been associated with higher risk of suicide among cannabis users with preexisting psychiatric disorders. The impact of cannabis on patients with psychiatric disorders needs to be studied in more detail.

In several of the studies reviewed, patients reported punitive or financial concerns. They were concerned that if they reported that CBD hemp extract or cannabis was effective for pain control, their access to opioid prescriptions may diminish. This may lead patients to not being able to afford their pain control methods due to a fear that opiate prescriptions were likely covered by insurance but cannabis products were not (Capano et al., 2020). Patients also reported concerns that they may fail mandatory drug tests in their place of work. Additionally, cannabis use is not generally accepted as a modality for back pain control. At least in the United States, many pain patients sign pain contracts stating that they will not take additional pain control adjuvants. Cannabis use may nullify some pain contracts and lead to pain patients being dismissed from pain management practices. These patient concerns are very valid within society currently, further outlining the importance of current legislation regarding medical cannabinoids to adapt in light of emerging science.

Conclusion

Cannabis—Low side-effect profile

Cannabis has been shown to have a very low side-effect profile. Most common minor side effects from cannabis use are dizziness, drowsiness, fatigue, appetite changes, or psychosis. There have been more severe side effects noted in patients with preexisting mental health disorders, and thus these patients are not candidates for treatment with medical marijuana (Lowe et al., 2019). Unfortunately, most of the side-effect reports have been from recreational cannabis use rather than medical cannabis use. This will be an important aspect of future research. When compared to opioids, the risks of morbidity and mortality with cannabis are astronomically less significant. Marijuana does not have the ability to cause lethal overdose in the traditional sense. Reports of patients taking over 30 times the "accepted" limit of THC (10mg) presented with symptoms such as anxiety, panic attacks, hallucinations, psychotic thinking, rapid heart rate, and high blood pressure. While this state may result in death indirectly, such as operating a motor vehicle during this time, the effects of cannabis themselves have not been shown to be lethal.

Cannabis with mental health diagnoses

Generally, literature has provided high-quality evidence to support the low side-effect profile of cannabis for medical intervention (Ashton & Milligan, 2008). It also has no overdose levels (Choo et al., 2016). While cannabis has been proven to be a safe alternative or adjunct to opiates for patients with chronic back pain, the side effects do need to be considered on a patient-to-patient basis. Acute intoxication with cannabis can interfere with users' perception of time and memory and can also alter motor function. Use of cannabis, especially long term, can worsen coexisting anxiety or mood disorders, as well as increase the likelihood of developing an anxiety or mood disorder in a predisposed person (Lowe et al., 2019). Cannabis has also been shown to hasten the development of psychotic disorders in those with a genetic predisposition for these conditions (Halladay et al., 2020). Cannabis use in adolescence can permanently change the brain biochemistry of the developing brain, eventually leading to a lower density of white matter in the corpus callosum and the hypothalamus, shown in diffusion weighted magnetic resonance imaging (MRI) (Zalesky et al., 2012). Early and regular use of cannabis over time, shown in a longitudinal study, decreased intelligence quotient (IQ) points by 8 (Hill & Palastro, 2017) These effects should all be taken seriously into consideration when physicians are prescribing cannabis to adolescent patients. This should also not be considered as a pain management solution for patients with pre-existing mental health diagnoses.

Much of what is known regarding the adverse effects of cannabis is known from recreational studies which is low quality evidence due to the lack of standardization, use of self-reporting, and potential use of multiple substances.

Cannabis for back pain

In an extensive systematic review using both natural and synthetic cannabinoid products, the average number of chronic pain candidates who reported a reduction in pain of greater than or equal to 30% was greater with the cannabinoid products than with placebo (Whiting et al., 2015).

In vitro studies have shown that the antiinflammatory properties of THC have two times the potency of hydrocortisone and twenty times the potency of aspirin (Temple, 2016). The effects of THC used concomitantly with COX-2 inhibitors have been shown to have synergistic effects on pain. Cannabis use has also been shown to benefit sleep and relaxation of users, a very common complaint of back pain patients seen in clinic, thus leading to reduced daytime functioning and societal impacts.

Numerous randomized controlled trials have shown that compared to placebo, cannabinoids directly reduce pain (Cunningham et al., 2020). In 2017, an extensive review concluded that there was substantial evidence supporting cannabis use being effective in treating chronic pain.

In a study involving patients with intractable pain after failed back surgery who reported no improvement after spinal cord stimulation, addition of oral cannabinoids led to a significant improvement in not only pain but also sleep and mood (Mondello et al., 2018). In a single-center study of 30 patients with noncancer spine pain unresponsive to nonsteroidal antiinflammatory drugs (NSAIDs), patients receiving nabilone treatment reported significant improvements in spine pain (Pinsger, 2012). Another study looking at back pain patients reported statistically significant decreases in pain intensity in patients self-administering dronabinol when compared to placebo (Narang et al., 2008). These clinical studies have been able to demonstrate the role that cannabinoids may play in the treatment of back pain. However, the lack of studies along with the small sample sizes of existing studies limits the ability to more widely generalize the data. This will be an important focus of future research regarding cannabis use for back pain management.

Reducing opiate use

High-quality evidence has shown that cannabis use for chronic back pain can have a major opiate-sparing effect, leading to a reduced need for opioid use and prescribing. The current state of the opiate crisis provides solid evidence that other non-addictive and less harmful substances need to be utilized for chronic pain control.

In animal studies, it has been shown that rodents' responses to noxious stimuli were dulled when THC was also used, suggesting that less MMEs would be needed to reduce pain if used concomitantly with THC (Smith et al., 1998). Up to 75% of medical cannabis dispensary users reported a reduction in their need for opiates after utilizing medical marijuana (Piper et al., 2017). When examining records of a prescription drug monitoring program, it showed that participants were 17 times more likely to eliminate need of controlled substances for pain control after beginning a medical cannabis regimen (Stith et al., 2018). This is also in line with the Medicare reports of states that have legalized medical cannabis having lower expenditures for prescription pain medications.

Forthcoming research

The future for medicinal cannabinoid use research is bright. There are proposals for studies regarding medical cannabis use for autoimmune disorders, fibromyalgia, many chronic cancer, and noncancer pain regimens. There are proposals for studies regarding the long-term adverse effects of medical cannabis rather than recreational cannabis. There are also studies underway regarding the legislation of cannabis and the racial stereotypes that have affected the justice system regarding marijuana use. These are just a few of the many examples of what the future of medical cannabis research holds, but the science is coming. Like cannabis, ketamine will also be on the rise as a study subject for alternative pain management options. The upside to this future research is that scientists will not only be learning about the many benefits of medical cannabis but will also learn how the drug is currently being misused (Krystal, 2018). Further research will highlight the growing list of medical conditions that can not only benefit from cannabinoid use but also highlight which conditions should limit or eliminate the use of marijuana, psychiatric disorders being a clear example.

Cannabis has been shown to be a safe modality for control of back pain. In fact, it may be a viable option to reduce opiate use as well. There has been promising research demonstrating its benefits on pain management while also maintaining a low side-effect profile and nonaddictive potential. Due to the ever-growing opioid crisis, discussion regarding cannabinoid pain management is coming to light. Further studies are needed to progress knowledge on cannabis benefits on different diagnoses as well as studying long-term side effects of using the substance.

Applications to other areas

This chapter exemplifies how cannabis use can be applied to back pain patients. This research can also be applied to other chronic pain patients, for example: fibromyalgia, cancer pain, and patients with autoimmune disorders. The substance of this article can also be applied to patients with sleeping issues and sleep disorders, as the research studied here was also shown to hold value on sleep. The chapter discusses at length the beneficial effects cannabis use can have on patients who suffer with opioid dependence and addiction, it is not able to be concluded at this time whether this can be applied to patients with other substance use dependencies and addictions.

Mini-dictionary of terms

Corticosteroid injection—An injection of local anesthetic and steroid medication, commonly into the sacroiliac joints or facet joints for back pain. To reduce local pain and inflammation. First injection is diagnostic of location, second injection is confirmatory.

Faber test—Faber stands for flexion, abduction, and external rotation. These three movements can provoke pain to diagnose pathology in the hip, lumbar spine, and sacroiliac regions.

Gabapentin—A prescription medication that is a gamma aminobutyric acid (GABA) analog. It reduces the excitability of nerve cells. It is frequently prescribed for chronic back pain especially when there is a neuropathic or radicular component to the pain.
Hoffmans sign—A neurological examination done by holding the patient's finger and flicking the nail downward, looking for a reflexive flick upward. This tests potential spinal cord compression in the cervical spine region.
Limbic system—A collection of structures in the brain, the hippocampus, amygdala, and hypothalamus. This system is involved in the processing of emotion, body response to stress, and is associated with the autonomic nervous system and endocrine system.
Milligram morphine—The amount of morphine (mg) that is equivalent to the strength of the prescribed opioid dose.
Radiofrequency ablation—A procedure that uses heat and radio waves to destroy nerves. The radiofrequency needle creates a heat lesion on the pain-transmitting nerve to prevent pain signal transduction to the brain. This is done after diagnostic and confirmatory corticosteroid injections have taken place.
Straight leg raise—When performed passively, is called Lasegues test, is used to determine if a patient presenting with low back pain has nerve root sensitivity in L5.

Key facts of opioid addiction

- Most opioid medications work by binding to G-protein-coupled receptors leading to cellular hyperpolarization. They mostly bind to MOP receptors in the nervous system to provide pain relief.
- Opioid overprescription and overuse has led to dangerous outcomes. Due to their pharmacological effects, opioids are highly addictive. Tolerance is achieved within days, and the withdrawal is severe.
- Opioid drug addiction has led to an epidemic in the United States and worldwide.
- Elements of opiate use disorder include, but are not limited to, continued use of opioids despite consequences, increasing tolerance or dose, excessive time spent using or acquiring substance, interferes with responsibilities, wishing to cut down use, strong desire to use, reducing or eliminating important activities due to use, withdrawal when the dose is decreased.
- Opioid morbidity and mortality have been on the rise (Fig. 2) which is why alternative pain management therapies that have less addictive potential are a key part of current and future research.

Summary points

- The opiate epidemic is a very serious problem worldwide that has increased the need to find pain management therapies that are nonaddictive and have a lower side-effect profile.
- There are many different types of back pain, showing that this is a multifaceted issue which has been shown to be responsive to the effects of cannabis.
- Cannabis legislation has become more accepting of cannabinoid use for medical purposes in recent years, now being legal in 37 US states and 30 countries around the globe. Europe is the most progressive leader in medicinal cannabis use and awareness.
- Internationally, cannabis use is widely accepted for medicinal purposes, pain management being one of the most well-researched uses for medicinal cannabinoids.
- Cannabis has two cannabinoids that are most commonly studied—THC and CBD. They work on the CB1 and CB2 receptors to produce analgesic effects.
- High-quality evidence has shown that cannabinoids are effective therapies for modulating back pain while also reducing opiate use in opiate-dependent patients.
- Cannabis has been shown to be a safe pain treatment modality with a low side-effect profile and low addictive potential. These benefits give it great possibility of reducing opiate use, thus improving the opiate epidemic globally.

References

Ashton, J. C., & Milligan, E. D. (2008). Cannabinoids for the treatment of neuropathic pain: Clinical evidence. *Current Opinion in Investigational Drugs*, *9* (1), 65–75. 18183533.

Assistant Secretary of Public Affairs (ASPA). (n.d.). What is the U.S. opioid epidemic? HHS.gov. Retrieved July 19, 2022, from https://www.hhs.gov/opioids/about-the-epidemic/index.html.

Aviram, J., Pud, D., Gershoni, T., et al. (2021). Medical cannabis treatment for chronic pain: Outcomes and prediction of response. *European Journal of Pain*, *25*, 359–374. https://doi-org.proxy.library.rcsi.ie/10.1002/ejp.1675.

Bahji, A., & Stephenson, C. (2019). International perspectives on the implications of cannabis legalization: A systematic review & thematic analysis. *International Journal of Environmental Research and Public Health*, *16*(17), 3095. https://doi.org/10.3390/ijerph16173095. PMID: 31454942.

Bradford, A. C., Bradford, W. D., Abraham, A., et al. (2018). Association between US state medical cannabis laws and opioid prescribing in the medicare part D population. *JAMA Internal Medicine*, *178*, 667–672.

Busse, J. W., Vankrunkelsven, P., Zeng, L., Heen, A. F., Merglen, A., Campbell, F., Granan, L. P., Aertgeerts, B., Buchbinder, R., Coen, M., Juurlink, D., Samer, C., Siemieniuk, R. A. C., Kumar, N., Cooper, L., Brown, J., Lytvyn, L., Zeraatkar, D., Wang, L., … Agoritsas, T. (2021). Medical cannabis or cannabinoids for chronic pain: a clinical practice guideline. *BMJ*, *374*, n2040. https://doi.org/10.1136/bmj.n2040. 34497062.

Capano, A., Weaver, R., & Burkman, E. (2020). Evaluation of the effects of CBD hemp extract on opioid use and quality of life indicators in chronic pain patients: a prospective cohort study. *Postgraduate Medicine*, *132*(1), 56–61. https://doi.org/10.1080/00325481.2019.1685298. Epub 2019 Nov 12 31711352.

Casiano, V. E., Sarwan, G., Dydyk, A. M., & Varacallo, M. (2022). Back pain. In *StatPearls*. Treasure Island (FL): StatPearls Publishing. 30844200.

Choo, E. K., Feldstein Ewing, S. W., & Lovejoy, T. I. (2016). Opioids out, cannabis in: Negotiating the unknowns in patient care for chronic pain. *JAMA*, *316*(17), 1763–1764. https://doi.org/10.1001/jama.2016.13677. 27802551. PMCID: PMC5332546.

Coussens, N. P., Sittampalam, G. S., Jonson, S. G., Hall, M. D., Gorby, H. E., Tamiz, A. P., McManus, O. B., Felder, C. C., & Rasmussen, K. (2019). The opioid crisis and the future of addiction and pain therapeutics. *The Journal of Pharmacology and Experimental Therapeutics*, *371*(2), 396–408. https://doi.org/10.1124/jpet.119.259408. PMID: 31481516.

Cunningham, C. O., Starrels, J. L., Zhang, C., Bachhuber, M. A., Sohler, N. L., Levin, F. R., Minami, H., Slawek, D. E., & Arnsten, J. H. (2020). Medical Marijuana and Opioids (MEMO) Study: Protocol of a longitudinal cohort study to examine if medical cannabis reduces opioid use among adults with chronic pain. *BMJ Open*, *10*(12), e043400. https://doi.org/10.1136/bmjopen-2020-043400. PMID: 33376181.

Degenerative cervical spine disease. (2022). Degenerative cervical spine disease—Symptoms, diagnosis and treatment. *BMJ Best Practice US*. https://bestpractice.bmj.com/topics/en-us/577. Accessed July 19, 2022.

Dydyk, A. M., Khan, M. Z., & Singh, P. (2021). Radicular back pain. In *StatPearls*. Treasure Island (FL): StatPearls Publishing. Available from: https://www.ncbi.nlm.nih.gov/books/NBK546593/.

Eisenberg, E., Morlion, B., Brill, S., & Häuser, W. (2022). Medicinal cannabis for chronic pain: The bermuda triangle of low-quality studies, countless meta-analyses and conflicting recommendations. *European Journal of Pain*, *26*, 1183–1185. https://doi-org.proxy.library.rcsi.ie/10.1002/ejp.1946.

First, L., Douglas, W., Habibi, B., Singh, J. R., & Sein, M. T. (2020). Cannabis use and low-back pain: A systematic review. *Cannabis and Cannabinoid Research*, *5*(4), 283–289. https://doi.org/10.1089/can.2019.0077. PMID: 33381642; PMCID: PMC7759283.

Fitzcharles, M. A., Baerwald, C., Ablin, J., & Häuser, W. (2016). Efficacy, tolerability and safety of cannabinoids in chronic pain associated with rheumatic diseases (fibromyalgia syndrome, back pain, osteoarthritis, rheumatoid arthritis): A systematic review of randomized controlled trials. *Schmerz*, *30*(1), 47–61. https://doi.org/10.1007/s00482-015-0084-3. PMID: 26767993.

Foster, N. E., Anema, J. R., Cherkin, D., Chou, R., Cohen, S. P., Gross, D. P., Ferreira, P. H., Fritz, J. M., Koes, B. W., Peul, W., Turner, J. A., Maher, C. G., & Lancet Low Back Pain Series Working Group. (2018). Prevention and treatment of low back pain: Evidence, challenges, and promising directions. *Lancet*, *391*(10137), 2368–2383. https://doi.org/10.1016/S0140-6736(18)30489-6. PMID: 29573872.

Greis, A., Renslo, B., Wilson-Poe, A. R., et al. (2022). Medical cannabis use reduces opioid prescriptions in patients with chronic back pain. *Cureus*, *14*(1), e21452. https://doi.org/10.7759/cureus.21452.

Halladay, J. E., MacKillop, J., Munn, C., Jack, S. M., & Georgiades, K. (2020). Cannabis use as a risk factor for depression, anxiety, and suicidality: Epidemiological associations and implications for nurses. *Journal of Addictions Nursing*, *31*(2), 92–101. https://doi.org/10.1097/JAN.0000000000000334. PMID: 32487935.

Hill, K. P., & Palastro, M. D. (2017). Medical cannabis for the treatment of chronic pain and other disorders: Misconceptions and facts. *Polish Archives of Internal Medicine*, *127*(11), 785–789. https://doi.org/10.20452/pamw.4123. Epub 2017 Oct 25. PMID: 29067992.

Kim, T. E., Townsend, R. K., Branch, C. L., Romero-Sandoval, E. A., & Hsu, W. (2020). Cannabinoids in the treatment of back pain. *Neurosurgery*, *87*(2), 166–175. https://doi.org/10.1093/neuros/nyz573. 32097466.

Krystal, H. (2018). The misclassification of medical marijuana. *The Journal of the American Academy of Psychiatry and the Law*, *46*(4), 472–479. https://doi.org/10.29158/JAAPL.003790-18. PMID: 30593477.

Lavonas, E. J., & Dezfulian, C. (2020). impact of the opioid epidemic. *Critical Care Clinics*, *36*(4), 753–769. https://doi.org/10.1016/j.ccc.2020.07.006. PMID: 32892827.

Lowe, D. J. E., Sasiadek, J. D., Coles, A. S., & George, T. P. (2019). Cannabis and mental illness: A review. *European Archives of Psychiatry and Clinical Neuroscience*, *269*(1), 107–120. https://doi.org/10.1007/s00406-018-0970-7. Epub 2018 Dec 19 30564886.

Meng, H., Johnston, B., Englesakis, M., et al. (2017). Selective cannabinoids for chronic neuropathic pain: A systematic review and meta-analysis. *Anesthesia and Analgesia*, *125*, 1638–1652.

Mondello, E., Quattrone, D., Cardia, L., et al. (2018). Cannabinoids and spinal cord stimulation for the treatment of failed back surgery syndrome refractory pain. *Journal of Pain Research*, *11*(6), 1761–1767.

Monshouwer, K., Van Laar, M., & Vollebergh, W. A. (2011). Buying cannabis in 'coffee shops'. *Drug and Alcohol Review*, *30*(2), 148–156. https://doi.org/10.1111/j.1465-3362.2010.00268.x. PMID: 21375615.

Mücke, M., Phillips, T., Radbruch, L., Petzke, F., & Häuser, W. (2018). Cannabis-based medicines for chronic neuropathic pain in adults. *Cochrane Database of Systematic Reviews*, *3*(3), CD012182. https://doi.org/10.1002/14651858.CD012182.pub2. 29513392.

Murray, M., Stone, A., Pearson, V., & Treisman, G. (2019). Clinical solutions to chronic pain and the opiate epidemic. *Preventive Medicine, 118*, 171–175. https://doi.org/10.1016/j.ypmed.2018.10.004. Epub 2018 Oct 11. PMID: 30315848.

Narang, S., Gibson, D., Wasan, A. D., et al. (2008). Efficacy of dronabinol as an adjuvant treatment for chronic pain patients on opioid therapy. *The Journal of Pain, 9*(3), 254–264.

O'Connell, M., Sandgren, M., Frantzen, L., Bower, E., & Erickson, B. (2019). Medical cannabis: Effects on opioid and benzodiazepine requirements for pain control. *The Annals of Pharmacotherapy, 53*(11), 1081–1086. https://doi.org/10.1177/1060028019854221. Epub 2019 May 25. PMID: 31129977.

Pinsger, M. (2012). Benefit of an add-on-treatment with a synthetic cannabinomimeticum on patients with chronic back pain—A randomized controlled trial. *European Spine Journal, 21*(11), 2366.

Pinsger, M., Schimetta, W., Volc, D., Hiermann, E., Riederer, F., & Pölz, W. (2006). Benefits of an add-on treatment with the synthetic cannabinomimetic nabilone on patients with chronic pain—A randomized controlled trial. *Wiener Klinische Wochenschrift, 118*(11–12), 327–335.

Piper, B. J., DeKeuster, R. M., Beals, M. L., et al. (2017). Substitution of medical cannabis for pharmaceutical agents for pain, anxiety, and sleep. *Journal of Psychopharmacology, 31*, 569–575. https://doi.org/10.1177/0269881117699616.

Price, R. L., Charlot, K. V., Frieler, S., Dettori, J. R., Oskouian, R., & Chapman, J. R. (2022). The efficacy of cannabis in reducing back pain: A systematic review. *Global Spine Journal, 12*(2), 343–352. https://doi.org/10.1177/21925682211065411. 35128969.

Ren, Y., Whittard, J., Higuera-Matas, A., et al. (2009). Cannabidiol, a nonpsychotropic component of cannabis, inhibits cue-induced heroin seeking and normalizes discrete mesolimbic neuronal disturbances. *The Journal of Neuroscience, 29*(47), 14764–14769.

Schmitz, N., & Richert, L. (2003). Pharmacists and the future of cannabis medicine. *Journal of the American Pharmacists Association, 60*(1), 207–211. https://doi.org/10.1016/j.japh.2019.11.007. Epub 2019 Dec 20 31870860.

Smith, F. L., Cichewicz, D., Martin, Z. L., et al. (1998). The enhancement of morphine antinociception in mice by delta9-tetrahydrocannabinol. *Pharmacology, Biochemistry, and Behavior, 60*, 559–566. PMID: 9632241.

Stith, S. S., Vigil, J. M., Adams, I. M., et al. (2018). Effects of legal access to cannabis on scheduled II-V drug prescriptions. *Journal of the American Medical Directors Association, 19*. https://doi.org/10.1016/j.jamda.2017.07.017. 59-64.e1.

Takakuwa, K. M., et al. (2020). The impact of medical cannabis on intermittent and chronic opioid users with back pain: How cannabis diminished prescription opioid usage. *Cannabis and Cannabinoid Research, 5*(3), 263–270.

Temple, L. M. (2016). Medical marijuana and pain management. *Disease-a-Month, 62*(9), 346–352. https://doi.org/10.1016/j.disamonth.2016.05.014. Epub 2016 Jun 27. PMID: 27364691.

Upp, L. A., & Waljee, J. F. (2020). The opioid epidemic. *Clinics in Plastic Surgery, 47*(2), 181–190. https://doi.org/10.1016/j.cps.2019.12.005. PMID: 32115045.

What Is Degenerative Disc Disease?n.d.What is degenerative disc disease? www.spine-health.com/conditions/degenerative-disc-disease/what-degenerative-disc-disease. Accessed July 19, 2022.

Whiting, P. F., Wolff, R. F., Deshpande, S., Di Nisio, M., Duffy, S., Hernandez, A. V., Keurentjes, J. C., Lang, S., Misso, K., Ryder, S., Schmidlkofer, S., Westwood, M., & Kleijnen, J. (2015). Cannabinoids for medical use: A systematic review and meta-analysis. *Journal of the American Medical Association, 313*(24), 2456–2473. https://doi.org/10.1001/jama.2015.6358. PMID: 26103030.

Wiese, B., & Wilson-Poe, A. R. (2018). Emerging evidence for cannabis' role in opioid use disorder. *Cannabis and Cannabinoid Research, 3*, 179–189.

Will, J. S., Bury, D. C., & Miller, J. A. (2018). Mechanical low back pain. *American Family Physician, 98*(7), 421–428. PMID: 30252425.

Zalesky, A., Solowij, N., Yücel, M., Lubman, D. I., Takagi, M., Harding, I. H., Lorenzetti, V., Wang, R., Searle, K., Pantelis, C., & Seal, M. (2012). Effect of long-term cannabis use on axonal fibre connectivity. *Brain, 135*(Pt 7), 2245–2255. https://doi.org/10.1093/brain/aws136. Epub 2012 Jun 4. PMID: 22669080.

Part III

Pharmacological and physiological aspects

Chapter 14

Cannabis, cannabinoids, and receptor responses

Yi Yang[a,b,c], Roya Derwish[a,c], and Lakshmi P. Kotra[a,b,c]
[a]*Krembil Research Institute, University Health Network, Toronto, ON, Canada,* [b]*Department of Pharmaceutical Sciences, Leslie Dan Faculty of Pharmacy, University of Toronto, ON, Canada,* [c]*Centre for Cannabinoid Therapeutics, Toronto, ON, Canada*

Abbreviations

11-OH-THC	11-hydroxy-Δ^9-THC
2-AG	2-arachidonoylglycerol
AAI	aminoalkylindole
AEA	*N*-arachidonoylethanolamine (anandamide)
API	active pharmaceutical ingredient
BCS	biopharmaceutical classification system
CB_1R	cannabinoid 1 receptor
CB_2R	cannabinoid 2 receptor
CBC	cannabichromene
CBCA	cannabichromenic acid
CBCA	cannabichromenic acid
CBD	cannabidiol
CBDA	cannabidiolic acid
CBDB	cannabidibutol
CBDV	cannabidivarin
CBE	cannabielsoin
CBEA-A	cannabielsoic acids A
CBEA-B	cannabielsoic acids B
CBG	cannabigerol
CBGA	cannabigerolic acid
CBGVA	cannabigerovarinic acid
CBL	cannabicyclol
CBN	cannabinol
CBND	cannabinodiol
CNS	central nervous system
DA	divarinic acid
GOT	geranylpyrophosphate: olivetolate geranyltransferase
GPCRs	G protein-coupled receptors
GPP	geranyl pyrophosphate
NADA	*N*-arachidonoyl dopamine
OA	olivetolic acid
$t_{1/2}$	half-life
THC-COOH	11-nor-9-carboxy-Δ^9-THC
Δ^9-THC	Δ^9-tetrahydrocannabinol
Δ^9-THCA	Δ^9-tetrahydrocannabinolic acid
Δ^9-THCV	Δ^9-tetrahydrocannabivarin

Cannabis Use, Neurobiology, Psychology, and Treatment. https://doi.org/10.1016/B978-0-323-89862-1.00025-8

Introduction

Cannabis and the cannabinoids

Cannabis spp. is one of the oldest botanicals known to man (Pertwee, 2014). An annual flowering herb from the *Cannabaceae* family of plants, this genus is primarily represented by *Cannabis sativa* (cannabis), although two additional species *Cannabis indica* and *Cannabis ruderalis* are sometimes recognized (McPartland & Guy, 2017). Cannabis is cultivated for many human purposes: aside from being a source of fabric and industrial fiber, cannabis seeds and their derivative oils are commonly consumed as food and health supplements while cannabis flowers are used as both medicine and a recreational drug (Pertwee, 2014). Varieties of the plant bred for food and fiber are colloquially known as hemp, while strains grown for medicinal purposes have been called marijuana, among other names.

Cannabis possesses rich phytochemistry, and to date more than 500 different compounds have been isolated from the plant (Pertwee, 2014). These span a diverse range of classes including terpenes, flavonoids, fatty acids, and steroids (Lewis et al., 2017; Pertwee, 2014). Among these phytochemicals are a distinct group of secondary metabolites known as the cannabinoids. Members of this compound class synthesized by the cannabis plant are known as phytocannabinoids and will be henceforth referred to as simply *cannabinoids*. In general, cannabinoids can be described as phenolic polycycles constituted by carbon, hydrogen, and oxygen (Fig. 1).

Within the plant, cannabinoids are biosynthesized by way of a complex, multibranched pathway, utilizing a combination of enzyme catalysis, oxidation, photochemistry, and heat to produce the more than 100 members identified to date (Lewis et al., 2017; Pertwee, 2014). These compounds exhibit a wide array of physiological interactions in animals

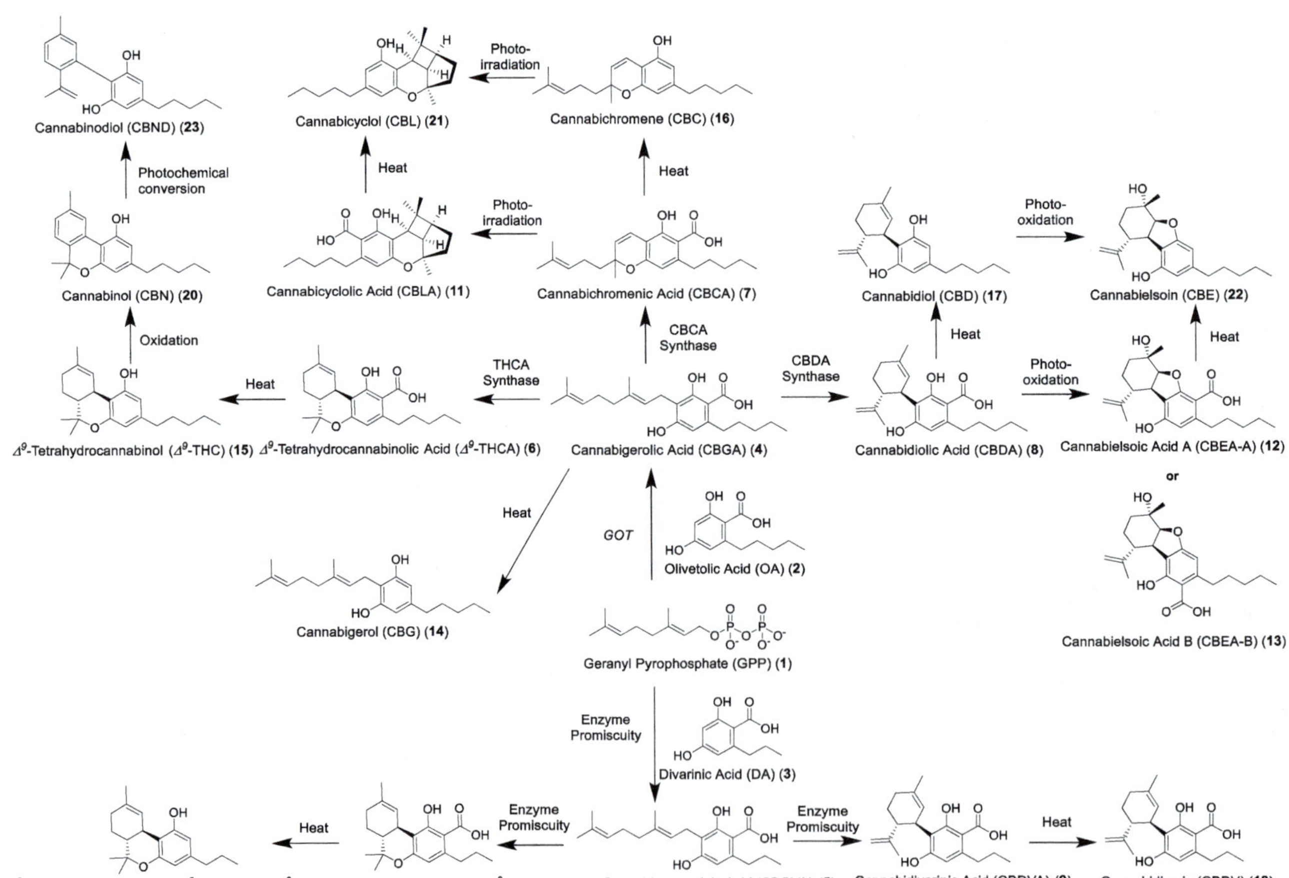

FIG. 1 Structures and biosynthesis of C_5 and C_3 cannabinoids (Fellermeier & Zenk, 1998; de Meijer & Hammond, 2016; Lewis et al., 2017; Luo et al., 2019).

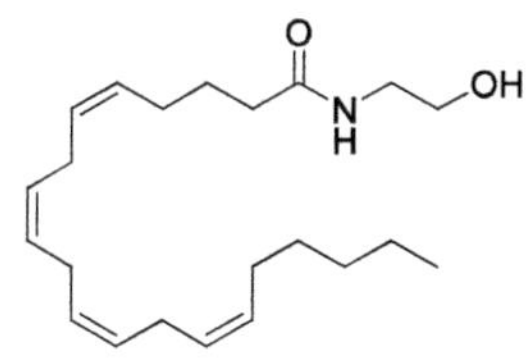

N-Arachidonoylethanolamine (AEA) (**24**)

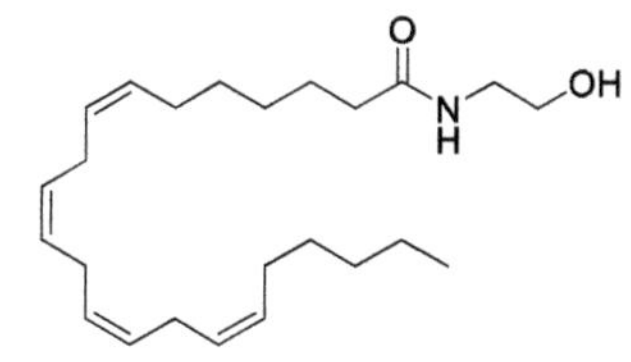

N-Docosotetraenoylethanolamine (**25**)

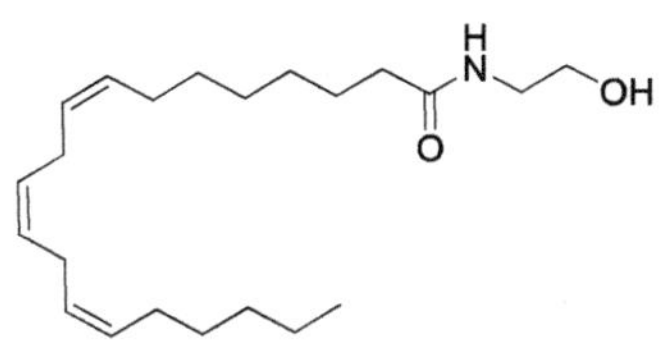

N-Homo-γ-linolenoylethanolamine (**26**)

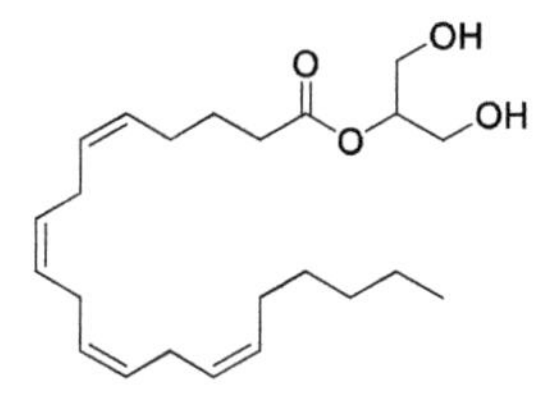

2-Arachidonoylglycerol (2-AG) (**27**)

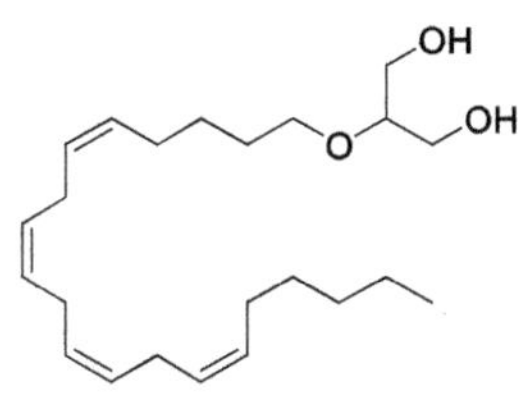

Noladin Ether (**28**)

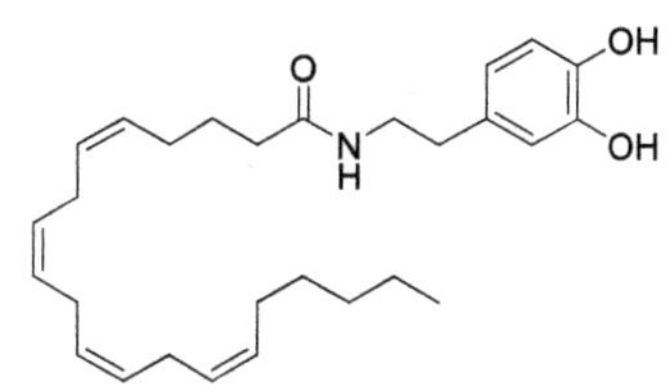

N-Arachidonoyl dopamine (NADA) (**29**)

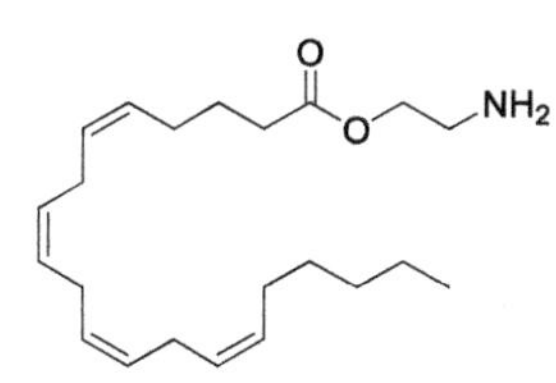

Virodhamine (**30**)

FIG. 2 Structures of known endocannabinoids.

including humans and are thought to be largely responsible for the medicinal effects of the cannabis plant. Accordingly, the focus of this chapter will be on the chemistry, biochemistry, and receptor activities of these cannabinoids.

Additionally, research into cannabinoid pharmacodynamics led to the discovery of endogenous cannabinoids (i.e., the endocannabinoids), which participate in signaling in both the central and peripheral nervous systems. To date, several endocannabinoids have been identified: the unsaturated fatty acid ethanolamides, AEA, *N*-homo-γ-linolenoylethanolamine, and *N*-docosotetraenoylethanolamine; the glycerol ester 2-AG and its ether analog 2-arachidonyl glyceryl ether (Noladin ether); the dopamine derivative, NADA, as well as the nonclassical eicosanoid virodhamine (Fig. 2, Reggio, 2010). Apart from AEA and 2-AG, the presence of some of these compounds in mammalian tissue has been debated. Nonetheless, endocannabinoids share many of the same molecular targets as phytocannabinoids and regulate a wide range of physiological processes including mood, appetite, immune function, and inflammatory responses (Reggio, 2010).

Furthermore, over the past 50 years, a plethora of non-natural cannabinoid analogs have been designed and synthesized. These range from nonclassical cannabinoids such as CP-55940 and HU-210, which share the same basic structural core as phytocannabinoids, to analogues of diverse structural classes such as rimonabant (SR141716), naphthoylindole derivatives, and AAI derivatives (Fig. 3, Le Boisselier et al., 2017; Reggio, 2010). Engineered with specific molecular targets in mind, synthetic cannabinoids generally exhibit higher selectivity, potency, or sometimes both, compared to endo- and phytocannabinoids at their site(s) of action. Several synthetic cannabinoids currently see routine usage as receptor probes and reference compounds for biomedical research, while others are under investigation as experimental therapeutics (Le Boisselier et al., 2017; Reggio, 2010).

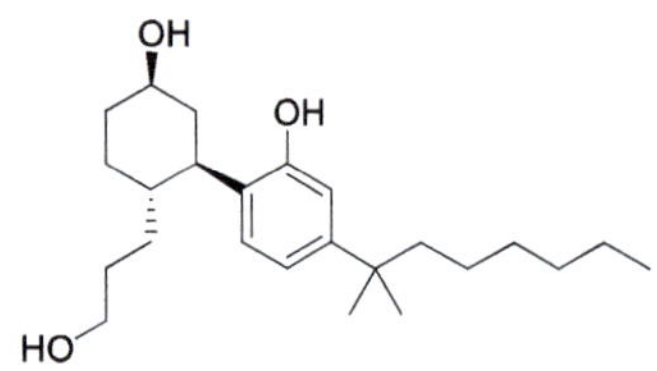

CP 55,940 (**31**)

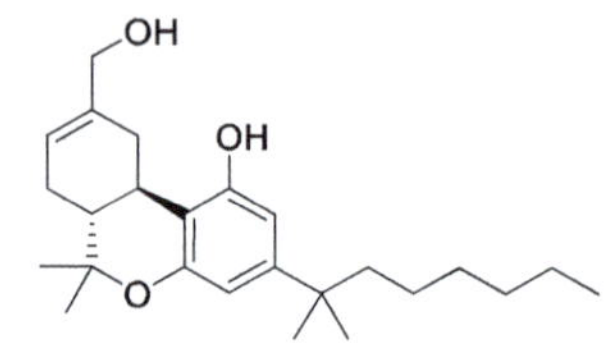

HU-210 (**32**)

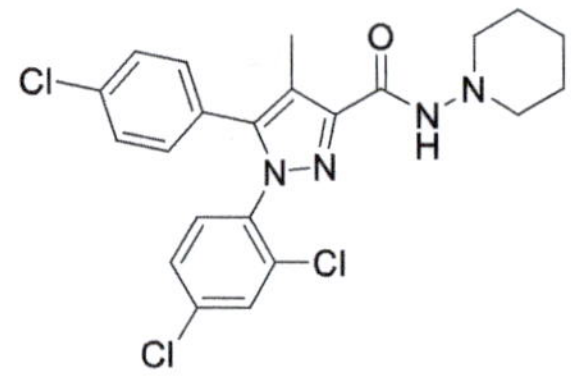

Rimonabant (**33**)

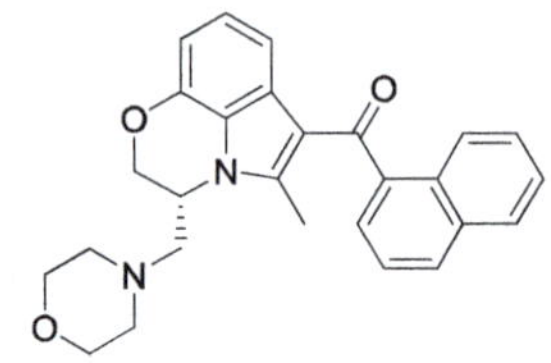

WIN 55,212-2 (**34**)

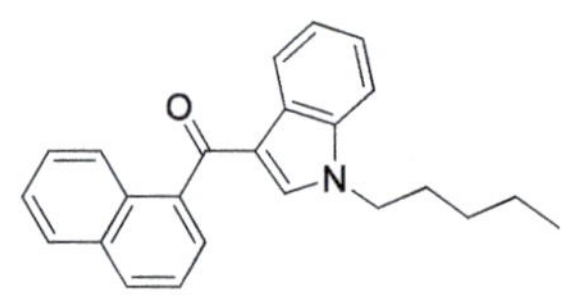

JWH-018 (**35**)

FIG. 3 Structures of selected synthetic cannabinoids. WIN 55,212-2 is an aminoalkyl indole derivative and JWH-018 belongs to the naphthoyl indole family of compounds.

Cannabinoid receptors

With the medicinal effects of cannabis well documented throughout history, the search for the molecular targets of the active constituents of cannabis resulted in the discovery of cannabinoid receptors and later the endocannabinoid system in the 1990s (Matsuda et al., 1990; Munro et al., 1993). Cannabinoid receptors are trans-membrane GPCRs and at present two subtypes, CB_1R and CB_2R, are well known, with moderate sequence similarity between them (Munro et al., 1993).

In mammals, the CB_1 receptor is primarily expressed in the CNS, particularly the brain, and is thus thought to mediate the psychoactive effects of cannabis consumption (Katona et al., 2001; Matsuda et al., 1990; Tsou et al., 1998). This is at least partially brought about by the modulation of GABAergic neurons to which CB_1R is localized within the CNS (Tsou et al., 1998). In rat amygdala nuclei, CB_1R activation was shown to depress presynaptic GABA signaling (Katona et al., 2001). In the periphery, CB_1R is found at lower levels in the eye, reproductive organs, endocrine organs, as well as in the digestive tract (Lee et al., 2016; Nielsen et al., 2019; Straiker et al., 1999). Notably, the potential to control gastrointestinal motility, appetite, and energy metabolism has made CB_1R modulation the subject of interest for managing obesity-related diseases (Pagotto et al., 2006). CB_1R exhibits relatively broad ligand specificity and serves as the main target of the endocannabinoid AEA as well as the phytocannabinoid Δ^9-THC, among others (Pertwee, 2014; Reggio, 2010).

On the other hand, the CB_2 receptor is generally localized to the peripheral tissues, especially that of the immune system. Aside from organs such as the spleen, tonsils, and bone marrow, CB_2R is also highly expressed in hematopoietic cell lines such as blood leukocytes (Gonsiorek et al., 2006; Graham et al., 2010). Interestingly, the CB_2R signaling pathway has downstream effects on a myriad of immune cell types and as such this receptor is currently under investigation as a potential target against several neurodegenerative and inflammatory conditions (Basu & Dittel, 2011; Ladha et al., 2020). Despite a long-standing belief that CB_2R was exclusive to the peripheral tissues, CB_2R was eventually found to be functionally expressed in the brain, albeit at much lower concentrations compared to CB_1R (Onaivi, 2006). Abnormalities in neuronal CB_2R signaling have been linked to depression and substance abuse in humans (Onaivi, 2006). Additionally, CB_2R is expressed in the gastrointestinal tract, where it is thought to regulate gut motility (a function shared with CB_1R) and intestinal inflammation (Wright et al., 2008). Like CB_1R, CB_2R is also the site of action for several different cannabinoids. Endogenously, it is the main target of 2-AG, although AEA as well as phytocannabinoids such as Δ^9-THC and CBD also bind to CB_2R with reasonable affinity (Basu & Dittel, 2011).

Recently, the receptor GPR55 has garnered attention as a putative third cannabinoid receptor. Originally, an orphan GPCR, GPR55 was shown by Ryberg et al. (2007) to bind to and be activated by a variety of natural and synthetic cannabinoids *in vitro* at high potency. Interestingly, these compounds included the atypical cannabinoids, abnormal CBD and O-1602, which showed no activity at CB_1R and CB_2R but were known to affect vascular tone and blood pressure (Ryberg et al., 2007). Thus, it was theorized that GPR55 may function as a vasodilating cannabinoid receptor; however, further research demonstrated that while atypical cannabinoids indeed stimulated GPR55, there was little evidence of the receptor mediating the downstream vasodilation (Johns et al., 2007).

In all, great strides have been made in the past three decades in exploring the endocannabinoid system. However, the molecular mechanisms behind the physiological effects of cannabis and cannabinoids remain relatively poorly understood and presently investigations continue not only towards CB_1R and CB_2R pharmacology but also in the search for novel cannabinoid receptors.

C_5 cannabinoids

The C_5 cannabinoid family is distinguished by their *n*-pentyl moieties, which are derived from the intermediate phytochemical, OA. During biosynthesis, GPP is combined with OA through the action of GOT to form the precursor C_5 cannabinoid, CBGA (Fellermeier & Zenk, 1998). Further catalysis by separate synthases then converts CBGA into either Δ^9-THCA, CBDA, or CBCA, with each conversion differing in the cyclization of the geranyl group. These C_5 cannabinoid acids are in turn transformed into other C_5 cannabinoids through nonenzymatic means (Fig. 1, Lewis et al., 2017). Among all cannabinoids, the C_5 family is synthesized in the highest concentration, with Δ^9-THC and CBD being the two most abundant cannabinoids found in most cannabis plants (Yang et al., 2020). Accordingly, most of the physiological effects of cannabis consumption have been attributed to one or more of the C_5 cannabinoids.

Major C_5 cannabinoids

Δ^9-THC

The tricyclic Δ^9-THC is the most abundant cannabinoid in most medicinal/drug varieties of cannabis and is perhaps the most studied member of the cannabinoid family. The cyclohexenyl ring of Δ^9-THC possesses two chiral centers at the neighboring 6a and 10a positions; in naturally synthesized Δ^9-THC these take on the levorotatory *trans* conformation, i.e., (−)-*trans*-(6a*R*,10a*R*)-Δ^9-THC (Fig. 4). Additionally, the structural isomer Δ^8-THC, in which the cyclohexenyl double bound is found at the C8 position, is also biosynthesized in minor quantities. The chemical structure and absolute stereochemistry of Δ^9-THC were elucidated in 1964 and confirmed in 1970 using nuclear magnetic resonance studies (Archer et al., 1970; Gaoni & Mechoulam, 1964).

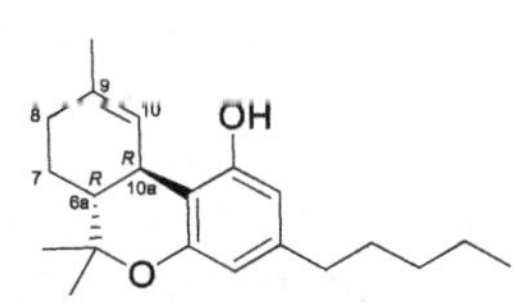

(-)-*trans*-(6a*R*,10a*R*)-Δ^9-THC (**15**)

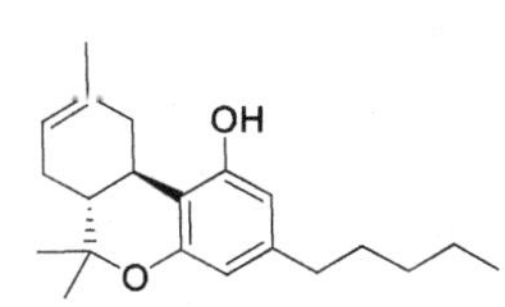

Δ^8-THC (**36**)

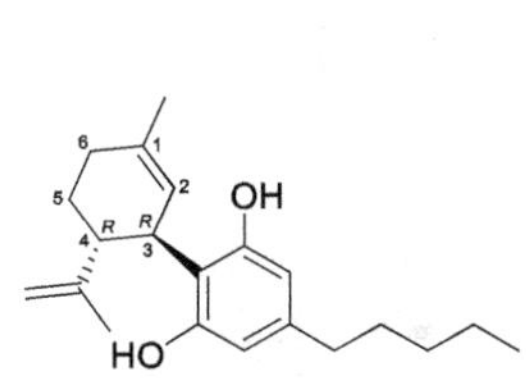

(-)-(3*R*,4*R*)-CBD (**17**)

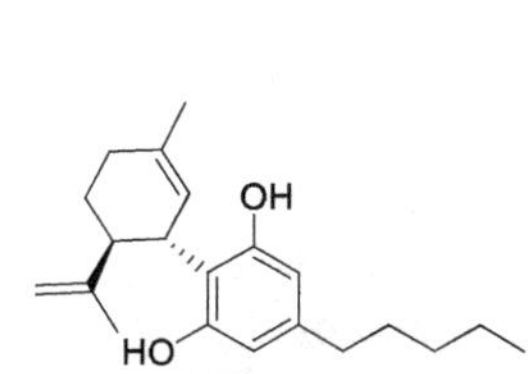

(+)-(3*S*,4*S*)-CBD (**37**)

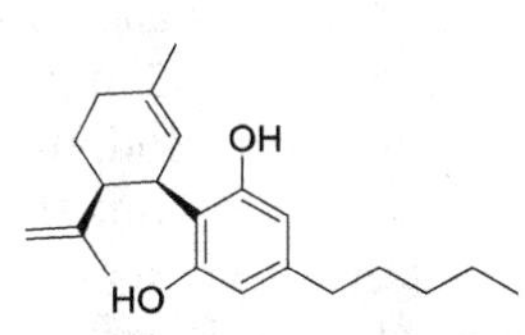

(3*R*,4*S*)-CBD (**38**)

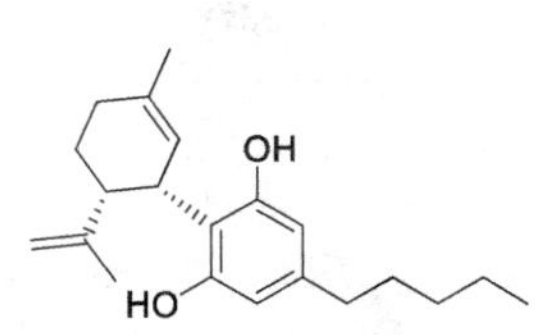

(3*S*,4*R*)-CBD (**39**)

FIG. 4 Isomers of Δ^9-THC and CBD.

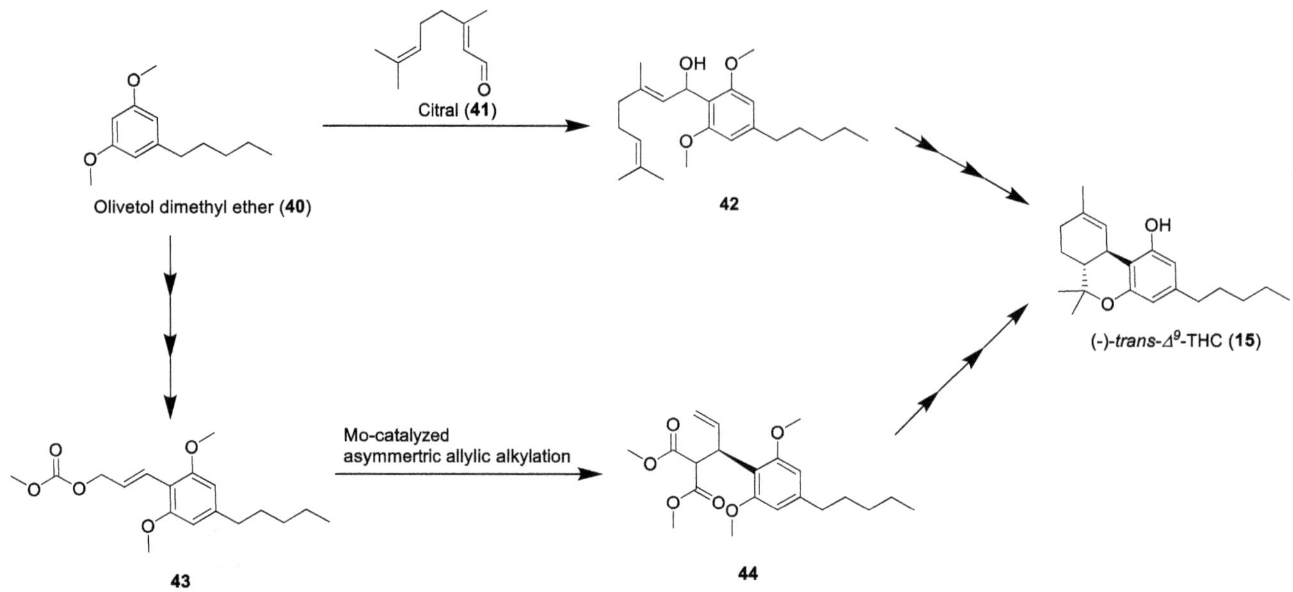

FIG. 5 Synthetic strategies for (−)-*trans*-Δ^9-THC.

Within the cannabis plant, Δ^9-THC is generally found as its acidic precursor Δ^9-THCA, which itself is derived from CBGA via THCA synthase. The loss of the carboxylic acid group from Δ^9-THCA from exposure to heat during either storage or combustion of plant material liberates the neutral Δ^9-THC (Lewis et al., 2017). Synthetic (−)-*trans*-Δ^9-THC, also known as dronabinol, was first produced in 1965 by Mechoulam and Gaoni. Employing a bio-mimetic approach, a methylated olivetol derivative was coupled to citral A (geranial) and cyclized to produce CBD dimethyl ether. The ether was then *O*-demethylated, and the resulting (±)-CBD was boiled in ethanolic hydrochloric acid to close the oxane ring (Mechoulam & Gaoni, 1965). However, being a nonstereoselective route Δ^9-THC was obtained as a racemic mixture of the (−) and (+) stereoisomers. Following this, Trost and Dogra (2007) successfully completed the stereoselective synthesis of (−)-*trans*-Δ^9-THC via an enantioselective allylic alkylation brought about by a molybdenum catalyst (Fig. 5).

In the licensed therapeutics sphere, Δ^9-THC is a major component of the oromucosal spray Nabiximols, which contains a standardized cannabis extract that delivers approximately equal quantities of Δ^9-THC and CBD per dose (Russo, 2008). Marketed as Sativex, this formulation from GW Pharmaceuticals is approved for neuropathic and cancer-related pain in Canada, as well as for multiple sclerosis-related spasticity in several European countries (Russo, 2008). Additionally, synthetic Δ^9-THC (dronabinol) formulated as Marinol capsules (Solvay Pharmaceuticals) has been FDA approved for the alleviation of chemotherapy-associated nausea and appetite loss in HIV/AIDS patients since 1985 and 1992, respectively (Russo, 2008). On a related note, nabilone, a synthetic derivative of Δ^9-THC, is also approved under the trade name Cesamet as a chemotherapy antiemetic in the United States (Russo, 2008).

CBD

In contrast to Δ^9-THC, CBD is a bicyclic compound comprising of a terpenoid ring coupled to a resorcinol moiety (Fig. 4). With chiral centers at the 3 and 4 positions of the terpenoid ring, CBD exhibits stereoisomerism and has 4 isomeric forms, where each chiral center takes either the R or S configuration (Fig. 4). Due to their optical activities, the levorotatory (3*R*,4*R*) stereoisomer is commonly known as (−)-CBD, while the dextrorotatory (3*S*,4*S*) stereoisomer is known as (+)-CBD. The naturally occurring and more therapeutically relevant (−)-CBD was discovered by Adams et al. in 1940. Its chemical structure and absolute stereochemistry were elucidated in 1963 and 1967, respectively, by Mechoulam and colleagues (Mechoulam & Gaoni, 1967; Pertwee, 2006). (+)-CBD is not found in the cannabis plant but has been synthesized and characterized as early as 1982 (Leite et al., 1982). On the other hand, the non-natural *cis* CBD stereoisomers (i.e., (3*R*,4*S*)-CBD and (3*S*,4*R*)-CBD) have yet to be isolated in their pure forms (Fig. 6).

During biosynthesis, CBDA synthase cyclizes CBGA into CBDA, which upon exposure to heat undergoes decarboxylation to yield CBD (Lewis et al., 2017). The first total synthesis of CBD was performed by Mechoulam and Gaoni (1965) as the final intermediate in the total synthesis of Δ^9-THC. Following this, improved synthetic strategies were developed to directly couple a terpenoid moiety onto substituted resorcinols to obtain enantiomerically pure CBD (Baek et al., 1985).

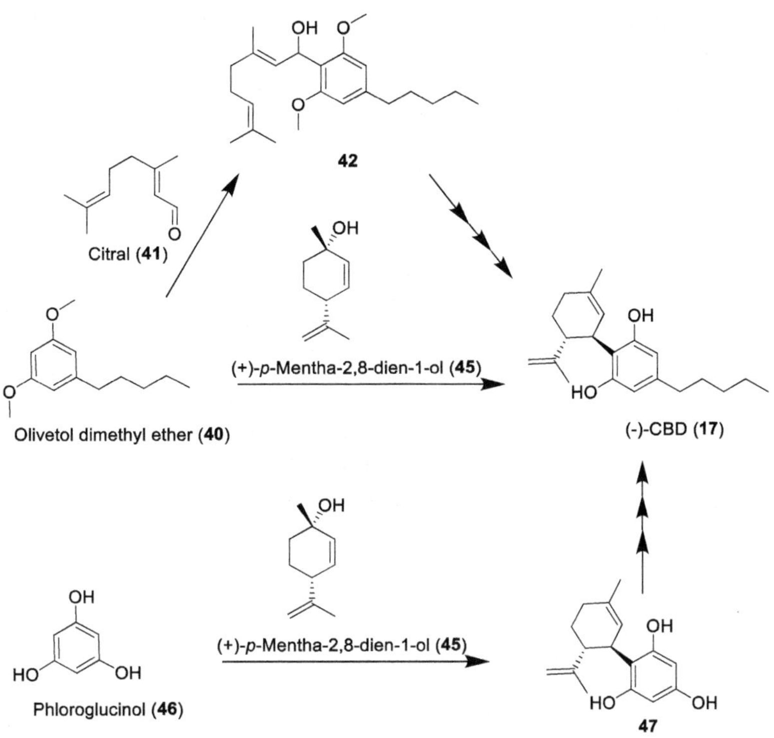

FIG. 6 Synthetic strategies for (−)-CBD.

Recently, Gong et al. (2020) reported a new synthetic route that coupled (+)-*p*-mentha-2,8-dien-1-ol to phloroglucinol rather than a modified resorcinol to eliminate regioisomer formation at this stage. Compared to classical routes, this method also provides easy access to the analogs of CBD (Fig. 5).

Presently, licensed therapeutics incorporating (−)-CBD as an API include the CBD/Δ^9-THC-containing oral spray Nabiximols (Sativex) and Epidiolex, an orally administered CBD solution which recently approved by the FDA to treat forms of childhood epilepsy (Medical Letter Inc., 2018). Notably, the CBD found in these products are extracted from the cannabis plant rather than synthetically produced.

Minor C_5 cannabinoids

As part of the same biosynthetic pathway for Δ^9-THC and CBD, cannabis also produces several other C_5 cannabinoids, albeit in significantly smaller quantities (Lewis et al., 2017). Some, such as CBC and CBG, are direct derivatives of CBGA, produced through enzyme catalysis by CBCA synthase followed by heat-induced decarboxylation for the former and straight decarboxylation for the latter (Fig. 1). The remaining members are secondary derivatives of existing C_5 cannabinoids, produced through a combination of oxidation, photo-irradiation, and heating. CBL is derived from CBC, CBE from CBD, and CBN from Δ^9-THC (Fig. 1). Through photochemistry, CBN can be further transformed into CBND, a fully aromatized analog of CBD (Fig. 1). Interestingly, most of these conversions can take place in both the acidic and neutral forms, providing multiple routes to arrive at the derivatized cannabinoids (Lewis et al., 2017).

Notably, CBN was the first cannabinoid ever isolated from the cannabis plant in the late 1800s. The elucidation of its structure and total synthesis in 1933 and 1940, respectively, laid the chemical groundwork for the subsequent identification of Δ^9-THC (Pertwee, 2006). Following its isolation from cannabis, CBND was first synthesized in 1976, which allowed for the conclusive elucidation of its structure (Lousberg et al., 1977). The first total synthesis of CBG was performed in 1968 through the coupling of a geranyl moiety to olivetol (Razdan, 1981). CBC was isolated from cannabis and identified shortly after Δ^9-THC in 1966, with the first total synthesis performed in 1968 through the dehydrogenation of CBG (Pollastro et al., 2018; Razdan, 1981). Interestingly, an alternative route for CBC synthesis based on the heating of citral and olivetol under basic conditions also led to the formation of CBL, allowing for its eventual structural elucidation in 1971 (Pollastro et al., 2018; Razdan, 1981). Finally, CBE, in the form its acid precursors CBEA-A and CBEA-B, was isolated and synthesized in 1974, and was also found as a CBD metabolite in mammals (Morales et al., 2017a, b).

Biological activities of C_5 cannabinoids

Δ⁹-THC

Δ^9-THC suffers from a low oral bioavailability of approximately 10%–20%, in part due to significant first-pass effects (Huestis, 2007). When inhaled, however, bioavailability increases to as high as 56% due to improved absorption and evasion of first-pass metabolism (Huestis, 2007). This has led to inhalation being the preferred route of administration for medical and recreational cannabis products. The metabolism of Δ^9-THC is mainly carried out by the liver cytochromes CYP2C9, CYP2C19, and CYP3A4, with 11-hydroxylation, 8β-hydroxylation, and (9α,10α)-epoxidation being the major phase I transformations (Vozella et al., 2019; Watanabe et al., 2007). Among these metabolites, 11-OH-THC, which is itself bioactive, is produced in the largest quantity and can be further oxidized into THC-COOH. Both compounds then undergo phase II metabolism via uridine diphosphate-glucuronosyltransferase (UGT) enzymes prior to renal excretion (Vozella et al., 2019). Clearance of Δ^9-THC is thought to be rapid, with half-lives of less than 2 hour (h) generally observed for the inhalation route (Mørland & Bramness, 2020). However, recent data suggest that due to its lipophilicity Δ^9-THC readily compartmentalizes into vascular and adipose tissue, leading to a short "initial" $t_{1/2}$ followed by a long "terminal" $t_{1/2}$ of up to 21.5 h that reflects the gradual release of accumulated Δ^9-THC back into the bloodstream (Mørland & Bramness, 2020).

Pharmacodynamically, Δ^9-THC is best known as an exogenous ligand of the endocannabinoid system. At CB_1R and CB_2R, it acts as a moderately potent partial agonist with K_i values of <100 nM at both receptors *in vitro* (Pertwee, 1999). The pharmacology of this cannabinoid appears to be quite promiscuous, and over the years, other putative targets have emerged including the orphan receptor GPR55, serotonin receptors, opioid receptors, and nuclear receptors (e.g., PPARγ), TRPV channels, as well as melastatin receptors (Morales et al., 2017a, b; Pertwee, 2014).

Exposure to Δ^9-THC has been well documented to cause neurological and motor impairment, producing a characteristic "high" in users of cannabis (Pertwee, 2014). Aside from psychoactivity, Δ^9-THC also exhibits several therapeutically useful properties, most prominently the modulation of appetite (Le Foll et al., 2013). In rats, Δ^9-THC was shown to increase feeding motivation in a CB_1R-dependant manner (Solinas & Goldberg, 2005) and anecdotally, cannabis users report increases in appetite upon cannabis consumption, a phenomenon colloquially known as the "munchies" (Le Foll et al., 2013). However, Le Foll et al. (2013) reported that the obesity rate among US adults was in fact lower in cannabis users compared to nonusers and that the difference could not be explained by variables such as sex, age, and tobacco usage. Meanwhile, a recent systematic review of fifteen randomized clinical trials by Spanagel and Bilbao (2021) found dronabinol and nabilone to have no significant effect on appetite. Interestingly, the short-lived obesity drug Rimonabant (SR141716) also utilized the CB_1R pathway to modulate metabolism, albeit as a synthetic inverse agonist designed to suppress appetite in dieting individuals (Le Foll et al., 2013).

Another therapeutically exploited property of Δ^9-THC is its antiemetic effect. In a least shrew (*Cryptotis parva*) model, a 10 mg/kg dose of Δ^9-THC provided up to 80% protection against cisplatin-induced emesis with ID_{50} values between 0.3 and 1.8 mg/kg, an effect that was potentiated by coadministration with an ultralow dose of a TRPV1 agonist (Darmani et al., 2014). Similar results were seen in primates, where pretreatment with 0.1 mg/kg of Δ^9-THC provided 100% and 75% protection against emesis induced by nicotine and lithium chloride, respectively, that were reversed with CB_1R inverse agonism (Wooldridge et al., 2020). In humans, however, the efficacy of Δ^9-THC appears to be limited, and a 2016 review by May and Glode (May & Glode, 2016) did not find sufficient evidence to support the universal use of dronabinol for chemotherapy-induced emesis. Similarly, a study by Kleine-Brueggeney et al. (2015) found IV Δ^9-THC to be potentially less effective than other prophylactic drugs for postoperative nausea and vomiting while exhibiting significant psychotropic side effects.

Furthermore, Δ^9-THC has also been investigated for analgesic and anxiolytic properties. Despite showing antinociceptive effects in animals, Δ^9-THC did not reduce pain in healthy human subjects and in fact induced hyperalgesia under certain conditions (Naef et al., 2003). However, further studies with combination Δ^9-THC/CBD formulations (e.g., Sativex) as well as whole cannabis in patients with chronic pain demonstrated therapeutic efficacy, suggesting that the analgesic effect may be brought about synergistically with other cannabinoids and/or phytochemicals in cannabis (van de Donk et al., 2019). Regarding anxiety, the effect of Δ^9-THC appears to be biphasic in animals, generating anxiolytic responses at low doses and anxiogenic responses at high doses, with evidence for dual modulation by the endocannabinoid and opioid systems (Berrendero & Maldonado, 2002). However, clinical studies in humans have generally shown an anxiogenic response to Δ^9-THC, particularly at high dosages, and as such the anxiolytic properties of whole cannabis have been attributed to CBD instead (Sharpe et al., 2020).

Due to the extensive distribution of CB_1R in the CNS and the GI tract, Δ^9-THC is believed to exert its psychoactive and therapeutic effects primarily through CB_1R activation (Le Foll et al., 2013; Morales et al., 2017a, b; Zagzoog et al., 2020). Lately, discoveries have been made regarding CB_2R-mediated effects of Δ^9-THC, such as the *in vitro* reversal of airway epithelial cell permeability and the attenuation of MDMA neurotoxicity in mice (Shang et al., 2016; Tourino et al., 2010). Continuing studies in this field will no doubt shed more light on the roles of CB_2R as well as non-endocannabinoid system targets in Δ^9-THC pharmacology.

CBD

Like Δ^9-THC, CBD undergoes extensive first-pass metabolism and exhibits a low oral bioavailability of 13%–19%, which improves to 11%–45% in the inhaled route of administration (Scuderi et al., 2009). Upon consumption, CBD is rapidly distributed and is primarily metabolized in the liver by cytochrome P450 enzymes from the 2C and 3A subfamilies (Scuderi et al., 2009). Observed half-lives for this cannabinoid include 1.4–10.9h with buccal administration, 24h with IV administration, 31h with inhalation, and 2–5days with chronic oral administration (Millar et al., 2018).

The non-psychoactive CBD binds to a variety of physiological targets which fall broadly into two categories: (1) modulators of antiinflammatory/antioxidation pathways and (2) modulators of motivational disorders (Burstein, 2015; Zlebnik & Cheer, 2016).

CBD has been shown to exert antiinflammatory effects through cannabinoid receptors, adenosine receptors, TRPV channels, and GPR55 receptors. Here, activation of the adenosine A_{2A} and TRPV1 signaling pathways, along with deactivation of the CB_2 and GPR55 receptor pathways by CBD, influence downstream events such as gene expression which in turn lead to the lowering of cytokine levels (Gonca & Darıcı, 2015; Pellati et al., 2018). Interestingly, CB_2R/5-$HT_{1A}R$ heterodimers have recently been observed in a human cell line, and a study on the neuroprotective effects of CBD in a hypoxic-ischemic newborn pig model suggested a role for serotonin receptors in the anti-inflammatory effects of CBD as well (Pazos et al., 2013).

CBD is also thought to improve psychological disorders including depression, psychosis, schizophrenia, and anxiety (Campos et al., 2013; Jenny et al., 2009; Leweke et al., 2012; Resstel et al., 2009). One of the proposed mechanisms for this is the augmentation of the endocannabinoid system tone, which has been shown to modulate behaviors related to anxiety as well as psychotic symptoms of schizophrenia (Campos et al., 2013; Leweke et al., 2012). Meanwhile, CBD agonism at the serotonin receptor 5-HT1A has been shown to induce anxiolytic effects in rodents, and at micromolar concentrations, CBD suppressed the degradation of tryptophan in vitro which potentially upregulates serotonin biosynthesis via increasing tryptophan availability (Resstel et al., 2009; Jenny et al., 2009).

Additionally, CBD has shown efficacy against neuropathic pain in animal models and even exhibits anticancer activity by inhibiting the growth of several tumor cell lines (Costa et al., 2007; Ligresti et al., 2006). However, much of the compound's pharmacology remains to be uncovered. For example, GPR55 receptors and TRPV1 channels are thought to play a role in the modulation of anxiety, while CB_1R, CB_2R, and TRPV1 antagonists reduced the reinstatement of drug-seeking behavior (i.e., relapse) in rats (Adamczyk et al., 2012; Rahimi et al., 2015; Terzian et al., 2009). CBD is known to interact with all these targets, but mechanistic links between CBD and these modulative pathways have yet to be established.

Minor C_5 cannabinoids

In contrast to Δ^9-THC and CBD, investigations into the pharmacology of the minor cannabinoids have been limited. In terms of abundance, CBG and CBC generally rank the highest among the minor cannabinoids, albeit at concentrations as low as 1/10 that of Δ^9-THC and CBD (Lewis et al., 2019; Pollastro et al., 2018). CBC did not show strong affinity to CB_1R and CB_2R (i.e., $K_i > 200$nM) but is a known potent TRPV1 agonist (Pollastro et al., 2018; Rosenthaler et al., 2014). In animal models, CBC exhibited anti-inflammatory and antinociceptive effects that could not be explained by TRPV1 agonism alone, suggesting undiscovered targets for this cannabinoid (Pollastro et al., 2018). Meanwhile, CBG demonstrates low-to-moderate affinity for the cannabinoid receptors with higher binding at CB_2R compared to CB_1R, and other putative targets include adenosine receptors, serotonin receptors, TRP channels, as well as both human cyclooxygenases (Pollastro et al., 2018; Rosenthaler et al., 2014). CBG has shown potential as an anti-inflammatory for bowel diseases and may even be efficacious against mental ailments such as depression (Morales et al., 2017a, b). Additionally, both CBC and CBG have been recognized for their antibacterial properties, with CBC also outperforming other cannabinoids and cannabinoid homologs as an antifungal (Morales et al., 2017a, b; Pollastro et al., 2018).

Compared to CBC and CBG, other derivatized cannabinoids such as CBN, CBND, CBL, and CBE are present at much lower concentrations in cannabis. Largely overshadowed by the more abundant cannabinoids, few evaluations of their pharmacology have been conducted and the therapeutic significance, if any, of most of these compounds remain unknown. The weakly psychoactive CBN binds to CB_1R and CB_2R with affinities comparable to that of Δ^9-THC and is also active at TRP channels (Morales et al., 2017a, b; Rosenthaler et al., 2014). Known to be a weak CB_1R agonist, the precise CB_2 pharmacology of CBN remains under investigation (Morales et al., 2017a, b). Remarkably, CBN was reported to possess antiviral properties against SARS-CoV2 (Pitakbut et al., 2021). Although the mechanism of action is not yet understood, further experiments showed that CBN does not target the main viral proteases of SARS-CoV2 (Pitakbut et al., 2021). Although CBE is yet to be evaluated for bioactivity against the cannabinoid receptors, it was shown to inhibit the liver cytochrome CYP1A1 *in vitro* with low potency through a mixed mode mechanism (Yamaori et al., 2013). Recently, CBE was discovered to bind to the Wnt/β-catenin receptor with high affinity, suggesting neuroprotective properties for the cannabinoid against pain (Monroe et al., 2021). On the other hand, while CBND is believed to be another psychoactive cannabinoid, little is known about its receptor pharmacology at present and similarly, the receptor activities of CBL have not been reported so far (Morales et al., 2017a, b).

C_3 cannabinoids

Compared to the C_5 family, C_3 cannabinoids possess an *n*-propyl rather than an *n*-pentyl hydrocarbon tail. Biosynthesis of this compound family occurs on a separate branch of the pathway, although the mechanism mirrors that of C_5 cannabinoids. Like their C_5 counterparts, C_3 cannabinoids also originate from GPP, with the difference being coupling to DA rather than OA at the first step to arrive at CBGVA, which bears the *n*-propyl moiety distinctive to this class. Subsequent cyclization of the geranyl group followed by decarboxylation yields the neutral C_3 cannabinoids (Fig. 1). Interestingly, rather than a separate set of enzymes, it is believed that catalytic phases of C_3 cannabinoid biosynthesis are performed by the same synthases as those of the C_5 branch—an example of enzyme promiscuity in natural product genesis (Fig. 1, de Meijer & Hammond, 2016; Luo et al., 2019).

Major C_3 cannabinoids

In cannabis, C_3 cannabinoids are generally synthesized at much lower concentrations than C_5 cannabinoids, with Δ^9-THCV and CBDV the only members produced in significant amounts (de Meijer & Hammond, 2016; Luo et al., 2019).

First isolated from a cannabis extract in 1970, the structure of Δ^9-THCV was confirmed by synthesis in the same year (Gill et al., 1970). As the *n*-propyl analog of Δ^9-THC, Δ^9-THCV is presumably biosynthesized by THCA synthase from CBGVA and can accumulate in surprisingly high concentrations in certain cannabis varieties, occasionally rivaling that of Δ^9-THC (de Meijer & Hammond, 2016; Gill et al., 1970). Presently, the enantiomerically pure form of Δ^9-THCV is largely botanically derived, and explorations of the total synthesis of this cannabinoid have been limited; however, mammalian metabolites of Δ^9-THCV have been generated through both total synthesis and oxidation of the parent compound (Feng & ElSohly, 2002; Tinto et al., 2020).

Meanwhile, CBDV, the *n*-propyl analog of CBD, was unambiguously identified in 1969 by Vollner and colleagues (Vollner et al., 1969) through isolation from a cannabis extract. Like Δ^9-THCV, CBDV is believed to be produced via the same synthase as its C_5 counterpart, with little information in the literature pertaining to its total synthesis (de Meijer & Hammond, 2016). Recently, Chiurchiù et al. (2021) pioneered a continuous flow technique utilizing terpenoid-resorcinol coupling, which produced CBDV in good yield (56%) while minimizing the formation of Δ^9-THCV and abnormal coupling side products. Interestingly, this method could also be applied to the synthesis of other CBD-type molecules of varying *n*-alkyl tail lengths, such as the C_4 cannabinoid CBDB (Chiurchiù et al., 2021).

Biological activities of C_3 cannabinoids

Δ^9-*THCV*

The major mammalian metabolites of Δ^9-THCV include 11-OH-THCV and THCV-COOH, the latter of which has been proposed for use as a biomarker for cannabis consumption (Newmeyer et al., 2016; Tinto et al., 2020). The pharmacokinetics of Δ^9-THCV is not well understood, although significant distribution into the brain as well as elimination half lives on the order of a few hours were observed in rats and mice following oral administration (Deiana et al., 2012). *In vivo*, Δ^9-THCV behaved as an antagonist at CB_1R at low doses (<3 mg/kg), while at high doses (>10 mg/kg), this effect changed to agonism (Tinto et al., 2020). On the other hand, *in vitro*, Δ^9-THCV was shown to be a weak partial agonist at CB_1R and an agonist at CB_2R, binding to both cannabinoid receptors with potencies comparable to that of Δ^9-THC (Zagzoog et al., 2020). Additionally, TRPV channels were recently discovered to be novel targets of Δ^9-THCV, where this compound activates the thermosensitive TRPV1, TRPV3, and TRPV4, while inhibiting TRPV5 and TRPV6 (de Petrocellis et al., 2011; de Petrocellis et al., 2012; Janssens et al., 2018).

Due to the potential of Δ^9-THCV to activate CB_2R while inhibiting CB_1R, it has been hypothesized to possess anti-inflammatory, antioxidant, and neuroprotective effects. In human sebocytes, Δ^9-THCV reduced inflammation and normalized excess sebum production in a manner indicative of an anti-acne agent and in a mice model of Parkinson's disease, Δ^9-THCV protected tyrosine hydroxylase-positive neurons from 6-hydroxydopamine-induced neurodegeneration via the CB_2R pathway (García et al., 2011; Olah et al., 2016). Likewise, Δ^9-THCV was reported to produce antinociceptive effects in animal models, although this appears to be mediated by both CB_1R and CB_2R (Zagzoog et al., 2020). Furthermore, Δ^9-THCV has also been demonstrated to have antiepileptic, hypolocomotive, antiemetic, and anxiolytic effects *in vivo* (Rock et al., 2013; Zagzoog et al., 2020). Interestingly, unlike many existing CB_1 blockers Δ^9-THCV does not show inverse agonism at CB_1R, thus avoiding adverse effects such as nausea and depression that are characteristic of CB_1R inverse agonists such as Rimonabant (Rock et al., 2013).

CBDV

The pharmacokinetics of CBDV is also relatively underexplored. Little has been reported about the metabolic pathway of CBDV in humans, although a decarbonylated metabolite, corresponding to the loss of a carbon monoxide molecule from CDBV, was recently detected *in vitro* after incubation with human liver microsomes (Kraemer et al., 2019). In rats and mice, CBDV exhibited rapid absorption after oral administration and eliminations half lives of 3–4h were observed (Deiana et al., 2012). *In vitro*, CBDV showed little-to-no activity at CB_1R but was a moderately potent ligand of CB_2R, where it behaves as a full agonist (Navarro et al., 2020; Rosenthaler et al., 2014; Zagzoog et al., 2020). In accordance with its CB_1 inactivity, CBDV did not demonstrate inverse agonism at CB_1R (Rock et al., 2013). TRP channels represent the only other known targets of CBDV, where it was shown to stimulate and desensitize TRPV1, TRPV2, TRPV4, as well as TRPA1 (Arturo et al., 2014; De Petrocellis et al., 2012).

Like other CB_2R agents, CBDV has been investigated as an analgesic; however, a randomized controlled study found CBDV to be safe but ineffective against HIV-induced neuropathic pain (Eibach et al., 2021). On the other hand, the action of CBDV at TRP channels postulated its potential as an anticonvulsant, and indeed in animal models of epilepsy, CBDV exhibited significant efficacy against pentylenetetrazole-induced seizures in a CB_1R-independent manner (Hill et al., 2013). Additionally, CBDV demonstrated an ability to modulate glutamate levels in the brain through a hitherto unknown mechanism, a potentially useful effect for the treatment of autism spectrum disorder (Pretzsch et al., 2019). However, CBDV was also found to cause DNA damage in human cell lines even at low doses, demonstrating the need for caution in the translation of cannabinoid pharmaceutics (Russo et al., 2019).

Other C_3 cannabinoids

Due to their low abundance in cannabis, the only other C_3 cannabinoid to be pharmacologically evaluated so far is cannabigerovarin (CBGV), the neutral form of CBGVA and the *n*-propyl analog of CBG. CBGV has been shown to bind to both CB_1R and CB_2R with higher potencies than Δ^9-THCV and CBDV, behaving as a full agonist at both cannabinoid receptors (Navarro et al., 2020). As with the major C_3 cannabinoids, CBGV also acts at TRP channels, desensitizing TRPV3 to carvacol (de Petrocellis et al., 2011). However, unlike Δ^9-THCV and CBDV, CBGV showed proinflammatory effects by increasing basal sebum synthesis in human sebocytes (Olah et al., 2016).

Applications to other areas: Deconvoluting the chemical complexity of medical cannabis

An emerging direction in cannabis research is that in contrast to traditional botanical drugs such as tobacco and psilocybin mushrooms, cannabis should be investigated as a mixture of physiologically active molecules, where its effect on receptors varies significantly with chemical composition (Ladha et al., 2020). Accordingly, efforts have been made to deconvolute this complex chemistry. Recently, Yang et al. (2020) applied chemoinformatic techniques to correlate the chemical profiles of medical cannabis extracts and cannabis oils to their activities at cannabinoid receptors, generating meaningful prediction models for the *in vitro* potencies of extracts and oils based on their chemical compositions. Interestingly, the models predicted the activities of individual cannabinoids at CB_1R and CB_2R with reasonable accuracy and provided evidence that cannabinoids behaved differently when administered as part of an extract compared to individual administration (Yang et al., 2020). Additional high-level analyses incorporating not only *in vitro* data but also biomarkers and clinical markers would help bridge the gap between the pharmacology of cannabinoids and that of whole cannabis, and thus allow our current knowledge of cannabinoid pharmacology to better inform the decisions of healthcare professionals when prescribing medical cannabis to patients.

Key facts of cannabinoid receptors

- Phytocannabinoids are synthesized by plants; endocannabinoids are produced by various animals including humans; synthetic cannabinoids are organic molecules that interact with cannabinoid receptors.
- Cannabinoids are biosynthesized through a complex, multibranched pathway, utilizing a combination of enzyme catalysis, heat, oxidation, and photoreactions.
- In cannabis, C_5 cannabinoids are produced in the largest quantity followed by C_3 cannabinoids.
- The endocannabinoid system mainly comprises of the CB_1 and CB_2 GPCR pathways.
- CB_1R is primarily expressed in the CNS and mediates appetite, energy metabolism, as well as the psychoactive effects of cannabis.
- CB_2R is generally found in peripheral tissues, particularly in immune organs, and is thought to modulate inflammation.

Mini-dictionary of terms

Agonist: A ligand that activates a receptor upon binding to it. Ligands that elicit the maximum receptor response are known as *full agonists*, while ligands that produce partial receptor responses are known as *partial agonists*.
Analgesic: Reliving pain. Analgesic drugs are also known as antinociceptives.
Antagonist: A ligand that blocks a receptor upon binding and inhibits the agonistic receptor response.
Antiemetic: Preventing emesis, i.e., nausea and vomiting.
Anxiogenic: Increasing anxiety.
Anxiolytic: Lowering anxiety.
Central nervous system: Portion of the nervous system comprising of the brain and spinal cord.
Chiral center: An asymmetric carbon atom attached to four different functional groups. Molecules possessing chiral centers exhibit stereoisomerism.
Inverse Agonist: A ligand that upon binding inhibits the constitutive activity of a receptor, in addition to not producing any agonistic effects.
Peripheral nervous system: Portion of the nervous system excluding the brain and spinal cord.

Summary points

- This chapter gives an overview of the chemistry of cannabinoids, their receptor activities, as well as the applications of basic research to future explorations of medical cannabis.
- Cannabinoids are a distinct class of secondary metabolites produced by *Cannabis* spp. of plants and are present in cannabis and cannabis extracts in a significantly variable range.
- Cannabinoids target the endocannabinoid system as well as several other receptors, exhibiting different affinities and modes of binding at different targets thus leading to a myriad of physiological effects.
- Several cannabinoids are therapeutically useful and are under investigation as drugs against various psychological and neurological ailments.
- Cannabis and cannabis extracts behave as a mixture of physiologically active molecules, and further research is necessary to link the pharmacology of individual cannabinoids to that of whole cannabis.

References

Adamczyk, P., Miszkiel, J., McCreary, A. C., Filip, M., Papp, M., & Przegaliński, E. (2012). The effects of cannabinoid CB1, CB2 and vanilloid TRPV1 receptor antagonists on cocaine addictive behavior in rats. *Brain Research, 1444*, 45–54.

Archer, R. A., Boyd, D. B., Demarco, P. V., Tyminski, I. J., & Allinger, N. L. (1970). Structural studies of cannabinoids. Theoretical and proton magnetic resonance analysis. *Journal of the American Chemical Society, 92*(17), 5200–5206.

Arturo, I. F., Hill, C. L., Leo, A., Alhusaini, A., Soubrane, C., Mazzarella, E., Russo, E., Whalley, B. J., Di Marzo, V., & Stephens, G. J. (2014). Non-psychotropic plant cannabinoids, cannabidivarin (CBDV) and cannabidiol (CBD), activate and desensitize transient receptor potential vanilloid 1 (TRPV1) channels in vitro: Potential for the treatment of neuronal hyperexcitability. *ACS Chemical Neuroscience, 5*, 1131–1141.

Baek, S. H., Srebink, M., & Mechoulam, R. (1985). Borontrifluoride on alumina—A modified Lewis acid reagent. An improved synthesis of cannabidiol. *Tetrahedron Letters, 26*, 1083–1086.

Basu, S., & Dittel, B. N. (2011). Unraveling the complexities of cannabinoid receptor 2 (CB2) immune regulation in health and disease. *Immunologic Research, 51*, 26–38.

Berrendero, F., & Maldonado, R. (2002). Involvement of the opioid system in the anxiolytic-like effects induced by Δ^9-tetrahydrocannabinol. *Psychopharmacology, 163*, 111–117.

Burstein, S. (2015). *Cannabidiol (CBD) and its analogs: A review of their effects on inflammation. Bioorganic and medicinal chemistry. 1* (pp. 1377–1385). Elsevier Ltd April.

Campos, A. C., Ortega, Z., Palazuelos, J., Fogaça, M. V., Aguiar, D. C., Díaz-Alonso, J., Ortega-Gutiérrez, S., Vázquez-Villa, H., Moreira, F. A., Guzmán, M., Galve-Roperh, I., & Guimarães, F. S. (2013). The anxiolytic effect of cannabidiol on chronically stressed mice depends on hippocampal neurogenesis: Involvement of the endocannabinoid system. *The International Journal of Neuropsychopharmacology, 16*(6), 1407–1419.

Cannabidiol (Epidiolex) for Epilepsy. (2018). *Cannabidiol (Epidiolex) for epilepsy. Medical letter on drugs and therapeutics* (pp. 182–184). Medical Letter Inc.

Chiurchiù, E., Sampaolesi, S., Allegrini, P., Ciceri, D., Ballini, R., & Palmieri, A. (2021). A novel and practical continuous flow chemical synthesis of cannabidiol (CBD) and its CBDV and CBDB analogues. *European Journal of Organic Chemistry, 2021*(8), 1286–1289.

Costa, B., Trovato, A. E., Comelli, F., Giagnoni, G., & Colleoni, M. (2007). The non-psychoactive cannabis constituent cannabidiol is an orally effective therapeutic agent in rat chronic inflammatory and neuropathic pain. *European Journal of Pharmacology, 556*(1–3), 75–83.

Darmani, N. A., Chebolu, S., Zhong, W., Trinh, C., McClanahan, B., & Brar, R. S. (2014). Additive antiemetic efficacy of low-doses of the cannabinoid CB1/2 receptor agonist Δ9-THC with ultralow-doses of the vanilloid TRPV1 receptor agonist resiniferatoxin in the least shrew (*Cryptotis parva*). *European Journal of Pharmacology, 722*, 147–155.

de Meijer, E. D. M., & Hammond, K. M. (2016). The inheritance of chemical phenotype in Cannabis sativa L. (V): Regulation of the propyl-/pentyl cannabinoid ratio, completion of a genetic model. *Euphytica, 210*, 291–307.

de Petrocellis, L., Ligresti, A., Schiano, M. A., Allara, M., Bisogno, T., Petrosino, S., Stott, C. G., & Di Marzo, V. (2011). Effects of cannabinoids and cannabinoid-enriched Cannabis extracts on TRP channels and endocannabinoid metabolic enzymes. *British Journal of Pharmacology, 163*, 1479–1494.

de Petrocellis, L., Orlando, P., Schiano, M. A., Aviello, G., Stott, C., Izzo, A. A., & di Marzo, V. (2012). Cannabinoid actions at TRPV channels: Effects on TRPV3 and TRPV4 and their potential relevance to gastrointestinal inflammation. *Acta Physiologica, 204*, 255–266.

Deiana, S., Watanabe, A., Yamasaki, Y., Amada, N., Arthur, M., Fleming, S., Woodcock, H., Dorward, P., Pigliacampo, B., Close, S., Platt, B., & Reidel, G. (2012). Plasma and brain pharmacokinetic profile of cannabidiol (CBD), cannabidivarine (CBDV), Δ9-tetrahydrocannabivarin (THCV) and cannabigerol (CBG) in rats and mice following oral and intraperitoneal administration and CBD action on obsessive compulsive behaviour. *Psychopharmacology, 219*, 859–873.

Eibach, L., Scheffel, S., Cardebring, M., Lettau, M., Özgür, C. M., Morguet, A., Roehle, R., & Stein, C. (2021). Cannabidivarin for HIV-associated neuropathic pain: A randomized, blinded, controlled clinical trial. *Clinical Pharmacology & Therapeutics, 109*(4), 1055–1062.

Fellermeier, M., & Zenk, M. H. (1998). Prenylation of olivetolate by a hemp transferase yields cannabigerolic acid, the precursor of tetrahydrocannabinol. *FEBS Letters, 427*, 283–285.

Feng, S., & ElSohly, M. A. (2002). Synthesis of (±)-6,6-[$^2H_6$]dimethyl-11-nor-Δ^9-tetrahydrocannabivarin-9-carboxylic acid. *Journal of Labelled Compounds and Radiopharmaceuticals, 45*, 577–581.

Gaoni, Y., & Mechoulam, R. (1964). Isolation, structure, and partial synthesis of an active constituent of hashish. *Journal of the American Chemical Society, 86*(8), 1646–1647.

García, C., Palomo-Garo, C., García-Arencibia, M., Ramos, J. A., Pertwee, R. G., & Fernández-Ruiz, J. (2011). Symptom-relieving and neuroprotective effects of the phytocannabinoid Δ^9-THCV in animal models of Parkinson's disease. *British Journal of Pharmacology, 163*, 1495–1506.

Gill, E. W., Paton, W. D. M., & Pertwee, R. G. (1970). Preliminary experiments on the chemistry and pharmacology of cannabis. *Nature, 228*, 134–136.

Gonca, E., & Darıcı, F. (2015). The effect of cannabidiol on ischemia/reperfusion-induced ventricular arrhythmias: The role of adenosine A1 receptors. *Journal of Cardiovascular Pharmacology and Therapeutics, 20*(1), 76–83.

Gong, X., Sun, C., Abame, M. A., Shi, W., Xie, Y., Xu, W., Zhu, F., Zhang, Y., Shen, J., & Aisa, H. A. (2020). Synthesis of CBD and its derivatives bearing various C4′-side chains with a late-stage diversification method. *The Journal of Organic Chemistry, 85*, 2704–2715.

Gonsiorek, W., Hesk, D., Chen, S., Kinsley, D., Fine, J. S., Jackson, J. V., Bober, L. A., Deno, G., Bian, H., Fossetta, J., Lunn, C. A., Kozlowski, J. A., Lavey, B., Piwinski, J., Narula, S. K., Lundell, D. J., & Hipkin, R. W. (2006). Characterization of peripheral human cannabinoid receptor (hCB2) expression and pharmacology using a novel radioligand, [^{35}S]Sch225336. *The Journal of Biological Chemistry, 281*(38), 28143–28151.

Graham, E. S., Angel, C. E., Schwarcz, L. E., Dunbar, P. R., & Glass, M. (2010). Detailed characterization of CB2 receptor protein expression in peripheral blood immune cells from healthy human volunteers using flow cytometry. *International Journal of Immunopathology and Pharmacology, 23*(1), 25–34.

Hill, T. D. M., Cascio, M.-G., Romano, B., Duncan, M., Pertwee, R. G., Williams, C. M., Whalley, B. J., & Hill, A. J. (2013). Cannabidivarin-rich cannabis extracts are anticonvulsant in mouse and rat via a CB1 receptor-independent mechanism. *British Journal of Pharmacology, 170*, 679–692.

Huestis, M. A. (2007). Human cannabinoid pharmacokinetics. *Chemistry & Biodiversity, 4*(8), 1770–1804.

Janssens, A., Silvestri, C., Martella, A., Vanoevelen, J. M., Di Marzo, V., & Voets, T. (2018). Δ^9-Tetrahydrocannabivarin impairs epithelial calcium transport through inhibition of TRPV5 and TRPV6. *Pharmacological Research, 136*, 83–89.

Jenny, M., Santer, E., Pirich, E., Schennach, H., & Fuchs, D. (2009). Δ9-Tetrahydrocannabinol and cannabidiol modulate mitogen-induced tryptophan degradation and neopterin formation in peripheral blood mononuclear cells in vitro. *Journal of Neuroimmunology, 207*(1–2), 75–82.

Johns, D. G., Behm, D. J., Walker, D. J., Ao, Z., Shapland, E. M., Daniels, D. A., Riddick, M., Dowell, S., Staton, P. C., Green, P., Shabon, U., Bao, W., Aiyar, N., Yue, T.-L., Brown, A. J., Morrison, A. D., & Douglas, S. A. (2007). The novel endocannabinoid receptor GPR55 is activated by atypical cannabinoids but does not mediate their vasodilator effects. *British Journal of Pharmacology, 152*, 825–831.

Katona, I., Rancz, E. A., Acsady, L., Ledent, C., Mackie, K., Hajos, N., & Freund, T. F. (2001). Distribution of CB1 cannabinoid receptors in the amygdala and their role in the control of GABAergic transmission. *The Journal of Neuroscience*, *21*(23), 9506–9518.

Kleine-Brueggeney, M., Greif, R., Brenneisen, R., Urwyler, N., Stueber, F., & Theiler, L. G. (2015). Intravenous delta-9-tetrahydrocannabinol to prevent postoperative nausea and vomiting: A randomized controlled trial. *Anesthesia and Analgesia*, *121*(5), 1157–1164.

Kraemer, M., Broecker, S., Madea, B., & Hess, C. (2019). Decarbonylation: A metabolic pathway of cannabidiol in humans. *Drug Testing and Analysis*, *11*, 957–967.

Ladha, K. S., Ajrawat, P., Yang, Y., & Clarke, H. (2020). Understanding the medical chemistry of the cannabis plant is critical to guiding real world clinical evidence. *Molecules*, *25*, 4042.

Le Boisselier, R., Alexandre, J., Lelong-Boulouard, V., & Debruyne, D. (2017). Focus on cannabinoids and synthetic cannabinoids. *Clinical Pharmacology and Therapeutics*, *101*(2), 220–229.

Le Foll, B., Trigo, J. M., Sharkey, K. A., & Le Strat, Y. (2013). Cannabis and D9-tetrahydrocannabinol (THC) for weight loss. *Medical Hypotheses*, *80*, 564–567.

Lee, Y., Jo, J., Chung, H. Y., Pothoulakis, C., & Im, E. (2016). Endocannabinoids in the gastrointestinal tract. *American Journal of Physiology. Gastrointestinal and Liver Physiology*, *311*, G655–G666.

Leite, J. R., Carlini, E. A., Lander, N., & Mechoulam, R. (1982). Anticonvulsant effects of the (–) and (+) isomers of cannabidiol and their dimethylheptyl homologs. *Pharmacology*, *24*(3), 141–146.

Leweke, F. M., Piomelli, D., Pahlisch, F., Muhl, D., Gerth, C. W., Hoyer, C., Klosterkötter, J., Hellmich, M., & Koethe, D. (2012). Cannabidiol enhances anandamide signaling and alleviates psychotic symptoms of schizophrenia. *Translational Psychiatry*, *2*(3), e94.

Lewis, M. M., Yang, Y., Vyawahare, R., & Kotra, L. P. (2019). Extractions of medical cannabis cultivars and the role of decarboxylation in optimal receptor responses. *Molecules*, *4*(3), 183–194.

Lewis, M. M., Yang, Y., Wasilewski, E., Clarke, H. A., & Kotra, L. P. (2017). Chemical profiling of medical cannabis extracts. *ACS Omega*, *2*(9), 6091–6103.

Ligresti, A., Moriello, A. S., Starowicz, K., Matias, I., Pisanti, S., De Petrocellis, L., Laezza, C., Portella, G., Bifulco, M., & Di Marzo, V. (2006). Antitumor activity of plant cannabinoids with emphasis on the effect of cannabidiol on human breast carcinoma. *The Journal of Pharmacology and Experimental Therapeutics*, *318*(3), 1375–1387.

Lousberg, R. J. J. C., Bercht, C. A. L., van Ooyen, R., & Spronck, H. J. W. (1977). Cannabinodiol: Conclusive identification and synthesis of a new cannabinoid from *Cannabis Sativa*. *Phytochemistry*, *16*, 595–597.

Luo, X., Reiter, M. A., d'Espaux, L., Wong, J., Denby, C. M., Lechner, A., Zhang, Y., Grzybowski, A. T., Harth, S., Lin, W., Lee, H., Yu, C., Shin, J., Deng, K., Benites, V. T., Wang, G., Baidoo, E. E. K., Chen, Y., Dev, I., … Keasling, J. D. (2019). Complete biosynthesis of cannabinoids and their unnatural analogues in yeast. *Nature*, *567*, 123–126.

Matsuda, L. A., Lolait, S. J., Brownstein, M. J., Young, A. C., & Bonner, T. I. (1990). Structure of a cannabinoid receptor and functional expression of the cloned cDNA. *Nature*, *346*, 561–564.

May, M. B., & Glode, A. E. (2016). Dronabinol for chemotherapy-induced nausea and vomiting unresponsive to antiemetics. *Cancer Management and Research*, *8*, 49–55.

McPartland, J. M., & Guy, G. W. (2017). Models of *Cannabis* taxonomy, cultural bias, and conflicts between scientific and vernacular names. *The Botanical Review*, *83*, 327–381.

Mechoulam, R., & Gaoni, Y. (1965). A total synthesis of dl-Δ1-tetrahydrocannabinol, the active constituent of hashish. *Journal of the American Chemical Society*, *87*(14), 3273–3275.

Mechoulam, R., & Gaoni, Y. (1967). The absolute configuration of delta-1-tetrahydrocannabinol, the major active constituent of hashish. *Tetrahedron Letters*, *8*(12), 1109–1111.

Pertwee, R. G. E. (Ed.). (2014). *Handbook of cannabis*. Oxford: Oxford University Press.

Reggio, P. H. (2010). Endocannabinoid binding to the cannabinoid receptors: What is known and what remains unknown. *Current Medicinal Chemistry*, *17* (14), 1468–1486.

Munro, S., Thomas, K. L., & Abu-Shaar, M. (1993). Molecular characterization of a peripheral receptor for cannabinoids. *Nature*, *365*, 61–65.

Tsou, K., Brown, S., Sanudo-Pena, M. C., Mackie, K., & Walker, J. M. (1998). Immunohistochemical distribution of cannabinoid CB1 receptors in the rat central nervous system. *Neuroscience*, *83*(2), 393–411.

Straiker, A. J., Maguire, G., Mackie, K., & Lindsey, J. (1999). Localization of cannabinoid CB1 receptors in the human anterior eye and retina. *Investigative Ophthalmology and Visual Science*, *40*(10), 2442–2448.

Nielsen, J. E., Rolland, A. D., Rajpert-De, M. E., Janfelt, C., Jorgensen, A., Winge, S. B., Kristensen, D. M., Juul, A., Chalmel, F., Jegou, B., & Skakkebaek, N. E. (2019). Characterisation and localisation of the endocannabinoid system components in the adult human testis. *Scientific Reports*, *9*, 12866.

Pagotto, U., Mariscano, G., Cota, D., Lutz, B., & Pasquali, R. (2006). The Emerging Role of the Endocannabinoid System in Endocrine Regulation and Energy Balance. *Endocrine Reviews*, *27*(1), 73–100.

Onaivi, E. S. (2006). Neuropsychobiological evidence for the functional presence and expression of cannabinoid CB2 receptors in the brain. *Neuropsychobiology*, *54*, 231–246.

Wright, K. L., Duncan, M., & Sharkey, K. A. (2008). Cannabinoid CB_2 receptors in the gastrointestinaltract: A regulatory system in states of inflammation. *British Journal of Pharmacology*, *153*, 263–270.

Ryberg, E., Larsson, N., Sjogren, S., Hjorth, S., Hermansson, N.-O., Leonova, J., Elebring, T., Nilsson, K., Drmota, T., & Greasley, P. J. (2007). The orphan receptor GPR55 is a novel cannabinoid receptor. *British Journal of Pharmacology*, *152*, 1092–1101.

Yang, Y., Vyawahare, R., Lewis-Bakker, M., Clarke, H. A., Wong, A. H. C., & Kotra, L. P. (2020). Bioactive chemical composition of cannabis extracts and cannabinoid receptors. *Molecules*, *25*, 3466.

Trost, B. M., & Dogra, K. (2007). Synthesis of (-)-Δ^9-*trans*-tetrahydrocannabinol: Stereocontrol via Mo-catalyzed asymmetric allylic alkylation reaction. *Organic Letters*, *9*(5), 861–863.

Russo, E. B. (2008). Cannabinoids in the management of difficult to treat pain. *Therapeutics and Clinical Risk Management*, *4*(1), 245–259.

Pertwee, R. G. (2006). Cannabinoid pharmacology: The first 66 years. *British Journal of Pharmacology*, S163.

Razdan, R. K. (1981). The total synthesis of cannabinoids. In *The total synthesis of natural products* (pp. 185–262).

Pollastro, F., Caprioglio, D., Del Prete, D., Rogati, F., Minassi, A., Taglialatela-Scafati, O., Munoz, E., & Appendino, G. (2018). Cannabichromene. *Natural Product Communications*, *13*(9), 1189–1194.

Morales, P., Reggio, P. H., & Jagerovic, N. (2017). An overview on medicinal chemistry of synthetic and natural derivatives of cannabidiol. *Frontiers in Pharmacology*, *8*, 422.

Watanabe, K., Yamaori, S., Funahashi, T., Kimura, T., & Yamamoto, I. (2007). Cytochrome P450 enzymes involved in the metabolism of tetrahydro-cannabinols and cannabinol by human hepatic microsomes. *Life Sciences*, *80*, 1415–1419.

Vozella, V., Zibardi, C., Ahmed, F., & Piomelli, D. (2019). Fast and sensitive quantification of Δ^9-tetrahydrocannabinol and its main oxidative metabolites by liquid chromatography/tandem mass spectrometry. *Cannabis and Cannabinoid Research*, *4*(2), 110–123.

Mørland, J., & Bramness, J. G. (2020). Δ9-Tetrahydrocannabinol (THC) is present in the body between smoking sessions in occasional non-daily cannabis users. *Forensic Science International*, *309*, 110188.

Pertwee, R. G. (1999). Pharmacology of cannabinoid receptor ligands. *Current Medicinal Chemistry*, *6*, 635–664.

Morales, P., Hurst, D. P., & Reggio, P. H. (2017). Molecular targets of the phytocannabinoids-A complex picture. *Progress in the Chemistry of Organic Natural Products*, *103*, 103–131.

Solinas, M., & Goldberg, S. R. (2005). Motivational effects of cannabinoids and opioids on food reinforcement depend on simultaneous activation of cannabinoid and opioid systems. *Neuropsychopharmacology*, *30*, 2035–2045.

Spanagel, R., & Bilbao, A. (2021). Approved cannabinoids for medical purposes—Comparative systematic review and meta-analysis for sleep and appetite. *Neuropharmacology*, *196*, 108680.

Wooldridge, L. M., Ji, L., Liu, Y., Nikas, S. P., Makriyannis, A., Bergman, J., & Kangas, B. D. (2020). Antiemetic effects of cannabinoid agonists in nonhuman primates. *The Journal of Pharmacology and Experimental Therapeutics*, *374*, 462–468.

Naef, M., Curatolo, M., Petersen-Felix, S., Arendt-Nielsen, L., Zbinden, A., & Brenneisen, R. (2003). The analgesic effect of oral delta-9-tetrahydrocannabinol (THC), morphine, and a THC-morphine combination in healthy subjects under experimental pain conditions. *Pain*, *105*, 79–88.

van de Donk, T., Neisters, M., Kowal, M. A., Olofsen, E., Dahan, A., & van Velzen, M. (2019). An experimental randomized study on the analgesic effects of pharmaceutical-grade cannabis in chronic pain patients with fibromyalgia. *Pain*, *160*, 860–869.

Sharpe, L., Sinclair, J., Kramer, A., de Manincor, M., & Sarris, J. (2020). Cannabis, a cause for anxiety? A critical appraisal of the anxiogenic and anxiolytic properties. *Journal of Translational Medicine*, *18*, 374.

Zagzoog, A., Mohamed, K. A., Kim, H. J. J., Kim, E. D., Frank, C. S., Black, T., Jadhav, P. D., Holbrook, L. A., & Laprairie, R. B. (2020). In vitro and in vivo pharmacological activity of minor cannabinoids isolated from Cannabis sativa. *Scientific Reports*, *10*, 20405.

Shang, V. C. M., Kendall, D. A., & Roberts, R. E. (2016). Δ^9-Tetrahydrocannabinol reverses TNFα-induced increase in airway epithelial cell permeability through CB_2 receptors. *Biochemical Pharmacology*, *120*, 63–71.

Tourino, C., Zimmer, A., & Valverde, O. (2010). THC prevents MDMA neurotoxicity in mice. *PLoS One*, *5*(2), e9143.

Scuderi, C., Filippis, D. D., Luvone, T., Blasio, A., Steardo, A., & Esposito, G. (2009). Cannabidiol in medicine: A review of its therapeutic potential in CNS disorders. *Phytotherapy Research*, *23*, 597–602.

Millar, S. A., Stone, N. L., Yates, A. S., & O'Sullivan, S. E. (2018). A systematic review on the pharmacokinetics of cannabidiol in humans. *Frontiers in Pharmacology*, *9*, 1365.

Zlebnik, N. E., & Cheer, J. F. (2016). Beyond the CB1 receptor: Is cannabidiol the answer for disorders of motivation? *Annual Review of Neuroscience*, *39*, 1–17.

Pellati, F., Borgonetti, V., Brighenti, V., Biagi, M., Benvenuti, S., & Corsi, L. (2018). Cannabis sativa L. and nonpsychoactive cannabinoids: Their chemistry and role against oxidative stress, inflammation, and cancer. *BioMed Research International*, 1–15.

Pazos, M. R., Mohammad, N., Lafuente, H., Santos, M., Martinez-Pinilla, E., Moreno, E., Valdizan, E., Romero, J., Pazos, A., Franco, R., Hillard, C. J., Alvarez, F. J., & Martinez-Orgado, J. (2013). Mechanisms of cannabidiol neuroprotection in hypoxic–ischemic newborn pigs: Role of 5HT1A and CB2 receptors. *Neuropharmacology*, *71*, 282–291.

Resstel, L. B. M., Tavares, R. F., Lisboa, S. F. S., Joca, S. R. L., Corrêa, F. M. A., & Guimarães, F. S. (2009). 5-HT 1A receptors are involved in the cannabidiol-induced attenuation of behavioural and cardiovascular responses to acute restraint stress in rats. *British Journal of Pharmacology*, *156*(1), 181–188.

Rahimi, A., Hajizadeh Moghaddam, A., & Roohbakhsh, A. (2015). Central administration of GPR55 receptor agonist and antagonist modulates anxiety-related behaviors in rats. *Fundamental & Clinical Pharmacology*, *29*(2), 185–190.

Terzian, A. L. B., Aguiar, D. C., Guimarães, F. S., & Moreira, F. A. (2009). Modulation of anxiety-like behaviour by transient receptor potential vanilloid type 1 (TRPV1) channels located in the dorsolateral periaqueductal gray. *European Neuropsychopharmacology*, *19*(3), 188–195.

Rosenthaler, S., Pöhn, B., Kolmanz, C., Huu, C. N., Krewenka, C., Huber, A., Kranner, B., Rausch, W.-D., & Moldzio, R. (2014). Differences in receptor binding affinity of several phytocannabinoids do not explain their effects on neural cell cultures. *Neurotoxicology and Teratology*, *46*, 49–56.

Pitakbut, T., Nguyen, G.-N., & Kayser, O. (2021). Activity of THC, CBD, and CBN on human ACE2 and SARS-CoV1/2 main protease to understand antiviral defense mechanism. *Planta Medica.*

Yamaori, S., Okushima, Y., Masuda, K., Kushihara, M., Katsu, T., Narimatsu, S., Yamamoto, I., & Watanabe, K. (2013). Structural requirements for potent direct inhibition of human cytochrome P450 1A1 by cannabidiol: Role of pentylresorcinol moiety. *Biological & Pharmaceutical Bulletin*, *36*(7), 1197–1203.

Monroe, A. Z., Gordon, W. H., Wood, J. S., Martin, G. E., Morgan, J. B., & Williamson, R. T. (2021). Structural revision of a Wnt/b-catenin modulator and confirmation of cannabielsoin constitution and configuration. *Chemical Communications*, *57*, 5658–5661.

Tinto, F., Villano, R., Kostrzewa, M., Ligresti, A., Straker, H., & Manzo, E. (2020). Synthesis of the major mammalian metabolites of THCV. *Journal of Natural Products*, *83*, 2060–2065.

Vollner, L., Bieniek, D., & Korte, F. (1969). Hashish. XX. Cannabidivarin, a new hashish constituent. *Tetrahedron Letters*, *3*, 145–147.

Newmeyer, M. N., Swortwood, M. J., Barnes, A. J., Abulseoud, O. A., Scheidweiler, K. B., & Huestis, M. A. (2016). Free and glucuronide whole blood cannabinoids' pharmacokinetics after controlled smoked, vaporized, and oral cannabis administration in frequent and occasional cannabis users: Identification of recent cannabis intake. *Clinical Chemistry*, *62*(12), 1579–1592.

Olah, A., Markovics, A., Szabo-Papp, J., Timea, S. P., Stott, C., Zouboulis, C. C., & Biro, T. (2016). Differential effectiveness of selected non-psychotropic phytocannabinoids on human sebocyte functions implicates their introduction in dry/seborrhoeic skin and acne treatment. *Experimental Dermatology*, *25*, 701–707.

Rock, E. M., Sticht, M. A., Duncan, M., Stott, C., & Parker, L. A. (2013). Evaluation of the potential of the phytocannabinoids, cannabidivarin (CBDV) and Δ^9-tetrahydrocannabivarin (THCV), to produce CB1 receptor inverse agonism symptoms of nausea in rats. *British Journal of Pharmacology*, *170*, 671–678.

Navarro, G., Varani, K., Lillo, A., Vincenzi, F., Rivas-Santisteban, R., Raïch, I., Reeyes-Resina, I., Ferreiro-Vera, C., Andrea, B. P., Sánchez de Medina, V., Nadal, X., & Franco, R. (2020). Pharmacological data of cannabidiol-and cannabigerol-type phytocannabinoids acting on cannabinoid CB1, CB2 and CB1/CB2 heteromer receptors. *Pharmacological Research*, *159*, 104940.

Pretzsch, C. M., Voinescu, B., Lythgoe, D., Horder, J., Mendez, M. A., Wichers, R., Ajram, L., Ivin, G., Heasman, M., Edden, R. A. E., Williams, S., Murphy, D. G. M., Daly, E., & McAlonan, G. M. (2019). Effects of cannabidivarin (CBDV) on brain excitation and inhibition systems in adults with and without autism spectrum disorder (ASD): A single dose trial during magnetic resonance spectroscopy. *Translational Psychiatry*, *9*, 313.

Russo, C., Ferk, F., Mišík, M., Ropek, N., Nersesyan, A., Mejri, D., Holzmann, K., Lavorgna, M., Isidori, M., & Knasmüller, S. (2019). Low doses of widely consumed cannabinoids (cannabidiol and cannabidivarin) cause DNA damage and chromosomal aberrations in human-derived cells. *Archives of Toxicology*, *93*, 179–188.

Chapter 15

Cannabidiol (CBD): Abuse and physical dependence potential

Kenneth William Sommerville[a] and Beatrice Setnik[b]
[a]Sommerville LLC, Hampstead, NC, United States, [b]Cabin Creek Ln, Raleigh, NC, United States

Abbreviations

AE	adverse event
CB_1	cannabinoid-1-receptor
CB_2	cannabinoid-2-receptor
CBD	cannabidiol
CNS	central nervous system
CPP	conditioned place preference
CSA	Controlled Substances Act
C-SSRS	Columbia-Suicide Severity Rating Scale
CWS	Cannabinoid Withdrawal Syndrome or Scale
DEA	Drug Enforcement Agency
DSM-5	Diagnostic and Statistical Manual of Mental Disorders, Fifth Edition
E_{max}	maximum (peak) effect
ESS	Epworth Sleepiness Scale
FDA	Food and Drug Administration
GABA	gamma amino butyric acid
HAM-D	Hamilton Depression Scale
HAP	human abuse potential
iv	intravenous
NRS	Numeric Rating Scale
PWC-20	Penn Physician Withdrawal Checklist
SD	standard deviation
THC	tetrahydrocannabinol
VAS	Visual Analog Scale

Introduction

Cannabidiol (CBD) is a cannabinoid that has been developed as a plant-derived pharmaceutical-grade medication and labeled as Epidiolex in USA and Epidyolex in Europe. It was approved by the FDA in 2018 for the epileptic syndromes of Dravet and Lennox-Gastaut, for ages one and above for the control of seizures associated with those conditions at doses of 10 and 20 mg/kg/day (Epidiolex Product label, 2021). It was also approved in 2020 for the treatment of seizures associated with the syndrome of Tuberous Sclerosis, for ages one and above at a dose of 25 mg/kg/day (Epidiolex Product label, 2021).

Doses of CBD much lower than approved in the epilepsy syndromes have been used as food supplements for a variety of conditions without FDA approval (Bon-Miller et al., 2017). Epidiolex was developed for epilepsy originally as a Schedule 1 entity, which required extensive special precautions and monitoring for abuse, overdose, and misuse of the compound in the clinical centers participating in the trials for epilepsy. The clinical development program did not report evidence of any abuse, misuse, or overdose. Epidiolex was originally approved as Schedule 5, the least regulated category for controlled substances in the United States. This was based on an abuse liability trial that is reviewed in this chapter (See below in Abuse Potential section). It was subsequently descheduled in 2020 in the Farm Bill that deregulated hemp, which included CBD and prohibited the presence of tetrahydrocannabinol (THC) at amounts greater than 0.3% (U.S. Department of

Cannabis Use, Neurobiology, Psychology, and Treatment. https://doi.org/10.1016/B978-0-323-89862-1.00015-5

Agriculture, 2018). A controlled study of withdrawal was conducted prior to de-scheduling and is reviewed later in this chapter. Despite the known abuse potential of THC, CBD has consistently shown a lack of abuse potential and any association with the Cannabis Withdrawal Syndrome (CWS). These aspects of CBD are the focus of this chapter. CBD in this chapter refers to the formulation of Epidiolex that is FDA-approved for epilepsy syndromes including Dravet, Lennox-Gastaut, and Tuberous Sclerosis. References to other CBD formulations will have a specific notation.

Due to its origins from the cannabis plant and its ability to cross the blood brain barrier, CBD was evaluated for its abuse and dependence potential. In the United States, drugs that have the potential for abuse are typically scheduled and controlled as per the Controlled Substances Act (CSA) of 1978. Scheduled drugs are strictly controlled in terms of the requirements for drug distribution, storage, accountability, and prescribing practices. Scheduling determinations are based on whether a drug has approved medical use, its potential to be abused, and its potential for physical dependency. In this context, physical dependency is a physiological state of adaptation that occurs in response to chronic administration of a drug. When a drug is abruptly discontinued, symptoms of drug withdrawal may occur. Scheduling is tiered from Schedule I, which includes drugs with abuse potential that have no approved medical use, to Schedules II–V, which include drugs with approved medical use in the United States that have varying degrees of relative abuse and dependence potential risk. Medically approved drugs with the highest risk of abuse and dependence potential are ranked as Schedule II and those with lesser risk fall into Schedules III-V. Drugs with no/negligible risk of abuse potential are generally not scheduled under the CSA.

Several agencies in the United States, including the Food and Drug Administration (FDA) and Drug Enforcement Agency, are pivotal to reviewing and determining the schedule status for new drugs approved for medical use. In the determination of scheduling, specific preclinical and clinical studies are required to inform the abuse potential of a new drug in development. In vitro testing determines the receptor-binding profile of a drug and identifies if a drug may be binding to receptors that are commonly associated with abuse potential (e.g., dopamine, opioid, GABA, and cannabinoid receptors). The chemical structure of the drug is also examined to determine if it is a precursor or can be converted to a drug with known abuse potential. In animals, general behavioral studies (e.g., Irwin test, motor performance test, and Tetrad test) are often conducted to determine whether a drug produces abuse-related signals such as hyperactivity indicative of stimulation or sedative effects. Preclinical abuse-related studies include those that evaluate the rewarding or reinforcing properties of a drug (self-administration and conditioned place preference studies) or the drug discrimination study, which evaluates whether a drug has similar effects to a drug with known abuse potential.

Physical dependency, or the ability of a drug to produce withdrawal symptoms when abruptly discontinued, is evaluated in preclinical physical dependency studies. In the clinical setting, the human abuse potential study is used to determine the positive and reinforcing effects of a drug in comparison to relevant drugs with known abuse potential in a face-valid population of recreational drug users who are not physically dependent on drugs. The data collected throughout the drug development program, including the adverse events collected in clinical studies, are analyzed for abuse and dependence potential. Evaluation of physical dependency is also evaluated in the clinical setting by evaluating signs and symptoms of withdrawal following study drug discontinuation in clinical trials where drug is administered for at least 28 consecutive days. During the review of a new drug application, the Controlled Substances Staff at the FDA uses a medical and scientific analysis of the available in vitro, preclinical, clinical, and post-marketing data to assess eight factors relating to the degree and nature of the risks of abuse and physical dependence (Food and Drug Administration., 2017).

Prior to the approval of Epiodolex for medical use, cannabidiol was a schedule 1 entity that required extensive special precautions and monitoring for abuse, overdose, and misuse of the compound in the clinical centers participating in the trials for epilepsy. The preclinical and clinical development program did not find evidence of any abuse, misuse, or overdose and cannabidiol was initially approved as Schedule 5, the lightest category for drug abuse. This was based on several studies, including the HAP study (Schoedel, Stockis, & Sellers, 2018) that will be later reviewed in this chapter. It was subsequently de-scheduled in July 2020 (Epidiolex Product label, 2021). A study of withdrawal was conducted prior to de-scheduling and will be reviewed later in this chapter. Despite the known abuse potential of THC, CBD has consistently shown a lack of abuse potential and any association with the CWS. These aspects of CBD are the focus of this chapter.

Preclinical abuse potential evaluation of CBD

The abuse potential of CBD has been evaluated with various nonclinical studies. In receptor-binding studies, CBD did not demonstrate affinity to any receptor sites currently associated with abuse potential, including cannabinoid CB_1 and CB_2 sites. In tests of general behavior, CBD produced some slight signals of CNS activity; however, such behavioral changes were transient and seen only at higher doses (e.g., 100 mg/kg showed reduced locomotor activity) (CDER, 2018).

CBD was evaluated in the Tetrad Test (Study #GWOR1211 and GWPP1370); a series of four behavioral tests (ambulation and rearing, immobility, hypothermia, and antinociception) that are known to produce positive results in rodents when receiving a drug that has cannabinoid agonist activity. Mice were evaluated for ambulation and rearing (activity meter test), immobility (bar test), hypothermia (rectal temperature), and antinociception (hot plate test), following ip administration of THC, CBD, THC:CBD (1:1 mixture), or one of three CBD metabolites (6-OHCBD, 7-OH-CBD, and 11-OH-CBD) at doses of 1, 10, 50, and 100 mg/kg. At CBD doses of 1, 10, and 50 mg/kg, no effects on behavior were observed. At 100 mg/kg, there was a decrease compared to vehicle in the number of rearings (but not the number of crossings) and decreased rectal temperature. However, the latency to remove forelimbs in the bar test or the foot-licking latency in the hot plate test were not affected (CDER, 2018).

The CBD metabolite, 6-OH-CBD, did not produce any changes in the four behavioral tests at 1, 10, 50, and 100 mg/kg. Another CBD metabolite, 7-OH-CBD, did not significantly affect the number of crossings and rearings at 1, 10, 50, and 100 mg/kg; however, significant increases in the latency to remove forelimbs and decreased the rectal temperature were seen at 50 and 100 mg/kg. No effects were observed on the foot-licking latency. This study suggested that CBD and two of its metabolites (7-OH THC and 11-OH THC) have slight cannabinoid effects at very high doses (CDER, 2018).

Drug discrimination studies determine whether a test drug produces physical and behavioral responses that are similar to a training drug (active control). When the control drug has known abuse potential, this model serves as a relevant method to determine if a test drug will have similar abuse potential. In this model, animals learn to press one bar when they receive the training drug (active control) and another bar when they receive placebo. Once responding is stable, a challenge session with the test drug is administered. A test drug displays "full generalization" to the training drug when the test drug produces bar pressing >75% on the bar associated with the training drug. In a drug discrimination study with oral CBD (Study #GWTX1554), rats were trained to discriminate THC (3 mg/kg i.p.) from vehicle. CBD did not produce a mean generalization of ≥80% to THC at any of the doses of CBD tested (20, 75, and 150 mg/kg) (CDER, 2018). In a study with both CBD and its three metabolites (tetrahydrocannabivarin, cannabidivarin, and cannabigerol), neither CBD nor its metabolites (1, 3, and 10 mg/kg) generalized to THC. In drug discrimination studies in rats, intravenous administration of CBD did not produce full generalization to either the THC interoceptive cue or the midazolam interoceptive cue. This showed CBD did not produce sensations similar to a cannabinoid or a depressant (CDER, 2018). When the training drug was changed to midazolam (0.5 mg/kg) (GW Study #15110), CBD did not produce generalization to the midazolam interoceptive cue at the oral doses tested (20, 75, and 150 mg/kg) (CDER, 2018).

In a study by Vann et al. (2008), CBD and THC were assessed alone and in combination using a 2-lever THC discrimination procedure (Long-Evans rats) and a conditioned place preference/aversion model in ICR mice. This study concluded that CBD alone did not produce effects in either model; however, when administered in combination with THC at therapeutically relevant ratios, aversive effects (e.g. dysphoria) induced by THC alone may be ameliorated (Vann et al., 2008). Complementary findings were observed in another study that evaluated the reinforcing properties of CBD (15, 30, and 60 mg/kg; intraperitoneal) in C57BL/6J mice using the conditioned place preference (CPP) model and an oral CBD self-administration test (50 mg/kg; CBD water soluble 1.2 mg/mL) in comparison to vehicle (water). This study concluded that CBD failed to induce CPP and did not show signs of reinforcing effects (Viudez-Martínez et al., 2019).

Self-administration studies evaluate whether or not a test drug has rewarding properties that produce reinforcement (i.e., repeated self-administration after an initial exposure to a drug). Animals are trained to press a bar in the test cage in order to obtain a food reward using a schedule that typically increases the number of presses needed from 1 to 10. Once established, the animals are trained to perform this task using an intravenous (iv) dose of a control drug with known abuse potential. Animals are also tested with vehicle (placebo) to confirm that bar pressing is not maintained for administration of a substance without abuse potential. Once bar pressing is stabilized for the control drug, they are then allowed to self-administer iv doses of the test drug. If the drug produces a high level of self-administration relative to the vehicle (placebo), this may be indicative of abuse potential (Food and Drug Administration, 2017). A series of self-administration studies were conducted in rats trained with cocaine (0.32 mg/kg/infusion, iv; Study GWTX1551) and heroin (0.015 mg/kg/injection, iv; Study GWTX1663), and in Rhesus monkeys trained with midazolam (0.01 mg/kg infusion in three monkeys and 0.032 mg/kg/infusion in two monkeys; Study #GWT1664). CBD did not produce self-administration in animals trained to self-administer cocaine, midazolam, or heroin (CDER, 2018).

Clinical abuse potential evaluation of CBD

The abuse potential of CBD was evaluated in a randomized, double-blind, placebo, and active controlled, single dose, 7-way crossover study (Schoedel, Szeto, et al., 2018). These types of studies are generally referred to as human abuse potential (HAP) or liability studies. The abuse potential of single oral, ascending does of CBD (750, 1500, and

4500 mg) were compared to placebo and two active controls; alprazolam (2 mg) and dronabinol (10 mg and 30 mg), a synthetic version of delta-9-tetrahydrocannibonol (delta-9-THC). Doses of the active controls were selected from prior HAP studies (Schoedel et al., 2012), whereas the doses of CBD were determined with a previous dose ranging study and confirmed to be well tolerated (Schoedel, Stockis, & Sellers, 2018). Doses of CBD tested included a range of doses from therapeutic to supratherapeutic, as often drugs are abused at much higher doses than the intended therapeutic dose. The investigational drug was a plant-derived pharmaceutical formulation of highly purified CBD (Ataken, 2012).

The study was conducted in 43 healthy, recreational, male and female polydrug users (aged 18–55 years; body mass index between 19.0 and 30.0 kg/m^2; minimum weight 50 kg) who provided informed consent, were not physically dependent on drugs or alcohol and met study inclusion criteria. Prior recreational (nontherapeutic) drug use was a study requirement ($\geq$10 uses each with central nervous system (CNS) depressants and cannabinoids in their lifetime; $\geq$1 CNS depressant and cannabinoid use in the 12 weeks prior to screening; $\geq$1 use of another class of drugs of abuse in their lifetime).

The study consisted of screening, qualification, treatment, and follow-up phases (Fig. 1). The primary endpoint to assess abuse potential was the maximum effect (E_{max}) on Drug-Liking visual analog scale (VAS). Other measurements included E_{max} on Overall Drug-Liking VAS, Take Drug Again VAS, positive and negative effects, other subjective effects, and Drug Similarity VAS. Cognitive and psychomotor functions were assessed using the Divided Attention Test, the Hopkins Verbal Learning Test-Revised, and the Digit-Symbol Substitution Task. Pharmacokinetic parameters were determined for CBD and its major metabolites. Standard safety measures and adverse events were assessed. Following the screening visit, subjects were randomized to a 7-day, double-blind, crossover qualification phase to confirm if they were able to distinguish the subjective effects of the positive control drugs (single doses of 2 mg alprazolam, 20 mg dronabinol) from placebo and were able to tolerate the control drugs. Only eligible subjects were randomized to the Treatment Phase.

Fig. 2 shows the mean Drug-Liking scores for placebo over the 12-h post-dose period for all treatments administered. As expected, scores for placebo within the neutral range (near 50 on the bipolar scale), whereas the highest responses were seen for both positive controls alprazolam and dronabinol. Drug Liking E_{max} was significantly higher for alprazolam and for both dronabinol doses relative to placebo, confirming study validity. Drug Liking E_{max} for CBD (750 mg) was not significantly different than placebo, whereas the higher doses of CBD (1500 and 4500 mg) were significantly greater than placebo (P=0.04 and 0.002, respectively). However, the net differences were less than a 10 point difference on the scale, indicating numerically low scores on the VAS. All three doses of CBD showed significantly lower Drug Liking E_{max} relative to alprazolam and both doses of dronabinol.

Other key measures of Overall Drug Liking and Take Drug Again showed a similar trend with alprazolam and dronabinol doses showing significantly higher E_{max} scores relative to placebo. CBD did not differentiate from placebo for all

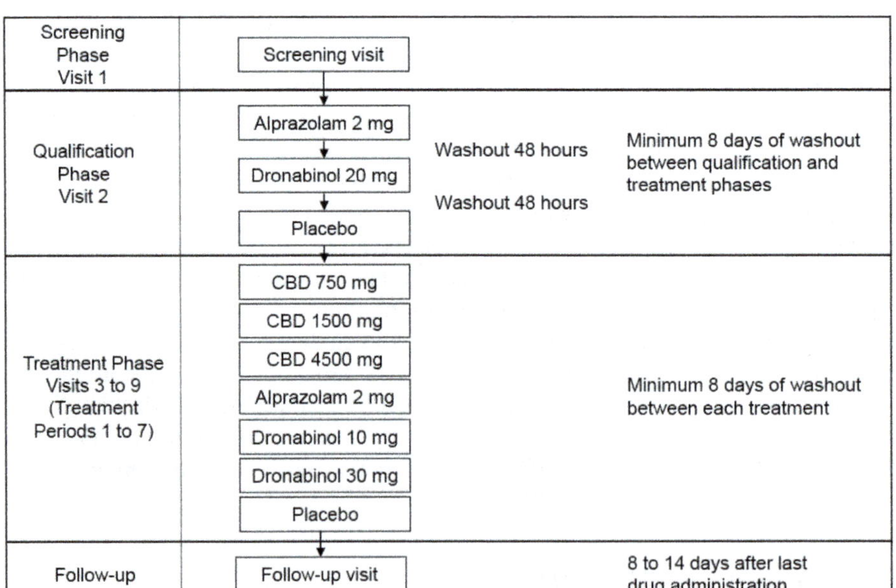

FIG. 1 Study design overview. Treatments were administered in a randomized fashion in both the Qualification and Treatment Phases. SNIP. *From Schoedel, K.A., Szeto, I., Setnik, B., Sellers, E.M., Levy-Cooperman, N., Mills, C., Etges, T., Sommerville, K. (2018). Abuse potential assessment of cannabidiol (CBD) in recreational polydrug users: A randomized, double-blind, controlled study. Epilepsy and Behavior 88, 162–171. with permission.*

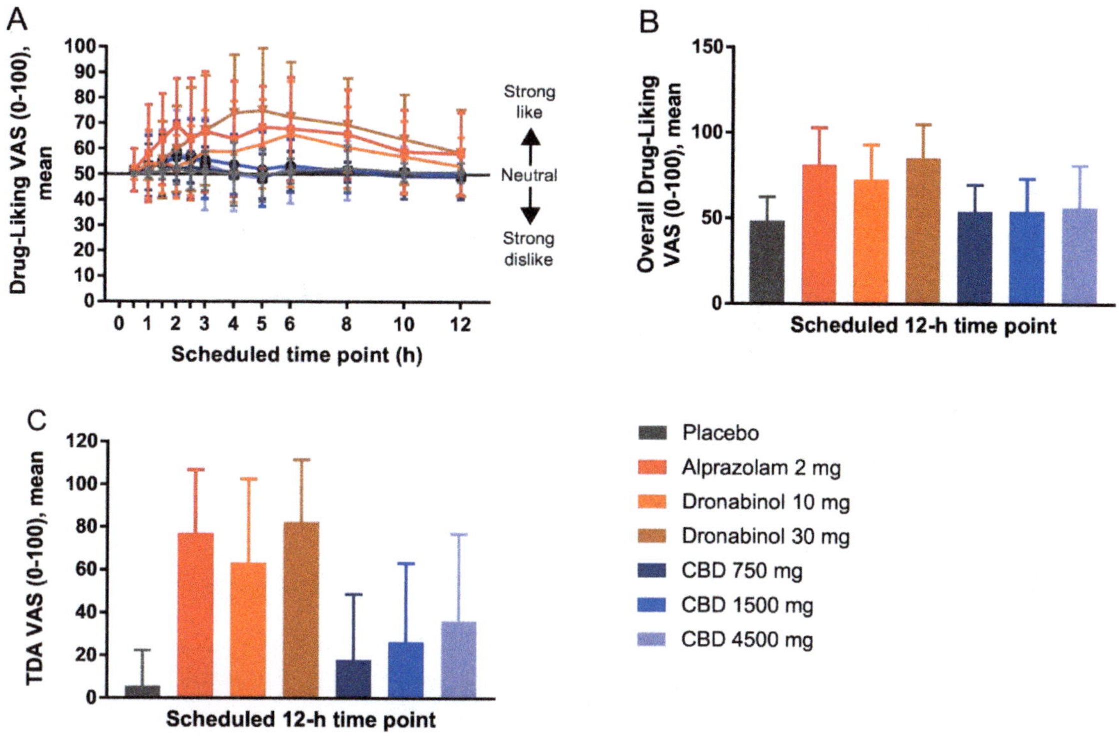

FIG. 2 Drug-Liking, Overall Drug-Liking, and Take Drug Again VAS scores for CBD, alprazolam, dronabinol, and placebo. (A) Time-response profiles of Drug-Liking VAS scores (mean ± standard deviation), (B) Overall Drug-Liking VAS scores (mean±standard deviation), (C) E_{max} values (mean±standard deviation) for Take Drug Again. CBD=cannabidiol; E_{max}=maximum effects; TDA=Take Drug Again; VAS=visual analog scale. Placebo; Alprazolam 2mg; Dronabinol 10mg; Dronabinol 30mg; CBD 750mg; CBD 1500mg; CBD 4500mg. SNIP. *From Schoedel, K.A., Szeto, I., Setnik, B., Sellers, E.M., Levy-Cooperman, N., Mills, C., Etges, T., Sommerville, K. (2018). Abuse potential assessment of cannabidiol (CBD) in recreational polydrug users: A randomized, double-blind, controlled study. Epilepsy and Behavior 88, 162–171. with permission.*

three doses for Overall Drug Liking and showed significantly higher values on Take Drug Again E_{max} only for the higher doses of 1500 and 4500mg. In relation to the positive controls, all three doses of CBD showed significantly decreased E_{max} on both Overall Drug Liking and Take Drug Again. Other positive subjective effects (Good Drug Effects, High, and Stoned VAS) also showed a similar trend to those observed with the above measures. In conclusion, CBD at a therapeutic dose of 750mg showed significantly decreased abuse potential in recreational polydrug users. Although the supratherapeutic doses of 1500 and 4500mg showed some differentiation from placebo, these effects were significantly lower compared to the positive controls alprazolam and dronabinol (Fig. 2).

These findings were supported by additional studies that examined the abuse potential of CBD.

In a study that examined the role of CBD on the reinforcing effects of cannabis, Haney et al conducted a randomized crossover study in nontreatment seeking, healthy cannabis smokers (N=31) (Haney et al., 2016). Study participants were pretreated with single, increasing doses of CBD (0, 200, 400, and 800mg p.o.) followed 90min later by smoked cannabis (0.01% (inactive), 5.30%–5.8% THC). During the smoking session, half of a cannabis cigarette was smoked followed by an ad lib session to examine the reinforcing effects of cannabis. This session occurred 150min later and gave subjects the opportunity to purchase up to three additional 5s puffs of the cannabis sampled earlier at a cost of $0.50 per puff (taken from study earnings). The behavioral and cardiovascular effects of cannabis were measured at baseline and throughout the sessions. Under placebo CBD conditions, self-administration of active cannabis was consumed by significantly more participants than placebo cannabis, and produced significantly higher, time dependent increases in subjective ratings of "High," "Good effects," ratings of the cannabis cigarette (e.g. strength, liking) and heart rate, compared to inactive cannabis. In this study, it was found that cannabis self-administration and subjective effects did not vary as a function of CBD dose relative to placebo CBD and therefore did not suggest that oral CBD reduces the reinforcing, physiological or positive subjective effects of smoked cannabis (Fig. 3).

Adverse events (AEs) collected during the clinical development program of Epidiolex were summarized. In the phase I studies conducted with CBD, none of the individuals reported experiencing "euphoria" related AEs. Higher rates of somnolence were noted in two of the phase I pharmacokinetic studies. In one study, 750 and 1500mg CBD produced

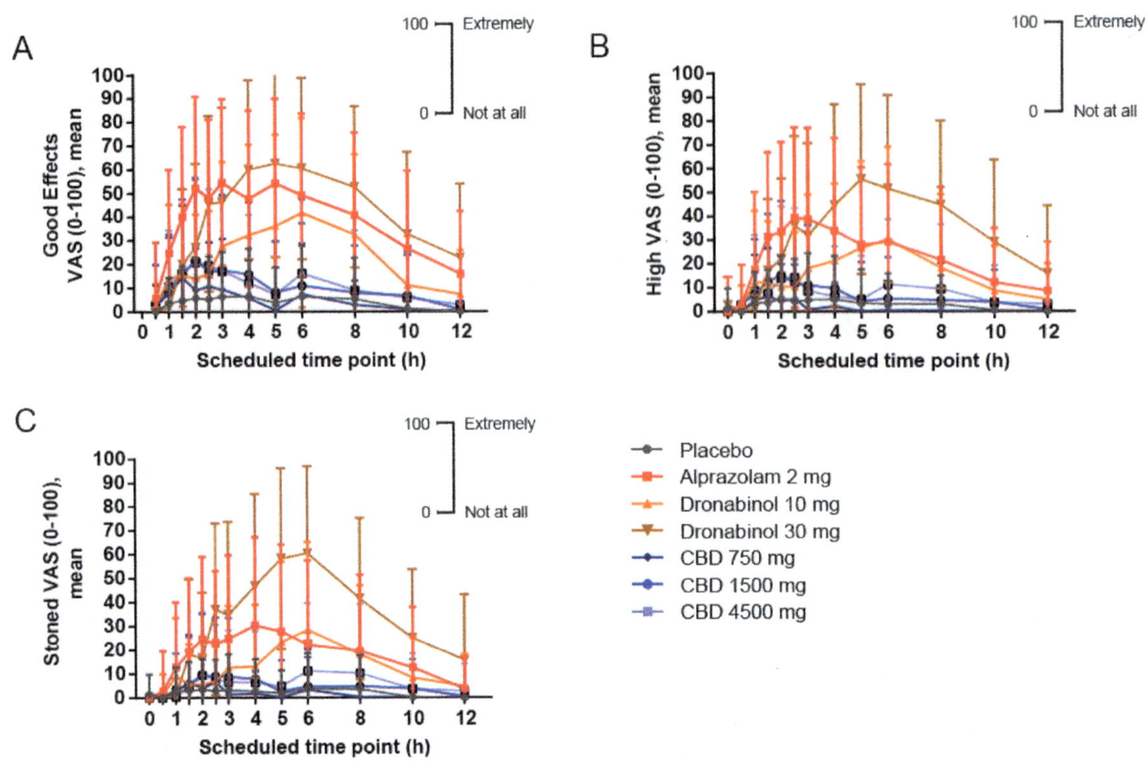

FIG. 3 Time-response profiles for positive effects of CBD, alprazolam, dronabinol, and placebo. (A) Good effects VAS scores (mean±SD), (B) high VAS scores (mean±SD), (C) stoned VAS scores (mean±SD). CBD=cannabidiol; VAS=visual analog scale. Placebo; Alprazolam 2mg; Dronabinol 10mg; Dronabinol 30mg; CBD 750mg; CBD 1500mg; CBD 4500mg. SNIP. *From Schoedel, K.A., Szeto, I., Setnik, B., Sellers, E.M., Levy-Cooperman, N., Mills, C., Etges, T., Sommerville, K. (2018). Abuse potential assessment of cannabidiol (CBD) in recreational polydrug users: A randomized, double-blind, controlled study. Epilepsy and Behavior 88, 162–171. with permission.*

"somnolence" in 2 and 4 of 9 subjects (22–44%), respectively, compared to 2 of 9 subjects (22%) from placebo. In the other study, 750 and 4500mg CBD produced "somnolence" in 5 and 11 of 49 subjects (10–22%), respectively, compared to 4 of 50 subjects (8%) from placebo (CDER, 2018).

In the human abuse potential study in 40 recreational drug users, the incidence of reported "euphoria" was considerably less following receipt of CBD doses of 750, 1500, and 4500mg (5%, 5%, and 8%, respectively) compared to the positive controls alprazolam 2mg (98%), and dronabinol 10 and 30mg (31% and 63%, respectively) (Schoedel, Stockis, & Sellers, 2018). In phase 2/3 studies, abuse-related adverse events related to CBD were not possible to evaluate due to the underlying neurological impairment of patients and confounding effects of other medications (CDER, 2018).

In summary, CBD was not shown to have signals of abuse potential in clinical studies.

Physical dependence potential of CBD

Cannabidiol was developed to treat certain epilepsy syndromes especially associated with children. Originally classified as a Schedule 1 compound during development, it was approved as Schedule 5, the least regulated of scheduled compounds based on the abuse liability study as described earlier in this chapter (Schoedel, Stockis, & Sellers, 2018). As a scheduled compound at the time of approval, there was concern that there might be withdrawal effects despite having little or no abuse potential. There was special concern about use of the compound in children. A specific study to test the potential of physical withdrawal was designed to evaluate this possibility. The likelihood of a withdrawal syndrome was thought to be small since the withdrawal observed with THC is likely to be from an interaction with the CB_1 receptor which CBD lacks (Sim-Selley, 2013).

CBD has been shown to have no interaction with CB_1 receptors (Sim-Selley, 2013) and has even been successfully used as a treatment for cannabis withdrawal (Pokorski et al., 2017). One preclinical study showed CBD attenuated the withdrawal effects with discontinuation of CP-55,940, a potent cannabinoid receptor agonist, in motor signs and behavior in mice (Navarron et al., 2016). It was routine in the clinical trials of epilepsy to withdraw CBD gradually to avoid possible withdrawal seizures for which all anti-epilepsy drugs are labeled, including Epidiolex (Epidiolex Product label). A Cannabis Withdrawal Syndrome (Allsop et al., 2011) has been well described and is listed in the DSM-5 but refers to discontinuation of THC (Livne et al., 2019). Estimates of the incidence of cannabis withdrawal vary but are in the range of 47% among regular users of THC (Bahji et al., 2020). Based on this and other reports of cannabis withdrawal, it was thought to be important to assess possible withdrawal secondary to CBD.

The withdrawal study of CBD was conducted at a research facility in the Netherlands (Taylor et al., 2020). This was an exploratory study that was not formally powered and did not have a hypothesis. The study recruited 30 healthy subjects without concomitant medical conditions including any history of drug abuse or misuse. Further details of the study are in the publication (Taylor et al., 2020) and are summarized here. Part 1 of study design placed all the subjects on a therapeutic dose of CBD (750 mg bid which is approximately 20 mg/kg/day for adults weighing about 75 kg) for epilepsy. The study design was two phases as shown in Fig. 4. In Part 1, subjects were dosed in a single-blind method with CBD for 28 days to allow extended exposure and enough time to enable a degree of physical dependence if it is present. On Day 29, the subjects who were previously randomized during the study baseline in a double-blind manner were placed on either CBD continuing at 750 mg bid or placebo for two additional weeks until Day 42. Subjects were then assessed on Days 31, 35, and 42 as a comparison between CBD and placebo and to the baseline. Subjects were assessed for withdrawal by the Cannabis Withdrawal Scale which was developed specifically to measure symptoms of withdrawal from cannabis. The cannabis withdrawal scale lists nineteen symptoms of cannabis withdrawal measured by two factors. One is the intensity of the symptom in the last 24 h and the other is the impact on the quality of life. Both aspects are measured on a 1–10 Numeric Rating Scale (NRS) with 10 being the highest intensity or interference and 0 not present. The total scores range from 0 to 190 for each of the two items. The DSM-5 lists seven key symptoms of cannabis withdrawal that must be present to diagnose cannabis withdrawal syndrome (CWS). Three symptoms should be present within 3 days of reduced or discontinued use (Bahji et al., 2020). The seven symptoms are (1) irritability, anger, or aggression, (2) nervousness or anxiety, (3) sleep disturbances, (4) appetite or weight disturbance, (5) restlessness, (6) depressed mood, and (7) somatic disturbances (headaches, sweating, nausea, vomiting, and abdominal pain). The Penn Physician Withdrawal Checklist-20 (PWC-20) (Rickels et al., 2008) was developed to measure anxiety symptoms from withdrawal of benzodiazepines and was included since anxiety is a key symptom of cannabinoid withdrawal. The total score is in the range of 0–60. It was also administered on Days-1, 21, 28, 31, 35, and 42. The testing on those days enabled sampling at baseline Day 1 and then over 2 weeks following the abrupt discontinuation on Day 29 which served as the baseline to compare changes on the CWS and PWC-20.

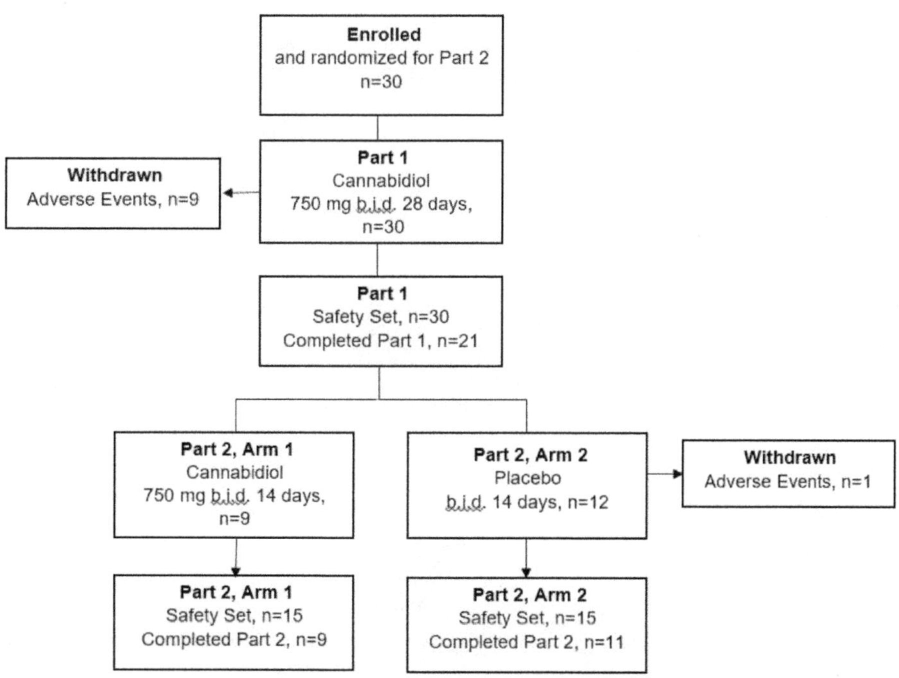

FIG. 4 The figure shows the design of the CBD Withdrawal Trial. Patients were all given CBD (Epidiolex) for 4 weeks at a therapeutic dose before abrupt discontinuation on Day 28 when the groups were randomized to continue CBD or to abruptly stop and were followed for a further 2 weeks to Day 42 to compare whether symptoms or signs of Cannabinoid withdrawal syndrome appeared in the group on placebo. SNIP. *From Taylor, L., Crockett, J., Tayo, B., Checketts, D., Sommerville, K. (2020). Abrupt withdrawal of cannabidiol (CBD): A randomized trial. Epilepsy and Behavior 104(Part A). with permission.*

Sleep disturbances are also a key symptom of cannabinoid withdrawal and were assessed with the Epworth Sleepiness Scale (ESS) (Johns, 1991), The Hamilton Depression Scale (Hamilton, 1960), and the Columbia Suicide Rating Scale (C-SSRS) (Posner et al., 2007) assessed symptoms of possible depression and suicidality. CBD concentrations were also checked on Days 1 (predose), 7, 14, 21, 28, and 35. In addition, electrocardiograms, routine labs, physical examinations, and AEs were checked throughout the study (Taylor et al., 2020).

The withdrawal study recruited healthy adult volunteers to avoid any issues with seizures that could occur with abrupt discontinuation and drug users, abusers, and subjects with a history of any drug abuse or dependence who could have confounding symptoms during withdrawal (Taylor et al., 2020). The volunteers (n=30) were 90% white at ages with a mean of 25.3 (SD=6.7) and body mass index with a mean of 23.8 (2.2). Subjects took CBD at therapeutic doses for epilepsy for four weeks so any possible physical dependence could develop in a clinically relevant setting. This was considered adequate as CBD has a half-life of approximately 60h (Taylor et al., 2020). An adequate test group would need to abruptly withdraw CBD while a control group would continue in parallel at the same dose. This allowed the adverse event profile after abrupt withdrawal to be compared to a group continuing at the chosen therapeutic dose for epilepsy to distinguish adverse events of withdrawal from those with continuing CBD therapy. Fig. 5 confirms that the group on placebo had a reduction in CBD levels while the control group continued to show adequate CBD levels. This confirms that CBD was withdrawn in the placebo group.

The results showed no evidence of physical or mental withdrawal. The change in CWS and PWS-20 scores are shown in Fig. 6A–C. There is a large overlap of the standard deviations across the two groups of the CWS and the total PWC-20 score following the abrupt discontinuation of CBD. Day 28 was the baseline when CBD was abruptly withdrawn from the placebo group. Scores on Day 28 were compared to Days 31, 35, and 42. A negative score on these days was indicative of improvement. The change in CWS score for severity of symptoms in the last 24h in Part 2 showed small negative changes in both groups. The negative scores on daily activities in the CWS and in the change from baseline in the PWC-20 were consistent in showing little difference between the groups.

All other scales were also unremarkable and similar between the groups. The ESS showed no disruption of sleep. The HAM-D scores greater than 7 which is needed to diagnose depression. The C-SRSS had zero scores for all questions at all timepoints meaning there was no evidence of suicidality.

The adverse events listed in the CWS as required to diagnose physical withdrawal did not occur during the second phase of the study where CBD was compared to placebo except for nightmares reported from a single subject in the group continuing CBD. The reported adverse events showed no evidence of irritability, restlessness, sleep disorders, anxiety, depression, suicidality, weight changes or other somatic disturbances. Headaches were common with a greater number 7/12 (58.3%) with placebo compared to 2/9 (22.2%) of those remaining on Epidiolex in Part 2 of the study. Although headaches are a symptom of cannabis withdrawal, there was a 15/30 (50%) incidence in Part 1 when there was no reduction in

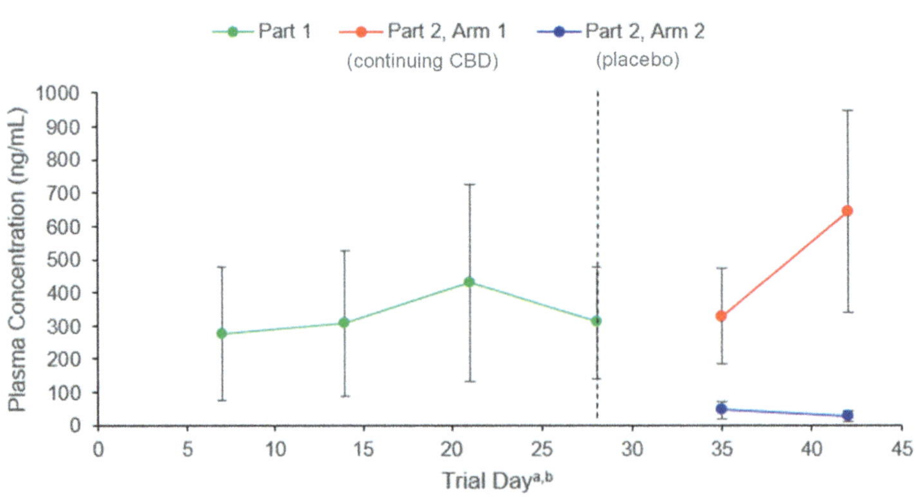

FIG. 5 The figure shows the concentrations of CBD during the trial. The *green* line is from Part 1 of the study when all volunteers were receiving CBD. Part 2 shows the group continuing CBD in *red* and the placebo group in *blue*. Note the very low concentrations in the placebo group as expected after abrupt discontinuation on Day 28, confirming this trial compared ongoing CBD compared to the discontinued group on placebo. Part 1 (*green*); Part 2, Arm 1 (continuing CBD) (*red*); Part 2, Arm 2 (placebo) (*blue*). SNIP. *From Taylor, L., Crockett, J., Tayo, B., Checketts, D., Sommerville, K. (2020). Abrupt withdrawal of cannabidiol (CBD): A randomized trial. Epilepsy and Behavior 104(Part A). with permission.*

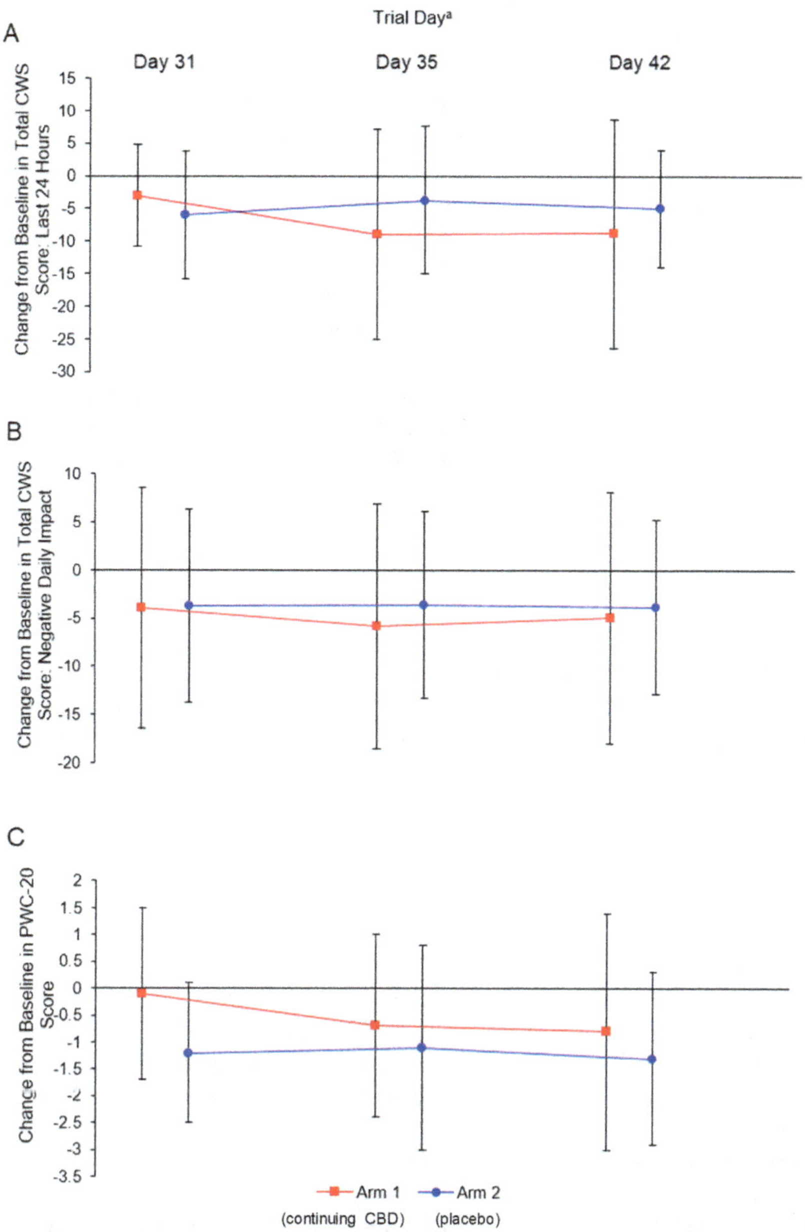

FIG. 6 (A–C) All figures show scores after abrupt CBD withdrawal on Day 28. There is a large overlap of the standard deviations showing no significant differences between the withdrawal group on placebo and those continuing on CBD. There is no evidence of physical withdrawal in the cannabis withdrawal scale or the PWC-20 assessments. A negative score indicates improvement on all figures. Figure A shows the overall score for the impact of 19 symptoms in the last 24 h. Figure B shows the negative daily impact in the 19 symptoms. Figure C shows the change in scores in the PWC-20 assessment. Arm1 (continuing CBD) (*red*); Arm 2 (placebo) (*blue*). SNIP. *From Taylor, L., Crockett, J., Tayo, B., Checketts, D., Sommerville, K. (2020). Abrupt withdrawal of cannabidiol (CBD): A randomized trial. Epilepsy and Behavior 104(Part A). with permission.*

the dose of Epidiolex. No subject had 3 or more of the key symptoms listed in the CWS to fulfill the criteria for cannabis withdrawal in DSM-5.

There were no serious adverse events during the trial. There were 10 discontinuations for adverse events including 7 rashes, one increased transaminases, one increased transaminases with eosinophilia, and one muscle injury unrelated to the study. The rashes were all generalized and had no systemic symptoms and resolved after discontinuation of CBD and were assessed as related to CBD. Nine (9) discontinued in Part 1 and one discontinued on the first day of Part 2 for adverse events starting in Part 1. All of these events had onset during the first part of the study and were unrelated to withdrawal. All other routine labs and safety evaluations were unremarkable except for the rashes noted on physical examinations. It was concluded that the study showed no evidence of a physical withdrawal syndrome.

This study demonstrated class 2 evidence that there is no withdrawal syndrome associated with the abrupt discontinuation of CBD in the formulation of Epidiolex. The lack of interaction of CBD with the CB_1 receptor, the preclinical data suggesting CBD may attenuate withdrawal in animal models, the lack of clear evidence of abuse liability, and the otherwise negative results of the controlled withdrawal study all suggest CBD has minimal or no abuse potential and no evidence of a withdrawal syndrome. The strengths of the study were that it was placebo -controlled at adequate doses to be used in epilepsy with abrupt and confirmed discontinuation of CBD with extended assessment for signs of withdrawal over two weeks. Limitations included a small sample size and 10/30 discontinuations prior to withdrawal. There is also the possibility that higher doses taken over a longer period might lead to a withdrawal syndrome although the long half-life of CBD (range of 60h) allows a gradual reduction of serum concentration over days as was noted in this study (Fig. 2).

In conclusion, CBD showed little or no evidence of abuse potential in preclinical models or in human abuse liability studies. A controlled study of volunteers abruptly withdrawing from CBD had no signs of physical addiction or tendency stay on doses therapeutic in some forms of epilepsy. Limited data suggest CBD may even be a treatment for cannabis withdrawal. Taken together, the preclinical and clinical data show CBD has minimal or no abuse potential and no tendency to physical addiction.

Applications to other areas

The immediate application of CBD to other areas can now be done without concern about physical dependence or withdrawal. CBD is not a controlled substance and may be studied for other potential medical uses including relief in spasticity associated with multiple sclerosis for which there is approval in several countries outside of the United States. There is also possible use for various types of pain including chronic syndromes that would require dosing over long periods. This is also the case with use in Parkinson's disease, also a chronic condition. There have been suggestions that CBD may also actually treat withdrawal symptoms from other forms of cannabis. CBD may be tested for any promising condition without concern that it could lead to dependence or be associated with withdrawal after prolonged use.

Mini-dictionary of terms

- **Abuse Potential**. The likelihood that a drug may be abused (see Drug Abuse).
- **Cannabis Withdrawal Scale (CWS)**. This is a diagnostic tool for clinical assessment of whether a patient may have or have had withdrawal from any substance in the cannabis family. It has 19 symptoms that a patient must assess as a form that asks the patient to rate each symptom on a 0–10 NRS for intensity and interference with daily life in the last 24h. Each item is given points 0–190.
- **Columbia Suicide Rating Scale (C-SRSS)**. This is an assessment of suicidality and is used extensively by the FDA to assess the possibility of suicidal symptoms.
- **Dravet Syndrome**. A genetic condition manifested as multiple refractory seizures and delayed cognitive development, usually starting in the first year of life. The seizures may be multiple types and highly resistant to treatment and provoked by fever, infection, or increased temperature. The seizures typically worsen in early childhood, and there is marked intellectual impairment. Most antiseizure medicines are of limited value and those with sodium-blocking properties may make the seizures worse. Many patients survive into adulthood but are seriously disabled.
- **Drug Abuse**. The intentional, nontherapeutic use of a drug product or substance for the purposes of obtaining a desired psychological or physiological effect. Most commonly abused drugs are CNS-active (e.g., opioids, stimulants, marijuana, hallucinogens, and sedative/hypnotics).
- **Epworth Sleepiness Scale (ESS)**. This scale estimates the degree of somnolence experienced by a patient by assigning a rating determined by the patient.
- **Hamilton Depression Inventory (HAM-D)**. This is a standard clinical assessment of depression assigning numbers to a scale assessing depressive symptoms.
- **Lennox-Gastaut Syndrome**. This a severe condition of epilepsy that starts at an early age or infancy and manifested as multiple seizure types that evolve from infantile spasms or brief seizures to motor seizures that are refractory to most antiseizure medications. Most afflicted individuals are severely disabled and intellectually impaired although most survive into adulthood. The seizures typically may be major bodily shaking or are manifested as drop attacks where

the person suddenly loses consciousness and falls, frequently sustaining multiple injuries. Certain surgeries on the brain may bring some relief, but the prognosis is otherwise generally poor.

- **Penn Physician Withdrawal Checklist (PWC-20).** A clinical checklist developed to assess symptoms of anxiety associated with discontinuation of benzodiazepines but also applied to assessment of withdrawal of cannabis.
- **Physical Dependency**. A physiological adaptation that occurs in response to repeated or chronic exposure to a drug such that when a drug is abruptly discontinued or exposure to drug is suddenly reduced, signs and symptoms of drug withdrawal may manifest.
- **Self-Administration**. A model that examines the reinforcing effects of a drug or substance.
- **Tetrad Test**. A series of four behavioral tests (ambulation and rearing, immobility, hypothermia, and antinociception) that are conducted in animal studies.
- **Tuberous sclerosis**. This is an inherited disorder manifested as multiple skin lesions, tumors of the nervous system, and in many cases severe epilepsy with multiple seizure types. The tumors may appear at any time throughout life and may require surgical excision. The seizures occur in the majority of cases. The cause is a genetic abnormality that increases production of proteins in the nervous system. Patients often live to adulthood but have increases risk for impairment or even death from epilepsy and the tumors.

Key facts of CBD

- Cannabidiol (CBD) is a compound of the cannabis family that includes tetrahydrocannabinol (THC) which is known to induce dependence and has a syndrome of physical withdrawal (CWS).
- CBD was developed for syndromes of epilepsy and approved by the FDA for the Dravet, Lennox-Gastaut, and tuberous sclerosis syndromes that occur predominantly in children, raising concerns there may be dependence and/or physical withdrawal with immediate discontinuation.
- A controlled human abuse potential study with polydrug users showed CBD had little or no propensity to nonmedical use (abuse) compared to alprazolam and a synthetic formulation of THC (dronabinol).
- A study of sudden withdrawal of CBD after one month on a therapeutic dose of CBD for epilepsy showed no evidence of a physical withdrawal syndrome among healthy volunteers without a history of drug use.
- CBD was originally listed as a category I compound, meaning there was a high potential for abuse and no known medical benefit. It was approved and listed as category V, the most mild category of a controlled substance. CBD was later completely de-scheduled and is no longer a controlled substance.

Summary points

- CBD is defined as being in the cannabis family where there are other compounds such as THC, which has abuse liability and a physical withdrawal syndrome.
- CBD was in Category I of controlled substances, meaning there was a high degree of potential abuse and no known therapeutic value.
- CBD was extensively studied in controlled trials of three syndromes with epilepsy as a prominent complication of each syndrome, Dravet, Lennox-Gastaut, and tuberous sclerosis.
- To address concerns that CBD may have properties of abuse and a physical withdrawal syndrome, controlled studies of potential abuse and possible physical dependence were performed.
- The human abuse liability study compared CBD to a benzodiazepine (alprazolam) and a synthetic formulation of THC (dronabinol) in a human abuse liability study among nondependent polydrug users. CBD was found to have little or no abuse potential compared to the two controlled substances known to have abuse potential. CBD had a closer profile to placebo that the active controls.
- CBD was approved for the three syndromes of epilepsy and placed in category V, the mildest category for a controlled substance. CBD was subsequently de-scheduled and remains an uncontrolled substance.
- CBD was then studied for the presence of a withdrawal syndrome found in other cannabis compounds in a trial where healthy volunteers were given a therapeutic dose of CBD for one month and then compared half of the volunteers who were discontinued to the other half continued on the same dose of CBD.
- The withdrawal trial showed no evidence of a withdrawal syndrome with CBD.

References

Allsop, D. J., Norberg, M. M., Copeland, J., Fu, S., & Budney, A. J. (2011). The cannabis withdrawal scale development: Patterns and predictors of cannabis withdrawal and distress. *Drug and Alcohol Dependence*, *119*(1-2), 123–129.

Ataken, Z. (2012). Cannabis, a complex plant: Different compounds and different effects on individuals. *Therapeutic Advances in Psychopharmacology*, *2*(6), 241–254.

Bahji, A., Stephenson, C., Tyo, R., Hawken, E. R., & Seitz, D. P. (2020). Prevalence of cannabis withdrawal symptoms among people with regular or dependent use of cannabinoids: A systemic review. *JAMA Network Open*, *3*(4), e202370.

Bon-Miller, M. O., Loflin, M. J. E., & Thomas, B. F. (2017). Labeling accuracy of cannabidiol extracts sold online. *Journal of the American Medical Association*, *318*, 1708–1709.

CDER. (2018). *Other review(s)*. Application number. 210365Orig1s000 https://www.accessdata.fda.gov/drugsatfda_docs/nda/2018/210365Orig1s000 OtherR.pdf.

Epidiolex Product label. (2021). *Greenwich Biosciences, Inc., Carlsbad, CA, USA.*

Food and Drug Administration. (2017). *Assessment of abuse potential of drugs-guidance for industry*. U.S. Department of Health and Human Services. Center for Drug Evaluation and Research.

Hamilton, M. (1960). A rating scale for depression. *Journal of Neurology, Neurosurgery, and Psychiatry*, *231*, 56–62.

Haney, M., Malcolm, R. J., Babalonis, S., Nuzzo, P. A., Cooper, Z. D., Bedi, G., Gray, K. M., McRae-Clark, A., Lofwall, M. R., Sparenborg, S., & Walsh, S. L. (2016). Oral cannabidiol does not alter the subjective, reinforcing or cardiovascular effects of smoked cannabis. *Neuropsychopharmacology*, *41* (8), 1974–1982.

Johns, M. W. (1991). A new method for measuring daytime sleepiness: The Epworth sleepiness scale. *Sleep*, *14*, 540–545.

Livne, O., Shamulewitz, D., Lev-Ran, S., & Hasin, D. (2019). DSM-5 cannabis withdrawal syndrome: Demographic and clinical correlates in US adults. *Drug and Alcohol Dependence*, *195*, 170–177.

Navarron, C. M., Aracil-Fernandez, A., Navarretic, R. F., Bianco-Gandia, M. C., Ding, Y. S., & Manzanares, J. (2016). The administration of cannabidiol reduced the spontaneous withdrawal syndrome induced by CP-55,940. *European Psychopharmacology*, *26*(Suppl. 2), S685–S686.

Pokorski, I., Clement, N., Phung, N., Weltman, M., Fu, S., & Copeland, J. (2017). Cannabidiol in the management of in-patient cannabis withdrawal. A clinical case series. *Future Neurology*, *12*(3).

Posner, K., Oquendo, M., Gould, M., Stanley, B., & Davies, M. (2007). Columbia classification algorithm of suicide assessment (C-CASA): Classification of suicidal events in the FDA's pediatric suicidal risk analysis of antidepressants. *The American Journal of Psychiatry*, *165*, 1035–1043.

Rickels, K., Garcia-Espana, F., Mandos, L. A., & Case, G. W. (2008). *Journal of Clinical Psychopharmacology*, *28*(4), 447–451.

Schoedel, K. A., Addy, C., Chakraborty, B., Rosko, K., Dunbar, S., Maes, A., Chen, N., Stoch, S. A., Wagner, J., Chodakewitz, J., & Sellers, E. M. (2012). Human abuse potential and cognitive effects of taranabant, a cannabinoid 1 receptor inverse agonist: A randomized, double-blind, placebo- and active-controlled, crossover study in recreational polydrug users. *Journal of Clinical Psychopharmacology*, *32*(4), 492–502.

Schoedel, K. A., Stockis, A., & Sellers, E. M. (2018). Human abuse potential of brivaracetam in healthy recreational central nervous system depressant users. *Epilepsy & Behavior*, *78*, 194–201.

Schoedel, K. A., Szeto, I., Setnik, B., Sellers, E. M., Levy-Cooperman, N., Mills, C., Etges, T., & Sommerville, K. (2018). Abuse potential assessment of cannabidiol (CBD) in recreational polydrug users: A randomized, double-blind, controlled study. *Epilepsy and Behavior*, *88*, 162–171.

Sim-Selley, L. J. (2013). Regulation of cannabinoid CB_1 receptors in the central nervous system by chronic cannabinoids. *Critical Reviews in Neurobiology*, *15*(2), 91–119.

Taylor, L., Crockett, J., Tayo, B., Checketts, D., & Sommerville, K. (2020). Abrupt withdrawal of cannabidiol (CBD): A randomized trial. *Epilepsy and Behavior*, *104*(Part A), 106938.

U.S. Department of Agriculture. (2018). https://www.usda.gov/farmbill.

Vann, R. E., Gamage, T. F., Warner, J. A., Marshall, E. M., Taylor, N. L., Martin, B. R., & Wiley, J. L. (2008). Divergent effects of cannabidiol on the discriminative stimulus and place conditioning effects of Delta(9)-tetrahydrocannabinol. *Drug and Alcohol Dependence*, *94*(1-3), 191–198.

Viudez-Martínez, A., García-Gutiérrez, M. S., Medrano-Relinque, J., Navarrón, C. M., Navarrete, F., & Manzanares, J. (2019). Cannabidiol does not display drug abuse potential in mice behavior. *Acta Pharmacologica Sinica*, *40*(3), 358–364.

Chapter 16

4,5-Dihydro-1*H*-pyrazole/3,4-Diarylpyrazoline class of cannabinoid-1 (CB1R) receptor antagonists and their potential in medicinal applications

Malliga R. Iyer

National Institute on Alcohol Abuse and Alcoholism, National Institutes of Health, Rockville, MD, United States

Abbreviations

2-AG	2-arachidonoylglycerol
ADME	absorption, distribution, metabolism, excretion
AEA	*N*-arachidonoylethanolamine
BBB	blood-brain barrier
CB1R	cannabinoid-1 receptor
CB2R	cannabinoid-2 receptor
CNS	central nervous system
DIO	diet-induced obesity
HFD	high-fat diet
PR	peripherally restricted
TNFα	tumor necrosis factor α
TPSA	total or theoretical polar surface area

Introduction

The endocannabinoid system came into being around late last century when the receptors were discovered to determine the binding target of the main psychoactive constituent of hashish/marijuana/*Cannabis sativa*, (−)-delta 9-tetrahydrocannabinol (Δ^9-THC) (Devane et al., 1992; Gaoni & Mechoulam, 1964; Howlett et al., 2002; Rodríguez de Fonseca et al., 2005). The CB1 receptor was cloned and characterized leading to the acceptance of the presence of endogenous cannabinoid system (ECS) in the mammalian system with its highest presence in the brain but also substantial presence in the adipose tissues, liver, GI tract, pancreas, and kidneys (Howlett et al., 2002) (De Petrocellis et al., 2004). Endogenous ligands or endocannabinoids namely anandamide (AEA) and 2-arachidonoylglycerol (2-AG) that activate the CB1 receptor were discovered and determined to be synthesized from arachidonic acid. Soon the presence of an additional receptor CB2 that are activated by these endogenous ligands was discovered along with the enzymes like fatty acid amide hydrolase (FAAH) and monoacylglycerol lipase (MAGL) that degrade the endocannabinoids (Mackie, 2008).

The two main cannabinoid receptors, CB1 and CB2, have now been well characterized with many exogenous agonists and antagonists acting together as well as selectively on these receptors. The CB2 receptor has limited presence in the brain but largely found in the immune system where it plays an important role in modulating inflammatory responses in various physiological and pathological conditions (Mackie, 2008; Pacher et al., 2006). The CB1 receptor, however, has gained considerable focus over the past thirty years due to its role in regulating food intake, behavior, and de novo lipogenesis (Black, 2004; Patel & Pathak, 2007; Pertwee, 2000). Clearly, "marijuana-munchies" and increased appetite associated with

Cannabis Use, Neurobiology, Psychology, and Treatment. https://doi.org/10.1016/B978-0-323-89862-1.00041-6

smoking marijuana showed that the functional interaction of THC in hypothalamic CB1 receptors induces an increase in food intake. Hence, appetitive behavior could be regulated by targeting brain CB1 receptors and blockade or antagonism of them could suppress food intake, paving for a paradigm to treat obesity via development of an antiobesity medication.

Targeting the ECS CB1R with brain-penetrant CB1 blockers

At the turn of this century, a large part of the first decade involved efforts from the pharmaceutical industry to utilize the paradigm of CB1 blockade by developing small-molecule drugs that could be validated in limiting caloric restriction and its associated metabolic benefits. Thus, these efforts resulted in a number of chemical agents antagonists/inverse agonists from diverse scaffolds and different groups to validate the rationale of antagonizing CB1 receptors (Janero & Makriyannis, 2009; Sharma et al., 2014). Despite all the efforts, rimonabant/Acomplia/Zimulti (**1**) was the sole drug to be approved as a first-in-class antiobesity medication in Europe (Després, Golay, Sjöström, & Rimonabant in Obesity-Lipids Study Group, 2005; Patel & Pathak, 2007). However, the rise of rimonabant was short lived as it was withdrawn from the market soon due to the increasing cases of neuropsychiatric incidences like depression, anxiety, and suicidal ideation. These incidents were attributed to blocking CB1 receptors in the brain. Consequently, development of other brain-penetrant antagonists, such as ibipinabant (**2**), taranabant (**3**), and otenabant (**4**), was also halted immediately (Fig. 1).

The abandonment of this approach by Pharma led to a lull period where redux strategies to resurrect this paradigm and generation of safer CB1 antagonists were eagerly sought (Kunos et al., 2009; Ward & Raffa, 2011). The focus of many companies was mainly on brain-penetrant agents as the dogma was that central engagement of CB1R were required for appetite suppression, weight loss, and related metabolic improvements. By beginning of the last decade, however, alternate CB1 blockade-mediated mechanisms without CNS involvement for weight loss and resolution of metabolic abnormalities started appearing (Tam et al., 2010). A large number of studies now provide a strong rationale for continued investment and effort to exploit the therapeutic potential of this paradigm (Chorvat, 2013; Cinar et al., 2020). These revelations largely relied on extricating favorable effects resulting from peripherally limited receptor blockade to achieve safer CB1-antagonist therapies. Detailed studies have now appeared where previous brain-penetrant molecular scaffolds have been modified to brain-impaired entities that allow to harness the yet fully untapped potential of CB1 blockade.

Peripherally-restricted (PR) CB1R blockers

By the end of 2008, literature evidence started appearing where the demarcation between the role of peripheral CB1 mechanism and central CB1 inputs showed distinct results. Kunos and co-workers investigated the role of hepatic versus extra-hepatic CB_1 receptors and metabolic effects of a high-fat diet (HFD) using liver-specific CB1 knockout ($LCB1^{-/-}$) mice. $LCB1^{-/-}$ mice fed a HFD developed a similar degree of obesity as wild-type mice, but like $CB1^{-/-}$ mice, had less steatosis, hyperglycemia, dyslipidemia, and insulin/leptin resistance than did wild-type mice on a HFD (Osei-Hyiaman et al., 2008). These studies indicated that in rodents hepatic CB1 receptors can contribute to the diet-induced steatosis and metabolic consequences independent of the weight gain resulting from HFD thereby providing one of the earliest proof of selectively utilizing peripheral CB1 receptors for the treatment of dyslipidemia, fatty liver, and restoring glucose homeostasis.

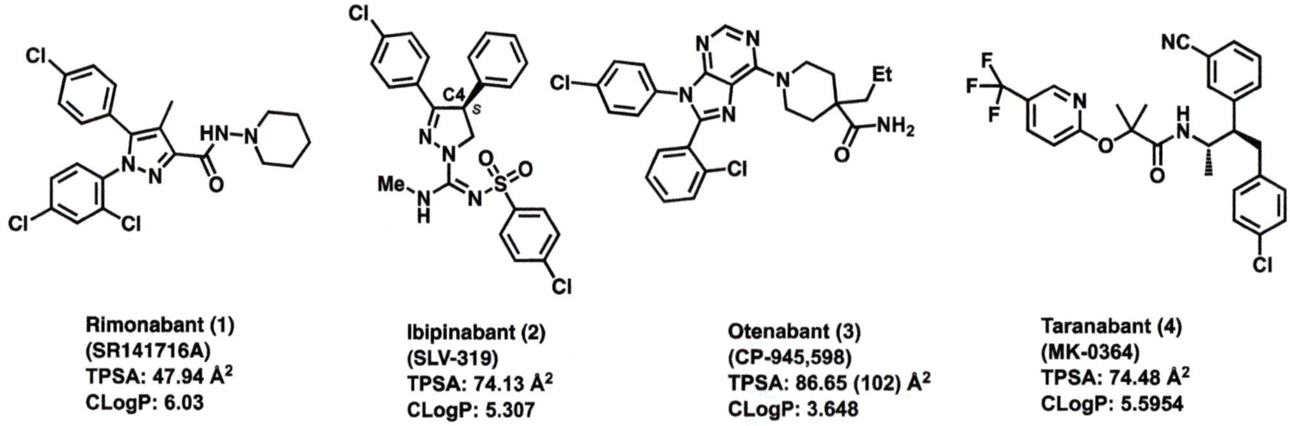

FIG. 1 Select CNS penetrant CB1R antagonist/inverse agonists that were in clinical development.

Results from clinical trials of rimonabant showed significant amelioration of triglycerides and cholesterol and such improvement in cardiometabolic risk profile required a combination of central effects as well as peripheral engagement of CB1 receptors. Additionally, the leptinergic system and its modulation by endocannabinoid system can also act as an ally for peripherally restricted cannabinoid blockers (Di Marzo et al., 2001; Tam et al., 2012). More recently, the ability of peripheral CB1 blockers to act via sympathetic nervous system may also contribute to their ability to induce hypophagic effects via direct, noncentral mechanisms (Bellocchio et al., 2013). As more results started appearing in the literature on the role of peripheral CB1 blockers as useful therapeutics, the field has seen renewed focus on developing various novel approaches to harness the peripheral CB1 blockade paradigm for mitigating the global health concern of diabesity and underlying metabolic comorbidities (Chorvat, 2013; Cinar et al., 2020; Kunos et al., 2009).

Designing PR CB1 blockers

In light of the mounting evidence for efficacy of PR blockers in positively altering metabolic profiles, many chemical agents were designed and tested in rodent models to assess the correlation of metabolic parameter improvement with impaired access in the CNS. From earlier studies, brain-penetrant compounds ibipinabant and otenabant with total polar surface area (TPSA), 78 and 102 $Å^2$, respectively, offered attractive scaffolds and higher propensity to generate nonbrain penetrant compounds based on the physicochemical principles. For molecules to penetrate the BBB in general, the PSA is generally found to be <80 (Ertl et al., 2000; Kelder et al., 1999). Hence, molecules with a PSA between 80 and 120 may combine ideal features for nonbrain penetrant yet orally bioavailable molecules (Lipinski et al., 2001). Otenabant, however, turned out to be an exception where molecular features offered higher PSA yet an intramolecular H-bond facilitated higher than expected brain access resulting in its halted clinical development from similarly probable neuropsychiatric liabilities (Griffith et al., 2009). A large amount of research is now available on modifications of rimonabant core and scaffold hopping strategies (Janero & Makriyannis, 2009; Lange, van Stuivenberg, Coolen, et al., 2005; Srivastava et al., 2007). Similar manipulations on otenabant core to disrupt the intramolecular H-bonding to offer brain-impaired compounds have also been extensively studied (Fulp et al., 2012, 2013). This review will focus on the design strategies utilizing the ibipinant core to deliver second generation and entry into novel, third-generation peripherally restricted safer CB1 antagonist-based therapeutics.

Chemistry of 4,5-1*H*-dihydropyrazole (3,4-diarylpyrazoline) scaffold-based CB1 antagonists

Solvay group

The 3,4-diarylpyrazoline scaffold was utilized by Solvay about two decades ago in their initial foray into developing CB1 antagonists/inverse agonist. One of the well-characterized compound/clinical candidate (Ibipinabant/SLV-319) showed potential in ameliorating CB1R-mediated pathologies (Lange et al., 2004). This compound's development was waylaid due to rimonabant's unfortunate turn of events. One of the earlier reports to modify the lipophilic nature and improve the solubility of CB1 antagonists came from Lange and co-workers at Solvay group where they utilized the 4,5-1*H*-dihydropyrazole scaffold to generate compounds of the type **A** (Table 1AA) (Lange, van Stuivenberg, Veerman, et al., 2005). While there was no explicit mention of peripheral restriction or increase in polar surface areas, this approach to modify the physicochemical properties of CB1 receptor to yield more soluble compounds seemed reasonable in comparison to the parent SLV-319/326 compounds. The *N*-diethyl analog **5** had a CB1-binding affinity of 30 nM as a racemate. The thiomorpholine analog **6** displayed the highest CB1-binding affinity; however, conversion of thio group to its SO_2 congener **7** resulted in considerable lowering of CB1-binding affinity. Another compound in the same series, a piperidinyl anaolg, **8** was studied in somewhat in detail (de Bruin et al., 2011, 2010). The *S*-enantiomer (**8a**) of **8** exhibited a 13-fold higher affinity than *R* enantiomer **8b** for the CB_1 receptor.

In vivo studies on **5**, **9**, **8a** were conducted in the CB1 agonist-induced rat hypotensive model and the hypothermia mouse model. The pyridyl analog **9** was inactive in the hypotension test, which was due not only because of its low CB1 functional potency but it was also identified as a P-glycoprotein substrate in vitro, which reduces CNS penetrability. Orally administered **5** showed CB1-antagonism-mediated effects in vivo in the (CNS-mediated) hypothermia mouse model, which was in line with its in vitro functional potencies (Lange, van Stuivenberg, Veerman, et al., 2005). The $\log P$ values of compounds **5**, **9**, and **8a** were lower than rimonabant and SLV-319 and **9** displayed the lowest $\log P$ value due to the presence of 3-pyridyl moiety in the pyrazoline core. Compound **8a** (SLV-330) was effective in reducing the ethanol and nicotine self-administration along with reinstatement of ethanol/nicotine seeking in Wistar rats

TABLE 1A Compounds with reduced lipophilic and improved aqueous solubility from Solvay group.

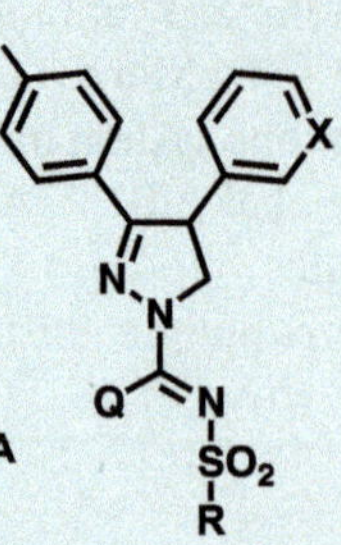

A

Code	R/X/Q=$NHCH_3$	*Ki* CB1 (nM)	*p*A2 (CB1)	*Ki* CB2 (nM)	PGP	log *P*
1	–	25.0±15	8.6±0.1	1580±150	1.2	5.5
2	–	7.8±1.4	9.9±0.6	7493±126	1.4	5.1
5	$N(C_2H_5)_2$/C	30±14	8.6±0.2	3270±1208	1.5	4.8
6	$N(CH_2)_2S(CH_2)_2$/C	24±14	8.8±0.2	3526±1015	NA	NA
7	$N(CH_2)_2S(O_2)(CH_2)_2$/C	832±168	8.1±0.2	NA	NA	NA
8	$N(CH_2)_5$/C	152±68	8.7±0.3	1321±264	NA	NA
8a	(*S*)-(−) Piperidinyl/C	58±19	8.7±0.2	3495±968	1.4	4.8
9	$N(C_2H_5)_2$/N	141±48	7.5±0.3	9077±923	3.4	3.2

at lowest effective dose of 10mg/kg (p.o) (de Bruin et al., 2011). This compound was also found to improve memory and cognitive improvement in preclinical models (de Bruin et al., 2010) attesting to CB1R antagonism-mediated effects.

In a subsequent study, Lange and co-workers designed compounds with the goal of improving aqueous solubilities to circumvent practical issues with pharmacological testing as well compound formulations. The compounds for, e.g., **10–13** were designed to produce CB1 receptor antagonists with improved aqueous solubilities, via the presence of either a polar moiety (**10** and **11**) or a nitrogen atom in their carboxamidine side chain **Q** (Formula A), which could be protonated (**12–13**) (Lange et al., 2010).

Compound **12** with the *N,N*-dimethylaminopropyl appendage elicited the highest CB1 receptor affinity (K*i* 3.1 nM) in this series. The compounds **10–13** also showed high CB1/CB2 receptor selectivity, which was similar to the parent compound **2**. The compounds **12–13**, which had an ionizable nitrogen atom in their pendant side chain, had lower CB1 antagonistic potencies (about ~10-fold) even though their binding affinities were higher compared to compounds **10** and **11,** which lacked ionizable nitrogen atom. As expected, the compounds in these series exhibited improved aqueous solubilities under acidic conditions (Table 1B).

TABLE 1B (R=4-ClPhenyl, X=C).

Code	Q	*Ki* CB1 (nM)	*p*A2 (CB1)	*Ki* CB2 (nM)	Aqueous solubility, μM pH 7.4/pH 2
10	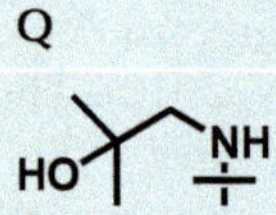	43±11	9.0±0.1	>1000	4/4
11	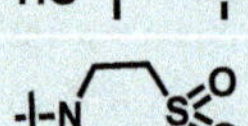	31±17	8.6±0.2	571±215	<3/<7
12	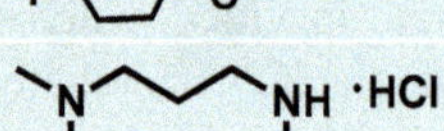	3.1±0.9	7.8±0.1	>1000	38/>100
13	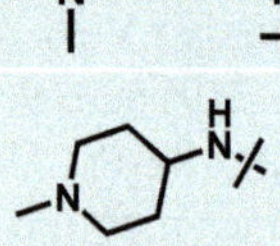	19±7	7.8±0.1	>1000	11/65

Jenrin group

The blueprint to generate non-CNS penetrant compounds are well known (Chen et al., 2003). The Solvay work likely provided impetus for the Jenrin group (currently Corbus) to utilize ibipinabant as a focal point for developing PR CB1 receptor blockers. In their approach, Chorvat, McElroy and coworkers found that modifications at the imidoyl *N*-methyl group of SLV-319 could be shuttled with polar pendants arising from natural and un-natural amino acid derivatives to provide analogs that adhered to physicochemical properties for impaired BBB penetrance via increased PSA, decreased lipophilicity, as well as retaining the high-binding CB1 affinity. In a detailed study on SAR, it was found that incorporation of a variety of more hydrophilic, hydrogen bonding substituents in place of the chloro group at the 4-position of the phenylsulfonyl ring was detrimental for the CB1 receptor. Retaining the diarylpyrazole pharmacophore, amino ester adducts at the central (*N*-methyl) region of the SLV-319 elicited high CB1R binding affinity though their easily ionizable acids possessed significantly diminished affinity for the CB1 receptor (Chorvat et al., 2012). Esters have a propensity to hydrolyze readily to low-affinity acids in vivo. Hence, their potential as drug candidates was limited. As opposed to esters, carboxamido group was determined to provide properties that were amenable to reduce brain penetration resulting from a reduction in passive diffusion, or being P-glycoprotein (PgP) substrates and efflux transporters (Chen et al., 2003; Trapa et al., 2019) and affording attracting physicochemical properties like lower lipophilicity, high polar surface area, and an increase in number of H-bond donors (Ertl et al., 2000; Lipinski, 2004; Lipinski et al., 2001).

The two main compounds from this approach, **14** and **15** (the *S*-enantiomer at C4 carbon), JD5006 **14a** (PSA 117, 3 HBDs, cLog *P* 4.64) and JD5037 **15a**, (PSA 117 Å^2, 3 HBDs, cLog *P* 6.02) included an appended N-substituted glycinamide and *L*-valinamide, respectively. The glycinamide analog **14** showed binding affinity (IC_{50} 18 nM) comparable to the parent compound **2**, whereas the L-valinamide analog **15** had an improved affinity for the CB1 receptor (IC_{50} 2 nM) (Fig. 2). Compound **14** required a chiral HPLC to allow for component enantiomer separation and characterization to yield the *S* eutomer JD5006 (**14a**). In case of **15**, L-amino acid carboxamide pendant afforded a pair of diastereomers, where the two components could be isolated via simple flash chromatography and upon recrystallization afforded the potent diastereomer labeled JD5037 (**15a**). Of the compounds evaluated in MDR-MDCK cell monolayer assay, **14** and **15** showed potentially low brain exposure based on their Papp A–B coefficients (1.8 and 1.1, respectively) and high efflux ratios (34 and 15, respectively). They were then evaluated for plasma and brain exposure, brain CB1-receptor occupancy (RO), and therapeutic efficacy in pharmacological assays in vivo for ameliorating metabolic dysfunction (Chorvat et al., 2012).

Racemic **14** was profiled as an inverse agonist in cAMP assays with its IC_{50} of 46 nM and was comparable to rimonabant, IC_{50} = 51 nM and better than for SLV-319, IC_{50} = 139 nM.

CB1-RO studies in mouse brain were conducted for both JD5006 and JD5037. Both compounds showed negligible occupancy in the cerebellum and less than 3% in hippocampus at 30 mg/kg as opposed to SLV-319 and rimonabant which were approaching 87%–100% at the same dose.

Blood to plasma ratios of **14a** and **15a** in chow fed mice showed high plasma levels of drug when analysis was performed 1 h after administration in doses up to 100 mg/kg p.o. over 3 days. While the plasma levels of JD5006 and JD5037

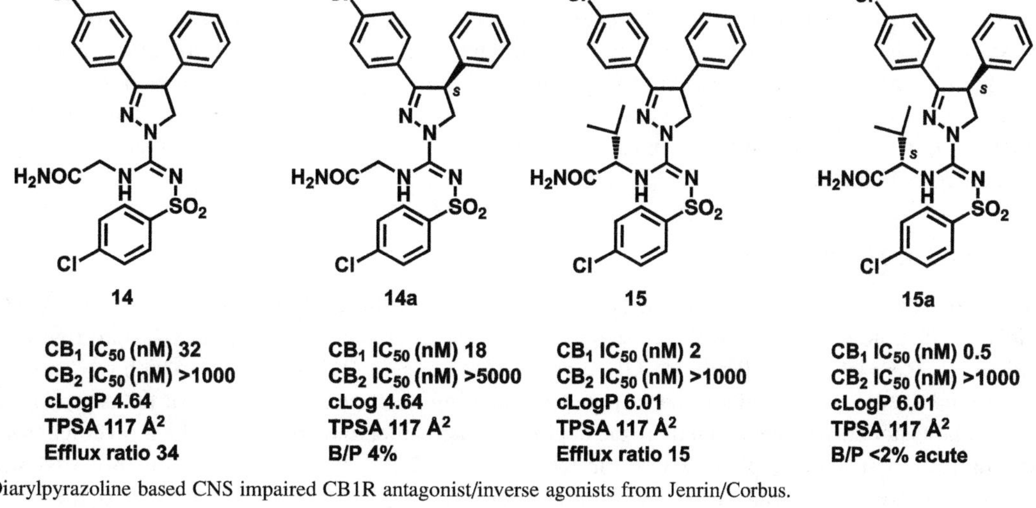

FIG. 2 3,4-Diarylpyrazoline based CNS impaired CB1R antagonist/inverse agonists from Jenrin/Corbus.

were greater than SLV-319 at this dose, the total brain levels remained negligible or below detection level for both compounds.

Despite the lack of direct brain receptor engagement, **14** (rac) was shown to have a beneficial effect in glucose tolerance test and enhance insulin sensitivity, normalized triglyceride levels, liver mass at doses ranging from 3 to 30 mg/kg in DIO mice. These results supported the hypothesis that antagonizing peripheral CB1 receptors may be adequate to produce various beneficial metabolic effects seen in case of global CB1 inverse agonists. Thus, PR CB1R inverse agonists with limited to even negligible brain access may offer a safer choice to highly brain-penetrant antagonists for the treatment of metabolic syndrome disorders.

JD5037 (**15a**) was further studied extensively for its effects arising from peripheral restriction(Tam et al., 2012). Absence of brain presence was verified through various assays like the MDCK cell assay, tissue distribution studies, behavioral paradigms, ex vivo RO studies, and PET imaging. The absence of in vivo brain CB1R occupancy following acute and chronic administration of **15a** was also reflected by PET imaging in mice. Minor displacement of a CB1 receptor PET ligand was indicated when **15a** was given acute or chronically either before or after the radioligand. CB1-receptor-mediated catalepsy or increased locomotor activity was not seen from behavioral assays at 3 mg/kg. In DIO mouse model to test food intake and weight loss after oral dosing for up to 28 days JD5037 was efficacious, suggesting that potent that peripheral inverse agonism at CB1 receptors and not central sites alone mediate the appetitive effects (Gómez et al., 2002). Leptin's role is further underlined in this process by the abrogation of these effects in DIO mice pretreated with a leptin antagonist. JD5037 also had beneficial effects on glucose tolerance, insulin resistance, fasting triglyceride levels, liver weight and liver enzymes, alanine transaminase, and aspartate transaminase (ALTs and ASTs). In primary cultured mouse hepatocytes and HepG2 cells, JD5037 treatment reduced the AEA, increased ceramide synthesis, and chronic treatment of DIO mice with JD5037 reversed hepatic insulin resistance, which was verified by euglycemic/hyperinsulinemic clamp (Cinar et al., 2014). Furthermore, JD5037 has shown a positive effect in mediating body weight related to a rare genetic disorder known as Prader Willi Syndrome (PWS), which is characterized by obesity and hyperphagia. It has been shown that PWS patients had elevated circulating levels of 2-AG, which manifested in increased eCBs and upregulated CB1R, indicative of increased adiposity and reduced energy expenditure. In mouse models of PWS (obese *Magel2*-null mice), JD5037 (3 mg/kg/d for 28 days) attenuated body weight, reversed hyperphagia, and improved metabolic parameters in accordance with obese phenotype, thereby offering potential for peripheral CB1 antagonism as a treatment option in PWS patients (Knani et al., 2016). JD5037 also could be a potential treatment in urinary disorders and CYP-induced cystitis. Treatment with JD5037in mice normalized the imbalance in CYP-mediated bladder ECS tone imbalance by affecting the CYP-induced inflammatory response via diminishing levels of the proinflammatory cytokine and tumor necrosis factor alpha (TNFα) (Hinden et al., 2022). JD5037 treatment was also shown to be effective in reducing hepatic steatosis in HFD-fed wild-type and liver-specific SIRT1 knockout (Sirt1-LKO) mice by increasing fatty acid β-oxidation and activating 5′adenosine monophosphate-activated protein kinase (AMPK) through liver kinase B1 (LKB1) (Liu et al., 2019).

Tan et al have shown that blockade of CB1 receptors with JD5037-attenuated CB1 receptor-regulated hepatic stellate cell (HSC) activation and ameliorated liver fibrosis by suppressing β-arrestin1/Akt signaling (Tan et al., 2020).

Recently, JD5037 was also shown to reduce ethanol drinking in wild-type mice that was abolished in mice lacking CB_1R, ghrelin peptide, or growth hormone secretagogue receptor (GHS-R_{1A}) (Godlewski et al., 2019). This was attributed to blocking CB1R in ghrelin-producing cells by inhibiting the formation of active acylated ghrelin and modulating gastric vagal afferent signaling. Thus, peripheral CB1R blockade may be a potential target in the treatment of alcoholism without perturbing CNS signaling directly. JD5037 administered by oral gavage at doses up to 150 mg/kg/day and for up to 34 days in Sprague Dawley rats, and up to 75 mg/kg/day for 28 consecutive days in Beagle dogs showed that the compound was well-tolerated with no-observed adverse effect levels (NOAELs) (Kale et al., 2019). JD5037/CRB-4001 is undergoing chronic pharmacokinetic studies in primates to measure brain exposure. JD5037 has achieved an IND status, and clinical trials with this compound are pending.

National Institutes of Health (NIH)

Kunos and co-workers were involved in meticulous characterization of the CB1-mediated effects of JD5037 in its early stages that fortified the role of peripheral CB1 receptors and their therapeutic potential. 3,4-Diarylpyrazoline scaffold was further utilized by them in the development of third generation of new CB1 antagonists, which were now hybrid target, peripherally restricted agents. The novel approach included attaching polar pendants in lieu of the central *N*-Methyl group such that these pendants not only provided an increase in polar surface amenable to keeping the molecules from penetrating BBB but also implicate a secondary pharmacophoric target that will be ally to CB1 antagonism. Attaching of a polar amidine moiety like, for, e.g., acetamidine or *S*-methylisothiourea known for its inducible nitric oxide synthase (iNOS)

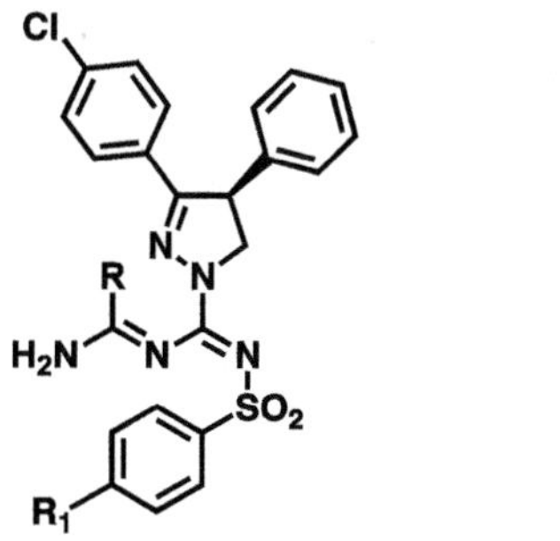

MRI-1867 R= Me, R_1 = CF_3 (16)
MRI-1569 R= Me, R_1 = Cl (17)
Compound R= SMe, R_1 = Cl (18)
MRI-1891 R= NHAc, R_1 = CF_3 (19)

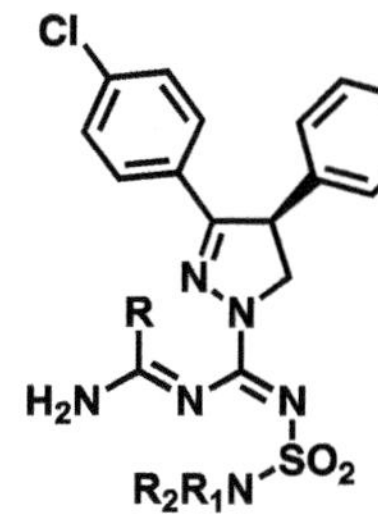

Compound R= Me, R_1, R_2= Et (20)
Compound R= Me, R_1, R_2= -$(CH_2)_5$- (21)
Compound R= SMe, R_1, R_2= -$(CH_2)_2CH(CF_3)(CH_2)_2$- (22)

FIG. 3 3,4-Diarylpyrazoline based peripherally restricted CB1R antagonist/inverse agonists from NIH.

inhibitory (Southan et al., 1995) activity to the core chemical scaffold would increase the polar surface area of the hybrid structure thereby limiting its brain penetrance. The hybrid compound could further serve as a prodrug and or a carrier for the secondary target/iNOS inhibitory moiety yet retain potent CB1-binding affinity/functional potency. Two compounds (R=CF_3; MRI-1867, CB1R *Ki*=2.3 nM, B/P ratio <3.1% acute, TPSA 100.48 Å^2 and R= Cl; MR-1569, CB1R *Ki*=7.9 nM, B/P ratio <3% acute, TPSA 100.48 Å^2) were well characterized, and both were orally bioavailable. In C57Bl6/J mice with HFD-induced obesity (DIO), chronic treatment with both compounds caused dose-dependent reductions in body weight, hepatic steatosis, food intake, and improved glucose tolerance and insulin sensitivity. The compounds thus showed beneficial effects in ameliorating metabolic deficiencies in line with peripheral CB1 blockade paradigm. Both compounds showed superior antifibrotic efficacy in CCl_4 liver fibrosis model and bile-duct ligation models (Cinar et al., 2016; Iyer, Cinar, Katz, et al., 2017). Compound MRI-1569 showed clean profile; however, MRI-1867 offered attractive druggable properties and has now advanced further. An additional compound (**18**) showed potent iNOS inhibitory and acceptable CB1R-binding affinity (60 nM) (Iyer, Cinar, Katz, et al., 2017). The compounds in these series acted as CB1R inverse agonists with negligible CNS penetration (Fig. 3).

MRI-1867 was studied extensively and showed superior antifibrotic efficacy in liver (Cinar et al., 2016), lung (Cinar et al., 2017), skin (Zawatsky et al., 2021), and kidney fibrosis (Udi et al., 2020) models. MRI-1867 also reduced hepatic steatosis, hepatic VLDL secretion, and diminished the circulating levels of proprotein convertase subtilisin/kexin type 9 (PCSK9). The decrease in VLDL secretion could be attributed to CB1R blockade, while the reduction of PCSK9 levels and the related increase in LDLR were due to iNOS inhibitory properties of the compound attesting to superior in vivo effects from dual-targeting of CB1R and iNOS (Roger et al., 2020). Furthermore, the hybrid CB1R and iNOS engagement by MRI-1867 provided greater antifibrotic efficacy than inhibiting either target alone in pale ear mice (HPS1$^{ep/ep}$) mice model of bleomycin-induced pulmonary fibrosis induced from Hermansky Pudlak Sydrome (HPS), a rare genetic disorder (Cinar et al., 2021). This compound, MRI-1867/INV-101 (Zevaquenabant), has undergone Phase I clinical trial in Canada to evaluate the safety, tolerability, and PK following single oral doses (NCT04531150).

Kunos, Cinar, Iyer, and colleagues at NIH have characterized another novel, peripherally restricted candidate MRI-1891 (**19**), which was shown to be a potent antagonist of CB1 receptor. This compound is the first example of biased, orthosteric CB1R antagonist (Liu et al., 2021). MRI-1891 is a potent, functionally selective, β-arrestin-2-biased antagonist of CB1 (IC_{50} 21 pM) with about 300-fold separation over G-protein signaling inhibition (IC_{50} 6 nM).

In tissue distribution studies, MRI-1891 showed a brain/plasma ratio (7%) compared to ibipinabant (22%) or rimonabant (100%) following acute oral doses of 3 mg/kg in male, wild-type mice. Acute treatment at doses of 1 or 10 mg/kg of MRI-1891 did not result in significant CB1R occupancy in the brain, as determined by CB1R positron emission tomography (PET). The compound also did not cause anxiogenic behavior at the high dose of 30 mg/kg. A moderate increase in the brain/plasma ratio of MRI-1891 was seen following 28 days of chronic administration at 3 mg/kg/day (19%). The higher brain CB1R occupancy following a 10 mg/kg/day but not a low dose (1 mg/kg/day) indicated that ABC transporters-based efflux was not efficient for MRI-1891 relative to MRI-1867 as indicated by the smaller increase in brain levels of MRI-1891 compared to MRI-1867 in *Mdr1a/b* double-knockout mice than in wild-type mice (Fig. 3).

Molecular dynamics (MD) simulations with a set of conformers of MRI-1891 was carried out using the X-ray structure of CB1R cocrystallized with taranabant and compared to the unbiased antagonist rimonabant and ibipinabant (Shao et al., 2016). It was shown that each arm plays a distinct role and interacts with a different region of the receptor. MRI-1891 is unique in that combines in a single scaffold the arms the spans the interactions for both agonists and antagonists, which may contribute to the biased property. It was postulated that S123 (TMH1) which interaction with the CF_3 group of MRI-1891 that is missing in Arm 3 ibipinabant and rimonabant do not interact with this residue. Further confirmation was obtained from S123A mutation that resulted in a fourfold reduction of inhibitory potency of MRI-1891, but not of rimonabant, toward βArr2 signaling mutation. Additionally, this mutation did not alter either CB1R binding affinity or CB1R inhibitory potency toward G protein signaling of either MRI-1891 or rimonabant. Significantly, it was found that this bias results in functional selectivity that modulates anxiogenic behavior, body weight, appetitive behavior and hepatic glucose production mainly via G protein activation, whereas CB1R regulation of muscle insulin sensitivity occurs predominantly via βArr2 signaling.

This novel paradigm can be further harnessed by developing signaling-selective molecules for higher efficacy and lesser side effects. MRI-1891/INV-202 is undergoing a Phase 1B study to examine the safety and tolerability as well as pharmacokinetics (PK) pharmacodynamic (PD) effects in subjects with metabolic syndrome over 28 days (NCT 05282446).

In continuation with the development of novel, peripherally restricted CB1 antagonists, dihydropyrazoline scaffold was utilized and designed to compounds with limited brain penetrance and decreased lipophilicity (for, e.g., **20–22**). Promising compounds with potent CB1R activity were evaluated in tissue distribution and pharmacokinetic studies. Three compounds (**20**, **21** and **22**) showed limited brain penetrance (<4%). This study also reinforced that CB1 potency in the 4,5-1*H*-dihydropyrazole resides with the *S*-enantiomer. While the compounds in this series behaved as potent inverse agonist at CB1, they had negligible iNOS inhibitor activity and limited oral bioavailability. It was observed that having an ionizable nitrogen in lieu of the phenyl group enhanced the CB1-binding affinities, yet it was detrimental to iNOS inhibitory properties in this series. Compounds **20**, **21,** and **22** at 1 hour after intraperitoneal injections showed good systemic exposure and liver distribution though they were highly plasma protein bound and had low metabolic stability. The tested compounds reduced food intake and body weight similarly at 3 mg/kg dose for **21**, 10 mg/kg dose for **20** and **22** in DIO mice. Body weight reduction was about 8% for the three compounds tested that was attributable to peripheral CB1 antagonism. Food-intake reduction though was modest. The lower in vivo efficacy was mainly due to the short-half life and once-daily dose (Fig. 3) (Iyer et al., 2022).

General synthesis schemes for dihydropyrazoline-type compounds

The general synthesis method for this series of compounds involves generating pyrazolines (e.g., 3-(4-chlorophenyl)-4-phenyl-2-pyrazoline, **23**) or appropriately substituted diphenyl pyrazolines as the first step (Grosscurt et al., 1979; Lange et al., 2004). The pyrazolines are then coupled with sulfonyl carbamates or sulfonyl isocyanates to obtain sulfonylureas of the type **24**. The sulfonylureas are then treated with a chlorinating agent like PCl_5 or $POCl_3$/base to yield the intermediate imidoyl chloride **25**. The imidoyl chloride could then be reacted with appropriate amino pendants to yield the target compounds as racemic or diastereomeric analogs (Fig. 4A). The racemates can be separated into component enantiomers via chiral SFC methods. In case of JD5037, because *L*-Valinamide pendant was used, a diastereomeric mixture was obtained and the *S,S*-diastereomer could be recrystallized as the eutomer. It is worth noting that in this series the CB1R binding is stereoselective and resides with the *S*-enantiomer at the C4 position of the pyrazoline ring (Chorvat et al., 2012; Iyer, Cinar, Katz, et al., 2017). An alternative approach was proposed by Solvay and co-workers via the generation of thiourea **27** and thiomethyl analogs **28** which could be displaced by amino pendants. Appropriate sulfonamides (**29**) serve as the starting points for generation of the carbamates (**31**), thiocyantes (**33**), and thioalkylated intermediates (**34**) (Fig. 4B) (Lange et al., 2011; Lange, van Stuivenberg, Veerman, et al., 2005).

Stable isotope-labeled compound synthesis and PET probes

Stable isotope standards are valuable entities for ADME studies. In this regard, stable isotope standards for JD5037 and MRI-1867 have been synthesized. In case of JD5037, the synthesis of the deuterated version was accomplished using D8-*L*-Valinamide that was synthesized from D8-*L*-Valine to afford D8-JD5037 (**35**) (Iyer, Cinar, Coffey, et al., 2017) The stable isotope standard for MRI-1867 was generated using $^{13}C_6$-chlorobenzene to give ^{13}C-MRI-1867 as a racemic compound that was then separated into component enantiomers to afford the stable isotope labeled eutomer (**36**) (Iyer et al., 2018) (Fig. 5). Imaging probes based on the SLV-319 without the central *N*-methyl group were generated by Donohue et al (Donohue et al., 2008). They synthesized compound **37** referred to as **12a** ((−)-3-(4-chlorophenyl)-*N′*-[(4-cyanophenyl) sulfonyl]-4-phenyl-4,5-dihydro-1*H*-pyrazole-1-carboxamidine), which was found to have high affinity and selectivity

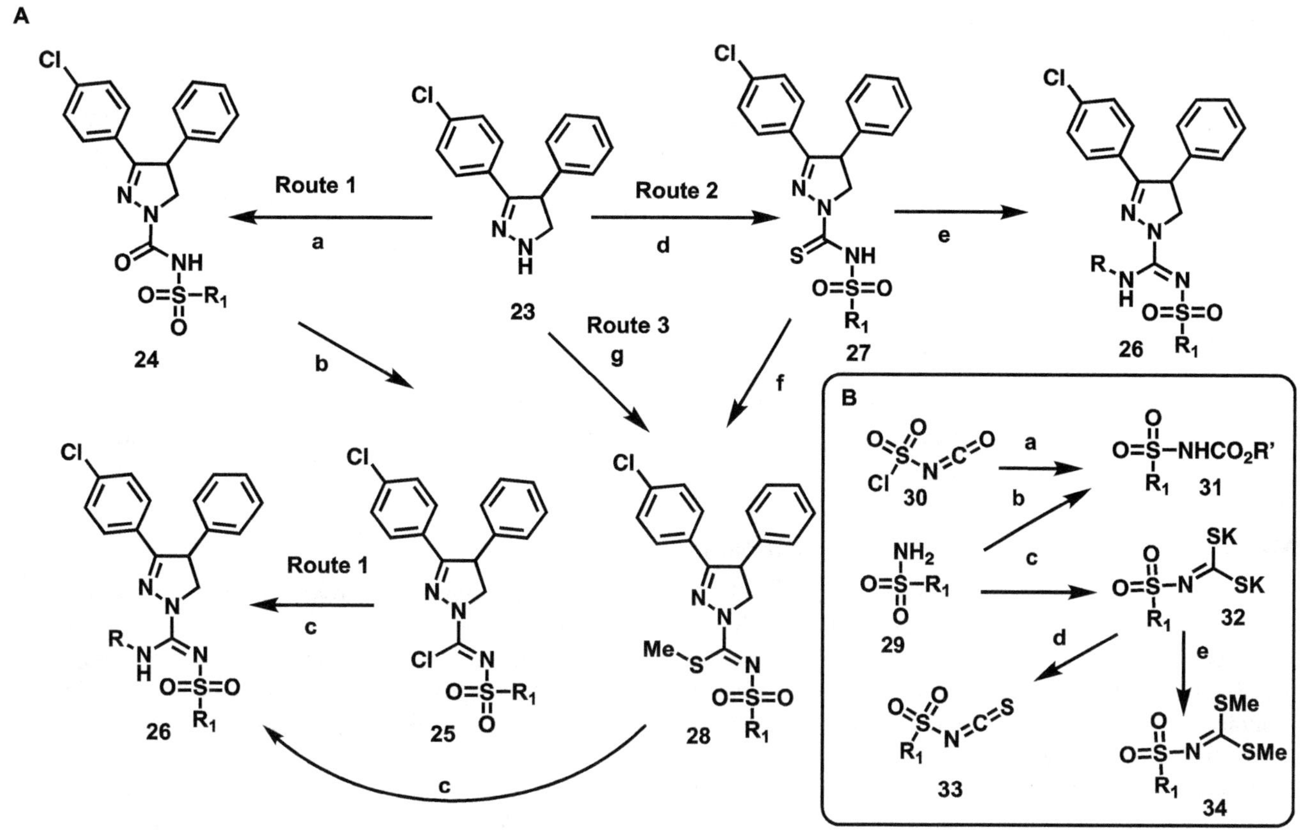

A. Reagents and conditions: (a) $ArylSO_2NHCO_2CH_3$ (**32**), toluene, reflux, 3– 5h; (b) PCl_5, chlorobenzene, reflux, 2 h or $POCl_3$, DMAP CH_2Cl_2 reflux 5 h; (c) RNH_2, Et_3N, CH_2Cl_2-MeOH 12-24 h; (d) **33**, CH_2Cl_2, rt, 2 h; (e) RNH_2, $HgCl_2$,CH_3CN, rt, 4 h; (f) CH_3I, Et_3N, acetone, rt, 16 h; (g) **34**, pyridine 100 C, 40 h.
B. Reagents and conditions: (a) (i) t-BuOH CH_2Cl_2, rt, 2 h; (ii) $R_1 = NHR^1R^2$ Et_3N, CH_2Cl_2 24 h; (b) $ClCO_2Me$, Et_3N, CH_3CN, 0 °C-rt 24 h (R_1 = Aryl); (c) CS_2, KOH, DMF, rt, 4 h; (d) diphosgene, CH_2Cl_2, rt, 20 h; (e) CH_3I, DMF, 0 C, 2 h.

FIG. 4 General scheme for synthesis of 3,4-Diarylpyrazoline CB1R antagonist/inverse agonists.

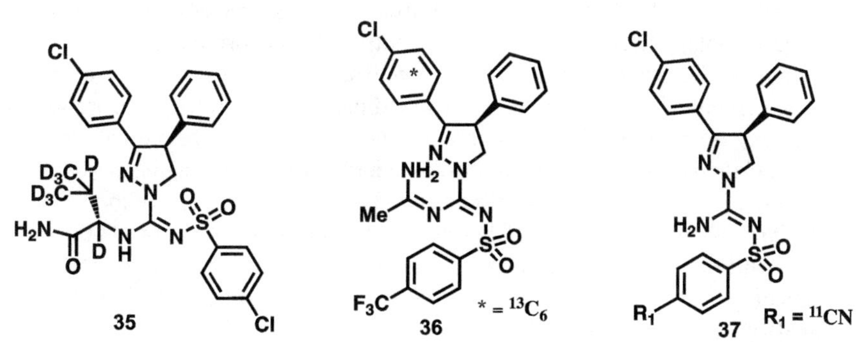

[$^2H_8$]-JD5037 *S*-**[$^{13}C_6$]-MRI-1867** **(-)-[^{11}C]-12a**

FIG. 5 Stable-isotope compounds and PET ligand based on 3,4-Diarylpyrazoline template.

for binding to CB_1 receptors. (−)-**37** and its lower affinity enantiomer ((+)-**12a**) was labeled with carbon-11 ($t_{1/2}$ = 20.4 min) using [^{11}C]cyanide ion as labeling agent and evaluated as PET radioligands in cynomolgus monkey. [^{11}C](−)-**12a** (**37**) showed high uptake and retention of radioactivity in the brain where the CB1 receptor densities were the highest. Radioligand [^{11}C](−)-**12a** was deemed to be a promising imaging agent for the study of brain CB1 receptors (Donohue et al., 2008) (Fig. 5).

Computational models

Four-arm antagonist

The prototypical brain-penetrant antagonist rimonabant and some of its derived analogs have been termed as 3-arm antagonists (Hua et al., 2017, 2016). SLV-319 with its *N*-methylamino arm can constitute four-arm antagonist. The short/abbreviated *N*-methyl fourth arm of SLV-319 while important for CB1R inverse agonism, however, is not known to impart functional selectivity in signal transduction pathways. In case of the unique compound, MRI-1891 (**19**) polar groups of Arm 4 are stabilized by water, and the NH_2 group of guanidine forms putative H-bond with S383 of TMH7. Importantly, groups of Arm 4 interact with several N-terminal hydrophobic residues, including F102, M103, and with F108. These interactions presumably contribute to the signaling bias of MRI-1891. Arm 4 of ibipinabant (SLV-319) lacks the acyl guanidine group which prevents contact points with these residues that shows in its negligible bias (G/βArr≈2) (Liu et al., 2021). While cocrystal structures of four-arm CB1 antagonist based on 3,4-diarylpyrazoline scaffold are yet to be available, docking and MD simulations of some of the four-arm antagonists have been carried out (Liu et al., 2018). The implications of the fourth arm in contributing to signaling bias may offer a new avenue to utilize the peripheral CB1-antagonism for mechanistic gain and functional cure.

Application to other areas and summary

In this chapter, we have reviewed the chemistry and biology of 4,5-1*H*-dihydropyrazole/3,4-diarylpyrazoline, peripherally restricted cannabinoid-1 receptor antagonists, and their potential in treating various diseases. Many compounds arising from this scaffold have attractive druggable properties. New as well as robust synthetic routes to develop chiral compounds on this scaffold have also been developed. This has enabled in-depth structure-activity studies and detailed biological investigations. The scaffold was originally utilized for brain-penetrant CB1 inverse agonist clinical candidate Ibipinabant. Subsequently, the scaffold was modified in a way where the polar surface areas (PSAs) and other physicochemical properties impaired BBB permeability allowing for the agents to act solely through the noninvolvement of the brain CB1 receptors. Of these second-generation antagonists that were nonbrain penetrant compounds, JD5037 and JD5006 were studied. These compounds, particularly JD5037 showed potential in treating various diseases in experimental rodent models. Interestingly, human CB1 receptor isoforms may also play a role in glucose metabolism disorders. Egan and coworkers identified an isoform (CB1b) that finds its specific presence in β-cells and hepatocytes but was not detected in brain (González-Mariscal et al., 2016). CB1b showed higher affinity for JD5037 than for rimonabant compared to CB1 full length and CB1b in peripheral metabolic tissues is a strong regulator of adenylyl cyclase activity, which can significantly enhance insulin secretion in β-cells. Hence, isoform-selective peripheral targeting can be an additional mode of treatment for metabolic disorders.

Further research showed that the 4,5-1*H*-dihydropyrazole scaffold was amenable to further modifications allowing entry into novel four-arm CB1 antagonists. Through an innovative design principle, which was further validated by proof-of-concept studies, some of these agents had not only potent CB1 antagonism but also selectively engaged a beneficial secondary target. Engagement of a secondary target resulted in superior efficacy in fibrotic disorders in comparison with the prototypical, single-target antagonist like rimonabant. These compounds hence have potential in treating multifactorial conditions. Most importantly, all these benefits could be had without causing neuropsychiatric liabilities making them potentially valuable agents for treating various pathologies. The most advanced compound in the second-generation CB1 antagonist include JD5037, which had received IND status and awaiting Phase I trial. In the third-generation series of compounds, two compounds with the most favorable profile have undergone Phase I trial (INV-101 and INV-202) indicating promising development prospects for this class of compounds belonging to the privileged 3,4-diarylpyrazoline/4,5-1*H*-dihydropyrazole scaffold.

Mini-dictionary of terms

- Biased signaling: Functional selectivity caused by a ligand to selectively modulate one signal transduction pathway over another.
- Inverse agonist: A compound that binds to the same receptor site as an agonist but produces opposite pharmacological response to an agonist.
- Inducible nitric oxide synthase: An enzyme that catalyzes the production of nitric oxide from L-arginine.
- Peripherally restricted: Compounds that do not enter the blood brain barrier.
- Stable isotope-labeled compound: Introduction of deuterium or carbon-13 atom in the compound to enable bioanalytical studies.

Key facts of 3,4-diarylpyrazolines CB1R modulators

- Metabolic syndrome is a multifactorial disease with limited treatment options.
- Cannabinoid receptor-1 antagonism is a validated target for treating many of the conditions affected by the metabolic syndrome.
- Globally acting CB1R antagonists/inverse agonists were in development to treat such conditions, however, because of CNS side effects they fell out of favor and further clinical development.
- Compared to the global antagonists, peripheral antagonists based on 3,4-diarylpyrazoline scaffold offer similar therapeutic benefits without CNS liabilities.
- Novel approaches to CNS impaired compounds utilizing the 3,4-diarylpyrazoline template and new modalities to improve efficacy and safety has been addressed.

Summary points

- The endocannabinoid system plays a significant role in many pathophysiological processes in the human body.
- The CB1 receptor is involved in the modulation of caloric intake, behavior, and lipogenesis.
- Globally acting CB1 antagonists are currently considered as unsafe because of their ability to induce anxiogenic effects leading to neuropsychiatric side effects.
- CNS sparing CB1R antagonists/inverse agonists retain many of the beneficial metabolic effects without causing CNS side effects.
- Dual targeting and biased signaling-based CB1 antagonists offer additional options to improve safety and efficacy for treating metabolic syndrome.

References

Bellocchio, L., Soria-Gómez, E., Quarta, C., Metna-Laurent, M., Cardinal, P., Binder, E., & Marsicano, G. (2013). Activation of the sympathetic nervous system mediates hypophagic and anxiety-like effects of CB_1 receptor blockade. *Proceedings of the National Academy of Sciences of the United States of America*, *110*(12), 4786–4791. https://doi.org/10.1073/pnas.1218573110.

Black, S. C. (2004). Cannabinoid receptor antagonists and obesity. *Current Opinion in Investigational Drugs*, *5*(4), 389–394.

Chen, C., Hanson, E., Watson, J. W., & Lee, J. S. (2003). P-glycoprotein limits the brain penetration of nonsedating but not sedating H1-antagonists. *Drug Metabolism and Disposition: The Biological Fate of Chemicals*, *31*(3), 312–318. https://doi.org/10.1124/dmd.31.3.312.

Chorvat, R. J. (2013). Peripherally restricted CB1 receptor blockers. *Bioorganic & Medicinal Chemistry Letters*, *23*(17), 4751–4760. https://doi.org/10.1016/j.bmcl.2013.06.066.

Chorvat, R. J., Berbaum, J., Seriacki, K., & McElroy, J. F. (2012). JD-5006 and JD-5037: Peripherally restricted (PR) cannabinoid-1 receptor blockers related to SLV-319 (Ibipinabant) as metabolic disorder therapeutics devoid of CNS liabilities. *Bioorganic & Medicinal Chemistry Letters*, *22*(19), 6173–6180. https://doi.org/10.1016/j.bmcl.2012.08.004.

Cinar, R., Gochuico, B. R., Iyer, M. R., Jourdan, T., Yokoyama, T., Park, J. K., & Kunos, G. (2017). Cannabinoid CB1 receptor overactivity contributes to the pathogenesis of idiopathic pulmonary fibrosis. *Journal of Clinical Investigation Insight*, *2*(8). https://doi.org/10.1172/jci.insight.92281.

Cinar, R., Godlewski, G., Liu, J., Tam, J., Jourdan, T., Mukhopadhyay, B., & Kunos, G. (2014). Hepatic cannabinoid-1 receptors mediate diet-induced insulin resistance by increasing de novo synthesis of long-chain ceramides. *Hepatology*, *59*(1), 143–153. https://doi.org/10.1002/hep.26606.

Cinar, R., Iyer, M. R., & Kunos, G. (2020). The therapeutic potential of second and third generation CB1R antagonists. *Pharmacology & Therapeutics*, *208*, 107477. https://doi.org/10.1016/j.pharmthera.2020.107477.

Cinar, R., Iyer, M. R., Liu, Z., Cao, Z., Jourdan, T., Erdelyi, K., & Kunos, G. (2016). Hybrid inhibitor of peripheral cannabinoid-1 receptors and inducible nitric oxide synthase mitigates liver fibrosis. *Journal of Clinical Investigation Insight*, *1*(11). https://doi.org/10.1172/jci.insight.87336.

Cinar, R., Park, J. K., Zawatsky, C. N., Coffey, N. J., Bodine, S. P., Abdalla, J., & Malicdan, M. C. V. (2021). CB1 R and iNOS are distinct players promoting pulmonary fibrosis in Hermansky-Pudlak syndrome. *Clinical and Translational Medicine*, *11*(7), e471. https://doi.org/10.1002/ctm2.471.

de Bruin, N. M. W. J., Lange, J. H. M., Kruse, C. G., Herremans, A. H., Schoffelmeer, A. N. M., van Drimmelen, M., & De Vries, T. J. (2011). SLV330, a cannabinoid CB(1) receptor antagonist, attenuates ethanol and nicotine seeking and improves inhibitory response control in rats. *Behavioural Brain Research*, *217*(2), 408–415. https://doi.org/10.1016/j.bbr.2010.11.013.

de Bruin, N. M. W. J., Prickaerts, J., Lange, J. H. M., Akkerman, S., Andriambeloson, E., de Haan, M., & Kruse, C. G. (2010). SLV330, a cannabinoid CB1 receptor antagonist, ameliorates deficits in the T-maze, object recognition and Social Recognition Tasks in rodents. *Neurobiology of Learning and Memory*, *93*(4), 522–531. https://doi.org/10.1016/j.nlm.2010.01.010.

De Petrocellis, L., Cascio, M. G., & Di Marzo, V. (2004). The endocannabinoid system: A general view and latest additions. *British Journal of Pharmacology*, *141*(5), 765–774. https://doi.org/10.1038/sj.bjp.0705666.

Després, J.-P., Golay, A., Sjöström, L., & Rimonabant in Obesity-Lipids Study Group. (2005). Effects of rimonabant on metabolic risk factors in overweight patients with dyslipidemia. *The New England Journal of Medicine*, *353*(20), 2121–2134. https://doi.org/10.1056/NEJMoa044537.

Devane, W. A., Hanus, L., Breuer, A., Pertwee, R. G., Stevenson, L. A., Griffin, G., & Mechoulam, R. (1992). Isolation and structure of a brain constituent that binds to the cannabinoid receptor. *Science*, *258*(5090), 1946–1949. https://doi.org/10.1126/science.1470919.

Di Marzo, V., Goparaju, S. K., Wang, L., Liu, J., Bátkai, S., Járai, Z., & Kunos, G. (2001). Leptin-regulated endocannabinoids are involved in maintaining food intake. *Nature*, *410*(6830), 822–825. https://doi.org/10.1038/35071088.

Donohue, S. R., Pike, V. W., Finnema, S. J., Truong, P., Andersson, J., Gulyás, B., & Halldin, C. (2008). Discovery and labeling of high-affinity 3,4-diarylpyrazolines as candidate radioligands for in vivo imaging of cannabinoid subtype-1 (CB1) receptors. *Journal of Medicinal Chemistry*, *51*(18), 5608–5616. https://doi.org/10.1021/jm800329z.

Ertl, P., Rohde, B., & Selzer, P. (2000). Fast calculation of molecular polar surface area as a sum of fragment-based contributions and its application to the prediction of drug transport properties. *Journal of Medicinal Chemistry*, *43*(20), 3714–3717. https://doi.org/10.1021/jm000942e.

Fulp, A., Bortoff, K., Zhang, Y., Seltzman, H., Mathews, J., Snyder, R., & Maitra, R. (2012). Diphenyl purine derivatives as peripherally selective cannabinoid receptor 1 antagonists. *Journal of Medicinal Chemistry*, *55*(22), 10022–10032. https://doi.org/10.1021/jm301181r.

Fulp, A., Bortoff, K., Zhang, Y., Snyder, R., Fennell, T., Marusich, J. A., & Maitra, R. (2013). Peripherally selective diphenyl purine antagonist of the CB1 receptor. *Journal of Medicinal Chemistry*, *56*(20), 8066–8072. https://doi.org/10.1021/jm401129n.

Gaoni, Y., & Mechoulam, R. (1964). Isolation, structure, and partial synthesis of an active constituent of Hashish. *Journal of the American Chemical Society*, *86*(8), 1646–1647. https://doi.org/10.1021/ja01062a046.

Godlewski, G., Cinar, R., Coffey, N. J., Liu, J., Jourdan, T., Mukhopadhyay, B., & Kunos, G. (2019). Targeting peripheral CB1 receptors reduces ethanol intake via a gut-brain axis. *Cell Metabolism*, *29*(6), 1320–1333.e8. https://doi.org/10.1016/j.cmet.2019.04.012.

Gómez, R., Navarro, M., Ferrer, B., Trigo, J. M., Bilbao, A., Del Arco, I., & Rodríguez de Fonseca, F. (2002). A peripheral mechanism for CB1 cannabinoid receptor-dependent modulation of feeding. *The Journal of Neuroscience*, *22*(21), 9612–9617. https://doi.org/10.1523/JNEUROSCI.22-21-09612.2002.

González-Mariscal, I., Krzysik-Walker, S. M., Doyle, M. E., Liu, Q.-R., Cimbro, R., Santa-Cruz Calvo, S., & Egan, J. M. (2016). Human CB1 Receptor Isoforms, present in Hepatocytes and β-cells, are Involved in Regulating Metabolism. *Scientific Reports*, *6*, 33302. https://doi.org/10.1038/srep33302.

Griffith, D. A., Hadcock, J. R., Black, S. C., Iredale, P. A., Carpino, P. A., DaSilva-Jardine, P., & Scott, D. O. (2009). Discovery of 1-[9-(4-chlorophenyl)-8-(2-chlorophenyl)-9H-purin-6-yl]-4-ethylaminopiperidine-4-carboxylic acid amide hydrochloride (CP-945,598), a novel, potent, and selective cannabinoid type 1 receptor antagonist. *Journal of Medicinal Chemistry*, *52*(2), 234–237. https://doi.org/10.1021/jm8012932.

Grosscurt, A. C., Van Hes, R., & Wellinga, K. (1979). 1-Phenylcarbamoyl-2-pyrazolines, a new class of insecticides. 3. Synthesis and insecticidal properties of 3,4-diphenyl-1-phenylcarbamoyl-2-pyrazolines. *Journal of Agricultural and Food Chemistry*, *27*(2), 406–409. https://doi.org/10.1021/jf60222a061.

Hinden, L., Ludyansky, R., Leidershnaider, S., Harris, Y., Nemirovski, A., Gofrit, O. N., & Hidas, G. (2022). Peripheral cannabinoid-1 receptor blockade ameliorates cystitis severity. *Cannabis and Cannabinoid Research*. https://doi.org/10.1089/can.2022.0077.

Howlett, A. C., Barth, F., Bonner, T. I., Cabral, G., Casellas, P., Devane, W. A., & Pertwee, R. G. (2002). International union of pharmacology. XXVII. Classification of cannabinoid receptors. *Pharmacological Reviews*, *54*(2), 163,164. https://doi.org/10.1124/pr.54.2.161.

Hua, T., Vemuri, K., Nikas, S. P., Laprairie, R. B., Wu, Y., Qu, L., & Liu, Z.-J. (2017). Crystal structures of agonist-bound human cannabinoid receptor CB1. *Nature*, *547*(7664), 468–471. https://doi.org/10.1038/nature23272.

Hua, T., Vemuri, K., Pu, M., Qu, L., Han, G. W., Wu, Y., & Liu, Z.-J. (2016). Crystal structure of the human cannabinoid receptor CB1. *Cell*, *167*(3), 750–762.e14. https://doi.org/10.1016/j.cell.2016.10.004.

Iyer, M. R., Cinar, R., Coffey, N. J., Chorvat, R. J., & Kunos, G. (2017). Synthesis of S-2-((S)-3-(4-chlorophenyl)-N'-((4-chlorophenyl)sulfonyl)-4-phenyl-4,5-dihydro-1H-pyrazole-1-carboximidamido)-3-(methyl-d3)butanamide-d5, octadeuterated JD5037. *Journal of Labelled Compounds & Radiopharmaceuticals*, *60*(10), 460–465. https://doi.org/10.1002/jlcr.3521.

Iyer, M. R., Cinar, R., Coffey, N. J., & Kunos, G. (2018). Synthesis of 13 C6 -labeled, dual-target inhibitor of cannabinoid-1 receptor (CB1 R) and inducible nitric oxide synthase (iNOS). *Journal of Labelled Compounds and Radiopharmaceuticals*. https://doi.org/10.1002/jlcr.3639.

Iyer, M. R., Cinar, R., Katz, A., Gao, M., Erdelyi, K., Jourdan, T., & Kunos, G. (2017). Design, synthesis, and biological evaluation of novel, non-brain-penetrant, hybrid cannabinoid CB1R Inverse agonist/inducible nitric oxide synthase (iNOS) inhibitors for the treatment of liver fibrosis. *Journal of Medicinal Chemistry*, *60*(3), 1126–1141. https://doi.org/10.1021/acs.jmedchem.6b01504.

Iyer, M. R., Cinar, R., Wood, C. M., Zawatsky, C. N., Coffey, N. J., Kim, K. A., & Lee, Y.-S. (2022). Synthesis, biological evaluation, and molecular modeling studies of 3,4-diarylpyrazoline series of compounds as potent, nonbrain penetrant antagonists of cannabinoid-1 (CB1R) receptor with reduced lipophilicity. *Journal of Medicinal Chemistry*, *65*(3), 2374–2387. https://doi.org/10.1021/acs.jmedchem.1c01836.

Janero, D. R., & Makriyannis, A. (2009). Cannabinoid receptor antagonists: Pharmacological opportunities, clinical experience, and translational prognosis. *Expert Opinion on Emerging Drugs*, *14*(1), 43–65. https://doi.org/10.1517/14728210902736568.

Kale, V. P., Gibbs, S., Taylor, J. A., Zmarowski, A., Novak, J., Patton, K., & Terse, P. S. (2019). Preclinical toxicity evaluation of JD5037, a peripherally restricted CB1 receptor inverse agonist, in rats and dogs for treatment of nonalcoholic steatohepatitis. *Regulatory Toxicology and Pharmacology*, *109*, 104483. https://doi.org/10.1016/j.yrtph.2019.104483.

Kelder, J., Grootenhuis, P. D., Bayada, D. M., Delbressine, L. P., & Ploemen, J. P. (1999). Polar molecular surface as a dominating determinant for oral absorption and brain penetration of drugs. *Pharmaceutical Research*, *16*(10), 1514–1519. https://doi.org/10.1023/a:1015040217741.

Knani, I., Earley, B. J., Udi, S., Nemirovski, A., Hadar, R., Gammal, A., & Tam, J. (2016). Targeting the endocannabinoid/CB1 receptor system for treating obesity in Prader-Willi syndrome. *Molecular Metabolism*, *5*(12), 1187–1199. https://doi.org/10.1016/j.molmet.2016.10.004.

Kunos, G., Osei-Hyiaman, D., Bátkai, S., Sharkey, K. A., & Makriyannis, A. (2009). Should peripheral CB(1) cannabinoid receptors be selectively targeted for therapeutic gain? *Trends in Pharmacological Sciences*, *30*(1), 1–7. https://doi.org/10.1016/j.tips.2008.10.001.

Lange, J. H. M., Coolen, H. K. A. C., van der Neut, M. A. W., Borst, A. J. M., Stork, B., Verveer, P. C., & Kruse, C. G. (2010). Design, synthesis, biological properties, and molecular modeling investigations of novel tacrine derivatives with a combination of acetylcholinesterase inhibition and cannabinoid CB1 receptor antagonism. *Journal of Medicinal Chemistry*, *53*(3), 1338–1346. https://doi.org/10.1021/jm901614b.

Lange, J. H. M., Coolen, H. K. A. C., van Stuivenberg, H. H., Dijksman, J. A. R., Herremans, A. H. J., Ronken, E., & Kruse, C. G. (2004). Synthesis, biological properties, and molecular modeling investigations of novel 3,4-diarylpyrazolines as potent and selective CB(1) cannabinoid receptor antagonists. *Journal of Medicinal Chemistry*, *47*(3), 627–643. https://doi.org/10.1021/jm031019q.

Lange, J. H. M., Sanders, H. J., & van Rheenen, J. (2011). An expedient atom-efficient synthesis of the cannabinoid CB1 receptor inverse agonist ibipinabant. *Tetrahedron Letters*, *52*(12), 1303–1305. https://doi.org/10.1016/j.tetlet.2011.01.068.

Lange, J. H. M., van Stuivenberg, H. H., Coolen, H. K. A. C., Adolfs, T. J. P., McCreary, A. C., Keizer, H. G., & Kruse, C. G. (2005). Bioisosteric replacements of the pyrazole moiety of rimonabant: Synthesis, biological properties, and molecular modeling investigations of thiazoles, triazoles, and imidazoles as potent and selective CB1 cannabinoid receptor antagonists. *Journal of Medicinal Chemistry*, *48*(6), 1823–1838. https://doi.org/10.1021/jm040843r.

Lange, J. H. M., van Stuivenberg, H. H., Veerman, W., Wals, H. C., Stork, B., Coolen, H. K. A. C., & Kruse, C. G. (2005). Novel 3,4-diarylpyrazolines as potent cannabinoid CB1 receptor antagonists with lower lipophilicity. *Bioorganic & Medicinal Chemistry Letters*, *15*(21), 4794–4798. https://doi.org/10.1016/j.bmcl.2005.07.054.

Lipinski, C. A. (2004). Lead- and drug-like compounds: The rule-of-five revolution. *Drug Discovery Today: Technologies*, *1*(4), 337–341. https://doi.org/10.1016/j.ddtec.2004.11.007.

Lipinski, C. A., Lombardo, F., Dominy, B. W., & Feeney, P. J. (2001). Experimental and computational approaches to estimate solubility and permeability in drug discovery and development settings. *Advanced Drug Delivery Reviews*, *46*(1–3), 3–26. https://doi.org/10.1016/S0169-409X(00)00129-0.

Liu, J., Godlewski, G., Jourdan, T., Liu, Z., Cinar, R., Xiong, K., & Kunos, G. (2019). Cannabinoid-1 receptor antagonism improves glycemic control and increases energy expenditure through sirtuin-1/mechanistic target of rapamycin complex 2 and 5'adenosine monophosphate-activated protein kinase signaling. *Hepatology*, *69*(4), 1535–1548. https://doi.org/10.1002/hep.30364.

Liu, Q.-R., Huang, N. S., Qu, H., O'Connell, J. F., Gonzalez-Mariscal, I., Santa-Cruz-Calvo, S., & Egan, J. M. (2018). Identification of novel mouse and rat CB1R isoforms and in silico modeling of human CB1R for peripheral cannabinoid therapeutics. *Acta Pharmacologica Sinica*, *40*(3), 387–397. https://doi.org/10.1038/s41401-018-0152-1.

Liu, Z., Iyer, M. R., Godlewski, G., Jourdan, T., Liu, J., Coffey, N. J., & Cinar, R. (2021). Functional selectivity of a biased cannabinoid-1 receptor (CB1R) antagonist. *ACS Pharmacology & Translational Science*, *4*(3), 1175–1187. https://doi.org/10.1021/acsptsci.1c00048.

Mackie, K. (2008). Cannabinoid receptors: Where they are and what they do. *Journal of Neuroendocrinology*, *20*(Suppl 1), 10–14. https://doi.org/10.1111/j.1365-2826.2008.01671.x.

Osei-Hyiaman, D., Liu, J., Zhou, L., Godlewski, G., Harvey-White, J., Jeong, W., & Kunos, G. (2008). Hepatic CB1 receptor is required for development of diet-induced steatosis, dyslipidemia, and insulin and leptin resistance in mice. *The Journal of Clinical Investigation*, *118*(9), 3160–3169. https://doi.org/10.1172/JCI34827.

Pacher, P., Bátkai, S., & Kunos, G. (2006). The endocannabinoid system as an emerging target of pharmacotherapy. *Pharmacological Reviews*, *58*(3), 389–462. https://doi.org/10.1124/pr.58.3.2.

Patel, P. N., & Pathak, R. (2007). Rimonabant: A novel selective cannabinoid-1 receptor antagonist for treatment of obesity. *American Journal of Health-System Pharmacy*, *64*(5), 481–489. https://doi.org/10.2146/060258.

Pertwee, R. G. (2000). Cannabinoid receptor ligands: Clinical and neuropharmacological considerations, relevant to future drug discovery and development. *Expert Opinion on Investigational Drugs*, *9*(7), 1553–1571. https://doi.org/10.1517/13543784.9.7.1553.

Rodríguez de Fonseca, F., Del Arco, I., Bermudez-Silva, F. J., Bilbao, A., Cippitelli, A., & Navarro, M. (2005). The endocannabinoid system: Physiology and pharmacology. *Alcohol and Alcoholism*, *40*(1), 2–14. https://doi.org/10.1093/alcalc/agh110.

Roger, C., Buch, C., Muller, T., Leemput, J., Demizieux, L., Passilly-Degrace, P., & Jourdan, T. (2020). Simultaneous inhibition of peripheral CB1R and iNOS mitigates obesity-related dyslipidemia through distinct mechanisms. *Diabetes*, *69*(10), 2120–2132. https://doi.org/10.2337/db20-0078.

Shao, Z., Yin, J., Chapman, K., Grzemska, M., Clark, L., Wang, J., & Rosenbaum, D. M. (2016). High-resolution crystal structure of the human CB1 cannabinoid receptor. *Nature*, *540*(7634), 602–606. https://doi.org/10.1038/nature20613.

Sharma, M. K., Murumkar, P. R., Kanhed, A. M., Giridhar, R., & Yadav, M. R. (2014). Prospective therapeutic agents for obesity: Molecular modification approaches of centrally and peripherally acting selective cannabinoid 1 receptor antagonists. *European Journal of Medicinal Chemistry*, *79*, 298–339. https://doi.org/10.1016/j.ejmech.2014.04.011.

Southan, G. J., Szabó, C., & Thiemermann, C. (1995). Isothioureas: Potent inhibitors of nitric oxide synthases with variable isoform selectivity. *British Journal of Pharmacology*, *114*(2), 510–516. https://doi.org/10.1111/j.1476-5381.1995.tb13256.x.

Srivastava, B. K., Joharapurkar, A., Raval, S., Patel, J. Z., Soni, R., Raval, P., & Patel, P. R. (2007). Diaryl dihydropyrazole-3-carboxamides with significant in vivo antiobesity activity related to CB1 receptor antagonism: Synthesis, biological evaluation, and molecular modeling in the homology model. *Journal of Medicinal Chemistry*, *50*(24), 5951–5966. https://doi.org/10.1021/jm061490u.

Tam, J., Cinar, R., Liu, J., Godlewski, G., Wesley, D., Jourdan, T., & Kunos, G. (2012). Peripheral cannabinoid-1 receptor inverse agonism reduces obesity by reversing leptin resistance. *Cell Metabolism*, *16*(2), 167–179. https://doi.org/10.1016/j.cmet.2012.07.002.

Tam, J., Vemuri, V. K., Liu, J., Bátkai, S., Mukhopadhyay, B., Godlewski, G., & Kunos, G. (2010). Peripheral CB1 cannabinoid receptor blockade improves cardiometabolic risk in mouse models of obesity. *The Journal of Clinical Investigation*, *120*(8), 2953–2966. https://doi.org/10.1172/JCI42551.

Tan, S., Liu, H., Ke, B., Jiang, J., & Wu, B. (2020). The peripheral CB1 receptor antagonist JD5037 attenuates liver fibrosis via a CB1 receptor/β-arrestin1/Akt pathway. *British Journal of Pharmacology*, *177*(12), 2830–2847. https://doi.org/10.1111/bph.15010.

Trapa, P. E., Troutman, M. D., Lau, T. Y., Wager, T. T., Maurer, T. S., Patel, N. C., & Liras, J. L. (2019). In vitro-in vivo extrapolation of key transporter activity at the blood-brain barrier. *Drug Metabolism and Disposition: The Biological Fate of Chemicals*, *47*(4), 405–411. https://doi.org/10.1124/dmd.118.083279.

Udi, S., Hinden, L., Ahmad, M., Drori, A., Iyer, M. R., Cinar, R., & Tam, J. (2020). Dual inhibition of cannabinoid CB1 receptor and inducible NOS attenuates obesity-induced chronic kidney disease. *British Journal of Pharmacology*, *177*(1), 110–127. https://doi.org/10.1111/bph.14849.

Ward, S. J., & Raffa, R. B. (2011). Rimonabant redux and strategies to improve the future outlook of CB1 receptor neutral-antagonist/inverse-agonist therapies. *Obesity*, *19*(7), 1325–1334. https://doi.org/10.1038/oby.2011.69.

Zawatsky, C. N., Park, J. K., Abdalla, J., Kunos, G., Iyer, M. R., & Cinar, R. (2021). Peripheral hybrid CB1R and iNOS antagonist MRI-1867 displays anti-fibrotic efficacy in bleomycin-induced skin fibrosis. *Frontiers in Endocrinology*, *12*, 744857. https://doi.org/10.3389/fendo.2021.744857.

Part IV

Neurobiology of cannabis

Chapter 17

Neurobiology of the endocannabinoid system and relevance to cannabis use

Deepak Kumar[a], **Bareera Khan**[b], **Aditi Sharma**[a], **and Rohit Goyal**[a]

[a]*Department of Neuropharmacology, School of Pharmaceutical Sciences, Shoolini University, Solan, HP, India,* [b]*Faculty of Applied Sciences and Biotechnology, Shoolini University of Biotechnology and Management Sciences, Solan, HP, India*

Introduction

The endocannabinoid system (ECS) is an essential neuromodulatory system required for the normal functioning of the central and peripheral nervous system (CNS and PNS). In addition, it appears to neuromodulate multiple vital physiologies such as circadian rhythms (CR), memory, brain homeostasis, and neuroprotection. For example, the ECS protects the suprachiasmatic nucleus (SCN), a central control center for the orchestration of endogenous CRs, from light-induced glutamatergic excitotoxicity by decreasing its ability to become entrained to the light stimulus (Acuna-Goycolea et al., 2010). Further, ECS also plays an essential role in protecting CNS and PNS from external insults such as lifestyle-mediated oxidative stress and related neurodegeneration (Parolaro et al., 2010; Ruehle et al., 2012).

The ECS comprises the cannabinergic receptors (cannabinoid and noncannabinoid receptors), endocannabinoids (eCBs) and enzymes regulating the biosynthesis and degradation of eCBs. In addition, exocannabinoids, such as synthetic and phytocannabinoids, also activate ECS. For instance, *Cannabis sativa* (*C. sativa*) was never new to mankind and was mentioned in the earliest Indian text known to humans, i.e., *Vedas* (Bapat, 2015). In Book 11, Hymn 6, Verse 15 of Atharva Veda-Samhita, it says:

पञ्;च राज्;यानि वीरुधां सोमश्;रेष्;ठानि ब्;रूमः।
दर्;भो भङ्;गो यवः सह ते नो मुञ्;चन्;त्;व्; अंहसः॥

This means "To the five kingdoms of the plants which Soma rules as Lord we speak. Darbha, hemp, barley, mighty power: may these deliver us from woe." (Ferrara, 2016; Yadav, 2022). One another shloka from Bhavaprakash, an ayurvedic book, suggests its uses as anti-phlegmatic, increases digestion, acid and bile-affecting, aphrodisiac, analgesic, anti-diarrheal, and so on (Yadav, 2022). In addition, aside from numerous ancient Indian books, other civilizations across the globe, such as the Chinese, Egyptians, Greece, and Roman empire, were also aware of the physio-psychological effects of *C. sativa* (Crocq, 2020). However, due to its psychotropic effects, plant use is mainly restricted to its medicinal use. The main phytochemical of *C. sativa*, i.e., Δ^9-THC, exerts its action by activating the cannabinoid receptors (CBR1 and CBR2) of the ECS, producing numerous pharmacological effects (Carlini, 2004).

In the present chapter, we ought to shed light on the neurobiology of ECS and the effects of various agonists and antagonists of the system, as numerous studies have depicted the direct/indirect role of ECS in distinct pathologies. For instance, dysfunctional ECs have been reported in both neurodegenerative (Alzheimer's disease, Parkinson's disease) and neuropsychological (depression and anxiety) disorders (Parolaro et al., 2010; Tiller, 2013).

Endocannabinoid system

ECS is primarily composed of two cannabinoid receptors, viz., cannabinoid receptors 1, 2 (CBR1, CBR2), endogenous cannabinoids N-arachidonoylethanolamine (or anandamide, where "ananda" means bliss in Sanskrit), and 2-arachidonoyl glycerol (2-AG). ECS also comprises the enzyme system responsible for the synthesis and degradation of eCBs, including NAPE-phospholipase-D (PLD), diacylglycerol lipase (DAGL), fatty acid amide hydrolase (FAAH), and monoacylglycerol lipase (MAGL) (Lu & Mackie, 2016). The ECS is a vital and highly complex neuromodulatory system whose components can be traced throughout the CNS and PNS (Lowe et al., 2021). The brief repertoire of ECS is attributed to the brain

Cannabis Use, Neurobiology, Psychology, and Treatment. https://doi.org/10.1016/B978-0-323-89862-1.00003-9

functions (e.g., brain homeostasis, cognition, and memory), gastrointestinal tract (GIT), cardiovascular system, endocrine system, immune system, etc. The detailed description is given in "Neuromodulation by endocannabinoids" section.

General pharmacology (Fig. 1A)

Unlike typical neurotransmitters such as glutamate, eCBs are synthesized and released on demand. Also, despite sharing an arachidonic acid chain, the synthesis and degradation pathways of eCBs are separate and require different enzymes (Felder & Glass, 1998). The synthesis of anandamide begins with an increase in intracellular Ca^{2+}. The activation of certain G-protein-coupled receptors (GPCRs), such as vanilloid receptors, also increases the intracellular levels of Ca^{2+} and induces the synthesis of anandamide. Initially, calcium-dependent N-acyltransferase (NAT) transfers an arachidonoyl group to a membrane-bound phospholipid phosphatidylethanolamine, resulting in N-arachidonoyl phosphatidylethanolamine (NAPE). Later, PLD catalyzes these NAPEs into anandamide (Felder & Glass, 1998; Snider et al., 2010). This synthesis pathway for anandamide is commonly accepted, however, it appears not exclusive. For example, a study by Leung et al. (2006) showed preserved, but five times reduced conversion of NAPEs to *N*-acyl ethanolamines (NAEs) in NAPE-PLD gene knockout mice. The other proposed pathways include PLC-mediated conversion of NAPEs into phospho-NAE species, which are then converted into anandamide by Protein tyrosine phosphatase non-receptor type 22 (PTPN22) or SH2-containing inositol-5-phosphatase (SHIP1). In addition, NAPEs are also catalyzed by the soluble form of phospholipase A2 (sPLA2) or serine hydrolase α/β-hydrolase 4 (ABHD4) into lyso-NAPE, which are then converted by ABHD4 into GP-NAE species. Finally, phosphodiesterase GDE1 facilitates the conversion of GP-NAE into anandamide (Snider et al., 2010). On the contrary, the degradation of anandamide occurs through three pathways out of which the FAAH-mediated hydrolysis of anandamide into arachidonic acid and ethanolamine appears to be the dominant one (Cravatt et al., 1996). The second pathway includes N-acylethanolamine-hydrolyzing acid amidase (NAAA), which, like FAAH, converts anandamide into arachidonic acid and ethanolamine (Tsuboi et al., 2005). Further, the third pathway is the cyclooxygenase-2 (COX-2)-mediated conversion of anandamide into prostamides (prostaglandin-ethanolamides) (Woodward et al., 2008).

The synthesis of 2-AG is rather more straightforward than anandamide. However, more enzymes are involved in its degradation. Like anandamide, synthesis of the 2-AG also initiates with a rise in intracellular Ca^{2+} levels. However,

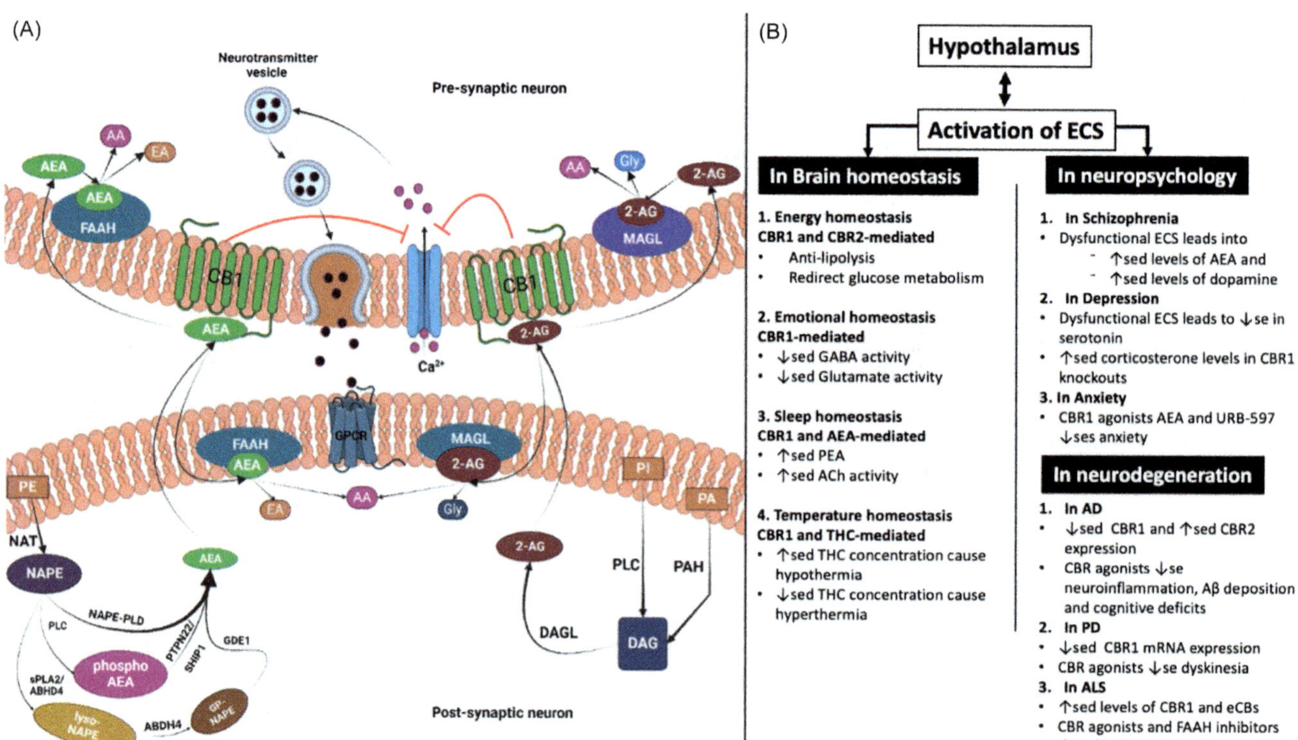

FIG. 1 General pharmacology and neuromodulation of ECS in brain physio/pathology. *AEA*, anandamide; *AA*, arachidonic acid; *EA*, ethanolamine; *Gly*, glycerol; *AD*, Alzheimers disease; *PD*, Parkinson's disease; *ALS*, amyotrophic lateral sclerosis.

the process appears independent of anandamide's synthesis (Piomelli, 2013). The activation of phospholipase C (PLC, PLCβ) results in the conversion of membrane-bound phospholipid phosphatidylinositol (PI; PIP2) into diacylglycerol (DAG), which ultimately is converted into 2-AG by DAGL. Interestingly, two isoforms of DAGL, i.e., DAGLα and DAGLβ can generate 2-AG in vitro, though DAGLα appears to be the important one, as shown by the knockout studies whereas DAGLβ is essential in autaptic neurons' physiology (Jain et al., 2013). Further, another phospholipid, phosphatidic acid (PA), also acts as a precursor for DAG, where phosphatidic acid hydrolase (PAH) acts as a catalytic enzyme (Lu & Mackie, 2016). The degradation of 2-AG is primarily acheived by MAGL, which hydrolyzes it into arachidonic acid and glycerol. Studies suggest that about 85% of 2-AG is hydrolyzed by MAGL across majorityof the brain area, including the hippocampus, with maximum localization at stratum oriens (Ludányi et al., 2011). The remaining 15% of 2-AG is hydrolyzed into arachidonic acid and glycerol by ABHD6 and ABHD12 (Blankman et al., 2007). Interestingly, FAAH is also reported to hydrolyze 2-AG in certain conditions. In another pathway, COX-2-mediated degradation of 2-AG generates an oxidative metabolite of 2-AG, i.e., PGE2-glycerol ester (PGE2-GE), which potentiates synaptic transmission and hyperalgesia (Sang et al., 2006).

Receptor targets for endocannabinoids

In the mid-1980s, Pfizer's development of multiple potent cannabinoids potentiated in vitro studies and suggested that psychotropic cannabinoids can inhibit adenylate cyclase through $G_{i/o}$ proteins (Pertwee, 2006). Later, the first successful cloning of CBR1 and CBR2 in early 1990 (Matsuda et al., 1990) and 1993 (Munro et al., 1993), respectively, led to the discovery of their endogenous ligands anandamide in 1992 (Devane et al., 1992) and 2-AG in 1995 (Mechoulam et al., 1995). CBR1 and CBR2 receptors are the primary targets of eCBs, phytocannabinoid, and synthetic cannabinoids. However, 2-AG has been found to be a more potent agonist than anandamide. The predominant expression of CBR1 is found abundantly in the CNS. Specifically, they are highly expressed in the hippocampus, amygdala, cortex, basal ganglia, cerebellum, and the external projection of CNS, i.e., retina. In addition, CBR1 is also expressed in the reproductive system (such as testis, sperm cells, uterus, and ovary), adipocytes, lungs, heart, adrenal glands, prostate, and colon (Reggio, 2010). CBR2 expression, on the other hand, is not as prominent in the brain as compared to CBR1 and appears to be more prevalent in the central and peripheral immune systems. CBR2 was initially cloned from the human promyelocytic leukemia (HL60) cell line and was found to be highly expressed in the microglia, natural killer cells, B-cells, macrophages, and to some extent in blood vessels. Further, in CNS, CBR2 expression is found in the basal ganglia, hippocampus, cerebellum, and brain stem (Fox & Faust-Socher, 2017). Apart from CBRs, *endo*- and exocannabinoids were also found to interact with several other GPCRs, such as GPR18, GPR55, and GPR119 (Morales & Reggio, 2017; Moriconi et al., 2010). Laun and Song (2017) also showed phytocannabinoid cannabidiol as an inverse agonist of class A orphan GPCRs, i.e., GPR3 and GPR6 through β-arrestin2 recruitment assay. Interestingly, eCBs, such as anandamide, has also been shown to modulate some other GPCRs as well, e.g., muscarinic (M1 and M4) and serotonin (5-HT_{1A} and 5-HT_{2A}) receptors (Morales & Reggio, 2017).

Neuromodulation by endocannabinoids

When CBRs are activated by endo/exogenous cannabinoids, they couple to the second messenger protein $G_{i/o}$, which inhibits adenylate cyclase and causes a decline in cAMP levels as there is not enough adenylate cyclase available to convert ATP into cAMP. Activation of Gi/o protein also prevents the influx of Ca^{2+}ions by blocking the calcium channels, thus alleviating the release of neurotransmitters in the synaptic region. In addition, $G_{i/o}$ activation also increases the outward flow of K^+ ions, causing alleviation of action potentials and synaptic transmission (Ashton et al., 2005). Besides typical CBRs, recent studies have also shown cannabinoids mediated modulation of various Transient receptor potential (TRP) channels which is suggestive of their role as ionotropic cannabinoid receptors. TRP channels are known to modulate ion entry inside the cells, thus mediating neural signaling in the functioning of multiple physiologies, such as vision, nociception, temperature, pressure, taste, and smell. TRPV1–4, TRPA1, and TRPM8 are some TRP channels modulated by cannabinoids (Muller et al., 2018). In addition, cannabinoids are also reported to modulate various other ion channels, including Ca^{2+}, Na^+, K^+ channels, and ligand-gated ion channels, such as GABA, glycine, and serotonin 5-HT_3 receptor-ion channel in a CBR-independent way (De Petrocellis et al., 2017).

By modulating the above-mentioned targets, ECS affects various physio/pathological processes and may contribute as a therapeutic target. For instance, in GIT, the eCBs are present ubiquitously. Also, the immunoreactivity of CBR1 has been found in the cholinergic neurons of the enteric nervous system, vasoactive intestinal peptide-expressing ganglion cells of submucosa, dorsovagal complex, and at the terminal ends of the vagus nerve, thus regulating intestinal motility, hunger, gastroesophageal reflux, and GIT secretions, such as gastric acid and intestinal secretion (Coutts & Izzo, 2004). Further,

CBR2 mRNA is expressed in the esophagus, stomach, ileum, and various immune cells, e.g., activated macrophages and plasma cells of lamina propria (Wright et al., 2008). The presence of ECS in GIT suggests a scope of various cannabinoid-based therapeutics in multiple pathologies, for instance, paralytic ileus (Mascolo et al., 2002), immunomodulation (Acharya et al., 2017), and obesity (Klein, 2005). In fact, plant-based cannabinoids have already been used since ancient times as a treatment for Crohn's disease, nausea, vomiting, anorexia, hyperemesis, and weight loss (Goyal et al., 2017). Besides, eCBs signaling is also known to suppress the activity of the hypothalamic-pituitary-adrenal (HPA) axis in stressful conditions. Hill et al. (2009) demonstrated that exposing lab animals to stressful situations can escalate the FAAH-mediated degradation of eCBs in the amygdala, which negatively correlates with the serum corticosterone levels, suggesting an augmented HPA-axis response to stress. These findings were corroborated by Newsom et al. (2012) in their study depicting the increased levels of adrenocorticotropic hormone and corticosterone in the plasma of Sprague-Dawley rats exposed to loud noise-mediated stressful conditions. Further, activation of CB2 receptors is also cardioprotective and exerts anti-inflammatory and antifibrosis effects (Kaschina, 2016).

Physiological outcomes of endocannabinoids

Brain homeostasis (Fig. 1B)

The brain is the master controller of the body and is responsible for various activities such as temperature, sleep, and drinking. These activities, in turn, are responsible for maintaining the body's homeostasis or being in a state of stability. Homeostasis is a self-managing mechanism that maintains physiological changes and ensures the stability of biological systems by adapting to the hostile variations of the external environment (Billman, 2020). Hence, this concept justifies how an individual's internal conditions can parallelly and constantly be monitored with the external environment. In humans, the complex process of homeostasis is controlled through chemical messengers such as hormones and regulatory mechanisms, comprising of three components: a receptor, an effector, and a control center. The regulatory mechanism works in a loop and can result in positive feedback leading to stimulation of a certain process or negative feedback leading to decelerating a certain process. The term homeostasis was coined by the scientist Walter Cannon and gave fundamental concepts on the topic, which are widely accepted in various fields of medicine, neuroscience, and endocrinology (Davies, 2016). The hypothalamus is the master regulator of the homeostatic modulation for crucial activities such as sleep, controlling appetite or energy homeostasis, body temperature mechanism, and emotional homeostasis (Cacioppo & Berntson, 2011).

Energy homeostasis

Energy homeostasis refers to a state where energy intake matches the energy spent. In cases where the body fails to attain so, conditions like obesity or anorexia may occur. The hypothalamus, which is the key structure and homeostasis regulator, pertains this information through various hormones secreted mainly from the white adipose tissues and gastrointestinal tract and via different pathways that can detect the body's energy status (Cristino et al., 2014). The neurohormonal signaling mechanism is a complex and integrated network system that receives information through an orexigenic pathway in case of increased food intake and through an anorexigenic pathway when energy intake is reduced. The two major peripheral signals the CNS receives are adipocity signals, such as insulin and leptin, which inform the brain about the stored energy levels and the satiety signals. In return, the brain delivers a homeostatic feedback response to the periphery (Bermudez-Silva et al., 2012). The ECS plays a vital role in energy storage and human obesity.

The first evidence of ECS use in the control of food intake was reported through a study that facilitated eating by systemically administering Δ^9-THC (Trojniar & Wise, 1991). Other studies have also reported that peripheral and intra-hypothalamic administration of CB1 agonists and eCBs can increase animal food intake. Thus, it has been observed that ECS signaling via CBR1 directs an individual to eat through general anabolic effects, such as (a) promoting food intake as means of motivation or pleasure to eat, (b) redirecting lipid metabolism to anti-lipolytic or pro-lipogenic processes, and (c) redirecting glucose metabolism. Therefore, modulation of the ECS can be studied as a therapeutic strategy for weight loss in obese. Currently, increased interest in the market is reported for selective CBR1 antagonists such as AM6545, an analog of earlier used rimonabant and CBR1 inverse agonist such as JD5037 (Kunos et al., 2008).

Emotional homeostasis

As the intensity of negative emotions such as fear and anxiety exceeds a disproportional number or becomes chronic and irreversible, they can result in increased impairment of psychological functions. Such emotions can result from neuropsychiatric disorders like stress disorder and generalized anxiety. The ECS plays a pivotal role in the brain's emotional development and motivational status. In fact, anxiety, fear, stress and reward-driven behaviors are importantly monitored by the ECS. It has been reported that CB1 receptors are highly distributed in areas that govern emotional activity, thereby controlling behavior and mood (Rusconi et al., 2020). They are highly expressed in high densities in areas such as limbic, frontal regions, hypothalamus, and midbrain monoaminergic nuclei and hence are responsible for modulating neurotransmitter release by activating CB1 receptors (Stampanoni Bassi et al., 2018).

Further, 2-AG has a vital role in the stress reduction phenomenon. When excitatory responses are evoked due to stress conditions, the ECS comes into the picture, limiting the release of glutamate via retrograde activation of the CBR1 receptor at the synapse and through 2-AG stress-induced synthesis. It is evident from studies that both glutamatergic and GABAergic transmissions are responsible for anxiety, and due to the presence of CB1R on axon terminals of both these neurotransmitters, CBR1 and the ECS can portray its mechanism in controlling anxiety. Therefore, it can be indicated that 2-AG inhibits stress-induced neurotransmitter release by suppressing excited glutamate release and inhibiting GABA release. However, the role of another ligand of the ECS, i.e., anandamide, is a quiet paradox because it exhibits dual nature of anxiolytic nature on CBR1 and anxiogenic nature on the TRPV1 channel (Ruehle et al., 2012).

Sleep homeostasis

The sleep-wake cycle is a complex process that involves various molecules and brain circuits. Different neurochemicals such as dopamine, acetylcholine, noradrenaline, and hypocretin interplay and regulates the neuroanatomical system (Vaughn et al., 2010). Sleep is induced by sleep neurons in the anterior hypothalamus that utilizes GABA neurotransmitters to inhibit the wakefulness of the brainstem and hypothalamus regions. It has been known, and many studies also have reported the use of cannabinoids such as marijuana and Δ^9-THC to promote sleep. The most promising link of cannabinoids inducing sleep is established due to the expression of CBR1 in the sleep-promoting areas. CBR1 is distributed in the basal forebrain's neurons, and pons activates cholinergic neurons localized in the same area. Further, when the CBR1 gets activated, they enhance the release of acetylcholine, and this sleep-promoting activity can be correlated with previous studies stating that acetylcholine level from the basal forebrain and brainstem is found to be higher during sleep (Murillo-Rodríguez, 2008).

Therefore, it can be interpreted that CBR1 expressed in cholinergic neurons increases sleep due to the release of acetylcholine, and this activation triggers the neuronal activity in the thalamus to potentiate cortical desynchronization (Zant et al., 2016). Other studies have also demonstrated the role of a fatty acid amide named palmitoylethanolamide (PEA) in increasing the levels of anandamide and thereby regulating sleep homeostasis. PEA is an endogenous compound that stimulates the production of cannabinoids in our body and promotes restful sleep again through the activation of CBR1; hence, the ECS can be targeted as an emerging approach in the treatment of sleep-related disturbances such as excessive diurnal somnolence and insomnia (Lu & Mackie, 2016).

Body temperature regulation

Thermoregulation is a mechanism through which our body regulates its temperature and responds to external changes. Neurotransmitters such as dopamine, opioid peptides, and GABA can produce effects like hypothermia when our temperature homeostasis is not regulated. Therefore, monitoring body temperature is one of the primary roles of brain homeostasis. Many reports give evident data stating the considerable effect of CBR1 on body temperature regulation (Vaughn et al., 2010). Exogenous administration of THC has portrayed significant hypothermia, while lower dose administration can lead to hyperthermia conditions. However, this regulation in the body temperature is a result of dose-dependent administration. The body temperature of individuals exhibits a circadian rhythm on a daily basis which dominates during its active phase, and this rhythm is regulated by the released melatonin from the pituitary gland. The melatonin release is higher in the inactive phase and is likely to drive lower body temperature, whereas if we correlate this with THC, its concentration is maximum in the inactive phase, especially during the second half and minimum in the active phase. Hence, it could be suggested that THC potentiates melatonin secretion and contributes to regulating body temperature (Wenger & Moldrich, 2002).

Circadian rhythms

To synchronize the physio-behavioral activities with the light and dark cycle of our planet, life forms on Earth have developed an endogenous time-keeping mechanism called CR. Situated in the hypothalamus, the keeper of this robust physiology is a master clock known as the SCN, which is made up of specialized neurons (Acuna-Goycolea et al., 2010). These neurons receive environmental cues from different sensory pathways, such as the retino-hypothalamic (RHT) pathway, and orchestrate daily activities by expressing specific proteins known as period (PER) and cryptochrome (CRY) proteins. According to the literature, ECS stays under the influence of the master clock and is a vital component in maintaining its robustness (Valenti et al., 2004) for review (Vaughn et al., 2010). Furthermore, there have been reports depicting the role of eCB signaling in preventing light-evoked phase shifts of locomotor activity in laboratory animals. Interestingly, CBR agonists have been reported as the inhibitor of both light-induced phase advances and delays by modulating the chief neurotransmitter of master clock, i.e., GABA (Acuna-Goycolea et al., 2010; Sanford et al., 2008). Although this field of research is new and in its primary state, the homeostatic role of ECS in circadian machinery seems promising and needs an array of investigation.

Memory and cognition

Memory formation is a multistep and dynamic procedure comprising information acquisition, consolidation, and retrieval. ECS appears to be one vital physiological system in this case, as it exists in almost all the critical memory-regulating regions of the brain, such as the hippocampus, amygdala, and cerebellum. Furthermore, various brain areas that participate in the cognitive processes perform distinct roles in forming different types of memories. For instance, the hippocampus is mainly indulged in the processing of declarative memories, e.g., episodic and semantic memories (Roediger et al., 2008). Further, fear memories are primarily processed through the amygdala (Josselyn, 2010). On the other hand, implicit memories (procedural memory, classical conditioning, and motor learning) are regulated by the cerebellum (Cooke et al., 2004).

Multiple studies suggest that the ECS is critical in memory consolidation. For instance, in the hippocampus, eCBs signaling is essential in memory updation, consolidation of fear memories, and behavioral flexibility (Lunardi et al., 2020). In the amygdala, arachidonylcyclopropylamide, a CB1 agonist, was shown to rescue the impaired memory consolidation induced by scopolamine (Nedaei et al., 2016). In the dorsal striatum, glucocorticoid-mediated recruitment of eCBs signaling increases retention avoidance in inhibitory avoidance training (Siller-Pérez et al., 2019). Interestingly, despite the negative effect of CBR agonists on memory acquisition, certain studies suggest no effect on its consolidation (Alarcon et al., 2020).

In addition, eCBs signaling is also necessary for memory and plasticity in the CA1 region of memory (Misner & Sullivan, 1999). Sullivan (1999) demonstrated that eCBs signaling can decrease the synaptic transmission of pyramidal neurons in the hippocampus by inhibiting glutamate's release. Pyramidal cells of the CA1 region and subiculum area of the hippocampus are responsible for transferring cognitive information throughout the brain which they generate by processing spatial, contextual, and emotional information and creating a cognitive map (Sullivan, 1999). Further, studies suggest that hybrid offspring of APP/PS1 transgenic mice and CB1 knockout mice displays less survival and dyssynchronized neurotransmission in hippocampus, resulting in memory impairement (Aso et al., 2018).

Evidence of endocannabinoids in neuropathology (Fig. 1B)

ECS in mood disorders

Mood disorders chiefly comprise conditions including anxiety, depression, and schizophrenia. However, these conditions cannot be looked at in isolation since their pathologies coincide. For instance, it has been reported that almost 85% of patients with depression experience anxiety and 90% of those with anxiety experience a depressive state of mind. In addition, patients with schizophrenia also display depression as one of their symptoms (Tiller, 2013). Pathologically, in both anxiety and depression, cerebrospinal fluid (CSF) levels of corticotropin-releasing factor (Tiller, 2013) and glucocorticoids (Rajkowska & Miguel-Hidalgo, 2007) were noted to be increased. Besides, treatment is also common for both anxiety and depression, involving antidepressants and rectifying endogenous circadian rhythm disruption, a recent treatment approach (Tiller, 2013). Schizophrenia, on the other hand, is a severe mental condition signified by delusions, hallucinations, social withdrawal, and cognitive deficits. In addition, numerous genetic and environmental factors cause dysfunction of cortical circuits, further resulting in dysregulation of NMDA receptors, glutamatergic signaling, and subcortical dopamine levels. The treatment options for schizophrenia currently include blockers of dopamine D_2-receptors (McCutcheon et al., 2020).

ECS in schizophrenia and depression

Literature suggests dysfunctional ECS in mood disorders. In schizophrenia, reports suggest polymorphisms in the gene encoding CBR1, i.e., cannabinoid-1 receptor (CNR1) gene. Further, the anandamide levels are significantly higher in patients with schizophrenia compared to healthy individuals, and pharmacological evaluations of CB1R antagonists in pre-clinical and clinical trials have shown positive results in schizophrenia (Ashton & Moore, 2011). Besides, GPCRs, such as NMDA receptors, are of interest in the case of mood disorders, particularly schizophrenia and depression. There is a crosstalk between the functioning of NMDA receptors with other GPCRs, which occurs through the NR1 C1 subunits and the tandem histidine triad nucleotide-binding protein 1 (HINT1). The levels of HINT1 were found to decrease in schizophrenia and vice-versa in the case of depression (Rodríguez-Muñoz et al., 2017). Interestingly, eCBs regulate the crosstalk between these receptors through the NR1 C1 subunits and negatively regulate the NMDA receptors (Ashton & Moore, 2011). Therefore, any dysfunction in the eCB control results in different pathologies. For instance, a dysfunctional ECS leads to increased dopamine signaling in schizophrenia and decreased serotonin activity in depression (Rodríguez-Muñoz et al., 2017). In addition, the ECS needs to be boosted via CBR1. Various studies have, hence, shown the antidepressive effects of CB1R agonists. One such study in rats reported that stimulation of CBR1 produced antidepression effects when analyzed in a forced swimming test (Ashton & Moore, 2011). Another study used CBR1 knockout mice and reported that these mice were more vulnerable to the effects of chronic stress. They also reported high corticosterone levels in knockout mice compared to the wild type, which is representative of a hyperactive HPA-axis (Parolaro et al., 2010). Similar results were observed after using URB-597, an inhibitor of FAAH that hydrolyzes anandamide. However, impairment of the CBR1, on the other hand, was found to be responsible for depression-like symptoms (Ashton & Moore, 2011).

ECS in anxiety

The involvement of ECS in anxiety has been correlated with euphoric feelings in recreational cannabis users, followed by reduced anxiety and increased sociability (Parolaro et al., 2010). The components of ECS, including ligand, anandamide, and receptors, CBR1 and TRPV1, are responsible for exerting their effects on anxiety (Rubino et al., 2008). It has been reported that at high doses, anandamide activates the CBR1 resulting in anxiolytic effects. A study also reports that when low doses of anandamide and URB-597 were given separately, anxiolytic effects were observed; however, in combination, they exerted the opposite, i.e., anxiogenic effects (Scherma et al., 2008). Besides, a series of studies using CBR1 antagonists and CBR1 knockout mice showed inconsistent observations, as some studies reported anxiogenic effects, while others showed no effects on anxiety (Parolaro et al., 2010).

ECS in neurodegenerative disorders

ECS in Alzheimer's disease

The most common form of dementia is Alzheimer's disease (AD), a neurodegenerative disorder involving the gradual loss of selective neurons in certain brain areas involved in learning and memory, such as the hippocampus and cortex. The disease is a combined pathology involving two proteins, amyloid beta (Aβ) and tau, and is associated with a decline in cognition and memory (Scheltens et al., 2016). The pathology of the disease involves the presence of amyloid plaques and neurofibrillary tangles (NFTs), which are the hallmarks of the disease (DeTure & Dickson, 2019). Plaques and NFTs hamper nutrient flow within the neurons, causing their death in a particular region. Furthermore, neuronal death leads to a cascade of events resulting in oxidative stress, which damages the cell organelles and proteins, thereby triggering the activation of microglia and astrocytes and the release of proinflammatory cytokines, which further leads to neuroinflammation. Therefore, both oxidative stress and inflammation are promising targets for the treatment of AD (Galasko & Montine, 2010).

The involvement of the ECS in AD is now an established fact. Various human and animal studies have reported that the accumulation of amyloid plaques activates surrounding microglia and astrocytes, which express CB1 and CB2 receptors on their surfaces (Ramírez et al., 2005). Besides, the expression of CBRs and their endogenous ligands is also affected in case of AD. For instance, the expression of CB1 receptors is reported to be reduced in AD patients and, in contrast, the CB2 receptor expression is reported to be increased (Solas et al., 2013). In AD, there is also and increase in CBR2 expression on microglia surrounding the plaques, whereas CB1 expression is reduced in the AD brain (Ramírez et al., 2005). There are also reports of enhanced activity of DAGL (a synthetic enzyme of 2-AG) and MAGL (a catabolic enzyme of 2-AG) in the hippocampus of the AD brain (Farooqui et al., 1988). In addition, elevated FAAH (catabolic enzyme of AEA) activity is also reported in astrocytes surrounding the plaques (Benito et al., 2003). Therefore, levels of these eCBs are consistently high in AD cases, especially that of 2-AG, whereas that of AEA may vary, and in some cases, no change is detected in the levels. Further, some studies have shown that activation of CB2 receptors inhibits TNF-α and nitric oxide production by microglia and macrophages and stimulates the phagocytosis of Aβ peptide by these cells (Tolón et al., 2009).

Numerous in -vitro and in vivo studies have reported the efficacy of cannbainods in ameliorating Aβ-induced neurotoxicity, and microglial activation. Martín-Moreno et al. (2012) reported that long-term administration of cannabinoids in Tg APP mice resulted in attenuation of microglila activation, reduction in TNF-α levels, COX-2 expression, and cortical Aβ deposition along with amelioration of cognitive deficits.

ECS in Parkinson's disease

Parkinson's disease (PD) is a neurodegenerative disease marked by the loss of dopaminergic neurons in the brain's substantia nigra, leading to insufficient dopamine production (Scotter et al., 2010). It is a proteinopathy involving the aggregation of α-synuclein (Bloem et al., 2021). The hallmarks of the disease include cell loss and Lewy bodies which are immunoreactive inclusions of synuclein (Davie, 2008). The disease is characterized by tremors, slow movements, and cognitive dysfunction (Scotter et al., 2010). Evidence suggests that, in PD, there is mitochondrial dysfunction along with increased activation of microglial cells. Both pathologies result in oxidative stress due to mitochondrial dysfunction and inflammation due to microglial activation (Beal, 2003).

There are lot of data inconsistency related to the changes in the levels of eCBs and associated receptors in PD (Scotter et al., 2010). CBR1 mRNA levels have found to be decreased in PD (Hurley et al., 2003). In PD patients, the CSF levels of AEA were more than twice as compared to controls (Pisani et al., 2005). In support of this, a study reported that levodopa treatment does not elevate AEA levels (Giuffrida et al., 1999) but upregulates the expression of CBR1 (Zeng et al., 1999). Similarly, in a human double-blinded placebo-controlled trial, CBR agonist nabilone, was found to reduce levodopa-induced dyskinesia symptoms (Sieradzan et al., 2001). However, a similar trial using cannabis extracts through the oral route showed no effect on the symptoms of dyskinesia in PD treatment with levodopa (Carroll et al., 2004). This inconsistency in clinical trial results cannot deny the role of ECS in PD and suggests a need for more extensive studies to establish consistent data.

ECS in amyotrophic lateral sclerosis

Amyotrophic lateral sclerosis (ALS), also known as Lou Gehrig's disease, is a neurodegenerative, crippling, and ultimately fatal motor neuron disease (Brown & Al-Chalabi, 2017). Excitotoxicity, oxidative damage, and neuroinflammation are thought to be the underlying causes of neuronal death in the cortical and spinal cord regions. Plasma and CSF levels of glutamate were significantly higher in ALS patients. These patients also showed a loss of function of Excitatory Amino Acid Transporter-2 (EAAT2), a major glutamate transporter expressed by glia (Scotter et al., 2010).

As per the literature, the role of CBR2 receptors has already been established in ALS due to their presence on microglia in regions of neuronal damage (Yiangou et al., 2006). In addition, CBR1 mRNA and protein levels were also found to be increased during disease progression (Zhao et al., 2008), and AEA and 2-AG are reported to accumulate in ALS.

Nevertheless, it has been shown that compounds that act on the ECS have the potential to alleviate excitotoxicity, oxidative damage, and inflammation associated with ALS. For instance, the administration of cannabinoid agonist, WIN55, 212-2, significantly delayed the disease pathology in mice and appeared beneficial in controlling the disease progression (Bilsland et al., 2006). Similarly, AM1241, a CB2 agonist, also slowed the disease progression on administration after disease onset (Kim et al., 2006). Further, Bilsland et al. (2006) demonstrated an exciting association between ECS and ALS, where they showed a significance of WIN55, 212-2 treatment and genetic ablation of FAAH but no effect of CBR1 ablation on the disease manifestation. Finally, the team concluded their study by signifying the role of cannabinoids by non-CBR mechanisms.

Conclusion

With the discovery of anandamide, 2-AG, CBR1, and CBR2 in the CNS and PNS, the ECS establishes its role in brain functions and can be considered an intriguing target for brain diseases and neurobiology. The evidence from various studies suggests that eCBs and cannabinergic receptors, majorly CBR1, exhibit interesting and potential targets for brain homeostasis and neurodegenerative diseases. Further, phytocannabinoids have been known in the Indian system of medicine since antiquity, and Ayurveda has mentioned *C. sativa* as one of the five sacred plants. The interaction of synthetic and phytocannabinoids with ECS plays a major role in attenuating the dyskinesia in the PD, neuroinflammation, and Aβ deposition in AD.

Summary points

- The ECS, with its ligands and receptors, is a complex neuromodulatory system throughout the central and peripheral nervous system that controls various physiologies, including brain functions.
- The gastrointestinal, immune, and endocrine systems also stay under the influence of the ECS.
- The dysfunctional ECS has been attributed to various neuropsychological and neurodegenerative disorders. For instance, the antagonists of CBR1 have shown the potential in decreasing schizophrenia.
- CBR1 knockouts have reported increased levels of corticosterone that were alleviated by the administration of CBR1 agonists in depression.
- ECS dysfunction is also reported in neurodegenerative conditions as well.
- CBR agonists are found to decrease neuroinflammation in Alzheimers' disease, decrease dyskinesia in Parkinson's disease, and delay disease progression in amyotrophic lateral sclerosis.

References

Acharya, N., Penukonda, S., Shcheglova, T., Hagymasi, A. T., Basu, S., & Srivastava, P. K. (2017). Endocannabinoid system acts as a regulator of immune homeostasis in the gut. *Proceedings of the National Academy of Sciences of the United States of America*, *114*, 5005–5010.

Acuna-Goycolea, C., Obrietan, K., & van den Pol, A. N. (2010). Cannabinoids excite circadian clock neurons. *The Journal of Neuroscience: The Official Journal of the Society for Neuroscience*, *30*, 10061–10066.

Alarcon, T. A., Areal, L. B., Herlinger, A. L., Paiva, K. K., Cicilini, M. A., Martins-Silva, C., & Pires, R. G. W. (2020). The cannabinoid agonist WIN-2 affects acquisition but not consolidation of a spatial information in training and retraining processes: Relation with transcriptional regulation of the endocannabinoid system? *Behavioural Brain Research*, *377*, 112231.

Ashton, C. H., & Moore, P. B. (2011). Endocannabinoid system dysfunction in mood and related disorders. *Acta Psychiatrica Scandinavica*, *124*, 250–261.

Ashton, C. H., Moore, P. B., Gallagher, P., & Young, A. H. (2005). Cannabinoids in bipolar affective disorder: A review and discussion of their therapeutic potential. *Journal of Psychopharmacology (Oxford, England)*, *19*, 293–300.

Aso, E., Andres-Benito, P., & Ferrer, I. (2018). Genetic deletion of CB1 cannabinoid receptors exacerbates the Alzheimer-like symptoms in a transgenic animal model. *Biochemical Pharmacology*, *157*, 210–216.

Bapat, S. (2015). Cannabis: The forgotten sacred plant of India. *Journal of Ayurveda and Holistic Medicine*, *3*, 92–96.

Beal, M. F. (2003). Mitochondria, oxidative damage, and inflammation in Parkinson's disease. *Annals of the New York Academy of Sciences*, *991*, 120–131.

Benito, C., Nunez, E., Tolon, R. M., Carrier, E. J., Rabano, A., Hillard, C. J., & Romero, J. (2003). Cannabinoid CB2 receptors and fatty acid amide hydrolase are selectively overexpressed in neuritic plaque-associated glia in Alzheimer's disease brains. *The Journal of Neuroscience: The Official Journal of the Society for Neuroscience*, *23*, 11136–11141.

Bermudez-Silva, F. J., Cardinal, P., & Cota, D. (2012). The role of the endocannabinoid system in the neuroendocrine regulation of energy balance. *Journal of Psychopharmacology*, *26*, 114–124.

Billman, G. E. (2020). Homeostasis: The underappreciated and far too often ignored central organizing principle of physiology. *Frontiers in Physiology*, *11*, 200.

Bilsland, L. G., Dick, J. R., Pryce, G., Petrosino, S., Di Marzo, V., Baker, D., & Greensmith, L. (2006). Increasing cannabinoid levels by pharmacological and genetic manipulation delay disease progression in SOD1 mice. *FASEB Journal : Official Publication of the Federation of American Societies for Experimental Biology*, *20*, 1003–1005.

Blankman, J. L., Simon, G. M., & Cravatt, B. F. (2007). A comprehensive profile of brain enzymes that hydrolyze the endocannabinoid 2-arachidonoylglycerol. *Chemistry & Biology*, *14*, 1347–1356.

Bloem, B. R., Okun, M. S., & Klein, C. (2021). Parkinson's disease. *The Lancet*, *397*, 2284–2303.

Brown, R. H., & Al-Chalabi, A. (2017). Amyotrophic lateral sclerosis. *The New England Journal of Medicine*, *377*, 162–172.

Cacioppo, J. T., & Berntson, G. G. (2011). *The brain, homeostasis, and health: Balancing demands of the internal and external milieu* (pp. 121–137). See Friedman.

Carlini, E. A. (2004). The good and the bad effects of (-) trans-delta-9-tetrahydrocannabinol (Delta 9-THC) on humans. *Toxicon: Official Journal of the International Society on Toxinology*, *44*, 461–467.

Carroll, C. B., Bain, P. G., Teare, L., Liu, X., Joint, C., Wroath, C., Parkin, S. G., Fox, P., Wright, D., Hobart, J., et al. (2004). Cannabis for dyskinesia in Parkinson disease: A randomized double-blind crossover study. *Neurology*, *63*, 1245–1250.

Cooke, S. F., Attwell, P. J., & Yeo, C. H. (2004). Temporal properties of cerebellar-dependent memory consolidation. *The Journal of Neuroscience: The Official Journal of the Society for Neuroscience*, *24*, 2934–2941.

Coutts, A. A., & Izzo, A. A. (2004). The gastrointestinal pharmacology of cannabinoids: An update. *Current Opinion in Pharmacology*, *4*, 572–579.

Cravatt, B. F., Giang, D. K., Mayfield, S. P., Boger, D. L., Lerner, R. A., & Gilula, N. B. (1996). Molecular characterization of an enzyme that degrades neuromodulatory fatty-acid amides. *Nature*, *384*, 83–87.

Cristino, L., Becker, T., & Di Marzo, V. (2014). Endocannabinoids and energy homeostasis: An update. *BioFactors*, *40*, 389–397.

Crocq, M. A. (2020). History of cannabis and the endocannabinoid system. *Dialogues in Clinical Neuroscience*, *22*, 223–228.

Davie, C. A. (2008). A review of Parkinson's disease. *British Medical Bulletin, 86*, 109–127.

Davies, K. J. (2016). Adaptive homeostasis. *Molecular Aspects of Medicine, 49*, 1–7.

De Petrocellis, L., Nabissi, M., Santoni, G., & Ligresti, A. (2017). Actions and regulation of ionotropic cannabinoid receptors. *Advances in Pharmacology (San Diego, Calif), 80*, 249–289.

DeTure, M. A., & Dickson, D. W. (2019). The neuropathological diagnosis of Alzheimer's disease. *Molecular Neurodegeneration, 14*, 1–18.

Devane, W. A., Hanus, L., Breuer, A., Pertwee, R. G., Stevenson, L. A., Griffin, G., Gibson, D., Mandelbaum, A., Etinger, A., & Mechoulam, R. (1992). Isolation and structure of a brain constituent that binds to the cannabinoid receptor. *Science (New York, N.Y.), 258*, 1946–1949.

Farooqui, A. A., Liss, L., & Horrocks, L. A. (1988). Stimulation of lipolytic enzymes in Alzheimer's disease. In *Lipid storage disorders* (pp. 689–698). Springer.

Felder, C., & Glass, M. (1998). Cannabinoid receptors and their endogenous agonists. *Annual Review of Pharmacology and Toxicology, 38*, 179–200.

Ferrara, M. S. (2016). *Sacred bliss: A spiritual history of cannabis.* Rowman & Littlefield.

Fox, S. H., & Faust-Socher, A. (2017). Cannabinoids☆. In *Reference module in neuroscience and biobehavioral psychology* Elsevier.

Galasko, D., & Montine, T. J. (2010). Biomarkers of oxidative damage and inflammation in Alzheimer's disease. *Biomarkers in Medicine, 4*, 27–36.

Giuffrida, A., Parsons, L. H., Kerr, T. M., Rodriguez de Fonseca, F., Navarro, M., & Piomelli, D. (1999). Dopamine activation of endogenous cannabinoid signaling in dorsal striatum. *Nature Neuroscience, 2*, 358–363.

Goyal, H., Singla, U., Gupta, U., & May, E. (2017). Role of cannabis in digestive disorders. *European Journal of Gastroenterology & Hepatology, 29*, 135–143.

Hill, M. N., McLaughlin, R. J., Morrish, A. C., Viau, V., Floresco, S. B., Hillard, C. J., & Gorzalka, B. B. (2009). Suppression of amygdalar endocannabinoid signaling by stress contributes to activation of the hypothalamic-pituitary-adrenal axis. *Neuropsychopharmacology: Official Publication of the American College of Neuropsychopharmacology, 34*, 2733–2745.

Hurley, M. J., Mash, D. C., & Jenner, P. (2003). Expression of cannabinoid CB1 receptor mRNA in basal ganglia of normal and parkinsonian human brain. *Journal of Neural Transmission (Vienna), 110*, 1279–1288.

Jain, T., Wager-Miller, J., Mackie, K., & Straiker, A. (2013). Diacylglycerol lipaseα (DAGLα) and DAGLβ cooperatively regulate the production of 2-arachidonoyl glycerol in autaptic hippocampal neurons. *Molecular Pharmacology, 84*, 296–302.

Josselyn, S. A. (2010). Continuing the search for the engram: Examining the mechanism of fear memories. *Journal of Psychiatry & Neuroscience, 35*, 221–228.

Kaschina, E. (2016). Cannabinoid CB1/CB2 receptors in the heart: Expression, regulation, and function. In *Cannabinoids in Health and Disease [Internet]* Intechopen.

Kim, K., Moore, D. H., Makriyannis, A., & Abood, M. E. (2006). AM1241, a cannabinoid CB2 receptor selective compound, delays disease progression in a mouse model of amyotrophic lateral sclerosis. *European Journal of Pharmacology, 542*, 100–105.

Klein, T. W. (2005). Cannabinoid-based drugs as anti-inflammatory therapeutics. *Nature Reviews Immunology, 5*, 400–411.

Kunos, G., Osei-Hyiaman, D., Liu, J., Godlewski, G., & Batkai, S. (2008). Endocannabinoids and the control of energy homeostasis. *Journal of Biological Chemistry, 283*, 33021–33025.

Laun, A. S., & Song, Z. H. (2017). GPR3 and GPR6, novel molecular targets for cannabidiol. *Biochemical and Biophysical Research Communications, 490*, 17–21.

Leung, D., Saghatelian, A., Simon, G. M., & Cravatt, B. F. (2006). Inactivation of N-acyl phosphatidylethanolamine phospholipase D reveals multiple mechanisms for the biosynthesis of endocannabinoids. *Biochemistry, 45*, 4720–4726.

Lowe, H., Toyang, N., Steele, B., Bryant, J., & Ngwa, W. (2021). The endocannabinoid system: A potential target for the treatment of various diseases. *International Journal of Molecular Sciences, 22*, 9472.

Lu, H. C., & Mackie, K. (2016). An introduction to the endogenous cannabinoid system. *Biological Psychiatry, 79*, 516–525.

Ludányi, A., Hu, S. S., Yamazaki, M., Tanimura, A., Piomelli, D., Watanabe, M., Kano, M., Sakimura, K., Maglóczky, Z., Mackie, K., et al. (2011). Complementary synaptic distribution of enzymes responsible for synthesis and inactivation of the endocannabinoid 2-arachidonoylglycerol in the human hippocampus. *Neuroscience, 174*, 50–63.

Lunardi, P., de Souza, L. W., dos Santos, B., Popik, B., & de Oliveira Alvares, L. (2020). Effect of the endocannabinoid system in memory updating and forgetting. *Neuroscience, 444*, 33–42.

Martín-Moreno, A. M., Brera, B., Spuch, C., Carro, E., García-García, L., Delgado, M., … de Ceballos, M. L. (2012). Prolonged oral cannabinoid administration prevents neuroinflammation, lowers β-amyloid levels and improves cognitive performance in Tg APP 2576 mice. *Journal of Neuroinflammation, 9*, 8.

Mascolo, N., Izzo, A. A., Ligresti, A., Costagliola, A., Pinto, L., Cascio, M. G., Maffia, P., Cecio, A., Capasso, F., & Di Marzo, V. (2002). The endocannabinoid system and the molecular basis of paralytic ileus in mice. *FASEB Journal : Official Publication of the Federation of American Societies for Experimental Biology, 16*, 1973–1975.

Matsuda, L. A., Lolait, S. J., Brownstein, M. J., Young, A. C., & Bonner, T. I. (1990). Structure of a cannabinoid receptor and functional expression of the cloned cDNA. *Nature, 346*, 561–564.

McCutcheon, R. A., Marques, T. R., & Howes, O. D. (2020). Schizophrenia—An overview. *JAMA Psychiatry, 77*, 201–210.

Mechoulam, R., Ben-Shabat, S., Hanus, L., Ligumsky, M., Kaminski, N. E., Schatz, A. R., Gopher, A., Almog, S., Martin, B. R., Compton, D. R., et al. (1995). Identification of an endogenous 2-monoglyceride, present in canine gut, that binds to cannabinoid receptors. *Biochemical Pharmacology, 50*, 83–90.

Misner, D. L., & Sullivan, J. M. (1999). Mechanism of cannabinoid effects on long-term potentiation and depression in hippocampal CA1 neurons. *The Journal of Neuroscience: The Official Journal of the Society for Neuroscience*, *19*, 6795–6805.

Morales, P., & Reggio, P. H. (2017). An update on non-CB(1), non-CB(2) cannabinoid related G-protein-coupled receptors. *Cannabis and Cannabinoid Research*, *2*, 265–273.

Moriconi, A., Cerbara, I., Maccarrone, M., & Topai, A. (2010). GPR55: Current knowledge and future perspectives of a purported "Type-3" cannabinoid receptor. *Current Medicinal Chemistry*, *17*, 1411–1429.

Muller, C., Morales, P., & Reggio, P. H. (2018). Cannabinoid ligands targeting TRP channels. *Frontiers in Molecular Neuroscience*, *11*, 487.

Munro, S., Thomas, K. L., & Abu-Shaar, M. (1993). Molecular characterization of a peripheral receptor for cannabinoids. *Nature*, *365*, 61–65.

Murillo-Rodríguez, E. (2008). The role of the CB1 receptor in the regulation of sleep. *Progress in Neuro-Psychopharmacology*, *32*, 1420–1427.

Nedaei, S. E., Rezayof, A., Pourmotabbed, A., Nasehi, M., & Zarrindast, M.-R. (2016). Activation of endocannabinoid system in the rat basolateral amygdala improved scopolamine-induced memory consolidation impairment. *Behavioural Brain Research*, *311*, 183–191.

Newsom, R. J., Osterlund, C., Masini, C. V., Day, H. E., Spencer, R. L., & Campeau, S. (2012). Cannabinoid receptor type 1 antagonism significantly modulates basal and loud noise induced neural and hypothalamic-pituitary-adrenal axis responses in male Sprague-Dawley rats. *Neuroscience*, *204*, 64–73.

Parolaro, D., Realini, N., Vigano, D., Guidali, C., & Rubino, T. (2010). The endocannabinoid system and psychiatric disorders. *Experimental Neurology*, *224*, 3–14.

Pertwee, R. G. (2006). Cannabinoid pharmacology: The first 66 years. *British Journal of Pharmacology*, *147*(Suppl 1), S163–S171.

Piomelli, D. (2013). Endocannabinoids. In W. J. Lennarz, & M. D. Lane (Eds.), *Encyclopedia of biological chemistry* (2nd ed., pp. 194–196). Waltham: Academic Press.

Pisani, A., Fezza, F., Galati, S., Battista, N., Napolitano, S., Finazzi-Agro, A., Bernardi, G., Brusa, L., Pierantozzi, M., Stanzione, P., et al. (2005). High endogenous cannabinoid levels in the cerebrospinal fluid of untreated Parkinson's disease patients. *Annals of Neurology*, *57*, 777–779.

Rajkowska, G., & Miguel-Hidalgo, J. J. (2007). Gliogenesis and glial pathology in depression. *CNS & Neurological Disorders Drug Targets*, *6*, 219–233.

Ramírez, B. G., Blázquez, C., del Pulgar, T. G., Guzmán, M., & de Ceballos, M. L. (2005). Prevention of Alzheimer's disease pathology by cannabinoids: Neuroprotection mediated by blockade of microglial activation. *Journal of Neuroscience*, *25*, 1904–1913.

Reggio, P. H. (2010). Endocannabinoid binding to the cannabinoid receptors: What is known and what remains unknown. *Current Medicinal Chemistry*, *17*, 1468–1486.

Rodríguez-Muñoz, M., Sánchez-Blázquez, P., Callado, L. F., Meana, J. J., & Garzón-Niño, J. (2017). Schizophrenia and depression, two poles of endocannabinoid system deregulation. *Translational Psychiatry*, *7*, 1–12.

Roediger, H. L., Zaromb, F. M., & Goode, M. K. (2008). 1.02—A typology of memory terms. In J. H. Byrne (Ed.), *Learning and memory: A comprehensive reference* (pp. 11–24). Oxford: Academic Press.

Rubino, T., Realini, N., Castiglioni, C., Guidali, C., Vigano, D., Marras, E., Petrosino, S., Perletti, G., Maccarrone, M., Di Marzo, V., et al. (2008). Role in anxiety behavior of the endocannabinoid system in the prefrontal cortex. *Cerebral Cortex*, *18*, 1292–1301.

Ruehle, S., Rey, A. A., Remmers, F., & Lutz, B. (2012). The endocannabinoid system in anxiety, fear memory and habituation. *Journal of Psychopharmacology*, *26*, 23–39.

Rusconi, F., Rubino, T., & Battaglioli, E. (2020). Endocannabinoid-epigenetic cross-talk: A bridge toward stress coping. *International Journal of Molecular Sciences*, *21*, 6252.

Sanford, A. E., Castillo, E., & Gannon, R. L. (2008). Cannabinoids and hamster circadian activity rhythms. *Brain Research*, *1222*, 141–148.

Sang, N., Zhang, J., & Chen, C. (2006). PGE2 glycerol ester, a COX-2 oxidative metabolite of 2-arachidonoyl glycerol, modulates inhibitory synaptic transmission in mouse hippocampal neurons. *The Journal of Physiology*, *572*, 735–745.

Scheltens, P., Blennow, K., Breteler, M. M., De Strooper, B., Frisoni, G. B., Salloway, S., & Van der Flier, W. M. (2016). Alzheimer's disease. *Lancet*, *388*, 505–517.

Scherma, M., Medalie, J., Fratta, W., Vadivel, S. K., Makriyannis, A., Piomelli, D., Mikics, E., Haller, J., Yasar, S., & Tanda, G. (2008). The endogenous cannabinoid anandamide has effects on motivation and anxiety that are revealed by fatty acid amide hydrolase (FAAH) inhibition. *Neuropharmacology*, *54*, 129–140.

Scotter, E. L., Abood, M. E., & Glass, M. (2010). The endocannabinoid system as a target for the treatment of neurodegenerative disease. *British Journal of Pharmacology*, *160*, 480–498.

Sieradzan, K., Fox, S., Hill, M., Dick, J., Crossman, A., & Brotchie, J. (2001). Cannabinoids reduce levodopa-induced dyskinesia in Parkinson's disease: A pilot study. *Neurology*, *57*, 2108–2111.

Siller-Pérez, C., Fuentes-Ibañez, A., Sotelo-Barrera, E. L., Serafín, N., Prado-Alcalá, R. A., Campolongo, P., Roozendaal, B., & Quirarte, G. L. (2019). Glucocorticoid interactions with the dorsal striatal endocannabinoid system in regulating inhibitory avoidance memory. *Psychoneuroendocrinology*, *99*, 97–103.

Snider, N. T., Walker, V. J., & Hollenberg, P. F. (2010). Oxidation of the endogenous cannabinoid arachidonoyl ethanolamide by the cytochrome P450 monooxygenases: Physiological and pharmacological implications. *Pharmacological Reviews*, *62*, 136–154.

Solas, M., Francis, P. T., Franco, R., & Ramirez, M. J. (2013). CB2 receptor and amyloid pathology in frontal cortex of Alzheimer's disease patients. *Neurobiology of Aging*, *34*, 805–808.

Stampanoni Bassi, M., Gilio, L., Maffei, P., Dolcetti, E., Bruno, A., Buttari, F., Centonze, D., & Iezzi, E. (2018). Exploiting the multifaceted effects of cannabinoids on mood to boost their therapeutic use against anxiety and depression. *Frontiers in Molecular Neuroscience*, *424*.

Sullivan, J. M. (1999). Mechanisms of cannabinoid-receptor-mediated inhibition of synaptic transmission in cultured hippocampal pyramidal neurons. *Journal of Neurophysiology*, *82*, 1286–1294.

Tiller, J. W. (2013). Depression and anxiety. *The Medical Journal of Australia*, *199*, S28–S31.

Tolón, R. M., Núñez, E., Pazos, M. R., Benito, C., Castillo, A. I., Martínez-Orgado, J. A., & Romero, J. (2009). The activation of cannabinoid CB2 receptors stimulates in situ and in vitro beta-amyloid removal by human macrophages. *Brain Research Bulletin*, *1283*, 148–154.

Trojniar, W., & Wise, R. A. (1991). Facilitory effect of Δ9-tetrahydrocannabinol on hypothalamically induced feeding. *Psychopharmacology*, *103*, 172–176.

Tsuboi, K., Sun, Y. X., Okamoto, Y., Araki, N., Tonai, T., & Ueda, N. (2005). Molecular characterization of N-acylethanolamine-hydrolyzing acid amidase, a novel member of the choloylglycine hydrolase family with structural and functional similarity to acid ceramidase. *The Journal of Biological Chemistry*, *280*, 11082–11092.

Valenti, M., Vigano, D., Casico, M. G., Rubino, T., Steardo, L., Parolaro, D., & Di Marzo, V. (2004). Differential diurnal variations of anandamide and 2-arachidonoyl-glycerol levels in rat brain. *Cellular and Molecular Life Sciences: CMLS*, *61*, 945–950.

Vaughn, L. K., Denning, G., Stuhr, K. L., de Wit, H., Hill, M. N., & Hillard, C. J. (2010). Endocannabinoid signalling: Has it got rhythm? *British Journal of Pharmacology*, *160*, 530–543.

Wenger, T., & Moldrich, G. (2002). The role of endocannabinoids in the hypothalamic regulation of visceral function. *Prostaglandins, Leukotrienes, and Essential Fatty Acids*, *66*, 301–307.

Woodward, D. F., Liang, Y., & Krauss, A. H. (2008). Prostamides (prostaglandin-ethanolamides) and their pharmacology. *British Journal of Pharmacology*, *153*, 410–419.

Wright, K. L., Duncan, M., & Sharkey, K. A. (2008). Cannabinoid CB2 receptors in the gastrointestinal tract: A regulatory system in states of inflammation. *British Journal of Pharmacology*, *153*, 263–270.

Yadav, A. (2022). *Legalize me*. The Sacred Plant of India.

Yiangou, Y., Facer, P., Durrenberger, P., Chessell, I. P., Naylor, A., Bountra, C., Banati, R. R., & Anand, P. (2006). COX-2, CB2 and P2X7-immunoreactivities are increased in activated microglial cells/macrophages of multiple sclerosis and amyotrophic lateral sclerosis spinal cord. *BMC Neurology*, *6*, 12.

Zant, J. C., Kim, T., Prokai, L., Szarka, S., McNally, J., McKenna, J. T., Shukla, C., Yang, C., Kalinchuk, A. V., McCarley, R. W., et al. (2016). Cholinergic neurons in the basal forebrain promote wakefulness by actions on neighboring non-cholinergic neurons: An opto-dialysis study. *The Journal of Neuroscience: The Official Journal of the Society for Neuroscience*, *36*, 2057–2067.

Zeng, B. Y., Dass, B., Owen, A., Rose, S., Cannizzaro, C., Tel, B. C., & Jenner, P. (1999). Chronic L-DOPA treatment increases striatal cannabinoid CB1 receptor mRNA expression in 6-hydroxydopamine-lesioned rats. *Neuroscience Letters*, *276*, 71–74.

Zhao, P., Ignacio, S., Beattie, E. C., & Abood, M. E. (2008). Altered presymptomatic AMPA and cannabinoid receptor trafficking in motor neurons of ALS model mice: Implications for excitotoxicity. *The European Journal of Neuroscience*, *27*, 572–579.

Chapter 18

Neurobiology of the kynurenine pathway and cannabis misuse: A new narrative

Nuria Morales-Puerto[a,b,c,d] and Esther O'Shea[a,b,c,d]

[a] *Departamento de Farmacología y Toxicología, Facultad de Medicina, Universidad Complutense, Madrid, Spain,* [b] *Instituto de Investigación Sanitaria, Hospital 12 de Octubre, Madrid, Spain,* [c] *Red de Trastornos Adictivos del Instituto de Salud Carlos III, Madrid, Spain,* [d] *Instituto Universitario de Investigación Neuroquímica (IUIN), Facultad de Medicina, Universidad Complutense, Madrid, Spain*

Abbreviations

AhR	aryl hydrocarbon receptor
AMPA	α-amino-3-hydroxy-5-methyl-4-isoxazolepropionic acid
CA	cinnabarinic acid
CBR	cannabinoid receptors
CDB	cannabidiol
CNS	central nervous system
GluN1	subunit 1 of NMDA receptor
GPR35	astrocytic G-protein coupled receptor 35
3-HAA	3-hydroxyanthranilic acid
3-HK	3-hydroxykynurenine
5-HT	serotonin
IDO	indolamine 2,3-dioxygenase
IFN	interferon
KAT	kynurenine aminotransferases
KYN	kynurenine
KYNA	kynurenic acid
KMO	kynurenine monooxygenase
LPS	lipopolysaccharide
mGluR	metabotropic glutamate receptor
α7nAChR	α7 nicotinic acetylcholine receptor
NMDA	N-methyl-D-aspartate
PBMC	peripheral blood mononuclear cells
QA	quinolinic acid
TDO	tryptophan dioxygenase
THC	Δ^9-tetrahydrocannabinol
TRP	tryptophan
XA	xanthurenic acid

Introduction

Kynurenine (KYN) is the central metabolite of the metabolic pathway through which most of the essential amino acid tryptophan (TRP) is degraded. While the best-known TRP subproduct is serotonin (5-HT), the synthesis of this mediator along with melatonin, tryptamine, and proteins accounts for only 5% of dietary TRP. The remaining 95% of ingested TRP undergoes degradation via the kynurenine pathway (Badawy, 2017; Leklem, 1971). The biomolecules originating from this route, which will be described in the following sections, are involved in diverse metabolic processes, such as the regulation of inflammatory, proliferative, and immune responses. Additionally, since some KYN derivates are modulators of

Cannabis Use, Neurobiology, Psychology, and Treatment. https://doi.org/10.1016/B978-0-323-89862-1.00033-7

glutamatergic signaling, the main excitatory neurotransmission in the central nervous system (CNS), the kynurenine pathway has drawn attention in several fields of neuroscience, from neurodegenerative to mental pathologies, including drug addiction.

The kynurenine pathway

The first step of the route is the conversion of TRP into KYN, a reaction carried out by tryptophan dioxygenase (TDO), found predominantly in the liver, or indolamine 2,3-dioxygenase (IDO), mainly present in the CNS and immune cells (Savitz, 2020) (Fig. 1). Although the presence of the enzyme in the CNS permits its local synthesis, in physiological conditions the majority of KYN found in the brain is obtained from the periphery thanks to the action of the large amino acid transporter-1, which facilitates the crossing of KYN through the blood-brain barrier (Fukui et al., 1991; Kita et al., 2002).

At this point, the kynurenine pathway diverges into two branches. The first branch leads to the formation of kynurenic acid (KYNA), a neuroprotective metabolite synthetized by the enzymes kynurenine aminotransferases (KAT). The synthesis of KYNA in the CNS takes place mainly in astrocytes, since it is in these cells where KATs are found (Guidetti et al., 2007; Schwarcz et al., 2012).

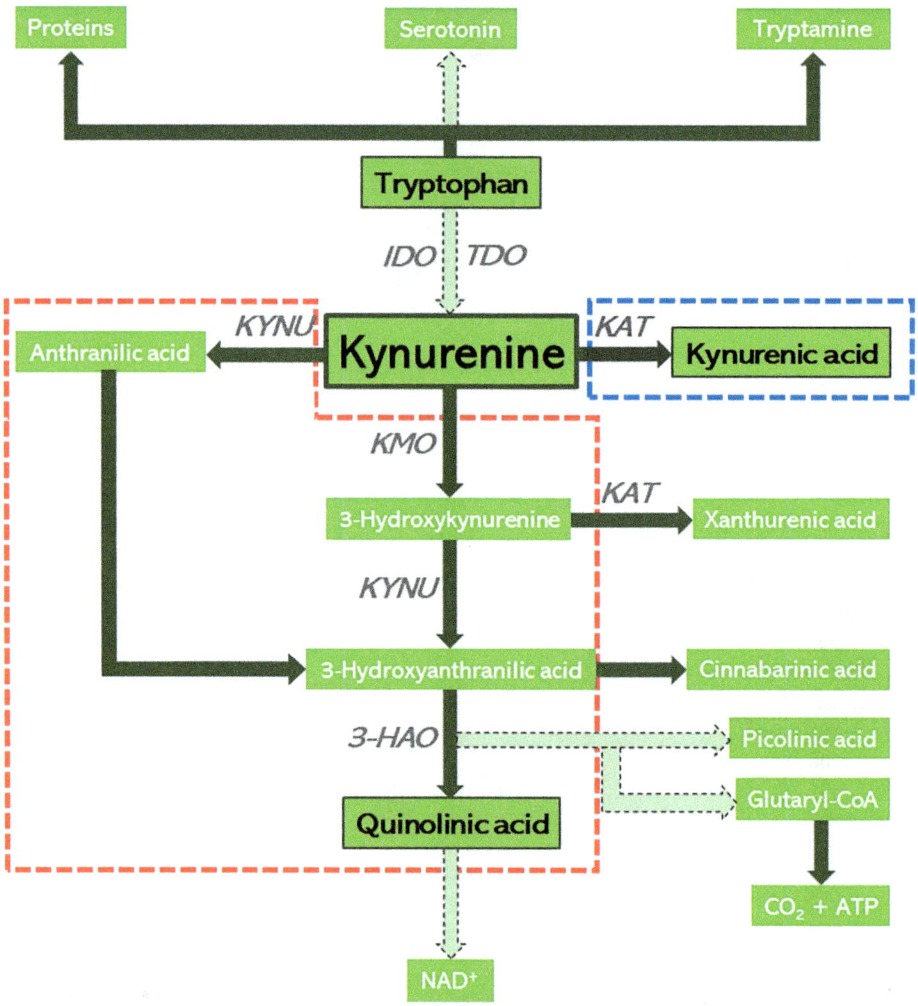

FIG. 1 The kynurenine pathway. The main route of tryptophan degradation takes place after the conversion of tryptophan into kynurenine. From the central metabolite of the pathway, two irreversible branches emerge: the neuroprotective branch (marked in *blue*) leads to the formation of kynurenic acid through the action of KATs, while the neurotoxic branch (marked in *red*), started by KMO, is composed of several kynurenine metabolites, eventually ending in the synthesis of quinolinic acid. 3-HAO: 3-hydroxyanthranilic acid oxygenase; IDO: indoleamine-2,3-dioxygenase; KAT: kynurenine amino transferase; KMO: kynurenine 3-monooxygenase; KYNU: kynureninase; TDO: tryptophan-2,3-dioxygenase. Dotted arrows depict multi-step transformations.

The second branch of the kynurenine pathway finds its source of production in the microglia, where the participating enzymes of the implicated reactions are located (Campbell et al., 2014). This branch is initiated with the conversion of KYN to 3-hydroxykynurenine (3-HK) by kynurenine monooxygenase (KMO), which is subsequently converted to 3-hydroxyanthranilic acid (3-HAA) by kynureninase. This molecule can alternatively be formed by spontaneous hydroxylation of anthranilic acid, which is the product of the action of kynureninase on KYN. The last step of the route entails the conversion, by the enzyme 3-hydroxyanthranilic acid oxygenase, of 3-HAA to quinolinic acid (QA), the source of nicotinamide adenine dinucleotide, a crucial cofactor involved in electron transfer for redox reactions but also a neurotoxic metabolite which gives this branch of the kynurenine pathway its neurotoxic denomination. Additionally, in this neurotoxic branch, several side-products emerge from each metabolite: xanthurenic acid (XA) is a subproduct synthesized by KAT from 3-HK; cinnabarinic acid (CA) is generated from 3-HAA following oxidative self-dimerization, and picolinic acid is the result of a decarboxylation from an intermediate molecule between 3-HAA and QA (Badawy, 2017; Campbell et al., 2014).

Targets of kynurenine pathway metabolites

Over the years, various receptors have been identified as targets of several of the kynurenine pathway metabolites. Most studies have focused on KYN, KYNA, and QA with only a small number having evaluated the remaining metabolites of the pathway. The most important findings are briefly outlined below.

KYN appears to have as its main target the aryl hydrocarbon receptor (AhR), at which it acts as an agonist (DiNatale et al., 2010). AhR is a transcription factor and cytoplasmic receptor first known for its role in xenobiotic detoxification and later also identified as a regulator of immune activity and cell differentiation and survival, which makes AhR an important factor in terms of drug use (Suzuki et al., 2020).

KYNA, similar to KYN, is an agonist of AhR (DiNatale et al., 2010). However, KYNA also acts on several other receptors. First, its neuroprotective actions emerge mainly from its role as a glutamatergic signaling modulator. KYNA is an endogenous competitive antagonist of the ionotropic glutamatergic kainate receptor, α-amino-3-hydroxy-5-methyl-4-isoxazolepropionic acid (AMPA) receptor and, especially, of the *N*-methyl-D-aspartate (NMDAR) receptor. At the latter, KYNA binds with great affinity to the glycine site on the receptor complex, blocking its activity (Perkins & Stone, 1982). Additionally, it has been suggested, by several different researchers, that KYNA inhibits the α7 nicotinic acetylcholine receptor (α7nAChR) (Alkondon et al., 2011; Hilmas et al., 2001), but this interaction remains unclear and in need of further characterization (Albuquerque & Schwarcz, 2013; Morales-Puerto et al., 2021; Stone, 2020). The association between KYNA and NMDA receptor and/or α7nAChR potentially provides a target for pharmacological treatment against cannabis addiction that will be discussed in detail in later sections. Moreover, it has been recently proposed that KYNA also acts on the astrocytic G-protein-coupled receptor 35 (GPR35), which seems to be involved in the regulation of glutamatergic neurotransmission (Wang et al., 2006). The structural and functional similarities between the GPR35 and the cannabinoid receptors (CBR) may be indicative of a possible site of action for KYNA on CBR. Nonetheless, this hypothesis has been ruled out in a recent study (Zádor et al., 2020), as explained in the following section.

Contrary to the effects of KYNA, the neurotoxic properties of QA are due to its activity as an NMDA and AMPA receptors agonist, which is known to induce neurodegeneration and aberrant electric activity in the brain leading to convulsions. Other mechanisms of toxicity by QA include its action as a gliotoxin, proinflammatory mediator, and precursor of reactive oxygen species (Guillemin, 2012; Savitz, 2020).

Thus, QA is implicated in the pathogenesis of multiple mental and neurodegenerative diseases.

Due to their numerous actions, KYNA and QA are the most studied metabolites of the kynurenine pathway, but other biomolecules of this route deserve to be reviewed, likewise for their involvement in glutamatergic neurotransmission. XA regulates glutamatergic signaling through two mechanisms. First, XA inhibits the vesicular glutamate transporter, which leads to a decrease in glutamate release in the synapse. Also, XA possibly acts as an allosteric agonist of metabotropic glutamate receptors (mGluR) 2 and 3, presynaptic receptors that block glutamate release (Bartlett et al., 1998; Kenny et al., 2005). Finally, CA acts as a partial agonist of mGluR4 (Fazio et al., 2012) and also activates AhR (Lowe et al., 2014).

Interaction of cannabinoids with the kynurenine pathway targets

In the context of cannabis use, it is relevant to mention that the activation state of CBRs may modulate the tone of the glutamatergic system and since the glutamate receptors are important targets of some of the kynurenine pathway metabolites exposure to cannabinoids may indirectly affect the downstream actions of the kynurenine pathway. Specifically, the CB1R is implicated in the regulation of NMDAR activation protecting nervous cells from glutamatergic excitotoxicity.

Thus, NMDA receptor hyperactivation induces the synthesis of endocannabinoids, such as anandamide which upon binding to the CB1R triggers the formation of the CB1R-NMDA receptor complex, resulting in its internalization and subsequent reduction in availability of functional NMDA receptors (Marsicano et al., 2003; Rodríguez-Muñoz et al., 2018; Sánchez-Blázquez et al., 2014).

In line with this, similar results were obtained in a study in 2017 where administration of WIN 55,212-2 (a synthetic nonselective CBR agonist) and URB597 (an inhibitor of fatty acid amide hydroxylase, enzyme responsible for anandamide catabolism) decreased the expression of NMDA receptor subunit 1 (GluN1) in the substantia nigra and striatum of mice, whether administered before or both before and after an excitotoxic insult with 1-methyl-4-phenyl-1,2,3,6-tetrahydropyridine, a mitochondrial toxin used to recreate a Parkinson disease model in rodents (Escamilla-Ramírez et al., 2017).

An example of how this target alteration can modulate the downstream effects of kynurenine metabolites is evidenced by a study in rats that analyzed the protective properties of URB597 against the excitotoxic damage caused by striatal QA administration. The use of the named inhibitor prevented QA-induced body weight decrease, motor alterations, GABAergic neuronal loss, and oxidative damage to protein and lipids. Furthermore, QA modified the expression and localization of both the CB1R and GluN1 subunit of NMDA receptor resulting in an increase in the GluN1/CB1R ratio, indicating a favorable scenario for excitotoxic conditions. Notably, administration of URB597 to rats that had received QA produced a decrease in the GluN1/CB1R ratio, hence diminishing the probability of excitotoxic episodes (Aguilera-Portillo et al., 2019).

Interaction of kynurenine pathway metabolites with CBR

Thus far, there is no evidence in the literature to document a direct binding or site of action of any of the kynurenine pathway metabolites on either of the cannabinoid receptors. Nonetheless, there are a few studies that account for the existence of interactions between both systems.

An in vivo study with the QA model of Huntington disease with Wistar rats showed, using small animal PET imaging and receptor-binding techniques, that the QA lesion produced a decrease in CB1R levels in the caudate-putamen, which agrees with findings in ex vivo models of the disease (Casteels et al., 2010).

A recent work assessed the direct interaction between KYNA and CB1R in rat brain for the first time. Competition-binding assays, using the nonselective synthetic cannabinoid [3H]WIN 55,212-2 as radioligand, revealed that KYNA lacked specific affinity toward CB1R and did not function as an allosteric modulator. However, when administered chronic and systemically, KYNA treatment increased the number of CB1R-coupled G-proteins, the basal activity level and the amount of maximally stimulated CB1R-coupled G-proteins. Also, KYNA treatment augmented the maximum binding capacity. All these results suggest that chronic, systemic KYNA induces an increase in the concentration of functional CB1R in the rat whole brain by a mechanism that does not involve a direct interaction between the metabolite and the receptor.

Interestingly, when performing these experiments in isolated brain regions, long-term KYNA treatment produced the same changes in the hippocampus, but not in cortex or brainstem, revealing a region-specific effect (Zádor et al., 2020).

Animal studies on the effects of cannabinoids on the kynurenine pathway

Other than a couple of studies carried out in the second half of the past century, research on the interaction of cannabis metabolites, kynurenine pathway subproducts, and related enzymes, whether in animal models, cell culture, or human subjects, is relatively recent (Table 1). Nevertheless, evidence on how modulation of the endocannabinoid tone may directly interfere with the actions of kynurenine metabolites is accumulating.

Δ^9-Tetrahydrocannabinol (THC)

The early studies were both published in the 1970s by the same research group and demonstrated for the first time that cannabis metabolites could alter the activity of rat liver TDO. Initially, they found that TDO activity was increased 6 hours after intraperitoneal injection of both cannabis extract and pure THC (Poddar & Ghosh, 1972). Two years later they reported that, while cannabidiol (CDB) did not induce any alteration on TDO activity, it potentiated the effect exerted by THC when administered in combination with it (Poddar et al., 1974).

TABLE 1 **Effects of exposure to cannabinoids on the kynurenine metabolites.**

Cannabinoid	Animal/sample type		Effects on kynurenine pathway	Reference
THC	Albino rat	Liver	↑ TDO activity	Poddar and Ghosh (1972)
	Sprague-Dawley rat	Nucleus accumbens shell	↓ [KYNA]	Secci et al. (2019)
	Wistar rat	Prefrontal cortex (adult offspring after prenatal exposure)	↑ [KYNA]	Beggiato et al. (2020)
			↑ KAT I mRNA	
	Rhesus monkey	Gingival epithelium (SIV-infected monkey)	↓ IDO1 expression	McDew-White et al. (2022)
		Plasma (SIV-infected monkey)	↓ [KYN], [KYNA] and [QA]	
	Human	Human peripheral mononuclear cells	*Nanomolar doses on mitogen-induced cells:*	Jenny et al. (2008)
			↑ IDO activity	
			↑ KYN/TRP ratio	
			Low micromolar doses on mitogen-induced cells:	
			Supression of IDO activity	
			Unstimulated cells:	
			Concentration-dependent ↓ KYN/TRP ratio	
	Human	Primary human gingival epithelium	↑ IFN-γ-induced IDO1 expression	McDew-White et al. (2022)
CBD	Albino rat	Liver	Potentiation of THC-induced ↑ TDO activity	Poddar et al. (1974)
	Rat	Hypothalamic Hypo-E22 cell line	↓ 3-HK/KYNA ratio	di Giacomo, Chiavaroli, Orlando, et al. (2020)
	Sprague-Dawley rat	Prefrontal cortex	↓ Upregulated [3-HK]	di Giacomo, Chiavaroli, Recinella, et al. (2020)
			↑ Downregulated [KYNA]	
	NMRI mouse	Cortex	↓ KYN/TRP ratio	Florensa-Zanuy et al. (2021)
		Hippocampus	↓ KYN/5-HT ratio	
			↓ IDO1 activity	
	Human	Human peripheral mononuclear cells	*Nanomolar doses on mitogen-induced cells:*	Jenny et al. (2008)
			↑ IDO activity	
			↑ KYN/TRP ratio	
			Low micromolar doses on mitogen-induced cells:	
			Supression of IDO activity	
			Unstimulated cells:	
			Concentration-dependent ↓ KYN/TRP ratio	
	Human	Hepatoma HepG2 cells	↑ AhR-mediated CYP1A1 expression	Yamaori et al. (2015)

Continued

TABLE 1 Effects of exposure to cannabinoids on the kynurenine metabolites—cont'd

Cannabinoid	Animal/sample type		Effects on kynurenine pathway	Reference
WIN 55,212-2	Wistar rat	Corticostriatal region	Inhibition of QA-induced:	Pintor et al. (2006)
			↑ $[Glutamate]_{ext}$	
			↓ Field potentials	
	Wistar rat	Striatal neurons	Protection against QA-induced loss of:	Rangel-López et al. (2015)
			Cell viability	
			CB1R-NeuN positive cells	
		Synaptosomes	Protection against QA-induced:	
			Mitochondrial dysfunction	
			Lipid peroxidation	
			ROS formation	
5F-APINAC	Zebrafish	Larvae	[XA] and [PA] alterations	Markin et al. (2021)
	Rabbit	Plasma	Concentration-dependent effects:	Shestakova et al. (2021)
			↓ [TRP], ↑ [KYN], ↑ [XA], ↑ [QA], ↑ [KYNA], ↓[AA]	

3-HK: 3-hydroxykynurenine; 5-HT: serotonin; AA: anthranilic acid; CB1R: cannabinoid receptor type 1; CBD: cannabidiol; IDO: indoleamine-2,3-dioxygenase; KAT: kynurenine amino transferase; KMO: kynurenine 3-monooxygenase; KYN: kynurenine; KYNA: kynurenic acid; KYNU: kynureninase; PA: picolinic acid; QA: quinolinic acid; ROS: reactive oxygen species; TDO: tryptophan-2,3-dioxygenase; THC: tetrahydrocannabinol; TRP: tryptophan; XA: xanthurenic acid.

Secci et al recently found that THC reduced the extracellular KYNA concentration in the nucleus accumbens shell of Sprague-Dawley rats measured by in vivo microdialysis, an effect that was blocked by treatment with the CB1R antagonist rimonabant (Secci et al., 2019).

It is well known that cannabis use during pregnancy has consequences on the correct development of the offspring's nervous system. A recent study analyzed the long-term consequences of prenatal THC exposure. Adult offspring showed increased KYNA concentration and augmented levels of KAT I mRNA in the prefrontal cortex.

Given that, as previously mentioned, KYNA is an antagonist of glutamate receptors, this result could be related to low performance on a short-term working memory task (Beggiato et al., 2020).

Another recent study examined the interaction of cannabinoids and the kynurenine pathway in a model of periodontal disease in simian immunodeficiency virus-infected Rhesus monkeys. Treatment with THC succeeded in diminishing IDO1 protein expression in the gingiva epithelium of infected monkeys compared to untreated infected animals. In consequence, long-term low-dose THC treatment decreased KYN, KYNA and QA plasma concentrations in infected monkeys. In addition, they analyzed the effects of THC exposure on the expression of IDO1 in primary human gingival epithelium cells. In contrast to the result with animals, THC slightly augmented interferon (IFN)-γ-induced IDO1 expression. Subsequently, using AM251 and AM630 as antagonists of CB1R and CB2R, respectively, only CB2R blockade was shown to exacerbate the above-mentioned IDO1 increase, suggesting that the ability of THC to downregulate IDO1 levels is mediated by its interaction with CB2R (McDew-White et al., 2022).

CBD

Recently, studies with different CNS-derived cells have aimed to elucidate the effects of CBD treatment after exposure to oxidative and excitotoxic stimulus. First, in experiments with rat hypothalamic cell line Hypo-E22 exposed to H_2O_2, as a form of oxidative stress, treatment with CBD resulted in a decrease of the 3-HK/KYNA ratio.

This result highlighted the neuroprotective properties of CBD, given that a decrease in the mentioned ratio implied an inhibition of the neurotoxic branch of the kynurenine pathway. Similar results were obtained when cells were treated with cannabigerol (di Giacomo, Chiavaroli, Orlando, et al., 2020). After that, the same researchers evaluated the neuroprotective effects of CBD in cultured prefrontal cortex against a neurotoxic stimulus triggered by external K^+ administration at a concentration equivalent to an excitotoxic depolarizing stimulus. CBD treatment was effective in preventing upregulated 3-HK and downregulated KYNA levels (di Giacomo, Chiavaroli, Recinella, et al., 2020).

The kynurenine pathway has been lately related to the pathogenesis of depression, with the neurotoxic branch of the pathway being upregulated. A group of researchers evaluated the effects of CBD on a model of lipopolysaccharide (LPS)-induced neuroinflammation in mice, given that it has been widely shown that depressive patients show altered inflammatory markers. Using the KYN/TRP ratio as a marker of IDO activity, the study revealed that CBD administration prevented an increase in the KYN/TRP ratio both in cortex and in hippocampus, suggesting a diminished activity of IDO. Also, the KYN/5-HT ratio was measured to assess the balance between the two main directions of TRP catabolism, finding that the KYN/5-HT ratio was shifted toward the formation of KYN in the animals treated with LPS, which confirmed the increase in IDO activation. Nonetheless, the administration of CBD prevented this effect, again in both brain regions. Altogether, this study manifests that CBD antidepressant effects are achieved through IDO1 activity downregulation (Florensa-Zanuy et al., 2021).

Synthetic cannabinoids

Synthetic cannabinoids are a group of new psychoactive substances, usually with more potent properties than natural cannabinoids, which work by mimicking the effects of THC through activation of cannabinoid receptors. Although they were first synthetized in the 1980s as prospective pharmaceutical tools and have been in the market ever since, the possible interactions between these compounds and the kynurenine pathway remain widely unexplored (Schifano et al., 2015; Alves et al., 2020).

WIN 55,212-2

In 2006, it was confirmed for the first time that cannabinoids could influence the deleterious consequences of a QA administration to the brain, in an animal model of Huntington's disease. Treatment with the synthetic CBR agonist WIN 55,212-2 inhibited the QA-induced increase in striatal extracellular glutamate levels and reduction of field potentials in corticostriatal slices. These findings were solely attributable to the CB1R, since pretreatment with a selective antagonist blocked the effects of WIN 55,212-2 (Pintor et al., 2006).

A decade later, another research group demonstrated that, after incubation with QA, rat striatal neurons cultured with WIN 55,212-1 were protected against a set of various toxic markers. Preincubation of striatal neurons with the mentioned cannabinoid prior to QA exposure prevented the loss of cell viability and preserved the number of NeuN-CB1R colocalizing neurons. Moreover, pretreatment of rat brain synaptosomes with WIN 55,212-2 before incubation with QA slightly preserved mitochondrial function, thwarted lipid peroxidation, and avoided reactive oxygen species formation (Rangel-López et al., 2015).

5F-APINAC

A study on the past year characterized the alterations in different neurotransmission systems, including the kynurenine pathway, induced by the exposure to the recently discovered synthetic cannabinoid 5F-APINAC in rabbit plasma. After administering a low, medium, and high dose of 5F-APINAC (0.1, 1, and 2 mg/kg, respectively), the kynurenine pathway was the most-affected system of those analyzed in the study. Levels of TRP diminished after the low and medium dose but remained unaltered with the high dose. KYN levels increased after the low dose, with no alterations with the higher doses. XA, QA, and KYNA concentrations were augmented after the three doses. Anthranilic acid is the only KYN metabolite that showed reduced levels after the 5F-APINAC exposure, while no regular trend was detected for the concentration of picolinic acid. The authors conclude that potential neurotoxic effects of 5F-APINAC may be related to the significant alterations on the kynurenine pathway (Shestakova et al., 2021). On the contrary, a previous work by the same group on zebrafish larvae, in which they administered increasing doses to the micromolar level of 5F-APINAC on acute and chronic exposure, did not find as many changes in the kynurenine pathway, with only XA and picolinic acid showing alterations after chronic administration of the drug (Markin et al., 2021).

Humans

To this day, there are no studies on humans about the actions of cannabis abuse on the metabolism of KYN. Likewise, preclinical research with human cell lines on the effects of phytocannabinoids on the kynurenine pathway is scarce (Table 1).

In 2008, Jenny et al. showed a concentration-dependent dual effect of THC and CBD on TRP catabolism in human peripheral blood mononuclear cells (PBMC) obtained from healthy subjects. After treating cells with nanomolar concentrations of THC or CBD, mitogen-stimulated IDO activity on human PBMC was slightly enhanced, as reflected by an increase in the KYN/TRP ratio. On the contrary, low micromolar doses of the cannabinoids suppressed mitogen-induced IDO activity. The inhibition of TRP degradation by IDO was 2–4 times stronger with CBD compared to THC. Furthermore, the biphasic effect also differs in receptor involvement: nanomolar doses increase in IDO activity was CB1- and CB2-receptor dependent, while low micromolar inhibition did not need the participation of the before mentioned receptors. On the other hand, unstimulated cells showed a concentration-dependent decrease in the KYN/TRP ratio (Jenny et al., 2008; Jenny et al., 2010).

A later study revealed the involvement of AhR, KYN target receptor, on the induction of CYP1A1 expression by phytocannabinoids. Human hepatoma HepG2 cells were used to assess the impact of THC, CBD, and cannabinol on the expression of human CYP1A1. The results showed that THC and CBD, but not cannabinol, induced the expression of CYP1A1 in a dose-dependent manner, with CBD being more powerful than that of THC. Moreover, CBD-induced CYP1A1 expression was mediated by AhR, given that, after transfection with a small interfering RNA against AhR, the mRNA levels for CYP1A1 after treatment with CBD were reduced (Yamaori et al., 2015).

Effects of modulation of the kynurenine pathway on cannabis misuse

The kynurenine pathway has drawn attention in the last few years due to its potential as a pharmacological target for the treatment of addiction to several drugs of abuse, including alcohol, nicotine, cannabis, amphetamines, and cocaine. A popular strategy consists of shifting the metabolism of KYN toward the neuroprotective branch using Ro 61-8048, a KMO inhibitor, and thus increasing the concentration of KYNA in addiction-related brain areas (Fig. 2). As mentioned previously, KYNA can act both through NMDA receptor and, most certainly, α7nAChR, leading to the modulation of the glutamatergic signaling that ultimately triggers the dopaminergic projections in the mesolimbic pathway, responsible for addictive behavior. In the case of cannabis misuse, a few studies have shown promising results.

A preclinical study revealed that treatment with Ro 61-8048 decreased self-administration of THC or the synthetic cannabinoid WIN 55,212-2 in squirrel monkeys and rats, respectively (Fig. 3). The inhibition of KMO by systemic Ro 61-8048 administration produced an increase in KYNA levels in the ventral tegmental area and the nucleus accumbens, brain regions involved in the reward circuit. This resulted in a blockade of dopamine release in these mesolimbic areas, thus depleting the effects of THC. Additionally, relapse to THC following reexposure to either THC itself or WIN 55,212-2 was prevented with Ro 61-8048 treatment. Although the blockade of NMDA receptor cannot be ruled out, this research group showed that the effect of Ro 61-8048 was attributable to the inhibition of α7nAChR, since pretreatment with PNU-120596, a α7nAChR-positive allosteric modulator, avoided the described results (Justinova et al., 2013).

In fact, further studies in rats by this research group showed that Ro 61-8048 blocked glutamate release caused by THC exposure in the ventral tegmental area, the nucleus accumbens, and the medial prefrontal cortex, thus explaining the decrease in dopamine signaling in the previous study (Secci et al., 2019). In addition, the results suggested a possible role for cortical astrocytic α7nAChR in the control of glutamate release. Thus, modulation of the kynurenine pathway may exert its effects at multiple areas of the brain reward pathway and act at more than one target.

Applications to other areas

After so many decades, witnessing the devastating development of this problem and despite the accumulated evidence on their harmful effects, drug abuse and addiction continues killing people, directly or indirectly, and the drug markets do not cease to expand and grow worldwide. The World Health Organization, in the World Drug Report 2021 of the Office on Drugs and Crime, stated that almost half a million people suffered a drug-related death in 2019 and around 275 million people declared having used drugs in the past year, that is, a 22% increase compared to the data from 2010 (United Nations Office on Drugs Crime (UNODC), 2021).

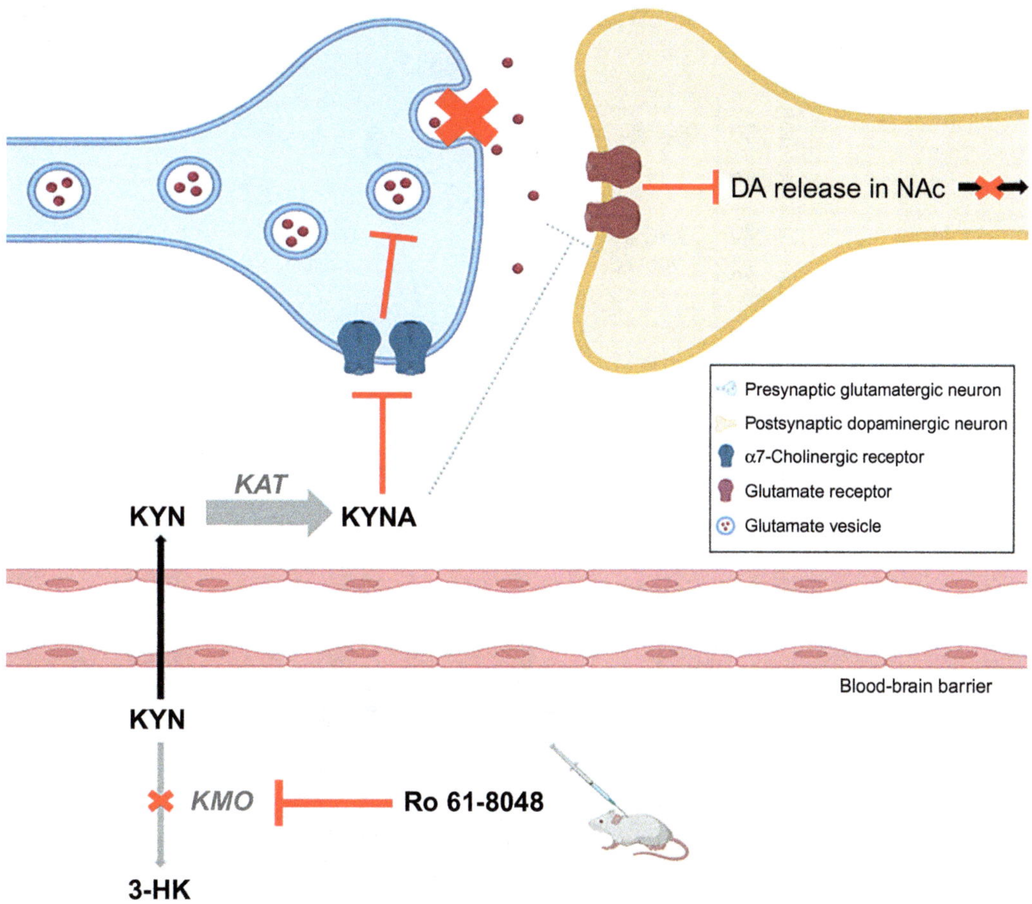

FIG. 2 Mechanism of dopaminergic signaling blockade by Ro 61-8048. A series of independent studies with different drugs of abuse have succeeded in abolishing drug seeking-like behaviors through KMO inhibition using Ro 61-8048. After blocking the conversion of KYN to 3-HK, the formation of KYNA by KATs is promoted, thus elevating the levels of this metabolite in the brain. KYNA inhibits the activity of α7nAChR, causing an interruption of glutamatergic neurotransmission, which, subsequently, results in the interruption of dopaminergic impulses to the nucleus accumbens. Since this brain area is essential for the development of drug addiction, the inhibition in the activity of the circuit due to Ro 61-8048 treatment would explain the abolishment of drug-related conducts (for more detail see 8. Effects of modulation of the kynurenine pathway on cannabis misuse and Applications to other areas). It is worth mentioning that, although the referred studies have explicitly demonstrated the involvement of α7nAChR, KYNA direct inhibition of glutamate receptors cannot be ruled out as an alternative explanation for the observed outcomes. 3-HK: 3-hydroxykynurenine; DA: dopamine; KAT: kynurenine amino transferase; KMO: kynurenine 3-monooxygenase; KYN: kynurenine; KYNA: kynurenic acid; NAc: nucleus accumbens.

Among illicit drugs, cannabis continues to be the most consumed drug and cannabis cultivation and trafficking has spread among almost all countries. A troubling fact: even though the potency of diverse cannabis products has increased, the risk perception of regular use of the drug has declined, especially among adolescent and young adults.

This is particularly serious if we consider that it has been widely reported that cannabis use during this period of life may exert impairing and long-term effects on neurodevelopment and cognitive functions. Several factors contribute to this crisis, including legalization for recreational purposes in some regions, presence on social media and the dynamic nature of illegal markets. All these data point to the necessity of promoting social awareness about the unsafe effects of indiscriminate cannabis use, while continuing research on the beneficial properties of cannabis derivates for medical purposes, such as moderate pain control or some forms of epilepsy.

The consequences of drug use have a huge impact on societies and economic systems, as reflected, for instance, in the global disease burden, with over 18 million years of healthy life lost. Unfortunately, pharmacological treatment and psychological therapies for drug addiction are far from successfully maintaining abstinence, not to mention that some drug types still remain orphan from pharmacological help.

Therefore, it is becoming imperative to develop new pharmacological approaches to fight against the pandemic of drug addiction.

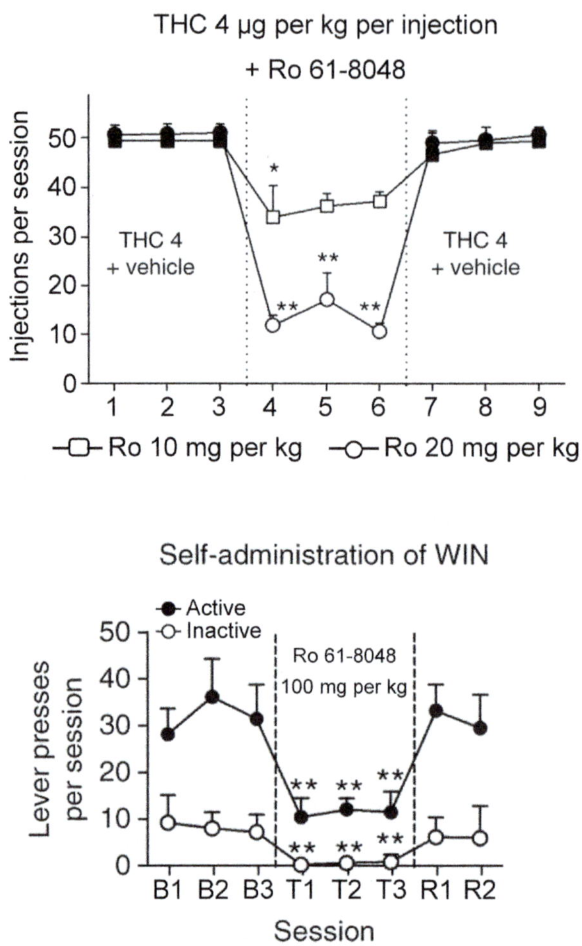

FIG. 3 Effects of Ro 61-8048 on THC and WIN 55-212,2 self-administration. Upper panel: Ro 61-8048 (10 and 20 mg per kg i.m.) significantly decreased the number of THC injections self-administered during one-h sessions by squirrel monkeys under a fixed-ratio ten (FR10) schedule at a THC dose of 4 µg per kg per injection $*P<0.05$, $**P<0.01$, post-hoc vs. the last session with vehicle pretreatment (session 3); Bonferroni test. Lower panel: Ro 61-8048 (sessions T1-T3), but not its vehicle (sessions B1-B3; recovery sessions R1-R2), decreased the number of WIN injections (12.5 µg per kg; 1 injection per active-lever response) self-administered by rats over 2-h sessions. Presses on active and inactive levers per session are shown. $*P<0.05$, $**P<0.01$, post-hoc vs. average of sessions B1-B3; Tukey test. For both panels: all points represent the means± s.e.m.; $n=5$. *From Justinova, Z., Mascia, P., Wu, H.-Q., Secci, M. E., Redhi, G. H., Panlilio, L. V, Scherma, M., Barnes, C., Parashos, A., Zara, T., Fratta, W., Solinas, M., Pistis, M., Bergman, J., Kangas, B. D., Ferré, S., Tanda, G., Schwarcz, R., & Goldberg, S. R. (2013). Reducing cannabinoid abuse and preventing relapse by enhancing endogenous brain levels of kynurenic acid. Nature Neuroscience, 16(11), 1652–1661. doi: 10.1038/nn.3540. Copyright permission obtained from Springer Nature.*

In the last decade, studies have emerged on a new prospective pharmacological tool for substance use disorder. Ro 61-8048 is an inhibitor of KMO; thus, it shifts the kynurenine pathway toward the synthesis of KYNA. As detailed in the chapter, preclinical research on cannabis misuse using Ro 61-8048 as treatment has shown efficacy on decreasing self-administration, avoiding reinstatement and blocking neurotransmission in the brain reward circuit. Two receptors have been proposed as targets for the elevated levels of KYNA following Ro 61-8048 treatment: the presynaptic α7nAChR, which would block the release of glutamate, inhibiting communication between regions of the mesolimbic pathway; and the postsynaptic NMDAR, which would stop responding to released glutamate.

The encouraging results on studies with Ro 61-8048 are not restricted just to cannabis abuse: experiments with models of alcohol and nicotine consumption have yielded similar outcomes as those of cannabis. In the case of alcohol, Ro 61-8048 successfully abolished cue-induced reinstatement (Vengeliene et al., 2016) and reduced ethanol consumption, both in binge-like (Giménez-Gómez et al., 2018) and chronic models (Gil de Biedma-Elduayen et al., 2022). In the nicotine studies, Ro 61-8048 managed to reduce nicotine self-administration and relapse-like effects after reexposure (Secci et al., 2017). For both drugs, the mechanism of action of Ro 61-8048 involved α7nAChR modulation, although NMDA receptor involvement cannot be ruled out.

The data resulting from Ro 61-8048 research are promising and further work shedding light on the effects of this new compound on models of addiction for other drugs should be carried out as soon as possible. It is especially interesting in the case of opioids: although most overdose deaths are attributable to opioids as a consequence of the dramatically high prevalence of opioid misuse, an important number of cases originate from a medical need. Several studies have linked the modulation of NMDA receptor activity by KYNA with prevention of allodynia, decreasing the hyperalgesia and suppression of aversive response to noxious stimuli (Knyihár-Csillik et al., 2007: Pineda-Farias et al., 2013; Zhou et al., 2018). Furthermore, activation of GPR35, another KYNA target, has been associated with an analgesic effect (Cosi et al., 2011). Taken together, since opioids are mostly prescribed for pain relief, modulation of the kynurenine pathway by Ro 61-8048 could, in the right scenario, provide a safer alternative to opioids, hence diminishing the chances of addiction problems.

Most importantly, these findings suggest that Ro 61-8048, or alternative means of increasing KYNA concentration in the brain such as administration of KYN itself (Giménez-Gómez et al., 2018), is an exciting novel strategy against addiction that may offer a general pharmacological treatment for substance use disorders.

Key facts of the kynurenine pathway

- The kynurenine pathway is the main metabolic route of TRP degradation, accounting for 95% of TRP catabolism.
- The kynurenine pathway was first identified as responsible for xenobiotic detoxification, but it is also involved in cellular differentiation and immune response.
- The kynurenine pathway is implicated in a great number of pathologies, including neurodegenerative diseases, like Huntington's disease, or mental disorders, such as schizophrenia or drug addiction.
- From KYN, the pathway splits into two branches: the neurotoxic branch, toward the synthesis of QA; and the neuroprotective branch, toward the formation of KYNA.
- While neuroprotective KYNA is an antagonist of glutamatergic NMDA and AMPA receptors, neurotoxic QA can activate them.
- Currently, there is no evidence of any of the kynurenine metabolites having direct interaction with the cannabinoid receptors.
- Cannabis metabolites, such as THC and CBD, as well as synthetic cannabinoids, can alter the activity of several enzymes of the kynurenine pathway.
- Recently, pharmacological diversion of the pathway to the neuroprotective branch has shown positive results on different aspects of addiction and with several drugs.

Mini-dictionary of terms

- **Cannabis misuse**. a mental disorder leading to an uncontrolled consumption of cannabis for nonmedical purposes.
- **Tryptophan**. an essential amino acid, precursor of the neurotransmitter serotonin. It is degraded mainly through the kynurenine pathway.
- **Kynurenine**. central metabolite of the kynurenine pathway, generated by the enzyme indolamine 2,3-dioxygenase.
- **Kynurenic acid**. neuroprotective metabolite of the kynurenine pathway, generated by the enzyme kynurenine aminotransferase. It acts as an NMDA receptor antagonist.
- **Quinolinic acid**. neurotoxic metabolite of the kynurenine pathway, generated after the degradation of kynurenine by kynurenine monooxygenase. It is an agonist of NMDA receptor and a source of NAD^+.
- **KYN/TRP ratio**. a measure of indolamine 2,3-dioxygenase activity.
- **3-HK/KYN ratio**. a measure of kynurenine monooxygenase activity.
- **Aryl hydrocarbon receptor**. main target of kynurenine, where it acts as an agonist. It is involved in xenobiotic detoxification, immune function, and cell proliferation.
- **Ro 61-8048**. inhibitor of kynurenine monooxygenase, diverts the metabolism of kynurenine toward the formation of kynurenic acid, promoting the neuroprotective branch.

Summary points

- The present chapter describes the interactions between the major compounds of cannabis, namely CBD and THC, and some synthetic cannabinoids with the main route of TRP degradation, i.e., the kynurenine pathway.

- Currently, there is no evidence of a direct interaction between any of the metabolites of the kynurenine pathway and the cannabinoid receptors.
- Nonetheless, several studies have reported alterations on the pathway after exposure to various exogenous cannabinoids.
- In animal models, THC can affect both IDO and KAT activity. Furthermore, in studies with human cells, effects on IDO activity after THC treatment have also been reported.
- CBD has shown beneficial properties in several studies both in animal and human cell culture studies, measured as a diminished 3-HK/KYNA or KYN/TRP ratio.
- Novel synthetic cannabinoid 5F-APINAC altered the concentration of many of the KYN metabolites in rabbit plasma after receiving an acute dose.
- Modulation of the kynurenine pathway through KMO inhibition with Ro 61-8048 is a potential tool for drug addiction treatment.
- Ro 61-8048 diminished WIN 55,212-2 self-administration and relapse to THC in research with monkeys and rats.
- The actions of Ro 61-8048 may be attributable to various mechanisms of action, including inhibition of α7nAChR and blockade of glutamate release in the mesolimbic pathway.

References

Aguilera-Portillo, G., Rangel-López, E., Villeda-Hernández, J., Chavarría, A., Castellanos, P., Elmazoglu, Z., Karasu, Ç., Túnez, I., Pedraza, G., Königsberg, M., & Santamaría, A. (2019). The pharmacological inhibition of fatty acid amide hydrolase prevents excitotoxic damage in the rat striatum: Possible involvement of CB1 receptors regulation. *Molecular Neurobiology*, *56*(2), 844–856. https://doi.org/10.1007/s12035-018-1129-2.

Albuquerque, E. X., & Schwarcz, R. (2013). Kynurenic acid as an antagonist of α7 nicotinic acetylcholine receptors in the brain: Facts and challenges. *Biochemical Pharmacology*, *85*(8), 1027–1032. https://doi.org/10.1016/j.bcp.2012.12.014.

Alkondon, M., Pereira, E. F. R., & Albuquerque, E. X. (2011). Endogenous activation of nAChRs and NMDA receptors contributes to the excitability of CA1 stratum radiatum interneurons in rat hippocampal slices: Effects of kynurenic acid. *The Biochemist*, *82*(8), 842–851. https://doi.org/10.1016/j.bcp.2011.06.004.

Alves, V. L., Gonçalves, J. L., Aguiar, J., Teixeira, H. M., & Câmara, J. S. (2020). The synthetic cannabinoids phenomenon: From structure to toxicological properties. A review. *Critical Reviews in Toxicology*, *50*(5), 359–382. https://doi.org/10.1080/10408444.2020.1762539.

Badawy, A. A. B. (2017). Kynurenine pathway of tryptophan metabolism: Regulatory and functional aspects. *International Journal of Tryptophan Research*, *10*(1). https://doi.org/10.1177/1178646917691938.

Bartlett, R. D., Esslinger, C. S., Thompson, C. M., & Bridges, R. J. (1998). Substituted quinolines as inhibitors of L-glutamate transport into synaptic vesicles. *Neuropharmacology*, *37*(7), 839–846. https://doi.org/10.1016/S0028-3908(98)00080-X.

Beggiato, S., Ieraci, A., Tomasini, M. C., Schwarcz, R., & Ferraro, L. (2020). Prenatal THC exposure raises kynurenic acid levels in the prefrontal cortex of adult rats. *Progress in Neuro-Psychopharmacology and Biological Psychiatry*, *100*(January), 109883. https://doi.org/10.1016/j.pnpbp.2020.109883.

Campbell, B. M., Charych, E., Lee, A. W., & Möller, T. (2014). Kynurenines in CNS disease: Regulation by inflammatory cytokines. *Frontiers in Neuroscience*, *8*(8 Feb), 1–22. https://doi.org/10.3389/fnins.2014.00012.

Casteels, C., Martinez, E., Bormans, G., Camon, L., De Vera, N., Baekelandt, V., Planas, M., & Van Laere, K. (2010). Type 1 cannabinoid receptor mapping with [18F]MK- 9470 PET in the rat brain after quinolinic acid lesion: A comparison to dopamine receptors and glucose metabolism. *European Journal of Nuclear Medicine and Molecular Imaging*, *37*(12), 2354–2363. https://doi.org/10.1007/S00259-010-1574-2.

Cosi, C., Mannaioni, G., Cozzi, A., Carl, V., Sili, M., Cavone, L., Maratea, D., & Moroni, F. (2011). G-protein coupled receptor 35 (GPR35) activation and inflammatory pain: Studies on the antinociceptive effects of kynurenic acid and zaprinast. *Neuropharmacology*, *60*(7–8), 1227–1231. https://doi.org/10.1016/j.neuropharm.2010.11.014.

di Giacomo, V., Chiavaroli, A., Orlando, G., Cataldi, A., Rapino, M., di Valerio, V., Leone, S., Brunetti, L., Menghini, L., Recinella, L., & Ferrante, C. (2020). Neuroprotective and neuromodulatory effects induced by cannabidiol and cannabigerol in rat hypo-E22 cells and isolated hypothalamus. *Antioxidants*, *9*(1), 71. https://doi.org/10.3390/antiox9010071.

di Giacomo, V., Chiavaroli, A., Recinella, L., Orlando, G., Cataldi, A., Rapino, M., Di Valerio, V., Ronci, M., Leone, S., Brunetti, L., Menghini, L., Zengin, G., Gunes, A., Abdallah, H. H., & Ferrante, C. (2020). Antioxidant and neuroprotective effects induced by cannabidiol and cannabigerol in rat CTX-TNA2 astrocytes and isolated cortexes. *International Journal of Molecular Sciences*, *21*(10). https://doi.org/10.3390/ijms21103575.

DiNatale, B. C., Murray, I. A., Schroeder, J. C., Flaveny, C. A., Lahoti, T. S., Laurenzana, E. M., Omiecinski, C. J., & Perdew, G. H. (2010). Kynurenic acid is a potent endogenous aryl hydrocarbon receptor ligand that synergistically induces interleukin-6 in the presence of inflammatory signaling. *Toxicological Sciences*, *115*(1), 89–97. https://doi.org/10.1093/toxsci/kfq024.

Escamilla-Ramírez, A., García, E., Palencia-Hernández, G., Colín-González, A. L., Galván- Arzate, S., Túnez, I., Sotelo, J., & Santamaría, A. (2017). URB597 and the cannabinoid WIN55,212-2 reduce behavioral and neurochemical deficits induced by MPTP in mice: Possible role of redox modulation and NMDA receptors. *Neurotoxicity Research*, *31*(4), 532–544. https://doi.org/10.1007/s12640-016-9698-1.

Fazio, F., Lionetto, L., Molinaro, G., Bertrand, H. O., Acher, F., Ngomba, R. T., Notartomaso, S., Curini, M., Rosati, O., Scarselli, P., Di Marco, R., Battaglia, G., Bruno, V., Simmaco, M., Pin, J. P., Nicoletti, F., & Goudet, C. (2012). Cinnabarinic acid, an endogenous metabolite of the kynurenine pathway, activates type 4 metabotropic glutamate receptors. *Molecular Pharmacology*, *81*(5), 643–656. https://doi.org/10.1124/mol.111.074765.

Florensa-Zanuy, E., Garro-Martínez, E., Adell, A., Castro, E., Díaz, Á., Pazos, Á., Mac- Dowell, K. S., Martín-Hernández, D., & Pilar-Cuéllar, F. (2021). Cannabidiol antidepressant-like effect in the lipopolysaccharide model in mice: Modulation of inflammatory pathways. *Biochemical Pharmacology*, *185*(December 2020), 1–11. https://doi.org/10.1016/j.bcp.2021.114433.

Fukui, S., Schwarcz, R., Rapoport, S. I., Takada, Y., & Smith, Q. R. (1991). Blood-brain barrier transport of kynurenines: Implications for brain synthesis and metabolism. *Journal of Neurochemistry*, *56*(6), 2007–2017. https://doi.org/10.1111/j.1471-4159.1991.tb03460.x.

Gil de Biedma-Elduayen, L., Giménez-Gómez, P., Morales-Puerto, N., Vidal, R., Calle, C. N., Gutiérrez-López, M. D., O'Shea, E., & Colado, M. I. (2022). Influx of kynurenine into the brain is involved in the reduction of ethanol consumption induced by Ro61-8048 after chronic intermittent ethanol in mice. *British Journal of Pharmacology*, 1–16. https://doi.org/10.1111/bph.15825.

Giménez-Gómez, P., Pérez-Hernández, M., Gutiérrez-López, M. D., Vidal, R., Abuin- Martínez, C., O'Shea, E., & Colado, M. I. (2018). Increasing kynurenine brain levels reduces ethanol consumption in mice by inhibiting dopamine release in nucleus accumbens. *Neuropharmacology*, *135*, 581–591. https://doi.org/10.1016/J.NEUROPHARM.2018.04.016.

Guidetti, P., Hoffman, G. E., Melendez-Ferro, M., Albuquerque, E. X., & Schwarcz, R. (2007). Astrocytic localization of kynurenine aminotransferase II in the rat brain visualized by immunocytochemistry. *Glia*, *55*, 78–92. https://doi.org/10.1002/glia.

Guillemin, G. J. (2012). Quinolinic acid, the inescapable neurotoxin. *FEBS Journal*, *279*(8), 1356–1365. https://doi.org/10.1111/j.1742-4658.2012.08485.x.

Hilmas, C., Pereira, E. F. R., Alkondon, M., Rassoulpour, A., Schwarcz, R., & Albuquerque, E. X. (2001). The brain metabolite kynurenic acid inhibits α7 nicotinic receptor activity and increases non-α7 nicotinic receptor expression: Physiopathological implications. *Journal of Neuroscience*, *21*(19), 7463–7473. https://doi.org/10.1523/jneurosci.21-19-07463.2001.

Jenny, M., Santer, E., Pirich, E., Schennach, H., & Fuchs, D. (2008). Δ9-Tetrahydrocannabinol and cannabidiol modulate mitogen-induced tryptophan degradation and neopterin formation in peripheral blood mononuclear cells in vitro. *Journal of Neuroimmunology*, *207*, 75–82. https://doi.org/10.1016/j.jneuroim.2008.12.004.

Jenny, M., Schröcksnadel, S., Überall, F., & Fuchs, D. (2010). The potential role of cannabinoids in modulating serotonergic signaling by their influence on tryptophan metabolism. *Pharmaceuticals*, *3*(8), 2647–2660. https://doi.org/10.3390/ph3082647.

Justinova, Z., Mascia, P., Wu, H.-Q., Secci, M. E., Redhi, G. H., Panlilio, L. V., Scherma, M., Barnes, C., Parashos, A., Zara, T., Fratta, W., Solinas, M., Pistis, M., Bergman, J., Kangas, B. D., Ferré, S., Tanda, G., Schwarcz, R., & Goldberg, S. R. (2013). Reducing cannabinoid abuse and preventing relapse by enhancing endogenous brain levels of kynurenic acid. *Nature Neuroscience*, *16*(11), 1652–1661. https://doi.org/10.1038/nn.3540.

Kenny, P. J., Boutrel, B., Gasparini, F., Koob, G. F., & Markou, A. (2005). Metabotropic glutamate 5 receptor blockade may attenuate cocaine self-administration by decreasing brain reward function in rats. *Psychopharmacology*, *179*(1), 247–254. https://doi.org/10.1007/s00213-004-2069-2.

Kita, T., Morrison, P. F., Heyes, M. P., & Markey, S. P. (2002). Effects of systemic and central nervous system localized inflammation on the contributions of metabolic precursors to the L-kynurenine and quinolinic acid pools in brain. *Journal of Neurochemistry*, *82*(2), 258–268. https://doi.org/10.1046/j.1471-4159.2002.00955.x.

Knyihár-Csillik, E., Toldi, J., Mihály, A., Krisztin-Péva, B., Chadaide, Z., Németh, H., Fenyo, R., & Vécsei, L. (2007). Kynurenine in combination with probenecid mitigates the stimulation-induced increase of c-fos immunoreactivity of the rat caudal trigeminal nucleus in an experimental migraine model. *Journal of Neural Transmission*, *114*(4), 417–421. https://doi.org/10.1007/s00702-006-0545-z.

Leklem, J. E. (1971). Quantitative aspects of tryptophan metabolism in humans and other species: A review. *The American Journal of Clinical Nutrition*, *24*(6), 659–672. https://doi.org/10.1093/ajcn/24.6.659.

Lowe, M. M., Mold, J. E., Kanwar, B., Huang, Y., Louie, A., Pollastri, M. P., Wang, C., Patel, G., Franks, D. G., Schlezinger, J., Sherr, D. H., Silverstone, A. E., Hahn, M. E., & McCune, J. M. (2014). Identification of cinnabarinic acid as a novel endogenous aryl hydrocarbon receptor ligand that drives IL-22 production. *PLoS One*, *9*(2), e87877. https://doi.org/10.1371/journal.pone.0087877.

Markin, P. A., Brito, A., Moskaleva, N. E., Tagliaro, F., La Frano, M. R., Savitskii, M. V., & Appolonova, S. A. (2021). Short- and long-term exposures of the synthetic cannabinoid 5F-APINAC induce metabolomic alterations associated with neurotransmitter systems and embryotoxicity confirmed by teratogenicity in zebrafish. *Comparative Biochemistry and Physiology, Part C: Toxicology and Pharmacology*, *243*(October 2020). https://doi.org/10.1016/j.cbpc.2021.109000.

Marsicano, G., Goodenough, S., Monory, K., Hermann, H., Eder, M., Cannich, A., Azad, S. C., Cascio, M. G., Ortega-Gutiérrez, S., Van der Stelt, M., López-Rodríguez, M. L., Casanova, E., Schütz, G., Zieglgänsberger, W., Di Marzo, V., Behl, C., & Lutz, B. (2003). CB1 cannabinoid receptors and on-demand defense against excitotoxicity. *Science*, *302*(5642), 84–88. https://doi.org/10.1126/science.1088208.

McDew-White, M., Lee, E., Alvarez, X., Sestak, K., Ling, B. J., Byrareddy, S. N., Okeoma, C. M., & Mohan, M. (2022). Cannabinoid control of gingival immune activation in chronically SIV-infected rhesus macaques involves modulation of the indoleamine- 2,3-dioxygenase-1 pathway and salivary microbiome. *eBioMedicine*, *75*(103769). https://doi.org/10.1016/j.ebiom.2021.103769j.

Morales-Puerto, N., Giménez-Gómez, P., Pérez-Hernández, M., Abuin-Martínez, C., Gil de Biedma-Elduayen, L., Vidal, R., Gutiérrez-López, M. D., O'Shea, E., & Colado, M. I. (2021). Addiction and the kynurenine pathway: A new dancing couple? *Pharmacology and Therapeutics*, *223*, 107807. https://doi.org/10.1016/j.pharmthera.2021.107807.

Perkins, M. N., & Stone, T. W. (1982). An iontophoretic investigation of the actions of convulsant kynurenines and their interaction with the endogenous excitant quinolinic acid. *Brain Research*, *247*(1), 184–187. https://doi.org/10.1016/0006- 8993(82)91048-4.

Pineda-Farias, J. B., Pérez-Severiano, F., González-Esquivel, D. F., Barragán-Iglesias, P., Bravo-Hernández, M., Cervantes-Durán, C., Aguilera, P., Ríos, C., & Granados-Soto, V. (2013). The L-kynurenine-probenecid combination reduces neuropathic pain in rats. *European Journal of Pain (United Kingdom)*, *17*(9), 1365–1373. https://doi.org/10.1002/j.1532-2149.2013.00305.x.

Pintor, A., Tebano, M. T., Martire, A., Grieco, R., Galluzzo, M., Scattoni, M. L., Pèzzola, A., Coccurello, R., Felici, F., Cuomo, V., Piomelli, D., Calamandrei, G., & Popoli, P. (2006). The cannabinoid receptor agonist WIN 55,212-2 attenuates the effects induced by quinolinic acid in the rat striatum. *Neuropharmacology*, *51*(5), 1004–1012. https://doi.org/10.1016/j.neuropharm.2006.06.013 Poddar.

Poddar, M. K., Bhattacharyya, K. C., & Ghosh, J. J. (1974). Potentiating effect of cannabidiol on Δ9-tetrahydrocannabinol-induced changes in hepatic enzymes. *Biochemical Pharmacology*, *23*(3), 758–759. https://doi.org/10.1016/0006-2952(74)90643-1.

Poddar, M. K., & Ghosh, J. J. (1972). Effect of cannabis extract, Δ9-tetrahydrocannabinol and lysergic acid diethylamide on rat liver enzymes. *Biochemical Pharmacology*, *21*(24), 3301–3303. https://doi.org/10.1016/0006-2952(72)90094-9.

Rangel-López, E., Colín-González, A. L., Paz-Loyola, A. L., Pinzón, E., Torres, I., Serratos, I. N., Castellanos, P., Wajner, M., Souza, D. O., & Santamaría, A. (2015). Cannabinoid receptor agonists reduce the short-term mitochondrial dysfunction and oxidative stress linked to excitotoxicity in the rat brain. *Neuroscience*, *285*, 97–106. https://doi.org/10.1016/j.neuroscience.2014.11.016.

Rodríguez-Muñoz, M., Onetti, Y., Cortés-Montero, E., Garzón, J., & Sánchez-Blázquez, P. (2018). Cannabidiol enhances morphine antinociception, diminishes NMDA-mediated seizures and reduces stroke damage via the sigma 1 receptor. *Molecular Brain*, *11*(1), 1–12. https://doi.org/10.1186/s13041-018-0395-2.

Sánchez-Blázquez, P., Rodríguez-Muñoz, M., & Garzón, J. (2014). The cannabinoid receptor 1 associates with NMDA receptors to produce glutamatergic hypofunction: Implications in psychosis and schizophrenia. *Frontiers in Pharmacology*, *4*, 1–10. https://doi.org/10.3389/fphar.2013.00169.

Savitz, J. (2020). The kynurenine pathway: A finger in every pie. *Molecular Psychiatry*, *25*(1), 131–147. https://doi.org/10.1038/s41380-019-0414-4.

Schifano, F., Orsolini, L., Papanti, G. D., & Corkery, J. M. (2015). Novel psychoactive substances of interest for psychiatry. *World Psychiatry*, *14*, 15–26. https://doi.org/10.1002/wps.20174.

Schwarcz, R., Bruno, J. P., Muchowski, P. J., & Wu, H. Q. (2012). Kynurenines in the mammalian brain: When physiology meets pathology. *Nature Reviews Neuroscience*, *13*(7), 465–477. https://doi.org/10.1038/nrn3257.KYNURENINES.

Secci, M. E., Auber, A., Panlilio, L. V., Redhi, G. H., Thorndike, E. B., Schindler, C. W., Schwarcz, R., Goldberg, S. R., & Justinova, Z. (2017). Attenuating nicotine reinforcement and relapse by enhancing endogenous brain levels of kynurenic acid in rats and squirrel monkeys. *Neuropsychopharmacology*, *42*(8), 1619–1629. https://doi.org/10.1038/npp.2017.21.

Secci, M. E., Mascia, P., Sagheddu, C., Beggiato, S., Melis, M., Borelli, A. C., Tomasini, M. C., Panlilio, L. V., Schindler, C. W., Tanda, G., Ferré, S., Bradberry, C. W., Ferraro, L., Pistis, M., Goldberg, S. R., Schwarcz, R., & Justinova, Z. (2019). Astrocytic mechanisms involving kynurenic acid control Δ9-tetrahydrocannabinol-induced increases in glutamate release in brain reward-processing areas. *Molecular Neurobiology*, *56*(5), 3563–3575. https://doi.org/10.1007/s12035-018-1319-y.

Shestakova, K. M., Mesonzhnik, N. V., Markin, P. A., Moskaleva, N. E., Nedorubov, A. A., Brito, A., Appolonova, E. G., Kuznetsov, R. M., Bochkareva, N. L., Kukharenko, A., Lyundup, A. V., Tagliaro, F., & Appolonova, S. A. (2021). Pharmacokinetic properties of the novel synthetic cannabinoid 5F-APINAC and its influence on metabolites associated with neurotransmission in rabbit plasma. *Pharmaceuticals*, *14*(7), 668. https://doi.org/10.3390/PH14070668.

Stone, T. W. (2020). Does kynurenic acid act on nicotinic receptors? An assessment of the evidence. *Journal of Neurochemistry*, *152*(6), 627–649. https://doi.org/10.1111/jnc.14907.

Suzuki, T., Hidaka, T., Kumagai, Y., & Yamamoto, M. (2020). Environmental pollutants and the immune response. *Nature Immunology*, *21*, 1486–1495.

United Nations Office on Drugs Crime (UNODC). (2021). *World drug report 2021*. https://www.unodc.org/res/wdr2021/field/WDR21_Booklet_4.pdf, 2021. https://www.unodc.org/res/wdr2021/field/WDR21_Booklet_3.pdf.

Vengeliene, V., Cannella, N., Takahashi, T., & Spanagel, R. (2016). Metabolic shift of the kynurenine pathway impairs alcohol and cocaine seeking and relapse. *Psychopharmacology*, *233*(18), 3449–3459. https://doi.org/10.1007/s00213-016-4384-9.

Wang, J., Simonavicius, N., Wu, X., Swaminath, G., Reagan, J., Tian, H., & Ling, L. (2006). Kynurenic acid as a ligand for orphan G protein-coupled receptor GPR35. *Journal of Biological Chemistry*, *281*(31), 22021–22028. https://doi.org/10.1074/jbc.M603503200.

Yamaori, S., Kinugasa, Y., Jiang, R., Takeda, S., Yamamoto, I., & Watanabe, K. (2015). Cannabidiol induces expression of human cytochrome P450 1A1 that is possibly mediated through aryl hydrocarbon receptor signaling in HepG2 cells. *Life Sciences*, *136*, 87–93. https://doi.org/10.1016/j.lfs.2015.07.007.

Zádor, F., Nagy-Grócz, G., Dvorácskó, S., Bohár, Z., Cseh, E. K., Zádori, D., Párdutz, Á., Szűcs, E., Tömböly, C., Borsodi, A., Benyhe, S., & Vécsei, L. (2020). Long-term systemic administration of kynurenic acid brain region specifically elevates the abundance of functional CB1 receptors in rats. *Neurochemistry International*, *138*(April), 104752. https://doi.org/10.1016/j.neuint.2020.104752.

Zhou, H., Zhang, Q., Martinez, E., Dale, J., Hu, S., Zhang, E., Liu, K., Huang, D., Yang, G., Chen, Z., & Wang, J. (2018). Ketamine reduces aversion in rodent pain models by suppressing hyperactivity of the anterior cingulate cortex. *Nature Communications*, *9*(1). https://doi.org/10.1038/s41467-018-06295-x.

Chapter 19

Central nervous system cancers and use of cannabis and related products: An updated narrative

J. Eduardo Rodriguez Almaraz[a,c] and Nicholas Butowski[a,b,d]

[a]Neurosurgery Department, Division of Neuro-Oncology, University of California, San Francisco, CA, United States, [b]Department of Molecular Science, Director of Translational Research, Neurosurgery Department, Division of Neuro-Oncology, University of California, San Francisco, CA, United States, [c]Department of Epidemiology & Biostatistics, University of California, San Francisco, CA, United States, [d]Department of Molecular Science, University of California, San Francisco, CA, United States

Abbreviations

2-AG	2-arachidonoylglycerol
Acadvl	acyl-CoA dehydrogenase, very long chain
Akt	protein kinase B
ACC1	acetyl-CoA carboxylase 1
AEA	arachidonoylethanolamide
Agpat1	1-acylglycerol-3-phosphate O-acyltransferase 1
CBD	cannabidiol
CBG	cannabigerol
CNS	central nervous system
COX-2	cyclooxygenase-2
CINV	chemotherapy-induced nausea and vomiting
FDA	Food and Drug Administration
GBM	glioblastoma (IDH-wildtype gliogblastoma)
IDH	isocitrate dehydrogenase
JCC	Joint Commission of Cancer
MAPK	mitogen-activated protein kinase
NASEM	National Academies of Sciences, Engineering and Medicine
PI3K	phophoinositide 3-kinase
Ppar-γ	peroxisome proliferator-activated receptor-γ
QOL	quality of life
Scd1	stearoyl-CoA desaturase 1
Srebp-1c	sterol regulatory element-binding protein 1c
TAG	triacylglycerol
THC	delta-9-tetrahydrocannabinol
TMZ	temozolomide
TNFα	tumor necrose factor α
QOL	quality of life
WHO	World Health Organization

Overview

High grade or malignant central nervous system (CNS) tumors are among the most fatal cancers. Between malignant and nonmalignant CNS tumors, there are over one hundred subtypes with different behavioral, epidemiology, morphology, and

Cannabis Use, Neurobiology, Psychology, and Treatment. https://doi.org/10.1016/B978-0-323-89862-1.00016-7

clinical characteristics (Miller et al., 2021). Due to these characteristics, CNS tumors differ in prognostic, treatment, clinical outcomes, and survival across histological subtypes (Burt Nabors et al., 2021).

While cancer treatments often appeal to dubious claimants with equally dubious cure for cancer statements often drawn from anecdotal evidence, in recent years there has been an increased number of studies that explore the mechanisms of antitumor effects of cannabinoids. In this chapter, we provide an overview on the signal transduction pathways related to the activation of cannabinoid receptors as well as the biological interaction between cannabinoids, their receptors, and their effects that can be clinically important in the context of primary brain malignancies. Additionally, we provide with the overall context of the status of the disease and how it distributes in populations.

Global incidence, prevalence, and survival of gliomas

The overall incidence of malignant brain tumors varies across different regions of the globe. In 2016, it was estimated that there were 330,000 (95% CI: 299,000–349,000) new cases (4.63 per 100,00 person-rates [4.17–4.90]) with 227,000 deaths (95% CI: 205,000–241,000) attributed to CNS cancer globally. The regions with higher incidence per 100,000 person-years include east Asia (98,000–122,000), western Europe (37,000–54,000), and south Asia (29,000–37,000). The countries with the highest number of incident cases are China (53,264–66,813), the United States (14,745–17,756), and India (18,847–25,993) (Patel et al., 2019).

In 2021, it is estimated that more than 84,000 people were diagnosed with a primary brain tumor in the United States of America (US). Of these, nearly one-third (29%) of brain and CNS tumors are malignant (Ostrom et al., 2019). Overall global incidence of these tumors was highest in adults older than 40 years old (Leece et al., 2017).

The survival rate at five years for all malignant brain tumors combined has increased due to development of new therapies, being 36% in the period observed between 2009 and 2015 compared to 23% from 1975 to 1977. However, the specific five-year survival rates vary across populations.

Glioblastoma (GBM) is the most commonly occurring malignant CNS tumor (14.3% of all tumors and 49.1% of malignant tumors) (Ostrom et al., 2019) and the most aggressive with a five-year survival rate of 7% and median survival time of 15 months. It affects more often older ($\geq$64 years of age), Caucasian (2.0 times higher in Caucasians), and males (1.6 times higher). GBM presents the lowest incidence among Asians and Native-Americans (Ostrom et al., 2019).

The incidence rate for other malignant CNS tumors (including spine, cranial nerves, and other locations) varies depending on geographical location and population diversity.

Given that the most common tumors are adult-type diffuse gliomas (often referred simply as gliomas), the remaining of the present chapter will focus on these types of tumors.

Classification of gliomas

The classification and reporting of CNS tumors have rapidly changed in recent years in parallel with expanding molecular and genetic understanding of the tumor biology and advances in diagnostic tools (Hoshide & Jandial, 2016).

Most cancers are staged according to the Joint Commission of Cancer (AJCC) collaborative Staging schema. However, scientists and clinicians follow the World Health Organization's (WHO) Classification of Tumors of the Central Nervous system guidelines. These guidelines were first published in 2000 (Kleihues et al., 2002), and the grading system assigned a grade (I–IV) based on clinical behavior. In 2016, these guidelines were updated to reflect diagnosis including some molecular features of tumors (Louis et al., 2016). Most recently, in 2021 (Louis et al., 2021), the 5th edition of the WHO classification of tumors of the CNS was published reflecting the advances made in our understanding of tumor biology by incorporating further molecular, histological, and immunohistochemical features while keeping a seemingly parsimonious classification (Louis et al., 2021) (Fig. 1).

The importance of this classification lays in the fact that it provides a great deal of guidance to the clinician regarding the course of treatment and prognosis (Wen & Packer, 2021) regardless of not been fully implemented yet.

Glioma risks factors

In general, there is no single risk factor that has been identified to be associated with CNS tumors. However, the most consistently identified exogenous risk factor associated with brain tumor development is exposure to either therapeutic or high-dose ionizing radiation. Other factors that have been associated with higher risk of glioma include decreased susceptibility to allergy, immune factors, and immune genes, as well as some single-nucleotide polymorphisms detected by genomic analysis (Hu et al., 1998).

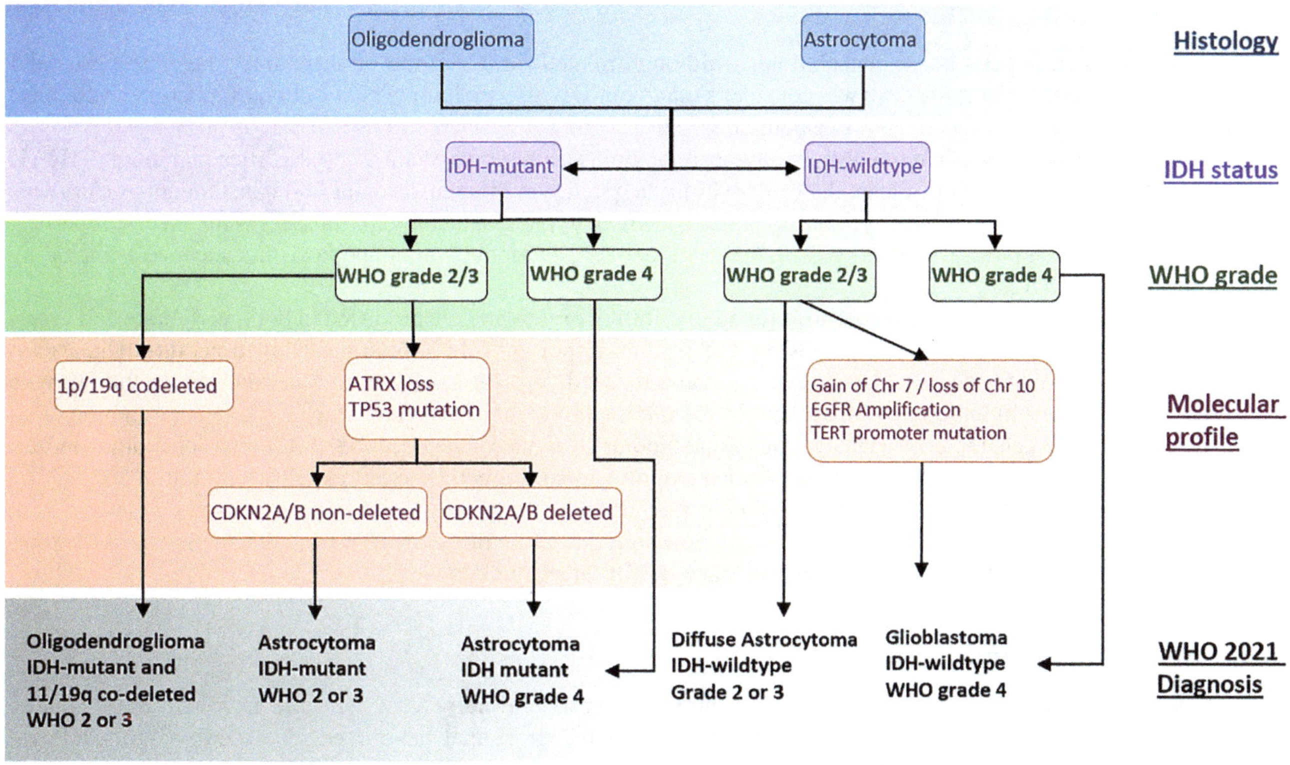

FIG. 1 Algorithm based on the 2021 WHO classification of gliomas. The use of the suffix NOS (not otherwise specified) is to denote that either the necessary molecular testing is lacking or the results are noninformative or inconclusive.

Additionally, there are a few rare, inherited genetic syndromes that are associated with brain tumors, including neurofibromatosis 1 (NF1 gene), neurofibromatosis2 (NF2 gene), Turcot syndrome (APC gene), Gorlin syndrome (PTCH gene), tuberous sclerosis (TSC1 and TSC2 genes), and Li-Fraumeni syndrome (TP53 gene).

Despite the fact that the association of these genes with CNS tumors is well established (Rice et al., 2016), only ~5% of persons with brain tumors have a family history of a brain tumor. The vast majority of CNS tumors appear no to be part of inherited genetic syndromes. While there are some hereditary similarities in glioma tumors between family members, there is not a statistically significant difference between families having tumors with similar hereditary features. Also, in families with more than one glioma, the tumors tend to have the same molecular markers (Molinaro et al., 2019).

Another controversial and potential risk factor that has been proposed is the impact on nonionizing radiation from mobile phones; however, this exposure and brain cancer have not been well established. Similarly, industrial chemicals have long been suspected as a cause of glioma due to their ability to cross the blood brain barrier (BBB) (Yang et al., 2017), but further research is necessary to establish a causal association.

Treatment and management of gliomas

Treatment of all CNS tumors is complex, initially consisting of maximally safe surgical resection. Treatment of glioblastoma is followed by radiation therapy and concurrent temozolomide (TMZ) chemotherapy followed by adjuvant chemotherapy At recurrence, the course of treatment depends on a number of factors that include the molecular features of the tumor as well as the clinical status of the patient and their personal preferences toward repeat surgery, more radiation, chemotherapies, or clinical trials.

All of the above factors make it a challenge to manage the symptoms associated with the tumor as well as the side effects of the therapies used to treat disease. Additionally, the symptoms of gliomas are varied depending on the size, location, and classification. These may include headaches, vision impairment, hearing impairment speech difficulty, and balance difficulty (Davies & Clarke, 2004). As the cancer in the brain grows and begins to compress adjacent brain areas other complications might be present such as fatigue, depression, memory or personality changes, decline in mental function or brain capacity, seizures, and muscle paralysis.

Cannabinoids and gliomas

As it is stated elsewhere in this book, cannabinoids—although related—are a group of structurally heterogeneous and complex compounds. For our purposes, we will consider cannabinoids within their three main subtypes: plant-derived (phytocannabinoids), synthetic, and endogenous cannabinoids.

Although the pharmacology of the majority of phytocannabinoids is unknown, delta-9-tetrahydrocannabinoid (THC) is recognized as the most potent out of those identified within the plant, yet there are around 600 constituents in *Cannabis sativa* (Giacoppo et al., 2014). Recently, studies have shown that phytocannabinoids possess anticancer properties, including inhibition of cell proliferation, migration and angiogenesis, and induction of apoptosis in cancer cells including glioma (Velasco et al., 2012).

In contrast, endocannabinoids are ubiquitously found in different organs and are produced in both, vertebrate and invertebrate tissues. Signaling that, endocannabinoids have played an important role in biological functions throughout evolution. One of the important processes that researchers have focused their efforts is the role of cannabis in the immune response. As such, two arachidonic acid derivatives, 2-arachidonoylglycerol (2-AG) and arachidonoylethanolamide (anadamide, AEA), are the best characterized endogenous cannabinoids (Katona & Freund, 2008). Although some other amides and esters of long-chain polyunsaturated fatty acids also exhibit cannabimimetic properties (Howlett et al., 2002).

Endocannabinoid production and release by glial cells may underlie the neuroprotective properties of cannabinoids in experimental models (Sánchez & García-Merino, 2012); however, due to the difficulty of testing this in patients with diagnosis of glioma, its clinical and physiological significance is still largely unknown.

Cannabinoid receptors and glioma

As it is known, cannabinoids elicit a wide range of central and peripheral nervous system effects, which are mediated through cannabinoid receptors. There are two specific types of receptors that have been identified so far: CB1 and CB2. These receptors are transmembrane $G_{i/o}$-protein-coupled receptors that interact with effector proteins to elicit its effects. Overall, the similarities between CB1 and CB2 are rather scarce due to significant disparities in the transmembrane domains. Additionally, they differ in their amino acid sequence, tissue distribution and expression pattern (Howlett et al., 2002). As a result, different compounds that activate CB1 and CB2 receptors (selectively and nonselectively) play distinct physiological roles and produce different effects.

CB1 is of particular interest in neurooncology due to their abundant presence in specific areas of the brain, such as the cortex, hippocampus, basal ganglia, and cerebellum, as well as in peripheral nerve terminals. CB1 mediates inhibition of neurotransmitter release and is involved in the control of motor activity, memory, cognition, appetite, and sensory perception (Howlett & Abood, 2017). Similarly, nonneuronal (glial cells) have been shown to express CB1 receptors (Devane et al., 1992), although their precise function in astrocytes, oligodendrocytes, and microglia has been only partly described.

By contrast, the CB2 receptor is predominantly expressed in cells and organs of the immune system. Although the role of peripheral CB2 receptor activation under physiologic conditions is not well defined, it is thought that CB2 signaling is involved in B-cell differentiation and modulation of immune response (Turcotte et al., 2016). Until recently, it was thought that healthy brain tissue lacked of CB2 receptor expression; however, its expression has been detected in brain macrophages and may be involved in central nervous system immune activation (Cabral et al., 2008).

Cannabinoid system in gliomas

A growing body of evidence suggests that alterations of a balance in the cannabinoid system between the levels of endogenous ligands and their receptors occur during malignant transformation in various types of cancer, including gliomas (Peeri et al., 2021). While nontransformed astrocytes express only the CB1 receptor, both functional cannabinoid receptors have been identified in several human glioblastoma cell lines, as well as in primary cultures derived from glioblastoma (Kaminska & Ellert-Miklaszewska, 2015). In previous immunohistochemical analysis of surgical specimens of human glioma (low and high grade), it was revealed that the expression of CB2 receptor was increased in tumoral cells, as well as in invading macrophages and endothelial vessel cells of the tumor (Ellert-Miklaszewska et al., 2007). Additionally, Sanchez and collaborators observed a positive correlation between the tumor malignancy grade and levels of CB2 receptor expression (Velasco et al., 2012). Interestingly, as observed by the same group, the immunoreactivity to the CB2 receptor markedly prevailed over detected CB1 receptor levels in IDH-wildtype glioblastomas grade 4 (as per WHO 2021 classification). These observations imply that high expression levels of CB1 receptor in the tumor make gliomas particularly susceptible to the antitumor effects of circulating endocannabinoids like AEA, resulting in tumor involution (Sredni et al., 2016). The increased level of cannabinoid receptor expression in malignant versus healthy tissues suggests a possible important role of the endocannabinoid system in the immune stimulation and the sustained suppression of cell division as an internal mechanism to attempt to control cancer growth.

In support of this hypothesis, inhibitors of endocannabinoid transport and degradation have been shown to suppress tumor progression in some types of cancer (Oláh et al., 2017). Thus, a promising field to explore is not only to increase the expression of CB1 and CB2 receptors but also to prolong the bioavailability of endocannabinoids by increasing their biosynthesis and uptake as well as minimize their degradation.

Antitumoral action of cannabinoids in glioma cells

Cannabinoids as agents to control glioma growth and proliferation

The inhibition of adenylyl cyclase via $G_{i/o}$ via activation of CB1 and CB2 activates different metabolic pathways such as mitogen-activated protein kinase (MAPK), phophoinositide 3-kinase (PI3K), and cyclooxygenase-2 pathways (COX-2), as well as accumulation of ceramide, modulation of protein kinase B (Akt), and ion channels (Valverde et al., 2005). In addition to its neuromodulatory function, endocannabinoids have been shown to play other important functions such as control of energy metabolism, immunity, cardiovascular tone, and reproduction (Katona & Freund, 2008). As such, cannabinoids can induce significant inhibition of cell growth in tumor cells, via to modulation of proteins and nuclear factors involved in the control of cell survival, transformation, and cell death (Aguado et al., 2007). Apoptosis of glioma cells promoted by cannabinoid treatment was first described by Manuel Guzman in 1998 (Sánchez et al., 1998). Guzman showed that THC is able to inhibit the growth of rat glioma cells in vitro and induces apoptosis. Later, Ellert-Miklaszewska and collaborators showed apoptotic activity triggered by a mixed CB1/CB2 synthetic agonist (WIN 55,212-2) in rat glioma cells (Ellert-Miklaszewska et al., 2005) adding to the body of evidence that cannabinoids exert antitumor effects that lead to a significant regression of malignant gliomas in cannabinoid-treated animal models. Specifically, local administration of THC or WIN55, 212-2 reduced the size of tumors generated by intracranial inoculation of C6 glioma cell lines in rats, leading to complete eradication of gliomas and increased survival in one-third of the treated rats (Galve-Roperh et al., 2000).

Cannabinoids as antiangiogenic agents

To meet the metabolic demand of a rapidly growing cell population tumors need to generate a new vascular supply, blocking this process constitutes one of the most important antitumoral approaches currently available. At first, gliomas acquire their blood supply by co-opting existing normal brain vessels to meet the metabolic demand of rapidly dividing cells. However, as gliomas progress, they become hypoxic triggering robust angiogenesis via the VEGF pathway and angiopoietin-2 (Vajkoczy et al., 2002). Cannabinoids have shown to impair the VEGF pathway by blunting VEGF production and signaling as well as VEGFR-2 (the most prominent VEGF receptor in glioma) (Bla´zquez et al., 2008).

Matrix metalloproteinase-2 (MMP2) is another important protein that plays a role in glioma evolution by promoting tissue breakdown and remodeling during the formation of neovessels. Local administration of THC reduces MMP2 activity and expression, in gliomas generated in mice (Blázquez et al., 2008; Velasco et al., 2007).

Cannabinoids as proapoptotic agents

Several signal transduction pathways are triggered by activation of the CB1 and CB2 receptors and participate in the cannabinoid-induced cell death in various tumor cells (Guzmán, 2003).

One mechanism by which cannabinoids induce glioma cell death involves the sustained accumulation of sphingolipid ceramide (a proapoptotic lipid), which modulates signaling pathways that are crucial in the control of tumor cell growth and cell survival (Guzmán et al., 2001). Galve-Roperh and collaborators postulated that increase ceramide levels induced by cannabinoids leads to prolonged activation of the Raf-1/MEK/ERK signaling cascade mediating the arrest of glioma cell cycle and inducing apoptosis (Galve-Roperh et al., 2000). Further, Aguado and collaborators later demonstrated that human glioma cell lines treated with cannabinoids enter apoptosis using the same mitochondria-dependent pathway (Aguado et al., 2007), validating this mechanism as an important contributor to the apoptosis induction. Thus, the proapoptotic action of cannabinoids relies on the disruption of signaling pathways crucial for the regulation of cellular proliferation and differentiation (Salazar et al., 2009).

Furthermore, phytocannabinoids have been shown to trigger glioma apoptosis via different signal transduction pathways, including cell cycle arrest, oxidative stress, endoplasmic reticulum (ER)-stress and autophagy (Sánchez et al., 1998). More specifically, THC has been shown to alter the balance between ceramides and dihydroceramides and subsequently activate ER-stress that results in apoptotic cell death. In turn, THC-induced ER-stress promotes autophagy, which it is a process upstream of apoptosis, and the activation of this pathway was essential for the in vivo antitumor action of THC in glioma xenografts (Kaminska & Ellert-Miklaszewska, 2015).

THC also induced $G_{0/1}$ phase arrest in glioblastoma cells. Interestingly, THC and CBD showed synergistic inhibition of cell proliferation in glioblastoma cell lines. Further, CBD was found to inhibit the invasiveness of glioblastoma cells (Galanti et al., 2009). Recently, cannabigerol (CBG) has shown to reduce glioblastoma cell viability and invasion in vitro, especially with a synergistic effect with THC (Zogopoulos et al., 2015). However, it has been observed that sublethal concentrations of THC leads to cell proliferation in glioblastoma cell lines. This proliferation is dependent on the activity of metalloprotease, endothelial growth factor (EGFR), and tumor necrosis factor α-converting enzyme (TACE/ADAM17) that mediates EGFR transactivation. This apparent "reversal" of the expected effect of THC seems to be critically dependent on drug concentrations and EGFR transactivation (Hart et al., 2004). Importantly, this proliferative effect at low concentrations should be taken into consideration, as it raises concerns for clinical use: Given that the benefit of treating patients with THC might be reversed at lower doses, continuous high doses of THC might be needed to ensure cell death rather than cell proliferation in patients (Peeri et al., 2021). Additionally, the clinical implications as well as the side effects and interaction with other drugs when using THC at high doses must be considered when treating with THC.

Despite the evidence of unveiling the potential apoptotic properties of cannabinoids, alternative receptor-dependent pathways cannot be ruled out and are in need of further investigation and human clinical trials need to be conducted to validate the role of cannabinoids in control of tumor growth.

Potential therapeutic applications of cannabinoids in gliomas

Antitumor

While standard therapies for glioma (chemotherapy, radiation, and surgery) have improved in efficacy and safety profile over the last decade, they are not free of side effects that range from nausea, vomiting, liver damage, hematologic anomalies, pulmonary damage, cardiac diseases, asthenia, mucosal damage, as well as neurological damage (Fernandes et al., 2017). In contrast, cannabinoids have shown a favorable safety profile in preclinical studies (in vivo and in vitro) (Velasco et al., 2012), as well as from pilot clinical trials in the context of recurrent glioblastoma and may counteract side effects from radiation and chemotherapy (Lynch et al., 2014; Short & Little, 2017). Furthermore, cannabinoids promote survival of healthy glial cells and neurons in different models of injury, suggesting ta potential neuroprotective effect of cannabinoids during radiation (Guzmán, 2003).

Cannabinoids as radiation sensitizers

Experiments conducted by Dumitru and collaborators showed that pretreating glioma cells with a mixture of THC and CBD increased cell sensitivity to irradiation (Dumitru et al., 2018), a mechanism similar to temozolomide, a chemotherapeutic agent commonly used in the treatment of glioblastoma. These promising preclinical observations then led to a phase II clinical trial investigating the treatment with a THC:CBD mixture that showed a higher survival rate in the intervention group compared to the control group treated with dose-intense temozolomide (clinical trial NCT01812603).

Neuroprotective effect

One possible mechanism of antitumor and neuro-protective effect of cannabinoid compounds is the signaling differential expression of endocannabinoid receptors (specifically, CB2) between glioma cells and healthy neural tissue (Howlett et al., 2002). In contrast to proapoptotic action of THC and WIN55 on transformed glial cells, treatment of primary cultured normal astrocytes with these CB1/CB2-activating cannabinoids does not trigger ceramide generation de novo, and induction of endoplasmic reticulum (ER) stress-related genes or apoptosis (Galve-Roperh et al., 2000). Thus, the apparent lack or low expression of CB2 receptor expression in the normal brain and its abundance in high-grade gliomas seems to confer a relative safety of CB2-selective agonists for targeted glioma therapy. Additionally, the CB2 selective compounds lack the undesirable psychotropic side effects that are often of concern for clinicians (C. Howlett et al., 2002).

Clinical experience of cannabis in gliomas

In recent years, the medicinal properties of the *cannabis* genus have received increased attention around the world. This attention combined with lack of therapeutic choices and side effect management from such therapies has greatly increased the interest of cannabis in brain tumor patients due to its safety profile.

Current evidence suggests medicinal cannabis may inhibit chemotherapy-induced nausea and vomiting, stimulate appetite, reduce pain, and decrease inflammation and cancer cell proliferation and survival (Abrams, 2016). In this context, cannabinoids have not only shown to possess properties that may help minimize the polypharmacy effect that chemotherapeutic agents entail but to be a potential co-adjuvant to these therapies.

Cannabinoid therapies, including marijuana and CBD products, may be a useful adjunct to steroids and palliative therapies to improve symptom management.

Side effect management

Although there has been advancement in the research of cannabinoids as symptom management in different cancers, the literature is limited in the context of gliomas.

However, some proposed properties of cannabis studied in other fields can be used in glioma patients such as antiinflammatory (Mukhopadhyay et al., 2011), anticonvulsive (Jones et al., 2010), antipsychotic (Zuardi et al., 2006), anxiolytic (Bergamaschi et al., 2011), antidepressant (Zanelati et al., 2010), hypnotic, sedative, antitumor, analgesic, and antiemetic activity.

Chemotherapy-induced nausea and vomiting (CINV)

Nausea and vomiting, which can present in different ways in patients undergoing chemotherapy (acute, subacute, delayed, and anticipatory), is one of the most common toxicities and is considered by patients as the most stressful and cumbersome side effect (Reblin et al., 2019). This relative unpredictability of the severity of CINV has the potential to cause depression, anxiety, and impact in mood given the potential disruption to the daily life of patients undergoing chemotherapy treatment (Rodriguez Almaraz et al., unpublished data). Cannabinoids are well known to exert palliative effects in cancer patients (Pacher, 2006). As such, one of the better established effects is as an inhibitor of chemotherapy-induced nausea and vomiting (Guzmán, 2003). Specifically, dronabinol, first approved in 1986 for the treatment of CINV, has shown to have a better acute antiemetic efficacy than conventional antiemetic drugs on cancer patients treated with potentially emesis-inducing chemotherapeutic agents. In contrast, nabilone and levonantradol did not show superior acute antiemetic efficacy (Machado Rocha et al., 2008). However, dronabinol (Marinol) and its synthetic analog, nabiolone (Cesamet) are currently approved for use as anti CINV. Given the widespread knowledge of this effect, observational studies have shown that CINV is the main reason for use of cannabinoids (including medical marijuana) among cancer patients (Reblin et al., 2019; Rodriguez-Almaraz et al., 2019).

Cancer cachexia control

It is estimated that approximately 50%–80% of cancer patients suffer from cancer-cachexia syndrome, and it has been implicated in 25% of cancer deaths (Barton, 2017). The basis for this syndrome to be as severe is due to abnormalities in protein and amino acid metabolism and the appearance of energy balance disorder. Subsequently, malnutrition due to possible side effects of chemotherapy and radiotherapy may further lead to a number of metabolic unbalances that negatively impact the overall clinical outcomes and quality of life of patients. In recent years, there has been a gradual but steady progress in unveiling the benefits of cannabinoid usage as supportive therapy and in palliative medicine (Dumitru et al., 2018). In 2011, Brisbois and collaborators demonstrated that THC improved chemosensory perception, altered macronutrient preference, appeal of savory foods, and appetite in patients with advanced cancer that had chemosensory perception (Brisbois et al., 2011). Additionally, dronabinol has been approved by the FDA to increase appetite as antiemetic in patients diagnosed with AIDS (Badowski & Yanful, 2018). Altogether, this information signals that cannabinoids play an important role in the control of anorexia induced by the disease itself as well as therapies for disease control such as chemoradiation and that by improving the patient's nutritional intake the overall health status and quality of life might improve as well.

Seizure control

Another symptom that is often present in glioma patients are seizures. Depending on the location size and shape of the tumor, seizures can be often difficult to treat leading to a significant detriment in patients' quality of life (Davies & Clarke, 2004). Additionally, seizures can accelerate the neurocognitive deterioration, which can have an additive effect resulting from the adverse effects from antiepileptic drugs themselves (Vecht et al., 2014). Management of gliomas with surgery, radiotherapy, and chemotherapy may contribute to seizure control, but tumor-related epilepsy is often refractory despite adequate treatment with standard antiepileptic medications (You et al., 2012). Research in this domain led to the approval of cannabidiol (Epidiolex) in oral solution for the treatment of seizures associated with two rare and severe forms of epilepsy in children (FDA News Release, 2018).

In the context of glioma, only anecdotal and observational data have been collected (Hsu et al., 2020; Rodriguez-Almaraz et al., unpublished data). However, the body of evidence in other domains suggests that cannabinoids may play an important role in seizure control in patients with glioma with few side effects.

Patient experience

The body of scientific evidence is important to guide clinicians and health professionals who help patients navigate the therapy for gliomas and concomitant medications used for symptom management. Despite successful efforts by groups of interest that include patients, clinicians, and private companies to move forward legislation that allows the legalization of cannabis and its derivatives for medicinal purposes, these efforts have been outpaced by the recreational use by the general public hampering the accurate systematic description of its effects in different diseases, including glioma. However, there have been efforts to describe these effects in the observational setting. On an observational study conducted at the Moffitt Cancer Center and Research Institute in Tampa, FL (US), only 20% of the patients were recommended to use cannabis by a health professional, and about half of those surveyed obtained a prescription. The main reason for use was pain relief (58%), nausea relief (54%), appetite stimulant (48%), relief through relaxation (52%), ability to cope emotionally (39%), decrease in spasm or tremors (22%), and sleep aid (8.7%) (Reblin et al., 2019). These results are consistent with our won observations at the University of California, San Francisco where we identified 4 greater domains that patients use cannabinoids for: symptom management (33%), tumor treatment (40%), recreational (20%), and use and other reason (7%). Under symptom management patients more often used cannabinoids for anxiety, insomnia, nausea, and pain control (Fig. 2). Of note, patients who reported cannabinoid use for "tumor treatment" was self-perceived with no direction from a clinician. Interestingly, our observations have shown that, while one of the worries for cannabis use is cognitive decline, cannabinoid users cognition remained the same or slightly improved over time (Rodriguez-Almaraz, et al.; Unpublished data). Even though there are strong limitations to observational studies, these efforts serve as basis for understanding the impact of cannabinoid use, patient preferences, and believes around the use of cannabinoids. Additionally, they serve as further confirmation on observations made in preclinical work and in other cancers, paving the pathway to further conduct randomized double-blinded clinical trials in patients with glioma.

Clinical trials

The increasing need for evidence-based arguments about potential harms and benefits of cannabis use paralleled a surge in empirical studies investigating the health impact of cannabis use, the majority of which are observational and focused upon negative health effects (Cousijn et al., 2018). However, there have been some interventional studies to try to understand the impact of cannabinoids in gliomas. The first clinical study to assess cannabinoid antitumoral action in gliomas was conducted by Guzman and collaborators in 2003 in which nine patients diagnosed with recurrent GBM were treated with THC intratumorally (intracranial). After 15 days of administration, THC did not show clear differences in patient survival. However, THC decreased tumor cell proliferation and increased tumor cell apoptosis. Interestingly, even though THC was delivered directly in the tumor cavity patients showed little no psychotropic symptoms related to THC or to the delivery method (Guzman et al., 2006).

Although the risks and benefits of cannabinoid use in patients with established glioma have not been studied in large scale, a placebo-controlled trial of nabiximols (n=27) conducted by Twelves et al. showed that, in general, cannabinoids are well tolerated and that their safety profile is fairly wide (Twelves et al., 2021). Additionally, this study showed a significant improvement of survival at year 1 for the naboximols (an oromucosal spray) group (83%) compared to the placebo group (44%; $p>0.05$) signaling that there is a clear benefit of cannabinoids in the treatment of glioma. These results were

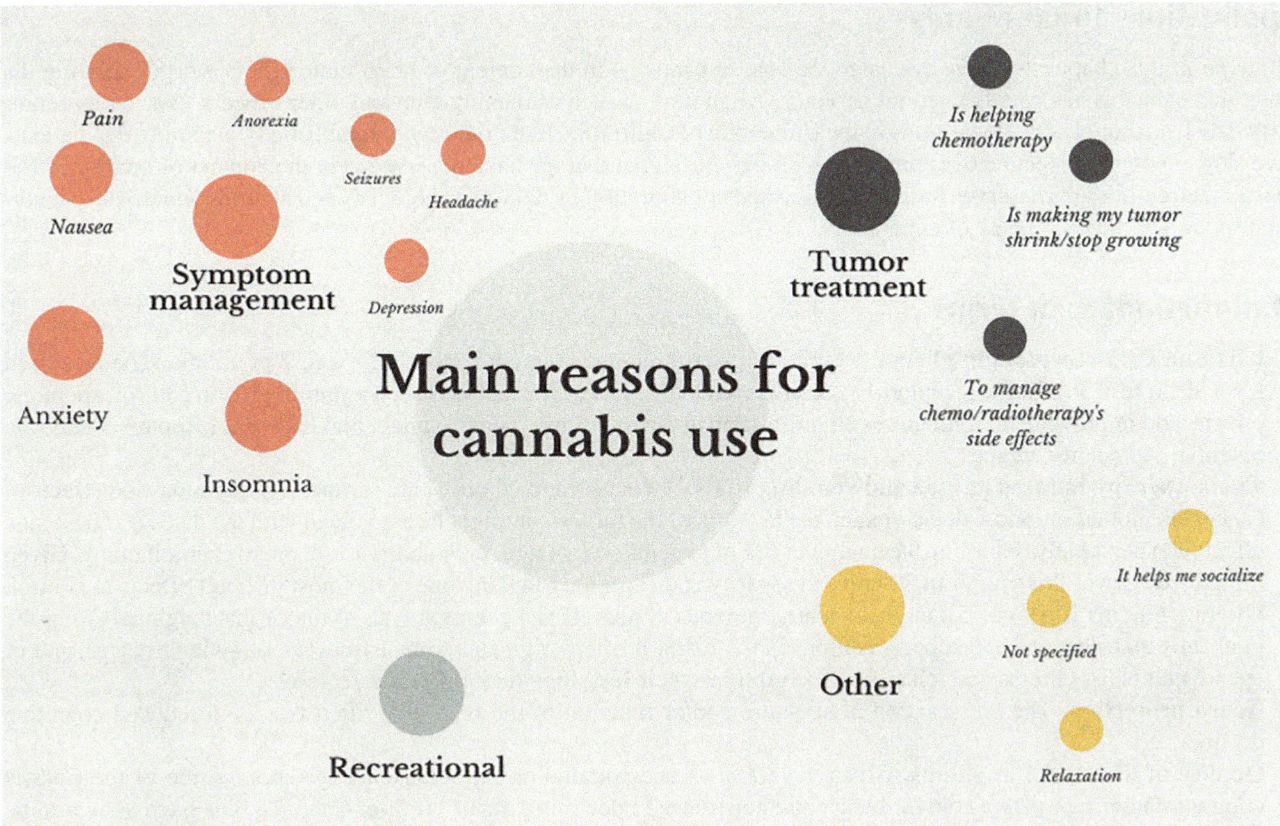

FIG. 2 Main reasons for cannabis use among patients with diagnosis of glioma from the Neuro oncology clinic at the University of California, San Francisco. The size of the circle indicates the frequency of the endorsed uses. Of note, the "tumor treatment" domain is self-perceived. This is, no clinician has prescribed cannabinoids for tumor treatment. Rather, it is believed by the patients that cannabinoids help with tumor growth.

aggregated on a systematic review and meta-analysis exploring the literature that included clinical level data where we showed that, although limited, cannabis improves survival among treated patients (J. E. Rodriguez-Almaraz et al., 2020).

More recently, Schloss and collaborators conducted a relatively large phase 2 randomized blinded clinical trial to assess different ratios of medicinal cannabis in patients with high-grade gliomas. The study included 88 patients with recurrent or inoperable high-grade glioma who were treated with a single dose per day of oil-based organic plant extracts of cannabis based on a 1:1 and 4:1 ratio of THC:CBD. Even though the treatment was well tolerated, the main side effects included muscle spasms and hallucinations (3.4%), which are well-known side effects of cannabinoids. Similar to our observations, the authors found improving quality of life scores among the treated group improving sleep, functional well-being, and contentment with quality of life (Schloss et al., 2021).

Altogether, this evidence suggests that medical cannabis and cannabinoids could become a therapy of choice in modern neuro oncology; however, further studies are needed to establish not only safe but also efficacious doses.

Potential negative effects

While the safety profile of cannabinoids is favorable with few side effects, this does not mean that cannabinoids are entirely innocuous. In 2017, the WHO and the National Academies of Sciences, Engineering and Medicine (NASEM) release their respective reports summarizing a large literature review and issued conclusions regarding potential health benefits and risks. The report was limited to 11 domains of which, an increased risk in cancer (nonseminoma-type testicular germ cell tumors and lung cancer), coronary events, myocardial infarctions, and strokes in young users, bronchitis, acute increases in the risk of traffic injuries, adolescent cognitive changes, and depression and suicidal ideation among adolescent and young adults were the most notable side effects (Cousijn et al., 2018). Unironically, these side effects of cannabinoids are similar to those effects that cannabinoids are trying to overcome, however, the majority of side effects are dose dependent (Cousijn et al., 2018) in chronic users and most prevalent in adolescent and young adults (Bridgeman & Abazia, 2017).

Applications to other areas

Although in this chapter we have discussed the role of cannabis in the context of brain tumors, it is worth exploring the effect that cannabis has on other central nervous system tumors such as meningioma and other cancers. Based on preliminary data from our observational study at the University of California, San Francisco (unpublished), we imply that patients have slower cognitive decline over time. Even though the signal that we have observed is in the context of quality-of-life questionnaires, this area deserves further attention and collaboration by neuropsychologists and neuropsychiatrists to incorporate accurate measurements of cognition.

Mini-dictionary of terms

- **CB1 and CB2 receptors in glioma.** Despite of prior thoughts, both CB1 and CB2 express in glioma. The increased level of expression of these receptors has been associated with beneficial effects such as tumor growth control, apoptotic effects, and improvement of immune cell infiltration in the tumor microenvironment making these receptors a valuable potential therapeutic target.
- **Chemotherapy-induced nausea and vomiting (CINV).** Nausea and vomiting are serious and common side effects of cancer chemotherapy and—depending on the location of the tumor—it might be associated with the disease. These side effects can negatively impact on the quality of life of patients and, in turn, their ability to adhere to chemotherapy. Given the subjectivity of this symptom, it varies in severity and chronicity making one of the most difficult effects to control.
- **Glioma.** Glioma is a type of malignant central nervous system. It is a common type of tumor that originates from the glial cells that surround and support neurons in the brain. It often originates from astrocytes, oligodendrocytes, and/or ependymal cells. Gliomas are classified according to their histology and molecular features.
- **Neuro protection.** The preservation of structure and/or function of the brain that often results in reduced cognitive decline.
- **Quality of life (QoL) in glioma.** Given the effect of chemoradiation on organs and systems outside of the CNS, a common factor that play a role in therapeutic adherence is declining quality of life. Cannabinoids have shown to be a safe therapeutic option not only to improve a number of symptoms inherent to the disease but also side effects from therapy, improving so QoL and therapeutic adherence leading to better clinical outcomes.

Key facts of central nervous system cancers and use of cannabis and related products

- Gliomas are a rare but often lethal disease.
- The etiology of gliomas is largely unknown, but there are well-identified risk factors.
- Different types of glioma have different frequency of incidence across populations.
- The treatment of gliomas is complex and noncurative (often limited to palliation) and might have a detrimental effect in quality of life of patients.
- Alternative therapies for the treatment of the disease and symptoms associated with the disease and side effects of chemoradiation are urgently needed.
- Cannabinoids present as an alternative with a rather safe profile as a coadjutant in the treatment of gliomas.
- Further high-quality clinical trials are needed to cement their use in humans beyond recreational use.
- Clinicians and scientists must advocate for legislation to enable high-quality research.

Summary points

(1) Cannabis use and research is of greater interest due to changing social and legal landscape.
(2) CBD may play a role in symptoms management.
(3) CBD can play a role in therapeutic strategies including RT sensitization and antitumor effect.
(4) Clinical research is in its infancy but growing.
(5) There are possible negative effects of use both on tumor and on QOL—needs greater study.

References

Abrams, D. I. (2016). Integrating cannabis into clinical cancer care. *Current Oncology*, *23*(2), S8–S14. https://doi.org/10.3747/co.23.3099.

Aguado, T., Carracedo, A., Julien, B., Velasco, G., Milman, G., Mechoulamluis, R., Alvarez, L., Guzmán, M., & Galve-Roperh, I. (2007). Cannabinoids induce glioma stem-like cell differentiation and inhibit gliomagenesis. *The Journal of Biological Chemistry*, *282*(9), 6854–6862.

Badowski, M. E., & Yanful, P. K. (2018). Dronabinol oral solution in the management of anorexia and weight loss in AIDS and cancer. *Therapeutics and Clinical Risk Management*, *14*, 643.

Barton, M. K. (2017). Cancer cachexia awareness, diagnosis, and treatment are lacking among oncology providers. *CA: A Cancer Journal for Clinicians*, *67*(2), 91–92.

Bergamaschi, M. M., Queiroz, R. H. C., Chagas, M. H. N., de Oliveira, D. C. G., de Martinis, B. S., Kapczinski, F., Quevedo, J., Roesler, R., Schröder, N., Nardi, A. E., Martín-Santos, R., Hallak, J. E. C., Zuardi, A. W., & Crippa, J. A. S. (2011). Cannabidiol reduces the anxiety induced by simulated public speaking in treatment-naïve social phobia patients. *Neuropsychopharmacology*, *36*(6), 1219.

Blázquez, C., Salazar, M., Carracedo, A., Lorente, M., Egia, A., González-Feria, L., Haro, A., Velasco, G., & Guzmán, M. (2008). Cannabinoids inhibit glioma cell invasion by down-regulating matrix metalloproteinase-2 expression. *Cancer Research*, *68*(6), 1945–1952.

Bla´zquez, C., Salazar, M., Carracedo, A., Lorente, M., Egia, A., Gonza´lez-Feria, L., … Guzma´n, M. (2008). Cannabinoids inhibit glioma cell invasion by down-regulating matrix metalloproteinase-2 expression. *Cancer Research*, *68*(6), 1945–1952. https://doi.org/10.1158/0008-5472.CAN-07-5176.

Bridgeman, M. B., & Abazia, D. T. (2017). Medicinal cannabis: history, pharmacology, and implications for the acute care setting. *Pharmacy and Therapeutics*, *42*(3), 180.

Brisbois, T. D., de Kock, I. H., Watanabe, S. M., Mirhosseini, M., Lamoureux, D. C., Chasen, M., MacDonald, N., Baracos, V. E., & Wismer, W. V. (2011). Delta-9-tetrahydrocannabinol may palliate altered chemosensory perception in cancer patients: results of a randomized, double-blind, placebo-controlled pilot trial. *Annals of Oncology*, *22*(9), 2086–2093.

Burt Nabors, L., Portnow, J., Baehring, J., Brem, S., Butowski, N., Campian, J. L., Cannon, D. M., Chao, S., Chheda, M. G., Clark, S. W., Fabiano, A. J., Forsyth, P., Susan Darlow, N., Anne Bergman Pierre Gigilio, M., Hattangadi-Gluth, J., Holdhoff, M., Horbinski, C., Lurie, R. H., Junck, L., & Willmarth, N. E. (2021). *NCCN Guidelines Version 2.2021 Central Nervous System Cancers Continue NCCN Guidelines Panel Disclosures.*

Cabral, G. A., Raborn, E. S., Griffin, L., Dennis, J., & Marciano-Cabral, F. (2008). CB2 receptors in the brain: role in central immune function. *British Journal of Pharmacology*, *153*(2), 240.

Cousijn, J., Núñez, A. E., & Filbey, F. M. (2018). Time to acknowledge the mixed effects of cannabis on health: a summary and critical review of the NASEM 2017 report on the health effects of cannabis and cannabinoids. *Addiction*, *113*(5), 958–966.

Davies, E., & Clarke, C. (2004). Early symptoms of brain tumours. *Journal of Neurology, Neurosurgery, and Psychiatry*, *75*(8), 1205.

Devane, W. A., Hanuš, L., Breuer, A., Pertwee, R. G., Stevenson, L. A., Griffin, G., Gibson, D., Mandelbaum, A., Etinger, A., & Mechoulam, R. (1992). Isolation and structure of a brain constituent that binds to the cannabinoid receptor. *Science (New York, N.Y.)*, *258*(5090), 1946–1949.

Dumitru, C. A., Sandalcioglu, I. E., & Karsak, M. (2018). Cannabinoids in glioblastoma therapy: new applications for old drugs. *Frontiers in Molecular Neuroscience*, *11*, 159. https://doi.org/10.3389/fnmol.2018.00159.

Ellert-Miklaszewska, A., Grajkowska, W., Gabrusiewicz, K., Kaminska, B., & Konarska, L. (2007). Distinctive pattern of cannabinoid receptor type II (CB2) expression in adult and pediatric brain tumors. *Brain Research*, *1137*(1), 161–169.

Ellert-Miklaszewska, A., Kaminska, B., & Konarska, L. (2005). Cannabinoids down-regulate PI3K/Akt and Erk signalling pathways and activate proapoptotic function of Bad protein. *Cellular Signalling*, *17*(1), 25–37.

FDA News Release. (2018). *FDA Approves First Drug Comprised of an Active Ingredient Derived from Marijuana to Treat Rare, Sever Forms of Epilepsy.* FDA. https://www.fda.gov/news-events/press-announcements/fda-approves-first-drug-comprised-active-ingredient-derived-marijuana-treat-rare-severe-forms. (Accessed 18 March 2022).

Fernandes, C., Costa, A., Osório, L., Lago, R. C., Linhares, P., Carvalho, B., & Caeiro, C. (2017). Current standards of care in glioblastoma therapy. *Glioblastoma*, 197–241.

Galanti, G., Fisher, T., Kventsel, I., Shoham, J., Gallily, R., Mechoulam, R., Lavie, G., Amariglio, N., Rechavi, G., & Toren, A. (2009). Δ9-Tetrahydrocannabinol inhibits cell cycle progression by downregulation of E2F1 in human glioblastoma multiforme cells. *Acta Oncology*, *47*(6), 1062–1070.

Galve-Roperh, I., Sánchez, C., Cortés, M. L., del Pulgar, T. G., Izquierdo, M., & Guzmán, M. (2000). Anti-tumoral action of cannabinoids: Involvement of sustained ceramide accumulation and extracellular signal-regulated kinase activation. *Nature Medicine*, *6*(3), 313–319.

Giacoppo, S., Molino, G., Galuppo, M., & Mazzon, P. B. E. (2014). Cannabinoids: new promising agents in the treatment of neurological diseases. *Molecules*, *19*(11), 18781.

Guzmán, M. (2003). Cannabinoids: potential anticancer agents. *Nature Reviews Cancer*, *3*(10), 745–755.

Guzman, M., Duarte, M. J., Blazquez, C., Ravina, J., Rosa, M. C., Galve-Roperh, I., Sanchez, C., Velasco, G., & Gonzalez-Feria, L. (2006). A pilot clinical study of Delta9-tetrahydrocannabinol in patients with recurrent glioblastoma multiforme. *British Journal of Cancer*, *95*(2), 197–203.

Guzmán, M., Galve-Roperh, I., & Sánchez, C. (2001). Ceramide: A new second messenger of cannabinoid action. *Trends in Pharmacological Sciences*, *22* (1), 19–22.

Hart, S., Fischer, O. M., & Ullrich, A. (2004). Cannabinoids induce cancer cell proliferation via tumor necrosis factor α-converting enzyme (TACE/ADAM17)-mediated transactivation of the epidermal growth factor receptor. *Cancer Research*, *64*(6), 1943–1950.

Howlett, A. C., Barth, F., Bonner, T. I., Cabral, G., Casellas, P., Devane, W. A., Felder, C. C., Herkenham, M., Mackie, K., Martin, B. R., Mechoulam, R., & Pertwee, R. G. (2002). Classification of cannabinoid receptors. *Pharmacological Reviews*, *54*(2), 161–202.

Hoshide, R., & Jandial, R. (2016). 2016 World Health Organization classification of central nervous system tumors: an era of molecular biology. *World Neurosurgery*, *94*, 561–562.

Howlett, A. C., & Abood, M. E. (2017). CB1 & CB2 Receptor Pharmacology. *Advances in Pharmacology (San Diego, CA)*, *80*, 169.

Hsu, K., Whitham, E., & Kichenadasse, G. (2020). Potential role of cannabidiol for seizure control in a patient with recurrent glioma. *Journal of Clinical Neuroscience: Official Journal of the Neurosurgical Society of Australasia*, *71*, 275–276.

Hu, J., Johnson, K. C., Mao, Y., Guo, L., Zhao, X., Jia, X., Bi, D., Huang, G., & Liu, R. (1998). Risk factors for glioma in adults: a case-control study in northeast China. *Cancer Detection and Prevention*, *22*(2), 100–108.

Jones, N. A., Hill, A. J., Smith, I., Bevan, S. A., Williams, C. M., Whalley, B. J., & Stephens, G. J. (2010). Cannabidiol displays antiepileptiform and antiseizure properties in vitro and in vivo. *The Journal of Pharmacology and Experimental Therapeutics*, *332*(2), 569.

Kaminska, B., & Ellert-Miklaszewska, A. (2015). Cannabinoid signalling in glioma cells. *Springerplus*, *4*(Suppl. 1), 1–32.

Katona, I., & Freund, T. F. (2008). Endocannabinoid signaling as a synaptic circuit breaker in neurological disease. *Nature Medicine*, *14*(9), 923–930.

Kleihues, P., Louis, D. N., Scheithauer, B. W., Rorke, L. B., Reifenberger, G., Burger, P. C., & Cavenee, W. K. (2002). The WHO classification of tumors of the nervous system. *Journal of Neuropathology and Experimental Neurology*, *61*(3), 215–225.

Leece, R., Xu, J., Ostrom, Q. T., Chen, Y., Kruchko, C., & Barnholtz-Sloan, J. S. (2017). Global incidence of malignant brain and other central nervous system tumors by histology, 2003–2007. *Neuro-Oncology*, *19*(11), 1553.

Louis, D. N., Perry, A., Reifenberger, G., von Deimling, A., Figarella-Branger, D., Cavenee, W. K., Ohgaki, H., Wiestler, O. D., Kleihues, P., & Ellison, D. W. (2016). The 2016 World Health Organization Classification of Tumors of the Central Nervous System: a summary. *Acta Neuropathologica*, *131*(6), 803–820.

Louis, D. N., Perry, A., Wesseling, P., Brat, D. J., Cree, I. A., Figarella-Branger, D., Hawkins, C., Ng, H. K., Pfister, S. M., Reifenberger, G., Soffietti, R., von Deimling, A., & Ellison, D. W. (2021). The 2021 WHO classification of tumors of the central nervous system: a summary. *Neuro-Oncology*, *23*(8), 1231–1251.

Lynch, M. E., Cesar-Rittenberg, P., & Hohmann, A. G. (2014). A double-blind, placebo-controlled, crossover pilot trial with extension using an oral mucosal cannabinoid extract for treatment of chemotherapy-induced neuropathic pain. *Journal of Pain and Symptom Management*, *47*(1), 166–173.

Machado Rocha, F. C., Stéfano, S. C., de Cássia Haiek, R., Rosa Oliveira, L. M. Q., & da Silveira, D. X. (2008). Therapeutic use of Cannabis sativa on chemotherapy-induced nausea and vomiting among cancer patients: systematic review and meta-analysis. *European Journal of Cancer Care*, *17*(5), 431–443.

Miller, K. D., Ostrom, Q. T., Kruchko, C., Patil, N., Tihan, T., Cioffi, G., Fuchs, H. E., Waite, K. A., Jemal, A., Siegel, R. L., & Barnholtz-Sloan, J. S. (2021). Brain and other central nervous system tumor statistics, 2021. *CA: A Cancer Journal for Clinicians*, *71*(5), 381–406.

Molinaro, A. M., Taylor, J. W., Wiencke, J. K., & Wrensch, M. R. (2019). Genetic and molecular epidemiology of adult diffuse glioma. *Nature Reviews. Neurology*, *15*(7), 405.

Mukhopadhyay, P., Rajesh, M., Horváth, B., Bátkai, S., Park, O., Tanchian, G., Gao, R. Y., Patel, V., Wink, D. A., Liaudet, L., Haskó, G., Mechoulam, R., & Pacher, P. (2011). Cannabidiol protects against hepatic ischemia/reperfusion injury by attenuating inflammatory signaling and response, oxidative/nitrative stress, and cell death. *Free Radical Biology & Medicine*, *50*(10), 1368.

Oláh, A., Szekanecz, Z., & Bíró, T. (2017). Targeting cannabinoid signaling in the immune system: "High"-ly exciting questions, possibilities, and challenges. *Frontiers in Immunology*, *8*, 1487. https://doi.org/10.3389/fimmu.2017.01487.

Ostrom, Q. T., Cioffi, G., Gittleman, H., Patil, N., Waite, K., Kruchko, C., & Barnholtz-Sloan, J. S. (2019). CBTRUS Statistical Report: Primary Brain and Other Central Nervous System Tumors Diagnosed in the United States in 2012–2016. *Neuro-Oncology*, *21*(Suppl. 5), v1.

Pacher, P. (2006). The endocannabinoid system as an emerging target of pharmacotherapy. *Pharmacological Reviews*, *58*(3), 389–462.

Patel, A. P., Fisher, J. L., Nichols, E., Abd-Allah, F., Abdela, J., Abdelalim, A., Abraha, H. N., Agius, D., Alahdab, F., Alam, T., Allen, C. A., Anber, N. H., Awasthi, A., Badali, H., Belachew, A. B., Bijani, A., Bjørge, T., Carvalho, F., Catalá-López, F., … Fitzmaurice, C. (2019). Global, regional, and national burden of brain and other CNS cancer, 1990–2016: a systematic analysis for the Global Burden of Disease Study 2016. *The Lancet Neurology*, *18*(4), 376–393.

Peeri, H., Shalev, N., Vinayaka, A. C., Nizar, R., Kazimirsky, G., Namdar, D., … Koltai, H. (2021). Specific compositions of *Cannabis sativa* compounds have cytotoxic activity and inhibit motility and colony formation of human glioblastoma cells *in vitro*. *Cancers*, *13*(7), 1720. https://doi.org/10.3390/cancers13071720.

Reblin, M., Sahebjam, S., Peeri, N. C., Martinez, Y. C., Thompson, Z., & Egan, K. M. (2019). Medical cannabis use in glioma patients treated at a comprehensive cancer center in Florida. *Journal of Palliative Medicine*, *22*(10), 1202–1207. https://doi.org/10.1089/jpm.2018.0528.

Rice, T., Lachance, D. H., Molinaro, A. M., Eckel-Passow, J. E., Walsh, K. M., Barnholtz-Sloan, J., Ostrom, Q. T., Francis, S. S., Wiemels, J., Jenkins, R. B., Wiencke, J. K., & Wrensch, M. R. (2016). Editor's choice: Understanding inherited genetic risk of adult glioma—a review. *Neuro-Oncology Practice*, *3*(1), 10.

Rodriguez-Almaraz, E., Butowski, N., Chang, S., Oberheim-Bush, N. A., Clarke, J., Taylor, J., Buerki, R., & Berger, M. (2019). QOLP-13. Impact of Cannabis use on quality of life in patients with central nervous system tumors. *Neuro-Oncology*, *21*(Suppl._6), vi200.

Rodriguez-Almaraz, J. E., Chang, S., Clarke, J., Oberheim-Bush, N. A., Taylor, J., Buerki, R., Berger, M., Zablotska, L., Lobach, I., & Butowski, N. (2020). A systematic review and meta-analysis examining the effects of cannabis and its derivatives in adults with malignant CNS tumors. *Neuro-Oncology Practice*, *7*(4), 376–383.

Salazar, M., Carracedo, A., Salanueva, I. J., et al. (2009). Cannabinoid action induces autophagy-mediated cell death through stimulation of ER stress in human glioma cells. *Journal of Clinical Investigation*, *119*(5), 1359–1372. https://doi.org/10.1172/jci37948.

Sánchez, A. J., & García-Merino, A. (2012). Neuroprotective agents: cannabinoids. *Clinical Immunology (Orlando, Fla.)*, *142*(1), 57–67.

Sánchez, C., Galve-Roperh, I., Canova, C., Brachet, P., & Guzmán, M. (1998). Δ9-Tetrahydrocannabinol induces apoptosis in C6 glioma cells. *FEBS Letters*, *436*(1), 6–10.

Schloss, J., Lacey, J., Sinclair, J., Steel, A., Sughrue, M., Sibbritt, D., & Teo, C. (2021). A phase 2 randomised clinical trial assessing the tolerability of two different ratios of medicinal cannabis in patients with high grade gliomas. *Frontiers in Oncology*, *11*, 1687.

Short, S. C., & Little, C. (2017). A 2-part safety and exploratory efficacy randomised double-blind placebo-controlled study of a 1:1 ration of cannabidiol and THC plus dose-intense temzolomide in patients with recurrent glioblastoma multiforme. *Neuro-Oncology*, *19*(6), vi13.

Sredni, S. T., Huang, C. C., Suzuki, M., Pundy, T., Chou, P., & Tomita, T. (2016). Spontaneous involution of pediatric low-grade gliomas: high expression of cannabinoid receptor 1 (CNR1) at the time of diagnosis may indicate involvement of the endocannabinoid system. *Child's Nervous System, 32*(11), 2061–2067.

Turcotte, C., Blanchet, M. R., Laviolette, M., & Flamand, N. (2016). The CB2 receptor and its role as a regulator of inflammation. *Cellular and Molecular Life Sciences, 73*(23), 4449.

Twelves, C., Sabel, M., Checketts, D., Miller, S., Tayo, B., Jove, M., … Short, S. C. (2021). A phase 1b randomised, placebo-controlled trial of nabiximols cannabinoid oromucosal spray with temozolomide in patients with recurrent glioblastoma. *British Journal of Cancer, 124*(8), 1379–1387. https://doi.org/10.1038/s41416-021-01259-3.

Vajkoczy, P., Farhadi, M., Gaumann, A., Heidenreich, R., Erber, R., Wunder, A., Tonn, J. C., Menger, M. D., & Breier, G. (2002). Microtumor growth initiates angiogenic sprouting with simultaneous expression of VEGF, VEGF receptor-2, and angiopoietin-2. *The Journal of Clinical Investigation, 109*(6), 777.

Valverde, O., Karsak, M., & Zimmer, A. (2005). Analysis of the endocannabinoid system by using CB cannabinoid receptor knockout mice. *Handbook of Experimental Pharmacology, 168*, 117–145.

Vecht, C. J., Kerkhof, M., & Duran-Pena, A. (2014). Seizure prognosis in brain tumors: new insights and evidence-based management. *The Oncologist, 19*(7), 751.

Velasco, G., Carracedo, A., Blázquez, C., Lorente, M., Aguado, T., Haro, A., Sánchez, C., Galve-Roperh, I., & Guzmán, M. (2007). Cannabinoids and gliomas. *Molecular Neurobiology, 36*(1), 60–67.

Velasco, G., Sánchez, C., & Guzmán, M. (2012). Towards the use of cannabinoids as antitumour agents. *Nature Reviews. Cancer, 12*(6), 436–444.

Wen, P. Y., & Packer, R. J. (2021). The 2021 WHO classification of tumors of the central nervous system: clinical implications. *Neuro-Oncology, 23*(8), 1215–1217.

Yang, M., Guo, W. W., Yang, C. S., Tang, J. Q., Huang, Q., Feng, S. X., … Jiang, G. (2017). Mobile phone use and glioma risk: A systematic review and meta-analysis. *PLoS One, 12*(5), e0175136. https://doi.org/10.1371/journal.pone.0175136.

You, G., Sha, Z. Y., Yan, W., Zhang, W., Wang, Y. Z., Li, S. W., Sang, L., Wang, Z., Li, G. L., Li, S. W., Song, Y. J., Kang, C. S., & Jiang, T. (2012). Seizure characteristics and outcomes in 508 Chinese adult patients undergoing primary resection of low-grade gliomas: a clinicopathological study. *Neuro-Oncology, 14*(2), 230.

Zanelati, T. V., Biojone, C., Moreira, F. A., Guimarães, F. S., & Joca, S. R. L. (2010). Antidepressant-like effects of cannabidiol in mice: possible involvement of 5-HT1A receptors. *British Journal of Pharmacology, 159*(1), 122.

Zogopoulos, P., Korkolopoulou, P., Patsouris, E., & Theocharis, S. (2015). The antitumor action of cannabinoids on glioma tumorigenesis. *Histology and Histopathology, 30*(6), 629–645.

Zuardi, A. W., Hallak, J. E. C., Dursun, S. M., Morais, S. L., Sanches, R. F., Musty, R. E., & Crippa, J. A. S. (2006). Cannabidiol monotherapy for treatment-resistant schizophrenia. *Journal of Psychopharmacology, 20*(5), 683–686.

Chapter 20

The relationship between glutamate and resting-state connectivity in chronic cannabis users

Sharlene D. Newman
Alabama Life Research Institute, The University of Alabama, Tuscaloosa, AL, United States

Abbreviations

RSfc	resting-state functional connectivity
Glu	glutamate
ACC	anterior cingulate
dACC	dorsal anterior cingulate
pgACC	pregenual anterior cingulate cortex
aMCC	anterior-mid cingulate cortex
NAc	nucleus accumbens
CB	cannabis
THC	Δ9-tetrahydrocannabinol
CBD	cannabidiol
MRS	magnetic resonance spectroscopy
fMRI	functional magnetic resonance imaging
Gln	glutamine
CB1	cannabinoid type-1 (receptors)
DMN	default-mode network

Introduction

The use of cannabis has increased over the past decade in the United States, and past-year prevalence of cannabis (CB) use exceeds 10% (Grucza et al., 2016) with few users seeking treatment (Brown et al., 2003). While the primary psychoactive component of CB is Δ9-tetrahydrocannabinol (THC), there are over 120 cannabinoids found in the plant *Cannabis sativa* (ElSohly & Gul, 2014). In addition to THC, cannabidiol (CBD) is the other most abundant and widely studied cannabinoid. CBD has been found to be neuroprotective and acts as an antiinflammatory and antioxidant (Maroon & Bost, 2018). A recent review has suggested that some of the lesser-known cannabinoids also may have neuroprotective properties (Stone et al., 2020). Even though there is increased use and legalization of CB, there is still limited understanding of its effect on human brain functioning, structure, and neurochemistry. In this chapter we review the current literature examining how CB use disrupts the relationship between one of the most abundant neurotransmitters, glutamate, and resting-state functional connectivity.

Glutamate and addiction

Neurotransmitters are the brain's chemical messengers; they transmit signals from neuron to neuron. The neurotransmitter most closely linked to addiction is dopamine. When presented with a motivationally salient stimulus, dopamine is released throughout the reward network from the ventral tegmental area. This dopamine signal is thought to have two functions in the reward network. First, it promotes learning when exposed to a novel salient stimulus; therefore, it is involved in creating

Cannabis Use, Neurobiology, Psychology, and Treatment. https://doi.org/10.1016/B978-0-323-89862-1.00031-3

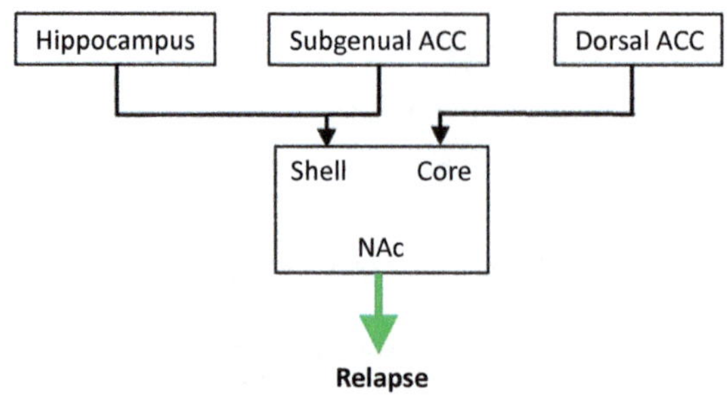

FIG. 1 Model of the network that results in relapse in animal models. The *arrows* denote the projection of glutamate *(black)* and GABA *(green)* release.

the initial drive to use drugs. Second, after learning has occurred, it provides an alert to the pending appearance of a familiar motivationally relevant stimulus and thus keeps the person in the addictive cycle. Although dopamine is an important neurotransmitter in addiction, it does not work alone. Glutamate (Glu) is one of the brain's primary excitatory neurotransmitters whose concentration in the extracellular space is tightly controlled due to its potential toxic properties. Dopamine is thought to potentiate or inhibit the capacity of Glu to depolarize neurons. Therefore, dopamine and Glu act together to modify the reward network. In sum, Glu plays as significant a role in addiction as dopamine.

The neural network that has been strongly linked to addiction in animal models is the reward network that includes the prefrontal cortex, the nucleus accumbens (NAc) in the ventral striatum, and the hippocampus (see Fig. 1, taken from Kalivas & Volkow, 2011). The NAc plays a primary role in addiction and has been strongly associated with mediating the expression of learned behaviors in response to a motivationally relevant stimulus (Kalivas & Volkow, 2005). The prefrontal cortex is one region that exerts some control over the NAc. The signaling pathway between prefrontal cortex and NAc is glutamatergic. Additionally, different regions of medial prefrontal cortex are connected to the shell and core of the NAc with the dorsal anterior cingulate (dACC) projecting to the core, while the ventral/subgenual ACC projects to the shell. The core and shell of the NAc have been linked to different functions. For example, phasic dopamine release in the core, but not the shell, has been observed following reward-predictive cues (Sunsay & Rebec, 2008) and that dopamine release is related to the subjective reward value of the cue (Saddoris et al., 2015; Sugam et al., 2012). Conversely, the shell appears to play an important role in processing motivating, hedonic stimulus properties (Rebec et al., 1997; Saddoris et al., 2015; Wheeler et al., 2011). These differential roles of the shell and the core parallel those of the ventral and dorsal ACC with ventral ACC being linked to emotion and reward associated behaviours (Etkin et al., 2011), whereas the dACC is linked to inhibitory control and appraisal (Botvinick et al., 2001; Venkatraman & Huettel, 2012). The ACC (Tsou et al., 1998) and NAc (Moldrich & Wenger, 2000) have also been found to have high cannabinoid type-1 (CB1) receptor density suggesting that CB is likely to have an effect on the glutamatergic signaling between these two regions.

It should also be noted that unlike dopamine, Glu can be measured in vivo in humans with the use of a standard magnetic resonance spectroscopy (MRS) acquisition sequence, making the examination of the role of Glu in substance abuse feasible in humans. MRS is a noninvasive technique that allows for the measurement of a number of molecules including Glu. Glutamate levels in humans have been reliably reported using a magnetic field strength of 3T (Cohen-Gilbert et al., 2014; Hurd et al., 2004; Yasen et al., 2017). While sophisticated 3D MRS sequences are available, single voxel MRS allows for a focus on discreet regions with the higher spatial and spectral resolution necessary for regions with susceptibility issues related to field inhomogeneities like those close to the sinuses (e.g., the ventral striatum; Cohen-Gilbert et al., 2014). The measurement of Glu is complicated by glutamine (Gln) because their signals overlap, particularly at lower magnetic field strengths. There is some debate as to whether Glu can be reliably separated from Gln at 3T (Ende, 2015; Mayer & Spielman, 2005; Schubert et al., 2004; Wijtenburg & Knight-Scott, 2011). However, with specialized editing and analysis techniques, the separation of molecules is possible, allowing for more accurate quantification of individual Glu levels (Cheng et al., 2021; Jensen et al., 2009; Mescher et al., 1998). Therefore, MRS technology has advanced to the point that Glu can be reliably measured in humans making it a powerful tool in the study of addiction.

Glutamate and cannabis

The primary psychoactive component of CB is THC. Cannabinoid receptors are highly expressed in axons and terminals of glutamatergic neurons and act to inhibit synaptic transmission or reduce neuronal excitability through depolarization-induced suppression of inhibition or excitation. Preclinical models have shown that THC depresses endocannabinoid-mediated Glu transmission and affects Glu release, enzyme activity, and the expression and activity of receptors and transporters (Colizzi et al., 2016). THC has been found to reduce the release and uptake of Glu in a dose-dependent manner in rat striatal slices (Brown et al., 2003); reduce Glu transmission in mice (Straiker & Mackie, 2005); and exposure in utero results in reductions in basal extracellular Glu in adolescent rats (Castaldo et al., 2007).

In addition to THC, CBD is the other most abundant and widely studied cannabinoid. CBD has been found to be neuroprotective and acts as an antiinflammatory and antioxidant (Maroon & Bost, 2018). CBD, like THC, also appears to interact with the glutamatergic system (Campos et al., 2016; Linge et al., 2016; O'Neill et al., 2021). For example, in a study examining its potential to treat heroin-seeking behavior, Ren et al. (2009) found that CBD normalized disturbances in Glu and CB1 receptor expression in the NAc that are associated with cue-induced drug seeking. Additionally, CBD was found to increase hippocampal Glu in humans and decrease psychosis symptom severity (O'Neill et al., 2021). Together, the evidence is accumulating to support the hypothesis that CB interacts with the glutamatergic signaling pathway.

Resting-state connectivity and cannabis

The process of addiction involves the reorganization of neural connectivity in the brain as well as altered neurochemistry. Resting-state functional connectivity (RSfc) is a measure of intrinsic brain connectivity, which has been used to characterize large-scale neural networks including the default-mode (DMN), the salience and sensorimotor networks (Smith et al., 2009), and the reward network (Barnes et al., 2010; Di Martino et al., 2008). Additionally, these resting-state networks have been found to be disrupted in substance-use disorders (Fedota & Stein, 2015; Müller-Oehring et al., 2015; Sutherland et al., 2012). In a recent study examining whole-brain RSfc, Ramaekers et al. (2022) found that chronic CB users showed hyperconnectivity compared to occasional users. These results are consistent with a previous study that also found stronger RSfc for chronic CB users compared to nonusers (Cheng et al., 2014). The Cheng study used multivoxel pattern analysis to predict whether the brain data corresponded to a CB user or control at a significantly high rate, demonstrating that the differences in RSfc are reliable and predictive. When examining the cerebellum, a region with a high CB1 receptor density (Schnakenberg Martin et al., 2021), both hyper- and hypo connectivity patterns were observed when comparing chronic CB users to controls. The hypo-connectivity was observed between the nonmotor regions of the cerebellum and prefrontal regions. These results demonstrate that CB use is associated with disruptions of RSfc and that the observation of hyper- or hypo-connectivity varies as a function of brain region.

Glutamate and resting-state connectivity

Glutamatergic and GABAergic function have been implicated as a driver of neuronal and glial activity (Hyder et al., 2006) with recent studies suggesting that Glu drives the cortico-cortical functional connections that are sustained during the resting state in normal functioning adults (Duncan et al., 2014; Moeller et al., 2016). Much of the work in this area has focused on specific regions of interest with Duncan et al. (2014) arguing that within the DMN, the relationship between Glu and connectivity strength is positively correlated. For example, it has been shown that the higher the Glu concentration in the posterior cingulate cortex (PCC; a primary hub in the DMN), the higher was the connectivity between the PCC and the pregenual ACC (members of the DMN; Duncan et al., 2013; Hu et al., 2013; Kapogiannis et al., 2013). However, this positive correlation does not seem to hold outside of the DMN (Li et al., 2022; Wang et al., 2020). Wagner et al. (2016) demonstrated that the connectivity between the hippocampus and the perigenual anterior cingulate cortex (pgACC), which is located anterior to the dACC, was negatively correlated with Glu in healthy adults.

Most studies have focused on the connectivity between a priori-defined regions of interest. A recent study examined the relationship between whole-brain RSfc and Glu using two ACC regions as seeds (MRS was measured from the regions and they were used as the seed for the RSfc analysis)—the pgACC and the anterior mid-cingulate cortex (aMCC) that differ in cytoarchitecture, chemoarchitecture, and function (Dou et al., 2013; Li et al., 2022; Menon, 2015; van Heukelum et al., 2020). The pgACC and aMCC are also members of different resting-state networks with pgACC being linked to the DMN and aMCC the salience network. The relationship between Glu and RSfc strength varied with the seed region. The pgACC Glu was positively correlated with whole-brain connectivity strength, while aMCC was negatively correlated. This pattern matched previous studies and suggests that the connectivity strength is affected by Glu, the direction of the relationship varies by brain region.

Cannabis, glutamate, and resting-state connectivity

Because chronic CB use has been associated with modulations in Glu levels, it is expected that the relationship between Glu and RSfc is also affected by CB use. To date, there are two studies that have examined these relationships. A study conducted by Mason et al. (2019) examined the *acute* effects of CB with a focus on the ACC and NAc. They found that acute exposure to CB resulted in increases in Glu in the NAc but not the ACC. CB also resulted in decreased connectivity between

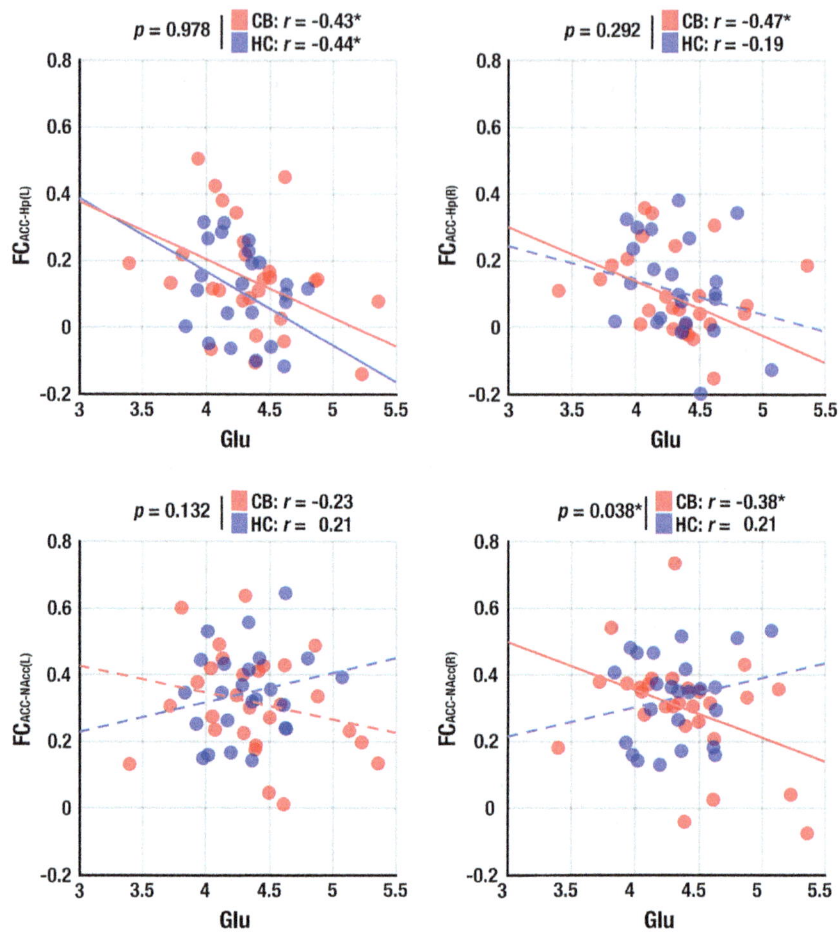

FIG. 2 The correlation between dACC Glu levels and RSfc between the dACC and the hippocamps (top) and NAc (bottom) for CB users *(red)* and nonusers *(blue)*.

the NAc and prefrontal regions. Finally, Mason and colleagues found a positive correlation between changes in Glu and change in connectivity strength. The second study was conducted by Newman et al. (2020). There were differences between the two studies: (1) *chronic* CB use was examined instead of acute effects; (2) Glu was measured only in the dACC; and (3) the dACC was the seed for the connectivity analysis. No statistically significant differences in Glu or RSfc strength were observed between the CB user and nonuser groups. However, for the CB user group, there was a significant correlation between connectivity between the dACC and NAc and dACC Glu levels that was not significant for the nonuser group (see Fig. 2). Additionally, a regression analysis that included an interaction between Glu and monthly CB use found that the interaction was significant and that monthly use has an influence on the relationship between Glu and RSfc. While CB use was associated with differences in connectivity between the dACC and NAc, it did not affect the connectivity between the dACC and hippocampus suggesting that the relationship is region dependent.

These two studies demonstrate that CB use, both the effects of acute CB intoxication and chronic CB use, interacts with the relationship between Glu and connectivity. However, they also raise important research questions that require further investigation. For example, the interaction with monthly CB use is interesting but difficult to control for in humans. The concentration of THC or CBD being consumed by chronic users is almost impossible to quantify given current laws. The results suggest that there may be a threshold of use whereby if using only occasionally there are no significant effects. More controlled studies with larger sample sizes are required to properly address this question.

Implications

The previous research demonstrates that Glu's effect on connectivity varies depending on the brain regions involved. A potential explanation for this difference in effect may be related to whether the regions are receiving or sending glutamatergic inputs. For example, the dACC sends glutamatergic inputs to the NAc. In a study examining alcohol-use disorder, Glu was measured in both the dACC and NAc while participants viewed alcohol cues or neutral cues (Cheng et al., 2018) (Cheng et al.

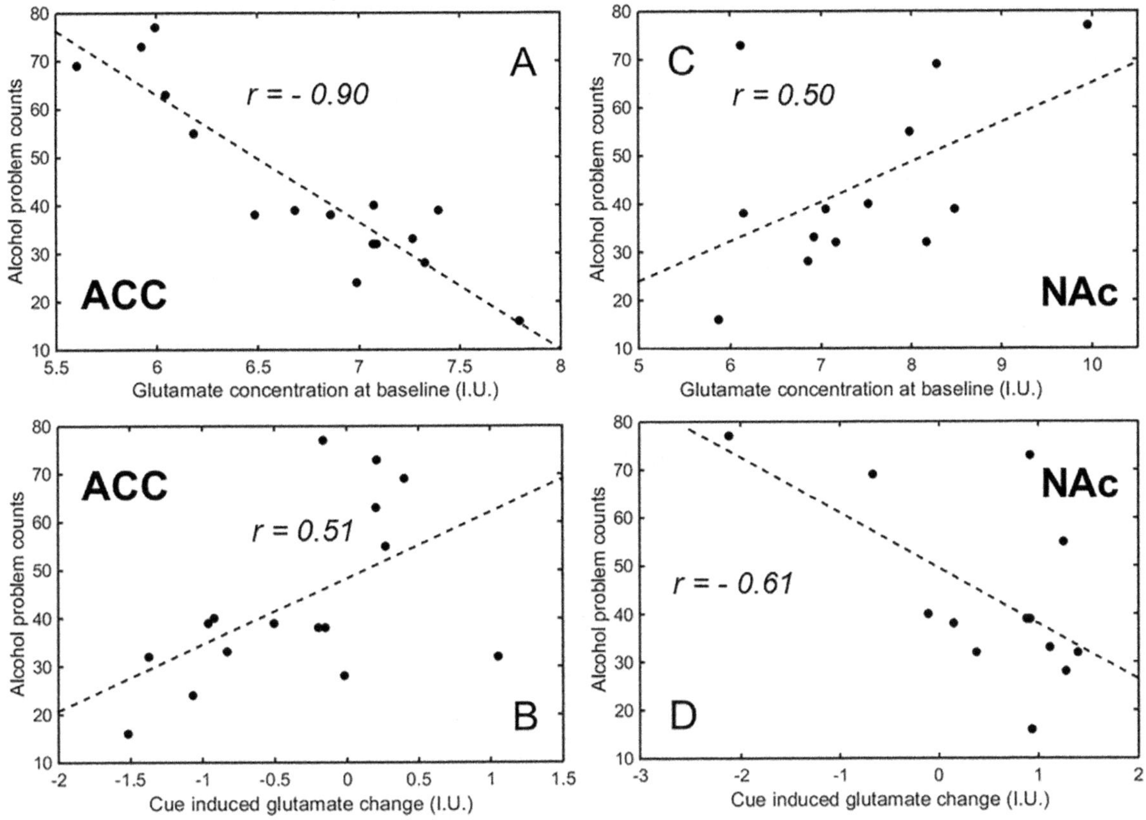

FIG. 3 Prelim data: (A) and (B) show the ACC response, while (C) and (D) show the NAc response. The top shows the correlation between problem counts and baseline glutamate, while the bottom shows the correlation with cue-induced glutamate change.

did not include the NAc data, only dACC data). Glu levels associated with each viewing condition were correlated with alcohol problem counts. Results showed a negative correlation between Glu in the dACC and problem counts such that participants with higher alcohol problem counts had lower basal levels of Glu ($r = -0.9$; see Fig. 3). In contrast, a positive correlation between NAc Glu levels and problem counts was observed, supporting the reciprocal relationship between dACC and NAc. These results suggest that it is important to account for the functioning of the region but also potentially for whether the region is sending or receiving glutamatergic inputs. This likely also affects the relationship between Glu and resting-state connectivity. For example, Li et al. (2022) found that the aMCC Glu, which overlaps with the dACC, had a negative correlation with global connectivity (the average connectivity between the region and the rest of the brain). Our work, on the other hand, found that a similar region's Glu level (dACC) was negatively correlated with connectivity with the hippocampus but positively correlated with connectivity with the NAc (for controls although not significantly, see Fig. 2). These results, like those described above, demonstrate that the relationship between Glu and RSfc is region dependent.

Examining the association between Glu and connectivity is important as it is directly related to human behavior. For example, Schmaal et al. (2012) performed a path analysis to show that the relationship between dACC Glu and impulsive decision-making is partially mediated by the RSfc between the dACC and ventral striatum. Drugs of addiction have been associated with increased risk-taking behaviours and impulsive decision-making (Stephan et al., 2017); however, there is evidence that CB may act differently than other drugs in that it may not be associated with increases in risky decision-making (Raymond et al., 2020). Therefore, it is important to understand how substance use, including CB, interacts with the relationship between Glu, RSfc, and behaviour.

Applications to other areas

In this chapter, we reviewed the relationship between glutamate, resting-state functional connectivity and CB use. Glutamate has been hypothesized to drive resting-state connectivity and CB use is thought to affect glutamate and connectivity. Other drugs of abuse, including alcohol, are expected to result in similar changes to glutamate and resting-state connectivity. The examination of these relationships is important and may potentially lead to pharmaceutical targets to disrupt the addiction cycle.

Mini-dictionary of terms

- **Connectivity**—the level of coactivation between brain regions.
- **Resting-state connectivity**—the temporal correlation of spontaneous fMRI signal between brain regions while not engaged in an explicit task or exposed to external stimuli.
- **Resting-state networks**—discernible functional communities that show within-community, high-level functional coupling with lower or intermittent coupling between communities (Fricchione & Beach, 2019). Examples include the default-mode network and the salience network.
- **Impulsive decision-making**—risky decisions, motivated more by immediate reward rather than the potential long-term negative consequences, suggesting greater sensitivity to reward and/or reduced sensitivity to negative outcomes (Ainslie, 1975).
- **Glutamatergic signaling pathway**—a series of molecular signals initiated by the binding of glutamate to a glutamate receptor on the surface of a target cell and ending with regulation of downstream cellular processes (Smith et al., 2019).

Key facts of resting-state functional connectivity

- Resting-state networks provide useful information about the functioning in the healthy and disordered brain independent of cognitive tasks and external stimuli.
- Resting-state connectivity is reproducible within and across individuals.
- Disrupted resting-state connectivity has been observed in a number of clinical conditions (e.g., substance-use disorders, depression, and schizophrenia) and appears to change with age.
- Resting-state network fluctuations occur both during tasks and during rest meaning that the network is coherent both at rest and during task performance.

Key facts of glutamate

- Glutamate is the most abundant free amino acid in the brain.
- It is the most abundant excitatory neurotransmitter in the nervous system.
- Excess extracellular glutamate is toxic and leads to neuronal damage and cell death.
- Glutamate has been associated with a number of psychiatric conditions including depression and schizophrenia.

Summary points

- CB use, chronic and acute, affects glutamate levels in the brain.
- Resting-state connectivity is disrupted in CB users.
- Resting-state connectivity is affected by glutamate levels and glutamate is thought to "drive" connectivity.
- CB use interferes with the relationship between glutamate and resting-state connectivity.
- The mechanism that links glutamate to resting-state connectivity is unclear.

References

Ainslie, G. (1975). Specious reward: A behavioral theory of impulsiveness and impulse control. *Psychological Bulletin*, *82*(4), 463.

Barnes, K. A., Cohen, A. L., Power, J. D., Nelson, S. M., Dosenbach, Y. B., Miezin, F. M., … Schlaggar, B. L. (2010). Identifying basal ganglia divisions in individuals using resting-state functional connectivity MRI. *Frontiers in Systems Neuroscience*, *4*, 7–11.

Botvinick, M. M., Braver, T. S., Barch, D. M., Carter, C. S., & Cohen, J. D. (2001). Conflict monitoring and cognitive control. *Psychological Review*, *108*(3), 624.

Brown, T. M., Brotchie, J. M., & Fitzjohn, S. M. (2003). Cannabinoids decrease corticostriatal synaptic transmission via an effect on glutamate uptake. *Journal of Neuroscience*, *23*(35), 11073–11077.

Campos, A. C., Fogaça, M. V., Sonego, A. B., & Guimarães, F. S. (2016). Cannabidiol, neuroprotection and neuropsychiatric disorders. *Pharmacological Research*, *112*, 119–127.

Castaldo, P., Magi, S., Gaetani, S., Cassano, T., Ferraro, L., Antonelli, T., … Cuomo, V. (2007). Prenatal exposure to the cannabinoid receptor agonist WIN 55,212-2 increases glutamate uptake through overexpression of GLT1 and EAAC1 glutamate transporter subtypes in rat frontal cerebral cortex. *Neuropharmacology*, *53*(3), 369–378.

Cheng, H., Kellar, D., Lake, A., Finn, P., Rebec, G. V., Dharmadhikari, S., … Newman, S. (2018). Effects of alcohol cues on MRS glutamate levels in the anterior cingulate. *Alcohol and Alcoholism*, *53*(3), 209–215.

Cheng, H., Skosnik, P. D., Pruce, B. J., Brumbaugh, M. S., Vollmer, J. M., Fridberg, D. J., … Newman, S. D. (2014). Resting state functional magnetic resonance imaging reveals distinct brain activity in heavy cannabis users–a multi-voxel pattern analysis. *Journal of Psychopharmacology*, *28*(11), 1030–1040.

Cheng, H., Wang, A., Newman, S., & Dydak, U. (2021). An investigation of glutamate quantification with PRESS and MEGA-PRESS. *NMR in Biomedicine*, *34*(2), e4453.

Cohen-Gilbert, J. E., Jensen, J. E., & Silveri, M. M. (2014). Contributions of magnetic resonance spectroscopy to understanding development: Potential applications in the study of adolescent alcohol use and abuse. *Development and Psychopathology*, *26*(02), 405–423.

Colizzi, M., McGuire, P., Pertwee, R. G., & Bhattacharyya, S. (2016). Effect of cannabis on glutamate signalling in the brain: A systematic review of human and animal evidence. *Neuroscience & Biobehavioral Reviews*, *64*, 359–381.

Di Martino, A., Scheres, A., Margulies, D. S., Kelly, A. M. C., Uddin, L. Q., Shehzad, Z., … Milham, M. P. (2008). Functional connectivity of human striatum: A resting state FMRI study. *Cerebral Cortex*, *18*(12), 2735–2747.

Dou, W., Palomero-Gallagher, N., van Tol, M. J., Kaufmann, J., Zhong, K., Bernstein, H. G., … Walter, M. (2013). Systematic regional variations of GABA, glutamine, and glutamate concentrations follow receptor fingerprints of human cingulate cortex. *Journal of Neuroscience*, *33*(31), 12698–12704.

Duncan, N. W., Wiebking, C., & Northoff, G. (2014). Associations of regional GABA and glutamate with intrinsic and extrinsic neural activity in humans—A review of multimodal imaging studies. *Neuroscience & Biobehavioral Reviews*, *47*, 36–52.

Duncan, N. W., Wiebking, C., Tiret, B., Marjańska, M., Hayes, D. J., Lyttleton, O., … Northoff, G. (2013). Glutamate concentration in the medial prefrontal cortex predicts resting-state cortical-subcortical functional connectivity in humans. *PLoS One*, *8*(4), e60312.

ElSohly, M., & Gul, W. (2014). Constituents of Cannabis sativa. In R.G. Pertwee (Ed). *Handbook of cannabis* (pp. 3–22), Oxford, UK, Oxford University Press.

Ende, G. (2015). Proton magnetic resonance spectroscopy: Relevance of glutamate and GABA to neuropsychology. *Neuropsychology Review*, *25*(3), 315–325.

Etkin, A., Egner, T., & Kalisch, R. (2011). Emotional processing in anterior cingulate and medial prefrontal cortex. *Trends in Cognitive Sciences*, *15*(2), 85–93.

Fedota, J. R., & Stein, E. A. (2015). Resting-state functional connectivity and nicotine addiction: prospects for biomarker development. *Annals of the New York Academy of Sciences*, *1349*(1), 64–82.

Fricchione, G., & Beach, S. (2019). Cingulate-basal ganglia-thalamo-cortical aspects of catatonia and implications for treatment. In *Vol. 166. Handbook of clinical neurology* (pp. 223–252).

Grucza, R. A., Agrawal, A., Krauss, M. J., Cavazos-Rehg, P. A., & Bierut, L. J. (2016). Recent trends in the prevalence of marijuana use and associated disorders in the United States. *JAMA Psychiatry*, *73*(3), 300–301.

Hu, Y., Chen, X., Gu, H., & Yang, Y. (2013). Resting-state glutamate and GABA concentrations predict task-induced deactivation in the default mode network. *Journal of Neuroscience*, *33*(47), 18566–18573.

Hurd, R., Sailasuta, N., Srinivasan, R., Vigneron, D. B., Pelletier, D., & Nelson, S. J. (2004). Measurement of brain glutamate using TE-averaged PRESS at 3T. *Magnetic Resonance in Medicine*, *51*(3), 435–440.

Hyder, F., Patel, A. B., Gjedde, A., Rothman, D. L., Behar, K. L., & Shulman, R. G. (2006). Neuronal–glial glucose oxidation and glutamatergic–GABAergic function. *Journal of Cerebral Blood Flow & Metabolism*, *26*(7), 865–877.

Jensen, J. E., Licata, S. C., Öngür, D., Friedman, S. D., Prescot, A. P., Henry, M. E., & Renshaw, P. F. (2009). Quantification of J-resolved proton spectra in two-dimensions with LCModel using GAMMA-simulated basis sets at 4 Tesla. *NMR in Biomedicine*, *22*(7), 762–769.

Kalivas, P. W., & Volkow, N. D. (2005). The neural basis of addiction: A pathology of motivation and choice. *American Journal of Psychiatry*, *162*(8), 1403–1413.

Kalivas, P. W., & Volkow, N. D. (2011). New medications for drug addiction hiding in glutamatergic neuroplasticity. *Molecular Psychiatry*, *16*(10), 974–986.

Kapogiannis, D., Reiter, D. A., Willette, A. A., & Mattson, M. P. (2013). Posteromedial cortex glutamate and GABA predict intrinsic functional connectivity of the default mode network. *NeuroImage*, *64*, 112–119.

Li, M., Danyeli, L. V., Colic, L., Wagner, G., Smesny, S., Chand, T., … Sen, Z. D. (2022). The differential association between local neurotransmitter levels and whole-brain resting-state functional connectivity in two distinct cingulate cortex subregions. *Human Brain Mapping*, *43*, 2833–2844.

Linge, R., Jiménez-Sánchez, L., Campa, L., Pilar-Cuéllar, F., Vidal, R., Pazos, A., … Díaz, Á. (2016). Cannabidiol induces rapid-acting antidepressant-like effects and enhances cortical 5-HT/glutamate neurotransmission: role of 5-HT1A receptors. *Neuropharmacology*, *103*, 16–26.

Maroon, J., & Bost, J. (2018). Review of the neurological benefits of phytocannabinoids. *Surgical Neurology International*, *9*.

Mason, N. L., Theunissen, E. L., Hutten, N. R., Desmond, H. Y., Toennes, S. W., Stiers, P., & Ramaekers, J. G. (2019). Cannabis induced increase in striatal glutamate associated with loss of functional corticostriatal connectivity. *European Neuropsychopharmacology*, *29*(2), 247–256.

Mayer, D., & Spielman, D. M. (2005). Detection of glutamate in the human brain at 3 T using optimized constant time point resolved spectroscopy. *Magnetic Resonance in Medicine*, *54*(2), 439–442.

Menon, V. (2015). Salience network. In A. W. Toga (Ed.), *Brain mapping* (pp. 597–611). Waltham: Academic Press.

Mescher, M., Merkle, H., Kirsch, J., Garwood, M., & Gruetter, R. (1998). Simultaneous in vivo spectral editing and water suppression. *NMR in Biomedicine*, *11*, 266–272. EPFL-ARTICLE-177509.

Moeller, S. J., London, E. D., & Northoff, G. (2016). Neuroimaging markers of glutamatergic and GABAergic systems in drug addiction: Relationships to resting-state functional connectivity. *Neuroscience & Biobehavioral Reviews*, *61*, 35–52.

Moldrich, G., & Wenger, T. (2000). Localization of the CB 1 cannabinoid receptor in the rat brain. An immunohistochemical study. *Peptides*, *21*(11), 1735–1742.

Müller-Oehring, E. M., Jung, Y. C., Pfefferbaum, A., Sullivan, E. V., & Schulte, T. (2015). The resting brain of alcoholics. *Cerebral Cortex*, *25*(11), 4155–4168.

Newman, S. D., Cheng, H., Kim, D. J., Schnakenberg-Martin, A., Dydak, U., Dharmadhikari, S., … O'Donnell, B. (2020). An investigation of the relationship between glutamate and resting state connectivity in chronic cannabis users. *Brain Imaging and Behavior*, *14*(5), 2062–2071.

O'Neill, A., Annibale, L., Blest-Hopley, G., Wilson, R., Giampietro, V., & Bhattacharyya, S. (2021). Cannabidiol modulation of hippocampal glutamate in early psychosis. *Journal of Psychopharmacology*, *35*(7), 814–822.

Ramaekers, J. G., Mason, N. L., Toennes, S. W., Theunissen, E. L., & Amico, E. (2022). Functional brain connectomes reflect acute and chronic cannabis use. *Scientific Reports*, *12*(1), 1–12.

Raymond, D. R., Paneto, A., Yoder, K. K., O'Donnell, B. F., Brown, J. W., Hetrick, W. P., & Newman, S. D. (2020). Does chronic cannabis use impact risky decision-making: An examination of fMRI activation and effective connectivity? *Frontiers in Psychiatry*, *1340*.

Rebec, G. V., Christensen, J. R., Guerra, C., & Bardo, M. T. (1997). Regional and temporal differences in real-time dopamine efflux in the nucleus accumbens during free-choice novelty. *Brain Research*, *776*(1), 61–67.

Ren, Y., Whittard, J., Higuera-Matas, A., Morris, C. V., & Hurd, Y. L. (2009). Cannabidiol, a nonpsychotropic component of cannabis, inhibits cue-induced heroin seeking and normalizes discrete mesolimbic neuronal disturbances. *Journal of Neuroscience*, *29*(47), 14764–14769.

Saddoris, M. P., Cacciapaglia, F., Wightman, R. M., & Carelli, R. M. (2015). Differential dopamine release dynamics in the nucleus accumbens core and shell reveal complementary signals for error prediction and incentive motivation. *The Journal of Neuroscience*, *35*(33), 11572–11582.

Schmaal, L., Goudriaan, A. E., van der Meer, J., van den Brink, W., & Veltman, D. J. (2012). The association between cingulate cortex glutamate concentration and delay discounting is mediated by resting state functional connectivity. *Brain and Behavior: A Cognitive Neuroscience Perspective*, *2*(5), 553–562.

Schnakenberg Martin, A. M., Kim, D. J., Newman, S. D., Cheng, H., Hetrick, W. P., Mackie, K., & O'Donnell, B. F. (2021). Altered cerebellar-cortical resting-state functional connectivity in cannabis users. *Journal of Psychopharmacology*, *35*(7), 823–832.

Schubert, F., Gallinat, J., Seifert, F., & Rinneberg, H. (2004). Glutamate concentrations in human brain using single voxel proton magnetic resonance spectroscopy at 3 Tesla. *NeuroImage*, *21*(4), 1762–1771.

Smith, S. M., Fox, P. T., Miller, K. L., Glahn, D. C., Fox, P. M., Mackay, C. E., … Beckmann, C. F. (2009). Correspondence of the brain's functional architecture during activation and rest. *Proceedings of the National Academy of Sciences*, *106*(31), 13040–13045.

Smith, C. M., Hayamizu, T. F., Finger, J. H., Bello, S. M., McCright, I. J., Xu, J., … Ringwald, M. (2019). The mouse gene expression database (GXD): 2019 update. *Nucleic Acids Research*, *47*(D1), D774–D779.

Stephan, R. A., Alhassoon, O. M., Allen, K. E., Wollman, S. C., Hall, M., Thomas, W. J., et al. (2017). Meta-analyses of clinical neuropsychological tests of executive dysfunction and impulsivity in alcohol use disorder. *American Journal of Drug and Alcohol Abuse*, *43*, 24–43. https://doi.org/10.1080/00952990.2016.1206113.

Stone, N. L., Murphy, A. J., England, T. J., & O'Sullivan, S. E. (2020). A systematic review of minor phytocannabinoids with promising neuroprotective potential. *British Journal of Pharmacology*, *177*(19), 4330–4352.

Straiker, A., & Mackie, K. (2005). Depolarization-induced suppression of excitation in murine autaptic hippocampal neurones. *The Journal of Physiology*, *569*(2), 501–517.

Sugam, J. A., Day, J. J., Wightman, R. M., & Carelli, R. M. (2012). Phasic nucleus accumbens dopamine encodes risk-based decision-making behavior. *Biological Psychiatry*, *71*(3), 199–205.

Sunsay, C., & Rebec, G. V. (2008). Real-time dopamine efflux in the nucleus accumbens core during Pavlovian conditioning. *Behavioral Neuroscience*, *122*(2), 358.

Sutherland, M. T., McHugh, M. J., Pariyadath, V., & Stein, E. A. (2012). Resting state functional connectivity in addiction: lessons learned and a road ahead. *NeuroImage*, *62*(4), 2281–2295.

Tsou, K., Brown, S., Sanudo-Pena, M. C., Mackie, K., & Walker, J. M. (1998). Immunohistochemical distribution of cannabinoid CB1 receptors in the rat central nervous system. *Neuroscience*, *83*(2), 393–411.

van Heukelum, S., Mars, R. B., Guthrie, M., Buitelaar, J. K., Beckmann, C. F., Tiesinga, P. H., … Havenith, M. N. (2020). Where is cingulate cortex? A cross-species view. *Trends in Neurosciences*, *43*(5), 285–299.

Venkatraman, V., & Huettel, S. A. (2012). Strategic control in decision-making under uncertainty. *European Journal of Neuroscience*, *35*(7), 1075–1082.

Wagner, G., Gussew, A., Koehler, S., de la Cruz, F., Smesny, S., Reichenbach, J. R., & Baer, K. J. (2016). Resting state functional connectivity of the hippocampus along the anterior–posterior axis and its association with glutamatergic metabolism. *Cortex*, *81*, 104–117.

Wang, K., Smolker, H. R., Brown, M. S., Snyder, H. R., Hankin, B. L., & Banich, M. T. (2020). Association of γ-aminobutyric acid and glutamate/glutamine in the lateral prefrontal cortex with patterns of intrinsic functional connectivity in adults. *Brain Structure and Function*, *225*(7), 1903–1919.

Wheeler, R. A., Aragona, B. J., Fuhrmann, K. A., Jones, J. L., Day, J. J., Cacciapaglia, F., … Carelli, R. M. (2011). Cocaine cues drive opposing context-dependent shifts in reward processing and emotional state. *Biological Psychiatry*, *69*(11), 1067–1074.

Wijtenburg, S. A., & Knight-Scott, J. (2011). Very short echo time improves the precision of glutamate detection at 3T in 1H magnetic resonance spectroscopy. *Journal of Magnetic Resonance Imaging*, *34*(3), 645–652.

Yasen, A. L., Smith, J., & Christie, A. D. (2017). Reliability of glutamate and GABA quantification using proton magnetic resonance spectroscopy. *Neuroscience Letters*, *643*, 121–124.

Chapter 21

Brain metabolic responses to cannabis use in people with multiple sclerosis: Insights from [^{18}F]-FDG positron emission tomography and functional MRI

Thorsten Rudroff and Craig D. Workman
Department of Health and Human Physiology, University of Iowa, Iowa City, IA, United States

Abbreviations

5-HT1A	5-hydroxytryplamine
AD	Azheimer's disease
AEA	N-arachidonylethanolamine
BOLD	blood oxygenation level dependent
CB	cannabinoid receptor
CBD	cannabidiol
CBF	cerebral blood flow
CNS	central nervous system
CT	computed tomography
FAAH	fatty acid amide hydrolase
FDA	Food and Drug Administration
FDG	fluorodeoxyglucose
fMRI	functional magnetic resonance imaging
GlyR	glycine receptor
GPCRs	G protein-coupled receptors
GU	glucose uptake
HD	Huntington's disease
MSFC	multiple sclerosis functional composite
NU	nonusers
PASAT	Paced Auditory Serial Addition Test
PDDS	patient determined disease steps
PET	positron emission tomography
PPARs	peroxisome proliferator-activated receptors
PwMS	people with multiple sclerosis
SDMT	Symbol Digit Modalities Test
SUV	standardized uptake values
THC	delta-9-tetrahydrocannabinol
TRPV	transient receptor potential vanilloid

Cannabis

Cannabis, colloquially called marijuana, refers to the dried leaves, flowers, stems, and seeds from the *Cannabis sativa* or *Cannabis indica* plant and is the most commonly used illicit substance in the world (NIDA, 2018). Cannabis includes 60–500 different compounds, including a class of chemicals known as cannabinoids (Bloomfield et al., 2019; Wycoff et al., 2018), of which delta-9-tetrahydrocannabinol (THC) and cannabidiol (CBD) are the most frequently investigated.

Cannabis Use, Neurobiology, Psychology, and Treatment. https://doi.org/10.1016/B978-0-323-89862-1.00026-X

THC-dominant strains with low CBD content have been associated with intoxicating, psychotropic effects including euphoria, enhancement of sensory perception, and memory impairment. In contrast, CBD-dominant strains with low THC content are considered nonpsychotropic and might contribute to many of the purported therapeutic benefits (Ferrari et al., 2013; Schreiner & Dunn, 2012; Walsh, McKinney, & Holmes, 2021).

THC acts centrally predominantly via the cannabinoid receptor 1 (CB1) (Wycoff et al., 2018), which attaches the main active ingredient of cannabis, 19-tetrahydrocannabinol (19-THC) and mediates most of the central nervous system (CNS) effects of 19-THC (Zimmer et al., 1999). CB1 is heterogeneously distributed throughout the human brain, with particularly high levels in the neocortex, basal ganglia, hippocampus, brainstem, and cerebellum (Herkenham et al., 1991). CB1 can also be found on peripheral nerve terminals and in extra-neural sites like the testis, vascular endothelium, eye, and spleen. CB1 has connections to the pathophysiology of many CNS disorders, including neurodegenerative disorders like multiple sclerosis (MS), Huntington's disease (HD), and Alzheimer's disease (AD) (Di Marzo et al., 2014; Zimmer et al., 1999). On the other hand, the cannabinoid receptor 2 (CB2) exhibits a more defined brain expression pattern and is mostly found in immune system cells and tissues (Di Marzo et al., 2014). In the CNS, CB2 expression is correlated with inflammation and is primarily localized to resident CNS macrophages (i.e., microglia) (Mackie, 2006, 2008).

In contrast to THC, CBD has little binding affinity for either CB1 or CB2 but can antagonize these receptors in the presence of THC (Palazuelos et al., 2009). CBD has been described as a noncompetitive negative allosteric modulator of CB1 and reduces the efficacy and potency of THC (Thomas et al., 2007). CBD also controls the perception of pain by affecting the activity of a significant number of other targets, including noncannabinoid G protein-coupled receptors (GPCRs) (e.g., 5-HT1A), ion channels (TRPV1, TRPA1 and TPRM8, GlyR), and peroxisome proliferator-activated receptors (PPARs), while also inhibiting the uptake of N-arachidonylethanolamine (AEA) and weakly inhibiting AEA hydrolysis via fatty acid amide hydrolase (FAAH) (Burstein, 2015; Laprairie et al., 2015). CBD displays an entourage effect and is able of enhancing the tolerability and safety of THC by reducing the possibility of psychoactive effects and antagonizing several other adverse effects (e.g., tachycardia, sedation, and anxiety) (Morales et al., 2017; Russo & Guy, 2006). Although differences in THC and CBD concentrations of cannabis reflect distinctions in the effects of different strains, this plant diversity might present unique avenues of study. For example, even though CBD as a monotherapy for pain treatment has not been clinically evaluated, the anti-inflammatory (Abrams & Guzman, 2015) and anti-epileptic benefits, in addition to its agreeable safety profile, suggest that CBD could be an effective and safe analgesic (Ko et al., 2016).

Multiple sclerosis

Multiple sclerosis (MS) is a chronic disorder characterized by peri-venular inflammatory lesions leading to demyelinating plaques, oligodendrocyte injury, and irreversible damage of axons within the gray and white brain matter (Dobson & Giovannoni, 2019); the result is a progressive loss of function across the neurological spectrum. MS is typically characterized into four disease courses: relapsing-remitting (~85% of MS diagnoses), primary progressive (~10% of MS diagnoses), secondary progressive (~50% of primary progressive patients), and progressive relapsing (~5% of MS diagnoses) (Bishop & Rumrill, 2015). The most frequent resulting symptoms affect both physical (e.g., pain, fatigue, gait impairment, spasticity, and paresthesia) and cognitive (e.g., impaired executive function, memory, information processing, and depression) function (Haussleiter, Brune, & Juckel, 2009; Kister et al., 2013). Interestingly, the majority of PwMS (50%–70%) (Koppel et al., 2014; Zajicek, Fox, & Sanders, 2003; Zajicek, Hobart, Slade, Barnes, & Mattison, 2012), including those early in the disease (Amato, Portaccio, & Zipoli, 2006; Amato, Zipoli, & Portaccio, 2006; Drew et al., 2008; Feinstein et al., 2013; Rao, 2004; Wallin et al., 2006), suffer from cognitive impairment, which often manifests as impaired executive function. Executive function encompasses the higher order cognitive skills essential to formulate and execute goals, which enables autonomous engagement in purposeful, self-serving activities (Lezak et al., 2004). Cognitive impairment is frequently accompanied by difficulties with emotional processing (i.e., the ability to perceive and use emotional information (Henry et al., 2017)). For example, PwMS experience emotional dysregulation (e.g., anxiety and depression) at a prevalence that is three times greater than the general population (Minden & Schiffer, 1990; Patten et al., 2005; Sadovnick et al., 1996); These patients also face considerable unpredictability and uncertainty inherent to the erratic nature of their disease (e.g., risks of relapses, hospital admissions, or progressing in disability) (Mullins et al., 2001). Moreover, because MS is currently incurable and usually diagnosed in early adulthood, many patients live with these debilitating symptoms and uncertain disease courses for many years (Simmons, 2010). All things considered, it is not surprising that fear and anxiety disorders have been reported in 36%–54% of PwMS, with approximately 30% experiencing symptoms consistent with generalized anxiety disorder (Hartoonian et al., 2015; Jones et al., 2012; Wood et al., 2012). In addition, more than half of PwMS experience depression during their illness (Merkelbach et al., 2003; Viner et al., 2014).

Despite the availability of disease-modifying treatments that reduce relapse rates by suppressing CNS inflammation (Bishop & Rumrill, 2015), most symptoms are not satisfactorily remedied. As a result, many people with MS (PwMS) seek alternative therapies for symptom management. For example, as many as 60% of PwMS currently use cannabis and up to 90% would consider using cannabis if the legal climate was more favorable (Rudroff & Honce, 2017). Interestingly, one study reported that 59% of PwMS reduced their use of prescription medication after starting cannabis (Kindred et al., 2017), which implies symptom amelioration. However, it is well accepted that cannabis, particularly products with a high THC content, negatively impairs cognitive and executive function in healthy adults, which may increase the risk of schizophrenic-like psychosis or other severe mental illnesses (Hindocha et al., 2015). Although some evidence suggests that CBD minimizes some of the negative side effects and enhances the therapeutic efficacy of THC (Arkell et al., 2019; Bhattacharyya et al., 2010; Martin-Santos et al., 2012; Russo & Guy, 2006), a recent study (Hocevar et al., 2019) found that a combination of THC and CBD increased plasma concentrations of THC metabolites and slightly increased cognitive impairment more than THC alone. These results may have significant consequences for PwMS that use cannabis products containing both THC and CBD. Significantly, PwMS already have an elevated risk of cognitive impairment and executive dysfunction. Thus, the physiological effects of chronic cannabis use may further increase that risk and lead to a higher prevalence of psychosis and severe mental illness. Understanding how cannabis use influences executive function is also important for clinicians because patients who use cannabis may have cognitive deficits that might make adhering to treatment, following medical advice, and experiencing successful disease-related outcomes more challenging.

On this point, the American Academy of Neurology summarized the outcomes of clinical studies that investigated the effects of cannabis in PwMS with neuropathic and spasticity-related pain and spasms (Koppel et al., 2014). The summary reported that the current evidence suggests that THC (Rog et al., 2005; Zajicek et al., 2003; Lakhan & Rowland, 2019) is probably effective for treating MS-related pain or painful spasms; however, the efficacy of combustible (smoked) cannabis for reducing pain is unclear (Corey-Bloom et al., 2012; Greenberg et al., 1994). A more recent systematic review (Riera et al., 2021) aimed to map the evidence of the therapeutic effects of cannabis derivatives and synthetic cannabis analogs. This overview, based on systematic reviews of disparate methodological quality, identified no benefit from these substances for spasticity-related pain in MS. For all other outcomes, the certainty of the evidence was very low, or not evaluated by the identified reviews, preventing recommendations for or against the routine use of these substances (Amato et al., 2016, 2017; Fu et al., 2018; Iskedjian et al., 2007; Lakhan & Rowland, 2019; Mills et al., 2007; Rovare et al., 2017; Shakespeare et al., 2003; Torres-Moreno et al., 2018).

Unfortunately, most studies (see Nielsen et al., 2018 for an overview) did not focus on concerns of trial quality and included various drugs, durations, doses, conditions, and outcomes. Thus, the efficacy of cannabis use to treat symptoms in MS remains largely unresolved. Furthermore, the safety and potential long-term effects of cannabis on PwMS have not been satisfactorily evaluated (Rudroff, 2019). In particular, the MS studies have only investigated short-term effects, with durations ranging from 6 to 15 weeks. Accordingly, the long-term risks and rare, but potentially severe, side effects cannot be captured in these short-duration trials. For example, cannabis specifically products with high THC content, negatively impairs cognition and executive function in healthy adults, and may increase the risk of schizophrenic-like psychosis or other severe mental illnesses (Bhattacharyya et al., 2012; Bloomfield et al., 2019). This has significant consequences for PwMS because this population is already more likely to experience impaired cognition and executive dysfunction (Bishop & Rumrill, 2015; Kister et al., 2013). Importantly, how chronic cannabis use, of any THC:CBD ratio, impacts cognitive and executive function in PwMS is currently unknown; thus, the effects of chronic cannabis use in PwMS might further increase their risk of psychological impairment. In addition, even CBD-dominant use is not risk-free because CBD has both adverse event and drug interaction potential (Huestis et al., 2019). Furthermore, because CBD is an effective anticonvulsant therapy, the FDA is concerned that its use might cause suicidal ideation (White, 2019). On the other hand, a recent review (Bergamaschi et al., 2011) substantiated and expanded the findings of an earlier review (Iffland & Grotenhermen, 2017) about the favorable safety profile of CBD. However, various aspects of CBD safety research still need to be explored, and longer-term safety data are critically needed to fully appreciate CBD's risk to benefit ratio (Rudroff, 2019).

In vivo neuroimaging

Neuroimaging has offered effective tools to study the in vivo effects of cannabis on brain function. These effects can be directly investigated in experimental settings following the administration of THC and CBD or indirectly by comparing subjects with and without a history of cannabis use. This chapter focuses on human neuroimaging studies, specifically positron emission tomography (PET) and functional magnetic resonance imaging (fMRI), investigating the effects of THC and CBD in PwMS.

Positron emission tomography

PET is a functional imaging modality that describes a given functional process (e.g., glucose metabolism, cerebral blood flow [CBF], receptor density and binding, pathological burden [amyloid, tau]) contingent on the radiopharmaceutical employed. The most commonly used PET radiotracer is the glucose analog [^{18}F]fluoro-deoxyglucose (FDG). Because the brain uses glucose as a primary fuel source, mapping FDG uptake provides insight into brain activity. Additionally, given the fact that the brain is exceptionally metabolically active, "activation" or "deactivation" of a particular brain region implies respective increases or decreases in glucose metabolism of that area. These changes are accompanied by corresponding alterations in CBF via [^{15}O]water. [^{15}O]water is a freely diffusible PET radiotracer that leverages the fact that water constitutes a large proportion of blood. Accordingly, FDG provides a means of mapping glucose metabolism, and [^{15}O]water provides a means of assessing changes in CBF in the cerebrum. Thus, the "metabolic trapping" properties of FDG and diffusivity of [^{15}O]water PET tracers allow the in vivo examination of the neuronal mechanisms underlying the effects of cannabis in unconstrained conditions (i.e., in realistic training scenarios outside the scanner environment) (Rudroff et al., 2020). Additionally, FDG-PET studies indicate that regular heavy alcohol use (Volkow et al., 2015), cannabis (Volkow et al., 1996), and other illegal drug use (Kim et al., 2009; Moreno-Lopez et al., 2012) might result in lower resting nervous system glucose uptake (GU). It has been suggested that lower cerebral GU might be driving for some of the cognitive and physical impairments observed in these substance abuse populations (Kim et al., 2009; Volkow et al., 1996). Importantly, lower brain GU has been correlated with clinical symptoms in PwMS, like fatigue (Roelcke et al., 1997), cognitive ability (Blinkenberg et al., 1999), and walking speed (Kindred et al., 2015) in PwMS. Thus, brain GU may be an effective biomarker for tracking disability and/or disease progression in PwMS.

Functional magnetic resonance imaging

fMRI is a commonly used noninvasive technique to study the neuronal mechanisms of CNS function and to assess how a disease disrupts brain activation patterns. fMRI assesses changes in the blood oxygenation level-dependent (BOLD) response, which, in turn, influence the transverse magnetization relaxation time (either T2* in a gradient-echo sequence, or T2 in a spin-echo sequence). These alterations are attributed to differences in deoxyhemoglobin concentrations resulting from variations in neuronal activity (Bandettini et al., 1992; Ogawa et al., 1990, 1992). The correlation between local deoxyhemoglobin levels and neuronal activity is thought to reflect changes in CBF, cerebral blood volume, and oxygen extraction, all of which are altered with neuronal activity (Bandettini et al., 1992; Ogawa et al., 1990, 1992).

fMRI is frequently used to study brain function in patients with various neurological conditions, including PwMS. For example, previous studies have assessed the role of brain plasticity in functional recovery after neurological injury and to characterize associations of neuroanatomy with cognitive decline and worsening disability in PwMS. Moreover, cross-sectional MRI studies have reported significant correlations between MS-associated impairment and several MRI measures, including cortical and subcortical atrophy, increased lesion load, and disruption of white matter tracts. A review of such studies (Rocca et al., 2015) found significant cross-sectional associations between MRI measures and cognitive impairment.

PET, cannabis, and MS

As indicated above, recent estimates indicate that 20%–66% of PwMS currently use cannabis (Bhattacharyya et al., 2012; Henry et al., 2017) despite little empirical evidence of efficacy and, importantly, the underlying neural mechanisms. Accordingly, how cannabis use affects cerebral GU in PwMS is largely unknown. Indeed, only two PET studies (Kindred et al., 2018; Workman et al., 2020) have investigated brain activity alterations in PwMS using cannabis to alleviate debilitating symptoms like spasticity and neuropathic pain. Kindred et al. (2018) examined whether PwMS using cannabis had altered resting brain GU and clinical test results compared with nonusing PwMS. Sixteen PwMS, eight cannabis users, and eight age/sex-matched nonusers (NU), undertook clinical testing and FDG-PET/computed tomography (CT) imaging. Only PwMS using cannabis for at least 6 months (i.e., chronic users) were included in the user group. Participants refrained from cannabis for $\geq$8 h (using group) and fasted for $\geq$4 h (both groups) before the experiments. Disability levels were assessed using the patient determined disease steps (PDDS) and clinical testing consisted of the following: Beck's Depression Inventory, Activities of Balance Confidence, Pain Effects Scale, Fatigue Severity Scale, Multiple Sclerosis Functional Composite (MSFC), Numerical Rating Scale measure of spasticity, handgrip strength testing, balance testing, and timed up-and-go. The authors reported that the subject demographics, scores on the questionnaires, and all but one of the clinical variables (paced auditory serial addition test [PASAT], part of the MSFC) were similar

($p \geq 0.12$). For the PASAT, the reported scores were lower for cannabis users than for nonusers ($p = 0.02$). In addition, functional brain analysis showed areas throughout the brain, primarily within the frontal and parietal lobes, in which cannabis users had greater GU, although these results did not persist when the authors accounted for disease duration. Additionally, mean standardized uptake values (SUVs) for the groups/regions were not different and PASAT scores did not correlate with regional SUVs ($R < -0.35$, $p > 0.18$).

Several studies have shown reduced GU in gray and white matter in PwMS (i.e., lower global brain activity), which might be related to the direct or indirect effects of CNS lesions in MS (Iskedjian et al., 2007; Lakhan et al., 2019; Mills et al., 2007; Torres-Moreno et al., 2018). However, in the study by Kindred et al. (2018), lower global GU in cannabis users was not detected. This suggests a different neuroadaptive process in the MS brains compared with non-MS brains. One factor integral to nervous system GU is circulating lactate, which was relevantly highlighted by Smith et al. (2003) who found that infusing peripheral lactate lowered GU in healthy individuals. However, PwMS have high lactate levels which might stem from increased inflammation and neurodegeneration within the CNS (Amorini et al., 2014). Given that cannabis might reduce CNS inflammation via the CB1 and CB2 pathways described above (Di Marzo et al., 2014; Klein, 2005; Mackie, 2006, 2008; Zimmer et al., 1999), cannabis use in PwMS might lead to lowered inflammation/neurodegeneration (Iuvone et al., 2009; Saito et al., 2012). This reduced inflammation might have an offsetting effect between speculated CB1 downregulation and reduced nervous system inflammation, leading to a net-zero change in GU. Furthermore, the Kindred et al. study (Kindred et al., 2018) found no correlation between resting brain activity and PASAT scores and the correlations performed for each group did not reveal an association between the variables. Still, their findings suggest that cannabis use may have adverse effects on cognitive performance, as evidenced by the lower PASAT scores; however, the mechanism behind this dysfunction is still unknown.

A conspicuous limitation of the Kindred et al. study (2018) is that the specific effects of THC-dominant and CBD-dominant cannabis were not distinguished. As discussed above, THC is psychoactive, may cause acute psychosis and negatively impact executive function (Bloomfield et al., 2019). However, evidence implies that THC may be antiepileptic (Saito et al., 2012), anti-spastic (Friedman & Devinsky, 2015), and analgesic for neuropathic pain (Collin et al., 2007). On the other hand, CBD is not intoxicating and might be anxiolytic, anti-inflammatory, antiepileptic, and neuroprotective (Wilsey et al., 2008); CBD might also counteract the negative effects of THC (Ebbert et al., 2018). In healthy cannabis users, a Bloomfield et al. (2019) review reported differing metabolic and CBF responses to THC and CBD in the brain. Specifically, acute THC consumption was associated with glucose hypermetabolism and increased global and regional blood flow in assorted brain areas, while CBD tended to mediate these effects (Bloomfield et al., 2019). On the other hand, acute CBD administration has been associated with decreased blood flow in limbic and paralimbic areas (e.g., hypothalamus, amygdala) (Crippa et al., 2004). In contrast, chronic THC use was associated with reduced blood flow and hypometabolism of various brain areas (Bloomfield et al., 2019). However, chronic CBD studies are rare, and only one study (Workman et al., 2020) investigated brain function alterations after chronic CBD, and THC, use. The hypothesis of the Workman et al. (2020) study was that PwMS who chronically use THC and CBD would demonstrate altered and divergent relative regional cerebral glucose metabolism than NU. Ten cannabis users (6 THC, 4 CBD; use $\geq$6 months), and 9 NU, underwent FDG-PET/CT imaging. Cannabis consumption was via both smoking and edible forms. Their results revealed different effects of chronic use of THC and CBD. Fig. 1A shows a 3D representation of brain areas with significantly increased (THC users >NU) and decreased (CBD users <NU; Fig. 1B) glucose metabolism from archetypal subjects from the respective groups. Fig. 2 displays significant metabolism differences between the groups. Specifically, hypermetabolism in the left superior temporal gyrus posterior part, the right posterior temporal lobe, and the left inferolateral remainder of the parietal lobe in THC-users compared with NU. CBD-users had significant hypometabolism in the left parahippocampal and ambient gyri, the left amygdala, the left hippocampus, the left fusiform gyrus, and the right nucleus accumbens compared with NU. There were no differences between THC users and CBD users, but there were trends for the right subcallosal area, the right nucleus accumbens, the left fusiform gyrus, and the left insula, with THC users > CBD users in all of these regions.

The Workman et al. (2020) findings suggest that, compared to NU, chronic THC and CBD use by PwMS might be associated with respective increases and decreases in cerebral glucose metabolism in PwMS, primarily in the left temporal areas. The contrasting effects (i.e., hyper- vs. hypometabolism) in this study emphasizes the necessity of differentiating between THC and CBD use in future cannabis research. Although, THC has previously improved self-reported spasticity and pain (Hocevar et al., 2019), THC-induced hypermetabolism of left temporal areas has also been associated with psychosis and executive dysfunction (Crippa et al., 2004). This is a relevant finding because these are extant problems in PwMS (Atakan et al., 2013) that might be exacerbated by THC use. CBD-induced hypometabolism of the left medial temporal areas (e.g., amygdala/hippocampus) was associated with reduced feelings of anxiety and fear (Fusar-Poli et al., 2009), which might ameliorate anxiety and depression symptoms common in PwMS (Atakan et al., 2013).

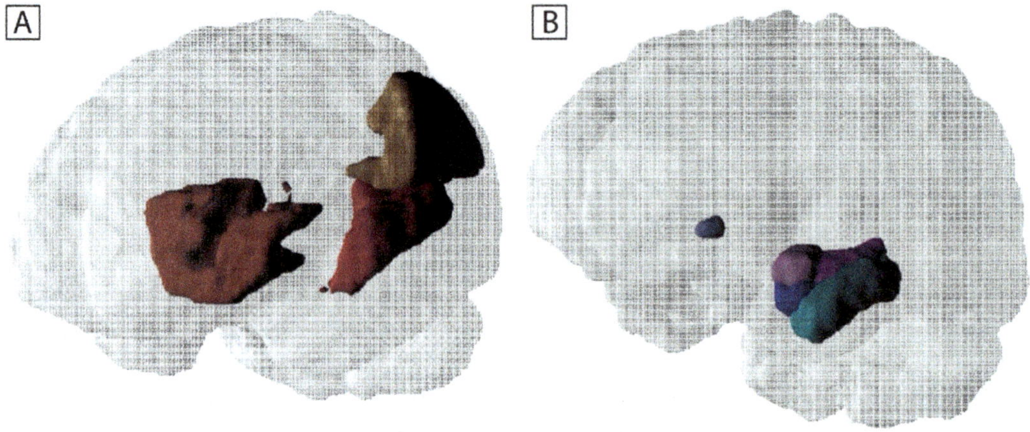

FIG. 1 Three-dimensional representation of brain areas with altered cerebral glucose metabolism (i.e., compared with nonusers) associated with chronic cannabis use. The brain regions (Hammers N30R83 atlas) are modeled with a nearly frontal view (i.e., left side of the brain is on the right side of the figure and the central sulcus can be visualized on the lateral left side of each figure) on a representative subject brain from the respective cannabis group. (A) Increased cerebral glucose metabolism in THC-users (*red* = left superior temporal gyrus posterior part; *brown* = left inferiolateral remainder of parietal lobe; *orange* = right posterior temporal lobe). (B) Decreased cerebral glucose metabolism in CBD-users (*blue* = left parahippocampal and ambient gyri; *light-purple* = left amygdala; *pink* = left hippocampus; *turquoise* = left fusiform gyrus; *dark-purple* = right nucleus accumbens) (Workman et al., 2020).

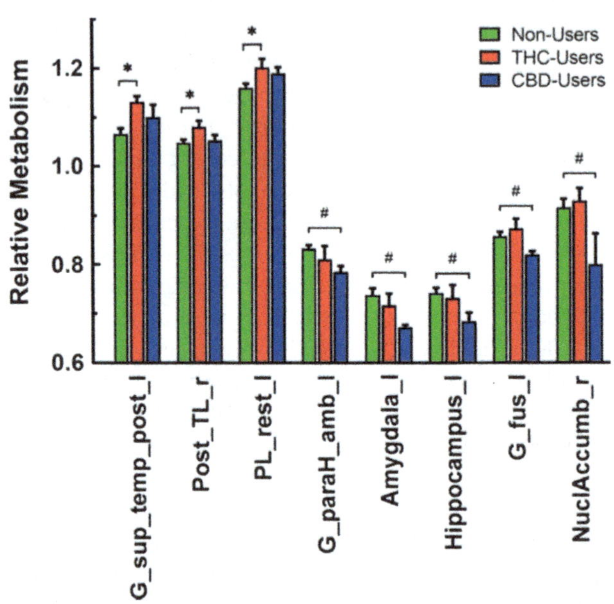

FIG. 2 Comparison of cannabis nonusers (NU), THC-users, and CBD-users. Data are mean ± SEM. *, THC users significantly different ($p < 0.05$, uncorrected) than NU; #, CBD users significantly different ($p < 0.05$, uncorrected) than NU. There were no differences between THC users and CBD users. Brain regions are from Hammers N30R83 atlas. Amygdala_l, left amygdala; CBD, cannabidiol; G_fus_l, left fusiform gyrus; G_paraH_amb_l, left parahippocampal and ambient gyri; G_sup_temp_post_l, left superior temporal gyrus posterior part; Hippocampus_l, left hippocampus; NuclAccumb_r, right nucleus accumbens; PL_rest_l, left inferiolateral remainder of parietal lobe; Post_TL_r, right posterior temporal lobe; THC, Δ-9-tetrahydrocannabinol (Workman et al., 2020).

Summary and future directions

Kindred et al. (2018) did not detect any changes in resting brain GU of chronic cannabis users, which is a marker for clinical MS disability. Furthermore, the mechanisms of cognitive dysfunction have yet to be clarified. Based on the existing FDG-PET findings, it is questionable if resting brain metabolism/GU sufficiently account for differences in cognitive ability, assessed by the PASAT (Banwell et al., 2016). Importantly, the study by Workman et al. (2020) indicated distinct effects of THC and CBD, which might explain, at least in part, the lack of correlations between GU and PASAT scores reported in Kindred et al. (2018), who did not stratify users into subgroups. Furthermore, these studies provide a foundation for future cross-sectional and longitudinal PET studies that should aim to differentiate between the cerebral effects of different THC:CBD ratios (e.g., THC-dominant vs. CBD-dominant vs. THC=CBD) and the subsequent influences on cognition and function.

fMRI, cannabis, and MS

Previous cross-sectional studies have investigated the correlation between functional and structural neuroimaging and cognitive dysfunction in PwMS who use cannabis (Pavisian et al., 2014, 2015; Romero et al., 2015). Pavisian et al. (2014) studied 20 PwMS who smoked cannabis and 19 matched PwMS who did not use cannabis (nonusers; NU). Resting-state fMRI and structural MRI data (i.e., lesion and normal-appearing brain tissue volumes, diffusion tensor imaging metrics) were collected, while the subjects were completing the n-back working memory test. Moreover, neuropsychological data on verbal (Selective Reminding Test Revised) and visual (10/36 Spatial Recall Test) memory, attention (Word List Generation), and information processing speed (PASAT [2- and 3-s versions] and Symbol Digit Modalities Test) were obtained. The results indicated that, compared to the NU, cannabis users displayed more diffuse cerebral activation during n-back trials, including increased activation in the anterior cingulate and parietal regions, which are involved in working memory. The authors concluded that cannabis use might further compromise the already weakened brain compensatory mechanisms in PwMS. The imaging performed in this studies increases the construct validity of the neuropsychological findings and serve as a cautionary note to both PwMS who use cannabis and potential cannabis prescribers.

In a follow-up study, Romero et al. (2015) investigated, in the same PwMS (Pavisian et al., 2014), whether structural brain changes were correlated with impaired cognition. The partial least squares analysis returned significant correlations between cognitive scores and both gray matter and white matter volume in users and NU. Specifically, gray matter volume in the thalamus, medial temporal, medial prefrontal regions, and basal ganglia, and white matter volume in the fornix were associated with cognitive deficits. Critically, the analysis also found that brain volume declines were associated with more general cognitive impairment in the cannabis-using group compared with the NU group (Fig. 3). These results indicate that

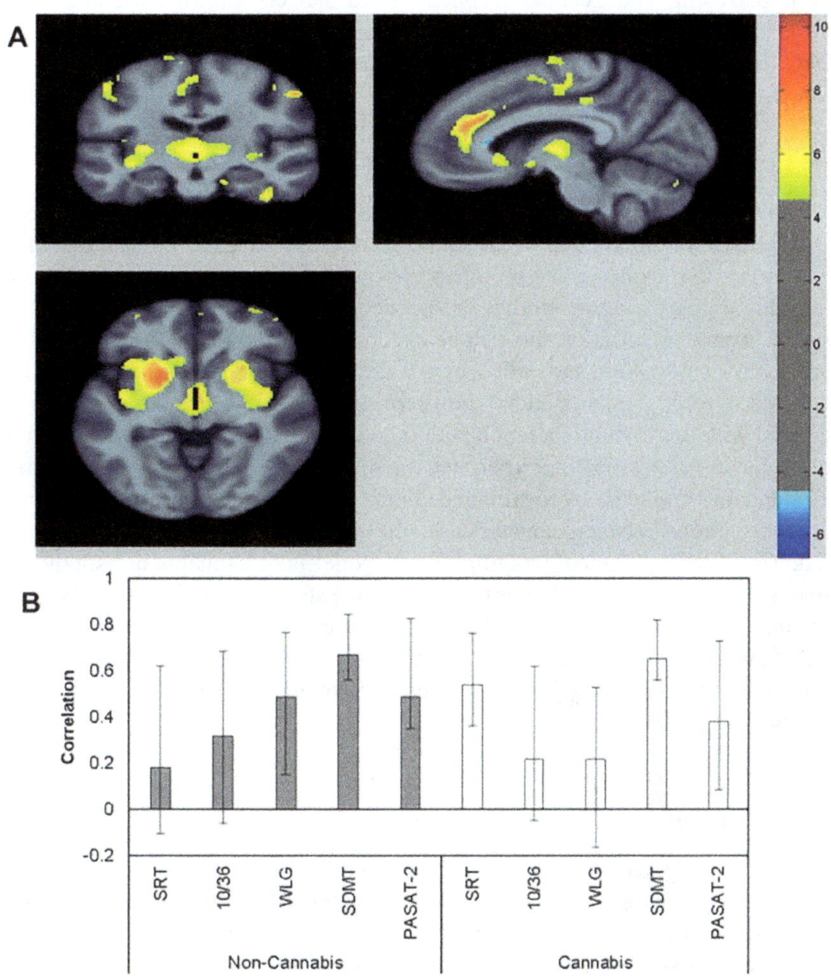

FIG. 3 Pattern of whole brain covariance of gray matter changes with neuropsychological test scores in MS patients who do or do not smoke cannabis (Romero et al., 2015).

cannabis use in MS results in more widespread cognitive deficits, which correlated with tissue volume in medial temporal, subcortical, and prefrontal regions.

A second study by Pavisian et al. (2015) investigated the functional brain correlates of information processing speed in PwMS who smoke cannabis ($n = 20$) and drug naïve ($n = 19$) subjects. They found that the cannabis group responded slower in nine of 11 blocks of the Symbol Digit Modalities Test (SDMT), indicating a trend toward a slower response time, but the groups did not differ in the accuracy of their responses. Both groups demonstrated activation in a prefrontal cortex-parietal network associated with information processing speed. However, when compared to the cannabis naïve group, users had lower activation in the right and left thalami and greater activation in the anterior cingulate cortex. The authors concluded that regular cannabis use in PwMS was associated with slower information processing speed and a pattern of cerebral activity that differed from cannabis naïve patients, most notably a reduction in bilateral thalamic activity.

Summary and future directions

Pavisian et al. (2014) found greater impairment of information processing speed (PASAT) and visual memory in cannabis users compared to NU. The pattern of cerebral activation also varied between these groups in two ways: First, greater activation was observed in the neural circuits underpinning the n-back task, especially in the anterior cingulate and posterior parietal regions, in the cannabis users. Second, the cannabis group showed activation in brain regions superfluous to the n-back paradigm. The second study by Pavisian et al. (2015) also found slower processing speed (i.e., on the SDMT) in the cannabis users, accompanied by excessive activation in paralimbic regions and reduced activation in the bilateral thalami. Finally, Romero et al. (2015) found structural MRI brain volume reductions in cannabis users which were associated with more extensive cognitive impairment. Thus, cannabis appears to further compromise the brain's structural integrity relevant to cognitive abilities. However, none of these studies (Pavisian et al., 2014, 2015; Romero et al., 2015) discriminated between the different THC:CBD ratios of the cannabis products used by their subjects.

Conclusion

A better understanding of how different THC:CBD ratios might alter brain activity in PwMS—and whether these effects have persistent deleterious effects or simply represent transient neuroadaptation—will have paramount importance to future and ongoing clinical trials that evaluate the therapeutic potential of cannabis and the new class of drugs that act on cannabinoid receptors. Future neuroimaging studies are needed to clarify the acute (i.e., the first few months of use) and longitudinal effects (i.e., from use initiation through chronic use [$\geq$6 months]) of different THC:CBD ratios. Establishing the time course of cognitive, symptom, and brain activity alterations, and their associations with functional (physical and cognitive) outcomes, is also strongly recommended. Furthermore, polysubstance use is frequent in MS (Hocevar et al., 2019) and has been associated with detrimental mental health outcomes in the absence of cannabis use. Thus, if other substance use is not controlled for in future studies, either statistically or via inclusion/exclusion criteria, the effects (both positive and negative) of cannabis might be overestimated. Finally, most studies did not control for baseline illness or symptom severity; hence, associations between cannabis use and symptoms that might motivate cannabis use (e.g., pain) should also be considered. The potential clinical or symptomatic benefits of cannabis use should be carefully considered against the negative impacts on brain health and cognition, the full extent of which is only beginning to emerge. These findings also generate the intriguing question of the sufficiency of use for these effects to manifest, or whether cognitive deficits are reversible with cannabis cessation, as suggested by studies in neurologically healthy populations (Schreiner & Dunn, 2012). However, considering the imaging findings presented here, it is questionable if this will be the case for PwMS; nevertheless, this is still a worthy topic of future study.

Applications to other areas

PwMS are not the only patients seeking to use cannabis for symptom relief or management; indeed, people with Parkinson's disease have reported an increased cannabis use rate. Other neurological populations, especially those with spasticity (e.g., stroke and spinal cord injury), are also prone to using cannabis. Given that the baseline brain activities of different neurological groups are diverse, neuroimaging studies like those discussed above are essential to illuminating the mechanisms of purported effects of cannabis in these populations as well.

Mini-dictionary of terms

Anticonvulsant Therapy Anticonvulsant medications are FDA approved to treat epileptic seizures and many other medical conditions unrelated to seizure disorders. In 2008, the Food and Drug Administration (FDA) issued a warning that any and all antiepileptic drugs (AEDs) might increase the risk of suicidal ideation, suicide attempt, and completed suicide.
Beck's Depression Inventory (BDI) The BDI is a 21-item self-reporting questionnaire for evaluating the severity of depression in normal and psychiatric populations.
Blood Oxygenation Level-Dependent (BOLD) BOLD imaging is a technique that is commonly used for measuring brain activity in humans using magnetic resonance imaging (MRI). Blood supplies oxygen to brain cells. When these cells are active, there is an increase in blood flow and blood oxygen in the surrounding area.
Demyelination A demyelinating disease is any condition that causes damage to the protective covering (myelin sheath) that surrounds nerve fibers in your brain, the nerves leading to the eyes (optic nerves) and spinal cord.
Determined Disease Steps (PDDS) The PDDS is a patient-reported outcome (PRO) of disability in multiple sclerosis.
Executive Function The capacity to plan and meet goals, display self-control, follow multiple-step directions even when interrupted, and stay focused despite distractions, among others.
Hypermetabolism Metabolism at an increased or excessive rate.
Hypometabolism Hypometabolism, characterized by decreased brain glucose consumption, is a common feature of many neurodegenerative diseases.
Multiple Sclerosis A potentially disabling disease of the central nervous system. The immune system attacks the protective sheath (myelin) that covers nerve fibers and causes communication problems between your brain and the rest of your body.
Multiple Sclerosis Functional Composite (MSFC) The MSFC is a multidimensional, three-component performance scale to assess the degree of impairment in MS patients.
Spasticity A condition in which muscles stiffen or tighten, preventing normal fluid movement. The muscles remain contracted and resist being stretched, thus affecting movement, speech and gait.
Standardized Uptake Value (SUV) SUV is a semiquantitative measurement of uptake in tissue. Many factors affect the SUV, including accuracy of dose calibration, time between injection and imaging (dose to scan time), patient weight (changes are common in oncology patients), motion artifacts, and blood glucose levels.
Timed up-and-go (TUG) The TUG test assesses mobility, balance, walking ability, and fall risk.

Key facts of neuroimaging

- PET is a functional imaging modality that characterizes a given functional process (e.g., metabolism).
- FDG PET/CT can be a useful tool to investigate brain function.
- Alterations of brain glucose metabolism with this technique may be a marker of the systemic process of MS.
- fMRI measures brain activity by detecting changes associated with blood flow.
- fMRI relies on the fact that cerebral blood flow and neuronal activation are coupled.

Summary points

- The use of cannabis as a symptom treatment of MS symptom results in alterations in brain function and structure.
- Cannabis appears to further compromise the brain's structural and functional integrity relevant to cognitive abilities.
- Cannabis use further compromises brain compensatory mechanisms, which are already weakened in MS.
- Future neuroimaging studies are needed to clarify the distinct acute and longitudinal effects of different THC:CBD ratios.
- The time course of cognitive, symptom, and brain activity alterations, and their associations with functional (physical and cognitive) outcomes require further investigation.

References

Abrams, D. I., & Guzman, M. (2015). Cannabis in cancer care. *Clinical Pharmacology and Therapeutics*, *97*, 575–586.

Amato, L., Davoli, M., Minozzi, S., Mitrova, Z., Parmelli, E., Saulle, R., & Vecchi, S. (2016). *Systematic reviews on therapeutic efficacy and safety of Cannabis (including extracts and tinctures) for patients with multiple sclerosis, chronic neuropathic pain, dementia and Tourette syndrome, HIV/AIDS and cancer receiving chemotherapy* (pp. 1–109). Rome, Italy: World Health Organization and ASL ROMA 1 - Department of Epidemiology Lazio Region.

Amato, L., Minozzi, S., Mitrova, Z., Parmelli, E., Saulle, R., Cruciani, F., & Davoli, M. (2017). Systematic review of safeness and therapeutic efficacy of cannabis in patients with multiple sclerosis, neuropathic pain, and in oncological patients treated with chemotherapy. *Epidemiologia e Prevenzione*, *41*(5–6), 279–293 (in Italian).

Amato, M. P., Portaccio, E., & Zipoli, V. (2006). Are there protective treatments for cognitive decline in MS? *Journal of the Neurological Sciences*, *245* (1–2), 183–186.

Amato, M. P., Zipoli, V., & Portaccio, E. (2006). Multiple sclerosis-related cognitive changes: A review of cross-sectional and longitudinal studies. *Journal of the Neurological Sciences*, *245*, 41–46.

Amorini, A. M., Nociti, V., Petzold, A., et al. (2014). Serum lactate as a novel potential biomarker in multiple sclerosis. *Biochimica et Biophysica Acta*, *1842*, 1137–1143.

Arkell, T. R., Lintzeris, N., Kevin, R. C., Ramaekers, J. G., Vandrey, R., Irwin, C. J., Haber, P. S., & McGregor, I. S. (2019). Cannabidiol (CBD) content in vaporized cannabis does not prevent tetrahydrocannabinol (THC)-induced impairment of driving and cognition. *Psychopharmacology*, *236*, 2713–2724.

Atakan, Z., Bhattacharyya, S., Allen, P., Martin-Santos, R., Crippa, J. A., Borgwardt, S. J., et al. (2013). Cannabis affects people differently: Inter-subject variation in the psychotogenic effects of delta-9-tetrahydrocannabinol: A functional magnetic resonance imaging study with healthy volunteers. *Psychological Medicine*, *43*(6), 1255–1267.

Bandettini, P. A., Wong, E. C., Hinks, R. S., et al. (1992). Time course EPI of human brain function during task activation. *Magnetic Resonance in Medicine*, *25*, 390–397.

Banwell, E., Pavisian, B., Lee, L., et al. (2016). Attitudes to cannabis and patterns of use among Canadians with multiple sclerosis. *Multiple Sclerosis and Related Disorders*, *10*, 123–126.

Bergamaschi, M. M., Queiroz, R. H., Zuardi, A. W., & Crippa, J. A. (2011). Safety and side effects of cannabidiol, a *Cannabis sativa* constituent. *Current Drug Safety*, *6*, 237–249.

Bhattacharyya, S., Atakan, Z., Martin-Santos, R., Crippa, J. A., & McGuire, P. K. (2012). Neural mechanisms for the cannabinoid modulation of cognition and affect in 202 man: A critical review of neuroimaging studies. *Current Pharmaceutical Design*, *18*(32), 5045–5054.

Bhattacharyya, S., Morrison, P. D., Fusar-Poli, P., et al. (2010). Opposite effects of delta-9-tetrahydrocannabinol and cannabidiol on human brain function and psychopathology. *Neuropsychopharmacology*, *35*, 764–774.

Bishop, M., & Rumrill, P. D. (2015). Multiple sclerosis: Etiology, symptoms, incidence and prevalence, and implications for community living and employment. *Work*, *52*(4), 725–734.

Blinkenberg, M., Jensen, C. V., Holm, S., et al. (1999). A longitudinal study of cerebral glucose metabolism, MRI, and disability in patients with MS. *Neurology*, *53*, 149–153.

Bloomfield, M. A. P., Hindocha, C., Green, S. F., Wall, M. B., Lees, R., Petrilli, K., et al. (2019). The neuropsychopharmacology of cannabis: A review of human imaging studies. *Pharmacology & Therapeutics*, *195*, 132–161.

Burstein, S. (2015). Cannabidiol (CBD) and its analogs: A review of their effects on inflammation. *Bioorganic & Medicinal Chemistry*, *23*, 1377–1385.

Collin, C., Davies, P., Mutiboko, I. K., Ratcliffe, S., & Sativex Spasticity in MS Study Group. (2007). Randomized controlled trial of cannabis-based medicine in spasticity caused by multiple sclerosis. *European Journal of Neurology*, *14*(3), 290–296.

Corey-Bloom, J., Wolfson, T., Gamst, A., et al. (2012). Smoked cannabis for spasticity in multiple sclerosis: A randomized, placebo-controlled trial. *CMAJ*, *184*, 1143–1150.

Crippa, J. A., Zuardi, A. W., Garrido, G. E., Wichert-Ana, L., Guarnieri, R., Ferarri, L., et al. (2004). Effects of cannabidiol (CBD) on regional cerebral blood flow. *Neuropsychopharmacology*, *29*(2), 417–426.

Di Marzo, V., Stella, N., & Zimmer, A. (2014). Endocannabinoid signaling and the deteriorating brain. *Nature Reviews. Neuroscience*, *16*, 30–42.

Dobson, R., & Giovannoni, G. (2019). Multiple sclerosis—A review. *European Journal of Neurology*, *26*, 27–40.

Drew, M., Tippett, L. J., Starkey, N. J., & Isler, R. B. (2008). Executive dysfunction and cognitive impairment in a large community-based sample with multiple sclerosis from New Zealand: A descriptive study. *Archives of Clinical Neuropsychology*, *23*, 1–19.

Ebbert, J. O., Scharf, E. L., & Hurt, R. T. (2018). Medical cannabis. *Mayo Clinic Proceedings*, *93*(12), 1842–1847.

Feinstein, A., DeLuca, J., Baune, B. T., Filippi, M., & Lassman, H. (2013). Cognitive and neuropsychiatric disease manifestations in MS. *Multiple Sclerosis and Related Disorders*, *2*, 4–12.

Ferrari, A. J., Charlson, F. J., Norman, R. E., Patten, S. B., Freedman, G., Murray, C. J., et al. (2013). Burden of depressive disorders by country, sex, age, and year: Findings from the global burden of disease study 2010. *PLoS Medicine*, *10*(11), e1001547. https://doi.org/10.1371/journal.pmed.1001547.

Friedman, D., & Devinsky, O. (2015). Cannabinoids in the treatment of epilepsy. *The New England Journal of Medicine*, *373*(11), 1048–1058.

Fu, X., Wang, Y., Wang, C., Wu, H., Li, J., Li, M., & Yang, W. (2018). A mixed treatment comparison on efficacy and safety of treatments for spasticity caused by multiple sclerosis: A systematic review and network meta-analysis. *Clinical Rehabilitation*, *32*(6), 713–721.

Fusar-Poli, P., Crippa, J. A., Bhattacharyya, S., Borgwardt, S. J., Allen, P., Martin-Santos, R., et al. (2009). Distinct effects of {delta}9-tetrahydrocannabinol and cannabidiol on neural activation during emotional processing. *Archives of General Psychiatry*, *66*(1), 95–105.

Greenberg, H. S., Werness, S. A., Pugh, J. E., Andrus, R. O., Anderson, D. J., & Domino, E. F. (1994). Short-term effects of smoking marijuana on balance in patients with multiple sclerosis and normal volunteers. *Clinical Pharmacology and Therapeutics*, *55*, 324–328.

Hartoonian, N., Terrill, A. L., Beier, M. L., Turner, A. P., Day, M. A., Alschuler, K. N., et al. (2015). Predictors of anxiety in multiple sclerosis. *Rehabilitation Psychology*, *60*, 91–98.

Haussleiter, I. S., Brune, M., & Juckel, G. (2009). Psychopathology in multiple sclerosis: Diagnosis, prevalence and treatment. *Therapeutic Advances in Neurological Disorders*, *2*(1), 13–29.

Henry, A., Tourbah, A., Chaunu, M. P., Bakchine, S., & Montreuil, M. (2017). Social cognition abilities in patients with different multiple sclerosis subtypes. *Journal of the International Neuropsychological Society*, *23*, 653–664.

Herkenham, M., Lynn, A. B., Johnson, M. R., Melvin, L. S., de Costa, B. R., & Rice, K. C. (1991). Characterization and localization of cannabinoid receptors in rat brain: A quantitative in vitro autoradiographic study. *The Journal of Neuroscience*, *11*, 563–583.

Hindocha, C., Freeman, T. P., Schafer, G., et al. (2015). Acute effects of delta-9-tetrahydrocannabinol, cannabidiol and their combination on facial emotion recognition: A randomized, double-blind, placebo-controlled study in cannabis users. *European Neuropsychopharmacology*, *25*, 325–334.

Hocevar, K., Ristic, S., & Peterlin, B. (2019). Pharmacogenomic of multiple sclerosis: A systematic review. *Frontiers in Neurology*, *10*, 34.

Huestis, M. A., Solimini, R., Pichini, R., Pacificy, R., Carlier, J., & Busardo, F. F. (2019). Cannabidiol adverse effects and toxicity. *Current Neuropharmacology*, *17*(10), 974–989.

Iffland, K., & Grotenhermen, F. (2017). An update on safety and side effects of cannabidiol: A review of clinical data and relevant animal studies. *Cannabis and Cannabinoid Research*, *2*(1), 139–154.

Iskedjian, M., Bereza, B., Gordon, A., Piwko, C., & Einarson, T. R. (2007). Meta-analysis of cannabis based treatments for neuropathic and multiple sclerosis-related pain. *Current Medical Research and Opinion*, *23*(1), 17–24.

Iuvone, T., Esposito, G., De Filippis, D., et al. (2009). Cannabidiol: A promising drug for neurodegenerative disorders? *CNS Neuroscience & Therapeutics*, *15*, 65–75.

Jones, K. H., Ford, D. V., Jones, P. A., John, A., Middleton, R. M., Lockhart-Jones, H., et al. (2012). A large-scale study of anxiety and depression in people with multiple sclerosis: A survey via the web portal of the UK MS register. *PLoS One*, *7*, 1–10.

Kim, Y. T., Lee, S. W., Kwon, D. H., et al. (2009). Dose-dependent frontal hypometabolism on FDG-PET in methamphetamine abusers. *Journal of Psychiatric Research*, *43*, 1166–1170.

Kindred, H., Honce, J. M., Kwak, J. J., & Rudroff, T. (2018). Multiple sclerosis, cannabis use, and clinical disability: A preliminary [^{18}F]-Fluorodeoxyglucose positron emission tomography study. *Cannabis and Cannabinoid Research*, *3*, 1.

Kindred, J. H., Li, K., Ketelhut, N. B., Proessl, F., Fling, B. W., Honce, J. M., et al. (2017). Cannabis use in people with Parkinson's disease and multiple sclerosis: A web-based investigation. *Complementary Therapies in Medicine*, *33*, 99–104.

Kindred, J. H., Tuulari, J. J., Bucci, M., et al. (2015). Walking speed and brain glucose uptake are uncoupled in patients with multiple sclerosis. *Frontiers in Human Neuroscience*, *9*, 84.

Kister, I., Bacon, T. E., Chamot, E., Salter, A. R., Cutter, G. R., Kalina, J. T., et al. (2013). Natural history of multiple sclerosis symptoms. *International Journal of MS Care*, *15*(3), 146–158.

Klein, T. W. (2005). Cannabinoid-based drugs as anti-inflammatory therapeutics. *Nature Reviews. Immunology*, *5*, 400–411.

Ko, G. D., Bober, S. L., Mindra, S., & Moreau, J. M. (2016). Medical cannabis—The Canadian perspective. *Journal of Pain Research*, *9*, 735–744.

Koppel, B. S., Brust, J. C., Fife, T., et al. (2014). Systematic review: Efficacy and safety of medical marijuana in selected neurologic disorders: Report of the Guideline Development Subcommittee of the American Academy of Neurology. *Neurology*, *82*(17), 1556–1563.

Lakhan, S. E., & Rowland, M. (2019). Whole plant cannabis extracts in the treatment of spasticity in multiple sclerosis: A systematic review. *BMC Neurology*, *9*, 59.

Laprairie, R. B., Bagher, A. M., Kelly, M. E., & Denovan-Wright, E. M. (2015). Cannabidiol is a negative allosteric modulator of the cannabinoid CB1 receptor. *British Journal of Pharmacology*, *172*, 4790–4805.

Lezak, M. D., Howieson, D. B., & Loring, D. W. (2004). *Neuropsychological assessment* (4th ed.). New York, NY: Oxford University Press.

Mackie, K. (2006). Mechanisms of CB1 receptor signaling: Endocannabinoid modulation of synaptic strength. *International Journal of Obesity*, *30*, S19–S23.

Mackie, K. (2008). Cannabinoid receptors: Where they are and what they do. *Journal of Neuroendocrinology*, *20*, 10–14.

Marsicano, G., & Kuner, R. (2008). Anatomical distribution of receptors, ligands and enzymes in the brain and in the spinal cord: Circuitries and neurochemistry. In A. Köfalvi (Ed.), *Cannabinoids and the brain* (pp. 161–201). Boston, MA: Springer US.

Martin-Santos, R., Crippa, J. A., Batalla, A., et al. (2012). Acute effects of a single, oral dose of Δ9-etrahydrocannabinol (THC) and cannabidiol (CBD) administration in healthy volunteers. *Current Pharmaceutical Design*, *18*, 4966–4979.

Merkelbach, S., Konig, J., & Sittinger, H. (2003). Personality traits in multiple sclerosis (MS) patients with and without fatigue experience. *Acta Neurologica Scandinavica*, *107*, 195–201.

Mills, R. J., Yap, L., & Young, C. A. (2007). Treatment for ataxia in multiple sclerosis. *Cochrane Database of Systematic Reviews*, *1*, CD005029.

Minden, S. L., & Schiffer, R. B. (1990). Affective disorders in multiple sclerosis. Review and recommendations for clinical research. *Archives of Neurology*, *47*, 98–104.

Morales, P., Hurst, D. P., & Reggio, P. H. (2017). Molecular targets of the phytocannabinoids—A complex picture. *Progress in the Chemistry of Organic Natural Products*, *103*, 103–131.

Moreno-Lopez, L., Stamatakis, E. A., Fernandez-Serrano, M. J., et al. (2012). Neural correlates of the severity of cocaine, heroin, alcohol, MDMA and cannabis use in polysubstance abusers: A resting-PET brain metabolism study. *PLoS One*, *7*, e39830.

Mullins, L. L., Cote, M. P., Fuemmeler, B. F., Jean, V. M., Beatty, W. W., & Paul, R. H. (2001). Illness intrusiveness, uncertainty, and distress in individuals with multiple sclerosis. *Rehabilitation Psychology*, *46*, 139–153.

NIDA. (2018). *Media guide*. Retrieved from https://archives.drugabuse.gov/publications/mediaguide.

Nielsen, S., Germanos, R., Weier, M., Pollard, J., Degenhardt, L., Hall, W., et al. (2018). The use of cannabis and cannabinoids in treating symptoms of multiple sclerosis: A systematic review of reviews. *Current Neurology and Neuroscience Reports*, *18*, 8.

Ogawa, S., Lee, T. M., Kay, A. R., et al. (1990). Brain magnetic resonance imaging with contrast dependent on blood oxygenation. *Proceedings of the National Academy of Sciences of the United States of America*, *87*, 9868–9872.

Ogawa, S., Tank, D. W., Menon, R., et al. (1992). Intrinsic signal changes accompanying sensory stimulation: Functional brain mapping with magnetic resonance imaging. *Proceedings of the National Academy of Sciences of the United States of America*, *89*, 5951–5955.

Palazuelos, J., Aguado, T., Pazos, M. R., Julien, B., Carrasco, C., Resel, E., et al. (2009). Microglial CB2 cannabinoid receptors are neuroprotective in Huntington's disease excitotoxicity. *Brain*, *132*, 3152–3164.

Patten, S. B., Svenson, L. W., & Metz, L. M. (2005). Descriptive epidemiology of affective disorders in multiple sclerosis. *CNS Spectrums*, *10*, 365–371.

Pavisian, B., MacIntosh, B. J., Szilagyi, G., Staines, R. W., O'Connor, P., & Feinstein, A. (2014). Effects of cannabis on cognition in patients with MS. *Neurology*, *82*, 1879–1887.

Pavisian, B., Staines, R. W., & Feinstein, A. (2015). Cannabis-induced alterations in brain activation during a test of information processing speed in patients with MS. *Multiple Sclerosis Journal - Experimental, Translational and Clinical*, *1*, 1–10.

Rao, S. M. (2004). Cognitive function in patients with multiple sclerosis: Impairment and treatment. *International Journal of MS Care*, *1*, 9–22.

Riera, R., Pacheco, R. L., Bagattini, A. M., & Martimbianco, A. L. C. (2021). Efficacy and safety of therapeutic use of cannabis derivates and their synthetic analogs: Overview of systematic reviews. *Phytotherapy Research*, 1–17.

Rocca, M. A., et al. (2015). Clinical and imaging assessment of cognitive dysfunction in multiple sclerosis. *Lancet Neurology*, *14*, 302–317.

Roelcke, U., Kappos, L., Lechner-Scott, J., et al. (1997). Reduced glucose metabolism in the frontal cortex and basal ganglia of multiple sclerosis patients with fatigue: A ^{18}F-fluorodeoxyglucose positron emission tomography study. *Neurology*, *48*, 1566–1571.

Rog, D. J., Nurmikko, T. J., Friede, T., & Young, C. A. (2005). Randomized, controlled trial of cannabis-based medicine in central pain in multiple sclerosis. *Neurology*, *65*, 812–819.

Romero, K., Pavisian, B., Staines, W. R., & Feinstein, A. (2015). Multiple sclerosis, cannabis, and cognition: A structural MRI study. *Neuroimage: Clinical*, *8*, 140–147.

Rovare, V. P., Magalhaes, G. P. A., Jardini, G. D. A., Beraldo, M. L., Gameiro, M. O., Agarwal, A., & El Dib, R. (2017). Cannabinoids for spasticity due to multiple sclerosis or paraplegia: A systematic review and meta-analysis of randomized clinical trials. *Complementary Therapies in Medicine*, *34*, 170–185.

Rudroff, T. (2019). Cannabis for neuropathic pain in multiple sclerosis—High expectations, poor data. *Frontiers in Pharmacology*, *10*, 1239.

Rudroff, T., & Honce, J. M. (2017). Cannabis and multiple sclerosis—The way forward. *Frontiers in Neurology*, *8*, 299.

Rudroff, T., Workman, C. D., Fietsam, A. C., & Ponto, A. C. (2020). Imaging transcranial direct current stimulation (tDCS) with positron emission tomography (PET). *Brain Sciences*, *10*(4), 236.

Russo, E., & Guy, G. W. (2006). A tale of two cannabinoids: The therapeutic rationale for combining tetrahydrocannabinol and cannabidiol. *Medical Hypotheses*, *66*, 234–246.

Sadovnick, A. D., Remick, R. A., Allen, J., Swartz, E., Yee, I. M., Eisen, K., et al. (1996). Depression and multiple sclerosis. *Neurology*, *46*, 628–632.

Saito, V. M., Rezende, R. M., & Teixeira, A. L. (2012). Cannabinoid modulation of neuroinflammatory disorders. *Current Neuropharmacology*, *10*, 159–166.

Schreiner, A. M., & Dunn, M. E. (2012). Residual effects of cannabis use on neurocognitive performance after prolonged abstinence: A meta-analysis. *Experimental and Clinical Psychopharmacology*, *20*(5), 420–429. https://doi.org/10.1037/a0029117.

Shakespeare, D. T., Boggild, M., & Young, C. (2003). Anti-spasticity agents for multiple sclerosis. *Cochrane Database of Systematic Reviews*, *4*, CD001332.

Simmons, R. D. (2010). Life issues in multiple sclerosis. *Nature Reviews. Neurology*, *6*, 603–610.

Smith, D., Pernet, A., Hallett, W. A., et al. (2003). Lactate: A preferred fuel for human brain metabolism in vivo. *Journal of Cerebral Blood Flow and Metabolism*, *23*, 658–664.

Thomas, A., Baillie, G. L., Phillips, A. M., Razdan, R. K., Ross, R. A., & Pertwee, R. G. (2007). Cannabidiol displays unexpectedly high potency as an antagonist of CB1 and CB2 receptor agonists in vitro. *British Journal of Pharmacology*, *150*, 613–623.

Torres-Moreno, M. C., Papaseit, E., Torrens, M., & Farré, M. (2018). Assessment of efficacy and tolerability of medicinal cannabinoids in patients with multiple sclerosis: A systematic review and meta-analysis. *JAMA Network Open*, *1*(6), e183485.

Viner, R., Fiest, K. M., Bulloch, A. G., Williams, J. V., Lavorato, D. H., Berzins, S., et al. (2014). Point prevalence and correlates of depression in a national community sample with multiple sclerosis. *General Hospital Psychiatry*, *36*, 352–354.

Volkow, N. D., Gillespie, H., Mullani, N., et al. (1996). Brain glucose metabolism in chronic marijuana users at baseline and during marijuana intoxication. *Psychiatry Research*, *167*, 29–38.

Volkow, N. D., Wang, G. J., Shokri Kojori, E., et al. (2015). Alcohol decreases baseline brain glucose metabolism more in heavy drinkers than controls but has no effect on stimulation-induced metabolic increases. *The Journal of Neuroscience*, *35*, 3248–3255.

Wallin, M. T., Wilken, J. A., & Kane, R. (2006). Cognitive dysfunction in multiple sclerosis: Assessment, imaging and risk factors. *Journal of Rehabilitation Research and Development*, *43*(1), 63–72.

Walsh, K. B., McKinney, A. E., & Holmes, A. E. (2021). Minor cannabinoids: Biosynthesis, molecular pharmacology and potential therapeutic uses. *Frontiers in Pharmacology*, *12*, 777804.

White, C. M. (2019). Review of human studies assessing cannabidiol's (CBD) therapeutic actions and potential. *Journal of Clinical Pharmacology*, *313* (24), 1–12.

Wilsey, B., Marcotte, T., Tsodikov, A., Millman, T. A., Bentley, H., & Gouaux, B. (2008). A randomized, placebo-controlled, crossover trial of cannabis cigarettes in neuropathic pain. *The Journal of Pain*, *9*(6), 506–521.

Wood, B., van der Mei, I. A., Ponsonby, A. L., Pittas, F., Quinn, S., Dwyer, T., et al. (2012). Prevalence and concurrence of anxiety, depression and fatigue over time in multiple sclerosis. *Multiple Sclerosis*, *J19*, 217–224.

Workman, C. D., Kindred, J. H., Boles Ponto, L. L., Kamholz, J., & Rudroff, T. (2020). The effects of chronic Δ-9-tetrahydrocannabinol (THC) and cannabidiol (CBD) use on cerebral glucose metabolism in multiple sclerosis: A pilot study. *Applied Physiology, Nutrition, and Metabolism*, *45*, 450–452.

Wycoff, A. M., Metrik, J., & Trull, T. J. (2018). Affect and cannabis use in daily life: A review and recommendations for future research. *Drug and Alcohol Dependence*, *191*, 223–233.

Zajicek, J. P., Fox, P., Sanders, H., et al. (2003). Cannabinoids for treatment of spasticity and other symptoms related to multiple sclerosis (CAMS study): Multicentre randomised placebo-controlled trial. *Lancet*, *362*, 1517–1526.

Zajicek, J. P., Hobart, J. C., Slade, A., Barnes, D., Mattison, P. G., & MUSEC Research Group. (2012). Multiple sclerosis and extract of cannabis: Results of the MUSEC trial. *Journal of Neurology, Neurosurgery, and Psychiatry*, *83*, 1125–1132.

Zimmer, A., Zimmer, A. M., Hohmann, A. G., Herkenham, M., & Bonner, T. I. (1999). Increased mortality, hypoactivity and hypoalgesia in cannabinoid CB1 receptor knockout mice. *Proceedings of the National Academy of Sciences of the United States of America*, *96*, 5780–5785.

Chapter 22

Genetic variation and acute responses to cannabis

Zoe Bourgault[a,b], Justin Matheson[a], and Bernard Le Foll[a,b,c,d,e,f]

[a]*Translational Addiction Research Laboratory, Campbell Family Mental Health Research Institute, Centre for Addiction and Mental Health, Toronto, ON, Canada,* [b]*Department of Pharmacology and Toxicology, University of Toronto, Toronto, ON, Canada,* [c]*Department of Psychiatry, Faculty of Medicine, University of Toronto, Toronto, ON, Canada,* [d]*Department of Family and Community Medicine, University of Toronto, Toronto, ON, Canada,* [e]*Institute of Medical Science, University of Toronto, Toronto, ON, Canada,* [f]*Waypoint Research Institute, Waypoint Centre for Mental Health Care, Penetanguishene, ON, Canada*

Abbreviations

2-AG	2-arachidonoylglycerol
AEA	*N*-arachidonoylethanolamide (anandamide)
ATP	adenosine triphosphate
CAPE	community assessment of psychic experiences
CB1	cannabinoid receptor type 1
CB2	cannabinoid receptor type 2
CEQ	cannabis experiences questionnaire
COMT	catechol-o-methyltransferase
CUD	cannabis use disorder
DAGL	diacylglycerol lipase
FAAH	fatty acid amide hydrolase
GABA	gamma-aminobutyric acid
GPCR	G-protein coupled receptor
GWAS	genome wide association study
MAGL	monoacylglycerol lipase
NAc	nucleus accumbens
PANSS	positive and negative syndrome scale
POMS	profile of mood states
SNP	single nucleotide polymorphism
THC	delta-9 tetrahydrocannabinol
UTR	untranslated region
VAS	visual analog scale
VNTR	variable number of tandem repeats
VTA	ventral tegmental area

Introduction

The use of cannabis has become increasingly prevalent in North America over the past decade, with an estimated 12.4% of the population reporting past-year use in 2018 (Connor et al., 2021). Cannabis produces a wide range of effects on mood, cognition, and perception. Positive subjective responses to cannabis can include enhanced affect, euphoria, and sedation, which motivate further consumption and underlie its misuse potential (Drennan et al., 2021). In addition to rewarding effects, people who use cannabis may experience anxiety, paranoia, or psychosis during intoxication (Mason et al., 2009). While anxiogenic and psychotomimetic responses are typically associated with high doses, certain individuals appear to be at a higher risk of experiencing them at any dose and at higher intensities (Okey & Meier, 2020). The acute effects of THC also involve cognitive impairments such deficits in executive function and reduced psychomotor control (Broyd et al., 2016). Sensitivity to cannabis and the nature of subjective experiences vary across individuals.

Cannabis Use, Neurobiology, Psychology, and Treatment. https://doi.org/10.1016/B978-0-323-89862-1.00045-3

The heritability of responses was initially determined by twin studies, which estimated that approximately 25% of the variance in subjective effects is genetically mediated (Lyons et al., 1997).

A growing body of research has focused on identifying genes associated with cannabis use and acute responses. Understanding the genetic factors that underlie response variability has significant implications, as subjective drug effects can predict clinical phenotypes such as abuse liability and risk of substance use disorder. For instance, greater rewarding effects of THC are associated with heavier use and progression to dependence (Le Strat et al., 2009). The etiology of cannabis use disorder (CUD) likely involves complex interactions between a large number of genes and environmental factors. Consequently, genomic strategies have had limited success in identifying specific risk alleles associated with the diagnostic category (Agrawal & Lynskey, 2009; Palmer et al., 2019). Instead, laboratory studies measuring responses to acute drug administration may provide more insight on the genetic basis of cannabis use and related problems. As such, in this review, we discuss current knowledge obtained from laboratory studies on genetic variants found to mediate acute responses to cannabis. We also review physiologically relevant candidate genes that warrant investigation in future research.

Genetics of the endocannabinoid system

The major pharmacological targets of cannabinoids are two inhibitory G-protein-coupled receptors (GPCRs) within the endocannabinoid system. The cannabinoid receptor type 1 (CB1) is encoded by the *CNR1* gene (Fig. 1), located on chromosome 6q14–15, and is one of the most abundantly expressed GPCRs in the central nervous system (Hartman et al., 2009; Lu & Mackie, 2016). Its activation by THC mediates the psychoactive effects of cannabis (Onaivi et al., 2002). The *CNR2* gene located on chromosome 1p3611 encodes the cannabinoid receptor type 2 (CB2). CB2 is expressed in the immune cells of the peripheral nervous system and to a lesser extent in the astroglia of the brain. It appears to mediate the anti-inflammatory properties of cannabinoids rather than their psychoactive effects (Mechoulam & Parker, 2013). In the central nervous system, levels the of endogenous cannabinoid receptor ligands *N*-arachidonoyletholamine (AEA) and 2-arachidonoylglycerol (2-AG) influence CB1 receptor availability and are tightly regulated by three major enzymes. 2-AG is synthesized by DAGL and inactivated by MAGL (Basavarajappa, 2007). Fatty acid amide hydrolase (FAAH) terminates the action of AEA (Lu & Mackie, 2016). These enzymes are encoded by the *DAGLA*, *MGLL*, and *FAAH* genes, respectively (Basavarajappa, 2007; Bisogno et al., 2003; López-Moreno et al., 2012). The synthesis of AEA is more complex and involves multiple enzymatic steps (Lu & Mackie, 2016).

Behavioral pharmacology of cannabinoids

CB1 is expressed in areas of the reward system such as the ventral tegmental area (VTA) and nucleus accumbens (NAc), where its activation by endogenous or exogenous cannabinoids modulates the activity of the major neurotransmitter systems (e.g., GABA, glutamate, dopamine) (Lu & Mackie, 2016). The rewarding properties of exogenous cannabinoids are mediated by their effects on dopamine transmission (Wenzel & Cheer, 2018). In the striatum, dopaminergic signaling functions as a prediction-error system that controls the cognitive, affective, and motor processes associated with reward-driven behavior (Salamone & Correa, 2002; Schultz et al., 1997). CB1 activation enhances extracellular dopamine levels in the VTA and NAc (Tanda et al., 1997). This occurs through an indirect mechanism involving the GABAergic and glutamatergic inputs (Wenzel & Cheer, 2018). Afferent GABAergic neurons exert tonic inhibitory control over VTA dopamine cells, which is relieved by presynaptic CB1 activation (Cheer et al., 2000; Maldonado et al., 2006) (Fig. 1). Exogenous cannabinoid administration disinhibits dopamine cells in the reward system thereby mediating reward and associated behaviors (Wenzel & Cheer, 2018).

The pharmacological effects of cannabinoids are further modulated by the endogenous opioid system due to functional interactions between the endocannabinoid system and opioid receptors (Wenzel & Cheer, 2018). The mu opioid receptor is of particular interest due to its co-localization with CB1 receptors in the reward system where it is thought to facilitate dopamine release in response to cannabinoids (Wenzel & Cheer, 2018). Mu opioid receptor antagonism has been shown to attenuate the rewarding effects of CB1 activation in primates (Justinova et al., 2004). Variability in endogenous opioid function may therefore affect subjective responses to cannabis.

A large body of epidemiological and clinical research has demonstrated an association between cannabis use and increased risk of psychotic disorders (Kuepper et al., 2010; Moore et al., 2007). Acute cannabis exposure is known to induce transient psychotic responses in certain individuals during and/or following intoxication (Bhattacharyya et al., 2009; Mason et al., 2009). The presence of psychosis-related experiences in only a subset of people who use cannabis suggests that they are caused by genetic predisposition. Given that increases in synaptic dopamine transmission are common to both cannabis intoxication and psychosis, genetic factors influencing dopaminergic function may affect individual vulnerability to psychotomimetic responses to THC (Bhattacharyya et al., 2009).

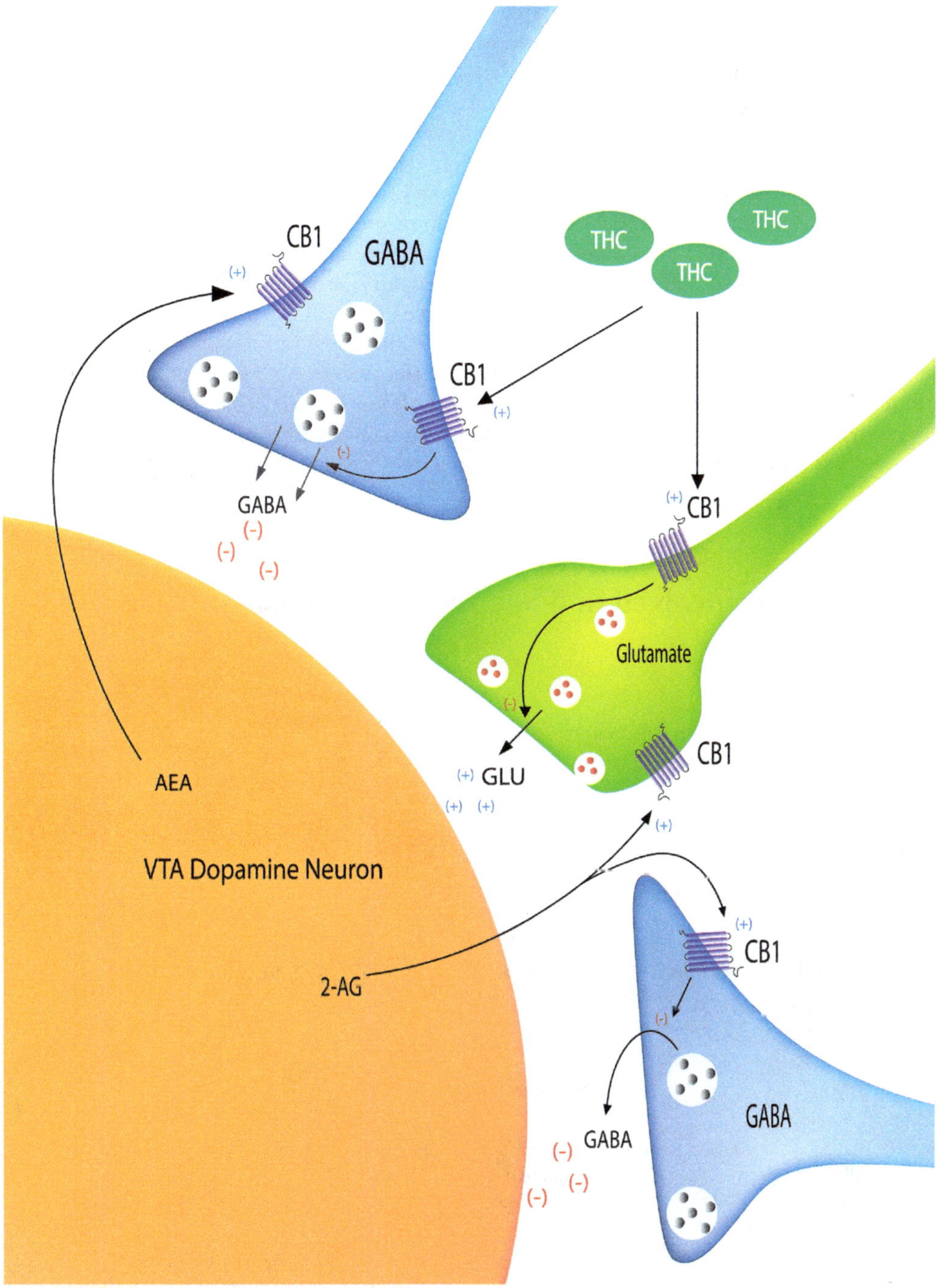

FIG. 1 THC mechanism of action in the reward system. THC binds presynaptic CB1 receptors at GABAergic and glutamatergic terminals in the VTA, which inhibits their signaling to dopaminergic neurons. This also disrupts the activity of the endocannabinoids AEA and 2-AG.

Experimental measures

Acute drug effects

Laboratory studies investigating the acute effects of cannabis generally use within-subjects, placebo-controlled, double-blind study designs. Participants are administered the drug in laboratory and subsequently assessed for behavioral and physiological responses. Validated instruments used to measure subjective effects can consist of clinician-interviewed or self-report questionnaires and scales (Carter & Griffiths, 2009). For example, the Profile of Mood States (POMS) is a widely used self-report measure that can capture changes in subjective and affective states (McNair et al., 1971).

Although the scale was not specifically developed to capture acute drug reactions, scores can be used to characterize subjective responses and quantify their intensity. Drug-related experiences such as subjective high and drug-liking are often measured using visual analog scales (VAS). Participants indicate the intensity of VAS items on a line scale typically ranging from 0 to 100 points (Fischman & Foltin, 1991).

The Positive and Negative Syndrome Scale (PANSS) was developed to measure both positive and negative symptoms in patients with schizophrenia (Kay et al., 1987). In laboratory studies, it is often adapted to measure only the intensity of positive symptoms as an indicator of psychotomimetic reactions to THC in both healthy and clinical populations. The Community Assessment of Psychic Experiences (CAPE) is a self-report questionnaire that assesses psychosis-related experiences. It can also be used to determine pre-existing psychosis liability and identify at-risk individuals (Mossaheb et al., 2012). Finally, the Psychotomimetic States Inventory was specifically designed to measure psychotomimetic responses associated with the use of cannabis and ketamine. It includes subscales of Delusory Thinking, Perceptual Distortions, Cognitive Disorganization, Anhedonia, and Paranoia (Mason et al., 2008).

Impairment of cognition is one of the most consistent adverse effects associated with acute cannabis intoxication (Broyd et al., 2016), and thus many laboratory studies of acute cannabis effects include measures of cognitive performance. The most consistently reported cognitive effect of cannabis is an impairment of verbal recall performance, though cannabis has effects on most cognitive domains, including attention, memory, psychomotor control, and executive functioning (Dellazizzo et al., 2022; Matheson & Le Foll, 2022).

Genetic polymorphisms involved in acute responses to cannabis

Subjective effects

CNR1

A small number of genetic association studies have investigated the effects of *CNR1* polymorphisms on subjective responses to THC. Findings from genetic association studies on acute subjective effects of cannabis are summarized in Table 1.

TABLE 1 Genetic association studies of acute subjective responses to cannabis.

Gene	Polymorphisms	Sample	Subjective measures	Main findings	Reference
CNR1	Rs1049353	Adults who use cannabis (*n*=48)	State satiety assessed by the Bodily Symptoms Scale	C/C genotype associated with greater state satiety following THC exposure	Hindocha, Freeman, et al. (2020), Hindocha, Quattrone, et al. (2020)
	Rs1049353 Rs2023239	Adults who use cannabis (*n*=52)	VAS of subjective effects and POMS	Rs1049353 T allele associated with elevated subjective effects Rs2023239 C allele associated with elevated subjective effects	Murphy et al. (2021)
	Rs2023239	Adults who use cannabis (*n*=118)	POMS	Elevated subjective effects in C allele carriers	Palmer et al. (2019)
FAAH	Rs32440 (C385A)	Adults who use cannabis (*n*=37)	POMS	C/C genotype associated with nonsignificant trend for increased happiness following THC exposure	Schacht et al. (2009)
OPRM1	Rs510769 Rs1799971	Adults who use cannabis (*n*=52)	VAS of subjective drug effects	Elevated positive subjective effects in rs510769 T-allele carriers. No effect of rs1799971	Bourgault et al. (2022)

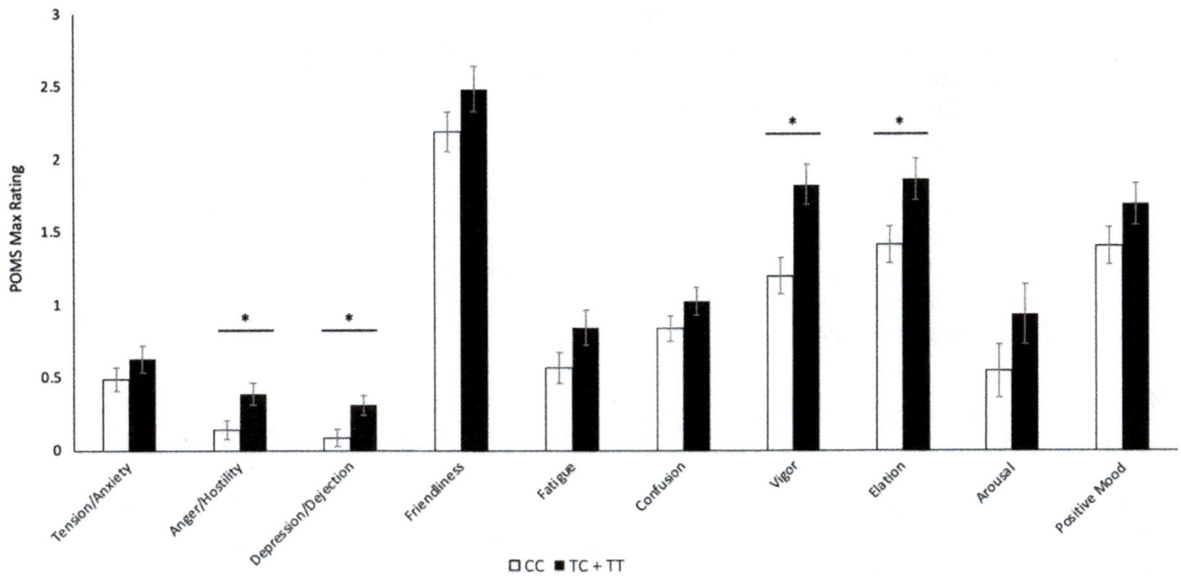

FIG. 2 *CNR1* rs1049353 aggregate profile of mood states maximum values. Unadjusted maximum Profile of Mood States (POMS) questionnaire results (measured on a scale of 0–4), measured at baseline and 1 h post-cannabis exposure of $n = 52$ subjects, separated by rs1049353 genotype. *(From Murphy, T., Matheson, J., Mann, R. E., Brands, B., Wickens, C. M., Tiwari, A. K., ... Le Foll, B. (2021). Influence of cannabinoid receptor 1 genetic variants on the subjective effects of smoked cannabis.* International Journal of Molecular Sciences, 22 *(14). https://doi.org/10.3390/ijms22147388.)*

The rs1049353 SNP is a silent mutation in *CNR1* resulting in a C/T substitution within the coding region of the gene. The C/C genotype has been associated with higher rates of cannabis consumption and symptoms of dependence (Gerra et al., 2018; Hartman et al., 2009). Work from our group (Murphy et al., 2021) recently found elevated subjective effects of smoked cannabis in T-allele carriers compared to C/C homozygotes. This included significantly higher ratings on a visual analog scale of drug-liking, as well as higher scores on the *Vigor*, *Depression/Dejection*, and *Anger/Hostility* dimensions of the POMS (Fig. 2). The T-allele was also associated with longer smoking duration during the ad libitum THC exposure session, and consequently higher blood-THC levels. This may indicate a compensatory increase in consumption in this genotype group resulting from reduced rewarding effects. Lower sensitivity to THC in T-allele carriers may also explain why the C/C genotype has been associated with greater risk of cannabis use and dependence. This is further supported by findings reported by Hindocha, Freeman, et al. (2020), Hindocha, Quattrone, et al. (2020). Following the administration of an oral THC capsule, which was dosed equally across participants, those carrying the C/C genotype exhibited greater state satiety and reduced attentional bias toward THC cues compared to T-allele carriers. The reduction in motivational salience of THC cues accompanied by satiety responses observed in C-allele homozygotes could be indicative of greater subjective reward in this group.

Another *CNR1* variant of interest is rs2023239: a noncoding T/C substitution located in intron 2. The minor C allele has been associated with increased CB1 expression and ligand binding (Hirvonen et al., 2013; Hutchison et al., 2008; Ketcherside et al., 2017). Cannabis-related phenotypes associated with the C-allele include increased cannabis consumption, increased severity of withdrawal symptoms, and heightened neural activation in the reward system following exposure to THC-related cues (Filbey et al., 2010; Haughey et al., 2008). There is recent evidence supporting a role of rs2023239 in subjective responses to cannabis. Elevations in the *Anger/hostility*, *Fatigue*, *Tension/Anxiety*, and *Vigor* dimensions of the POMS have been reported in C-allele carriers compared to those with the T/T genotype (Palmer et al., 2019). This effect on POMS items was not observed in a subsequent study; however, increased subjective responses assessed by visual analog scales were reported. The intensity of the statements *I like the drug* and *It feels like cannabis* in response to cannabis exposure were significantly higher in C-allele carriers (Murphy et al., 2021). Taken together, these findings indicate greater subjective effects of THC in C-allele carriers, which may be either rewarding or adverse in nature. This is consistent with the higher CB1 expression and behavioral phenotypes found in carriers of the variant.

FAAH

The majority of studies investigating variation in the *FAAH* gene have focused on the rs32440 (C385A) polymorphism. The C/A single nucleotide substitution results in a proline to threonine exchange in exon 3 (López-Moreno et al., 2012). The enzymatic activity of the mutant protein is reduced by 30%, resulting in elevated levels of its endocannabinoid substrate

AEA. The effects of C385A on subjective responses to cannabis have been investigated by one study that found no significant differences in POMS scores between A-allele carriers and C-allele homozygotes (Schacht et al., 2009). However, genotype was associated with a trend in happiness ratings over time. After controlling for baseline scores, individuals with the C/C genotype reported increased happiness following cannabis exposure. Interestingly, the opposite trend was seen in A-allele carriers, who reported decreasing levels of happiness. The A-allele has also been associated with greater attentional bias toward THC stimuli following exposure, as well as reduced severity of withdrawal symptoms during abstinence (Hindocha, Freeman, et al., 2020; Hindocha, Quattrone, et al., 2020). These findings suggest that the C-allele is associated with both increased positive responses during intoxication and worsened affect during withdrawal. Lower CB1 receptor availability resulting from the heightened endocannabinoid tone associated with the A allele may reduce sensitivity to exogenous cannabinoid administration. Increased endocannabinoid signaling may also counteract withdrawal symptoms, thus improving affect during abstinence.

OPRM1

We recently found that genetic variation in the mu opioid receptor, encoded by *OPRM1* (chromosome 6q24-q25), was associated with subjective responses to smoked cannabis (Bourgault et al., 2022). *OPRM1* rs510769 is a noncoding C/T substitution located in intron 1. The minor T allele was associated with increased sensitivity to positive subjective responses to cannabis assessed by visual analog scales (Fig. 3). T-allele carriers also exhibited higher blood THC levels despite consuming amounts of THC comparable to C/C individuals. The effect of this SNP may thus be mediated by pharmacokinetic factors. Rs1799971, a widely studied functional SNP in *OPRM1* associated with substance dependence, was found to have no effect on subjective responses to cannabis (Bourgault et al., 2022).

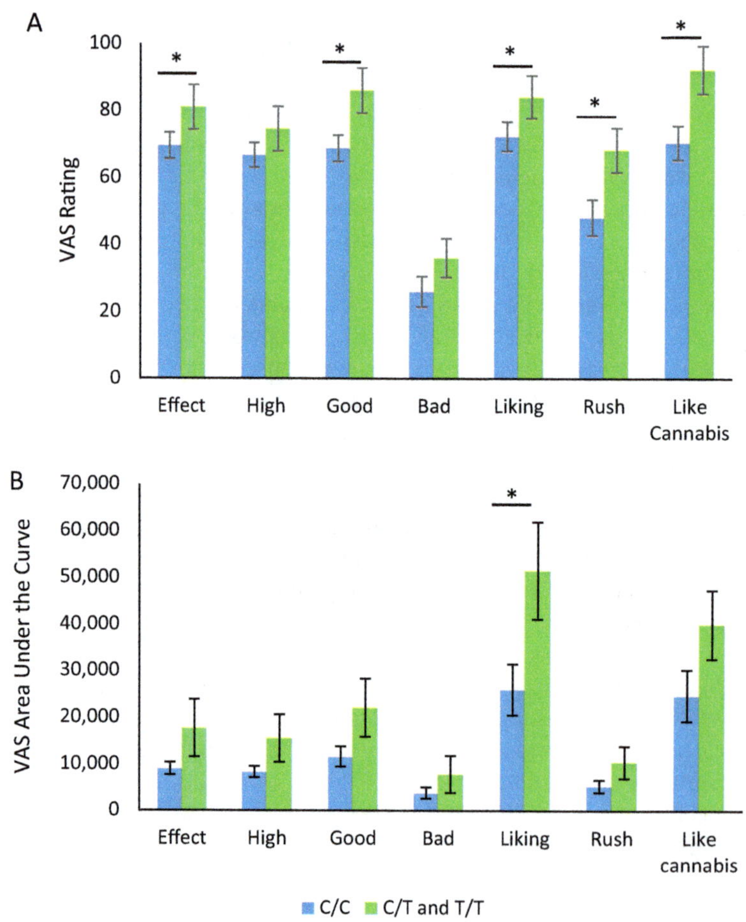

FIG. 3 Effects of rs510769 genotype on subjective responses to cannabis measured by visual analog scales. Data expressed as mean ± SEM of (A) maximum VAS ratings by rs510769 genotype (C/C: $N=38$; C/T and T/T: $N=13$) and (B) area under the curve of VAS ratings over time (C/C: $N=34$; C/T and T/T: $N=11$). C/C genotype compared to C/T and T/T genotypes using independent sample *t*-tests. *(From Bourgault, Z., Matheson, J., Mann, R. E., Brands, B., Wickens, C. M., Tiwari, A. K., ... Le Foll, B. (2022). Mu opioid receptor gene variant modulates subjective response to smoked cannabis.* American Journal of Translational Research, 14*(1), 623–632.)*

Psychotomimetic effects

Hyperdopaminergic signaling in the striatum is a known physiological characteristic of psychotic disorders (Bhattacharyya et al., 2009). Similarly, acute THC exposure is thought to exert its subjective effects via the indirect augmentation of striatal dopamine (Bloomfield et al., 2016; Kuepper et al., 2010). Polymorphisms related to dopamine regulation and signaling have therefore been investigated as candidate genes for psychotomimetic responses to THC. Findings from these studies are summarized in Table 2.

TABLE 2 Genetic association studies of acute psychotomimetic responses to cannabis.

Gene	Polymorphisms	Sample	Psychotomimetic measures	Main findings	Reference
COMT	Rs4680 (Val158Met)	Healthy participants with past-month paranoia experience ($n=78$)	CAPE positive dimension	No significant effect	Tunbridge et al. (2015)
	Rs4680 (Val158Met)	Healthy participants ($n=74$)	PANSS	No significant effect	Ranganathan et al. (2019)
	Rs4680 (Val158Met)	First episode psychosis patients ($n=143$), young adults who use cannabis ($n=485$), and aged-matched controls ($n=92$)	Cannabis experiences questionnaire (CEQ)	No significant effect	Hindocha, Freeman, et al. (2020), Hindocha, Quattrone, et al. (2020)
	Rs4680 (Val158Met)	Patients with a psychotic disorder diagnosis ($n=30$), relatives of patients with a psychotic disorder ($n=12$), and healthy controls ($n=32$)	CAPE-positive dimension and CAPE-trait	No overall effect of genotype Val allele associated with elevated psychotomimetic responses in individuals with high CAPE-trait ratings	Henquet et al. (2006)
AKT1	Rs2494732	People who use cannabis ($n=442$)	Psychotomimetic States Inventory	C allele associated with elevated psychotomimetic effects	Morgan et al. (2016)
	Rs2494732	First-episode psychosis patients ($n=143$), young adults who use cannabis ($n=485$), and aged-matched controls ($n=92$)	Cannabis Experiences Questionnaire	No significant effect	Hindocha, Freeman, et al. (2020), Hindocha, Quattrone, et al. (2020)
	Rs1130223	Healthy volunteers with minimal previous cannabis exposure ($n=35$)	PANSS	G/G genotype associated with nonsignificant trend for increased psychotic symptoms	Bhattacharyya et al. (2012)
DAT1	3′UTR VNTR		PANSS	9-repeat allele associated with increased psychotic symptoms	Bhattacharyya et al. (2012)
CNR1	rs202329, rs1535255, rs806379 haplotype block	Students who had used cannabis at least once ($n=544$)	Cannabis Subjective Experience Questionnaire	AAA haplotype associated with reduced psychotomimetic effects	Krebs et al. (2014)

COMT

Catechol-*O*-methyltransferase, encoded by *COMT*, is an enzyme involved in the extracellular breakdown of prefrontal dopamine. There is significant variability in the activity of COMT, the majority of which results from the rs4680 (Val158Met) polymorphism (Farrell et al., 2012). The Val allele is associated with increased COMT expression in the prefrontal cortex, consequently reducing synaptic dopamine availability in this area (Mier et al., 2010). Due to the inverse relationship between prefrontal and striatal dopamine levels, Val allele carriers also exhibit elevated striatal dopamine levels, which may increase susceptibility to psychosis.

Overall, laboratory studies investigating the effects of Val158Met on cannabis-induced psychotic experiences have found no direct association between the polymorphism and behavioral outcomes (Hindocha, Freeman, et al., 2020; Hindocha, Quattrone, et al., 2020; Ranganathan et al., 2019; Tunbridge et al., 2015). However, one study demonstrated a potential influence of the variant in the presence of additional risk factors. Henquet et al. (2006) used an adapted version of the CAPE (community assessment of psychotic experiences) to examine acute responses to THC as well as pre-existing psychosis liability. CAPE-trait ratings measured baseline predisposition to psychosis, whereas an adapted version of the CAPE-positive dimension checklist captured transient psychotic responses to drug administration. Val158Met genotype did not affect acute CAPE scores following THC exposure. However, in individuals with high CAPE-trait ratings, cannabis was associated with elevated psychotomimetic responses in Val allele carriers. These findings indicate that vulnerability to cannabis-induced psychosis is likely polygenic and implicate the Val158Met polymorphism as one of multiple causative factors.

AKT1

The AKT1 protein, encoded by *AKT1* on chromosome 14q32, is a member of the protein kinase B family. It acts as a serine/threonine kinase with widespread physiological functions. In the central nervous system, AKT1 has an important role in the dopamine signaling cascade, acting through the beta-arrestin pathway downstream of the D2 receptor (Ozaita et al., 2007). Loss of function of AKT1 is associated with altered responses to D2 receptor stimulation (Arguello & Gogos, 2008). THC activates AKT1, thereby affecting dopaminergic signaling independently of D2 receptor stimulation (Ozaita et al., 2007).

The *AKT1* rs2494732 polymorphism has been implicated in a gene X environment interaction that associates cannabis use with increased risk developing of psychotic disorders in individuals with the C/C genotype. One study has investigated the impact of rs2494732 on acute psychotic responses to cannabis (Morgan et al., 2016). The C allele was associated with increased psychotic symptoms during intoxication assessed by the Psychotomimetic States Inventory. Hindocha and colleagues (Hindocha, Freeman, et al., 2020; Hindocha, Quattrone, et al., 2020) also examined the interaction between *AKT1* genotype and cannabis. In contrast to Morgan et al. and previous case-control studies linking *AKT1* with cannabis use and psychotic disorders, genotype did not affect psychotic experiences. The contrasting results may be explained by methodological differences between studies. Psychotomimetic symptoms were assessed using the Cannabis Experiences Questionnaire (CEQ), a retrospective self-report of lifetime experiences, rather than during acute intoxication.

Rs1130223 is a silent mutation in AKT1, causing a synonymous single nucleotide exchange (G/A). The A-allele has been associated with reduced AKT1 expression (Blest-Hopley et al., 2021). In a sample of healthy individuals, exposure to THC was associated with a nonsignificant trend of increased psychotic symptoms assessed by the PANSS in G allele homozygotes (Bhattacharyya et al., 2012).

DAT1

The dopamine transporter, encoded by *DAT1*, mediates the reuptake of extracellular dopamine into the presynaptic cell. The 3′ untranslated region (UTR) of *DAT1* contains a polymorphic variable number of tandem repeats (VNTR). The most common alleles are 9 and 10-repeat polymorphisms (Bhattacharyya et al., 2012). The 9-repeat allele is associated with reduced DAT1 expression (Fuke et al., 2001) and increased extracellular dopamine in the striatum (Heinz et al., 2000), which has been associated with cannabis-induced psychosis. Compared to 10-repeat homozygotes, individuals carrying the 9-repeat allele report significantly elevated psychotic symptoms assessed by the PANSS in response to cannabis exposure (Bhattacharyya et al., 2012). In addition, a significant interaction was found between *DAT1* 3′UTR VNTR and *AKT1* rs1130233 genotypes. Psychotomimetic effects of cannabis were greater in individuals carrying both polymorphisms compared to those carrying only one. This suggests that polymorphisms associated with hyperdopaminergic signaling in the striatum may have cumulative effects on susceptibility to cannabis-induced psychosis and related reactions.

CNR1

A haplotype block comprises of *CNR1* rs202329, rs1535255, and rs806379 has been associated with the psychotomimetic effects of cannabis (Krebs et al., 2014). Individuals with the AAA haplotype reported reduced psychotomimetic effects compared to all other haplotypes. Ratings of psychotomimetic experiences were obtained with a retrospective self-report questionnaire rather than during acute intoxication.

Cognitive effects

AKT1

Only a single study has investigated the role of AKT1 genotype in moderating acute cognitive effects of cannabis. Morgan et al. (2016) did not find a significant effect of the rs2494732 polymorphism on spatial working memory performance (using the spatial n-back task) after naturalistic smoking of cannabis in a large sample ($n=442$) of young people (aged 16 to 23 years).

CNR1

Hindocha, Freeman, et al. (2020), Hindocha, Quattrone, et al. (2020) examined associations between two CNR1 polymorphisms (rs1049353 and rs806378) and salience of appetitive cues (both cannabis- and food-related cues) in a sample of 48 healthy volunteers. The rs1049353 G/G genotype group showed reduced cue salience after exposure to vaporized THC, while A-allele carriers did not vary in cue salience after exposure, suggesting increased vulnerability to cannabis misuse (i.e., since their attention bias to appetitive cannabis cues does not decrease even while intoxicated). The rs806378 C/C genotype group showed greater cue salience compared to T-allele carriers.

COMT

Five studies have evaluated the role of the COMT Val158Met polymorphism in moderating acute cognitive effects of cannabis. In a mixed sample of patients with a psychotic disorder ($n=30$), their relatives ($n=12$), and a group of healthy volunteers ($n=32$), Henquet et al. (2006) found poorer performance in delayed recognition in a visual verbal learning task and increased reaction time in a sustained attention task (continuous performance task) in Val/Val carriers after smoking cannabis. Similarly, Tunbridge et al. (2015) found an association between the Val/Val genotype and impaired performance in working memory performance (Digit Span Backwards task) after infusion of intravenous THC in a sample of 78 healthy participants who were vulnerable to paranoia. Ranganathan et al. (2019) also found a significant association between Val/Val genotype and poorer performance in an attention task (Rapid Visual Information Processing test) after infusion of intravenous THC in a sample of 74 healthy volunteers. In contrast, two studies failed to find a significant association between COMT genotype and cognitive effects after acute exposure to cannabis. In line with their AKT1 results, Morgan et al. (2016) failed to find a significant association between COMT Val158Met genotype and working memory performance during cannabis intoxication. Spronk et al. (2016) also failed to find an association between COMT Val158Met genotype and performance in a reversal learning task after vaporized cannabis.

DRD2

In addition to the COMT Val158Met polymorphism, Spronk et al. (2016) also examined the DRD2 Taq1A polymorphism (rs1800497) and found that genotype did not influence performance in the reversal learning task after vaporized cannabis.

FAAH

Hindocha, Freeman, et al. (2020), Hindocha, Quattrone, et al. (2020) found that FAAH rs324420 A-allele carriers demonstrated a greater attentional bias to appetitive cues after exposure to vaporized THC.

Findings from genetic association studies on the acute cognitive effects of cannabis are summarized in Table 3.

TABLE 3 Genetic association studies of acute cognitive effects of cannabis.

Gene	Polymorphisms	Sample	Cognitive tasks	Main findings	Reference
AKT1	Rs2494732	People who use cannabis aged 16–23 years (n=442)	Spatial n-back	No significant effect	Morgan et al. (2016)
CNR1	Rs1049353 and rs806378	Adults who use cannabis (n=48)	Dot-probe task (salience of cannabis- and food-related stimuli)	rs1049353 G/G group had lower attentional bias after THC administration rs806378 C/C group had greater attentional bias	Hindocha, Freeman, et al. (2020), Hindocha, Quattrone, et al. (2020)
COMT	Rs4680 (Val158Met)	Patients with a psychotic disorder (n=30), their relatives (n=12), and healthy controls (n=32)	Visual Verbal Learning Test; Abstract Visual Pattern Learning; Continuous performance task; Stroop Color-Word test; Digit Symbol Substitution Test	Val/val associated with impaired delayed recognition performance and increased reaction time in CPT	Henquet et al. (2006)
	Rs4680 (Val158Met)	People who use cannabis aged 16–23 years (n=442)	Spatial n-back	No significant effect	Morgan et al. (2016)
	Rs4680 (Val158Met)	Healthy participants (n=74)	CANTAB Rapid Visual Information Processing test (RVIP), Spatial Working Memory (SWM), and RAVLT	Val/val had greater latency in RVIP	Ranganathan et al. (2019)
	Rs4680 (Val158Met)	Healthy participants with experience of paranoia in the past month (n=78)	Digit span backwards task	Val/val associated with poorer performance	Tunbridge et al. (2015)
DRD2	DRD2 Taq1A (rs1800497)	Adults aged 18–40 reporting poly-drug use (n=61)	Reversal learning task	No significant difference	Spronk et al. (2016)
FAAH	Rs32440 (C385A)	Adults who use cannabis (n=48)	Dot-probe task (salience of cannabis- and food-related stimuli)	C/C genotype associated with lower attentional bias after THC	Hindocha, Freeman, et al. (2020), Hindocha, Quattrone, et al. (2020)

Candidate genes for future research

MGLL *and* DAGLA

Similarly to *FAAH*, polymorphisms in *MGLL* and *DAGLA*, which encode MAGL and DAGL, may also contribute to variation in endocannabinoid levels and responses to cannabis. Two studies have implicated *MGLL* in cannabis use disorder (Agrawal & Lynskey, 2009). To our knowledge, *DAGLA* has not been associated with cannabis use phenotypes.

GABRA2

CB1 receptor activation modulates GABAergic signaling in the reward system. The ionotropic GABA-A receptor contains an alpha subunit encoded by *GABRA2*. One study found an association between cannabis dependence and rs279858, a synonymous polymorphism in *GABRA2* (Agrawal et al., 2006). However, a subsequent study reported no effect of the SNP on lifetime cannabis use frequency (Lind et al., 2008).

ABCB1

ABCB1 encodes the first member of the B superfamily of ATP-binding cassette transporter proteins. The protein is expressed in cells of the blood brain barrier where it mediates the efflux of certain drugs from the brain. Evidence from preclinical studies suggests that ABCB1 limits the accumulation of THC in the central nervous system. Rs1045642 is a silent mutation in *ABCB1* which has been associated with cannabis dependence. The C allele has been associated with increased risk of cannabis dependence (Benyamina et al., 2009) and increased blood THC concentrations in patients with cannabis dependence (Kebir et al., 2018).

SLC6A4

5-HTTLPR is a polymorphism in the serotonin transporter gene *SLC6A4*. The short (s) allele is associated with reduced serotonin transporter expression. One study found a gene X environment interaction between the polymorphism and cannabis use. s/s genotype was associated with poorer cognitive performance in participants with a history of cannabis use but not in controls (Verdejo-García et al., 2013). No studies have investigated effects of 5-HTTLPR genotype on acute cognitive effects of cannabis.

Applications to other areas

Historically, the *Cannabis sativa* plant has been widely used for its medicinal properties. Commonly recognized conditions known to benefit from cannabinoid treatment include epilepsy, multiple sclerosis, pain, nausea associated with chemotherapy, and anorexia. Cannabis is also commonly used to treat symptoms of psychiatric disorders including depression, anxiety, and posttraumatic stress disorder (Committee on the Health Effects of Marijuana: An Evidence Review and Research Agenda et al., 2017). The medical use of cannabinoids remains controversial due to the potential for side effects and adverse reactions in certain individuals. These concerns could potentially be alleviated through the implementation of pharmacogenetic testing strategies. Genomic screening for risk alleles associated with adverse effects such as cognitive impairment and psychotomimetic responses could identify patients at risk of negative reactions. One study evaluated the feasibility of a pilot pharmacogenetic testing program for cannabis use (Papastergiou et al., 2020). The program was successfully implemented in two community pharmacies, which offered consultations to guide the safe use of cannabis based on genetic testing. Risk alleles in *COMT* and/or *AKT1* associated with cognitive impairment and cannabis-induced psychosis were identified in over half of the sample, suggesting that similar programs may benefit a large proportion of patients.

Further research into the genetic basis of the favorable effects of medical cannabinoids could also be applied to personalized medicine. If identified, genetic polymorphisms associated with improved analgesia, anxiolytic or antidepressant effects, or other treatment outcomes could be implemented into pharmacogenetic screening.

Mini-dictionary of terms

- **Allele:** the version of a given gene that an organism possesses. Humans inherit one copy of each gene from each parent and thus have two alleles at each genetic locus. For a given gene, an individual can inherit the same allele from both parents (homozygous), or a different allele from each parent (heterozygous).
- **Genetic polymorphism:** the presence of multiple versions of a genetic sequence across individuals, with the least common variant occurring in at least 1% of the population. Single nucleotide polymorphisms are common genetic variants characterized by a single nucleotide change at a specific position within a gene.
- **Genotype:** the identity of the two alleles that an individual carries at a given genetic locus.
- **Pharmacogenetics:** the study of how an individual's genes affect their response to drugs or medications.
- **Phenotype:** observable physical or behavioral trait of an organism resulting from its genes and environmental influences.

Key facts of candidate gene association studies

- Candidate gene studies investigate associations between a small number of preselected genes and a trait or disease. Genes of interest are selected based on the role of the proteins they encode in the expression of a phenotype. These are typically proteins involved in signaling pathways or physiological mechanisms related to a disease state.
- Hardy-Weinberg Equilibrium is a principle of population genetics which states that the frequencies of different alleles of a given gene should remain predictable within a population over time. Candidate gene association analyses involve quality control procedures, which include testing whether the polymorphisms being studied are in Hardy-Weinberg

Equilibrium within the study sample. This ensures that the frequencies of genetic alleles do not differ from expected frequencies in the general population

- An important consideration in genetic association studies is controlling for population stratification. There is considerable variation in the frequencies of different alleles across ancestry groups. Statistical analyses must account for population structure to ensure that results are reliable. Failure to correct for population stratification can confound associations between genotype, ancestry, and the trait being studied, leading to false-positive or false-negative findings.

Summary points

- We have reviewed current knowledge on genetic variants that modulate the acute effects of cannabis obtained in human laboratory studies and identified candidate genes that warrant investigation in future research.
- Polymorphisms affecting components of the endocannabinoid system such as the CB1 receptor and FAAH have been associated with variation in subjective responses to THC. Genes unrelated to the endocannabinoid system have not been studied for their impact on subjective effects with the exception of *OPRM1* which was shown to affect positive subjective responses.
- Genes that encode components of the dopamine system modulate the psychotomimetic effects of cannabis. Genetic predisposition to hyperdopaminergic signaling in the striatum appears to increase the risk of psychotic-like responses to cannabis in a cumulative manner.
- Evidence for genetic influences on acute cognitive effects of cannabis is sparse, but preliminary evidence suggests an association between the *COMT* Val158Met polymorphism and performance in memory and attention tasks.
- There are many genes that encode proteins involved in cannabinoid responses, which should be investigated in future research. These include ECS-specific genes as well as genes related to other neurotransmitter systems such as the GABAergic and serotonergic systems and transporter proteins.

Conflict of interest statement

Dr. Le Foll has obtained funding from Pfizer (GRAND Awards, including salary support) for investigator-initiated projects. Dr. Le Foll has some in-kind donation of cannabis product from Aurora and medication donation from Pfizer and Bioprojet and was provided a coil for TMS study from Brainsway. Dr. Le Foll has obtained industry funding from Canopy (through research grants handled by CAMH or University of Toronto), Bioprojet, ACS and Alkermes. Dr. Le Foll has received in kind donations of nabiximols from GW Pharma for past studies funded by CIHR and NIH. He has been consultant for Shionogi. He is supported by CAMH and a clinician-scientist award from the department of Family and Community Medicine of the University of Toronto and an Addiction Psychiatry Chair from the department of Psychiatry of the University of Toronto.

References

Agrawal, A., Edenberg, H. J., Foroud, T., Bierut, L. J., Dunne, G., Hinrichs, A. L., … Dick, D. M. (2006). Association of GABRA2 with drug dependence in the collaborative study of the genetics of alcoholism sample. *Behavior Genetics*, *36*(5), 640–650. https://doi.org/10.1007/s10519-006-9069-4.

Agrawal, A., & Lynskey, M. T. (2009). Candidate genes for cannabis use disorders: Findings, challenges and directions. *Addiction*, *104*(4), 518–532. https://doi.org/10.1111/j.1360-0443.2009.02504.x.

Arguello, P. A., & Gogos, J. A. (2008). A signaling pathway AKTing up in schizophrenia. *The Journal of Clinical Investigation*, *118*(6), 2018–2021. https://doi.org/10.1172/jci35931.

Basavarajappa, B. S. (2007). Critical enzymes involved in endocannabinoid metabolism. *Protein and Peptide Letters*, *14*(3), 237–246. https://doi.org/10.2174/092986607780090829.

Benyamina, A., Bonhomme-Faivre, L., Picard, V., Sabbagh, A., Richard, D., Blecha, L., … Reynaud, M. (2009). Association between ABCB1 C3435T polymorphism and increased risk of cannabis dependence. *Progress in Neuro-Psychopharmacology & Biological Psychiatry*, *33*(7), 1270–1274. https://doi.org/10.1016/j.pnpbp.2009.07.016.

Bhattacharyya, S., Atakan, Z., Martin-Santos, R., Crippa, J. A., Kambeitz, J., Prata, D., … McGuire, P. K. (2012). Preliminary report of biological basis of sensitivity to the effects of cannabis on psychosis: AKT1 and DAT1 genotype modulates the effects of δ-9-tetrahydrocannabinol on midbrain and striatal function. *Molecular Psychiatry*, *17*(12), 1152–1155. https://doi.org/10.1038/mp.2011.187.

Bhattacharyya, S., Fusar-Poli, P., Borgwardt, S., Martin-Santos, R., Nosarti, C., O'Carroll, C., … McGuire, P. (2009). Modulation of mediotemporal and ventrostriatal function in humans by Delta9-tetrahydrocannabinol: A neural basis for the effects of Cannabis sativa on learning and psychosis. *Archives of General Psychiatry*, *66*(4), 442–451. https://doi.org/10.1001/archgenpsychiatry.2009.17.

Bisogno, T., Howell, F., Williams, G., Minassi, A., Cascio, M. G., Ligresti, A., … Doherty, P. (2003). Cloning of the first sn1-DAG lipases points to the spatial and temporal regulation of endocannabinoid signaling in the brain. *The Journal of Cell Biology*, *163*(3), 463–468. https://doi.org/10.1083/jcb.200305129.

Blest-Hopley, G., Colizzi, M., Prata, D., Giampietro, V., Brammer, M., McGuire, P., & Bhattacharyya, S. (2021). Epigenetic mediation of AKT1 rs1130233's effect on Delta-9-tetrahydrocannabinol-induced medial temporal function during fear processing. *Brain Sciences*, *11*(9). https://doi.org/10.3390/brainsci11091240.

Bloomfield, M. A. P., Ashok, A. H., Volkow, N. D., & Howes, O. D. (2016). The effects of Δ9-tetrahydrocannabinol on the dopamine system. *Nature*, *539* (7629), 369–377. https://doi.org/10.1038/nature20153.

Bourgault, Z., Matheson, J., Mann, R. E., Brands, B., Wickens, C. M., Tiwari, A. K., … Le Foll, B. (2022). Mu opioid receptor gene variant modulates subjective response to smoked cannabis. *American Journal of Translational Research*, *14*(1), 623–632.

Broyd, S. J., van Hell, H. H., Beale, C., Yucel, M., & Solowij, N. (2016). Acute and chronic effects of cannabinoids on human cognition—A systematic review. *Biological Psychiatry*, *79*(7), 557–567. https://doi.org/10.1016/j.biopsych.2015.12.002.

Carter, L. P., & Griffiths, R. R. (2009). Principles of laboratory assessment of drug abuse liability and implications for clinical development. *Drug and Alcohol Dependence*, *105*(Suppl 1), S14–S25. https://doi.org/10.1016/j.drugalcdep.2009.04.003.

Cheer, J. F., Marsden, C. A., Kendall, D. A., & Mason, R. (2000). Lack of response suppression follows repeated ventral tegmental cannabinoid administration: An in vitro electrophysiological study. *Neuroscience*, *99*(4), 661–667. https://doi.org/10.1016/S0306-4522(00)00241-4.

Committee on the Health Effects of Marijuana: An Evidence Review and Research Agenda, Board on Population Health and Public Health Practice, Health and Medicine Division, & National Academies of Sciences, Engineering, and Medicine. (2017). *The health effects of cannabis and cannabinoids: The current state of evidence and recommendations for research. A report of the National Academies of Sciences, Engineering, and Medicine*. Washington, DC: The National Academies Press.

Connor, J. P., Stjepanović, D., Le Foll, B., Hoch, E., Budney, A. J., & Hall, W. D. (2021). Cannabis use and cannabis use disorder. *Nature Reviews. Disease Primers*, *7*(1), 16. https://doi.org/10.1038/s41572-021-00247-4.

Dellazizzo, L., Potvin, S., Giguère, S., & Dumais, A. (2022). Evidence on the acute and residual neurocognitive effects of cannabis use in adolescents and adults: A systematic meta-review of meta-analyses. *Addiction*. https://doi.org/10.1111/add.15764.

Drennan, M. L., Karoly, H. C., Bryan, A. D., Hutchison, K. E., & Bidwell, L. C. (2021). Acute objective and subjective intoxication effects of legal-market high potency THC-dominant versus CBD-dominant cannabis concentrates. *Scientific Reports*, *11*(1), 21744. https://doi.org/10.1038/s41598-021-01128-2.

Farrell, S. M., Tunbridge, E. M., Braeutigam, S., & Harrison, P. J. (2012). COMT Val(158)Met genotype determines the direction of cognitive effects produced by catechol-O-methyltransferase inhibition. *Biological Psychiatry*, *71*(6), 538–544. https://doi.org/10.1016/j.biopsych.2011.12.023.

Filbey, F. M., Schacht, J. P., Myers, U. S., Chavez, R. S., & Hutchison, K. E. (2010). Individual and additive effects of the CNR1 and FAAH genes on brain response to marijuana cues. *Neuropsychopharmacology*, *35*(4), 967–975. https://doi.org/10.1038/npp.2009.200.

Fischman, M. W., & Foltin, R. W. (1991). Utility of subjective-effects measurements in assessing abuse liability of drugs in humans. *British Journal of Addiction*, *86*(12), 1563–1570. https://doi.org/10.1111/j.1360-0443.1991.tb01749.x.

Fuke, S., Suo, S., Takahashi, N., Koike, H., Sasagawa, N., & Ishiura, S. (2001). The VNTR polymorphism of the human dopamine transporter (DAT1) gene affects gene expression. *The Pharmacogenomics Journal*, *1*(2), 152–156. https://doi.org/10.1038/sj.tpj.6500026.

Gerra, M. C., Jayanthi, S., Manfredini, M., Walther, D., Schroeder, J., Phillips, K. A., … Donnini, C. (2018). Gene variants and educational attainment in cannabis use: Mediating role of DNA methylation. *Translational Psychiatry*, *8*(1), 23. https://doi.org/10.1038/s41398-017-0087-1.

Hartman, C. A., Hopfer, C. J., Haberstick, B., Rhee, S. H., Crowley, T. J., Corley, R. P., … Ehringer, M. A. (2009). The association between cannabinoid receptor 1 gene (CNR1) and cannabis dependence symptoms in adolescents and young adults. *Drug and Alcohol Dependence*, *104*(1–2), 11–16. https://doi.org/10.1016/j.drugalcdep.2009.01.022.

Haughey, H. M., Marshall, E., Schacht, J. P., Louis, A., & Hutchison, K. E. (2008). Marijuana withdrawal and craving: Influence of the cannabinoid receptor 1 (CNR1) and fatty acid amide hydrolase (FAAH) genes. *Addiction*, *103*(10), 1678–1686. https://doi.org/10.1111/j.1360-0443.2008.02292.x.

Heinz, A., Goldman, D., Jones, D. W., Palmour, R., Hommer, D., Gorey, J. G., … Weinberger, D. R. (2000). Genotype influences in vivo dopamine transporter availability in human striatum. *Neuropsychopharmacology*, *22*(2), 133–139. https://doi.org/10.1016/s0893-133x(99)00099-8.

Henquet, C., Rosa, A., Krabbendam, L., Papiol, S., Fananás, L., Drukker, M., … van Os, J. (2006). An experimental study of catechol-o-methyltransferase Val158Met moderation of delta-9-tetrahydrocannabinol-induced effects on psychosis and cognition. *Neuropsychopharmacology*, *31*(12), 2748–2757. https://doi.org/10.1038/sj.npp.1301197.

Hindocha, C., Freeman, T. P., Schafer, G., Gardner, C., Bloomfield, M. A. P., Bramon, E., … Curran, H. V. (2020). Acute effects of cannabinoids on addiction endophenotypes are moderated by genes encoding the CB1 receptor and FAAH enzyme. *Addiction Biology*, *25*(3), e12762. https://doi.org/10.1111/adb.12762.

Hindocha, C., Quattrone, D., Freeman, T. P., Murray, R. M., Mondelli, V., Breen, G., … Di Forti, M. (2020). Do AKT1, COMT and FAAH influence reports of acute cannabis intoxication experiences in patients with first episode psychosis, controls and young adult cannabis users? *Translational Psychiatry*, *10*(1), 143. https://doi.org/10.1038/s41398-020-0823-9.

Hirvonen, J., Zanotti-Fregonara, P., Umhau, J. C., George, D. T., Rallis-Frutos, D., Lyoo, C. H., … Heilig, M. (2013). Reduced cannabinoid CB1 receptor binding in alcohol dependence measured with positron emission tomography. *Molecular Psychiatry*, *18*(8), 916–921. https://doi.org/10.1038/mp.2012.100.

Hutchison, K. E., Haughey, H., Niculescu, M., Schacht, J., Kaiser, A., Stitzel, J., … Filbey, F. (2008). The incentive salience of alcohol: Translating the effects of genetic variant in CNR1. *Archives of General Psychiatry*, *65*(7), 841–850. https://doi.org/10.1001/archpsyc.65.7.841.

Justinova, Z., Tanda, G., Munzar, P., & Goldberg, S. R. (2004). The opioid antagonist naltrexone reduces the reinforcing effects of Delta 9 tetrahydrocannabinol (THC) in squirrel monkeys. *Psychopharmacology*, *173*(1–2), 186–194. https://doi.org/10.1007/s00213-003-1693-6.

Kay, S. R., Fiszbein, A., & Opler, L. A. (1987). The positive and negative syndrome scale (PANSS) for schizophrenia. *Schizophrenia Bulletin*, *13*(2), 261–276. https://doi.org/10.1093/schbul/13.2.261.

Kebir, O., Lafaye, G., Blecha, L., Chaumette, B., Mouaffak, F., Laqueille, X., & Benyamina, A. (2018). ABCB1 C3435T polymorphism is associated with tetrahydrocannabinol blood levels in heavy cannabis users. *Psychiatry Research*, *262*, 357–358. https://doi.org/10.1016/j.psychres.2017.09.006.

Ketcherside, A., Noble, L. J., McIntyre, C. K., & Filbey, F. M. (2017). Cannabinoid receptor 1 gene by cannabis use interaction on CB1 receptor density. *Cannabis and Cannabinoid Research*, *2*(1), 202–209. https://doi.org/10.1089/can.2017.0007.

Krebs, M. O., Morvan, Y., Jay, T., Gaillard, R., & Kebir, O. (2014). Psychotomimetic effects at initiation of cannabis use are associated with cannabinoid receptor 1 (CNR1) variants in healthy students. *Molecular Psychiatry*, *19*(4), 402–403. https://doi.org/10.1038/mp.2013.188.

Kuepper, R., Morrison, P. D., van Os, J., Murray, R. M., Kenis, G., & Henquet, C. (2010). Does dopamine mediate the psychosis-inducing effects of cannabis? A review and integration of findings across disciplines. *Schizophrenia Research*, *121*(1–3), 107–117. https://doi.org/10.1016/j.schres.2010.05.031.

Le Strat, Y., Ramoz, N., Horwood, J., Falissard, B., Hassler, C., Romo, L., … Gorwood, P. (2009). First positive reactions to cannabis constitute a priority risk factor for cannabis dependence. *Addiction*, *104*(10), 1710–1717. https://doi.org/10.1111/j.1360-0443.2009.02680.x.

Lind, P. A., Macgregor, S., Agrawal, A., Montgomery, G. W., Heath, A. C., Martin, N. G., & Whitfield, J. B. (2008). The role of GABRA2 in alcohol dependence, smoking, and illicit drug use in an Australian population sample. *Alcoholism, Clinical and Experimental Research*, *32*(10), 1721–1731. https://doi.org/10.1111/j.1530-0277.2008.00768.x.

López-Moreno, J. A., Echeverry-Alzate, V., & Bühler, K.-M. (2012). The genetic basis of the endocannabinoid system and drug addiction in humans. *Journal of Psychopharmacology*, *26*(1), 133–143. https://doi.org/10.1177/0269881111416689.

Lu, H. C., & Mackie, K. (2016). An introduction to the endogenous cannabinoid system. *Biological Psychiatry*, *79*(7), 516–525. https://doi.org/10.1016/j.biopsych.2015.07.028.

Lyons, M. J., Toomey, R., Meyer, J. M., Green, A. I., Eisen, S. A., Goldberg, J., … Tsuang, M. T. (1997). How do genes influence marijuana use? The role of subjective effects. *Addiction*, *92*(4), 409–417.

Maldonado, R., Valverde, O., & Berrendero, F. (2006). Involvement of the endocannabinoid system in drug addiction. *Trends in Neurosciences*, *29*(4), 225–232. https://doi.org/10.1016/j.tins.2006.01.008.

Mason, O., Morgan, C. J. A., Dhiman, S. K., Patel, A., Parti, N., Patel, A., & Curran, H. V. (2009). Acute cannabis use causes increased psychotomimetic experiences in individuals prone to psychosis. *Psychological Medicine*, *39*(6), 951–956. https://doi.org/10.1017/s0033291708004741.

Mason, O. J., Morgan, C. J., Stefanovic, A., & Curran, H. V. (2008). The psychotomimetic states inventory (PSI): Measuring psychotic-type experiences from ketamine and cannabis. *Schizophrenia Research*, *103*(1–3), 138–142. https://doi.org/10.1016/j.schres.2008.02.020.

Matheson, J., Le Foll, B., Martin, Patel, & Preedy. (2022). Acute and chronic impact of cannabis on human cognition. In *Cannabis use, neurobiology, psychology, and treatment* Elsevier.

McNair, D. M., Lorr, M., & Droppleman, L. F. (1971). *Manual for the profile of mood states*. San Diego, CA: Educational and Industrial Testing Services.

Mechoulam, R., & Parker, L. A. (2013). The endocannabinoid system and the brain. *Annual Review of Psychology*, *64*, 21–47. https://doi.org/10.1146/annurev-psych-113011-143739.

Mier, D., Kirsch, P., & Meyer-Lindenberg, A. (2010). Neural substrates of pleiotropic action of genetic variation in COMT: A meta-analysis. *Molecular Psychiatry*, *15*(9), 918–927. https://doi.org/10.1038/mp.2009.36.

Moore, T. H., Zammit, S., Lingford-Hughes, A., Barnes, T. R., Jones, P. B., Burke, M., & Lewis, G. (2007). Cannabis use and risk of psychotic or affective mental health outcomes: A systematic review. *Lancet*, *370*(9584), 319–328. https://doi.org/10.1016/s0140-6736(07)61162-3.

Morgan, C. J., Freeman, T. P., Powell, J., & Curran, H. V. (2016). AKT1 genotype moderates the acute psychotomimetic effects of naturalistically smoked cannabis in young cannabis smokers. *Translational Psychiatry*, *6*(2), e738. https://doi.org/10.1038/tp.2015.219.

Mossaheb, N., Becker, J., Schaefer, M. R., Klier, C. M., Schloegelhofer, M., Papageorgiou, K., & Amminger, G. P. (2012). The community assessment of psychic experience (CAPE) questionnaire as a screening-instrument in the detection of individuals at ultra-high risk for psychosis. *Schizophrenia Research*, *141*(2–3), 210–214. https://doi.org/10.1016/j.schres.2012.08.008.

Murphy, T., Matheson, J., Mann, R. E., Brands, B., Wickens, C. M., Tiwari, A. K., … Le Foll, B. (2021). Influence of cannabinoid receptor 1 genetic variants on the subjective effects of smoked cannabis. *International Journal of Molecular Sciences*, *22*(14). https://doi.org/10.3390/ijms22147388.

Okey, S. A., & Meier, M. H. (2020). A within-person comparison of the subjective effects of higher vs. lower-potency cannabis. *Alcohol Dependence*, *216*. https://doi.org/10.1016/j.drugalcdep.2020.108225.

Onaivi, E. S., Leonard, C. M., Ishiguro, H., Zhang, P. W., Lin, Z., Akinshola, B. E., & Uhl, G. R. (2002). Endocannabinoids and cannabinoid receptor genetics. *Progress in Neurobiology*, *66*(5), 307–344. https://doi.org/10.1016/s0301-0082(02)00007-2.

Ozaita, A., Puighermanal, E., & Maldonado, R. (2007). Regulation of PI3K/Akt/GSK-3 pathway by cannabinoids in the brain. *Journal of Neurochemistry*, *102*(4), 1105–1114. https://doi.org/10.1111/j.1471-4159.2007.04642.x.

Palmer, R. H. C., McGeary, J. E., Knopik, V. S., Bidwell, L. C., & Metrik, J. M. (2019). CNR1 and FAAH variation and affective states induced by marijuana smoking. *The American Journal of Drug and Alcohol Abuse*, *45*(5), 514–526. https://doi.org/10.1080/00952990.2019.1614596.

Papastergiou, J., Li, W., Sterling, C., & van den Bemt, B. (2020). Pharmacogenetic-guided cannabis usage in the community pharmacy: Evaluation of a pilot program. *Journal of Cannabis Research*, *2*(1), 24. https://doi.org/10.1186/s42238-020-00033-1.

Ranganathan, M., De Aquino, J. P., Cortes-Briones, J. A., Radhakrishnan, R., Pittman, B., Bhakta, S., & D'Souza, D. C. (2019). Highs and lows of cannabinoid-dopamine interactions: Effects of genetic variability and pharmacological modulation of catechol-O-methyl transferase on the acute response to delta-9-tetrahydrocannabinol in humans. *Psychopharmacology*, *236*(11), 3209–3219. https://doi.org/10.1007/s00213-019-05273-5.

Salamone, J. D., & Correa, M. (2002). Motivational views of reinforcement: Implications for understanding the behavioral functions of nucleus accumbens dopamine. *Behavioural Brain Research*, *137*(1–2), 3–25. https://doi.org/10.1016/s0166-4328(02)00282-6.

Schacht, J. P., Selling, R. E., & Hutchison, K. E. (2009). Intermediate cannabis dependence phenotypes and the FAAH C385A variant: An exploratory analysis. *Psychopharmacology*, *203*(3), 511–517. https://doi.org/10.1007/s00213-008-1397-z.

Schultz, W., Dayan, P., & Montague, P. R. (1997). A neural substrate of prediction and reward. *Science*, *275*(5306), 1593–1599. https://doi.org/10.1126/science.275.5306.1593.

Spronk, D. B., Van der Schaaf, M. E., Cools, R., De Bruijn, E. R., Franke, B., van Wel, J. H., … Verkes, R. J. (2016). Acute effects of cocaine and cannabis on reversal learning as a function of COMT and DRD2 genotype. *Psychopharmacology*, *233*(2), 199–211. https://doi.org/10.1007/s00213-015-4141-5.

Tanda, G., Pontieri, F. E., & Di Chiara, G. (1997). Cannabinoid and heroin activation of mesolimbic dopamine transmission by a common mu1 opioid receptor mechanism. *Science*, *276*(5321), 2048–2050. https://doi.org/10.1126/science.276.5321.2048.

Tunbridge, E. M., Dunn, G., Murray, R. M., Evans, N., Lister, R., Stumpenhorst, K., … Freeman, D. (2015). Genetic moderation of the effects of cannabis: Catechol-O-methyltransferase (COMT) affects the impact of Δ9-tetrahydrocannabinol (THC) on working memory performance but not on the occurrence of psychotic experiences. *Journal of Psychopharmacology*, *29*(11), 1146–1151. https://doi.org/10.1177/0269881115609073.

Verdejo-García, A., Fagundo, A. B., Cuenca, A., Rodriguez, J., Cuyás, E., Langohr, K., … de la Torre, R. (2013). COMT val158met and 5-HTTLPR genetic polymorphisms moderate executive control in cannabis users. *Neuropsychopharmacology*, *38*(8), 1598–1606. https://doi.org/10.1038/npp.2013.59.

Wenzel, J. M., & Cheer, J. F. (2018). Endocannabinoid regulation of reward and reinforcement through interaction with dopamine and endogenous opioid signaling. *Neuropsychopharmacology*, *43*(1), 103–115. https://doi.org/10.1038/npp.2017.126.

Chapter 23

Cell signaling of the CB_1 cannabinoid receptor via β-arrestins, cannabinoid receptor interacting protein (CRIP1a) and other regulatory proteins

Allyn C. Howlett[a], Christopher S. Breivogel[b], and Khalil Eldeeb[c,d]

[a]*Department of Physiology and Pharmacology, Wake Forest University School of Medicine, Winston-Salem, NC, United States,* [b]*Department of Pharmaceutical & Clinical Sciences, Campbell University College of Pharmacy & Health Sciences, Buies Creek, NC, United States,* [c]*Department of Pharmacology, Campbell University School of Osteopathic Medicine, Lillington, NC, United States,* [d]*Department of Pharmacology, Al Azhar Faculty of Medicine, Al Azhar University, New Damietta, Egypt*

Abbreviations

2-AG	2-arachidonoylglycerol
AP2	clathrin adaptor protein 2
BiP	binding protein; also known as glucose-regulated protein78 (GRP78) or heat shock protein a5 (HSPa5)
BRET	bioluminescence resonance energy transfer
CB_1R	CB_1 cannabinoid receptor
CB_2R	CB_2 cannabinoid receptor
DAMGO	[D-Ala^2, N-MePhe^4,Gly-ol^5]-enkephalin
ERK1/2	extracellular signal-regulated kinases1 and 2
FAAH	fatty acid amide hydrolase
GABA	γ-aminobutyric acid
GAP	GTPase-accelerating protein
GASP1	G protein-coupled receptor-associated sorting protein1
GPCR	G protein-coupled receptor
GRK	G protein-coupled receptor kinase
MOR	Mu opioid receptor
RGS proteins	regulators of G protein signaling proteins
SGIP1	Src homology 3 (SH3)-domain growth factor receptor-bound 2 (GRB)-like (endophilin) interacting protein1
Δ^9-THC	Δ^9-tetrahydrocannabinol
WLS	Wntless, also known as GPR177, Evi, or Sprinter

Introduction

In addition to stimulation by endocannabinoid agonist ligands, CB_1 and CB_2 cannabinoid receptors (CB_1R and CB_2R) are regulated by cellular proteins within their interactome (Mattheus et al., 2016; Molina-Holgado et al., 2021; Njoo et al., 2015). This chapter expands upon previous overviews of CB_1R-associated proteins (Fletcher-Jones et al., 2020; Howlett et al., 2010; Smith et al., 2010) to highlight interactions with proteins that are pertinent to cellular signaling or to trafficking of the receptor at various stages in the CB_1R life cycle. We also seek to provide physiological relevance of the associated protein interactions with the CB_1R, and in some cases with the CB_2R where known.

Cannabis Use, Neurobiology, Psychology, and Treatment. https://doi.org/10.1016/B978-0-323-89862-1.00029-5

Proteins that regulate signal transduction pathways

Heterotrimeric G proteins

CB_1R and CB_2R couple to heterotrimeric G proteins that lead to signaling pathways regulated primarily by Gαi/o subtypes or Gβγ dimers (Fig. 1) (Howlett et al., 2002; Kendall & Yudowski, 2016; Turu & Hunyady, 2010). G protein subunits are released in response to endocannabinoid agonists, 2-arachidonoylglycerol (2-AG) or anandamide (N-arachidonylethanolamide), or exogenous agonists such as Δ^9-tetrahydrocannabinol (Δ^9-THC) and its classical and nonclassical analogs (e.g., CP55940), as well as cannabimimetic aminoalkylindoles (e.g., WIN55212-2). Although most cellular signaling responses are pertussis toxin-sensitive indicating that they are mediated by Gi/o proteins (Howlett & Abood, 2017; Howlett et al., 2002), Gq-mediated Ca^{2+} mobilization has also been reported (Lauckner et al., 2005). CB_1Rs located at presynaptic terminals in both γ-aminobutyric acid (GABA)ergic and glutamatergic neurons modulate neurotransmission via feedback mechanisms that govern Ca^{2+} and K^+ channel activity (Mackie, 2008). However, a wide range of signaling via both Gαi/o and Gβγ proteins can be initiated by receptors located at neuronal soma and other cell types (Howlett & Abood, 2017; Leo & Abood, 2021; Lutz, 2020).

RGS proteins in complex with CB_1R and CB_2R

The CB_1R signals through heterotrimeric G proteins by promoting GDP-to-GTP exchange on the Gα subunit, resulting in dissociation of Gα-GTP from Gβγ subunits and activation of their specific effectors. Gα subunits hydrolyze GTP through their intrinsic GTPase activity, which leads to G protein inactivation. However, this process is very slow and does not correspond to the rapid turn-off of signaling events in cells (Sethakorn et al., 2010). In the past 20 years, regulators of G protein signaling (RGS) proteins have emerged as central control points of GPCR signaling and promising targets for drug development (Fig. 2) (Roman & Traynor, 2011). RGS proteins are perceived as crucial modulators of the magnitude and duration of G protein-mediated signals triggered by extracellular stimuli. RGS proteins reduce the amplitude and duration of G protein signaling by binding to activated Gα subunits and dramatically increasing their GTPase activity (Fig. 2). This GTPase-accelerating protein (GAP) activity of RGS proteins is limited mainly to Gα subunits of the Gq and Gi/o families (Sethakorn et al., 2010; Sjogren et al., 2010; Sjogren & Neubig, 2010; Stewart et al., 2012). More than 30 known mammalian RGS proteins have strong GAP activity at Gα subunits, and they are divided into several subfamilies based on sequence homology (R4, R7, R12, and RZ). Common to all RGS proteins is a highly conserved region, the RGS box, comprised of ~130 amino acids, which is sufficient and critical for GAP activity (Ishii & Kurachi, 2003; Roman & Traynor, 2011; Sjogren & Neubig, 2010).

CB_1Rs are integral to the physiological and therapeutic interaction with the dopaminergic systems in neurological disorders like Parkinson's disease (Van der Stelt & Di Marzo, 2003). While the complete machinery of endocannabinoid neurotransmission is highly represented in striatal spine modules, the ability of CB_1R to heterodimerize with dopamine, adenosine, and opioid receptors, which are differentially expressed in different types of striatal spine modules, can

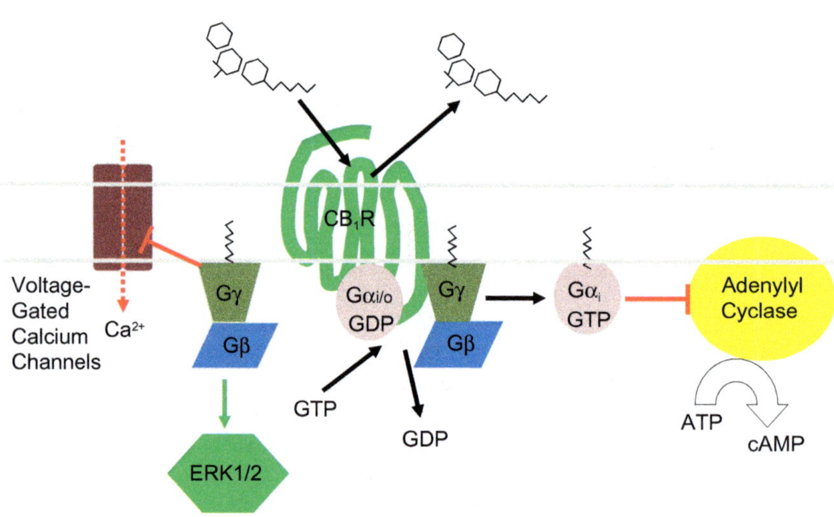

FIG. 1 Cannabinoid receptors couple to and activate G protein subtypes. Gαi/o proteins inhibit adenylyl cyclase (AC) to slow the formation of cAMP from ATP. Gβγ inhibits Ca^{2+} channels and opens GIRK channels and can work in concert with β-arrestins to activate other pathways like ERK1/2.

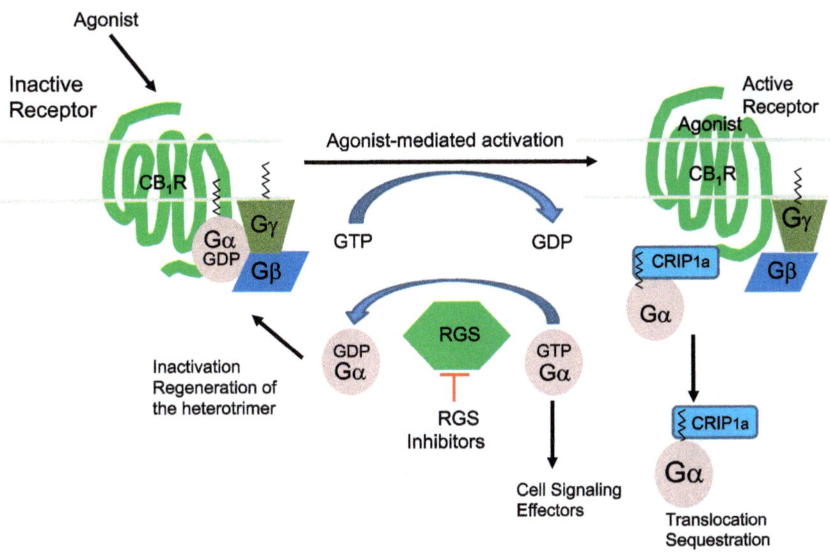

FIG. 2 Mechanisms of action of regulators of G protein signaling (RGS) proteins and cannabinoid receptor interacting protein 1a (CRIP1a). RGS proteins negatively modulate GPCR activity by increasing the rate of inactivation of Gα-GTP. The diverse expression of RGS proteins suggests that their inhibitors may have therapeutic potential. CRIP1a interacts with the CB_1R to bind the myristoyl moiety of Gαi to allow translocation or sequestration in the cytosol.

represent a relevant fine-tuning of neurotransmission. In a study by Lerner and Kreitzer to characterize the mechanisms linking D_2 dopamine and A_{2A} adenosine receptor signaling to the mobilization of endocannabinoids, RGS4 was found to be a crucial link between dopamine signaling, synaptic plasticity, and motor behavior (Lerner & Kreitzer, 2012). RGS4 inhibitors with suitable characteristics for clinical use could be potential pharmacotherapeutics for Parkinson's disease.

Specific induction of RGS4 after peripheral nerve injury can be considered a disease-modifying strategy. The pharmacological inhibition of spinal RGS4 counteracted the silencing of the spinal CB_1R signaling analgesic system and effectively attenuated neuropathic pain and reduced hyperalgesia during the treatment period (Bosier et al., 2015).

The endogenous cannabinoid system in the gastrointestinal tract plays a vital role in regulating energy balance and food intake. Increased expression of RGS4 occurs in several pathological conditions in the gut, such as ulcerative colitis. Stimulation of endocannabinoid signaling by inhibition of RGS4 is a potential strategy for treating inflammatory bowel disease that is limited to the colon (Salaga et al., 2016). This therapeutic strategy may be further improved by combined treatment with a fatty acid amide hydrolase (FAAH) blocker, which decreases breakdown of anandamide by FAAH, thereby increasing endogenous cannabinoid agonist (Salaga et al., 2016).

RGS17, a member of the RGS-RZ subfamily that commonly targets GTP-bound $G\alpha i_{1\text{-}3}$, Gαo, Gαz, and Gαq has been studied for its role in inhibiting cannabinoid receptor signaling. The Ramkumar lab showed that the CB_2R agonist JWH-015 protected against cisplatin-induced hearing loss (Ghosh et al., 2018). Cisplatin increased the expression of RGS17 in the cochlea, which was abolished by pretreatment with the CB_2R agonist. Interestingly, knockdown of RGS17 in the cochlea significantly increased CB_2R expression, and overexpression of RGS17 led to significant reduction in CB_2R expression, suggesting a reciprocal regulation, disruption of which might produce hearing loss (Dhukhwa et al., 2021). The ototoxic effect observed with the knockdown of CB_2R may be attributed to the induction of RGS17. These findings suggest a close association between the regulation of CB_2R, G proteins, and RGS17 in cisplatin toxicity in the cochlea (Dhukhwa et al., 2021).

Cannabinoid receptor signaling through β-arrestins

β-Arrestins were initially characterized as negative regulators of the G protein-coupled β2 adrenergic receptor (Attramadal et al., 1992) and then of GPCRs in general (Lefkowitz, 2013). β-Arrestins block G protein coupling to, and mediate internalization of, GPCRs by coupling them to clathrin and its adaptor protein 2 (AP2) (Lefkowitz, 2013) and thus they are widely held to be important mediators of receptor desensitization and downregulation as mechanisms underlying tolerance (Fig. 3). In addition to arresting CB1R-G protein signaling, β-arrestins link GPCRs to activating intracellular signaling pathways such as ERK1/2 (DeFea et al., 2000; Luttrell et al., 2001; Ma & Pei, 2007), JNK, p38, and Src (Delgado-Peraza et al., 2016). Gβγ subunits can induce rapid and short-lived activation followed by slower and more prolonged

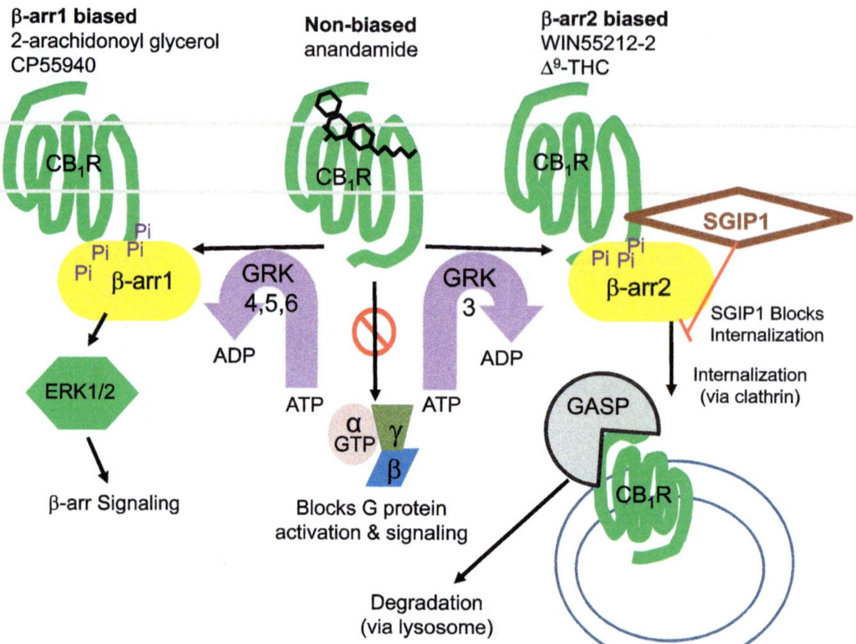

FIG. 3 Cannabinoid receptor regulation and signaling via β-arrestins and effects of SGIP1 and GASP. Agonist-occupied CB_1R is a substrate for G protein receptor kinases (GRKs), and phosphorylated-CB_1R binds β-arrestins with high affinity. β-arrestin1 promotes slow, longer-duration ERK1/2 activation. β-arrestin2 mediates desensitization and internalization of CB_1R. The internalization of the CB_1R-β-arrestin2 can be blocked by SGIP1. Once internalized, GASP may direct the receptor to the lysosome for degradation.

activation by β-arrestins (Flores-Otero et al., 2014; Nogueras-Ortiz & Yudowski, 2016), underlying the multiple waves of GPCR signaling (Nogueras-Ortiz & Yudowski, 2016).

β-Arrestin binding to the agonist-induced conformation of a GPCR involves phosphorylation of an intracellular domain by one of the seven members of the G protein-coupled receptor kinase (GRK) family. Recruitment of β-arrestin1 to the CB_1R depends on GRK4, 5, and/or 6 and results in signaling, whereas recruitment of β-arrestin2 depends on GRK3 and results in desensitization and internalization. Although the CB_1R is internalized by continued exposure to some agonists (including Δ^9-THC), CB_1R agonists can activate ERK1/2 whether at the plasma membrane or internalized (Daigle et al., 2008).

GRK3 phosphorylation of serines 426 and 430 on the CB_1R C-terminus is essential for β-arrestin2-mediated desensitization (Jin et al., 1999), and phosphorylation of serine 465 and/or threonine 466 near the distal CB_1R C-terminus are important for the recruitment of β-arrestin2 to the membrane and CB_1R internalization. There have been a number of studies utilizing a mutant CB_1R with serines 426 and 430 replaced by alanine (S426A/S430A) to examine the effects of preventing GRK3/β-arrestin2-mediated actions. Delgado-Peraza and colleagues (Delgado-Peraza et al., 2016) showed that a slow phase of activation of ERK1/2 by WIN55212-2 was dependent on GRK4, 5, and/or 6 and β-arrestin1, and that desensitization of this activity was mediated by β-arrestin2. Morgan and colleagues (Morgan et al., 2014) showed that mice expressing the S426A/S430A mutant CB_1R showed greater acute effects and delayed tolerance to Δ^9-THC, and impaired desensitization of the CB_1R mutant in cultured neurons from these mice. Nealon and colleagues (Nealon et al., 2019) found that S426A/S430A mice exhibited impaired tolerance to Δ^9-THC and WIN55212-2, but not to CP55940. Effects of mutations in the DRY motif of CB_1R, have been reported (Gyombolai et al., 2015). The aspartic acid and arginine residues within the DRY motif facilitated agonist-induced interaction with β-arrestin1 and suppressed interaction with β-arrestin2. The tyrosine residue facilitated agonist-induced coupling of CB_1R with β-arrestin2 (Gyombolai et al., 2015).

The phytocannabinoid Δ^9-THC was shown to be a low efficacy partial agonist for G protein activation by the CB_1R (Sim et al., 1996). However, for inducing CB_1R to interact with β-arrestin2, Δ^9-THC was as efficacious as arachidonyl-2'-chloroethylamide (Navarro et al., 2018) or CP55940 (Raich et al., 2021) as determined by bioluminescence resonance energy transfer (BRET) in transfected HEK293 or STHdh cells. Δ^9-THC was as efficacious as 2-AG and CP55940 for inducing CB_1 to interact with β-arrestin1, and Δ^9-THC and CP55940 were more efficacious for inducing internalization than 2-AG (Laprairie et al., 2014). Δ^9-THC showed low efficacy for inducing β-arrestin1 or β-arrestin2 translocation to plasma membrane CB_1Rs (2-AG $\geq$ WIN55212-2 > CP55940 > anandamide and Δ^9-THC) (Finlay et al., 2019; Ibsen et al., 2019; Patel et al., 2020), and for inducing CB_1R internalization (WIN55212-2 > CP55940 > Δ^9-THC) (Finlay et al., 2019). For CB_1R exogenously expressed in AtT20 cells, WIN55212-2, CP55940, and HU210 all induced rapid internalization of high fractions of plasma membrane CB_1Rs, but Δ^9-THC and methanandamide were less effective (Hsieh et al., 1999) as expected for Δ^9-THC's low efficacy to recruit β-arrestin2. In contrast, in mice, Δ^9-THC caused a greater degree and more rapid tolerance than CP55940 (Breivogel et al., 2008; Breivogel & Vaghela, 2015).

In animal model studies comparing β-arrestin2$^{(-/-)}$ (knock out) versus wild-type mice, the absence of β-arrestin2 enhanced the acute hypothermic and antinociceptive effects of Δ^9-THC, and no difference was found in the rate of tolerance to either effect of Δ^9-THC (Breivogel et al., 2008). We (Breivogel et al., 2013) found no differences in the overall density of CB_1Rs or β-arrestin1 in brain homogenates of drug-naïve wild-type versus β-arrestin2$^{(-/-)}$ mice. Nguyen and colleagues (Nguyen et al., 2012) confirmed these findings for acute Δ^9-THC in the β-arrestin2$^{(-/-)}$. They also reported that tolerance to Δ^9-THC treatment was attenuated for antinociception but enhanced for catalepsy, perhaps due to mixed effects on G protein activation in different brain regions (Nguyen et al., 2012). Comparing β-arrestin1$^{(-/-)}$ versus wild-type mice, the absence of β-arrestin1 decreased the acute antinociceptive and hypothermic response to CP55940 but not to Δ^9-THC (Breivogel & Vaghela, 2015). No difference in the tolerance to either agonist was found.

Taken together, these studies are beginning to yield a clearer picture of the interactions of β-arrestins with CB_1Rs. Generally, β-arrestin1 interaction with CB_1R is mediated by GRK4/5/6 and is associated with signaling via kinases like ERK1/2, and β-arrestin2 is mediated by GRK3 and is associated with terminating signaling and CB_1R internalization. CP55940 and 2-AG appear to preferentially recruit β-arrestin1, and Δ^9-THC and WIN55212-2 appear to prefer β-arrestin2. Ligands biased toward G proteins rather than β-arrestins are expected to exhibit reduced levels of tolerance.

Cannabinoid receptor interacting protein 1a (CRIP1a) and CB_1Rs

Cannabinoid receptor interacting protein1a (CRIP1a) is a protein that interacts with the distal C-terminus of CB_1R but not CB_2R (Fig. 2) (Booth et al., 2019; Oliver et al., 2020). CRIP1a was discovered in a yeast two-hybrid screen for CB_1R C-terminus-binding proteins and was confirmed by co-immunoprecipitation with CB_1Rs from rat brain and cultured neuroblastoma cells and pull-down assays with a glutathione S-transferase (GST)-CB_1R C-terminal peptide (Niehaus et al., 2007). CRIP1a functions as a broad negative regulator of CB_1R signaling. CRIP1a significantly attenuated tonic inhibition of voltage-gated Ca^{2+} channels by CB_1R (Niehaus et al., 2007). CRIP1a inhibited both constitutive and agonist-stimulated CB_1R-mediated G protein activation (Smith et al., 2015) and facilitated a Gi/o protein subtype preference for Gi1 and Gi2, accompanied by an overall suppression of G protein-mediated signaling (Blume et al., 2015). CRIP1a attenuated the agonist-driven CB_1R internalization from the cell surface (Blume et al., 2016), in competition with β-arrestins for interaction with C-terminal CB_1R domains (Blume et al., 2017). CRIP1a was also identified as a negative regulator for endocytosis of both the presynaptic CB_1R and the mGlu8 metabotropic glutamate receptor (Mascia et al., 2017). CRIP1a binds N-terminally myristoylated Gαi peptide (Fig. 2), providing a mechanism by which CRIP1a binding to Gαi proteins can regulate CB_1R cell signaling (Booth et al., 2021).

The mRNA of both CRIP1a and CB_1R was co-localized in significant amounts in the forebrain and pyramidal neurons and interneurons of the hippocampal formation. Based on the reported CB_1R-mediated neuroprotection against excitotoxicity-induced cell death and the CRIP1a co-expression with CB_1R in excitatory glutamatergic neurons, Stauffer and colleagues reported that in primary neuronal cortical cultures, CRIP1a disrupts agonist-induced neuroprotection and confers antagonist-induced neuroprotection (Stauffer et al., 2011). CRIP1a overexpression extended the depression of excitatory currents by cannabinoid agonists in pyramidal neurons of the hippocampus and diminished the severity of chemically induced acute epileptiform seizures (Guggenhuber et al., 2016). CRIP1a modulation has a neuroprotective role by attenuating neuronal damage and hyperactivity caused by ischemic damage (Kwon et al., 2020). These data suggest that expression of CRIP1a may reduce neuronal damage induced by different mechanisms and support targeting CRIP1a for drug development.

Outside the nervous system, the presence of CRIP1a in the vertebrate retina and a close juxtaposition of CRIP1a and CB_1R in cone terminals was reported (Hu et al., 2010). CRIP1a was detected in uterine tissue with a lower level in uterine tissue exposed to labor (Kozakiewicz et al., 2021). These findings provide a basis for future investigations on the function of CRIP1a in health and disease.

Proteins that regulate CB_1R trafficking

Binding protein (BiP) chaperone of CB_1R

Binding protein (BiP), also known as glucose-regulated protein (GRP78) or heat shock protein a5 (HSPa5), is a member of the HSP70 family of chaperone proteins that appeared in a yeast two-hybrid screen of a human cDNA library for targets that interact with the CB_1R C-terminus (Costas-Insua et al., 2021). Direct interaction of CB_1R with the BiP C-terminus interacting domain was confirmed using fluorescence polarization binding assays and co-immunoprecipitations from lysates of HEK293T cells exogenously expressing tagged CB_1R and BiP proteins and mutants (Costas-Insua et al., 2021). The CB_2R failed to interact with the BiP interaction region. The CB_1R C-terminal distal 23 amino acids were the target for the BiP interaction region (Fig. 4), and most C-terminal domain phosphomimetic mutants substituting aspartate for each serine or threonine were able to promote the association. The only exception was the S452D mutant, which failed to interact in the yeast two-hybrid assay or in WIN55212-2-stimulated cellular mass redistribution in HEK293-CB_1R cells (Costas-Insua et al., 2021). Cellular signaling via Gi/o protein activation (GTPγS binding) and inhibition of cAMP production in HEK293-CB_1R cells was not altered by co-expression of the BiP-interacting region (Costas-Insua et al., 2021). However, the BiP-interacting region reduced Gq/11-activation (Fig. 4) and attenuated WIN55212-2-stimulated Akt/mTORC1 pathway proteins, ERK1/2, and CREB phosphorylation (Costas-Insua et al., 2021).

BiP expression throughout the mouse brain occurred predominantly in GABAergic rather than glutamatergic neurons. BiP protein was localized at the endoplasmic reticulum lumen and cytosol, but could co-localize with the CB_1R as punctae on GABAergic terminals and in synaptosomes isolated from hippocampus, striatum, and cortex (Costas-Insua et al., 2021). In mice, $BiP^{(+/+)}$ wild-type versus $BiP^{(+/-)}$ heterozygous mice were not different in their response to Δ^9-THC in the tetrad of cannabinoid behaviors or the development of tolerance. However, Δ^9-THC-evoked measures of anxiety were augmented and failed to exhibit tolerance in $BiP^{(+/-)}$ mice expressing reduced BiP levels (Costas-Insua et al., 2021). Thus, there may be a special role for BiP chaperoning of CB_1Rs associated with Gq-coupled signaling in GABAergic terminals in promoting anxiolytic responses to cannabinoid agonists.

In a mouse model of experimental amyotrophic lateral sclerosis (ALS), $BiP^{(+/-)}$ mice exhibited a greater disease-related deterioration of neurological scores and survival times compared with their $BiP^{(+/+)}$ ALS controls (Gómez-Almería et al., 2021). This pathology was manifest as greater activation of microglia and astroglia at the regions of spinal motor neuron degeneration. These findings suggest that neuroprotective effects of CB_1R might be promoted by BiP.

SGIP1 and CB_1R internalization

The nervous system protein "Src homology 3 (SH3)-domain growth factor receptor-bound 2 (GRB)-like (endophilin) interacting protein 1" (SGIP1) was identified in a yeast two-hybrid screen for targets that interact with the CB_1R C-terminal peptide (Hájková et al., 2016). Interaction with CB_1R was confirmed by co-immunoprecipitation with CB_1R from rodent brain extracts, and BRET in HEK293 cells transiently expressing tagged SGIP1 and CB_1R (Hájková et al., 2016). In cultured neurons from rat embryonic brain cortex, endogenous SGIP1 was found in axonal neurites, and co-localized with CB_1R at presynaptic terminals. SGIP1 expression in HEK293 cells diminished internalization of CB_1Rs desensitized by WIN55212-2, concurrent with the prolonged β-arrestin2-CB_1R association at the plasma membrane and reduced

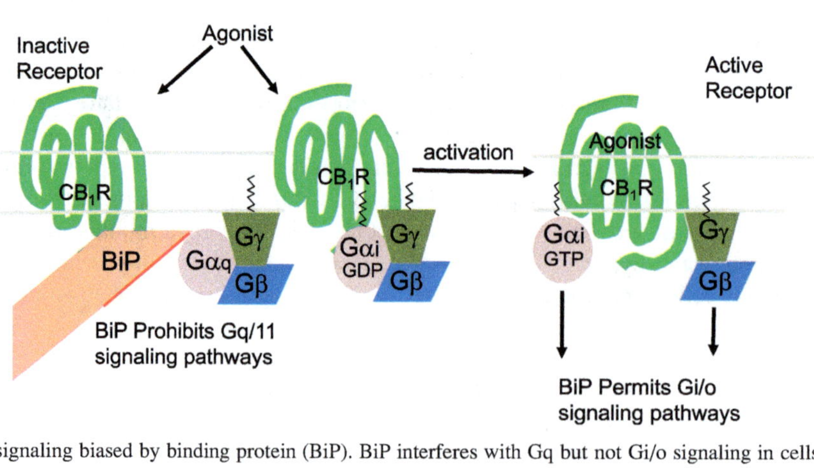

FIG. 4 Cannabinoid receptor signaling biased by binding protein (BiP). BiP interferes with Gq but not Gi/o signaling in cells.

internalization-dependent ERK1/2 phosphorylation (Fig. 3) (Hájková et al., 2016). One ramification of SGIP1's disrupting clathrin-mediated endocytosis may be that desensitized CB_1Rs accumulate where SGIP1 is located at neuronal axons.

One physiological impact of SGIP1 is found in models of obesity in rodents, which identified SGIP1 gene expression as directly related to food intake and body weight (Trevaskis et al., 2005). A human population genetic association study found that polymorphisms in the SGIP1 gene were associated with fat mass index (Cummings et al., 2012). SGIP1's effects may be associated with CB_1Rs in multiple brain regions associated with food intake, or at peripheral organs associated with energy storage, as the endocannabinoid system is a well-recognized regulator of energy balance (DiPatrizio, 2021; Pucci et al., 2021).

In SGIP1$^{(-/-)}$ mice, Δ^9-THC-evoked hypothermia and thermal antinociception, but not catalepsy (Dvorakova et al., 2021). Both cannabinoid and opioid antinociception in the mouse tail immersion model was accentuated in SGIP1$^{(-/-)}$ mice, suggesting that multiple receptor types are regulated by SGIP1 (Dvorakova et al., 2021). Tolerance to Δ^9-THC developed for these responses in SGIP1$^{(-/-)}$ mice, but the rate of tolerance was reduced for only the immobility response. SGIP1$^{(-/-)}$ mice spent more time in the open arms of the elevated plus maze as a measure of antianxiety and exhibited greater resilience to the tail suspension response to an inescapable stress (Dvorakova et al., 2021). These findings suggest that endogenous SGIP1 in wild-type mice may be associated with anxiogenic behavior and less effective emotional coping mechanisms. Other tests for cognitive function, sensorimotor gating, working memory, exploration, ambulation, and others did not show a distinct SGIP1$^{(-/-)}$ cannabinoid phenotype. Thus, CB_1R regulation by SGIP1 may differ with neuronal subtypes (glutamatergic versus GABAergic), neurotransmission strength or thresholds, CB_1R densities or cellular locations, endocannabinoid tone, or other regulatory factors in retrograde endocannabinoid signaling.

Wntless (WLS/GPR177) as CB_1R- and CB_2R-interacting proteins

Wnt is a protein with a complex synthesis, distribution, and receptor signaling mechanism that regulates mammalian embryonic cell polarity in blastocyst formation, cell proliferation, cell fate, as well as adult cell regulation (Huang et al., 2019; Nusse & Clevers, 2017). Wntless (WLS), also known as mammalian GPR177, Evi, or Sprinter, is a 7-transmembrane protein that functions to transport palmitoylated Wnt proteins from their site of synthesis and modification at the endoplasmic reticulum/Golgi complex to the plasma membrane for secretion (Routledge & Scholpp, 2019; Zhong et al., 2021). In a WLS C-terminal yeast two-hybrid screen, SGIP1 was identified and verified using GST-WLS C-terminus pull-down (Petko et al., 2018). WLS C-terminus could be pulled down by a GST-mu opioid receptor (MOR) intracellular loop 2 domain, and MOR antibodies could immunoprecipitate WLS from HEK293-MOR cells, PC12 cells, and rat brain homogenates (Jin et al., 2010). This complex of WLS with a GPCR generated the notion that other GPCRs might function as WLS-interacting proteins. Although not found in the yeast two-hybrid screen, both HA-tagged CB_1R and CB_2R were co-immunoprecipitated by Flag-WLS exogenously expressed in HEK293 cells (Petko et al., 2018).

The nature of the interaction between WLS and CB_1R or CB_2R is not understood. Perhaps endocytic trafficking of WLS and CB_1Rs may result from a complex with SGIP1. WLS interacted with MOR in neurons in the rat striatum, co-localizing in dendritic plasma membrane and cytoplasm domains (Jin et al., 2010). MOR and WLS co-localization in HEK293-MOR cells was augmented at the plasma membrane after treatment with morphine, but internalized after DAMGO ([D-Ala^2-, N-MePhe^4,Gly-ol^5]-enkephalin) (Jin et al., 2010). Wnt secretion by WLS was attenuated by morphine but not DAMGO in HEK293-MOR cells exogenously expressing myc-Wnt2; this attenuation could be rescued by overexpression of WLS (Jin et al., 2010). Thus, the retention of WLS at the membrane by either SGIP1 or MOR could alter WLS recycling which is necessary for continued Wnt secretion. Whether (or how) CB_1R might intervene in these processes is not established. Nevertheless, common functions of Wnt signaling and the endocannabinoid system are found in neurodevelopment and synaptic plasticity (Gaffuri et al., 2012; Galve-Roperh et al., 2013; Jin et al., 2010; Petko et al., 2018; Prenderville et al., 2015; Rodrigues et al., 2019).

G protein receptor-associated sorting protein 1 (GASP1) and CB_1R lysosomal degradation

G protein receptor-associated sorting protein 1 (GASP1) was identified to be a CB_1 receptor-associated protein by pull-down assays using a GST-CB_1R distal C-terminal 14-mer peptide (Martini et al., 2007). The association was confirmed by co-immunoprecipitation of GASP1 from N-terminally tagged-CB_1R-expressing HEK293 cell lysates or from rat brain synaptosomes using anti-CB_1R antibodies (Martini et al., 2007; Tappe-Theodor et al., 2007). GASP family proteins include a highly conserved 15 amino acid, "GASP motif" containing the conserved SWFW core sequence, which is repeated multiple times (Bornert et al., 2013). The GPCR-GASP association requires the conserved "FR" amino acid sequence within the

GPCR C-terminal helix 8 motif (Simonin et al., 2004; Thompson et al., 2007). Codes for internalization versus transit to lysosomes are distinct: CB_1R distal C-terminal valines 460 and 464 and the segment of the receptor between these sites are important for signaling internalization (Hsieh et al., 1999). Truncation at either of these valines (which removes the 14-mer peptide that originally pulled down GASP1 (Martini et al., 2007)) failed to disrupt the ability of CB_1R to co-immunoprecipitate GASP (Martini et al., 2007; Tappe-Theodor et al., 2007), lending support to the idea that CB_1R-GASP1 interaction occurs at multiple C-terminal sites (Fig. 3).

As for multiple GPCRs, GASP1 directs internalized CB_1Rs to the lysosome for degradation (Fig. 3) (Martini et al., 2007; Tappe-Theodor et al., 2007). In HEK293 cells, WIN55212-2 stimulated enhanced green fluorescent protein-CB_1Rs co-localization with GASP1 and with a lysosomal marker (Tappe-Theodor et al., 2007). In neonatal rat spinal cord neuronal cultures, WIN55212-2-stimuated downregulation of CB_1Rs was attenuated by expression of a dominant-negative cGASP1 (Tappe-Theodor et al., 2007). In vivo, the reduction in CB_1Rs in the mouse spinal dorsal horn after WIN55212-2 treatment was attenuated threefold by dominant-negative cGASP1 expression, consistent with attenuation of analgesic tolerance (Tappe-Theodor et al., 2007).

Conclusions

CB_1R trafficking and cellular signaling mechanisms can be promoted or mitigated by the expression of these CB_1R-associated proteins. Dysregulation of these associated proteins can result in pathological responses to CB_1R stimulation by agonists. Thus, it is important to consider the CB_1R-associated proteins as determinants of cell-specific responses to endocannabinoid and phytocannabinoid stimulation.

Key facts

Key facts of yeast two-hybrid screen

- Is a method of mapping a protein's interactome by expressing a cDNA fusion construct for one segment of an enzyme to be fused with the "bait" protein, along with the cDNA construct for the remaining segment of the enzyme fused to constructs from a cDNA library derived from a relevant cell system (referred to as "prey").
- When expressed in yeast, the enzyme becomes active only if both segments are assembled by a bait-prey protein-protein interaction.
- The enzyme then functions as the detection system. The prey is subsequently identified.

Key facts of co-immunoprecipitation and western blotting

- Is a method to detect the protein(s) that are associated in a stable complex with a target protein in solution (detergent-solubilized if from cell membranes).
- The target protein interacts with a specific antibody bound to a solid phase (e.g., agarose bead coupled to proteins that bind antibodies).
- The "bead-antibody-target protein-associated protein(s)" complex is separated from other proteins by sedimentation and dissociated by low pH or other means; proteins are then separated based upon size, shape, and charge using polyacrylamide gel electrophoresis.
- The proteins on the gel are iontophoretically transferred onto a blotting membrane, and identified by "western blot" with "primary" antibodies to the target protein and to predicted associated proteins, followed by "secondary" antibodies that recognize the primary antibody's species-specific epitope.
- The secondary antibodies are covalently bound to a fluorescent dye or an enzyme for detection.

Key facts of bioluminescence resonance energy transfer (BRET)

- BRET is a method to detect protein-protein associations within living cells based upon expression in a host cell (e.g., HEK293 fibroblasts) of fusion proteins created by joining the proteins of interest with fluorescent proteins.
- The "donor" typically is a fusion of the first protein of interest and luciferase, which in the presence of a substrate emits fluorescence at a wavelength that can excite an "acceptor" fusion protein of the second protein of interest with a fluorescent protein.
- When the two fusion proteins are in close proximity (<10 nm), the recipient emission is detected.

Mini-dictionary of terms

- **Heterotrimeric G proteins**. The G protein heterotrimer consists of an inactive Gα protein bound to GDP and associated with a Gβγ dimer. When the G protein-coupled receptor (GPCR) is activated by an agonist, the Gα-GDP releases the GDP, and the Gα protein binds GTP, which is abundant in the cytosol. Gα-GTP and Gβγ dimer are released from the GPCR to interact with effector proteins (e.g., adenylyl cyclase, phospholipase Cβ). The Gα deactivates via its GTPase activity that hydrolyzes bound GTP to GDP. The inactive Gα-GDP is able to couple to a Gβγ dimer, and the heterotrimer re-associates with a GPCR to continue another activation cycle.
- **CP55940**. A nonclassical analog of 11-hydroxy-tetrahydrocannabinol that is a full agonist at CB_1 and CB_2 cannabinoid receptors.
- **WIN55212-2**. A cannabimimetic indole structure that was originally developed from a nonsteroidal antiinflammatory lead compound and is a full agonist at CB_1 and CB_2 cannabinoid receptors.

Summary points

- CB_1 and CB_2 cannabinoid receptors are G protein-coupled receptors that regulate primarily Gi/o family of proteins to promote initial signaling pathways. Regulators of G protein signaling (RGS) proteins accelerate the inactivation of Gα-GTP proteins and have been associated with cannabinoid actions in abrogating neuropathic pain and inflammatory bowel disease.
- Phosphorylation of CB_1 receptor C-terminal serine and threonine residues by G protein receptor kinases (GRKs) allows binding of β-arrestins1 and 2.
- β-Arrestins1 and 2 initiate alternative cellular signaling such as by extracellular signal regulated kinases (ERK1/2) phosphorylation and promote clathrin-mediated endocytosis, respectively. β-Arrestins alter rates of tolerance to cannabinoid responses.
- Cannabinoid receptor interacting protein 1a (CRIP1a) attenuates CB_1 receptor-mediated G protein signaling probably by binding Gαi/o proteins, and competitively with β-arrestins1 and 2 binding to CB_1 receptors. CRIP1a has been associated with epileptiform seizures and visual system signaling.
- Binding protein (BiP), a member of the heat shock protein 70 chaperone family, binds to CB_1 receptors to alter Ca^{2+}-mobilization signaling and is associated with antianxiety behaviors and neuroprotection in amyotrophic lateral sclerosis in animal models.
- Src homology3-domain growth factor receptor-bound2-like (endothelin) interacting protein (SGIP1) blocks endocytosis of the CB_1 receptor-β-arrestin2 complex at the plasma membrane. SGIP1 expression is associated with an obesity phenotype and suppression of the cannabinoid-mediated analgesia and antianxiety responses in animal models.
- Wntless (WLS)/GPR177 binds to the CB_1 receptor and to SGIP1 and might be involved in the well-recognized neurodevelopment and synaptic plasticity responses to CB_1 receptor stimulation.
- G protein receptor-associated sorting protein1 (GASP1) shuttles CB_1 receptors to the lysosome for degradation and is associated with analgesic tolerance to cannabinoid agonists.

References

Attramadal, H., Arriza, J. L., Aoki, C., Dawson, T. M., Codina, J., Kwatra, M. M., … Lefkowitz, R. J. (1992). Beta-arrestin2, a novel member of the arrestin/beta-arrestin gene family. *The Journal of Biological Chemistry*, *267*(25), 17882–17890.

Blume, L. C., Eldeeb, K., Bass, C. E., Selley, D. E., & Howlett, A. C. (2015). Cannabinoid receptor interacting protein (CRIP1a) attenuates CB1R signaling in neuronal cells. *Cellular Signalling*, *27*(3), 716–726. https://doi.org/10.1016/j.cellsig.2014.11.006.

Blume, L. C., Leone-Kabler, S., Luessen, D. J., Marrs, G. S., Lyons, E., Bass, C. E., … Howlett, A. C. (2016). Cannabinoid receptor interacting protein suppresses agonist-driven CB(1) receptor internalization and regulates receptor replenishment in an agonist-biased manner. *Journal of Neurochemistry*, *139*(3), 396–407. https://doi.org/10.1111/jnc.13767.

Blume, L. C., Patten, T., Eldeeb, K., Leone-Kabler, S., Ilyasov, A. A., Keegan, B. M., … Howlett, A. L. (2017). Cannabinoid receptor interacting protein 1a competition with β-arrestin for CB1 receptor binding sites. *Molecular Pharmacology*, *91*(2), 75–86. https://doi.org/10.1124/mol.116.104638.

Booth, W. T., Clodfelter, J. E., Leone-Kabler, S., Hughes, E. K., Eldeeb, K., Howlett, A. C., & Lowther, W. T. (2021). Cannabinoid receptor interacting protein 1a interacts with myristoylated Gα(i) N terminus via a unique gapped β-barrel structure. *The Journal of Biological Chemistry*, *297*(3), 101099. https://doi.org/10.1016/j.jbc.2021.101099.

Booth, W. T., Walker, N. B., Lowther, W. T., & Howlett, A. C. (2019). Cannabinoid receptor interacting protein 1a (CRIP1a): Function and structure. *Molecules*, *24*(20). https://doi.org/10.3390/molecules24203672.

Bornert, O., Møller, T. C., Boeuf, J., Candusso, M. P., Wagner, R., Martinez, K. L., & Simonin, F. (2013). Identification of a novel protein-protein interaction motif mediating interaction of GPCR-associated sorting proteins with G protein-coupled receptors. *PLoS One*, *8*(2), e56336. https://doi.org/10.1371/journal.pone.0056336.

Bosier, B., Doyen, P. J., Brolet, A., Muccioli, G. G., Ahmed, E., Desmet, N., … Deumens, R. (2015). Inhibition of the regulator of G protein signalling RGS4 in the spinal cord decreases neuropathic hyperalgesia and restores cannabinoid CB1 receptor signalling. *British Journal of Pharmacology*, *172*(22), 5333–5346. https://doi.org/10.1111/bph.13324.

Breivogel, C. S., Lambert, J. M., Gerfin, S., Huffman, J. W., & Razdan, R. K. (2008). Sensitivity to delta9-tetrahydrocannabinol is selectively enhanced in beta-arrestin2-/- mice. *Behavioural Pharmacology*, *19*(4), 298–307. https://doi.org/10.1097/FBP.0b013e328308f1e6.

Breivogel, C. S., Puri, V., Lambert, J. M., Hill, D. K., Huffman, J. W., & Razdan, R. K. (2013). The influence of beta-arrestin2 on cannabinoid CB1 receptor coupling to G-proteins and subcellular localization and relative levels of beta-arrestin1 and 2 in mouse brain. *Journal of Receptor and Signal Transduction Research*, *33*(6), 367–379. https://doi.org/10.3109/10799893.2013.838787.

Breivogel, C. S., & Vaghela, M. S. (2015). The effects of beta-arrestin1 deletion on acute cannabinoid activity, brain cannabinoid receptors and tolerance to cannabinoids in mice. *Journal of Receptor and Signal Transduction Research*, *35*(1), 98–106. https://doi.org/10.3109/10799893.2014.1003659.

Costas-Insua, C., Moreno, E., Maroto, I. B., Ruiz-Calvo, A., Bajo-Grañeras, R., Martín-Gutiérrez, D., … Guzmán, M. (2021). Identification of BiP as a CB(1) receptor-interacting protein that fine-tunes cannabinoid signaling in the mouse brain. *The Journal of Neuroscience*, *41*(38), 7924–7941. https://doi.org/10.1523/jneurosci.0821-21.2021.

Cummings, N., Shields, K. A., Curran, J. E., Bozaoglu, K., Trevaskis, J., Gluschenko, K., … Jowett, J. B. (2012). Genetic variation in SH3-domain GRB2-like (endophilin)-interacting protein 1 has a major impact on fat mass. *International Journal of Obesity*, *36*(2), 201–206. https://doi.org/10.1038/ijo.2011.67.

Daigle, T. L., Kearn, C. S., & Mackie, K. (2008). Rapid CB1 cannabinoid receptor desensitization defines the time course of ERK1/2 MAP kinase signaling. *Neuropharmacology*, *54*(1), 36–44.

DeFea, K. A., Vaughn, Z. D., O'Bryan, E. M., Nishijima, D., Dery, O., & Bunnett, N. W. (2000). The proliferative and antiapoptotic effects of substance P are facilitated by formation of a beta-arrestin-dependent scaffolding complex. *Proceedings of the National Academy of Sciences of the United States of America*, *97*(20), 11086–11091. https://doi.org/10.1073/pnas.190276697.

Delgado-Peraza, F., Ahn, K. H., Nogueras-Ortiz, C., Mungrue, I. N., Mackie, K., Kendall, D. A., & Yudowski, G. A. (2016). Mechanisms of biased beta-arrestin-mediated signaling downstream from the cannabinoid 1 receptor. *Molecular Pharmacology*, *89*(6), 618–629. https://doi.org/10.1124/mol.115.103176.

Dhukhwa, A., Al Aameri, R. F. H., Sheth, S., Mukherjea, D., Rybak, L., & Ramkumar, V. (2021). Regulator of G protein signaling 17 represents a novel target for treating cisplatin induced hearing loss. *Scientific Reports*, *11*(1), 8116. https://doi.org/10.1038/s41598-021-87387-5.

DiPatrizio, N. V. (2021). Endocannabinoids and the gut-brain control of food intake and obesity. *Nutrients*, *13*(4). https://doi.org/10.3390/nu13041214.

Dvorakova, M., Kubik-Zahorodna, A., Straiker, A., Sedlacek, R., Hajkova, A., Mackie, K., & Blahos, J. (2021). SGIP1 is involved in regulation of emotionality, mood, and nociception and modulates in vivo signalling of cannabinoid CB(1) receptors. *British Journal of Pharmacology*, *178*(7), 1588–1604. https://doi.org/10.1111/bph.15383.

Finlay, D. B., Manning, J. J., Ibsen, M. S., Macdonald, C. E., Patel, M., Javitch, J. A., … Glass, M. (2019). Do toxic synthetic cannabinoid receptor agonists have signature in vitro activity profiles? A case study of AMB-FUBINACA. *ACS Chemical Neuroscience*, *10*(10), 4350–4360. https://doi.org/10.1021/acschemneuro.9b00429.

Fletcher-Jones, A., Hildick, K. L., Evans, A. J., Nakamura, Y., Henley, J. M., & Wilkinson, K. A. (2020). Corrigendum: Protein interactors and trafficking pathways that regulate the cannabinoid type 1 receptor (CB1R). *Frontiers in Molecular Neuroscience*, *13*, 142. https://doi.org/10.3389/fnmol.2020.00142.

Flores-Otero, J., Ahn, K. H., Delgado-Peraza, F., Mackie, K., Kendall, D. A., & Yudowski, G. A. (2014). Ligand-specific endocytic dwell times control functional selectivity of the cannabinoid receptor 1. *Nature Communications*, *5*, 4589. https://doi.org/10.1038/ncomms5589.

Gaffuri, A. L., Ladarre, D., & Lenkei, Z. (2012). Type-1 cannabinoid receptor signaling in neuronal development. *Pharmacology*, *90*(1-2), 19–39. https://doi.org/10.1159/000339075.

Galve-Roperh, I., Chiurchiu, V., Diaz-Alonso, J., Bari, M., Guzman, M., & Maccarrone, M. (2013). Cannabinoid receptor signaling in progenitor/stem cell proliferation and differentiation. *Progress in Lipid Research*, *52*(4), 633–650. https://doi.org/10.1016/j.plipres.2013.05.004.

Ghosh, S., Sheth, S., Sheehan, K., Mukherjea, D., Dhukhwa, A., Borse, V., … Ramkumar, V. (2018). The endocannabinoid/cannabinoid receptor 2 system protects against cisplatin-induced hearing loss. *Frontiers in Cellular Neuroscience*, *12*, 271. https://doi.org/10.3389/fncel.2018.00271.

Gómez-Almería, M., Burgaz, S., Costas-Insua, C., Rodríguez-Cueto, C., Santos-García, I., Rodríguez-Crespo, I., … Fernández-Ruiz, J. (2021). BiP heterozigosity aggravates pathological deterioration in experimental amyotrophic lateral sclerosis. *International Journal of Molecular Sciences*, *22*(22). https://doi.org/10.3390/ijms222212533.

Guggenhuber, S., Alpar, A., Chen, R., Schmitz, N., Wickert, M., Mattheus, T., … Klugmann, M. (2016). Cannabinoid receptor-interacting protein Crip1a modulates CB1 receptor signaling in mouse hippocampus. *Brain Structure & Function*, *221*(4), 2061–2074. https://doi.org/10.1007/s00429-015-1027-6.

Gyombolai, P., Toth, A. D., Timar, D., Turu, G., & Hunyady, L. (2015). Mutations in the 'DRY' motif of the CB1 cannabinoid receptor result in biased receptor variants. *Journal of Molecular Endocrinology*, *54*(1), 75–89. https://doi.org/10.1530/JME-14-0219.

Hájková, A., Techlovská, Š., Dvořáková, M., Chambers, J. N., Kumpošt, J., Hubálková, P., … Blahos, J. (2016). SGIP1 alters internalization and modulates signaling of activated cannabinoid receptor 1 in a biased manner. *Neuropharmacology*, *107*, 201–214. https://doi.org/10.1016/j.neuropharm.2016.03.008.

Howlett, A. C., & Abood, M. E. (2017). CB(1) and CB(2) receptor pharmacology. *Advances in Pharmacology*, *80*, 169–206. https://doi.org/10.1016/bs.apha.2017.03.007.

Howlett, A. C., Barth, F., Bonner, T. I., Cabral, G., Casellas, P., Devane, W. A., … Pertwee, R. G. (2002). International Union of Pharmacology. XXVII. Classification of cannabinoid receptors. *Pharmacological Reviews*, *54*(2), 161–202. https://doi.org/10.1124/pr.54.2.161.

Howlett, A. C., Blume, L. C., & Dalton, G. D. (2010). CB(1) cannabinoid receptors and their associated proteins. *Current Medicinal Chemistry*, *17*(14), 1382–1393. https://doi.org/10.2174/092986710790980023.

Hsieh, C., Brown, S., Derleth, C., & Mackie, K. (1999). Internalization and recycling of the CB1 cannabinoid receptor. *Journal of Neurochemistry*, *73*(2), 493–501. https://doi.org/10.1046/j.1471-4159.1999.0730493.x.

Hu, S. S., Arnold, A., Hutchens, J. M., Radicke, J., Cravatt, B. F., Wager-Miller, J., … Straiker, A. (2010). Architecture of cannabinoid signaling in mouse retina. *Journal of Comparative Neurology*, *518*(18), 3848–3866. https://doi.org/10.1002/cne.22429.

Huang, P., Yan, R., Zhang, X., Wang, L., Ke, X., & Qu, Y. (2019). Activating Wnt/beta-catenin signaling pathway for disease therapy: Challenges and opportunities. *Pharmacology & Therapeutics*, *196*, 79–90. https://doi.org/10.1016/j.pharmthera.2018.11.008.

Ibsen, M. S., Finlay, D. B., Patel, M., Javitch, J. A., Glass, M., & Grimsey, N. L. (2019). Cannabinoid CB1 and CB2 receptor-mediated arrestin translocation: Species, subtype, and agonist-dependence. *Frontiers in Pharmacology*, *10*, 350. https://doi.org/10.3389/fphar.2019.00350.

Ishii, M., & Kurachi, Y. (2003). Physiological actions of regulators of G-protein signaling (RGS) proteins. *Life Sciences*, *74*(2–3), 163–171. https://doi.org/10.1016/j.lfs.2003.09.004.

Jin, W., Brown, S., Roche, J. P., Hsieh, C., Celver, J. P., Kovoor, A., … Mackie, K. (1999). Distinct domains of the CB1 cannabinoid receptor mediate desensitization and internalization. *The Journal of Neuroscience*, *19*(10), 3773–3780.

Jin, J., Kittanakom, S., Wong, V., Reyes, B. A., Van Bockstaele, E. J., Stagljar, I., … Levenson, R. (2010). Interaction of the mu-opioid receptor with GPR177 (Wntless) inhibits Wnt secretion: Potential implications for opioid dependence. *BMC Neuroscience*, *11*, 33. https://doi.org/10.1186/1471-2202-11-33.

Kendall, D. A., & Yudowski, G. A. (2016). Cannabinoid receptors in the central nervous system: Their signaling and roles in disease. *Frontiers in Cellular Neuroscience*, *10*, 294. https://doi.org/10.3389/fncel.2016.00294.

Kozakiewicz, M. L., Zhang, J., Leone-Kabler, S., Yamaleyeva, L. M., McDonald, A. G., Brost, B. C., & Howlett, A. C. (2021). Differential expression of CB(1) cannabinoid receptor and cannabinoid receptor interacting protein 1a in labor. *Cannabis and Cannabinoid Research*. https://doi.org/10.1089/can.2020.0107.

Kwon, H. J., Kim, D. S., Kim, W., Jung, H. Y., Yu, Y. H., Ju, Y. I., … Yoo, D. Y. (2020). Tat-cannabinoid receptor interacting protein reduces ischemia-induced neuronal damage and its possible relationship with 14-3-3eta. *Cells*, *9*(8). https://doi.org/10.3390/cells9081827.

Laprairie, R. B., Bagher, A. M., Kelly, M. E., Dupre, D. J., & Denovan-Wright, E. M. (2014). Type 1 cannabinoid receptor ligands display functional selectivity in a cell culture model of striatal medium spiny projection neurons. *The Journal of Biological Chemistry*, *289*(36), 24845–24862. https://doi.org/10.1074/jbc.M114.557025.

Lauckner, J. E., Hille, B., & Mackie, K. (2005). The cannabinoid agonist WIN55,212-2 increases intracellular calcium via CB1 receptor coupling to Gq/11 G proteins. *Proceedings of the National Academy of Sciences of the United States of America*, *102*(52), 19144–19149. https://doi.org/10.1073/pnas.0509588102.

Lefkowitz, R. J. (2013). Arrestins come of age: A personal historical perspective. *Progress in Molecular Biology and Translational Science*, *118*, 3–18. https://doi.org/10.1016/B978-0-12-394440-5.00001-2.

Leo, L. M., & Abood, M. E. (2021). CB1 cannabinoid receptor signaling and biased signaling. *Molecules*, *26*(17). https://doi.org/10.3390/molecules26175413.

Lerner, T. N., & Kreitzer, A. C. (2012). RGS4 is required for dopaminergic control of striatal LTD and susceptibility to parkinsonian motor deficits. *Neuron*, *73*(2), 347–359. https://doi.org/10.1016/j.neuron.2011.11.015.

Luttrell, L. M., Roudabush, F. L., Choy, E. W., Miller, W. E., Field, M. E., Pierce, K. L., & Lefkowitz, R. J. (2001). Activation and targeting of extracellular signal-regulated kinases by beta-arrestin scaffolds. *Proceedings of the National Academy of Sciences of the United States of America*, *98*(5), 2449–2454. https://doi.org/10.1073/pnas.041604898.

Lutz, B. (2020). Neurobiology of cannabinoid receptor signaling. *Dialogues in Clinical Neuroscience*, *22*(3), 207–222. https://doi.org/10.31887/DCNS.2020.22.3/blutz.

Ma, L., & Pei, G. (2007). Beta-arrestin signaling and regulation of transcription. *Journal of Cell Science*, *120*(Pt 2), 213–218. https://doi.org/10.1242/jcs.03338.

Mackie, K. (2008). Signaling via CNS cannabinoid receptors. *Molecular and Cellular Endocrinology*, *286*(1–2 Suppl 1), S60–S65. https://doi.org/10.1016/j.mce.2008.01.022.

Martini, L., Waldhoer, M., Pusch, M., Kharazia, V., Fong, J., Lee, J. H., … Whistler, J. L. (2007). Ligand-induced down-regulation of the cannabinoid 1 receptor is mediated by the G-protein-coupled receptor-associated sorting protein GASP1. *The FASEB Journal*, *21*(3), 802–811. https://doi.org/10.1096/fj.06-7132com.

Mascia, F., Klotz, L., Lerch, J., Ahmed, M. H., Zhang, Y., & Enz, R. (2017). CRIP1a inhibits endocytosis of G-protein coupled receptors activated by endocannabinoids and glutamate by a common molecular mechanism. *Journal of Neurochemistry*, *141*(4), 577–591. https://doi.org/10.1111/jnc.14021.

Mattheus, T., Kukla, K., Zimmermann, T., Tenzer, S., & Lutz, B. (2016). Cell Type-specific tandem affinity purification of the mouse hippocampal CB1 receptor-associated proteome. *Journal of Proteome Research*, *15*(10), 3585–3601. https://doi.org/10.1021/acs.jproteome.6b00339.

Molina-Holgado, E., Paniagua-Torija, B., Arevalo-Martin, A., Moreno-Luna, R., Esteban, P. F., Le, M. Q. U., … Garcia-Ovejero, D. (2021). Cannabinoid Receptor 1 associates to different molecular complexes during GABAergic neuron maturation. *Journal of Neurochemistry*, *158*(3), 640–656. https://doi.org/10.1111/jnc.15381.

Morgan, D. J., Davis, B. J., Kearn, C. S., Marcus, D., Cook, A. J., Wager-Miller, J., … Mackie, K. (2014). Mutation of putative GRK phosphorylation sites in the cannabinoid receptor 1 (CB1R) confers resistance to cannabinoid tolerance and hypersensitivity to cannabinoids in mice. *The Journal of Neuroscience*, *34*(15), 5152–5163. https://doi.org/10.1523/JNEUROSCI.3445-12.2014.

Navarro, G., Reyes-Resina, I., Rivas-Santisteban, R., Sanchez de Medina, V., Morales, P., Casano, S., … Franco, R. (2018). Cannabidiol skews biased agonism at cannabinoid CB1 and CB2 receptors with smaller effect in CB1-CB2 heteroreceptor complexes. *Biochemical Pharmacology*, *157*, 148–158. https://doi.org/10.1016/j.bcp.2018.08.046.

Nealon, C. M., Henderson-Redmond, A. N., Hale, D. E., & Morgan, D. J. (2019). Tolerance to WIN55,212-2 is delayed in desensitization-resistant S426A/S430A mice. *Neuropharmacology*, *148*, 151–159. https://doi.org/10.1016/j.neuropharm.2018.12.026.

Nguyen, P. T., Schmid, C. L., Raehal, K. M., Selley, D. E., Bohn, L. M., & Sim-Selley, L. J. (2012). Beta-arrestin2 regulates cannabinoid CB_1 receptor signaling and adaptation in a central nervous system region-dependent manner. *Biological Psychiatry*, *71*(8), 714–724. https://doi.org/10.1016/j.biopsych.2011.11.027.

Niehaus, J. L., Liu, Y., Wallis, K. T., Egertová, M., Bhartur, S. G., Mukhopadhyay, S., … Lewis, D. L. (2007). CB1 cannabinoid receptor activity is modulated by the cannabinoid receptor interacting protein CRIP 1a. *Molecular Pharmacology*, *72*(6), 1557–1566. https://doi.org/10.1124/mol.107.039263.

Njoo, C., Agarwal, N., Lutz, B., & Kuner, R. (2015). The cannabinoid receptor CB1 interacts with the WAVE1 complex and plays a role in actin dynamics and structural plasticity in neurons. *PLoS Biology*, *13*(10), e1002286. https://doi.org/10.1371/journal.pbio.1002286.

Nogueras-Ortiz, C., & Yudowski, G. A. (2016). The multiple waves of cannabinoid 1 receptor signaling. *Molecular Pharmacology*, *90*(5), 620–626. https://doi.org/10.1124/mol.116.104539.

Nusse, R., & Clevers, H. (2017). Wnt/beta-catenin signaling, disease, and emerging therapeutic modalities. *Cell*, *169*(6), 985–999. https://doi.org/10.1016/j.cell.2017.05.016.

Oliver, E. E., Hughes, E. K., Puckett, M. K., Chen, R., Lowther, W. T., & Howlett, A. C. (2020). Cannabinoid receptor interacting protein 1a (CRIP1a) in health and disease. *Biomolecules*, *10*(12). https://doi.org/10.3390/biom10121609.

Patel, M., Manning, J. J., Finlay, D. B., Javitch, J. A., Banister, S. D., Grimsey, N. L., & Glass, M. (2020). Signalling profiles of a structurally diverse panel of synthetic cannabinoid receptor agonists. *Biochemical Pharmacology*, *175*, 113871. https://doi.org/10.1016/j.bcp.2020.113871.

Petko, J., Tranchina, T., Patel, G., Levenson, R., & Justice-Bitner, S. (2018). Identifying novel members of the Wntless interactome through genetic and candidate gene approaches. *Brain Research Bulletin*, *138*, 96–105. https://doi.org/10.1016/j.brainresbull.2017.07.004.

Prenderville, J. A., Kelly, A. M., & Downer, E. J. (2015). The role of cannabinoids in adult neurogenesis. *British Journal of Pharmacology*, *172*(16), 3950–3963. https://doi.org/10.1111/bph.13186.

Pucci, M., Zaplatic, E., Micioni Di Bonaventura, M. V., Micioni Di Bonaventura, E., De Cristofaro, P., Maccarrone, M., … D'Addario, C. (2021). On the role of central type-1 cannabinoid receptor gene regulation in food intake and eating behaviors. *International Journal of Molecular Sciences*, *22*(1). https://doi.org/10.3390/ijms22010398.

Raich, I., Rivas-Santisteban, R., Lillo, A., Lillo, J., Reyes-Resina, I., Nadal, X., … Franco, R. (2021). Similarities and differences upon binding of naturally occurring Delta(9)-tetrahydrocannabinol-derivatives to cannabinoid CB1 and CB2 receptors. *Pharmacological Research*, *174*, 105970. https://doi.org/10.1016/j.phrs.2021.105970.

Rodrigues, R. S., Lourenco, D. M., Paulo, S. L., Mateus, J. M., Ferreira, M. F., Mouro, F. M., … Xapelli, S. (2019). Cannabinoid actions on neural stem cells: Implications for pathophysiology. *Molecules*, *24*(7). https://doi.org/10.3390/molecules24071350.

Roman, D. L., & Traynor, J. R. (2011). Regulators of G protein signaling (RGS) proteins as drug targets: Modulating G-protein-coupled receptor (GPCR) signal transduction. *Journal of Medicinal Chemistry*, *54*(21), 7433–7440. https://doi.org/10.1021/jm101572n.

Routledge, D., & Scholpp, S. (2019). Mechanisms of intercellular Wnt transport. *Development*, *146*(10). https://doi.org/10.1242/dev.176073.

Salaga, M., Storr, M., Martemyanov, K. A., & Fichna, J. (2016). RGS proteins as targets in the treatment of intestinal inflammation and visceral pain: New insights and future perspectives. *BioEssays*, *38*(4), 344–354. https://doi.org/10.1002/bies.201500118.

Sethakorn, N., Yau, D. M., & Dulin, N. O. (2010). Non-canonical functions of RGS proteins. *Cellular Signalling*, *22*(9), 1274–1281. https://doi.org/10.1016/j.cellsig.2010.03.016.

Sim, L. J., Hampson, R. E., Deadwyler, S. A., & Childers, S. R. (1996). Effects of chronic treatment with Δ^9-tetrahydrocannabinol on cannabinoid-stimulated [^{35}S]GTPγS autoradiography in rat brain. *The Journal of Neuroscience*, *16*, 8057–8066.

Simonin, F., Karcher, P., Boeuf, J. J., Matifas, A., & Kieffer, B. L. (2004). Identification of a novel family of G protein-coupled receptor associated sorting proteins. *Journal of Neurochemistry*, *89*(3), 766–775. https://doi.org/10.1111/j.1471-4159.2004.02411.x.

Sjogren, B., Blazer, L. L., & Neubig, R. R. (2010). Regulators of G protein signaling proteins as targets for drug discovery. *Progress in Molecular Biology and Translational Science*, *91*, 81–119. https://doi.org/10.1016/S1877-1173(10)91004-1.

Sjogren, B., & Neubig, R. R. (2010). Thinking outside of the "RGS box": New approaches to therapeutic targeting of regulators of G protein signaling. *Molecular Pharmacology*, *78*(4), 550–557. https://doi.org/10.1124/mol.110.065219.

Smith, T. H., Blume, L. C., Straiker, A., Cox, J. O., David, B. G., McVoy, J. R., … Selley, D. E. (2015). Cannabinoid receptor-interacting protein 1a modulates CB1 receptor signaling and regulation. *Molecular Pharmacology*, *87*(4), 747–765. https://doi.org/10.1124/mol.114.096495.

Smith, T. H., Sim-Selley, L. J., & Selley, D. E. (2010). Cannabinoid CB1 receptor-interacting proteins: Novel targets for central nervous system drug discovery? *British Journal of Pharmacology*, *160*(3), 454–466. https://doi.org/10.1111/j.1476-5381.2010.00777.x.

Stauffer, B., Wallis, K. T., Wilson, S. P., Egertova, M., Elphick, M. R., Lewis, D. L., & Hardy, L. R. (2011). CRIP1a switches cannabinoid receptor agonist/antagonist-mediated protection from glutamate excitotoxicity. *Neuroscience Letters*, *503*(3), 224–228. https://doi.org/10.1016/j.neulet.2011.08.041.

Stewart, A., Huang, J., & Fisher, R. A. (2012). RGS proteins in heart: Brakes on the vagus. *Frontiers in Physiology*, *3*, 95. https://doi.org/10.3389/fphys.2012.00095.

Tappe-Theodor, A., Agarwal, N., Katona, I., Rubino, T., Martini, L., Swiercz, J., … Kuner, R. (2007). A molecular basis of analgesic tolerance to cannabinoids. *The Journal of Neuroscience*, *27*(15), 4165–4177. https://doi.org/10.1523/jneurosci.5648-06.2007.

Thompson, D., Pusch, M., & Whistler, J. L. (2007). Changes in G protein-coupled receptor sorting protein affinity regulate postendocytic targeting of G protein-coupled receptors. *The Journal of Biological Chemistry*, *282*(40), 29178–29185. https://doi.org/10.1074/jbc.M704014200.

Trevaskis, J., Walder, K., Foletta, V., Kerr-Bayles, L., McMillan, J., Cooper, A., … Collier, G. R. (2005). Src homology 3-domain growth factor receptor-bound 2-like (endophilin) interacting protein 1, a novel neuronal protein that regulates energy balance. *Endocrinology*, *146*(9), 3757–3764. https://doi.org/10.1210/en.2005-0282.

Turu, G., & Hunyady, L. (2010). Signal transduction of the CB1 cannabinoid receptor. *Journal of Molecular Endocrinology*, *44*(2), 75–85. https://doi.org/10.1677/JME-08-0190.

Van der Stelt, M., & Di Marzo, V. (2003). The endocannabinoid system in the basal ganglia and in the mesolimbic reward system: Implications for neurological and psychiatric disorders. *European Journal of Pharmacology*, *480*(1–3), 133–150. https://doi.org/10.1016/j.ejphar.2003.08.101.

Zhong, Q., Zhao, Y., Ye, F., Xiao, Z., Huang, G., Xu, M., … Ma, D. (2021). Cryo-EM structure of human Wntless in complex with Wnt3a. *Nature Communications*, *12*(1), 4541. https://doi.org/10.1038/s41467-021-24731-3.

Chapter 24

Impact of cannabis on memory: From in utero to adulthood

Jennifer A. Willford[a], Meaghan E. Hazelet[a], Mohamad N. Khalaifa[a], Natacha De Genna[b], and Gale A. Richardson[b]

[a]*Department of Psychology, Slippery Rock University, Slippery Rock, PA, United States,* [b]*Department of Psychiatry, University of Pittsburgh School of Medicine, Pittsburgh, PA, United States*

Abbreviations

CBD	cannabidiol
CUD	Cannabis Use Disorder
EEG	electroencephalogram
fMRI	functional magnetic resonance imaging
IQ	intelligence quotient
LTM	long-term memory
MHPCD	Maternal Health Practices and Child Development Project
OPPS	Ottawa Prenatal Prospective Study
PCE	prenatal cannabis exposure
PFC	prefrontal cortex
SBIS	Stanford-Binet Intelligence Scale
SC	synthetic Cannabinoid
STM	short-term memory
THC	Δ9-tetrahydrocannabinol
WRAML	Wide Range Assessment of Memory and Learning
WRAT	Wide Range Achievement Test

Introduction

Approximately 12% of pregnant women (Volkow et al., 2019), 12.5% of adolescents between 12 and 17 years old (Substance Abuse and Mental Health Services Administration, 2019), and an estimated 200 million adults (World Drug Report, 2021) report cannabis use. Among users, exposure to cannabis occurs during important periods of brain development and may occur at multiple times across the lifespan. The 2020 report of the Monitoring the Future Study, a large-scale study of drug use patterns among adolescents in the United States, indicates that among 8th, 10th, and 12th grade high school students, daily marijuana use is at its highest levels since the early 1990s (Johnston et al., 2020). Further, adolescents are more likely to use vaping devices to increase the frequency and intensity of cannabis use, while decreasing the likelihood of being caught using the drug, which increases their vulnerability for addiction. In addition, the perceived risk of using marijuana has steadily decreased since the mid-2010s (Johnston et al., 2020).

In this chapter, the literature about the effects of cannabis exposure during pregnancy, adolescence, and adulthood on memory function is reviewed and then considered from the lifespan perspective. Polydrug exposure is also considered in this review as cannabis is often consumed in the presence of other drugs such as alcohol and tobacco. Neuroimaging studies provide further insights about potential neural compensation that may occur with cannabis use, as indicated by functional brain activation differences in the absence of behavioral differences. Lastly, the methodological limitations of current human studies of cannabis use and memory function are evaluated with an appraisal of the benefits of prospective,

Cannabis Use, Neurobiology, Psychology, and Treatment. https://doi.org/10.1016/B978-0-323-89862-1.00044-1

longitudinal study design. Importantly, the impact of cannabis exposure in a developmental context is evaluated to better understand the proximal and distal factors that influence the association between cannabis exposure and memory function.

Prenatal cannabis exposure and memory

There are two North American prospective longitudinal cohort studies that investigated the effects of prenatal cannabis exposure (PCE) on offspring development, including growth, behavior, cognition, and memory. The Ottawa Prenatal Prospective Study (OPPS) enrolled low-risk mothers who were predominantly white and middle class between 1978 and 1983. The Maternal Health Practices and Child Development Study (MHPCD) enrolled high-risk mothers (51.8% African American, 48.2% white) with low socioeconomic status (Day et al., 1994) between 1982 and 1985. The OPPS and MHPCD studies both examined pregnant women and their offspring from birth to young adulthood, statistically controlling for multiple sociodemographic (e.g., race, education, family income, and marital status), environmental (e.g., home environment), and child characteristics (e.g., age, gender, depression, and anxiety), maternal physical and mental health, and prenatal and current polysubstance use.

The effects of PCE on memory were assessed in both the OPPS and MHPCD at multiple developmental timepoints (Table 1). When the cohorts are compared, heavy PCE (>6 joints/week) was associated with memory deficits on the McCarthy Scales at age 4 in the OPPS (Fried & Watkinson, 1990). First and second trimester PCE at similar levels were associated with deficits in verbal reasoning and short-term memory (STM) function at age 3 (Day et al., 1994), STM and composite intelligence quotient (IQ) scores at age 6 (Goldschmidt et al., 2008), and design memory at age 10 (Richardson et al., 2002) in the MHPCD. Memory deficits emerged in both cohorts during a critical period of cognitive and language development. The lack of findings regarding memory performance before this age (i.e., during infancy) suggest that memory deficits associated with PCE may not have been reliably assessed at younger ages or may not be detectable until specific points of brain maturation. Furthermore, enhancing the child's environment (e.g., childcare/school attendance) may buffer the negative effects of PCE on memory during early childhood (Day et al., 1994). The OPPS and MHPCD do not report associations between PCE and behavioral measures of memory during adolescence (Willford, Richardson, Leech, & Day, 2004; Fried et al., 2003) and young adulthood (Smith, Fried, Hogan, & Cameron, 2006; Willford et al., 2021). Despite a lack of behavioral differences on memory tasks at these later phases of development, PCE was associated with slower processing speeds in adolescence (Fried et al., 2003) and altered functional brain activity was noted in young adults (Smith et al., 2010) in the OPPS.

In more recent cohort studies, the impact of PCE on memory can be ascertained within the context of increasing Δ9-tetrahydrocannabinol (THC) levels in cannabis. These studies include the Generation R, a prospective study of 9,778 pregnant women recruited from 2002 to 2006 in the Netherlands, and the Norwegian Mother and Child Cohort Study (MoBa), a population-based cohort of over 95,000 women recruited from 1999 to 2008, the Adolescent Brain Cognitive Development (ABCD) study of 11,875 American children born between 2005 and 2009, and the Lifestyle and Early Achievement in Families (LEAF), an ambidirectional cohort study that includes 356 women who gave birth between 2010 and 2016. The Generation R, MoBa, and ABCD studies have not yet assessed the association between PCE and memory function. The LEAF study reported no association between PCE and working and episodic memory at age 3 (Murnan et al., 2021), a result that is consistent with the OPPS, but not the MHPCD, which found lower STM at three years. In the more recent large cohort studies of PCE, no data are available on quantity or frequency of the exposure.

Adolescent cannabis use and memory

Adolescence is a sensitive period for the development of the brain, cognitive abilities, and emotional regulation (Casey et al., 2008; Sturman & Moghaddam, 2011). During adolescence, structural and functional brain development occur in a region-specific manner including the hippocampus, prefrontal cortex (PFC), amygdala, basal ganglia, and the nucleus accumbens, which are brain regions that contribute to the maturation of higher cognitive functioning (Casey et al., 2008). Across studies, there is evidence that cannabis use by adolescents is associated with deficits in multiple domains of memory, including episodic, working, verbal-auditory, and visual-spatial memories (Table 2). The timing of cannabis use onset is an important factor in understanding the association between adolescent cannabis use and memory.

There is an association between early onset (prior to age 16 or 17) cannabis use and memory compared to late onset use (age 16 or older). For example, deficits in working memory are associated with early onset use compared to late onset (Gruber et al., 2012) and deficits in verbal memory are associated with early onset compared to late onset and controls (Pope Jr et al., 2003). The impact of the timing of onset adolescent cannabis use is also associated with memory function in young adults. Schuster et al. (2016) reported that young adults with early-onset cannabis use (before age 16) demonstrated deficits in reduced overall learning and delayed recall compared to those with later onset and no use.

TABLE 1 Summary of the results of prenatal cannabis exposure on memory from the OPPS and MHPCD.

Developmental time period	Prenatal cohort study	
	OPPS	MHPCD
Infancy	***12 and 24 months:*** No significant association between PCE and cognitive or language development (Fried & Watkinson, 1988)	***9–18 months:*** At 9 months, third trimester PCE was associated with lower mental scores on Bayley Scales of Infant Development (BSID), which includes learning and memory. This result did not remain significant at the 18-month follow-up (Richardson, Day, & Goldschmidt, 1995)
Early childhood	***3 and 4 years:*** No association between PCE and memory at age 3, however, heavy prenatal exposure to cannabis (>1 joint/day) was negatively associated with verbal and memory function at age 4 (Fried & Watkinson, 1990) ***5 and 6 years:*** No relation between PCE and verbal or memory outcomes as measured by the McCarthy Scales of Child Abilities (Fried, O'Connell, & Watkinson, 1992)	***3 years:*** First trimester PCE predicted lower verbal reasoning scores while third trimester PCE predicted lower STM scores on the SBIS in children who did not attend preschool/childcare (Day et al., 1994) ***6 years:*** Heavy PCE (>1 joint/day) during the first trimester was associated with decreased verbal reasoning scores while second trimester PCE was associated with lower composite IQ and STM scores (Goldschmidt et al., 2008)
Adolescence	***9–12 years:*** No deficits found between PCE and working memory; however, it significantly affected visual analysis, an important aspect of visual working memory (Fried et al., 1998) ***13–16 years:*** No association between PCE and aspects of auditory and visual memory but slower processing speed was observed in exposed individuals (Fried et al., 2003)	***10 years:*** First trimester PCE was associated with poorer performance on the design memory subscale of the WRAML (Richardson et al., 2002) ***14 years:*** No association between PCE and verbal and visual spatial memory (Willford et al., 2004)
Young Adulthood	***18–22 years:*** Increased levels of PCE resulted in altered brain activation in PFC and parahippocampal gyrus during visuospatial working memory task (Smith et al., 2016) Prenatally exposed individuals required greater activation in left brain regions to perform working memory tasks (Smith et al., 2016)	***22 years:*** First trimester PCE indirectly predicted deficits in young adult memory function via childhood IQ at 6 years, memory deficits at age 10, and early onset marijuana use in adolescence (Willford et al., 2021)

Adult cannabis use and memory

Cannabis use in adults is associated with deficits in multiple domains of memory, including episodic, working, verbal-auditory, and visual-spatial (Table 2). Numerous meta-analytic and systematic review studies have attempted to summarize the impact of cannabis use on memory in adulthood. Across studies, there are reports of subtle effects of use on adult memory compared to non-users. Chronic use was associated with learning and memory deficits in encoding, storage, and memory retrieval (Figueiredo et al., 2020; Ganzer et al., 2016) with small to medium effect sizes (Duperrouzel et al., 2020) and linked to age of onset, dose, duration of use, and length of abstinence in one study that required 14 days of abstinence (Ganzer et al., 2016), factors that were not significant in a recent metaanalysis (Lovell et al., 2020). Key factors related to cannabis use including the duration, frequency, and dose of cannabis, periods of abstinence, and age of onset during adolescence are important for understanding cannabis related impairments in learning and memory in adults (Table 3).

Compared to young adulthood, less research has been done to study the impact of long-term cannabis use in middle age or later. Verbal learning and memory deficits have been reported in middle-aged people with cannabis use compared to

TABLE 2 Summary of the types of memory and associated deficits in adolescent and adult cannabis users described in recent studies.

Associations with adolescent cannabis use	Associations with adult cannabis use
(A) Episodic memory	
Increasing use of cannabis over a one-year span was associated with worsening, immediate episodic memory (Duperrouzel et al., 2019) Declarative memory, specifically the encoding of novel information, showed major improvements in those who abstained from cannabis use for a month compared to those who continued use (Schuster et al., 2018)	Current, young adult cannabis use (at least 4 times/week, 24 hours of abstinence before testing) was associated with verbal episodic memory (Gonzalez et al., 2012) Daily use (at least 25 times/month) with high THC metabolites in hair samples was associated with poorer verbal episodic memory (Morgan et al., 2012) Young adults meeting criteria for CUD showed deficits in verbal episodic memory (Selamoglu et al., 2021) Deficits in verbal episodic memory and alterations in hippocampal morphology were associated with a past history of CUD (Smith et al., 2015) Current, high-frequency cannabis users reported lower ratings for their subjective experiences of vividness and excitement (but not importance or enjoyment) for episodic memories relative to low and moderate frequency users (Sofis et al., 2021)
(B) Working memory	
Adolescents who reported regular cannabis use ($\geq$ four times/week for $\geq$6 months) showed impairments on short-term memory compared to nonusing adolescents (Dougherty et al., 2013) Adolescents who used cannabis performed worse than nonusers on a verbal working memory subtest, even after two weeks of abstaining (Hanson et al., 2010)	Acute Dose of THC (0, 29.3, 49.1, 69.4 mg/THC) was linearly associated with memory search and item recognition in males (Hunault et al., 2009) A small sample of frequent, but relatively moderate cannabis users, abstinent for 1 week prior to testing, showed no association with verbal working memory (Jager et al., 2006) Recreational, young adult cannabis users (at least 1 time/week). Compared to nonusers, cannabis use was associated with worse verbal working memory performance (Levar et al., 2018) Large sample of young adult nonusers, occasional users (monthly or weekly), and daily users at high risk for drinking showed an association between daily use and auditory working memory (Parlar et al., 2021) Young adults with CUD showed deficits in visual working memory (Selamoglu et al., 2021)
(C) Verbal-auditory memory	
Adolescent-onset cannabis users showed slower retention in verbal learning tasks compared to non-using adolescents (Blest-Hopley et al., 2021) Cannabis users between the ages of 15 and 21 years showed verbal memory deficits compared to nonusers (Takagi et al., 2011) Adolescents who used cannabis on a regular basis showed greater impairments in verbal learning, retention, and retrieval, compared to alcohol users and nonusers (Solowij et al., 2011) Early onset cannabis users learned fewer words overall compared to late onset users and nonusers (Schuster et al., 2016)	Among chronic and former cannabis users, both groups showed verbal memory deficits (Lorenzetti et al., 2021) Long-term cannabis users (mean 32.3 years of use) were compared to long-term tobacco users (mean 37.3 years of use). Only cannabis users showed deficits in verbal learning and recall. Lovell et al., 2018 Cannabis use in young adults (past 30 days and past-year) was associated with deficits in verbal learning and memory (Laspada et al., 2019) Verbal learning deficits were found in cannabis users after abstinence (THC metabolites were no longer detectable in urine) (Duperrouzel et al., 2020) Long-term users (4 or more joints/week for at least 10 years) showed overall verbal memory deficits in learning, retention, and retrieval of information compared to shorter term (4 or more joints/week for at least 5 years, but less than 10), and noncannabis users, assessed after a 24 hr period of abstinence (Messinis et al., 2006) Frequency and duration of cannabis use were negatively associated with verbal memory recall (Wagner et al., 2010)

TABLE 2 Summary of the types of memory and associated deficits in adolescent and adult cannabis users described in recent studies—cont'd

Associations with adolescent cannabis use	Associations with adult cannabis use
(D) Visual-spatial memory	
Adolescents who used cannabis on a regular basis (more than once a week) performed significantly worse on spatial working memory (Harvey et al., 2007) Adolescent cannabis use was associated with impaired spatial working memory compared to controls (Meier et al., 2018)	Among chronic and former cannabis users, only current users showed visual-spatial memory deficits (Lorenzetti et al., 2021) Young adults with CUD showed reduced accuracy on a visual memory task (Selamoglu et al., 2021) A small sample of heavy chronic cannabis users (smoked a minimum of 2500 times and 5 days in the past week) showed deficits in memory retention that corresponded to brain activation while performing a virtual Morris-Water maze task (Sneider et al., 2013)

TABLE 3 Factors that influence the association between memory performance and cannabis use.

Memory: – Types of memory (episodic, working, visual–spatial, verbal-auditory, false) – Variations in how memory is assessed (neuropsychological testing, neuroimaging).
User characteristics: – Drug expectancies – Titrating behavior
Use characteristics: – Age of onset – Acute vs. chronic use – Frequency, duration, and dose of cannabis – Route of administration (smoking, vaping, oral, IV consumption) – Polydrug use
Smoking related factors: – Rate of inhalation – Depth of puffs – Volume inhaled – Extent of breath holding after inhalation

controls (Pope Jr et al., 2001; Solowij et al., 2002), which may dissipate with abstinence from cannabis use (Pope Jr et al., 2001). In another study, those with frequent use (weekly or greater) showed deficits in immediate recall of verbal information, but results did not indicate an increased risk for cognitive decline (McKetin et al., 2016). Memory deficits in middle-aged individuals with cannabis use may reflect decreased memory performance that does not recover after a period of abstinence, deficits in memory that existed prior to cannabis exposure, or may develop because of lifetime cannabis exposure effects on educational and career attainment, protective factors related to natural cognitive decline.

Recent studies point to differences in the effect of cannabis use in adults on memory based on use of synthetic cannabis-related drug products such as "Spice" or "K2" (Cohen et al., 2020) that, although less prevalent than plant-derived cannabis products, are popular with a subset of users due to a more intense high, lack of detection in urine drug screens, and quasi-legal status as new formulations are developed to avoid regulation (Castaneto et al., 2014). Synthetic cannabinoid (SC) use is associated with decreased accuracy on a working memory task (Cohen et al., 2017, 2020) and poorer word recall on a verbal long-term memory task (Cohen et al., 2017) compared to nonsynthetic or no cannabis use. Several factors may explain the effects of SC on memory compared to nonsynthetic cannabis use. First, SC mimics endogenous cannabinoids and has a greater affinity for cannabinoid receptors (Castaneto et al., 2014), Second, SC contains other psychoactive substances that could impact memory function (Cohen & Weinstein, 2018). Lastly, the cannabidiol (CBD) found in plant-derived cannabis products may counteract the effects of THC as indicated by studies that dissociate effects of CBD vs. THC on memory (Curran et al., 2020; Morgan et al., 2012).

Effects of cannabis on memory across the lifespan

While many studies have characterized the proximal effect of cannabis on memory at key developmental time points, research shows that distal factors must also be considered including 1) the age of onset, 2) memory deficits that pre-date cannabis initiation, 3) the toxic effects of cannabis on the developing brain, particularly in circuits of the brain that underlie learning, memory, and cognition (e.g., PFC, hippocampal circuitry), 4) lack of academic achievement seen in early initiators, with ensuing long-term impact on subsequent learning, and 5) polysubstance use that often co-occurs with cannabis use across the lifespan. Methodologically, prospective, longitudinal studies are needed to adequately examine both proximal and distal factors related to cannabis exposure and memory function.

Longitudinal studies are useful for examining explanatory hypotheses for cannabis effects on memory across the lifespan. For example, Meier et al. (2012) found a dose-response relationship between persistent cannabis use over 20 years and neuropsychological decline and increased levels of third-party ratings of memory problems in daily living. The effects were detected across multiple cognitive domains, including memory, suggesting global neuropsychological impairment. Deficits in cognitive function were not mediated by academic achievement in this study, but were associated with it in another study (Castellanos-Ryan et al., 2017). Early onset cannabis use, however, predicted greater deficits in neuropsychological function that persisted after a year of abstinence (Meier et al., 2012). The results of this study were corroborated by other longitudinal studies showing an association between early-onset cannabis use (Castellanos-Ryan et al., 2017; Scott et al., 2017) and cognitive decline, and deficits in learning and memory from pre- to post-initiation in adolescents and young adults who transitioned from no cannabis use at one assessment phase to frequent use at a later phase (>6 uses/past year) (Ajmera et al., 2021). Fried et al. (2005) demonstrated, prospectively, after controlling for important explanatory factors including prenatal and concurrent substance exposure, that current, regular, heavy cannabis use (>5 joints/week) in young adults was associated with deficits in immediate and delayed memory. However, there were no memory deficits in former users with at least 3 months of abstinence (Fried et al., 2005). Longitudinal studies provide a great deal of clarity about the impact of cannabis on memory and highlight the importance of considering multiple explanatory pathways.

These pathways for cannabis-induced deficits in memory emerge when prenatal, adolescent, and adult cannabis use are examined together using statistical modeling. A recent study examined the complex association between exposures at all three developmental time points to better understand memory impairments in young adults with cannabis use (Willford et al., 2021). In this study, current cannabis use in adults was initially associated with verbal memory impairments. Similar to past studies (Ganzer et al., 2016; Gruber et al., 2012; Pope Jr et al., 2003), this study found an association between early-onset cannabis use and adult memory function, but also showed that the relation persisted when current use was controlled. When adolescent and adult use were both included in the models, only adolescent use predicted memory impairments in adulthood. Lastly, there were indirect effects of PCE on adult memory. PCE predicted intelligence at age six, memory function at age 10, and early-onset cannabis use, which was then related to adult memory. This study highlights the importance of understanding the association between cannabis exposure and memory in the context of exposures that occur at different developmental time points and antecedents of memory impairment in adults who use cannabis.

Polydrug use

Cannabis use often occurs concurrently with tobacco, alcohol, and other illicit drugs in adolescents and adults (World Health Organisation (WHO), 2016), but more research is needed on the effects of cannabis use on memory function in the context of polydrug use. Of the few reports available on adult samples, the data are mixed. Concurrent use of tobacco was reported to mitigate some of the negative effects of cannabis on verbal memory (Filbey et al., 2015), working memory (Schuster et al., 2016), and episodic memory (Schuster et al., 2015), but not in all studies (Hindocha et al., 2017; Wade et al., 2021). Winward et al. (2014) reported that concurrent alcohol use exacerbated the impact of cannabis use on working memory, but Schuster et al. (2016) did not replicate that finding. Changes in visual memory were reported in individuals with cannabis use, regardless of (illicit) polydrug use in one study (Binkowska, Jakubowska, Gaca, et al., 2021). However, oscillatory activity (rhythmic electrical brain activity important for the synchronization of neural activity) generated during working memory maintenance was different between regular users with and without polydrug use (Binkowska,

Jakubowska, Krystecka, et al., 2021). The EEG differences in oscillatory activity are thought to reflect compensatory brain activity during working memory. These initial data should be interpreted with caution as the operational definition of polydrug use varies greatly among studies and sample sizes tend to be small.

Compensatory brain function

There is evidence that individuals who use cannabis must put forth more effort to perform memory tasks, which leads to comparable behavioral performance but greater memory network activation in the brain. Functionally, several neuroimaging studies have shown that even in the absence of behavioural differences, brain regions associated with memory function are differentially activated in adults who use cannabis including visual-spatial memory (Jager et al., 2007; Smith et al., 2010) and working memory (Cousijn et al., 2014; Jager et al., 2006, 2007; Kroon et al., 2021). A recent meta-analysis concluded that cannabis use leads to decreased neural efficiency even after long periods of abstinence (Blest-Hopley, Giampietro, & Bhattacharyya, 2018). Similar results have emerged from studies in adolescents. For example, among adolescents, early onset cannabis use was associated with increased activation of the left superior parietal lobe, but accuracy on verbal working memory tests did not differ compared to late onset cannabis use (Becker et al., 2010). Similarly, another study found that after one month of abstinence, adolescents who used cannabis and those who were abstinent performed similarly on a spatial working memory task, but cannabis use was associated with increased brain activity in the right posterior parietal cortex and lower activity in the right dorsolateral PFC and occipital cortices (Schweinsburg et al., 2008). Several explanations are plausible for functional and structural brain differences associated with adolescent and adult cannabis use in the absence of behavioral differences, including differences in the short- vs. long-term effects of cannabis use, moderate vs. heavy use of cannabis, and the frequency of use across study samples. Taken together, however, studies suggest that there are compensatory mechanisms in which exposed individuals require increased effort, whether it be circuitry rewiring, recruitment of other regions, or increases/reductions in activation, to achieve the same behavioral outcomes as unexposed individuals.

Methodological limitations

A number of methodological issues and limitations must be considered. The prenatal longitudinal prospective cohort designs afford several benefits including large samples, the ability to prospectively measure dose, timing, and duration of cannabis exposure and other drugs, maternal psychosocial factors, home environment, and demographics, as well as assess memory function prior to the onset of cannabis use in adolescence and adulthood. Prenatal studies generally rely on self-reported substance use, which may be subject to recall bias and misreporting. However, with carefully constructed, detailed interviews and well-trained, nonjudgmental interviewers, accurate reports of substance use can be obtained. Goldschmidt et al. (2012) found that urine toxicology samples were consistent with self-reported use in the MHPCD cohort, demonstrating reporting accuracy.

Most studies of adolescent and adult cannabis use and memory function are cross-sectional and cannabis use is ascertained retrospectively and through self-report. Samples are not well-characterized, mainly derived from student populations or volunteers from the community, which introduces biases such as educational attainment into the data collection. In addition, studies have used samples that vary based on the frequency and dose of cannabis and periods of abstinence. Studies have failed to address whether withdrawal, tolerance, or residual carryover effects of THC are important variables in assessing cannabis effects on memory. Assessing the impact of cannabis use on memory function is further complicated by the varied types of memory testing paradigms examining different domains of function including episodic memory, working memory, verbal/auditory, and visual/spatial memory, in addition to neuropsychological testing and neuroimaging cognitive paradigms. The types of covariates and polydrug exposures are measured inconsistently across studies. Lastly, most, but not all, studies in adult samples have focused on young adults. Comparatively, there is less research that examines memory function in older adults, who may be at higher risk for memory impairment. Publication bias and lack of replication further contribute to the controversy and confusion surrounding cannabis' effects on memory in adolescents and adults

Concluding remarks

Cannabis exposure during pregnancy, adolescence, and adulthood coincides with the development of memory functions that reflect normative changes in information-processing systems that affect the efficiency of and access to memory. Our understanding of the association between cannabis exposure and memory function across the lifespan is much greater now than even a decade ago. Longitudinal studies have provided important insights about the associations between cannabis exposure across time when important covariates are considered, the timing, frequency, and dose of cannabis are measured, and pre-exposure function has been assessed (Willford et al., 2021). More recent cohort studies of the developmental impact of cannabis on memory and cognition are warranted, given the changes in perceived risk, legal status of and access to cannabis, and the increased concentrations of THC in cannabis compared to the exposures measured in the older cohort studies. In the future, research on epigenetic changes associated with cannabis exposure will provide additional clarity about interactions between development and environment interactions that underlie brain function and behavior.

Applications to other areas

In this chapter, we have reviewed the impact of prenatal, adolescent, and adult cannabis exposure on memory function. Cross-sectional studies demonstrate that cannabis exposures at key developmental timepoints are associated with behavioral and brain measures of memory function. Longitudinal studies show that the consideration of exposures earlier in development (prenatal and adolescent), the assessment of memory function prior to the onset of drug use, the age of onset during adolescence, and the duration of use over the lifetime are important in understanding the association between cannabis exposure/use and memory function. The methodological considerations that emerged from the review are also relevant to the literature on exposure to drugs such as alcohol, tobacco, and opioids. First, the measurement of key variables such as age, demographics, environment, polysubstance use, and length of abstinence are important covariates in determining the impact of drug exposure on cognitive functions. Second, conclusions that can be drawn about the association between drug exposure and cognitive function vary depending on the neuropsychological or neuroimaging paradigm used in the assessment. Third, the use of prospective, longitudinal research design allows for the evaluation of the impact of exposures over time, the dose, frequency, and timing of drug exposure, and the ability to determine whether pre-exposure deficits in cognitive function underlie the association between drug exposure and cognitive function. Lastly, given the current trends for decreased perceived risk of cannabis use, increasing levels of THC in cannabis, increasing availability of medical marijuana and unregulated synthetic cannabis products, and the widespread legalization of and subsequent increase in recreational cannabis use, the information in this chapter is applicable in public health, education, and clinical care settings including obstetrics and paediatrics.

Mini-dictionary of terms

- *Episodic memory*: A form of declarative memories, which are memories that can be recalled consciously and includes the capacity to remember details of an experience including sensory, affective, and contextual aspects of personal events.
- *Working memory*: A main component of executive function that reflects the capacity to store a limited amount of information to be manipulated and used for complex, goal-directed behavior It is sometimes referred to as short-term or immediate memory.
- *Verbal-auditory memory*: A subcategory of episodic memory that refers to the capacity to remember phonological information, including words and language narrative. Information is acquired through reading or hearing.
- *Visual-spatial memory*: A subcategory of episodic memory that refers to the capacity to remember visual-spatial information, including pictures, scenes, words, or other information that is presented visually.

Key facts of the impact of cannabis on memory: From in utero to adulthood

- *Key facts of development and memory*:

  Studies show effects of prenatal, adolescent, and adult cannabis exposure on episodic, working, visual-spatial, and verbal-auditory memory.
- *Key facts of methodological considerations*:

  The effects of developmental cannabis exposures vary based on methodological considerations that are best addressed by longitudinal, prospective research designs that allow for ascertainment of the dose, timing, and frequency of drug exposure/use, environmental covariates of the exposure, and the assessment of pre-exposure function.
- *Key facts of brain function*:

  Neuroimaging studies show cannabis-related changes in brain activation while performing memory tasks in the absence of behavioral deficits, suggesting compensatory brain function.

- *Key facts of adolescence*:
  The effects of cannabis exposure in adolescence on memory are related to early age of onset.
- *Key facts of adulthood*:
  The effects of adult cannabis exposure must be considered in the context of earlier developmental exposures including prenatal exposure, age of initiation, length of use, and periods of abstinence.

Summary points

The effects of prenatal cannabis exposure emerge with the development of memory and language during early childhood.

- Cannabis-related memory impairment in adolescence depends on factors including the age of onset, dose and duration of use, and periods of abstinence.
- In adults, cannabis use is associated with subtle effects on the encoding, storage, and retrieval of memories. This effect is dependent on age of onset, duration of use, periods of abstinence, and tolerance.
- At each developmental phases, studies show that prenatal, adolescent, and adult exposure to cannabis leads to compensatory or adaptive brain activation to perform memory tasks in the absence of behavioral deficits.
- When cannabis exposure is considered across the lifespan, longitudinal studies provide a method for evaluating multiple hypotheses about the effects of cannabis exposure at different developmental periods on memory function and the ability to statistically control for prior exposures and important covariates.
- Longitudinal studies show that the consideration of exposures earlier in development and the assessment of function prior to the onset of drug use are important in understanding the association between cannabis exposure/use and memory function.

References

Ajmera, N., Collins, P. F., Weiss, H., & Luciana, M. (2021). Initiation of moderately frequent cannabis use in adolescence and young adulthood is associated with declines in verbal learning and memory: A longitudinal comparison of pre- vs. post-initiation cognitive performance. *Journal of the International Neuropsychological Society: JINS*, *27*(6), 621–636. https://doi.org/10.1017/S1355617721000631.

Becker, B., Wagner, D., Gouzoulis-Mayfrank, E., Spuentrup, E., & Daumann, J. (2010). The impact of early-onset cannabis use on functional brain correlates of working memory. *Progress in Neuro-Psychopharmacology & Biological Psychiatry*, *34*(6), 837–845. https://doi.org/10.1016/j.pnpbp.2010.03.032.

Binkowska, A. A., Jakubowska, N., Gaca, M., Galant, N., Piotrowska-Cyplik, A., & Brzezicka, A. (2021). Not just a pot: Visual episodic memory in cannabis users and polydrug cannabis users: ROC and ERP preliminary investigation. *Frontiers in Human Neuroscience*, *15*, 677793. https://doi.org/10.3389/fnhum.2021.677793.

Binkowska, A. A., Jakubowska, N., Krystecka, K., Galant, N., Piotrowska-Cyplik, A., & Brzezicka, A. (2021). Theta and alpha oscillatory activity during working memory maintenance in long-term cannabis users: The importance of the polydrug use context. *Frontiers in Human Neuroscience*, *15*, 740277. https://doi.org/10.3389/fnhum.2021.740277.

Blest-Hopley, G., O'Neill, A., Wilson, R., Giampietro, V., & Bhattacharyya, S. (2021). Disrupted parahippocampal and midbrain function underlie slower verbal learning in adolescent-onset regular cannabis use. *Psychopharmacology*, *238*(5), 1315–1331. https://doi.org/10.1007/s00213-019-05407-9.

Blest-Hopley, G., Giampietro, V., & Bhattacharyya, S. (2018). Residual effects of cannabis use in adolescent and adult brains—A meta-analysis of fMRI studies. *Neuroscience & Biobehavioral Reviews*, *88*, 26–41. https://doi.org/10.1016/j.neubiorev.2018.03.008.

Casey, B. J., Jones, R. M., & Hare, T. A. (2008). The adolescent brain. *Annals of the New York Academy of Sciences*, *1124*, 111–126. https://doi.org/10.1196/annals.1440.010.

Castaneto, M. S., Gorelick, D. A., Desrosiers, N. A., Hartman, R. L., Pirard, S., & Huestis, M. A. (2014). Synthetic cannabinoids: Epidemiology, pharmacodynamics, and clinical implications. *Drug and Alcohol Dependence*, *144*, 12–41. https://doi.org/10.1016/j.drugalcdep.2014.08.005.

Castellanos-Ryan, N., Pingault, J. B., Parent, S., Vitaro, F., Tremblay, R. E., & Séguin, J. R. (2017). Adolescent cannabis use, change in neurocognitive function, and high-school graduation: A longitudinal study from early adolescence to young adulthood. *Development and Psychopathology*, *29*(4), 1253–1266. https://doi.org/10.1017/S0954579416001280.

Cohen, K., Kapitány-Fövény, M., Mama, Y., Arieli, M., Rosca, P., Demetrovics, Z., & Weinstein, A. (2017). The effects of synthetic cannabinoids on executive function. *Psychopharmacology*, *234*(7), 1121–1134. https://doi.org/10.1007/s00213-017-4546-4.

Cohen, K., Mama, Y., Rosca, P., Pinhasov, A., & Weinstein, A. (2020). Chronic use of synthetic cannabinoids is associated with impairment in working memory and mental flexibility. *Frontiers in Psychiatry*, *11*, 602. https://doi.org/10.3389/fpsyt.2020.00602.

Cohen, K., & Weinstein, A. (2018). The effects of cannabinoids on executive functions: Evidence from cannabis and synthetic cannabinoids—a systematic review. *Brain Sciences*, *8*(3), 40. https://doi.org/10.3390/brainsci8030040.

Cousijn, J., Wiers, R. W., Ridderinkhof, K. R., van den Brink, W., Veltman, D. J., & Goudriaan, A. E. (2014). Effect of baseline cannabis use and working-memory network function on changes in cannabis use in heavy cannabis users: A prospective fMRI study. *Human Brain Mapping*, *35*(5), 2470–2482. https://doi.org/10.1002/hbm.22342.

Curran, T., Devillez, H., YorkWilliams, S. L., & Bidwell, L. C. (2020). Acute effects of naturalistic THC vs. CBD use on recognition memory: A preliminary study. *Journal of Cannabis Research*, *2*(1), 28. https://doi.org/10.1186/s42238-020-00034-0.

Day, N. L., Richardson, G., Goldschmidt, L., Robles, N., Taylor, P. M., Stoffer, D., … Geva, D. (1994). Effect of PCE on the cognitive development of offspring at age three. *Neurotoxicology and Teratology*, *16*, 169–175. https://doi.org/10.1016/0892-0362(94)90114-7.

Dougherty, D. M., Mathias, C. W., Dawes, M. A., Furr, R. M., Charles, N. E., Liguori, A., … Acheson, A. (2013). Impulsivity, attention, memory, and decision-making among adolescent cannabis users. *Psychopharmacology*, *226*(2), 307–319. https://doi.org/10.1007/s00213-012-2908-5.

Duperrouzel, J. C., Granja, K., Pacheco-Colón, I., & Gonzalez, R. (2020). Adverse effects of cannabis use on neurocognitive functioning: A systematic review of meta-analytic studies. *Journal of Dual Diagnosis*, *16*(1), 43–57. https://doi.org/10.1080/15504263.2019.1626030.

Duperrouzel, J. C., Hawes, S. W., Lopez-Quintero, C., Pacheco-Colon, I., Coxe, S., Hayes, T., & Gonzalez, R. (2019). Adolescent cannabis use and its associations with decision-making and episodic memory: Preliminary results from a longitudinal study. *Neuropsychology*, *33*(5), 701–710. https://doi.org/10.1037/neu0000538.

Figueiredo, P. R., Tolomeo, S., Steele, J. D., & Baldacchino, A. (2020). Neurocognitive consequences of chronic cannabis use: A systematic review and meta-analysis. *Neuroscience & Biobehavioral Reviews*, *108*, 358–369. https://doi.org/10.1016/j.neubiorev.2019.10.014.

Filbey, F. M., McQueeny, T., Kadamangudi, S., Bice, C., & Ketcherside, A. (2015). Combined effects of marijuana and nicotine on memory performance and hippocampal volume. *Behavioural Brain Research*, *293*, 46–53. https://doi.org/10.1016/j.bbr.2015.07.029.

Fried, P. A., O'Connell, C. M., & Watkinson, B. (1992). 60- and 72-month follow-up of children prenatally exposed to marijuana, cigarettes, and alcohol: Cognitive and language assessment. *Journal of Developmental and Behavioral Pediatrics*, *13*(6), 383–391.

Fried, P. A., & Watkinson, B. (1988). 12- and 24-month neurobehavioural follow-up of children prenatally exposed to marihuana, cigarettes and alcohol. *Neurotoxicology. Teratology*, *10*(4), 305–313. https://doi.org/10.1016/0892-0362(88)90032-3.

Fried, P. A., & Watkinson, B. (1990). 36- and 48-month neurobehavioral follow-up of children prenatally exposed to cannabis, cigarettes, and alcohol. *Journal of Developmental and Behavioral Pediatrics*, *11*(2), 49–58. https://doi.org/10.1097/00004703-199004000-00003.

Fried, P. A., Watkinson, B., & Gray, R. (1998). Differential effects on cognitive functioning in 9- to 12-year olds prenatally exposed to cigarettes and marihuana. *Neurotoxicology and Teratology*, *20*(3), 293–306. https://doi.org/10.1016/s0892-0362(97)00091-3.

Fried, P. A., Watkinson, B., & Gray, R. (2003). Differential effects on cognitive functioning in 13- to 16-year-olds prenatally exposed to cigarettes and marihuana. *Neurotoxicology and Teratology*, *25*(4), 427–436. https://doi.org/10.1016/s0892-0362(03)00029-1.

Fried, P. A., Watkinson, B., & Gray, R. (2005). Neurocognitive consequences of marihuana—a comparison with pre-drug performance. *Neurotoxicology and Teratology*, *27*(2), 231–239. https://doi.org/10.1016/j.ntt.2004.11.003.

Ganzer, F., Broening, S., Kraft, S., Sack, P. M., & Thomasius, R. (2016). Weighing the evidence: A systematic review on long-term neurocognitive effects of cannabis use in abstinent adolescents and adults. *Neuropsychology Review*, *26*(2), 186–222. https://doi.org/10.1007/s11065-016-9316-2.

Goldschmidt, L., Richardson, G. A., Willford, J., & Day, N. L. (2008). PCE and intelligence test performance at age 6. *Journal of the American Academy of Child and Adolescent Psychiatry*, *47*(3), 254–263. https://doi.org/10.1097/CHI.0b013e318160b3f0.

Goldschmidt, L., Richardson, G. A., Willford, J. A., Severtson, S. G., & Day, N. L. (2012). School achievement in 14-year-old youths prenatally exposed to marijuana. *Neurotoxicology and Teratology*, *34*(1), 161–167. https://doi.org/10.1016/j.ntt.2011.08.009.

Gonzalez, R., Schuster, R. M., Mermelstein, R. J., Vassileva, J., Martin, E. M., & Diviak, K. R. (2012). Performance of young adult cannabis users on neurocognitive measures of impulsive behavior and their relationship to symptoms of cannabis use disorders. *Journal of Clinical and Experimental Neuropsychology*, *34*(9), 962–976. https://doi.org/10.1080/13803395.2012.703642.

Gruber, S. A., Sagar, K. A., Dahlgren, M. K., Racine, M., & Lukas, S. E. (2012). Age of onset of cannabis use and executive function. *Psychology of Addictive Behaviors*, *26*(3), 496–506. https://doi.org/10.1037/a0026269.

Hanson, K. L., Winward, J. L., Schweinsburg, A. D., Medina, K. L., Brown, S. A., & Tapert, S. F. (2010). Longitudinal study of cognition among adolescent cannabis users over three weeks of abstinence. *Addictive Behaviors*, *35*(11), 970–976. https://doi.org/10.1016/j.addbeh.2010.06.012.

Harvey, M. A., Sellman, J. D., Porter, R. J., & Frampton, C. M. (2007). The relationship between non-acute adolescent cannabis use and cognition. *Drug and Alcohol Review*, *26*(3), 309–319. https://doi.org/10.1080/09595230701247772.

Hindocha, C., Freeman, T. P., Xia, J. X., Shaban, N., & Curran, H. V. (2017). Acute memory and psychotomimetic effects of cannabis and tobacco both 'joint' and individually: A placebo-controlled trial. *Psychological Medicine*, *47*(15), 2708–2719. https://doi.org/10.1017/S0033291717001222.

Hunault, C. C., Mensinga, T. T., Böcker, K. B., Schipper, C. M., Kruidenier, M., Leenders, M. E., … Meulenbelt, J. (2009). Cognitive and psychomotor effects in males after smoking a combination of tobacco and cannabis containing up to 69 mg delta-9-tetrahydrocannabinol (THC). *Psychopharmacology*, *204*(1), 85–94. https://doi.org/10.1007/s00213-008-1440-0.

Jager, G., Kahn, R. S., Van Den Brink, W., Van Ree, J. M., & Ramsey, N. F. (2006). Long-term effects of frequent cannabis use on working memory and attention: An fMRI study. *Psychopharmacology*, *185*(3), 358–368. https://doi.org/10.1007/s00213-005-0298-7.

Jager, G., Van Hell, H. H., De Win, M. M., Kahn, R. S., Van Den Brink, W., Van Ree, J. M., & Ramsey, N. F. (2007). Effects of frequent cannabis use on hippocampal activity during an associative memory task. *Journal of the European College of Neuropsychopharmacology*, *17*(4), 289–297. https://doi.org/10.1016/j.euroneuro.2006.10.003.

Johnston, L.D., Miech, R.A., O'Malley, P.M., Bachman, J.G., Schulenberg, J.E., & Patrick, M.E. (2020) 2020 Overview Key Findings on Adolescent Drug Use, Monitoring the Future National Survey Results on Drug Use 1975-2020, Retrieved from https://files.eric.ed.gov/fulltext/ED611736.pdf.

Kroon, E., Kuhns, L., & Cousijn, J. (2021). Context-dependent differences in working memory-related brain activity in heavy cannabis users. *Psychopharmacology*. https://doi.org/10.1007/s00213-021-05956-y. Advance online publication.

Laspada, N., Delker, E., Blanco, E., Encina, P., Caballero, G., Delva, J., … Gahagan, S. (2019). Cannabis use associated with worse verbal learning and delayed recall in a sample of young adults. *Revista Médica de Chile*, *147*(2), 206–211. https://doi.org/10.4067/s0034-98872019000200206.

Levar, N., Francis, A. N., Smith, M. J., Ho, W. C., & Gilman, J. M. (2018). Verbal memory performance and reduced cortical thickness of brain regions along the uncinate fasciculus in young adult cannabis users. *Cannabis and Cannabinoid Research*, *3*(1), 56–65. https://doi.org/10.1089/can.2017.0030.

Lorenzetti, V., Takagi, M., van Dalen, Y., Yücel, M., & Solowij, N. (2021). Investigating the residual effects of chronic cannabis use and abstinence on verbal and visuospatial learning. *Frontiers in Psychiatry*, *12*, 663701. https://doi.org/10.3389/fpsyt.2021.663701.

Lovell, M. E., Akhurst, J., Padgett, C., Garry, M. I., & Matthews, A. (2020). Cognitive outcomes associated with long-term, regular, recreational cannabis use in adults: A meta-analysis. *Experimental and Clinical Psychopharmacology*, *28*(4), 471–494. https://doi.org/10.1037/pha0000326.

Lovell, M. E., Bruno, R., Johnston, J., Matthews, A., McGregor, I., Allsop, D. J., & Lintzeris, N. (2018). Cognitive, physical, and mental health outcomes between long-term cannabis and tobacco users. *Addictive Behaviors*, *79*, 178–188. https://doi.org/10.1016/j.addbeh.2017.12.009.

McKetin, R., Parasu, P., Cherbuin, N., Eramudugolla, R., & Anstey, K. J. (2016). A longitudinal examination of the relationship between cannabis use and cognitive function in mid-life adults. *Drug and Alcohol Dependence*, *169*, 134–140. https://doi.org/10.1016/j.drugalcdep.2016.10.022.

Meier, M. H., Caspi, A., Ambler, A., Harrington, H., Houts, R., Keefe, R. S., … Moffitt, T. E. (2012). Persistent cannabis users show neuropsychological decline from childhood to midlife. *Proceedings of the National Academy of Sciences of the United States of America*, *109*(40), E2657–E2664. https://doi.org/10.1073/pnas.1206820109.

Meier, M. H., Caspi, A., Danese, A., Fisher, H. L., Houts, R., Arseneault, L., & Moffitt, T. E. (2018). Associations between adolescent cannabis use and neuropsychological decline: A longitudinal co-twin control study. *Addiction*, *113*(2), 257–265. https://doi.org/10.1111/add.13946.

Messinis, L., Kyprianidou, A., Malefaki, S., & Papathanasopoulos, P. (2006). Neuropsychological deficits in long-term frequent cannabis users. *Neurology*, *66*(5), 737–739. https://doi.org/10.1212/01.wnl.0000201279.83203.c6.

Morgan, C. J. A., Gardener, C., Schafer, G., Swan, S., Demarchi, C., Freeman, T. P., … Wingham, G. (2012). Sub-chronic impact of cannabinoids in street cannabis on cognition, psychotic-like symptoms and psychological well-being. *Psychological Medicine*, *42*(2), 391–400. https://doi.org/10.1017/S0033291711001322.

Murnan, A. W., Keim, S. A., Yeates, K. O., Boone, K. M., Sheppard, K. W., & Klebanoff, M. A. (2021). Behavioral and cognitive differences in early childhood related to prenatal marijuana exposure. *Journal of Applied Developmental Psychology*, *77*. https://doi.org/10.1016/j.appdev.2021.101348.

Parlar, M., MacKillop, E., Petker, T., Murphy, J., & MacKillop, J. (2021). Cannabis use, age of initiation, and neurocognitive performance: Findings from a large sample of heavy drinking emerging adults. *Journal of the International Neuropsychological Society*, *27*(6), 533–545. https://doi.org/10.1017/S1355617721000618.

Pope, H. G., Jr., Gruber, A. J., Hudson, J. I., Cohane, G., Huestis, M. A., & Yurgelun-Todd, D. (2003). Early-onset cannabis use and cognitive deficits: What is the nature of the association? *Drug and Alcohol Dependence*, *69*(3), 303–310. https://doi.org/10.1016/s0376-8716(02)00334-4.

Pope, H. G., Jr., Gruber, A. J., Hudson, J. I., Huestis, M. A., & Yurgelun-Todd, D. (2001). Neuropsychological performance in long-term cannabis users. *Archives of General Psychiatry*, *58*(10), 909–915. https://doi.org/10.1001/archpsyc.58.10.909.

Richardson, G. A., Day, N. L., & Goldschmidt, L. (1995). Prenatal alcohol, marijuana, and tobacco use: Infant mental and motor development. *Neurotoxicology and Teratology*, *17*(4), 479–487. https://doi.org/10.1016/0892-0362(95)00006-d. 7565494.

Richardson, G. A., Ryan, C., Willford, J., Day, N. L., & Goldschmidt, L. (2002). Prenatal alcohol and cannabis exposure: Effects on neuropsychological outcomes at 10 years. *Neurotoxicology and Teratology*, *24*(3), 309–320. https://doi.org/10.1016/s0892-0362(02)00193-9.

Schuster, R. M., Crane, N. A., Mermelstein, R., & Gonzalez, R. (2015). Tobacco may mask poorer episodic memory among young adult cannabis users. *Neuropsychology*, *29*(5), 759–766. https://doi.org/10.1037/neu0000173.

Schuster, R. M., Gilman, J., Schoenfeld, D., Evenden, J., Hareli, M., Ulysse, C., … Evins, A. E. (2018). One month of cannabis abstinence in adolescents and young adults is associated with improved memory. *The Journal of Clinical Psychiatry*, *79*(6). https://doi.org/10.4088/JCP.17m11977.

Schuster, R. M., Hoeppner, S. S., Evins, A. E., & Gilman, J. M. (2016). Early onset cannabis use is associated with learning inefficiencies. *Neuropsychology*, *30*(4), 405–415. https://doi.org/10.1037/neu0000281.

Schweinsburg, A. D., Nagel, B. J., Schweinsburg, B. C., Park, A., Theilmann, R. J., & Tapert, S. F. (2008). Abstinent adolescent cannabis users show altered fMRI response during spatial working memory. *Psychiatry Research*, *163*(1), 40–51. https://doi.org/10.1016/j.pscychresns.2007.04.018.

Scott, J. C., Wolf, D. H., Calkins, M. E., Bach, E. C., Weidner, J., Ruparel, K., … Gur, R. C. (2017). Cognitive functioning of adolescent and young adult cannabis users in the Philadelphia Neurodevelopmental Cohort. *Psychology of Addictive Behaviors*, *31*(4), 423–434. https://doi.org/10.1037/adb0000268.

Selamoglu, A., Langley, C., Crean, R., Savulich, G., Cormack, F., Sahakian, B. J., & Mason, B. (2021). Neuropsychological performance in young adults with cannabis use disorder. *Journal of Psychopharmacology*, *35*(11), 1349–1355. https://doi.org/10.1177/02698811211050548.

Smith, A. M., Fried, P. A., Hogan, M. J., & Cameron, I. (2006). Effects of prenatal marijuana on visuospatial working memory: An fMRI study in young adults. *Neurotoxicology and Teratology*, *28*(2), 286–295. https://doi.org/10.1016/j.ntt.2005.12.008.

Smith, M. J., Cobia, D. J., Reilly, J. L., Gilman, J. M., Roberts, A. G., Alpert, K. I., … Csernansky, J. G. (2015). Cannabis-related episodic memory deficits and hippocampal morphological differences in healthy individuals and schizophrenia subjects. *Hippocampus*, *25*(9), 1042–1051. https://doi.org/10.1002/hipo.22427.

Smith, A. M., Longo, C. A., Fried, P. A., Hogan, M. J., & Cameron, I. (2010). Effects of cannabis on visuospatial working memory: An fMRI study in young adults. *Psychopharmacology*, *210*(3), 429–438. https://doi.org/10.1007/s00213-010-1841-8.

Smith, A. M., Mioduszewski, O., Hatchard, T., Byron-Alhassan, A., Fall, C., & Fried, P. A. (2016). PCE impacts executive functioning into young adulthood: An fMRI study. *Neurotoxicology and Teratology*, *58*, 53–59. https://doi.org/10.1016/j.ntt.2016.05.010.

Sneider, J. T., Gruber, S. A., Rogowska, J., Silveri, M. M., & Yurgelun-Todd, D. A. (2013). A preliminary study of functional brain activation among cannabis users during performance of a virtual water maze task. *Journal of Addiction*, *2013*, 461029. https://doi.org/10.1155/2013/461029.

Sofis, M. J., Lemley, S. M., & Budney, A. J. (2021). The effects of cannabis use frequency and episodic specificity training on the recall of specific and rewarding events. *Frontiers in Psychiatry*, *12*, 643819. https://doi.org/10.3389/fpsyt.2021.643819.

Solowij, N., Jones, K. A., Rozman, M. E., Davis, S. M., Ciarrochi, J., Heaven, P. C., ... Yücel, M. (2011). Verbal learning and memory in adolescent cannabis users, alcohol users and non-users. *Psychopharmacology*, *216*(1), 131–144. https://doi.org/10.1007/s00213-011-2203-x.

Solowij, N., Stephens, R. S., Roffman, R. A., Babor, T., Kadden, R., Miller, M., ... Marijuana Treatment Project Research Group. (2002). Cognitive functioning of long-term heavy cannabis users seeking treatment. *JAMA*, *287*(9), 1123–1131. https://doi.org/10.1001/jama.287.9.1123.

Sturman, D. A., & Moghaddam, B. (2011). The neurobiology of adolescence: Changes in brain architecture, functional dynamics, and behavioral tendencies. *Neuroscience and Biobehavioral Reviews*, *35*(8), 1704–1712. https://doi.org/10.1016/j.neubiorev.2011.04.003.

Substance Abuse and Mental Health Services Administration. (2019). *Key substance use and mental health indicators in the United States: Results from the 2018 national survey on drug use and health (HHS Publication No. PEP19–5068, NSDUH Series H-54)*. Rockville, MD: Center for Behavioral Health Statistics and Quality, Substance Abuse and Mental Health Services Administration.

Takagi, M., Yücel, M., Cotton, S. M., Baliz, Y., Tucker, A., Elkins, K., & Lubman, D. I. (2011). Verbal memory, learning, and executive functioning among adolescent inhalant and cannabis users. *Journal of Studies on Alcohol and Drugs*, *72*(1), 96–105. https://doi.org/10.15288/jsad.2011.72.96.

Volkow, N. D., Han, B., Compton, W. M., & McCance-Katz, E. F. (2019). Self-reported medical and nonmedical cannabis use among pregnant women in the United States. *JAMA*, *322*(2), 167–169. https://doi.org/10.1001/jama.2019.7982.

Wade, N. E., Baca, R., Courtney, K. E., McCabe, C. J., Infante, M. A., Huestis, M. A., & Jacobus, J. (2021). Preliminary evidence for cannabis and nicotine urinary metabolites as predictors of verbal memory performance and learning among young adults. *Journal of the International Neuropsychological Society*, *27*(6), 546–558. https://doi.org/10.1017/S1355617721000205.

Wagner, D., Becker, B., Gouzoulis-Mayfrank, E., & Daumann, J. (2010). Interactions between specific parameters of cannabis use and verbal memory. *Progress in Neuro-Psychopharmacology & Biological Psychiatry*, *34*(6), 871–876. https://doi.org/10.1016/j.pnpbp.2010.04.004.

Willford, J. A., Goldschmidt, L., De Genna, N. M., Day, N. L., & Richardson, G. A. (2021). A longitudinal study of the impact of cannabis on adult memory function: Prenatal, adolescent, and young adult exposures. *Neurotoxicology and Teratology*, *84*, 106958. https://doi.org/10.1016/j.ntt.2021.106958.

Willford, J. A., Richardson, G. A., Leech, S. L., & Day, N. L. (2004). Verbal and visuospatial learning and memory function in children with moderate prenatal alcohol exposure. *Alcoholism: Clinical and Experimental Research*, *28*(3), 497–507. https://doi.org/10.1097/01.alc.0000117868.97486.2d.

Winward, J. L., Hanson, K. L., Tapert, S. F., & Brown, S. A. (2014). Heavy alcohol use, cannabis use, and concomitant use by adolescents are associated with unique and shared cognitive decrements. *Journal of the International Neuropsychological Society*, *20*(8), 784–795. https://doi.org/10.1017/S1355617714000666.

World Drug Report. (2021). *United Nations Office on Drugs and Crim 2021, Booklet 3*. https://www.unodc.org/unodc/en/data-an-analysis/wdr2021.html.

World Health Organisation (WHO) (2016). The health and social effects of nonmedical cannabis use. Available online at: https://www.who.int/substance_abuse/publications/msbcannabis.pdf (Accessed 12 January 2022).

Chapter 25

Age-related considerations in the use of cannabinoids: Focus on drug interactions, physiology, and drug effects

Hanana AlTfaili and Brett C. Ginsburg
Department of Psychiatry and Behavioral Sciences, The University of Texas Health Science Center at San Antonio, San Antonio, TX, United States

Abbreviations

CB1	cannabinoid receptor type1
CBD	cannabidiol
DDI	drug-drug interaction
Δ9-THC	Δ9-tetrahydrocannabinol
P-gp	P-glycoprotein
ROS	reactive oxygen species

Introduction

Cannabis use has been increasing among the elderly recently, aligning with trends of increased cannabis product use in other age groups (Han et al., 2016; Han & Palamar, 2018). Several factors, including liberalization of cannabis laws, decreased social stigma, and awareness of potential therapeutic benefits of cannabis, may be responsible for this observation (Gallup, 2018). However, age-linked physiological changes and health issues could impact the consequences of cannabis use, so additional studies are needed to assess whether there are age-related changes in the metabolism or behavioral effects of cannabis, or alterations in endogenous cannabinoid receptor expression, function, or drug–receptor interactions with cannabinoids. Such studies should be conducted to provide evidence about the impacts of cannabis use in the aged to determine safe and effective conditions for cannabis use.

Uses of cannabis

Cannabinoids are generally used either for recreational purposes or medical reasons (D'Souza & Ranganathan, 2015). Some of these uses are ancient while others have been more recently identified (D'Souza & Ranganathan, 2015; Pisanti & Bifulco, 2018). Different cannabis products are currently in use or under consideration to treat various diseases (summarized in Table 1).

Medically, cannabinoids are used to treat several problems—some of which are more prevalent among the elderly such as chronic pain and insomnia—that are not treated sufficiently by other medications, or to avoid the intolerable side effects of the medications in current clinical use (Minerbi et al., 2018). Another reason cannabinoids may be preferred is the perception that cannabinoids are "natural" plant extracts and thus safer than pharmaceuticals (Minerbi et al., 2018). Examples of cannabinoid therapeutic uses include managing pain, treating constipation (Pisanti & Bifulco, 2018) and more recently, controlling spasticity from multiple sclerosis and managing Crohn's disease. However, the efficacy of cannabinoids to treat some of these conditions is not well established, despite the growing interest in using them for these indications (D'Souza & Ranganathan, 2015).

Cannabis Use, Neurobiology, Psychology, and Treatment. https://doi.org/10.1016/B978-0-323-89862-1.00039-8

TABLE 1 Potential uses of cannabis products.

Uses	Examples
Analgesic	Chronic pain, chemotherapy-related pain
Antiproliferation—cancer treatment	Lung adenocarcinoma, prostate cancer, colon cancer
Antineurodegeneration	Alzheimer's disease, Parkinson's disease
Substance use	Opioid dependence

Representative examples of some primary effects of cannabinoids that are used for their potential to treat and control the diseases and conditions listed.

Innovative uses of cannabinoids

A revolutionary use of cannabinoids that is undergoing preclinical study is their potential role as antiproliferative drugs that could hinder uncontrolled proliferation of cells in cancer tumors. This potential use is particularly relevant to aging individuals given the significant age-linked surge of cancer incidence (Pellati et al., 2018). Evidence supporting this potential use comes from several studies including the one conducted by Munson et al. (1975) that provides evidence for the ability of specific cannabinoids to retard the proliferation of lung adenocarcinoma tumor cells. Examples of cancer types that would be controlled through growth suppression and spread inhibition by various cannabinoids are many and include prostate cancer and colon cancer (Pisanti et al., 2009) both of which are more common in the elderly (Byrne & Carney, 1993). Moreover, cannabinoid use as adjuvant therapy to palliate cancer symptoms and adverse effects of chemotherapy such as pain, nausea, vomiting, and reduced appetite is promising as well (Pisanti et al., 2009).

Another innovative use of cannabinoids is as an antioxidant with possible neuroprotective effects that could be used for neurodegenerative diseases including dementias. The antioxidant effect of cannabinoids arises from their ability to interact with reactive oxygen species (ROS) that injure many cellular components (Hartsel et al., 2016). These antioxidation effects contribute to the neuroprotective potential of cannabinoids that would be employed in controlling neurodegenerative diseases including Alzheimer's disease, Parkinson's disease (Izzo et al., 2009), and Parkinson's disease dementia all of which are diseases prevalent among the elderly (Lopez & Kuller, 2019). Additional supporting evidence of the ability of these drugs to enhance cognitive ability in the aged comes from the experiment conducted by Bilkei-Gorzo et al. (2017), which shows that chronic administration of low dose cannabinoids to aging mice could reverse cognitive impairments caused by aging process. Additional evidence linking these findings to the antiinflammatory effect of cannabinoids is reported in which cannabidiol (CBD), through its antineuroinflammatory effect, is found to be able to slow down the development of Alzheimer's disease in a rodent model (Esposito et al., 2009). These potential uses of cannabis products could be utilized in humans, mainly the elderly, to maintain brain health and cognition and thus a better quality of life.

Cannabinoids could play a role in treating drug abuse such as opioid dependence. For instance, it has been observed that opioid-dependent patients on opioid agonist treatment who use cannabis have a remarkable decrease in their opioid agonist requirements without increasing the risk of withdrawal or cravings compared to noncannabis users (Bagra et al., 2018). Whereas the possible role of cannabinoid system in drug abuse treatment is also suggested by the work of Le Foll and Goldberg (2004) who utilized CB1 receptor antagonists Rimonabant for treating drug abuse including alcoholism, the work done by Ginsburg and Lamb (2006) shows that the same drug did not affect alcohol-seeking behavior or even food-seeking behavior in aged subjects with chronic alcohol exposure. This suggests that age-related cannabinoid system variation could hinder the effectiveness of cannabis in treating alcoholism in the elderly. Still, targeting alcohol use disorder among the elderly would be essential given the increasing number of people in this age group with drinking problems that can go unnoticed (Berks & McCormick, 2008).

Geriatric special considerations

Despite these promising initial findings, there are specific geriatric precautions to be considered before using a new drug in this population. These include the physiologic changes that occur as part of aging process, the prevalence of comorbidities in this age group, and thus potential drug-drug interactions (DDI) that could occur with the addition of a new drug.

TABLE 2 Physiologic changes linked to age potentially affecting cannabis.

Tissue/ organ characteristic	Change
Body fat content	Increase (35%)
Liver	Decrease mass, decrease blood flow
Gastric acidity	Decrease

Main organs/tissue/body factors and their specific physiologic changes that occur as part of aging and would affect cannabinoids metabolism and dosing in the human body.

Physiologic changes

Several physiologic changes (summarized in Table 2) occur in the body as part of aging process and could affect the pharmacodynamics or pharmacokinetics of cannabinoids (Hines & Murphy, 2011). Examples of these changes include a 35% rise in body fat, 17% drop in total volume of water in the body, 8% decline in plasma water, and 40% decrease in extracellular water in those above 65-year-old compared to young adults, which significantly affect the volume of distribution of both water-soluble and fat-soluble drugs; in addition, there is a decrease in hepatic mass and hepatic blood flow that would affect drug pharmacokinetics (Klotz, 2009). These variations significantly affect the bioavailability of various cannabinoids. For instance, THC undergoes extensive metabolism in the liver and only 4%–12% of the amount orally consumed reaches circulation and is taken up immediately by fatty tissue as it is a fat-soluble molecule (Chayasirisobhon, 2020). This implies that variation affecting the liver and an increase in body fat composition would alter cannabinoid elimination and distribution, dosing, and availability. Other age-related changes that could alter dosing, drug elimination, drug availability, and other important drug aspects in this age group include diminished renal clearance, reduced absorption, and decreased gastric acidity, which is relevant for orally administered cannabinoids as some of the THC consumed is destroyed by the acidity of the stomach (Grotenhermen, 2003; Klotz, 2009).

Comorbidities

Health-related problems linked to aging or that are more prevalent among the elderly could interact with drug dosing and effect (Hines & Murphy, 2011). The prevalence of multiple comorbidities increases with age (Salive, 2013), and these diseases can act as a confounding variable that would alter drug metabolism themselves or by contributing to age-related physiologic changes (Hines & Murphy, 2011). For instance, aging is a main risk factor for developing metabolic diseases which in turn might lead to liver steatosis that can progress to more serious hepatic conditions and disrupt various hepatic metabolic roles (Heindel et al., 2017). Given the significant role of hepatic metabolism of cannabinoids mentioned earlier (Chayasirisobhon, 2020), such effects would potentially affect cannabinoids dosing and effects. Although it might be challenging to set a guideline that consider comorbidities, there are successful examples and ongoing efforts to establish such guidelines that would set different priorities for the elderly who have a shorter life expectancy (Davies & O'Mahony, 2015).

Effects of other drugs (Table 3)

The third variable is the high likelihood of elderly individuals to be using several other medications to treat age-related diseases that could interact with a newly prescribed drug (Hines & Murphy, 2011). To control various comorbidities elderly people suffer from, they often take multiple medications daily, and the prevalence of probable interactions while on

TABLE 3 Drug-drug interaction potentially affecting cannabis action.

Factor	Effect
Cytochrome P450 enzyme	Metabolism of various medications
Transmembrane protein: P-glycoprotein	Pump drugs out of human cells

Examples of major factors and their mechanisms of action that are affected by and affect drugs metabolism and are part of drug-drug interactions that would potentially affect cannabinoids.

polypharmacy is not determined (Hines & Murphy, 2011). For example, a study conducted in Italy targeting possible drug-drug interaction (DDI) among the elderly shows that around half of the participants were at risk of a DDI that would be classified as a serious risk (Petrini et al., 2019). While other drugs could alter the metabolism of cannabinoids, there is also evidence that this is a reciprocal relationship and cannabinoids can alter metabolism of other drugs by interfering with cytochrome P450 enzyme that is involved in the metabolism of several medications (Chayasirisobhon, 2020).

Another cause of DDI can arise from the effect of some drugs on drug transporters that play an important role in the intake, dissemination, and elimination of drugs (DuBuske, 2005). An example of these transporters is the transmembrane protein P-glycoprotein (P-gp) that was initially found in cancer cells and plays a role multidrug resistance (Nigam, 2014), but it also plays a role in pumping drugs out of normal human cells (DuBuske, 2005). The level of activity of these transporters can be affected by other drugs, some of which are over the counter products (Wang et al., 2002). Moreover, the study conducted by Holland et al. (2006) provides evidence that extended exposure to cannabinoids might reduce the level of P-gp expression in cells. This complicates the ability to reach a therapeutic level yet avoid a toxic level of a newly prescribed drug.

Other factors

In addition to the factors mentioned earlier that are common to all aged people, aging also augments the variation between individuals in drug metabolism and side effects (Kinirons & O'Mahony, 2004). Moreover, guidelines on drug use are derived from studies that often do not include the elderly which make the side effects profile in this age group unpredictable (Davies & O'Mahony, 2015). This calls attention to the need for studying any drug to be used in the elderly population extensively and specifically in this age group before it can be approved and safely used. This need is augmented for drugs with well-known side effects like cannabinoids. These side effects include, but are not limited to, their psychoactive effect and their cardiovascular risks such as myocardial infarction, cannabis arteritis, and stroke (Cohen et al., 2019) which would be especially dangerous in aged people.

Age-related CB1 receptor expression variation

The consensus of studies targeting age linked variation of cannabinoid receptor type1 (CB1) expression is that their expression is generally reduced, but this might be brain region specific. For example, one of the few human postmortem studies conducted on this topic suggests that there might be an aged-related decrease in the expression of cannabinoid receptor in the frontal cortex of humans (Mato & Pazos, 2004). Ginsburg and Hensler (2022) study examined this phenomenon in several cortical and subcortical areas of young and aged rats, revealing that while medial prefrontal, temporal and entorhinal cortices of older subjects show a significant decrease in CB1 binding sites (via [^{3}H] CP55,940 cannabis agonist binding) in aged subjects compared to the young, other areas, including the subcortical ones, do not show age-related variations. However, another study targeting age linked variation of this receptor expression in and around the hippocampus area suggests a temporal cortex increase, post-rhinal cortex decrease, and no hippocampus variation in the level CB1 receptor expression in aged rats compared to younger subjects (Liu et al., 2003). Romero et al. (1998) showed extrapyramidal area-specific reduction in receptor expression in aged rats mainly in the lateral region of the caudate and putamen areas, similar to the results of Mailleux and Vanderhaeghen (1992) in which a significant drop in receptor expression in the striatum of aged rats was reported. Thus, there is some variability in reporting the areas that showed this reduction of expression, but the conditions in which the rats were kept are either not mentioned or varied between studies, and technical details of protein detection are different between studies as well. This would imply that these conditions might play a role in expression variation, and the techniques used to detect receptor availability on cell surface versus whole tissue expression might also be significant as cell surface expression might not be reflected in whole tissue expression (Ginsburg & Hensler, 2022). These variations (indicated in Table 4) would highlight the need for additional studies to investigate this difference under controlled conditions and determine the contribution of other variables to the reported variation.

Age-related CB1 receptor function variation

Decreases in receptor expression might be compensated for by a compensatory increase in their function in aging subjects. To address this, Ginsburg and Hensler also measured CB1 function using [^{35}S] GTPγS labeling in the presence of a synthetic cannabinoid agonist. Despite decreased receptor expression in some brain regions, the function of CB1 receptor

TABLE 4 Age-related variation in CB1 receptor expression in different brain areas.

Author/s (year)	Variation	Area studied
Ginsburg and Hensler (2022)	Decrease	Prefrontal, temporal, entorhinal cortices
	No change	Subcortical areas
Mato and Pazos (2004)	Decrease	Frontal cortex
Liu et al. (2003)	Increase	Temporal cortex
	Decrease	Post-rhinal cortex
	No change	Hippocampus
Romero et al. (1998)	Decrease	Lateral region of the caudate and putamen
Mailleux and Vanderhaeghen (1992)	Decrease	Striatum

Main studies targeting variation in the level of expression of cannabinoid receptor type 1 that is linked to aging and their area specific findings.

TABLE 5 Age-related variation of CB1 receptor function.

Author/s (year)	Variation	Area studied
Ginsburg and Hensler (2022)	No change	Cortical and subcortical areas
Wang et al. (2003)	Decrease	Limbic forebrain
Romero et al. (1998)	Decrease	Basal ganglia

Main studies targeting variation in the functional level of cannabinoid receptor type 1 that is linked to aging and their area specific findings.

signaling was not different between young and aged rats, suggesting that enhancement of receptor function can compensate for the net decrease in receptors and cannabinoid-binding sites (Ginsburg & Hensler, 2022). However, other animal studies reported a decrease in receptor function in various brain regions of aging adults. For instance, Romero et al. (1998) reported an age-related decrease in agonist stimulated GTP binding linked to area specific decrease in receptor expression. Wang et al. (2003) also reported age-linked region-specific decreases in agonist-stimulated signaling. This discrepancy (indicated in Table 5) in results necessitates further studies to determine functional variability and factors that might be affecting it.

Age-related cannabinoids behavioral effect variation

Ginsburg and Hensler (2022) utilized the active component of cannabis and CB1 agonist Δ9-tetrahydrocannabinol (THC) and the cannabinoid receptor type1 (CB1) antagonist rimonabant in young and old adult rats and measured its effects on decreased operant food responding. Although both the CB1 agonist and antagonist had a significant effect on behavioral responding, none of these effects were significantly different between age groups (Ginsburg & Hensler, 2022). However, food-maintained operant behavior is a relatively robust behavior and behavioral measures more sensitive to cannabinoid effects could reveal differences. For example, a survey-based study on memory effects of cannabis in human subjects found that younger participants reported being more forgetful and having more difficulty finding words than older participants (Sexton et al., 2019). Although one possible explanation for this would be an age-related decrease in cannabinoid system activity and drug effect, it is also possible that aged subjects related these symptoms to aging and the cognitive impairment they already have due to their age and thus did not report it as an effect of their cannabis use (Sexton et al., 2019). These results (summarized in Table 6) support the need for further studies targeting age-related variations in cannabinoid effects.

TABLE 6 Age-related variation of cannabinoids behavioral effects.

Author/s (year)	Behavior	Variation
Ginsburg and Hensler (2022)	Operant food responding	No change
Sexton et al. (2019)	Reported memory deficit	Decrease

Results of major studies exploring age linked variation in the effect of cannabinoids on targeted behaviors.

Conclusion

The scarcity of studies targeting how cannabinoids affect people in this age group calls attention to the importance of conducting highly controlled studies targeting this issue to improve our understanding of the mechanism of action of cannabinoids in aging individuals. This need is growing with the growing demand for medical use of cannabinoids, as there are multiple potential therapeutic benefits of this drug in the elderly. It would also give insight about the impact of recreational use of this drug among the aging.

This chapter provides an overview of issues relevant to the growing occurrence of cannabinoid use among the elderly, including the reasons for and prevalence of this increased use. It also covers ancient, current, and future potential uses of this group of drugs. In addition, we have discussed some of the considerations and special precautions that must be evaluated when introducing a new drug to the aged.

Studies targeting age-related variation of cannabinoid receptor expression and function, and behavioral effects of the drug are also reviewed. A measurable region-specific decrease in cannabis receptor expression in the elderly is discussed, although there is no consensus about the specific brain regions in which this occurs or the impact of these changes. Moreover, the results of the few studies targeting age-related variation of cannabinoids behavioral effects are reviewed and are inconsistent. Thus, additional studies with more consistent experimental conditions are warranted.

Applications to other areas

In this chapter, the CB1 receptor and its effect on various regions of the brain is reviewed. While this review has focused on the therapeutic potential of cannabinoids in aging individuals, some of the same considerations are relevant to younger adults. For example, chronic pain is not solely an affliction of the aged, nor is multiple sclerosis, cancer, or addiction. Drug-drug interactions are also relevant to younger individuals. Thus, these applications and considerations are relevant to adults of all ages, though age-dependent changes in the cannabinoid system become more important with advancing age.

Further, as recreational use of cannabis and cannabinoids increases, especially among older adults, clinical consideration of the impact of this use and potential interactions with other medications or the cognitive or physiological changes that occur with age become more important. Clinicians should ask about cannabis use and consider whether it might impact clinical care or medications.

Additionally, the CB2 receptor, predominantly found in the immune system, represents another important area of therapeutic development which may be subject to the same age-related changes, and it should be the target of further study as well. Thus, it is essential to conduct studies targeting different cannabinoids receptors to conclusively determine the extent and impact of age-linked variations in the cannabinoid system and drug effects to better guide their therapeutic use and provide better clinical counseling on their recreational use in aging patients.

Mini-dictionary of terms

CP55,940: a synthetic cannabinoid receptor agonist

Δ9-tetrahydrocannabinol: Δ9-THC is a natural cannabinoid extracted from cannabis plant.

Insomnia: sleeping problem that would include inability to fall or stay asleep.

Metabolic diseases: examples include diabetes, hepatic lipid diseases, and high blood lipid levels.

[^{35}S] GTPγS: a method used for measuring the level of G protein activation following stimulation of a G protein-coupled receptor by an agonist.

Steatosis: abnormal accumulation of lipids in tissues.

Key facts of drug-drug interaction

- Administering several drugs simultaneously has the risk of drugs interfering with each other's metabolism and mechanism of action.
- This risk can decrease the effect of the therapeutic dose of a drug to a subtherapeutic level or increase it to a toxic level.
- Main factors involved in this effect are Cytochrome P 450 enzymes which are a family of enzymes that are involved in metabolism of several drugs, and yet they get induced and inhibited by a set of drugs as well.
- Another possibility is having synergistic or additive effects of drugs, which would exaggerate their targeted effect.
- Polypharmacy has a risk of side effects addition especially in elderly, such as the sedative side effects of multiple drugs that can result in confusion.

Key facts of cannabinoid receptor function and expression

- Cannabinoid receptor is a transmembrane protein that belong to G protein-coupled receptors.
- It has a transmembrane domain that anchors it to the cell membrane, an extra cellular domain, and an intracellular domain.
- The extracellular domain binds to cannabinoids products or agonist and trigger conformational changes eventually affecting the coupled G protein.
- The activated G protein releases its G subunit that induces intracellular signaling.
- Intracellular signaling is propagated by the effect of G subunit on adenylate cyclase and its product cAMP.

Summary points

- Cannabinoid use is growing among the elderly.
- Some of this increased use is due to perceived or legitimate therapeutic effects.
- Age-related physiologic changes that would affect cannabinoids' function and availability are complex.
- It is difficult to predict how cannabinoids interact with comorbidities and chronic medications commonly used in the elderly or the implications of these interactions.
- Few studies targeted the variation of cannabinoids effect in this age group.
- CB1 receptor expression seems to have brain-area specific reduction in expression although results vary across studies.
- Studies targeting age-related variation in the function of this receptor also report contradictory results.
- Age-related behavioral changes in cannabinoid effects should be more carefully studied.

References

Bagra, I., Krishnan, V., Rao, R., & Agrawal, A. (2018). Does cannabis use influence opioid outcomes and quality of life among buprenorphine maintained patients? A cross-sectional comparative study. *Journal of Addiction Medicine*, *12*(4), 315–320. https://doi.org/10.1097/adm.0000000000000406.

Berks, J., & McCormick, R. (2008). Screening for alcohol misuse in elderly primary care patients: A systematic literature review. *International Psychogeriatrics*, *20*(06), 1090. https://doi.org/10.1017/s1041610208007497.

Bilkei-Gorzo, A., Albayram, O., Draffehn, A., Michel, K., Piyanova, A., Oppenheimer, H., Dvir-Ginzberg, M., Rácz, I., Ulas, T., Imbeault, S., Bab, I., Schultze, J. L., & Zimmer, A. (2017). A chronic low dose of Δ9-tetrahydrocannabinol (THC) restores cognitive function in old mice. *Nature Medicine*, *23*(6), 782–787. https://doi.org/10.1038/nm.4311.

Byrne, A., & Carney, D. N. (1993). Cancer in the elderly. *Current problems in cancer*, *17*(3), 145–218. https://doi.org/10.1016/0147-0272(93)90007-o.

Chayasirisobhon, S. (2020). Mechanisms of action and pharmacokinetics of cannabis. *The Permanente Journal*, *24*(5). https://doi.org/10.7812/tpp/19.200.

Cohen, K., Weizman, A., & Weinstein, A. (2019). Positive and negative effects of cannabis and cannabinoids on health. *Clinical Pharmacology & Therapeutics*, *105*(5). https://doi.org/10.1002/cpt.1381.

D'Souza, D. C., & Ranganathan, M. (2015). Medical marijuana. *JAMA*, *313*(24), 2431. https://doi.org/10.1001/jama.2015.6407.

Davies, E. A., & O'Mahony, M. S. (2015). Adverse drug reactions in special populations—The elderly. *British Journal of Clinical Pharmacology*, *80*(4), 796–807. https://doi.org/10.1111/bcp.12596.

DuBuske, L. M. (2005). The role of P-glycoprotein and organic anion-transporting polypeptides in drug interactions. *Drug Safety*, *28*(9), 789–801. https://doi.org/10.2165/00002018-200528090-00004.

Esposito, G., Scuderi, C., Savani, C., Steardo, L., Filippis, D., Cottone, P., Iuvone, T., Cuomo, V., & Steardo, L. (2009). Cannabidiol in vivo blunts β-amyloid induced neuroinflammation by suppressing IL-1β and iNOS expression. *British Journal of Pharmacology*, *151*(8), 1272–1279. https://doi.org/10.1038/sj.bjp.0707337.

Gallup. (2018). *In the news: Marijuana legalization (Online)*. http://news.gallup.com/poll/225017/news-marijuana-legalization.aspx.

Ginsburg, B. C., & Hensler, J. G. (2022). Age-related changes in CB1 receptor expression and function and the behavioral effects of cannabinoid receptor ligands. *Pharmacology Biochemistry and Behavior*, *213*, 173339. https://doi.org/10.1016/j.pbb.2022.173339.

Ginsburg, B. C., & Lamb, R. J. (2006). Cannabinoid effects on behaviors maintained by ethanol or food: A within-subjects comparison. *Behavioural Pharmacology*, *17*(3), 249–257. https://doi.org/10.1097/00008877-200605000-00006.

Grotenhermen, F. (2003). Pharmacokinetics and pharmacodynamics of cannabinoids. *Clinical Pharmacokinetics*, *42*(4), 327–360. https://doi.org/10.2165/00003088-200342040-00003.

Han, B. H., & Palamar, J. J. (2018). Marijuana use by middle-aged and older adults in the United States, 2015–2016. *Drug and Alcohol Dependence*, *191*, 374–381. https://doi.org/10.1016/j.drugalcdep.2018.07.006.

Han, B. H., Sherman, S., Mauro, P. M., Martins, S. S., Rotenberg, J., & Palamar, J. J. (2016). Demographic trends among older cannabis users in the United States, 2006-13. *Addiction*, *112*(3), 516–525. https://doi.org/10.1111/add.13670.

Hartsel, J. A., Eades, J., Hickory, B., & Makriyannis, A. (2016). Cannabis sativa and Hemp. In *Nutraceuticals: Efficacy, Safety and Toxicity* (pp. 735–754). Elsevier. https://doi.org/10.1016/B978-0-12-802147-7.00053-X.

Heindel, J. J., Blumberg, B., Cave, M., Machtinger, R., Mantovani, A., Mendez, M. A., Nadal, A., Palanza, P., Panzica, G., Sargis, R., Vandenberg, L. N., & vom Saal, F. (2017). Metabolism disrupting chemicals and metabolic disorders. *Reproductive Toxicology*, *68*, 3–33. https://doi.org/10.1016/j.reprotox.2016.10.001.

Hines, L. E., & Murphy, J. E. (2011). Potentially harmful drug–drug interactions in the elderly: A review. *The American Journal of Geriatric Pharmacotherapy*, *9*(6), 364–377. https://doi.org/10.1016/j.amjopharm.2011.10.004.

Holland, M. L., Panetta, J. A., Hoskins, J. M., Bebawy, M., Roufogalis, B. D., Allen, J. D., & Arnold, J. C. (2006). The effects of cannabinoids on P-glycoprotein transport and expression in multidrug resistant cells. *Biochemical Pharmacology*, *71*(8), 1146–1154. https://doi.org/10.1016/j.bcp.2005.12.033.

Izzo, A. A., Borrelli, F., Capasso, R., Di Marzo, V., & Mechoulam, R. (2009). Non-psychotropic plant cannabinoids: New therapeutic opportunities from an ancient herb. *Trends in Pharmacological Sciences*, *30*(10), 515–527. https://doi.org/10.1016/j.tips.2009.07.006.

Kinirons, M. T., & O'Mahony, M. S. (2004). Drug metabolism and ageing. *British Journal of Clinical Pharmacology*, *57*(5), 540–544.

Klotz, U. (2009). Pharmacokinetics and drug metabolism in the elderly. *Drug Metabolism Reviews*, *41*(2), 67–76. https://doi.org/10.1080/03602530902722679.

Le Foll, B., & Goldberg, S. R. (2004). Cannabinoid CB1 receptor antagonists as promising new medications for drug dependence. *Journal of Pharmacology and Experimental Therapeutics*, *312*(3), 875–883. https://doi.org/10.1124/jpet.104.077974.

Liu, P., Bilkey, D. K., Darlington, C. L., & Smith, P. F. (2003). Cannabinoid CB1 receptor protein expression in the rat hippocampus and entorhinal, perirhinal, postrhinal and temporal cortices: Regional variations and age-related changes. *Brain Research*, *979*(1–2), 235–239. https://doi.org/10.1016/s0006-8993(03)02872-5.

Lopez, O. L., & Kuller, L. H. (2019). Epidemiology of aging and associated cognitive disorders: Prevalence and incidence of Alzheimer's disease and other dementias. *Handbook of Clinical Neurology*, *167*, 139–148. https://doi.org/10.1016/B978-0-12-804766-8.00009-1.

Mailleux, P., & Vanderhaeghen, J.-J. (1992). Age-related loss of cannabinoid receptor binding sites and mRNA in the rat striatum. *Neuroscience Letters*, *147*(2), 179–181. https://doi.org/10.1016/0304-3940(92)90589-y.

Mato, S., & Pazos, A. (2004). Influence of age, postmortem delay and freezing storage period on cannabinoid receptor density and functionality in human brain. *Neuropharmacology*, *46*(5), 716–726. https://doi.org/10.1016/j.neuropharm.2003.11.004.

Minerbi, A., Häuser, W., & Fitzcharles, M.-A. (2018). Medical cannabis for older patients. *Drugs & Aging*, *36*(1), 39–51. https://doi.org/10.1007/s40266-018-0616-5.

Munson, A. E., Harris, L. S., Friedman, M. A., Dewey, W. L., & Carchman, R. A. (1975). Antineoplastic activity of cannabinoids2. *JNCI: Journal of the National Cancer Institute*, *55*(3), 597–602. https://doi.org/10.1093/jnci/55.3.597.

Nigam, S. K. (2014). What do drug transporters really do? *Nature Reviews Drug Discovery*, *14*(1), 29–44. https://doi.org/10.1038/nrd4461.

Pellati, F., Borgonetti, V., Brighenti, V., Biagi, M., Benvenuti, S., & Corsi, L. (2018). Cannabis sativa L. and nonpsychoactive cannabinoids: Their chemistry and role against oxidative stress, inflammation, and cancer. *BioMed Research International*, *2018*, 1–15. https://doi.org/10.1155/2018/1691428.

Petrini, E., Caviglia, G. P., Pellicano, R., Saracco, G. M., Morino, M., & Ribaldone, D. G. (2019). Risk of drug interactions and prescription appropriateness in elderly patients. *Irish Journal of Medical Science (1971)*, *189*(3), 953–959. https://doi.org/10.1007/s11845-019-02148-8.

Pisanti, S., & Bifulco, M. (2018). Medical cannabis: A plurimillennial history of an evergreen. *Journal of Cellular Physiology*, *234*(6), 8342–8351. https://doi.org/10.1002/jcp.27725.

Pisanti, S., Malfitano, A. M., Grimaldi, C., Santoro, A., Gazzerro, P., Laezza, C., & Bifulco, M. (2009). Use of cannabinoid receptor agonists in cancer therapy as palliative and curative agents. *Best Practice & Research Clinical Endocrinology & Metabolism*, *23*(1), 117–131. https://doi.org/10.1016/j.beem.2009.02.001.

Romero, J., Berrendero, F., Garcia-Gil, L., de la Cruz, P., Ramos, J. A., & Fernandez-Ruiz, J. J. (1998). Loss of cannabinoid receptor binding and messenger RNA levels and cannabinoid agonist-stimulated [35s]guanylyl-5′-O-(thio)-triphosphate binding in the basal ganglia of aged rats. *Neuroscience*, *84*(4), 1075–1083. https://doi.org/10.1016/s0306-4522(97)00552-6.

Salive, M. E. (2013). Multimorbidity in older adults. *Epidemiologic Reviews*, *35*(1), 75–83. https://doi.org/10.1093/epirev/mxs009.

Sexton, M., Cuttler, C., & Mischley, L. K. (2019). A survey of cannabis acute effects and withdrawal symptoms: Differential responses across user types and age. *The Journal of Alternative and Complementary Medicine*, *25*(3), 326–335. https://doi.org/10.1089/acm.2018.0319.

Wang, L., Liu, J., Harvey-White, J., Zimmer, A., & Kunos, G. (2003). Endocannabinoid signaling via cannabinoid receptor 1 is involved in ethanol preference and its age-dependent decline in mice. *Proceedings of the National Academy of Sciences of the United States of America*, *100*(3), 1393–1398. https://doi.org/10.1073/pnas.0336351100.

Wang, Z., Hamman, M. A., Huang, S.-M., Lesko, L. J., & Hall, S. D. (2002). Effect of St John's wort on the pharmacokinetics of fexofenadine*. *Clinical Pharmacology & Therapeutics*, *71*(6), 414–420. https://doi.org/10.1067/mcp.2002.124080.

Further reading

Berrendero, F., Romero, J., Garcı´a-Gil, L., Suarez, I., De la Cruz, P., Ramos, J. A., & Fernández-Ruiz, J. J. (1998). Changes in cannabinoid receptor binding and mRNA levels in several brain regions of aged rats. *Biochimica et Biophysica Acta (BBA)—Molecular Basis of Disease*, *1407*(3), 205–214. https://doi.org/10.1016/s0925-4439(98)00042-8.

Chapter 26

Cannabis, binge drinking, and memory in adolescents and young adults

Concepción Vinader-Caerols and Santiago Monleón
Department of Psychobiology, Faculty of Psychology, University of Valencia, Valencia, Spain

Abbreviations

AUDIT	alcohol use disorders identification test
AUDs	alcohol use disorders
BAC	blood alcohol concentration
BD	binge drinking
CAST	cannabis abuse screening test
IVM	immediate visual memory
NIAAA	National Institute on Alcohol Abuse and Alcoholism
SDU	standard drink unit
THC	Δ-9-tetrahydrocannabinol
WM	working memory

Introduction

Cannabis and binge drinking

Cannabis is the most prevalent illegal drug among adolescents and young adults. Thus, 28.6% of Spanish students between the ages of 14 and 18 have ever consumed it, and 22.2% of them admit to having used cannabis in the last year (OEDA, 2021).

On the other hand, alcohol is one of the most widely consumed psychoactive substances in the world, especially among adolescents and young adults (Chavez et al., 2011; Hibell et al., 2009). The prevalent pattern of alcohol consumption of this population in Western countries is binge drinking (BD), which has been defined by the National Institute on Alcohol Abuse and Alcoholism (NIAAA) as a pattern of drinking that raises a person's blood alcohol concentration (BAC) to 0.8 g/L or above (NIAAA, 2004). This pattern is characterized by the intake of large quantities of alcohol in short periods (about 2 hours), followed by a period of abstinence (Maurage et al., 2012; Petit et al., 2014; Vinader-Caerols & Monleón, 2019) (see Fig. 1). This repeated alternation between intoxications and abstinence is particularly detrimental in a still developing brain, such as the adolescent brain.

In recent years, there has been an increase in the BD consumption pattern during late adolescence and early youth. In Europe, the prevalence of BD in adolescents aged 15–19 years is 24.1% (WHO, 2018). In Spain, the prevalence of this pattern among adolescents is even higher: According to the latest Survey on Drug Use in Secondary Education, 27.9% of 14-18-year-old students admitted having carried out this consumption pattern in the last 30 days at the time of answering the survey (OEDA, 2021).

Furthermore, 13.4% of the Spanish adolescents who consumed alcohol during the last year also used cannabis; and 91.7% of those who used cannabis also claimed to drink alcohol (OEDA, 2019). This indicates that almost all cannabis users are cannabis and alcohol polyconsumers.

Often, the low perception of risk about these substances and their availability make that cannabis and alcohol are considered to be recreational drugs whose consumption is trivialized. But the truth is that the use and abuse of cannabis and alcohol is one of the most important public health problems worldwide due to the social and health consequences they cause. Besides, alcohol BD, alone or together with cannabis, when started at an early age is also a precipitant that can lead

Cannabis Use, Neurobiology, Psychology, and Treatment. https://doi.org/10.1016/B978-0-323-89862-1.00018-0

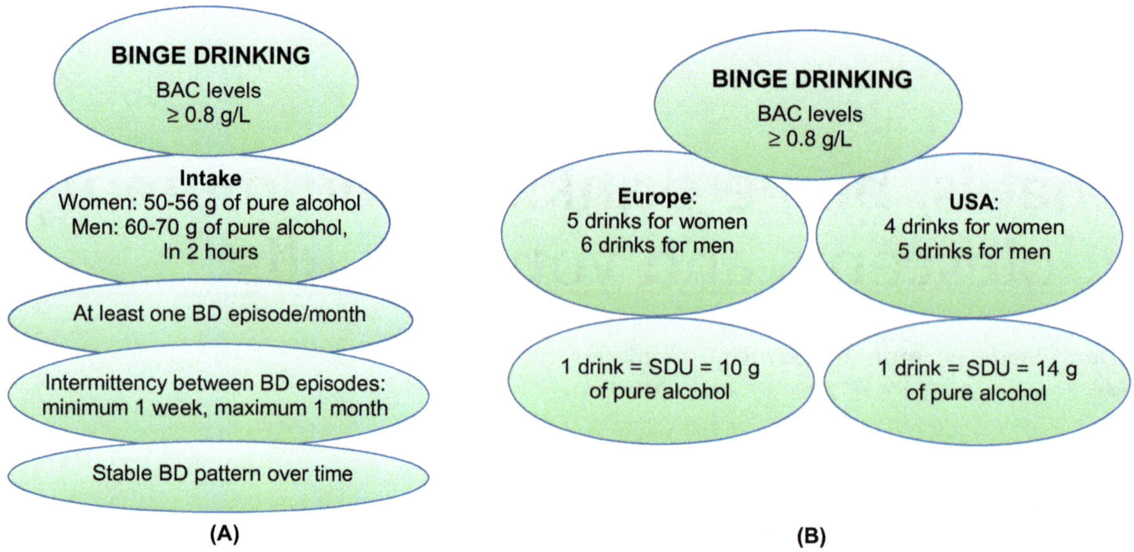

FIG. 1 (A) Binge drinking pattern criteria. Quantity: intake of 50–56 g of pure alcohol in women and 60–70 g in men, in 2 h. Frequency: at least one BD episode per month. Intermittency: abstinence between BD episodes over time (minimum 1 week, maximum 1 month). (B) Number of drinks (1 drink = SDU, standard drink unit) in the Europe and USA for binge drinking's BAC levels.

to the development of dependence and other alcohol use disorders (AUDs). Consequently, in this risk population, the combined effects of cannabis and alcohol BD should be studied with special attention due to their wide range of negative consequences, including their detrimental effects on memory.

Adolescents and young adults

Adolescence (10–19 years) and young adulthood (20–24 years) are unique transitional periods associated with age-related neural and cognitive changes (Lees et al., 2019). These maturational changes could contribute to a young person's increased propensity to experiment and engage in risk-taking behaviors, including an increased propensity to drug abuse (for a review, see Lees et al., 2020). The adolescent brain is highly plastic and undergoes developmental and biological changes that may be required for proper behavioral and cognitive development. Unfortunately, due to this highly malleable state, the adolescent brain is also particularly vulnerable to the effects of drugs of abuse. Human and animal findings validate these concerns by demonstrating that exposure to alcohol or cannabis can lead to long-term cognitive dysfunction into adulthood, even after extended abstinence periods (Mooney-Leber & Gould, 2018). Therefore, it is important to further evaluate the effects of cannabis and alcohol on brain structure and cognitive function, especially during the adolescent period. Large longitudinal human cognitive and neuroimaging studies with more diverse samples are currently underway which will improve our understanding of the impact of polysubstance use, as well as the interactive effects of substance abuse on neurodevelopment and cognition.

Gender differences

Adolescent brain developmental trajectories tend to differ by sex, with female brains developing one to two years earlier than males (Giedd, 2004). Adolescent drug use may differentially impact male and female cognitive function. Thus, gender differences in the effects of alcohol have been reported, supporting the view that brains of male and female adolescents are differentially affected by alcohol use (Alfonso-Loeches et al., 2013). There is evidence suggesting that female adolescents are more vulnerable to the neurotoxic effects of alcohol on cognition (Alfonso-Loeches et al., 2013; Squeglia et al., 2011). For example, the cognitive tolerance effect of alcohol on immediate visual memory (IVM) develops in female binge drinkers but not in male binge drinkers (Vinader-Caerols, Talk, et al., 2017). However, there are no studies evaluating the cognitive tolerance phenomenon in cannabis and alcohol polyconsumers of both sexes. Several studies have also demonstrated the differential associations of cannabis and alcohol use concerning the effects on brain structure and neurocognitive functioning, showing a different pattern of brain alterations (e.g., Sullivan et al., 2020) and neurocognitive impairment (e.g., Noorbakhsh et al., 2020) in men and women. In the light of these data, it seems crucial including both sexes in this research field, carrying out more longitudinal studies on sex differences in different cognitive domains known to be affected by cannabis and/or alcohol use. Furthermore, it has been proposed that longitudinal studies may provide insight into the cause-effect relationships between structural and functional brain differences, neurocognitive deficits, and BD in adolescents and young adults (Lees et al., 2019).

Effects of cannabis on memory (in healthy adolescents with recreational use)

Recreational cannabis use is understudied, especially during the adolescent period, when neural maturation may make users particularly vulnerable to the effects of Δ-9-tetrahydrocannabinol (THC) on brain structure. However, the differences in brain structure between cannabis consumers and nonconsumers, if present, are subtle, and may vary as a function of age, cumulative cannabis exposure, and sex in young people (Allick et al., 2021). The effects of withdrawal are unclear, although there are indications of cognitive recovery, but not brain recovery, even after weeks of abstinence (Harris et al., 2022; Sullivan et al., 2020).

However, it should be noted that, since cannabis users are mostly also alcohol users, there are no studies in healthy adolescents and young adults investigating recreational effects in cannabis-only users.

The literature on the effects of cannabis on cognitive functions generally shows that different types of memory are affected by cannabis recreational use. Episodic/visual memory (VM), working memory (WM), face processing, verbal memory, and short-term memory are affected by cannabis consumption. Few studies note a lack of effect of cannabis on memory (Matheson et al., 2020; Ross et al., 2020). The age of onset and the amount of total consumption of this drug of abuse may have a strong influence on the cognitive effects of cannabis.

A summary of the main memory impairments produced by cannabis consumption is shown in Table 1.

TABLE 1 Summary of the main memory impairments by cannabis use.

References	Sample	Impaired memory
Lawn et al. (2022)	$n=76$ (16–17 years old)	Verbal episodic memory
Ajmera et al. (2021)	$n=86$ (9–23 years old)	Verbal memory
Blest-Hopley et al. (2021)	$n=42$ (adolescents)	Verbal memory
Selamoglu et al. (2021)	$n=50$ (21–30 years old)	Visual memory Working memory Episodic memory
Frolli et al. (2021)	$n=300$ (15–16 years old)	Working memory
Ellingson et al. (2021)	$n=1192$ (17–24 years old)	Verbal memory
Infante et al. (2020)	$n=175$ (12–15 years old)	Working memory
Chung et al. (2020)	$n=60$ (18–25 years old)	Working memory
Leiker et al. (2019)	$n=104$ (14–18 years old)	Face processing
Wade et al. (2019)	$n=79$ (16–26 years old)	Working memory
Tervo-Clemmens et al. (2018)	$n=75$ (14, 16, 22 years old)	Working memory
Levar et al. (2018)	$n=41$ (18–25 years old)	Verbal memory
Mokrysz et al. (2016)	$n=40$ (16–17, 24–28 years old)	Working memory Episodic memory
Becker et al. (2015)	$n=46$ (15–23 years old)	Verbal memory
Shollenbarger et al. (2015)	$n=65$ (18–25 years old)	Working memory
Winward et al. (2014)	$n=128$ (16–18 years old)	Verbal memory
Dougherty et al. (2013)	$n=93$ (14–17 years old)	Short-term memory

Effects of binge drinking on memory

Findings from human adolescent studies suggest that BD and heavy alcohol use are associated with poorer cognitive functioning on a broad range of neuropsychological assessments, including learning, memory, visuospatial functioning, psychomotor speed, attention, executive functioning, and impulsivity (Lees et al., 2020). When it comes to memory, the effects of BD on different kinds of memory have been investigated and impairments produced by alcohol BD consumption have been observed in IVM, episodic memory, WM, short-term memory, verbal memory, and long-term memory.

A summary of the main memory impairments produced by alcohol BD consumption is shown in Table 2.

Among these memory impairments, IVM, word fragment completion, and free recall appear to be the most sensitive ones, as they are affected by moderate doses of alcohol (BAC = 0.3–0.38 g/L), both in adolescents and in young adults (e.g., Söderlund et al., 2005; Vinader-Caerols, Talk, et al., 2017). However, higher doses of alcohol (BAC levels of BD, i.e., around 0.8 g/L) are necessary so that this significant impairment can be observed in other types of memory, such as WM (e.g., Vinader-Caerols, Duque, et al., 2017) or short-term memory (e.g., Brumback et al., 2007). A plausible explanation for the lack of effects reported with BACs under 0.8 g/L (e.g., Brumback et al., 2007; Rose & Duka, 2007; Spinola et al., 2017) is that the brain of binge drinkers employs compensatory mechanisms in additional brain areas in order to perform the tasks adequately and that these resources are undermined at higher BACs (e.g., Day et al., 2013).

Effects of polydrug use (cannabis + alcohol BD) on memory

Given the scarcity of studies assessing the recreational effects of poly-drug use (cannabis together with alcohol BD consumption), we examined the interaction between a history of BD, alone or with cannabis consumption, and the effects of acute alcohol exposure on IVM and WM in adolescents of both sexes (Vinader-Caerols & Monleón, 2021). In our studies, the third edition of the Wechsler Memory Scale (WMS-III; version adapted for the Spanish population) (Wechsler, 2004) was used to assess IVM (faces memory task, scenes memory task and IVM-IQ) and WM (letter-number task, spatial location task and WM-IQ). The IVM subscales require the respondent to recognize faces and remember scenes, while the WM subscales require the respondent to put letter-number sets in order and to reproduce visual-spatial sequences.

In addition, a problematic use of alcohol or cannabis among the BD subjects and co-users was assessed using the AUDIT and CAST tests, respectively. None of the subjects was found to be alcohol or cannabis-dependent (mean AUDIT: 8.35 ± 0.39 in men; 7.3 ± 0.28 in women; mean CAST: 2.83 ± 0.38 in men; 2.57 ± 0.28 in women).

TABLE 2 Summary of the main memory impairments by alcohol BD use.

References	Sample	Impaired memory
Vinader-Caerols and Monleón (2021)	$n = 290$ (18–19 years old)	Immediate visual memory
Carbia et al. (2020)	$n = 180$ (18–19 years old)	Episodic memory
Mahedy et al. (2018)	$n = 461$ (15–18 years old)	Working memory
Carbia et al. (2017)	$n = 155$ (18–23 years old)	Short-term memory
Vinader-Caerols, Duque, et al. (2017)	$n = 154$ (18–19 years old)	Immediate visual memory Working memory
Vinader-Caerols, Talk, et al. (2017)	$n = 172$ (18–19 years old)	Immediate visual memory
Nguyen-Louie (2016)	$n = 112$ (12–16 years old)	Verbal memory
Cash et al. (2015)	$n = 19$ (18–35 years old)	Short-term memory
Nguyen-Louie et al. (2015)	$n = 234$ (12–18 years old)	Verbal memory
Winward et al. (2014)	$n = 128$ (16–18 years old)	Verbal memory
Day et al. (2013)	$n = 91$ (18–20 years old)	Working memory
Mota et al. (2013)	$n = 89$ (18–20 years old)	Short-term memory Working memory Long-term memory
Brumback et al. (2007)	$n = 72$ (21–35 years old)	Short-term memory

The effects of cannabis and alcohol polyconsumption on memory have been less investigated and are more ambiguous than the effects of these substances alone. Thus, some authors (Cummings et al., 2019; Squeglia & Gray, 2016) have reported a synergistic effect of cannabis on the deteriorating effects of alcohol on memory. In contrast, other authors (Jacobus et al., 2009, 2013; Mahmood et al., 2010) consider that cannabis has a damping effect on the deteriorating effect of alcohol on cognitive functions. Nevertheless, with the exception of our research (Vinader-Caerols & Monleón, 2021), there are not experimental studies evaluating specifically the combined effects of the cannabis+BD pattern on memory in this risk population (adolescents and young adults). Therefore, a summarizing table of these combined effects is not included in this section.

In line with the last mentioned studies, we have observed, after a previous year of simultaneous consumption, a dampening effect of cannabis on the deteriorating effects of alcohol in the faces memory test when performed by adolescent men (see Fig. 2) but not by women (Vinader-Caerols & Monleón, 2021).

A cognitive tolerance effect to alcohol in scenes memory task and IVM-IQ was seen in female binge drinkers (see Figs. 3 and 4), which was not observed in men (Vinader-Caerols & Monleón, 2021). Then, we can conclude that exposure to alcohol, alone or with cannabis, during adolescence can trigger different cognitive effects that contribute to enduring cognitive deficits in adulthood.

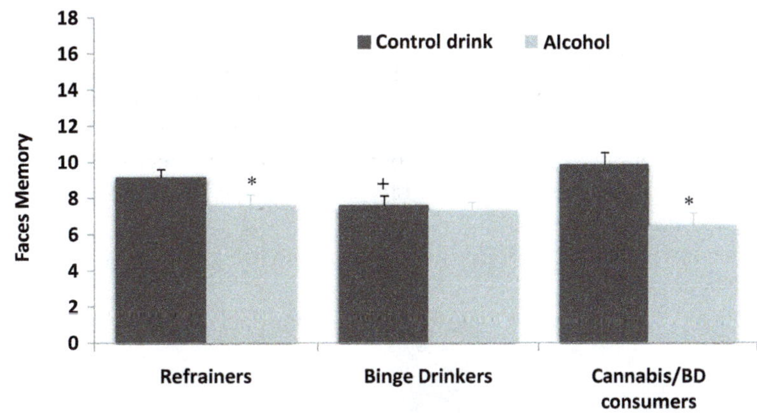

FIG. 2 Performance in faces memory task (mean+SEM) in men ($n = 118$) with a history of consumption: refrainers, binge drinkers, and cannabis+BD consumers. Subjects received an acute dose of control drink or alcohol (BAC: 0.36±0.008 g/L). *$p < 0.05$ *vs* control drink; +$p < 0.05$ *vs* refrainers-control drink and cannabis+BD consumers-control drink.

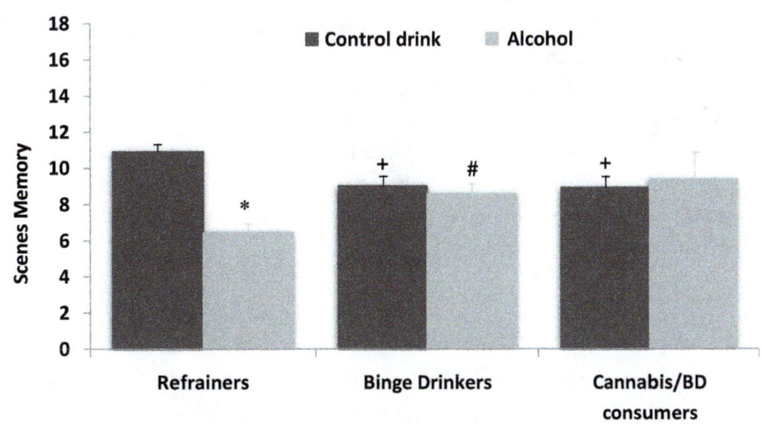

FIG. 3 Performance in scenes memory task (mean+SEM) in women ($n = 172$) with a history of consumption: refrainers, binge drinkers, and cannabis+BD consumers. Subjects received an acute dose of control drink or alcohol (BAC: 0.53±0.01 g/L). *$p < 0.05$ *vs* control drink; +$p < 0.05$ *vs* refrainers-control drink; #$p < 0.05$ *vs* refrainers-alcohol.

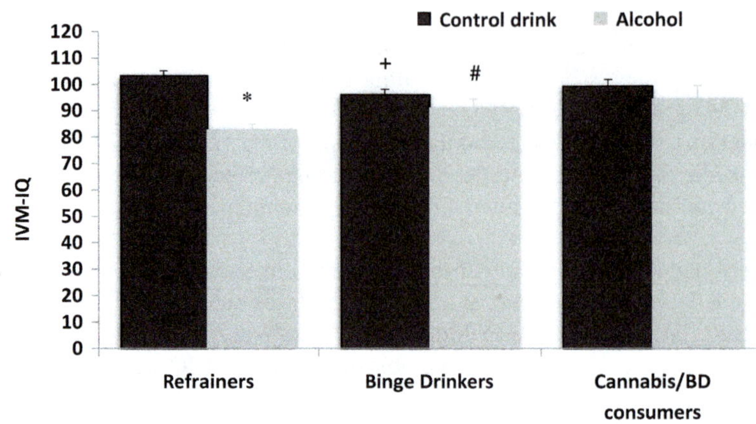

FIG. 4 Immediate Visual Memory-IQ (mean+SEM) in women ($n = 172$) with a history of consumption: refrainers, binge drinkers, and cannabis+BD consumers. Subjects received an acute dose of control drink or alcohol (BAC: 0.53 ± 0.01 g/L). *$p < 0.05$ *vs* control drink; +$p < 0.05$ *vs* refrainers-control drink; #$p < 0.05$ *vs* refrainers-alcohol.

Conclusions

The main findings of the published research show that:

- Cannabis and alcohol BD have differential effects on several types of memory, being IVM one of the most affected.
- Women are more vulnerable than men to the harmful effects of alcohol on memory. These cognitive differences between men and women seem to be overridden as the blood alcohol concentration progressively increases. A buffering effect of cannabis has been observed in men while a cognitive tolerance effect has been reported in women.
- It is essential including both sexes in this research field for evaluating sex differences in the effects of cannabis and/or alcohol BD on different cognitive functions.
- Exposure to alcohol, alone or with cannabis, during adolescence can trigger different cognitive effects that contribute to enduring cognitive deficits in adulthood.
- Longitudinal studies are needed to assess the long-term cognitive effects of these drugs of abuse whose consumption begins in adolescence.

Applications to other areas

In this chapter, we have reviewed the main effects of cannabis and alcohol BD consumption on memory in adolescents and young adults. Nevertheless, the memory impairment produced by these drugs could be extended to other periods of the lifespan, such as adulthood or elderly. These negative consequences may be also present in those individuals because of the maintenance of this consumption pattern. Even if they reduce or stop drug use, it is possible that this pattern induces sequelae over time. Also, the negative impact of cannabis and alcohol BD consumption is not limited to memory, but it could be extended to other cognitive functions and behavior.

A better understanding of the detrimental cognitive effects of cannabis and alcohol BD consumption in a high-risk population, such as adolescents and young adults, may promote a greater social awareness, involving society as a whole, as an active part. This increased social awareness will allow to delay the age of onset and decrease the consumption of cannabis and alcohol BD and will help to establish better (pharmacological, social, or educational) intervention strategies, necessary for the prevention (e.g., to prevent the development of other AUDs) and treatment of the observed damages, contributing to the achievement of more specific and effective action programs.

Mini-dictionary of terms

- **Short-term memory**. The capacity for holding a small amount of information in the mind and keep it readily available for a short period of time (seconds). Its capacity is considered to be seven (plus minus two) items. This type of memory is essential for daily functioning.
- **Working memory**. The capacity for using, manipulating, and applying temporarily limited information. This type of memory is important for reasoning and the guidance of decision making. WM has been considered synonymous with

short-term memory, but WM focus on the manipulation of information, whilst short-term memory refers to the short-term storage of information.

- **Visual memory**. Memory which preserves some characteristics of our visual experience. It consists in recalling something we have seen (objects, places, people…). VM occurs over a broad time range spanning from eye movements to years.
- **Verbal memory**. Storage of phonological information. Ability to remember words. This memory is essential for language.
- **Face processing**. Facial recognition. The capacity for recognizing an individual's identity by his/her face.
- **Long-term memory**. Storage of an unlimited amount of information for a long period of time (years) or indefinitely. Short-term memories can be stored into long-term memory.
- **Episodic memory**. Subtype of long-term memory which refers to memory for particular events situated in space and time, as well as supporting their formation and retrieval. It is the collection of past personal experiences that occurred at particular times and places.

Key facts of cannabis, binge drinking, and memory in adolescents and young adults

- Cannabis and alcohol are the most used drugs by adolescents and young adults.
- Almost all of young cannabis users are polyconsumers of cannabis and alcohol.
- The BD pattern is characterized by intermittent consumption of large amounts of alcohol in short periods of time.
- The developing adolescent brain is particularly vulnerable to drugs of abuse and their neurotoxic effects.
- The exposure to cannabis and/or alcohol during adolescence and youth can lead to long-term cognitive impairment into adulthood.

Summary points

- Adolescence is a particularly vulnerable neurodevelopmental period with high rates of engagement with drugs of abuse.
- Our experimental research work has evaluated specifically the combined effects of cannabis and BD in male and female adolescents.
- Immediate VM is one of the most affected memories by cannabis and BD.
- Women are more vulnerable than men to the effects of cannabis and BD on memory.
- A buffering effect of cannabis on the impairing effects of alcohol has been observed in men.
- Longitudinal studies are needed to assess the long-term cognitive effects of these drugs of abuse whose consumption begins in adolescence.

References

Ajmera, N., Collins, P., Weiss, H., & Luciana, M. (2021). Initiation of moderately frequent cannabis use in adolescence and young adulthood is associated with declines in verbal learning and memory: A longitudinal comparison of pre- versus post-initiation cognitive performance. *Journal of the International Neuropsychological Society*, *27*, 621–636.

Alfonso-Loeches, S., Pascual, M., & Guerri, C. (2013). Gender differences in alcohol-induced neurotoxicity and brain damage. *Toxicology*, *311*, 27–34.

Allick, A., Park, G., Kim, K., Vintimilla, M., Rathod, K., Lebo, R., Nanavati, J., & Hammond, C. J. (2021). Age- and sex-related cortical gray matter volume differences in adolescent cannabis users: A systematic review and meta-analysis of voxel-based morphometry studies. *Frontiers in Psychiatry*, *12*, 745193.

Becker, M. P., Collins, P. F., Lim, K. O., Muetzel, R. L., & Luciana, M. (2015). Longitudinal changes in white matter microstructure after heavy cannabis use. *Developmental Cognitive Neuroscience*, *16*, 23–35.

Blest-Hopley, G., O'Neill, A., Wilson, R., Giampietro, V., & Bhattacharyya, S. (2021). Disrupted parahippocampal and midbrain function underlie slower verbal learning in adolescent-onset regular cannabis use. *Psychopharmacology*, *238*, 1315–1331.

Brumback, T., Cao, D., & King, A. (2007). Effects of alcohol on psychomotor performance and perceived impairment in heavy binge social drinkers. *Drug and Alcohol Dependence*, *91*, 10–17.

Carbia, C., Cadaveira, F., Caamaño-Isorna, F., Rodríguez-Holguín, S., & Corral, M. (2017). Binge drinking during adolescence and young adulthood is associated with deficits in verbal episodic memory. *PLoS One*, *12*, e0171393.

Carbia, C., Corral, M., Caamaño-Isorna, F., & Cadaveira, F. (2020). Emotional memory bias in binge drinking women. *Drug and Alcohol Dependence*, *209*, 107888.

Cash, C., Peacock, A., Barrington, H., Sinnett, N., & Bruno, R. (2015). Detecting impairment: Sensitive cognitive measures of dose-related acute alcohol intoxication. *Journal of Psychopharmacology*, *29*, 436–446.

Chavez, P. R., Nelson, D. E., Naimi, T. S., & Brewer, R. D. (2011). Impact of a new gender-specific definition for binge drinking on prevalence estimates for women. *Am J Prev Med*, *40*, 468–471.

Chung, T., Bae, S. W., Mun, E. Y., Suffoletto, B., Nishiyama, Y., Jang, S., & Dey, A. K. (2020). Mobile assessment of acute effects of marijuana on cognitive functioning in young adults: Observational study. *JMIR mHealth and uHealth*, *8*, e16240.

Cummings, C., Beard, C., Habarth, J. M., Weaver, C., & Haas, A. (2019). Is the sum greater than its parts? Variations in substance-related consequences by conjoint alcohol-marijuana use patterns. *Journal of Psychoactive Drugs*, *51*, 351–359.

Day, A. M., Celio, M. A., Lisman, S. A., Johansen, G. E., & Spear, L. P. (2013). Acute and chronic effects of alcohol on trail making test performance among underage drinkers in a field setting. *Journal of Studies on Alcohol and Drugs*, *74*, 635–641.

Dougherty, D. M., Mathias, C. W., Dawes, M. A., Furr, R. M., Charles, N. E., Liguori, A., Shannon, E. E., & Acheson, A. (2013). Impulsivity, attention, memory, and decision-making among adolescent marijuana users. *Psychopharmacology*, *226*, 307–319.

Ellingson, J.M., Ross, J.M., Winiger, E., Stallings, M.C., Corley, R.P., Friedman, N.P., Hewitt, J.K., Tapert, S.F., Brown, S.A., Wall, T.L., & Hopfer, C.J. (2021). Familial factors may not explain the effect of moderate-to-heavy cannabis use on cognitive functioning inadolescents: A sibling-comparison study. Addiction Research Report, 116, 833–844.

Frolli, A., Ricci, M. C., Cavallaro, A., Lombardi, A., Bosco, A., Di Carmine, F., Operto, F. F., & Franzese, L. (2021). Cognitive development and cannabis use in adolescents. *Behavioral Sciences*, *11*, 37.

Giedd, J. N. (2004). Structural magnetic resonance imaging of the adolescent brain. *Annals of the New York Academy of Sciences*, *1021*, 77–85.

Harris, J. C., Wallace, A. L., Thomas, A. M., Wirtz, H. G., Kaiver, C. M., & Lisdahl, K. M. (2022). Disrupted resting state attentional network connectivity in adolescent and young adult cannabis users following two-weeks of monitored abstinence. *Brain Sciences*, *12*, 287.

Hibell, B., Guttormsson, U., Ahlström, S., Balakireva, O., Bjarnason, T., Kokkevi, A., & Kraus, L. (2009). *Substance use among students in 35 european countries (The 2007 European School Survery Project on Alcohol and other Drugs, ESPAD, Report)*. Stockholm: The Swedish Council for Information on Alcohol and other Drugs (CAN).

Infante, M. A., Nguyen-Louie, T. T., Worley, M., Courtney, K. E., Coronado, C., & Jacobus, J. (2020). Neuropsychological trajectories associated with adolescent alcohol and cannabis use: A prospective 14-year study. *Journal of the International Neuropsychological Society*, *26*, 480–491.

Jacobus, J., McQueeny, T., Bava, S., Schweinsburg, B. C., Frank, L. R., Yang, T. T., & Tapert, S. F. (2009). White matter integrity in adolescents with histories of marijuana use and binge drinking. *Neurotoxicology and Teratology*, *31*, 349–355.

Jacobus, J., Squeglia, L. M., Bava, S., & Tapert, S. F. (2013). White matter characterization of adolescent binge drinking with and without co-occurring marijuana use: A 3-year investigation. *Psychiatry Research*, *214*, 374–381.

Lawn, W., Fernandez-Vinson, N., Mokrysz, C., Hogg, G., Lees, R., Trinci, K., Petrilli, K., Borissova, A., Ofori, S., Waters, S., Michór, P., Wall, M. B., Freeman, T. P., & Curran, H. V. (2022). The CannTeen study: Verbal episodic memory, spatial working memory, and response inhibition in adolescent and adult cannabis users and age-matched controls. *Psychopharmacology*, *239*, 1629–1641.

Lees, B., Meredith, L. R., Kirkland, A. E., Bryant, B. E., & Squeglia, L. M. (2020). Effect of alcohol use on the adolescent brain and behavior. *Pharmacology, Biochemistry and Behavior*, *192*, 172906.

Lees, B., Mewton, L., Stapinski, L. A., Squeglia, L. M., Rae, C. D., & Teesson, M. (2019). Neurobiological and cognitive profile of young binge drinkers: A systematic review and meta-analysis. *Neuropsychology Reviews*, *29*, 357–385.

Leiker, E. K., Meffert, H., Thornton, L. C., Taylor, B. K., Aloi, J., Abdel-Rahim, H., Shah, N., Tyler, P. M., White, S. F., Blair, K. S., Filbey, F., Pope, K., Dobbertin, M., & Blair, R. J. R. (2019). Alcohol use disorder and cannabis use disorder symptomatology in adolescents are differentially related to dysfunction in brain regions supporting face processing. *Psychiatry Research Neuroimaging*, *292*, 62–71.

Levar, N., Francis, A. N., Smith, M. J., Ho, W. C., & Gilman, J. M. (2018). Verbal memory performance and reduced cortical thickness of brain regions along the uncinate fasciculus in young adult cannabis users. *Cannabis and Cannabinoid Research*, *8*, 56–65.

Mahedy, L., Field, M., Gage, S., Hammerton, G., Heron, J., Hickman, M., & Munafò, M. R. (2018). Alcohol use in adolescence and later working memory: Findings from a large population- based birth cohort. *Alcohol and Alcoholism*, *53*, 251–258.

Mahmood, O. M., Jacobus, J., Bava, S., Scarlett, A., & Tapert, S. F. (2010). Learning and memory performances in adolescent users of alcohol and marijuana: Interactive effects. *Journal of Studies on Alcohol and Drugs*, *71*, 885–894.

Matheson, J., Sproule, B., Di Ciano, P., Fares, A., Le Foll, B., Mann, R. E., & Brands, B. (2020). Sex differences in the acute effects of smoked cannabis: Evidence from a human laboratory study of young adults. *Psychopharmacology*, *237*, 305–316.

Maurage, P., Joassin, F., Speth, A., Modave, J., Philippot, P., & Campanella, S. (2012). Cerebral effects of binge drinking: Respective influences of global alcohol intake and consumption pattern. *Clinical Neurophysiology*, *123*, 892–901.

Mokrysz, C., Freeman, T. P., Korkki, S., Griffiths, K., & Curran, H. V. (2016). Are adolescents more vulnerable to the harmful effects of cannabis than adults? A placebo-controlled study in human males. *Translational Psychiatry*, *6*, e961.

Mooney-Leber, S. M., & Gould, T. J. (2018). The long-term cognitive consequences of adolescent exposure to recreational drugs of abuse. *Learning and Memory*, *25*, 481–491.

Mota, N., Parada, M., Crego, A., Doallo, S., Caamaño-Isorna, F., Holguin, S. R., Cadaveira, F., & Corral, M. (2013). Binge drinking trajectory and neuropsychological functioning among university students: A longitudinal study. *Drug and Alcohol Dependence*, *133*, 108–114.

National Institute of Alcohol Abuse and Alcoholism. (2004). NIAAA council approves definition of binge drinking. *NIAAA Newsletter*, *3*, 3.

Nguyen-Louie, T. T. (2016). Learning and memory in adolescent moderate, binge, and extreme-binge drinkers. *Alcoholism: Clinical and Experimental Research*, *40*, 1895–1904.

Nguyen-Louie, T. T., Castro, N., Matt, G. E., Squeglia, L. M., Brumback, T., & Tapert, S. F. (2015). Effects of emerging alcohol and marijuana use behaviors on adolescents' neuropsychological functioning over four years. *Journal of Studies on Alcohol and Drugs*, *76*, 738–748.

Noorbakhsh, S., Afzali, M. H., Boers, E., & Conrod, P. J. (2020). Cognitive function impairments linked to alcohol and cannabis use during adolescence: A study of gender differences. *Frontiers in Human Neuroscience*, *14*, 95.

Observatorio Español de las Drogas y las Adicciones (OEDA). (2019). *Alcohol, tabaco y drogas ilegales en España*. Retrieved from: https://pnsd.sanidad.gob.es/profesionales/sistemasInformacion/informesEstadisticas/pdf/2019OEDA-INFORME.pdf.

Observatorio Español de las Drogas y las Adicciones (OEDA). (2021). *Encuesta sobre Uso de Drogas en Enseñanzas Secundarias en España (ESTUDES) 1994-2021*. Observatorio Español de las Drogas y las Adicciones, Delegación del Gobierno para el Plan Nacional sobre Drogas, Secretaría de Estado de Servicios Sociales, Ministerio de Sanidad, Consumo y Bienestar Social.

Petit, G., Maurage, P., Kornreich, C., Verbanck, P., & Campanella, S. (2014). Binge drinking in adolescents: A review of neurophysiological and neuroimaging research. *Alcohol and Alcoholism*, *49*, 198–206.

Rose, A. K., & Duka, T. (2007). The influence of alcohol on basic motoric and cognitive disinhibition. *Alcohol and Alcoholism*, *42*, 544–551.

Ross, J. M., Ellingson, J. M., Rhee, S. H., Hewitt, J. K., Corley, R. P., Lessem, J. M., & Friedman, N. P. (2020). Investigating the causal effect of cannabis use on cognitive function with a quasi-experimental co-twin design. *Drug and Alcohol Dependence*, *206*, 107712.

Selamoglu, A., Langley, C., Crean, R., Savulich, G., Cormack, F., Sahakian, B. J., & Mason, B. (2021). Neuropsychological performance in young adults with cannabis use disorder. *Journal of Psychopharmacology*, *35*, 1349–1355.

Shollenbarger, S. G., Price, J., Wieser, J., & Lisdahl, K. (2015). Impact of cannabis use on prefrontal and parietal cortex gyrification and surface area in adolescents and emerging adults. *Developmental Cognitive Neuroscience*, *16*, 46–53.

Söderlund, H., Parker, E. S., Schwartz, B. L., & Tulving, E. (2005). Memory encoding and retrieval on the ascending and descending limbs of the blood alcohol concentration curve. *Psychopharmacology*, *182*, 305–317.

Spinola, S., Maisto, S. A., White, C. N., & Huddleson, T. (2017). Effects of acute alcohol intoxication on executive functions controlling self-regulated behavior. *Alcohol*, *61*, 1–8.

Squeglia, L. M., & Gray, K. M. (2016). Alcohol and drug use and the developing brain. *Current Psychiatry Reports*, *18*, 46.

Squeglia, L. M., Schweinsburg, A. D., Pulido, C., & Tapert, S. F. (2011). Adolescent binge drinking linked to abnormal spatial working memory brain activation: differential gender effects. *Alcoholism: Clinical and Experimental Research*, *35*, 1831–1841.

Sullivan, R. M., Wallace, A. L., Wade, N. E., Swartz, A. M., & Lisdahl, K. M. (2020). Assessing the role of cannabis use on cortical surface structure in adolescents and young adults: Exploring gender and aerobic fitness as potential moderators. *Brain Sciences*, *10*, 117.

Tervo-Clemmens, B., Simmonds, D., Calabro, F. J., Day, N. L., Richardson, G. A., & Luna, B. (2018). Adolescent cannabis use and brain systems supporting adult working memory encoding, maintenance, and retrieval. *Neuroimage*, *169*, 496–509.

Vinader-Caerols, C., Duque, A., Montañés, A., & Monleón, S. (2017). Blood alcohol concentration-related lower performance in immediate visual memory and working memory in adolescent binge drinkers. *Frontiers in Psychology*, *8*, 1720.

Vinader-Caerols, C., & Monleón, S. (2019). Binge drinking and memory in adolescents and young adults. In S. Palermo, & M. Bartoli (Eds.), *Inhibitory control training. A multidisciplinary approach* IntechOpen.

Vinader-Caerols, C., & Monleón, S. (2021). Binge drinking, alone or with cannabis, during adolescence triggers different effects on immediate visual memory in men and women. *Frontiers in Psychiatry*, *12*, 797221.

Vinader-Caerols, C., Talk, A., Montañés, A., Duque, A., & Monleón, S. (2017). Differential effects of alcohol on memory performance in adolescent men and women with a binge drinking history. *Alcohol and Alcoholism*, *52*, 610–616.

Wade, N. E., Wallace, A. L., Swartz, A. M., & Lisdahl, K. M. (2019). Aerobic fitness level moderates the association between cannabis use and executive functioning and psychomotor speed following abstinence in adolescents and young adults. *Journal of the International Neuropsychological Society: JINS*, *25*, 134–145.

Wechsler, D. (2004). *Wechsler adult intelligence scale-III*. San Antonio, EEUU: The Psychological Corporation.

Winward, J. L., Hanson, K. L., Tapert, S. F., & Brown, S. A. (2014). Heavy alcohol use, marijuana use, and concomitant use by adolescents are associated with unique and shared cognitive decrements. *Journal of the International Neuropsychological Society*, *20*, 784–795.

World Health Organization (WHO). (2018). *Global Status Report on Alcohol and Health*. Geneva.

Part V

Neurobiology of cannabinoids

Chapter 27

The role of ΔFosB in the plasticity of the cannabinoid system

Arturo Venebra-Muñoz[a] and Sara Mejía-Chávez[b]
[a]Neurobiology of Addiction and Brain Plasticity Laboratory, Faculty of Sciences, Autonomous University of the State of Mexico, Toluca, State of Mexico, Mexico, [b]Institute of Neurobiology, Autonomous National University of Mexico, Juriquilla, Queretaro, Mexico

Abbreviations

AC	adenylyl cyclase
ECBS	endocannabinoids
Gi	inhibitory G protein
PKA	protein kinase A
THC	tetrahydrocannabinol

Introduction

At present, derivatives obtained from the Cannabis plant are produced and used in a wide variety of ways, for recreational purposes, in the manufacture of fibers or for medicinal or therapeutic purposes. The legalization of the cultivation, processing, and commercialization of the plant in several countries greatly detonated its exploitation. In the United States, there are specialized crops where different varieties are cultivated for their variable concentrations of phytocannabinoids, originating different added values of the plants.

In Latin American countries, legalization issues of Cannabis plants are still being worked on for its cultivation, distribution, use, and commercialization; however, the issue of traffic and its implications makes it difficult to use it to its full potential. In the specialized scientific field, the effects of cannabinoids are discussed at different levels, from molecular to behavioral. In this chapter, we will discuss the impact that cannabinoids have on the generation of changes at the cellular level to induce neuronal plasticity, and proposals for mechanisms by which plastic changes can be generated, emphasizing the involvement of the transcription factor ΔFosB.

Cannabis and cannabinoids

Taxonomic descriptions of Cannabis continue to be provided, and the number of species within the genus is still disputed, although three species may be recognized: *Cannabis indica*, *Cannabis ruderalis*, and *Cannabis sativa*, the latter being the most used in the production of fibers, whereas *Cannabis indica* is mostly used to produce hasish.

It is considered that Cannabis can contain several dozens of phytocannabinoids (ElSohly, 2002), of which three have psychoactive properties; tetrahydrocannabinols Δ^9-THC and Δ^8-THC, and the propyl homolog of Δ9-Tetrahydrocannabidivarin, whose psychoactive properties is lesser compared to the other mentioned phytocannabinoids (Gaoni & Mechoulam, 1964; Gill et al., 1970; Grunfeld & Edery, 1969; Koob & Le Moal, 2006).

Some of the reported effects induced by cannabinoids are enhanced relaxation, senses become more perceptive, a state of deeper thought, happiness, dizziness, elation, hallucinations, enhanced appetite, the feeling of pleasantness, among others. An attractive effect is that subjects *think deeper*, meaning that cannabinoids have a disinhibiting effect, but when comparing them with other substances such as alcohol, besides their disinhibiting effect, they enhance cognitive activity (Koob & Le Moal, 2006). Another interesting effect is that subjects report that time *slows down*. These effects suggest that cannabinoids emulate in an exacerbated way the action of endogenous cannabinoids, endocannabinoids, in addition to their effects on the dopaminergic system inducing the emergence of a pleasant state and reinforcing sensation; that is, an effect of pleasure when deepening cognitively and interest in thought.

Cannabis Use, Neurobiology, Psychology, and Treatment. https://doi.org/10.1016/B978-0-323-89862-1.00019-2

Endocannabinoids

Endocannabinoids (ECBS) are derived from membrane fatty acids, produced on demand in response to increases in intracellular Ca^{2+} (Di Marzo et al., 1998; Mechoulam & Parker, n.d.). Endocannabinoids such as anandamide (arachidonoylethanolamide) and 2-AG (2-arachidonoyl-glycerol) act retrogradely on membrane receptors located in a presynaptic neuron (Di Marzo et al., 1998; Di Marzo et al., 2015; Patricio et al., 2020). Two types of endocannabinoid receptors have been described, CB1 and CB2, and both belong to the G protein coupled receptor family. These proteins cross the cell membrane 7 times, and once coupled to their ligand, they activate an inhibitory G protein (Gi) that hinders the release of neurotransmitter molecules into the synaptic cleft.

The cellular activity of endocannabinoids leads us to attribute them a modulatory role in neurotransmission processes, and their retrograde action has led various groups to study their role in time perception. Another highlight is the changes that endocannabinoids could induce at the molecular, cellular, and morphological level, meaning their influence on neuroplasticity.

When ECBS interact with their receptor, the Gi protein inhibits a adenylate cyclase adenylyl cyclase (AC), and thus inhibits the protein kinase A (PKA) and voltage-gated Ca^{2+} channels (Kano et al., 2009; Mackie & Stella, 2006), which could lead to changes in alternative signaling pathways through molecular switches, such as the protein ΔFosB.

ΔFOSB

ΔFosB is a transcription factor that dimerizes with proteins of the Jun family to form the AP1 activator complex. This transcription factor regulates the activity of genes involved in neuronal remodeling through the production of proteins such as the GluR2 subunit of AMPA glutamate receptors, which regulates the production of dynorphin, inducing an increase in the activity of dopaminergic neurons in the ventral tegmental area. Additionally, the overexpression of ΔFosB induces an increase in the expression of the Cdk5 gene, which is involved in the formation of dendritic spines. Cdk5 upregulates the G-protein alpha subunit ($G_{\alpha O}$), CaMKII, and protein kinase Cβ (McClung & Nestler, 2003; Nestler et al., 2001). These proteins are considered as molecular switches since they regulate different intracellular signaling pathways, and therefore, having an impact on neuronal activity.

The expression of ΔFosB also increases upon the intake of psychostimulants. For example, its overexpression has been observed in animals to which cocaine (Solinas et al., 2009), sugar (Wallace et al., 2008), and nicotine has been administered (Venebra-Muñoz et al., 2014), after the intake of artificial sweeteners such as stevia or sucralose, but also after the chronic administration of the endocannabinoid anandamide (Salaya-Velazquez et al., 2020; Salinas-Velarde et al., 2021) (Fig. 1).

An increase in the expression of ΔFosB is observed mainly in neurons that express the dopamine D1 receptor and has been related to addictive-compulsive behaviors. Therefore, it can be used as a marker of addiction-related behaviors but also as an indicator of neuronal plasticity. Additionally, its expression patterns give insights to the molecular mechanisms that could underlie decision-making and cognitive processes, since its expression increases notably in brain regions such as the prefrontal cortex (Lafragette et al., 2017; Lobo et al., 2013; Salinas-Velarde et al., 2021; Sikora et al., 2018; Stairs & Bardo, 2009; Venebra-Muñoz et al., 2014).

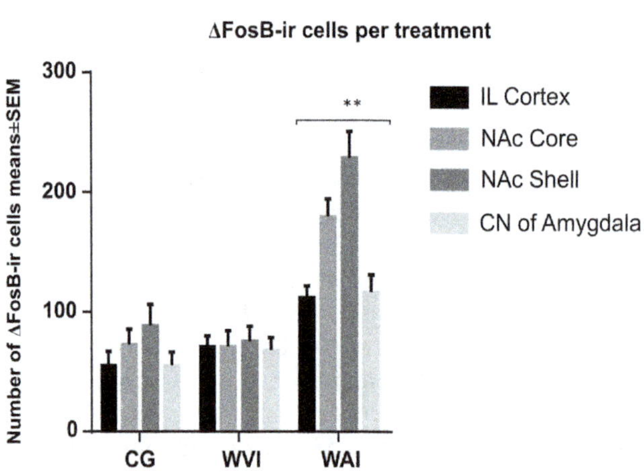

FIG. 1 Bars show the mean +/- standard error of ΔFosB immunoreactive neurons in different brain regions upon administration of anandamide, at a dose of 0.5 mg/kg injected daily from postnatal day 30 to 50. Infralimbic cortex. Nucleus accumbens core. Nucleus accumbens shell. Central Amygdala. CG, control group. WVI, group with vehicle administration. WAI, anandamide administration group. Asterisks indicate significant differences between groups. *(Modified with permission of Salaya-Velazquez, N. F., López-Muciño, L. A., Mejía-Chávez, S., Sánchez-Aparicio, P., Domínguez-Guadarrama, A. A., & Venebra-Muñoz, A. (2020). Anandamide and sucralose change ΔfosB expression in the reward system.* NeuroReport, 31(3), *240–244.)*

Interaction

Evidence indicates that the administration of most addictive substances, including cannabinoids, increase the expression of the transcription factor ΔFosB in regions of the dopaminergic system (Nestler, 2008; Salaya-Velazquez et al., 2020; Werme et al., 2002). However, the expression of ΔFosB increases in striatal regions or ventral tegmental area, but after repeated administration, its expression decreases in the prefrontal cortex (Solinas et al., 2009; Venebra-Muñoz et al., 2014). Similar results have also been shown for the case of the cannabinoid tetrahydrocannabinol (THC); the chronic administration of THC increases the expression of ΔFosB in the striatum but decreases the activation of CB1 receptors in medium spiny neurons and the desensitization of these receptors in dopaminergic regions (Lazenka et al., 2014).

As mentioned, ΔFosB together with proteins of the Jun family make up the AP1 activator complex, which intervenes in the expression of genes related to neuronal structural modifications, by integrating, for example, glutamate receptors and generating plastic processes. Furthermore, cannabinoids reduce desensitization in the striatum; however, the reduction in the expression of ΔFosB that occurs in the prefrontal cortex after its repeated administration could partially explain addictive processes and the decision-making repercussion observed in addict subjects. Therefore, studies on the interaction between the administration or intake of cannabinoids and ΔFosB expression in brain regions involved in cognitive processes are necessary to approach the underlying mechanisms of addiction-related behaviors. We propose that there may be cellular mechanisms by which, when faced with a maximum expression of this transcription factor, its elimination begins, perhaps through specialized proteases that maintain an optimal level of ΔFosB expression in each brain region. It should also be remembered that this transcription factor is accumulates and has a much longer intracellular half-life compared to other transcription factors such as c-Fos and c-Jun.

Conclusion

The impact of cannabinoids is relevant at all biological levels, and the mechanisms by which they can induce drug dependence, or the development of addiction-related behaviors are still being studied. A possible explanation for the establishment of a dependence on some addictive substances is through changes in receptor activity or through changes in the generation of new synapses or neurotransmitter release that influence the overexpression of transcription factors such as ΔFosB. Experiments could be carried out to assess the mechanisms by with this transcription factor is regulated upon the administration of cannabinoids, and the focus on brain regions such as the prefrontal cortex could uncode the mechanisms underlying decision-making in the development of addiction-related behaviors.

Applications to other areas

In this chapter, we review and discuss the action of cannabinoids with evidence derived from studies in animal models, without leaving aside some important effects reported in human subjects. The evidence presented at the cellular-molecular level is discussed, aiming to provide physiological explanations, and propose possible action pathways before the administration of cannabinoids, the activation of their receptors, the inhibition of intracellular protein targets, their role modulator at the synaptic level, and its relationship with the effects observed mainly with the expression of the transcription factor ΔFosB, which has been shown to be involved in the generation of neuronal plasticity through neuronal remodeling. By showing that cannabinoids induce molecular changes, explanations can be proposed for higher levels. The information discussed in this chapter can lead to hypotheses and study proposals, at the molecular and cellular, but also at higher levels, whether physiological or behavioral, and might be taken into account for therapeutic preclinical or clinical studies related to addiction, since it could partly explain the difficulty in reversing the effects caused by the intake of drugs such as cannabinoids, considering that the development of an addictive behavior not only has a behavioral story but also has cellular-molecular bases in the individuals who suffer from it.

Key facts

Key facts of ΔFosB

- ΔFosB is a protein that acts as a cellular transcription factor.
- Evidence shows that ΔFosB forms dimers with protein members of the Jun family.
- It has a longer half-life than other members of its family, such as c-Fos.
- Characteristically, the ΔFosB protein accumulates mainly in neurons belonging to the dopaminergic system.
- Its expression is induced by stimuli such as drug use or stressful events and has been related to neuronal plasticity processes.

Mini-dictionary of terms

- **Neuroplasticity.** Changes at the molecular, metabolic, and structural level, which occur upon neuronal stimulation.
- **Transcription factor.** Protein (s) involved in gene transcription.
- **Desensitization of receptors.** Mechanisms by which neurons reduce the activity or inactivate proteins that act as receptors, including downregulation or intracellular receptor inhibition.
- **Proteases.** Enzymes that catalyze proteolysis, breaking down proteins into smaller polypeptides or single amino acids, and spurring the formation of new protein products.
- **Dopaminergic System.** Nuclei or cell aggregates in the central nervous system whose activity mainly depends on the neurotransmitter dopamine and dopaminergic receptors.

Summary points

- Cannabinoids are currently of great interest in all areas due to their health, economic, and social implications, hence the relevance of their study.
- The effect of cannabinoids can be evaluated in animal models at different levels, from molecular to behavioral, and in different ways, at the cellular level, their effects can be related to protein expression.
- There are proteins called transcription factors, such as ΔFosB, that are indicators of neuronal activity. ΔFosB is involved in mechanisms of neuronal plasticity, so its expression upon the administration of cannabinoids may indicate changes at the neural level.
- The expression of ΔFosB in regions of the dopaminergic system can be induced by the administration of cannabinoids.
- Differences in ΔFosB expression patterns upon the administration of cannabinoids in regions such as the prefrontal cortex may play an important role in decision-making in addictive-related behavior processes.

References

Di Marzo, V., Bisogno, T., Sugiura, T., MelK, D., Petrocellis, D., & L. (1998). The novel endogenous cannabinoid 2 arachidonoylglycerol is inactivated by neural- and basophil-like cells: Connections with anandamide. *The Biochemical Journal*, *331*, 15–19.

Di Marzo, V., Stella, N., & Zimmer, A. (2015). Endocannabinoid signalling and the deteriorating brain. *Nature Reviews. Neuroscience*, *16*(1), 30–42.

ElSohly, M. A. (2002). Chemical constituents of cannabis. In F. Grotenhermen, & E. Russo (Eds.), *Cannabis and cannabinoids: Pharmacology, toxicology and therapeutic potential* (pp. 27–36). New York: Haworth Integrative Healing Press.

Gaoni, Y., & Mechoulam, R. (1964). Isolation, structure, and partial synthesis of an active constituent of hashish. *Journal of the American Chemical Society*, *86*, 1646–1647.

Gill, E. W., Paton, W. D., & Peterwee, R. G. (1970). Preliminary experiments on the chemistry and pharmacology of cannabis. *Nature*, *228*, 134–136.

Grunfeld, Y., & Edery, H. (1969). Psychopharmacological activity of the active constituents of hashish and some related cannabinoids. *Psychopharmacologia*, *14*, 200–210.

Kano, M., Ohno-Shosaku, T., Hashimotodani, Y., Uchigashima, M., & Watanabe, M. (2009). Endocannabinoid-mediated control of synaptic transmission. *The Psychological Review*, *89*(1), 309–380.

Koob, G. F., & Le Moal, M. (2006). *Neurobiology of addiction*. United Kingdom: Academic Press.

Lafragette, A., Bardo, M. T., Lardeux, V., Solinas, M., & Thiriet, N. (2017). Reduction of cocaine-induced locomotor effects by enriched environment is associated with Cell-specific accumulation of ΔFosB in striatal and cortical subregions. *International Journal of Neuropsychopharmacology*, *20*(3), 237–246.

Lazenka, M. F., David, B. G., Lichtman, A. H., Nestler, E. J., Selley, E. D., & Sim-Selley, L. J. (2014). Delta FosB and AP-1-mediated transcription modulate cannabinoid CB_1 receptor signaling and desensitization in striatal and limbic brain regions. *Biochemical Pharmacology*, *9*(3), 380–389.

Lobo, M. K., Zaman, S., Damez-Werno, D. M., Koo, J. W., Bagot, R. C., DiNieri, J. A., Nugent, A., Finkel, E., Chaudhury, D., Chandra, R., Riberio, E., Rabkin, J., Mouzon, E., Cachope, R., Cheer, J. F., Han, M. H., Dietz, D. M., Self, D. W., Hurd, Y. L., … Nestler, E. J. (2013). FosB induction in striatal medium spiny neuron subtypes in response to chronic pharmacological, emotional, and optogenetic stimuli. *Journal of Neuroscience*, *33*(47), 18381–18395.

Mackie, M., & Stella, N. (2006). Cannabinoid receptors and endocannabinoids: Evidence for new players. *The AAPS Journal*, *8*(2), E298–E306.

McClung, C. A., & Nestler, E. J. (2003). Regulation of gene expression and cocaine reward by CREB and DeltaFosB. *Nature Neuroscience*, *6*, 1208–1215.

Mechoulam, R., & Parker, L. Towards a better cannabis drug. British Journal of Pharmacology, 170 (7), 1363-1364.

Nestler, E. J. (2008). Transcriptional mechanisms of addiction: Role of DeltaFosB. *Philosophical Transactions of the Royal Society B: Biological Sciences*, *363*, 3245–3255.

Nestler, E. J., Barrot, M., & Self, D. W. (2001). DeltaFosB: A sustained molecular switch for addiction. *Proceedings of the National Academy of Sciences of the United States of America*, *98*, 11042–11046.

Patricio, F., Morales-Andrade, A. A., Patricio-Martínez, A., & Limón, I. D. (2020). Cannabidiol as a therapeutic target: Evidence of its neuroprotective and neuromodulatory function in Parkinson's disease. *Frontiers in Pharmacology*, *15*(11), 595635.

Salaya-Velazquez, N. F., López-Muciño, L. A., Mejía-Chávez, S., Sánchez-Aparicio, P., Domínguez-Guadarrama, A. A., & Venebra-Muñoz, A. (2020). Anandamide and sucralose change ΔfosB expression in the reward system. *NeuroReport*, *31*(3), 240–244.

Salinas-Velarde, I. D., Bernal-Morales, B., Pacheco-Cabrera, P., Sánchez-Aparicio, P., Pascual-Marhey, L. I., & Venebra-Muñoz, A. (2021). Lower ΔFosB expression in the dopaminergic system after stevia consumption in rats housed nder environmental enrichment conditions. *Brain Research Bulletin*, *177*, 172–180.

Sikora, M., Nicolas, C., Istin, M., Jaafari, N., Thiriet, N., & Solinas, M. (2018). Generalization of effects of environmental enrichment on seeking for different classes of drugs of abuse. *Behavioural Brain Research*, *341*, 109–113.

Solinas, M., Thiriet, N., Rawas, R., Lardeux, V., & Jaber, M. (2009). Environmental enrichment during early stages of life reduces the behavioral, neurochemical, and molecular effects of cocaine. *Neuropsychopharmacology*, *34*(5), 1102–1111.

Stairs, D. J., & Bardo, M. T. (2009). Neurobehavioral effects of environmental enrichment and drug abuse vulnerability. *Pharmacology Biochemistry and Behavior*, *92*(3), 377–382.

Venebra-Muñoz, A., Corona-Morales, A., Santiago-García, J., Melgarejo-Gutiérrez, M., Caba, M., & García-García, F. (2014). Enriched environment attenuates nicotine self-administration and induces changes in ΔFosB expression in the rat prefrontal cortex and nucleus accumbens. *NeuroReport*, *25*(9), 688–692.

Wallace, D. L., Vialou, V., Rios, L., Carle-Florence, T. L., Chakravarty, S., Kumar, A., Graham, D. L., Green, T. A., Kirk, A., Iñiguez, S. D., Perrotti, L. I., Barrot, M., DiLeone, R. J., Nestler, E. J., & Bolaños-Guzmán, C. A. (2008). The influence of ΔfosB in the nucleus accumbens on natural reward-related behavior. *Journal of Neuroscience*, *28*(41), 10272–10277.

Werme, M., Messer, C., Olson, L., Gilden, L., Thoren, P., Nestler, E. J., & Brené, S. (2002). ΔFosB regulates weel running. *The Journal of Neuroscience*, *22*(18), 8133–8138.

Chapter 28

Functional profile of synthetic cannabinoid receptor agonists: Exploring cannabinoid and noncannabinoid targets

Shivani Sachdev[a,b], Marina Santiago[a], and Chris Bladen[a]
[a]*Macquarie Medical School, Macquarie University, Sydney, NSW, Australia,* [b]*Laboratory of Bioorganic Chemistry, National Institute of Diabetes, Digestive, and Kidney Diseases, National Institutes of Health, Bethesda, MD, United States*

Abbreviations

SCRA	synthetic cannabinoid receptor agonists
NPS	new psychoactive substances
THC	Δ^9-tetrahydrocannabinol
GPCR	G protein-coupled receptor
AC	adenylyl cyclase
AtT20	pituitary adenoma cell line
BRET	bioluminescence energy resonance transfer
cAMP	3′-5′-cyclic adenosine monophosphate
CB1/CB2	cannabinoid type 1/2 receptor
Cav3.1-3.3	T-type calcium channel
EC_{50}	half maximal effective concentration
E_{max}	maximal effect
GIRK	G protein-activated inwardly rectifying potassium channels
GPR55	G protein-coupled receptor 55
TRP	transient receptor potential channel

Introduction

Synthetic cannabinoid receptor agonists (SCRAs) are new psychoactive substances (NPS) responsible for significant toxicity and death reported globally; an outcome that is very different from users of herbal cannabis and other forms of THC (Fantegrossi et al., 2014; Hermanns-Clausen et al., 2013). SCRAs are a group of structurally diverse molecules marketed as herbal incense blends initially branded as "Spice" (Europe) or "K2" (North America) (for detailed review, see Banister and Connor (2018)). The recreational use of SCRAs was first reported in the early 2000s and ever since, newer, unregulated compounds appear as soon as older compounds become controlled under state and national laws. They are traditionally sold over the internet and local headshops (Auwärter et al., 2009) and unlike their natural counterpart (THC), SCRAs produce a wide range of physiological and psychological effects, including psychosis, cardiac arrest, hyperthermia, cerebral ischemia, and seizures (reviewed in Cooper (2016)). Given that new variants of recreational SCRAs are constantly being developed and distributed, it is difficult to keep up with their extensive range of pharmacological and toxic effects.

Synthetic cannabinoids were originally developed as a pharmacological tool to probe the endocannabinoid system and to find a viable alternative to traditional medicine (Adams et al., 1940; Wiley et al., 2011). Several studies exist on the molecular pharmacology of early generation synthetic cannabinoids derived from different structural templates to investigate drug-receptor interactions at cannabinoid and noncannabinoid systems for potential medicinal effects. However, these are outside of the scope of this chapter and have been extensively reviewed elsewhere (Wiley et al., 2016). The focus of this chapter is to primarily review the pharmacological activity that may underlie the multifaceted toxic effects of unregulated synthetic cannabinoids SCRAs in comparison to the phytocannabinoid THC, and the cannabinoids most

Cannabis Use, Neurobiology, Psychology, and Treatment. https://doi.org/10.1016/B978-0-323-89862-1.00046-5

extensively used in research, CP55940 and WIN55212. While the last decade has seen incremental development in the current understanding of the pharmacology of SCRAs, particularly at cannabinoid type-1 (CB1) and type-2 (CB2) receptors (Chung et al., 2021; Sachdev, 2020), some evidence exists that SCRAs may act at targets other than CB1 and CB2. Other receptors such as TRPA1 (Pearson, 2017), GPR55 (Wang et al., 2020), $5HT_{1A}$ receptor (Yano et al., 2020), and T-type calcium channels (Bladen et al., 2021) have been identified as being modulated by SCRAs; however, these interactions have not yet been fully characterized.

The apparent greater susceptibility to abuse and complex toxicological profile of SCRAs, likely add to their ability to induce the plethora of signaling effects downstream from their targets. More research into the molecular pharmacology of SCRAs at cannabinoid and noncannabinoid targets is desperately needed in order to identify mechanisms underlying SCRA toxicity. This chapter therefore summarizes the current paradigms of SCRA signaling in terms of the intracellular transduction pathways across multiple receptor targets and utilizes these approaches to explain their ongoing impact across clinical aspects.

Molecular targets of synthetic cannabinoid receptor agonists (SCRAs)

Cannabinoid receptors

SCRA-induced effects in CB1 receptor

SCRAs were originally developed to assess the structural requirements of CB1 receptor binding in an effort to dissect the potential medicinal effects of cannabinoids from unwanted psychotropic effects (Devane et al., 1988; Matsuda et al., 1990). Many prior research articles have summarized the use of earlier generations of synthetic cannabinoids to explore the pharmacology and signal transduction of CB1 (Wiley et al., 2011; Wiley et al., 2016). This chapter focuses on exploring the multiple receptor mechanisms that may underlie the adverse effects of SCRAs synthetized specifically for recreational use. A line of evidence has shown that SCRAs, like THC, bind to CB1 receptor with appreciable affinity, and display varying intrinsic activity relative to THC, both in cellular assays and animal studies (Devane et al., 1988; Huestis et al., 2001; Mackie, 2005). The pharmacological profile of the first SCRA to be identified in early formulations of Spice, JWH-018, has been studied in a battery of in vitro assays: GTPγS binding (Wiley et al., 2012), cAMP signaling (Chin et al., 1999), ligand-binding assay (Huffman et al., 1994), electrophysiological recording from autaptic neurons (Atwood et al., 2010), MAPK activation, and receptor internalization assay (Atwood et al., 2010). Pharmacological assessment revealed that JWH-018 exhibits higher potency and efficacy at CB1 than THC in each of these signaling endpoints. The receptor saturating concentration of JWH-018 produced over fourfold greater G protein stimulation than a similar maximal concentration of THC in mice brain homogenates containing CB1 (Wiley et al., 2012). In radioligand-binding assay, JWH-018 demonstrated nanomolar-range affinity for CB1 with a K_i value of 9 nM—approximately 4.5 times higher than THC (Huffman et al., 1994). JWH-018 also exhibited nanomolar-range potency for CB1 in the cAMP inhibition assay (EC_{50}=14 nM), being approximately threefold more potent than the reference agonist CP55940 (Chin et al., 1999). Electrophysiological recordings showed that JWH-018 potently inhibited excitatory postsynaptic currents (IC_{50}=15 nM) when tested in autaptic hippocampal neurons, and that this effect was blocked by SR141716A (CB1 antagonist). These results suggest the adverse psychomimetic effects of JWH-018 are likely due to their interactions with CB1 receptors, but given the diverse physiological effects of SCRAs, research should continue exploring this complex interaction on additional molecular targets (Atwood et al., 2010). Consistent with the in vitro agonist profile of JWH-018 at CB1, the behavioral assessment of a cluster of four classical tetrad endpoints (analgesia, hypothermia, hypomotility, and catalepsy) was investigated in the rodent model following administration of JWH-018 (Wiley et al., 2012). The authors found that JWH-018 can produce the full spectrum of tetrad effects (ranging from 0.09 mg/kg for analgesia to 1.47 mg/kg for hypothermia) typical of cannabinoids, and these effects could conceivably account for some of the adverse effects associated with JWH-018 in humans (Huffman et al., 1994). Several reports have documented significant neurological effects, such as seizures, following recreational use of SCRAs in humans and laboratory animals (Breivogel et al., 2020; Malyshevskaya et al., 2017). A study conducted by Malyshevskaya et al. (2017) showed that the acute administration of JWH-018 (2.5 mg/kg), as well as THC (10 mg/kg), triggered electrographic seizures in mice in an AM-251 (CB1 antagonist) sensitive manner. However, in later studies, JWH-018 was demonstrated to potently and effectively elicit convulsions in mice, while THC failed to induce observable convulsant effects at any tested dose (30–100 mg/kg) (Wilson et al., 2019). This enhanced ability of JWH-018 to induce cannabimimetic effects in vivo compared to the low efficacy phytocannabinoid THC clearly implicates the adverse effects of SCRAs are at least partly due to potent CB1 receptor activation.

The discovery of the first-generation SCRAs was soon followed by the identification of tetramethylcyclopropyl ketone indoles—UR144 and XLR-11 (5″-fluoro-UR-144) as NPS (Kavanagh et al., 2013). Multiple cases have reported acute kidney injury (AKI) after smoking SCRA-product containing XLR-11 (Buser et al., 2014). The activity profiles of UR144 and XLR-11 were assessed at CB1 receptor in an assay of GTPγS binding in HEK cells and found that these SCRAs exhibit comparable potency, with THC approximately 80-fold less potent in the same pathway (Wiley et al., 2013). Subsequent studies investigated the cannabinoid tetrad effects following administration of UR-144 and XLR-11 (Wiley et al., 2013). The authors found that UR-144 and XLR-11 exhibited a complete profile of tetrad effects in mice, with potencies up to 15- and 17-fold greater than THC, respectively (Wiley et al., 2013). A later investigation reported the real-time measurement of body temperature and heart rate in rats following administration with UR144/XLR-11 (Banister et al., 2015). Both UR144/ XLR-11 dose-dependently induced hypothermia and reduced heart rate at a 10 mg/kg dose (Banister et al., 2015). Further, CB1 receptors appear to mediate in part the in vivo effects of these SCRAs as evidenced by attenuation of these effects by CB1 receptor antagonist rimonabant (Bosier et al., 2010; Janoyan et al., 2002) and in CB1 receptor knockout mice (Zimmer et al., 1999). Together, these studies show that many of SCRAs' effects are mediated through multiple molecular mechanisms of action, including CB1 receptor activation in the central and peripheral nervous system (Sim-Selley, 2003).

Several putative SCRAs continue to emerge on the market without previous characterization in the scientific literature (Banister & Connor, 2018). The most prevalent SCRAs now feature indole and indazole-3-carboxamide-based scaffolds with pendant amino acid amides or esters and encompass compounds like AB-CHMINACA, AB-PINACA, MDMB-FUBINACA, 5F-MDMB-PICA, MMB-4en-PICA, and 5F-CUMYL-PICA (Banister et al., 2016; Longworth et al., 2017). The pharmacological profile of these new generations of SCRAs were characterized in various pharmacological assays—GTPγS binding (Thomas et al., 2017), ligand-binding (Thomas et al., 2017), membrane potential assay (Banister et al., 2016; Longworth et al., 2017; Pike et al., 2021), and NanoBiT complementation assay for β-arrestins recruitment (Cannaert et al., 2016). The majority of the novel SCRAs tested illustrated high potency and full efficacy in these assays—for instance, a selection of SCRAs tested completely displaced [^{3}H]SR141716 binding in CB1-expressing HEK293 membranes and produced greater maximal effects than THC in GTPγS signaling; SCRAs tested also exhibited comparable efficacy to inhibit forskolin-stimulated cAMP production in CB1-expressing CHO cells, whereas THC exhibited approximately fourfold lower efficacy, consistent with partial agonism (Thomas et al., 2017). Similarly, these SCRAs possessed potencies greater than that of THC in fluorometric assay of membrane potential, which correlates with the hyperpolarization induced by CB1-mediated activation of the GIRK channel (Banister et al., 2016; Longworth et al., 2017). In the NanoBit complementation assay, all the SCRA tested displayed higher potency and efficacy for CB1-mediated recruitment of β-arrestin 2 and was, in many instances, much higher than that of JWH-018 (Cannaert et al., 2016). They found that ADB-FUBINACA was the most potent SCRA within the evaluated cohort with an EC_{50} below 1 nM for CB1 activation compared to its L-valinamide derivative AB-FUBINACA. Similarly, 5F-MDMB-PICA was found to be more potent than 5F-AMB-PICA (Cannaert et al., 2016). These results suggest that the subtle modifications of SCRAs could lead to significant changes in their profile or pattern of responses, which may have implications for toxicity in vivo. Additional studies further investigate the cannabimimetic psychoactive effects of the new generation of SCRAs using rodent behavioral and motor tests (Gatch & Forster, 2015). The authors found that all the SCRAs but not THC demonstrated decreased locomotor activity in the staircase test (Gatch & Forster, 2015). Another group examined the SCRAs for their discriminative stimulus properties in THC drug discrimination test to ascertain whether they produce interoceptive effects similar to that of cannabis (Gamage et al., 2018). They found that all of the SCRAs tested substituted for THC with significantly greater potency (Gamage et al., 2018). These effects may be indicative of CB1 agonism, a necessary prerequisite for psychotropic and abuse-related effects of SCRAs.

To gain further mechanistic insight into the wide-ranging adverse effects of SCRAs, it is important to understand the extent of CB1 receptor activation by distinct classes of SCRAs to allow comparisons between the phytocannabinoid THC, and classical research cannabinoid CP55940. In an effort to more accurately define efficacy, we used operational analysis approach combined with an irreversible antagonist of CB1, AM6544, to deplete receptor reserve while measuring the hyperpolarization induced by SCRAs in AtT20-CB1 cells (Sachdev et al., 2019). The depletion of spare receptors is a useful strategy to surmount the confounding influence on efficacy measurement that arises from the large receptor reserve in transfected cell lines. This study revealed that all the SCRAs tested showed substantially higher agonist activity at CB1 receptors than THC, with operational efficacy ranging between 28 and 233 for SCRAs and 0.9 for THC. The SCRAs 5F-MDMB-PICA and XLR-11 exhibited considerably higher efficacies compared to CP55940 (Sachdev et al., 2019). This suggests that the profile or pattern of responses produced by SCRAs differs from THC and traditional research agonists like CP55940, which may help explain their enhanced toxicity.

Recently, Patel et al. (2020) found that a panel of indole- and indazole-carboxamide SCRAs exhibited highly potent and efficacious agonist behavior at several CB1-mediated signaling pathways in HEK cells. Compared to THC, SCRAs induced highly potent and efficacious inhibition of cAMP, ranging from 0.11 nM (4-CN-CUMYL-BUTINACA) to 24.55 nM (XLR-11) (Patel et al., 2020). The functional activity of the same panel of SCRAs was also characterized in the

CB1-Gαs signaling pathway. The majority of SCRAs increased cAMP levels above those generated by forskolin, although with lower potencies and greater variation in efficacies when compared to the inhibition of cAMP (Patel et al., 2020). This aligned with our previous findings comparing SCRAs to THC, where SCRAs showed similar increase in cAMP levels, while THC appeared to exert no measurable effect on CB1-Gαs signaling pathway (Sachdev et al., 2020). Consistent with the effects on cAMP production, many emerging SCRAs demonstrated high efficacy at inducing β-arrestin 1 and β-arrestin 2 recruitment to CB1, ranging from 7.94 nM (4CN-CUMYL-BUTINACA) to 1.02 μM (WIN55,212-2) (Patel et al., 2020). This potent and robust β-arrestin recruitment by recently emerged SCRAs has also been established in other studies (reviewed in Patel et al., 2021), which may have implications for their in vivo toxicological impacts. In particular, THC has been shown to have very low or undetectable recruitment of either β-arrestin, aligning with the studies classifying THC as a low efficacy agonist of CB1 receptor (Kumar et al., 2019). This suggests the profile or pattern of differential signaling responses produced by SCRAs differs from THC (termed "biased agonism" and reviewed elsewhere (Patel et al., 2021)), which is associated with relatively low incidences of toxicity. Using a quantitative immunocytochemistry approach, this study also assessed the influence of SCRA on CB1 receptor internalization. The results revealed that a selection of SCRAs tested produced a complete loss of surface CB1, which differed to the reduced internalization observed with THC. The enhanced ability of SCRAs to induce robust receptor internalization suggests substantial loss of CB1 on cell surfaces for activation may be associated with the adverse effects of SCRAs, such as the development of tolerance and withdrawal symptoms upon drug cessation. However, the biological significance of SCRA-mediated differential coupling of CB1 to Gi/o, Gs, β-arrestin, and internalization pathway is not well understood, despite the increasing array of molecular pharmacology data. One of the greatest obstacles in this field of research is the difficulty in relating in vitro findings of signaling phenotype of cannabinoids to their physiological role and toxicity in humans.

SCRA-induced effects in CB2 receptor

In addition to SCRAs having high potency, affinity, and efficacy at CB1, which may contribute to these drugs psychoactive profile, many compounds also possess activity at CB2. In contrast to CB1, CB2 receptors are expressed at high levels outside the CNS on immune cells and possess a key role in immunomodulation (Atwood & Mackie, 2010). Although also present in relatively low numbers in the brain and/or spinal cord, increasing evidence suggests that activation of CB2 in the CNS seems to be involved in the effects produced by drugs of abuse, and more specifically in the field of addiction (Onaivi et al., 2008).

Several SCRAs have demonstrated equally high, if not higher intrinsic activity at CB2 than CB1 (Banister et al., 2016; Banister et al., 2019; Longworth et al., 2017; Wiley et al., 2015). Perhaps the most characterized, recreationally used SCRA, JWH-018, demonstrated canonical signaling effects at CB2, inducing Gi-mediated inhibition of adenylyl cyclase (AC) activity (Rajasekaran et al., 2013), GIRK activation (Banister et al., 2015), and recruitment of β-arrestin 2 (Zagzoog et al., 2021). Studies have also examined the molecular pharmacological properties of several phase I hydroxylated metabolites of JWH-018 at CB2. It was found that all but one of the metabolites of JWH-018 bind with high nanomolar affinity, activate G proteins and inhibit AC-activity via CB2 receptors (Rajasekaran et al., 2013). Furthermore, when compared to CP55940 and THC, a major omega hydroxyl metabolite of JWH-018 (JWH-018-M5) required significantly less CB2 receptor occupancy to produce equivalent levels of AC-activity regulation, indicating that these compounds may more efficiently couple to CB2 than well-characterized classical cannabinoid agonists (Rajasekaran et al., 2013).

New generations of SCRAs also act as potent and relatively high-affinity agonists at CB2 (Banister et al., 2016; Banister et al., 2019). SCRA designer drugs based on indole and indazole scaffold with amino acid-derived side chains, including AB-CHMINACA, AB-PINACA, 5F-CUMYL-PICA, 5F-CUMYL-PINACA, and 5F-ADB-PINACA were found to be potent, highly efficacious agonists in [^{35}S]GTPγS binding (Gamage et al., 2018), and fluorescence-based membrane potential assays (Banister et al., 2016; Longworth et al., 2017), as compared with the low efficacy agonist THC.

Zagzoog et al. (2021) further assessed the ability of SCRAs to elicit differential activation of cAMP inhibition or β-arrestin 2 recruitment and receptor subtype selectivity between CB1 and CB2 receptors. Indeed, all SCRAs assayed induced highly potent and efficacious inhibition of forskolin-stimulated cAMP at CB2, which contrasts with the partial efficacy of THC. A select few of the SCRAs displayed none or very weak agonist activity for the recruitment of β-arrestin 2 to CB2 relative to CP55940 (Zagzoog et al., 2021). The study observed that AB-PINACA N-(2-fluoropentyl) isomer, 3-chloro AB-PINACA, and 4-fluoro MDMB-BINACA exhibited bias toward CB2-mediated recruitment of β-arrestin 2 over inhibition of cAMP, when evaluated using the operational analysis. The authors also noted that all the SCRAs tested have similar activity profiles for both CB1 and CB2 receptors with no evidence of any selectivity for an individual receptor type (Zagzoog et al., 2021). In vivo studies in a rodent model revealed that the highly β-arrestin 2-biased ligand 4-fluoro MDMB-BINACA was the most potent and efficacious at inducing catalepsy and hypothermia among the cannabinoids tested (Zagzoog et al., 2021). For the cannabinoid receptors, it remains unclear whether β-arrestin 2 bias attributes to the on-target pharmacodynamic toxicities associated with SCRAs, warranting further investigation. Therefore, like CB1, it is critical to determine how SCRAs and CB2 receptors interact with each other and how the subsequent cell signaling events effect human physiological conditions.

HU-210, another compound identified in the SCRA products, is a high-affinity CB1 and CB2 receptor agonist reported to be highly potent in producing cannabinoid-like effects in rats and pigeons, including discriminative stimulus effects, decreased locomotor activity, rearing, and grooming, and long-lasting hypothermia of a greater magnitude than that produced by THC (Hruba & McMahon, 2014; Ovadia et al., 1995). In experiments performed with rhesus monkeys, HU-210 was compared with the discriminative stimulus effects of THC (Hruba & McMahon, 2014). When discriminative stimulus effects were examined over time, the duration of action of intravenous HU-210 was much longer (at least 24 h) than that of THC (less than 5 h), CP55940 (less than 3 h), and WIN55212-2 (less than 1 h) (Hruba & McMahon, 2014). This suggests that HU-210 can produce protracted withdrawal symptoms similar to what has been observed with the long-acting opioid agonist buprenorphine. The results of this study indicate yet another physiological outcome that may result from the differing activity of THC and SCRAs on CB receptors. Another interesting observation of this study is that at CB1, HU-210 appeared to be more potent in attenuating the effects of rimonabant as compared with CP55940, WIN55212-2, and THC, consistent with pseudo-irreversible binding of HU-210 to cannabinoid receptors (Hruba & McMahon, 2014). Although this study did not explicitly employ CB2 antagonist to investigate the SCRA-mediated effects associated with CB2 receptor subtype, it cannot be ruled out that some unanticipated adverse effects of SCRAs are caused by their effect on CB2.

Noncannabinoid G protein-coupled receptors

SCRA-induced effects in GPR55

G protein-coupled receptor 55 (GPR55) has been shown to be targeted by a range of endogenous, phyto and synthetic cannabinoids (see Marichal-Cancino chapter, this issue). Given that GPR55 and CB1 receptors are co-expressed in brain regions such as striatum, frontal cortex, hippocampus, and hypothalamus, it is critically important to understand whether GPR55 may also play a role in SCRA-mediated toxicity (Wu et al., 2013). Ryberg et al. (2007) showed that cannabinoids such as CP55940, AM-251, and THC can bind to and signal in cells heterologously expressing FLAG-tagged human GPR55. Like its activity at CB1 or CB2 receptors, CP55940 and AM-251 appear to function as a highly potent GPR55 agonists determined using GTPγS assay (EC_{50} values of 5 and 39 nM, respectively). Another study using β-galactosidase enzyme fragment complementation technology for β-arrestin recruitment demonstrated inhibitory effects of SCRAs such as EAM-2201, UR-144, and APICA on lysophosphatidylinositol (GPR55 agonist) induced activation of GPR55, with EAM-2201 being the most potent of this series with an IC_{50} value of 1.86 μM (Hess et al., 2016). While these studies themselves do not point to an unequivocal role of GPR55 in SCRA action, these findings suggest that GPR55 activation may also be involved in some of the pharmacological effects of SCRAs.

To examine the in vivo effects, a multiple GPR55/CB1 knockout mouse line combined with pharmacological approaches was used to study the mechanism underlying the effects produced by phytocannabinoids versus SCRAs (Wang et al., 2020). Not surprisingly, systemic administration of THC, WIN55212-2 or XLR11, produced dose-dependent tetrad effects in WT mice. Specifically, the genetic deletion of CB1 receptors, but not GPR55, blocked the tetrad effects produced by all three cannabinoids (Wang et al., 2020), consistent with the potential involvement of CB1 in this response. Moreover, they showed that the mice lacking GPR55 receptors caused enhanced tetrad responses to low doses of THC or WIN55212-2, but not to XLR11. These findings imply that distinct receptor mechanisms might be responsible for the effects produced by THC versus XLR-11 (Wang et al., 2020).

Serotonin 5-HT_{1A} receptors

Studies have shown that 5-HT_{1A} and CB1 may form heteromers in hippocampus and other brain regions related to memory formation (Viñals et al., 2015). Hence, coactivation of 5-HT_{1A} and CB1 receptors by cannabinoids may modulate each other's signaling pathways and therefore alter each other's functional properties. The idea that SCRAs might have modulatory effects at the serotonin 1A receptors 5-HT_{1A} was postulated by Yano et al. (2020). This study utilized bioluminescence resonance energy transfer (BRET)-based CB1R/Gi protein engagement assay to investigate "off-target" interactions of a series of cannabinoids, including SCRAs and THC with GPCRs other than CB1 and CB2—dopamine D2 receptors, serotonin receptor subtypes 5-HT_{1A}, 5-HT_{2A}, 5-HT_{2B}, and adrenergic α1 and α2 receptors (Yano et al., 2020). The authors found that the SCRAs tested were unable to interact with D2, 5-HT_{2A}, α1A, or α2A in either an orthosteric or an allosteric way, but selectively enhanced 5-HT_{1A}-mediated endogenous agonist 5-HT binding. Their results indicate that AM2201 and JWH-018 led to a more than 28% augmentation in the maximal effect of 5-HT (E_{max}) on Gi activation, suggesting allosteric potentiation of the 5-HT_{1A} receptor activation (Yano et al., 2020). Consistent with this, BRET-based cAMP assay demonstrated that AM2201 and JWH-018 both significantly increased the 5-HT-mediated inhibition of cAMP accumulation, most likely acting as a positive allosteric modulator, devoid of intrinsic agonist activity. To further reinforce the notion that SCRA is interacting with the allosteric site of 5-HT_{1A} receptors, the authors showed that in cells expressing the 5-HT_{1A} receptor, AM2201 enhanced the 5-HT-mediated displacement of [^{3}H]WAY100635 (selective 5-HT_{1A} antagonist) by 11.2% compared to vehicle control (Yano et al., 2020). This suggests that AM2201 acts at 5-HT_{1A} receptor via binding in an allosteric rather than an orthosteric site. Furthermore, using the whole-cell voltage-clamp experiments in dorsal raphe nucleus cells, AM2201 converted an ineffective dose of 5-HT into a substantial activator of GIRK-mediated hyperpolarization, supporting an allosteric activity for SCRAs at this receptor (Yano et al., 2020).

To explore the possibility that SCRA might exert allosteric activity at 5-HT_{1A} receptor in vivo, Yano et al. (2020) examined the effects of AM2201 administration on temperature responses to 8-OH-DPAT, a selective 5-HT_{1A} receptor agonist. Compared to the hypothermic effect elicited by 8-OH-DPAT alone, effects of a combined treatment with AM2201 (0.3 mg/kg) produced a significant enhancement in the magnitude and duration of 5-HT_{1A}-mediated hypothermia (Yano et al., 2020). These results were in agreement with a study reporting that repeated injections of JWH-018 produced a modest yet significant increase in the hypothermic effects induced by 8-OH-DPAT via the same 5-HTA_{1A} receptor mechanisms (Elmore & Baumann, 2018). In summary, there is compelling evidence to suggest the involvement of a noncannabinoid 5-HT_{1A} receptor mechanism could also underlie some of the untoward clinical effects of certain classes of SCRAs.

Ion channels

TRPA1

Recently, studies have shown that transient receptor potential (TRP) channels belonging to subfamily A type 1 (TRPA1) are also a potential target for SCRAs. Many cannabinoids have been reported to activate TRP channels, with evidence existing in the literature of THC agonist activity at the TRPA1 (reviewed by Muller et al., 2019). While research into SCRA activation of TRPA1 began in the context of better understanding of their structural relationships and pharmacological profile; Soethoudt et al. (2017) found that the synthetic cannabinoids WIN-55212-2, AM-251, HU308, and JWH-133 were able to evoke intracellular Ca^{2+} response in HEK cells expressing TRPA1. SCRAs found in new designer drug products sold to consumers have also been shown to activate TRPA1 (Pearson, 2017). The SCRAs MDMB-CHMICA, PB-22, 5F-PB-22, UR-144, XLR-11, and 5-OH-UR-144 produced a concentration-dependent elevation of intracellular calcium in HEK cells expressing TRPA1, as assessed by a fluorescent calcium indicator dye kit (Pearson, 2017). Although THC was shown to function as a more potent agonist of TRPA1 when compared to all other SCRAs tested (anywhere from 3- to 11-fold), the activity profile of recent SCRAs at TRPA1 cannot be ruled out. Considering many adverse effects to SCRA exposure include some form of respiratory or cardiovascular distress, it is reasonable to believe some of these effects may be, in part, mediated by TRPA1 activation (Adamowicz & Gieroń, 2016; Monte et al., 2014).

T-type Ca channels

Several studies have now explored the functional activity of cannabinoids on T-type calcium channels that are crucial to brain and heart rhythmicity and function (Bender et al., 2012; Zamponi et al., 2010). Although the adverse psychotomimetic effects seen in patients following SCRA consumption are likely due to their interactions with CB1 receptors, the reasons for more profound effects of SCRAs such as abnormal heart rate, hyperalgesia, and seizure activity are still not clear. Subsequent to the initial discovery of THC interaction with T-type calcium channels (Ross et al., 2008), a recent study has examined whether SCRAs alter T-type activity, and if these interactions may contribute to a physiological role in SCRA toxicity (Bladen et al., 2021). This study established that of the 16 SCRAs tested, two of the SCRAs, MDMB-CHMICA and AMB-CHMINACA, potently blocked Cav3.2 T-type calcium channel with IC_{50} values of 1.5 and 0.74 μM, respectively. Bladen et al. (2021) further assessed the ability of MDMB-CHMICA and AMB-CHMINACA to modulate the activation and inactivation states of human Cav3.2 expressed in HEK cells compared to phytocannabinoid THC, as assessed by patch clamp electrophysiology. The study observed that the SCRAs were not able to affect the steady-state inactivation state of human Cav3.2, in contrast to THC, which has been shown to significantly alter this parameter. However, for activation kinetics of Cav3.2, the significant negative shift caused by MDMB-CHMICA is similar to that of THC, whereas AMB-CHMINACA failed to significantly alter activation (Bladen et al., 2021). This suggests that the functional interaction

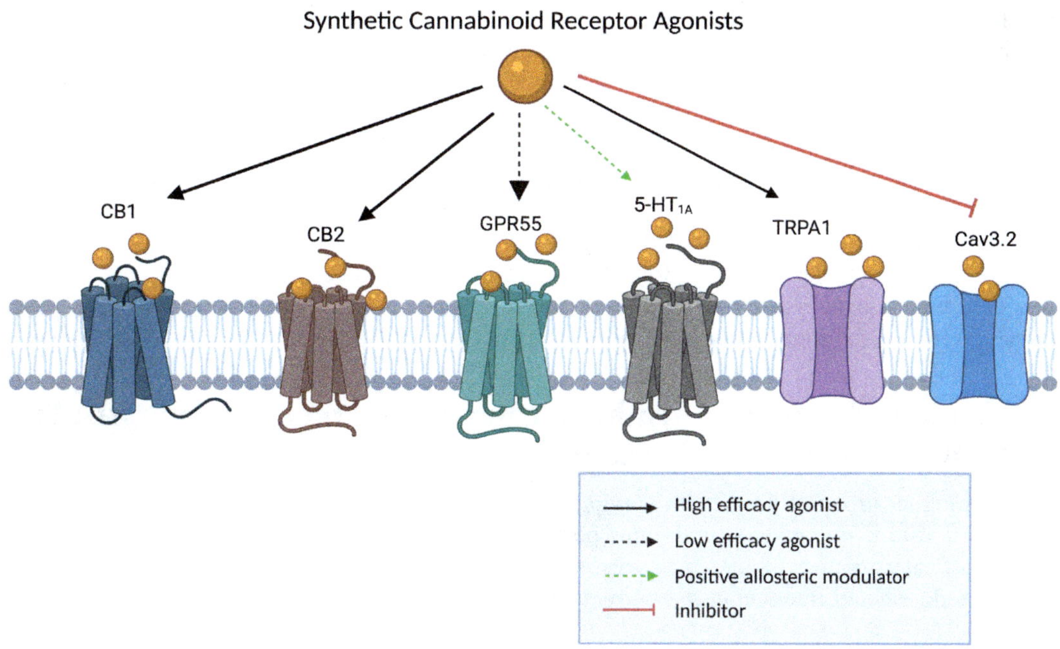

FIG. 1 Multiple molecular targets for SCRAs—synthetic cannabinoid receptor agonists have multiple targets within the cell. They behave like a high efficacy agonist of cannabinoid CB1 and CB2 receptors in canonical pathways of receptor activity, including cAMP inhibition, activation of G protein inwardly rectifying potassium channels, and arrestin recruitment, which may contribute to the toxicity and/or tolerance seen with SCRA consumption. Partial agonism of SCRAs at GPR55 and TRPA1 might account for the cardiotoxic effects of SCRAs. While the positive allosteric modulation of 5-HT_{1A} may represent a novel receptor mechanism underlying the complex profile of effects for certain SCRAs. SCRAs have an overall inhibitory effect on Cav3.2, raising the possibility that SCRA toxicity may be in part to Cav3.2 modulation. These interactions alone or in combination with cannabinoid receptors may explain some of their adverse effects.

of SCRA with Cav3.2 channels in relation to SCRA toxicity is multifaceted and structurally dependent on both "how and where" the drug binds to the channel. Nevertheless, this study supports the notion that T-type Cav3.2 channels may represent an additional mechanism implicated in the adverse effect of SCRAs (Fig. 1).

Conclusion

We have identified a large number of potential molecular targets of SCRAs that are likely to be relevant to many of the adverse effects associated with these compounds. While the increase in molecular pharmacology information has aided our understanding of potential mechanisms of toxicity within the SCRA family, it remains difficult to correlate in vitro findings with physiological relevant effects in humans and SCRA-related causes of toxicity and death have not yet been resolved. It is worth noting that SCRA-associated toxicity may result from a combination of factors including pharmacokinetic differences, pharmacological profile of thermolytic degradants and metabolites compared to their parent compounds, and their activity at multiple receptor targets. Many questions still remain unanswered regarding the etiology of SCRA toxicity-related manifestations at the tissue level? Could off-target noncannabinoid receptor effects of SCRAs account for the complex adverse effect profile associated with their use? Further studies will be needed in order to further delineate the role of traditional and nontraditional cannabinoid targets, their specific pathways, and their response to SCRAs.

Applications to other areas

Better understanding of how drugs work can inform not only the development of newer and more effective treatments but also safer drugs that help prevent serious adverse effects. In this chapter, we discuss multiple targets that are activated, or effected by SCRAs. These targets have many important physiological roles and are also involved in medical conditions such as chronic pain and epilepsy. SCRAs have very high efficacy and potency at the cannabinoid receptors that may also result in serious adverse effects and even death. This level of toxicity is not observed with the consumption of *Cannabis sativa* and at least partially explains the large difference in the pharmacological profile between the phytocannabinoid THC and these synthetic cannabinoids. Identifying these differences is important if we aim to develop drugs that have superior therapeutic benefits to THC but lower adverse effects. Also, identifying all the potential targets of these drugs will help support the treatment of SCRA-induced psychosis, toxicity, overdose, and dependence.

Mini-dictionary of terms

- Molecular pharmacology—a subcategory of pharmacological research that seeks to understand the mechanism of a drug at molecular level, with focus on cellular processes, structures and the macromolecules involved.
- Recreational drug use—use of drugs for leisure purposes (not as part of medical therapy).
- Molecular target—a macromolecule where drugs bind to for the effect to occur.
- Intracellular transduction pathways—signaling inside the cell that is 'turned on" by activation of targets.
- Cannabimimetic effects—substances that have similar pharmacological properties to those of cannabis.

Key facts of functional profile of synthetic cannabinoid receptor agonists: Exploring cannabinoid and noncannabinoid targets

- SCRA is a heterologous group of drugs with varying activities across multiple targets.
- Profiling SCRA in vitro is important to determine possible targets and signaling cascades activated or affected.
- SCRA are high efficacy agonists of CB1 receptors, while THC is a low efficacy agonist.
- SCRA targets are distributed throughout the body, which may explain adverse effects observed in multiple physiological systems.
- Although studies have provided a good insight on SCRA cellular effects, the correlation between activation of certain pathways and the overall physiological effects observed is not well understood.

Summary points

- SCRAs are recreational, psychoactive drugs that are generally far more toxic than THC.
- Multiple receptors can be effected by SCRA, including the cannabinoid receptors (CB1 and CB2), GPR55 receptors, 5-HT_{1A} receptors, TRP channels, and T-type calcium channels. Given these identified targets have very similar structures to other receptors, it is highly likely as yet unidentified receptors will also strongly interact with SCRAs, rendering it a challenge to discern the precise molecular mechanism by which a SCRA exerts its toxic effects.
- The current chapter has demonstrated a broad pharmacological profile of SCRAs, including the substantially high efficacy profile of SCRAs at activating cannabinoid CB1 and CB2 receptors, the sub-nanomolar range agonistic activity of SCRAs on GPR55, positive allosteric modulation of 5-HT_{1A} receptors, and an overall inhibitory effect of SCRAs on T-type calcium channels.
- SCRA potency and efficacy vary dramatically but are generally more efficacious and potent than THC on activating the cannabinoid receptors.
- The differing natures of SCRAs and THC effects in the activation of multiple targets may help identify some of the potential mechanisms implicated in the toxic effects of SCRAs.

References

Adamowicz, P., & Gieroń, J. (2016). Acute intoxication of four individuals following use of the synthetic cannabinoid mab-chminaca. *Clinical Toxicology (Philadelphia, Pa.)*, *54*(8), 650–654.

Adams, R., Cain, C., & Baker, B. (1940). Structure of cannabinol. II. Synthesis of two new isomers, 3-Hydroxy-4-n-amyl-and 3-Hydroxy-2-n-amyl 6, 6, 9-Trimethyl-6-dibenzopyrans1. *Journal of the American Chemical Society*, *62*(8), 2201–2204.

Atwood, B. K., Huffman, J., Straiker, A., & Mackie, K. (2010). JWH018, a common constituent of 'spice'herbal blends, is a potent and efficacious cannabinoid CB1 receptor agonist. *British Journal of Pharmacology*, *160*(3), 585–593.

Atwood, B. K., & Mackie, K. (2010). CB2: A cannabinoid receptor with an identity crisis. *British Journal of Pharmacology*, *160*(3), 467–479.

Auwärter, V., Dresen, S., Weinmann, W., Müller, M., Pütz, M., & Ferreirós, N. (2009). 'Spice'and other herbal blends: Harmless incense or cannabinoid designer drugs? *Journal of Mass Spectrometry*, *44*(5), 832–837.

Banister, S. D., Adams, A., Kevin, R. C., Macdonald, C., Glass, M., Boyd, R., et al. (2019). Synthesis and pharmacology of new psychoactive substance 5F-CUMYL-P7AICA, a scaffold-hopping analog of synthetic cannabinoid receptor agonists 5F-CUMYL-PICA and 5F-CUMYL-PINACA. *Drug Testing and Analysis*, *11*(2), 279–291.

Banister, S. D., & Connor, M. (2018). The chemistry and pharmacology of synthetic cannabinoid receptor agonists as new psychoactive substances: Origins. In *Handbook of experimental pharmacology* (pp. 165–190).

Banister, S. D., Longworth, M., Kevin, R., Sachdev, S., Santiago, M., Stuart, J., et al. (2016). Pharmacology of valinate and tert-leucinate synthetic cannabinoids 5F-AMBICA, 5F-AMB, 5F-ADB, AMB-FUBINACA, MDMB-FUBINACA, MDMB-CHMICA, and their analogues. *ACS Chemical Neuroscience*, *7*(9), 1241–1254.

Banister, S. D., Stuart, J., Kevin, R. C., Edington, A., Longworth, M., Wilkinson, S. M., et al. (2015). Effects of bioisosteric fluorine in synthetic cannabinoid designer drugs JWH-018, AM-2201, UR-144, XLR-11, PB-22, 5F-PB-22, APICA, and STS-135. *ACS Chemical Neuroscience*, *6*(8), 1445–1458.

Bender, K. J., Uebele, V. N., Renger, J. J., & Trussell, L. O. (2012). Control of firing patterns through modulation of axon initial segment t-type calcium channels. *The Journal of Physiology*, *590*(1), 109–118.

Bladen, C., Mirlohi, S., Santiago, M., Longworth, M., Kassiou, M., Banister, S., et al. (2021). Modulation of human T-type calcium channels by synthetic cannabinoid receptor agonists in vitro. *Neuropharmacology*, *187*, 108478.

Bosier, B., Sarre, S., Smolders, I., Michotte, Y., Hermans, E., & Lambert, D. M. (2010). Revisiting the complex influences of cannabinoids on motor functions unravels pharmacodynamic differences between cannabinoid agonists. *Neuropharmacology*, *59*(6), 503–510.

Breivogel, C. S., Wells, J. R., Jonas, A., Mistry, A. H., Gravley, M. L., Patel, R. M., et al. (2020). Comparison of the neurotoxic and seizure-inducing effects of synthetic and endogenous cannabinoids with Δ9-tetrahydrocannabinol. *Cannabis and Cannabinoid Research*, *5*(1), 32–41.

Buser, G., Gerona, R., Horowitz, B. Z., Vian, K., Troxell, M., Hendrickson, R., et al. (2014). Acute kidney injury associated with smoking synthetic cannabinoid. *Clinical Toxicology*, *52*(7), 664–673.

Cannaert, A., Storme, J., Franz, F., Auwärter, V., & Stove, C. P. (2016). Detection and activity profiling of synthetic cannabinoids and their metabolites with a newly developed bioassay. *Analytical Chemistry*, *88*(23), 11476–11485.

Chin, C. N., Murphy, J. W., Huffman, J. W., & Kendall, D. A. (1999). The third transmembrane helix of the cannabinoid receptor plays a role in the selectivity of aminoalkylindoles for CB2, peripheral cannabinoid receptor. *The Journal of Pharmacology and Experimental Therapeutics*, *291*(2), 837–844.

Chung, E. Y., Cha, H. J., Min, H. K., & Yun, J. (2021). Pharmacology and adverse effects of new psychoactive substances: Synthetic cannabinoid receptor agonists. *Archives of Pharmacal Research*, 1–12.

Cooper, Z. D. (2016). Adverse effects of synthetic cannabinoids: Management of acute toxicity and withdrawal. *Current Psychiatry Reports*, *18* (5), 52.

Devane, W. A., Dysarz, F. A., III, Johnson, M. R., Melvin, L. S., & Howlett, A. C. (1988). Determination and characterization of a cannabinoid receptor in rat brain. *Molecular Pharmacology*, *34*(5), 605–613.

Elmore, J. S., & Baumann, M. H. (2018). Repeated exposure to the "spice" cannabinoid JWH-018 induces tolerance and enhances responsiveness to 5-HT1A receptor stimulation in male rats. *Frontiers in Psychiatry*, *9*, 55.

Fantegrossi, W. E., Moran, J. H., Radominska-Pandya, A., & Prather, P. L. (2014). Distinct pharmacology and metabolism of K2 synthetic cannabinoids compared to Δ9-THC: Mechanism underlying greater toxicity? *Life Sciences*, *97*(1), 45–54.

Gamage, T. F., Farquhar, C. E., Lefever, T. W., Marusich, J. A., Kevin, R. C., McGregor, I. S., et al. (2018). Molecular and behavioral pharmacological characterization of abused synthetic cannabinoids MMB-and MDMB-FUBINACA, MN-18, NNEI, CUMYL-PICA, and 5-FLUORO-CUMYL-PICA. *The Journal of Pharmacology and Experimental Therapeutics*, *365*(2), 437–446.

Gatch, M. B., & Forster, M. J. (2015). Δ9-Tetrahydrocannabinol-like effects of novel synthetic cannabinoids found on the gray market. *Behavioural Pharmacology*, *26*(5), 460.

Hermanns-Clausen, M., Kneisel, S., Szabo, B., & Auwarter, V. (2013). Acute toxicity due to the confirmed consumption of synthetic cannabinoids: Clinical and laboratory findings. *Addiction*, *108*(3), 534–544.

Hess, C., Schoeder, C. T., Pillaiyar, T., Madea, B., & Müller, C. E. (2016). Pharmacological evaluation of synthetic cannabinoids identified as constituents of spice. *Forensic Toxicology*, *34*(2), 329–343.

Hruba, L., & McMahon, L. R. (2014). The cannabinoid agonist HU-210: Pseudo-irreversible discriminative stimulus effects in rhesus monkeys. *European Journal of Pharmacology*, *727*, 35–42.

Huestis, M. A., Gorelick, D. A., Heishman, S. J., Preston, K. L., Nelson, R. A., Moolchan, E. T., et al. (2001). Blockade of effects of smoked marijuana by the CB1-selective cannabinoid receptor antagonist SR141716. *Archives of General Psychiatry*, *58*(4), 322–328.

Huffman, J. W., Dai, D., Martin, B. R., & Compton, D. R. (1994). Design, synthesis and pharmacology of cannabimimetic indoles. *Bioorganic & Medicinal Chemistry Letters*, *4*(4), 563–566.

Janoyan, J. J., Crim, J. L., & Darmani, N. A. (2002). Reversal of SR141716A-induced head-twitch and ear-scratch responses in mice by Δ9-THC and other cannabinoids. *Pharmacology, Biochemistry, and Behavior*, *71*(1-2), 155–162.

Kavanagh, P., Grigoryev, A., Savchuk, S., Mikhura, I., & Formanovsky, A. (2013). UR-144 in products sold via the internet: Identification of related compounds and characterization of pyrolysis products. *Drug Testing and Analysis*, *5*(8), 683–692.

Kumar, K. K., Shalev-Benami, M., Robertson, M. J., Hu, H., Banister, S. D., Hollingsworth, S. A., et al. (2019). Structure of a signaling cannabinoid receptor 1-g protein complex. *Cell*, *176*(3), 448–458.e412.

Longworth, M., Banister, S. D., Boyd, R., Kevin, R. C., Connor, M., McGregor, I. S., et al. (2017). Pharmacology of Cumyl-carboxamide synthetic cannabinoid new psychoactive substances (nps) CUMYL-BICA, CUMYL-PICA, CUMYL-5F-PICA, CUMYL-5F-PINACA, and their analogues. *ACS Chemical Neuroscience*, *8*(10), 2159–2167.

Mackie, K. (2005). Distribution of cannabinoid receptors in the central and peripheral nervous system. In *Cannabinoids* (pp. 299–325). Springer.

Malyshevskaya, O., Aritake, K., Kaushik, M. K., Uchiyama, N., Cherasse, Y., Kikura-Hanajiri, R., et al. (2017). Natural (△ 9-THC) and synthetic (JWH-018) cannabinoids induce seizures by acting through the cannabinoid CB1 receptor. *Scientific Reports*, *7*(1), 1–8.

Matsuda, L. A., Lolait, S. J., Brownstein, M. J., Young, A. C., & Bonner, T. I. (1990). Structure of a cannabinoid receptor and functional expression of the cloned cDNA. *Nature*, *346*(6284), 561–564.

Monte, A. A., Bronstein, A. C., Cao, D. J., Heard, K. J., Hoppe, J. A., Hoyte, C. O., et al. (2014). An outbreak of exposure to a novel synthetic cannabinoid. *The New England Journal of Medicine*, *370*(4), 389.

Muller, C., Morales, P., & Reggio, P. H. (2019). Cannabinoid ligands targeting TRP channels. *Frontiers in Molecular Neuroscience*, *11*, 487.

Onaivi, E. S., Ishiguro, H., Gong, J.-P., Patel, S., Meozzi, P. A., Myers, L., et al. (2008). Functional expression of brain neuronal CB2 cannabinoid receptors are involved in the effects of drugs of abuse and in depression. *Annals of the New York Academy of Sciences*, *1139*, 434.

Ovadia, H., Wohlman, A., Mechoulam, R., & Weidenfeld, J. (1995). Characterization of the hypothermic effect of the synthetic cannabinoid HU-210 in the rat. Relation to the adrenergic system and endogenous pyrogens. *Neuropharmacology*, *34*(2), 175–180.

Patel, M., Finlay, D. B., & Glass, M. (2021). Biased agonism at the cannabinoid receptors—Evidence from synthetic cannabinoid receptor agonists. *Cellular Signalling*, *78*, 109865.

Patel, M., Manning, J. J., Finlay, D. B., Javitch, J. A., Banister, S. D., Grimsey, N. L., et al. (2020). Signalling profiles of a structurally diverse panel of synthetic cannabinoid receptor agonists. *Biochemical Pharmacology*, *175*, 113871.

Pearson, S. J. (2017). *Synthetic cannabinoid activation of hTRPA1 and naturally occurring channel variants*. Master of Research, Macquarie University.

Pike, E., Grafinger, K. E., Cannaert, A., Ametovski, A., Luo, J. L., Sparkes, E., et al. (2021). Systematic evaluation of a panel of 30 synthetic cannabinoid receptor agonists structurally related to MMB-4EN-PICA, MDMB-4EN-PINACA, ADB-4EN-PINACA, and MMB-4CN-BUTINACA using a combination of binding and different cb1 receptor activation assays: Part I—Synthesis, analytical characterization, and binding affinity for human CB1 receptors. *Drug Testing and Analysis*, *13*.

Rajasekaran, M., Brents, L. K., Franks, L. N., Moran, J. H., & Prather, P. L. (2013). Human metabolites of synthetic cannabinoids JWH-018 and JWH-073 bind with high affinity and act as potent agonists at cannabinoid type-2 receptors. *Toxicology and Applied Pharmacology*, *269*(2), 100–108.

Ross, H. R., Napier, I., & Connor, M. (2008). Inhibition of recombinant human T-type calcium channels by Δ9-tetrahydrocannabinol and cannabidiol. *Journal of Biological Chemistry*, *283*(23), 16124–16134.

Ryberg, E., Larsson, N., Sjorgen, S., Hjorth, S., Hermansson, N., Leonova, J., et al. (2007). The orphan receptor GPR55 is a novel cannabinoid receptor. *British Journal of Pharmacology*, *152*(1092), 124.

Sachdev, S. (2020). *Molecular pharmacology of cannabinoids: Beyond THC*. PhD Thesis by publication Macquarie University.

Sachdev, S., Banister, S. D., Santiago, M., Bladen, C., Kassiou, M., & Connor, M. (2020). Differential activation of G protein-mediated signaling by synthetic cannabinoid receptor agonists. *Pharmacology Research & Perspectives*, *8*(2), e00566.

Sachdev, S., Vemuri, K., Banister, S. D., Longworth, M., Kassiou, M., Santiago, M., et al. (2019). In vitro determination of the efficacy of illicit synthetic cannabinoids at CB1 receptors. *British Journal of Pharmacology*, *176*(24), 4653–4665.

Sim-Selley, L. J. (2003). Regulation of cannabinoid CB1 receptors in the central nervous system by chronic cannabinoids. *Critical Reviews in Neurobiology*, *15*(2).

Soethoudt, M., Grether, U., Fingerle, J., Grim, T. W., Fezza, F., De Petrocellis, L., et al. (2017). Cannabinoid CB2 receptor ligand profiling reveals biased signalling and off-target activity. *Nature Communications*, *8*(1), 1–14.

Thomas, B. F., Lefever, T. W., Cortes, R. A., Grabenauer, M., Kovach, A. L., Cox, A. O., et al. (2017). Thermolytic degradation of synthetic cannabinoids: Chemical exposures and pharmacological consequences. *The Journal of Pharmacology and Experimental Therapeutics*, *361*(1), 162–171.

Viñals, X., Moreno, E., Lanfumey, L., Cordomí, A., Pastor, A., de La Torre, R., et al. (2015). Cognitive impairment induced by Δ9-tetrahydrocannabinol occurs through heteromers between cannabinoid CB1 and serotonin 5-HT2A receptors. *PLoS Biology*, *13*(7), e1002194.

Wang, X. F., Galaj, E., Bi, G. H., Zhang, C., He, Y., Zhan, J., et al. (2020). Different receptor mechanisms underlying phytocannabinoid-versus synthetic cannabinoid-induced tetrad effects: Opposite roles of CB1/CB2 versus GPR55 receptors. *British Journal of Pharmacology*, *177*(8), 1865–1880.

Wiley, J. L., Lefever, T. W., Marusich, J. A., Grabenauer, M., Moore, K. N., Huffman, J. W., et al. (2016). Evaluation of first generation synthetic cannabinoids on binding at noncannabinoid receptors and in a battery of in vivo assays in mice. *Neuropharmacology*, *110*, 143–153.

Wiley, J. L., Marusich, J. A., Huffman, J. W., Balster, R. L., & Thomas, B. F. (2011). Hijacking of basic research: The case of synthetic cannabinoids. *Methods Report (RTI Press)*, *17971*.

Wiley, J. L., Marusich, J. A., Lefever, T. W., Antonazzo, K. R., Wallgren, M. T., Cortes, R. A., et al. (2015). AB-CHMINACA, AB-PINACA, and FUBIMINA: Affinity and potency of novel synthetic cannabinoids in producing δ9-tetrahydrocannabinol-like effects in mice. *The Journal of Pharmacology and Experimental Therapeutics*, *354*(3), 328–339.

Wiley, J. L., Marusich, J. A., Lefever, T. W., Grabenauer, M., Moore, K. N., & Thomas, B. F. (2013). Cannabinoids in disguise: Δ9-tetrahydrocannabinol-like effects of tetramethylcyclopropyl ketone indoles. *Neuropharmacology*, *75*, 145–154.

Wiley, J. L., Smith, V. J., Chen, J., Martin, B. R., & Huffman, J. W. (2012). Synthesis and pharmacology of 1-alkyl-3-(1-naphthoyl) indoles: Steric and electronic effects of 4-and 8-halogenated naphthoyl substituents. *Bioorganic & Medicinal Chemistry*, *20*(6), 2067–2081.

Wilson, C. D., Tai, S., Ewing, L., Crane, J., Lockhart, T., Fujiwara, R., et al. (2019). Convulsant effects of abused synthetic cannabinoids JWH-018 and 5F-AB-PINACA are mediated by agonist actions at CB1 receptors in mice. *The Journal of Pharmacology and Experimental Therapeutics*, *368*(2), 146–156.

Wu, C.-S., Chen, H., Sun, H., Zhu, J., Jew, C. P., Wager-Miller, J., et al. (2013). GPR55, a G-protein coupled receptor for lysophosphatidylinositol, plays a role in motor coordination. *PLoS One*, *8*(4), e60314.

Yano, H., Adhikari, P., Naing, S., Hoffman, A. F., Baumann, M. H., Lupica, C. R., et al. (2020). Positive allosteric modulation of the 5-HT1A receptor by indole-based synthetic cannabinoids abused by humans. *ACS Chemical Neuroscience*, *11*(10), 1400–1405.

Zagzoog, A., Brandt, A. L., Black, T., Kim, E. D., Burkart, R., Patel, M., et al. (2021). Assessment of select synthetic cannabinoid receptor agonist bias and selectivity between the type 1 and type 2 cannabinoid receptor. *Scientific Reports*, *11*(1), 1–18.

Zamponi, G. W., Lory, P., & Perez-Reyes, E. (2010). Role of voltage-gated calcium channels in epilepsy. *Pflügers Archiv - European Journal of Physiology*, *460*(2), 395–403.

Zimmer, A., Zimmer, A. M., Hohmann, A. G., Herkenham, M., & Bonner, T. I. (1999). Increased mortality, hypoactivity, and hypoalgesia in cannabinoid CB1 receptor knockout mice. *Proceedings of the National Academy of Sciences*, *96*(10), 5780–5785.

Chapter 29

Linking the G-protein-coupled receptor 55 (GPR55) to the cannabinoid receptors (CB_1 and CB_2): A new narrative

Paula Morales[a], Raquel Guerrero-Alba[b], and Bruno A. Marichal-Cancino[b]

[a]*Instituto de Química Médica, CSIC, Madrid, Spain,* [b]*Departamento de Fisiología y Farmacología, Centro de Ciencias Básicas, Universidad Autónoma de Aguascalientes, Ciudad Universitaria, Aguascalientes, Mexico*

Abbreviations

2-AG	2-arachidonoylglycerol
Δ^9-THC	Δ^9-tetrahydrocannabinol
AArG	sn-1acyl-2-arachidonoylglycerols
ABHD4	α,β-hydrolase-4
AEA	anandamide
CB_1	cannabinoid receptor type 1
CB_2	cannabinoid receptor type 2
CB_3	cannabinoid receptor type 3
ECS	endocannabinoid system
ECs	endocannabinoids
GPCRs	G-protein-coupled receptors
GPR55	G-coupled protein receptor 55
CNS	central nervous system
CRIP1a	cannabinoid receptor interacting protein 1a
FAAH	fatty acid amide hydrolase
FABPs	fatty acid-binding proteins
GDE1	glycerophosphodiesterase-1
GPR55	G-protein-coupled receptor 55
GRKs	G-protein-coupled receptor kinases
IUPHAR	International Union of Basic and Clinical Pharmacology
LPI	lysophsophatidylinositol
NADA	*N*-arachidonoyldopamine
NArPE	*N*-arachidonoylphosphatidylethanolamine
NK	natural killer cells
PAG	periaqueductal gray
PLCβ	phospholipase Cβ
PPARs	peroxisome proliferator-activated receptors
TRPV1	transient receptor potential cation channel type 1

Introduction

The recent discovery of cosmic ethanolamine (Rivilla et al., 2021), an abundant group of phospholipid polar heads in membranes, but also a precursor of the *N*-acylethanolamines (an ancient group of cellular mediators), may provide a new perspective to the origin and evolution of lipid mediators, for example, the endocannabinoids (Fig. 1). The endocannabinoids characterize a ubiquitous system of eukaryote cellular control which has been getting complexity on its way to the mammal

Cannabis Use, Neurobiology, Psychology, and Treatment. https://doi.org/10.1016/B978-0-323-89862-1.00024-6

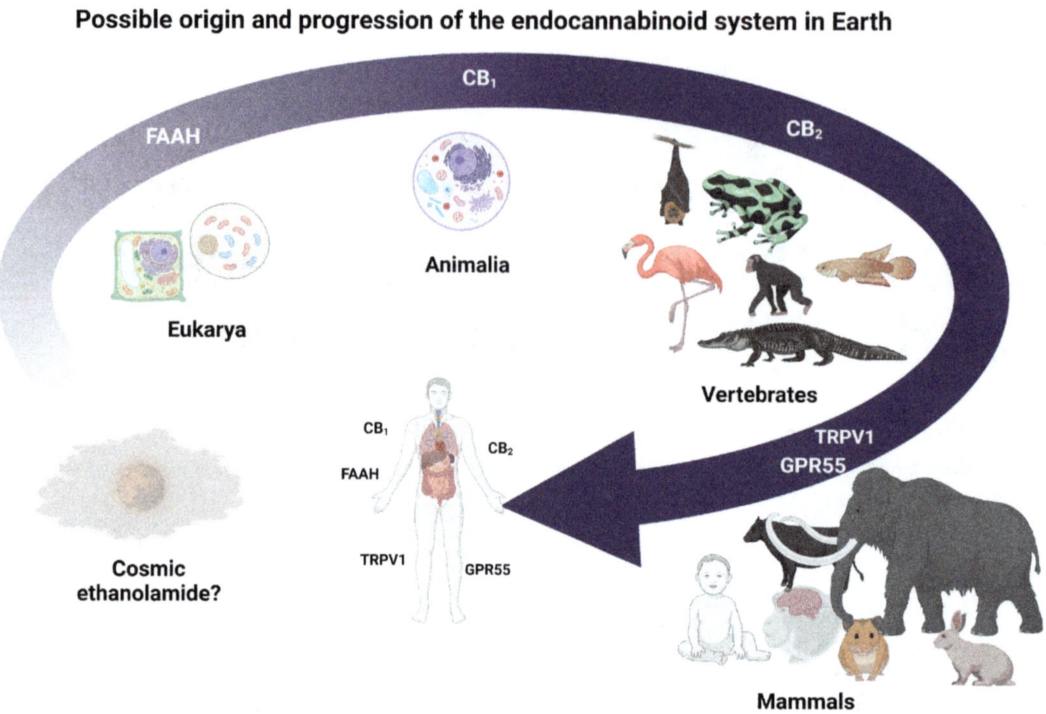

FIG. 1 Timeline of the possible origin and progression of endocannabinoid system in Earth. The sequence for the origin of the endocannabinoid system was taken from McPartland et al. (2006), the existence of cosmic ethanolamide was a recent discovery from Rivilla et al. (2021). Note that the link between cosmic ethanolamide and the origin of the endocannabinoiod system is merely speculative. This figure was created with Biorender with license.

physiology (McPartland et al., 2006). Endocannabinoid targets include the transient receptor potential cation channel type 1 (TRPV1) and the orphan G-coupled protein receptor 55 (GPR55), which are mammalian receptors; whereas a cannabinoid type 2 (CB_2) receptor is expressed in the vertebrates, the cannabinoid receptor 1 (CB_1) is ubiquitous all over *Animalia* and the fatty acid amide hydrolase (FAAH), a catabolic enzyme which interrupts the signaling pathways of the *N*-acylethanolamines, may be found extensively distributed in *Eukarya* (Aziz & Chapman, 2020; McPartland et al., 2006) (Fig. 1). Thus, the endocannabinoid system may be seen as a master controlling system for complex organisms in Earth (and perhaps out of Earth if life is finally found).

In a chronological succession of events, upon the discovery of the first endocannabinoid (i.e., anandamide; AEA) (Devane et al., 1992), and the CB_1-DNA sequencing report (Matsuda et al., 1990), the study of the endocannabinoid system became naturally biased to the AEA structure and CB_1 as the prototypical cannabinoid receptor. Few years later, a closely related receptor was cloned and highly detected out of the central nervous system (CNS), the CB_2 receptor (Munro et al., 1993). Since those discoveries, the actions of cannabinoids, which include phytocannabinoids, endocannabinoids, and synthetic cannabinoids, have been mainly related to CB_1 and CB_2 receptors. Nevertheless, cannabinoids can interact with several other targets including nuclear receptors (e.g., peroxisome proliferator-activated receptors; PPARs), metabotropic non-CB_1/CB_2 receptors (e.g., GPR55 and GPR18), ionotropic receptors (e.g., TRPV1 and 5-HT_3), and ion channels (e.g., Cav3.3) (Demuth & Molleman, 2006; Pertwee et al., 2010). Hence, cannabinoid-based medicines should be understood as multitarget bullets with an outstanding therapeutic potential, but also likely to produce undesired outcomes due to off-target effects. And thus, both pharmacokinetic and toxicologic studies are mandatory to establish the security ranges.

On the other hand, if the endothelial GPR55 receptors, which are also targeted by several cannabinoids, had been initially identified as the first AEA physiological target, the current cannabinoid receptors classification could be quite different. The International Union of Basic and Clinical Pharmacology (IUPHAR) current classification of cannabinoid receptors (Pertwee et al., 2010), as any well thought classification, remains functional, but it might be imprecise (in part due its inherently biased origin), and thus, future knowledge may guide subtle changes to improve it. The IUPHAR-proposed criteria (Pertwee et al., 2010) to a type 3 cannabinoid receptor (CB_3) are shown in Table 1.

By strictly following the IUPHAR-proposed criteria, GPR55 could be classified as a cannabinoid member. Moreover, stimulation of GPR55 by lysophosphatidylinosiol (LPI) (Oka et al., 2007), a non-cannabinoid lipid, suggests a complex biochemistry for this receptor. However, as discussed in this chapter, the great interaction among the current members of the endocannabinoid system and GPR55 is functionally strong and clinically relevant.

TABLE 1 Criteria proposed by IUPAHR to classify the CB_3 receptor (Pertwee et al., 2010).

IUPHAR-proposed criteria for a CB_3 receptor
1. It should be activated at its orthosteric site and with significant potency by an established CB_1/CB_2 receptor ligand
2. It should be activated by at least one established endogenous CB_1/CB_2 receptor agonist at "physiologically relevant" concentrations
3. If it is a GPCR, it should display significant amino acid sequence similarity with the CB_1 or the CB_2 receptor, which are members of the α group of rhodopsin-type GPCRs
4. It should not be a "well established" non-CB_1/CB_2 receptor or channel, especially if there is already strong evidence that: (1) this is activated endogenously by a non-CB_1/CB_2 receptor ligand with appropriate potency and relative intrinsic activity and (2) this is not activated endogenously by any endocannabinoid with appropriate potency and relative intrinsic activity
5. It should be expressed by mammalian cells that are known to be exposed to concentrations of endogenously released endocannabinoid molecules capable of eliciting a response

The endocannabinoid system

Extracts from the marijuana plant, *Cannabis sativa*, have been therapeutically used as analgesics, appetite stimulants, anti-inflammatories, and other purposes for centuries, but its medicinal use has been greatly subsided due to the lack of knowledge of the molecular pharmacology of its substances, its abuse for recreational purpose and additional undesirable effects, such as impairments in locomotor activity and thermoregulation (Crean et al., 2011), harms in executive functions (Crean et al., 2011), and memory consolidation (Ranganathan & D'Souza, 2006). However, the identification of the major psychoactive component Δ^9-tetrahydrocannabinol (Δ^9-THC) (Gaoni & Mechoulam, 1964), and the subsequent isolation of cannabinoid receptors with high expression levels in the nervous system and in the periphery (Matsuda et al., 1990; Munro et al., 1993), led to a detonation of studies exploring the existence of an endogenous cannabinoid system (ECS) and its regulatory functions in health and disease.

The ECS consists of endogenous cannabinoids ligands (e.g., AEA and 2-arachidonoylglycerol; 2-AG), the CB_1 and CB_2 receptors, the enzymes responsible for synthesis and degradation of endocannabinoids and the corresponding genes (K. Mackie, 2008). As mentioned in the preamble, other receptors have been reported to be activated by cannabinoids and related molecules, this includes the class A orphan GPCRs GPR18, GPR55 and GPR119, transient receptor potential channels such as TRPV1, TRPA1, or TPRM8 and nuclear receptors, as well as several nonreceptor targets (Pertwee et al., 2010). Their ability to target such a wide variety of proteins explains their implication in numerous pathophysiological processes.

Endocannabinoids (ECs) biosynthesis and metabolism

ECs are so-called because they activate the same receptors as the major psychoactive components of *Cannabis sativa*, Δ^9-THC. The first EC isolated in the brain, AEA, was named based on the Sanskrit word for bliss, *Ananda* (Devane et al., 1992). AEA is a fatty acid neurotransmitter derived from the nonoxidative metabolism of arachidonic acid. The second endocannabinoid identified was the 2-AG (Mechoulam et al., 1995; Sugiura et al., 1995), an ester formed from arachidonic acid and glycerol. As the search for THC-like compounds continued, other bioactive lipids were extracted from animal tissues. These include noladin ether (Hanuš et al., 2001), virodhamine (Porter et al., 2002), and *N*-arachidonoyldopamine (NADA) (Huang et al., 2002). However, the most widely investigated ECs are the AEA and 2-AG; levels of AEA in brain tissue have been shown to be about 170-fold lower than 2-AG (Stella et al., 1997).

ECs are unconventional neurotransmitters or neuromodulators because they are not stored in synaptic vesicles and are not released from presynaptic terminals via exocytotic mechanism. In fact, they are synthesized and released "on-demand" from cell membrane phospholipids depending on intracellular calcium. However, this conviction has been lately discussed because of the discovery of intracellular carriers and storage in specific reservoirs that might use as potential platforms for EC trafficking and accumulation (e.g., in adiposomes) (Kaczocha et al., 2010; Maccarrone et al., 2010; Oddi et al., 2008, 2009).

The biosynthesis of AEA and other NAEs include the enzymatic hydrolysis of the *N*-arachidonoylphosphatidylethanolamine (NArPE), which is formed from phosphatidylethanolamine by calcium-dependent *N*-acyl transferase and is then converted through at least four alternative metabolic pathways into AEA. The most studied route for such as conversion involves the *N*-acylphosphatidylethanolamine-specif phospholipase D (NAPE-PLD), into directly converts NArPE into AEA. Alternative yet relevant pathways engage α,β-Hydrolase-4 (ABHD4) and glycerophosphodiesterase-1 (GDE1), which catalyze the conversion of NArPE into lyso-NArPE first, then glycerophosphoanandamide and, finally, AEA. Formation of lyso-NArPE has also been suggested to occur through the action of a soluble phospholipase A2, followed by direct conversion into AEA by a lyso-phospholipase D. Finally, an as-yet-unidentified phospholipase C, followed by the action of phosphatases (such as protein tyrosine phosphatase N22 or SH2 domain-containing inositol phosphatase), converts NArPE first into phosphoanandamide and then AEA (Chiurchiù et al., 2015; Di Marzo, 2011).

On the other hand, the most important biosynthetic precursor of 2-AG are the sn-1acyl-2-arachidonoylglycerols (AArG), which are mostly produced by phospholipase Cβ (PLCβ) acting on membrane phosphatidylinositol, and then converted to 2-AG by the action of either of two isoforms of the same enzyme, the sn-1-diacylglycerol lipases α and β (DAGLα and DAGLβ) (Bisogno et al., 2003; Di Marzo, 2011). A less-characterized pathway for 2-AG involves the generation of 2-AG-3-phosphatase, which is an arachidonic acid-containing lysophosphatidic acid (Nakane et al., 2002). After their biosynthesis, AEA and 2-AG are immediately released into the extracellular medium by an unknown mechanism, but it is thought that is via the same putative membrane transporter proposed to facilitate its cellular uptake (Chicca et al., 2012; Ligresti et al., 2004). The movements of AEA and 2-AG across the cell membrane remain to be elucidated and are subject to ongoing controversy (Fowler, 2012). Into the cell, a group of proteins, the fatty acid-binding proteins (FABPs), are involved in AEA catabolism by transporting it to the FAAH enzymes (Elmes et al., 2015).

ECs are inactivated by intracellular hydrolysis. AEA is principally hydrolyzed to arachidonic acid and ethanolamine by FAAH, which also catalyzes the hydrolysis of other long-chain NAEs (McKinney & Cravatt, 2005). The hydrolysis of 2-AG to arachidonic acid and glycerol can be performed by FAAH (Goparaju et al., 1998), though its main hydrolase is a monoacylglycerol lipase (MAGL) in the mouse brain. Other enzymes, such as α/β-hydrolase domain (ABHD)-6 and ABHD-12 also hydrolyze to 2-AG (Blankman et al., 2007). ABHD-6 plays a key role in 2-AG hydrolysis in the microglial cell line BV-2 that does not express MAGL (Muccioli et al., 2007). Other enzymes, such as carboxylesterase 1 and 2, are involved in 2-AG hydrolysis in cultured human monocytic leukemia cell line (THP-1) (Xie et al., 2010).

Cannabinoid receptors CB_1 and CB_2

Although several proteins are target for cannabinoids, only CB_1 and CB_2 are the cannabinoid classified receptors (Pertwee et al., 2010). These receptors belong to the class A superfamily of GPCRs. Both CB_1 and CB_2 receptors primarily signal through the inhibitory G proteins G_i and G_o (Howlett et al., 2002). However, CB_1 receptor may interact with G_s and $G_{q/11}$ under certain conditions and with specific agonists (Mackie, 2005, 2008). CB_1 is negatively coupled to adenylate cyclase through $G_{i/o}$ proteins (Caulfield & Brown, 1992; Howlett et al., 2002). Activation of these receptors activates inward rectifying potassium channels and A-type potassium channels and inhibits N-type and P/Q type calcium channels (Demuth & Molleman, 2006). CB_2 receptor is also negatively coupled to adenylate cyclase but is not coupled to calcium channels (Felder et al., 1995). CB_1 and CB_2 receptors also act to regulate the phosphorylation and activation of p42/p44 mitogen-activated- protein kinase (MAPK), p38 MAPK, and Jun N terminal kinase (JNK) as signaling to regulate nuclear transcription factors related to cell proliferation, motility, adhesion, and apoptosis (Howlett, 2005). Besides G protein signaling, cannabinoid receptors have shown to interact with other effectors that regulate crucial downstream signal transduction. In this context, β-arrestin recruitment has lately gained a lot of attention in the cannabinoid field (Manning et al., 2021). Arrestins are scaffolding proteins recruited upon phosphorylation of the cytoplasmic domain of the receptor by G-protein-coupled receptor kinases (GRKs). They are well known by their role in G-protein-mediated signaling arrest via desensitization or internalization (Ferguson, 2001). More recently, their ability to trigger β-arrestin signaling through G protein-independent pathways has opened novel therapeutic avenues in the search of functionally selective cannabinoid modulators (Patel et al., 2021; Wouters et al., 2019). While CB_2 remains to be further explored, CB_1-β-arrestin1-mediated signaling has shown to regulate ERK1/2 phosphorylation, p38, JNK, or gene transcription (Wouters et al., 2019). Moreover, other cannabinoid-interacting partners such as the cannabinoid receptor interacting protein 1a (CRIP1a) (Fletcher-Jones et al., 2020) or the intracellular protein BiP (Costas-Insua et al., 2021) have been discovered but their therapeutic relevance has not been fully unraveled yet.

The CB_1 receptor is highly expressed in the allocortex, cerebellum, and associational cortical regions of the neocortex (Glass et al., 1997). Peripherally, CB_1 receptor was expressed in the dorsal horn (Sañudo-Peña et al., 1999) and dorsal root ganglion cells of heterogenous size (Hohmann & Herkenham, 1999). Also, in sympathetic nerves terminals (Ishac et al., 1996) and a very low levels in adrenal gland, thymus, heart, bone marrow, tonsils, prostate, uterus, ovary, and lung (Galiègue et al., 1995). Few studies have found CB_2 immunoreactivity expression in glial and neuronal cells in some areas of the rodent brain (Gong et al., 2006; Onaivi et al., 2006). Moreover, several studies have shown upexpression of the

TABLE 2 CB_1/CB_2 **receptor protein expression in some human peripheral blood immune cells.**

Immune cell type	CB_1/CB_2 receptors	References
B lymphocyte	CB_2	(Carayon et al., 1998; Galiègue et al., 1995; Graham et al., 2010)
T4 lymphocyte	CB_2	(Carayon et al., 1998; Galiègue et al., 1995; Graham et al., 2010)
T8 lymphocyte	CB_2	(Carayon et al., 1998; Galiègue et al., 1995; Graham et al., 2010)
Macrophages	CB_1/CB_2	(Ramirez et al., 2013)
Microglia	CB_2	(Klegeris et al., 2003)
Monocytes	CB_1/CB_2	(Castaneda et al., 2013; Galiègue et al., 1995; Graham et al., 2010)
Mast cells	CB_1/CB_2	(Samson et al., 2003)
Dendritic cells	CB_1 /CB_2	(Matias et al., 2002)
Natural Killer (NK) cells	CB_2	(Carayon et al., 1998; Galiègue et al., 1995; Graham et al., 2010)
Neutrophils	CB_2	(Galiègue et al., 1995; Graham et al., 2010)
Platelets	CB_1/CB_2	(Signorello et al., 2013)

CB_2 receptor in microglia and neurons with neural injury and inflammation (Beltramo & Piomelli, 2000; Hsieh et al., 2011). Clearly, CB_2 receptor is mostly expressed on cells of the immune system and the spleen (Marzo et al., 2004), where it is commonly associated with the regulation of different immune functions (Basu & Dittel, 2011). In a study using flow-cytometry, CB_2 protein levels from healthy human donors were analyzed, and it was found that natural killer (NK) cells, B lymphocytes, and monocytes express higher levels of CB_2 than CD4+ and CD8+ T lymphocytes or neutrophils. NK cells had the greatest variation in CB_2 expression levels, whereas for each of the other cell types levels was relatively similar between subjects (Graham et al., 2010).

Extensive literature has been focused on the exploration of the multiple pathophysiological actions of CB_1 and CB_2 receptors [revised in (Morales & Reggio, 2021)]. Both CB_1 and CB_2 are powerful negative modulators of immune processes as they are expressed in the main cell types (Table 2). It is possible that, at least a good part of the popularity of cannabinoid-based medication is due to the anti-inflammatory action by CB receptors. After all, most pathological phenomena include inflammation as an important cause of signs and symptoms during the start, progression, or culmination of the disease (e.g., cancer, epilepsy, obesity, etc.). Table 2 summarizes studies that have documented the presence of CB receptors in human immune cells at the protein levels.

Linking GPR55

Both AEA and 2-AG are considered endocannabinoids that activate CB_1/CB_2 cannabinoid receptors; AEA shows preferential affinity for CB_1 receptor *versus* CB_2 receptor (K_i=89 and 371 nM for CB_1 and CB_2 receptors, respectively) (Gauldie et al., 2001; Stella et al., 1997), whereas 2-AG is a full agonist at both receptors (Sugiura & Waku, 2000). Interestingly, both AEA and 2-AG were reported to activate GPR55 with higher potency than CB_1/CB_2 cannabinoid receptors (Table 3).

Since its discovery, GPR55 receptor has been reported to be involved in several actions in the CNS (Marichal-Cancino et al., 2017). GPR55 stimulation in dorsal striatum seems to facilitate the formation of procedural memories in rats, whereas its pharmacological blockade impaired the efficacy of rats to solve the T-maze (Marichal-Cancino et al., 2016). Hippocampally, GPR55 modulates glutamate release (Sylantyev et al., 2013), and it is associated with hippocampal synaptic plasticity (Hurst et al., 2017) and memory formation mechanisms (Marichal-Cancino et al., 2018). In the periacueductal gray (PAG), GPR55 has pronociceptive functions and its stimulation produced anxiolytic-like actions in alcohol pre-exposed rats, whereas its blockade induced aggressive defensive behaviors (Vázquez-León, Miranda-Páez, Calvillo-Robledo, & Marichal-Cancino, 2021). The PAG structure has been associated with reinforcing-guided behaviors related with food and drugs of abuse (Vázquez-León, Miranda-Páez, Chávez-Reyes, et al., 2021). Interestingly, in a very recent report, Liu et al. (Liu et al., 2021) showed high expression of GPR55 in the nucleus accumbens (NAc), a key brain region in the brain reward system and reported as a possible preventive action of nicotine reinforcing effects in the conditional place

TABLE 3 Reported functionality of cannabinoids and related molecules at GPR55, CB_1, and CB_2 receptors.

Ligand	GPR55	CB_1	CB_2
Endogenous ligands			
AEA	7.7[a]pEC50	7.5[a] pEC50	7.5[a] pEC50
2-AG	8.5[a] pEC50	6.3[a]pEC50	6.2[a] pEC50
Noladin ether	8.1[a] pEC50	7.5[a] pEC50	<4.5[a]pEC50
Virhodamide	7.9[a] pEC50	5.5[a] pEC50	6.4[a] pEC50
PEA	8.4[a]pEC50	<4.5[a] pEC50	<4.5[a] pEC50
LPI	5.9[b]–7.3[c] pEC50	<4.5[b] pEC50	N.D.
Phytocannabinoids and related compounds			
Δ^9-THC	8.1[a] pEC50	8.2[a] pEC50	9.4[a] pEC50
Abnormal cannabidiol	5.5–8.6[a]pEC50	<4.5[a] pEC50	<4.5[a] pEC50
Cannabidiol	6.3[a]pIC50	<4.5[a]pEC50	<4.5[a]pEC50
Synthetic chemotypes			
AM251	5.5[d]–7.4[a] pEC50	7.5[b] pEC50	5.5[a] pEC50
Rimonabant	5.5[d] pEC50	7.9[e]–8.6[d]pIC50	N.D.
CID16020046	6.1–6.7[f]pIC50	N.D.	N.D.
ML-193	6.7[g] pIC50	4.6[g] pIC50	N.D.

AEA, anandamide; *PEA*, palmitoylethanolamide; *N.D.*, not determined.
[a]*Data taken from Ryberg et al. (2007).*
[b]*Data taken from Kapur et al. (2009).*
[c]*Data taken from Henstridge et al. (2010).*
[d]*Data taken from Brown et al. (2011).*
[e]*Data taken from Felder et al. (1995).*
[f]*Data taken from Kargl et al. (2013).*
[g]*Data taken from Heynen-Genel et al. (2010).*

preference paradigm with the systemic administration of O-1602 (a GPR55 receptor agonist). Thus, the participation of GPR55 in the brain reward system represents an issue that deserves further investigation. Specially, considering that substances of abuse such the Δ^9-THC and palmitoylethanolamide (PEA) (which is a component of some medicaments) are full GPR55 agonists (Table 3). Knowledge about the neurobiology of GPR55 has constantly increased in the last years demonstrating its important role in the CNS of mammals. To support the above notion, García-Gutiérrez et al. (2018) reported a possible link between suicide victims and low gene expression of GPR55 in the prefrontal cortex. GPR55 is expressed in the striatum, hippocampus, forebrain, cortex and cerebellum (Wu et al., 2013), while at the peripheral level, it is found abundantly in the gut (Ryberg et al., 2007) and found in cells of the immune system, including microglia in brain (Henstridge et al., 2011) as well as in osteoclasts and osteoblasts (Whyte et al., 2009) and other tissues such as the pancreatic beta cells where it regulates the insulin secretion (Ramirez-Orozco et al., 2019). In the vasculature, GPR55 is expressed on endothelial cells and its activation drives to nitric oxide production. Hence, GPR55 is a member of the vasorelaxant mechanisms that produce hypotension and vasodilatation (Marichal-Cancino et al., 2013) and may be activated by PEA and other cannabinoid-related molecules (Marichal-Cancino et al., 2020). Hence, GPR55 is a ubiquitous receptor linked with a plethora of physiological actions (like CB_1 receptors). Moreover, it is interesting to note that GPR55 is able to form heterodimers with CB_1 (Martínez-Pinilla et al., 2020) and CB_2 (Balenga et al., 2014) receptors. Although their pathophysiological relevance remains to be further studied, this demonstrates the potential crosstalk and synergistic effects between GPR55 and the cannabinoid receptors.

One of the main facts against the cannabinoid nature of GPR55 is its activation by the non-cannabinoid lipid LPI (Yamashita et al., 2013). However, the lack of consistent in vivo data forced to retain its orphan status (Alexander & Irving, 2019).

Considerations on GPR55 as a potential therapeutic target: Pathophysiologic and toxicologic issues

The clinical potential that GPR55 could have as a therapeutic target remains eclipsed by its involvement in important issues included: (i) its proliferative and protumoral activity (ii) its possible participation during the neurodevelopment, and (iii) its poorly explored participation in the brain reward system and suicide behaviors.

Cancer

GPR55 has been linked with proliferative and protumoral actions. In fact, there is a GPR55/LPI axis that appears to be significant in the cancer disease (Falasca & Ferro, 2016) (see below). Whereas a biased endocannabinoids/GPR55 could exert antitumoral actions (Akimov et al., 2021). Thus, any GPR55-based medication should be carefully developed. The GPR55/LPI axis has been reported as overexpressed in different types of cancer including breast (Andradas et al., 2011; Ford et al., 2010), prostate (Piñeiro et al., 2011), glioblastoma (Andradas et al., 2011), colon (Kargl et al., 2016), ovarian (Hofmann et al., 2015), and skin cancer (Pérez-Gómez et al., 2013), among others (Andradas et al., 2011). Therefore, this receptor and its endogenous lipid have been proposed as potential biomarkers for specific types of cancer. Importantly, via a biased agonism, its stimulation by LPI seems to be responsible for the procancerogenic actions (e.g., in breast cancer cells) (Zhou et al., 2018), whereas its activation by endocannabinoids such as AEA and *N*-arachidonoyl dopamine could drive to protective anticancer effects (Akimov et al., 2021) via mechanisms that include Ca^{2+} increase and overstimulation of the neuronal nitric oxide synthase (nNOS) (Akimov et al., 2017; Zhou et al., 2018).

Neuronal development

Interestingly, GPR55 could have an important function during the neurodevelopment. To support the above notion, stimulation of GPR55 by LPI produced the retraction of neurites in a mechanism that requires the RhoA signaling pathway, which in turn suggested the involvement of a G-coupled protein $_{13}$ (G_{13}) (Obara et al., 2011). Nevertheless, the mutant ice that lack of GPR55 exerted normal behavior in multiple tasks with only minor alterations in motor skills (Wu et al., 2013). Thus, both the physiological and pharmacological roles of GPR55 require further investigation.

Brain reward system and destructive behaviors

The participation of GPR55 in emotional behaviors is limited to some observations regarding its modulatory role in anxiety. Nevertheless, a recent study by Liu et al. (Liu et al., 2021) reported a high expression of GPR55 in the accumbens nucleus. Moreover, the possible stimulation of GPR55 by systemic O-1602 seems to prevent the reinforcing actions of substances of abuse such as nicotine. To support the above observation, pharmacological manipulation of central GPR55 also prevented the reinforcing actions of nicotine in the conditioning place preference paradigm (Díaz-Barba et al., 2022). As Δ^9-THC is a full agonist at GPR55 (Table 3), its participation in the marihuana intake phenomena remains to be elucidated. If an important participation of GPR55 in the reward brain system modulation is confirmed, the dogma of CB_1 receptors as responsible of the whole reinforcing actions of marihuana must be revisited.

On the other hand, the report by García Gutierrez et al. (García-Gutiérrez et al., 2018) about the possible participation of GPR55 in the neurobiology of suicide was highly disturbing. According with this report, victims of suicide had lower levels of GPR55 in the dorsolateral prefrontal cortex (García-Gutiérrez et al., 2018). Moreover, Vázquez-León, Miranda-Páez, Calvillo-Robledo, and Marichal-Cancino (2021) reported that blockade of GPR55 in periaqueductal gray of alcohol-pre-exposed rats during alcoholic abstinence developed aggressive behaviors.

Conclusions

GPR55 is target of lipid mediators including some endocannabinoids and LPI. CB_1-GPR55 and CB_2-GPR55 heterodimers are expressed into the CNS adding complexity to the understanding of this cannabinoid-related receptor. Some limitations for its potential as a therapeutic target include its pro-cancerogenic actions when the GPR55/LPI axis is overexpressed, its possible link with the brain reward system and its involvement during the neurodevelopment. Clearly, these limitations could be used also in the clinic once we have unraveled its complex pathophysiology. Currently, its applications may include the modulation of anxiety, pain, inflammation, and insulin secretion. The ability of numerous endo-, phyto-, and synthetic cannabinoids to modulate GPR55 clearly shows its relevance in the search of cannabinoid-based medicines.

Indeed, pharmacological characterization of any novel cannabinoid should require its assessment at GPR55. Independently of its classification as the CB_3 or not, GPR55 is a receptor highly related with the ECS which will help us to understand the functions of this fantastic and ancestral system.

Applications to other areas

In this chapter, we have reviewed the main physiological processes that involve GPR55 receptors. Moreover, it was discussed the great interaction that exists between cannabinoids and GPR55 receptors and the potential risks and therapeutic reach of these receptors. This information should motivate further research focused on decoding the specific participation of GPR55 in the brain reward system and human neurodevelopment. But also, it should provide a global view of the GPR55 in the context of the endocannabinoid system and the cannabis-based medicine.

Mini-dictionary of terms

- **GPR55 heterodimers.** The GPR55 receptor is an integral protein embedded in the cell membrane that has been reported to conform transmembrane adhesion with other protein receptors included CB_1 and CB_2.
- **Cosmic ethanolamine.** The detection of ethanolamine in the cosmos without mediation of living systems represents an advance in the understanding of life materials. Endocannabinoids such as AEA has ethanolamine as a structural molecule. Thus, this discovery may be interesting for those which study the origin of lipid mediators.
- **CB_3.** A cannabinoid receptor type 3 (CB_3) has been expected for a while. The GPR55 was proposed as the potential CB_3. But its activation by lysophophatidylinositol (LPI), a non-cannabinoid lipid, obstructed its cannabinoid classification.
- **GPR55/LPI axis.** In some types of cancers, both GPR55 expression and LPI levels are high. Thus, this axis has been proposed as a cancer biomarker.

Key facts of GPR55

- GPR55 is a cellular receptor that seems to be exclusive to mammals.
- The cannabinoid receptors and GPR55 are closely related both functional and pharmacologically.
- Even thought still classified as orphan by the IUPHAR, the bioactive lipid LPI has been proposed as GPR55 main endogenous ligand. GPR55 is a potential target for developing novel therapeutics.
- GPR55 is linked to proliferative and pro-cancerogenic actions. Thus, the development of pharmacological tools that interact at GPR55 should be cautious.

Summary points

- GPR55 interacts with cannabinoids and LPI. The interaction with cannabinoids requires further in vivo investigation, whereas a GPR55/LPI axis is overexpressed in some type of cancers.
- The relationship between the cannabinoid receptors and GPR55 is not related to their genetic homology (which is low). GPR55 forms heterodimers with CB_1 and CB_2 cannabinoid receptors, but also it seems to be involved in similar actions in the body (e.g., modulations of cardiovascular, metabolic, immune, and cognitive phenomena).
- Classification of GPR55 as a cannabinoid receptor requires further pharmacological in vivo reports. Currently, the endogen ligand of GPR55 seems to be LPI.
- GPR55 may be target of phytocannabinoids such as Δ^9-THC and medicaments which contains PEA. However, the repercussions of this interaction remain to be analyzed.

References

Akimov, M. G., Ashba, A. M., Gretskaya, N. M., & Bezuglov, V. V. (2017). N-acyl dopamines induce apoptosis in PC12 cell line via the GPR55 receptor activation. *Doklady. Biochemistry and Biophysics*, *474*(1), 155–158. https://doi.org/10.1134/s1607672917030012.

Akimov, M. G., Gamisonia, A. M., Dudina, P. V., Gretskaya, N. M., Gaydaryova, A. A., Kuznetsov, A. S., & Bezuglov, V. V. (2021). GPR55 receptor activation by the N-acyl dopamine family lipids induces apoptosis in cancer cells via the nitric oxide synthase (nNOS) over-stimulation. *International Journal of Molecular Sciences*, *22*(2). https://doi.org/10.3390/ijms22020622.

Alexander, S., & Irving, A. (2019). GPR18, GPR55 and GPR119 (version 2019.4) in the IUPHAR/BPS guide to pharmacology database. In *IUPHAR/BPS guide to pharmacology CITE, 2019*. https://doi.org/10.2218/gtopdb/F114/2019.4.

Andradas, C., Caffarel, M. M., Pérez-Gómez, E., Salazar, M., Lorente, M., Velasco, G., & Sánchez, C. (2011). The orphan G protein-coupled receptor GPR55 promotes cancer cell proliferation via ERK. *Oncogene*, *30*(2), 245–252. https://doi.org/10.1038/onc.2010.402.

Aziz, M., & Chapman, K. D. (2020). Fatty acid amide hydrolases: An expanded capacity for chemical communication? *Trends in Plant Science*, *25*(3), 236–249. https://doi.org/10.1016/j.tplants.2019.11.002.

Balenga, N. A., Martínez-Pinilla, E., Kargl, J., Schröder, R., Peinhaupt, M., Platzer, W., & Franco, R. (2014). Heteromerization of GPR55 and cannabinoid CB2 receptors modulates signalling. *British Journal of Pharmacology*, *171*(23), 5387–5406. https://doi.org/10.1111/bph.12850.

Basu, S., & Dittel, B. N. (2011). Unraveling the complexities of cannabinoid receptor 2 (CB2) immune regulation in health and disease. *Immunologic Research*, *51*(1), 26–38. https://doi.org/10.1007/s12026-011-8210-5.

Beltramo, M., & Piomelli, D. (2000). Carrier-mediated transport and enzymatic hydrolysis of the endogenous cannabinoid 2-arachidonylglycerol. *Neuroreport*, *11*(6). Retrieved from: https://journals.lww.com/neuroreport/Fulltext/2000/04270/Carrier_mediated_transport_and_enzymatic.18.aspx.

Bisogno, T., Howell, F., Williams, G., Minassi, A., Cascio, M. G., Ligresti, A., & Doherty, P. (2003). Cloning of the first sn1-DAG lipases points to the spatial and temporal regulation of endocannabinoid signaling in the brain. *Journal of Cell Biology*, *163*(3), 463–468. https://doi.org/10.1083/jcb.200305129.

Blankman, J. L., Simon, G. M., & Cravatt, B. F. (2007). A comprehensive profile of brain enzymes that hydrolyze the endocannabinoid 2-arachidonoylglycerol. *Chemistry & Biology*, *14*(12), 1347–1356. https://doi.org/10.1016/j.chembiol.2007.11.006.

Brown, A. J., Daniels, D. A., Kassim, M., Brown, S., Haslam, C. P., Terrell, V. R., & Dowell, S. J. (2011). Pharmacology of GPR55 in yeast and identification of GSK494581A as a mixed-activity glycine transporter subtype 1 inhibitor and GPR55 agonist. *The Journal of Pharmacology and Experimental Therapeutics*, *337*(1), 236–246. https://doi.org/10.1124/jpet.110.172650.

Carayon, P., Marchand, J., Dussossoy, D., Derocq, J. M., Jbilo, O., Bord, A., & Casellas, P. (1998). Modulation and functional involvement of CB2 peripheral cannabinoid receptors during B-cell differentiation. *Blood*, *92*(10), 3605–3615.

Castaneda, J. T., Harui, A., Kiertscher, S. M., Roth, J. D., & Roth, M. D. (2013). Differential expression of intracellular and extracellular CB2 cannabinoid receptor protein by human peripheral blood leukocytes. *Journal of Neuroimmune Pharmacology*, *8*(1), 323–332. https://doi.org/10.1007/s11481-012-9430-8.

Caulfield, M. P., & Brown, D. A. (1992). Cannabinoid receptor agonists inhibit Ca current in NG108–15 neuroblastoma cells via a Pertussis toxin-sensitive mechanism. *British Journal of Pharmacology*, *106*(2), 231–232. https://doi.org/10.1111/j.1476-5381.1992.tb14321.x.

Chicca, A., Marazzi, J., Nicolussi, S., & Gertsch, J. (2012). Evidence for bidirectional endocannabinoid transport across cell membranes. *Journal of Biological Chemistry*, *287*(41), 34660–34682. https://doi.org/10.1074/jbc.M112.373241.

Chiurchiù, V., Battistini, L., & Maccarrone, M. (2015). Endocannabinoid signalling in innate and adaptive immunity. *Immunology*, *144*(3), 352–364. https://doi.org/10.1111/imm.12441.

Costas-Insua, C., Moreno, E., Maroto, I. B., Ruiz-Calvo, A., Bajo-Grañeras, R., Martín-Gutiérrez, D., & Guzmán, M. (2021). Identification of BiP as a CB_1 receptor-interacting protein that fine-tunes cannabinoid signaling in the mouse brain. *The Journal of Neuroscience*, *41*(38), 7924. https://doi.org/10.1523/JNEUROSCI.0821-21.2021.

Crean, R. D., Crane, N. A., & Mason, B. J. (2011). An evidence based review of acute and long-term effects of cannabis use on executive cognitive functions. *Journal of Addiction Medicine*, *5*(1), 1–8. https://doi.org/10.1097/ADM.0b013e31820c23fa.

Demuth, D. G., & Molleman, A. (2006). Cannabinoid signalling. *Life Sciences*, *78*(6), 549–563. https://doi.org/10.1016/j.lfs.2005.05.055.

Devane, W. A., Hanus, L., Breuer, A., Pertwee, R. G., Stevenson, L. A., Griffin, G., & Mechoulam, R. (1992). Isolation and structure of a brain constituent that binds to the cannabinoid receptor. *Science*, *258*(5090), 1946–1949. https://doi.org/10.1126/science.1470919.

Di Marzo, V. (2011). Endocannabinoid signaling in the brain: Biosynthetic mechanisms in the limelight. *Nature Neuroscience*, *14*(1), 9–15. https://doi.org/10.1038/nn.2720.

Díaz-Barba, A., Calvillo-Robledo, A., Vázquez-León, P., Gallegos-Vieyra, E., Quintanar, J. L., & Marichal-Cancino, B. A. (2022). Central GPR55 may prevent nicotine reinforcing actions: a preliminary study. *Acta Neurobiologiae Experimentalis*, *82*(3), 304–314. https://doi.org/10.55782/ane-2022-029.

Elmes, M. W., Kaczocha, M., Berger, W. T., Leung, K., Ralph, B. P., Wang, L., & Deutsch, D. G. (2015). Fatty acid-binding proteins (FABPs) are intracellular carriers for Δ9-tetrahydrocannabinol (THC) and cannabidiol (CBD). *The Journal of Biological Chemistry*, *290*(14), 8711–8721. https://doi.org/10.1074/jbc.M114.618447.

Falasca, M., & Ferro, R. (2016). Role of the lysophosphatidylinositol/GPR55 axis in cancer. *Advances in Biological Regulation*, *60*, 88–93. https://doi.org/10.1016/j.jbior.2015.10.003.

Felder, C. C., Joyce, K. E., Briley, E. M., Mansouri, J., Mackie, K., Blond, O., & Mitchell, R. L. (1995). Comparison of the pharmacology and signal transduction of the human cannabinoid CB1 and CB2 receptors. *Molecular Pharmacology*, *48*(3), 443–450.

Ferguson, S. S. G. (2001). Evolving concepts in G protein-coupled receptor endocytosis: The role in receptor desensitization and signaling. *Pharmacological Reviews*, *53*(1), 1. Retrieved from: http://pharmrev.aspetjournals.org/content/53/1/1.abstract.

Fletcher-Jones, A., Hildick, K. L., Evans, A. J., Nakamura, Y., Henley, J. M., & Wilkinson, K. A. (2020). Protein interactors and trafficking pathways that regulate the cannabinoid type 1 receptor (CB1R). *Frontiers in Molecular Neuroscience*, *13*. https://doi.org/10.3389/fnmol.2020.00108.

Ford, L. A., Roelofs, A. J., Anavi-Goffer, S., Mowat, L., Simpson, D. G., Irving, A. J., & Ross, R. A. (2010). A role for L-alpha-lysophosphatidylinositol and GPR55 in the modulation of migration, orientation and polarization of human breast cancer cells. *British Journal of Pharmacology*, *160*(3), 762–771. https://doi.org/10.1111/j.1476-5381.2010.00743.x.

Fowler, C. J. (2012). Anandamide uptake explained? *Trends in Pharmacological Sciences*, *33*(4), 181–185. https://doi.org/10.1016/j.tips.2012.01.001.

Galiègue, S., Mary, S., Marchand, J., Dussossoy, D., Carrière, D., Carayon, P., & Casellas, P. (1995). Expression of central and peripheral cannabinoid receptors in human immune tissues and leukocyte subpopulations. *European Journal of Biochemistry*, *232*(1), 54–61. https://doi.org/10.1111/j.1432-1033.1995.tb20780.x.

Gaoni, Y., & Mechoulam, R. (1964). Isolation, structure, and partial synthesis of an active constituent of Hashish. *Journal of the American Chemical Society*, *86*(8), 1646–1647. https://doi.org/10.1021/ja01062a046.

García-Gutiérrez, M. S., Navarrete, F., Navarro, G., Reyes-Resina, I., Franco, R., Lanciego, J. L., & Manzanares, J. (2018). Alterations in gene and protein expression of cannabinoid CB(2) and GPR55 receptors in the dorsolateral prefrontal cortex of suicide victims. *Neurotherapeutics: The journal of the American Society for Experimental NeuroTherapeutics*, *15*(3), 796–806. https://doi.org/10.1007/s13311-018-0610-y.

Gauldie, S. D., McQueen, D. S., Pertwee, R., & Chessell, I. P. (2001). Anandamide activates peripheral nociceptors in normal and arthritic rat knee joints. *British Journal of Pharmacology*, *132*(3), 617–621. https://doi.org/10.1038/sj.bjp.0703890.

Glass, M., Dragunow, M., & Faull, R. L. (1997). Cannabinoid receptors in the human brain: A detailed anatomical and quantitative autoradiographic study in the fetal, neonatal and adult human brain. *Neuroscience*, *77*(2), 299–318. https://doi.org/10.1016/s0306-4522(96)00428-9.

Gong, J.-P., Onaivi, E. S., Ishiguro, H., Liu, Q.-R., Tagliaferro, P. A., Brusco, A., & Uhl, G. R. (2006). Cannabinoid CB2 receptors: Immunohistochemical localization in rat brain. *Brain Research*, *1071*(1), 10–23. https://doi.org/10.1016/j.brainres.2005.11.035.

Goparaju, S. K., Ueda, N., Yamaguchi, H., & Yamamoto, S. (1998). Anandamide amidohydrolase reacting with 2-arachidonoylglycerol, another cannabinoid receptor ligand. *FEBS Letters*, *422*(1), 69–73. https://doi.org/10.1016/S0014-5793(97)01603-7.

Graham, E. S., Angel, C. E., Schwarcz, L. E., Dunbar, P. R., & Glass, M. (2010). Detailed characterisation of CB2 receptor protein expression in peripheral blood immune cells from healthy human volunteers using flow cytometry. *International Journal of Immunopathology and Pharmacology*, *23*(1), 25–34. https://doi.org/10.1177/039463201002300103.

Hanuš, L., Abu-Lafi, S., Fride, E., Breuer, A., Vogel, Z., Shalev, D. E., & Mechoulam, R. (2001). 2-Arachidonyl glyceryl ether, an endogenous agonist of the cannabinoid CB_1 receptor. *Proceedings of the National Academy of Sciences*, *98*(7), 3662. https://doi.org/10.1073/pnas.061029898.

Henstridge, C. M., Balenga, N. A., Kargl, J., Andradas, C., Brown, A. J., Irving, A., & Waldhoer, M. (2011). Minireview: Recent developments in the physiology and pathology of the lysophosphatidylinositol-sensitive receptor GPR55. *Molecular Endocrinology*, *25*(11), 1835–1848. https://doi.org/10.1210/me.2011-1197.

Henstridge, C. M., Balenga, N. A., Schröder, R., Kargl, J. K., Platzer, W., Martini, L., & Irving, A. J. (2010). GPR55 ligands promote receptor coupling to multiple signalling pathways. *British Journal of Pharmacology*, *160*(3), 604–614. https://doi.org/10.1111/j.1476-5381.2009.00625.x.

Heynen-Genel, S., Dahl, R., Shi, S., Milan, L., Hariharan, S., Sergienko, E., & Abood, M. E. (2010). Screening for selective ligands for GPR55—Antagonists. In *Probe reports from the NIH molecular libraries program*. Bethesda (MD): National Center for Biotechnology Information (US).

Hofmann, N. A., Yang, J., Trauger, S. A., Nakayama, H., Huang, L., Strunk, D., & Graier, W. F. (2015). The GPR 55 agonist, L-α-lysophosphatidylinositol, mediates ovarian carcinoma cell-induced angiogenesis. *British Journal of Pharmacology*, *172*(16), 4107–4118. https://doi.org/10.1111/bph.13196.

Hohmann, A. G., & Herkenham, M. (1999). Localization of central cannabinoid CB1 receptor messenger RNA in neuronal subpopulations of rat dorsal root ganglia: A double-label in situ hybridization study. *Neuroscience*, *90*(3), 923–931. https://doi.org/10.1016/S0306-4522(98)00524-7.

Howlett, A. C. (2005). Cannabinoid receptor signaling. In *Vol. 168*. *Handbook of experimental pharmacology* (pp. 53–79). https://doi.org/10.1007/3-540-26573-2_2.

Howlett, A. C., Barth, F., Bonner, T. I., Cabral, G., Casellas, P., Devane, W. A., & Pertwee, R. G. (2002). International Union of Pharmacology. XXVII. Classification of cannabinoid receptors. *Pharmacological Reviews*, *54*(2), 161. https://doi.org/10.1124/pr.54.2.161.

Hsieh, G. C., Pai, M., Chandran, P., Hooker, B. A., Zhu, C. Z., Salyers, A. K., & Meyer, M. D. (2011). Central and peripheral sites of action for CB2 receptor mediated analgesic activity in chronic inflammatory and neuropathic pain models in rats. *British Journal of Pharmacology*, *162*(2), 428–440. https://doi.org/10.1111/j.1476-5381.2010.01046.x.

Huang, S. M., Bisogno, T., Trevisani, M., Al-Hayani, A., De Petrocellis, L., Fezza, F., & Di Marzo, V. (2002). An endogenous capsaicin-like substance with high potency at recombinant and native vanilloid VR1 receptors. *Proceedings of the National Academy of Sciences*, *99*(12), 8400. https://doi.org/10.1073/pnas.122196999.

Hurst, K., Badgley, C., Ellsworth, T., Bell, S., Friend, L., Prince, B., & Edwards, J. G. (2017). A putative lysophosphatidylinositol receptor GPR55 modulates hippocampal synaptic plasticity. *Hippocampus*, *27*(9), 985–998. https://doi.org/10.1002/hipo.22747.

Ishac, E. J. N., Jiang, L., Lake, K. D., Varga, K., Abood, M. E., & Kunos, G. (1996). Inhibition of exocytotic noradrenaline release by presynaptic cannabinoid CB1 receptors on peripheral sympathetic nerves. *British Journal of Pharmacology*, *118*(8), 2023–2028. https://doi.org/10.1111/j.1476-5381.1996.tb15639.x.

Kaczocha, M., Glaser, S. T., Chae, J., Brown, D. A., & Deutsch, D. G. (2010). Lipid droplets are novel sites of N-acylethanolamine inactivation by fatty acid amide hydrolase-2. *Journal of Biological Chemistry*, *285*(4), 2796–2806. https://doi.org/10.1074/jbc.M109.058461.

Kapur, A., Zhao, P., Sharir, H., Bai, Y., Caron, M. G., Barak, L. S., & Abood, M. E. (2009). Atypical responsiveness of the orphan receptor GPR55 to cannabinoid ligands. *The Journal of Biological Chemistry*, *284*(43), 29817–29827. https://doi.org/10.1074/jbc.M109.050187.

Kargl, J., Andersen, L., Hasenöhrl, C., Feuersinger, D., Stančić, A., Fauland, A., & Schicho, R. (2016). GPR55 promotes migration and adhesion of colon cancer cells indicating a role in metastasis. *British Journal of Pharmacology*, *173*(1), 142–154. https://doi.org/10.1111/bph.13345.

Kargl, J., Brown, A. J., Andersen, L., Dorn, G., Schicho, R., Waldhoer, M., & Heinemann, A. (2013). A selective antagonist reveals a potential role of G protein-coupled receptor 55 in platelet and endothelial cell function. *The Journal of Pharmacology and Experimental Therapeutics*, *346*(1), 54–66. https://doi.org/10.1124/jpet.113.204180.

Klegeris, A., Bissonnette, C. J., & McGeer, P. L. (2003). Reduction of human monocytic cell neurotoxicity and cytokine secretion by ligands of the cannabinoid-type CB2 receptor. *British Journal of Pharmacology*, *139*(4), 775–786. https://doi.org/10.1038/sj.bjp.0705304.

Ligresti, A., Morera, E., van der Stelt, M., Monory, K., Lutz, B., Ortar, G., & Di Marzo, V. (2004). Further evidence for the existence of a specific process for the membrane transport of anandamide. *Biochemical Journal*, *380*(1), 265–272. https://doi.org/10.1042/bj20031812.

Liu, Q., Yu, J., Li, X., Guo, Y., Sun, T., Luo, L., & Yang, Q. (2021). Cannabinoid receptor GPR55 activation blocks nicotine use disorder by regulation of AMPAR phosphorylation. *Psychopharmacology*, *238*(11), 3335–3346. https://doi.org/10.1007/s00213-021-05949-x.

Maccarrone, M., Dainese, E., & Oddi, S. (2010). Intracellular trafficking of anandamide: New concepts for signaling. *Trends in Biochemical Sciences*, *35*(11), 601–608. https://doi.org/10.1016/j.tibs.2010.05.008.

Mackie, K. (2005). Cannabinoid receptor homo- and heterodimerization. *Life Sciences*, *77*(14), 1667–1673. https://doi.org/10.1016/j.lfs.2005.05.011.

Mackie, K. (2008). Cannabinoid receptors: Where they are and what they do. *Journal of Neuroendocrinology*, *20*(s1), 10–14. https://doi.org/10.1111/j.1365-2826.2008.01671.x.

Manning, J. J., Green, H. M., Glass, M., & Finlay, D. B. (2021). Pharmacological selection of cannabinoid receptor effectors: Signalling, allosteric modulation and bias. *Neuropharmacology*, *193*, 108611. https://doi.org/10.1016/j.neuropharm.2021.108611.

Marichal-Cancino, B. A., Fajardo-Valdez, A., Ruiz-Contreras, A. E., Mendez-Diaz, M., & Prospero-Garcia, O. (2017). Advances in the physiology of GPR55 in the central nervous system. *Current Neuropharmacology*, *15*(5), 771–778. https://doi.org/10.2174/1570159X14666160729155441.

Marichal-Cancino, B. A., Fajardo-Valdez, A., Ruiz-Contreras, A. E., Mendez-Diaz, M., & Prospero-Garcia, O. (2018). Possible role of hippocampal GPR55 in spatial learning and memory in rats. *Acta Neurobiologiae Experimentalis (Wars)*, *78*(1), 41–50. Retrieved from: https://www.ncbi.nlm.nih.gov/pubmed/29694340.

Marichal-Cancino, B. A., González-Hernández, A., MaassenVanDenBrink, A., Ramírez-San Juan, E., & Villalón, C. M. (2020). Potential mechanisms involved in palmitoylethanolamide-induced vasodepressor effects in rats. *Journal of Vascular Research*. https://doi.org/10.1159/000506158.

Marichal-Cancino, B. A., Manrique-Maldonado, G., Altamirano-Espinoza, A. H., Ruiz-Salinas, I., Gonzalez-Hernandez, A., Maassenvandenbrink, A., & Villalon, C. M. (2013). Analysis of anandamide- and lysophosphatidylinositol-induced inhibition of the vasopressor responses produced by sympathetic stimulation or noradrenaline in pithed rats. *European Journal of Pharmacology*, *721*(1-3), 168–177. https://doi.org/10.1016/j.ejphar.2013.09.039.

Marichal-Cancino, B. A., Sanchez-Fuentes, A., Mendez-Diaz, M., Ruiz-Contreras, A. E., & Prospero-Garcia, O. (2016). Blockade of GPR55 in the dorsolateral striatum impairs performance of rats in a T-maze paradigm. *Behavioural Pharmacology*, *27*(4), 393–396. https://doi.org/10.1097/FBP.0000000000000185.

Martínez-Pinilla, E., Rico, A. J., Rivas-Santisteban, R., Lillo, J., Roda, E., Navarro, G., & Lanciego, J. L. (2020). Expression of cannabinoid CB(1) R-GPR55 heteromers in neuronal subtypes of the Macaca fascicularis striatum. *Annals of the New York Academy of Sciences*, *1475*(1), 34–42. https://doi.org/10.1111/nyas.14413.

Marzo, V. D., Bifulco, M., & Petrocellis, L. D. (2004). The endocannabinoid system and its therapeutic exploitation. *Nature Reviews Drug Discovery*, *3*(9), 771–784. https://doi.org/10.1038/nrd1495.

Matias, I., Pochard, P., Orlando, P., Salzet, M., Pestel, J., & Di Marzo, V. (2002). Presence and regulation of the endocannabinoid system in human dendritic cells. *European Journal of Biochemistry*, *269*(15), 3771–3778. https://doi.org/10.1046/j.1432-1033.2002.03078.x.

Matsuda, L. A., Lolait, S. J., Brownstein, M. J., Young, A. C., & Bonner, T. I. (1990). Structure of a cannabinoid receptor and functional expression of the cloned cDNA. *Nature*, *346*(6284), 561–564. https://doi.org/10.1038/346561a0.

McKinney, M. K., & Cravatt, B. F. (2005). Structure and function of fatty acid amide hydrolase. *Annual Review of Biochemistry*, *74*, 411–432. https://doi.org/10.1146/annurev.biochem.74.082803.133450.

McPartland, J. M., Matias, I., Di Marzo, V., & Glass, M. (2006). Evolutionary origins of the endocannabinoid system. *Gene*, *370*, 64–74. https://doi.org/10.1016/j.gene.2005.11.004.

Mechoulam, R., Ben-Shabat, S., Hanus, L., Ligumsky, M., Kaminski, N. E., Schatz, A. R., & Vogel, Z. (1995). Identification of an endogenous 2-monoglyceride, present in canine gut, that binds to cannabinoid receptors. *Biochemical Pharmacology*, *50*(1), 83–90. https://doi.org/10.1016/0006-2952(95)00109-D.

Morales, P., & Reggio, P. H. (2021). Emerging roles of cannabinoids and synthetic cannabinoids in clinical experimental models. *Advances in Experimental Medicine and Biology*, *1264*, 47–65. https://doi.org/10.1007/978-3-030-57369-0_4.

Muccioli, G. G., Xu, C., Odah, E., Cudaback, E., Cisneros, J. A., Lambert, D. M., & Stella, N. (2007). Identification of a novel endocannabinoid-hydrolyzing enzyme expressed by microglial cells. *The Journal of Neuroscience*, *27*(11), 2883. https://doi.org/10.1523/JNEUROSCI.4830-06.2007.

Munro, S., Thomas, K. L., & Abu-Shaar, M. (1993). Molecular characterization of a peripheral receptor for cannabinoids. *Nature*, *365*(6441), 61–65. https://doi.org/10.1038/365061a0.

Nakane, S., Oka, S., Arai, S., Waku, K., Ishima, Y., Tokumura, A., & Sugiura, T. (2002). 2-Arachidonoyl-sn-glycero-3-phosphate, an arachidonic acid-containing lysophosphatidic acid: Occurrence and rapid enzymatic conversion to 2-arachidonoyl-sn-glycerol, a cannabinoid receptor ligand, in rat brain. *Archives of Biochemistry and Biophysics*, *402*(1), 51–58. https://doi.org/10.1016/S0003-9861(02)00038-3.

Obara, Y., Ueno, S., Yanagihata, Y., & Nakahata, N. (2011). Lysophosphatidylinositol causes neurite retraction via GPR55, G13 and RhoA in PC12 cells. *PLoS One*, *6*(8), e24284. https://doi.org/10.1371/journal.pone.0024284.

Oddi, S., Fezza, F., Pasquariello, N., D'Agostino, A., Catanzaro, G., De Simone, C., & Maccarrone, M. (2009). Molecular identification of albumin and Hsp70 as cytosolic anandamide-binding proteins. *Chemistry & Biology*, *16*(6), 624–632. https://doi.org/10.1016/j.chembiol.2009.05.004.

Oddi, S., Fezza, F., Pasquariello, N., De Simone, C., Rapino, C., Dainese, E., & Maccarrone, M. (2008). Evidence for the intracellular accumulation of anandamide in adiposomes. *Cellular and Molecular Life Sciences*, *65*(5), 840. https://doi.org/10.1007/s00018-008-7494-7.

Oka, S., Nakajima, K., Yamashita, A., Kishimoto, S., & Sugiura, T. (2007). Identification of GPR55 as a lysophosphatidylinositol receptor. *Biochemical and Biophysical Research Communications*, *362*(4), 928–934. https://doi.org/10.1016/j.bbrc.2007.08.078.

Onaivi, E. S., Ishiguro, H., Gong, J.-P., Patel, S., Perchuk, A., Meozzi, P. A., & Uhl, G. R. (2006). Discovery of the presence and functional expression of cannabinoid CB2 receptors in brain. *Annals of the New York Academy of Sciences*, *1074*(1), 514–536. https://doi.org/10.1196/annals.1369.052.

Patel, M., Finlay, D. B., & Glass, M. (2021). Biased agonism at the cannabinoid receptors—Evidence from synthetic cannabinoid receptor agonists. *Cellular Signalling*, *78*, 109865. https://doi.org/10.1016/j.cellsig.2020.109865.

Pérez-Gómez, E., Andradas, C., Flores, J. M., Quintanilla, M., Paramio, J. M., Guzmán, M., & Sánchez, C. (2013). The orphan receptor GPR55 drives skin carcinogenesis and is upregulated in human squamous cell carcinomas. *Oncogene*, *32*(20), 2534–2542. https://doi.org/10.1038/onc.2012.278.

Pertwee, R. G., Howlett, A. C., Abood, M. E., Alexander, S. P. H., Di Marzo, V., Elphick, M. R., & Ross, R. A. (2010). International Union of Basic and Clinical Pharmacology. LXXIX. Cannabinoid receptors and their ligands: Beyond CB_1 and CB_2. *Pharmacological Reviews*, *62*(4), 588–631. https://doi.org/10.1124/pr.110.003004.

Piñeiro, R., Maffucci, T., & Falasca, M. (2011). The putative cannabinoid receptor GPR55 defines a novel autocrine loop in cancer cell proliferation. *Oncogene*, *30*(2), 142–152. https://doi.org/10.1038/onc.2010.417.

Porter, A. C., Sauer, J.-M., Knierman, M. D., Becker, G. W., Berna, M. J., Bao, J., & Felder, C. C. (2002). Characterization of a novel endocannabinoid, virodhamine, with antagonist activity at the CB1 receptor. *Journal of Pharmacology and Experimental Therapeutics*, *301*(3), 1020. https://doi.org/10.1124/jpet.301.3.1020.

Ramirez, S. H., Reichenbach, N. L., Fan, S., Rom, S., Merkel, S. F., Wang, X., & Persidsky, Y. (2013). Attenuation of HIV-1 replication in macrophages by cannabinoid receptor 2 agonists. *Journal of Leukocyte Biology*, *93*(5), 801–810. https://doi.org/10.1189/jlb.1012523.

Ramirez-Orozco, R. E., Garcia-Ruiz, R., Morales, P., Villalon, C. M., Villafan-Bernal, J. R., & Marichal-Cancino, B. A. (2019). Potential metabolic and behavioural roles of the putative endocannabinoid receptors GPR18, GPR55 and GPR119 in feeding. *Current Neuropharmacology*. https://doi.org/10.2174/1570159x17666190118143014.

Ranganathan, M., & D'Souza, D. C. (2006). The acute effects of cannabinoids on memory in humans: A review. *Psychopharmacology*, *188*(4), 425–444. https://doi.org/10.1007/s00213-006-0508-y.

Rivilla, V. M., Jiménez-Serra, I., Martín-Pintado, J., Briones, C., Rodríguez-Almeida, L. F., Rico-Villas, F., & Requena-Torres, M. A. (2021). Discovery in space of ethanolamine, the simplest phospholipid head group. *Proceedings of the National Academy of Sciences*, *118*(22), e2101314118. https://doi.org/10.1073/pnas.2101314118.

Ryberg, E., Larsson, N., Sjögren, S., Hjorth, S., Hermansson, N. O., Leonova, J., & Greasley, P. J. (2007). The orphan receptor GPR55 is a novel cannabinoid receptor. *British Journal of Pharmacology*, *152*(7), 1092–1101. https://doi.org/10.1038/sj.bjp.0707460.

Samson, M.-T., Small-Howard, A., Shimoda, L. M. N., Koblan-Huberson, M., Stokes, A. J., & Turner, H. (2003). Differential roles of CB1 and CB2 cannabinoid receptors in mast cells. *The Journal of Immunology*, *170*(10), 4953. https://doi.org/10.4049/jimmunol.170.10.4953.

Sañudo-Peña, M. C., Strangman, N. M., Mackie, K., Walker, J. M., & Tsou, K. (1999). CB1 receptor localization in rat spinal cord and roots, dorsal root ganglion, and peripheral nerve. *Zhongguo Yao Li Xue Bao*, *20*(12), 1115–1120.

Signorello, M. G., Giacobbe, E., Passalacqua, M., & Leoncini, G. (2013). The 2-arachidonoylglycerol effect on myosin light chain phosphorylation in human platelets. *Biochimie*, *95*(8), 1620–1628. https://doi.org/10.1016/j.biochi.2013.05.003.

Stella, N., Schweitzer, P., & Piomelli, D. (1997). A second endogenous cannabinoid that modulates long-term potentiation. *Nature*, *388*(6644), 773–778. https://doi.org/10.1038/42015.

Sugiura, T., Kondo, S., Sukagawa, A., Nakane, S., Shinoda, A., Itoh, K., & Waku, K. (1995). 2-Arachidonoylgylcerol: A possible endogenous cannabinoid receptor ligand in brain. *Biochemical and Biophysical Research Communications*, *215*(1), 89–97. https://doi.org/10.1006/bbrc.1995.2437.

Sugiura, T., & Waku, K. (2000). 2-Arachidonoylglycerol and the cannabinoid receptors. *Chemistry and Physics of Lipids*, *108*(1), 89–106. https://doi.org/10.1016/S0009-3084(00)00189-4.

Sylantyev, S., Jensen, T. P., Ross, R. A., & Rusakov, D. A. (2013). Cannabinoid- and lysophosphatidylinositol-sensitive receptor GPR55 boosts neurotransmitter release at central synapses. *Proceedings of the National Academy of Sciences of the United States of America*, *110*(13), 5193–5198. https://doi.org/10.1073/pnas.1211204110.

Vázquez-León, P., Miranda-Páez, A., Calvillo-Robledo, A., & Marichal-Cancino, B. A. (2021). Blockade of GPR55 in dorsal periaqueductal gray produces anxiety-like behaviors and evocates defensive aggressive responses in alcohol-pre-exposed rats. *Neuroscience Letters*, 136218. https://doi.org/10.1016/j.neulet.2021.136218.

Vázquez-León, P., Miranda-Páez, A., Chávez-Reyes, J., Allende, G., Barragán-Iglesias, P., & Marichal-Cancino, B. A. (2021). The periaqueductal gray and its extended participation in drug addiction phenomena. *Neuroscience Bulletin*. https://doi.org/10.1007/s12264-021-00756-y.

Whyte, L. S., Ryberg, E., Sims, N. A., Ridge, S. A., Mackie, K., Greasley, P. J., & Rogers, M. J. (2009). The putative cannabinoid receptor GPR55 affects osteoclast function in vitro and bone mass in vivo. *Proceedings of the National Academy of Sciences of the United States of America*, *106*(38), 16511–16516. https://doi.org/10.1073/pnas.0902743106.

Wouters, E., Walraed, J., Banister, S. D., & Stove, C. P. (2019). Insights into biased signaling at cannabinoid receptors: Synthetic cannabinoid receptor agonists. *Biochemical Pharmacology*, *169*, 113623. https://doi.org/10.1016/j.bcp.2019.08.025.

Wu, C. S., Chen, H., Sun, H., Zhu, J., Jew, C. P., Wager-Miller, J., & Lu, H. C. (2013). GPR55, a G-protein coupled receptor for lysophosphatidylinositol, plays a role in motor coordination. *PLoS One*, *8*(4), e60314. https://doi.org/10.1371/journal.pone.0060314.

Xie, S., Borazjani, A., Hatfield, M. J., Edwards, C. C., Potter, P. M., & Ross, M. K. (2010). Inactivation of lipid glyceryl ester metabolism in human THP1 monocytes/macrophages by activated organophosphorus insecticides: Role of carboxylesterases 1 and 2. *Chemical Research in Toxicology*, *23*(12), 1890–1904. https://doi.org/10.1021/tx1002194.

Yamashita, A., Oka, S., Tanikawa, T., Hayashi, Y., Nemoto-Sasaki, Y., & Sugiura, T. (2013). The actions and metabolism of lysophosphatidylinositol, an endogenous agonist for GPR55. *Prostaglandins & Other Lipid Mediators*, *107*, 103–116. https://doi.org/10.1016/j.prostaglandins.2013.05.004.

Zhou, X. L., Guo, X., Song, Y. P., Zhu, C. Y., & Zou, W. (2018). The LPI/GPR55 axis enhances human breast cancer cell migration via HBXIP and p-MLC signaling. *Acta Pharmacologica Sinica*, *39*(3), 459–471. https://doi.org/10.1038/aps.2017.157.

Chapter 30

Proteomic analysis of cannabinoids in human oligodendrocyte cells

Valéria de Almeida[a,b]

[a]*Laboratory of Neuroproteomics, Department of Biochemistry and Tissue Biology, Institute of Biology, University of Campinas (UNICAMP), Campinas, Brazil,* [b]*Department of Psychiatry and Psychotherapy, University Hospital, LMU Munich, Munich, Germany*

Abbreviations

THC	Δ^9-tetrahydrocannabinol
2-AG	2-arachidonoylglycerol
ABHD6	alpha/beta-hydrolase domain-containing 6
AEA	anandamide
Akt	serine/threonine-protein kinase
CB1	cannabinoid receptor, type 1
CB2	cannabinoid receptor, type 2
CBD	cannabidiol
CNS	central nervous system
CYP51A1	cytochrome P450 51A1
D2	dopamine receptor, type 2
DAGLA	diacylglycerol lipase alpha
DAGLB	diacylglycerol lipase beta
DEPs	differentially expressed proteins
EAE	experimental autoimmune encephalomyelitis
EIF2	eukaryotic initiation factor 2
FAAH	fatty acid amide hydrolase
FDR	false discovery rate
IPA	ingenuity pathway analysis
iPSC	induced stem cell
LC	liquid chromatography
MAGL	monoacylglycerol lipase
MAPK	mitogen-activated protein kinase
MBP	myelin-binding protein
MS	multiple sclerosis
mTOR	mammalian target of the rapamycin
mTORC1	mammalian target of the rapamycin complex 1
NAPE-PLD	*N*-acyl phosphatidylethanolamine phospholipase D
OPC	oligodendrocyte precursor cell
PIP3	phosphatidylinositol (3,4,5)-trisphosphate
TTRUST	transcriptional regulatory relationships unraveled by sentence-based text mining
TRPV1	transient receptor potential cation channel, subfamily V, member 1

Oligodendrocytes

Myelin ensures the fast and efficient transmission of electrical signals through axonal traits, provides energy support, and contributes to axonal integrity in the central nervous system (CNS) (Fünfschilling et al., 2012). Damage or loss of oligodendrocytes is involved in the pathology of demyelination diseases. Despite multiple sclerosis (MS) being the most well-known demyelinating disease, others, such as schizophrenia and bipolar disorders (Davis & Haroutunian, 2003), depression

Cannabis Use, Neurobiology, Psychology, and Treatment. https://doi.org/10.1016/B978-0-323-89862-1.00047-7

(Miyata et al., 2016), and Alzheimer's disease (McAleese et al., 2017) present white matter deficits due to oligodendrocyte dysfunctions. To investigate myelination, demyelination, and remyelinating processes, studies rely on in vitro (primary oligodendrocyte precursor cells (OPCs) and, immortalized oligodendroglial cells lines (e.g., MO.313), ex vivo (oligodendrocytes obtained from organotypic brain slices), and animal models (transgenic mouse, transgenic zebrafish, experimental autoimmune encephalomyelitis (EAE) and cuprizone-induced toxicity models). Recently, with the advance of induced stem cell (iPSC) technology, derived-OPCs or oligodendrocytes from patients and controls have been used to investigate oligodendrocyte biology (Kuhn et al., 2019).

Due to the importance of myelination in the brain, an increasing interest in the drugs able to repair myelin-related pathology has been observed. Mature or myelinating oligodendrocytes are generated from OPCs after proliferation, migration, and differentiation steps (Kuhn et al., 2019). Multiple intrinsic and extrinsic factors, including the endocannabinoids, drive these three steps of oligodendrocyte biology (Gomez et al., 2011; Molina-Holgado et al., 2002; Solbrig et al., 2010; Tomas-Roig et al., 2016). Thus, pharmacological modulators of the endocannabinoid signaling seem to represent a promising therapeutic tool to target proliferation, migration, and differentiation of OPCs, which can repair the loss of myelin observed in some diseases. However, efforts still are needed to comprehend how the endocannabinoid signaling and its pharmacological modulators can affect oligodendrocytes and myelin. Therefore, this chapter shows that proteomic analysis is an approach that can open new hypotheses and lead to the discovery of new targets.

Proteomics

Proteomics relies mainly on mass spectrometry analysis to evaluate the complexity of a proteome, which can indicate the biological processes and molecular function affected by a given sample under specific conditions, for instance, control vs. treatment, healthy vs. disease states, and others (Aebersold & Mann, 2003). In particular, the mass spectrometry-based proteomics associated with bioinformatics tools has been valuable to study of cell biology (Walther & Mann, 2010). Briefly, a shotgun proteomics workflow with cells consists of protein extraction, protein digestion with trypsin, separation, and ionization of peptides by liquid chromatography coupled to mass spectrometry, data process of the spectra for quantification and identification of proteins, statistical and bioinformatics analyses.

Taking into account that quantitative proteomics analysis determines the relative abundance of proteins in two different conditions, for instance, treatment vs. control, it has an important contribution to cell biology. Recently, we published proteome data (de Almeida et al., 2022)—Proteomic data available at ProteomeXchange with the identifier: PXD031923) of the effect of cannabinoids on the human oligodendrocyte cell line (MO3.13). This cell line has been used to investigate the characteristics of immature oligodendrocytes or OPCs (Iwata et al., 2013). Since we found that CB1 and CB2 receptors are expressed on MO3.13 cells (de Almeida et al., 2022), we used it as a model to evaluate the effects of cannabinoid. The four main pharmacological classes were investigated in our previous study: endocannabinoids (*N*-arachidonoyl-ethanolamine (anandamide, AEA) and 2-arachidonoylglycerol (2-AG)), CB1 and CB2 agonists (WIN55,212-2, ACEA, and HU308), CB1 and CB2 antagonists (AM251 and AM630), and the phytocannabinoid (cannabidiol—CBD). Despite our published papers showing a detailed analysis, for this chapter, we depicted a new analysis and compared the main findings with recent data from the literature.

Endocannabinoids

Gomez and colleagues showed the expression of enzymes involved in the synthesis (diacylglycerol lipase alpha (DAGLα) and beta (DAGLβ)) and degradation (monoacylglycerol lipase (MAGL)) of 2-AG in OPCs and mature oligodendrocytes (Gomez et al., 2010, 2015). Reinforcing these findings, a recent study demonstrated the expression of the 2-AG synthesizing and degrading enzymes in the spinal cord and cerebral cortex-derived OPCs and mature oligodendrocytes (Moreno-Luna et al., 2021). Several studies have shown that the treatment with 2-AG or with its metabolizing enzyme alpha/beta-hydrolase domain-containing 6 (ABHD6) has a key role in oligodendrocyte biology and therapeutic effect on the demyelination process both in vitro and in vivo experiments (Alpár et al., 2014; Arévalo-Martín et al., 2007; Bernal-Chico et al., 2015; Feliú et al., 2017; Ilyasov et al., 2018; Manterola et al., 2018; Molina-Holgado et al., 2002). In sum, evidence shows the potential of 2-AG on demyelination-related diseases. Contributing with the literature, we showed that 2-AG modulated proteins associated with other disorders that also present myelin pathology processes, such as amyotrophic lateral sclerosis, motor neuron, and neurodegenerative diseases (Fig. 1).

Notably, 2-AG is the most studied endocannabinoid in oligodendrocyte biology. However, studies should not exclude other endocannabinoids, such as AEA, since we observed that AEA had similar effects on the proteome (Fig. 2A) and the canonical pathways (Fig. 2B) compared to 2-AG treatment on MO3.13 oligodendrocytes. Clinical evidence also pointed to

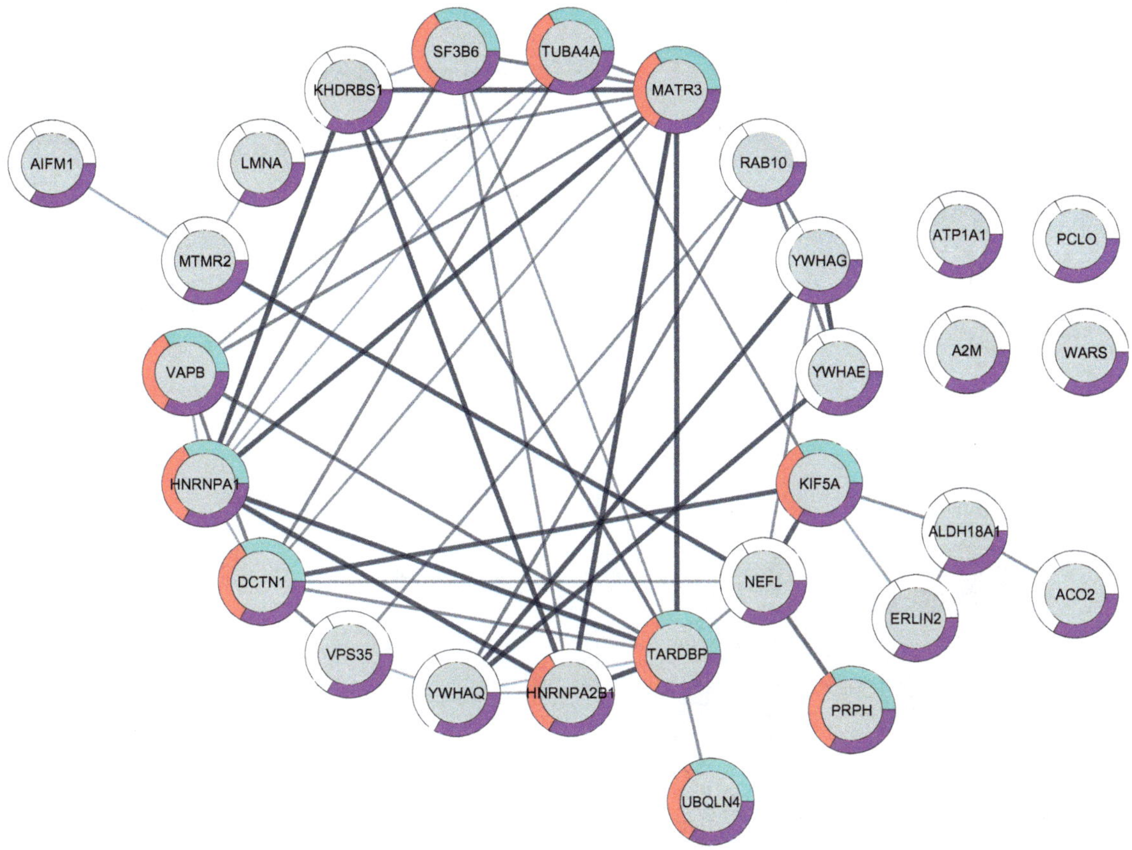

FIG. 1 Differentially expressed proteins associated with diseases that were modulated by 2-AG on MO3.13 cells. The inner circle shows the protein name. The outer ring segments represent the diseases: Amyotrophic lateral sclerosis in *green* (p-value=2.03E10−7), motor neuron disease in *orange* (p-value=8.08E10−7), and neurodegenerative disease in *purple*, p-value=2.48E10−6). Created in Cytoscape with the data available (de Almeida et al., 2022, Proteomic data available at ProteomeXchange with the identifier PXD031923).

the role of AEA in MS. These studies showed alterations in AEA levels and its enzyme FAAH in patients with MS (Centonze et al., 2007; Di Filippo et al., 2008; Jean-Gilles et al., 2009), reinforcing the importance of investigations of AEA treatment on myelin disorders.

Despite some similarities between AEA e 2-AG effects, our data pointed out a different profile of transcriptional regulatory circuitry according to the changes on the proteome (Transcriptional Regulatory Relationships Unraveled by Sentence-based Text mining tool (TTRUST)—Fig. 2C) (Han et al., 2018). Among those transcription factors predicted only for 2-AG, we highlight that histone deacetylase 1 (HDAC1) has a key role in oligodendrocyte differentiation (Egawa et al., 2019; Ye et al., 2009). Thus, the exploratory proteomic data may result in a new hypothesis about the effect of 2-AG and AEA in the biology of oligodendrocytes.

CB1 and CB2 agonists

The role of synthetic agonists of CB1 and CB2 receptors on the differentiation of OPCs was demonstrated by Gomez and colleagues (Gomez et al., 2011). The main mechanisms of ACEA (CB1 agonist), JWH133 (CB2 agonist), and HU210 (CB1/CB2 agonist) on the oligodendrocyte maturation were associated with the phosphorylation of Akt and activation of mTOR signaling. In contrast, according to our proteome data, the agonist CB2 (HU308) had no significant effect on MO3.13 oligodendrocytes, which leads us to suggest that the time of treatment with CB2 agonists (8 hours in our study and 48 h in Gomez's study) may impact the role of the CB2 signaling. Other differences between the experimental designs may explain the discrepancies in the results. On the another hand, we found that CB1 agonist (ACEA) and CB1/CB2 agonist (WIN55,212-2) induced a significant effect on the proteome of oligodendrocytes.

The potential effect of WIN55,212-2 on OPC differentiation and myelination was demonstrated both in vivo and in vitro studies (Molina-Holgado et al., 2002; Solbrig et al., 2010; Sun et al., 2013; Tomas-Roig et al., 2016; Tomas-Roig et al.,

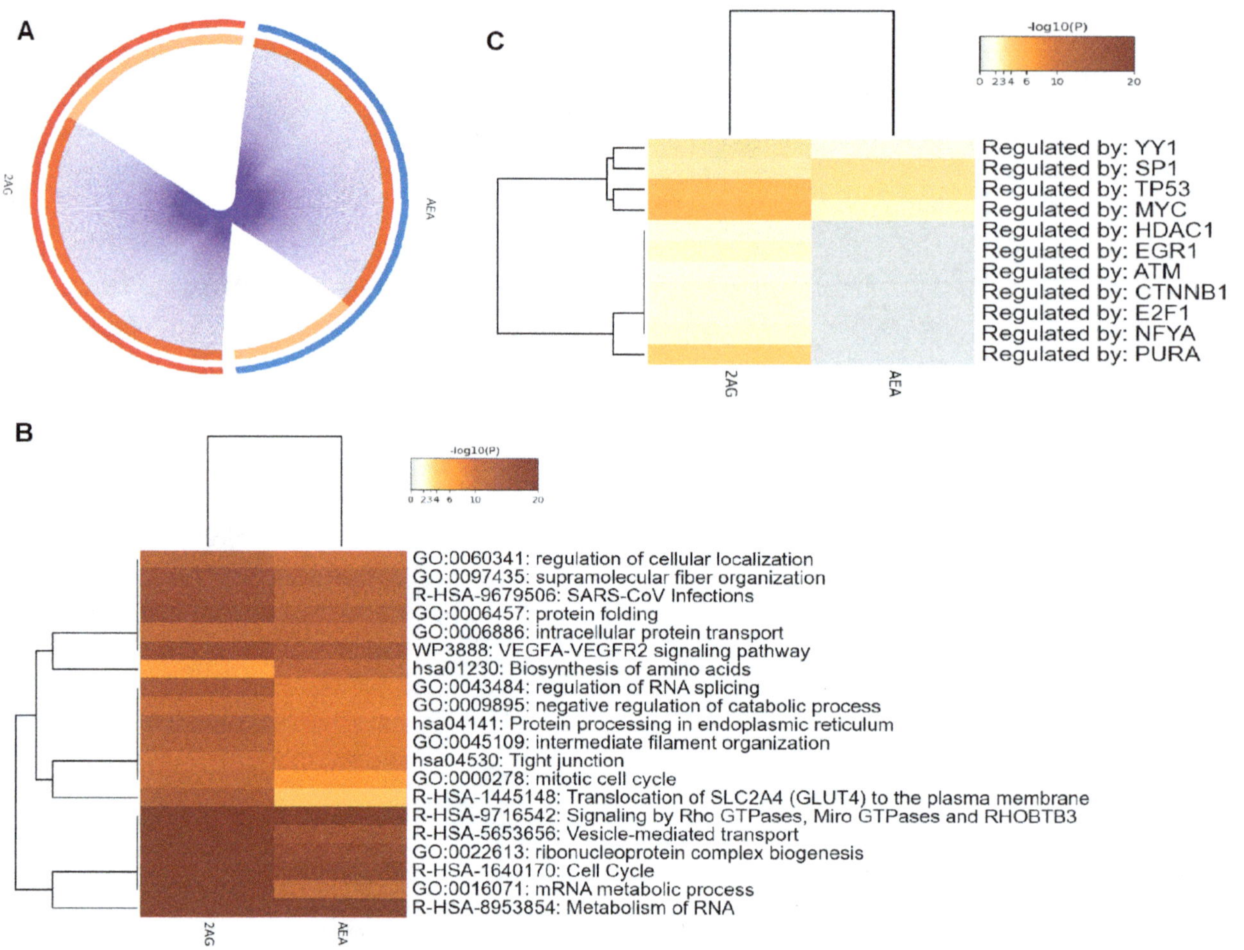

FIG. 2 (A) Overlap between protein lists of AEA and 2-AG treatment on MO3.13 oligodendrocytes (data available at PXD031923). *Purple* lines indicate shared genes between treatments. The outer circle indicates the number of proteins modulated by each treatment compared to controls. The inner-circle indicates shared (in *dark orange*) and unique proteins to each treatment (*light orange*). Analysis and figures generated from Metascape (Zhou et al., 2019). (B) Heatmap of enriched terms of diverse database and (C) Heatmap of enrichment analysis for transcriptor factors of TRRUST of MO3.13 cells treated with AEA or 2-AG compared to control. Created in Metascape with the data available (de Almeida et al., 2022, Proteomic data available at ProteomeXchange with the identifier PXD031923).

2020). These studies have also shown that activation of cannabinoid receptors on oligodendrocytes occurs due to activation of PI3K/Akt and the mTOR pathways. In contribution to these findings, we described the signaling pathways associated with WIN55,212-2 on oligodendrocyte cells (de Almeida et al., 2022). For this review, we highlight that WIN55,212-2 modulated several proteins associated with mitochondrial function (Fig. 3). Notably, the mitochondria have an important role in oligodendrocyte biology and myelin formation (Della-Flora Nunes et al., 2021; Rinholm et al., 2016), which leads us to reinforce the importance of new investigations to evaluate the therapeutic or toxic effect of this agonist and others on mitochondrial function. Despite the limitation of lacking functionality assays in our previous study, we found that WIN55,121-2 inhibited mitochondrial function signaling. As depicted in Fig. 3, WIN55,212-2 modulated the levels of mitochondria matrix and inner membrane proteins.

Regarding the selective CB1 agonist, few studies have shown the signaling pathways on oligodendrocytes. Similar to WIN55,21-2, ACEA promoted oligodendrocyte differentiation due to the PIP3/Akt/mTOR cascade (Gomez et al., 2011). In contrast, a study showed that ACEA did not induce the differentiation of neural stem cells (NSCs) into OPCs (Compagnucci et al., 2013). The cell type (OPCs vs NSCs) may explain the divergent results. Besides that, our analysis reported that ACEA-modulated proteins involved in key signaling pathways, for instance, EIF2, MAPK, protein kinase A signaling, and oxidative phosphorylation (de Almeida et al., 2022).

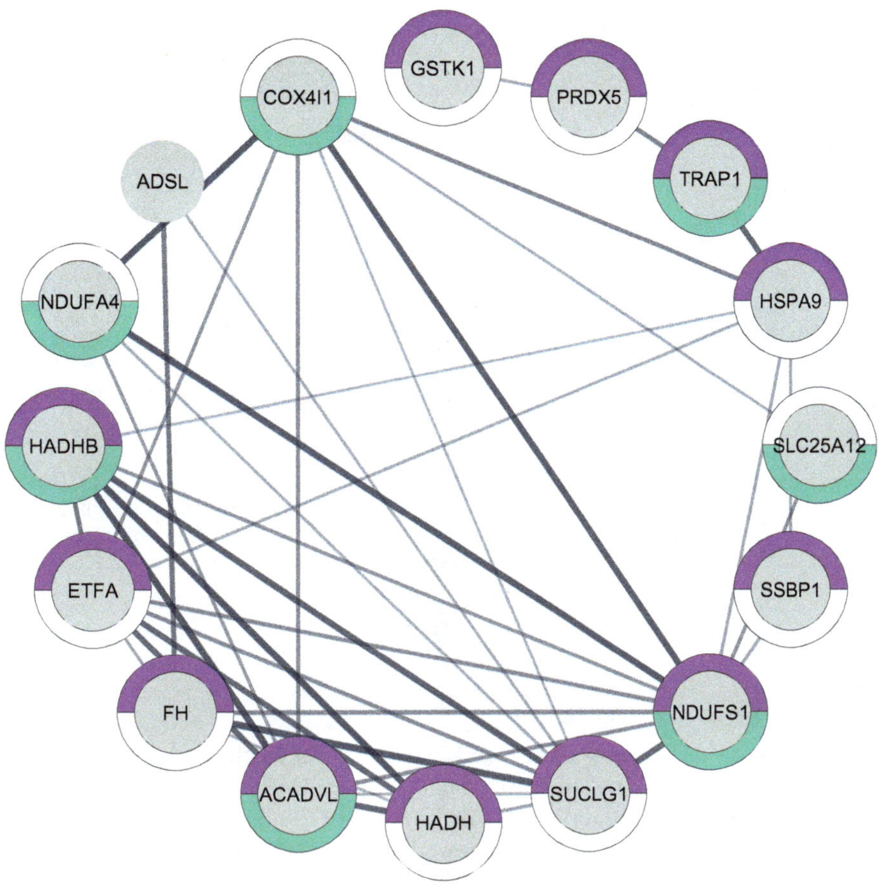

FIG. 3 Gene Ontology enrichment for cell compartment of proteins modulated by WIN55,212. The outer ring segments are colored according to the cell compartment of each protein: proteins of the mitochondrial matrix in purple (p-value=3.59E−06) and proteins of the mitochondrial inner membrane in green (p-value=1.9E−04). Created in Cytoscape with the data available (de Almeida et al., 2022, Proteomic data available at ProteomeXchange with the identifier PXD031923).

CB1 and CB2 antagonists

As reviewed, activation of cannabinoid receptors by endocannabinoids or exogenous agonists seems to play a critical role in oligodendrocyte proliferation, migration, and differentiation. Therefore, blocking the CB1 and CB2 receptors with antagonists may reveal important pathways of the basal endocannabinoid signaling in the oligodendrocyte biology. Several studies have evaluated the effect of antagonists to decipher the role of CB1 and CB2 receptors on myelination induced by exogenous, endogenous, or phytocannabinoids (Gomez et al., 2010; Huerga-Gómez et al., 2021). However, no data have shown the pharmacological or toxicity profile of the treatment with only antagonists on oligodendrocytes. In agreement with other studies, our previous data showed that AM251 and AM630 inhibited MAPK, PIP3/Akt, and mTOR signaling. Interestingly, with the proteomic data, we identified a significant inhibition of both antagonists in several signaling pathways that can play a role in oligodendrocyte biology. Besides those pathways described (de Almeida et al., 2022), here we highlighted some terms enriched, for instance, the amyotrophic lateral sclerosis, VERGFA-VEGFR2 signaling, and generation of precursor metabolites and energy, which are important to the oligodendrocyte function and pathological mechanisms involved in CNS diseases (Fig. 4B). Additionally, the antagonists of CB1 and CB2 receptors similarly induced proteome changes as can be observed in the overlap of the proteome changes (Fig. 4A) and the terms enrichment (Fig. 4B). As other studies have suggested, the crosstalk between CB1 and CB2 is possible in oligodendrocyte cells since the same effect was reported (Feliú et al., 2017; Gomez et al., 2011; Huerga-Gómez et al., 2021). Therefore, our data corroborate with those studies and can be further investigated for other groups (Proteomic data available at ProteomeXchange with the identifier PXD031923).

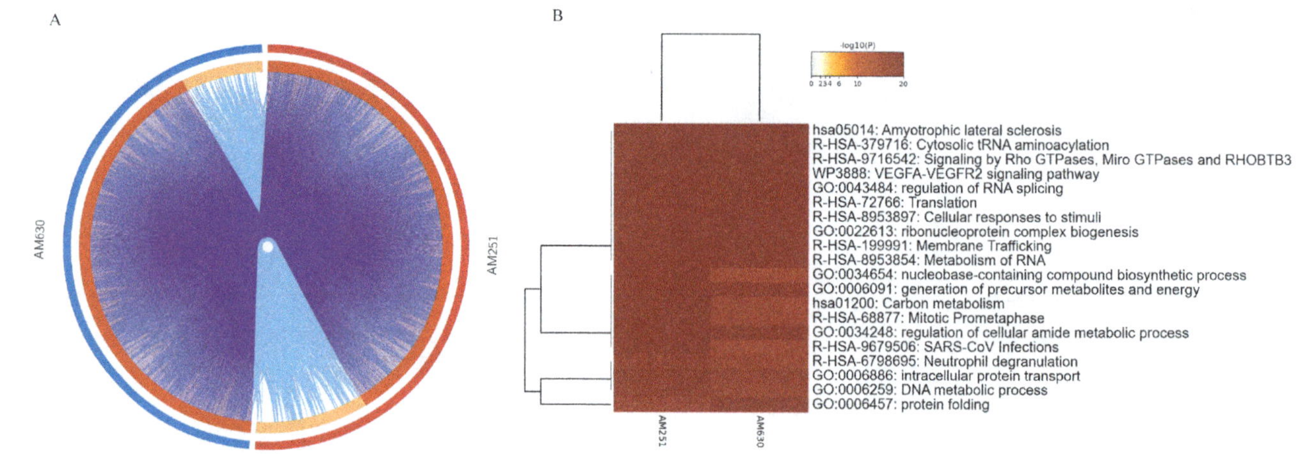

FIG. 4 (A) Overlap between protein lists, where *purple* curves link shared genes; the inner circle represents gene lists, where hits are arranged along the arc. Genes that hit both lists are colored in *dark orange*, and genes unique to a list are shown in *light orange*. *Blue* lines link the genes under the same ontology term. (B) Heatmap of enrichment analysis for transcriptor factors from TRRUST analysis of MO3.13 cells treated with AM251 or AM630 compared to control. Created in Metascape with the data available (de Almeida et al., 2022, Proteomic data available at ProteomeXchange with the identifier PXD031923).

Phytocannabinoids

The major two cannabinoids found in Cannabis—THC, and CBD—have been clinically used for myelin disorders, such as MS (Cristino et al., 2020). The therapeutic use of these cannabinoids on MS is the management of spasticity and neuropathic pain. However, clinical studies did not report mechanisms related to myelination due to limitations in studying the process in the human brain. Recently, preclinical studies showed that the THC increased myelination in vivo (Aguado et al., 2021; Huerga-Gómez et al., 2021). These studies have reported that THC induces oligodendrocyte differentiation by the activation of CB1 and the mammalian target of the rapamycin complex 1 (mTORC1) signaling pathway. Despite the promising result of THC on myelination, this phytocannabinoid has concerns to be considered, such as psychotomimetic property and effects on neurodevelopmental stages. In contrast, CBD presents a safety profile for several neurological and psychiatric disorders.

Evidence of the mechanism of CBD on oligodendrocytes has been reported. The protective effect of CBD against demyelination was observed in an animal model of MS (Rahimi et al., 2015) and hypomyelination in a model of neonatal hypoxia-ischemia in newborn rats (Ceprián et al., 2019). Moreover, in vitro study reported that CBD prevents apoptosis of OPCs (Mecha et al., 2012). However, Mato and colleagues showed the toxicity effect of CBD on optic nerve oligodendrocytes (Mato et al., 2010). Therefore, the molecular signaling and mechanisms specifically associated with the CBD effect on oligodendrocytes are not fully comprehended.

We have previously evaluated the effect of CBD on MO3.13 in two different experimental designs. First, we evaluated and compared the effect of CBD with drugs such as the antipsychotics haloperidol and clozapine and the muscarinic antagonist benztropine that present a protective effect in oligodendrocytes and myelination on a cuprizone-based toxicity model (Deshmukh et al., 2013; Patergnani et al., 2021). The main finding of our study is that CBD shares pathways with benztropine and clozapine, which leads us to hypothesize the protective effect of CBD on demyelination processes (Falvella et al., 2021). Second, investigating the effect of CBD on MO3.13 without cuprizone insult, we found glycolysis and EIF2 signaling were the main pathways affected by CBD (de Almeida et al., 2022). However, further investigations are demanded to confirm the potential of CBD in vivo myelinating models.

As well known, CBD has several pharmacological targets. Thus, we depicted the molecular function of proteins modulated by CBD on oligodendrocyte cells (Fig. 5). For instance, proteins associated with catalytic activity, such as the oxidation-reduction process and glycolysis, and with a structural role such as microtubule-associated proteins were modulated by CBD. These data suggest that CBD has different targets on oligodendrocytes as well known for other cell types (Ibeas Bih et al., 2015).

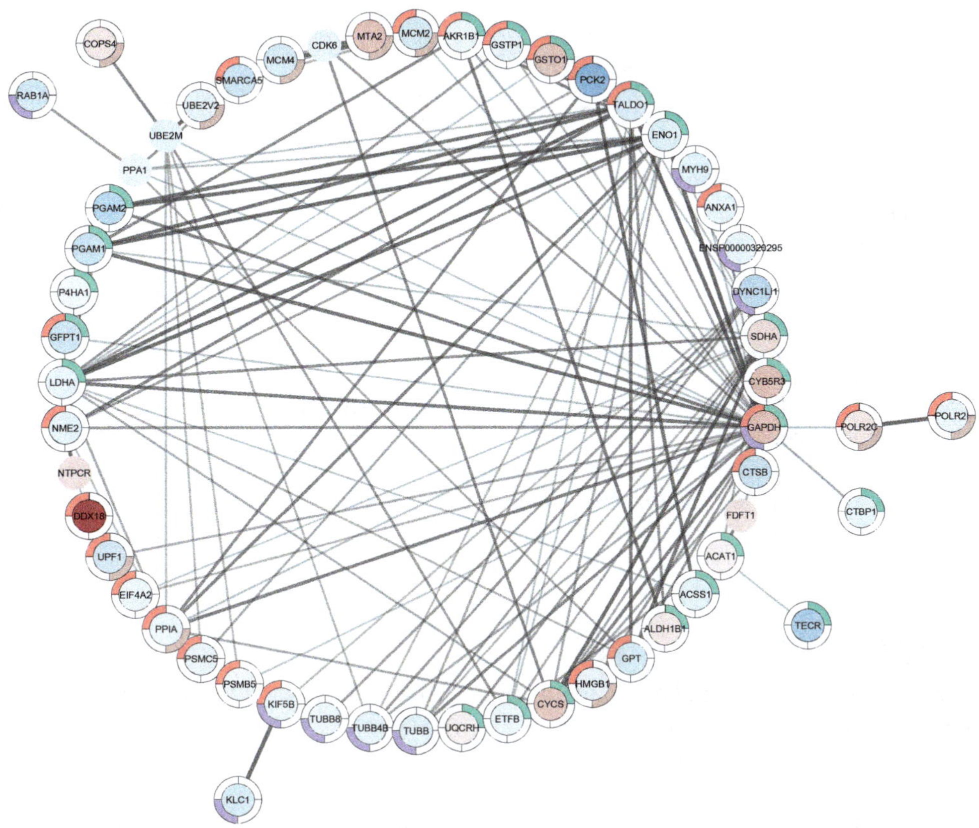

FIG. 5 Gene ontology enrichment for molecular function of proteins modulated by CBD. The inner circle contains the protein name and the colors *blue* and *red* represent the down- and upregulation of the protein by CBD, respectively. The outer ring segments are colored based on the molecular function of each protein: Oxidation-reduction process in green (p-value=2.11E−09), Cellular response to organic substance in *orange* (p-value=4.55E−05), Microtubule-based process in *purple* (p-value=0.012), and DNA metabolic process in *brown* (p-value=0.016). Created in Cytoscape with the data available (de Almeida et al., 2022, Proteomic data available at ProteomeXchange with the identifier PXD031923).

Conclusion

Proteomics can reveal, among other applications, novel targets for treatments. Data from literature reinforce the key role of the endocannabinoid signaling in oligodendrocytes. For this chapter, we have analyzed the proteomic data available and we encourage other researchers to use the proteomic data to investigate new targets in more complex models (Chanoumidou et al., 2020). In conclusion, according to literature, 2-AG and THC are the most potential myelinating drug, and CB1 receptor as main endocannabinoid target for signaling pathways associated with myelination. According to our proteomics, AEA and 2-AG modulated signaling pathways related to myelination (proliferation, migration, and differentiation of OPCs). Thus, AEA also should be considered as a target in further studies. Moreover, the antagonists for CB1 and CB2 receptor had a significant effect on the proteome of the MO3.13 cells, reinforcing the role of these receptors for the biology of the OPCs and OLs. Despite the therapeutic aspect discussed here, further studies should clarify the toxicicity of cannabinoid modulators on the biology of oligodendrocytes.

Applications to other areas

This chapter describes the potential role of endocannabinoid signaling in the biology of oligodendrocytes. Findings described here are related to the proteomic data from in vitro and other animal studies. These exploratory studies may lead to new hypotheses regarding the function of endocannabinoid signaling on myelination, and may contribute to select compounds for further clinic studies. The proteomic data used for the analyses in this chapter can be found in an open databank, which can help other investigators to clarify and search for new targets for oligodendrocyte biology.

Key fact

Key facts of oligodendrocyte biology and pathology

- Oligodendrocytes are the myelinating cells of the central nervous system responsible for efficient neurotransmission and neuron integrity.
- Oligodendrocyte pathologies are involved in neurodegenerative and neurodevelopmental diseases, such as multiple sclerosis, Alzheimer's disease, and schizophrenia.
- Damages on myelin are only partially replaced by the oligodendrocyte precursor cells.
- Currently, there are no treatments for oligodendrocytes and myelin pathologies.

Key facts of cannabinoid for myelin-related diseases

- The replacement of oligodendrocytes in myelin-injured areas requires the proliferation, migration, and differentiation of oligodendrocyte precursor cells.
- Several intrinsic and extrinsic factors can guide oligodendrocyte maturation, for instance, endocannabinoid signaling.
- The activation of the cannabinoid receptor (CB1) seems to be a potential target to stimulate myelin repair.
- Cannabinoid compounds may provide the future therapeutic solutions for oligodendrocyte pathology and myelin diseases.

Mini-dictionary of terms

- **Cytoscape**. An open-source software platform for visualizing complex networks and integrating these with any type of data, for instance, proteomics (Shannon et al., 2003).
- **Metascape**. A web-based portal designed to provide a comprehensive gene list annotation and analysis resource for experimental data, such as proteomics (Zhou et al., 2019)
- **Proteomics**. The large-scale study of proteins or the proteome in a given sample.
- **Proteome**. The set of proteins in a sample at a given condition and time.
- **String**. A web-based portal with a database of known and predicted protein-protein interactions (Szklarczyk et al., 2021).

Summary points

- OPCs generate mature oligodendrocytes that are responsible for myelination in the central nervous system.
- Oligodendrocyte pathology is involved in neurological and psychiatric disorders.
- Endocannabinoid signaling plays a role in oligodendrocyte biology.
- Proteomics contributes to elucidating the molecular pathways and biological process of cannabinoids on oligodendrocytes.

References

Aebersold, R., & Mann, M. (2003). Mass spectrometry-based proteomics. *Nature*, *422*(6928), 198–207. https://doi.org/10.1038/nature01511.

Aguado, T., Huerga-Gómez, A., Sánchez-de la Torre, A., Resel, E., Chara, J. C., Matute, C., Mato, S., Galve-Roperh, I., Guzman, M., & Palazuelos, J. (2021). Δ-Tetrahydrocannabinol promotes functional remyelination in the mouse brain. *British Journal of Pharmacology*, *178*(20), 4176–4192. https://doi.org/10.1111/bph.15608.

Alpár, A., Tortoriello, G., Calvigioni, D., Niphakis, M. J., Milenkovic, I., Bakker, J., Cameron, G. A., Hanics, J., Morris, C. V., Fuzik, J., Kovacs, G. G., Cravatt, B. F., Parnavelas, J. G., Andrews, W. D., Hurd, Y. L., Keimpema, E., & Harkany, T. (2014). Endocannabinoids modulate cortical development by configuring Slit2/Robo1 signalling. *Nature Communications*, *5*, 4421. https://doi.org/10.1038/ncomms5421.

Arévalo-Martín, Á., García-Ovejero, D., Rubio-Araiz, A., Gómez, O., Molina-Holgado, F., & Molina-Holgado, E. (2007). Cannabinoids modulate Olig2 and polysialylated neural cell adhesion molecule expression in the subventricular zone of post-natal rats through cannabinoid receptor 1 and cannabinoid receptor 2. *European Journal of Neuroscience*, *26*(6), 1548–1559. https://doi.org/10.1111/j.1460-9568.2007.05782.x.

Bernal-Chico, A., Canedo, M., Manterola, A., Victoria Sánchez-Gómez, M., Pérez-Samartín, A., Rodríguez-Puertas, R., Matute, C., & Mato, S. (2015). Blockade of monoacylglycerol lipase inhibits oligodendrocyte excitotoxicity and prevents demyelination in vivo. *Glia*, *63*(1), 163–176. https://doi.org/10.1002/glia.22742.

Centonze, D., Bari, M., Rossi, S., Prosperetti, C., Furlan, R., Fezza, F., De Chiara, V., Battistini, L., Bernardi, G., Bernardini, S., Martino, G., & Maccarrone, M. (2007). The endocannabinoid system is dysregulated in multiple sclerosis and in experimental autoimmune encephalomyelitis. *Brain: A Journal of Neurology*, *130*(Pt 10), 2543–2553. https://doi.org/10.1093/brain/awm160.

Ceprián, M., Vargas, C., García-Toscano, L., Penna, F., Jiménez-Sánchez, L., Achicallende, S., Elezgarai, I., Grandes, P., Hind, W., Pazos, M. R., & Martínez-Orgado, J. (2019). Cannabidiol administration prevents hypoxia-ischemia-induced hypomyelination in newborn rats. *Frontiers in Pharmacology*, *10*, 1131. https://doi.org/10.3389/fphar.2019.01131.

Chanoumidou, K., Mozafari, S., Baron-Van Evercooren, A., & Kuhlmann, T. (2020). Stem cell derived oligodendrocytes to study myelin diseases. *Glia*, *68* (4), 705–720. https://doi.org/10.1002/glia.23733.

Compagnucci, C., Di Siena, S., Bustamante, M. B., Di Giacomo, D., Di Tommaso, M., Maccarrone, M., Grimaldi, P., & Sette, C. (2013). Type-1 (CB1) cannabinoid receptor promotes neuronal differentiation and maturation of neural stem cells. *PLoS One*, *8*(1), e54271. https://doi.org/10.1371/journal.pone.0054271.

Cristino, L., Bisogno, T., & Di Marzo, V. (2020). Cannabinoids and the expanded endocannabinoid system in neurological disorders. *Nature Reviews. Neurology*, *16*(1), 9–29. https://doi.org/10.1038/s41582-019-0284-z.

Davis, K. L., & Haroutunian, V. (2003). Global expression-profiling studies and oligodendrocyte dysfunction in schizophrenia and bipolar disorder [Review of Global expression-profiling studies and oligodendrocyte dysfunction in schizophrenia and bipolar disorder]. *The Lancet*, *362*(9386), 758. https://doi.org/10.1016/S0140-6736(03)14297-3.

de Almeida, V., Seabra, G., Reis-de-Oliveira, G., Zuccoli, G. S., Rumin, P., Fioramonte, M., … Martins-de-Souza, D. (2022). Cannabinoids modulate proliferation, differentiation, and migration signaling pathways in oligodendrocytes. *European Archives of Psychiatry and Clinical Neuroscience*, *272*(7), 1311–1323. https://doi.org/10.1007/s00406-022-01425-5.

Della-Flora Nunes, G., Wilson, E. R., Hurley, E., He, B., O'Malley, B. W., Poitelon, Y., Wrabetz, L., & Feltri, M. L. (2021). Activation of mTORC1 and c-Jun by Prohibitin1 loss in Schwann cells may link mitochondrial dysfunction to demyelination. *eLife*, *10*. https://doi.org/10.7554/eLife.66278.

Deshmukh, V. A., Tardif, V., Lyssiotis, C. A., Green, C. C., Kerman, B., Kim, H. J., Padmanabhan, K., Swoboda, J. G., Ahmad, I., Kondo, T., Gage, F. H., Theofilopoulos, A. N., Lawson, B. R., Schultz, P. G., & Lairson, L. L. (2013). A regenerative approach to the treatment of multiple sclerosis. *Nature*, *502*(7471), 327–332. https://doi.org/10.1038/nature12647.

Di Filippo, M., Pini, L. A., Pelliccioli, G. P., Calabresi, P., & Sarchielli, P. (2008). Abnormalities in the cerebrospinal fluid levels of endocannabinoids in multiple sclerosis. *Journal of Neurology, Neurosurgery, and Psychiatry*, *79*(11), 1224–1229. https://doi.org/10.1136/jnnp.2007.139071.

Egawa, N., Shindo, A., Hikawa, R., Kinoshita, H., Liang, A. C., Itoh, K., Lok, J., Maki, T., Takahashi, R., Lo, E. H., & Arai, K. (2019). Differential roles of epigenetic regulators in the survival and differentiation of oligodendrocyte precursor cells. *Glia*, *67*(4), 718–728. https://doi.org/10.1002/glia.23567.

Falvella, A. C. B., Smith, B. J., Silva-Costa, L. C., Valença, A. G. F., Crunfli, F., Zuardi, A. W., Hallak, J. E., Crippa, J. A., de Almeida, V., & Martins-de-Souza, D. (2021). Cannabidiol displays proteomic similarities to antipsychotics in cuprizone-exposed human oligodendrocytic cell line MO3.13. *Frontiers in Molecular Neuroscience*, *14*, 673144. https://doi.org/10.3389/fnmol.2021.673144.

Feliú, A., del Río, I. B., Carrillo-Salinas, F. J., Hernández-Torres, G., Mestre, L., Puente, N., Ortega-Gutiérrez, S., López-Rodríguez, M. L., Grandes, P., Mecha, M., & Guaza, C. (2017). 2-Arachidonoylglycerol reduces proteoglycans and enhances remyelination in a progressive model of demyelination. *The Journal of Neuroscience*, *37*(35), 8385–8398. https://doi.org/10.1523/jneurosci.2900-16.2017.

Fünfschilling, U., Supplie, L. M., Mahad, D., Boretius, S., Saab, A. S., Edgar, J., Brinkmann, B. G., Kassmann, C. M., Tzvetanova, I. D., Möbius, W., Diaz, F., Meijer, D., Suter, U., Hamprecht, B., Sereda, M. W., Moraes, C. T., Frahm, J., Goebbels, S., & Nave, K.-A. (2012). Glycolytic oligodendrocytes maintain myelin and long-term axonal integrity. *Nature*, *485*(7399), 517–521. https://doi.org/10.1038/nature11007.

Gomez, O., Arevalo-Martin, A., Garcia-Ovejero, D., Ortega-Gutierrez, S., Cisneros, J. A., Almazan, G., Sánchez-Rodriguez, M. A., Molina-Holgado, F., & Molina-Holgado, E. (2010). The constitutive production of the endocannabinoid 2-arachidonoylglycerol participates in oligodendrocyte differentiation. *Glia*, *58*(16), 1913–1927. https://doi.org/10.1002/glia.21061.

Gomez, O., Sanchez-Rodriguez, A., Le, M., Sanchez-Caro, C., Molina-Holgado, F., & Molina-Holgado, E. (2011). Cannabinoid receptor agonists modulate oligodendrocyte differentiation by activating PI3K/Akt and the mammalian target of rapamycin (mTOR) pathways. *British Journal of Pharmacology*, *163*(7), 1520–1532. https://doi.org/10.1111/j.1476-5381.2011.01414.x.

Gomez, O., Sanchez-Rodriguez, M. A., Ortega-Gutierrez, S., Vazquez-Villa, H., Guaza, C., Molina-Holgado, F., & Molina-Holgado, E. (2015). A basal tone of 2-arachidonoylglycerol contributes to early oligodendrocyte progenitor proliferation by activating phosphatidylinositol 3-kinase (PI3K)/AKT and the mammalian target of rapamycin (MTOR) pathways. *Journal of Neuroimmune Pharmacology: The Official Journal of the Society on NeuroImmune Pharmacology*, *10*(2), 309–317. https://doi.org/10.1007/s11481-015-9609-x.

Han, H., Cho, J.-W., Lee, S., Yun, A., Kim, H., Bae, D., Yang, S., Kim, C. Y., Lee, M., Kim, E., Lee, S., Kang, B., Jeong, D., Kim, Y., Jeon, H.-N., Jung, H., Nam, S., Chung, M., Kim, J.-H., & Lee, I. (2018). TRRUST v2: An expanded reference database of human and mouse transcriptional regulatory interactions. *Nucleic Acids Research*, *46*(D1), D380–D386. https://doi.org/10.1093/nar/gkx1013.

Huerga-Gómez, A., Aguado, T., Sánchez-de la Torre, A., Bernal-Chico, A., Matute, C., Mato, S., Guzmán, M., Galve-Roperh, I., & Palazuelos, J. (2021). Δ-Tetrahydrocannabinol promotes oligodendrocyte development and CNS myelination in vivo. *Glia*, *69*(3), 532–545. https://doi.org/10.1002/glia.23911.

Ibeas Bih, C., Chen, T., Nunn, A. V. W., Bazelot, M., Dallas, M., & Whalley, B. J. (2015). Molecular targets of cannabidiol in neurological disorders. *Neurotherapeutics: The Journal of the American Society for Experimental NeuroTherapeutics*, *12*(4), 699–730. https://doi.org/10.1007/s13311-015-0377-3.

Ilyasov, A. A., Milligan, C. E., Pharr, E. P., & Howlett, A. C. (2018). The endocannabinoid system and oligodendrocytes in health and disease. *Frontiers in Neuroscience*, *0*. https://doi.org/10.3389/fnins.2018.00733.

Iwata, K., Café-Mendes, C. C., Schmitt, A., Steiner, J., Manabe, T., Matsuzaki, H., Falkai, P., Turck, C. W., & Martins-de-Souza, D. (2013). The human oligodendrocyte proteome. *Proteomics*, *13*(23-24), 3548–3553. https://doi.org/10.1002/pmic.201300201.

Jean-Gilles, L., Feng, S., Tench, C. R., Chapman, V., Kendall, D. A., Barrett, D. A., & Constantinescu, C. S. (2009). Plasma endocannabinoid levels in multiple sclerosis. *Journal of the Neurological Sciences*, *287*(1-2), 212–215. https://doi.org/10.1016/j.jns.2009.07.021.

Kuhn, S., Gritti, L., Crooks, D., & Dombrowski, Y. (2019). Oligodendrocytes in development, myelin generation and beyond. *Cell, 8*(11). https://doi.org/10.3390/cells8111424.

Manterola, A., Bernal-Chico, A., Cipriani, R., Canedo-Antelo, M., Moreno-García, Á., Martín-Fontecha, M., Pérez-Cerdá, F., Sánchez-Gómez, M. V., Ortega-Gutiérrez, S., Brown, J. M., Hsu, K.-L., Cravatt, B., Matute, C., & Mato, S. (2018). Deregulation of the endocannabinoid system and therapeutic potential of ABHD6 blockade in the cuprizone model of demyelination. *Biochemical Pharmacology, 157*, 189–201. https://doi.org/10.1016/j.bcp.2018.07.042.

Mato, S., Victoria Sánchez-Gómez, M., & Matute, C. (2010). Cannabidiol induces intracellular calcium elevation and cytotoxicity in oligodendrocytes. *Glia, 58*(14), 1739–1747. https://doi.org/10.1002/glia.21044.

McAleese, K. E., Walker, L., Graham, S., Moya, E. L. J., Johnson, M., Erskine, D., Colloby, S. J., Dey, M., Martin-Ruiz, C., Taylor, J.-P., Thomas, A. J., McKeith, I. G., De Carli, C., & Attems, J. (2017). Parietal white matter lesions in Alzheimer's disease are associated with cortical neurodegenerative pathology, but not with small vessel disease. *Acta Neuropathologica, 134*(3), 459–473. https://doi.org/10.1007/s00401-017-1738-2.

Mecha, M., Torrao, A. S., Mestre, L., Carrillo-Salinas, F. J., Mechoulam, R., & Guaza, C. (2012). Cannabidiol protects oligodendrocyte progenitor cells from inflammation-induced apoptosis by attenuating endoplasmic reticulum stress. *Cell Death & Disease, 3*, e331. https://doi.org/10.1038/cddis.2012.71.

Miyata, S., Taniguchi, M., Koyama, Y., Shimizu, S., Tanaka, T., Yasuno, F., Yamamoto, A., Iida, H., Kudo, T., Katayama, T., & Tohyama, M. (2016). Association between chronic stress-induced structural abnormalities in Ranvier nodes and reduced oligodendrocyte activity in major depression. *Scientific Reports, 6*, 23084. https://doi.org/10.1038/srep23084.

Molina-Holgado, E., Vela, J. M., Arévalo-Martín, A., Almazán, G., Molina-Holgado, F., Borrell, J., & Guaza, C. (2002). Cannabinoids promote oligodendrocyte progenitor survival: Involvement of cannabinoid receptors and phosphatidylinositol-3 kinase/Akt signaling. *The Journal of Neuroscience: The Official Journal of the Society for Neuroscience, 22*(22), 9742–9753. https://www.ncbi.nlm.nih.gov/pubmed/12427829.

Moreno-Luna, R., Esteban, P. F., Paniagua-Torija, B., Arevalo-Martin, A., Garcia-Ovejero, D., & Molina-Holgado, E. (2021). Heterogeneity of the endocannabinoid system between cerebral cortex and spinal cord oligodendrocytes. *Molecular Neurobiology, 58*(2), 689–702. https://doi.org/10.1007/s12035-020-02148-1.

Patergnani, S., Bonora, M., Ingusci, S., Previati, M., Marchi, S., Zucchini, S., Perrone, M., Wieckowski, M. R., Castellazzi, M., Pugliatti, M., Giorgi, C., Simonato, M., & Pinton, P. (2021). Antipsychotic drugs counteract autophagy and mitophagy in multiple sclerosis. *Proceedings of the National Academy of Sciences of the United States of America, 118*(24). https://doi.org/10.1073/pnas.2020078118.

Rahimi, A., Faizi, M., Talebi, F., Noorbakhsh, F., Kahrizi, F., & Naderi, N. (2015). Interaction between the protective effects of cannabidiol and palmitoylethanolamide in experimental model of multiple sclerosis in C57BL/6 mice. *Neuroscience, 290*, 279–287. https://doi.org/10.1016/j.neuroscience.2015.01.030.

Rinholm, J. E., Vervaeke, K., Tadross, M. R., Tkachuk, A. N., Kopek, B. G., Brown, T. A., Bergersen, L. H., & Clayton, D. A. (2016). Movement and structure of mitochondria in oligodendrocytes and their myelin sheaths. *Glia, 64*(5), 810–825. https://doi.org/10.1002/glia.22965.

Shannon, P., Markiel, A., Ozier, O., Baliga, N. S., Wang, J. T., Ramage, D., Amin, N., Schwikowski, B., & Ideker, T. (2003). Cytoscape: A software environment for integrated models of biomolecular interaction networks. *Genome Research, 13*(11), 2498–2504. https://doi.org/10.1101/gr.1239303.

Solbrig, M. V., Fan, Y., Hermanowicz, N., Morgese, M. G., & Giuffrida, A. (2010). A synthetic cannabinoid agonist promotes oligodendrogliogenesis during viral encephalitis in rats. *Experimental Neurology, 226*(1), 231–241. https://doi.org/10.1016/j.expneurol.2010.09.003.

Sun, J., Fang, Y., Chen, T., Guo, J., Yan, J., Song, S., Zhang, L., & Liao, H. (2013). WIN55, 212-2 promotes differentiation of oligodendrocyte precursor cells and improve remyelination through regulation of the phosphorylation level of the ERK 1/2 via cannabinoid receptor 1 after stroke-induced demyelination. *Brain Research, 1491*, 225–235. https://doi.org/10.1016/j.brainres.2012.11.006.

Szklarczyk, D., Gable, A. L., Nastou, K. C., Lyon, D., Kirsch, R., Pyysalo, S., Doncheva, N. T., Legeay, M., Fang, T., Bork, P., Jensen, L. J., & von Mering, C. (2021). The STRING database in 2021: Customizable protein-protein networks, and functional characterization of user-uploaded gene/measurement sets. *Nucleic Acids Research, 49*(D1), D605–D612. https://doi.org/10.1093/nar/gkaa1074.

Tomas-Roig, J., Agbemenyah, H. Y., Celarain, N., Quintana, E., Ramió-Torrentà, L., & Havemann-Reinecke, U. (2020). Dose-dependent effect of cannabinoid WIN-55,212-2 on myelin repair following a demyelinating insult. *Scientific Reports, 10*(1), 590. https://doi.org/10.1038/s41598-019-57290-1.

Tomas-Roig, J., Wirths, O., Salinas-Riester, G., & Havemann-Reinecke, U. (2016). The cannabinoid CB1/CB2 agonist WIN55212.2 promotes oligodendrocyte differentiation in vitro and neuroprotection during the cuprizone-induced central nervous system demyelination. *CNS Neuroscience & Therapeutics, 22*(5), 387–395. https://doi.org/10.1111/cns.12506.

Walther, T. C., & Mann, M. (2010). Mass spectrometry-based proteomics in cell biology. *The Journal of Cell Biology, 190*(4), 491–500. https://doi.org/10.1083/jcb.201004052.

Ye, F., Chen, Y., Hoang, T., Montgomery, R. L., Zhao, X.-H., Bu, H., Hu, T., Taketo, M. M., van Es, J. H., Clevers, H., Hsieh, J., Bassel-Duby, R., Olson, E. N., & Lu, Q. R. (2009). HDAC1 and HDAC2 regulate oligodendrocyte differentiation by disrupting the beta-catenin-TCF interaction. *Nature Neuroscience, 12*(7), 829–838. https://doi.org/10.1038/nn.2333.

Zhou, Y., Zhou, B., Pache, L., Chang, M., Khodabakhshi, A. H., Tanaseichuk, O., Benner, C., & Chanda, S. K. (2019). Metascape provides a biologist-oriented resource for the analysis of systems-level datasets. *Nature Communications, 10*(1), 1523. https://doi.org/10.1038/s41467-019-09234-6.

Chapter 31

Cannabinoids and endocannabinoid signaling at the basal forebrain cholinergic system

Iker Bengoetxea de Tena, Marta Moreno-Rodríguez, Alberto Llorente-Ovejero, Iván Manuel, and Rafael Rodríguez-Puertas

Department of Pharmacology, Faculty of Medicine and Nursing, University of the Basque Country (UPV/EHU), Vizcaya, Spain

Abbreviations

2-AG	2-arachidonoylglycerol
THC	Δ-9-Tetrahidrocannabinol
ACh	acetylcholine
ACEA	arachidonyl-2-chloroethylamide
AD	Alzheimer's disease
AEA	anandamide
B	*nucleus basalis magnocellularis*
BFCN	basal forebrain cholinergic neurons
BLA	basolateral amygdala
CBD	cannabidiol
ChAT	choline acetyltransferase
CNS	central nervous system
eCB	endocannabinoid
EMA	European Medicines Agency
FAAH	fatty acid amide hydrolase
FDA	Food and Drug Administration
GABA	γ-aminobutyric acid
HDB	horizontal limb nuclei of diagonal band of Broca
LC	*locus coeruleus*
MS	medial septal nuclei
mAChR	muscarinic acetylcholine receptors
MAGL	monoacylglycerol lipase
NADPHd	nicotinamide adenine dinucleotide phosphate diaphorase
NGFr	high affinity nerve growth factor receptor
$p75^{NTR}$	low affinity p75 nerve growth factor receptor
VDB	vertical limb nuclei of the diagonal band of Broca
VTA	ventral tegmental area

Introduction: The basal forebrain cholinergic system

The various nuclei of the basal forebrain system are, collectively, the main source of cholinergic innervations to brain areas involved in the most essential learning and memory processes, including the neocortex, the hippocampus, and the amygdala. The cholinergic neurons present in these nuclei project distant axons to cortical areas, providing them with their main source of acetylcholine (ACh) (Ballinger et al., 2016). The basal forebrain includes the medial septum nuclei (MS), the vertical (VDB) and horizontal limb nuclei of diagonal band of Broca (HDB), and the *nucleus basalis magnocellularis* (B). MS, VDB, and HDB project mainly to the hippocampus, while B provides the main source of cholinergic innervations

Cannabis Use, Neurobiology, Psychology, and Treatment. https://doi.org/10.1016/B978-0-323-89862-1.00034-9

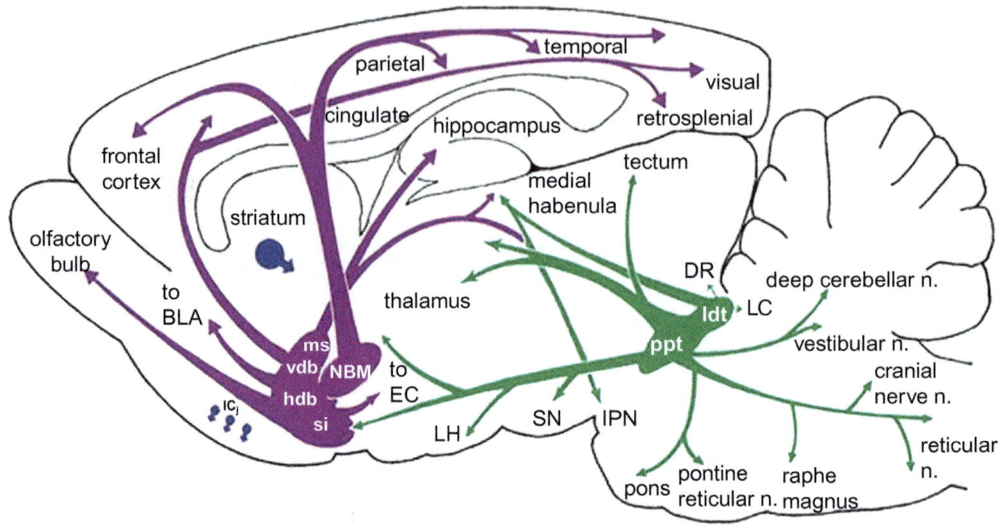

FIG. 1 Central cholinergic system. Outline of the mammalian central cholinergic system. Cholinergic neurons of the basal forebrain project to multiple brain areas, including the cortex, hippocampus and amygdala, and are implicated in the most essential learning and memory processes (*purple*). *From Woolf, N. J. (1997). A Possible Role for Cholinergic Neurons of the Basal Forebrain and Pontomesencephalon in Consciousness. Consciousness and Cognition 6(4), 574–596. The publication is available at Science Direct through: https://www.sciencedirect.com/science/article/pii/S1053810097903196?via%3Dihub.*

to the cerebral cortex and the amygdala (Mesulam et al., 1983) (Fig. 1). These latter projections arising from B are particularly relevant in learning and memory and suffer early degeneration in Alzheimer's disease (AD), which leads to the characteristic symptoms of dementia in these patients (Arendt et al., 1985).

B is formed by a collection of magnocellular and isodendritic neurons. In rodents, these neurons are located medial and ventral to the *globus pallidus* (Adams, 1997). In humans, this area is divided into four major sectors of cholinergic cells designated as Ch1-Ch4, where B is located at the intermediate Ch4 region, which is one of the most physiologically relevant nuclei of cholinergic neurons in the mammal brain, but it contains non-cholinergic neurons too. Cholinergic neurons in this area express choline acetyltransferase (ChAT), the vesicular ACh transporter, AChE, calbindin-d28k, the high affinity nerve growth factor receptor (NGFr) trkA, and the low affinity p75 NGFr ($p75^{NTR}$). A minority of these cholinergic neurons are NGFr-negative and, in rat, they project preferentially to the amygdala. It also contains a complex mosaic of non-cholinergic neurons that are nicotinamide adenine dinucleotide phosphate diaphorase (NADPHd)-positive, γ-aminobutyric acid (GABA)ergic, peptidergic, or tyrosine hydroxylase-positive (Gritti et al., 1994; Mesulam & Geula, 1988; Mufson et al., 2003).

Immunohistochemical studies in rodents reveal a substantial number of catecholaminergic afferents to the basal forebrain from the ventral tegmental area (VTA) and the *substantia nigra pars compacta*, serotonergic projections from the dorsal raphe nucleus and VTA, and cholinergic innervations from the pedunculopontine nucleus and latero-dorsal tegmental area. In rat, all of these nuclei project to Ch4 region (Jones & Cuello, 1989). Dendrites at the Ch4 region receive innervations from cholinergic (ChAT-positive), GABAergic and dopaminergic neurons (Smiley & Mesulam, 1999). Galanin, glutamate and estrogen receptors have also been described at the human Ch4 sector as postsynaptic components (Mufson et al., 2003).

The specific depletion of basal forebrain cholinergic neurons (BFCN) in the Ch4 region of rodents and primates significantly reduces the cortical densities of ACh and AChE (McGeer & McGeer, 1986; Wenk et al., 1994). Those lesion animal models have constituted an important tool to demonstrate that the cholinergic innervation of the cerebral cortex plays a central role in the modulation of arousal, attention, learning, and memory (Mesulam, 1995).

Degeneration of the basal forebrain cholinergic system: Role in AD

Basal cholinergic denervation in the brain is widely accepted as a key feature in the development of AD and related clinical symptoms of cognitive impairment (Geula & Mesulam, 1989). During the progression of the disease, there is a clear reduction in the density of muscarinic acetylcholine receptors (mAChR) in several areas of the brain, including the hippocampus and the entorhinal cortex, as well as in ChAT activity (Davies, 1979). In addition, it has long been reported that the acute administration of selective antagonists for muscarinic receptors, such as scopolamine, induces transient amnesia and cognitive deficits, mimicking the symptoms observed in AD patients (Drachman & Leavitt, 1974).

Although a decrease in other neurotransmitter systems has been reported, such as the noradrenergic, including degeneration of the *locus coeruleus* (LC) (Slater & Wang, 2021), the cholinergic loss is the most consistent with the different stages of the disease (Minger et al., 2000; Perry et al., 1981).

These and other data led to the formulation, in the 1980s, of the "the cholinergic hypothesis of geriatric memory dysfunction" (Bartus et al., 1982), according to which, in aged and demented patients, there is a significant dysfunction of the cholinergic system which correlates with memory loss, as well as with some neuropsychiatric symptoms of dementia, including, but not limited to, agitation, anxiety and impaired motor behaviors (Cummings et al., 2016). This hypothesis was the base for the development of cholinergic enhancement therapies for the treatment of AD, using inhibitors of the AChE with the aim of increasing the availability of ACh at the synaptic cleft. In spite of the limited clinical efficacy of AChE inhibitors, namely galantamine, rivastigmine, and donepezil, they still remain today together with memantine (NMDA receptor antagonist), as the only approved drugs for dementia (Aducanumab and Lecanemab, for blocking β-amyloid have only been approved in the USA and the clinical trials are presenting serious concerns). All these evidences indicate the relevance of cholinergic neurotransmission in the control of learning and memory.

The dysfunction of cholinergic neurotransmission is specific to some brain areas. Cholinergic inputs from striatal interneurons to the striatum and from the brainstem to the thalamus, for instance, remain relatively unaffected in AD until the very advanced stages, but innervations from B to the cortex are severely damaged from the prodromal stages (Mesulam, 2004). The reasons behind the specific vulnerability of this basal forebrain cholinergic pathway are currently unknown.

The maintenance of the synaptic tone communication between presynaptic and postsynaptic cells in neuronal circuits, including cholinergic ones, requires feedback mechanisms to control the release of neurotransmitters from the presynaptic terminal to the synaptic cleft. Among the neural systems that contribute to this key physiological function, the endocannabinoid (eCB) system is of particular relevance, given the abundant expression of its receptors and enzymatic machinery throughout the central nervous system (CNS). The release of eCB, namely anandamide (AEA) and 2-arachidonoylglycerol (2-AG), from postsynaptic neurons has been described as a mean of retrograde synaptic signaling that modulates neurotransmission (Kim et al., 2002; Kreitzer & Regehr, 2002).

The endocannabinoid system in the basal forebrain cholinergic pathways

The literature regarding the distribution, expression, and physiological role of the eCB system in the mammalian central cholinergic system is rather scarce and controversial (Harkany et al., 2003). This fact is particularly true for B, which is the main cholinergic source for cortical areas among the different nuclei that form the basal forebrain. The immunohistochemical studies describe that multiple nuclei surrounding B, including the *globus pallidus* and the basolateral/central amygdaloid nuclei, exhibit strong CB_1 receptor immunoreactivity. In fact, the *globus pallidus* is the brain area with the highest density of CB_1 receptors (Herkenham et al., 1990). BFCN are devoid of CB_1 receptors but receive dense CB_1 receptor innervation. The AEA degrading enzyme, the fatty acid amide hydrolase (FAAH), is expressed in the cholinergic neurons of B, which lack CB_1 receptors, suggesting that they may employ eCB for the retrograde control of neurotransmission (Harkany et al., 2003). In the septal nuclei, CB_1 and muscarinic M2 receptor subtypes are expressed in GABAergic interneurons. Nyiri et al. described that one type of septal cholinergic cells express GABAB and CB_1 receptor subtypes, and these cells project to the hippocampus (Nyíri et al., 2005). In the amygdala, GABAergic interneurons may also modulate cholinergic signaling by CB_1 receptor activation (Marsicano & Lutz, 1999).

Interaction between cholinergic and endocannabinoid systems

The eCB system modulates synaptic plasticity of both excitatory and inhibitory synapses via CB_1 receptors. These receptors, which constitute the predominant receptor subtype in the CNS and particularly in neurons, are localized in presynaptic terminals and preterminal axon segments as demonstrated in GABAergic basket cells (Katona et al., 1999). CB_1-mediated signaling decreases GABA release in most of the brain areas, including the cortex, hippocampus, amygdala, and basal forebrain (Hrabovszky et al., 2012). This CB_1 inhibition on GABA release may be responsible to a great extent of the cannabinoid-mediated modulation of cholinergic neurotransmission. Both CB_1 and cholinergic M_2 receptors are preferentially presynaptic and inhibitory, and both of them are coupled to the same type of $G_{i/0}$ proteins.

In fact, the eCB system may directly regulate the release of ACh itself, at least in the hippocampus (Gifford & Ashby, 1996). In vitro studies show that, in hippocampal and cortical synaptosomes, cannabinoid agonist WIN 55,212-2 inhibits ACh release (Gifford et al., 2000), and in CB_1-deficient mice, this release of ACh is also altered in the hippocampus, but not in the striatum (Kathmann et al., 2001). The modulation of ACh release by the eCB system is biphasic and dependent on the dose of the cannabinoid agonists, e.g., high doses of WIN 55,212-2 (5 mg/kg, i.p) inhibit hippocampal ACh release, while a dose ten times lower (0.5 mg/kg, i.p) promotes it. This opposite effect has been attributed to a differential functional association of anatomically distinct subpopulations of $G_{i/0}$-coupled CB_1 receptors to neurotransmitter systems (Tzavara et al., 2003).

At the basal forebrain cholinergic nuclei, this interaction between eCB and cholinergic systems has demonstrated that the activity of CB_1 receptors in B correlates with freezing behavior, an indicative of fear response, in an aversive memory task as a consequence of the increase in the activity of both CB_1 and M2/M4 muscarinic receptors in B, but not at innervated areas, suggesting that both systems are involved in the modulation of fear response (Bengoetxea de Tena et al., 2022) (Fig. 2).

Overall, the heterogeneous neural input from predominantly limbic structures, combined with the massive cholinergic output to the almost entire neocortical mantle, makes the basal forebrain cholinergic system the principal candidate for the modulation of several aspects of complex behaviors such as learning, executive functions, and working memory, many of which are modulated by the eCB system in physiological and pathological conditions (Gratwicke et al., 2013; Mesulam et al., 1987).

Cholinergic-cannabinoid crosstalk in animal models of genetic AD

The eCB system is involved in the regulation of different learning and memory processes, many of which are controlled by the cholinergic system. CB_1 receptors are expressed in several brain areas involved in the regulation of these brain functions, especially in the pathways from BFCN to amygdala, hippocampus, and cortical areas. Unsurprisingly, the literature is plenty of evidence reporting modulation of the eCB system in neurodegenerative disorders, including AD, in which many of the above-mentioned areas are critically affected already in the prodromal stages (Manuel et al., 2016; Westlake et al., 1994). Both in vitro and in vivo studies, as well as studies performed in *postmortem* human samples from patients, have demonstrated that pharmacological or genetic modulations of the eCB system affect cognition and histopathological and biochemical markers of AD, including those related with the cholinergic system.

Studies in animal models of genetic AD, such as transgenic 3xTg-AD mice, indicate that the activity of CB_1 receptors is increased in some areas such as the anterior thalamus, but slightly downregulated in the basal forebrain (Manuel et al., 2016). In an additional genetic mouse model of genetic AD, 5xFAD, that develops alterations in anxiety and memory-related parameters, CB_1 receptor expression is reduced in the hippocampus (Medina-Vera et al., 2020). In line with these findings, a treatment using low doses of arachidonyl-2-chloroethylamide (ACEA), an agonist of CB_1 receptors, improved cognition in another transgenic mice model of genetic AD when it was administered during the prodromal stages of the disease (Aso et al., 2012).

Beyond CB_1, the other main receptor subtype of the eCB system, CB_2, is also modulated in neuritic plaques-associated neuronal and glial cells (Benito et al., 2003; Solas et al., 2013).

All these findings further highlight the key role of the eCB system in the modulation of learning and memory under pathological conditions related with neurodegeneration, as well as in transgenic animal models of genetic AD. Among them, the above-mentioned 3xTg-AD mice is the most used one.

The 3xTg-AD mice model has consistently been used to evaluate the effects of some of the mechanisms observed in AD that are related to the genetic forms of the disorder. The triple transgenic model harbors the human transgenes APP_{Swe}, $PS1_{M146V}$, and tau_{P301L}, and it consequently emulates the development of the main histopathological characteristics of AD, namely β-amyloid deposits and the hyperphosphorylation of the tau protein (Oddo et al., 2003). This model progressively develops these typical neuropathological markers following a temporal and regional pattern similar to the one described in some AD patients, starting at 4 months of age (equivalent to the prodromal stages of the disease), when intraneuronal β-amyloid is already detectable and there are very slight signs of cognitive deterioration (Billings et al., 2005; Cañete et al., 2015; Giménez-Llort et al., 2007).

At 6 months, the presence of extracellular β-amyloid plaques becomes evident in different areas of the hippocampus and amygdala, which receive afferences from the basal forebrain cholinergic system. This pathological hallmark seems to be intimately linked to cognitive impairment and behavioral changes (Blázquez et al., 2014). At 12 months and beyond, there is a massive presence of extraneuronal β-amyloid in the hippocampus and neocortex, and from that moment on, these accumulations are accompanied by tangles of hyperphosphorylated tau protein (Oddo et al., 2003).

While the histopathological evolution and the development of cognitive impairment of this model are widely described, data regarding the modulation of different neurotransmitter systems is surprisingly scarce. This is particularly true for the basal forebrain cholinergic system and for cholinergic neurotransmission in general. The few studies that have explored this

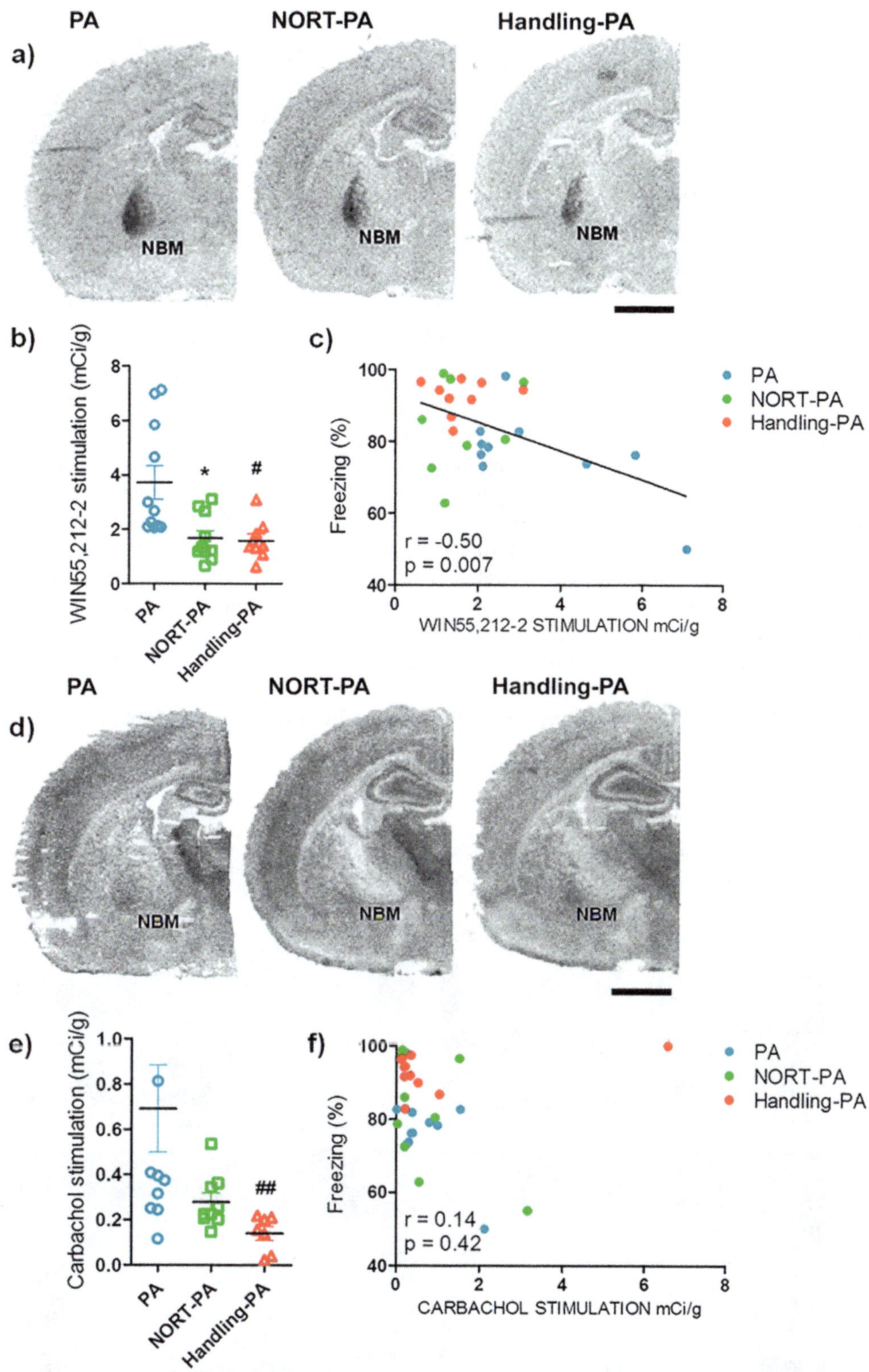

FIG. 2 Freezing behavior correlates with CB_1 receptor activity in the *nucleus basalis magnocellularis*. In rats that have performed an aversive memory task (passive avoidance test, PA) preceded by either novel object recognition test (NORT) or animal handling, freezing, an indicative of fear response, correlates with CB_1 receptor activity at B (NBM). (A) Representative autoradiograms of PA, NORT-PA, and Handling-PA groups in rat coronal sections that show CB_1 receptor activity. Scale bar: 5 mm. (B) CB_1 receptor activity (evoked by the cannabinoid agonist WIN55,212-2) in B. Rats that had performed NORT or being handled before PA have reduced CB_1 receptor activity in B. (C) Correlation between CB_1 receptor activity in B and freezing time. (D) Representative autoradiograms of PA, NORT-PA, and Handling-PA groups in rat coronal sections that show M2/M4 muscarinic receptors activity. Scale bar: 5 mm. (E) M2/M4 receptors activity (evoked by the muscarinic agonist carbachol) in B. Rats that had performed NORT or being handled before PA had reduced M2/M4 receptor activity, in line with the reduction observed in CB_1 activity, suggesting that both systems were key in the modulation of the observed fear response. (F) Correlation between M2/M4 receptors activity in B and freezing time. In this case, there is no correlation. *From Bengoetxea de Tena, I., Moreno-Rodríguez, M., Llorente-Ovejero, A., Monge-Benito, S., Martínez-Gardeazabal, J., Onandia-Hinchado, I., Manuel, I., Giménez-Llort, L., & Rodríguez-Puertas, R. (2022). Handling and novel object recognition modulate fear response and endocannabinoid signaling in* nucleus basalis magnocellularis. *European Journal of Neuroscience 55(6), 1532–1546. The publication is available at Wiley Online Library through: https://onlinelibrary.wiley.com/doi/10.1111/ejn.15642.*

relevant aspect of AD pathogenesis indicate that the basal forebrain cholinergic system begins to be affected very early, when β-amyloid is still intracellular (Perez et al., 2011). In B, there is a slight decrease in the M2/M4 muscarinic receptor activity at 4 months, but not at 15 months (Manuel et al., 2016). In other basal forebrain nuclei, such as the medial septum, a decrease in ChAT-positive cells has been described in 3xTg-AD animals over 18 months. In a subcortical projection area innervated by the basal forebrain, the hippocampus, there is a decrease in ChAT activity in relatively middle stages of the disease in these mice (Perez et al.).

Cholinergic-cannabinoid crosstalk in the 3xTg-AD mice model of genetic AD

With respect to the eCB system, high levels of CB_1 mRNA and functional protein have been described in 6-7-month-old 3xTg-AD mice in memory and fear-related areas innervated by the basal forebrain, such as the hippocampus, prefrontal cortex, and basolateral amygdala (BLA) (Bedse et al., 2014; Llorente-Ovejero et al., 2018). A study performed using specific radioligands for cannabinoid receptors reported an increase in receptor density in the dorsal hippocampus, among other areas not specifically related to memory, like the *substantia nigra* and the cingulate cortex (González de San Román et al., 2021). In terms of receptor activity, there is a decrease in the posterior amygdala and in the motor cortex at 6 months of age and a slight decrease in their activity in B at 15 months (Manuel et al., 2016). This diminution in CB_1 signaling in 3xTg-AD animals has been detected in areas directly innervated by cholinergic cells of the basal forebrain, e.g., amygdala and the innermost layers of the motor cortex. Therefore, a modulation of cannabinoid precursors to counteract the decreased signaling could be expected. There is evidence revealing that potential lipid precursor of eCB containing arachidonic acid (20:4), such as phosphatidylcholine (16:0/20:4) or phosphatidylinositol (16:0/20:4), are modulated in the same brain areas innervated by BFCN where CB_1 activity was decreased. This suggests a compensatory mechanism to overcome this cannabinoid deficit by synthesizing new eCB molecules (González de San Román et al., 2021) (Fig. 3).

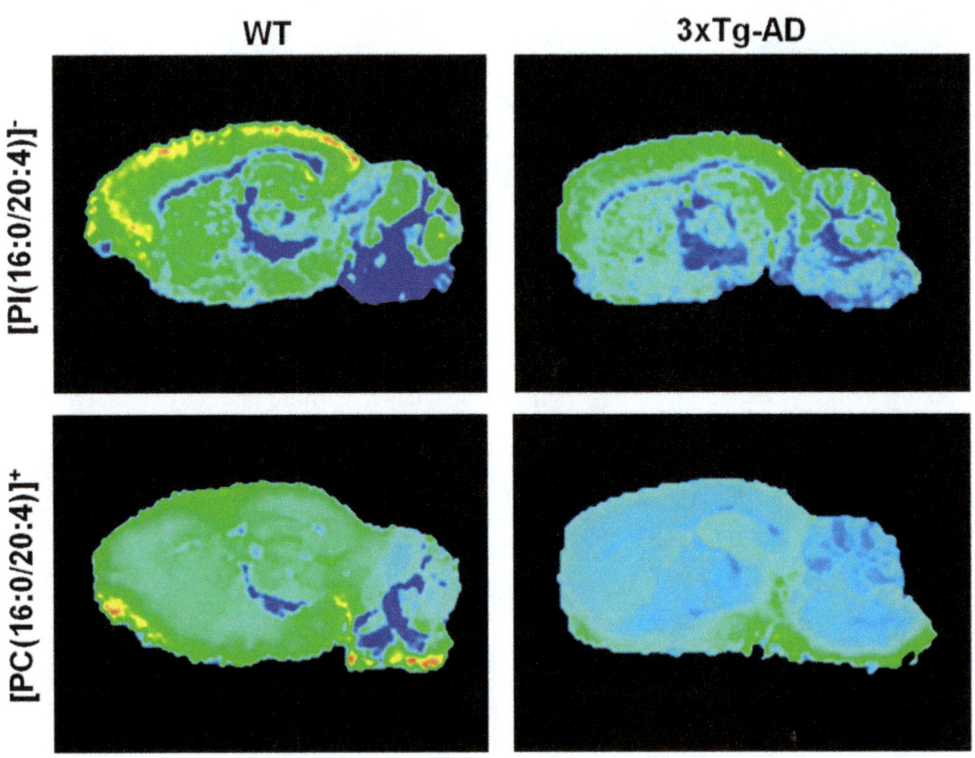

FIG. 3 Mass spectrometry imaging of phosphatidylinositol (16:0/20:4) and phosphatidylcholine (16:0/20:4) in wild-type and 3xTg-AD mice. Representative images obtained by mass spectrometry imaging (MSI) showing the distribution of phosphatidylinositol (16:0/20:4) and phosphatidylcholine (16:0/20:4) from sagittal brain sections from wild-type and 3xTg-AD mice. A decrease in phosphatidylcholine (16:0/20:4) and phosphatidylinositol (16:0/20:4), which are presumable eCB precursors, is observed in areas innervated by basal forebrain cholinergic neurons, like the amygdala and the motor cortex. *Modified from González de San Román, E., Llorente-Ovejero, A., Martínez-Gardeazabal, J., Moreno-Rodríguez, M., Giménez-Llort, L., Manuel, I., & Rodríguez-Puertas, R. (2021). Modulation of neurolipid signaling and specific lipid species in the triple transgenic mouse model of alzheimer's disease. International Journal of Molecular Sciences 22(22), 1–22. The publication is available at MDPI through: https://www.ncbi.nlm.nih.gov/pmc/articles/PMC8620566/.*

The potential therapeutic benefits of a pharmacological modulation of the eCB system in these mice have also been investigated. In 7-month-old mice, the evidence indicates neuroanatomical and neurochemical modifications related to the efficiency of the eCB and muscarinic cholinergic signaling in different brain areas associated with learning and memory functions. The subchronic administration of low doses of WIN 55,212-2 decreased cannabinoid receptor density in the BLA. Indirect stimulation of cannabinoid receptors via inhibition of the monoacylglycerol lipase (MAGL) with JZL184 caused a dramatic reduction in CB_1 receptor density in the hippocampus, cortex, amygdala, and B. Referring to cannabinoid receptor activity, no alteration was observed following WIN 55,212-2 treatment, but in animals treated with JZL184, a decrease in the activity of cannabinoid receptors in the amygdala was reported, probably related to innervation from B. This regulation, evoked by a repeated administration of cannabinoid compounds, seems to be associated with the improved performance of these animals in passive avoidance test, by which aversive learning and memory are evaluated under stressful conditions. The close relationship between the cholinergic and the eCB systems, which involves the upregulation of the muscarinic system in the BLA and the hippocampus, could be responsible for the observed improvement in memory tests (Llorente-Ovejero et al., 2018) (Fig. 4).

While transgenic animal models of AD, like 3xTg-AD, offer an interesting approach for the study of the eCB modulatory role in dementia and its crosstalk with the cholinergic system, alternative models with a more pronounced cholinergic degeneration exist, such as those based on stereotaxic lesions to specifically eliminate cholinergic neurons in the basal forebrain. Given the relevance of the cholinergic-eCB crosstalk and to better understand its role in the neurochemical mechanisms of dementia, such animals with a cholinergic lesion provide a specific model of basalo-cortical cholinergic impairment to identify new druggable targets for AD-type dementia.

Lesion of the *nucleus basalis magnocellularis*: Animal models of dementia associated to cholinergic impairment

The first lesion models targeting BFCN used excitotoxin-based approaches. In these surgical procedures, excitotoxins such as ibotenic acid and kainic acid (Reed & de Belleroche, 1990) were stereotaxically infused into the basal forebrain, with the aim of causing the death of cholinergic neurons and the subsequent denervation of their subcortical and, mainly, cortical pathways. However, this lesion models also caused cell death in the GABAergic neurons surrounding BFCN, among other cell types, leading to divergent results depending on the toxin used (Wenk et al., 1992), thus making the model unreliable.

In 1991, Wiley et al. (1991) described a new lesion model to study the cognitive functions of the BFCN. In this model, a ribosome inactivating protein, saporin, bound to a 192IgG monoclonal antibody directed to $p75^{NTR}$, specifically expressed in BFCN and very few other cell phenotypes (Woolf et al., 1989).

Following a bilateral intraparenchymal infusion at B, the toxin is transported to the cell soma, causing specific cell death of BFCN. The selective cholinergic denervation of rat neocortex and hippocampus can be achieved by selective deletion p75 NGF receptor expressing neurons in B (Heckers et al., 1994). Extensive evidence indicates that 192IgG-saporin-induced lesion causes severe cognitive deficits in rodents in various memory-related tasks, including impairments in novelty-seeking-related recognition memory (novel object recognition task), spatial memory (Morris water maze), and aversive memory (passive avoidance) (Berger-Sweeney et al., 2001). The observed cognitive damage in rodents highlighted the suitability of this cholinergic lesion model for the study of learning and memory deficits and, consequently, to analyze the neuromodulatory role of the eCB system in these processes and its cholinergic relationship.

Modulation of the eCB system in the 192IgG-saporin cholinergic lesion model

The 192IgG-saporin cholinergic lesion model has been consistently used throughout the past three decades for studies exploring new treatments for cognitive and behavioral deficits (Dobryakova et al., 2021; Ramezani et al., 2022). Regarding the modulation of the eCB system following this lesion and to explore the effect of cannabinoid treatments on cognition and memory in this model, Llorente-Ovejero et al. (2017) analyzed the eCB system in rats following infusion of 192IgG-saporin in B, which led to cognitive impairment in the animals, as evaluated in passive avoidance test seven days after the surgical procedure. As a consequence of the lesion, approximately 75% of the BFCN were eliminated (Fig. 5).

Using an autoradiographic approach to explore the density and activity of cannabinoid receptors reported an increase in CB_1 receptor density and activity in B following the lesion. In the cortex, which is the main projection area of BFCN, the clear reduction in AChE-positive stained fibers as a consequence of the lesion was not accompanied by an equivalent decrease in CB_1 receptor density. This is in line with previous studies indicating that cortical cholinergic terminals do not express significant levels of CB_1 receptors (Harkany et al., 2003) (Fig. 6).

In a subsequent study, Llorente-Ovejero et al. (2022) described the glial response after this cholinergic lesion. The intraparenchymal infusion of 192IgG-saporin induced a microglial response at the lesion site, as well as in cortical projection areas. Remarkably, the increase in microglial density was not accompanied by a similar increase on the presence of

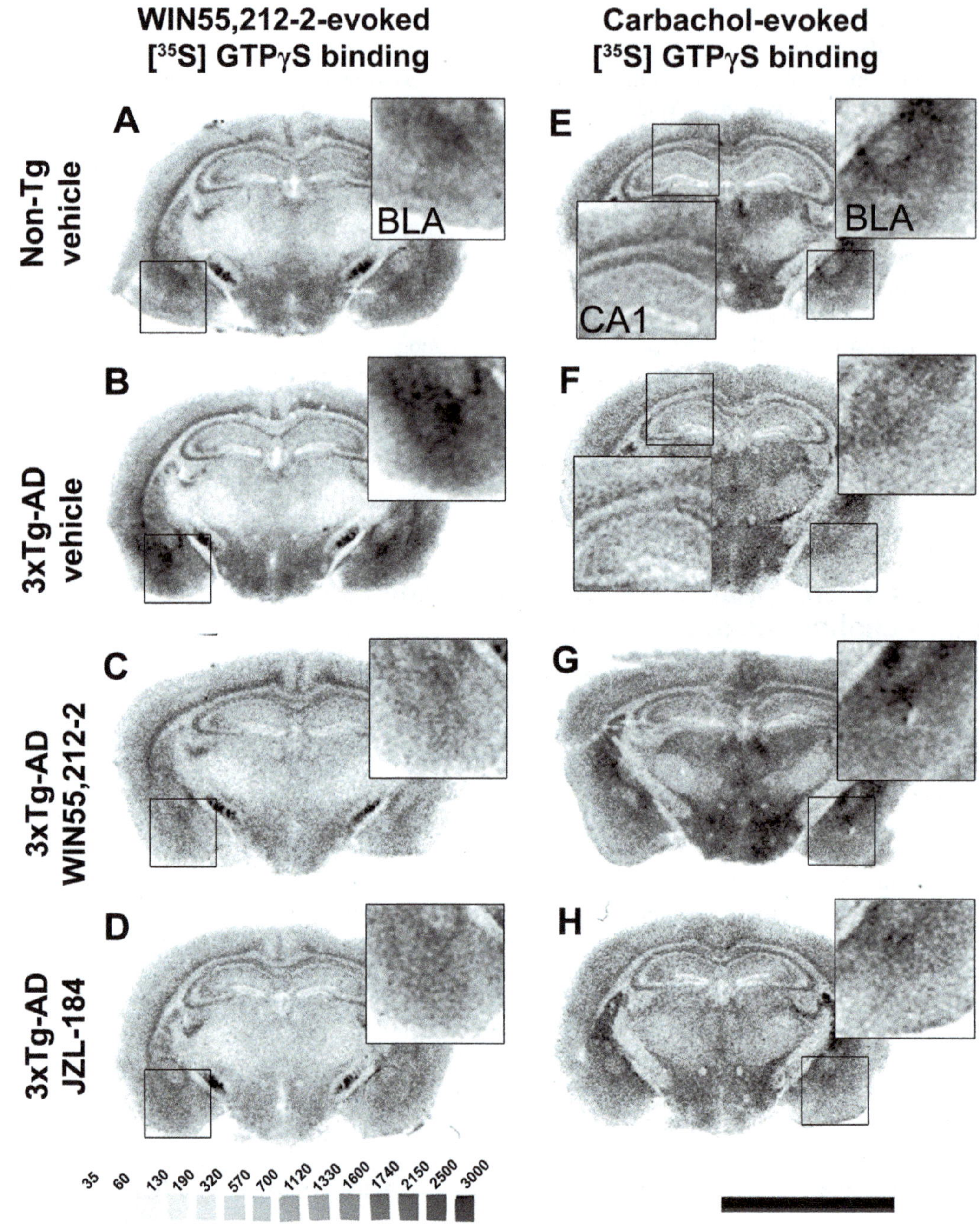

FIG. 4 CB_1 and M2/M4 receptor activity in wild-type and transgenic 3xTg-AD mice model of genetic AD. Representative autoradiograms of CB_1 (A–D) and M2/M4 muscarinic (E–H) receptors activity in coronal brain sections from wild-type and 3xTg-AD mice treated with vehicle and cannabinoid agonists. CB_1 activity in the amygdala is upregulated in 3xTg-AD mice (B) and reduced following the treatments with cannabinoid direct (WIN55,212-2) or indirect (JZL-184) agonists. M2/M4 cholinergic receptor activity in 3xTg-AD mice is lower compare to WT mice at the latero-basolateral region and in the pyramidal layer of the hippocampal CA1 region (boxed areas), indicating that both the eCB and the cholinergic systems are altered in this mouse model of AD. *From Llorente-Ovejero, A., Manuel, I., Lombardero, L., Giralt, M. T., Ledent, C., Giménez-Llort, L., & Rodríguez-Puertas, R. (2018). Endocannabinoid and Muscarinic Signaling Crosstalk in the 3xTg-AD Mouse Model of Alzheimer's Disease. Journal of Alzheimer's Disease 64(1), 117–136. The publication is available at IOS Press through: https://content.iospress.com/articles/journal-of-alzheimers-disease/jad180137.*

astrocytes, which almost disappear from the lesion site. The role of astrocytes in the BFCN is not clear; their lost could be a secondary consequence of the deletion of the BFCN, or may arise from the lesion itself since astrocytes could express $p75^{NTR}$ in response to damage (Cragnolini et al., 2009).

The possible involvement of the eCB system in those glial regulations was also analyzed. The contribution of the two main receptor subtypes of the eCB, CB_1, and CB_2, was analyzed in areas where microglial density increased as a

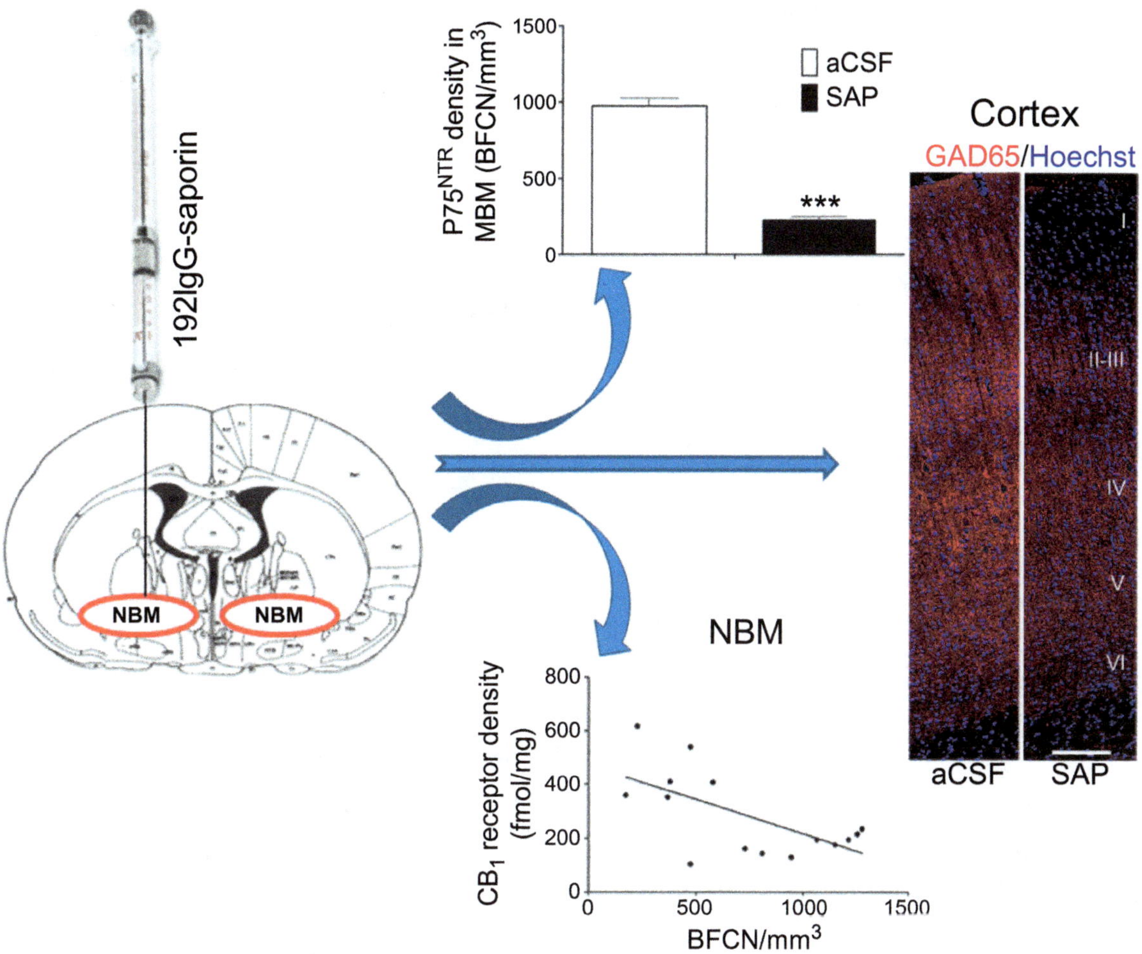

FIG. 5 192IgG-saporin cholinergic lesion in rat. Following the 192IgG-saporin cholinergic lesion, approximately 75% of the cholinergic neurons in B were eliminated and cortical cholinergic denervation was observed. This denervation as a consequence of the lesion induced a decrease in GAD65 GABAergic neuron marker, causing a deregulation of GABAergic neurotransmission. CB_1 receptor density at the lesion site inversely correlated with the number of cholinergic neurons in B, suggesting an interaction between the eCB and cholinergic systems following the lesion. *From Llorente-Ovejero, A., Manuel, I., Giralt, M. T., & Rodríguez-Puertas, R. (2017). Increase in cortical endocannabinoid signaling in a rat model of basal forebrain cholinergic dysfunction. Neuroscience 362, 206–218. The publication is available at Science Direct through: https://www.sciencedirect.com/science/article/pii/S0306452217305663?via%3Dihub.*

consequence of the lesion, including the lesion site and the cortex. Most of the measured activity corresponds to CB_1, with very limited contribution of CB_2. This minor CB_2 activity was restricted to the lesion site. This fact evidences that the microglial cells present in areas affected by the lesion mainly express CB_1 receptors over CB_2. However, previous studies performed in transgenic mice models of genetic AD reported CB_2 upregulation in areas affected by β-amyloid-triggered neuroinflammation (López et al., 2018).

Applications to other areas

The described preclinical evidence suggests that modulation of the eCB system can have beneficial effects in improving cognitive impairment affecting learning and memory processes. In accordance with that, data from a case report described that, in a 75-year-old patient with mild-stage Alzheimer's disease, cannabinoid extract in micro-doses ameliorated mnemonic and non-mnemonic symptoms (Ruver-Martins et al., 2022). However, nowadays, there are no available data from large-scale and monitorized clinical trials performed in demented patients supporting this hypothesis. As a consequence, institutions such as the US Food and Drug Administration (FDA) or the European Medicines Agency (EMA) have not approved, so far, any cannabinoid-based compound for the treatment of dementia or AD.

The limited data testing the use of Δ-9-tetrahidrocannabinol (THC) (dronabinol and nabilone) or cannabidiol (CBD) does indicate that such compounds improve some neuropsychiatric symptoms, such as agitation and aggression in AD

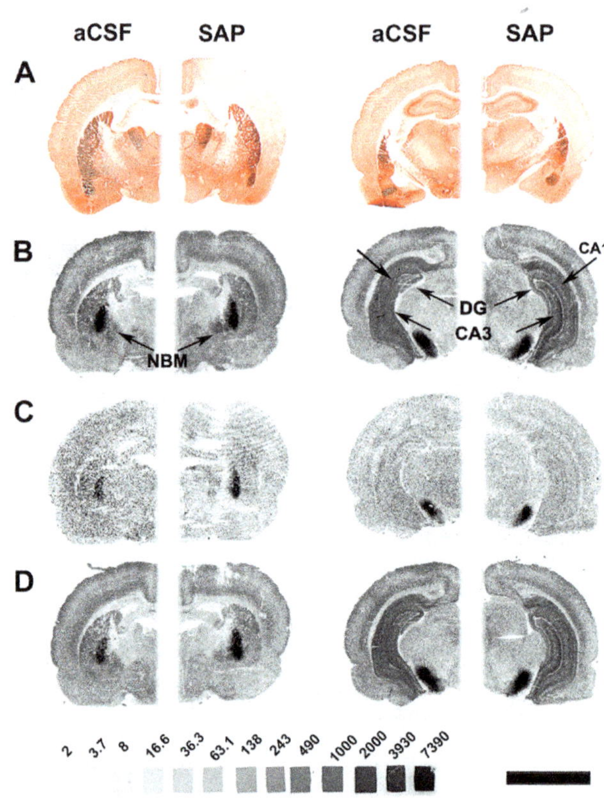

FIG. 6 Cholinergic denervation and modulation of the eCB system following the 192IgG-saporin lesion. Representative hemibrain sections from control (aCSF) and 192IgG-saporin infused at B (SAP) rats as a model of cholinergic cognitive impairment. (A) Acetylcholinesterase (AChE) staining; note the marked decrease in AChE in the cortex following the lesion. (B) Cannabinoid receptor density ([^{3}H]CP55,940 binding); note the differences in cannabinoid receptor density in B and in hippocampal regions following the lesion. [^{3}H]CP55,940 binding was blocked in the presence of CB_1 receptor antagonist, SR141716A (C), but not in the presence of CB_2 antagonist SR144528 (D), indicating that the observed changes in binding following the lesion correspond to CB_1 receptor for the most part. *From Llorente-Ovejero, A., Manuel, I., Giralt, M. T., & Rodríguez-Puertas, R. (2017). Increase in cortical endocannabinoid signaling in a rat model of basal forebrain cholinergic dysfunction. Neuroscience 362, 206–218. The publication is available at Science Direct through: https://www.sciencedirect.com/science/article/pii/S0306452217305663?via%3Dihub.*

patients, but not cognitive performance (Herrmann et al., 2019). A clinical trial in 24 patients with dementia receiving a daily dose of oral THC for three weeks reported no significant beneficial and/or side effects (van den Elsen et al., 2015).

These contradictory results between preclinical and clinical studies may arise from a wide range of reasons. For instance, it has been widely reported that, in AD patients, dementia symptoms lag years, or even decades, behind the first molecular and cellular neurodegenerative processes. The potential neuroprotective role of modulating the eCB system may thus require very long and early treatments, which are extremely difficult to carry out in the context of a clinical trial and such data are not yet available. Such studies would also raise important ethical considerations, as they would likely include treating people with no obvious signs of dementia. Moreover, identifying an effective dose of eCB for such treatments is challenging.

The eCB system, with its widespread distribution throughout the whole CNS and its varied and potent pharmacological effects, is a potentially good target for the amelioration or treatment of different pathological conditions, including dementia. The available data clearly endorse the continued study of the pharmacological modulation of the eCB system, but further preclinical research is required to support the performance of wide clinical trials.

Mini-dictionary of terms

- **Basal forebrain cholinergic system**. a brain area formed by a number of nuclei that provide cholinergic innervation to several brain structures, including the cortex, the hippocampus, and the amygdala. The basal forebrain cholinergic system includes nuclei such as the medial septum nuclei, the vertical and horizontal limb nuclei of diagonal band of Broca, and the *nucleus basalis magnocellularis*.

- ***Nucleus basalis magnocellularis***. one of the nuclei that are part of the basal forebrain, it is formed by a collection of magnocellular and isodendritic neurons. It provides the main source of cholinergic innervations to the cerebral cortex and the amygdala and is prematurely degenerated in AD, leading to learning and memory deficits.
- **Dementia**. degeneration or loss of cognition, affecting learning, memory, language and social abilities, among other functions. It is the most characteristic symptom of AD, which is the most common neurodegenerative disorder worldwide, but can also result from other diseases such as Parkinson's or Huntington's disease.
- **3xTg-AD mice**. one of the most commonly used animal models for the study of genetic AD, these transgenic mice contain three mutations associated with the familial form of the disease (APP_{Swe}, $PS1_{M146V}$, and tau_{P301L}). As a consequence, they show the characteristic amyloid-β and tau pathology observed in human patients and impaired cognitive functions.
- **192IgG-saporin cholinergic lesion model**. a rat model of dementia associated with cholinergic degeneration. Through the intraparenchymal infusion of 192 IgG-saporin toxin, which is directed at $p75^{NTR}$-positive cells, at the *nucleus basalis magnocellularis*, cholinergic neurons are targeted and deleted, impairing the basalo-cortical cholinergic pathway and producing cognitive impairment.
- **Autoradiography**. a pharmacological radioligand technique using tissue slices to obtain microscopical anatomical information of the density and activity of G protein-coupled receptors.
- **Imaging mass spectrometry**. a technique that uses tissue slices to obtain the spatial distribution of unmarked biomolecules, such as proteins and lipids, identifying them by their molecular mass.

Key facts of the endocannabinoid system in cognitive impairment

- The eCB system regulates the release of neurotransmitters such as acetylcholine from the presynaptic terminal to the synaptic cleft, regulating both excitatory and inhibitory neurotransmission via CB_1 receptors.
- The eCB system is modulating the cholinergic signaling at the basal forebrain and innervated areas.
- In neurodegenerative disorders, including AD, the eCB system is modulated in areas such as the amygdala, the hippocampus, the basal forebrain cholinergic system, and its cortical innervation areas.
- In transgenic rodent models of genetic AD, such as 3xTg-AD and 5xFAD mice, CB_1 receptor expression is altered in several brain areas.
- In a rat lesion model of basal forebrain cholinergic degeneration, CB_1 receptor density and activity increased in the basal forebrain following the lesion, suggesting a crosstalk between both systems.
- Pharmacological modulation of the eCB system is arising as a therapeutic approach for the treatment of dementia.

Summary points

- The basal forebrain cholinergic system is the main source of acetylcholine to brain areas responsible for learning and memory, including the cortex, the hippocampus, and the amygdala.
- The degeneration of the different nuclei that form the basal forebrain cholinergic system occurs prematurely in AD and leads to cognitive impairment.
- The eCB system modulates learning and memory under pathological conditions, both in animal models of dementia and in *postmortem* human samples from AD patients.
- In the 3xTg-AD transgenic mice model of genetic AD, the upregulation of the CB_1 receptor activity in the BLA is related to fear response, which is reversed by indirect cannabinoid agonists.
- In the cholinergic lesion rat model of cognitive impairment, CB_1 receptor density, and activity are increased at the basal forebrain to compensate the cholinergic damage.

References

Adams, N. C. (1997). Complexities in the thalamocortical and corticothalamic pathways. *European Journal of Neuroscience*, *9*(2), 204–209.

Arendt, T., Bigl, V., Tennstedt, A., & Arendt, A. (1985). Neuronal loss in different parts of the nucleus basalis is related to neuritic plaque formation in cortical target areas in alzheimer's disease. *Neuroscience*, *14*(1), 1–14.

Aso, E., Palomer, E., Juvés, S., Maldonado, R., Muoz, F. J., & Ferrer, I. (2012). CB1 agonist ACEA protects neurons and reduces the cognitive impairment of AβPP/PS1 mice. *Journal of Alzheimer's Disease*, *30*(2), 439–459.

Ballinger, E. C., Ananth, M., Talmage, D. A., & Role, L. W. (2016). Basal forebrain cholinergic circuits and signaling in cognition and cognitive decline. *Neuron*, *91*(6), 1199–1218.

Bartus, R. T., Dean, R. L., Beer, B., & Lippa, A. S. (1982). The cholinergic hypothesis of geriatric memory dysfunction. *Science*, *217*(4558), 408–417.

Bedse, G., Romano, A., Cianci, S., Lavecchia, A. M., Lorenzo, P., Elphick, M. R., Laferla, F. M., Vendemiale, G., Grillo, C., Altieri, F., Cassano, T., & Gaetani, S. (2014). Altered expression of the CB1 cannabinoid receptor in the triple transgenic mouse model of alzheimer's disease. *Journal of Alzheimer's Disease*, *40*(3), 701–712.

Bengoetxea de Tena, I., Moreno-Rodríguez, M., Llorente-Ovejero, A., Monge-Benito, S., Martínez-Gardeazabal, J., Onandia-Hinchado, I., Manuel, I., Giménez-Llort, L., & Rodríguez-Puertas, R. (2022). Handling and novel object recognition modulate fear response and endocannabinoid signaling in nucleus basalis magnocellularis. *European Journal of Neuroscience*, *55*(6), 1532–1546.

Benito, C., Núñez, E., Tolón, R. M., Carrier, E. J., Rábano, A., Hillard, C. J., & Romero, J. (2003). Cannabinoid CB2 receptors and fatty acid amide hydrolase are selectively overexpressed in neuritic plaque-associated glia in Alzheimer's disease brains. *Journal of Neuroscience*, *23*(35), 11136–11141.

Berger-Sweeney, J., Stearns, N. A., Murg, S. L., Floerke-Nashner, L. R., Lappi, D. A., & Baxter, M. G. (2001). Selective immunolesions of cholinergic neurons in mice: Effects on neuroanatomy, neurochemistry, and behavior. *Journal of Neuroscience*, *21*(20), 8164–8173.

Billings, L. M., Oddo, S., Green, K. N., McGaugh, J. L., & LaFerla, F. M. (2005). Intraneuronal Aβ causes the onset of early Alzheimer's disease-related cognitive deficits in transgenic mice. *Neuron*, *45*(5), 675–688.

Blázquez, G., Cañete, T., Tobeña, A., Giménez-Llort, L., & Fernández-Teruel, A. (2014). Cognitive and emotional profiles of aged Alzheimer's disease (3×TgAD) mice: Effects of environmental enrichment and sexual dimorphism. *Behavioural Brain Research*, *268*, 185–201.

Cañete, T., Blázquez, G., Tobeña, A., Giménez-Llort, L., & Fernández-Teruel, A. (2015). Cognitive and emotional alterations in young Alzheimer's disease (3xTgAD) mice: Effects of neonatal handling stimulation and sexual dimorphism. *Behavioural Brain Research*, *281*, 156–171.

Cragnolini, A. B., Huang, Y., Gokina, P., & Friedman, W. J. (2009). Nerve growth factor attenuates proliferation of astrocytes via the p75 neurotrophin receptor. *Glia*, *57*(13), 1386–1392.

Cummings, J., Lai, T. J., Hemrungrojn, S., Mohandas, E., Yun Kim, S., Nair, G., & Dash, A. (2016). Role of donepezil in the management of neuropsychiatric symptoms in Alzheimer's disease and dementia with lewy bodies. *CNS Neuroscience and Therapeutics*, *22*(3), 159–166.

Davies, P. (1979). Neurotransmitter-related enzymes in senile dementia of the alzheimer type. *Brain Research*, *171*, 319–327.

Dobryakova, Y. V., Spivak, Y. S., Zaichenko, M. I., Koryagina, A. A., Markevich, V. A., Stepanichev, M. Y., & Bolshakov, A. P. (2021). Intrahippocampal adeno-associated virus-mediated overexpression of nerve growth factor reverses 192IgG-Saporin-induced impairments of hippocampal plasticity and behavior. *Frontiers in Neuroscience*, *15*(November), 1–14.

Drachman, D. A., & Leavitt, J. (1974). Human memory and the cholinergic system: A relationship to aging? *Archives of Neurology*, *30*(2), 113–121.

Geula, C., & Mesulam, M. M. (1989). Cortical cholinergic fibers in aging and Alzheimer's disease: A morphometric study. *Neuroscience*, *33*(3), 469–481.

Gifford, A. N., & Ashby, C. R. (1996). Electrically evoked acetylcholine release from hippocampal slices is inhibited by the cannabinoid receptor agonist, WIN 55212-2, and is potentiated by the cannabinoid antagonist, SR 141716A. *Journal of Pharmacology and Experimental Therapeutics*, *277*(3), 1431–1436.

Gifford, A. N., Bruneus, M., Gatley, S. J., & Volkow, N. D. (2000). Cannabinoid receptor-mediated inhibition of acetylcholine release from hippocampal and cortical synaptosomes. *British Journal of Pharmacology*, *131*(3), 645–650.

Giménez-Llort, L., Blázquez, G., Cañete, T., Johansson, B., Oddo, S., Tobeña, A., LaFerla, F. M., & Fernández-Teruel, A. (2007). Modeling behavioral and neuronal symptoms of Alzheimer's disease in mice: A role for intraneuronal amyloid. *Neuroscience and Biobehavioral Reviews*, *31*(1), 125–147.

González de San Román, E., Llorente-Ovejero, A., Martínez-Gardeazabal, J., Moreno-Rodríguez, M., Giménez-Llort, L., Manuel, I., … Rodríguez-Puertas, R. (2021). Modulation of neurolipid signaling and specific lipid species in the triple transgenic mouse model of alzheimer's disease. *International Journal of Molecular Sciences*, *22*(22), 1–22.

Gratwicke, J., Kahan, J., Zrinzo, L., Hariz, M., Limousin, P., Foltynie, T., & Jahanshahi, M. (2013). The nucleus basalis of Meynert: A new target for deep brain stimulation in dementia? *Neuroscience and Biobehavioral Reviews*, *37*(10), 2676–2688.

Gritti, I., Mainville, L., & Jones, B. E. (1994). Projections of GABAergic and cholinergic basal forebrain and GABAergic preoptic-anterior hypothalamic neurons to the posterior lateral hypothalamus of the rat. *Journal of Comparative Neurology*, *339*(2), 251–268.

Harkany, T., Härtig, W., Berghuis, P., Dobszay, M. B., Zilberter, Y., Edwards, R. H., Mackie, K., & Ernfors, P. (2003). Complementary distribution of type 1 cannabinoid receptors and vesicular glutamate transporter 3 in basal forebrain suggests input-specific retrograde signalling by cholinergic neurons. *European Journal of Neuroscience*, *18*(7), 1979–1992.

Heckers, S., Ohtake, T., Wiley, R. G., Lappi, D. A., Geula, C., & Mesulam, M. (1994). Complete and selective cholinergic denervation of rat neocortex and hippocampus but not amygdala by an immunotoxin against the p75 NGF receptor. *Journal of Neuroscience*, *14*(3 Pt 1), 1271–1289.

Herkenham, M., Lynn, A. B., Little, M. D., Johnson, M. R., Melvin, L. S., De Costa, B. R., & Rice, K. C. (1990). Cannabinoid receptor localization in brain. *Proceedings of the National Academy of Sciences of the United States of America*, *87*(5), 1932–1936.

Herrmann, N., Ruthirakuhan, M., Gallagher, D., Verhoeff, N. P. L. G., Kiss, A., Black, S. E., & Lanctôt, K. L. (2019). Randomized placebo-controlled trial of nabilone for agitation in Alzheimer's disease. *American Journal of Geriatric Psychiatry*, *27*(11), 1161–1173.

Hrabovszky, E., Wittmann, G., Kalló, I., Füzesi, T., Fekete, C., & Liposits, Z. (2012). Distribution of type 1 cannabinoid receptor-expressing neurons in the septal-hypothalamic region of the mouse: Colocalization with GABAergic and glutamatergic markers. *Journal of Comparative Neurology*, *520*(5), 1005–1020.

Jones, B. E., & Cuello, A. C. (1989). Afferents to the basal forebrain cholinergic cell area from pontomesencephalic-catecholamine, serotonin, and acetylcholine-neurons. *Neuroscience*, *31*(1), 37–61.

Kathmann, M., Weber, B., Zimmer, A., & Schlicker, E. (2001). Enhanced acetylcholine release in the hippocampus of cannabinoid CB1 receptor-deficient mice. *British Journal of Pharmacology*, *132*(6), 1169–1173.

Katona, I., Sperlágh, B., Sík, A., Käfalvi, A., Vizi, E. S., Mackie, K., & Freund, T. F. (1999). Presynaptically located CB1 cannabinoid receptors regulate GABA release from axon terminals of specific hippocampal interneurons. *Journal of Neuroscience*, *19*(11), 4544–4558.

Kim, J., Isokawa, M., Ledent, C., & Alger, B. E. (2002). Activation of muscarinic acetylcholine receptors enhances the release of endogenous cannabinoids in the hippocampus. *Journal of Neuroscience*, *22*(23), 10182–10191.

Kreitzer, A. C., & Regehr, W. G. (2002). Retrograde signaling by endocannabinoids. *Current Opinion in Neurobiology*, *12*(3), 324–330.

Llorente-Ovejero, A., Bengoetxea de Tena, I., Martínez-Gardeazabal, J., Moreno-Rodríguez, M., Lombardero, L., Manuel, I., & Rodríguez-Puertas, R. (2022). Cannabinoid receptors and glial response following a basal forebrain cholinergic lesion. *ACS Pharmacology & Translational Science*, *5*(9), 791–802.

Llorente-Ovejero, A., Manuel, I., Giralt, M. T., & Rodríguez-Puertas, R. (2017). Increase in cortical endocannabinoid signaling in a rat model of basal forebrain cholinergic dysfunction. *Neuroscience*, *362*, 206–218.

Llorente-Ovejero, A., Manuel, I., Lombardero, L., Giralt, M. T., Ledent, C., Giménez-Llort, L., & Rodríguez-Puertas, R. (2018). Endocannabinoid and muscarinic signaling crosstalk in the 3xTg-AD mouse model of Alzheimer's disease. *Journal of Alzheimer's Disease*, *64*(1), 117–136.

López, A., Aparicio, N., Pazos, M. R., Grande, M. T., Barreda-Manso, M. A., Benito-Cuesta, I., Vázquez, C., Amores, M., Ruiz-Pérez, G., García-García, E., Beatka, M., Tolón, R. M., Dittel, B. N., Hillard, C. J., & Romero, J. (2018). Cannabinoid CB 2 receptors in the mouse brain: Relevance for Alzheimer's disease. *Journal of Neuroinflammation*, *15*(1), 1–11.

Manuel, I., Lombardero, L., LaFerla, F. M., Giménez-Llort, L., & Rodríguez-Puertas, R. (2016). Activity of muscarinic, galanin and cannabinoid receptors in the prodromal and advanced stages in the triple transgenic mice model of Alzheimer's disease. *Neuroscience*, *329*(May), 284–293.

Marsicano, G., & Lutz, B. (1999). Expression of the cannabinoid receptor CB1 in distinct neuronal subpopulations in the adult mouse forebrain. *European Journal of Neuroscience*, *11*(12), 4213–4225.

McGeer, E., & McGeer, P. L. (1986). Possible destruction of cholinoceptive neurons by overstimulation of cholinergic tracts. *Canadian Journal of Physiology and Pharmacology*, *64*(3), 363–368.

Medina-Vera, D., Rosell-Valle, C., López-Gambero, A. J., Navarro, J. A., Zambrana-Infantes, E. N., Rivera, P., Santín, L. J., Suarez, J., & de Fonseca, F. R. (2020). Imbalance of endocannabinoid/lysophosphatidylinositol receptors marks the severity of alzheimer's disease in a preclinical model: A therapeutic opportunity. *Biology*, *9*(11), 1–22.

Mesulam, M. (1995). The cholinergic contribution to neuromodulation in the cerebral cortex. *Seminars in the Neurosciences*, *7*(5), 297–307.

Mesulam, M. (2004). The cholinergic lesion of Alzheimer's disease: Pivotal factor or side show? *Learning and Memory*, *11*(1), 43–49.

Mesulam, M., & Geula, C. (1988). Nucleus basalis (Ch4) and cortical cholinergic innervation in the human brain: Observations based on the distribution of acetylcholinesterase and choline acetyltransferase. *Journal of Comparative Neurology*, *275*(2), 216–240.

Mesulam, M., Geula, C., & Asuncion Morán, M. (1987). Anatomy of cholinesterase inhibition in Alzheimer's disease: Effect of physostigmine and tetrahydroaminoacridine on plaques and tangles. *Annals of Neurology*, *22*(6), 683–691.

Mesulam, M., Mufson, E. J., Wainer, B. H., & Levey, A. I. (1983). Central cholinergic pathways in the rat: An overview based on an alternative nomenclature (Ch1-Ch6). *Neuroscience*, *10*(4), 1185–1201.

Minger, S. L., Esiri, M. M., McDonald, B., Keene, J., Carter, J., Hope, T., & Francis, P. T. (2000). Cholinergic deficits contribute to behavioral disturbance in patients with dementia. *Neurology*, *55*(10), 1460–1467.

Mufson, E. J., Ginsberg, S. D., Ikonomovic, M. D., & DeKosky, S. T. (2003). Human cholinergic basal forebrain: Chemoanatomy and neurologic dysfunction. *Journal of Chemical Neuroanatomy*, *26*(4), 233–242.

Nyíri, G., Szabadits, E., Cserép, C., Mackie, K., Shigemoto, R., & Freund, T. F. (2005). GABAB and CB1 cannabinoid receptor expression identifies two types of septal cholinergic neurons. *European Journal of Neuroscience*, *21*(11), 3034–3042.

Oddo, S., Caccamo, A., Kitazawa, M., Tseng, B. P., & LaFerla, F. M. (2003). Amyloid deposition precedes tangle formation in a triple transgenic model of Alzheimer's disease. *Neurobiology of Aging*, *24*(8), 1063–1070.

Perez, S. E., He, B., Muhammad, N., Oh, K. J., Fahnestock, M., Ikonomovic, M. D., & Mufson, E. J. (2011). Cholinotrophic basal forebrain system alterations in 3xTg-AD transgenic mice. *Neurobiology of Disease*, *41*(2), 338–352.

Perry, E. K., Tomlinson, B. E., Blessed, G., Perry, R. H., Cross, A. J., & Crow, T. T. (1981). Noradrenergic and cholinergic systems in senile dementia of Alzheimer type. *The Lancet*, *318*(8238), 149.

Ramezani, F., Salehian, S., Hosseinzadeh, S., Mahjour, Z., Babajani, T., Ghorbanian, D., Feizi, F., & Pourbagher, R. (2022). Serotonin-1A receptor activation in the median raphe nucleus improves response learning-based strategy in 192IgG saporin-induced cognitive impairments. *European Journal of Pharmacology*, *918*(September 2021), 174774.

Reed, L. J., & de Belleroche, J. (1990). Induction of ornithine decarboxylase in cerebral cortex by excitotoxin lesion of nucleus basalis: Association with postsynaptic responsiveness and N-methyl-d-aspartate receptor activation. *Journal of Neurochemistry*, *55*(3), 780–787.

Ruver-Martins, A. C., Bicca, M. A., Soares de Araujo, F., de Noronha, L., Sales Maia, B. H., Pamplona, F. A., & Gomes da Silva, E. (2022). Cannabinoid extract in microdoses ameliorates mnemonic and nonmnemonic Alzheimer's disease symptoms: A case report. *Journal of Medical Case Reports*, 1–7.

Slater, C., & Wang, Q. (2021). Alzheimer's disease: An evolving understanding of noradrenergic involvement and the promising future of electroceutical therapies. *Clinical and Translational Medicine*, *11*(4).

Smiley, J. F., & Mesulam, M. M. (1999). Cholinergic neurons of the nucleus basalis of Meynert receive cholinergic, catecholaminergic and GABAergic synapses: An electron microscopic investigation in the monkey. *Neuroscience*, *88*(1), 241–255.

Solas, M., Francis, P. T., Franco, R., & Ramirez, M. J. (2013). CB2 receptor and amyloid pathology in frontal cortex of Alzheimer's disease patients. *Neurobiology of Aging*, *34*(3), 805–808.

Tzavara, E. T., Wade, M., & Nomikos, G. G. (2003). Biphasic effects of cannabinoids on acetylcholine release in the hippocampus: Site and mechanism of action. *Journal of Neuroscience*, *23*(28), 9374–9384.

van den Elsen, G. A. H., Ahmed, A. I. A., Verkes, R.-J., Kramers, C., Feuth, T., Rosenberg, P. B., van der Marck, M. A., Olde Rikkert, M. G. M., & Aiaa, T. (2015). Tetrahydrocannabinol for neuropsychiatric symptoms in dementia: A randomized controlled trial. *Neurology*, *84*(23), 2338–2346.

Wenk, G. L., Harrington, C. A., Tucker, D. A., Rance, N. E., & Walker, L. C. (1992). Basal forebrain neurons and memory: A biochemical, histological, and behavioral study of differential vulnerability to ibotenate and quisqualate. *Behavioral Neuroscience*, *106*(6), 909–923.

Wenk, G. L., Stoehr, J. D., Quintana, G., Mobley, S., & Wiley, R. G. (1994). Behavioral, biochemical, histological, and electrophysiological effects of 192 IgG-saporin injections into the basal forebrain of rats. *Journal of Neuroscience*, *14*(10), 5986–5995.

Westlake, T. M., Howlett, A. C., Bonner, T. I., Matsuda, L. A., & Herkenham, M. (1994). Cannabinoid receptor binding and messenger RNA expression in human brain: An in vitro receptor autoradiography and in situ hybridization histochemistry study of normal aged and Alzheimer's brains. *Neuroscience*, *63*(3), 637–652.

Wiley, R. G., Oeltmann, T. N., & Lappi, D. A. (1991). Immunolesioning: Selective destruction of neurons using immunotoxin to rat NGF receptor. *Brain Research*, *562*(1), 149–153.

Woolf, N. J., Gould, E., & Butcher, L. L. (1989). Nerve growth factor receptor is associated with cholinergic neurons of the basal forebrain but not the pontomesencephalon. *Neuroscience*, *30*(1), 143–152.

Chapter 32

Cannabidiol (the major nonpsychotropic cannabinoid in the cannabis plant) and the PI3K/AKT signaling pathway in the CNS

Nadja Schröder
Departamento de Fisiologia, Instituto de Ciências Básicas da Saúde, Universidade Federal do Rio Grande do Sul (UFRGS), Porto Alegre, Brazil

Abbreviations

Aβ	amyloid-beta
ACEA	arachidonyl-2'-chloroethylamide
AD	Alzheimer's disease
AEA	*N*-arachidonoyl ethanolamine (Anandamide)
AMPA	α-amino-3-hydroxy-5-methyl-4-isoxazolepropionic acid
APP/PS1	amyloid precursor protein/presenilin 1
BAD	BCL-2-associated agonist of cell death
BDNF	brain-derived neurotrophic factor
CBD	cannabidiol
CB1R	cannabinoid receptor 1
CB2R	cannabinoid receptor 2
CNS	central nervous system
CPP	conditioned place preference
CUS	chronic unpredictable stress
DRG	dorsal root ganglia
EAE	experimental autoimmune encephalomyelitis
FAAH	fatty acid amide hydrolase
FITC	fluorescein isothiocyanate
FoxO	forkhead box protein O
GPCRs	G-protein-coupled receptors
GSK-3	glycogen synthase kinase-3
HS	hemorrhagic shock
HU210	(−)-1,1-dimethylheptyl-11-hydroxy-delta-8-tetrahydrocannabinol
JWH133	dimethylbutyl-deoxy-delta-8-tetrahydrocannabinol
MS	multiple sclerosis
MOG	myelin oligodendroglial glycoprotein
MPP+	1-methyl-4-phenylpyridinium
MPTP	1-methyl-4-phenyl-1,2,3,6-tetrahydropyridine
MSCs	mesenchymal stem cells
mTORC1	mammalian target of rapamycin complex 1
mTORC2	mammalian target of rapamycin complex 2
NFT	neurofibrillary tangles
NGF	nerve growth factor
NLRP3	NOD-, LRR-, and pyrin domain-containing protein 3
NMDA	*N*-methyl-D-aspartate
NS	nervous system
NT-3	neurotrophin 3
NT-4	neurotrophin 4

Cannabis Use, Neurobiology, Psychology, and Treatment. https://doi.org/10.1016/B978-0-323-89862-1.00012-X

PD	Parkinson's disease
PDK1	phosphoinositide-dependent kinase-1
PH	pleckstrin homology
PIPN	paclitaxel-induced peripheral neuropathy
PI3K	phosphatidylinositol-3-kinase
PIP2	phosphoinositide 4,5-biphosphate
PIP3	phosphatidylinositol (3,4,5)-triphosphate
PKB	protein kinase B
RTKs	receptor tyrosine kinases
TGF-β	transforming growth factor beta
TH	tyrosine hydroxylase
THC	tetrahydrocannabinol (or delta-9-tetrahydrocannabinol)
THCV	tetrahydrocannabivarin
TNF-α	tumor necrosis factor α
Trks	tropomyosin receptor kinases

Introduction

Cannabinoids, particularly cannabidiol (CBD), are being increasingly used for medical purposes. CBD has been shown to reduce seizures and is currently used in the treatment of drug-resistant epilepsy. Sativex, which is a *Cannabis sativa* plant extract that contains CBD and tetrahydrocannabinol (THC) in equivalent amounts, has been approved in several countries to treat spasticity associated to multiple sclerosis (MS) (Legare et al., 2022). Clinical and preclinical studies indicate that CBD possesses antidepressant, anxiolytic, anti-inflammatory, and neuroprotective effects (Crippa et al., 2018).

In terms of pharmacological profile, CBD has low binding affinity for cannabinoid receptor type 1 (CB1R) and cannabinoid receptor type 2 (CB2R), and thus other mechanisms are likely involved in mediating CBD effects (Legare et al., 2022). CBD binds to serotonin receptors, other noncannabinoid G-protein coupled receptors, and the ligand-gated ion channels TRPV1 and TRPV2. Moreover, CBD modulates the tone of the endocannabinoid system by inhibiting uptake of the endocannabinoid arachidonoylethanolamide (AEA), increasing the levels of AEA, which will interact with cannabinoid receptors. CBD is a highly lipophilic compound and may act intracellularly on other targets such as ion channels and enzymes (Silvestro et al., 2020). Despite all this knowledge, the precise biochemical and molecular mechanisms of action of CBD have not been completely elucidated.

The Phosphatidylinositol-3-kinase/AKT (PI3K/AKT) signaling pathway is essential in the regulation of a multitude of biological processes, including cell metabolism, growth, proliferation, and survival. In the central nervous system (CNS) the PI3K/AKT pathway plays a vital role in healthy physiological functioning, while its disruption has been implicated in the pathophysiology of several neurological disorders (Rai et al., 2019).

Recent evidence indicates that CBD modulates the PI3K/AKT signaling pathway, a mechanism that may contribute with CBD's anti-inflammatory, prosurvival, and neuroprotective effects. Also, by modulating the PI3K/AKT pathway, CBD has beneficial effects on behavioral functional endpoints, including memory amelioration, anxiety attenuation, and decrease in epileptic seizures. In the present chapter, we discuss current knowledge that supports the understanding that the modulation of the PI3K/AKT signaling pathway by CBD may underlie its therapeutic effects on neuropsychiatric disorders. This knowledge may contribute to the advance of its therapeutic applications, as well as to the comprehension of its potential undesirable effects.

PI3K/AKT signaling pathway

The PI3K/AKT signaling pathway, described in the 1980s, involves multiple signaling molecules and is essential in the regulation of a multitude of biological processes, including cell metabolism, growth, proliferation, survival, autophagy, and angiogenesis (Jin et al., 2022; Walkowski et al., 2022). In the nervous system, the PI3K/AKT signaling pathway has been implicated in neuronal survival, differentiation, proliferation, motility, nerve regeneration, neuronal migration, and myelination as well as the development of neuronal polarity. Moreover, activation of the PI3K/AKT pathway plays critical roles in signal transmission, synaptic plasticity, memory consolidation, and stress response (Yang, Du, Zhao, Wu, & Yu, 2022).

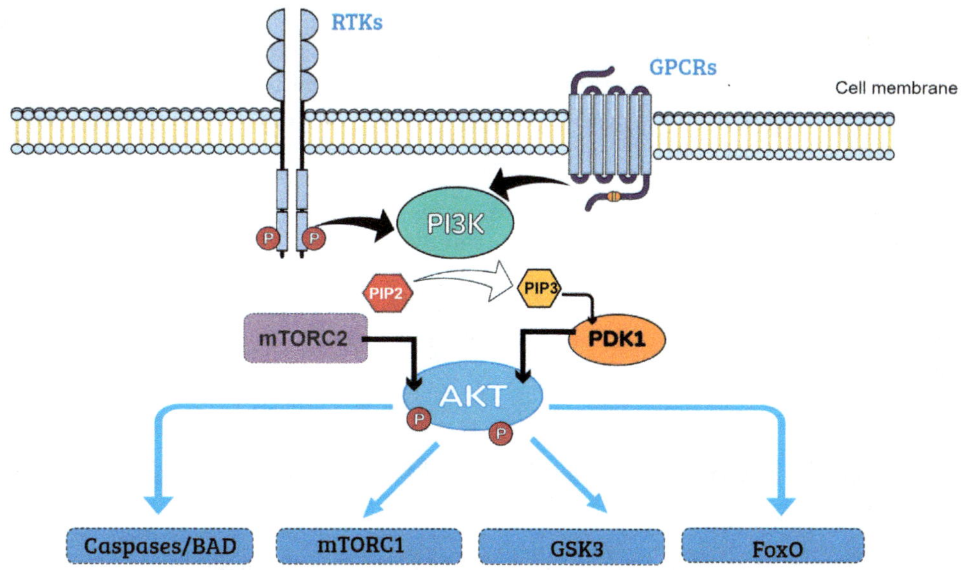

FIG. 1 Schematic representation of the main components and activation of the PI3K/AKT signaling pathway. The PI3K/AKT signaling pathway may be activated by membrane receptors, including G-protein-coupled receptors (GPCRs) and receptor tyrosine kinases (RTKs). Upon the activation of membrane receptors by their specific ligands, which can be hormones, growth factors, neurotransmitters, PI3K is activated and catalyzes the conversion of PIP2 to PIP3. PIP3, in turn, recruits AKT to the cell membrane. PIP3 also induces PDK1 to phosphorylate AKT, and the complete activation of AKT is achieved by its phosphorylation by mTORC2. Once activated, AKT phosphorylates specific target substrates, enhancing or inhibiting their activity. Downstream effectors of AKT, such as FoxO, caspases, mTORC1, and GSK-3, are key components of numerous cell functions, including metabolism, autophagy, cell survival, proliferation, and apoptosis.

PI3K/AKT signaling can be activated upon interaction of ligands with their receptors on the cell surface, including receptor tyrosine kinases (RTKs), cytokine receptors, B cell and T cell receptors, integrins, and G-protein-coupled receptors (GPCRs) (Camaya et al., 2022; Hausott et al., 2022) (Fig. 1). After the interaction of ligands with their specific receptors, PI3K, through its regulatory domain P85, interacts with the receptor either directly, or indirectly, through adapter molecules, which leads to PI3K activation (Camaya et al., 2022).

PI3Ks are a family of lipid kinases comprising three classes (I, II, III), in which class I consists of subclass IA PI3K (α, β, δ) and subclass IB PI3K (γ). PI3K subclasses IA and IB are activated by RTKs or GPCRs, respectively (Rinne et al., 2021). For instance, in the CNS, growth factors, such as nerve growth factor (NGF), brain-derived neurotrophic factor (BDNF), neurotrophin 3 (NT-3), and NT-4, stimulate the PI3K/AKT pathway by interacting with their tropomyosin receptor kinases (Trks) (Hausott et al., 2022). Activated PI3K catalyzes the phosphorylation of the 3' position of phosphoinositide 4,5-biphosphate (PIP2), generating phosphatidylinositol (3,4,5)-triphosphate (PIP3). The buildup of PIP3 interacts with AKT (also called Protein kinase B, PKB) via the pleckstrin homology (PH) domain, recruiting it to the cell membrane (Hausott et al., 2022). PIP3 also binds to phosphoinositide-dependent kinase-1 (PDK1), enabling PDK1 to interact with AKT (Walkowski et al., 2022). PDK1 activates AKT through the phosphorylation at Threonine-308 site, while the full activation of AKT is achieved when the mammalian target of rapamycin complex 2 (mTORC2) complex, phosphorylates it at the Serine-473 site (Rinne et al., 2021).

AKT kinases, the main downstream target of PI3K, consist of a family of three isoforms of serine-threonine kinases, AKT1/PKBα, AKT2/PKBβ, and AKT3/PKBγ (Khezri et al., 2021; Walkowski et al., 2022). AKT isoforms share 85% homology and the structure of three domains, a central catalytic domain, an N-terminal PH domain, and a regulatory domain at the C-terminal (Yang, Du, Zhao, Wu, & Yu, 2022). Once AKT becomes activated at the cell membrane, it can be transported to other cell compartments where it phosphorylates its substrates. The phosphorylation of specific target substrates by AKT may enhance or inhibit their activity, thereby modulating cellular processes. Downstream effectors of AKT include key components of numerous cell functions. For instance, AKT phosphorylates Threonine-24, Serine-256, and Serine-319 of the Forkhead box protein O (FoxO), inactivating it and preventing its proapoptotic and catabolic activities (Camaya et al., 2022; Chen et al., 2022). AKT also phosphorylates cysteine-aspartic proteases (caspases) such as caspase-3, caspase-7, caspase-9 (at Ser196), and BCL-2 associated agonist of cell death (BAD), thus suppressing the apoptotic cascade (Walkowski et al., 2022). Other critical downstream targets of AKT include glycogen synthase kinase-3 (GSK-3) and mTOR. While mTORC2 is an upstream regulator of AKT, mammalian target of rapamycin complex 1 (mTORC1), crucially involved in controlling autophagy and protein synthesis, is a downstream target of AKT (Chen et al., 2022; Querfurth & Lee, 2021). mTORC1 exerts a repressive control of the autophagic cascade.

GSK-3 is a highly conserved serine/threonine protein kinase, which presents two isoforms, GSK-3α and GSK-3β. GSK-3 is constitutively active under basal conditions, and its activity is mainly controlled by inhibition (Marosi et al., 2022). In general, AKT-mediated phosphorylation at the amino-terminal Serine-9 residue of GSK-3β, and Serine-21 of GSK-3α suppresses its activity (Camaya et al., 2022; Hausott et al., 2022). Considerable evidence indicates that GSK-3 hyperactivity may be implicated in psychiatric conditions, cognitive dysfunction, and neurodegeneration (Marosi et al., 2022; Ruiz & Eldar-Finkelman, 2022). Thus, great efforts are being made to discover GSK-3 inhibitors with potential application in treatment of neuropsychiatric diseases.

The following sections discuss current findings that relate the modulation of PI3K/AKT signaling by CBD with its therapeutic effects. Considering the myriad of cellular processes regulated by the PI3K/AKT pathway, this may represent a relevant mechanism underlying CBD effects.

PI3K/AKT signaling pathway is involved in CBD-induced neuroprotection: Evidence from animal models

Experimental multiple sclerosis (MS) can be modeled by inducing experimental autoimmune encephalomyelitis (EAE), in which mice are immunized with myelin oligodendroglial glycoprotein peptide (MOG)35-55. (MOG)35–55-immunized mice exhibit similar clinical features of the human disease, such as limbs paralysis and muscle mass loss, resulting in a high clinical score of disease. This is accompanied by pathological alterations that resemble MS in humans, such as demyelination and infiltration of inflammatory cells in the CNS. Treatment with CBD for 14 days in EAE mice significantly reduced clinical score and proinflammatory cytokines in EAE. Moreover, findings indicated that EAE downregulated PI3K/AKT/mTOR pathway, while CBD was able to restore it, significantly increasing the phosphorylation of PI3K, AKT, and mTOR in the spinal cord, when compared to the nontreated EAE group. These findings suggest that the therapeutic effects of CBD involve the activation of PI3K/AKT/mTOR signaling, together with reduction of proinflammatory mediators (Giacoppo et al., 2017). Corroborating these findings, in vitro experiments demonstrated that the PI3K/AKT/mTOR signaling pathway may play a role in mediating the effects of cannabinoids on oligodendrocytes. CB1R and CB2R agonists, arachidonyl-2′-chloroethylamide (ACEA), dimethylbutyl-deoxy-delta-8-tetrahydrocannabinol (JWH133), and (−)-1,1-dimethylheptyl-11-hydroxy-Delta-8-tetrahydrocannabinol (HU210), enhanced differentiation of oligodendrocyte progenitor cells and induced a time-dependent phosphorylation of AKT and mTOR. In addition, the inhibition of PI3K/AKT and mTOR with LY294002 and rapamycin, respectively, blocked cannabinoids effects on oligodendrocyte proliferation and kinase activation (Gomez et al., 2011). Moreover, trophic support deprivation in oligodendrocyte progenitor cells, downregulates AKT activity, leading to apoptosis. Exposure to CB1R agonist ACEA and the nonselective cannabinoid agonists HU210 and (+)-Win-55212-2 increased AKT phosphorylation and cell survival, and these effects were abolished by PI3K inhibition (Molina-Holgado et al., 2002).

The effects of CBD were also investigated in an experimental model of brain injury induced by hemorrhagic shock (HS) followed by reperfusion in rats. CBD, when administered intravenously during the 90-min HS, attenuated neuronal apoptosis and increased expression of crucial proteins in autophagy, Beclin-1 and LC3II. Levels of AKT phosphorylation (at Serine-473 and Threonine-308) increased in response to CBD treatment in HS-challenged rats compared with those treated with vehicle, indicating that CBD upregulated the PI3K/AKT signaling pathway. Interestingly, CBD-induced activation of autophagy was suppressed by a PI3K inhibitor, demonstrating that PI3K/AKT signaling is involved in CBD regulation of autophagy. This study also showed that the use of a selective CB1R antagonist, but not of a CB2R antagonist, prevented CBD-induced increase in AKT phosphorylation, suggesting that CBD stimulates the PI3K/AKT signaling pathway via CB1R activation (Lin et al., 2020). Confirming these findings, the CB1R agonist WIN55,212-2 and the fatty acid amide hydrolase (FAAH) inhibitor, URB597 ameliorated cognitive impairment associated with chronic cerebral hypoperfusion, while increasing AKT, GSK-3β, and CREB phosphorylation, as well as BDNF expression in the hippocampus and frontal cortex of rats. These effects were blocked by co-treatment with LY294002, a PI3K/AKT inhibitor (Su et al., 2016), suggesting that the activation of PI3K/AKT pathway by cannabinoids is critical for their neuroprotective and memory-rescuing effects. ACEA, a highly selective CB1R agonist improved neurological behavior, reduced infarct volume, and inhibited apoptosis in an animal model of cerebral ischemia, and these effects were accompanied by increased GSK-3β phosphorylation (Bai et al., 2017). In this line, Blázquez et al. (2015) demonstrated that CB1R agonists, delta-9 tetrahydrocannabinol (THC), and HU-210 increase cell viability and phosphorylation of AKT and ribosomal S6 protein, a canonical substrate of the AKT/mTORC1 pathway in STHdhQ7/Q7 mouse striatal neuroblasts. Moreover, THC and HU-210 rescued cells from

N-methyl-D-aspartate (NMDA)-induced death, and incubation with wortmannin, a PI3K inhibitor, Akti-1/2, a highly selective AKT inhibitor, or rapamycin, an mTORC1 inhibitor, abolished cannabinoid-evoked cytoprotection. The neuroprotective effects of CB1R agonists against excitotoxity were also investigated in an in vivo retinal model of AMPA (α-amino-3-hydroxy-5-methyl-4-isoxazolepropionic acid)-induced excitotoxicity. The endocannabinoid N-arachidonoyl ethanolamine (anandamide/AEA) and the synthetic cannabinoid HU-210 induced phosphorylation of AKT, and the use of selective CB1R antagonist AM251 and wortmannin reversed the cannabinoid-induced neuroprotection (Kokona & Thermos, 2015). These findings suggest the involvement of CB1 receptors in the activation of PI3K/AKT pathway and neuroprotection induced by cannabinoids.

Recently, there has been increasing interest in the use of CBD in the treatment of epilepsy. Many studies have demonstrated the antiseizure effects of CBD in animal models and in humans. However, the exact mechanism by which CBD yields anticonvulsant effects remains unclear. In a recent study, Lima et al. (2020) suggested that the anticonvulsant effects, involve at least in part, the PI3K/AKT pathway. CBD treatment increased latency and reduced severity of pilocarpine-induced seizures in wild-type mice. CBD also prevented microgliosis, astrocytosis, and neuronal cell death in the hippocampus of mice injected with pilocarpine. These effects were absent in mice lacking PI3K (PI3K$\gamma^{-/-}$). In addition, in primary cultures of hippocampal neurons, PI3Kγ inhibition or deficiency prevented CBD-induced neuroprotection against glutamatergic neurotoxicity. Together, these findings suggest that PI3K/AKT pathway may be involved in the anticonvulsant and neuroprotective effects of CBD. Interestingly, one study described the effects of PI3K/AKT signaling following CB2R activation in pilocarpine-induced status epilepticus in rats. The levels of p-AKT and PI3K 110α were significantly decreased in rats with epilepsy compared with the control group, while the use of CB2R agonist, JWH133, significantly upregulated p-AKT and PI3K 110α expression compared with rats treated with pilocarpine only (Cao et al., 2018).

CBD reversed the memory-impairing effects of ovariectomy, an experimental model used to investigate the effects of menopause on CNS functions. Menopause, characterized by a decline in estrogen levels, may affect diverse CNS functions, including synaptic plasticity, survival, proliferation, and protection against age-associated deterioration in women (Bustamante-Barrientos et al., 2021). The memory-ameliorating effects of CBD in this model were accompanied by a reversion of ovariectomy-induced downregulation of AKT and GSK-3β phosphorylation in the hippocampus. Moreover, CBD also increased the levels of the antiapoptotic protein, Bcl2. Thus, PI3K/AKT signaling activation may result in protection against apoptosis and preserve cognitive functions (Corrê et al., 2022).

Clinical and preclinical studies have demonstrated that CBD displays anxiolytic properties, and a study by Campos et al. (2013) showed that the anxiolytic effects of CBD in a chronic unpredictable stress (CUS) model involve adult hippocampal neurogenesis. A recent study from the same research group revealed that CBD given to animals subjected to CUS increased hippocampal neurogenesis and spine density in the hippocampal dentate gyrus. CBD also increased phosphorylated GSK-3β expression in stressed animals. The anxiolytic effects and the effects on spine density were blocked by pretreatment with a CB1R (AM251) or CB2R (AM630) antagonist, while the effects on p-GSK-3β expression were attenuated by AM630 (Fogaça et al., 2018). GSK-3β phosphorylation inhibits its enzymatic activity, and it has been suggested that aberrant GSK-3β activity could be a risk factor for neuropsychiatric and neurodegenerative disorders (Marosi et al., 2022).

CBD has been proposed as a potential treatment for inflammatory- and neuropathic-associated pain. Thus, a recent study by Kumar Kalvala et al. (2022) investigated the mechanisms involved in CBD protection against paclitaxel-induced peripheral neuropathy (PIPN). CBD and tetrahydrocannabivarin (THCV) reduced thermal and mechanical hyperalgesia in mice treated with placlitaxel. Rimonabant, a CB1R antagonist, and AM630, CB2R antagonist, prevented the effects of CBD on mechanical and thermal hypersensitivity. PIPN increased the expression of PI3K, p-AKT, p-P38 MAP kinase, BAX, TGF-β (Transforming growth factor beta), NLRP3 (NOD-, LRR-, and pyrin domain-containing protein 3) inflammasome, and caspase 3 in dorsal root ganglia (DRG) homogenates, while CBD and THCV reversed these effects. To further characterize the effects of CBD and THCV, in vitro experiments indicated that CBD and THCV ameliorated placlitaxel-induced reduction in neurite outgrowth and percentage of neurite-bearing cells in primary DRG neurons. The activation of PI3K/AKT signaling pathway may be involved in the progression of neuropathic pain, previous evidence indicate that it regulates inflammatory markers, such as TNF-α (tumor necrosis factor α) and NLRP3 inflammasome. Thus, it is possible that CBD and THCV, by reversing placlitaxel-induced effects on PI3K/AKT signaling pathway, thus modulating inflammatory pathways, may reduce neuropathic pain.

Yang and collaborators (Yang et al., 2020) investigated the involvement of AKT/GSK-3β signaling pathway in the therapeutic effects of CBD on conditioned place preference (CPP) induced by methamphetamine, a highly addictive psychostimulant. CBD suppressed methamphetamine-induced CPP in a dose dependent fashion. Moreover, methamphetamine increased the protein levels of AKT and p-AKT in the prefrontal cortex, nucleus accumbens, and p-AKT levels were elevated in the prefrontal cortex, nucleus accumbens, hippocampus, and ventral tegmental area. p-GSK-3β and p-GSK-3β/GSK-3β ratio were also increased in the prefrontal cortex, nucleus accumbens, hippocampus and ventral tegmental area, compared with controls. Pretreatment with CBD before methamphetamine administration, mainly attenuated p-AKT/AKT ratios and reduced the expression levels of p-GSK-3β and p-GSK-3β/GSK-3β ratio. These findings suggest that the imbalance in the PI3K/AKT/GSK-3β signaling pathway may play a critical role on drug-induced neurotoxicity. CBD attenuated methamphetamine-induced behavioral alterations related to drug abuse, while restoring PI3K/AKT/GSK-3β signaling pathway in limbic regions related to drug reward system. Cocaine-induced neurotoxicity was reversed using a FAAH inhibitor, URB597, or the CB1R selective agonist, ACEA. Pretreatment with these compounds inhibited cocaine-induced seizures and epileptiform electroencephalographic activity in mice, while URB597 also inhibited cocaine-induced hippocampal cell death both in vivo and in primary cell culture. URB597 injection enhanced ERK and AKT phosphorylation in the hippocampus, and its neuroprotective effects in the hippocampus were abrogated by the PI3K inhibitor, LY294002, suggesting that the activation of PI3K/AKT signaling pathway plays a critical role in protecting against cocaine intoxication models (Vilela et al., 2015).

Table 1 summarizes the animal studies demonstrating that the modulation of the PI3K/AKT pathway underly, at least in part, CBD therapeutic effects. Experimental findings also show that the blockade of this pathway abolishes the beneficial effects of CBD.

TABLE 1 Evidence of the effects of CBD, acting through the PI3K/AKT signaling pathway, in in vivo models.

Experimental model	Dose/regimen of treatment of CBD	Effects of CBD	Reference
In vivo			
Experimental model of MS (C57BL/6 mice)	10 mg/kg for 14 days	↑ Phosphorylation of PI3K, AKT, and mTOR in spinal cord	Giacoppo et al. (2017)
Paclitaxel-induced peripheral neuropathy (PIPN) (C57BL/6 mice)	10 mg/kg CBD (i.p.) twice a week for six weeks following the previous PTX injection	CBD reversed Paclitaxel-induced increase in PI3K e p-AKT in dorsal root ganglia	Kumar Kalvala et al. (2022)
Pilocarpine-induced behavioral seizures (C57BL/6 wild type or PI3K$\gamma^{-/-}$ mice)	30, 60 or 90 mg/kg (i.p.), 60 min before Pilocarpine (single dose)	CDB anticonvulsant effect was abolished in PI3K$\gamma^{-/-}$ mice	Lima et al. (2020)
Rat model of hemorrhagic shock (HS)-induced brain injury	5 mg/kg administered via the intravenous route following the 90-mins HS, after which reinfusion of the shed blood was performed	↑ Phosphorylation of AKT (Ser473 and Thr308) in cerebral cortex ↑ Autophagy and ↓ apoptosis in cerebral cortex, these effects were blocked by the PI3K inhibitor, LY294002	Lin et al. (2020)
Methamphetamine (METH)-induced conditioned place preference (CPP) in rats	10, 20, 40, and 80 mg/kg CBD administered (i.p.) 1 h prior to the injection of saline or METH	CBD dose-dependently attenuated p-AKT/AKT ratio in prefrontal cortex, nucleus accumbens, hippocampus, and ventral tegmental area compared to the METH-only group ↓ p-GSK3 and p-GSK3/GSK3 ratio	Yang et al. (2020)
Anxiogenic effects induced by chronic unpredictable stress (CUS) in male C57BL/6 mice	30 mg/kg concomitant with CUS procedure for 14 days	↑ p-GSK3 levels in comparison to stressed and nonstressed vehicle animals ↑ Hippocampal neurogenesis and spine density in the dentate gyrus of the hippocampus	Fogaça et al. (2018)
Memory deficits associated to ovariectomy	10 mg/kg for 14 days	CBD reversed ovariectomy-induced memory deficits ↑ p-AKT/AKT ratio ↑ p-GSK3/GSK3 ratio ↑ Antiapoptotic protein Bcl2	Corrê et al. (2022)

Studies investigating the effects of CBD using animal models of nervous system injury, neurodegeneration, neuropathic pain, epilepsy, anxiety, and behavioral deficits. CBD doses and regimen of administration are detailed, as well as main outcomes related to the PI3K/AKT pathway modulation by CBD.

Use of in vitro models to elucidate the involvement of PI3k/AKT signaling pathway in the neuroprotective effects of Cannabidiol

Alzheimer's disease (AD), the most prevalent neurodegenerative disorder worldwide, is characterized by memory loss and progressive cognitive decline. The neuropathological hallmarks of AD are the presence of extracellular aggregates of amyloid-beta (Aβ), known as amyloid plaques, and neurofibrillary tangles (NFT), constituted of intracellular aggregates of hyperphosphorylated tau protein (Scheltens et al., 2021). GSK-3β has been implicated in the pathogenesis of AD, because it can phosphorylate tau. In a cellular model, in which PC12 cells were exposed to Aβ(1–42), CBD increased cell viability and reversed Aβ-induced GSK-3β increased phosphorylation, in a concentration-dependent manner. CBD also protected against Aβ-induced tau hyperphosphorylation (Esposito et al., 2006). The investigation of the effects of CBD on mesenchymal stem cells (MSCs) indicated that CBD downregulated genes related to AD, including genes coding for kinases involved with aberrant tau phosphorylation, such as GSK-3β, and secretases involved in the process of amyloidogenesis, such as the β-(BACE-1) and γ-secretases. Moreover, CBD upregulated genes coding for PI3K subunits and for AKT1. When examining the activation of the PI3K/AKT signaling pathway, results confirmed that CBD increased cytoplasmatic expression of p-PI3K, AKT, and p-AKT, which resulted in a reduction of GSK-3β and p-GSK-3β. The administration of a vanilloid receptor subtype transient receptor potential (TRPV1) antagonist, Capsazepine, but not CB1R or CB2R antagonists, counteracted the effects of CBD on GSK-3β, PI3K, and AKT (Libro et al., 2016).

Recently, other cellular mechanisms associated to CBD neuroprotective effects against Aβ have been investigated. Primary microglial cells in culture were exposed to fluorescein isothiocyanate (FITC)-labeled Aβ to examine whether the engulfment of the fluorescent Aβ peptides by microglia would be stimulated by CBD. CBD treatment significantly increased the efficiency of FITC-Aβ uptake by microglial cells. The use of a TRPV2 channel antagonist, tranilast, or TRPV2 siRNA inhibited CBD-induced Aβ uptake by microglial cells, indicating that TRPV2 activation is critical for CBD-associated Aβ phagocytosis. Moreover, CBD enhanced AKT phosphorylation and the expression of TRPV2 in microglial cells exposed to Aβ. Levels of autophagy-related proteins, LC3B-II, p62, and Beclin-1, were also upregulated by CBD, while the use of tranilast lessened CBD-induced AKT phosphorylation, p62 and Beclin-1. The authors concluded that the modulation of autophagy mediated by AKT phosphorylation in crucial for TRPV2-mediated Aβ phagocytosis and clearance. Interestingly, TRPV2 was significantly reduced in the entorhinal cortex and hippocampal samples of AD patients compared with the controls, and in the cortex and hippocampus of the AD transgenic APP/PS1 mouse brain compared with the levels in the wild-type littermates (Yang, Du, Zhao, Tang, et al., 2022). A previous study has also showed that CBD enhances the phagocytic properties of microglial cells (Hassan et al., 2014). CBD increased fluorescently labeled latex beads by cultured microglial cells, and this effect was also mediated by TRPV2. CBD induced the translocation of TRPV2 to the membrane, which was abolished with the use of wortmannin, a PI3K inhibitor. Wortmannin also prevented CBD-induced phagocytosis, suggesting that the PI3K/AKT pathway mediates CBD effects of increasing microglial phagocytosis. Consistently with these findings, Reichhart and collaborators (Reichhart et al., 2015) demonstrated that CBD stimulated TRPV2 surface expression and consequent membrane conductance in the retinal pigment epithelium. Interestingly, this effect was blocked in the presence of the PI3K inhibitor, LY294002. Taken together, these findings support the use of CBD in treating neuroinflammatory conditions.

Parkinson's disease (PD) is a motor neurodegenerative disorder, with progressive loss of dopaminergic neurons in the nigrostriatal pathway. PD can be modeled in vivo and in vitro by using the neurotoxin 1-methyl-4-phenyl-1,2,3,6-tetrahydropyridine (MPTP) or its active toxic metabolite, 1-methyl-4-phenylpyridinium (MPP+), which is selectively carried through dopamine transporters and accumulates in dopaminergic neurons (Dionísio et al., 2021). The neuroprotective effects of CBD were evaluated in retinoic acid differentiated SH-SY5Y cells exposed to MPP+. In this cellular model, CBD reversed cell viability loss, apoptosis, and reduction of tyrosine hydroxylase (TH) induced by MPP+. MPP+ also decreased the levels of pERK, pAKT, and p-mTOR, while CBD treatment restored pAKT and p-mTOR levels. Remarkably, the use of AKT1/2 inhibitor or mTOR inhibitor, rapamycin, suppressed CBD neuroprotective effects. These findings suggest that the AKT/mTOR pathway, by modulating autophagy, is directly involved in neuroprotective effects induced by CBD (Gugliandolo et al., 2020). The selective CB2R agonist AM1241 reversed motor deficits in MPTP-treated mice, increased striatal dopamine and serotonin levels, and nigral TH-positive cells. These neuroprotective effects against the dopaminergic neurotoxin were related to PI3K/AKT/MEK activation in the substantia nigra and hippocampus in these animals. Accordingly, AM1241 increased pAKT and p-PI3K and upregulated Parkin/PINK in PD mice in comparison to untreated controls, confirming that the activation of PI3K/AKT is involved in dopaminergic neurons protection (Shi et al., 2017).

The effects of CBD on the PI3K/AKT/autophagy pathway were investigated in human neuroblastoma SH-SY5Y and murine astrocyte cell lines (Vrechi et al., 2021). CBD stimulated the expression of LC3II, indicating the accumulation of autophagosomes, without reducing cell viability. The effects of CBD in promoting autophagic flux was prevented by pre-incubation with CB1R, CB2R, and TRPV1 antagonists, AM251, AM630, and CPZ, respectively. Remarkably, CBD increased ERK1/2 phosphorylation, while decreasing pAKT levels. Although this study was done in the same cell line used in the study by Gugliandolo and coworkers, cells were not challenged with neurotoxins, such as MPP+, suggesting that CBD effects may diverge, depending on the conditions to which cells are exposed to.

Findings from cellular studies are outlined in Table 2. The regulation of PI3K/AKT/GSK3-β and autophagy are implicated in the neuroprotective effects of CBD, investigated with the use of cellular models of neurodegenerative diseases, such as AD and PD.

TABLE 2 Evidence of the effects of CBD, acting through the PI3K/AKT signaling pathway, in in vitro models.

Experimental model	CBD concentration	Effects of CBD	Reference
In vitro			
PC12 neuronal cells treated with β-amyloid (1–42) for 24 h	10^{-7}–10^{-5} M	CBD inhibited Aβ-induced p-GSK-3β levels	Esposito et al. (2006)
Primary cultures of microglial cells	5–10 μmol pretreatment for 12-24 h before A-β(1–42) exposure	↑ Microglial β-amyloid phagocytosis through TRPV2 activation, this effect was blocked by Akt inhibitor ↑ Phosphorylation of Akt in microglia cells ↑ Levels of autophagy related proteins	Yang, Du, Zhao, Tang, et al. (2022)
BV-2 microglial cells using two sorts of latex (amino-modified polystyrene) beads with or without fluorescent as phagocytosis targets	100 nM, 1 μM, 10 μM for 24 h	↑ Microglial phagocytosis through TRP channels ↑ Translocation of TRPV2 to the membrane and this effect was blocked by wortmannin, a PI3K inhibitor ↑ Phosphorylation of Akt in microglia cells ↑ Levels of autophagy related proteins	Hassan et al. (2014)
Mesenchymal stem cells (MSCs) derived from gingiva	5 μM for 24 h	↑ p-PI3K, PI3K, p-Akt, and Akt levels ↑ Expression of genes coding PI3K subunits (PIK3CA and PIK3CB) and Akt1 ↓ GSK3 and p-GSK3	Libro et al. (2016)
Retinal pigment epithelium	15 μM for 10 min	↑ Current density ↑ TRPV2 surface expression PI3K inhibitor LY294002 abolished these effects	Reichhart et al. (2015)
Human neuroblastoma SH-SY5Y and murine astrocyte cell lines	10 μM CBD for 5, 10, 30 min, 1 and 2 h	↑ Autophagy (LC3-II levels) ↑ ERK1/2 phosphorylation ↓ AKT phosphorylation at Ser473	Vrechi et al. (2021)
In vitro PD model (retinoic acid differentiated SH-SY5Y cells exposed to 1-methyl-4-phenylpyridinium (MPP+))	Pre-treatment with 10 μM CBD for 24 h followed by 1 mM MPP+, with or without 10 μM CBD, or 10 μM CBD alone for 48 h	↑ AKT phosphorylation ↑ Increased cell viability	Gugliandolo et al. (2020)

Beneficial effects of CBD mediated by the modulation of the PI3K/AKT signaling on in vitro models of neurodegeneration, including AD and PD, and effects of CBD on the PI3K/AKT signaling using mesenchymal stem cells, retinal epithelium, SH-SY5Y, and murine astrocyte cell lines.

Concluding remarks

The findings discussed in this chapter deepen the current knowledge on the cellular and molecular mechanisms underlying the therapeutic effects of cannabinoids, particularly CBD, on diseases affecting the nervous system. Evidence from animal and in vitro models indicates that CBD acts through the PI3K/AKT and its main downstream targets, GSK3 and autophagy, promoting neurogenesis, anti-inflammatory, antiapoptotic, and survival effects. Moreover, the studies using animal models also indicate that CBD has beneficial behavioral functional effects that might be related to the modulation of the PI3K/AKT signaling. CBD has its use approved in many countries for the treatment of drug-resistant epilepsy. Considering the current knowledge of the involvement of the PI3K/AKT signaling in neuropsychiatric disorders, and the potential of CBD in modulating it, the therapeutic uses of CBD may be expanded to other CNS diseases.

Applications to other areas

In this chapter, we have reviewed evidence that suggest that beneficial effects of CBD on neuropsychiatric disorders may be mediated, ate least in part, by the modulation of the PI3K/AKT signaling and its downstream targets. Evidence presented come from animal models of multiple sclerosis, neuropathic pain, epilepsy, drug abuse, anxiety/depression, and memory deficits. Also, the use of cellular in vitro models allowed the investigation of the neuroprotective effects of CBD on neurodegenerative diseases such as AD and PD. In humans, previous studies suggest that PI3K/AKT signaling disturbances may underlie the pathogenesis of neuropsychic disorders. A recent study of exome-sequenced samples of 4600 subjects with AD found an association of increased susceptibility to late-onset AD with damaging coding variants in the *PI3K* gene, suggesting that increased activity of the PI3K/AKT signaling pathway is protective against AD (Curtis et al., 2020). Previously, a post-mortem study with brains from AD patients showed that active GSK-3β, one of the main downstream targets of PI3K/AKT, is colocalized with the progressive accumulation of hyperphosphorylated tau and neurofibrillary tangles, following the Braak staging (Pei et al., 1999). A post-mortem study also showed that total and pAKT are reduced in midbrain fractions of PD patients compared to non-PD controls. Double immunofluorescence microscopy indicated that AKT is highly expressed in nigral dopaminergic neurons in control brains, thus relating a marked neuronal loss of AKT in PD (Timmons et al., 2009). Additionally, post-mortem studies have also demonstrated a lowered activity of the PI3K/AKT signaling pathway in the ventral prefrontal cortex in depression and suicidal patients (Karege et al., 2007, 2011).

Notably, there is a significant concordance between the findings from animal and cellular models with findings from human studies, concerning the involvement of aberrant PI3K/AKT signaling in neuropsychiatric disorders. Thus, drugs that modulate this pathway, such as CBD and possibly other cannabinoids, may serve as candidates for novel therapeutic strategies for diseases affecting the nervous system.

Mini-dictionary of terms

- **Chronic unpredictable stress (CUS) model.** CUS model is an experimental model that aims to induce depressive-like phenotypes in rodents, through chronical exposure to interchanging unpredictable stressors, such as wet cage, dampened sawdust, tilted cage, empty cage, social stress, restraint, and disruption of light/dark cycle.
- **Conditioned place preference (CPP).** CPP is a model used to investigate the role of context associations in reward-induced behaviors, such as those induced by drugs of abuse. CPP comprises the association of a particular environment with drug treatment.
- **G-protein-coupled receptors (GPCRs).** G-protein-coupled receptors (GPCRs) constitute the largest family of membrane receptors, activated by a multitude of ligands, including ions, odorants, hormones, neurotransmitters, and other signaling molecules. GPCRs are formed by seven transmembrane domains, and their carboxi terminus extends intracellularly where interacts with heterotrimeric G proteins, which activate intracellular signaling cascades.
- **Microgliosis.** microgliosis refers to the activation of microglial cells in response to central nervous system injury. Microglial activation includes morphological changes and cellular hypertrophy.
- **Protein phosphorylation.** a reversible post-translational modification, achieved by the addition of a phosphate group to amino acid residues, catalyzed by protein kinases. Phosphorylation induces conformational changes that can activate or inhibit target proteins.
- **Receptor tyrosine kinases (RTKs).** Receptor tyrosine kinases (RTK) are a broad family of ubiquitously distributed membrane receptors, which respond to extracellular signals, including neurotransmitter, hormones, and growth factors. Upon the binding of their specific ligands, RTKs undergo dimerization, triggering mutual phosphorylation of tyrosine residues located in intracellular domains, which leads to receptor activation.

Key facts

Key facts of apoptosis

- Apoptosis is a type of programmed cell death that occurs through energy-consuming biochemical processes.
- Apoptosis is recognized as an essential physiological process, particularly during development of the nervous system.
- Apoptotic cell death is characterized by morphological alterations that include cell shrinkage, chromatin condensation, plasma membrane blebbing, followed by separation of cell fragments forming apoptotic bodies, which are phagocytosed by macrophages.
- There are two apoptotic pathways: the extrinsic pathway, which is initiated by activation of death receptors at the cell surface, and the intrinsic pathway, which is mainly regulated by mitochondria.
- Biochemical modifications include protein degradation, achieved by activation of proteases called caspases, and DNA fragmentation.
- Aberrant apoptosis has been implicated in excessive cell death linked to neurodegenerative diseases and hypoxia/ischemia-associated injury.

Key facts of amyloidogenesis

- Amyloidogenesis refers to the formation of amyloid structures, with specific cross-β sheet conformation, which tend to aggregate into insoluble fibrils that accumulate in disease states, including many neurodegenerative disorders.
- Amyloidogenesis has been implicated in the pathogenesis of Alzheimer's disease (AD), where β-secretase-induced cleavage of the amyloid precursor protein (APP) followed by a second cleavage catalyzed by γ-secretase forms amyloid β peptide (Aβ) that aggregate.
- APP is an integral membrane glycoprotein, expressed in a variety of neural and non-neural cells, and can be processed in a nonamyloidogenic pathway.
- Extracellular aggregates of Aβ constitute amyloid plaques, also known as senile plaques, a pathological hallmark of AD.
- Amyloid plaques were observed for the first time in the brain in 1892, by Alois Alzheimer.

Key facts of autophagy

- Autophagy is a highly conserved cellular recycling process, which plays an important role in cell survival and maintenance.
- Cytoplasmic organelles, proteins, and other macromolecules are degraded, and the breakdown products are reutilized in biosynthetic pathways or in energy production.
- Cells may respond to a variety of extra- and intracellular signals of stress, such as nutrient starvation, hypoxia, and increased cytosolic Ca^{2+} concentration, by inducing autophagy.
- Autophagy may be also viewed as a cellular quality control mechanism, being responsible for the clearance of malformed proteins, which may result in protein aggregates found in neurodegeneratives disorders.
- Autophagy is a complex multistep cascade, mainly under repressive control of mTORC1.
- Autophagy and its regulators have attracted interest as potential therapeutic targets for the treatment of neurodegerative diseases.

Key facts of neuropathic pain

- Neuropathic pain is a chronic condition that results from damage to the peripheral and/or central somatosensory system.
- Injury of the somatosensory nervous system may result in aberrant neural transmission of sensory signals into the spinal cord and brain.
- Neuropathic pain is associated with multiple causes, but increased incidence has been related to aging, increased incidence of diabetes mellitus, and survival from cancer after chemotherapy.
- Peripheral neuropathy can be induced by chemotherapeutic drugs and constitutes a major limitating factor during cancer treatment.
- There are no preventive treatments for chemotherapy-induced peripheral neuropathic pain.

Summary points

- CBD mechanisms of action have not been completely elucidated.
- The phosphatidylinositol-3-kinase/AKT (PI3K/AKT) signaling pathway is essential in the regulation of various essential biological processes, including cell metabolism, growth, proliferation, and survival.
- Recent evidence indicates that CBD modulates the PI3K/AKT signaling pathway.
- Regulation of the PI3K/AKT signaling pathway may underlie, at least in part, CBD's anti-inflammatory, pro survival, and neuroprotective effects.
- Modulation of the PI3K/AKT signaling pathway by CBD has been shown to be associated with recovery of behavioral functional endpoints, such as anxiety relief, memory deficits, pain sensibility, and severity of seizures.

References

Bai, F., Guo, F., Jiang, T., Wei, H., Zhou, H., Yin, H., Zhong, H., Xiong, L., & Wang, Q. (2017). Arachidonyl-2-Chloroethylamide alleviates cerebral ischemia injury through glycogen synthase kinase-3β-mediated mitochondrial biogenesis and functional improvement. *Molecular Neurobiology*, *54*, 1240–1253. https://doi.org/10.1007/s12035-016-9731-7.

Blázquez, C., Chiarlone, A., Bellocchio, L., Resel, E., Pruunsild, P., García-Rincón, D., Sendtner, M., Timmusk, T., Lutz, B., Galve-Roperh, I., & Guzmán, M. (2015). The CB_1 cannabinoid receptor signals striatal neuroprotection via a PI3K/Akt/mTORC1/BDNF pathway. *Cell Death and Differentiation*, *22*, 1618–1629. https://doi.org/10.1038/cdd.2015.11.

Bustamante-Barrientos, F. A., Méndez-Ruette, M., Ortloff, A., Luz-Crawford, P., Rivera, F. J., Figueroa, C. D., Molina, L., & Bátiz, L. F. (2021). The impact of estrogen and estrogen-like molecules in neurogenesis and neurodegeneration: Beneficial or harmful? *Frontiers in Cellular Neuroscience*, *15*, 636176. https://doi.org/10.3389/fncel.2021.636176.

Camaya, I., Donnelly, S., & O'Brien, B. (2022). Targeting the PI3K/Akt signaling pathway in pancreatic β-cells to enhance their survival and function: An emerging therapeutic strategy for type 1 diabetes. *Journal of Diabetes*, *14*, 247–260. https://doi.org/10.1111/1753-0407.13252.

Campos, A. C., Ortega, Z., Palazuelos, J., Fogaca, M. V., Aguiar, D. C., Diaz-Alonso, J., Ortega-Gutierrez, S., Vazquez-Villa, H., Moreira, F. A., Guzman, M., Galve-Roperh, I., & Guimaraes, F. S. (2013). The anxiolytic effect of cannabidiol on chronically stressed mice depends on hippocampal neurogenesis: Involvement of the endocannabinoid system. *International Journal of Neuropsychopharmacology*, *16*, 1407–1419. https://doi.org/10.1017/S1461145712001502.

Cao, Q., Liu, X., Yang, F., & Wang, H. (2018). CB2R induces a protective response for epileptic seizure via the PI3K 110α-AKT signaling pathway. *Experimental and Therapeutic Medicine*, *16*, 4784–4790. https://doi.org/10.3892/etm.2018.6788.

Chen, K. Q., Wei, B. H., Hao, S. L., & Yang, W. X. (2022). The PI3K/AKT signaling pathway: How does it regulate development of Sertoli cells and spermatogenic cells? *Histology and Histopathology*, *18457*. https://doi.org/10.14670/HH-18-457.

Corrê, M. D. S., de Freitas, B. S., Machado, G. D. B., Pires, V. N., Bromberg, E., Hallak, J. E. C., Zuardi, A. W., Crippa, J. A. S., & Schröder, N. (2022). Cannabidiol reverses memory impairments and activates components of the Akt/GSK3β pathway in an experimental model of estrogen depletion. *Behavioural Brain Research*, *417*, 113555. https://doi.org/10.1016/j.bbr.2021.113555.

Crippa, J. A., Guimarães, F. S., Campos, A. C., & Zuardi, A. W. (2018). Translational investigation of the therapeutic potential of cannabidiol (CBD): Toward a new age. *Frontiers in Immunology*, *9*, 1–16. https://doi.org/10.3389/fimmu.2018.02009.

Curtis, D., Bakaya, K., Sharma, L., & Bandyopadhyay, S. (2020). Weighted burden analysis of exome-sequenced late-onset Alzheimer's cases and controls provides further evidence for a role for PSEN1 and suggests involvement of the PI3K/Akt/GSK-3β and WNT signalling pathways. *Annals of Human Genetics*, *84*, 291–302. https://doi.org/10.1111/ahg.12375.

Dionísio, P. A., Amaral, J. D., & Rodrigues, C. M. P. (2021). Oxidative stress and regulated cell death in Parkinson's disease. *Ageing Research Reviews*, *67*, 101263. https://doi.org/10.1016/j.arr.2021.101263.

Esposito, G., De Filippis, D., Carnuccio, R., Izzo, A. A., & Iuvone, T. (2006). The marijuana component cannabidiol inhibits beta-amyloid-induced tau protein hyperphosphorylation through Wnt/beta-catenin pathway rescue in PC12 cells. *Journal of Molecular Medicine (Berlin)*, *84*, 253–258. https://doi.org/10.1007/s00109-005-0025-1.

Fogaça, M. V., Campos, A. C., Coelho, L. D., Duman, R. S., & Guimarães, F. S. (2018). The anxiolytic effects of cannabidiol in chronically stressed mice are mediated by the endocannabinoid system: Role of neurogenesis and dendritic remodeling. *Neuropharmacology*, *135*, 22–33. https://doi.org/10.1016/j.neuropharm.2018.03.001.

Giacoppo, S., Pollastro, F., Grassi, G., Bramanti, P., & Mazzon, E. (2017). Target regulation of PI3K/Akt/mTOR pathway by cannabidiol in treatment of experimental multiple sclerosis. *Fitoterapia*, *116*, 77–84. https://doi.org/10.1016/j.fitote.2016.11.010.

Gomez, O., Sanchez-Rodriguez, A., Le, M., Sanchez-Caro, C., Molina-Holgado, F., & Molina-Holgado, E. (2011). Cannabinoid receptor agonists modulate oligodendrocyte differentiation by activating PI3K/Akt and the mammalian target of rapamycin (mTOR) pathways. *British Journal of Pharmacology*, *163*, 1520–1532. https://doi.org/10.1111/j.1476-5381.2011.01414.x.

Gugliandolo, A., Pollastro, F., Bramanti, P., & Mazzon, E. (2020). Cannabidiol exerts protective effects in an in vitro model of Parkinson's disease activating AKT/mTOR pathway. *Fitoterapia*, *143*, 104553. https://doi.org/10.1016/j.fitote.2020.104553.

Hassan, S., Eldeeb, K., Millns, P. J., Bennett, A. J., Alexander, S. P. H., & Kendall, D. A. (2014). Cannabidiol enhances microglial phagocytosis via transient receptor potential (TRP) channel activation. *British Journal of Pharmacology*, *171*, 2426–2439. https://doi.org/10.1111/bph.12615.

Hausott, B., Glueckert, R., Schrott-Fischer, A., & Klimaschewski, L. (2022). Signal transduction regulators in axonal regeneration. *Cell, 11*, 1537. https://doi.org/10.3390/cells11091537.

Jin, T., Zhang, Y., Botchway, B. O. A., Zhang, J., Fan, R., Zhang, Y., & Liu, X. (2022). Curcumin can improve Parkinson's disease via activating BDNF/PI3k/Akt signaling pathways. *Food and Chemical Toxicology, 164*, 113091. https://doi.org/10.1016/j.fct.2022.113091.

Karege, F., Perroud, N., Burkhardt, S., Fernandez, R., Ballmann, E., La Harpe, R., & Malafosse, A. (2011). Alterations in phosphatidylinositol 3-kinase activity and PTEN phosphatase in the prefrontal cortex of depressed suicide victims. *Neuropsychobiology, 63*, 224–231. https://doi.org/10.1159/000322145.

Karege, F., Perroud, N., Burkhardt, S., Schwald, M., Ballmann, E., La Harpe, R., & Malafosse, A. (2007). Alteration in kinase activity but not in protein levels of protein kinase B and glycogen synthase kinase-3beta in ventral prefrontal cortex of depressed suicide victims. *Biological Psychiatry, 61*, 240–245. https://doi.org/10.1016/j.biopsych.2006.04.036.

Khezri, M. R., Jafari, R., Yousefi, K., & Zolbanin, N. M. (2021). The PI3K/AKT signaling pathway in cancer: Molecular mechanisms and possible therapeutic interventions. *Experimental and Molecular Pathology, 127*, 104787. https://doi.org/10.1016/j.yexmp.2022.104787.

Kokona, D., & Thermos, K. (2015). Synthetic and endogenous cannabinoids protect retinal neurons from AMPA excitotoxicity in vivo, via activation of CB1 receptors: Involvement of PI3K/Akt and MEK/ERK signaling pathways. *Experimental Eye Research, 136*, 45–58. https://doi.org/10.1016/j.exer.2015.05.007.

Kumar Kalvala, A., Bagde, A., Arthur, P., Kumar Surapaneni, S., Ramesh, N., Nathani, A., & Singh, M. (2022). Role of cannabidiol and tetrahydrocannabivarin on paclitaxel-induced neuropathic pain in rodents. *International Immunopharmacology, 107*, 108693. https://doi.org/10.1016/j.intimp.2022.108693.

Legare, C. A., Raup-Konsavage, W. M., & Vrana, K. E. (2022). Therapeutic potential of Cannabis, Cannabidiol, and Cannabinoid-based pharmaceuticals. *Pharmacology, 107*, 131–149. https://doi.org/10.1159/000521683.

Libro, R., Diomede, F., Scionti, D., Piattelli, A., Grassi, G., Pollastro, F., Bramanti, P., Mazzon, E., & Trubiani, O. (2016). Cannabidiol modulates the expression of Alzheimer's disease-related genes in mesenchymal stem cells. *International Journal of Molecular Sciences, 18*, 26. https://doi.org/10.3390/ijms18010026.

Lima, I. V. A., Bellozi, P. M. Q., Batista, E. M., Vilela, L. R., Brandão, I. L., Ribeiro, F. M., Moraes, M. F. D., Moreira, F. A., & de Oliveira, A. C. P. (2020). Cannabidiol anticonvulsant effect is mediated by the PI3Kγ pathway. *Neuropharmacology, 176*, 108156. https://doi.org/10.1016/j.neuropharm.2020.108156.

Lin, B., Gao, Y., Li, Z., Zhang, Z., Lin, X., & Gao, J. (2020). Cannabidiol alleviates hemorrhagic shock-induced neural apoptosis in rats by inducing autophagy through activation of the PI3K/AKT pathway. *Fundamental & Clinical Pharmacology, 34*, 640–649. https://doi.org/10.1111/fcp.12557.

Marosi, M., Arman, P., Aceto, G., D'Ascenzo, M., & Laezza, F. (2022). Glycogen synthase kinase 3: Ion channels, plasticity, and diseases. *International Journal of Molecular Sciences, 23*, 4413. https://doi.org/10.3390/ijms23084413.

Molina-Holgado, E., Vela, J. M., Arévalo-Martín, A., Almazán, G., Molina-Holgado, F., Borrell, J., & Guaza, C. (2002). Cannabinoids promote oligodendrocyte progenitor survival: Involvement of cannabinoid receptors and phosphatidylinositol-3 kinase/Akt signaling. *Journal of Neuroscience, 22*, 9742–9753. https://doi.org/10.1523/JNEUROSCI.22-22-09742.2002.

Pei, J. J., Braak, E., Braak, H., Grundke-Iqbal, I., Iqbal, K., Winblad, B., & Cowburn, R. F. (1999). Distribution of active glycogen synthase kinase 3beta (GSK-3beta) in brains staged for Alzheimer disease neurofibrillary changes. *Journal of Neuropathology and Experimental Neurology, 58*, 1010–1019. https://doi.org/10.1097/00005072-199909000-00011.

Querfurth, H., & Lee, H. (2021). Mammalian/mechanistic target of rapamycin (mTOR) complexes in neurodegeneration. *Molecular Neurodegeneration, 16*, 44. https://doi.org/10.1186/s13024-021-00428-5.

Rai, S. N., Dilnashin, H., Birla, H., Singh, S. S., Zahra, W., Rathore, A. S., Singh, B. K., & Singh, S. P. (2019). The role of PI3K/Akt and ERK in neurodegenerative disorders. *Neurotoxicity Research, 35*, 775–795. https://doi.org/10.1007/s12640-019-0003-y.

Reichhart, N., Keckeis, S., Fried, F., Fels, G., & Strauss, O. (2015). Regulation of surface expression of TRPV2 channels in the retinal pigment epithelium. *Graefe's Archive for Clinical and Experimental Ophthalmology, 253*, 865–874. https://doi.org/10.1007/s00417-014-2917-7.

Rinne, N., Christie, E. L., Ardasheva, A., Kwok, C. H., Demchenko, N., Low, C., Tralau-Stewart, C., Fotopoulou, C., & Cunnea, P. (2021). Targeting the PI3K/AKT/mTOR pathway in epithelial ovarian cancer, therapeutic treatment options for platinum-resistant ovarian cancer. *Cancer Drug Resistance, 4*, 573–595. https://doi.org/10.20517/cdr.2021.05.

Ruiz, S. M. A., & Eldar-Finkelman, H. (2022). Glycogen synthase kinase-3 inhibitors: Preclinical and clinical focus on CNS—A decade onward. *Frontiers in Molecular Neuroscience, 14*, 792364. https://doi.org/10.3389/fnmol.2021.792364.

Scheltens, P., De Strooper, B., Kivipelto, M., Holstege, H., Chételat, G., Teunissen, C. E., Cummings, J., & van der Flier, W. M. (2021). Alzheimer's disease. *Lancet, 397*, 1577–1590. https://doi.org/10.1016/S0140-6736(20)32205-4.

Shi, J., Cai, Q., Zhang, J., He, X., Liu, Y., Zhu, R., & Jin, L. (2017). AM1241 alleviates MPTP-induced Parkinson's disease and promotes the regeneration of DA neurons in PD mice. *Oncotarget, 8*, 67837–67850. https://doi.org/10.18632/oncotarget.18871.

Silvestro, S., Schepici, G., Bramanti, P., & Mazzon, E. (2020). Molecular targets of Cannabidiol in experimental models of neurological disease. *Molecules, 25*, 5186. https://doi.org/10.3390/molecules25215186.

Su, S., Wang, Y., Wu, Y., Wang, D., Lin, Q., & Hai, J. (2016). Cannabinoid receptor agonist WIN55,212-2 and fatty acid amide hydrolase inhibitor URB597 may protect against cognitive impairment in rats of chronic cerebral hypoperfusion via PI3K/AKT signaling. *Behavioural Brain Research, 313*, 334–344. https://doi.org/10.1016/j.bbr.2016.07.009.

Timmons, S., Coakley, M. F., Moloney, A. M., & O'Neill, C. (2009). Akt signal transduction dysfunction in Parkinson's disease. *Neuroscience Letters, 467*, 30–35. https://doi.org/10.1016/j.neulet.2009.09.055.

Vilela, L. R., Gobira, P., Viana, T. G., Medeiros, D. C., Ferreira-Vieira, T. H., Doria, J. G., Rodrigues, F., Aguiar, D. C., Pereira, G. S., Massessini, A. R., Ribeiro, F. M., de Oliveira, A. C. P., Moraes, M. F. D., & Moreira, F. A. (2015). Enhancement of endocannabinoid signaling protects against cocaine-induced neurotoxicity. *Toxicology and Applied Pharmacology*, *286*, 178–187. https://doi.org/10.1016/j.taap.2015.04.013.

Vrechi, T. A. M., Leão, A. H. F. F., Morais, I. B. M., Abílio, V. C., Zuardi, A. W., Hallak, J. E. C., Crippa, J. A., Bincoletto, C., Ureshino, R. P., Smaili, S. S., & Pereira, G. J. S. (2021). Cannabidiol induces autophagy via ERK1/2 activation in neural cells. *Scientific Reports*, *11*, 5434. https://doi.org/10.1038/s41598-021-84879-2.

Walkowski, B., Kleibert, M., Majka, M., & Wojciechowska, M. (2022). Insight into the role of the PI3K/Akt pathway in ischemic injury and post-infarct left ventricular remodeling in normal and diabetic heart. *Cell*, *11*, 1553. https://doi.org/10.3390/cells11091553.

Yang, S., Du, Y., Zhao, X., Tang, Q., Su, W., Hu, Y., & Yu, P. (2022). Cannabidiol enhances microglial beta-Amyloid peptide phagocytosis and clearance via Vanilloid Family Type 2 Channel activation. *International Journal of Molecular Sciences*, *23*, 5367. https://doi.org/10.3390/ijms23105367.

Yang, S., Du, Y., Zhao, X., Wu, C., & Yu, P. (2022). Reducing PDK1/Akt activity: An effective therapeutic target in the treatment of Alzheimer's disease. *Cell*, *11*, 1735. https://doi.org/10.3390/cells11111735.

Yang, G., Liu, L., Zhang, R., Li, J., Leung, C., Huang, J., Li, Y., Shen, B., Zeng, X., & Zhang, D. (2020). Cannabidiol attenuates methamphetamine-induced conditioned place preference via the Sigma1R/AKT/GSK-3β/CREB signaling pathway in rats. *Toxicology Research*, *9*, 202–211. https://doi.org/10.1093/toxres/tfaa021.

Chapter 33

Effects of cannabidiol and other phytocannabinoids on voltage- and ligand-gated ion channels

Murat Oz[a], Keun-Hang Susan Yang[b], Mohamed Mahgoub[c], Georg Petroianu[d], and Dietrich E. Lorke[e]

[a]*Department of Pharmacology and Therapeutics, Faculty of Pharmacy, Kuwait University, Safat, Kuwait,* [b]*Department of Biological Sciences, Schmid College of Science and Technology, Chapman University, Orange, CA, United States,* [c]*Department of Health and Medical Sciences, Khawarizmi International College, Abu Dhabi, United Arab Emirates,* [d]*Department of Pharmacology, College of Medicine and Health Sciences, Khalifa University, Abu Dhabi, United Arab Emirates,* [e]*Department of Anatomy and Cellular Biology, College of Medicine and Health Sciences, Khalifa University, Abu Dhabi, United Arab Emirates*

Introduction

The *Cannabis sativa* plant (marijuana) contains more than 500 chemical compounds, and about 120 of them are structurally related phytocannabinoids, which are terpenophenolic compounds chemically related to the terpenes with their ring structure derived from a geranyl pyrophosphate (ElSohly et al., 2017). Among these phytocannabinoids, Δ^9-tetrahydrocannabinol (THC) and cannabidiol (CBD) are major psychoactive and nonpsychoactive components of marijuana, respectively (Izzo et al., 2009).

In recent years, the lack of psychoactive effects of some phytocannabinoids, e.g., CBD, increased access to cannabis products due to decriminalization of medical use of cannabis in several western countries, and new approvals for the clinical use of cannabinoid-based compounds have revitalized the interest in pharmacological effects of phytocannabinoids, specifically CBD. In clinical practice, Nabilone (Cesamet), a synthetic cannabinoid, and Dronabinol (Marinol), a synthetic (−) enantiomer of THC, have been used for the treatment of anorexia and chemotherapy-induced nausea and vomiting in the United States, Canada, and United Kingdom (Pagano et al., 2022; Wright & Guy, 2014). Nabiximols (Sativex), an oral spray containing CBD and THC in a 1:1 ratio, has been approved in several countries, including United Kingdom, European Union, and Canada, for the treatment of multiple sclerosis-associated spasticity (Ortiz et al., 2022; Pagano et al., 2022). CBD has been recently approved in the United States and the European Union as an add-on antiepileptic drug (Epidiolex) for the treatment of patients affected by refractory epilepsy, such as Dravet and Lennox–Gastaut syndromes (Ortiz et al., 2022; Pagano et al., 2022).

Among more than 120 phytocannabinoids found in cannabis, to date, beside THC, only cannabinol, (−)-trans-Δ^8-tetrahydrocannabinol, (−)-trans-Δ^9-tetrahydrocannabivarin, and the sesquiterpene (E)-β-caryophyllene have been shown to bind and activate cannabinoid receptors (Pertwee & Cascio, 2014). The majority of these phytocannabinoids and phenolic terpenes also interact with various ion channels (Oz et al., 2015), suggesting that these compounds can also potentially contribute to allosteric effects of cannabis through cannabinoid receptor-independent mechanisms. While this review focuses on the direct cannabinoid receptor-independent effects of CBD and some other phytocannabinoids on the functions of ion channels, the effects of these compounds on various G-protein-coupled receptors, enzymes, and neurotransmitter transporters have been reviewed in earlier reports (Cifelli et al., 2020; Senn et al., 2020; Vitale et al., 2021).

Effects of cannabidiol on voltage-gated sodium channels

Voltage-gated sodium channels (VGSCs) are mainly involved in the initiation of action potentials in neuronal, muscle, and other excitable cells. They are activated by alterations in membrane potential, which causes the movement of gating particles in the transmembrane domains of the channels and leads to opening of the pore allowing sodium influx (Catterall et al., 2020).

In earlier studies, low concentrations (10 nM) of 11-OH-THC, a principal metabolite of THC, were reported to inhibit Na^+ action potentials in squid giant axon (Brady & Carbone, 1973). In subsequent investigations, CBD (Turkanis & Karler,

Cannabis Use, Neurobiology, Psychology, and Treatment. https://doi.org/10.1016/B978-0-323-89862-1.00038-6

1986a), THC (Turkanis & Karler, 1983), and 11-hydroxy-THC (Turkanis & Karler, 1984) were shown to increase firing thresholds and suppress the amplitudes of action potentials in cat motoneurons and frog sartorius muscle (Kumbaraci & Nastuk, 1980). In subsequent patch-clamp studies, THC (Turkanis, Partlow, & Karler, 1991) and 11-hydroxy-THC (Turkanis, Karler, & Partlow, 1991) inhibited VGSCs in mouse neuroblastoma cells with IC_{50} values in the range of 0.3–2 μM. In recent years, approval of CBD for the treatment of Dravet syndrome and Lennox–Gastaut syndrome, following a series of successful randomized controlled trials (Ortiz et al., 2022; Pagano et al., 2022), has caused resurgence of interest in the pharmacology of phytocannabinoids and CBD in particular. The results of studies investigating the effects of CBD and other phytocannabinoids on the functional properties of VGSCs are summarized in Table 1 and have been reviewed in a recent report (Ghovanloo & Ruben, 2021). Notably, CBD-dominant hemp-derived products appeared to inhibit VGSCs more potently based on their CBD content than purified CBD, suggesting a synergistic (entourage) effect among different phytocannabinoids and terpenes found in the cannabis plant (Milligan et al., 2022). Notably, considering

TABLE 1 Effects of phytocannabinoids on voltage-gated sodium channels.

Preparation/recording method	Effect and conclusion	Reference
Rat CA1 hippocampal neurons, human SH-SY5Y cells; mouse cortical neurons, CHO cells/patch-clamp technique and fluorescence-based assay	CBD (10 μM) inhibits Nav in cell cultures. Inhibition of human Nav1.1., 1.2, and 1.5 with IC_{50}s of 33, 39, and 27 μM, respectively	Hill et al. (2014)
Rat cerebral cortex synaptosomes/[^{14}C]-guanidine flux assay and [^{12}H]-batrachotoxin-B binding	CBD inhibits [^{14}C]-guanidine flux and [^{12}H]-batrachotoxin-B binding with IC_{50}s of 0.9 μM and 3–4 μM, respectively	Pryce et al. (2015)
HEK-293 cells transfected with human Nav1.6 encoding for voltage-gated Na^+ channel subunit and mice striatal neurons/patch-clamp technique	Inhibition of resurgent human Nav1.6 current and Na^+ currents in striatal neurons by 1 μM CBD	Patel et al. (2016)
Small molecule neural precursor cells/patch-clamp technique	Inhibition of Na^+ current and action potential amplitudes by 10 μM THC	Stanslowsky et al. (2017)
HEK-293 cells transfected with human Nav1.1–1.7 channels/patch-clamp technique	Inhibition of Human Nav1.1–1.7 currents, with CBD, IC_{50}s of 1.9–3.8 μM	Ghovanloo et al. (2018)
COS cells transfected with Nav1.5 α-subunit/ patch-clamp technique	Inhibition of Nav1.5 currents by 1–5 μM CBD	Fouda et al. (2020)
Rat cultured vagal afferent neurons/patch-clamp study	CBD (30 μM) inhibited (60%–70%) Na^+ current in capsaicin-insensitive A fibers	Kowalski et al. (2020)
HEK-293 cells transfected with human Nav1.5 α-subunits/patch-clamp technique	Inhibition of peak Nav1.5 current by CBD with IC_{50} of 7.4 μM	Le Marois et al. (2020)
HEK-293 cells transfected with human Nav1.2 α-subunits/patch-clamp technique	Preferential inhibition of resurgent Na^+ current by 1 μM CBD	Mason and Cummins (2020)
Rabbit ventricular cardiomyocytes/patch-clamp technique	Inhibition of tetrodotoxin sensitive Na^+ current by 10 μM CBD	Orvos et al. (2020)
CHOK1 cells transfected with prokaryotic NavMs α and β-subunits/patch-clamp technique	CBD inhibition of NavMs current with IC_{50} of 17.8 μM	Sait et al. (2020)
CHOK1 cells transfected with Nav1.4 α and β-subunits/patch-clamp technique	Inhibition of Nav1.4 (skeletal muscle type) by 10 μM CBD	Ghovanloo et al. (2021)
CHOK1 cells transfected with Nav1.4 α-subunits/ patch-clamp technique	State-dependent CBD inhibition of Nav1.4 current with Kd 117- 51 μM	Huang et al. (2021)
Mouse dorsal root ganglion neurons/patch-clamp technique	Inhibition of tetrodotoxin-sensitive and resistant Na^+ currents by 2–5 μM CBD	Zhang and Bean (2021)
HEK-293 cells transfected with human Nav1.7 and rat dorsal root ganglion neurons/patch-clamp technique	Cannabigerol inhibited NaV1.7 current in a state-dependent manner with IC_{50} ranging from 4.6 to 18.8 μM. Inhibition (50%) of TTX-resistant Na^+ current by 10 μM cannabigerol in neurons.	Ghovanloo et al. (2022)
Rabbit ventricular cardiomyocytes/patch-clamp technique	Inhibition of peak Na^+ current by CBD with IC_{50} of 5.4 μM	Isaev et al. (2022)

that the majority (80%) of VGSC (*SCN1A* gene) mutations are loss of function mutations in Dravet syndrome patients, the facilitation rather than inhibition of VGSCs has been proposed to be beneficial. In line with this proposal, in Dravet-syndrome-patient-derived pluripotent stem cells, CBD (50–500 nM) attenuated the excitability of excitatory neurons and potentiated the excitability of inhibitory neurons by increasing the number of action potentials in response to submaximal current injections, thereby resulting in reestablishment of normal network activity in this model (Sun & Dolmetsch, 2018).

Effects of cannabidiol on voltage-gated calcium channels

Voltage-gated Ca^{2+} channels (VGCCs) activate in response to membrane depolarization and conduct Ca^{2+} into the cell, where it initiates several physiological responses, including secretion, contraction, and gene transcription (Catterall et al., 2020). Structurally, VGCCs are heteromultimeric complexes consisting of a central pore-forming α1 subunit surrounded by several auxiliary subunits (α2δ1-4, β1-4, and γ1-8). Broadly, VGCCs have been classified as L-type (dihydropyridine sensitive), N-type neuronal, and T-type channels (Catterall et al., 2020).

In an earlier study, THC (10 μM) was shown to be ineffective on the function of L-type VGCCs in T-tubule membranes of rabbit skeletal muscle (Oz, Tchugunova, & Dinc, 2004). However, both THC and CBD have been shown to inhibit the functions of various T-type and cardiac L-type VGCCs (Ali et al., 2015; Isaev et al., 2022; Ross et al., 2008). The results of studies investigating the effects of CBD and THC on the functional properties of VGCCs are summarized in Table 2.

TABLE 2 Effects of phytocannabinoids on voltage-gated calcium channels.

Preparation/recording method	Effect and conclusion	Reference
HEK-293 cells transfected with CaV3.1, 3.2, and 3.3 encoding for voltage-gated Ca^{2+} channel subunits and mouse trigeminal ganglion neurons/patch-clamp technique	Inhibition of CaV3.1, 3.2, and 3.3 by CBD with IC_{50}s of 820 nM, 780 nM, and 3.7 μM, respectively, and by THC with IC_{50}s of 1.5, 1.3, and 4.2 μM, respectively.	Ross et al. (2008)
Rat isolated aorta/contraction force measurement	Inhibition of concentration–response curves to $CaCl_2$ by 1–30 μM CBD	O'Sullivan et al. (2009)
Rat ventricular cardiomyocyte/patch-clamp technique	Inhibition (50%) of peak Ca^{2+} current by 1 μM CBD	Ali et al. (2015)
HEK-293T cells transfected with human α1B-1 and α2bδ−1 subunits/patch-clamp technique	Complete inhibition of peak CaV2.2 current by 10 μM CBD	Mendis et al. (2019)
COS cells transfected with CaV1.2/β2/α2δ1 subunits/patch-clamp technique	Inhibition of peak CaV1.2 current by CBD with IC_{50} of 9.6 μM	Le Marois et al. (2020)
Human induced pluripotent stem cells/single-cell Ca^{2+} imaging	Inhibition of KCl-induced intracellular Ca^{2+} responses by 10 μM THC and CBD	Miranda et al. (2020)
Rabbit ventricular cardiomyocytes/patch-clamp technique	Frequency-dependent inhibition of Ca^{2+} current by 10 μM CBD	Orvos et al. (2020)
HEK-293 cells transfected with CaV3.1, 3.2, and 3.3/patch-clamp technique	CBD and THC preferentially inhibited channels in slow-inactivated conformation	Bladen et al. (2021)
Rat isolated mesenteric arteries/contraction force measurement	Inhibition of high-K^+ (80 mM) and BayK-8644-induced contractures by 1–10 μM CBD	Mazeh et al. (2021)
Rabbit ventricular cardiomyocytes/patch-clamp technique	Inhibition of peak Ca^{2+} current by CBD with IC_{50} of 4.8 μM	Isaev et al. (2022)
HEK-293 cells transfected with human CaV3.1, 3.2, and 3.3/fluorescence imaging plate reader	CBD (10 μM) inhibited CaV3.1, 3.2, and 3.3 by 40%, 35%, and 8%, respectively	Mirlohi, Bladen, Santiago, Arnold, et al. (2022)
HEK-293 cells stably expressing human CaV3.1, 3.2, and 3.3/patch-clamp technique	THC and Δ^9-tetrahydrocannabinolic acid inhibited the peak current amplitude CaV3.1 with pEC_{50}s of 5.6.and 6, respectively	Mirlohi, Bladen, Santiago, and Connor (2022)
tsA-201 cells expressing either Cav3.2 or Cav2.2/Patch-clamp technique	CBD inhibited CaV3.2 with IC_{50} of 4.06 μM in cannabinoid receptor 1 independent manner and showed no effect on CaV2.2	Harding et al. (2023)

Effects of cannabidiol on voltage-gated potassium channels

Voltage-gated K^+ channels are integral membrane proteins that are activated upon changes in membrane potential to hyperpolarize the cell membrane by allowing passage of K^+ ions down its electrochemical gradient across the membrane, playing important roles in various physiological and pathological conditions. Structurally, they are assembled in the lipid bilayer as tetramers and demonstrate a considerable subtype diversity, due to the large number of different genes present, auxiliary β-subunits and metabolic regulation (Latorre et al., 2013).

In earlier studies, THC (30 μM) has been shown to reversibly inhibit the falling phase of the action potential, which is due to activation of K^+ conductance, in frog sartorius muscle (Kumbaraci & Nastuk, 1980). The results of studies investigating the effects of CBD on the functional properties of K^+ channels are summarized in Table 3. Recently, in addition to K^+ channels, THC and CBD were shown to modulate hyperpolarization-activated cyclic nucleotide-gated (HCN) channels, which are structurally related to K^+ channels. THC inhibited HCN1 channels expressed in *Xenopus* oocytes with an IC_{50} of 28.9 μM, whereas CBD-activated HCN1 channel with an EC_{50} of 28.5 μM (Mayar et al., 2022).

Effects of cannabidiol on ligand-gated ion channels

Cys-loop ligand-gated ion channels (LGICs) are integral membrane proteins, including the excitatory cation-selective nicotinic acetylcholine (nACh), 5-HT_3, and ionotropic glutamate receptors as well as the inhibitory anion-selective $GABA_A$ and glycine receptors. Typical LGICs are composed of multiple subunits in homo- or hetero-pentameric structure with a pore-forming α subunit allowing the regulated flow of ions across the plasma membrane in response to binding of endogenous or exogenous molecules. Structurally, an α-subunit of an LGIC contains a large extracellular N-terminus, four transmembrane (TM) domains, a short extracellular C-terminus, and a large cytoplasmic domain between TM3 and TM4 (Plested, 2016).

nACh receptors are prototypical LGIC members that comprise of α1-10, β1-6, δ, and γ subunit combinations and mediate fast cholinergic neurotransmission in the central and peripheral nervous systems. In earlier studies, CBD was reported to decrease the amplitudes of miniature end-plate potentials at the frog neuromuscular junction, suggesting an effect on postsynaptic nACh receptors (Turkanis & Karler, 1986b). Later studies indicated that CBD inhibits human

TABLE 3 Effects of phytocannabinoids on voltage-gated potassium channels.

Preparation/recording method	Effect and conclusion	Reference
Murine fibroblasts (B82 cell line) stably transfected with Shaker-related Kv1.2 encoding for voltage-gated K^+ channel subunits/patch-clamp technique	Inhibition of K^+ currents by THC with IC_{50} of 2.4 μM	Poling et al. (1996)
Small molecule neural precursor cells/patch-clamp technique	Inhibition of K^+ current amplitudes by 10 μM THC	Stanslowsky et al. (2017)
HEK-293 cells transfected with Kv2.1 and human iPSC/patch-clamp technique	Inhibition of Kv2.1 and IK (in human iPSC) by CBD with IC_{50}s of 3.8 μM and 1 μM, respectively.	Ghovanloo et al. (2018)
COS cells transfected with Kv11.1 (IKr); Kv7.1/minK (IKs); Kv4.3/KChIP2.2 (Ito)/patch-clamp technique	Inhibition of Kv11.1, Kv7.1/mink, and Kv4.3/KChIP2.2 by CBD with IC_{50}s of 15 μM; 2.7 μM, and 7.3 μM, respectively.	Le Marois et al. (2020)
Rabbit ventricular cardiomyocytes and HEK-293 cells transfected with hERG/patch-clamp technique	Inhibition of hERG and IKr by CBD with IC_{50} of 2.1 μM and 6.5 μM, respectively.	Orvos et al. (2020)
Rabbit ventricular cardiomyocytes/patch-clamp technique	Inhibition of IKr and IKs by CBD with IC_{50}s of 4.9 μM and 3.1 μM, respectively.	Topal et al. (2021)
Rabbit ventricular cardiomyocytes/patch-clamp technique	Inhibition of IKr and IKs by CBD with IC_{50}s of 2.4 μM and 2.1 μM, respectively.	Isaev et al. (2022)
CHO cells coexpressing human Kv7.2 and Kv7.3 channels; mouse superior cervical ganglion and rat hippocampal neurons/fluorescence-based thallium flux and patch-clamp technique	Activation of Kv7.2/Kv7.3 current HEK-293 cells and M-current native neurons by 30 nM to 1 μM CBD.	Zhang et al. (2022)

α7-nACh receptors with an IC_{50} of 11.3 μM (Mahgoub et al., 2013; Oz et al., 2014) and suppresses choline-induced inward currents in rat hippocampal interneurons. CBD has also been shown to inhibit nicotine-induced [^{3}H]-norepinephrine release in rat hippocampal slices (Mahgoub et al., 2013) and reduce withdrawal symptoms in nicotine-dependent rats (Smith et al., 2021), further suggesting an interaction between CBD and nicotinic receptors. Notably, the inhibitory effect of CBD was time-dependent with 50% inhibition occurring in 2.1 min sustained CBD application, voltage-independent, and noncompetitive, i.e., while the EC_{50} of acetylcholine (ACh) remained unaltered, maximal ACh responses were significantly decreased (Mahgoub et al., 2013). In line with these findings, specific [^{125}I] α-bungarotoxin binding was not affected by CBD. However, THC and cannabinol up to 10 μM were ineffective (10%–20% inhibition) on α7-nACh receptors expressed in *Xenopus* oocytes (Mahgoub et al., 2013; Oz, Zhang, et al., 2004). Similarly, THC (1 μM) did not affect the amplitudes of currents through α4β2-nACh receptors (Spivak et al., 2007).

The 5-HT_3 receptor is another member of the LGIC superfamily. Although five subtypes (A-E) of 5-HT_3 receptors have been identified to date, the 5-HT_{3A} subunit can form a homo- or heteromeric functional 5-HT_3 receptor with expected pharmacological properties. Both THC and CBD were shown to directly interact with 5-HT_3 receptors expressed in HEK-293 cells (Barann et al., 2002; Xiong, Koo, et al., 2011) and *Xenopus* oocytes (Xiong, Koo, et al., 2011;Yang, Galadari, et al., 2010; Yang, Isaev, et al., 2010). Depending on the expression levels of the receptor, THC and CBD inhibited human 5-HT_{3A} receptors with respective IC_{50} values of 1.2 μM and 0.6 μM in *Xenopus* oocytes (Yang, Galadari, et al., 2010; Yang, Isaev, et al., 2010) and 119 nM and 329 nM in HEK-293 cells (Xiong, Koo, et al., 2011). Inhibition by these cannabinoids was voltage-independent, noncompetitive, and increased with sustained (continuous) application (preincubation) time. In line with the noncompetitive nature of their effect, THC and CBD did not alter the specific [^{3}H]GR65630 binding to 5-HT_{3A} receptors (Yang, Galadari, et al., 2010; Yang, Isaev, et al., 2010). Importantly, slowing the receptor desensitization by pharmacological agents, such as 5-hydroxyindole and nocodazole, or mutations, such as R427L, markedly decreased the extent of 5-HT_3 receptor inhibition by CBD (Xiong, Koo, et al., 2011). Similarly, increasing the expression level and density of 5-HT_3 receptors on the cell membrane caused a noticeable decrease in receptor desensitization and in the potency of THC on the 5-HT_3 receptor (Yang, Isaev, et al., 2010). In in vivo studies, CBD has been reported to inhibit the reflex bradycardia induced by the 5-HT_3 receptor agonist phenylbiguanide in spontaneously hypertensive rats (Kossakowski et al., 2019).

In addition to nACh and 5-HT_3 receptors, the AMPA (α-amino-3-hydroxy-5-methyl-4-isoxazole-propionic acid)-glutamate receptor is another excitatory LGIC, and CBD was shown to inhibit recombinant GluA1 currents with an IC_{50} of 22.5 μM and suppress AMPA receptor-mediated currents in hippocampal neurons (Yu et al., 2020).

Glycine receptors, inhibitory members of LGICs, are formed as pentameric homomeric chloride channels of α1-4 subunits or as αβ heteromeric functional channels and are involved in several physiological and pathological processes, including neuromotor activity, antinociception, muscle relaxation, anxiety, and reward mechanisms (Lynch, 2009). CBD and THC have been shown to potentiate glycine receptor-mediated currents in a cannabinoid receptor-independent manner in *Xenopus* oocytes expressing homomeric α1 and heteromeric α1β glycine receptors and in acutely isolated CNS neurons (Hejazi et al., 2006; Xiong, Cheng, et al., 2011). CBD was reported to act as a positive allosteric modulator of α1 and α1β1 glycine receptors expressed in HEK-293 cells with EC_{50} values of 12.3 μM and 18.1 μM, respectively (Ahrens et al., 2009). Notably, at a higher concentration range, CBD directly acts as an agonist of α1 and α1β1 glycine receptors with respective EC_{50} values of 132.4 μM and 144.3 μM. Mutation of the α1 subunit TM2 serine (267) residue to isoleucine abolished both coactivation and direct activation of the glycine receptor by CBD (Foadi et al., 2010). However, THC-induced potentiation was not affected when S267Q mutant α1 glycine receptors were expressed in *Xenopus* oocytes (Hejazi et al., 2006). In subsequent studies, it was shown that continuous application, but not co-application, of THC (1 μM) for about 5 min caused a gradually developing marked potentiation (up to 1200%) of glycine-activated currents in HEK-293 cells expressing α1, α3, and α1β1, but not α2, subunits of the glycine receptor (Xiong, Cheng, et al., 2011). Notably, the S296 residue of the α1 and the S307 residue of the α3 subunits have been shown to be critical for the potentiation of glycine receptors by continuous but not simultaneous (THC plus glycine) THC application (Xiong, Cheng, et al., 2011; Xiong, Wu, et al., 2012). Conversely, substitution of the corresponding residue, Ala303 of the α2 subunit, with serine converted the α2 subunit from a low to a high THC sensitivity. Similar to THC, CBD, dehydroxyl-CBD, and dehydroxyl-THC have also been shown to potentiate glycine-activated currents of α1 and α3 subunits (Xiong, Cheng, et al., 2011; Xiong, Cui, et al., 2012; Xiong, Wu, et al., 2012). Further studies indicated that CBD and dehydroxyl-CBD suppress persistent inflammatory and neuropathic pain by potentiating α3 glycine receptor-mediated currents in HEK-293 cells and cultured spinal neurons, and CBD suppression of neuropathic pain was reversed in α3-knock-out mice (Xiong, Cui, et al., 2012). In subsequent studies, dehydroxyl-CBD was reported to suppress inflammatory pain and potentiate α1-glycine receptors by interacting with the S296 amino acid residue. Accordingly, the analgesic effect of dehydroxyl-CBD was reversed in transgenic mice carrying an S296A mutation, which also blocks dehydroxyl-CBD-induced potentiation of glycine-activated currents in neurons of the spinal cord dorsal horn in mutant mice (Lu et al., 2018).

The $GABA_A$ receptor, another inhibitory member of the LGIC family, exhibits high receptor subtype heterogeneity, assembling from a combination of subunits derived from a family of 19 genes, which code for α1–6, β1–3, γ1–3, δ, π, ε, and θ subunits. These subunits coassemble to form hetero-pentamers that allow the conductance of chloride ions through a central pore (Plested, 2016). CBD has been shown to act as a positive allosteric modulator, with EC_{50} values ranging from 0.9 μM to 16.1 μM and magnitudes of potentiation in the range of 72%–332% at αβγ2 receptor combinations, with higher levels of potentiation on α2 containing subtype combinations (Bakas et al., 2017). The greatest levels of enhancements by CBD were reported as 332% on α2β2γ2L and 752% on α4β2δ subunit combinations, and the classical benzodiazepine-binding site located at the α-γ2L interface does not seem to be involved in the CBD interaction with $GABA_A$ receptors. Furthermore, CBD potentiation of α2β2γ2L was significantly decreased by the α2β2(V436T)γ2L mutation on transmembrane domain 4. Notably, CBD potentiation of $GABA_A$ receptors was significantly decreased with increasing GABA concentrations, resulting in decreased GABA EC_{50} values with no apparent change in the E_{max}. In a recent study, CBD (2 μM) induced a 30%–50% potentiation of GABA (1–10 μM) activated currents in *Xenopus* oocytes expressing human α1β2γ2, α1β2, and α2β2γ2, α2β2 subunit combinations of $GABA_A$ receptors (Ruffolo et al., 2018). In another investigation, CBD (10 μM) caused a 106% potentiation, with an EC_{50} of 2.4 μM, of GABA (15 μM)-activated currents in *Xenopus* oocytes expressing an α1β2γ2 subunit combination of $GABA_A$ receptors (Anderson et al., 2019). In addition to CBD, THC (3 μM) has also been shown to potentiate (about 100%) α1β2γ2 $GABA_A$ receptors activated by low concentrations (EC_2) of GABA (Yao, Wells, et al., 2020). In another study, the run-down of GABA (500 μM) activated currents in *Xenopus* oocytes transplanted with hippocampal membranes from epileptic patients was significantly reversed by 2-hour-pretreatment with 50 nM cannabidivarin, a nonpsychoactive homolog of CBD (Morano et al., 2016). In a recent study, cannabigerolic acid, the biosynthetic precursor to both THC and CBD, has been shown to increase GABA (15 μM) activated currents to 271% of controls with an EC_{50} of 910 nM in *Xenopus* oocytes expressing human α1β2γ2 $GABA_A$ receptors (Anderson et al., 2021). Collectively, these results indicate that both CBD and THC act as positive allosteric modulators of glycine and $GABA_A$ receptors, especially at low agonist concentrations in a subunit-specific manner.

Effects of cannabidiol on transient receptor potential channels

Transient receptor potential (TRP) channels are nonselective Ca^{2+} permeable cation channels involved in response to a broad range of chemical and physical stimuli acting as molecular integrators of these stimuli. The mammalian TRP superfamily consists of 26 channels within six subfamilies, including vanilloid (TRPV), canonical (TRPC), melastatin (TRPM), ankyrin (TRPA), polycystin (TRPP), and mucolipin (TRPML) channels. Six of these channels have been shown to be activated by a variety of endogenous, phytogenic, and synthetic cannabinoids. These six TRP channels, TRPV1-TRPV4, TRPA1, and TRPM8, are also termed the ionotropic cannabinoid receptors.

In HEK-293 cells transfected with human TRPV1, both (-)-CBD and (+)-CBD activated human TPRV1 with respective EC_{50}s of 3.5 and 3.2 μM and inhibited the binding of [^{3}H]-Resiniferatoxin at the same concentration range (Bisogno et al., 2001). In line with these findings, THC, CBD, and cannabigerol elicited Ca^{2+} influx and caused desensitization of capsaicin responses in a concentration-dependent manner in rat dorsal root ganglion neurons (Anand et al., 2020, 2021). In addition to TRPV1, CBD, THC, and cannabinol activated the rat TRPV2 channel with EC_{50} values of 3.7, 14, and 77.7 μM, respectively (Qin et al., 2008). Similarly, CBD and tetrahydrocannabivarin, but not THC, stimulated rat TRPV3-mediated Ca^{2+} increases, whereas Δ^9-tetrahydrocannabidivarin, not CBD and THC, increased intracellular Ca^{2+} responses mediated by rat TRPV4 (De Petrocellis et al., 2012). In addition to TRPV1-4 channels, CBD and THC activated rat TRPA1 with EC_{50}s of 32.3 and 81.4 μM, respectively (Qin et al., 2008), confirming earlier results on the THC activation of human and rat TRPA1 channels (Jordt et al., 2004). In another study, while THC, CBD, and cannabichromene increased rat TRPA1-mediated intracellular Ca^{2+} levels, THC and CBD, but not cannabichromene, inhibited menthol-induced intracellular Ca^{2+} responses in a concentration-dependent manner (De Petrocellis et al., 2008). Cannabichromene, CBD, and cannabinol stimulated rat TRPA1 channels with EC_{50}s of 90, 110, and 180 nm, respectively, but antagonized TRPM8 channels with respective IC_{50}s of 40.7 μM, 60 nM, and 210 nM (De Petrocellis et al., 2011). In this study, THC, while found to be ineffective on human TRPV1 channels, stimulated rat TRPV2 with an EC_{50} of 650 nM. On the other hand, CBD-activated human TRPV1 and rat TRPV2 with EC_{50}s of 1 and 1.2 μM, respectively. Similarly, in patch-clamp studies, CBD and cannabidivarin dose-dependently activated and rapidly desensitized TRPV1, as well as TRPV2 and TRPA1 channels (Iannotti et al., 2014). Pharmacology of cannabinoid effects on TRP channels is highly complex and involves ligand-, species-, and cell-dependent effects of these compounds and has recently been reviewed in detail (De Petrocellis et al., 2017; Etemad et al., 2022; Muller et al., 2019).

Conclusion

CBD is a highly lipophilic compound with a LogP (octanol-water partition coefficient) value ranging between 6 and 8 (Lodzki et al., 2003). Thus, it is likely that CBD first dissolves into the lipid cell membrane and then diffuses into a nonannular lipid space to inhibit the ion channels. Consistent with this idea, the effect of CBD on ion channels usually reaches a maximal level within several minutes (5–10 min) of application. Notably, functional properties of LGICs have been shown to be affected by the activation of second messenger pathways (Hoffman et al., 1994; Nishizaki & Sumikawa, 1998; Siara et al., 1990; Zhang et al., 1995). However, agents modulating cAMP and protein kinase C pathways and chelation of intracellular Ca^{2+} do not alter the effects of endocannabinoids on LGICs (Oz et al., 2003; Oz, Zhang, & Morales, 2002), but the involvement of second messengers in phytocannabinoid modulation of ion channels has not yet been studied. In addition to phytocannabinoids, actions of several lipophilic modulators, such as capsaicin (Alzaabi et al., 2019; Lundbaek et al., 2005; Nebrisi et al., 2020), endocannabinoids (Oz, Tchugunova, & Dinc, 2004; Spivak et al., 2007), general anesthetics (Jackson et al., 2008; Zhang et al., 1997), and steroids (Oz, Zhang, & Spivak, 2002), on various ion channels require 5–10 min application times to reach their maxima, suggesting that their binding site(s) is/are located inside the lipid membrane and the effect requires a relatively slow equilibrium time. From this aspect, it appears that one or a combination of two mechanisms can describe the lipophilic actions of CBD (Ghovanloo & Ruben, 2021; Oz, 2006). First, CBD, like other lipophilic molecules, partitions into the lipid bilayer and alters the biophysical membrane properties by reducing stiffness, changing phase preference, membrane curvature, fluidity, and cellular elasticity (Bach et al., 1976; Ghovanloo et al., 2021; Hillard et al., 1985; James et al., 2022). Second, CBD can bind directly to transmembrane domains of ion channels embedded in the cell membrane (Ghovanloo et al., 2021;Lu et al., 2018; Xiong, Cheng, et al., 2011; Xiong, Wu, et al., 2012). In fact, residue S296 in the transmembrane domain of the α1 glycine receptor has been reported to interact with cannabinoids and mediate their potentiating effect on these receptors (Lu et al., 2018; Xiong, Cheng, et al., 2011; Xiong, Wu, et al., 2012). Interestingly, cholesterol seems to provide structural support to CBD-binding site on glycine receptors. Removal of membrane cholesterol by cyclodextrin (Yao, Wells, et al., 2020) or anticholesterol drugs, such as simvastatin (Yao, Liu, et al., 2020), markedly reversed THC or dehydroxyl-CBD potentiation of α1 and α3 glycine receptors by interacting with the S296 residue, suggesting that cholesterol directly interacts with CBD on this binding pocket. Interestingly, another recent study indicates that membrane orientation of CBD and its effect on water permeability are significantly altered in the presence of cholesterol, suggesting an interaction between CBD and cholesterol in lipid membranes (Perez et al., 2022). As a result of these mechanisms, it is likely that CBD and THC, similar to endocannabinoids (Spivak et al., 2007), affect the energy requirements for gating-related conformational changes and modulate the functional properties of ion channels.

Plasma concentrations of CBD following intraperitoneal, oral, and intravenous administrations have been studied previously (Deiana et al., 2012; Lodzki et al., 2003; Varvel et al., 2006). Commonly used doses of CBD (3–30 mg/kg I.V.; mouse) are found to result in mean plasma CBD levels of 0.42–11.8 μM, respectively (Varvel et al., 2006). In another study, following oral and intraperitoneal CBD administration (120 mg/kg, rat), maximal plasma levels of CBD were 6.4 μM and 8.3 μM, respectively (Deiana et al., 2012). Furthermore, due to its high lipophilicity, the membrane concentration of CBD is expected to be considerably higher than its blood levels (Deiana et al., 2012). In an earlier study, it has been shown that perfusion of isolated rat hearts with buffer containing [^{3}H]-CBD results in strong accumulation of radioactivity in the tissue (Smiley et al., 1976). Therefore, the functional modulation of ion channels by the concentration ranges of CBD and phytocannabinoids described in this review are expected to be pharmacologically relevant.

Summary points

- Among more than 100 phytocannabinoids, cannabidiol, and Δ^9-tetrahydrocannabinol are major nonpsychotropic and psychotropic cannabinoids.
- Both cannabidiol and Δ^9-tetrahydrocannabinol have been reported to affect several molecular targets independent of cannabinoid receptors.
- Notably, functions of both voltage- and ligand-gated ion channels have been shown to be directly affected by cannabidiol and Δ^9-tetrahydrocannabinol.

References

Ahrens, J., Demir, R., Leuwer, M., de la Roche, J., Krampfl, K., Foadi, N., … Haeseler, G. (2009). The nonpsychotropic cannabinoid cannabidiol modulates and directly activates alpha-1 and alpha-1-Beta glycine receptor function. *Pharmacology*, *83*(4), 217–222. https://doi.org/10.1159/000201556.

Ali, R. M., Al Kury, L. T., Yang, K. H., Qureshi, A., Rajesh, M., Galadari, S., … Oz, M. (2015). Effects of cannabidiol on contractions and calcium signaling in rat ventricular myocytes. *Cell Calcium*, *57*(4), 290–299. https://doi.org/10.1016/j.ceca.2015.02.001.

Alzaabi, A. H., Howarth, L., El Nebrisi, E., Syed, N., Susan Yang, K. H., Howarth, F. C., & Oz, M. (2019). Capsaicin inhibits the function of α(7)-nicotinic acetylcholine receptors expressed in Xenopus oocytes and rat hippocampal neurons. *European Journal of Pharmacology*, *857*, 172411. https://doi.org/10.1016/j.ejphar.2019.172411.

Anand, U., Jones, B., Korchev, Y., Bloom, S. R., Pacchetti, B., Anand, P., & Sodergren, M. H. (2020). CBD effects on TRPV1 signaling pathways in cultured DRG neurons. *Journal of Pain Research*, *13*, 2269–2278. https://doi.org/10.2147/jpr.S258433.

Anand, U., Oldfield, C., Pacchetti, B., Anand, P., & Sodergren, M. H. (2021). Dose-related inhibition of capsaicin responses by cannabinoids CBG, CBD, THC and their combination in cultured sensory neurons. *Journal of Pain Research*, *14*, 3603–3614. https://doi.org/10.2147/jpr.S336773.

Anderson, L. L., Absalom, N. L., Abelev, S. V., Low, I. K., Doohan, P. T., Martin, L. J., ... Arnold, J. C. (2019). Coadministered cannabidiol and clobazam: Preclinical evidence for both pharmacodynamic and pharmacokinetic interactions. *Epilepsia*, *60*(11), 2224–2234. https://doi.org/10.1111/epi.16355.

Anderson, L. L., Heblinski, M., Absalom, N. L., Hawkins, N. A., Bowen, M. T., Benson, M. J., ... Arnold, J. C. (2021). Cannabigerolic acid, a major biosynthetic precursor molecule in cannabis, exhibits divergent effects on seizures in mouse models of epilepsy. *British Journal of Pharmacology*, *178*(24), 4826–4841. https://doi.org/10.1111/bph.15661.

Bach, D., Raz, A., & Goldman, R. (1976). The interaction of hashish compounds with planar lipid bilayer membranes (BLM). *Biochemical Pharmacology*, *25*(11), 1241–1244. https://doi.org/10.1016/0006-2952(76)90084-8.

Bakas, T., van Nieuwenhuijzen, P. S., Devenish, S. O., McGregor, I. S., Arnold, J. C., & Chebib, M. (2017). The direct actions of cannabidiol and 2-arachidonoyl glycerol at GABA(A) receptors. *Pharmacological Research*, *119*, 358–370. https://doi.org/10.1016/j.phrs.2017.02.022.

Barann, M., Molderings, G., Brüss, M., Bönisch, H., Urban, B. W., & Göthert, M. (2002). Direct inhibition by cannabinoids of human 5-HT3A receptors: probable involvement of an allosteric modulatory site. *British Journal of Pharmacology*, *137*(5), 589–596. https://doi.org/10.1038/sj.bjp.0704829.

Bisogno, T., Hanus, L., De Petrocellis, L., Tchilibon, S., Ponde, D. E., Brandi, I., ... Di Marzo, V. (2001). Molecular targets for cannabidiol and its synthetic analogues: effect on vanilloid VR1 receptors and on the cellular uptake and enzymatic hydrolysis of anandamide. *British Journal of Pharmacology*, *134*(4), 845–852. https://doi.org/10.1038/sj.bjp.0704327.

Bladen, C., Mirlohi, S., Santiago, M., Longworth, M., Kassiou, M., Banister, S., & Connor, M. (2021). Modulation of human T-type calcium channels by synthetic cannabinoid receptor agonists in vitro. *Neuropharmacology*, *187*, 108478. https://doi.org/10.1016/j.neuropharm.2021.108478.

Brady, R. O., & Carbone, E. (1973). Comparison of the effects of 9-tetrahydrocannabinol, 11-hydroxy-9-tetrahydrocannabinol, and ethanol on the electrophysiological activity of the giant axon of the squid. *Neuropharmacology*, *12*(6), 601–605. https://doi.org/10.1016/0028-3908(73)90010-5.

Catterall, W. A., Lenaeus, M. J., & Gamal El-Din, T. M. (2020). Structure and pharmacology of voltage-gated sodium and calcium channels. *Annual Review of Pharmacology and Toxicology*, *60*, 133–154. https://doi.org/10.1146/annurev-pharmtox-010818-021757.

Cifelli, P., Ruffolo, G., De Felice, E., Alfano, V., van Vliet, E. A., Aronica, E., & Palma, E. (2020). Phytocannabinoids in neurological diseases: Could they restore a physiological GABAergic transmission? *International Journal of Molecular Sciences*, *21*(3). https://doi.org/10.3390/ijms21030723.

De Petrocellis, L., Ligresti, A., Moriello, A. S., Allarà, M., Bisogno, T., Petrosino, S., ... Di Marzo, V. (2011). Effects of cannabinoids and cannabinoid-enriched Cannabis extracts on TRP channels and endocannabinoid metabolic enzymes. *British Journal of Pharmacology*, *163*(7), 1479–1494. https://doi.org/10.1111/j.1476-5381.2010.01166.x.

De Petrocellis, L., Nabissi, M., Santoni, G., & Ligresti, A. (2017). Actions and regulation of ionotropic cannabinoid receptors. *Advances in Pharmacology*, *80*, 249–289. https://doi.org/10.1016/bs.apha.2017.04.001.

De Petrocellis, L., Orlando, P., Moriello, A. S., Aviello, G., Stott, C., Izzo, A. A., & Di Marzo, V. (2012). Cannabinoid actions at TRPV channels: effects on TRPV3 and TRPV4 and their potential relevance to gastrointestinal inflammation. *Acta Physiologica (Oxford, England)*, *204*(2), 255–266. https://doi.org/10.1111/j.1748-1716.2011.02338.x.

De Petrocellis, L., Vellani, V., Schiano-Moriello, A., Marini, P., Magherini, P. C., Orlando, P., & Di Marzo, V. (2008). Plant-derived cannabinoids modulate the activity of transient receptor potential channels of ankyrin type-1 and melastatin type-8. *The Journal of Pharmacology and Experimental Therapeutics*, *325*(3), 1007–1015. https://doi.org/10.1124/jpet.107.134809.

Deiana, S., Watanabe, A., Yamasaki, Y., Amada, N., Arthur, M., Fleming, S., ... Riedel, G. (2012). Plasma and brain pharmacokinetic profile of cannabidiol (CBD), cannabidivarine (CBDV), Δ^9-tetrahydrocannabivarin (THCV) and cannabigerol (CBG) in rats and mice following oral and intraperitoneal administration and CBD action on obsessive-compulsive behaviour. *Psychopharmacology*, *219*(3), 859–873. https://doi.org/10.1007/s00213-011-2415-0.

ElSohly, M. A., Radwan, M. M., Gul, W., Chandra, S., & Galal, A. (2017). Phytochemistry of Cannabis sativa L. *Progress in the Chemistry of Organic Natural Products*, *103*, 1–36. https://doi.org/10.1007/978-3-319-45541-9_1.

Etemad, L., Karimi, G., Alavi, M. S., & Roohbakhsh, A. (2022). Pharmacological effects of cannabidiol by transient receptor potential channels. *Life Sciences*, *300*, 120582. https://doi.org/10.1016/j.lfs.2022.120582.

Foadi, N., Leuwer, M., Demir, R., Dengler, R., Buchholz, V., de la Roche, J., ... Ahrens, J. (2010). Lack of positive allosteric modulation of mutated alpha(1)S267I glycine receptors by cannabinoids. *Naunyn-Schmiedeberg's Archives of Pharmacology*, *381*(5), 477–482. https://doi.org/10.1007/s00210-010-0506-9.

Fouda, M. A., Ghovanloo, M. R., & Ruben, P. C. (2020). Cannabidiol protects against high glucose-induced oxidative stress and cytotoxicity in cardiac voltage-gated sodium channels. *British Journal of Pharmacology*, *177*(13), 2932–2946. https://doi.org/10.1111/bph.15020.

Ghovanloo, M. R., Choudhury, K., Bandaru, T. S., Fouda, M. A., Rayani, K., Rusinova, R., ... Ruben, P. C. (2021). Cannabidiol inhibits the skeletal muscle Nav1.4 by blocking its pore and by altering membrane elasticity. *The Journal of General Physiology*, *153*(5). https://doi.org/10.1085/jgp.202012701.

Ghovanloo, M. R., Estacion, M., Higerd-Rusli, G. P., Zhao, P., Dib-Hajj, S., & Waxman, S. G. (2022). Inhibition of sodium conductance by cannabigerol contributes to a reduction of dorsal root ganglion neuron excitability. *British Journal of Pharmacology*. https://doi.org/10.1111/bph.15833.

Ghovanloo, M. R., & Ruben, P. C. (2021). Cannabidiol and sodium channel pharmacology: General overview, mechanism, and clinical implications. *The Neuroscientist*. https://doi.org/10.1177/10738584211017009.

Ghovanloo, M. R., Shuart, N. G., Mezeyova, J., Dean, R. A., Ruben, P. C., & Goodchild, S. J. (2018). Inhibitory effects of cannabidiol on voltage-dependent sodium currents. *The Journal of Biological Chemistry*, *293*(43), 16546–16558. https://doi.org/10.1074/jbc.RA118.004929.

Harding, E. K., Souza, I. A., Gandini, M. A., Gadotti, V. M., Ali, M. Y., Huang, S., … Zamponi, G. W. (2023). Differential regulation of Cav 3.2 and Cav 2.2 calcium channels by CB1 receptors and cannabidiol. *British Journal of Pharmacology*. https://doi.org/10.1111/bph.16035.

Hejazi, N., Zhou, C., Oz, M., Sun, H., Ye, J. H., & Zhang, L. (2006). Delta9-tetrahydrocannabinol and endogenous cannabinoid anandamide directly potentiate the function of glycine receptors. *Molecular Pharmacology*, *69*(3), 991–997. https://doi.org/10.1124/mol.105.019174.

Hill, A. J., Jones, N. A., Smith, I., Hill, C. L., Williams, C. M., Stephens, G. J., & Whalley, B. J. (2014). Voltage-gated sodium (NaV) channel blockade by plant cannabinoids does not confer anticonvulsant effects per se. *Neuroscience Letters*, *566*, 269–274. https://doi.org/10.1016/j.neulet.2014.03.013.

Hillard, C. J., Harris, R. A., & Bloom, A. S. (1985). Effects of the cannabinoids on physical properties of brain membranes and phospholipid vesicles: Fluorescence studies. *The Journal of Pharmacology and Experimental Therapeutics*, *232*(3), 579–588.

Hoffman, P. W., Ravindran, A., & Huganir, R. L. (1994). Role of phosphorylation in desensitization of acetylcholine receptors expressed in Xenopus oocytes. *The Journal of Neuroscience*, *14*(7), 4185–4195. https://doi.org/10.1523/jneurosci.14-07-04185.1994.

Huang, C. W., Lin, P. C., Chen, J. L., & Lee, M. J. (2021). Cannabidiol selectively binds to the voltage-gated sodium channel Na(v)1.4 in its slow-inactivated state and inhibits sodium current. *Biomedicines*, *9*(9). https://doi.org/10.3390/biomedicines9091141.

Iannotti, F. A., Hill, C. L., Leo, A., Alhusaini, A., Soubrane, C., Mazzarella, E., … Stephens, G. J. (2014). Nonpsychotropic plant cannabinoids, cannabidivarin (CBDV) and cannabidiol (CBD), activate and desensitize transient receptor potential vanilloid 1 (TRPV1) channels in vitro: potential for the treatment of neuronal hyperexcitability. *ACS Chemical Neuroscience*, *5*(11), 1131–1141. https://doi.org/10.1021/cn5000524.

Isaev, D., Shabbir, W., Dinc, E. Y., Lorke, D. E., Petroianu, G., & Oz, M. (2022). Cannabidiol inhibits multiple ion channels in rabbit ventricular cardiomyocytes. *Frontiers in Pharmacology*, *13*, 821758. https://doi.org/10.3389/fphar.2022.821758.

Izzo, A. A., Borrelli, F., Capasso, R., Di Marzo, V., & Mechoulam, R. (2009). Non-psychotropic plant cannabinoids: New therapeutic opportunities from an ancient herb. *Trends in Pharmacological Sciences*, *30*(10), 515–527. https://doi.org/10.1016/j.tips.2009.07.006.

Jackson, S. N., Singhal, S. K., Woods, A. S., Morales, M., Shippenberg, T., Zhang, L., & Oz, M. (2008). Volatile anesthetics and endogenous cannabinoid anandamide have additive and independent inhibitory effects on alpha(7)-nicotinic acetylcholine receptor-mediated responses in Xenopus oocytes. *European Journal of Pharmacology*, *582*(1–3), 42–51. https://doi.org/10.1016/j.ejphar.2007.12.023.

James, T. R., Richards, A. A., Lowe, D. A., Reid, W. A., Watson, C. T., & Pepple, D. J. (2022). The in vitro effect of delta-9-tetrahydrocannabinol and cannabidiol on whole blood viscosity, elasticity and membrane integrity. *J Cannabis Res*, *4*(1), 15. https://doi.org/10.1186/s42238-022-00126-z.

Jordt, S. E., Bautista, D. M., Chuang, H. H., McKemy, D. D., Zygmunt, P. M., Högestätt, E. D., … Julius, D. (2004). Mustard oils and cannabinoids excite sensory nerve fibres through the TRP channel ANKTM1. *Nature*, *427*(6971), 260–265. https://doi.org/10.1038/nature02282.

Kossakowski, R., Schlicker, E., Toczek, M., Weresa, J., & Malinowska, B. (2019). Cannabidiol affects the Bezold-Jarisch reflex via TRPV1 and 5-HT(3) receptors and has peripheral sympathomimetic effects in spontaneously hypertensive and normotensive rats. *Frontiers in Pharmacology*, *10*, 500. https://doi.org/10.3389/fphar.2019.00500.

Kowalski, C. W., Ragozzino, F. J., Lindberg, J. E. M., Peterson, B., Lugo, J. M., McLaughlin, R. J., & Peters, J. H. (2020). Cannabidiol activation of vagal afferent neurons requires TRPA1. *Journal of Neurophysiology*, *124*(5), 1388–1398. https://doi.org/10.1152/jn.00128.2020.

Kumbaraci, N. M., & Nastuk, W. L. (1980). Effects of delta 9-tetrahydrocannabinol on excitable membranes and neuromuscular transmission. *Molecular Pharmacology*, *17*(3), 344–349.

Latorre, R., Morera, F. J., & Zaelzer, C. (2013). Voltage-dependent K+ channels. In W. D. Lennarz, & M. D. Lane (Eds.), *Encyclopedia of biological chemistry* (pp. 555–559). MA: Waltham Academic Press.

Le Marois, M., Ballet, V., Sanson, C., Maizières, M. A., Carriot, T., Chantoiseau, C., … Bohme, G. A. (2020). Cannabidiol inhibits multiple cardiac ion channels and shortens ventricular action potential duration in vitro. *European Journal of Pharmacology*, *886*, 173542. https://doi.org/10.1016/j.ejphar.2020.173542.

Lodzki, M., Godin, B., Rakou, L., Mechoulam, R., Gallily, R., & Touitou, E. (2003). Cannabidiol-transdermal delivery and anti-inflammatory effect in a murine model. *Journal of Controlled Release*, *93*(3), 377–387. https://doi.org/10.1016/j.jconrel.2003.09.001.

Lu, J., Fan, S., Zou, G., Hou, Y., Pan, T., Guo, W., … Xiong, W. (2018). Involvement of glycine receptor α1 subunits in cannabinoid-induced analgesia. *Neuropharmacology*, *133*, 224–232. https://doi.org/10.1016/j.neuropharm.2018.01.041.

Lundbaek, J. A., Birn, P., Tape, S. E., Toombes, G. E., Søgaard, R., Koeppe, R. E., 2nd, … Andersen, O. S. (2005). Capsaicin regulates voltage-dependent sodium channels by altering lipid bilayer elasticity. *Molecular Pharmacology*, *68*(3), 680–689. https://doi.org/10.1124/mol.105.013573.

Lynch, J. W. (2009). Native glycine receptor subtypes and their physiological roles. *Neuropharmacology*, *56*(1), 303–309. https://doi.org/10.1016/j.neuropharm.2008.07.034.

Mahgoub, M., Keun-Hang, S. Y., Sydorenko, V., Ashoor, A., Kabbani, N., Al Kury, L., … Oz, M. (2013). Effects of cannabidiol on the function of α7-nicotinic acetylcholine receptors. *European Journal of Pharmacology*, *720*(1–3), 310–319. https://doi.org/10.1016/j.ejphar.2013.10.011.

Mason, E. R., & Cummins, T. R. (2020). Differential inhibition of human Nav1.2 resurgent and persistent sodium currents by cannabidiol and GS967. *International Journal of Molecular Sciences*, *21*(7). https://doi.org/10.3390/ijms21072454.

Mayar, S., Memarpoor-Yazdi, M., Makky, A., Eslami Sarokhalil, R., & D'Avanzo, N. (2022). Direct regulation of hyperpolarization-activated cyclic-nucleotide gated (HCN1) channels by cannabinoids. *Frontiers in Molecular Neuroscience*, *15*, 848540. https://doi.org/10.3389/fnmol.2022.848540.

Mazeh, A. C., Angus, J. A., & Wright, C. E. (2021). Cannabidiol selectively inhibits the contraction of rat small resistance arteries: Possible role for CGRP and voltage-gated calcium channels. *European Journal of Pharmacology*, *891*, 173767. https://doi.org/10.1016/j.ejphar.2020.173767.

Mendis, G. D. C., Berecki, G., Morrisroe, E., Pachernegg, S., Li, M., Varney, M., … Petrou, S. (2019). Discovering the pharmacodynamics of conolidine and cannabidiol using a cultured neuronal network based workflow. *Scientific Reports*, *9*(1), 121. https://doi.org/10.1038/s41598-018-37138-w.

Milligan, C. J., Anderson, L. L., Bowen, M. T., Banister, S. D., McGregor, I. S., Arnold, J. C., & Petrou, S. (2022). A nutraceutical product, extracted from Cannabis sativa, modulates voltage-gated sodium channel function. *Journal of Cannabis Research*, *4*(1), 30. https://doi.org/10.1186/s42238-022-00136-x.

Miranda, C. C., Barata, T., Vaz, S. H., Ferreira, C., Quintas, A., & Bekman, E. P. (2020). hiPSC-based model of prenatal exposure to cannabinoids: Effect on neuronal differentiation. *Frontiers in Molecular Neuroscience*, *13*, 119. https://doi.org/10.3389/fnmol.2020.00119.

Mirlohi, S., Bladen, C., Santiago, M. J., Arnold, J. C., McGregor, I., & Connor, M. (2022). Inhibition of human recombinant T-type calcium channels by phytocannabinoids in vitro. *British Journal of Pharmacology*. https://doi.org/10.1111/bph.15842.

Mirlohi, S., Bladen, C., Santiago, M., & Connor, M. (2022). Modulation of recombinant human T-type calcium channels by Δ(9)-tetrahydrocannabinolic acid in vitro. *Cannabis Cannabinoid Research*, *7*(1), 34–45. https://doi.org/10.1089/can.2020.0134.

Morano, A., Cifelli, P., Nencini, P., Antonilli, L., Fattouch, J., Ruffolo, G., … Giallonardo, A. T. (2016). Cannabis in epilepsy: From clinical practice to basic research focusing on the possible role of cannabidivarin. *Epilepsia Open*, *1*(3–4), 145–151. https://doi.org/10.1002/epi4.12015.

Muller, C., Morales, P., & Reggio, P. H. (2019). Cannabinoid ligands targeting TRP channels. *Frontiers in Molecular Neuroscience*, *11*, 487. https://doi.org/10.3389/fnmol.2018.00487.

Nebrisi, E. E., Prytkova, T., Lorke, D. E., Howarth, L., Alzaabi, A. H., Yang, K. S., … Oz, M. (2020). Capsaicin is a negative allosteric modulator of the 5-HT(3) receptor. *Frontiers in Pharmacology*, *11*, 1274. https://doi.org/10.3389/fphar.2020.01274.

Nishizaki, T., & Sumikawa, K. (1998). Effects of PKC and PKA phosphorylation on desensitization of nicotinic acetylcholine receptors. *Brain Research*, *812*(1–2), 242–245. https://doi.org/10.1016/s0006-8993(98)00836-1.

Ortiz, Y. T., McMahon, L. R., & Wilkerson, J. L. (2022). Medicinal cannabis and central nervous system disorders. *Frontiers in Pharmacology*, *13*, 881810. https://doi.org/10.3389/fphar.2022.881810.

Orvos, P., Pászti, B., Topal, L., Gazdag, P., Prorok, J., Polyák, A., … Csupor, D. (2020). The electrophysiological effect of cannabidiol on hERG current and in guinea-pig and rabbit cardiac preparations. *Scientific Reports*, *10*(1), 16079. https://doi.org/10.1038/s41598-020-73165-2.

O'Sullivan, S. E., Sun, Y., Bennett, A. J., Randall, M. D., & Kendall, D. A. (2009). Time-dependent vascular actions of cannabidiol in the rat aorta. *European Journal of Pharmacology*, *612*(1–3), 61–68. https://doi.org/10.1016/j.ejphar.2009.03.010.

Oz, M. (2006). Receptor-independent actions of cannabinoids on cell membranes: Focus on endocannabinoids. *Pharmacology & Therapeutics*, *111*(1), 114–144. https://doi.org/10.1016/j.pharmthera.2005.09.009.

Oz, M., Al Kury, L., Keun-Hang, S. Y., Mahgoub, M., & Galadari, S. (2014). Cellular approaches to the interaction between cannabinoid receptor ligands and nicotinic acetylcholine receptors. *European Journal of Pharmacology*, *731*, 100–105. https://doi.org/10.1016/j.ejphar.2014.03.010.

Oz, M., Lozon, Y., Sultan, A., Yang, K. H., & Galadari, S. (2015). Effects of monoterpenes on ion channels of excitable cells. *Pharmacology & Therapeutics*, *152*, 83–97. https://doi.org/10.1016/j.pharmthera.2015.05.006.

Oz, M., Ravindran, A., Diaz-Ruiz, O., Zhang, L., & Morales, M. (2003). The endogenous cannabinoid anandamide inhibits alpha7 nicotinic acetylcholine receptor-mediated responses in Xenopus oocytes. *The Journal of Pharmacology and Experimental Therapeutics*, *306*(3), 1003–1010. https://doi.org/10.1124/jpet.103.049981.

Oz, M., Tchugunova, Y., & Dinc, M. (2004). Differential effects of endogenous and synthetic cannabinoids on voltage-dependent calcium fluxes in rabbit T-tubule membranes: Comparison with fatty acids. *European Journal of Pharmacology*, *502*(1–2), 47–58. https://doi.org/10.1016/j.ejphar.2004.08.052.

Oz, M., Zhang, L., & Morales, M. (2002). Endogenous cannabinoid, anandamide, acts as a noncompetitive inhibitor on 5-HT3 receptor-mediated responses in Xenopus oocytes. *Synapse*, *46*(3), 150–156. https://doi.org/10.1002/syn.10121.

Oz, M., Zhang, L., Ravindran, A., Morales, M., & Lupica, C. R. (2004). Differential effects of endogenous and synthetic cannabinoids on alpha7-nicotinic acetylcholine receptor-mediated responses in Xenopus Oocytes. *The Journal of Pharmacology and Experimental Therapeutics*, *310*(3), 1152–1160. https://doi.org/10.1124/jpet.104.067751.

Oz, M., Zhang, L., & Spivak, C. E. (2002). Direct noncompetitive inhibition of 5-HT(3) receptor-mediated responses by forskolin and steroids. *Archives of Biochemistry and Biophysics*, *404*(2), 293–301. https://doi.org/10.1016/s0003-9861(02)00279-5.

Pagano, C., Navarra, G., Coppola, L., Avilia, G., Bifulco, M., & Laezza, C. (2022). Cannabinoids: Therapeutic use in clinical practice. *International Journal of Molecular Sciences*, *23*(6). https://doi.org/10.3390/ijms23063344.

Patel, R. R., Barbosa, C., Brustovetsky, T., Brustovetsky, N., & Cummins, T. R. (2016). Aberrant epilepsy-associated mutant Nav1.6 sodium channel activity can be targeted with cannabidiol. *Brain*, *139*(Pt 8), 2164–2181. https://doi.org/10.1093/brain/aww129.

Perez, E., Ceja-Vega, J., Krmic, M., Gamez Hernandez, A., Gudyka, J., Porteus, R., & Lee, S. (2022). Differential interaction of cannabidiol with biomembranes dependent on cholesterol concentration. *ACS Chemical Neuroscience*, *13*(7), 1046–1054. https://doi.org/10.1021/acschemneuro.2c00040.

Pertwee, R. G., & Cascio, M. G. (2014). Known pharmacological actions of delta-9-tetrahydrocannabinol and of four other chemical constituents of cannabis that activate cannabinoid receptors. In R. G. Pertwee (Ed.), *Handbook of cannabis* (pp. 115–136). Oxford, UK: Oxford University Press.

Plested, A. J. (2016). Structural mechanisms of activation and desensitization in neurotransmitter-gated ion channels. *Nature Structural & Molecular Biology*, *23*(6), 494–502. https://doi.org/10.1038/nsmb.3214.

Poling, J. S., Rogawski, M. A., Salem, N., Jr., & Vicini, S. (1996). Anandamide, an endogenous cannabinoid, inhibits Shaker-related voltage-gated K+ channels. *Neuropharmacology*, *35*(7), 983–991. https://doi.org/10.1016/0028-3908(96)00130-x.

Pryce, G., Riddall, D. R., Selwood, D. L., Giovannoni, G., & Baker, D. (2015). Neuroprotection in experimental autoimmune encephalomyelitis and progressive multiple sclerosis by cannabis-based cannabinoids. *Journal of Neuroimmune Pharmacology*, *10*(2), 281–292. https://doi.org/10.1007/s11481-014-9575-8.

Qin, N., Neeper, M. P., Liu, Y., Hutchinson, T. L., Lubin, M. L., & Flores, C. M. (2008). TRPV2 is activated by cannabidiol and mediates CGRP release in cultured rat dorsal root ganglion neurons. *The Journal of Neuroscience*, *28*(24), 6231–6238. https://doi.org/10.1523/jneurosci.0504-08.2008.

Ross, H. R., Napier, I., & Connor, M. (2008). Inhibition of recombinant human T-type calcium channels by Delta9-tetrahydrocannabinol and cannabidiol. *The Journal of Biological Chemistry*, *283*(23), 16124–16134. https://doi.org/10.1074/jbc.M707104200.

Ruffolo, G., Cifelli, P., Roseti, C., Thom, M., van Vliet, E. A., Limatola, C., … Palma, E. (2018). A novel GABAergic dysfunction in human Dravet syndrome. *Epilepsia*, *59*(11), 2106–2117. https://doi.org/10.1111/epi.14574.

Sait, L. G., Sula, A., Ghovanloo, M. R., Hollingworth, D., Ruben, P. C., & Wallace, B. A. (2020). Cannabidiol interactions with voltage-gated sodium channels. *eLife*, *9*. https://doi.org/10.7554/eLife.58593.

Senn, L., Cannazza, G., & Biagini, G. (2020). Receptors and channels possibly mediating the effects of phytocannabinoids on seizures and epilepsy. *Pharmaceuticals (Basel)*, *13*(8). https://doi.org/10.3390/ph13080174.

Siara, J., Ruppersberg, J. P., & Rüdel, R. (1990). Human nicotinic acetylcholine receptor: the influence of second messengers on activation and desensitization. *Pflügers Archiv*, *415*(6), 701–706. https://doi.org/10.1007/bf02584008.

Smiley, K. A., Karler, R., & Turkanis, S. A. (1976). Effects of cannabinoids on the perfused rat heart. *Research Communications in Chemical Pathology and Pharmacology*, *14*(4), 659–675.

Smith, L. C., Tieu, L., Suhandynata, R. T., Boomhower, B., Hoffman, M., Sepulveda, Y., … George, O. (2021). Cannabidiol reduces withdrawal symptoms in nicotine-dependent rats. *Psychopharmacology*, *238*(8), 2201–2211. https://doi.org/10.1007/s00213-021-05845-4.

Spivak, C. E., Lupica, C. R., & Oz, M. (2007). The endocannabinoid anandamide inhibits the function of alpha4beta2 nicotinic acetylcholine receptors. *Molecular Pharmacology*, *72*(4), 1024–1032. https://doi.org/10.1124/mol.107.036939.

Stanslowsky, N., Jahn, K., Venneri, A., Naujock, M., Haase, A., Martin, U., … Wegner, F. (2017). Functional effects of cannabinoids during dopaminergic specification of human neural precursors derived from induced pluripotent stem cells. *Addiction Biology*, *22*(5), 1329–1342. https://doi.org/10.1111/adb.12394.

Sun, Y., & Dolmetsch, R. E. (2018). Investigating the therapeutic mechanism of cannabidiol in a human induced pluripotent stem cell (iPSC)-based model of dravet syndrome. *Cold Spring Harbor Symposia on Quantitative Biology*, *83*, 185–191. https://doi.org/10.1101/sqb.2018.83.038174.

Topal, L., Naveed, M., Orvos, P., Pászti, B., Prorok, J., Bajtel, Á., … Jost, N. (2021). The electrophysiological effects of cannabidiol on action potentials and transmembrane potassium currents in rabbit and dog cardiac ventricular preparations. *Archives of Toxicology*, *95*(7), 2497–2505. https://doi.org/10.1007/s00204-021-03086-0.

Turkanis, S. A., & Karler, R. (1983). Effects of delta 9-tetrahydrocannabinol on cat spinal motoneurons. *Brain Research*, *288*(1–2), 283–287. https://doi.org/10.1016/0006-8993(83)90104-x.

Turkanis, S. A., & Karler, R. (1984). Influence of 11-hydroxy-delta-9-tetrahydrocannabinol on spinal motoneurons in cat. *Neuropharmacology*, *23*(12a), 1435–1439. https://doi.org/10.1016/0028-3908(84)90086-8.

Turkanis, S. A., & Karler, R. (1986a). Cannabidiol-caused depression of spinal motoneuron responses in cats. *Pharmacology, Biochemistry, and Behavior*, *25*(1), 89–94. https://doi.org/10.1016/0091-3057(86)90235-2.

Turkanis, S. A., & Karler, R. (1986b). Effects of delta-9-tetrahydrocannabinol, 11-hydroxy-delta-9-tetrahydrocannabinol and cannabidiol on neuromuscular transmission in thc frog. *Neuropharmacology*, *25*(11), 1273–1278. https://doi.org/10.1016/0028-3908(86)90147-4.

Turkanis, S. A., Karler, R., & Partlow, L. M. (1991). Differential effects of delta-9-tetrahydrocannabinol and its 11-hydroxy metabolite on sodium current in neuroblastoma cells. *Brain Research*, *560*(1–2), 245–250. https://doi.org/10.1016/0006-8993(91)91239-w.

Turkanis, S. A., Partlow, L. M., & Karler, R. (1991). Delta-9-tetrahydrocannabinol depresses inward sodium current in mouse neuroblastoma cells. *Neuropharmacology*, *30*(1), 73–77. https://doi.org/10.1016/0028-3908(91)90045-d.

Varvel, S. A., Wiley, J. L., Yang, R., Bridgen, D. T., Long, K., Lichtman, A. H., & Martin, B. R. (2006). Interactions between THC and cannabidiol in mouse models of cannabinoid activity. *Psychopharmacology*, *186*(2), 226–234. https://doi.org/10.1007/s00213-006-0356-9.

Vitale, R. M., Iannotti, F. A., & Amodeo, P. (2021). The (poly)pharmacology of cannabidiol in neurological and neuropsychiatric disorders: Molecular mechanisms and targets. *International Journal of Molecular Sciences*, *22*(9). https://doi.org/10.3390/ijms22094876.

Wright, S., & Guy, G. (2014). Licensed cannabis-based medicines: Benefits and risks. In R. G. Pertwee (Ed.), *Handbook of cannabis* (pp. 373–392). Oxford, UK: Oxford University Pres.

Xiong, W., Cheng, K., Cui, T., Godlewski, G., Rice, K. C., Xu, Y., & Zhang, L. (2011). Cannabinoid potentiation of glycine receptors contributes to cannabis-induced analgesia. *Nature Chemical Biology*, *7*(5), 296–303. https://doi.org/10.1038/nchembio.552.

Xiong, W., Cui, T., Cheng, K., Yang, F., Chen, S. R., Willenbring, D., … Zhang, L. (2012). Cannabinoids suppress inflammatory and neuropathic pain by targeting α3 glycine receptors. *The Journal of Experimental Medicine*, *209*(6), 1121–1134. https://doi.org/10.1084/jem.20120242.

Xiong, W., Koo, B. N., Morton, R., & Zhang, L. (2011). Psychotropic and nonpsychotropic cannabis derivatives inhibit human 5-HT(3A) receptors through a receptor desensitization-dependent mechanism. *Neuroscience*, *184*, 28–37. https://doi.org/10.1016/j.neuroscience.2011.03.066.

Xiong, W., Wu, X., Li, F., Cheng, K., Rice, K. C., Lovinger, D. M., & Zhang, L. (2012). A common molecular basis for exogenous and endogenous cannabinoid potentiation of glycine receptors. *The Journal of Neuroscience*, *32*(15), 5200–5208. https://doi.org/10.1523/jneurosci.6347-11.2012.

Yang, K. H., Galadari, S., Isaev, D., Petroianu, G., Shippenberg, T. S., & Oz, M. (2010). The nonpsychoactive cannabinoid cannabidiol inhibits 5-hydroxytryptamine3A receptor-mediated currents in Xenopus laevis oocytes. *The Journal of Pharmacology and Experimental Therapeutics*, *333*(2), 547–554. https://doi.org/10.1124/jpet.109.162594.

Yang, K. H., Isaev, D., Morales, M., Petroianu, G., Galadari, S., & Oz, M. (2010). The effect of Δ9-tetrahydrocannabinol on 5-HT3 receptors depends on the current density. *Neuroscience*, *171*(1), 40–49. https://doi.org/10.1016/j.neuroscience.2010.08.044.

Yao, L., Liu, C., Wang, N., Du, F., Fan, S., Guo, Y., … Xiong, W. (2020). Cholesterol regulates cannabinoid analgesia through glycine receptors. *Neuropharmacology*, *177*, 108242. https://doi.org/10.1016/j.neuropharm.2020.108242.

Yao, L., Wells, M., Wu, X., Xu, Y., Zhang, L., & Xiong, W. (2020). Membrane cholesterol dependence of cannabinoid modulation of glycine receptor. *The FASEB Journal*, *34*(8), 10920–10930. https://doi.org/10.1096/fj.201903093R.

Yu, Y., Yang, Z., Jin, B., Qin, X., Zhu, X., Sun, J., … Zhang, W. (2020). Cannabidiol inhibits febrile seizure by modulating AMPA receptor kinetics through its interaction with the N-terminal domain of GluA1/GluA2. *Pharmacological Research*, *161*, 105128. https://doi.org/10.1016/j.phrs.2020.105128.

Zhang, H. B., & Bean, B. P. (2021). Cannabidiol inhibition of murine primary nociceptors: tight binding to slow inactivated states of Na(v)1.8 channels. *The Journal of Neuroscience*, *41*(30), 6371–6387. https://doi.org/10.1523/jneurosci.3216-20.2021.

Zhang, H. B., Heckman, L., Niday, Z., Jo, S., Fujita, A., Shim, J., … Bean, B. P. (2022). Cannabidiol activates neuronal Kv7 channels. *eLife*, *11*. https://doi.org/10.7554/eLife.73246.

Zhang, L., Oz, M., Stewart, R. R., Peoples, R. W., & Weight, F. F. (1997). Volatile general anaesthetic actions on recombinant nACh alpha 7,5-HT3 and chimeric nACh alpha 7-5-HT3 receptors expressed in Xenopus oocytes. *British Journal of Pharmacology*, *120*(3), 353–355. https://doi.org/10.1038/sj.bjp.0700934.

Zhang, L., Oz, M., & Weight, F. F. (1995). Potentiation of 5-HT3 receptor-mediated responses by protein kinase C activation. *Neuroreport*, *6*(10), 1464–1468. https://doi.org/10.1097/00001756-199507100-00025.

Chapter 34

Cannabis, neurodevelopment, and the "two-hit" hypothesis

Elisa Guma[a,b,c], Lani Cupo[a,b,c], and M. Mallar Chakravarty[a,b,c,d,e]

[a]*Computational Brain Anatomy Laboratory, Cerebral Imaging Centre, Douglas Mental Health University Institute, Montreal, QC, Canada,* [b]*Integrated Program in Neuroscience, McGill University, Montreal, QC, Canada,* [c]*Cerebral Imaging Centre, Douglas Mental Health University Institute, Montreal, QC, Canada,* [d]*Department of Psychiatry, McGill University, Montreal, QC, Canada,* [e]*Department of Biomedical Engineering, McGill University, Montreal, QC, Canada*

Abbreviations

AEA	anandamide
AMPA	α-amino-3-hydroxy-5-methyl-4-isoxazolepropionic acid
CB	cannabinoid
CBD	cannabidiol
COMT	catechol-o-methyltransferase
FAAH	fatty acid amide hydrolase
GABA	gamma-amino butyric acid
GD	gestational day
MAM	methylazoxymethanol
MIA	maternal immune activation
MRI	magnetic resonance imaging
OFC	orbitofrontal cortex
PND	postnatal day
PFC	prefrontal cortex
SNP	single nucleotide polymorphism
THC	delta-9-tetrahydrocannabinol

Introduction

Adolescence is a dynamic developmental period characterized by social, behavioral, and neurobiological changes (Keshavan et al., 2014). The brain undergoes substantial neuroanatomical, functional, and chemical maturation leading to the refinement of functional circuits. Many of the brain's neurotransmitter systems, such as the dopaminergic, glutamatergic, and gamma-aminobutyric acid (GABA)-ergic systems play important roles in guiding the organization of these circuits. Endocannabinoids play a critical role in brain maturation and are linked to numerous neurodevelopmental processes in the maturing brain, including neuronal differentiation, axonal pathfinding, and gliogenesis (Harkany et al., 2008). Interestingly, the endocannabinoid system, through the endogenous or exogenous activation of cannabinoid (CB)-1 receptors, can influence the developmental trajectory of some of these neurotransmitter systems. One potential exogenous stimulus includes the main psychoactive component of cannabis, delta-9-tetrahydrocannabinol (THC), which binds to the CB-1 (and to some extent CB-2) receptors in the brain. Therefore, cannabis use in adolescence could interfere with the development of the endocannabinoid system as well as other related neurotransmitter systems, potentially leading to neuroanatomical, functional, and behavioural changes later in development.

Cannabis is among the most frequently used psychoactive drugs, particularly in adolescents and young adults. Use in adolescence has been associated with impairments in executive function and impulse control (Morin et al., 2019). Converging lines of evidence also point to use in this developmental window conferring risk for psychiatric illness such as

Cannabis Use, Neurobiology, Psychology, and Treatment. https://doi.org/10.1016/B978-0-323-89862-1.00022-2

schizophrenia (Arseneault et al., 2002), particularly if use begins before 16 years of age (French et al., 2015), which aligns with a peak in CB-1 receptor expression in the brain (Meyer et al., 2018). The adolescent peak of CB-1 receptor expression has been confirmed in rodents (~postnatal day [PND] 30), followed by a steady decline toward adulthood (~PND 70) (Ellgren et al., 2008). Importantly, CB-1 receptors are highly concentrated in the prefrontal cortex (PFC) and striatum (Heng et al., 2011), two regions which undergo developmental changes during adolescence, and whose abnormal development has been associated with numerous behavioral alterations. Initiation of cannabis use in concurrence with the peak in CB-1 receptor expression may derail aspects of brain development. Of further concern is the increasing concentration of THC in both the cannabis plant and in synthetic products (ElSohly et al., 2016), as THC is thought to be responsible for negative outcomes of cannabis use on neurodevelopment. The known risks associated with THC use, the high adolescent consumption of cannabis, and rising psychoactive potency of this drug represent an increasing global public health concern.

Cannabis consumption is a risk factor for several major mental illnesses, but response to cannabis consumption is variable as there are individuals with a range of susceptibilities (i.e., not all individuals who consume cannabis develop psychiatric illness). This has led researchers to formulate the "two-hit hypothesis" (or dual-hit hypothesis) (Maynard et al., 2001) which suggests that prenatal genetic or environmental "first-hits" (i.e., early life neglect, exposure to maternal stress or infection (Guma, Bordeleau, et al., 2022; Guma et al., 2019, 2021; Meyer, 2014), etc.) disrupt certain aspects of neurodevelopment, increasing vulnerability to "second-hits" experienced later in the lifespan, such as adolescent cannabis use (Fig. 1). In this chapter, we will discuss the susceptibility of the adolescent brain to cannabis use, the mechanisms of action of THC in the brain, the evidence for the association between cannabis use and major mental illness, and how cannabis use fits into the "two-hit" hypothesis.

Adolescence is a sensitive developmental window

Adolescence is a time of dramatic physical and behavioral change characterized by the onset of sexual maturation and the acquisition of adult social roles. During this developmental window, an increased desire for novelty seeking and impulsive behaviors occurs in parallel with the continued development of cognitive abilities. Furthermore, adolescence marks the onset of many major mental illnesses, which can interfere with academic achievement, family relationships, and further increases risk-taking behavior and substance misuse (Keshavan et al., 2014). Dynamic changes are also occurring in the

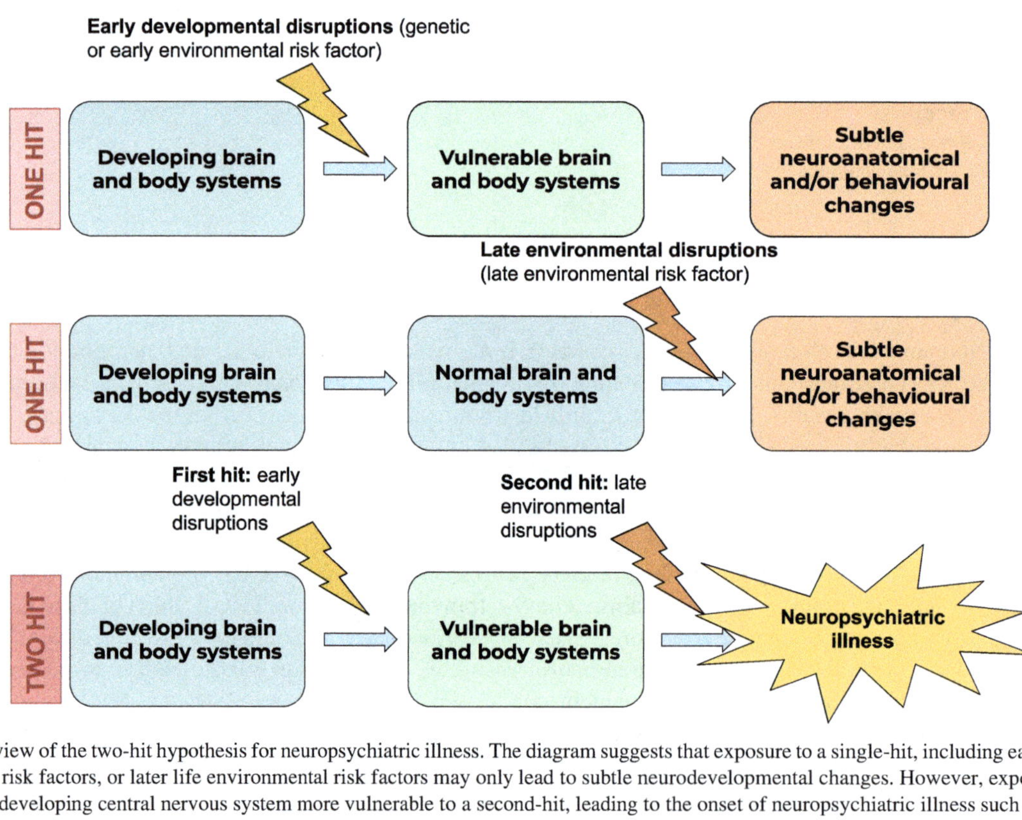

FIG. 1 Overview of the two-hit hypothesis for neuropsychiatric illness. The diagram suggests that exposure to a single-hit, including early life genetic or environmental risk factors, or later life environmental risk factors may only lead to subtle neurodevelopmental changes. However, exposure to a first-hit may make the developing central nervous system more vulnerable to a second-hit, leading to the onset of neuropsychiatric illness such as schizophrenia.

brain. Gray matter volume exhibits a steady, near-linear decline (following ~6 years of age), while white matter volume is still growing (peaking at ~19 years of age) (Bethlehem et al., 2022). Importantly, subcortical gray matter and white matter volume peak in adolescence (~14 years of age) (Bethlehem et al., 2022), which might make these regions, and their connections to cortical regions, more vulnerable to risk factors experienced during this time. Understanding the neurobiological factors underlying dynamic adolescent neurodevelopment is crucial to better understanding vulnerability to disease onset. Furthermore, understanding the role that genetic or environmental risk factors, such as adolescent cannabis use, play in shaping these changes may provide further insight into the mechanisms leading to the presentation of disease.

The endocannabinoid system and the developing brain

Endocannabinoid signaling is critically involved in guiding prenatal brain development. It supports the proliferation and differentiation of progenitor cells, axon guidance, neuronal migration, and outgrowth, as well as the positioning of cortical interneurons (Harkany et al., 2008). Indeed, disruption of endocannabinoid signaling via blockade of the CB-1 receptor during gestation, or via THC exposure, leads to impaired neuronal proliferation, cortical lamination, corticogenesis, and inhibitory neuron signaling (Mulder et al., 2008). While the endocannabinoid system plays an important role in shaping the earliest phases of brain development, it is also involved in adolescent brain development (Meyer et al., 2018). Expression of the CB-1 receptor has been shown to increase dramatically throughout the lifespan in corticolimbic regions, in both human and rodent studies (Mato et al., 2003).

The PFC has received a great deal of attention in the context of adolescent brain development, where the endocannabinoid system also plays an important role in shaping its development and function. This region is also central to the pathology of numerous neuropsychiatric disorders, including schizophrenia (Selemon & Zecevic, 2015). CB-1 receptors are present on GABAergic neurons in the PFC. Activation of these receptors via endogenous cannabinoids or exogenous cannabis consumption can decrease GABA release, which decreases the inhibitory break on excitatory glutamatergic pyramidal neurons as well as on dopaminergic neurons leading to an increase in either neurotransmitter at the synapse (Fig. 2) (Eggan et al., 2010). These states of hyperglutamatergia or hyperdopaminergia may affect cognitive function and increase risk for abnormal circuit development underlying psychosis symptoms (Eggan et al., 2010).

Exogenous activation of the endocannabinoid system via cannabis consumption has been associated with accelerated cortical thinning in adolescence (Albaugh et al., 2021; French et al., 2015). Interestingly, there is a dose-dependent relationship between the amount of cannabis consumed and the frequency to which cannabis is consumed, predominantly in prefrontal regions. Importantly, there is convergence between regions of cortical thinning and expression of CB-1 receptors (as identified by positron emission tomography mapping (Albaugh et al., 2021)). While better characterization of these dynamic changes via longitudinal investigation is required at a more region and age specific level, the endocannabinoid system is critically involved in adolescent brain development. Disruption of this system during this sensitive developmental window may, therefore, have long-term impact on adult brain structure and function.

Mechanisms of action of cannabis

THC is the main active compound of the cannabis plant. It binds to CB-1 (and CB-2) receptors in the central nervous system and activates the endocannabinoid system. The human body also produces its own endocannabinoids, which include anandamide (AEA) and 2-arachidonoyl glycerol (2-AG) (Lu & Mackie, 2021). AEA is a high-affinity, partial agonist of CB-1 receptors and has low-affinity binding for CB-2 receptors, while 2-AG is a full agonist at both receptors. Interestingly, THC is roughly equivalent to AEA in its binding properties (Mackie, 2006).

CB-1 receptors are among the most abundantly expressed G-protein coupled receptors in the brain, with peak concentrations in numerous cortical regions, the cerebellum, hippocampus, amygdala, striatum, and globus pallidus (Herkenham et al., 1991). Within these brain regions, they are expressed on glutamate, GABA, dopamine, serotonin, and acetylcholine releasing neurons, which play a role in the pathology of neuropsychiatric illnesses (Freund et al., 2003). They are expressed presynaptically, where they tune neurotransmitter release. Endocannabinoids such as AEA and 2-AG are synthesized and released postsynaptically and can retrogradely traverse to bind to presynaptic receptors and influence synaptic plasticity (Shrivastava et al., 2014). The mechanisms used to break down and recycle endogenous endocannabinoids do not work on THC resulting in sustained high levels of the compound which may have undesired effects (Keimpema et al., 2011). Indeed, activation of CB-1 receptors, via exogenous THC, can suppress neurotransmitter release to the synapse (Malone et al., 2010). This suppression, particularly if it occurs during adolescence when the system is maturing, may interfere with the development of both glutamatergic pyramidal cells, as well as GABA-ergic parvalbumin-containing basket cells. This may result in an excitatory-inhibitory signaling imbalance, thought to underlie some of the disruptions observed in

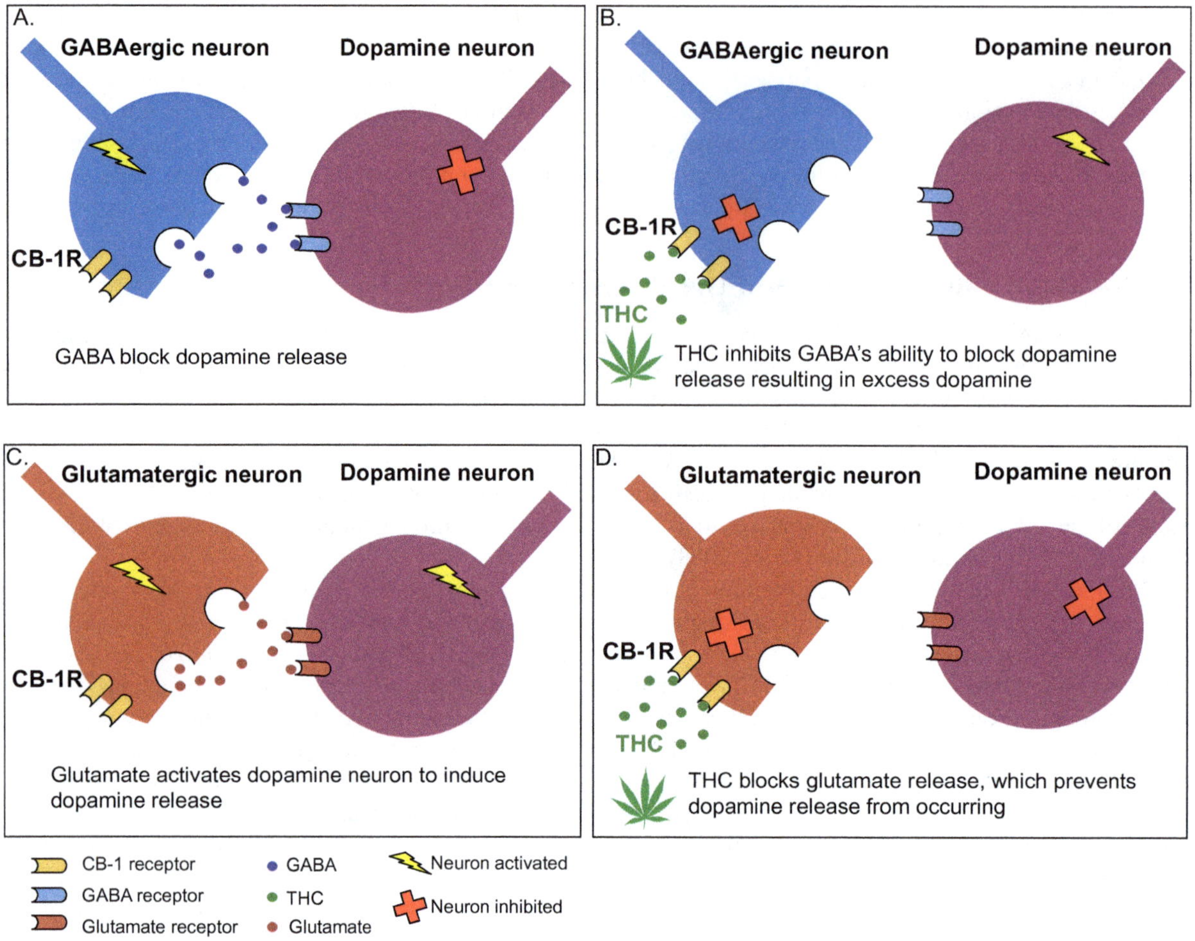

FIG. 2 Mechanism of action of THC on GABAergic, dopaminergic, and glutamatergic neurons. (A) GABA release from GABAergic neurons bind to GABA A or B receptors on dopamine neurons to inhibit the neurons and suppress dopamine release. (B) THC binds to CB-1 receptors on the GABAergic neurons to suppress GABA release. This in turn prevents inhibition of the dopamine neurons, leading to excess dopamine release. (C) Glutamatergic neurons release glutamate, which binds to glutamate neurons on dopamine cells (NMDA or AMPA), which activate the dopamine neuron, signalling the dopamine cell to release dopamine. (D) THC binds to the CB-1 receptors on the glutamatergic neurons, which inhibit glutamate release into the synapse, interfering with activation of the dopamine neurons, and inhibiting dopamine release. GABA = γ-aminobutyric acid, CB-1 = cannabinoid receptor 1, THC = tetrahydrocannabinol, NMDA = N-methyl-D-aspartate, AMPA = α-amino-3-hydroxy-5-methyl-4-isoxazole propionic acid.

psychotic symptoms (Keshavan et al., 2014). CB-1 activation can also induce acute dopamine release, associated with positive psychotic symptoms such as hallucinations or delusions (Keshavan et al., 2014). While CB-1 receptors are mainly expressed in neurons, they have also been identified on microglia (Cabral et al., 2008). CB-2 receptors, to which THC may also bind, are predominantly found in peripheral tissues but are also expressed on microglia in the central nervous system (Cabral et al., 2008). These glial cells play an interesting role in synaptic remodelling during adolescence and may be implicated in the pathology of neuropsychiatric illness.

Differential effects of THC and CBD

While THC receives much academic attention as a potential agent modifying development, cannabis contains a milieu of other compounds, both psycho-active and not. In addition to THC, cannabidiol (CBD) is the other main biologically active fatty cannabinoid found in cannabis. Relative to THC, CBD is widely considered nonpsychoactive and of high potential as a therapeutic agent, especially among the lay press (White, 2019). Compared to THC, CBD binds more weakly to CB-1 and CB-2 receptors (Peng et al., 2022). CBD contrasts the mechanism of action of THC, as it is an antagonist and THC is a partial agonist of CB-1 and -2 receptors (Pertwee, 2008). It exhibits a diversity of molecular effects as it binds to other receptors (transient receptor potential vanilloid 1 channel and 5HT1-alpha); this is speculated to contribute to hypothesised therapeutic effects including anti-inflammatory effects, antiepileptic effects, and pain relief (Peng et al., 2022; White,

2019). Nevertheless, there is a dearth of well-designed, randomized, placebo-controlled, double-blind clinical trials with diverse populations, and many of the findings remain speculatory (Peng et al., 2022; White, 2019). While more research is required, there is much enthusiasm (perhaps premature) for the potential use of CBD as an antipsychotic.

Evidence for cannabis use and risk for psychosis

Human studies focusing on THC

Cannabis use in humans has been associated with acute side effects that resemble the symptoms of psychosis (Marconi et al., 2016). Participants exposed to THC and then assessed with scales commonly used in assessing psychotic episodes (Positive and Negative Syndrome Scale, Brief Psychiatric Rating Scale, State Social Paranoia Scale, and self-report) reliably demonstrate temporarily increased symptom profiles (Colizzi et al., 2019). Whether chronic exposure to cannabis contributes to the emergence of psychotic disorders such as schizophrenia is more widely debated in the scientific literature.

There are several major challenges to assessing a causal relationship between cannabis use and the emergence of psychosis in humans. The potential of cannabis use as a risk for psychosis has been investigated for over 40 years. One of the confounding factors that complicates this relationship is that there is evidence that there is shared genetic predisposition for both cannabis use and psychosis (Verweij et al., 2017). Additionally, there is evidence for a dose-response relationship between the amount of high-potency cannabis used and psychosis-related outcomes including everything from self-reported symptoms to formal diagnosis (Marconi et al., 2016). It can be challenging to assess the dosage or timing of cannabis use among a population, especially as cannabis has only recently been decriminalized or legalized in many countries. The rising potency (concentration of THC) of cannabis also makes it difficult to interpret older evidence of associations between cannabis use and emergence of psychosis in the context of currently available strains and preparations (Wilson, Freeman, & Mackie, 2019).

There is plentiful evidence that the timing of initiation of cannabis use is important to the emergence of psychosis. Psychotic disorders typically emerge in late adolescence, 21–25 years of age for men and 25–30 for women (with a second peak around menopause) (Li et al., 2016). Epidemiological evidence suggests that early use of high potency THC is associated with the onset of psychosis in a dose-dependent manner (Bagot et al., 2015). Interestingly, age of onset of cannabis use has been correlated with age of onset of psychosis (Di Forti et al., 2014; Large et al., 2011). Additionally, longitudinal, prospective studies can provide evidence that cannabis use is not only correlated with the emergence of psychoses, but actually precedes it (Arsenault et al., 2014; Volkow et al., 2016). Such studies can also account for confounding factors, such as polydrug use, demonstrating that individuals with a cannabis dependence also had higher rates of psychotic symptoms (van Os et al., 2002; Volkow et al., 2016).

In human subjects, there are also many challenges to accurately assessing the neuroanatomical impact of chronic or acute exposure to cannabis with neuroimaging, including small sample sizes, heterogeneity in study design, differences in image acquisition, and inaccurate self-report (Cupo et al., 2021). Nevertheless, early meta-analyses revealed a smaller hippocampus and orbitofrontal cortex in nonpsychotic individuals who chronically use cannabis (Lorenzetti et al., 2019; Rocchetti et al., 2013). Reduced hippocampal (and to a lesser extent orbitofrontal cortex) volume, is an extremely consistent finding in studies of patients with schizophrenia (Roeske et al., 2021), as well as other psychiatric disorders such as bipolar disorder or major depressive disorder (Brosch et al., 2022).

While the exact mechanism by which cannabis increases risk of psychosis is unknown, it has been hypothesized that the dopamine system is involved, with exposure to CB-1 agonists (either endocannabinoids or exogenous cannabinoids) modulating both inhibitory and excitatory inputs to dopaminergic neurons (Luzi et al., 2008). They may increase striatal dopamine levels, mirroring an effect seen in psychosis (Howes et al., 2009; Luzi et al., 2008). It is important to note that cannabis use is neither necessary for the onset of psychosis (people develop psychosis without ever using cannabis) nor sufficient for it (most people who use cannabis do not develop a psychosis) (Volkow et al., 2016). Instead, it seems to interact with previous risk factors to contribute to the development of pathological psychiatric symptoms, explored more fully in the section below.

Animal studies focusing on THC

While animal models do not suffer from psychosis, many are thought to exhibit behaviours relevant to psychiatric disorders in humans. When properly modeled, the causal association of the impact of chronic cannabis exposure can be studied in a manner relevant to but not possible in humans. In humans, symptoms of psychoses and schizophrenia fall into three categories: positive (such as hallucinations and delusions), negative (such as avolition and amotivation), and cognitive symptoms (such as disordered speech). Additionally, psychoses are often accompanied by certain core features that are not typically involved in diagnosis, but can be assessed more readily in animal models, such as memory impairments and altered integration of sensory stimuli. Between the symptoms and features, behavioral paradigms have been established in rodents to assess aspects of all three categories (Malone et al., 2010).

Cognitive behaviours, such as learning and memory, are some of the more widely accessible psychiatric symptoms in animal models. In rodents, adolescent exposure to both phytocannabinoids and synthetic cannabinoids has been associated with disruptions in cognition (Rubino & Parolaro, 2016; Rubino, Realini, Braida, Alberio, et al., 2009; Rubino, Realini, Braida, Guidi, et al., 2009). In addition to memory, in rodent models, THC, but not WIN 55,212-2, a synthetic cannabinoid and CB-1-agonist, has been found to interfere with decision-making behaviors in a delay-based choice task (Silveira et al., 2017). Further research is required to establish whether the difference between phytocannabinoids and synthetic cannabinoids is maintained.

As hallucinations and delusions are uniquely human, two categories of behaviors are used to assess other aspects of positive symptoms: hyperlocomotion and disruptions to sensory motor gating (Rubino & Parolaro, 2014). Prepulse inhibition (PPI) is a task that assesses sensory motor gating and is widely accepted as indicative of a schizophrenia-like phenotype. The test has a high degree of translatability, as it can be investigated in humans (and other species) (Rubino & Parolaro, 2014). Chronic pubertal, but not adult, treatment with WIN 55,212-2 in rats has been associated with impaired PPI in adulthood (Schneider & Koch, 2003). This suggests the peripubertal period may represent a sensitive time for the effects of cannabis.

Finally, negative symptoms can be assessed with certain social deficits that may mirror human social withdrawal as well as decreases in exploratory behaviors or movement (hypolocomotion). While some studies still include only male animals, sex differences have been documented following THC exposure, with low doses of acute THC exposure altering anxiety and exploratory behaviours in female but not male rats, while high doses of THC accompanied alterations in both (Ruiz et al., 2021). While this study investigates the acute effects of THC, females show increased susceptibility to chronic THC treatment is some studies (Zamberletti et al., 2014), while others document no differences (Bruijnzeel et al., 2019; Ginder et al., 2021).

Findings of brain volume following exposure to cannabis in adolescent-aged rodents differ. Some studies report no significant differences in whole-brain or regional volume (Coleman et al., 2022), while others report increased volume in the hippocampus of female rats (Keeley et al., 2015), or transient volume changes in adolescence (following treatment), which resolve in adulthood (Guma, Cupo, et al., 2022). Together the findings fail to replicate the robust effect in human studies, potentially drawing attention to confounding factors in epidemiological data. There are a host of confounding factors that provide potential explanations for the variation in results in the preclinical literature. The species used, age of administration, age at testing, assays, dosage, compound, and route of administration must all be selected, introducing variance to studies. Additionally, the statistical approaches to analysing data may differ, as well as the sample size used, impacting the power of individual studies.

Animal models can allow scientists to assess potential mechanisms underlying the effects of chronic cannabis exposure. Repeat exposure to cannabis can lead to deficits in long-term synaptic plasticity, potentially underlying the cognitive deficits observed (Prini et al., 2020). Decreased release of hippocampal glutamate in females and greater release of GABA in both sexes have been observed in adult rodents exposed to THC in adolescence (Higuera-Matas et al., 2012). In contrast, in the PFC, there are reports of lower basal levels of GABA in adulthood following adolescent exposure (Zamberletti et al., 2014). Chronic exposure to THC was also associated with a downregulation in CB-1 receptors in both sexes, which could contribute to the disinhibition of synapses, potentially increasing neurotransmitter release where the receptors are located (Rubino & Parolaro, 2016). Interestingly, while the human literature focuses on the dopaminergic system as a potential mechanism explaining the alterations to behavior, the preclinical literature focuses more on the glutamatergic/GABAergic excitatory/inhibitory imbalance.

The "two-hit" hypothesis

Definition of the concept

As mentioned previously not all individuals exposed to cannabis in adolescence develop psychosis, suggesting that preexisting vulnerabilities may be present. Some evidence points to genetic susceptibility (Caspi et al., 2005; Estrada et al., 2011), which would classify as a "first-hit," priming individuals to become more responsive to environmental insults experienced later in life ("second-hit"). Genes regulating dopamine synthesis, glutamate receptors, calcium channels, and the immune system have all been associated with schizophrenia via genome-wide association studies (Schizophrenia Working

Group of the Psychiatric Genomics Consortium, 2014). While there is no causal link between these risk genes and schizophrenia, several risk genes together may increase risk for disease onset (Guerrin et al., 2021).

In addition to genetic risk factors, several environmental risk factors have been associated with increased risk for developing a psychosis spectrum disorder. Some of these include having a compromised prenatal environment due to maternal infection, malnutrition, drug abuse, or psychosocial stress (Cattane et al., 2020; Guma et al., 2019.; Selemon & Zecevic, 2015). Furthermore, obstetric complications, such as at delivery (caesarean section), preeclampsia, diabetes, or excessive bleeding (Cannon et al., 2002), have also been associated with increased risk of developing psychosis. Other early life risk factors include winter birth, growing up in an urban environment, as well as childhood adversity, including neglect, abuse, trauma, and bullying (van Os et al., 2010). Later life risk factors include discrimination and bullying, drug abuse and heavy cannabis use, social isolation, immigration, and poverty (van Os et al., 2010). While each of these factors alone may not be sufficient to lead to disease onset, a combination of genetic, early, and later life risk factors may have additive effects, which contribute to disease onset. A subset of these risk factors is highlighted in Fig. 3; however, we have simply illustrated a few examples of genetic and environmental risk factors as many more have been identified.

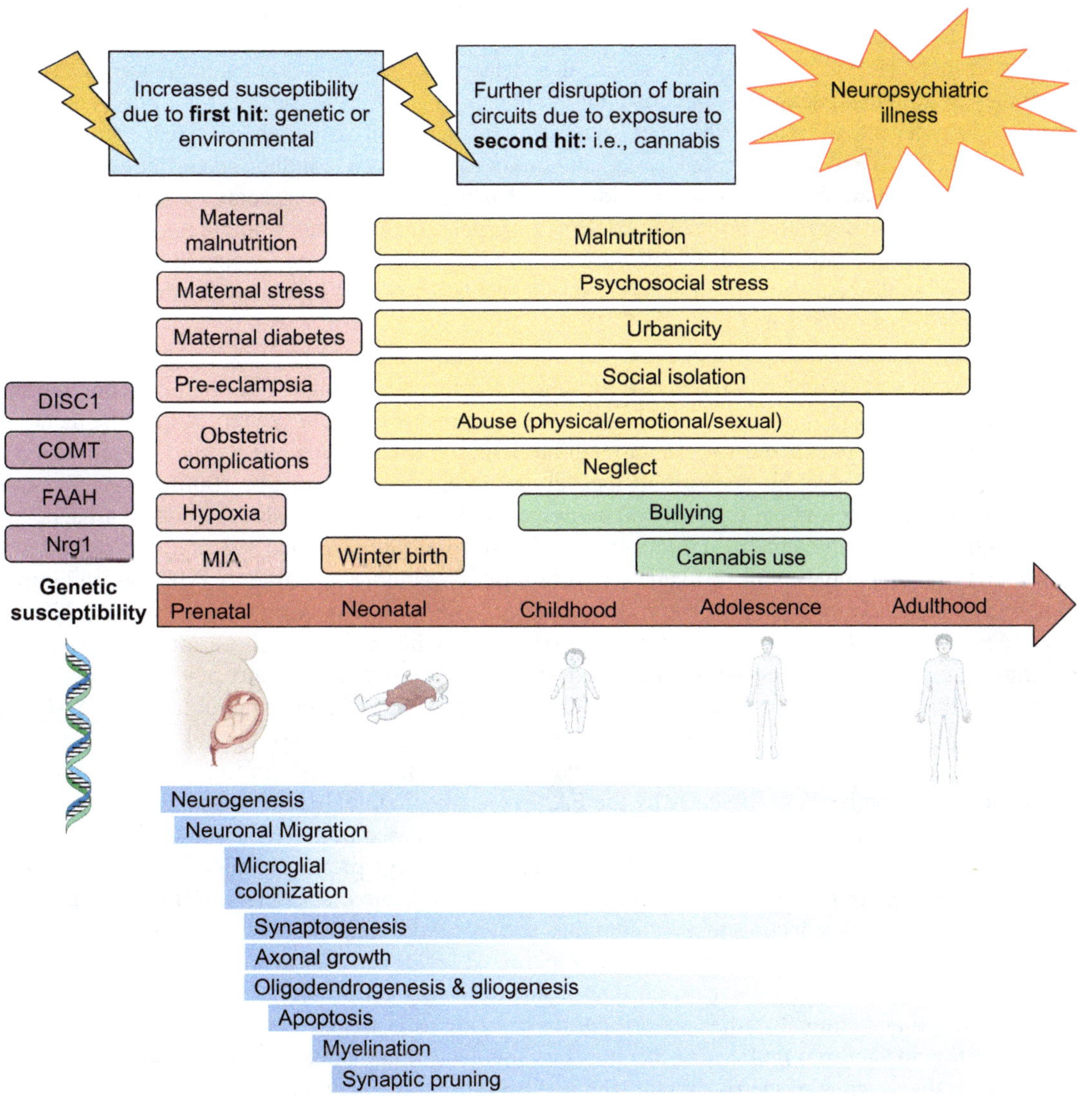

FIG. 3 Risk factors identified in human and animal studies to be associated with the two-hit hypothesis. Risk factors are color coded based on "type," with genetic risk factors in purple, prenatal risk factors in pink, neonatal risk factors in orange, neonatal to early adult risk factors in yellow, and adolescent risk factors in green. The timing of important neurodevelopmental events is highlighted in the blue boxes below the developmental timeline. DISC1 = disrupted-in-schizophrenia, COMT = catechol-o-methyltransferase, FAAH = fatty acid amide hydrolase, Nrg1 = neuregulin 1, MIA = maternal immune activation.

Evidence in humans

Outside of cannabis use, there is significant evidence that the accumulation of exposures across sensitive periods of development may elevate risk for neuropsychiatric disorders. Numerous studies have investigated whether the effect of accumulated environmental risk factor exposure impacts the age of schizophrenia onset. In general, risk factors that directly affect brain maturation, such as neurotrauma or drug use, tend to decrease the age of disease onset (Stepniak et al., 2014), while risk factors that increase psychosocial stress, such as urbanicity, have not been found to directly decrease age of onset, but may contribute to accumulated risk more indirectly (Stepniak et al., 2014). One study conducted only in males found that an accumulation of environmental insults (4 or more) resulted in an earlier age of onset (8 years earlier) compared to individuals exposed to only 1 risk factor (Stepniak et al., 2014). Gestational exposure to infection has been associated with an increased risk for developing schizophrenia in several epidemiological studies, including a very large Danish birth cohort of almost 1,000,000 individuals, of whom close to 10,000 had a schizophrenia diagnosis; in this cohort, the increased risk was observed in females but not males. In that same cohort, both males and females who had been exposed to peripubertal trauma were at increased risk for developing schizophrenia, while exposure to both maternal infection and psychosocial stress increased risk in males more than females. These findings reflect interesting sex-dependent synergistic interactions between risk factors, which require further investigation (Debost et al., 2017).

While these environmental risk factors are of interest, the focus of this chapter is on cannabis use as a risk factor. Not many human studies exist investigating the effect of adolescent cannabis use as a "second-hit"; however, several studies have identified a dose-dependent relationship between cannabis use and schizophrenia risk (as discussed above) (Manrique-Garcia et al., 2012) and have reported heavy adolescent cannabis use to induce an earlier age of onset for developing psychosis (Donoghue et al., 2014). In male individuals with a high polygenic risk score for schizophrenia, adolescent cannabis use has been shown to decrease cortical thickness at a steeper rate than peers who did not smoke cannabis (French et al., 2015). These findings indicate that individuals with a high polygenic risk score for schizophrenia may be more susceptible to the effects of adolescent cannabis use, based on measures of cortical thickness, a morphometric brain feature often associated with schizophrenia (van Erp et al., 2018).

Rather than focusing on more broad polygenic risk scores, others have investigated specific gene polymorphisms identified in large GWAS studies as conferring greater risk for schizophrenia. One example is a functional polymorphism of the Catechol-o-methyltransferase (COMT) gene (COMT valine), an enzyme involved in synaptic dopamine metabolism. Individuals who carry this polymorphism and use cannabis in adolescence seem to have increased risk for developing psychosis, pointing to a potential dual-hit scenario (Caspi et al., 2005; Henquet et al., 2006). This gene-environment interaction was observed in a longitudinal study of 800 participants in New Zealand with the COMT valine polymorphism, wherein those who had adolescent cannabis use had increased risk for adult psychotic symptoms (Caspi et al., 2005). A different study found that individuals with this gene polymorphism were more sensitive to THC-induced impairments in memory and attention indicative of possible gene by environment effects on cognition (Henquet et al., 2006). However, these results were not replicated in a study with 500 UK participants with schizophrenia (Zammit et al., 2007).

Another common single nucleotide polymorphism (SNP) to consider is in the gene fatty acid amide hydrolase (FAAH) (C385A; rs324420), which codes for the principal catabolic enzyme for lipid endocannabinoids. This SNP is carried by 28% of individuals of European descent (1000 Genomes Project Consortium et al., 2012), and renders the FAAH protein more vulnerable to degradation. This results in increased levels of anandamide (AEA), an endogenous endocannabinoid, in the brain as it cannot be catabolized as efficiently by the FAAH enzyme. While this SNP has not been directly associated with increased risk for psychosis, humans carrying this SNP display increased impulsivity and connectivity between their amygdala and PFC (Dincheva et al., 2015). Importantly, these regions undergo protracted adolescent development and play a central role in the pathology of neuropsychiatric illness, which may indicate that carriers of this SNP may be more vulnerable to additional risk factors affecting these vulnerable regions. Additionally, they are more likely to try cannabis (Tyndale et al., 2007). If they do, it is possible that having increased levels of AEA in addition to exogenous THC would interfere with brain development to a greater extent than if this SNP was not present.

While epidemiological and clinical studies are critical for drawing associations between risk factors and clinical outcomes, they provide less insight into underlying mechanisms involved in the development of major mental illness. It is difficult to account for the severity of the risk factor, the precise timing, the actual number of stressors, and their effects on disease outcomes. To probe some of these more detailed, mechanistic questions, researchers can, and do, turn to animal models. These models allow for more controlled experimental manipulation of the type, timing, and severity of the risk factor exposure, as well as a more controlled characterization of the effects throughout rodent development. These studies will be discussed in the next section.

Evidence in model organisms

Several rodent models have been used to test the two-hit hypothesis, using various combinations of environmental risk factors, with different severities, at different ages (prenatal, neonatal, adolescent, or adult) (Stepniak et al., 2014). Some of the genes associated with increased risk for developing schizophrenia have also been investigated as "first-hits'' in conjunction with adolescent cannabis use. For example, genetic knockout of the COMT gene in mice was found to enhance effects of adolescent THC administration; exposure to both risk factors induced greater impairments in spatial working memory, and larger increases in exploratory activity (O'Tuathaigh et al., 2010). COMT knockout mice exposed to THC also had increased dopaminergic cell size in the ventral tegmental area, increased CB-1 receptor expression in the hippocampus, and increased parvalbumin cell size in the PFC relative to "single-hit" controls (Behan et al., 2012). Others have explored the synergistic effects between neuregulin (NRG1) knockout mice and adolescent THC and found mixed results; in heterozygous mice, adolescent THC induced sex-specific deficits in social behavior, wherein females were not impaired (Long et al., 2010).

Several groups have investigated the "two-hit" hypothesis by using exposure to maternal immune activation (MIA) as a "first-hit'' (a well characterized risk factor for neurodevelopmental disorders), and adolescent cannabis exposure as the "second-hit." In rats prenatally exposed to MIA in late gestation (gestational day [GD] 16), and then postnatally exposed to daily injections of a synthetic cannabinoid (HU210) during adolescence (postnatal day [PND] 35–49), only subtle changes in gene expression were observed for a single risk factor exposure (either MIA or cannabinoid exposure) (Hollins et al., 2016). However, dramatic and persistent alterations of small noncoding microRNA in the entorhinal cortex associated with transcriptional networks enriched with genes associated with neurotransmission, cellular signaling, and schizophrenia were observed in rats exposed to both risk factors (Hollins et al., 2014, 2016). Work from the same group also found a synergistic effect between MIA-exposure and adolescent THC on serotonin receptor binding in the hippocampus, such that binding was only increased in the group exposed to both risk factors (Dalton et al., 2012). Conversely, others have found that adolescent THC exposure (PND 45–55) attenuated several MIA-induced (GD15) deficits, such as normalising the MIA-induced hyperactivity of ventral tegmental area dopamine cells (Lecca et al., 2019). This latter study was only conducted in male rodents and requires replication in females. Putative additive effects of exposure to MIA in early gestation (GD 9) and chronic adolescent THC exposure (PND 28–45) have been investigated in both male and female offspring (Guma, Cupo, et al., 2022). In this work, exposure to a single risk factor was not sufficient to induce lasting changes in adult brain anatomy or behavior, although transient brain anatomy changes were observed in the adolescent and early adult period in response to adolescent THC exposure. Mice exposed to both risk factors did have lasting neuroanatomical changes in adulthood in some regions but not others, as well as some subthreshold exploratory and sensorimotor gating alterations. This suggests that some brain regions and some behaviors may be susceptible to the dual-hit, but not all. Interestingly, in brain regions showing neuroanatomical remodeling due to MIA and THC exposure, exposure to prenatal MIA, adolescent THC, or both caused a significant decrease in the expression of CB-1 and CB-2 receptors (Guma, Cupo, et al., 2022). This indicates that the endocannabinoid system may play a role in modulating effects due to both prenatal and postnatal risk factors for neurodevelopment.

Adolescent cannabinoid exposure (synthetic cannabinoid WIN 55,212-2) has also been investigated in conjunction with a different prenatal risk factor, maternal exposure to methylazoxymethanol (MAM; GD 17); this compound is a DNA methylating agent shown to disrupt neurodevelopment in offspring consistent with those observed in schizophrenia (Gomes, Guimarães, & Grace, 2014). Pubertal WIN 55,212-2 treatment in normal animals and prenatal MAM exposure led to schizophrenia-like deficits in adulthood, including deficits in behavioural flexibility, augmented locomotor response to amphetamine, and increased spontaneous activity in ventral tegmental area dopamine neurons. Contrary to expectations, however, pubertal WIN 55,212-2 in conjunction with prenatal MAM did not exacerbate deficits, but rather normalized amphetamine-induced hyperlocomotion (Gomes, Guimarães, & Grace, 2014). There is evidence for both attenuating and potentiating effects due to multirisk factor exposure; there may be interacting effects that remain to be elucidated (Gomes et al., 2016).

CBD as a second-hit

While accumulating evidence points to THC exposure as a risk factor for psychosis, there is also some promising evidence pointing to the potential antipsychotic effects of CBD. In humans, much of the neuroimaging literature cited for the opposite effects of THC and CBD involves small, homogenous populations, and should be viewed as preliminary (Cupo et al., 2021). Recently, the potential use of CBD as a novel treatment for psychiatric disorders has emerged as a burgeoning field of research. There is limited work investigating its role as a "second-hit" or protective factor. In humans, the putative antipsychotic properties of CBD have been explored in individuals at clinical high risk for psychosis wherein CBD administration attenuated abnormal brain activity in the insula (amongst other regions) during a reward processing task (Wilson, Bossong, et al., 2019). This may be interpreted as a "protective hit" for individuals who are at risk but have not yet developed full blown symptoms. The data are mixed in individuals with schizophrenia, for whom CBD treatment has improved cognitive deficits (Leweke et al., 2021) or not (Hallak et al., 2010).

Investigations of CBD in animal models can provide causal evidence for the long-term impact that is confounded in the human literature. Most animal model studies first induce a phenotype mimicking clinical presentations of psychotic-like symptoms, which can be interpreted as a "first-hit." Some common rodent models include prenatal MIA exposure, repeated injections of MK-801 to disrupt the glutamate system via N-methyl-d-aspartate (NMDA) receptors, exposure to ketamine or amphetamine to mimic symptoms on an acute scale, or acute and repeat exposure to THC. In rodents prenatally exposed to MIA, early adult CBD exposure (3 weeks starting from PND 56) was associated with a reduction in MIA-induced cognitive deficits, both in male and female offspring. CBD treatment also increased parvalbumin levels and glutamate decarboxylase 67 levels, the enzyme that converts glutamate to GABA, in the hippocampus. While CBD seemed to counteract negative effects in the MIA-exposed offspring, it did lead to a decrease in social interaction and reduced CB-1 and NMDA receptor binding in the PFC in otherwise healthy rats (Osborne et al., 2017, 2019). This suggests that CBD may be useful in buffering the effects of a previous risk factor exposure, particularly if that risk factor alters immune function, but that by itself, it may also cause both behavioural and neuroanatomical changes which may not be ideal. This challenges the translation into clinical practice, as it may be difficult to know for whom CBD is successful or not. In mice repeatedly injected with MK-801 only high doses of CBD were sufficient to attenuate memory deficits included by the MK-801 treatment, but not low doses (Gomes, Issy, et al., 2014). The studies provide mixed evidence for the restorative impact of CBD behavioral abnormalities induced by exposure to a "first-hit" (Almeida et al., 2013; Peres et al., 2018). While the results do suggest CBD may have anxiolytic properties and could help moderate some of the symptoms associated with psychosis, they do not provide convincing evidence of CBD as an antipsychotic. Overall, the research on synergistic effects between either genetic or environmental risk factors and CBD treatment is in its infancy. There is some promising evidence for its potential therapeutic use, but also numerous null reports. This necessitates more rigorous research to better understand its mechanisms in the brain and interaction with preexisting vulnerabilities.

Conclusion

Adolescence represents a period of dynamic refinement in the brain. Therefore, altering signaling in the endocannabinoid system, such as with the introduction of exogenous cannabinoids like THC or CBD, can cause downstream effects on these other systems, impacting brain development. Importantly, disruptions in these same systems are often implicated in psychiatric disorders, such as psychosis spectrum disorder and schizophrenia. Furthermore, the acute effects of heavy THC exposure mimic the symptoms observed in these disorders. Substantial evidence from nonhuman animals and epidemiological studies indicate that heavy adolescent cannabis exposure is a risk factor for developing such disorders; however, many individuals expose themselves to cannabinoids without serious mental health consequences. Rather, cannabis exposure may interact with a milieu of other risk factors or "hits" that serve as disease primers, ultimately culminating in psychosis spectrum disorders. Both human and nonhuman animal studies provide evidence for this "two-hit" hypothesis, linking "first-hits" such as MIA or early-life stress with the "second-hit" of cannabis exposure. The "two-hit" hypothesis has emerged as a predominant theory explaining the role of the environment influencing mental health outcomes today. Further investigation of the timing and dosage of various hits may ultimately help prevent the emergence of these disorders.

Applications to other areas

In this chapter, we have reviewed the role of cannabis as a risk factor for psychosis and other neuropsychiatric disorders, specifically when its exposure occurs in combination with previous exposure to genetic or early life environmental risk factors that make the developing central nervous system more vulnerable. The concepts explored here could be applied to other adolescent risk factors, such as bullying, psychosocial stress, abuse, neglect, and exposure to different drugs of abuse. In general, our framework of the two-hit hypothesis should be extended to a multihit hypothesis to include exposure to more than just two risk factors. Additionally, many of the ways in which exogenous cannabinoids such as THC interfere with adolescent brain development may also be applicable to other neurodevelopmental stages where the endocannabinoid system plays a crucial role. This includes fetal brain development; investigating the effects of prenatal THC exposure in the context of the two-hit hypothesis, or as a general risk factor for mental illness should be of great interest to the field. While it can be challenging to separate risk factors in humans, the hypothesis increasingly extends into nonhuman animal models, where risk factors can be experimentally manipulated as variables to understand how they compound and interact.

Mini-dictionary of terms

- **Cannabidiol:** An active, but not psychoactive, ingredient of the cannabis plant, investigated for its possible antipsychotic effects.
- **Delta-9-tetrahydrocannabinol:** The main psychoactive compound of the cannabis (marijuana) plant that selectively binds to CB-1 and CB-2 receptors in the brain.
- **Environmental risk factor:** Environmental exposures associated with increased risk for adverse health outcomes.
- **Genetic risk factor:** Genetic mutations or polymorphisms that may predispose individuals to a certain disease (ex: DISC1 is associated with schizophrenia, FMR1 is associated with intellectual disability).
- **Phytocannabinoid:** Naturally occurring chemical constituents of the cannabis plant
- **Schizophrenia:** A serious neuropsychiatric condition characterized by an abnormal perception of reality, characterized by a combination of hallucinations, delusions, disordered thinking, and behavior that may interfere with daily function.
- **Two-hit hypothesis:** A hypothesis explaining the convergence between genetic and environmental risk factors in the emergence of psychotic disorders. Focuses on the role of very early life (prenatal to childhood) risk factors ("first-hits") and adolescent risk factors ("second-hits").

Key facts of cannabis and the "two-hit" hypothesis

- The "two-hit hypothesis" (or dual-hit hypothesis) suggests that prenatal genetic or environmental "first-hits" (i.e., exposure to maternal infection, early life neglect, etc) disrupt certain aspects of neurodevelopment, increasing vulnerability to the effect of "second-hits" experienced later in the lifespan, such as adolescent cannabis use.
- THC is the main active compound of the cannabis plant that activates the endocannabinoid system by binding to CB-1 (and CB-2) receptors in the central nervous system.
- CB-1 receptors are expressed on glutamate, GABA, dopamine, serotonin, and acetylcholine-releasing neurons.
- During adolescence brain circuits undergo significant refinement involving the maturation of many neurotransmitter systems with which the endocannabinoid system interacts, such as the dopaminergic, GABAergic, and glutamatergic systems. This makes adolescence an extremely sensitive period to external insults.
- Altering signaling in the endocannabinoid system, such as with the introduction of exogenous cannabinoids like THC or CBD, can cause downstream effects on these other systems, impacting brain development.
- Evidence from animal models substantiates the epidemiological evidence examined in humans with causal relationships between adolescent exposure to cannabinoids and altered brain and behavioral effects.
- There is evidence for a dose-response relationship between the amount of high-potency cannabis used and psychosis-related outcomes.
- Very limited evidence suggests CBD may serve as a protective second-hit; however, more data are required to support this finding.

Summary points

- Heavy adolescent cannabis use has been identified as a risk factor for neuropsychiatric illness, including schizophrenia.
- Not all individuals who consume cannabis will go on to develop a major psychiatric illness, which suggests that cannabis may act as a "second-hit," incurring additional damage to a vulnerable system
- Activation of CB-1 receptors via exogenous THC can suppress neurotransmitter release, such as GABA, to the synapse, which may interfere with the development of both glutamatergic pyramidal cells, as well as GABA-ergic parvalbumin-containing basket cells, which may disrupt excitatory-inhibitory signaling imbalance, thought to underlie some of the disruptions observed in psychotic symptoms.
- In humans, evidence suggests that cannabis consumption may decrease the age of onset for emergence of psychosis, and that this may occur in a dose-dependent manner. Chronic cannabis smoking has also been associated with further reductions in certain brain regions, and to increase the severity of certain symptoms
- In humans, individuals who carry risk genes for schizophrenia, either multiple (based on a polygenic risk score) or individual genetic mutations (ex: COMT gene polymorphism) tend to be more susceptible to the effects of cannabis use on neuroanatomy and psychosis spectrum symptoms.
- There are significant challenges in assessing a potential causal relationship between multiple hits and disease progression in humans, including inaccurate self-report, confounding variables, and small sample sizes.

- In animals, adolescent cannabis exposure has been studied in combination with several prenatal risk factors such as maternal immune activation, genetic risk factors, or increased DNA methylation. Findings are mixed, although there is evidence for altered neuroanatomy, behavior, and gene expression in mice exposed to two risk factors relative to those exposed to a single risk factor.
- Of the cannabinoids, THC has been highlighted as the most likely risk-factor for psychosis, with CBD affording potential protective effects.

References

Albaugh, M. D., Ottino-Gonzalez, J., Sidwell, A., Lepage, C., Juliano, A., Owens, M. M., Chaarani, B., Spechler, P., Fontaine, N., Rioux, P., Lewis, L., Jeon, S., Evans, A., D'Souza, D., Radhakrishnan, R., Banaschewski, T., Bokde, A. L. W., Quinlan, E. B., Conrod, P., … IMAGEN Consortium. (2021). Association of cannabis use during adolescence with neurodevelopment. *JAMA Psychiatry*. https://doi.org/10.1001/jamapsychiatry.2021.1258.

Almeida, V., Levin, R., Peres, F. F., Niigaki, S. T., Calzavara, M. B., Zuardi, A. W., Hallak, J. E., Crippa, J. A., & Abílio, V. C. (2013). Cannabidiol exhibits anxiolytic but not antipsychotic property evaluated in the social interaction test. *Progress in Neuro-Psychopharmacology & Biological Psychiatry*, *41*, 30–35.

Arsenault, D., St-Amour, I., Cisbani, G., Rousseau, L.-S., & Cicchetti, F. (2014). The different effects of LPS and poly I:C prenatal immune challenges on the behavior, development and inflammatory responses in pregnant mice and their offspring. *Brain, Behavior, and Immunity*, *38*, 77–90.

Arseneault, L., Cannon, M., Poulton, R., Murray, R., Caspi, A., & Moffitt, T. E. (2002). Cannabis use in adolescence and risk for adult psychosis: Longitudinal prospective study. *BMJ*, *325*(7374), 1212–1213.

Bagot, K. S., Milin, R., & Kaminer, Y. (2015). Adolescent initiation of cannabis use and early-onset psychosis. *Substance Abuse: Official Publication of the Association for Medical Education and Research in Substance Abuse*, *36*(4), 524–533.

Behan, A. T., Hryniewiecka, M., O'Tuathaigh, C. M. P., Kinsella, A., Cannon, M., Karayiorgou, M., Gogos, J. A., Waddington, J. L., & Cotter, D. R. (2012). Chronic adolescent exposure to delta-9-tetrahydrocannabinol in COMT mutant mice: Impact on indices of dopaminergic, endocannabinoid and GABAergic pathways. *Neuropsychopharmacology: Official Publication of the American College of Neuropsychopharmacology*, *37*(7), 1773–1783.

Bethlehem, R. A. I., Seidlitz, J., White, S. R., Vogel, J. W., Anderson, K. M., Adamson, C., Adler, S., Alexopoulos, G. S., Anagnostou, E., Areces-Gonzalez, A., Astle, D. E., Auyeung, B., Ayub, M., Bae, J., Ball, G., Baron-Cohen, S., Beare, R., Bedford, S. A., Benegal, V., … Alexander-Bloch, A. F. (2022). Brain charts for the human lifespan. *Nature*, *604*(7906), 525–533.

Brosch, K., Stein, F., Schmitt, S., Pfarr, J.-K., Ringwald, K. G., Thomas-Odenthal, F., Meller, T., Steinsträter, O., Waltemate, L., Lemke, H., Meinert, S., Winter, A., Breuer, F., Thiel, K., Grotegerd, D., Hahn, T., Jansen, A., Dannlowski, U., Krug, A., … Kircher, T. (2022). Reduced hippocampal gray matter volume is a common feature of patients with major depression, bipolar disorder, and schizophrenia spectrum disorders. *Molecular Psychiatry*. https://doi.org/10.1038/s41380-022-01687-4.

Bruijnzeel, A. W., Knight, P., Panunzio, S., Xue, S., Bruner, M. M., Wall, S. C., Pompilus, M., Febo, M., & Setlow, B. (2019). Effects in rats of adolescent exposure to cannabis smoke or THC on emotional behavior and cognitive function in adulthood. *Psychopharmacology*, *236*(9), 2773–2784.

Cabral, G. A., Raborn, E. S., Griffin, L., Dennis, J., & Marciano-Cabral, F. (2008). CB2 receptors in the brain: Role in central immune function. *British Journal of Pharmacology*, *153*(2), 240–251.

Cannon, M., Jones, P. B., & Murray, R. M. (2002). Obstetric complications and schizophrenia: Historical and meta-analytic review. *The American Journal of Psychiatry*, *159*(7), 1080–1092.

Caspi, A., Moffitt, T. E., Cannon, M., McClay, J., Murray, R., Harrington, H., Taylor, A., Arseneault, L., Williams, B., Braithwaite, A., Poulton, R., & Craig, I. W. (2005). Moderation of the effect of adolescent-onset cannabis use on adult psychosis by a functional polymorphism in the catechol-o-methyltransferase gene: Longitudinal evidence of a Gene X environment interaction. *Biological Psychiatry*, *57*(10), 1117–1127. https://doi.org/10.1016/j.biopsych.2005.01.026.

Cattane, N., Richetto, J., & Cattaneo, A. (2020). Prenatal exposure to environmental insults and enhanced risk of developing Schizophrenia and Autism Spectrum Disorder: Focus on biological pathways and epigenetic mechanisms. *Neuroscience & Biobehavioral Reviews*, *117*, 253–278. https://doi.org/10.1016/j.neubiorev.2018.07.001.

Coleman, J. R., Madularu, D., Ortiz, R. J., Athanassiou, M., Knudsen, A., Alkislar, I., Cai, X., Kulkarni, P. P., Cushing, B. S., & Ferris, C. F. (2022). Changes in brain structure and function following chronic exposure to inhaled vaporised cannabis during periadolescence in female and male mice: A multimodal MRI study. *Addiction Biology*, *27*(3), e13169.

Colizzi, M., Weltens, N., McGuire, P., Van Oudenhove, L., & Bhattacharyya, S. (2019). Descriptive psychopathology of the acute effects of intravenous delta-9-tetrahydrocannabinol administration in humans. *Brain Sciences*, *9*(4). https://doi.org/10.3390/brainsci9040093.

Cupo, L., Plitman, E., Guma, E., & Chakravarty, M. M. (2021). A systematic review of neuroimaging and acute cannabis exposure in age-of-risk for psychosis. *Translational Psychiatry*, *11*(1), 217.

Dalton, V. S., Verdurand, M., Walker, A., Hodgson, D. M., & Zavitsanou, K. (2012). Synergistic effect between maternal infection and adolescent cannabinoid exposure on serotonin 5HT1A receptor binding in the hippocampus: Testing the "two hit" hypothesis for the development of schizophrenia. *ISRN Psychiatry*, *2012*, 451865.

Debost, J.-C. P. G., Larsen, J. T., Munk-Olsen, T., Mortensen, P. B., Meyer, U., & Petersen, L. (2017). Joint effects of exposure to prenatal infection and peripubertal psychological trauma in schizophrenia. *Schizophrenia Bulletin*, *43*(1), 171–179.

Di Forti, M., Sallis, H., Allegri, F., Trotta, A., Ferraro, L., Stilo, S. A., Marconi, A., La Cascia, C., Reis Marques, T., Pariante, C., Dazzan, P., Mondelli, V., Paparelli, A., Kolliakou, A., Prata, D., Gaughran, F., David, A. S., Morgan, C., Stahl, D., … Murray, R. M. (2014). Daily use, especially of high-potency cannabis, drives the earlier onset of psychosis in cannabis users. *Schizophrenia Bulletin*, *40*(6), 1509–1517.

Dincheva, I., Drysdale, A. T., Hartley, C. A., Johnson, D. C., Jing, D., King, E. C., Ra, S., Megan Gray, J., Yang, R., DeGruccio, A. M., Huang, C., Cravatt, B. F., Glatt, C. E., Hill, M. N., Casey, B. J., & Lee, F. S. (2015). FAAH genetic variation enhances fronto-amygdala function in mouse and human. *Nature Communications*, *6*(1). https://doi.org/10.1038/ncomms7395.

Donoghue, K., Doody, G. A., Murray, R. M., Jones, P. B., Morgan, C., Dazzan, P., Hart, J., Mazzoncini, R., & Maccabe, J. H. (2014). Cannabis use, gender and age of onset of schizophrenia: Data from the ÆSOP study. *Psychiatry Research*, *215*(3), 528–532.

Eggan, S. M., Mizoguchi, Y., Stoyak, S. R., & Lewis, D. A. (2010). Development of cannabinoid 1 receptor protein and messenger RNA in monkey dorsolateral prefrontal cortex. *Cerebral Cortex*, *20*(5), 1164–1174. https://doi.org/10.1093/cercor/bhp179.

Ellgren, M., Artmann, A., Tkalych, O., Gupta, A., Hansen, H. S., Hansen, S. H., Devi, L. A., & Hurd, Y. L. (2008). Dynamic changes of the endogenous cannabinoid and opioid mesocorticolimbic systems during adolescence: THC effects. *European Neuropsychopharmacology: The Journal of the European College of Neuropsychopharmacology*, *18*(11), 826–834.

ElSohly, M. A., Mehmedic, Z., Foster, S., Gon, C., Chandra, S., & Church, J. C. (2016). Changes in cannabis potency over the last 2 decades (1995–2014): Analysis of current data in the United States. *Biological Psychiatry*, *79*(7), 613–619.

Estrada, G., Fatjó-Vilas, M., Muñoz, M. J., Pulido, G., Miñano, M. J., Toledo, E., Illa, J. M., Martín, M., Miralles, M. L., Miret, S., Campanera, S., Bernabeu, C., Navarro, M. E., & Fañanás, L. (2011). Cannabis use and age at onset of psychosis: Further evidence of interaction with COMT Val158Met polymorphism. *Acta Psychiatrica Scandinavica*, *123*(6), 485–492. https://doi.org/10.1111/j.1600-0447.2010.01665.x.

French, L., Gray, C., Leonard, G., Perron, M., Pike, G. B., Richer, L., Séguin, J. R., Veillette, S., Evans, C. J., Artiges, E., Banaschewski, T., Bokde, A. W. L., Bromberg, U., Bruehl, R., Buchel, C., Cattrell, A., Conrod, P. J., Flor, H., Frouin, V., … Paus, T. (2015). Early cannabis use, polygenic risk score for schizophrenia and brain maturation in adolescence. *JAMA Psychiatry*, *72*(10), 1002–1011.

Freund, T. F., Katona, I., & Piomelli, D. (2003). Role of endogenous cannabinoids in synaptic signaling. *Physiological Reviews*, *83*(3), 1017–1066.

Genomes Project Consortium, Abecasis, G. R., Auton, A., Brooks, L. D., DePristo, M. A., Durbin, R. M., Handsaker, R. E., Kang, H. M., Marth, G. T., & McVean, G. A. (2012). An integrated map of genetic variation from 1,092 human genomes. *Nature*, *491*(7422), 56–65.

Ginder, D. E., Wright, H. R., & McLaughlin, R. J. (2021). Chapter Four—The stoned age: Sex differences in the effects of adolescent cannabinoid exposure on prefrontal cortex structure and function in animal models. In R. L. Bell, & S. Rahman (Eds.), *161. International review of neurobiology* (pp. 121–145). Academic Press.

Gomes, F. V., Guimarães, F. S., & Grace, A. A. (2014). Effects of pubertal cannabinoid administration on attentional set-shifting and dopaminergic hyper-responsivity in a developmental disruption model of schizophrenia. *The International Journal of Neuropsychopharmacology/Official Scientific Journal of the Collegium Internationale Neuropsychopharmacologicum*, *18*(2). https://doi.org/10.1093/ijnp/pyu018.

Gomes, F. V., Issy, A. C., Ferreira, F. R., Viveros, M.-P., Del Bel, E. A., & Guimarães, F. S. (2014). Cannabidiol attenuates sensorimotor gating disruption and molecular changes induced by chronic antagonism of NMDA receptors in mice. *The International Journal of Neuropsychopharmacology/Official Scientific Journal of the Collegium Internationale Neuropsychopharmacologicum*, *18*(5). https://doi.org/10.1093/ijnp/pyu041.

Gomes, F. V., Rincón-Cortés, M., & Grace, A. A. (2016). Adolescence as a period of vulnerability and intervention in schizophrenia: Insights from the MAM model. *Neuroscience and Biobehavioral Reviews*, *70*, 260–270.

Guerrin, C. G. J., Doorduin, J., Sommer, I. E., & de Vries, E. F. J. (2021). The dual hit hypothesis of schizophrenia: Evidence from animal models. *Neuroscience and Biobehavioral Reviews*, *131*, 1150–1168.

Guma, E., Bordeleau, M., González Ibáñez, F., Picard, K., Snook, E., Desrosiers-Grégoire, G., Spring, S., Lerch, J. P., Nieman, B. J., Devenyi, G. A., Tremblay, M.-E., & Chakravarty, M. M. (2022). Differential effects of early or late exposure to prenatal maternal immune activation on mouse embryonic neurodevelopment. *Proceedings of the National Academy of Sciences of the United States of America*, *119*(12), e2114545119.

Guma, E., Cupo, L., Ma, W., Gallino, D., Moquin, L., Gratton, A., Devenyi, G. A., & Chakravarty, M. M. (2022). Investigating the "two-hit hypothesis": Effects of prenatal maternal immune activation and adolescent cannabis use on neurodevelopment in mice. *bioRxiv*. https://doi.org/10.1101/2022.03.21.485178.

Guma, E., do Bordignon, P. C., Devenyi, G. A., Gallino, D., Anastassiadis, C., Cvetkovska, V., Barry, A. D., Snook, E., Germann, J., Greenwood, C. M. T., Misic, B., Bagot, R. C., & Chakravarty, M. M. (2021). Early or late gestational exposure to maternal immune activation alters neurodevelopmental trajectories in mice: An integrated neuroimaging, behavioral, and transcriptional study. *Biological Psychiatry*, *90*(5), 328–341.

Guma, E., Plitman, E., & Chakravarty, M. M. (2019). The role of maternal immune activation in altering the neurodevelopmental trajectories of offspring: A translational review of neuroimaging studies with implications for autism spectrum disorder and schizophrenia. *Neuroscience and Biobehavioral Reviews*, *104*, 141–157.

Hallak, J. E. C., Machado-de-Sousa, J. P., Crippa, J. A. S., Sanches, R. F., Trzesniak, C., Chaves, C., Bernardo, S. A., Regalo, S. C., & Zuardi, A. W. (2010). Performance of schizophrenic patients in the Stroop Color Word Test and electrodermal responsiveness after acute administration of cannabidiol (CBD). *Revista Brasileira de Psiquiatria (Sao Paulo, Brazil: 1999)*, *32*(1), 56–61.

Harkany, T., Keimpema, E., Barabás, K., & Mulder, J. (2008). Endocannabinoid functions controlling neuronal specification during brain development. *Molecular and Cellular Endocrinology*, *286*(1–2 Suppl 1), S84–S90.

Heng, L., Beverley, J. A., Steiner, H., & Tseng, K. Y. (2011). Differential developmental trajectories for CB1 cannabinoid receptor expression in limbic/associative and sensorimotor cortical areas. *Synapse*, *65*(4), 278–286.

Henquet, C., Rosa, A., Krabbendam, L., Papiol, S., Fananás, L., Drukker, M., Ramaekers, J. G., & van Os, J. (2006). An experimental study of catechol-o-methyltransferase Val158Met moderation of delta-9-tetrahydrocannabinol-induced effects on psychosis and cognition. *Neuropsychopharmacology: Official Publication of the American College of Neuropsychopharmacology*, *31*(12), 2748–2757.

Herkenham, M., Lynn, A. B., Johnson, M. R., Melvin, L. S., de Costa, B. R., & Rice, K. C. (1991). Characterization and localization of cannabinoid receptors in rat brain: A quantitative in vitro autoradiographic study. *The Journal of Neuroscience: The Official Journal of the Society for Neuroscience*, *11*(2), 563–583.

Higuera-Matas, A., Miguéns, M., Coria, S. M., Assis, M. A., Borcel, E., del Olmo, N., & Ambrosio, E. (2012). Sex-specific disturbances of the glutamate/GABA balance in the hippocampus of adult rats subjected to adolescent cannabinoid exposure. *Neuropharmacology*, *62*(5-6), 1975–1984.

Hollins, S. L., Zavitsanou, K., Walker, F. R., & Cairns, M. J. (2014). Alteration of imprinted Dlk1-Dio3 miRNA cluster expression in the entorhinal cortex induced by maternal immune activation and adolescent cannabinoid exposure. *Translational Psychiatry*, *4*, e452.

Hollins, S. L., Zavitsanou, K., Walker, F. R., & Cairns, M. J. (2016). Alteration of transcriptional networks in the entorhinal cortex after maternal immune activation and adolescent cannabinoid exposure. *Brain, Behavior, and Immunity*, *56*, 187–196.

Howes, O. D., Montgomery, A. J., Asselin, M.-C., Murray, R. M., Valli, I., Tabraham, P., Bramon-Bosch, E., Valmaggia, L., Johns, L., Broome, M., McGuire, P. K., & Grasby, P. M. (2009). Elevated striatal dopamine function linked to prodromal signs of schizophrenia. *Archives of General Psychiatry*, *66*(1), 13–20.

Keeley, R. J., Trow, J., & McDonald, R. J. (2015). Strain and sex differences in puberty onset and the effects of THC administration on weight gain and brain volumes. *Neuroscience*, *305*, 328–342.

Keimpema, E., Mackie, K., & Harkany, T. (2011). Molecular model of cannabis sensitivity in developing neuronal circuits. *Trends in Pharmacological Sciences*, *32*(9), 551–561.

Keshavan, M. S., Giedd, J., Lau, J. Y. F., Lewis, D. A., & Paus, T. (2014). Changes in the adolescent brain and the pathophysiology of psychotic disorders. *The Lancet. Psychiatry*, *1*(7), 549–558.

Large, M., Sharma, S., Compton, M. T., Slade, T., & Nielssen, O. (2011). Cannabis use and earlier onset of psychosis: A systematic meta-analysis. *Archives of General Psychiatry*, *68*(6), 555–561.

Lecca, S., Luchicchi, A., Scherma, M., Fadda, P., Muntoni, A. L., & Pistis, M. (2019). Δ9-Tetrahydrocannabinol during adolescence attenuates disruption of dopamine function induced in rats by maternal immune activation. *Frontiers in Behavioral Neuroscience*, *13*, 202.

Leweke, F. M., Rohleder, C., Gerth, C. W., Hellmich, M., Pukrop, R., & Koethe, D. (2021). Cannabidiol and amisulpride improve cognition in acute schizophrenia in an explorative, double-blind, active-controlled, randomized clinical trial. *Frontiers in Pharmacology*, *12*, 614811.

Li, R., Ma, X., Wang, G., Yang, J., & Wang, C. (2016). Why sex differences in schizophrenia? *Journal of Translational Neuroscience*, *1*(1), 37–42.

Long, L. E., Chesworth, R., Arnold, J. C., & Karl, T. (2010). A follow-up study: Acute behavioural effects of Delta(9)-THC in female heterozygous neuregulin 1 transmembrane domain mutant mice. *Psychopharmacology*, *211*(3), 277–289.

Lorenzetti, V., Chye, Y., Silva, P., Solowij, N., & Roberts, C. A. (2019). Does regular cannabis use affect neuroanatomy? An updated systematic review and meta-analysis of structural neuroimaging studies. *European Archives of Psychiatry and Clinical Neuroscience*, *269*(1), 59–71.

Lu, H.-C., & Mackie, K. (2021). Review of the Endocannabinoid System. *Biological Psychiatry. Cognitive Neuroscience and Neuroimaging*, *6*(6), 607–615.

Luzi, S., Morrison, P. D., Powell, J., di Forti, M., & Murray, R. M. (2008). What is the mechanism whereby cannabis use increases risk of psychosis? *Neurotoxicity Research*, *14*(2–3), 105–112.

Mackie, K. (2006). Mechanisms of CB1 receptor signaling: Endocannabinoid modulation of synaptic strength. *International Journal of Obesity*, *30*(Suppl. 1), S19–S23.

Malone, D. T., Hill, M. N., & Rubino, T. (2010). Adolescent cannabis use and psychosis: Epidemiology and neurodevelopmental models. *British Journal of Pharmacology*, *160*(3), 511–522.

Manrique-Garcia, E., Zammit, S., Dalman, C., Hemmingsson, T., Andreasson, S., & Allebeck, P. (2012). Cannabis, schizophrenia and other non-affective psychoses: 35 years of follow-up of a population-based cohort. *Psychological Medicine*, *42*(6), 1321–1328.

Marconi, A., Di Forti, M., Lewis, C. M., Murray, R. M., & Vassos, E. (2016). Meta-analysis of the association between the level of cannabis use and risk of psychosis. *Schizophrenia Bulletin*, *42*(5), 1262–1269.

Mato, S., Del Olmo, E., & Pazos, A. (2003). Ontogenetic development of cannabinoid receptor expression and signal transduction functionality in the human brain. *European Journal of Neuroscience*, *17*(9), 1747–1754. https://doi.org/10.1046/j.1460-9568.2003.02599.x.

Maynard, T. M., Sikich, L., Lieberman, J. A., & LaMantia, A.-S. (2001). Neural development, cell-cell signaling, and the "two-hit" hypothesis of schizophrenia. *Schizophrenia Bulletin*, *27*(3), 457–476. https://doi.org/10.1093/oxfordjournals.schbul.a006887.

Meyer, U. (2014). Prenatal poly (i:C) exposure and other developmental immune activation models in rodent systems. *Biological Psychiatry*, *75*(4), 307–315.

Meyer, H. C., Lee, F. S., & Gee, D. G. (2018). The role of the endocannabinoid system and genetic variation in adolescent brain development. *Neuropsychopharmacology: Official Publication of the American College of Neuropsychopharmacology*, *43*(1), 21–33.

Morin, J.-F. G., Afzali, M. H., Bourque, J., Stewart, S. H., Séguin, J. R., O'Leary-Barrett, M., & Conrod, P. J. (2019). A population-based analysis of the relationship between substance use and adolescent cognitive development. *The American Journal of Psychiatry*, *176*(2), 98–106.

Mulder, J., Aguado, T., Keimpema, E., Barabás, K., Ballester Rosado, C. J., Nguyen, L., Monory, K., Marsicano, G., Di Marzo, V., Hurd, Y. L., Guillemot, F., Mackie, K., Lutz, B., Guzmán, M., Lu, H.-C., Galve-Roperh, I., & Harkany, T. (2008). Endocannabinoid signaling controls pyramidal cell specification and long-range axon patterning. *Proceedings of the National Academy of Sciences of the United States of America*, *105*(25), 8760–8765.

O'Tuathaigh, C. M. P., Hryniewiecka, M., Behan, A., Tighe, O., Coughlan, C., Desbonnet, L., Cannon, M., Karayiorgou, M., Gogos, J. A., Cotter, D. R., & Waddington, J. L. (2010). Chronic adolescent exposure to Δ-9-tetrahydrocannabinol in COMT mutant mice: Impact on psychosis-related and other phenotypes. *Neuropsychopharmacology: Official Publication of the American College of Neuropsychopharmacology*, *35*(11), 2262–2273.

Osborne, A. L., Solowij, N., Babic, I., Huang, X.-F., & Weston-Green, K. (2017). Improved social interaction, recognition and working memory with cannabidiol treatment in a prenatal infection (poly I:C) rat model. *Neuropsychopharmacology: Official Publication of the American College of Neuropsychopharmacology*, *42*(7), 1447–1457.

Osborne, A. L., Solowij, N., Babic, I., Lum, J. S., Huang, X.-F., Newell, K. A., & Weston-Green, K. (2019). Cannabidiol improves behavioural and neurochemical deficits in adult female offspring of the maternal immune activation (poly I:C) model of neurodevelopmental disorders. *Brain, Behavior, and Immunity*, *81*, 574–587.

Peng, J., Fan, M., An, C., Ni, F., Huang, W., & Luo, J. (2022). A narrative review of molecular mechanism and therapeutic effect of cannabidiol (CBD). *Basic & Clinical Pharmacology & Toxicology*, *130*(4), 439–456.

Peres, F. F., Diana, M. C., Levin, R., Suiama, M. A., Almeida, V., Vendramini, A. M., Santos, C. M., Zuardi, A. W., Hallak, J. E. C., Crippa, J. A., & Abílio, V. C. (2018). Cannabidiol administered during peri-adolescence prevents behavioral abnormalities in an animal model of schizophrenia. *Frontiers in Pharmacology*, *9*, 901.

Pertwee, R. G. (2008). The diverse CB1 and CB2 receptor pharmacology of three plant cannabinoids: Delta9-tetrahydrocannabinol, cannabidiol and delta9-tetrahydrocannabivarin. *British Journal of Pharmacology*, *153*(2), 199–215.

Prini, P., Zamberletti, E., Manenti, C., Gabaglio, M., Parolaro, D., & Rubino, T. (2020). Neurobiological mechanisms underlying cannabis-induced memory impairment. *European Neuropsychopharmacology: The Journal of the European College of Neuropsychopharmacology*, *36*, 181–190.

Rocchetti, M., Crescini, A., Borgwardt, S., Caverzasi, E., Politi, P., Atakan, Z., & Fusar-Poli, P. (2013). Is cannabis neurotoxic for the healthy brain? A meta-analytical review of structural brain alterations in non-psychotic users. *Psychiatry and Clinical Neurosciences*, *67*(7), 483–492.

Roeske, M. J., Konradi, C., Heckers, S., & Lewis, A. S. (2021). Hippocampal volume and hippocampal neuron density, number and size in schizophrenia: A systematic review and meta-analysis of postmortem studies. *Molecular Psychiatry*, *26*(7), 3524–3535.

Rubino, T., & Parolaro, D. (2014). Cannabis abuse in adolescence and the risk of psychosis: A brief review of the preclinical evidence. *Progress in Neuro-Psychopharmacology & Biological Psychiatry*, *52*, 41–44.

Rubino, T., & Parolaro, D. (2016). The impact of exposure to cannabinoids in adolescence: Insights from animal models. *Biological Psychiatry*, *79*(7), 578–585.

Rubino, T., Realini, N., Braida, D., Alberio, T., Capurro, V., Viganò, D., Guidali, C., Sala, M., Fasano, M., & Parolaro, D. (2009). The depressive phenotype induced in adult female rats by adolescent exposure to THC is associated with cognitive impairment and altered neuroplasticity in the prefrontal cortex. *Neurotoxicity Research*, *15*(4), 291–302.

Rubino, T., Realini, N., Braida, D., Guidi, S., Capurro, V., Viganò, D., Guidali, C., Pinter, M., Sala, M., Bartesaghi, R., & Parolaro, D. (2009). Changes in hippocampal morphology and neuroplasticity induced by adolescent THC treatment are associated with cognitive impairment in adulthood. *Hippocampus*, *19*(8), 763–772.

Ruiz, C. M., Torrens, A., Lallai, V., Castillo, E., Manca, L., Martinez, M. X., Justeson, D. N., Fowler, C. D., Piomelli, D., & Mahler, S. V. (2021). Pharmacokinetic and pharmacodynamic properties of aerosolized ("vaped") THC in adolescent male and female rats. *Psychopharmacology*, *238*(12), 3595–3605.

Schizophrenia Working Group of the Psychiatric Genomics Consortium. (2014). Biological insights from 108 schizophrenia-associated genetic loci. *Nature*, *511*(7510), 421–427.

Schneider, M., & Koch, M. (2003). Chronic pubertal, but not adult chronic cannabinoid treatment impairs sensorimotor gating, recognition memory, and the performance in a progressive ratio task in adult rats. *Neuropsychopharmacology: Official Publication of the American College of Neuropsychopharmacology*, *28*(10), 1760–1769.

Selemon, L. D., & Zecevic, N. (2015). Schizophrenia: A tale of two critical periods for prefrontal cortical development. *Translational Psychiatry*, *5*, e623.

Shrivastava, A., Johnston, M., Terpstra, K., & Bureau, Y. (2014). Cannabis and psychosis: Neurobiology. *Indian Journal of Psychiatry*, *56*(1), 8–16.

Silveira, M. M., Arnold, J. C., Laviolette, S. R., Hillard, C. J., Celorrio, M., Aymerich, M. S., & Adams, W. K. (2017). Seeing through the smoke: Human and animal studies of cannabis use and endocannabinoid signalling in corticolimbic networks. *Neuroscience and Biobehavioral Reviews*, *76*(Pt B), 380–395.

Stepniak, B., Papiol, S., Hammer, C., Ramin, A., Everts, S., Hennig, L., Begemann, M., & Ehrenreich, H. (2014). Accumulated environmental risk determining age at schizophrenia onset: A deep phenotyping-based study. *The Lancet. Psychiatry*, *1*(6), 444–453.

Tyndale, R. F., Payne, J. I., Gerber, A. L., & Sipe, J. C. (2007). The fatty acid amide hydrolase C385A (P129T) missense variant in cannabis users: Studies of drug use and dependence in Caucasians. *American Journal of Medical Genetics. Part B, Neuropsychiatric Genetics: The Official Publication of the International Society of Psychiatric Genetics*, *144B*(5), 660–666.

van Erp, T. G. M., Walton, E., Hibar, D. P., Schmaal, L., Jiang, W., Glahn, D. C., Pearlson, G. D., Yao, N., Fukunaga, M., Hashimoto, R., Okada, N., Yamamori, H., Bustillo, J. R., Clark, V. P., Agartz, I., Mueller, B. A., Cahn, W., de Zwarte, S. M. C., Hulshoff Pol, H. E., … Turner, J. A. (2018). Cortical brain abnormalities in 4474 individuals with schizophrenia and 5098 control subjects via the enhancing neuro imaging genetics through meta analysis (ENIGMA) Consortium. *Biological Psychiatry*, *84*(9), 644–654.

van Os, J., Bak, M., Hanssen, M., Bijl, R. V., de Graaf, R., & Verdoux, H. (2002). Cannabis use and psychosis: A longitudinal population-based study. *American Journal of Epidemiology*, *156*(4), 319–327.

van Os, J., Kenis, G., & Rutten, B. P. F. (2010). The environment and schizophrenia. *Nature*, *468*(7321), 203–212.

Verweij, K. J. H., Abdellaoui, A., Nivard, M. G., Sainz Cort, A., Ligthart, L., Draisma, H. H. M., Minică, C. C., International Cannabis Consortium, Gillespie, N. A., Willemsen, G., Hottenga, J.-J., Boomsma, D. I., & Vink, J. M. (2017). Short communication: Genetic association between schizophrenia and cannabis use. *Drug and Alcohol Dependence*, *171*, 117–121.

Volkow, N. D., Swanson, J. M., Evins, A. E., DeLisi, L. E., Meier, M. H., Gonzalez, R., Bloomfield, M. A. P., Curran, H. V., & Baler, R. (2016). Effects of cannabis use on human behavior, including cognition, motivation, and psychosis: A review. *JAMA Psychiatry*, *73*(3), 292–297.

White, C. M. (2019). A review of human studies assessing cannabidiol's (CBD) therapeutic actions and potential. *Journal of Clinical Pharmacology*, *59*(7), 923–934.

Wilson, R., Bossong, M. G., Appiah-Kusi, E., Petros, N., Brammer, M., Perez, J., Allen, P., McGuire, P., & Bhattacharyya, S. (2019). Cannabidiol attenuates insular dysfunction during motivational salience processing in subjects at clinical high risk for psychosis. *Translational Psychiatry*, *9*(1), 203.

Wilson, J., Freeman, T. P., & Mackie, C. J. (2019). Effects of increasing cannabis potency on adolescent health. *The Lancet. Child & Adolescent Health*, *3*(2), 121–128.

Zamberletti, E., Beggiato, S., Steardo, L., Jr., Prini, P., Antonelli, T., Ferraro, L., Rubino, T., & Parolaro, D. (2014). Alterations of prefrontal cortex GABAergic transmission in the complex psychotic-like phenotype induced by adolescent delta-9-tetrahydrocannabinol exposure in rats. *Neurobiology of Disease*, *63*, 35–47.

Zammit, S., Spurlock, G., Williams, H., Norton, N., Williams, N., O'Donovan, M. C., & Owen, M. J. (2007). Genotype effects of CHRNA7, CNR1 and COMT in schizophrenia: Interactions with tobacco and cannabis use. *The British Journal of Psychiatry: the Journal of Mental Science*, *191*, 402–407.

Part VI

New and synthetic cannabinoids

Chapter 35

Impact of Δ9-tetrahydrocannabinol (THC) and synthetic "spice" cannabinoids on nicotine use and abuse

Angeline J. Dukes, Malia Bautista, and Christie D. Fowler
Department of Neurobiology and Behavior, University of California Irvine, Irvine, CA, United States

Abbreviations

CB1R	cannabinoid 1 receptor
CB2R	cannabinoid 2 receptor
ENDS	electronic nicotine delivery system
NAc	nucleus accumbens
nAChRs	nicotinic acetylcholine receptors
NRT	nicotine replacement therapies
THC	Δ9-tetrahydrocannabinol
VTA	ventral tegmental area
WIN	WIN 55,212-2

Substance use disorder is typically characterized by excessive and compulsive drug-seeking behaviors, in which the individual continues to use the drug despite harmful consequences. Worldwide, nicotine is the most abused substance, and cannabis is the most abused illicit substance (SAMHSA, 2019). Following initial experimentation, continued use may lead to future patterns of abuse. Thus, tobacco use disorder and cannabis use disorder represent the consequences of dependence on either nicotine or an exogenous cannabinoid, respectively.

Tobacco and nicotine

Over 1.3 billion people around the world report using tobacco products, which includes cigarettes, cigars, e-cigarettes, and smokeless tobacco (WHO, 2021). Nicotine is the main psychoactive component derived from the leaves of the tobacco plant, *Nicotiana tabacum* (Fig. 1). Nicotine use in humans can lead to multiple positive effects (Gipson & Fowler, 2020), including mild euphoria, decreased appetite, reduced stress/anxiety, and improvements in memory and concentration. In addition to nicotine, tobacco smoke contains many other toxic chemicals, such as ammonia, arsenic, formaldehyde, acetaldehyde, and tar (US_DHHS, 2010). Thus, it is perhaps not surprising that tobacco smoking has been causally linked to multiple types of cancer, stroke, coronary heart disease, lung disease, chronic obstructive pulmonary disease, and periodontitis (US_DHHS, 2014). In the United States alone, the combined direct cost of healthcare to treat smoking-related disease and the human capital losses amass to more than $300 billion each year (Xu et al., 2021).

In the past decade, there has been the emergence of a potentially safer alternative to tobacco cigarettes, known as electronic nicotine delivery systems (ENDS) (Fig. 2). ENDS heat liquid to produce an aerosol that users inhale, and like tobacco smoke, allows for nicotine absorption through the lungs. In recent years, tobacco cigarette use has been declining, but the use of ENDS has conversely increased (Creamer et al., 2019). In 2011, it was estimated that there were seven million ENDS users worldwide, but by 2018, the number dramatically increased to 41 million. ENDS were developed to assist individuals in reducing the number of tobacco cigarettes smoked as a potential cessation strategy. For instance, in one study, over 80% of former tobacco smokers reported that ENDS helped them quit smoking tobacco cigarettes (Mayer et al., 2020). However, ENDS use by individuals who have never smoked tobacco cigarettes remains a major concern, due to the increased potential of developing nicotine dependence, later tobacco cigarette use, and unknown long-term health consequences

Cannabis Use, Neurobiology, Psychology, and Treatment. https://doi.org/10.1016/B978-0-323-89862-1.00014-3

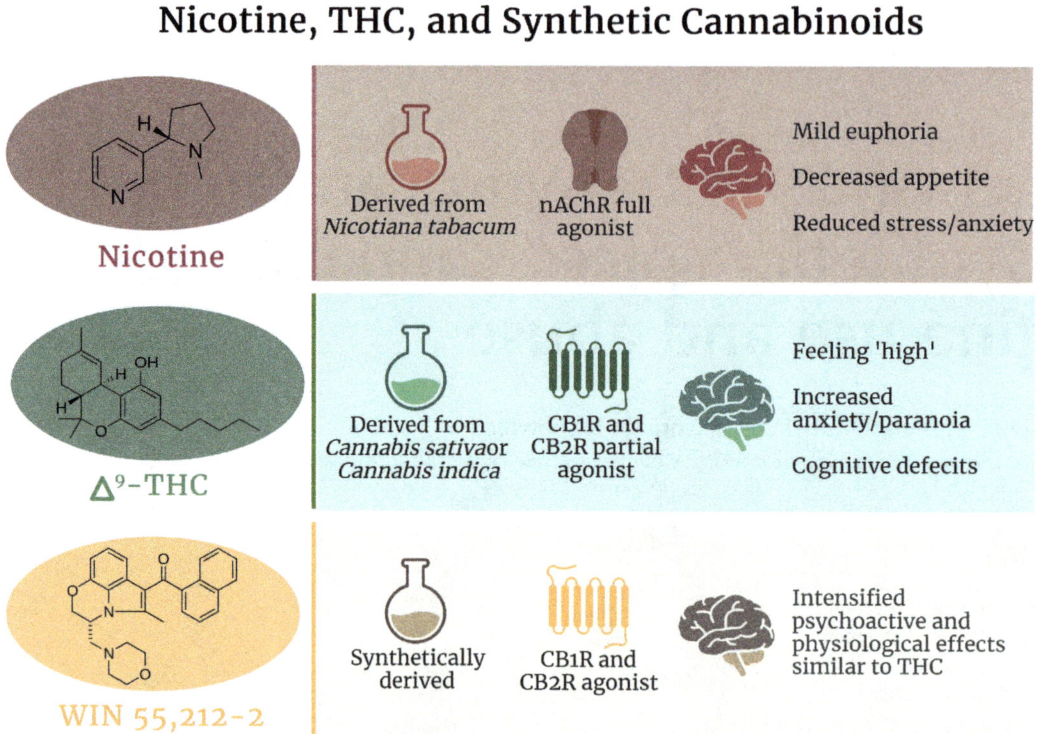

FIG. 1 Nicotine, THC, and synthetic cannabinoids. Summary of the origins, pharmacological actions, and psychological effects of nicotine, THC, and the synthetic cannabinoid WIN 55,212-2 (Castaneto et al., 2014; D'Souza et al., 2004; Gipson & Fowler, 2020; Zou & Kumar, 2018). Created with Bio Render.com.

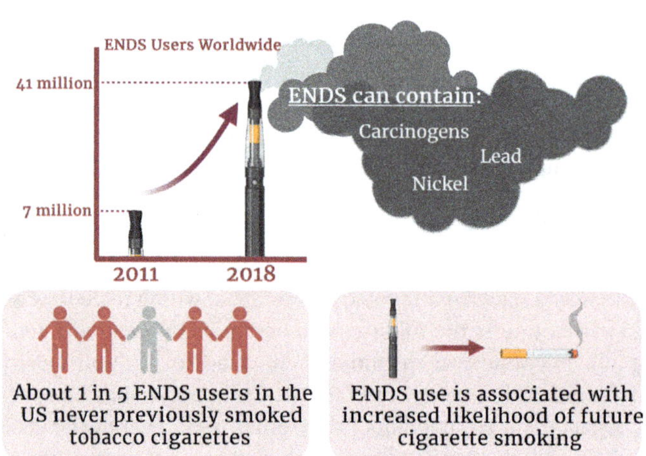

FIG. 2 Electronic nicotine delivery system (ENDS). Illustration of current trends of ENDS use from 2011 to 2018 (data derived from Mayer et al. (2020) and US_DHHS (2016)). Created with BioRender.com.

of vapor exposure. Indeed, from that same study, of the estimated 5.6 million US adults who currently used ENDS, 1.3 million of them were never smokers (Mayer et al., 2020). Moreover, of those that were dual users of both tobacco cigarettes and ENDS, 70% of them reported trying to use ENDS to quit smoking, which then led to dual use (Mayer et al., 2020). Although ENDS are considered to be safer than tobacco cigarettes, ENDS can emit harmful constituents, including carcinogens, nickel, and lead (US_DHHS, 2016). Further, to appeal to youth, additives in the ENDS solutions can provide a sensory perception of candy or fruit, but these flavoring chemicals may lead to adverse health consequences when inhaled, which needs further investigation.

The vast majority of adult smokers began smoking during adolescence (US_DHHS, 2016). Each day, approximately 2000 youth smoke their first cigarette and over 300 become daily smokers (SAMHSA, 2017). Previous studies have shown that the younger people are when they begin using tobacco products, the more likely they are to develop nicotine dependence and other substance use disorders (Hanna & Grant, 1999). It has also been shown that adolescents can become nicotine dependent very quickly, even after occasional intermittent use (DiFranza et al., 2000). Furthermore, a meta-analysis of longitudinal studies confirms that ENDS use is associated with an increased likelihood of future cigarette smoking (Soneji et al., 2017). Although the percentage of teen ENDS use had been consistently increasing over the years, the percentage of high school seniors vaping nicotine actually decreased from 34.5% in 2020 to 26.6% in 2021 during the COVID-19 pandemic in the United States (Johnston et al., 2022). This decrease in adolescent drug use was surprisingly consistent across many substances, including alcohol and opioids. This phenomenon could possibly be attributed to limited access to drugs during government "stay at home" orders, reduced in-person peer pressure, increased messaging of the harmful effects of drug products targeted at youth, and/or potential survey response biases as adolescents are likely responding in the presence of a parent at home during the pandemic.

Cannabis, THC, and spice

According to the World Health Organization, cannabis, also known as weed or marijuana, is the most abused illicit drug (UN, 2021). As of 2021, only eight countries have at least partially legalized cannabis for recreational use, but approximately 200 million people report using cannabis worldwide (UN, 2021). Cannabis is derived from the *Cannabis sativa* or *Cannabis indica* plants that contain over 100 compounds called phytocannabinoids, with Δ^9-tetrahydrocannabinol (THC) and cannabidiol being the most well characterized (Fig. 1). Cannabis use may induce sensations of euphoria, altered sense of time, distorted sensory/body perception, and mood changes (D'Souza et al., 2004). THC is the main psychoactive component in cannabis and can result in feeling "high," along with increased anxiety and paranoia, altered perception, impaired working memory, slower movements, and cognitive deficits, depending on the dose and drug use history (D'Souza et al., 2004; Martin-Santos et al., 2012). Among youth, smoking cannabis appears to be becoming somewhat less prevalent, but this downward trend has been paralleled by an upward trend in consumption of edibles and use of THC vapes (Patrick et al., 2020). Of concern, high-frequency adolescent cannabis use has been linked to deficits in attention, learning, and memory (Broyd et al., 2016), as well as mental health issues including increased depression, anxiety, suicidal ideation, and schizophrenia (Gobbi et al., 2019; Patton et al., 2002). Similar to nicotine, however, in correlation to the COVID-19 pandemic, there was a self-reported decrease in the percentage of adolescents who self-reported using cannabis in 2021 (Johnston, 2022). Interestingly, in older humans (age 65 and above), short-term low-dose cannabis consumption does not seem to have adverse effects on cognition and can aid in pain management (Vacaflor et al., 2020). THC can be also used to help counteract weight loss in HIV/AIDS patients and to alleviate nausea for patients undergoing chemotherapy (NIH, 2017). In older mice, there is also evidence that low doses of THC may help restore cognitive function (Bilkei-Gorzo et al., 2017).

Synthetic cannabinoids are those created in a laboratory and fall under the drug classification "New Psychoactive Drugs." Examples include WIN 55,212-2 (WIN) and CP-55,940—which are compounds that can be used in the street drug termed "spice" (Fig. 1). Some forms of spice are created by spraying synthetic cannabinoids onto shredded plant material, with users smoking the resulting combination like a joint. However, these synthetic cannabinoids may induce more intense psychoactive and physiological effects compared to THC (Castaneto et al., 2014). Synthetic spice cannabinoids were initially created for pharmaceutical research and were not intended for human consumption. However, because spice has similar physiological effects as THC and was not federally illegal, spice has been available in US and European street drug markets since the early 2000s. Further, these synthetic cannabinoids are not typically tested for during routine drug screens and often remain undetected when law enforcement or hospitalization is involved. Spice induces many more intense effects than merely altering one's mood and perceptions, such as tachycardia, psychosis, hallucinations, respiratory distress, and in some cases, death (Muller et al., 2016; Yeruva et al., 2019). In 2012, 26 different formulations of spice were banned with the Synthetic Drug Abuse Prevention Act in the US, classifying them as Schedule I drugs. Nevertheless, as different versions of these synthetic compounds become illegal, new modified versions created by underground chemists consistently reappear in the market.

Drug co-use is also of concern, given the potential synergistic effects on the user. For instance, a study in youth aged 12–17 years found that those who used cannabis were more likely to use nicotine products at the same time, or initiate nicotine consumption within 1 year (Weinberger et al., 2021). Daily cannabis use has also been associated with co-use of opiates, cocaine, and/or inhalants (Tzilos et al., 2014), and approximately half of young cannabis users report simultaneously consuming alcohol and cannabis (Yurasek et al., 2017). Moreover, young adults co-using tobacco and cannabis were more likely to use nicotine ENDS, cocaine, and greater amounts of cannabis than those that just consumed cannabis alone (Reboussin et al., 2021). In older adults, using cannabis is associated with an increased likelihood of being diagnosed with a substance use disorder for either nicotine, alcohol, or cannabis (Vacaflor et al., 2020). Thus, the drug co-use condition represents a significant health concern among various age groups.

Brain mechanisms underlying drug use

When inhaled, nicotine readily enters the bloodstream through the alveoli in the lungs and becomes absorbed into the brain within seconds. Nicotine selectively binds to nicotinic acetylcholine receptors (nAChRs), which are pentameric ligand-gated ion channels located on either the presynaptic or postsynaptic membrane (Gipson & Fowler, 2020). The nAChR subtype may be heteromeric or homomeric, containing a combination of α and β subunits or containing all the same subunit, respectively. The α subunits present in nine different types, α2–α10; whereas the β subunits present in three different types, β2–β4. Various combinations of these subunits result in diverse effects on the pharmacokinetics of the receptor with ligand binding. The α7, α4, and β2 subunits are the most prevalent in the central nervous system. The homomeric α7 nAChR has a relatively lower affinity for nicotine and is important for modulating inflammation (Bagdas et al., 2018). The α4 and β2 subunits combine to form a functional heteromeric receptor with a high affinity for nicotine; the α4β2 nAChR is involved in mediating nicotine's reinforcing and rewarding properties through receptor localization in the mesolimbic circuit (Picciotto et al., 1998). Further, receptors containing the α5, α3, and β4 subunits have been shown to modulate aversive signaling that limits nicotine intake and aspects of the withdrawal syndrome via receptor localization in the habenulo-interpeduncular circuit (Gipson & Fowler, 2020).

Like nicotine, when cannabis is inhaled, active phytocannabinoids pass from the lungs into the bloodstream and are carried throughout the brain and body. THC acts on the cannabinoid 1 receptor (CB1R) and cannabinoid 2 receptor (CB2R) (Zou & Kumar, 2018). In the brain, CB1Rs are mainly localized in neurons and astrocytes, whereas CB2Rs are primarily found on immune cells. In humans and rodents, CB1Rs are highly expressed on axon terminals of neurons in the neocortex, hippocampus, amygdala, cerebellum, and basal ganglia (Castillo et al., 2012; Zou & Kumar, 2018). Since endogenous cannabinoids engage in retrograde signaling from the cell body to the presynaptic axon terminal, activation of CB1Rs results in the inhibition of presynaptic neurotransmitter release (Zou & Kumar, 2018). CB1Rs are also located on postsynaptic membranes and astrocytes, where they function to modulate cellular activity. Of note, THC is a partial agonist of the CB1Rs and CB2Rs, whereas synthetic cannabinoids are typically full agonists. Cannabinoids can also act on other receptors, including GPR55 and TRPV1, although the functional role of these receptors is lesser known.

When individuals co-use nicotine and cannabinoids, one would expect activity at both the nAChRs and CB1Rs (Fig. 3). Indeed, both receptor types exhibit overlapping expression patterns in drug addiction-associated brain regions, such as the prefrontal cortex, nucleus accumbens (NAc), ventral tegmental area (VTA), and amygdala (Renard et al., 2014; Tuesta et al., 2011). In particular, the mesolimbic dopamine pathway, a circuit from the VTA to the NAc, controls reward

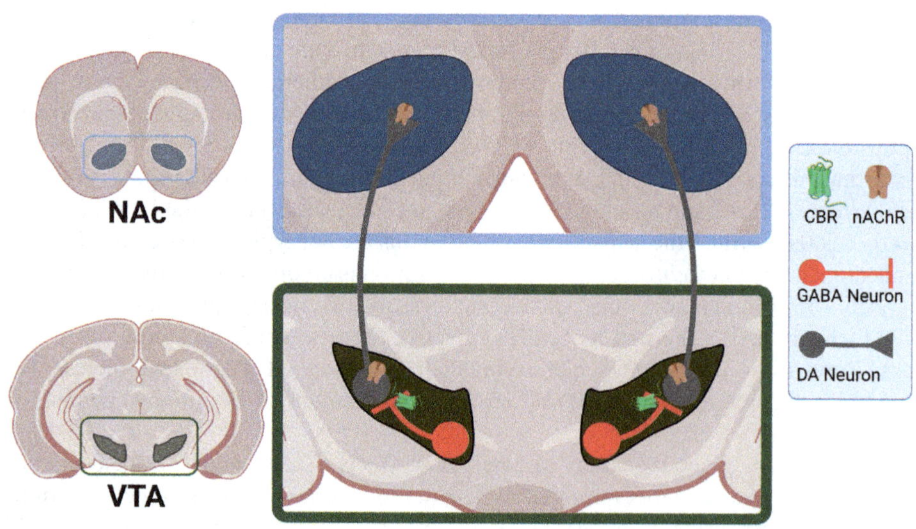

FIG. 3 Illustration of one type of interaction between cannabinoid receptors (CBRs) and nicotinic acetylcholine receptors (nAChRs) in the mesolimbic dopamine pathway. Both CBRs and nAChRs are expressed in areas involved in the mesolimbic dopaminergic pathway. While nicotine directly regulates dopamine release into the nucleus accumbens (NAc) via nAChRs expressed on the ventral tegmental area (VTA) neurons, CBRs expressed on GABAergic terminals in the VTA indirectly influence dopamine release into the NAc (for further information see Renard et al., 2014; Tuesta et al., 2011; Covey et al., 2017; Oleson & Cheer, 2012; Picciotto & Kenny, 2013). Created with BioRender.com.

processing and the reinforcement of natural rewards and most drugs of abuse. Mechanistically, as nicotine and cannabinoids bind to nAChRs and CB1Rs in the VTA-NAc circuit, dopamine levels increase in the NAc, which is subsequently thought to reinforce drug intake. Nicotine acts on the α4β2-containing nAChRs to mediate dopamine signaling via their localization on both dopaminergic and GABAergic neurons in the VTA and neuron terminals in the NAc (Picciotto & Kenny, 2013). In contrast, cannabinoids act on this circuit indirectly via the CB1Rs (Covey et al., 2017). Through retrograde mechanisms, cannabinoid binding to presynaptic CB1Rs expressed on GABAergic presynaptic terminals decrease GABAergic inhibition, thereby increasing dopamine release in the NAc (Covey et al., 2017; Oleson & Cheer, 2012). In addition, it is important to acknowledge the involvement of other neurotransmitters, such as acetylcholine, glutamate, and serotonin, in consideration of the intricate complexity of the projections among brain regions. Thus, nicotinic and cannabinoid signaling within reward-related brain regions may lead to interactions among signaling mechanisms that modulate various aspects of drug-taking behaviors.

Impact of cannabinoids on cognition, behavior, and nicotine use

Approximately 20%–30% of people who experiment with cigarettes will meet criteria for nicotine use disorder during their lifetime (AAAP, 1995). Around 30% of cannabis users are anticipated to develop some degree of cannabis use disorder (Hasin et al., 2015). Individuals who begin using cannabis prior to 18 years of age have a four to seven times increased likelihood of developing cannabis use disorder (Winters & Lee, 2008). Moreover, as noted above, the co-use of both substances is quite frequent. Around 60% of cigarette smokers reported ever using cannabis, and 90% of cannabis users reported ever smoking cigarettes in their lifetime (Agrawal et al., 2012). The rates of daily cannabis and nicotine co-use have doubled from 2002 to 2014 in the United States (Goodwin et al., 2018). While individual drug use has its own potential to develop into a substance use disorder, co-users of both substances are at an increased risk of developing both nicotine and cannabis dependence (Agrawal et al., 2012; Ream et al., 2008). This increased risk is of further concern because adults with co-occurring cannabis and nicotine use disorder are more likely to be diagnosed with bipolar, anxiety, or personality disorders (Peters et al., 2014).

Among youth and young adults, the concern of these co-occurring substance use disorders is also prevalent. Teens who vape nicotine or smoke hookah are four times more likely to start smoking cannabis within 2 years (Audrain-McGovern et al., 2018), and current adolescent tobacco smokers who frequently use cannabis are more likely to report nicotine dependence (Rubinstein et al., 2014). Likewise, adolescents who use cannabis are more likely to become daily cigarette smokers and develop nicotine dependence (Patton et al., 2005; Timberlake et al., 2007). Additionally, a longitudinal study examining the trajectories of nicotine and cannabis vaping from adolescence into early adulthood revealed that those who frequently vape were also very likely to be users of both substances (Lanza et al., 2020). Further, adolescent and young adult cannabis and tobacco cigarette co-users exhibit increased cannabis use disorder symptoms including continued use despite negative consequences, tolerance, and inability to reduce cannabis use, in comparison to those who only use cannabis (Dierker et al., 2018). Finally, young adults who co-use cannabis and nicotine together report consuming more cannabis and nicotine in the past year than those who only use one of the drugs (Tucker et al., 2019). In sum, across age ranges, nicotine and cannabis co-use conditions increase the risk of individual drug use and the development of substance use disorders.

Sex differences have emerged with the prevalence and patterns of substance use disorders in the population (Fig. 4). Adult men are more likely to initiate drug use, but adult women develop substance use disorders more rapidly (UN, 2018). Interestingly, during adolescence, drug use is initiated at similar rates between sexes, but boys appear to escalate their drug use faster than girls (Kuhn, 2015). According to a recent report, men are more likely to smoke cigarettes, vape nicotine, use smokeless tobacco, and be daily cannabis users (Schulenberg et al., 2019). Women with a history of cannabis use are four times more likely to become regular cigarette smokers and almost three times as likely to develop nicotine dependence (Agrawal et al., 2008). Moreover, women report experiencing more intense nicotine withdrawal symptoms (O'Dell & Torres, 2014). While the above findings suggest an important role of age or sex in nicotine and cannabis co-use, many individual, societal, and familial factors also influence drug consumption in humans, and as such, animal models are important to delineate the precise effects of such factors on drug-taking behaviors.

Findings from rodent models have established that cannabinoids and nicotine exert unique effects on drug-related behaviors based on the duration and timing of exposure, in addition to the animal's sex. Interestingly, adolescent male and female rats will self-administer greater amounts of nicotine than adults (Levin et al., 2003, 2007), and nicotine exposure during adolescence can lead to increased self-administration of other drugs of abuse, including alcohol, methamphetamine, and cocaine (Dao et al., 2011). Drug exposure during adolescence can also induce long-lasting effects on the animal into adulthood. For instance, our studies have shown that mid- to late adolescent exposure to nicotine and/or the synthetic

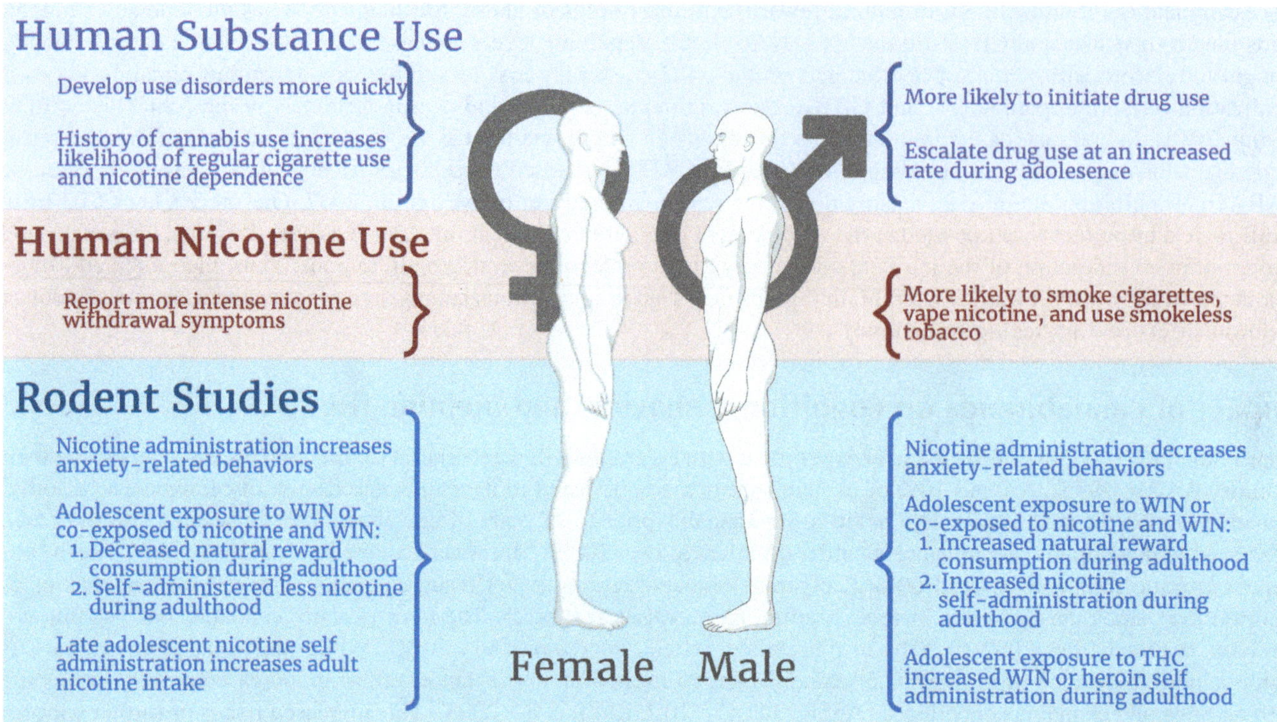

FIG. 4 Sex differences in drug use in humans and rodents. Both human and rodent studies reveal sex differences in substance use. In some cases, male and female rodent studies demonstrate differential effects following drug exposure, thus emphasizing the need to investigate drug action within each sex (Dao et al., 2011; Dukes et al., 2020; Elliott et al., 2004; O'Dell & Torres, 2014; Schulenberg et al., 2019; UN, 2018). Created with BioRender.com.

cannabinoid WIN across 14 days can alter nicotine self-administration in adulthood in a sex-specific manner (Dukes et al., 2020). Specifically, male mice exposed to WIN or co-exposed to nicotine and WIN during adolescence were found to self-administer higher levels of nicotine during adulthood (Dukes et al., 2020). Conversely, female mice adolescently exposed to WIN or co-exposed to nicotine and WIN self-administered significantly less nicotine in adulthood (Dukes et al., 2020). In these studies, the findings were dose-specific, as differences were not found with an adolescent history of a low dose of WIN, or nicotine alone (Dukes et al., 2020). Furthermore, exposure to the plant-derived cannabinoid THC can also induce long-term changes in the brain leading to altered adult drug-taking behavior. Adolescent male rats exposed to THC were found to self-administer higher levels of the synthetic cannabinoid WIN or heroin in adulthood (Ellgren et al., 2007; Scherma et al., 2016), and such adolescent exposure was further shown to induce changes in endogenous opioid signaling mechanisms (Ellgren et al., 2007). One may note that the stage of developmental exposure also appears to be important. For instance, female rats that were permitted to self-administer nicotine beginning in later adolescence exhibited higher levels of nicotine intake compared to those that initiated self-administration in adulthood (Levin et al., 2003).

Cannabinoid and nicotine co-exposure in adulthood can mediate one another given their actions on similar brain mechanisms. For instance, administering the synthetic cannabinoid WIN was shown to decrease nicotine self-administration in both male and female mice, following both acute exposure and with multiple chronic pairings (Dukes et al., 2020). However, WIN's nicotine lowering effects were interestingly dependent on prior drug history. Adult WIN treatment in nicotine self-administering mice did not alter nicotine intake if the mice had an adolescent history of nicotine or cannabinoid drug exposure (Dukes et al., 2020). Of further interest, while WIN exposure decreased nicotine self-administration in adult male rats at a moderate nicotine dose, this effect was reversed when the level of effort required to obtain drug infusions was increased under a progressive ratio schedule of reinforcement (Gamaleddin et al., 2012). Similarly, under operant conditions requiring high levels of behavioral effort, a brief history of THC administration in adulthood increased subsequent nicotine self-administration in male rats (Panlilio et al., 2013). Finally, cannabinoid signaling may also be involved in cue-associated nicotine seeking, a behavior representative of drug relapse; male rats administered WIN prior to a cue-induced reinstatement session exhibited increased nicotine-seeking behavior compared to controls (Gamaleddin et al., 2012), suggesting acute cannabinoid receptor activation potentiated the incentive salience of the cue to drive lever pressing behavior. Taken together, these studies highlight the importance of prior drug history at varying developmental stages and level of effort required on the effectiveness of cannabinoids in modulating nicotine reinforcement.

Nicotine and/or cannabinoid use may also alter cognitive and emotion-associated behaviors, which are often found correlated with substance use disorders. Acute cannabinoid or nicotine exposure has been shown to induce either anxiolytic or anxiogenic effects dependent on dose, age, or sex (Balerio et al., 2006; Renard et al., 2014; Tirado Munoz et al., 2020). For example, nicotine decreased anxiety-associated behaviors in adolescent male rats, but paradoxically increased anxiety-associated behaviors in females (Elliott et al., 2004). Further, male and female adolescent rats exposed to cannabinoids exhibited a decrease in short-term and spatial working memory but an increase in depressive-like behaviors (Renard et al., 2014). In a study assessing chronic co-exposure of nicotine and the synthetic cannabinoid CP 55,940, both male and female adolescent rats developed increased anxiety-like behavior that was further reflected physiologically by elevated corticosterone, a stress-associated hormone (Marco et al., 2006). In contrast, in adult mice, chronic co-exposure to both nicotine and THC decreased anxiety-like behaviors (Valjent et al., 2002). Similarly, nicotine treatment can reduce some of the anxiogenic effects of acute THC exposure, and THC treatment can attenuate the anxiogenic effects of acute nicotine exposure (Balerio et al., 2006; Manwell et al., 2019).

Finally, nicotine and/or cannabinoids can induce a significant developmental impact on cognitive outcomes when consumed by the pregnant dam or during adolescence. Chronic in utero exposure to nicotine, THC, or co-exposure to both drugs has been associated with long-term effects into adolescence (Lallai et al., 2022). Specifically, adolescent male and female rats exposed prenatally to THC exhibited deficits in short-term memory (Lallai et al., 2022). Interestingly, the adolescent male rats with a prenatal history of nicotine and THC co-exposure also exhibited deficits in short-term memory, as well as a deficit in pre-pulse inhibition (Lallai et al., 2022), a behavioral outcome associated with schizophrenia symptomology. It is worthwhile to note that the nicotine and THC prenatal co-exposure condition only induced memory-related effects in the males but not females (Lallai et al., 2022), suggesting that nicotine may have buffered the effects of THC on the developing female brain, although this contention requires further investigation. Exposure during adolescence has also been shown to alter cognitive processing later in adulthood. For instance, adult male mice with a history of adolescent WIN, or nicotine and WIN co-exposure exhibited an increase in natural reward consumption, a decrease in anxiety-associated behaviors, and enhanced cognitive flexibility (Pushkin et al., 2019). Conversely, adult females with a history of adolescent WIN exposure showed less natural reward consumption, but the female WIN and nicotine co-exposure condition did not result in any differences in these behaviors (Pushkin et al., 2019). Together, these findings indicate that nicotine and cannabinoids induce complex interactions on the brain across various stages of development (Fig. 5).

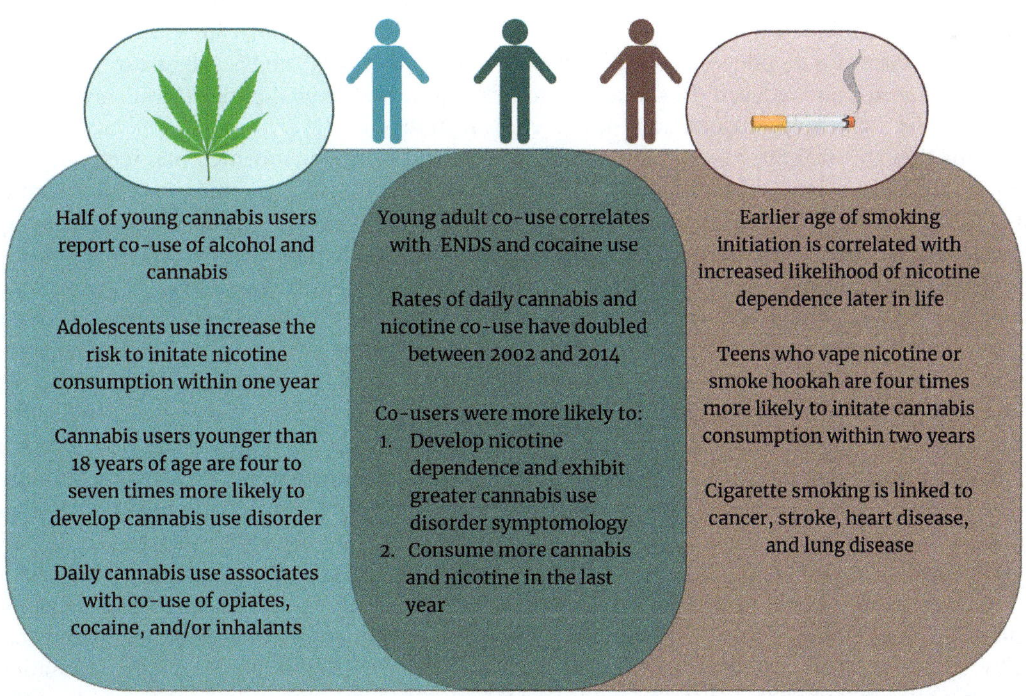

FIG. 5 Cannabis, nicotine, and co-use conditions. During adolescence, cannabis, nicotine, and co-use have different implications for substance abuse later in life (Dierker et al., 2018; Goodwin et al., 2018; Hanna & Grant, 1999; Audrain-McGovern et al., 2018; Reboussin et al., 2021; Tucker et al., 2019; Tzilos et al., 2014; US_DHHS, 2014; Weinberger et al., 2021; Winters & Lee, 2008; Yurasek et al., 2017). Created with BioRender.com.

Nicotine and cannabis cessation

Less than 10% of those who want to quit smoking cigarettes are successful in the long term (Babb et al., 2017). Most people attempt to quit "cold-turkey," without the help of any nicotine replacement therapies (NRT), other pharmacotherapies, or behavioral support programs. Unfortunately, this cold-turkey approach induces significant nicotine withdrawal symptoms, such as cravings, irritability, difficulty concentrating, headaches, and insomnia, which can promote relapse as the user attempts to alleviate these effects with drug re-exposure. By using NRTs, such as nicotine patches, lozenges, or gum, the success of quitting increases to 50%–60% at the 6-month time point (Hartmann-Boyce et al., 2018). This is likely due to smokers being able to obtain nicotine from a source other than cigarettes, thereby reducing withdrawal symptoms and easing the transition to abstinence. ENDS were also developed as a type of NRT for adult smokers. It was proposed that this method of administration may be more successful given that the same physical and sensory cues are present as with smoking cigarettes, such as raising the hand to the mouth and inhaling/exhaling smoke. In 2014, 4% of adults in the United States reported using ENDS for cigarette cessation, but by 2018, the percentage decreased to 3.2% (Creamer et al., 2019). Furthermore, about half of adults who vape nicotine also smoke tobacco cigarettes, a behavior known as "dual use" (CDC, 2016). Surprisingly, a recent study found that people who quit smoking for more than a year have an increased risk of relapse if they vape nicotine during that time (Dai & Leventhal, 2019). Additionally, there is increasing evidence of cannabis use in vape pens among teens and adults. People who use THC vapes report a high incidence of tobacco product use as well (Nicksic et al., 2020). As such, the absolute effectiveness of ENDS for tobacco cessation remains to be determined. Regardless of the smoking cessation tools implemented, high rates of nicotine relapse remain prevalent.

Modulation of the cannabinoid receptor has also been employed as a novel approach for smoking cessation. In rat models, CB1R antagonists were shown to decrease nicotine self-administration and reduce nicotine-induced dopamine release in the NAc (Cohen et al., 2002), which then led to the progression of these compounds along the drug development pipeline. Two different CB1R antagonists, rimonabant and taranabant, underwent clinical trials for smoking cessation and were found to be marginally effective (Cahill et al., 2013; Cahill & Ussher, 2011). However, both drugs have now been withdrawn from the market due to adverse psychological side effects in humans, including increased anxiety and depression. Nevertheless, rodent studies have shown that both acute and chronic pretreatment with the synthetic cannabinoid WIN significantly reduces nicotine intake in male and female mice; however, if either sex has been previously exposed to nicotine or cannabinoids during adolescence, this effect does not occur (Dukes et al., 2020). Thus, previous drug history may be a mediating factor in the effectiveness of pharmacological cessation treatments and needs to be considered in the clinical realm going forward. Cannabidiol, a CB1R and GPR55 antagonist and CB2R reverse agonist, has also been assessed as a modulator for nicotine withdrawal symptoms in a pre-clinical study. Administration of cannabidiol following chronic nicotine exposure reduced the somatic signs of nicotine withdrawal in rats (Smith et al., 2021), suggesting that cannabidiol may be a potential therapeutic in future clinical studies.

Individuals with cannabis use disorder exhibit similar withdrawal symptoms as nicotine, including increased irritability, aggression and depression, sleep difficulty, and physical symptoms (e.g., tremors, chills, and/or headaches) (Gilbert et al., 2020). Indeed, daily cannabis users who attempted to quit report similar withdrawal symptom severity as daily cigarette smokers attempting to quit (Budney et al., 2008). A preliminary study has shown that synthetic cannabinoids, such as nabilone, may be used to attenuate these withdrawal symptoms, which was demonstrated in a small sample of cannabis users in a clinical setting (Haney et al., 2013). Targeting nAChRs may also be effective as a treatment for cannabis use disorder. A preclinical study in rats showed that blocking α7 nAChRs with a selective antagonist, methyllacotinine, reduced self-administration of the synthetic cannabinoid WIN and prevented THC from increasing dopamine in the NAc shell (Solinas et al., 2007). This is quite promising because this putative therapeutic did not result in any depressant or toxic effects (Solinas et al., 2007). More recently, nicotine patches have been examined for alleviation of cannabis withdrawal symptoms. A low-dose nicotine patch was shown to reduce negative affective withdrawal symptoms in subjects that were not heavy tobacco users, but a side effect of nausea was also observed (Gilbert et al., 2020). Importantly, in consideration of the co-use condition, adult tobacco smokers who also smoke cannabis were twice as likely as non-cannabis users to continue smoking tobacco even years later (Ford et al., 2002). This could be due to the cannabinoids enhancing the effects of nicotine-associated cues in reinstating the drug-seeking behavior after a quit attempt (Gamaleddin et al., 2012). However, one study found that people attempting to quit or reduce cannabis intake also report using less tobacco on abstinent days (Hughes et al., 2016). Thus, research on effective cessation methods for co-users is heavily understudied and needs to be conducted to aid in the smoking cessation of people suffering from co-occurring cannabis and nicotine use disorders.

Applications to other areas

Nicotine and cannabis use disorders represent a significant health concern for individuals across age groups. While the implications of either drug alone represent a major concern, more pressing issues include single drug use leading to co-use and the subsequent unique effects occurring from drug co-exposure. This will need to be further explored by defining how nicotine, THC, synthetic cannabinoids, or a combination of them induces neurological changes across the lifespan, in addition to further investigation into the co-use of other substances of abuse, such as opioids or alcohol. Additionally, assessment of trends in human subjects reveals that co-occurring nicotine and cannabis use disorders are linked with an increased risk of other mood disorders and psychiatric diagnoses (Degenhardt et al., 2003; Peters et al., 2014; Tucker et al., 2019), and as such, understanding the similar neurobiological mechanisms common to these disorders may identify unique targets for future therapeutic development.

Key facts

- Tobacco and cannabis products are commonly co-used across age groups.
- Spice is a term used to refer to a broad class of synthetic cannabinoids.
- Single and co-use nicotine and cannabinoid consumption leads to unique effects on cognition and drug-taking behaviors, with sex-specific and developmental impacts.
- Novel approaches targeting either the nicotinic and/or cannabinoid systems may lead to therapeutic development for substance use disorders.

Mini-dictionary of terms

Cessation: Following chronic drug use, an individual stops using the substance, which can result in adverse effects due to drug withdrawal.

ENDS: Electronic devices used to deliver aerosolized nicotine, which include e-cigarettes, vapes, vape pens, and JUULs. Modified versions can deliver THC or other drugs of abuse.

Mesolimbic pathway: A brain circuit extending from the VTA to the NAc, which releases dopamine and has been linked to the reinforcing properties of drugs.

Self-administration: A process by which an animal makes a behavioral response, usually a nose poke or lever press, to receive a drug infusion.

Substance use disorder: A chronic disease characterized by excessive and compulsive drug seeking behaviors, in which the individual continues to use the drug despite harmful consequences. This term is preferred to the label "addiction" and may be applied to nicotine, cannabis, or other drugs of abuse.

Summary points

- Nicotine use patterns have varied across time due to the availability of novel products and marketing to select groups.
- Cannabinoid products can contain the cannabis plant (e.g., weed or marijuana), extracts of phytocannabinoids (e.g., THC edibles or vape solutions), or synthetic "spice" chemicals.
- Increased prevalence of nicotine and cannabinoid co-use has emerged in recent years.
- Drug co-use has been found to exert unique effects on neurodevelopmental processes, leading to altered behaviors across the lifespan, as supported by animal models.
- Nicotine and/or cannabinoid use has been associated with an increased comorbidity of mental health disorders.

References

AAAP. (1995). *Nicotine dependence*. Accessible at: American Academy of Addiction Psychiatry Board of Directors. https://www.aaap.org/wp-content/uploads/2014/05/Nicotine-Dependence.pdf.

Agrawal, A., Budney, A. J., & Lynskey, M. T. (2012). The co-occurring use and misuse of cannabis and tobacco: A review. *Addiction*, *107*, 1221–1233.

Agrawal, A., Madden, P. A., Bucholz, K. K., Heath, A. C., & Lynskey, M. T. (2008). Transitions to regular smoking and to nicotine dependence in women using cannabis. *Drug and Alcohol Dependence*, *95*, 107–114.

Audrain-McGovern, J., Stone, M. D., Barrington-Trimis, J., Unger, J. B., & Leventhal, A. M. (2018). Adolescent E-cigarette, hookah, and conventional cigarette use and subsequent marijuana use. *Pediatrics*, *142*.

Babb, S., Malarcher, A., Schauer, G., Asman, K., & Jamal, A. (2017). Quitting smoking among adults - United States, 2000-2015. *MMWR. Morbidity and Mortality Weekly Report*, *65*, 1457–1464.

Bagdas, D., Gurun, M. S., Flood, P., Papke, R. L., & Damaj, M. I. (2018). New insights on neuronal nicotinic acetylcholine receptors as targets for pain and inflammation: A focus on alpha7 nAChRs. *Current Neuropharmacology*, *16*, 415–425.

Balerio, G. N., Aso, E., & Maldonado, R. (2006). Role of the cannabinoid system in the effects induced by nicotine on anxiety-like behaviour in mice. *Psychopharmacology*, *184*, 504–513.

Bilkei-Gorzo, A., Albayram, O., Draffehn, A., Michel, K., Piyanova, A., Oppenheimer, H., Dvir-Ginzberg, M., Racz, I., Ulas, T., Imbeault, S., Bab, I., Schultze, J. L., & Zimmer, A. (2017). A chronic low dose of delta(9)-tetrahydrocannabinol (THC) restores cognitive function in old mice. *Nature Medicine*, *23*, 782–787.

Broyd, S. J., van Hell, H. H., Beale, C., Yucel, M., & Solowij, N. (2016). Acute and chronic effects of cannabinoids on human cognition-a systematic review. *Biological Psychiatry*, *79*, 557–567.

Budney, A. J., Vandrey, R. G., Hughes, J. R., Thostenson, J. D., & Bursac, Z. (2008). Comparison of cannabis and tobacco withdrawal: Severity and contribution to relapse. *Journal of Substance Abuse Treatment*, *35*, 362–368.

Cahill, K., Stevens, S., Perera, R., & Lancaster, T. (2013). Pharmacological interventions for smoking cessation: an overview and network meta-analysis. *Cochrane Database of Systematic Reviews*, CD009329.

Cahill, K., & Ussher, M. H. (2011). Cannabinoid type 1 receptor antagonists for smoking cessation. *Cochrane Database of Systematic Reviews*, CD005353.

Castaneto, M. S., Gorelick, D. A., Desrosiers, N. A., Hartman, R. L., Pirard, S., & Huestis, M. A. (2014). Synthetic cannabinoids: Epidemiology, pharmacodynamics, and clinical implications. *Drug and Alcohol Dependence*, *144*, 12–41.

Castillo, P. E., Younts, T. J., Chavez, A. E., & Hashimotodani, Y. (2012). Endocannabinoid signaling and synaptic function. *Neuron*, *76*, 70–81.

CDC. (2016). Cigarette smoking status among current adult E-cigarette users, by age group - National Health Interview Survey, United States, 2015. *MMWR. Morbidity and Mortality Weekly Report*, *65*, 1177.

Cohen, C., Perrault, G., Voltz, C., Steinberg, R., & Soubrie, P. (2002). SR141716, a central cannabinoid (CB(1)) receptor antagonist, blocks the motivational and dopamine-releasing effects of nicotine in rats. *Behavioural Pharmacology*, *13*, 451–463.

Covey, D. P., Mateo, Y., Sulzer, D., Cheer, J. F., & Lovinger, D. M. (2017). Endocannabinoid modulation of dopamine neurotransmission. *Neuropharmacology*, *124*, 52–61.

Creamer, M. R., Wang, T. W., Babb, S., Cullen, K. A., Day, H., Willis, G., Jamal, A., & Neff, L. (2019). Tobacco product use and cessation indicators among adults - United States, 2018. *MMWR. Morbidity and Mortality Weekly Report*, *68*, 1013–1019.

Dai, H., & Leventhal, A. M. (2019). Association of electronic cigarette vaping and subsequent smoking relapse among former smokers. *Drug and Alcohol Dependence*, *199*, 10–17.

Dao, J. M., McQuown, S. C., Loughlin, S. E., Belluzzi, J. D., & Leslie, F. M. (2011). Nicotine alters limbic function in adolescent rat by a 5-HT1A receptor mechanism. *Neuropsychopharmacology*, *36*, 1319–1331.

Degenhardt, L., Hall, W., & Lynskey, M. (2003). Exploring the association between cannabis use and depression. *Addiction*, *98*, 1493–1504.

Dierker, L., Braymiller, J., Rose, J., Goodwin, R., & Selya, A. (2018). Nicotine dependence predicts cannabis use disorder symptoms among adolescents and young adults. *Drug and Alcohol Dependence*, *187*, 212–220.

DiFranza, J. R., Rigotti, N. A., McNeill, A. D., Ockene, J. K., Savageau, J. A., St Cyr, D., & Coleman, M. (2000). Initial symptoms of nicotine dependence in adolescents. *Tobacco Control*, *9*, 313–319.

D'Souza, D. C., Perry, E., MacDougall, L., Ammerman, Y., Cooper, T., Wu, Y. T., Braley, G., Gueorguieva, R., & Krystal, J. H. (2004). The psychotomimetic effects of intravenous delta-9-tetrahydrocannabinol in healthy individuals: Implications for psychosis. *Neuropsychopharmacology*, *29*, 1558–1572.

Dukes, A. J., Fowler, J. P., Lallai, V., Pushkin, A. N., & Fowler, C. D. (2020). Adolescent cannabinoid and nicotine exposure differentially alters adult nicotine self-administration in males and females. *Nicotine & Tobacco Research*, *22*, 1364–1373.

Ellgren, M., Spano, S. M., & Hurd, Y. L. (2007). Adolescent cannabis exposure alters opiate intake and opioid limbic neuronal populations in adult rats. *Neuropsychopharmacology*, *32*, 607–615.

Elliott, B. M., Faraday, M. M., Phillips, J. M., & Grunberg, N. E. (2004). Effects of nicotine on elevated plus maze and locomotor activity in male and female adolescent and adult rats. *Pharmacology, Biochemistry, and Behavior*, *77*, 21–28.

Ford, D. E., Vu, H. T., & Anthony, J. C. (2002). Marijuana use and cessation of tobacco smoking in adults from a community sample. *Drug and Alcohol Dependence*, *67*, 243–248.

Gamaleddin, I., Wertheim, C., Zhu, A. Z., Coen, K. M., Vemuri, K., Makryannis, A., Goldberg, S. R., & Le Foll, B. (2012). Cannabinoid receptor stimulation increases motivation for nicotine and nicotine seeking. *Addiction Biology*, *17*, 47–61.

Gilbert, D. G., Rabinovich, N. E., & McDaniel, J. T. (2020). Nicotine patch for cannabis withdrawal symptom relief: A randomized controlled trial. *Psychopharmacology*, *237*, 1507–1519.

Gipson, C. D., & Fowler, C. D. (2020). Nicotinic receptors underlying nicotine dependence: Evidence from transgenic mouse models. *Current Topics in Behavioral Neurosciences*, *45*, 101–121.

Gobbi, G., Atkin, T., Zytynski, T., Wang, S., Askari, S., Boruff, J., Ware, M., Marmorstein, N., Cipriani, A., Dendukuri, N., & Mayo, N. (2019). Association of cannabis use in adolescence and risk of depression, anxiety, and suicidality in young adulthood: A systematic review and meta-analysis. *JAMA Psychiatry*, *76*, 426–434.

Goodwin, R. D., Pacek, L. R., Copeland, J., Moeller, S. J., Dierker, L., Weinberger, A., Gbedemah, M., Zvolensky, M. J., Wall, M. M., & Hasin, D. S. (2018). Trends in daily cannabis use among cigarette smokers: United States, 2002-2014. *American Journal of Public Health*, *108*, 137–142.

Haney, M., Cooper, Z. D., Bedi, G., Vosburg, S. K., Comer, S. D., & Foltin, R. W. (2013). Nabilone decreases marijuana withdrawal and a laboratory measure of marijuana relapse. *Neuropsychopharmacology*, *38*, 1557–1565.

Hanna, E. Z., & Grant, B. F. (1999). Parallels to early onset alcohol use in the relationship of early onset smoking with drug use and DSM-IV drug and depressive disorders: Findings from the National Longitudinal Epidemiologic Survey. *Alcoholism, Clinical and Experimental Research*, *23*, 513–522.

Hartmann-Boyce, J., Chepkin, S. C., Ye, W., Bullen, C., & Lancaster, T. (2018). Nicotine replacement therapy versus control for smoking cessation. *Cochrane Database of Systematic Reviews*, *5*, CD000146.

Hasin, D. S., Saha, T. D., Kerridge, B. T., Goldstein, R. B., Chou, S. P., Zhang, H., Jung, J., Pickering, R. P., Ruan, W. J., Smith, S. M., Huang, B., & Grant, B. F. (2015). Prevalence of marijuana use disorders in the United States between 2001-2002 and 2012-2013. *JAMA Psychiatry*, *72*, 1235–1242.

Hughes, J. R., Naud, S., Budney, A. J., Fingar, J. R., & Callas, P. W. (2016). Attempts to stop or reduce daily cannabis use: An intensive natural history study. *Psychology of Addictive Behaviors*, *30*, 389–397.

Johnston, L. D., Miech, R. A., O'Malley, P. M., Bachman, J. G., Schulenberg, J. E., & Patrick, M. E. (2022). *Monitoring the future national survey results on drug use, 1975–2021: Overview, key findings on adolescent drug use* (p. 154). Ann Arbor: Institute for Social Research, The University of Michigan.

Kuhn, C. (2015). Emergence of sex differences in the development of substance use and abuse during adolescence. *Pharmacology & Therapeutics*, *153*, 55–78.

Lallai, V., Manca, L., Sherafat, Y., & Fowler, C. D. (2022). Effects of prenatal nicotine, THC or co-exposure on cognitive behaviors in adolescent male and female rats. *Nicotine & Tobacco Research*, *24*(8), 1150–1160.

Lanza, H. I., Barrington-Trimis, J. L., McConnell, R., Cho, J., Braymiller, J. L., Krueger, E. A., & Leventhal, A. M. (2020). Trajectories of nicotine and cannabis vaping and Polyuse from adolescence to young adulthood. *JAMA Network Open*, *3*, e2019181.

Levin, E. D., Lawrence, S. S., Petro, A., Horton, K., Rezvani, A. H., Seidler, F. J., & Slotkin, T. A. (2007). Adolescent vs. adult-onset nicotine self-administration in male rats: Duration of effect and differential nicotinic receptor correlates. *Neurotoxicology and Teratology*, *29*, 458–465.

Levin, E. D., Rezvani, A. H., Montoya, D., Rose, J. E., & Swartzwelder, H. S. (2003). Adolescent-onset nicotine self-administration modeled in female rats. *Psychopharmacology*, *169*, 141–149.

Manwell, L. A., Miladinovic, T., Raaphorst, E., Rana, S., Malecki, S., & Mallet, P. E. (2019). Chronic nicotine exposure attenuates the effects of Delta(9) - tetrahydrocannabinol on anxiety-related behavior and social interaction in adult male and female rats. *Brain and Behavior: A Cognitive Neuroscience Perspective*, *9*, e01375.

Marco, E. M., Llorente, R., Moreno, E., Biscaia, J. M., Guaza, C., & Viveros, M. P. (2006). Adolescent exposure to nicotine modifies acute functional responses to cannabinoid agonists in rats. *Behavioural Brain Research*, *172*, 46–53.

Martin-Santos, R., Crippa, J. A., Batalla, A., Bhattacharyya, S., Atakan, Z., Borgwardt, S., Allen, P., Seal, M., Langohr, K., Farre, M., Zuardi, A. W., & McGuire, P. K. (2012). Acute effects of a single, oral dose of d9-tetrahydrocannabinol (THC) and cannabidiol (CBD) administration in healthy volunteers. *Current Pharmaceutical Design*, *18*, 4966–4979.

Mayer, M., Reyes-Guzman, C., Grana, R., Choi, K., & Freedman, N. D. (2020). Demographic characteristics, cigarette smoking, and e-cigarette use among US adults. *JAMA*, *3*(10), e2020694.

Muller, H. H., Kornhuber, J., & Sperling, W. (2016). The behavioral profile of spice and synthetic cannabinoids in humans. *Brain Research Bulletin*, *126*, 3–7.

Nicksic, N. E., Do, E. K., & Barnes, A. J. (2020). Cannabis legalization, tobacco prevention policies, and Cannabis use in E-cigarettes among youth. *Drug and Alcohol Dependence*, *206*, 107730.

NIH. (2017). *The health effects of cannabis and cannabinoids: The current state of evidence and recommendations for research*. Washington, DC: The National Academies Collection: Reports funded by National Institutes of Health.

O'Dell, L. E., & Torres, O. V. (2014). A mechanistic hypothesis of the factors that enhance vulnerability to nicotine use in females. *Neuropharmacology*, *76*(Pt B), 566–580.

Oleson, E. B., & Cheer, J. F. (2012). A brain on cannabinoids: The role of dopamine release in reward seeking. *Cold Spring Harbor Perspectives in Medicine*, *2*.

Panlilio, L. V., Zanettini, C., Barnes, C., Solinas, M., & Goldberg, S. R. (2013). Prior exposure to THC increases the addictive effects of nicotine in rats. *Neuropsychopharmacology: Official Publication of the American College of Neuropsychopharmacology*, *38*, 1198–1208.

Patrick, M. E., Miech, R. A., Kloska, D. D., Wagner, A. C., & Johnston, L. D. (2020). Trends in marijuana vaping and edible consumption from 2015 to 2018 among adolescents in the US. *JAMA Pediatrics*, *174*(9), 900–902.

Patton, G. C., Coffey, C., Carlin, J. B., Degenhardt, L., Lynskey, M., & Hall, W. (2002). Cannabis use and mental health in young people: Cohort study. *BMJ*, *325*, 1195–1198.

Patton, G. C., Coffey, C., Carlin, J. B., Sawyer, S. M., & Lynskey, M. (2005). Reverse gateways? Frequent cannabis use as a predictor of tobacco initiation and nicotine dependence. *Addiction*, *100*, 1518–1525.

Peters, E. N., Schwartz, R. P., Wang, S., O'Grady, K. E., & Blanco, C. (2014). Psychiatric, psychosocial, and physical health correlates of co-occurring cannabis use disorders and nicotine dependence. *Drug and Alcohol Dependence*, *134*, 228–234.

Picciotto, M. R., & Kenny, P. J. (2013). Molecular mechanisms underlying behaviors related to nicotine addiction. *Cold Spring Harbor Perspectives in Medicine*, *3*, a012112.

Picciotto, M. R., Zoli, M., Rimondini, R., Lena, C., Marubio, L. M., Pich, E. M., Fuxe, K., & Changeux, J. P. (1998). Acetylcholine receptors containing the beta2 subunit are involved in the reinforcing properties of nicotine. *Nature*, *391*, 173–177.

Pushkin, A. N., Eugene, A. J., Lallai, V., Torres-Mendoza, A., Fowler, J. P., Chen, E., & Fowler, C. D. (2019). Cannabinoid and nicotine exposure during adolescence induces sex-specific effects on anxiety- and reward-related behaviors during adulthood. *PLoS One*, *14*, e0211346.

Ream, G. L., Benoit, E., Johnson, B. D., & Dunlap, E. (2008). Smoking tobacco along with marijuana increases symptoms of cannabis dependence. *Drug and Alcohol Dependence*, *95*, 199–208.

Reboussin, B. A., Wagoner, K. G., Ross, J. C., Suerken, C. K., & Sutfin, E. L. (2021). Tobacco and marijuana co-use in a cohort of young adults: Patterns, correlates and reasons for co-use. *Drug and Alcohol Dependence*, *227*, 109000.

Renard, J., Krebs, M. O., Le Pen, G., & Jay, T. M. (2014). Long-term consequences of adolescent cannabinoid exposure in adult psychopathology. *Frontiers in Neuroscience*, *8*, 361.

Rubinstein, M. L., Rait, M. A., & Prochaska, J. J. (2014). Frequent marijuana use is associated with greater nicotine addiction in adolescent smokers. *Drug and Alcohol Dependence*, *141*, 159–162.

SAMHSA. (2017). *2017 National survey on drug use and health: Detailed tables*. Rockville, MD: Department of Health and Human Services (SAMHSA), Center for Behavioral Health Statistics and Quality.

SAMHSA. (2019). *Key substance use and mental health indicators in the United States: Results from the 2018 National survey on drug use and health* (HHS Publication No. PEP19-5068 NSH-(ed)). Rockville, MD: Center for Behavioral Health Statistics and Quality, Substance Abuse and Mental Health Services Administration (SAMHSA).

Scherma, M., Dessi, C., Muntoni, A. L., Lecca, S., Satta, V., Luchicchi, A., Pistis, M., Panlilio, L. V., Fattore, L., Goldberg, S. R., Fratta, W., & Fadda, P. (2016). Adolescent Delta(9)-tetrahydrocannabinol exposure alters WIN55,212-2 self-Administration in adult rats. *Neuropsychopharmacology*, *41*, 1416–1426.

Schulenberg, J. E., Johnston, L. D., O'Malley, P. M., Bachman, J. G., Miech, R. A., & Patrick, M. E. (2019). *Monitoring the future national survey results on drug use, 1975–2018: Volume II, college students and adults ages 19–60*. Ann Arbor: Institute for Social Research, The University of Michigan.

Smith, L. C., Tieu, L., Suhandynata, R. T., Boomhower, B., Hoffman, M., Sepulveda, Y., Carrette, L. L. G., Momper, J. D., Fitzgerald, R. L., Hanham, K., Dowling, J., Kallupi, M., & George, O. (2021). Cannabidiol reduces withdrawal symptoms in nicotine-dependent rats. *Psychopharmacology*, *238*, 2201–2211.

Solinas, M., Scherma, M., Fattore, L., Stroik, J., Wertheim, C., Tanda, G., Fratta, W., & Goldberg, S. R. (2007). Nicotinic alpha 7 receptors as a new target for treatment of cannabis abuse. *The Journal of Neuroscience: The Official Journal of the Society for Neuroscience*, *27*, 5615–5620.

Soneji, S., Barrington-Trimis, J. L., Wills, T. A., Leventhal, A. M., Unger, J. B., Gibson, L. A., Yang, J., Primack, B. A., Andrews, J. A., Miech, R. A., Spindle, T. R., Dick, D. M., Eissenberg, T., Hornik, R. C., Dang, R., & Sargent, J. D. (2017). Association between initial use of e-cigarettes and subsequent cigarette smoking among adolescents and young adults: A systematic review and meta-analysis. *JAMA Pediatrics*, *171*, 788–797.

Timberlake, D. S., Haberstick, B. C., Hopfer, C. J., Bricker, J., Sakai, J. T., Lessem, J. M., & Hewitt, J. K. (2007). Progression from marijuana use to daily smoking and nicotine dependence in a national sample of U.S. adolescents. *Drug and Alcohol Dependence*, *88*, 272–281.

Tirado Munoz, J., Belen Lopez-Rodriguez, A., Fonseca, F., Farre, M., Torrens, M., & Viveros, M. P. (2020). Effects of cannabis exposure in the prenatal and adolescent periods: Preclinical and clinical studies in both sexes. *Frontiers in Neuroendocrinology*, *100841*.

Tucker, J. S., Pedersen, E. R., Seelam, R., Dunbar, M. S., Shih, R. A., & D'Amico, E. J. (2019). Types of cannabis and tobacco/nicotine co-use and associated outcomes in young adulthood. *Psychology of Addictive Behaviors*, *33*, 401–411.

Tuesta, L. M., Fowler, C. D., & Kenny, P. J. (2011). Recent advances in understanding nicotinic receptor signaling mechanisms that regulate drug self-administration behavior. *Biochemical Pharmacology*, *82*, 984–995.

Tzilos, G. K., Reddy, M. K., Caviness, C. M., Anderson, B. J., & Stein, M. D. (2014). Getting higher: Co-occurring drug use among marijuana-using emerging adults. *Journal of Addictive Diseases*, *33*, 202–209.

UN. (2018). *United Nations Office on Drugs and Crime, World Drug Report 2018, Vienna, Austria*. Accessed at: https://www.unodc.org/wdr2018/.

UN. (2021). *United Nations (UN) world drug report 2021 booklet 3: Drug market trends cannabis and opioids*.

US_DHHS. (2010). *How tobacco smoke causes disease: The biology and behavioral basis for smoking-attributable disease: A report of the surgeon general (publications and reports of the surgeon general)*. Atlanta, GA: U.S. Department of Health and Human Services.

US_DHHS. (2014). *The health consequences of smoking-50 years of progress: A report of the surgeon general (reports of the surgeon general)*. Atlanta, GA: U.S. Department of Health and Human Services.

US_DHHS. (2016). E-Cigarette use among youth and young adults: A report of the surgeon general. In *Services USDoHaH*. Atlanta, GA: Publications and Reports of the Surgeon General.

Vacaflor, B. E., Beauchet, O., Jarvis, G. E., Schavietto, A., & Rej, S. (2020). Mental health and cognition in older cannabis users: A review. *The Canadian Geriatrics Journal*, *23*, 242–249.

Valjent, E., Mitchell, J. M., Besson, M. J., Caboche, J., & Maldonado, R. (2002). Behavioural and biochemical evidence for interactions between delta 9-tetrahydrocannabinol and nicotine. *British Journal of Pharmacology*, *135*, 564–578.

Weinberger, A. H., Zhu, J., Lee, J., Xu, S., & Goodwin, R. D. (2021). Cannabis use and the onset of cigarette and E-cigarette use: A prospective, longitudinal study among youth in the United States. *Nicotine & Tobacco Research*, *23*, 609–613.

WHO. (2021). *World Health Organization report on the global tobacco epidemic 2021: Addressing new and emerging products, Geneva.*

Winters, K. C., & Lee, C. Y. (2008). Likelihood of developing an alcohol and cannabis use disorder during youth: Association with recent use and age. *Drug and Alcohol Dependence*, *92*, 239–247.

Xu, X., Shrestha, S. S., Trivers, K. F., Neff, L., Armour, B. S., & King, B. A. (2021). U.S. healthcare spending attributable to cigarette smoking in 2014. *Preventive Medicine*, *150*, 106529.

Yeruva, R. R., Mekala, H. M., Sidhu, M., & Lippmann, S. (2019). Synthetic cannabinoids-"spice" can induce a psychosis: A brief review. *Innovations in Clinical Neuroscience*, *16*, 31–32.

Yurasek, A. M., Aston, E. R., & Metrik, J. (2017). Co-use of alcohol and cannabis: A review. *Current Addiction Reports*, *4*, 184–193.

Zou, S., & Kumar, U. (2018). Cannabinoid receptors and the endocannabinoid system: Signaling and function in the central nervous system. *International Journal of Molecular Sciences*, *19*.

Chapter 36

Synthetic cannabinoid receptor agonists compared to Δ^9-tetrahydrocannabinol: Neurological effects and beyond

Christopher S. Breivogel[a] and Victor M. Pulgar[a,b]

[a]Department of Pharmaceutical & Clinical Sciences, Campbell University College of Pharmacy & Health Sciences, Buies Creek, NC, United States, [b]Department of Obstetrics and Gynecology, Wake Forest School of Medicine, Winston-Salem, NC, United States

Abbreviations

2-AG	2-arachidonoyl glycerol
ACEA	arachidonyl-2′-chloroethylamide
AEA	arachidonyl ethanolamide (AKA anandamide)
CBC	cannabichromene
CBD	cannabidiol
CBG	cannabigerol
CBV	cannabivarin
EC	endocannabinoid
EEG	electroencephalogram
mAEA	methanandamide
SCRA	synthetic cannabinoid receptor agonist
THC	Δ^9-tetrahydrocannabinol

Introduction

There are multiple meanings of the term “cannabinoid”. One refers to a group of related plant-based compounds found in *Cannabis sativa*, the origin of the term itself. A second meaning of the term “cannabinoid” refers to compounds that act at the CB_1 and/or CB_2 receptors and the system of endogenous signaling molecules and receptors. CB_1 receptors are found throughout the body and are found in especially high density in the central nervous system; so high that they are among the most prevalent types of receptor in the brain (Herkenham et al., 1991). CB_2 are mostly found on immune cells in the periphery and central nervous system (Pertwee, 1997). Also known as phytocannabinoids, cannabinoids found in *C. sativa* (Mechoulam et al., 1970) include Δ^9-tetrahydrocannabinol (THC), cannabidiol (CBD), cannabigerol (CBG), cannabichromene (CBC), cannabivarin (CBV), and dozens of other structurally related compounds. THC is also a “cannabinoid” in the sense that it is an agonist for (activates) CB_1 and CB_2 receptors, and CBD appears to be a negative allosteric modulator of CB_1 (Tham et al., 2019). Chemically synthesized THC (dronabinol) is also a medication approved by the United States Food and Drug Administration (FDA) for treating chemotherapy-induced nausea and vomiting and stimulating appetite. Cannabis-derived CBD is FDA-approved as an antiseizure medication for some rare childhood seizure disorders including Dravet’s and Lennox-Gastaut syndromes and tuberous sclerosis complex (Steele et al., 2019). Other phytocannabinoids have not been shown to have much affinity or exert effects via CB_1 or CB_2, but this is currently an area of active research. The endogenous cannabinoid receptor ligands, also termed endocannabinoids (ECs), are the arachidonic acid derivatives, arachidonoyl ethanolamide (anandamide or AEA) and 2-arachidonoyl glycerol (2-AG) (Mechoulam et al., 1996). Methanandamide (mAEA) is a metabolism-resistant methylated analog of anandamide that has similar pharmacology to anandamide as a partial agonist for CB_1 and CB_2 (Abadji et al., 1994). Arachidonyl-2′-chloroethylamide (ACEA) is a metabolism-resistant, CB_1-selective analog of anandamide (Coronado-Alvarez et al., 2021).

Cannabis Use, Neurobiology, Psychology, and Treatment. https://doi.org/10.1016/B978-0-323-89862-1.00021-0

The term "synthetic cannabinoid" or "synthetic cannabinoid receptor agonist" (SCRA) has come to be used to refer to agonists for CB_1 and/or CB_2 that are not phyto- or endocannabinoids. SCRAs belong to several chemical classes, and include close analogs of THC like HU-210, and more distantly related chemical classes like alkyl phenols (e.g., CP55940) and aminoalkylindoles (e.g., WIN55212-2, JWH-018, AM2201 or PB-22). Though additional targets have been identified for some compounds, most psychological, neurological, and physiological effects of THC, ECs, and SCRAs appear to be mediated by CB_1 receptors. Cannabis and THC have a wide margin of safety and few incidents of severe acute adverse events. In spite of having the same target for their primary mechanism of action, use of SCRAs carries a relative risk approximately 30-fold greater than cannabis for the need for emergency medical treatment (Winstock et al., 2015). Increased use of SCRAs, sold to the public under brand names like Spice, Incense, or K2, and their apparent toxicity (Armenian et al., 2018) has led to a number of in vivo and in vitro studies in recent years to try to understand the mechanisms of greater toxicity compared to THC (Breivogel et al., 2020).

In animals, activation of CB_1 produces hypothermia, antinociception, decreased spontaneous activity, and catalepsy (Adams & Martin, 1996). In humans, cannabinoid effects on the psyche and perception include fatigue, euphoria, anxiety or anxiolysis, increased appetite, decreased nausea, amnesia, depersonalization, and hallucinations (Grotenhermen, 2003). Cannabinoid somatic effects include tachycardia, analgesia, xerostomia, conjunctivitis, decreased intraocular pressure, and unsteady gait (Grotenhermen, 2003). Recreational use of SCRAs is reported to cause serious adverse events (Seely et al., 2012), especially related to central nervous system and cardiovascular systems and include (but not limited to): acute kidney injury (Centers for Disease Control and Prevention (CDC), 2013; Buser et al., 2014, Luciano & Perazella, 2014), respiratory depression (Alon & Saint-Fleur, 2017; Schmid et al., 2003), anxiety/delusions (Armenian et al., 2018), seizures (Wolfe et al., 2019), changes in blood pressure/heart rate (Andonian et al., 2017; Young et al., 2012), stroke (Bernson-Leung et al., 2014; Freeman et al., 2013), angina/myocardial infarction (Ibrahim et al., 2014; Mir et al., 2011), and dependence (Seely et al., 2012).

Central nervous system effects

While seizures had been documented as one of the adverse effects from SCRA use in humans (Wolfe et al., 2019) two studies in late 2017 and early 2018 were the first to show that cannabinoids can cause seizures in lab mice, so that they could be used as a model for understanding the underlying mechanism. The first by Malyshevskaya et al. (2017) quantified seizures visually and by electroencephalogram (EEG) and showed that THC induced fewer seizures than the aminoalkylindole SCRA, JWH-018, and that those seizures were prevented by pretreatment with the CB_1-selective antagonist, AM-251. The second study by Funada and Takebayashi-Ohsawa (2018) found that a second aminoalkylindole SCRA, AM-2201 produced seizures which were also prevented by AM-251 or a metabotropic glutamate receptor antagonist, but not CB_2 or vanilloid receptor antagonists. A third study by Breivogel et al. (2020) included a wider variety of measures and cannabinoids from a variety of chemical classes. This study compared the effect of THC to compounds representing SCRAs chemical classes including aminoalkylindoles JWH-073 and AM-2201 and found that all three compounds produced generally similar effects in mice. All 3 produced antinociception, albeit with different potencies that correspond to their relative affinities for the CB_1 receptor, and all three produced hypothermia, but the two SCRAs showed 1.5- to 2-fold greater decrease in body temperature than THC. They all produced tolerance to these effects after only 2–3 days of daily treatments with each drug, and they also produced similar signs of antagonist-precipitated withdrawal after 4 days of daily agonist treatments. This study also included microscopic evaluation of the brains of treated mice, but found no significant pathology from any treatment. The lack of pathologic changes from cannabinoids is consistent with other studies in which animals had minimal to no pathology or the pathologic changes were associated with factors such as route of administration (Cha et al., 2015; Chan et al., 1996; Fitzgerald et al., 2013; Thompson et al., 1974; Thompson, Mason, et al., 1973; Thompson, Rosenkrantz, et al., 1973). Where the compounds really differed was in their ability to induce seizures in mice. The assay to measure seizures and hypothermia also included three additional SCRA including the THC analog, HU-210; the alkyl phenol, CP55940; and an additional aminoalkylindole, PB-22. For comparison, methanandamide (an analog of the endocannabinoid anandamide) and the phytocannabinoid, CBD, were also assessed. All 5 of the SCRAs produced a 1.5- to 2-fold greater drop in body temperature than THC, and 4 of the 5 SCRAs produced between 4- and 6.5-fold as many seizures as THC. HU-210, a close analog of THC with much greater potency, produced almost exactly the same number of seizures as THC in spite of twofold greater hypothermic effect. Interestingly, methanandamide also produced a greater than twofold change in temperature but no seizures at all. Not surprisingly, CBD, now a drug approved as a treatment for seizures in 2018 (Steele et al., 2019), produced no change in temperature and no seizures. These studies also included confirmation that a CB_1, but not a CB_2, antagonist prevented seizures (Breivogel et al., 2020).

One possible explanation for the discrepancy between inducing hypothermia and seizures is that different signal transduction proteins may mediate each effect. Biased agonism at G-protein coupled receptors is a well-established phenomenon where different agonists at the same receptor activate different signaling pathways by activating G-proteins rather than beta-arrestins or vice-versa or even different subtypes of G-proteins or beta-arrestins. Since THC, HU-210, and mAEA are all known to activate G-proteins (Breivogel et al., 2001), and THC generally shows only weak activation of beta-arrestins (Finlay et al., 2019; Ibsen et al., 2019) (and activity for beta-arrestins is not known for HU-210 or mAEA), perhaps seizures induced by the SCRAs are related to increased signaling via beta-arrestins (Fig. 1).

In addition, the phytocannabinoid CBD has shown some benefits in models of experimental autoimmune encephalomyelitis (EAE), a preclinical murine model of Multiple Sclerosis (Elliott et al., 2018). Similarly, a role for CBD in models of Parkinson's disease (Zhao et al., 2022) and Huntington's disease (Pérez-Olives et al., 2021) have been proposed. Since CBD displays weak agonistic or antagonistic activity toward classical cannabinoid CB_1 and CB_2 receptors, the observed neuroprotective effects of CBD are probably driven by interactions with other oxidative and metabolic pathways (Fernández-Ruiz et al., 2013) and/or other receptors, enzymes, or transporters.

Interestingly, stimulation of the CB_2 receptor has shown promise in the treatment of brain damage such as subarachnoid hemorrhage (Fujii et al., 2014) and traumatic brain injury (Li et al., 2022) in rats. Different from CB_1 activation, agonism at CB_2 has shown anti-inflammatory, cardioprotective, and nephroprotective actions among other. Consequently, CB_2 activation by the highly selective SCRA JWH-133 displays great potential for the treatment of brain disorders (Hashiesh et al., 2021).

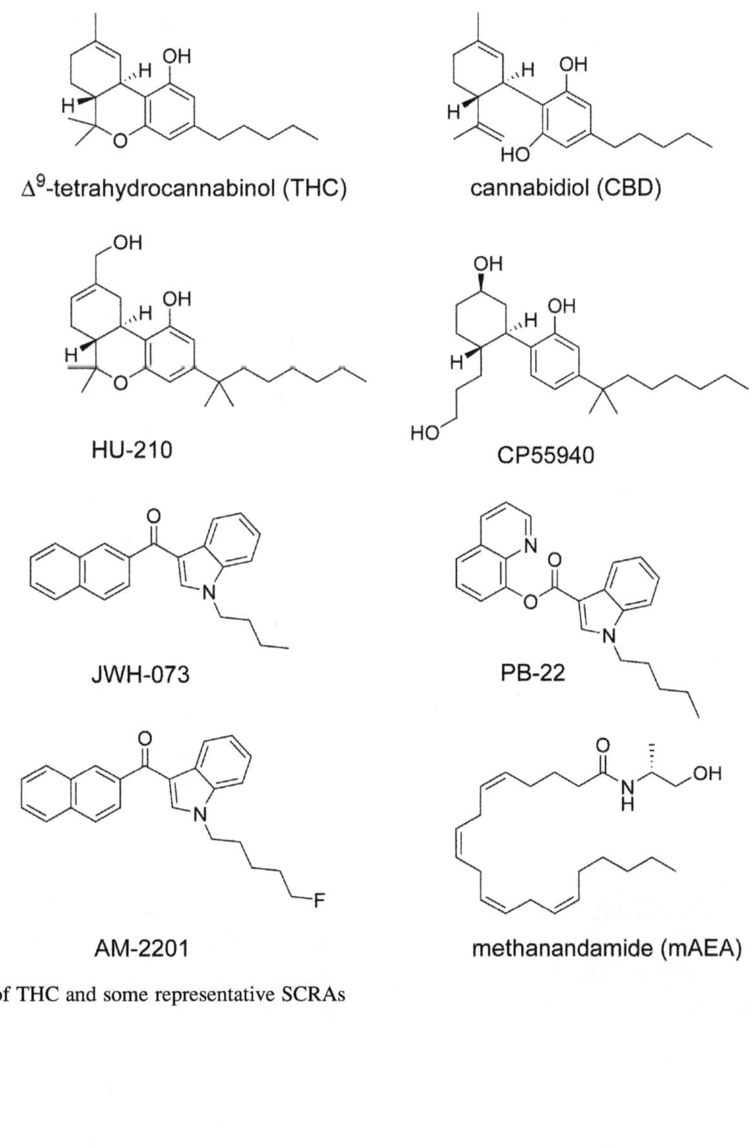

FIG. 1 Chemical structures of THC and some representative SCRAs

In addition to the well-described psychotropic effects of cannabis, the advent of more specific strategies to target components of the endocannabinoid system will provide therapeutic alternatives for the treatment of central nervous system pathologies.

Cardiovascular effects

Cannabis and THC can produce some changes in cardiovascular function like first increased, and then decreased heart rate, and changes in vascular contractility, but the effects are generally mild and not life-threatening (Pabon et al., 2021). In fact, numerous studies have found that cannabinoids including THC and SCRAs can decrease infarct volume from ischemia/reperfusion (Maslov et al., 2016) and mice lacking CB_1 in vascular smooth muscle experience greater infarct volume (Wang et al., 2022). In contrast, SCRAs have been reported to produce severe adverse events and even death in humans. A typical oral dose of THC (alone or in cannabis or cannabis product) for a heavy user is up to 1–2 mg/kg. Several toxicological studies dating back to the 1970s used extremely high doses of THC. THC given for 28 consecutive days in monkeys found no deaths up to 9000 mg/kg orally but did find cardiac failure resulting from respiratory arrest at doses of 128 mg/kg THC given intravenously. Some monkeys experienced nonlethal bradycardia at lower doses, but no damage observed in the heart or vasculature (Thompson et al., 1974). Rats given up to 500 mg/kg/day THC for up 119 consecutive days were found to exhibit changes in behavior, convulsions, hypothermia, and bradypnea during treatments. Post-mortem analysis found endocrine-related changes by no damage to the heart or vasculature (Thompson, Mason, et al., 1973). Another study involving even higher acute doses (up to 9000 mg/kg THC orally) in rats, dogs, and monkeys found no lethality or post-mortem histopathology (Thompson, Rosenkrantz, et al., 1973).

Previous studies using cultured H9C2 cardiac myocytes found that incubation with SCRAs HU-210 (Mukhopadhyay et al., 2007), JWH-030, JWH-210, or JWH-250 (Yun et al., 2016) up to 1 μM for 16-20 hours decreased H9C2 viability. In contrast, THC or JWH-073 at up to 3 μM for 24 hours had no effect on H9C2 cell viability (personal communication, Khalil Eldeeb and Allyn Howlett). JWH-081 and JWH-210 failed to show overt signs of cardiovascular toxicity.

Although early reviews of the cardiovascular effects of THC suggested that only pre-existing conditions predisposed to greater cardiac risks (Hollister, 1986), more recent studies indicate cardiovascular effects of marijuana involving increased catecholamine release with concomitant tachycardia and vasodilation, leading to orthostatic hypotension (Pacher et al., 2018). Additionally, an association between increased blood pressure, depending on the level of marijuana use, has also been reported (Vidot et al., 2019). The reported increases in respiratory burden and carboxyhemoglobin levels in cannabis users (Wu et al., 1988), are also of interest when analyzing cardiovascular effects.

Given their higher potency and efficacy toward cannabinoid receptors, an exacerbation of cardiovascular effects caused by SCRAs is expected. Although evidence is still limited, the contribution of SCRAs to death is being documented. For example, use of SCRAs have been associated with the occurrence of myocardial infarction (Tse et al., 2014) and cardiac death associated with arrythmias (Davis & Boddington, 2015; Ibrahim et al., 2014; Mir et al., 2011). Interestingly, when studying the effects of SCRAs use on the heart electrical conduction system through electrocardiogram (EKG) recordings, examinations in patients using SCRAs (smoking of K2) showed alterations including QT prolongation (Von Der Haar et al., 2016), ST elevation, lower T-wave amplitude, and extended QRS segment among others (Yeniocak et al., 2021). These alterations may generate arrythmia and potentially contribute to decreased systemic blood pressure.

Respiratory depression

While THC or cannabis has not been associated with respiratory depression, animal studies have shown that the SCRAs WIN55212-2 and CP55940 produce hypotension and bradycardia concomitantly with a decrease in respiratory rate, hypoxia, hypercapnia, and blood acidosis (Schmid et al., 2003). Similar effects on respiration have been reported in patients admitted to the Intensive Care Unit with acute respiratory failure after use of SCRAs (Alon & Saint-Fleur, 2017). These effects probably involve several different mechanisms of action, with CB_1-dependent suppression of excitatory as well as inhibitory neuronal activity at the level of chemo- and baroreceptors potentially playing a role.

Renal effects

Kidneys play vital roles to maintain body homeostasis, among them filtration, hormone synthesis, and blood pressure regulation. Consequently, renal diseases are associated with increased morbidity and mortality. Studies in humans have not found an association between cannabis use and progression of chronic kidney disease, CKD (Bundy et al., 2018). Although some benefits appear to exist in CKD patients, limited studies suggest the recommendation of non-SCRAs only for the treatment of chronic neuropathic pain in these patients (Ho et al., 2019). In animal models, CBD has been shown to have protective effects against cisplatin-induced nephrotoxicity in mice (Pan et al., 2009). In this model, activation of CB_1 promotes nephrotoxicity (Mukhopadhyay et al., 2010), whereas nephroprotection seems to be mediated by CB_2 activation (Mukhopadhyay et al., 2016).

Several reports have indicated an association between SCRAs and acute kidney injury (AKI). For example, it was reported that smoking SCRAs has been associated with AKI in patients with toxicologic syndrome (Centers for Disease Control and Prevention (CDC), 2013; Buser et al., 2014), and AKI and hyponatremia seem to be common effects of overuse of some designer drugs, including SCRAs (Luciano & Perazella, 2014).

Since a complete endocannabinoid system is present in the kidney, including expression of cannabinoid receptors in several renal structures, the understanding of the roles of nonsynthetic and SCRAs in renal physiology will continue to expand.

Cannabinoids in cancer treatment

In addition to the known antinociceptive and antiemetic properties of cannabinoid compounds, useful in cancer treatment, initial observations back in mid-1970s indicated that THC and CBD shared the ability to reduce lung adenocarcinoma tumor size in mice (Munson et al., 1975). Growing evidence also indicates additional roles for cannabinoid compounds as modulators of cell proliferation (Ligresti et al., 2006; Robson, 2005). THC inhibits growth of several cancer cell types, but its use in chemotherapy is limited given the psychotropic effects. Additional recent evidence suggests that CBD inhibits cell proliferation and migration in several cancer cell types (Valenti et al., 2022), and this effect has been recently shown in human colorectal cancer cells to be mediated by the CB_2 receptor (Lee et al., 2022). The importance of the CB_2 receptor in regulating proliferation of cancer cells has also prompted the development of novel SCRAs with CB_2 specificity for the treatment of neuroblastoma (Gado et al., 2022). This is another example where SCRAs show novel potential, as with the CB_2-selective SCRA JWH-015 in the treatment of bone cancer pain (Mao et al., 2019).

Overall, the SCRAs with higher CB_2 specificity seem to provide additional therapeutic anticancer strategies. The use of cannabinoids drugs as part of combination therapies appears as attractive alternatives, for example, THC and CBD are being formulated as part of potential cancer treatments in combination with known anticancer medications such as bortezomib and cisplatin (see Hinz & Ramer, 2022 for a recent review) (Fig. 2).

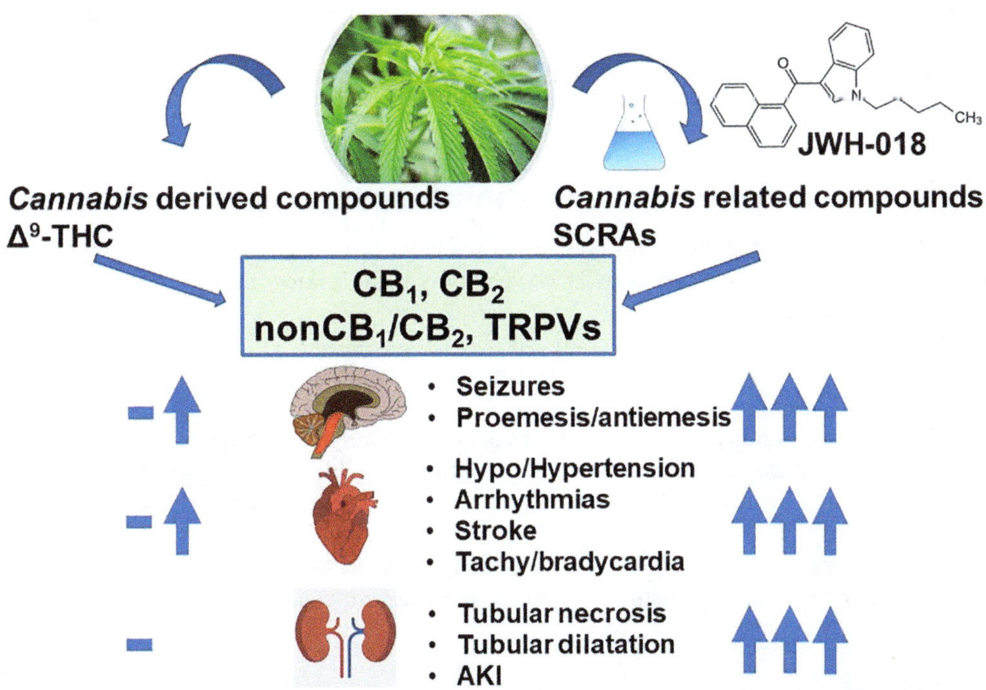

FIG. 2 Comparison of the effects of natural and synthetic cannabis-derived compounds on different physiological systems. SCRAs: synthetic cannabinoid receptors agonists. *CB_1, CB_2*, cannabinoids receptors 1 and 2; *TRPVs*, family of transient receptor potential cation channels.

Applications to other areas

THC and CBD have been the focus of most research on the clinical benefits of natural cannabinoid compounds, but there are more than 100 phytocannabinoids are found in the *Cannabis sativa* plant (Procaccia et al., 2022). Additionally, other compounds in cannabis such as flavonoids and terpenoids may also display therapeutic potential.

In this chapter, we have reviewed the main effects mediated by the nonsynthetic and synthetic cannabinoids receptor agonists (SCRAs) in different organ systems. Initial studies using THC aimed to understand the psychotropic effects of cannabis. Over the last several decades, a growing body of research has characterized the endocannabinoid system in the body, explaining the multiple physiological roles of endocannabinoids and the pharmacological effects of phytocannabinoids and SCRAs.

Although the initial studies focused on effects at the central nervous system, growing evidence shows the multiplicity of roles of cannabinoids, mostly but not exclusively, based on their interactions and modulation of the endogenous cannabinoid system. Thus, the important effects of these compounds in the cardiovascular system, kidney, liver, and on growth and proliferation of several cell types are becoming clearer. Another system in which cannabinoids play an important regulatory role is the immune system and thus modulation of inflammatory processes by cannabinoids. Similarly, the role of SCRAs as adjuvants in pain control, lowering the need for opioids drugs, for example, is highly relevant. Of great interest are the studies showing beneficial therapeutic effects of novel SCRAs with increased specificity for one of the cannabinoid receptors, as in the case of novel CB_2 agonists for cancer treatment. Through well-designed medicinal chemistry strategies and appropriate tests, the dissociation of CNS effects from the beneficial effects on cell proliferation and/or inflammation will provide new avenues for the clinical use of synthetic and phytocannabinoid compounds as anticancer drugs. Given the widespread distribution of the endocannabinoid system throughout the body, and the multiplicity of receptors cannabinoid compounds interact with, it is expected that roles and relevance of cannabis-derived chemicals and SCRAs will grow in the years to come.

Mini-dictionary of terms

- **Agonist**. A substance that binds to a receptor and activates it to produce a response in the cell expressing the receptor.
- **Antagonist** A substance that binds to a receptor, but does not activate it, and in doing so can block an agonist from activating the receptor.
- **Endogenous** A substance made naturally by the body.
- **Phytocannabinoid**. The chemical class of compounds including tetrahydrocannabinol that is derived from, and found only in, *Cannabis sativa.*
- **Synthetic cannabinoid (receptor agonist) (SCRA)**. Has come to mean any substance that can activate cannabinoid receptors that is not endogenous or derived from *C. sativa.*

Key facts of THC compared to synthetic cannabinoid receptor agonists

- Tetrahydrocannabinol (THC) is the main psychoactive constituent of *C. sativa.*
- THC is relatively nontoxic, with the worst acute effect generally being anxiety in some users, though it produces anti-anxiety effects in some users.
- Synthetic cannabinoids belong to a variety of chemical classes.
- Synthetic cannabinoids were originally developed as potential medicines or for research to better understanding the body's own cannabinoid system.
- Synthetic cannabinoids were initially not controlled substances, but were made illegal by federal and state governments after it become known that these substances were being sold to the public as "designer drugs."
- The most prevalent serious side effects of synthetic cannabinoids appear to be anxiety/panic attacks, seizures, stroke, and heart attacks.
- The difference in the extent to which each type of compound activates cannabinoid receptors appears to be the reason that synthetic cannabinoid receptor agonists are so much more hazardous than THC or cannabis.

Summary points

- Medicinal properties of the cannabis plant have been known for thousands of years with first reports appeared in the Chinese Emperor's Shen Nung's pharmacopeia around 2800 BC.

- Plant-derived and synthetic cannabinoid compounds with affinity for cannabinoids CB_1 and CB_2 receptors display actions on most organs studied to date.
- While THC is a weak activator of the body's cannabinoid CB_1 and CB_2 receptors, most synthetic cannabinoid receptor agonists (SCRAs) activate those receptors to a much higher degree.
- Higher activity of SCRAs toward cannabinoid and non-cannabinoid receptors may help explain different effects of natural and synthetic compounds.
- Novel SCRAs with enhanced cannabinoid receptor specificity, particularly toward the CB_2 receptor show promise as anticancer and/or anti-inflammatory drugs.

References

Abadji, V., Lin, S., Taha, G., Griffin, G., Stevenson, L. A., Pertwee, R. G., & Makriyannis, A. (1994). (R)-methanandamide: A chiral novel anandamide possessing higher potency and metabolic stability. *Journal of Medicinal Chemistry*, *37*, 1889–1893.

Adams, I. B., & Martin, B. R. (1996). Cannabis: Pharmacology and toxicology in animals and humans. *Addiction*, *91*, 1585–1614.

Alon, M. H., & Saint-Fleur, M. O. (2017). Synthetic cannabinoid induced acute respiratory depression: Case series and literature review. *Respiratory Medicine Case Reports*, *22*, 137–141. https://doi.org/10.1016/j.rmcr.2017.07.011.

Andonian, D. O., Seaman, S. R., & Josephson, E. B. (2017). Profound hypotension and bradycardia in the setting of synthetic cannabinoid intoxication—A case series. *American Journal of Emergency Medicine*, *35*, 940.e945–940.e946.

Armenian, P., Darracq, M., Gevorkyan, J., Clark, S., Kaye, B., & Brandehoff, N. P. (2018). Intoxication from the novel synthetic cannabinoids ab-pinaca and adb-pinaca: A case series and review of the literature. *Neuropharmacology*, *134*, 82–91.

Bernson-Leung, M. E., Leung, L. Y., & Kumar, S. (2014). Synthetic cannabis and acute ischemic stroke. *Journal of Stroke and Cerebrovascular Disesase*, *23*, 1239–1241.

Breivogel, C. S., Griffin, G., Di Marzo, V., & Martin, B. R. (2001). Evidence for a new g protein-coupled cannabinoid receptor in mouse brain. *Molecular Pharmacology*, *60*, 155–163.

Breivogel, C. S., Wells, J. R., Jonas, A., Mistry, A. H., Gravley, M. L., Patel, R. M., Whithorn, B. E., & Brenseke, B. M. (2020). Comparison of the neurotoxic and seizure-inducing effects of synthetic and endogenous cannabinoids with delta(9)-tetrahydrocannabinol. *Cannabis and Cannabinoid Research*, *5*, 32–41.

Bundy, J. D., Bazzano, L. A., Xie, D., Cohan, J., Dolata, J., Fink, J. C., Hsu, C. Y., Jamerson, K., Lash, J., Makos, G., Steigerwalt, S., Wang, X., Mills, K. T., Chen, J., He, J., & CRIC Study Investigators. (2018). Self-reported tobacco, alcohol, and illicit drug use and progression of chronic kidney disease. *Clinical Journal of the American Society of Nephrology*, *13*(7), 993–1001. https://doi.org/10.2215/CJN.11121017.

Buser, G. L., Gerona, R. R., Horowitz, B. Z., Vian, K. P., Troxell, M. L., Hendrickson, R. G., Houghton, D. C., Rozansky, D., Su, S. W., & Leman, R. F. (2014). Acute kidney injury associated with smoking synthetic cannabinoid. *Clinical Toxicology (Philadelphia, Pa.)*, *52*(7), 664–673.

Centers for Disease Control and Prevention (CDC). (2013). Acute kidney injury associated with synthetic cannabinoid use—Multiple states. *Morbidity and Mortality Weekly Report*, *62*(6), 93–98.

Cha, H. J., Seong, Y. H., Song, M. J., Jeong, H. S., Shin, J., Yun, J., Han, K., Kim, Y. H., Kang, H., & Kim, H. S. (2015). Neurotoxicity of synthetic cannabinoids JWH-081 and JWH-210. *Biomolecular Therapeutics (Seoul)*, *23*, 597–603.

Chan, P. C., Sills, R. C., Braun, A. G., Haseman, J. K., & Bucher, J. R. (1996). Toxicity and carcinogenicity of delta 9-tetrahydrocannabinol in fischer rats and b6c3f1 mice. *Fundamental and Applied Toxicology*, *30*, 109–117.

Coronado-Alvarez, A., Romero-Cordero, K., Macias-Triana, L., Tatum-Kuri, A., Vera-Barron, A., Budde, H., Machado, S., Yamamoto, T., Imperatori, C., & Murillo-Rodriguez, E. (2021). The synthetic CB1 cannabinoid receptor selective agonists: Putative medical uses and their legalization. *Progress in Neuro-Psychopharmacology and Biological Psychiatry*, *110*, 110301.

Davis, C., & Boddington, D. (2015). Teenage cardiac arrest following abuse of synthetic cannabis. *Heart, Lung & Circulation*, *24*, e162–e163.

Elliott, D. M., Singh, N., Nagarkatti, M., & Nagarkatti, P. S. (2018). Cannabidiol attenuates experimental autoimmune encephalomyelitis model of multiple sclerosis through induction of myeloid-derived suppressor cells. *Frontiers in Immunology*, *9*, 1782.

Fernández-Ruiz, J., Sagredo, O., Pazos, M. R., García, C., Pertwee, R., Mechoulam, R., & Martínez-Orgado, J. (2013). Cannabidiol for neurodegenerative disorders: Important new clinical applications for this phytocannabinoid? *British Journal of Clinical Pharmacology*, *75*(2), 323–333.

Finlay, D. B., Manning, J. J., Ibsen, M. S., Macdonald, C. E., Patel, M., Javitch, J. A., Banister, S. D., & Glass, M. (2019). Do toxic synthetic cannabinoid receptor agonists have signature in vitro activity profiles? A case study of amb-fubinaca. *ACS Chemical Neuroscience*, *10*, 4350–4360.

Fitzgerald, K. T., Bronstein, A. C., & Newquist, K. L. (2013). Marijuana poisoning. *Topics in Companion Animal Medicine*, *28*, 8–12.

Freeman, M. J., Rose, D. Z., Myers, M. A., Gooch, C. L., Bozeman, A. C., & Burgin, W. S. (2013). Ischemic stroke after use of the synthetic marijuana "spice". *Neurology*, *81*, 2090–2093.

Fujii, M., Sherchan, P., Krafft, P. R., Rolland, W. B., Soejima, Y., & Zhang, J. H. (2014). Cannabinoid type 2 receptor stimulation attenuates brain edema by reducing cerebral leukocyte infiltration following subarachnoid hemorrhage in rats. *Journal of the Neurological Sciences*, *342*(1-2), 101–106.

Funada, M., & Takebayashi-Ohsawa, M. (2018). Synthetic cannabinoid AM2201 induces seizures: Involvement of cannabinoid CB1 receptors and glutamatergic transmission. *Toxicology and Applied Pharmacology*, *338*, 1–8.

Gado, F., Ferrisi, R., Di Somma, S., Napolitano, F., Mohamed, K. A., Stevenson, L. A., Rapposelli, S., Saccomanni, G., Portella, G., Pertwee, R. G., Laprairie, R. B., Malfitano, A. M., & Manera, C. (2022). Synthesis and in vitro characterization of selective cannabinoid CB2 receptor agonists: Biological evaluation against neuroblastoma cancer cells. *Molecules (Basel, Switzerland)*, *27*(9), 3019.

Grotenhermen, F. (2003). Pharmacokinetics and pharmacodynamics of cannabinoids. *Clinical Pharmacokinetics*, *42*, 327–360.

Hashiesh, H. M., Jha, N. K., Sharma, C., Gupta, P. K., Jha, S. K., Patil, C. R., Goyal, S. N., & Ojha, S. K. (2021). Pharmacological potential of JWH133, a cannabinoid type 2 receptor agonist in neurodegenerative, neurodevelopmental and neuropsychiatric diseases. *European Journal of Pharmacology*, *909*, 174398.

Herkenham, M., Lynn, A. B., Johnson, M. R., Melvin, L. S., de Costa, B. R., & Rice, K. C. (1991). Characterization and localization of cannabinoid receptors in rat brain: A quantitative in vitro autoradiographic study. *Journal of Neuroscience*, *11*, 563–583.

Hinz, B., & Ramer, R. (2022). Cannabinoids as anticancer drugs: Current status of preclinical research. *British Journal of Cancer*. https://doi.org/10.1038/s41416-022-01727-4.

Ho, C., Martinusen, D., & Lo, C. (2019). A review of cannabis in chronic kidney disease symptom management. *Canadian Journal of Kidney Health and Disease*, *6*. https://doi.org/10.1177/2054358119828391.

Hollister, L. E. (1986). Health aspects of cannabis. *Pharmacological Reviews*, *38*, 1–20.

Ibrahim, S., Al-Saffar, F., & Wannenburg, T. (2014). A unique case of cardiac arrest following k2 abuse. *Case Reports in Cardiology*, 120607.

Ibsen, M. S., Finlay, D. B., Patel, M., Javitch, J. A., Glass, M., & Grimsey, N. L. (2019). Cannabinoid CB1 and CB2 receptor-mediated arrestin translocation: Species, subtype, and agonist-dependence. *Frontiers in Pharmacology*, *10*, 350.

Lee, H. S., Tamia, G., Song, H. J., Amarakoon, D., Wei, C. I., & Lee, S. H. (2022). Cannabidiol exerts anti-proliferative activity via a cannabinoid receptor 2-dependent mechanism in human colorectal cancer cells. *International Immunopharmacology*, *108*, 108865. https://doi.org/10.1016/j.intimp.2022.108865.

Li, L., Luo, Q., Shang, B., Yang, X., Zhang, Y., Pan, Q., Wu, N., Tang, W., Du, D., Sun, X., & Jiang, L. (2022). Selective activation of cannabinoid receptor-2 reduces white matter injury via PERK signaling in a rat model of traumatic brain injury. *Experimental Neurology*, *347*, 113899. https://doi.org/10.1016/j.expneurol.2021.113899.

Ligresti, A., Moriello, A. S., Starowicz, K., Matias, I., Pisanti, S., De Petrocellis, L., Laezza, C., Portella, G., Bifulco, M., & Di Marzo, V. (2006). Antitumor activity of plant cannabinoids with emphasis on the effect of cannabidiol on human breast carcinoma. *The Journal of Pharmacology and Experimental Therapeutics*, *318*(3), 1375–1387.

Luciano, R. L., & Perazella, M. A. (2014). Nephrotoxic effects of designer drugs: Synthetic is not better! *Nature Reviews. Nephrology*, *10*(6), 314–324.

Malyshevskaya, O., Aritake, K., Kaushik, M. K., Uchiyama, N., Cherasse, Y., Kikura-Hanajiri, R., & Urade, Y. (2017). Natural (delta(9)-THC) and synthetic (JWH-018) cannabinoids induce seizures by acting through the cannabinoid CB1 receptor. *Scientific Reports*, *7*, 10516.

Mao, Y., Huang, Y., Zhang, Y., Wang, C., Wu, H., Tian, X., Liu, Y., Hou, B., Liang, Y., Rong, H., Gu, X., & Ma, Z. (2019). Cannabinoid receptor 2-selective agonist JWH015 attenuates bone cancer pain through the amelioration of impaired autophagy flux induced by inflammatory mediators in the spinal cord. *Molecular Medicine Reports*, *20*(6), 5100–5110.

Maslov, L. N., Khaliulin, I., Zhang, Y., Krylatov, A. V., Naryzhnaya, N. V., Mechoulam, R., De Petrocellis, L., & Downey, J. M. (2016). Prospects for creation of cardioprotective drugs based on cannabinoid receptor agonists. *Journal of Cardiovascular Pharmacology and Therapeutics*, *21*, 262–272.

Mechoulam, R., Ben Shabat, S., Hanus, L., Fride, E., Vogel, Z., Bayewitch, M., & Sulcova, A. E. (1996). Endogenous cannabinoid ligands—Chemical and biological studies. *Journal of Lipid Mediators and Cell Signaling*, *14*, 45–49.

Mechoulam, R., Shani, A., Edery, H., & Grunfeld, Y. (1970). Chemical basis of hashish activity. *Science*, *169*, 611–612.

Mir, A., Obafemi, A., Young, A., & Kane, C. (2011). Myocardial infarction associated with use of the synthetic cannabinoid k2. *Pediatrics*, *128*, e1622–e1627.

Mukhopadhyay, P., Baggelaar, M., Erdelyi, K., Cao, Z., Cinar, R., Fezza, F., Ignatowska-Janlowska, B., Wilkerson, J., van Gils, N., Hansen, T., Ruben, M., Soethoudt, M., Heitman, L., Kunos, G., Maccarrone, M., Lichtman, A., Pacher, P., & Van der Stelt, M. (2016). The novel, orally available and peripherally restricted selective cannabinoid CB2 receptor agonist LEI-101 prevents cisplatin-induced nephrotoxicity. *British Journal of Pharmacology*, *173* (3), 446–458.

Mukhopadhyay, P., Batkai, S., Rajesh, M., Czifra, N., Harvey-White, J., Hasko, G., Zsengeller, Z., Gerard, N. P., Liaudet, L., Kunos, G., & Pacher, P. (2007). Pharmacological inhibition of CB1 cannabinoid receptor protects against doxorubicin-induced cardiotoxicity. *Journal of the American College of Cardiology*, *50*, 528–536.

Mukhopadhyay, P., Pan, H., Rajesh, M., Bátkai, S., Patel, V., Harvey-White, J., Mukhopadhyay, B., Haskó, G., Gao, B., Mackie, K., & Pacher, P. (2010). CB1 cannabinoid receptors promote oxidative/nitrosative stress, inflammation and cell death in a murine nephropathy model. *British Journal of Pharmacology*, *160*(3), 657–668.

Munson, A. E., Harris, L. S., Friedman, M. A., Dewey, W. L., & Carchman, R. A. (1975). Antineoplastic activity of cannabinoids. *Journal of the National Cancer Institute*, *55*(3), 597–602.

Pabon, E., Rockwood, F., Norman, G. J., & de Wit, H. (2021). Acute effects of oral delta-9-tetrahydrocannabinol (THC) on autonomic cardiac activity and their relation to subjective and anxiogenic effects. *Psychophysiology*, e13955.

Pacher, P., Steffens, S., Haskó, G., Schindler, T. H., & Kunos, G. (2018). Cardiovascular effects of marijuana and synthetic cannabinoids: The good, the bad, and the ugly. *Nature Reviews Cardiology*, *15*, 151–166.

Pan, H., Mukhopadhyay, P., Rajesh, M., Patel, V., Mukhopadhyay, B., Gao, B., Haskó, G., & Pacher, P. (2009). Cannabidiol attenuates cisplatin-induced nephrotoxicity by decreasing oxidative/nitrosative stress, inflammation, and cell death. *The Journal of Pharmacology and Experimental Therapeutics*, *328*(3), 708–714.

Pérez-Olives, C., Rivas-Santisteban, R., Lillo, J., Navarro, G., & Franco, R. (2021). Recent advances in the potential of cannabinoids for neuroprotection in Alzheimer's, Parkinson's, and Huntington's diseases. *Advances in Experimental Medicine and Biology*, *1264*, 81–92.

Pertwee, R. G. (1997). Pharmacology of cannabinoid CB1 and CB2 receptors. *Pharmacology and Therapeutics*, *74*, 129–180.

Procaccia, S., Lewitus, G. M., Lipson Feder, C., Shapira, A., Berman, P., & Meiri, D. (2022). Cannabis for medical use: Versatile plant rather than a single drug. *Frontiers in Pharmacology*, *13*, 894960. https://doi.org/10.3389/fphar.2022.894960.

Robson, P. (2005). Human studies of cannabinoids and medicinal cannabis. *Handbook of Experimental Pharmacology*, *168*, 719–756.

Schmid, K., Niederhoffer, N., & Szabo, B. (2003). Analysis of the respiratory effects of cannabinoids in rats. *Naunyn-Schmiedeberg's Archives of Pharmacology*, *368*(4), 301–308.

Seely, K. A., Lapoint, J., Moran, J. H., & Fattore, L. (2012). Spice drugs are more than harmless herbal blends: A review of the pharmacology and toxicology of synthetic cannabinoids. *Progress in Neuro-Psychopharmacology & Biological Psychiatry*, *39*, 234–243.

Steele, G., Arneson, T., & Zylla, D. (2019). A comprehensive review of cannabis in patients with cancer: Availability in the USA, general efficacy, and safety. *Current Oncology Reports*, *21*(1), 10.

Tham, M., Yilmaz, O., Alaverdashvili, M., Kelly, M. E. M., Denovan-Wright, E. M., & Laprairie, R. B. (2019). Allosteric and orthosteric pharmacology of cannabidiol and cannabidiol-dimethylheptyl at the type 1 and type 2 cannabinoid receptors. *British Journal of Pharmacology*, *176*, 1455–1469.

Thompson, G. R., Fleischman, R. W., Rosenkrantz, H., & Braude, M. C. (1974). Oral and intravenous toxicity of delta9-tetrahydrocannabinol in rhesus monkeys. *Toxicology and Applied Pharmacology*, *27*, 648–665.

Thompson, G. R., Mason, M. M., Rosenkrantz, H., & Braude, M. C. (1973). Chronic oral toxicity of cannabinoids in rats. *Toxicology and Applied Pharmacology*, *25*, 373–390.

Thompson, G. R., Rosenkrantz, H., Schaeppi, U. H., & Braude, M. C. (1973). Comparison of acute oral toxicity of cannabinoids in rats, dogs and monkeys. *Toxicology and Applied Pharmacology*, *25*, 363–372.

Tse, R., Kodur, S., Squires, B., & Collins, N. (2014). Sudden cardiac death complicating acute myocardial infarction following synthetic cannabinoid use. *Internal Medicine Journal*, *44*, 934–936.

Valenti, C., Billi, M., Pancrazi, G. L., Calabria, E., Armogida, N. G., Tortora, G., Pagano, S., Barnaba, P., & Marinucci, L. (2022). Biological effects of cannabidiol on human cancer cells: Systematic review of the literature. *Pharmacological Research*, 106267. https://doi.org/10.1016/j.phrs.2022.106267.

Vidot, D. C., Powers, M., Gonzalez, R., Jayaweera, D. T., Roy, S., Dong, C., & Messiah, S. E. (2019). Blood pressure and marijuana use: Results from a decade of NHANES data. *American Journal of Health Behavior*, *43*(5), 887–897.

Von Der Haar, J., Talebi, S., Ghobadi, F., Singh, S., Chirurgi, R., Rajeswari, P., … Hassan, G. W. (2016). Synthetic Cannabinoids and their effects on the cardiovascular system. *The Journal of Emergency Medicine*, *50*(2), 258–262. https://doi.org/10.1016/j.jemermed.2015.10.009.

Wang, L. N., Xing, M. D., Qu, W. T., Wang, C. B., Liu, Z. Q., Han, J., Ren, W., & Qiao, Y. N. (2022). Impaired vessel relaxation response and increased infarct size in smooth muscle cannabinoid receptor 1 knockout mice. *Microvascular Research*, *139*, 104263.

Winstock, A., Lynskey, M., Borschmann, R., & Waldron, J. (2015). Risk of emergency medical treatment following consumption of cannabis or synthetic cannabinoids in a large global sample. *Journal of Psychopharmacology*, *29*, 698–703.

Wolfe, C. E., Wood, D. M., Dines, A., Whatley, B. P., Yates, C., Heyerdahl, F., Hovda, K. E., Giraudon, I., & Dargan, P. I. (2019). Seizures as a complication of recreational drug use: Analysis of the euro-den plus data-set. *Neurotoxicology*, *73*, 183–187.

Wu, T. C., Tashkin, D. P., Djahed, B., & Rose, J. E. (1988). Pulmonary hazards of smoking marijuana as compared with tobacco. *The New England Journal of Medicine*, *318*(6), 347–351.

Yeniocak, S., Kalkan, A., Yamanoğlu, A., Öztürk, S., Söğüt, Ö., & Metiner, M. (2021). The effects of synthetic cannabinoids on the cardiovascular system: A case-control study. *Turkish Journal of Emergency Medicine*, *21*(4), 198–204.

Young, A. C., Schwarz, E., Medina, G., Obafemi, A., Feng, S. Y., Kane, C., & Kleinschmidt, K. (2012). Cardiotoxicity associated with the synthetic cannabinoid, k9, with laboratory confirmation. *American Journal of Emergency Medicine*, *30*(1320), e1325–e1327.

Yun, J., Yoon, K. S., Lee, T. H., Lee, H., Gu, S. M., Song, Y. J., Cha, H. J., Han, K. M., Seo, H., Shin, J., Park, H. K., Kim, H. S., & Kim, Y. H. (2016). Synthetic cannabinoid, JWH-030, induces QT prolongation through herg channel inhibition. *Toxicology Research*, *5*, 1663–1671.

Zhao, J., Gao, X., Zhao, L., Wang, Y., Zhang, J., & Liu, S. (2022). Effects of cannabidiol on Parkinson's disease in a transgenic mouse model by gut-brain metabolic analysis. *Evidence-based Complementary and Alternative Medicine*, *2022*, 1525113.

Further reading

Barbaccia, M. L., Affricano, D., Trabucchi, M., Purdy, R. H., Colombo, G., Agabio, R., et al. (1999). Ethanol markedly increases "GABAergic" neurosteroids in alcohol-preferring rats. *European Journal of Pharmacology*, *384*, R1–R2.

Beattie, M. C., Maldonado-Devincci, A. M., Porcu, P., O'Buckley, T. K., Daunais, J. B., Grant, K. A., et al. (2017). Voluntary ethanol consumption reduces GABAergic neuroactive steroid (3α,5α)3-hydroxypregnan-20-one (3α,5α-THP) in the amygdala of the cynomolgus monkey. *Addiction Biology*, *22*, 318–330.

Besheer, J., Lindsay, T. G., O'Buckley, T. K., Hodge, C. W., & Morrow, A. L. (2010). Pregnenolone and ganaxolone reduce operant ethanol self-administration in alcohol-preferring p rats. *Alcoholism: Clinical and Experimental Research*, *34*, 2044–2052.

Bukke, V. N., Archana, M., Villani, R., Serviddio, G., & Cassano, T. (2021). Pharmacological and toxicological effects of phytocannabinoids and recreational synthetic cannabinoids: Increasing risk of public health. *Pharmaceuticals (Basel, Switzerland)*, *14*(10), 965.

DeFilippis, E. M., Bajaj, N. S., Singh, A., Malloy, R., Givertz, M. M., Blankstein, R., Bhatt, D. L., & Vaduganathan, M. (2020). Marijuana use in patients with cardiovascular disease: JACC review topic of the week. *Journal of the American College of Cardiology*, *75*(3), 320–332.

Giorgetti, A., Busardò, F. P., Tittarelli, R., Auwärter, V., & Giorgetti, R. (2020). Post-mortem toxicology: A systematic review of death cases involving synthetic cannabinoid receptor agonists. *Frontiers in Psychiatry*, *11*, 464. https://doi.org/10.3389/fpsyt.2020.00464.

Kunos, G., Bátkai, S., Offertáler, L., Mo, F., Liu, J., Karcher, J., & Harvey-White, J. (2002). The quest for a vascular endothelial cannabinoid receptor. *Chemistry and Physics of Lipids*, *121*(1-2), 45–56. https://doi.org/10.1016/s0009-3084(02)00145-7.

Ozturk, H. M., Yetkin, E., & Ozturk, S. (2019). Synthetic cannabinoids and cardiac arrhythmia risk: Review of the literature. *Cardiovascular Toxicology*, *19*(3), 191–197.

Rein, J. L. (2020). The nephrologist's guide to cannabis and cannabinoids. *Current Opinion in Nephrology and Hypertension*, *29*(2), 248–257.

Richards, J. R. (2020). Mechanisms for the risk of acute coronary syndrome and arrhythmia associated with phytogenic and synthetic cannabinoid use. *Journal of Cardiovascular Pharmacology and Therapeutics*, *25*(6), 508–522.

Stanley, C., & O'Sullivan, S. E. (2014). Vascular targets for cannabinoids: Animal and human studies. *British Journal of Pharmacology*, *171*(6), 1361–1378.

Weresa, J., Pędzińska-Betiuk, A., Mińczuk, K., Malinowska, B., & Schlicker, E. (2022). Why do marijuana and synthetic cannabimimetics induce acute myocardial infarction in healthy young people? *Cell*, *11*(7), 1142.

Yu, D. H., Zhang, L., Eiselel, J. L., Bertrand, D., Changeux, J. P., & Weight, F. F. (1996). Ethanol inhibition of nicotinic acetylcholine type alpha-7 receptors involves the amino-terminal domain of the receptor. *Molecular Pharmacology*, *50*, 1010–1016.

Part VII

Biomarkers, analysis and screening

Chapter 37

The DSM: 5 Criteria of cannabis use disorder: Methods and applications

Mahadev Singh Sen[a], Siddharth Sarkar[b], and Yesh Chandra Singh[b]
[a]*Department of Psychiatry, Institute of Human Behaviour and Allied Sciences, New Delhi, India,* [b]*Department of Psychiatry and NDDTC, Room no 4096, Teaching Block, All India Institute of Medical Sciences, New Delhi, India*

Abbreviations

AUDADIS	alcohol use disorder and associated disabilities interview schedule
CUDIT-R	cannabis use disorders identification test—revised
DSM-5	diagnostic and statistical manual of mental disorders, 5th edition
ICD	international classification of diseases

Introduction

Cannabis is the most commonly abused illicit drug worldwide with approximately 10 million people suffering from cannabis dependence across the globe (Degenhardt et al., 2013; Merz, 2018). According to the world drug report 2017, cannabis remains the most widely used drug globally, with nearly 3.8 percent adults across the world using cannabis and there are more than 180 million users in past year (World Drug Report: Global Overview of Drug Demand and Supply, 2017). Even for those who are treated for other substance use disorders (like opioid use disorder), cannabis remains one of the most prominent substances being taken concomitantly. Although there are limited and variable treatment options, people seeking treatment for the first time at global level for cannabis use disorders remain high, with nearly 50% of those who are treated having cannabis use disorders. It has also been seen that more and more younger people are seeking treatment for cannabis use disorders. In fact, cannabis continues to be the most cultivated drug worldwide and the largest quantities of drugs seized globally were also of cannabis. High-income countries seem to be having highest prevalence of cannabis use, but the prevalence is gradually increasing in low- and middle-income countries as well. While the public health burden of cannabis may be lesser than other substances, the overall high prevalence of cannabis use translates into considerable number of individuals globally being consumers of cannabis (Connor et al., 2021), and some of them suffer from cannabis use disorders.

Different terminologies have been used in identifying problematic use of cannabis, viz., abuse, dependence, and use disorder. Various psychiatric classificatory systems have been devised for identifying problematic cannabis use, and they have been revised time and again as ongoing research provided new insights into the phenomenon. The two major diagnostic systems include the DSM and ICD. They provide classification systems that are relevant to clinical diagnosis, public health, research pursuits, and service provision. DSM is commonly used in the United States and various other countries across the world for various psychiatric illnesses including substance use disorders.

The first edition of DSM, i.e., DSM-I was conceived in 1952 which discouraged multiple diagnoses and used "reaction" for describing the illnesses. In 1968, DSM-II was launched which removed the word reaction and encouraged multiple diagnoses. The first time that inclusion and exclusion criteria were defined was in DSM-III in 1980. The fourth edition of DSM was launched in 1994 with subsequent text revision (DSM-IVTR) in 2000 with nearly 300 psychiatric diagnoses (Zimmerman & Spitzer, 2009). The latest edition is that of DSM-5, launched in 2013. The fifth edition of DSM also had Arabic number "5" replacing roman numeral "V" in its name. This change created a discontinuity from previous DSM naming conventions, conveying the significant changes made in this version of DSM (Surís et al., 2016). Of the various prominent changes from DSM-IV to DSM-5 was introduction of spectrum concept which also applied to the diagnosis of substance use disorders. The major change was elimination of distinction of abuse and dependence (Clark et al., 2017).

Cannabis Use, Neurobiology, Psychology, and Treatment. https://doi.org/10.1016/B978-0-323-89862-1.00008-8

Some of the notable changes in the diagnosis of substance use disorder include: deletion of recurrent legal problems criterion for abuse, addition of craving criteria, reducing the threshold from ≥3 criteria for dependence to 2 or more criteria for use disorder, addition of withdrawal syndromes in cannabis and caffeine, deletion of specifier for physiological subtype and polysubstance dependence, addition of specifiers "in a controlled environment" and "on maintenance therapy" (American Psychiatric Association, 2013).

Many users of cannabis do not fulfill the criteria for substance use disorder. It is thus important to understand and assess cannabis use and cannabis use disorders, and how DSM-5 criteria for cannabis use disorders may be applicable. In this chapter, we intend to discuss the changes in DSM-5 criteria for substance use, especially cannabis, the reasons behind these revisions of criteria for cannabis use disorder in DSM-5, and how it may be applicable in clinical and research settings. We will also throw some light upon the research comparing DSM-IV and DSM-5 criteria for cannabis use disorder, limitations in the current criteria, and controversies related to the new criteria for cannabis use disorders.

Inception of new DSM-5 criteria for cannabis use disorder

DSM is one of the established classification systems of mental disorders that is used for clinical, research, and policymaking in various regions across the globe (Hasin et al., 2013). It thus has widespread importance and influence on how psychiatric and substance use disorders are diagnosed, investigated, and treated. It was first published in 1952 and has undergone multiple revisions ever since incorporating our advancement of understanding of mental illnesses. Substance use criteria in DSM-IV were considered reliable and valid for establishing substance dependence but were majorly criticized for two reasons: divisions into dependence and abuse and not having a specifier for severity of dependence. Abuse could be diagnosed as per DMS-IV when dependence could not be established indicative of social and interpersonal consequences of heavy use (abuse was placed lower in hierarchy than dependence apparently). There were issues with validity and reliability of abuse, issues of diagnostic orphans, consideration of continuum for diagnostic purposes and threshold criteria (Compton et al., 2009; Grant et al., 2007; Hasin et al., 2006; Hasin et al., 2013; Hasin & Paykin, 1998; Martin et al., 2006; Muthén, 2006).

So, removing the major drawback of DSM-IV led to new set of criteria for a single diagnosis substance use disorders, including cannabis use disorder. Another major change was addition of withdrawal syndrome for cannabis. The criteria for cannabis use disorder in DMS-5 include an 11-item severity index. The 11 items include: consuming cannabis longer than intended, persistent desire/difficulty cutting down, more time consumed in consumption of or recovery from the effect of cannabis, craving/strong desire, recurrent substance consumption leading to failure of fulfilling major obligations at home/work/school, consumption despite substance causing social and interpersonal trouble, giving up on other recreational and pleasurable activities due to substance, getting into hurtful/unsafe situation because of substance, consumption despite physical/psychological harm, tolerance, withdrawal, preoccupation. The severity is graded into "mild," "moderate," "severe." Those who meet 2–3 criteria are classified into "mild" category, those who fulfill 4–5 criteria are classified as "moderate," while those with six or more criteria are classified as "severe" (American Psychiatric Association, 2013; Kopak, Proctor, & Hoffmann, 2014). The key changes from DSM-IV to DSM-5 are presented in Table 1.

TABLE 1 Major Changes in DSM-5 criteria specific to cannabis.

Major change	What led to those changes
Removal of abuse and dependence distinction	Abuse was not as well validated as dependence
Addition of single diagnosis of cannabis use disorder	Simple and parsimonious model, emphasizing that the quality of disorder is the same
Dropping off the criteria on legal problems	This criterion was found to be less prevalent in research/literature
Addition of withdrawal and craving criteria	Based on available literature showing higher prevalence
Change in course specifiers	Former course specifiers (DSM-IV) seldom used in practice
Grading severity based on the number of criteria met	SUDs considered on a continuum/as a unidimensional construct
Addition of cannabis withdrawal syndrome	Studies have shown validity of physical and physiological withdrawal symptoms in frequent and heavy cannabis users

DSM, diagnostic and statistical manual of mental disorders; *SUD*, substance use disorder.

DSM-5 workgroup on devising criteria for cannabis use disorder

Using a collection of 2006 reviews at the outset, the substance use disorder work group observed weaknesses, identified data sets to explore possible solutions to the issues in DSM-IV, and performed analyses to address knowledge gaps, monitored new research pertaining to the field and made interim recommendations for changes to DSM 5. The workgroup obtained input on the proposed changes through commentaries, expert advisers, the website of DSM-5, and presentations and symposiums at over thirty meetings (Hasin et al., 2013). The following issues were addressed (depicted in Table 1 as well):

(1) Are dependence and abuse two separate diagnoses?

The division of substance use into two disorders (abuse and dependence) was driven by the understanding that dependence formed one aspect of substance-related problems, while interpersonal and social consequences of heavy use were another (Rounsaville et al., 1986). Although these were supposed to be related, DSM-IV had seemingly placed dependence higher than abuse in the hierarchy. It suggested that if dependence was present, abuse should not be diagnosed. Some problems were noted in the classification. It was seen that when diagnosed by DSM-IV, the reliability and validity of dependence were much higher than those for abuse. Also, a syndrome is a constellation of more than one symptom, while most diagnoses of abuse were made with only one criterion (primarily hazardous use). Furthermore, although abuse is hierarchically below dependence, and assumed to be less severe, some of the abuse criteria are suggestive of severe problems (e.g., inability to fulfill major responsibilities). Additionally, some common assumptions were shown to be false, e.g., that abuse is simply a prodrome to dependence and that patients with dependence also fulfill criteria for abuse. Also, there were many cases of "diagnostic orphans," i.e., fulfilling only two criteria for dependence and not fulfilling any criteria for abuse (Hasin & Paykin, 1998). These cases could have significant complications arising out of substance use, but still unable to receive a diagnosis. Finally, factor analysis of the criteria of both abuse and dependence showed that it formed a single factor or two strongly correlated factors (Krueger et al., 2004; Martin et al., 2006). The workgroup examined thirty-nine studies (n>200,000) on the item response theory. Two main findings appeared, with similar results across various substances and demographics. First, all DSM-IV criteria for abuse and dependence were found to be of the same dimension except legal problems, indicating that both dependence and the rest of the abuse criteria, are indicative of a common underlying condition. Second, abuse and dependence criteria were always distributed across the severity spectrum. Thus, this large body of evidence backed combining abuse and dependence criteria.

(2) Should some criteria be added or dropped?

The group also planned to drop legal problems from the criteria due to meagre rates of occurrence in many samples, low discrimination, poor conformity with other criteria, and little added information in analyses of item response theory. Item response theory was also used to analyze craving and its relationship with other criteria. Adding craving to the diagnostic criteria added significant information in the case of cannabis, nicotine, alcohol, and opioid use (Keyes et al., 2011). Also craving was already a part of ICD-10 diagnostic criteria, and hence introducing it to DSM 5 criteria increased consistency between the two diagnostic systems.

(3) What should the diagnostic threshold be?

Presence of two or more criteria for the diagnosis of substance use disorder was agreed upon due to greater agreement, ranging from very good to excellent, between studies was observed.

(4) Specifiers

Physiological cases—This specifier was present in DSM-IV, but it had inconsistent predictive value. It was observed that this specifier was used only in the studies investigating its validity, not outside it, questioning the utility during large-scale application (Agrawal et al., 2014).

Course—The specifiers for time frame and completion of remission were complicated and seldom used. Partial remission was eliminated and the time frame was divided into two categories, sustained and early.

(5) Is cannabis withdrawal a distinct diagnostic entity?

The cannabis withdrawal syndrome has a short-term course after stoppage of cannabis use and pharmacological specificity. It has been reported by about one-third of regular cannabis users and by 50%–95% of heavy users. The importance of this withdrawal is seen by use of cannabis or other substances to alleviate it, its presence leading to difficulty quitting, and worse outcomes seen if withdrawal is of higher severity. In latent variable modeling, the addition of withdrawal to the diagnostic criteria of cannabis use disorder improved the model fit (Copersino et al., 2006; Hasin et al., 2013).

Concordance/compatibility of DSM-5 with DSM-IV and other classificatory systems

With the introduction of DSM-5, it was anticipated that the prevalence of diagnosable cases of substance use disorder may range from $<1\%$ to $>11\%$. Many individuals who met criteria for abuse without meeting dependence criteria may not receive any DSM-5 diagnosis because they would not meet the minimum threshold of two criteria for mild substance use disorder. Similarly, those who met abuse criteria with one or two dependence criteria in DSM-IV may receive a diagnosis of mild substance use disorder. Similarly, those with two or more abuse criteria and two or three dependence criteria, or those who met three dependence criteria with no abuse criteria may be classified under moderate substance use disorder (Kopak, Proctor, & Hoffmann, 2014).

A study was conducted on prison inmates to compare DMS-IV cannabis abuse and dependence designations with moderate and severe cannabis use disorder diagnoses of DSM-5 (Kopak et al., 2012). The authors also assessed the need for giving weightage to different criteria and whether any additional criteria, which is not a part of current classification, needed to be included. They observed that almost all those who were not diagnosed according to the DSM-IV could also not receive diagnosis under the proposed DSM-5 criteria, and of those who were diagnosed with cannabis dependence on DSM-IV, almost all met the criteria for severe cannabis use disorder under DSM-5. The study also indicated that all criteria do not have equal weightage as some may appear as important indicators of severe cannabis use disorder (e.g., withdrawal, relinquishing activities for substance use, and craving in severe cannabis use disorder) while others may appear more frequently among moderate cannabis use disorder (e.g., use in hazardous conditions and interpersonal conflicts). Kelly et al. (2014) assessed 525 adolescents for substance use disorders. Fifty-six (10.7%) subjects met the criteria for DSM-5 cannabis use disorder (Kelly et al., 2014). However, DSM-IV diagnosis of dependence was met by 43 (8.2%) participants. Of the 36 adolescents who could not be diagnosed as per DSM-IV cannabis dependence, 13 (36.1%) met criteria for DSM-5 cannabis use disorder. Only one adolescent who met criteria for DSM-IV abuse did not meet the criteria for DSM-5 cannabis use disorder due to discarding of the legal problems in DSM-5. Goldstein et al. 2016 examined the concordances between DSM IV and DSM 5 substance use disorders using the data from NSEARC III (Goldstein et al., 2015). Prevalence of any DSM-5 substance use disorders was 13% lower than those of DSM-IV dependence and abuse combined for cannabis. But prevalence of moderate to severe DSM 5 cannabis use disorder was 43% which was higher than DSM IV dependence. Concordances of DSM 5 moderate to severe use and DSM IV dependence were significant ($\kappa=0.78$). Concordance was moderate between DSM-IV dependence and any DSM 5 use disorder ($\kappa=0.48$).

A cross-sectional household survey was conducted to establish concordance in the diagnoses of alcohol use disorder ($n=12{,}182$ regular alcohol users) and cannabis use disorder ($n=1{,}788$ cannabis users) made using DSM-5, DSM-IV, International Classification of diseases—tenth edition (ICD-10), International Classification of diseases—eleventh edition (ICD-11) in 10 countries of the World Mental Health Survey Initiative (Degenhardt et al., 2019). Among cannabis users, nearly 9% met criteria for ICD-11 harmful use and approximately 3% for dependence. Excellent concordance was noted for both alcohol and cannabis diagnoses between ICD-11 with ICD-10 and DSM-IV ($\kappa \geq 0.9$). Concordance between ICD-11 and DSM-5 ranged from poor (for comparisons of harmful use and mild substance use disorders) to good (for substance use disorder and comparisons of dependence and severe substance use disorder). The authors concluded that the ICD-11 classifications for substance use disorder diagnoses were highly consistent with the ICD-10 and DSM-IV. Concordance between ICD 11 and DSM-5 was found to be variable, mainly due to low levels of agreement between DSM-5 mild use disorder and the ICD harmful use. Another study has shown good concordance between ICD-11 and DSM-5 for cannabis comparing dependence and moderate to severe substance use disorders but poor for ICD-11 harmful use and DSM-5 mild substance use disorder (Chung et al., 2017). Therefore, the prevalence of cannabis use disorder as per the proposed ICD-11 classification was lower as compared to DSM-5, but the prevalence of cannabis dependence as per proposed ICD-11 was higher than DSM-5. The authors concluded ICD-11 and DSM-5 substance use disorder diagnoses showed only moderate concordance.

Lago et al., 2016 collected data from the National Survey of Mental Health and Well-Being (NSMHWB). The prevalence was slightly lower in DSM 5 as compared to DSM IV (28.1% vs. 32.2%). Cannabis use disorder in DSM-5 and ICD-11 had only moderate agreement. ICD-11 dependence syndrome and DSM-5 moderate to severe cannabis use disorder also had moderate agreement. Concordance between DSM-IV dependence and DSM-5 severe or moderate severe use disorders was good ($\kappa=0.78$). It was found that DSM 5 as compared to other diagnostic systems was able to identify a greater number of cases but with lesser severity.

In India, a cross-sectional study was conducted on 100 out-patients in a tertiary care center with recent history of cannabis use in the past (Sarkar et al., 2020). The patients were evaluated for cannabis use disorder as per DSM-5 and cannabis dependence as per ICD-10. They were also evaluated for cannabis withdrawal. Fifty-eight patients met the criteria for cannabis dependence according to ICD-10, as against 74 patients who met DSM-5 criteria for cannabis use disorder.

So, to summarize, various studies have investigated concordance of DSM-5 with various other diagnostic criteria. The findings have been variable from moderate to good concordance with different criteria. In most studies, the using criteria for cannabis use disorder as per DSM-5 have diagnosed more individuals being studied with the disorder than while applying other criteria, say DSM-IV, ICD-10, or ICD-11.

Prevalence of various DSM-5 criteria for cannabis use disorder

Fig. 1 presents the schematic representation of the various criteria for cannabis use disorder, which can be further found in the text of the DSM-5. The prevalence of different criteria has been evaluated by certain studies.

National Epidemiologic Survey on Alcohol and Related Conditions-III (NESARC III) was conducted with a population of noninstitutionalized civilians. The diagnostic interview used was the NIAAA Alcohol Use Disorder and Associated Disabilities Interview Schedule-5 (AUDADIS-5). Psychiatric Research Interview for Substance and Mental Disorders, DSM-5 version (PRISM-5), a clinician administered instrument was used to assess procedural validity. The twelve-month prevalence of cannabis use disorder was 2.54% (n=972) and lifetime prevalence of cannabis use disorder as per DSM-5 was 6.27% (n=2242). The 12-month and lifetime prevalence of mild, moderate, and severe cannabis use disorder were 1.38%, 0.59%, and 0.57% and 2.85%, 1.42%, and 2.00%, respectively (Hasin et al., 2016).

In a study by Chung et al., 2018 (Chung et al., 2018), adolescents (ages 14–18) (n=339) who were admitted to community-based intensive outpatient (IOP) treatment for addictions were recruited to participate in research interviews for assessment of drug use. Structured clinical interview for DSM IV (SCID) was used for assessment. 82.8% participants fulfilled the criteria for any DSM-5 cannabis use disorder. The prevalence of mild, moderate, and severe cannabis use disorder was 27.1%, 26.8%, and 28.9%, respectively. In the past year, the most common symptoms were continuing despite problems in relationships (75.2%), followed by spending a lot of time (63.7%), tolerance (58.7%), role impairment (58.4%), use in hazardous situations (33.6%), withdrawal (31.3%), inability to quit/cut down (29.2), craving (27.4%), reducing other activities (13%), use in larger amounts/longer duration (10.6%), and psychological/physical problems (4.7% each).

Degenhardt et al. (2019) found the following prevalence for cannabis use disorder—impaired control (12.4%), use in larger amounts or longer duration (11.1%), centrality of drug use (8.0%), hazardous use (7.8%), tolerance/withdrawal (7.6%), continuous use despite harm (6.5%), interpersonal problems (5.6%), tolerance (5.5%), role impairment (5.4%), craving (5.2%), withdrawal (4.9%), and increased time spent (4.3%).

Mewton et al. (2013) compared dependence criteria in DSM IV with DSM 5 and reported symptom profiles of individuals with a cannabis use disorder in both DSM-IV and DSM-5. They reported most common symptom was Hazard Use 73%. Other criteria/symptoms assessed were tolerance (58%), major role impairment (47%), social problems (46%), withdrawal (45%), use for longer duration or larger amounts (76%), inability to cut down (15%), increased time spent (34%), giving up important activities (24%), continuation despite harmful consequences (40%), and craving (46%).

A study by Sarkar et al. (2020), recruited 100 purposively sampled adult participants who had used cannabis at least on one occasion in the past 6 months. The prevalence of mild, moderate, and severe cannabis use disorder was 21%, 21%, and 32%, respectively. The most common symptoms of cannabis use disorder were tolerance (76%), craving (71%), withdrawal (55%), hazardous use (47%), unable to quit/cut down (40%), use for larger/longer duration than intended (30%), time spent (21%), interpersonal problems (19%), role impairment (15%), and reduced activities and psychological/physical problems

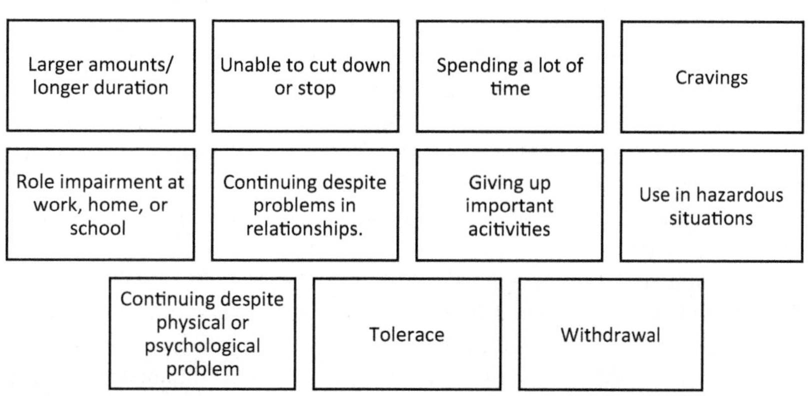

FIG. 1 Schematic representation of the DSM5 criteria of cannabis use disorder. Key terms related to the diagnosis are mentioned. The reader is referred to the text of DSM-5 (American Psychiatric Association, 2013) for details of the criteria.

TABLE 2 Different criteria of DSM5 diagnosis met in a few representative studies.

	Mewton et al. (2013)	Chung et al. (2018)	Degenhardt et al. (2019)	Sarkar et al. (2020)
Major role	47%	58%	5%	15%
Hazard	73%	34%	8%	47%
Social	46%	75%	8%	19%
Tolerance	58%	59%	6%	76%
Withdrawal	45%	31%	5%	55%
Use in larger amounts/for longer duration	76%	11%	11%	30%
Inability to cut down	15%	29%	12%	40%
Time spent	34%	64%	4%	21%
Craving	46%	27%	5%	71%

(10% each). The commonest withdrawal symptoms experienced were sleep difficulty (62%), decreased appetite or weight loss (59%), irritability, anger, or aggression (44%), restlessness (39%), depressed mood (29%), nervousness (25%), and at least one physical symptom causing significant discomfort (21%). Among 100, only 29 people experienced withdrawal symptoms that were severe enough to cause significant distress or impairment in social, occupational, or other important areas of functioning (Table 2).

Compton et al. (2013) while assessing agreement between DSM-IV and DSM-5 recruited nationally representative US adult sample. Past year users of cannabis (n=1622) were also assessed as a part of this study. They found hazardous use of cannabis to be the most common criteria being met followed by trying to stop and craving. Withdrawal was noted in less than 10% of individuals.

Different studies, as highlighted above, have shown differences in the prevalence of various criteria while diagnosing cannabis use disorder in a given population. These differences could be attributable to the form of cannabis being used, the context of cannabis use, differences in the settings, presence of comorbidities in the form of other substances or psychiatric illnesses, interviewer's bias, or simply cultural or regional variations.

Assessment instruments based on DSM-5 for cannabis use disorder

With the introduction of DSM-5 diagnostic criteria for cannabis use disorder including addition of cannabis withdrawal and craving, it was considered imperative to develop screening tools and other instruments according to the latest classification for identifying cannabis use disorder in clinics. Adamson et al., 2010 developed 8 item screening tool Cannabis Use Disorders Identification Test-Revised (CUDIT-R) to identify last 6 months of problematic cannabis use (Adamson et al., 2010). It had demonstrated good sensitivity and specificity when cannabis use disorders were identified by DSM IV criteria. When it was used along with DSM 5 criteria, increasing severity of the diagnosis was associated with greater CUDIT-R scores (Schultz et al., 2019). However, it is not considered ideal for application in busy clinical settings. Moreover, it was primarily developed in accordance with DSM-IV diagnostic criteria (which does not consider cannabis withdrawal and craving) (Bonn-Miller et al., 2016). Therefore, Bonn-Miller et al. (2016) aimed to develop a shortened version of CUDIT-R (CUDIT-SF) in accordance with DSM-5 criteria by utilizing item response theory for application in busy clinical settings (Bonn-Miller et al., 2016).

A completely structured computer-assisted diagnostic interview, the Alcohol Use Disorder and Associated Disabilities Interview Schedule (AUDADIS) has been designed for trained lay interviewers for assessment of substance use disorders. The AUDADIS-5 developed in 2011 (Grant et al., 2015) assesses diagnoses as per DSM-5 criteria, and was used in NSEARC-III. For its reappraisal, PRISM-5 was used. For the past one-year dimensional DSM-5 cannabis use disorder measures, intra class coefficients (ICC) indicated excellent AUDADIS-5/PRISM-5 concordance (ICC=0.79). For lifetime dimensional measures, excellent concordance was seen for cannabis use (ICC=0.78) (Hasin et al., 2015).

TABLE 3 Application of the DSM-5 criteria for cannabis use disorder.

Clinical application
• Diagnosis when cannabis use disorder presenting as the only condition
• Diagnosis when cannabis use disorder is present with additional substance use disorder(s)
• Diagnosis when cannabis use disorder is present with another psychiatric disorder (like schizophrenia)
Public health application
• Workplaces, to assess employees, and maybe coupled with laboratory investigations
• Educational institutions for monitoring students
• When recreational cannabis becomes problematic
Research application
• Assessing epidemiology of cannabis use disorder
• Important selection/outcome measure in cannabis therapeutic trials

Application of current criteria

The application of cannabis use disorder diagnostic criteria has relevance in clinical, public health, as well as research settings (summarized in Table 3). In clinical settings, cannabis use may be found among those coming for treatment of other substance use disorders, other psychiatric illnesses, or primarily cannabis use disorder. In addition to these, there may be individuals using cannabis for medical purposes (medical marijuana). Due to lowered threshold of diagnosing cannabis use disorder in DSM-5 as compared to dependence in previous classificatory systems, many more individuals may be diagnosed with a disorder. This was also seen in our study, where out of 100 individuals using cannabis 58 met diagnosis of dependence syndrome as per ICD-10 and 74 met criteria for cannabis use disorder as per DSM-5 (Sarkar et al., 2020). However, craving and tolerance in a medically ill patient taking medical marijuana, or an individual with psychiatric illness with cannabis use which may fulfil criteria for use despite psychological harm, and impairment in functioning may not be the candidates for focused treatment of cannabis use disorder, despite receiving a diagnosis of mild cannabis use disorder. This may impact the clinical workload of the said settings. On the contrary, even without meeting criteria for cannabis use disorder, cannabis use in a patient with psychiatric illness, like psychosis, or mood disorder, cannot be withheld the benefits of intervention.

It may also be difficult to comprehend if the same criteria can apply in different contexts, for example, recreational versus work. It has been seen that cannabis use is associated with certain arts like music (Pedersen, 2009). Some artists may prefer using cannabis before performing which may enhance their performance and work. It may be a challenge in applying criteria for cannabis use disorder as their functioning/work performance may actually be enhanced rather than impaired on cannabis intake. Similarly, it is also common to find adolescents using cannabis in school settings (Heyman et al., 1999). The prevalence of diagnosable cannabis use disorder in these populations may eventually affect the policies of the school or nation as any student with some decline in scholastic performance and craving may be labeled with cannabis use disorder.

In research settings, the change in criteria may affect the epidemiology of cannabis use disorder. After legalization of cannabis, coupled with modifications in DSM criteria over years for cannabis use disorder, a rising prevalence of cannabis use disorder may be seen (Cerdá et al., 2020), and it can lead to difficulty distinguishing the increase attributable to diagnostic systems or actual prevalence of cannabis use disorder. This may affect the laws and policies of the nation, as well as the clinical application of the research findings. For example, clinical trials being taken up on population with cannabis use disorder might need to be tailored to the severity of cannabis use disorder. Any drug development underway for cannabis dependence in the previous or ongoing trials may need replication in less than severe cannabis use disorder diagnosis as per DSM-5, as studies have shown that cannabis dependence as per DSM-IV mostly falls in the severe cannabis use disorder category in DSM-5 (Compton et al., 2013). In the absence of any significant evidence currently on pharmacological management for cannabis use disorder, it may be challenging for further designing research to explore this area.

Another challenge in application cannabis use disorder criteria of DSM-5 would be in different forms of cannabis preparation, e.g., edible or smoked. The potency of and potential dependence on a particular cannabis preparation may vary that may lead to discrepancies in applying the same criteria for all preparations. Application of the criteria may be muddled when there is a presence of a comorbid psychiatric or medical condition. For example, role impairment in social and occupational functioning can be due to cannabis use due to comorbid major depressive episode (a psychiatric disorder) or systemic lupus erythematosus (a medical illness).

The threshold of diagnosis is something that is of relevance while making a diagnosis of cannabis use disorder. Different clinicians can appraise the impairment criteria differently to make a diagnosis. There may be some subjectivity in how the appraisal of impairment occurs in the clinical setting and what is considered as impairment. For example, considering educational impairment with an individual using cannabis, whether missing a single class due to cannabis use would constitute impairment or whether the threshold should be missing several classes and/or with falling grades. This subjectivity is likely to remain while making a diagnosis

Cannabis withdrawal syndrome: Insights and controversies

Cannabis withdrawal also figures now in the DSM-5, which is closely aligned to cannabis use disorder. Cannabis withdrawal syndrome manifests three or more of the following symptoms are present: sleep difficulty, irritability, nervousness/anxiety, decreased appetite or weight loss, depressed mood, and one of the following physical symptoms such as shakiness/tremors, abdominal pain, sweating, chills, fever, or headache (American Psychiatric Association, 2013). Unlike other substances like alcohol and opioids, the withdrawal syndrome in cannabis is relatively less clear. Two types of cannabis withdrawal have been described in the literature: (i) Type A: peaks between 2nd and 6th day of abstinence (unmasked after the intoxication symptoms abate); (ii) Type B: continues to decrease with time after abstinence. It has been seen that naturalistic severity of cannabis withdrawal varies a lot depending upon the amount, and type (potency) of cannabis preparation, gender, environment, and other hereditary factors. Some other symptoms have also been studied and reported in cannabis withdrawal syndrome, which are not yet a part of DSM-5 criteria, like eating disorder-like picture, nausea and vomiting (Bonnet & Preuss, 2017). While some emphasize in-patient detoxification of cannabis use disorder patients who have moderate to severe dependence, certain countries like Germany do not advocate in-patient treatment of cannabis use disorder and consider cannabis withdrawal syndrome to be a inconsequential. Some studies have shown that presence of cannabis withdrawal syndrome does not impact the outcome of or abstinence from cannabis among patients dependent on cannabis (Arendt et al., 2007).

Some studies in the past, before the launch of DSM-5, have shown that nearly 90% patients with cannabis dependence experience cannabis withdrawal syndrome (Budney et al., 2004; Budney & Hughes, 2006; Greene & Kelly, 2014). However, more recent studies have variable findings with respect to prevalence of cannabis withdrawal syndrome amongst those seeking treatment for cannabis dependence, ranging from 35% to 75% (Allsop et al., 2011; Allsop et al., 2012; Gorelick et al., 2012). It has in fact been seen that presence of cannabis withdrawal syndrome has been considered a specific indicator of cannabis dependence (Greene & Kelly, 2014).

A study was done on 384 adults of cannabis smokers who were not seeking treatment to assess the onset, prevalence, and peak intensity of cannabis withdrawal symptoms (Gorelick et al., 2012). The authors found that nearly 41% individuals experience cannabis withdrawal symptoms (checklist of 39 symptoms derived from literature). Those with withdrawal symptoms had shorter abstinence duration. The prevalence was unchanged after dropping the physical symptom criteria of DSM-5 of cannabis withdrawal syndrome. The authors thus supported the concurrent and predictive validity of cannabis withdrawal syndrome, they proposed revising the list of symptoms diagnosing cannabis withdrawal syndrome in DSM-5 as well as the number of symptoms required to reach a diagnosis (Allsop et al., 2012).

Limitations in the current DSM-5 classification in the context of cannabis use disorder

While there are welcome changes in the current DSM-5 classification in the context of substance use disorders, most changes are not specific to cannabis. Data from Australian National Survey of Mental Health and Wellbeing showed that unlike other substances prevalence of cannabis use disorder reduced in the transition from DSM-IV to DSM-5, possibly due to high ratio of cannabis abuse to dependence when compared with other substances (Mewton et al., 2013).

A criticism that is common to the change in substance use disorder diagnosis in general is whether substance use disorder should be considered unidimensional disorders with all criteria being given equal weightage (Kopak, Proctor, & Hoffmann, 2014). This is particularly relevant to cannabis which comes in various forms with different potencies. Hence, the dependence potential may be different (Ambekar et al., 2019). There is much more variation in how milder cases will be addressed with DSM-5 as compared to DSM-IV (Kopak, Proctor, & Hoffmann, 2014). Moreover, the DSM-5 task force chose a simple parsimonious approach of grading severity into mild, moderate, and severe. Other ways of classifying severity were possible which may not have been considered, e.g., level of use, impairment, or comorbidity (Hasin et al., 2013).

The current criteria have many implications in treatment. It assumes that substance use disorders are qualitatively similar but vary in severity. While DSM-5 is considered an advancement in the diagnostic system, the answer to the question of how it will inform treatment is yet to be answered (Kopak, Proctor, & Hoffmann, 2014). Moreover, different studies have shown differences in the most commonly endorsed DSM-5 criteria. For example, a study reported that craving and tolerance were the most common while withdrawal was reported by only half of those with cannabis use disorder (Sarkar et al., 2020), while Compton et al. (2013) reported hazardous use to be the most common criterion while craving being third most common. A study was conducted from a nationally representative US adult sample assessing agreement between DSM-5 and DSM-IV for alcohol, opioid, cocaine, and cannabis (Compton et al., 2013). It was found that optimal concordance occurred for DSM-IV cocaine, alcohol and opioid dependence when four or more DSM-5 criteria were met, which corresponded to moderate DSM-5 severity threshold. However, for cannabis use disorder, concordance of DSM-5 cannabis use disorder and DSM-IV cannabis dependence occurred optimally when six or more criteria were met, corresponding to the threshold for severe DSM-5 disorder. Therefore, the DSM-5 substance use disorder workgroup while making recommendations for DSM-5 did acknowledge that the research must continue to identify the most effective indicators of substance use disorders including cannabis (Schuckit & Saunders, 2006). Studies have recently been conducted on the newly added entity of cannabis withdrawal syndrome and further changes have been proposed by various authors on improvising the criteria, for example, adding criterion of vivid, unpleasant dreams (Hesse & Thylstrup, 2013).

Future research

Future research may be undertaken in validating the criteria, comparing the impact of changing the criteria in cannabis versus other substances, understanding the clinical and research usage of this criteria in cultural context, for instance comparing nations which have legalized edible cannabis products (e.g., "Bhang" in India) versus those which allow smoked marijuana for recreational purposes (e.g., various states in the United States like Columbia, California, Virginia, etc.). With the launch of ICD-11, comparing the two diagnostic systems for understanding the epidemiology of the disorder and compatibility of the two systems may be worthwhile. Studies have also evaluated new criteria like preoccupation and consumption to relieve stress in alcohol use disorder (Kopak, Metze, & Hoffmann, 2014). Similar studies may be taken up in cannabis use disorder as well. Since there is limited evidence of effective evidence-based pharmacological treatment available for cannabis use disorders yet, it may be worthwhile in using the DSM-5 criteria with severity ratings as inclusion and outcome criteria for studies. The user perspective (mental health professionals) on utility and ease of application of the DSM-5 criteria for cannabis use disorders may also be studied.

Conclusion

The major revisions in DSM-5 criteria for cannabis included combining cannabis abuse and dependence into cannabis use disorder, with addition of cannabis withdrawal syndrome. The classification has moved from a dichotomy of apparently milder abuse category and severe dependence category in DSM-IV to dimensional category in DSM-5. While most changes are in positive direction confirming uniformity with other substance use disorders certain limitations in the current criteria continue to be there. Hence, the research must continue to identify most specific indicators of cannabis use disorders and improvise the current criteria. The current criteria may continue to change and be modified over time thus influencing policies, management guidelines, and laws. At the same time, applications of DSM-5 diagnostic criteria for cannabis use disorders should be rigorously explored in various clinical and nonclinical settings.

Applications to other areas

In this chapter, we have reviewed the methods and applications of DSM-5 diagnosis of cannabis use disorder. Cannabis use disorders in the clinical setting can be diagnosed using various means, i.e., by clinical interview, patient-reported self-assessment and other methods. DSM-5 was developed after refinement of the previous assessment criteria (the DSM-IV-TR). This suggests that the diagnostic systems and criteria may need updating from time to time, cognizant of the emerging literature in the field, and feedback from the clinical use scenarios. This has potential applicability for the International Classification of Diseases as well, which has largely retained the previous dependence and harmful use as cannabis use disorder diagnoses. Harmonization or diagnostic concurrence should be evaluated so that interpretations from different studies becomes possible. This has application for other substance use disorders as well, where the diagnostic criteria may apply slightly differently. Furthermore, variability in interpretation of diagnostic threshold has applicability for other psychiatric conditions, since what should be considered as impairment is somewhat subjective. Scale development processes for determining cannabis use disorder or problematic cannabis and other substance use disorder need to be cognizant of the issues and limitations of the present diagnostic system, so that the utility and applicability of the scale can be opined about.

Mini-dictionary of terms

- **Cannabis dependence**. The diagnostic syndromal construct of cannabis usage which is problematic and affects functioning of the individual.
- **Cannabis use disorder**. The diagnostic label for problematic cannabis use which has been introduced in the DSM-5.
- **Diagnostic orphan**. A situation where the individual having some of the features of a problematic substance use cannot be given a diagnostic label as he/she does not fulfill the criteria of diagnosis any disorder.
- **DSM-5**. A psychiatric diagnostic classificatory system developed by the American Psychiatric Association. There have been various versions, and the present is the fifth edition which was released in 2013.
- **ICD 10**. A psychiatric diagnostic classificatory system developed and endorsed by the World Health Organization.

Key facts of DSM-5

- DSM-5 is the psychiatric diagnostic system developed by the American Psychiatric Association.
- It was released in 2013.
- It replaced the DSM-IV-TR which was used before this version came.
- The document contains descriptions of psychiatric diagnoses for clinical and research use.
- The DSM-5 has been used frequently in research studies.

Summary points

- The DSM-5 criteria for cannabis use disorder subsumes the previous diagnoses of cannabis dependence and cannabis abuse.
- Diagnosis of cannabis use disorder requires at least 2 out of 11 criteria, along with impairment criteria.
- Development of the DSM-5 criteria of cannabis use disorder was through a rigorous process of the DSM-5 substance use disorder working group.
- Addition of cannabis withdrawal syndrome in DSM-5 is in consonance of recognition and formalization of withdrawal criteria.
- Several applications of DSM-5 cannabis use disorder exist, including in the clinical, public health, and research settings.

The views expressed in the chapter are author's own and do not necessarily reflect that of their institutions.

References

Adamson, S. J., Kay-Lambkin, F. J., Baker, A. L., Lewin, T. J., Thornton, L., Kelly, B. J., & Sellman, J. D. (2010). An improved brief measure of cannabis misuse: The cannabis use disorders identification test-revised (CUDIT-R). *Drug and Alcohol Dependence*, *110*(1–2). https://doi.org/10.1016/j.drugalcdep.2010.02.017.

Agrawal, A., Lynskey, M. T., Bucholz, K. K., Kapoor, M., Almasy, L., Dick, D. M., Edenberg, H. J., Foroud, T., Goate, A., Hancock, D. B., Hartz, S., Johnson, E. O., Hesselbrock, V., Kramer, J. R., Kuperman, S., Nurnberger, J. I., Schuckit, M., & Bierut, L. J. (2014). DSM-5 cannabis use disorder: A phenotypic and genomic perspective. *Drug and Alcohol Dependence*, *134*(1), 362–369. https://doi.org/10.1016/j.drugalcdep.2013.11.008.

Allsop, D. J., Copeland, J., Norberg, M. M., Fu, S., Molnar, A., Lewis, J., & Budney, A. J. (2012). Quantifying the clinical significance of cannabis withdrawal. *PLoS One*, *7*(9), e44864. https://doi.org/10.1371/journal.pone.0044864.

Allsop, D. J., Norberg, M. M., Copeland, J., Fu, S., & Budney, A. J. (2011). The cannabis withdrawal scale development: Patterns and predictors of cannabis withdrawal and distress. *Drug and Alcohol Dependence*, *119*(1–2), 123–129. https://doi.org/10.1016/j.drugalcdep.2011.06.003.

Ambekar, A., Agrawal, A., Rao, R., Mishra, A., Khandelwal, S., & Chadda, R. (2019). *Magnitude of substance use in India*. Ministry of Social Justice and Empowerment, Government of India.

American Psychiatric Association. (2013). *Diagnostic and statistical manual of mental disorders (DSM-5)* (5th ed.). Arlington, VA: American Psychiatric Publishing.

Arendt, M., Rosenberg, R., Foldager, L., Sher, L., & Munk-Jørgensen, P. (2007). Withdrawal symptoms do not predict relapse among subjects treated for cannabis dependence. *The American Journal on Addictions*, *16*(6), 461–467. https://doi.org/10.1080/10550490701640985.

Bonnet, U., & Preuss, U. W. (2017). The cannabis withdrawal syndrome: Current insights. *Substance Abuse and Rehabilitation*, *8*, 9. https://doi.org/10.2147/SAR.S109576.

Bonn-Miller, M. O., Heinz, A. J., Smith, E. V., Bruno, R., & Adamson, S. (2016). Preliminary development of a brief cannabis use disorder screening tool: The cannabis use disorder identification test short-form. *Cannabis and Cannabinoid Research*, *1*(1), 252–261. https://doi.org/10.1089/can.2016.0022.

Budney, A. J., & Hughes, J. R. (2006). The cannabis withdrawal syndrome. *Current Opinion in Psychiatry*, *19*(3), 233–238. https://doi.org/10.1097/01.yco.0000218592.00689.e5.

Budney, A. J., Hughes, J. R., Moore, B. A., & Vandrey, R. (2004). Review of the validity and significance of cannabis withdrawal syndrome. *The American Journal of Psychiatry*, *161*(11), 1967–1977. https://doi.org/10.1176/appi.ajp.161.11.1967.

Cerdá, M., Mauro, C., Hamilton, A., Levy, N. S., Santaella-Tenorio, J., Hasin, D., Wall, M. M., Keyes, K. M., & Martins, S. S. (2020). Association between recreational Marijuana legalization in the United States and changes in Marijuana use and cannabis use disorder from 2008 to 2016. *JAMA Psychiatry*, *77*(2), 165–171. https://doi.org/10.1001/jamapsychiatry.2019.3254.

Chung, T., Cornelius, J., Clark, D., & Martin, C. (2017). Greater prevalence of proposed ICD-11 alcohol and cannabis dependence compared to ICD-10, DSM-IV, and DSM-5 in treated adolescents. *Alcoholism: Clinical and Experimental Research*, *41*(9), 1584–1592.

Chung, T., Cornelius, J., Clark, D., & Martin, C. (2018). *Adolescents*, *41*(9), 1584–1592. https://doi.org/10.1111/acer.13441.Greater.

Clark, L. A., Cuthbert, B., Lewis-Fernández, R., Narrow, W. E., & Reed, G. M. (2017). Three Approaches to understanding and classifying mental disorder: ICD-11, DSM-5, and the National Institute of Mental Health's Research Domain Criteria (RDoC). *Psychological Science in the Public Interest*, *18*(2), 72–145. https://doi.org/10.1177/1529100617727266.

Compton, W. M., Dawson, D. A., Goldstein, R. B., & Grant, B. F. (2013). Crosswalk between DSM-IV dependence and DSM-5 substance use disorders for opioids, cannabis, cocaine and alcohol. *Drug and Alcohol Dependence*, *132*(1), 387–390. https://doi.org/10.1016/j.drugalcdep.2013.02.036.

Compton, W. M., Saha, T. D., Conway, K. P., & Grant, B. F. (2009). The role of cannabis use within a dimensional approach to cannabis use disorders. *Drug and Alcohol Dependence*, *100*(3), 221–227. https://doi.org/10.1016/j.drugalcdep.2008.10.009.

Connor, J. P., Stjepanović, D., Le Foll, B., Hoch, E., Budney, A. J., & Hall, W. D. (2021). Cannabis use and cannabis use disorder. *Nature Reviews. Disease Primers*, *7*(1), 1–24. https://doi.org/10.1038/s41572-021-00247-4.

Copersino, M. L., Boyd, S. J., Tashkin, D. P., Huestis, M. A., Heishman, S. J., Dermand, J. C., Simmons, M. S., & Gorelick, D. A. (2006). Cannabis withdrawal among non-treatment-seeking adult cannabis users. *American Journal on Addictions*, *15*(1). https://doi.org/10.1080/10550490500418997.

Degenhardt, L., Bharat, C., Bruno, R., Glantz, M. D., Sampson, N. A., Lago, L., Aguilar-Gaxiola, S., Alonso, J., Andrade, L. H., Bunting, B., Caldas-de-Almeida, J. M., Cia, A. H., Gureje, O., Karam, E. G., Khalaf, M., McGrath, J. J., Moskalewicz, J., Lee, S., Mneimneh, Z., & Behalf of the WHO World Mental Health Survey Collaborators. (2019). Concordance between the diagnostic guidelines for alcohol and cannabis use disorders in the draft ICD-11 and other classification systems: Analysis of data from the WHO's World Mental Health Surveys. *Addiction*, *114*(3), 534–552. https://doi.org/10.1111/add.14482.

Degenhardt, L., Ferrari, A. J., Calabria, B., Hall, W. D., Norman, R. E., McGrath, J., Flaxman, A. D., Engell, R. E., Freedman, G. D., & Whiteford, H. A. (2013). The global epidemiology and contribution of cannabis use and dependence to the global burden of disease: Results from the GBD 2010 study. *PLoS One*, *8*(10), e76635.

Goldstein, R. B., Patricia Chou, S., Smith, S. M., Jung, J., Zhang, H., Saha, T. D., Pickering, R. P., June Ruan, W., Huang, B., & Grant, B. F. (2015). Nosologic comparisons of DSM-IV and DSM-5 alcohol and drug use disorders: Results from the national epidemiologic survey on alcohol and related conditions-III. *Journal of Studies on Alcohol and Drugs*, *76*(3), 378–388. https://doi.org/10.15288/jsad.2015.76.378.

Gorelick, D. A., Levin, K. H., Copersino, M. L., Heishman, S. J., Liu, F., Boggs, D. L., & Kelly, D. L. (2012). Diagnostic criteria for cannabis withdrawal syndrome. *Drug and Alcohol Dependence*, *123*(1–3), 141–147. https://doi.org/10.1016/j.drugalcdep.2011.11.007.

Grant, B. F., Compton, W. M., Crowley, T. J., Hasin, D. S., Helzer, J. E., Li, T.-K., Rounsaville, B. J., Volkow, N. D., & Woody, G. E. (2007). Errors in assessing DSM-IV substance use disorders. *Archives of General Psychiatry*, *64*(3), 379–380. author reply 381-382. https://doi.org/10.1001/archpsyc.64.3.379.

Grant, B. F., Goldstein, R. B., Smith, S. M., Jung, J., Zhang, H., Chou, S. P., Pickering, R. P., Ruan, W. J., Huang, B., Saha, T. D., Aivadyan, C., Greenstein, E., & Hasin, D. S. (2015). The alcohol use disorder and associated disabilities interview schedule-5 (AUDADIS-5): Reliability of substance use and psychiatric disorder modules in a general population sample. *Drug and Alcohol Dependence*, *148*. https://doi.org/10.1016/j.drugalcdep.2014.11.026.

Greene, M. C., & Kelly, J. F. (2014). The prevalence of cannabis withdrawal and its influence on adolescents' treatment response and outcomes: A 12-month prospective investigation. *Journal of Addiction Medicine*, *8*(5), 359–367. https://doi.org/10.1097/ADM.0000000000000064.

Hasin, D. S., Greenstein, E., Aivadyan, C., Stohl, M., Aharonovich, E., Saha, T., Goldstein, R., Nunes, E. V., Jung, J., Zhang, H., & Grant, B. F. (2015). The alcohol use disorder and associated disabilities interview schedule-5 (AUDADIS-5): Procedural validity of substance use disorders modules through clinical re-appraisal in a general population sample. *Drug and Alcohol Dependence*, *148*, 40–46. https://doi.org/10.1016/j.drugalcdep.2014.12.011.

Hasin, D., Hatzenbuehler, M. L., Keyes, K., & Ogburn, E. (2006). Substance use disorders: Diagnostic and statistical manual of mental disorders, fourth edition (DSM-IV) and international classification of diseases, tenth edition (ICD-10). *Addiction (Abingdon, England)*, *101*(Suppl 1), 59–75. https://doi.org/10.1111/j.1360-0443.2006.01584.x.

Hasin, D. S., Kerridge, B. T., Saha, T. D., Huang, B., Pickering, R., Smith, S. M., Jung, J., Zhang, H., & Grant, B. F. (2016). Prevalence and correlates of DSM-5 cannabis use disorder, 2012-2013: Findings from the national epidemiologic survey on alcohol and related conditions-III. *American Journal of Psychiatry*, *173*(6), 588–599. https://doi.org/10.1176/appi.ajp.2015.15070907.

Hasin, D. S., O'Brien, C. P., Auriacombe, M., Borges, G., Bucholz, K., Budney, A., Compton, W. M., Crowley, T., Ling, W., Petry, N. M., Schuckit, M., & Grant, B. F. (2013). DSM-5 criteria for substance use disorders: Recommendations and rationale. *American Journal of Psychiatry*, *170*(8), 834–851. https://doi.org/10.1176/appi.ajp.2013.12060782.

Hasin, D., & Paykin, A. (1998). Dependence symptoms but no diagnosis: Diagnostic "orphans" in a community sample. *Drug and Alcohol Dependence*, *50* (1), 19–26. https://doi.org/10.1016/s0376-8716(98)00007-6.

Hesse, M., & Thylstrup, B. (2013). Time-course of the DSM-5 cannabis withdrawal symptoms in poly-substance abusers. *BMC Psychiatry*, *13*(1), 258. https://doi.org/10.1186/1471-244X-13-258.

Heyman, R. B., Anglin, T. M., Copperman, S. M., Joffe, A., McDonald, C. A., Rogers, P. D., Shah, R. Z., Armentano, M., Boyd, G. M., & Czechowicz, D. (1999). American Academy of Pediatrics. Committee on Substance Abuse. Marijuana: A continuing concern for pediatricians. *Pediatrics*, *104*(4 Pt 1), 982–985.

Kelly, S. M., Gryczynski, J., Mitchell, S. G., Kirk, A., O'Grady, K. E., & Schwartz, R. P. (2014). Concordance between DSM-5 and DSM-IV nicotine, alcohol, and cannabis use disorder diagnoses among pediatric patients. *Drug and Alcohol Dependence*, *140*. https://doi.org/10.1016/j.drugalcdep.2014.03.034.

Keyes, K. M., Krueger, R. F., Grant, B. F., & Hasin, D. S. (2011). Alcohol craving and the dimensionality of alcohol disorders. *Psychological Medicine*, *41*(3). https://doi.org/10.1017/S003329171000053X.

Kopak, A. M., Metze, A. V., & Hoffmann, N. G. (2014). Alcohol use disorder diagnoses in the criminal justice system: An analysis of the compatibility of current DSM-IV, proposed DSM-5.0, and DSM-5.1 diagnostic criteria in a correctional sample. *International Journal of Offender Therapy and Comparative Criminology*, *58*(6), 638–654. https://doi.org/10.1177/0306624X13485929.

Kopak, A. M., Proctor, S. L., & Hoffmann, N. G. (2012). An assessment of the compatibility of DSM-IV and proposed DSM-5 criteria in the diagnosis of cannabis use disorders. *Substance Use & Misuse*, *47*(12), 1328–1338. https://doi.org/10.3109/10826084.2012.714039.

Kopak, A. M., Proctor, S. L., & Hoffmann, N. G. (2014). The elimination of abuse and dependence in DSM-5 substance use disorders: What does this mean for treatment? *Current Addiction Reports*, *1*(3), 166–171. https://doi.org/10.1007/s40429-014-0020-0.

Krueger, R. F., Nichol, P. E., Hicks, B. M., Markon, K. E., Patrick, C. J., Iacono, W. G., & McGue, M. (2004). Using latent trait modeling to conceptualize an alcohol problems continuum. *Psychological Assessment*, *16*(2). https://doi.org/10.1037/1040-3590.16.2.107.

Lago, L., Bruno, R., & Degenhardt, L. (2016). Concordance of ICD-11 and DSM-5 definitions of alcohol and cannabis use disorders: A population survey. *The Lancet Psychiatry*, *3*(7), 673–684. https://doi.org/10.1016/S2215-0366(16)00088-2.

Martin, C. S., Chung, T., Kirisci, L., & Langenbucher, J. W. (2006). Item response theory analysis of diagnostic criteria for alcohol and cannabis use disorders in adolescents: Implications for DSM-V. *Journal of Abnormal Psychology*, *115*(4), 807–814. https://doi.org/10.1037/0021-843X.115.4.807.

Merz, F. (2018). United Nations Office on Drugs and Crime: World drug report 2017. *SIRIUS-Zeitschrift Für Strategische Analysen*, *2*(1), 85–86.

Mewton, L., Slade, T., & Teesson, M. (2013). An evaluation of the proposed DSM-5 cannabis use disorder criteria using Australian National Survey Data. *Journal of Studies on Alcohol and Drugs*, *74*(4), 614–621. https://doi.org/10.15288/jsad.2013.74.614.

Muthén, B. (2006). Should substance use disorders be considered as categorical or dimensional? *Addiction*, *101*(s1), 6–16. https://doi.org/10.1111/j.1360-0443.2006.01583.x.

Pedersen, W. (2009). Cannabis use: Subcultural opposition or social marginality?: A population-based longitudinal study. *Acta Sociologica*, *52*(2), 135–148. https://doi.org/10.1177/0001699309104001.

Rounsaville, B. J., Spitzer, R. L., & Williams, J. B. W. (1986). Proposed changes in DSM-III substance use disorders: Description and rationale. *American Journal of Psychiatry*, *143*(4). https://doi.org/10.1176/ajp.143.4.463.

Sarkar, S., Parmar, A., & Singh, A. (2020). An exploratory study of cannabis use pattern and treatment seeking in patients attending an addiction treatment facility. *Indian Journal of Psychiatry*, *62*(2), 145–151. https://doi.org/10.4103/psychiatry.IndianJPsychiatry_132_19.

Schuckit, M. A., & Saunders, J. B. (2006). The empirical basis of substance use disorders diagnosis: Research recommendations for the diagnostic and statistical manual of mental disorders, fifth edition (DSM-V). *Addiction*, *101*(s1), 170–173. https://doi.org/10.1111/j.1360-0443.2006.01611.x.

Schultz, N. R., Bassett, D. T., Messina, B. G., & Correia, C. J. (2019). Evaluation of the psychometric properties of the cannabis use disorders identification test—Revised among college students. *Addictive Behaviors*, *95*. https://doi.org/10.1016/j.addbeh.2019.02.016.

Surís, A., Holliday, R., & North, C. S. (2016). The evolution of the classification of psychiatric disorders. *Behavioral Science*, *6*(1), 5. https://doi.org/10.3390/bs6010005.

World Drug Report: Global Overview of Drug Demand and Supply. (2017). *United Nations Office on Drugs and Crime*. https://www.unodc.org/wdr2017/field/Booklet_2_HEALTH.pdf.

Zimmerman, M., & Spitzer, R. (2009). Psychiatric classification. In B. Sadock, V. Sadock, A. Virginia, & P. Ruiz (Eds.), *Kaplan & Sadock's comprehensive textbook of psychiatry* (pp. 1108–1139). Lippincott Williams & Wilkins.

Chapter 38

Cannabis and clinical high risk for psychosis: Monitoring and smartphones

Olga Santesteban-Echarri, Sara Jalali, and Jean Addington
Department of Psychiatry, Hotchkiss Brain Institute, University of Calgary, Calgary, AB, Canada

Abbreviations

APSS	attenuated positive symptoms syndrome
AUS/DUS	alcohol use scale/drug use scale
BIPS	brief intermittent psychotic syndrome
CAARMS	Comprehensive Assessment of the At-Risk Mental States
CBD	cannabidiol
CEQ	The Cannabis Experience Questionnaire
CEQ-I	The Cannabis Experiences Questionnaire-Intoxication effects checklist
COPS	Criteria of Psychosis-Risk Syndromes
DSM	Diagnostic and Statistical Manual of Mental Disorders
DFAQ-CU	The Daily Sessions, Frequency, Age of Onset, and Quantity of Cannabis Use Inventory
DUF	the drug use frequency
GRD	genetic risk and deterioration syndrome
SIPS	Structured Interview for Psychosis-Risk Syndromes
THC	tetrahydrocannabinol
uMARS	Mobile Application Rating Scale—user version

Introduction and orientation to the chapter

In the past decade, the increase in cannabis use has been a societal concern due to its potential impact on youth mental health, particularly the association of cannabis use with the development of psychosis. Therefore, within current work on the identification of youth at risk of psychosis and predictors of later development of psychosis, cannabis has been studied as a potential risk factor associated with an increased likelihood of transition to psychosis.

The first part of this chapter offers a brief overview of cannabis use in the general population and its impact on psychosis. This is followed by definitions of clinical high risk for psychosis and the role that cannabis may play in this young at-risk population. In the second part of the chapter, we address the issue of monitoring with a focus first on available assessment tools for cannabis use, and secondly on mobile-based applications. Finally, we present the features of an application specifically developed for assessing cannabis use and cannabis-related experiences for individuals at clinical high risk for developing psychosis.

General population studies on cannabis use and its impact on psychosis

In recent years, cannabis has been of relevance due to its increasing legalization in North America and its impact on youth. Overall, cannabis is the most used illicit substance in the world (United Nations Office on Drugs and Crime, 2019) and is certainly used by teenagers and young adults. In the general population, compared to individuals who do not use cannabis,

Cannabis Use, Neurobiology, Psychology, and Treatment. https://doi.org/10.1016/B978-0-323-89862-1.00011-8

those who do use have four to five times more probability of developing psychosis (Marconi et al., 2016; Mustonen et al., 2018). Moreover, the evidence suggests a dose-response with daily cannabis users and high potency cannabis users being two and four times more likely to develop psychosis, respectively (Marconi et al., 2016). Among those experiencing a first episode of psychosis, the average age of onset of psychosis was 6 years earlier for daily cannabis users compared to nonusers (Di Forti et al., 2014), and daily use of high-potency cannabis has been related to more severe psychotic symptoms (Quattrone et al., 2021). Importantly, cannabis use is associated with significant adverse effects among those who already have psychosis (Lowe et al., 2019), such as increased symptom severity, relapse rates, decreased treatment adherence, and poorer prognosis overall (Schoeler et al., 2016). Given the increasing availability of high-potency cannabis and the potential association between cannabis and the development of psychosis in the general population, there has been an effort to investigate whether cannabis use in those already at risk for psychosis is a risk factor for transition to psychosis.

Clinical high risk for developing psychosis

There has been a tremendous effort in the past 2 decades to prospectively identify individuals who may be at risk of developing psychosis and to understand the predictors that may lead those at high risk for psychosis to develop a full-blown psychotic illness (Addington et al., 2020). The criteria for being at high risk for developing psychosis were initially introduced in Australia and unified into the Comprehensive Assessment of the At-Risk Mental States (CAARMS; Yung et al., 2005). These criteria were modified by McGlashan to develop the Criteria of Psychosis-Risk Syndromes (COPS; McGlashan et al., 2010), which are assessed with The Structured Interview for Psychosis-risk Syndromes (SIPS). Both CAARMS and SIPS are widely used measures with excellent predictive validity, reliability, and agreement in identifying CHR individuals (Fusar-Poli et al., 2016). Both criteria propose three syndromes: i) genetic risk and deterioration syndrome (GRD), which involves having a first-degree relative with a psychotic illness plus a significant decline in functioning, ii) attenuated positive symptoms syndrome (APSS) where the individual experiences subthreshold psychotic symptoms, and iii) brief intermittent psychotic syndrome (BIPS) where psychotic symptoms are present but only for a very brief period and remit spontaneously. The literature uses different terms to refer to these criteria (e.g., psychosis-risk syndrome, at-risk mental state or at clinical high risk) or to describe the population (e.g., ultrahigh risk or clinical high risk). For this chapter, we will use the term clinical high risk (CHR).

CHR individuals are usually between 12- and 25-years old, and typically present with subthreshold psychotic symptoms such as unusual thought content, suspiciousness, grandiosity, perceptual abnormalities, and/or disorganized communication. However, CHR individuals are a heterogeneous group and often have ongoing difficulties with comorbid anxiety and depression, negative symptoms, or poor social and role functioning (Mittal & Addington, 2021). Current meta-analyses suggest that approximately 25% of CHR individuals will go on to develop psychosis within 3 years (Salazar De Pablo et al., 2021). However, a proportion of those who do not transition to psychosis, continue to have unremitting subthreshold psychotic symptoms, other symptoms and poor functioning (Beck et al., 2019).

Although the focus in this chapter is on cannabis, several other risk factors have been investigated in CHR studies as potential predictors of psychosis including environmental factors, trauma and stress, cognition, and social functioning, as well as a range of biomarkers (Addington et al., 2019).

Cannabis use in individuals at clinical high risk for psychosis

A recent review that included 36 CHR studies on cannabis use reported a prevalence of 25.8% for current cannabis use, 48.7% for lifetime cannabis use, and 14.9% with a DSM cannabis use disorder (Farris et al., 2020). When compared to healthy controls, cannabis use is more prevalent among CHR individuals. Moreover, users who develop psychosis before age 15 have been reported to have higher cannabis use prevalence and continue using more frequently compared to late users (Valmaggia et al., 2014).

The evidence regarding whether cannabis use among CHR individuals is associated with more severe APS remains contradictory. Schifani et al. (2020) reported more severe ratings on positive and disorganized symptom scores among CHR cannabis users compared to nonusers, and a meta-analysis reported more severe ratings on suspiciousness and unusual thought content (Carney et al., 2017). However, the most recent review showed no differences in APS between CHR users and nonusers (Farris et al., 2020).

The literature is also inconsistent with respect to negative symptoms. While several studies did not report significant differences for total negative symptoms between users and nonusers (Carney et al., 2017; Gupta et al., 2020; Schifani et al., 2020), others reported less severe negative symptoms associated with an increase in cannabis use (MacHielsen et al., 2010; Mizrahi et al., 2014; Nieman et al., 2016).

Cannabis and use and transition to psychosis

Initial meta-analyses suggested a dose-response association with transition to psychosis (Marconi et al., 2016) and an association between current cannabis use disorder among CHR individuals and transition to psychosis (Kraan et al., 2016). However, more recently other reviews are reporting no significant association between cannabis use and transition to psychosis (Farris et al., 2020; Mongan et al., 2020). The accumulating evidence seems to indicate that cannabis use may not distinctively contribute to the prediction of the development of psychosis in youth at CHR (Oliver et al., 2020). This is surprising given that the evidence from epidemiological studies in the general population shows an association between cannabis use and the development of psychosis. Thus, it would be reasonable to expect an increased probability of transition to psychosis for individuals at CHR who also use cannabis. However, one plausible explanation may be that CHR individuals with a more severe cannabis use may be less prone to join research studies and are, therefore, underrepresented in most CHR samples. Since we know that CHR youth have increased cannabis use relative to healthy controls, a second explanation might be that cannabis use has already had an impact in increasing their vulnerability to developing psychosis.

Cannabis use monitoring

There are more than 25 screening and diagnostic tools for assessing cannabis use in the general population. However, most are limited by single or discrete time-frames for screening, lack of succinctness, lack of validity for some populations (psychiatric patients, female users, or adolescents), and lack of relevant information to differentiate risky use from regular use (López-Pelayo et al., 2015). Particularly, the lack of quantitative measurement of consumption (grams of the cannabinoid content: tetrahydrocannabinol, THC, or cannabidiol, CBD) impedes comparability among data retrieved with different tools.

The majority of CHR studies have assessed cannabis use dichotomously, with a lack of longitudinal appraisal, and only a small percentage examined the frequency of cannabis use in detail (Farris et al., 2020). This is relevant because compared to daily use or several sessions per day, low exposure to cannabis in terms of frequency or potency may have minimal impact on an individual. Some CHR studies rated the frequency of cannabis use with the Drug and Alcohol Use Scale (AUS/DUS; Drake et al., 1996), which assesses cannabis use frequency and severity in the last month. This scale has been widely used in psychosis-spectrum populations, has high reliability, moderate concurrent consistency, and good convergent validity. Similarly, the Drug Use Frequency (DUF; O'Farrell et al., 2003) is a brief self-report tool that measures the frequency of drug and cannabis use in the past six months. Finally, Arseneault et al. (2002) have supplemented previous scales by collecting more relevant details on cannabis use, including being a past/current user, patterns of use, current and past frequency, and age at first cannabis use.

Since the cannabis legalization movement, the cannabis market has increased exponentially. Users have access to more precise information about the cannabis they consume and different modes of consumption as concentrates are becoming gradually popular. The Daily Sessions, Frequency, Age of Onset, and Quantity of Cannabis Use Inventory (DFAQ-CU; Cuttler & Spradlin, 2017) is a 33-item self-report inventory that echoes the new consumption trends. It measures cannabis use across six domains: frequency of cannabis use, number of cannabis use sessions per day, age of onset of cannabis use, and the quantity of cannabis, edibles, and concentrates consumed. The DFAQ-CU includes a picture of different amounts of cannabis in bud, loose, and joint form to ease the identification of the quantity of cannabis used and asks about a variety of methods, amounts personally used versus shared, and typical THC concentrations.

Given the association of cannabis use, psychotic-like experiences, psychosis proneness, and schizotypy, Barkus et al. (2006) developed the Cannabis Experience Questionnaire (CEQ) and the Cannabis Experiences Questionnaire-Intoxication effects checklist (CEQ-I; Quinn et al., 2017) to capture the cannabis-related experiences in detail. The CEQ is a 55-item questionnaire that assesses the subjective effects of cannabis. It contains three subscales that record pleasurable experiences (feeling happy, feeling laid back, able to understand the world better), psychosis-like experiences (delusional thinking, experiencing auditory hallucinations, and feeling paranoid), and after effects (feeling slowed down, loss of drive, and decreased attention).

Questionnaires like the DFAQ-CU and the CEQ provide a great avenue for capturing detailed data regarding potency, frequency, quantity, and methods of cannabis consumption, which could lead to the development of profiles of use (i.e., high potency concentrate users). However, regardless of the detail of the information gathered, studies using traditional paper-based measures rely on retrospective information, single time-point data, and cannot obtain a precise picture of the daily use of cannabis (Farris et al., 2020). Thus, mobile apps may improve symptom monitoring offering a 24-h accessible and more precise capture of current cannabis use in near real time.

Cannabis use tracking and mobile applications

There has been a proliferation of mobile-based mental health apps, often free, that offer portable, accessible, and cost-effective resources, such as tools for tracking symptoms. Most of the apps targeted to support substance users in recovery include some type of self-monitoring feature (Nesvåg & McKay, 2018). However, most of the apps do not support their claims with peer-reviewed research. To have a better picture of the apps available on the market, a search of English mobile-based applications with the search terms "cannabis" and "marijuana" was conducted in the Apple App Store. There were 246 apps focusing on cannabis, of which 20 (8.1%) were apps to track cannabis use and symptoms. Table 1 presents a description of the apps. Of these, Strainprint (https://strainprint.ca/app/) and Releaf App (https://releafapp.com/) have published peer-reviewed research.

Strainprint is a free smartphone application developed to prospectively track cannabis use for medicinal purposes including 279 physical and psychological conditions. The app tracks cannabis dose, varieties of cannabis products, administration mode, cannabinoid content, and cannabis effects on 46 symptoms. For each session, users log their general mood, cannabis product, mode of consumption, and the quantity used. The user receives a notification after the end of each session to log the main effects. The app comprises several features including (i) a calendar; (ii) "my stats;" (iii) a tracking page of sessions, products, symptoms, and effects; (iv) the "explore section" with information regarding cannabis products; and (v) a rewards section. Strainprint has been used to describe profiles of medical cannabis users (Kalaba & Ware, 2021).

Releaf App is an educational software application designed to track the effects of medical cannabis products used through naturalistic observation. Before beginning each session, the user identifies the condition to be treated, type of cannabis product, combustion method, cannabis subspecies, cannabinoid content, ratings of symptom severity levels (before and after the session), and side effects.

Some of the strengths of the studies that have used Releaf App and Strainprint include the naturalistic design and large sample sizes (n ranging from 252 to 3548), with the apps including extensive conditions and symptoms and examining perceived effects over time. However, these apps were intended to monitor cannabis use for medical purposes in adults (18–74 years) from the general population. Although "schizophrenia" can be selected as a condition in Strainprint, a user with a schizophrenia diagnosis may select "anxiety" as the condition to treat, making it difficult to study specific populations. Furthermore, due to the lengthy list of symptoms, some may be overlooked by users. Each user can select their symptoms, which is valuable for intra-subject comparisons but not for intersubject comparisons. Finally, despite including symptoms like visual or sound sensitivity, and paranoia, other psychotic symptoms such as hearing voices, or having visual hallucinations, were not addressed. Due to this gap in the field, an app, *LooseLeaf*, was developed to monitor cannabis use and cannabis-related experiences (psychotic-like, anxious, and euphoric experiences), which are pivotal to ascertaining the effect of cannabis use among those at CHR.

LooseLeaf, a mobile-application to track psychotic-like experiences

Based on the CEQ-I, *LooseLeaf* is a mobile-based application developed through participatory co-design that monitors cannabis use and cannabis-related experiences for youth at CHR (Santesteban-Echarri et al., 2020). The application comprises five main features: home screen, inventory, daily questions, two calculators, and the stats section. Through 11 daily questions, the application monitors (i) daily cannabis use: quantity of cannabis use, strain, cannabinoid content, and cannabis subspecies; method of consumption and type of cannabis product; the context of consumption (i.e., using with friends versus alone; in the mall versus outdoors); and subjective perception of the amount consumed; (ii) cannabis-related experiences: anxiety-paranoid experiences, enjoyable experiences, and psychotic-like experiences; (iii) other substance use; and (iv) perception of their day. Figs. 1 to 5 depict the application's features.

LooseLeaf demonstrated initial acceptability, safety, and feasibility to monitor cannabis use and cannabis-related experiences in a one-month pilot study with 14 CHR participants who reported current cannabis use (Santesteban-Echarri et al., 2020). The app obtained a 42.4% response rate, with nearly 60% of the participants logging in a third of the time. Active engagement (i.e., having 10 logins over the month) was high during the first 24 days. Within the month, a total of 180 cannabis sessions were logged. The application captured context of consumption (most occurring at night, and alone), strain category (evenly distributed between sativa, indica, and hybrid), average daily consumption (i.e., 0.96g), THC versus CBD content (17.8% and 5.8%, respectively), and cannabis-related experiences (mostly enjoyable experiences and some psychotic-like experiences or anxiety-paranoid experiences). Moreover, after the study, all participants rated *LooseLeaf* as safe. The application received a good total score in the Mobile Application Rating Scale user-version (uMARS) (Stoyanov et al., 2016) obtaining the highest score in information and functionality, followed by esthetics, engagement, subjective quality of the app, engagement, and perceived impact in user behavior,

TABLE 1 Characteristics of mobile-based applications for tracking cannabis use.

App name	What the App does
Blenddy	• Tracks products used, method of consumption, and feelings associated with cannabis strains
Buds	• Personalization of cannabis use • Tracks cannabis type and strain, method of consumption, experienced effects, and personalized notes
CannaBook—Medical Companion	• Journaling app for medical cannabis users • Allows journal entries, medicated recipes, and dosage calculator
CHAMPS	• Therapeutic application that tracks cannabis use practices • For young people who have had a psychotic episode and use cannabis • Objective: reducing negative consequences of use
Dependn'	• Quitting tracker for cannabis, tobacco, and alcohol • Monitors addictions and progress • Tracks consumption time and location and cravings intensity
Greencamp	• Discover strains best suited to alleviate medical symptoms • Has 2 tools: Strainblazer and "Which cannabis delivery method is best for you?" • Tracks conditions to be treated, time of consumption, THC and CBD %, desired effects, and flavor
Grounded	• Quitting tracker • Tracks consumption time, daily feelings of craving and withdrawal, money saved
HighBreed—Weed Collection	• Tracks photo of bud, breed name, subjective rating, breed type, and feelings
Highness—Cannabis diary	• Tracks experience with different cannabis strains
HowHi App	• Survey to track products used and physical effects
It is High Time	• Tracks type, strain, method of consumption, amount consumed, and effects produced and how they impact mood and symptoms for medical conditions
Marijuana Addiction Calendar	• Quitting tracker • Tracks reasons for quitting, time since last use, and time and money saved • Can be shared with an accountability partner • Provides distractions to keep mind off use
MyDx One	• Designed to be used with a portable cannabis potency and chemical analyzer device • Provides crowdsources predictions of how a specific cannabis strain might make user feel • Tracks cannabis strain, symptoms, and feelings
OnTracka—CBD & Cannabis App	• Tracks method of consumption and frequency of use for medical cannabis
PotBot	• For medical marijuana patients • Tracks condition wishing to be treated and provides information on cannabis strains suited for specific medical symptoms
Puff Count	• Tracks vaping habits • Tracks frequency of consumption, allows limit setting, and goal tracking
Releaf App	• Medical cannabis app • Tracks cannabis type and strain, method of consumption, THC and CBD %, and changes in symptom severity before and after cannabis use
Strainprint	• Medical cannabis app • Tracks cannabis type and strain, method of consumption, THC and CBD %, changes in symptom severity before and after cannabis use • Provides % efficacy data after tracking
Talk Medicine	• Tracks strain, dosage, method of consumption, desired effects, and feelings during each session
WeedUp	• Tracks amount of cannabis consumed and expenses accrued
Ween—Marijuana Usage Tracker	• Tracks cannabis usage • Provides usage trends

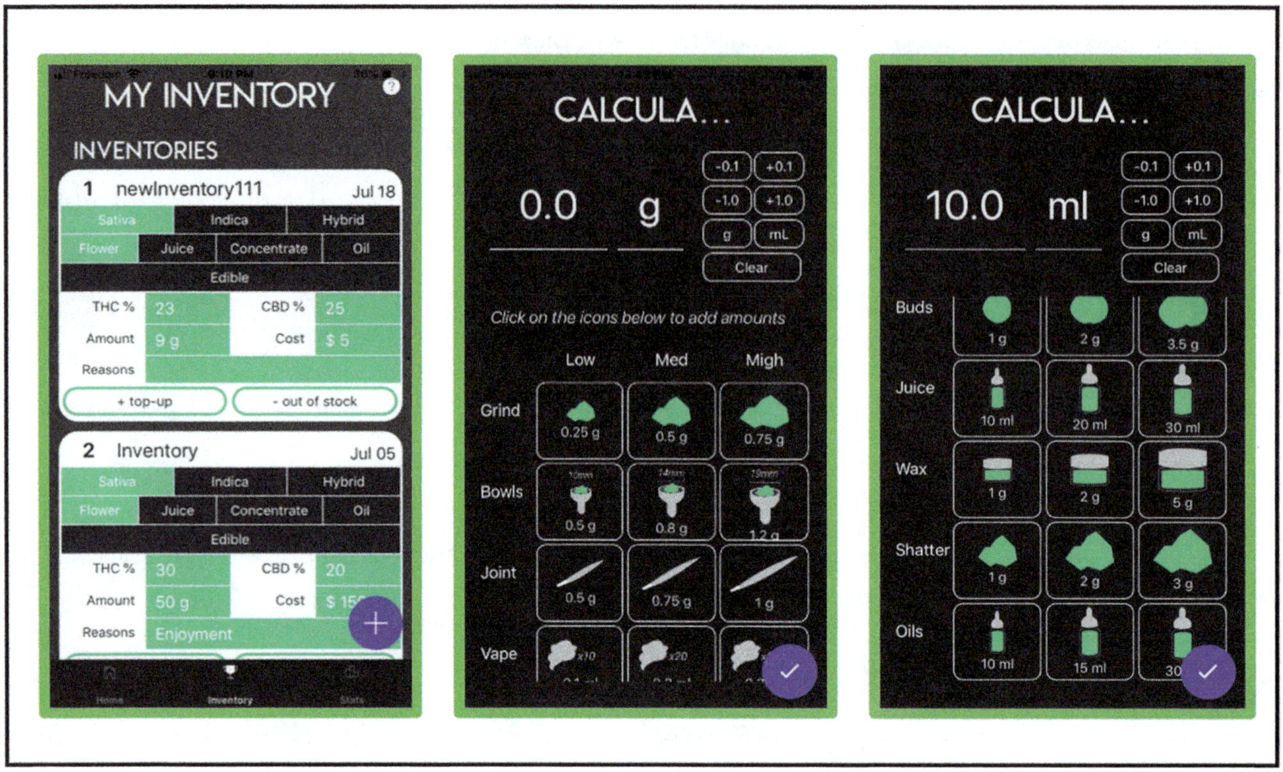

FIG. 1 Cannabis inventory and cannabis calculators.

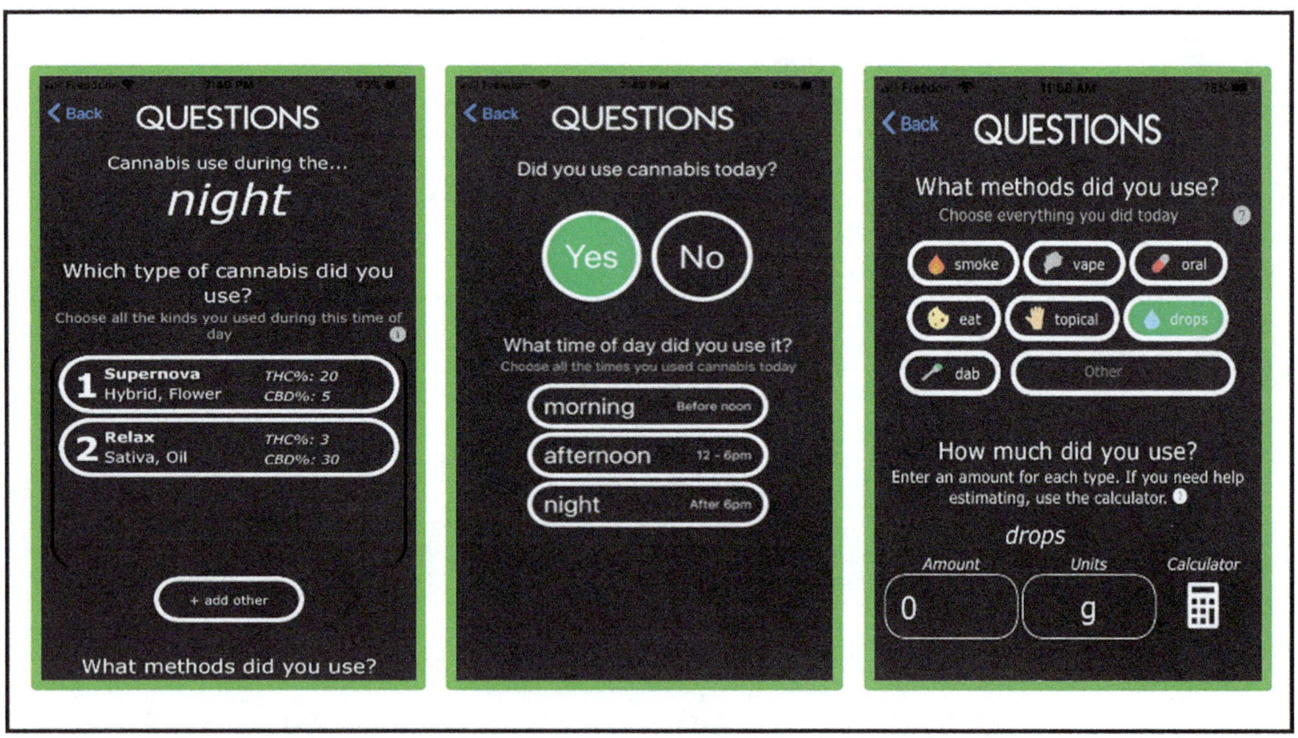

FIG. 2 Cannabis use questions.

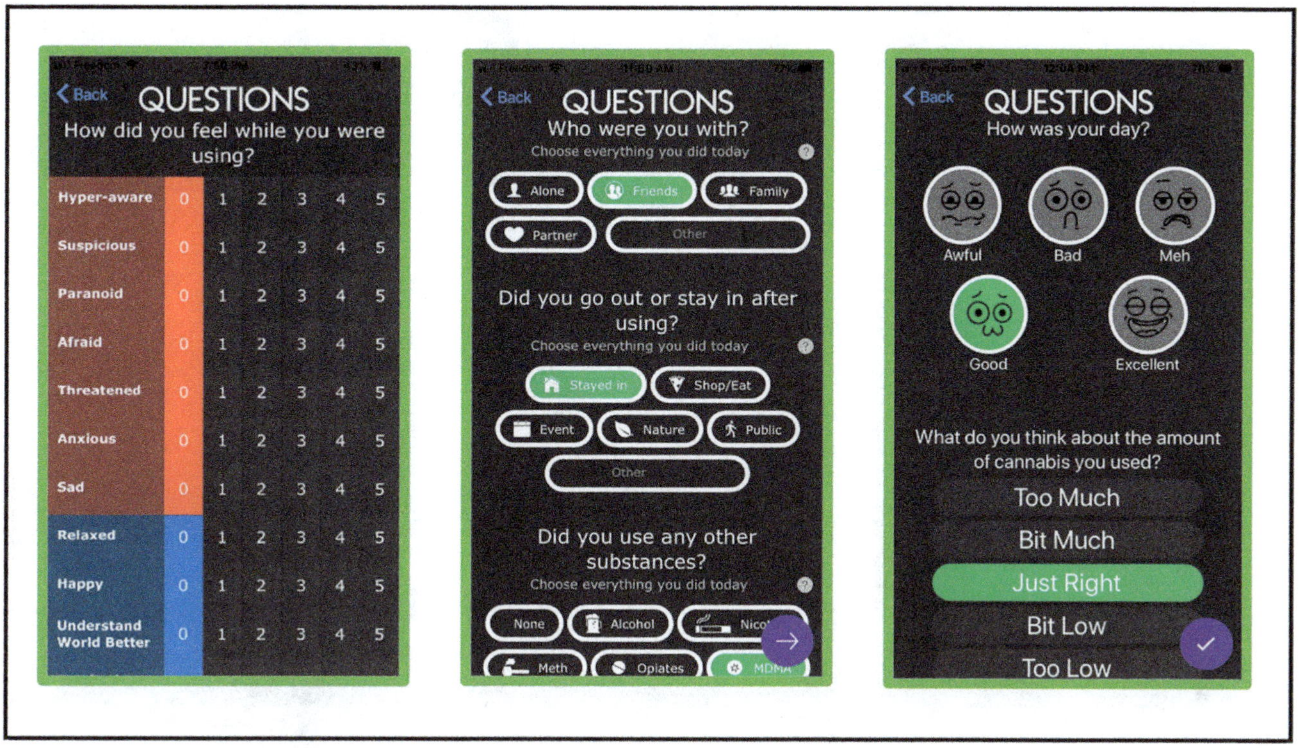

FIG. 3 Questions: cannabis-related experiences.

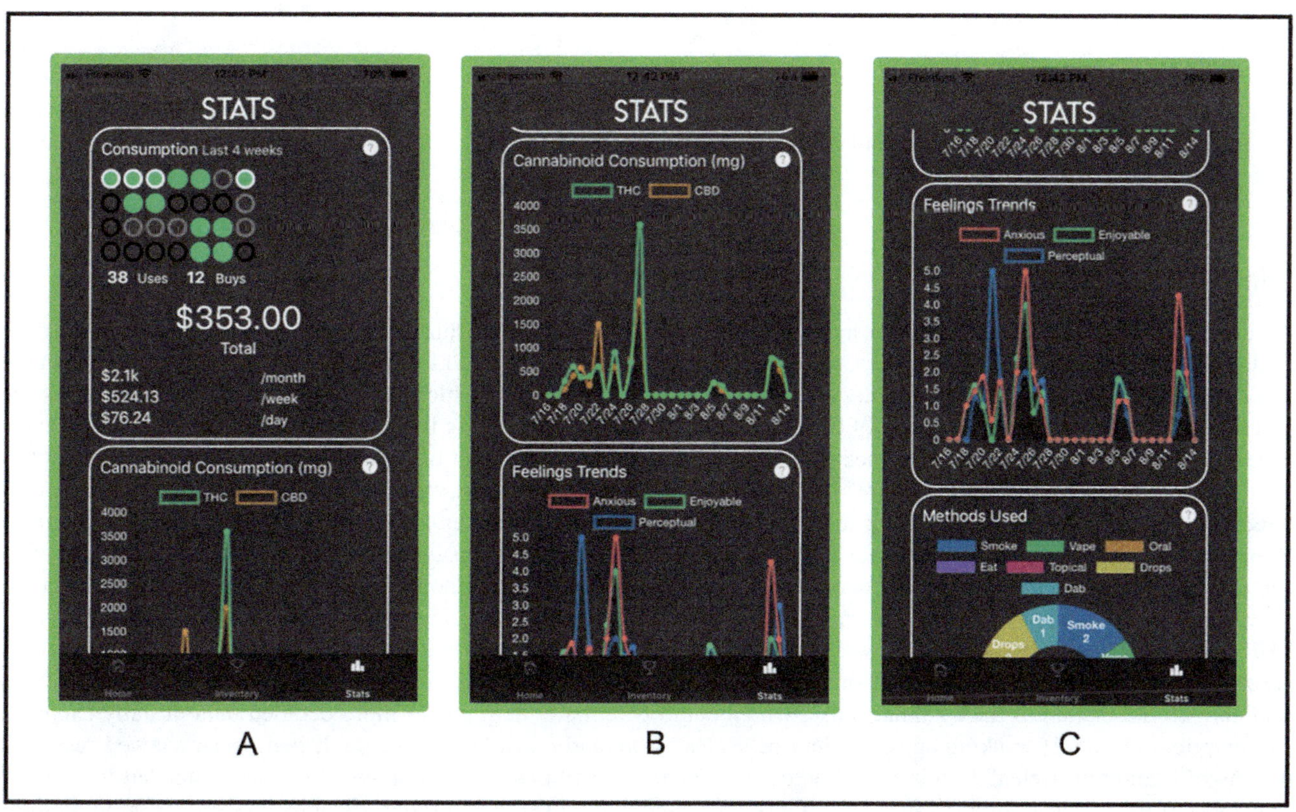

FIG. 4 Stats feature: (A) daily consumption, (B) cannabinoid consumption, and (C) feeling trends.

FIG. 5 Stats feature: (A) methods and form used, (B) context summary and context effects by feeling, and (C) feelings by cannabinoid concentration.

respectively. In conclusion, *LooseLeaf* seems to provide precise, thorough, context-sensitive, and ecologically valid data in near-real-time, which could not be assessed by traditional retrospective paper-based questionnaires. Moreover, an app like *LooseLeaf* may increase awareness of cannabis effects and usage among CHR youth, potentially leading to behavior change regarding consumption.

Summary

Regular and high-potency cannabis use is associated with an increased risk of schizophrenia among the general population. Additionally, 50% of CHR individuals report that they have tried cannabis in their life, which is significantly higher than the number using among age-matched healthy controls. A key CHR research question is whether cannabis use increases the severity of symptoms or the likelihood of developing psychosis. However, this is not supported in the literature. Nevertheless, it has been apparent there is a lack of accurate and detailed assessment of cannabis use patterns among this population. We examined 25 assessment tools for assessing cannabis use and most were limited by deficient time-frames for screening, lack of succinctness, validity for some populations, and relevant information to differentiate risky use from regular use. In fact, most CHR studies have assessed cannabis use dichotomously and at single-time points, with a small percentage examining cannabis use frequency or other cannabis details. Only the DFAQ-CU inquires about detailed cannabis use information. For additional information, the CEQ assesses cannabis-related experiences, particularly relevant for CHR individuals who use cannabis. However, these tools rely on retrospective and single time-point information which prevents obtaining a precise picture of the daily use of cannabis. Therefore, we examined mobile-based applications available on the market to track cannabis use, which could potentially help gather more detailed data of daily cannabis consumption. Of the 21 tracking apps, Strainprint, Releaf App, and *LooseLeaf*, have published peer-reviewed research. Although Strainprint, Releaf App included numerous general conditions and symptoms, they were intended to monitor cannabis use for medical purposes for adults. Only *LooseLeaf* was developed through participatory co-design to monitor cannabis use and cannabis-related experiences for individuals at CHR.

Applications to other areas

Alcohol use disorders are the most prevalent substance use disorder globally, contributing significantly to the global disease burden (Degenhardt et al., 2018). Similarly, after cannabis, alcohol is the most reported substance among CHR youth (Addington et al., 2014), with rates ranging from 17% to 44%. CHR individuals endorse similar or less alcohol use than their age-matched healthy controls counterparts (Buchy et al., 2015; Calvo et al., 2020). Alcohol is not related to the transition to psychosis among CHR individuals. However, cannabis and alcohol are usually consumed together (Dragt et al., 2012), leaving the unique contribution of these substances to clinical outcome unclear. It has been suggested that alcohol may be confounding the relationship between cannabis and transition to psychosis (Dragt et al., 2012), with a resulting weakened association when alcohol use is taken into account (Auther et al., 2012).

Similar to cannabis, most CHR studies report basic information on alcohol consumption, and most do not report amounts or frequencies (Addington et al., 2014). Although research has incorporated phone-based ecological momentary assessments to capture alcohol and cannabis consumption among youth (Comulada et al., 2016), this has not disentangled the contribution of the variables to clinical outcome among CHR individuals. Thus, to gather near-real-time information about comorbid substance use, there is a need to integrate in the tracking apps a detailed assessment of alcohol and cannabis consumption.

Mini-dictionary of terms

- *Attenuated psychotic symptoms (APS)*: Psychotic symptoms at subthreshold intensity (not severe enough) or subthreshold frequency (not occurring often enough) to meet a diagnosis of psychosis. APS include symptoms such as unusual thought content, suspiciousness or persecutory ideas, grandiose ideas or expansiveness, perceptual abnormalities, or disorganized speech.
- *Attenuated positive symptom syndrome (APSS)*: Requires the presence of attenuated positive symptoms but only at an attenuated or subthreshold level and should not be at a psychotic intensity.
- *Brief intermittent psychotic syndrome (BIPS)*: Here symptoms are present at a fully psychotic level but are very brief. These symptoms typically last for only a few minutes at a time or if longer for perhaps only a day. They remit spontaneously and are not sufficiently frequent or disorganizing to meet psychosis criteria.
- *Comprehensive Assessment of At-Risk Mental States (CAARMS)*: Is a diagnostic instrument developed at the PACE Clinic in Melbourne (Australia) to rate APS and determine whether one of the high-risk criteria (APS, BIPS, or GRD) were met.
- *Clinical high risk (CHR):* One of the terms used in the literature to refer to psychosis risk syndromes. It was coined by Correll et al. (2010) to describe characteristic symptoms and signs that are present in an attenuated form during the risk phase, compared to the acute phase of psychosis.
- *Genetic risk and deterioration syndrome (GRD)*: Defined by a combination of (a) meeting criteria for Schizotypal personality disorder or having a first-degree family member with a diagnosis of a schizophrenia spectrum disorder or mood disorder with psychotic features *and* (b) functional deterioration, which is met if there has been $\geq$30% decrease in global functioning during the last year.
- *Negative symptoms*: Refer to a diminution or an absence of normal "processes" and are defined as a reduction in emotion, motivation, and/or expressive behavior. Negative symptoms are a core feature of schizophrenia and the attenuated psychosis syndrome and include five core domains such as a sociality, avolition (i.e., lack of motivation), anhedonia (i.e., inability to feel pleasure), blunted affect (i.e., difficulty to express emotions), and alogia (i.e., poverty of speech).
- *The Structured Interview for Psychosis-risk Syndromes (SIPS)*: Developed by McGlashan et al. (2010), is a semi-structured interview and rating instrument intended to evaluate psychosis risk syndromes. The SIPS was developed to diagnose the risk syndromes, the presence of transition to psychosis, and to rate the severity of risk symptoms longitudinally.

Key facts of cannabis use for those at CHR

- Regular cannabis use (daily or near-daily use) is associated with an increased risk of schizophrenia and psychotic symptoms among the general population.
- The use of high-potency cannabis (high THC%) has been associated with an increased risk of psychosis.
- When cannabis use is initiated before the age of 16, the risk of developing psychosis may increase.

- Cannabis use is more prevalent among individuals at CHR than healthy controls.
- The acute effects of cannabis comprise euphoria, changes in perception, relaxation, time distortion, increased heart rate and blood pressure, decreased attention span and memory, and impaired motor functioning.

Summary points

- Cannabis use is more prevalent among individuals at CHR than healthy controls.
- Half of the individuals at CHR report having tried cannabis in their life.
- Twenty-five percent of CHR individuals report current cannabis use.
- Evidence regarding the association of cannabis use and an increase of attenuated positive and negative symptoms is mixed.
- Current meta-analyses do not support the notion that cannabis use among CHR individuals is associated with the transition to psychosis.
- The majority of CHR studies have assessed cannabis use dichotomously and in single-time points, with a small percentage examining cannabis use frequency or other cannabis details.
- Of the 246 cannabis apps available on the market, 8.1% track cannabis use and symptoms, of which Strainprint, Releaf App, and *LooseLeaf* have published peer-reviewed research.
- *LooseLeaf* is the only mobile-based application developed through participatory co-design that monitors cannabis use and cannabis-related experiences for youth at CHR.

References

Addington, J., Case, N., Saleem, M. M., Auther, A. M., Cornblatt, B. A., & Cadenhead, K. S. (2014). Substance use in clinical high risk for psychosis: A review of the literature. *Early Intervention in Psychiatry*, *8*(2), 104–112.

Addington, J., Farris, M. S., Devoe, D. J., & Metzak, P. (2020). Progression from being at-risk to psychosis: Next steps. *NPJ Schizophrenia*, *6*(1), 27.

Addington, J., Farris, M., Stowkowy, J., Santesteban-Echarri, O., Metzak, P., & Shakeel, M. K. (2019). Predictors of transition in individuals at clinical high risk. *Current Psychiatry Reports*, *21*(6), 39–49.

Arseneault, L., Cannon, M., Poulton, R., Murray, R., Caspi, A., Moffitt, T. E., Arseneault, L., Cannon, M., Murray, R., & Poulton, R. (2002). Cannabis use in adolescence and risk for adult psychosis: Longitudinal prospective study. *BMJ*, *325*, 1212–1213.

Auther, A. M., McLaughlin, D., Carrión, R. E., Nagachandran, P., Correll, C. U., & Cornblatt, B. A. (2012). Prospective study of cannabis use in adolescents at clinical high risk for psychosis: Impact on conversion to psychosis and functional outcome. *Psychological Medicine*, *42*, 2485–2497.

Barkus, E. J., Stirling, J., Hopkins, R. S., & Lewis, S. (2006). Cannabis-induced psychosis-like experiences are associated with high schizotypy. *Psychopathology*, *39*, 175–178.

Beck, K., Andreou, C., Studerus, E., Heitz, U., Ittig, S., Leanza, L., & Riecher-Rössler, A. (2019). Clinical and functional long-term outcome of patients at clinical high risk (CHR) for psychosis without transition to psychosis: A systematic review. *Schizophrenia Research*, *210*, 39–47.

Buchy, L., Cadenhead, K. S., Cannon, T. D., Cornblatt, B. A., McGlashan, T. H., Perkins, D. O., Seidman, L. J., Tsuang, M. T., Walker, E. F., Woods, S. W., Heinssen, R., Bearden, C. E., Mathalon, D., & Addington, J. (2015). Substance use in individuals at clinical high risk of psychosis. *Psychological Medicine*, *45*(11), 2275–2284.

Calvo, E., Mittal, V., Schiffman, J., & Ellman, L. (2020). Alcohol and cannabis use in individuals at clinical high-risk for psychosis—Relationship between symptom course and patterns of use. *Biological Psychiatry*, *87*, S247.

Carney, R., Cotter, J., Firth, J., Bradshaw, T., & Yung, A. R. (2017). Cannabis use and symptom severity in individuals at ultra high risk for psychosis: A meta-analysis. *Acta Psychiatrica Scandinavica*, *136*(1), 5–15.

Comulada, W. S., Swendeman, D., & Wu, N. (2016). Cell phone-based ecological momentary assessment of substance use context for Latino youth in outpatient treatment: Who, what, when and where. *Drug and Alcohol Dependence*, *167*, 207–213.

Correll, C. U., Hauser, M., Auther, A. M., & Cornblatt, B. A. (2010). Research in people with psychosis risk syndrome: A review of the current evidence and future directions. *Journal of Child Psychology and Psychiatry, and Allied Disciplines*, *51*(4), 390–431.

Cuttler, C., & Spradlin, A. (2017). Measuring cannabis consumption: Psychometric properties of the daily sessions, frequency, age of onset, and quantity of cannabis use inventory (DFAQ-CU). *PLoS One*, *12*(5), 1–14.

Degenhardt, L., Charlson, F., Ferrari, A., Santomauro, D., Erskine, H., Mantilla-Herrara, A., Whiteford, H., Leung, J., Naghavi, M., Griswold, M., Rehm, J., Hall, W., Sartorius, B., Scott, J., Vollset, S. E., Knudsen, A. K., Haro, J. M., Patton, G., Kopec, J., & Vos, T. (2018). The global burden of disease attributable to alcohol and drug use in 195 countries and territories, 1990–2016: A systematic analysis for the Global Burden of Disease Study 2016. *The Lancet Psychiatry*, *5*(12), 987–1012.

Di Forti, M., Sallis, H., Allegri, F., Trotta, A., Ferraro, L., Stilo, S. A., Marconi, A., La Cascia, C., Marques, T. R., Pariante, C., Dazzan, P., Mondelli, V., Paparelli, A., Kolliakou, A., Prata, D., Gaughran, F., David, A. S., Morgan, C., Stahl, D., & Murray, R. M. (2014). Daily use, especially of high-potency cannabis, drives the earlier onset of psychosis in cannabis users. *Schizophrenia Bulletin*, *40*(6), 1509–1517.

Dragt, S., Nieman, D. H., Schultze-Lutter, F., van der Meer, F., Becker, H., de Haan, L., Dingemans, P. M., Birchwood, M., Patterson, P., Salokangas, R. K. R., Heinimaa, M., Heinz, A., Juckel, G., Graf von Reventlow, H., French, P., Stevens, H., Ruhrmann, S., Klosterkötter, J., & Linszen, D. H. (2012). Cannabis use and age at onset of symptoms in subjects at clinical high risk for psychosis. *Acta Psychiatrica Scandinavica*, *125*(1), 45–53.

Drake, R. E., Mueser, K., & McHugo, G. (1996). Clinical rating scales: alcohol use scale (AUS), drug use scale (DUS) and substance abuse treatment scales (STATS). In L. I. Sederer, & B. Dickey (Eds.), *Outcomes assessment in clinical practice* (pp. 113–116). Williams and Wilkins.

Farris, M. S., Shakeel, M. K., & Addington, J. (2020). Cannabis use in individuals at clinical high-risk for psychosis: A comprehensive review. *Social Psychiatry and Psychiatric Epidemiology*, *55*(5), 527–537.

Fusar-Poli, P., Cappucciati, M., Rutigliano, G., Lee, T. Y., Beverly, Q., Bonoldi, I., Lelli, J., Kaar, S. J., Gago, E., Rocchetti, M., Patel, R., Bhavsar, V., Tognin, S., Badger, S., Calem, M., Lim, K., Kwon, J. S., Perez, J., & McGuire, P. (2016). Towards a standard psychometric diagnostic interview for subjects at ultra high risk of psychosis: CAARMS versus SIPS. *Psychiatry Journal*, *2016*, 1–11.

Gupta, T., Cowan, H. R., Strauss, G. P., Walker, E. F., & Mittal, V. A. (2020). Deconstructing negative symptoms in individuals at clinical high-risk for psychosis: Evidence for volitional and diminished emotionality subgroups that predict clinical presentation and functional outcome. *Schizophrenia Bulletin*, 1–10. In Press.

Kalaba, M., & Ware, M. A. (2021). Cannabinoid profiles in medical cannabis users: Effects of age, gender, symptoms, and duration of use. *Cannabis and Cannabinoid Research*, *X*(X), 1–12.

Kraan, T., Velthorst, E., Koenders, L., Zwaart, K., Ising, H. K., Van Den Berg, D., De Haan, L., & Van Der Gaag, M. (2016). Cannabis use and transition to psychosis in individuals at ultra-high risk: Review and meta-analysis. *Psychological Medicine*, *46*(4), 673–681.

López-Pelayo, H., Batalla, A., Balcells, M. M., Colom, J., & Gual, A. (2015). Assessment of cannabis use disorders: A systematic review of screening and diagnostic instruments. *Psychological Medicine*, *45*(6), 1121–1133.

Lowe, D. J. E., Sasiadek, J. D., Coles, A. S., & George, T. P. (2019). Cannabis and mental illness: A review. *European Archives of Psychiatry and Clinical Neuroscience*, *269*(1), 107–120.

MacHielsen, M., Van Der Sluis, S., & De Haan, L. (2010). Cannabis use in patients with a first psychotic episode and subjects at ultra high risk of psychosis: Impact on psychotic- and pre-psychotic symptoms. *Australian and New Zealand Journal of Psychiatry*, *44*(8), 721–728.

Marconi, A., Di Forti, M., Lewis, C. M., Murray, R. M., & Vassos, E. (2016). Meta-Analysis of the association between the level of cannabis use and risk of psychosis. *Schizophrenia Bulletin*, *42*(5), 1262–1269.

McGlashan, T., Walsh, B. C., & Woods, S. W. (2010). *The psychosis-risk syndrome: Handbook for diagnosis and follow-up*. Oxford University Press.

Mittal, V. A., & Addington, J. M. (2021). Embracing heterogeneity creates new opportunities for understanding and treating those at clinical-high risk for psychosis. *Schizophrenia Research*, *227*, 1–3.

Mizrahi, R., Kenk, M., Suridjan, I., Boileau, I., George, T. P., McKenzie, K., Wilson, A. A., Houle, S., & Rusjan, P. (2014). Stress-induced dopamine response in subjects at clinical high risk for schizophrenia with and without concurrent cannabis use. *Neuropsychopharmacology*, *39*(6), 1479–1489.

Mongan, D., Föcking, M., Healy, C., Susai, S. R., Heurich, M., Wynne, K., Nelson, B., McGorry, P. D., Amminger, G. P., Nordentoft, M., Krebs, M.-O., Riecher-Rössler, A., Bressan, R. A., Barrantes-Vidal, N., Borgwardt, S., Ruhrmann, S., Sachs, G., Pantelis, C., van der Gaag, M., … McGuire, P. (2020). Development of proteomic prediction models for transition to psychotic disorder in the clinical high-risk state and psychotic experiences in adolescence. *JAMA Psychiatry*, *78*.

Mustonen, A., Niemelä, S., Nordström, T., Murray, G. K., Mäki, P., Jääskeläinen, E., & Miettunen, J. (2018). Adolescent cannabis use, baseline prodromal symptoms and the risk of psychosis. *British Journal of Psychiatry*, *212*(4), 227–233.

Nesvåg, S., & McKay, J. R. (2018). Feasibility and effects of digital interventions to support people in recovery from substance use disorders: Systematic review. *Journal of Medical Internet Research*, *20*(8).

Nieman, D. H., Dragt, S., van Duin, E. D. A., Denneman, N., Overbeek, J. M., de Haan, L., Rietdijk, J., Ising, H. K., Klaassen, R. M. C., van Amelsvoort, T., Wunderink, L., van der Gaag, M., & Linszen, D. H. (2016). COMT Val158Met genotype and cannabis use in people with an at risk mental state for psychosis: Exploring gene x environment interactions. *Schizophrenia Research*, *174*(1–3), 24–28.

O'Farrell, T. J., Fals-Stewart, W., & Murphy, M. (2003). Concurrent validity of a brief self-report drug use frequency measure. *Addictive Behaviors*, *28*(2), 327–337.

Oliver, D., Reilly, T. J., Baccaredda Boy, O., Petros, N., Davies, C., Borgwardt, S., McGuire, P., & Fusar-Poli, P. (2020). What causes the onset of psychosis in individuals at clinical high risk? A meta-analysis of risk and protective factors. *Schizophrenia Bulletin*, *46*.

Quattrone, D., Ferraro, L., Tripoli, G., La Cascia, C., Quigley, H., Quattrone, A., Jongsma, H. E., Del Peschio, S., Gatto, G., Gayer-Anderson, C., Jones, P. B., Kirkbride, J. B., La Barbera, D., Tarricone, I., Berardi, D., Tosato, S., Lasalvia, A., Szöke, A., Arango, C., & Di Forti, M. (2021). Daily use of high-potency cannabis is associated with more positive symptoms in first-episode psychosis patients: The EU-GEI case-control study. *Psychological Medicine*, *51*(8), 1329–1337.

Quinn, C. A., Wilson, H., Cockshaw, W., Barkus, E., & Hides, L. (2017). Development and validation of the cannabis experiences questionnaire—Intoxication effects checklist (CEQ-I) short form. *Schizophrenia Research*, *189*, 91–96.

Salazar De Pablo, G., Radua, J., Pereira, J., Bonoldi, I., Arienti, V., Besana, F., Soardo, L., Cabras, A., Fortea, L., Catalan, A., Vaquerizo-Serrano, J., Coronelli, F., Kaur, S., Da Silva, J., Shin, J. I., Solmi, M., Brondino, N., Politi, P., McGuire, P., & Fusar-Poli, P. (2021). Probability of transition to psychosis in individuals at clinical high risk: An updated meta-analysis. *JAMA Psychiatry*, *78*(9), 970–978.

Santesteban-Echarri, O., Haffey, P., Kim, G. H., Tang, J., & Addington, J. (2020). A mobile-based app to monitor cannabis use among youth at clinical high risk (CHR) for psychosis: Feasibility and acceptability of LooseLeaf. *Schizophrenia Research*, *222*, 505–506.

Schifani, C., Tseng, H.-H., Rao, N., Wilson, A. A., Houle, S., Rusjan, P. M., Pruessner, J., Tagore, A., & Mizrahi, R. (2020). Stress-induced cortical dopamine response is altered in subjects at clinical high risk for psychosis using cannabis. *Addiction Biology*, *25*(4), e12812.

Schoeler, T., Monk, A., Sami, M. B., Klamerus, E., Foglia, E., Brown, R., Camuri, G., Altamura, A. C., Murray, R., & Bhattacharyya, S. (2016). Continued versus discontinued cannabis use in patients with psychosis: A systematic review and meta-analysis. *The Lancet Psychiatry*, *3*, 215–225.

Stoyanov, S. R., Hides, L., Kavanagh, D. J., & Wilson, H. (2016). Mobile application rating scale: User version (uMARS). *JMIR mHealth and uHealth*, *4*(2), e72.
United Nations Office on Drugs and Crime. (2019). *Cannabis and hallucinogens*. United Nations publication.
Valmaggia, L. R., Day, F. L., Jones, C., Bissoli, S., Pugh, C., Hall, D., Bhattacharyya, S., Howes, O., Stone, J., Fusar-Poli, P., Byrne, M., & McGuire, P. K. (2014). Cannabis use and transition to psychosis in people at ultra-high risk. *Psychological Medicine*, *44*(12), 2503–2512.
Yung, A. R., Yuen, H. P., McGorry, P. D., Phillips, L. J., Kelly, D., Dell'Olio, M., Francey, S. M., Cosgrave, E. M., Killackey, E., Stanford, C., Godfrey, K., & Buckby, J. (2005). Mapping the onset of psychosis: The comprehensive assessment of at-risk mental states. *Australian and New Zealand Journal of Psychiatry*, *39*(11–12), 964–971.

Part VIII

Impact on non-neurological systems

Chapter 39

The impact of cannabis and cannabinoids on the skin

Kimberly Shao[a], Jane M. Grant-Kels[a,b], and Campbell Stewart[a]
[a]*Department of Dermatology, University of Connecticut Health Center, Farmington, CT, United States,* [b]*Department of Dermatology, University of Florida, Gainesville, FL, United States*

Abbreviations

2-AG	2-arachidonoylglycerol
ACD	allergic contact dermatitis
AEA	*N*-arachidonoylethanolamide, anandamide
cAMP	cyclic adenosine monophosphate
CB1R	cannabinoid receptor 1
CB2R	cannabinoid receptor 2
CBD	cannabidiol
CCL	chemokine ligand
ECS	endocannabinoid system
FAAH	fatty acid amide hydrolase
GDP	adenosine diphosphate
GPR	G-protein-couple receptors
GTP	adenosine triphosphate
IFN	interferon
IL	interleukin
MAP	mitogen-activated protein
NMSC	nonmelanoma skin cancer
OEA	*N*-oleoylethanolamine
PEA	*N*-palmithylethanolamide
PKA	protein kinase A
PPAR	peroxisome-proliferators-activated receptors
SLE	systemic lupus erythematosus
TRPV-1	transient receptor potential vanilloid 1
UVB	ultraviolet B

Introduction

Cannabinoid studies imply their role in regulating pain, proliferation, and inflammation, giving them therapeutic potential for many dermatologic conditions. A survey of dermatologists reported that providers are interested in prescribing cannabinoids but have limited knowledge (Robinson et al., 2018). Herein we review the physiology of the cutaneous endocannabinoid system as well as potential therapeutic uses, current research, potential cutaneous adverse effects, and need for further research. Table 1 shows an overview of the dermatologic conditions we will discuss.

Cannabinoid receptors

Cannabinoid receptor 1 (CB1R) and cannabinoid receptor 2 (CB2R) are well-documented cannabinoid receptors, which function as G-protein-coupled receptors (GPRs) when activated (Elphick & Egertová, 2001). Activation of cannabinoid receptors causes a reduction in intracellular cyclic adenosine monophosphate (cAMP) (Basavarajappa, 2007). This reduction leads to suppression of protein kinase A (PKA) and an increase in intracellular calcium. The change in calcium

Cannabis Use, Neurobiology, Psychology, and Treatment. https://doi.org/10.1016/B978-0-323-89862-1.00006-4

TABLE 1 Potential role of cannabinoids in dermatologic conditions.

Cannabinoid effect	Dermatologic condition
Anti-proliferative	Basal cell carcinoma
	Squamous cell carcinoma
	Melanoma
	Psoriasis
Anti-inflammatory	Atopic dermatitis
	Allergic contact dermatitis
	Psoriasis
	Acne vulgaris
	Dermatomyositis
	Systemic sclerosis
	Lupus erythematosus
	Alopecia areata
Anti-pruritic	Pruritus
	Atopic dermatitis
	Allergic contact dermatitis
Increased hair growth	Alopecia areata
	Androgenetic Alopecia
Decreased hair growth	Hirsutism/Hypertrichosis
Anti-nociceptive	Pain
	Facial postherpetic neuralgia
	Epidermolysis bullosa
Wound healing	Wounds
	Epidermolysis bullosa
Anti-microbial	Atopic dermatitis
	Acne vulgaris
	Wounds

ion concentration causes downstream effects, simultaneously activating and inhibiting certain enzymes and the release of various interleukins, chemokines, and growth factors (Kupczyk et al., 2009).

The highest expression of CB1R appears in the nervous system, which includes cutaneous nerves. CB1R has also been observed on keratinocytes in the stratum spinosum and stratum granulosum and on epithelial cells of hair follicles within the infundibulum and inner hair root sheath. Macrophages and dermal mast cells also express CB1R (Ständer et al., 2005). CB2R expression is primarily concentrated within non-neural tissues especially cells of the immune system, including mast cells and macrophages (Klein et al., 2003; Ständer et al., 2005). CB2R has also been found on cutaneous nerves and undifferentiated cells of the hair follicular infundibulum, outer root sheath, and bulb (Ständer et al., 2005).

Non-cannabinoid receptors

Cannabinoids serve as ligands for non-cannabinoid receptors. Transient receptor potential vanilloid 1 (TRPV-1) can be activated by various inflammatory mediators, including endocannabinoids (Ahluwalia et al., 2003; Muller et al., 2018).

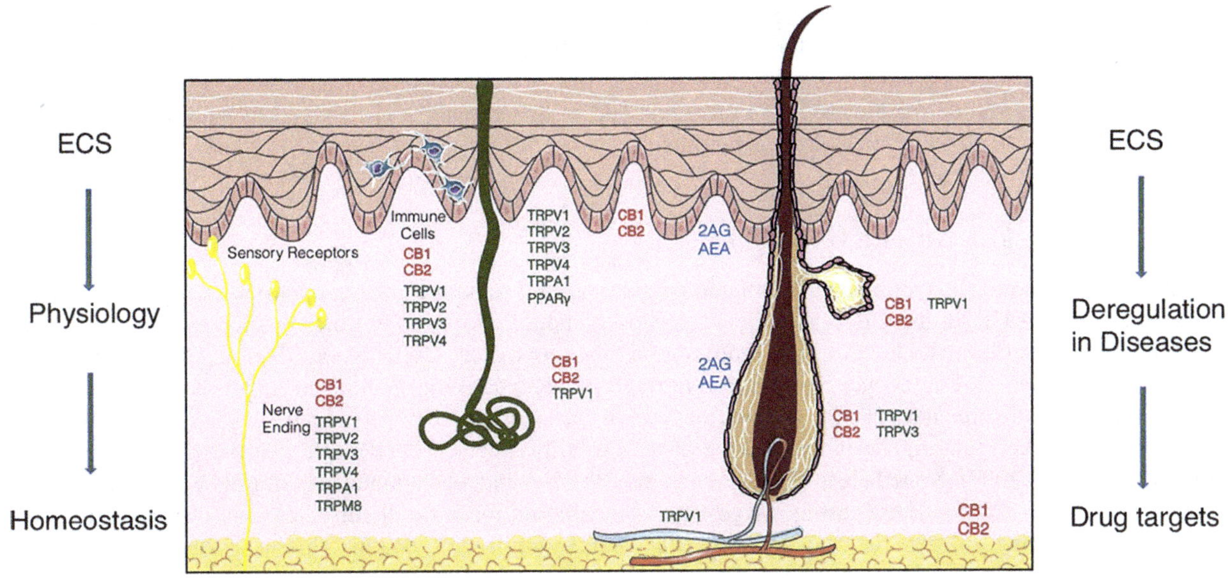

FIG. 1 Cannabinoid receptors within the skin. This figure shows locations of CB1R/CB2R and TRPV receptors within cellular departments of the skin, implying their potential use as targets for therapy. *From Tijani, A. O., Thakur, D., Mishra, D., Frempong, D., Chukwunyere, U. I., & Puri, A. (2021). Delivering therapeutic cannabinoids via skin: Current state and future perspectives. Journal of Controlled Release, 334, 427–451. https://doi.org/10.1016/j.jconrel.2021.05.005 with permission.*

TRPV-1 is an ion channel found on central neurons, perivascular and cutaneous sensory nerves, macrophages, Langerhans cells, endothelial cells, epithelial cells, epidermal and hair follicle keratinocytes, and smooth muscle cells (Movahed et al., 2005; Veronesi & Oortgiesen, 2006). When activated, TRPV-1 can induce nitric oxide synthesis and vasodilation, inhibit L-type calcium channels, and decrease cAMP (Kupczyk et al., 2009). Peroxisome-proliferators-activated receptors (PPAR) can be directly activated by endocannabinoids. PPARs affect lipid and glucose metabolism and inflammatory responses (Sun & Bennett, 2007).

Fig. 1 illustrates an overview of the endocannabinoid system (ECS) in the skin (Tijani et al., 2021).

Physiologic roles of cannabinoids in the skin

Two of the most well-studied endocannabinoids (referring to endogenous ligands for cannabinoid receptors) are *N*-arachidonoylethanolamine (anandamide, AEA) and 2-arachidonoylglycerol (2-AG) are endogenous agonists of CB1R and CB2R. In addition to the cannabinoid receptors, both endocannabinoids can activate TRPV1. AEA is an agonist for several PPARs (Hillard, 2018). Responses to these endocannabinoids help define the role cannabinoids play in the skin.

Cannabinoids may have a function in keratinocyte differentiation. In models of human epidermal keratinocytes exogenous application of AEA caused inhibition of both the formation of cornified envelopes and keratinocyte differentiation (Maccarrone et al., 2003). AEA also downregulated the gene transcription of keratin 1, 10, and involucrin (Paradisi et al., 2008).

Mouse models demonstrated exposure to AEA, 2-AG, and delta-9-tetrahydrocannabinol (delta-9-THC), a partial CB1R agonist, resulted in decreased proliferation of B-cells and macrophages (Salzet et al., 2000). Cannabinoids appeared to regulate Th1 and Th2 homeostasis, with activation of cannabinoid receptors leading to greater suppression of Th1. Proinflammatory cytokines like TNF-alpha, IL-1, IL-6, IL-10, and IL-12 as well as chemokines including chemokine ligand (CCL)-2, CCL5, C-X-C motif chemokine ligand (CXCL)-8, and CXCL10 are downregulated following the application of cannabinoids (Berdyshev et al., 1997; Kupczyk et al., 2009).

The antinociceptive effect of endocannabinoids is recognized, but their relationship with pain is complex. Multiple studies indicate that the activation of peripheral CB1Rs and CB2Rs decreases pain response (Johanek & Simone, 2004; LaBuda et al., 2005). A study of chronic hyperalgesia in rodent models, however, showed pain is attenuated by a selective CB1R blocker (Costa et al., 2005). The effects of endocannabinoids on pain may also involve other receptors such as TRPV-1 and PPAR-gamma (Costa et al., 2008).

Cannabinoids are also candidates for pruritus treatment. CB1R and CB2R agonists reduce itch via activation of receptors on cutaneous sensory nerve fibers, mast cells, and keratinocytes (Ständer et al., 2008). CB1R activation in the central nervous system decreases scratching behaviors, likely contributing to the antipruritic effects (Ward et al., 2009). Certain cannabinoids interact with TRPV-1 receptors and have been shown to reduce itch (Muller et al., 2018; Tóth et al., 2014).

Table 2 summarizes the dermatologic conditions reviewed in this chapter and the type of studies that were found regarding the potential therapeutic roles of cannabinoids.

Non-melanoma skin cancer (NMSC)

Malignant cells of keratinocyte carcinomas (basal cell carcinomas and squamous cell carcinomas) express CB1 and CB2 receptors. Activation of CB1R and CB2R in the epidermis may induce apoptosis of tumorigenic epidermal cells, impair vascularization, and interfere with keratinocyte proliferation (Casanova et al., 2003; Kupczyk et al., 2009). Endocannabinoids may further decrease keratinocyte proliferation through PPAR-γ activity (Wilkinson & Williamson, 2007).

Cannabinoid-based treatment for NMSC remains theoretical and findings are mostly in animal studies. Mice who lacked CB1R/CB2R had reduced tumorigenesis following exposure to ultraviolet B (UVB) light compared to mice with these receptors (Zheng et al., 2008). Nonetheless, the endocannabinoid AEA-induced apoptosis in murine cell lines with NMSC (Soliman & Van Dross, 2016). These results suggest that cannabinoid receptors themselves may be required for tumor development, but giving exogenous cannabinoids can induce tumor cell death. This is further complicated by

TABLE 2 Cannabinoids research for dermatologic conditions by study design.

Dermatologic condition	Study type
Non-melanoma skin cancer	In vitro
	In vivo: mouse injected
Melanoma	In vitro
	In vivo: mouse injected
Psoriasis	In vitro
	Human topical, randomized control trial (n=2)
Acne vulgaris/Seborrhea	In vitro
	Human inhaled, retrospective survey
	Human topical, split face study
	Human topical, phase 2 clinical trial
Alopecia/Hair growth	In vitro
	In vivo, mouse oral and topical
	Human inhaled, retrospective survey
	Human topical, case report/case series
Pruritus	In vitro
	In vivo: mouse oral and topical
	Human oral, case series
	Human topical, case report/case series
	Human topical, observational
	Human topical, randomized control trial (n=18)
Atopic dermatitis	In vitro
	In vivo: mouse oral and topical
	Human topical, observational

TABLE 2 Cannabinoids research for dermatologic conditions by study design—cont'd

Dermatologic condition	Study type
Allergic contact dermatitis	In vitro
	In vivo: mouse topical
	In vivo: mouse oral
	Human oral: phase 3 clinical trial
Dermatomyositis	Human oral: phase 3 clinical trial
Cutaneous lupus erythematosus	In vivo: mouse oral and topical
	Human oral: phase 2 clinical trial
Cutaneous pain	In vivo: mouse injection and topical
	Human topical, case report/case series
	Human oral (sublingual), case series
Wound healing and skin aging	In vitro
	In vivo: mouse injection
	Human topical, case series

concentration-dependent effects. Nanomolar concentrations of endocannabinoids stimulated tumor development while high concentrations diminished NMSC growth (Van Dross et al., 2013).

Melanoma

CB1R and CB2R are present on both keratinocytes and melanocytes (Pucci et al., 2012). Like NMSC, the relationship of cannabinoids to melanoma is complex. In a small study, melanomas were found to have upregulated CB2Rs compared to melanocytes in nevi (Zhao et al., 2012). Differences in endocannabinoid levels were also found. In melanoma patients, AEA levels were decreased while 2-AG levels were increased. AEA and 2-AG are both cannabinoid receptor agonists, but this suggests that the roles of AEA and 2-AG may differ in regard to tumorigenesis. Additionally, an endocannabinoid-like substance *N*-oleoylethanolamine (OEA), which downregulates AEA but not 2-AG, was found to be elevated in melanoma patients (Tóth et al., 2019).

In vitro studies on melanoma cell lines showed cannabinoid-induced apoptosis (Kupczyk et al., 2009), but the apoptosis may be dose dependent. While higher doses of AEA ($>5\,\mu M$) resulted in apoptosis, lower doses of AEA ($<3\,\mu M$) instead stimulated melanogenesis (Pucci et al., 2012). Two synthetic cannabinoids, also CB2R agonists, WIN-55,212-2 and JWH-133, decreased cell proliferation and caused cell cycle arrest in murine models with melanoma (Blázquez et al., 2006).THC inhibited cell growth and induced apoptosis in melanoma cell lines, even when compared to a chemotherapy control (Armstrong et al., 2015). This effect may be due to THC's ability to mitigate the proinflammatory microenvironment (Sánchez et al., 2003). A 1:1 compound of cannabidiol (CBD) and THC appeared to have a synergistic cytotoxic effect on melanoma cells (Armstrong et al., 2015). Agonists of CB2R also reduce angiogenesis and inhibit metastatic spread (Eagleston et al., 2018).

In contrast, researchers reported that activation of CB1R alone in vitro may incite tumor growth (Carpi et al., 2017; Zhou et al., 2015).

Importantly, cannabis use may lower the response rate of nivolumab, a programmed cell death protein (PD)-1 inhibitor used in the treatment of melanoma (Taha et al., 2019).

Psoriasis

Cannabinoids' ability to inhibit keratinocyte proliferation is advantageous for keratinocyte carcinomas and inflammatory disorders, like psoriasis (Kupczyk et al., 2009; Scheau et al., 2020; Sheriff et al., 2019). A synthetic CB1 agonist reduces the expression of keratin 6 and 16 and inhibits keratinocyte proliferation within human skin models (Ramot et al., 2013).

Cannabinoids may also act independently of CB1R and CB2R to affect psoriasis (Wilkinson & Williamson, 2007). Cannabinoids can interact with PPARs, which include retinoid and vitamin D receptors (O'Sullivan & Kendall, 2010). Cannabinoids also act on immune cells, another promising target for psoriasis (Kupczyk et al., 2009). In 2019, a patent was filed for a topical treatment of psoriasis containing 15% CBD and cannabigerol, a natural phytocannabinoid. The patent claimed that after 6 weeks of twice daily use two subjects had 16% to 33% improvement (Changoer & Anastassov, 2018). There is no clinical trial with published results evaluating cannabinoid-containing products and psoriasis.

Acne vulgaris and seborrhea

Human sebaceous glands express CB1R and CB2R (Ständer et al., 2005). Human sebocytes also express key enzymes that produce and breakdown endocannabinoids (Szöllősi et al., 2017). Hormones like testosterone result in increased lipid production and sebum secretion, associated with acne and seborrhea. Some cannabinoids may counteract the influence of these hormones through activation of CB2R (Dobrosi et al., 2008; Oláh et al., 2014; Oláh et al., 2016). Exogenous CBD and THC were found to have anti-acne and anti-sebocyte proliferative effects (Oláh et al., 2014; Oláh et al., 2016). CBD may also lower sebocyte production through TRPV-4 activation (Oláh et al., 2014). Bacteria, especially Propionibacterium acnes, impact the development of acne. Hemp seed hexane extracts have been shown to have antimicrobial activity, including against *P. acnes* (Appendino et al., 2008; Jin & Lee, 2018).

There are few human trials analyzing specific cannabinoid products' effect on acne and seborrhea. One split-face study showed that 3% topical cannabis seed extract decreases erythema and skin sebum (Ali & Akhtar, 2015). A phase 1b trial had participants with moderate or severe acne apply 5% topical synthetic CBD (BTX1503) twice daily. Patients reported no serious treatment-related adverse events (Spleman et al., 2018). A phase 2 clinical trial of this product has been completed, but results are not published.

Conversely, recreational synthetic cannabinoid addiction and smoking cannabis have been associated with both the development and worsening of acne (Inci et al., 2017; Wolkenstein et al., 2015). Some specific phytocannabinoids such as cannabigerol and cannabigerovarin were found to prolong sebocyte viability illustrating the complex relationship between cannabinoids and cutaneous disorders like acne vulgaris (Oláh et al., 2016).

Alopecia and hair growth disorders

CB1R are expressed within human hair follicles; stimulation of CB1R has been shown to hinder hair shaft elongation and induce premature hair follicle regression (Ständer et al., 2005). Both CB1R and TRP ion channel activation may stimulate the onset of catagen phase and/or maintain anagen phase when they are activated (Szabó et al., 2019). Cannabinoid dysregulation may therefore pay a role in hair growth disorders. A mutation in tyrosine phosphatase, an enzyme involved in the production of the endocannabinoid anandamide, was shown to be linked to alopecia areata through the inability to effectively suppress cytotoxic T cells suggesting CB1R or TRP antagonists may have therapeutic potential (Betz et al., 2008; Salinas-Santander et al., 2015). Agonists, conversely, may benefit hirsutism or hypertrichosis. Concentration may influence outcomes as low concentrations of CBD targeting TRPV-4 receptors were shown to increase hair shaft elongation, while high concentrations hinder hair shaft development and promote apoptosis (Szabó et al., 2019).

An oral CB1R antagonist promoted hair growth in mice. However, topical use demonstrated no change in hair density (Srivastava et al., 2009). A single case report observed hair growth in alopecia areata using 1.0% CBD ointment (Arkaki, 2015). A case series of androgenetic alopecia subjects reported daily use of topical hemp oil containing 3–4 mg of CBD per dose resulted in an average of 93.5% increase in hair after 6 months with no reported adverse effects (Gupta & Talukder, 2021). In contrast, hair loss was reported as a common complaint among users of synthetic cannabinoids (Inci et al., 2017).

Pruritus

Although pruritus is not always associated with a rash, the urge to scratch can initiate an itch-scratch cycle resulting in cutaneous inflammation and a secondary rash. There are multiple potential mechanisms by which cannabinoids may alleviate pruritus, which makes the relationship between cannabinoids and itch multifaceted (Mounessa et al., 2017). Cutaneous nerve fibers contain CB1R/CB2R. Stimulation of these receptors may dampen excitation of associated nerve fibers and decrease symptoms of pruritus (Dvorak et al., 2003). In addition to peripheral effects of cannabinoid receptors, stimulation of CB1R in the central nervous system may diminish the perception of itch and associated behaviors (Bilir et al., 2018; Schlosburg et al., 2011). At a cellular level, cannabinoids may inhibit mast cell degranulation and histamine release (Avila et al., 2020; Nam et al., 2016; Sugawara et al., 2012). In skin prick tests, an irritant called falcarinol acts as a CB1R antagonist. It caused histamine-induced edema and an increase in proinflammatory cytokines interleukin (IL)-8 and monocyte chemotactic protein-1 (MCP-1) (Leonti et al., 2010). Independent of cannabinoid receptors, cannabinoids may modulate pruritus through TRPV1, a capsaicin receptor whose blockage is known to reduce itch (Ambrosino et al., 2013; Muller et al., 2018).

Animal and human models also show mixed results. THC and AEA suppressed fatty acid amide hydrolase (FAAH), an enzyme involved in pruritus, and in mice models resulted in reduced scratching (Schlosburg et al., 2011). Another selective synthetic CB2R agonist also decreased scratching behaviors when taken orally (Odan et al., 2012). In contrast, a synthetic CB2R antagonist called JTE-907, similarly suppressed scratching in mice (Maekawa et al., 2006). In humans HU210, a synthetic cannabinoid receptor agonist with a similar structure to THC, relieved itch (Dvorak et al., 2003). Patients with various conditions associated with pruritus (prurigo nodularis, lichen simplex, atopic dermatitis, asteatotic eczema, and uremic pruritus) reported itch relief using a topical emollient containing the endocannabinoid palmitoylethanolamide (PEA) (Eberlein et al., 2008; Ständer et al., 2006; Szepietowski et al., 2005; Yuan et al., 2014). Oral dronabinol, a synthetic THC, relieved intractable pruritus in patients with chronic liver disease (Neff et al., 2002).

Atopic dermatitis

Cannabinoids' anti-pruritic and anti-inflammatory effects could be leveraged for the treatment of atopic dermatitis (Bíró et al., 2009; Klein, 2005). Mice without CB1R have elevated levels of IL-4 and chemokine ligand (CCL)-8, similar to patients with atopic dermatitis. They also had an impaired epidermal barrier (Gaffal et al., 2014). Topical THC in mouse models of atopic dermatitis lowered production of CCL2 by keratinocytes and interferon (IFN) gamma by T-cells (Wollenberg et al., 2014). Agonists of CB1R instead resulted in an enhanced epidermal barrier function in human models (Kim et al., 2015; Roelandt et al., 2012). Topical PEA resulted in reduced itch for patients with atopic dermatitis. PEA-containing emollient creams resulted in decreased use of topical steroids, less excoriations, improved dryness, and decreased loss of sleep related to scratching (Eberlein et al., 2008).

Allergic contact dermatitis

Endocannabinoids impact on itch and inflammation may be leveraged for treatment of allergic contact dermatitis (ACD). In vitro studies of human keratinocytes exposed to topical CBD reported less inflammation following polyinosinic:polycytidylic acid induced allergic dermatitis (Petrosino et al., 2018). In vivo studies in mice showed that topical THC inhibited proinflammatory mediators such as CCL8 and CXCL10 and reduced symptoms associated with ACD (Gaffal et al., 2013). Mice who did not have CB1R or CB2R were more likely to develop contact hypersensitivity when exposed to nickel or 2,4-dinitrofluorobenzene (Karsak et al., 2007). This effect may be dependent on duration of activation; initial CB2R stimulation lowered inflammation, while chronic CB2R blockade was proinflammatory (Karsak et al., 2007; Ueda et al., 2005).

Connective tissue disorders

Cannabinoids mitigate inflammation, which may have a downstream effect on preventing tissue thickening and fibrosis in sclerosing disorders. CB1R and CB2R are also found on dermal fibroblasts (Tóth et al., 2019). This implicates that endocannabinoids may both play a part in the pathogenesis of and have therapeutic potential for connective tissue disorders such as systemic sclerosis, dermatomyositis, and systemic lupus erythematosus (SLE).

For example, plasma levels of 2-AG were significantly higher in patients with SLE compared with healthy patients (Navarini et al., 2018). However, the highest levels of 2-AG were found in SLE patients who had lower disease activity, which illustrates the complex relationship the endocannabinoid system may have with lupus. In mice modes, WIN55,212-2 and VCE-004.8, both synthetic cannabinoids, decreased dermal fibrosis in models of systemic sclerosis (Balistreri et al., 2011; Servettaz et al., 2010). In animal models of cutaneous lupus, topical anandamide reportedly reversed symptoms (Chalmers et al., 2018).

JBT-101 (also known as lenabasum ajulemic acid, or anabasum) is a synthetic analog of THC that binds to the CB2R. Hypothetically, JBT-101 may inhibit expression of certain growth factors expression including TGF-β, platelet-derived growth factor, and connective tissue growth factor. JBT-101 has been studied for systemic sclerosis, dermatomyositis and SLE. In mice, JBT-101 lowered the risk of bleomycin-induced dermal fibrosis in mice (Gonzalez et al., 2012). Phase 2 data of patients with diffuse cutaneous systemic sclerosis reported that those who took oral lenabasum have improved symptom index scores (Spiera et al., 2017). However, phase 3 clinical trials (RESOLVE-1) failed to reach the primary endpoint (Spiera et al., 2021). Similarly, oral lenabasum saw promising results in a phase 2 trial of patients with dermatomyositis. Patients reported improved symptom severity index scores (Werth et al., 2018). The phase 3 data (DETERMINE) did not meet the primary endpoint (GlobalNewswire, 2021). Of note, the study enrolled patients with two major classifications of dermatomyositis: classic dermatomyositis and amyopathic dermatomyositis. While muscle weakness was heavily weighted within the composite primary endpoint score, one of the study's secondary endpoints assessing inflammatory skin involvement did show improvement in skin manifestations (GlobalNewswire, 2021). A phase II clinical trial of oral lenabasum and SLE is underway (ClinicalTrials.gov, 2017).

Pain

While there are reports of cannabinoids alleviating various types of pain, we will focus on pain associated with disruption of the skin barrier. In animal models, an intraplantar injection of a CB2R agonist alleviated capsaicin-induced pain (Katsuyama et al., 2013). A topical mixed CB1R/CB2R agonist was found to have dose-dependent antinociceptive effect, and even worked synergistically with topical morphine (Dogrul et al., 2003; Yesilyurt et al., 2003). There are anecdotal reports of pain relief using topical CBD oil in patients with epidermolysis bullosa (Chelliah et al., 2018). A small observational study of facial postherpetic neuralgia and topical PEA reported significant pain reduction (Phan et al., 2010). A case series of three patients with pyoderma gangrenosum found that topical cannabis oil resulted in reduced analgesic and opioid use (Maida & Corban, 2017).

Wound healing and skin aging

Cannabinoids may accelerate wound re-epithelization and scar formation (Kong et al., 2021). CB2R expression varies by cell type and time post-injury over the process of wound healing in mice (Zheng et al., 2012). Both CB2R agonists and antagonists affect inflammation in incised wounds of mouse skin by regulating macrophage infiltration (Du et al., 2018). CB2R activation also diminished neutrophil infiltration, fostered keratinocyte migration, promoted fibroblast accumulation, and stimulated fibroblast to myofibroblast transformation (Wang et al., 2016). In cases of epidermolysis bullosa, topical CBD resulted in reported faster wound healing and less blistering (Chelliah et al., 2018). Interestingly, mice without CB1R have less skin rejuvenation and earlier onset aging (Bilkei-Gorzo et al., 2012).

Cutaneous adverse effects

The potential applications of cannabinoids for dermatologic therapy are exciting but potential risks exist. Oral symptoms immediately following use of inhaled cannabis include xerostomia, anesthesia, and erythema due to mucosal irritation (Darling & Arendorf, 1993). Long-term inhaled use may be associated with chronic inflammation, hyperkeratosis, leukoplakia, gingivitis, dental caries, oral candidiasis, and oral carcinoma (Darling, 2003; Darling & Arendorf, 1993; Firth, 1997). The relationship between oral cancer and cannabis remains controversial. Studies have quantified a 1.2- to 2.6-fold greater risk of head and neck cancers in patients who smoked cannabis (Firth, 1997). Other reports have found low to a reduced risk of oral cancer with cannabis use (Aldington et al., 2008; Marks et al., 2014).

While cannabinoids could be used to treat certain skin conditions, the concentration, and type of usage may have deleterious effects. Cannabinoids could improve acne yet regular use of inhaled cannabis and synthetic cannabinoids was associated with acne flares (Inci et al., 2017; Wolkenstein et al., 2015). Hair loss, gray hair, and premature aging have all been reported with chronic cannabis systemic use (Inci et al., 2017). THC and anandamide inhibit hair shaft elongation and hair matrix keratinocytes proliferation via CB1R (Telek et al., 2007). Chronic cannabis may develop abnormally enlarged, node-shaped areas on their hair shafts under scanning electron microscopy (Turkmenoglu et al., 2015).

Allergy to cannabis presents as urticaria, pruritus, and angioedema (Majmudar et al., 2006; Stöckli & Bircher, 2007; Williams et al., 2008). A type I and type IV hypersensitivity reaction to cannabis can been confirmed by scratch, prick, and patch testing (Fig. 2) (Armentia et al., 2011; Dhadwal & Kirchhof, 2018; Foster et al., 2018). Additive ingredients in over-the-counter cannabis products can cause cutaneous reactions. A study of products in California dispensaries found 1 or more allergens defined by the North American Contact Dermatitis Group listed in 84% of the products (Adler & DeLeo, 2019).

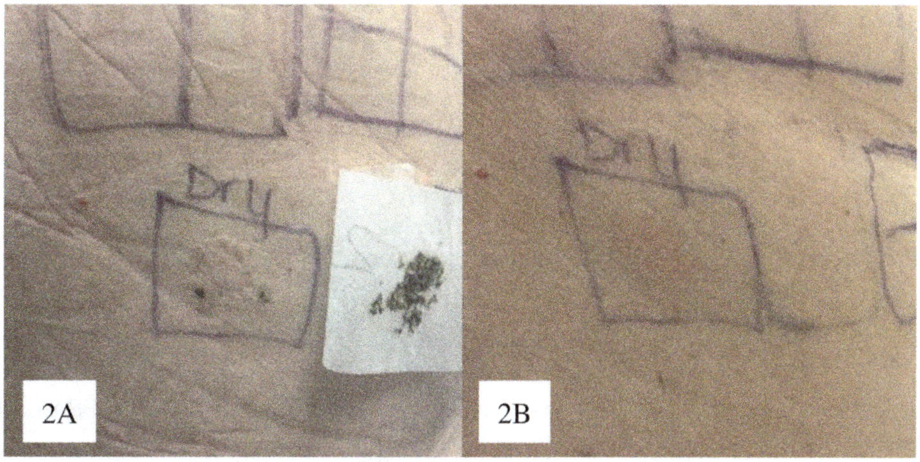

FIG. 2 Positive patch testing to cannabis. Figure A shows a positive patch test to dry cannabis 48-h after placement. Figure B shows a positive patch test read an additional 2 days after removal. *From Foster, E., Nguyen, C., & Norris, P. (2018). Contact buzz: Allergic contact dermatitis to cannabis. Dermatitis, 29(4), 223-224. https://doi.org/10.1097/DER.0000000000000378 with permission.*

Conclusion

Encouraging data indicate possible therapeutic roles for cannabinoids in dermatology. However, the interpretation of cannabinoids' therapeutic potential is complex. Cannabinoids bind to multiple cannabinoid and non-cannabinoid receptors with varying affinities, making their biological outcomes difficult to predict. The facts are complicated further by the various unregulated products that are available to consumers making claims that outpace clinical data. Additionally, cannabinoids have potential adverse effects.

Applications to other areas

Herein, we reviewed the endocannabinoid systems complex role in cutaneous disorders. Cannabinoids can affect nerve receptors and modulate immune function through cannabinoid receptors and non-cannabinoid receptors including TRP ion channels. Synthetic and natural cannabinoids could therefore be utilized to manage neural and immune-mediated conditions. The mechanism for how cannabinoids affect cutaneous pain could impact nociception in other tissues like joints and muscles. The impact cannabinoids have on cutaneous immune cells may be leveraged as potential targets for the therapy of disorders as asthma, allergic rhinitis, and autoimmune conditions. Applications of cannabinoids in treatment of cutaneous malignancies may relate to carcinomas elsewhere in the body. However, the majority of the data is in vitro or in vivo animal models with few human clinical trials which emphasizes the knowledge gap that exists regarding cannabinoids and skin. Therefore, our conclusions regarding clinical use of cannabinoids remain limited, especially since most available consumer products remain unregulated.

Mini-dictionary of terms

- **Dermatomyositis.** Inflammatory disease characterized by skeletal muscle weakness and cutaneous manifestations such as facial rash, photosensitivity, and nail changes.
- **Epidermolysis bullosa.** Group of inherited conditions characterized by blisters of the skin and mucous membranes.
- **Fibrosis.** Thickening of tissue by excess deposition of extracellular matrix (i.e., collagen) and increased fibroblast proliferation.
- **Lupus erythematosus.** Autoimmune connective tissue disorders that can range from localization to the skin to association with systemic lupus erythematosus and organ dysfunction.
- **Systemic sclerosis.** Autoimmune inflammatory disease that is associated with fibrosis that can affect the skin as well as extracutaneous tissues in the lungs, gastrointestinal tract, kidneys, and heart.

Key facts of inflammation in the skin

- The cutaneous inflammatory response is a complex process involving many different cells and initiating factors.
- Despite the variety of triggers, inflammation involves an upsurge of proinflammatory cytokines (i.e., interferons and interleukins) and can lead to a chronic process.
- There is also a link between inflammation and malignancy, such that multiple inflammatory chemokines, cytokines, and growth factors have roles in tumorigenesis.
- Targeting inflammation has resulted in antiproliferative effects.
- Upticks in inflammation can cause disruption in the nervous system further adding to symptoms of pain and pruritus.

Summary points

- Cannabinoid receptors, CB1R and CB2R, and non-cannabinoid receptors, like TRPV-1 and PPARs, are located on keratinocytes, immune cells, and on nerves within the skin and hair follicles.
- Studies reveal a complex relationship between cannabinoids and cutaneous malignancies, though overall may be favorable.
- Anecdotal evidence suggests that cannabinoids may lead to pain relief in dermatologic conditions, but more studies are needed.
- In vitro and in vivo studies implicate multiple mechanisms by which cannabinoids can ameliorate pruritus, which has been translated to a few human studies but there are no clinical trials.
- Cannabinoids can regulate immune cell infiltration and the activity of fibroblasts and may therefore accelerate wound healing.
- Based on their anti-inflammatory effect, cannabinoids could benefit inflammatory connective tissue disorders such as systemic sclerosis, dermatomyositis, and cutaneous lupus; however, clinical trials have had mixed results.
- Adverse cutaneous effects from cannabinoids may come from overuse as well as allergic reactions due to cannabis itself or additive ingredients in consumer products.

References

Adler, B. L., & DeLeo, V. A. (2019). Allergenic ingredients in commercial topical cannabinoid preparations. *Journal of the American Academy of Dermatology*, *81*(3), 847–848. https://doi.org/10.1016/j.jaad.2019.03.015.

Ahluwalia, J., Urban, L., Bevan, S., & Nagy, I. (2003). Anandamide regulates neuropeptide release from capsaicin-sensitive primary sensory neurons by activating both the cannabinoid 1 receptor and the vanilloid receptor 1 in vitro. *The European Journal of Neuroscience*, *17*(12), 2611–2618. https://doi.org/10.1046/j.1460-9568.2003.02703.x.

Aldington, S., Harwood, M., Cox, B., Weatherall, M., Beckert, L., Hansell, A., & The Cannabis And Respiratory Disease Research Group. (2008). Cannabis use and cancer of the head and neck: Case-control study. *Otolaryngology and Head and Neck Surgery*, *138*(3), 374–380. https://doi.org/10.1016/j.otohns.2007.12.002.

Ali, A., & Akhtar, N. (2015). The safety and efficacy of 3% Cannabis seeds extract cream for reduction of human cheek skin sebum and erythema content. *Pakistan Journal of Pharmaceutical Sciences*, *28*(4), 1389–1395.

Ambrosino, P., Soldovieri, M. V., Russo, C., & Taglialatela, M. (2013). Activation and desensitization of TRPV1 channels in sensory neurons by the PPARα agonist palmitoylethanolamide. *British Journal of Pharmacology*, *168*(6), 1430–1444. https://doi.org/10.1111/bph.12029.

Appendino, G., Gibbons, S., Giana, A., Pagani, A., Grassi, G., Stavri, M., & Rahman, M. M. (2008). Antibacterial cannabinoids from Cannabis sativa: A structure-activity study. *Journal of Natural Products*, *71*(8), 1427–1430. https://doi.org/10.1021/np8002673.

Arkaki, M. (2015). The effect of 0.1% cannabidiol ointment for intractable alopecia areata: A clinical case report. In *Cannabinoid international conference, Sestri Levante, Italy*. http://www.cannabinoidconference2015.org/media/2015/abstractbook.pdf.

Armentia, A., Castrodeza, J., Ruiz-Muñoz, P., Martínez-Quesada, J., Postigo, I., Herrero, M., & Guisantes, J. A. (2011). Allergic hypersensitivity to cannabis in patients with allergy and illicit drug users. *Allergol Immunopathol (Madr)*, *39*(5), 271–279. https://doi.org/10.1016/j.aller.2010.09.008.

Armstrong, J. L., Hill, D. S., McKee, C. S., Hernandez-Tiedra, S., Lorente, M., Lopez-Valero, I., & Lovat, P. E. (2015). Exploiting cannabinoid-induced cytotoxic autophagy to drive melanoma cell death. *The Journal of Investigative Dermatology*, *135*(6), 1629–1637. https://doi.org/10.1038/jid.2015.45.

Avila, C., Massick, S., Kaffenberger, B. H., Kwatra, S. G., & Bechtel, M. (2020). Cannabinoids for the treatment of chronic pruritus: A review. *Journal of the American Academy of Dermatology*, *82*(5), 1205–1212. https://doi.org/10.1016/j.jaad.2020.01.036.

Balistreri, E., Garcia-Gonzalez, E., Selvi, E., Akhmetshina, A., Palumbo, K., Lorenzini, S., & Distler, J. W. (2011). The cannabinoid WIN55, 212-2 abrogates dermal fibrosis in scleroderma bleomycin model. *Annals of the Rheumatic Diseases*, *70*(4), 695–699. https://doi.org/10.1136/ard.2010.137539.

Basavarajappa, B. S. (2007). Neuropharmacology of the endocannabinoid signaling system-molecular mechanisms, biological actions and synaptic plasticity. *Current Neuropharmacology*, *5*(2), 81–97. https://doi.org/10.2174/157015907780866910.

Berdyshev, E. V., Boichot, E., Germain, N., Allain, N., Anger, J. P., & Lagente, V. (1997). Influence of fatty acid ethanolamides and delta9-tetrahydrocannabinol on cytokine and arachidonate release by mononuclear cells. *European Journal of Pharmacology*, *330*(2-3), 231–240. https://doi.org/10.1016/s0014-2999(97)01007-8.

Betz, R. C., König, K., Flaquer, A., Redler, S., Eigelshoven, S., Kortüm, A. K., & Nöthen, M. M. (2008). The R620W polymorphism in PTPN22 confers general susceptibility for the development of alopecia areata. *The British Journal of Dermatology*, *158*(2), 389–391. https://doi.org/10.1111/j.1365-2133.2007.08312.x.

Bilir, K. A., Anli, G., Ozkan, E., Gunduz, O., & Ulugol, A. (2018). Involvement of spinal cannabinoid receptors in the antipruritic effects of WIN 55,212-2, a cannabinoid receptor agonist. *Clinical and Experimental Dermatology*, *43*(5), 553–558. https://doi.org/10.1111/ced.13398.

Bilkei-Gorzo, A., Drews, E., Albayram, Ö., Piyanova, A., Gaffal, E., Tueting, T., & Zimmer, A. (2012). Early onset of aging-like changes is restricted to cognitive abilities and skin structure in Cnr1⁻/⁻ mice. *Neurobiology of Aging*, *33*(1). https://doi.org/10.1016/j.neurobiolaging.2010.07.009. 200.e211–222.

Bíró, T., Tóth, B. I., Haskó, G., Paus, R., & Pacher, P. (2009). The endocannabinoid system of the skin in health and disease: Novel perspectives and therapeutic opportunities. *Trends in Pharmacological Sciences*, *30*(8), 411–420. https://doi.org/10.1016/j.tips.2009.05.004.

Blázquez, C., Carracedo, A., Barrado, L., Real, P. J., Fernández-Luna, J. L., Velasco, G., & Guzmán, M. (2006). Cannabinoid receptors as novel targets for the treatment of melanoma. *The FASEB Journal*, *20*(14), 2633–2635. https://doi.org/10.1096/fj.06-6638fje.

Carpi, S., Fogli, S., Polini, B., Montagnani, V., Podestà, A., Breschi, M. C., & Nieri, P. (2017). Tumor-promoting effects of cannabinoid receptor type 1 in human melanoma cells. *Toxicology In Vitro*, *40*, 272–279. https://doi.org/10.1016/j.tiv.2017.01.018.

Casanova, M. L., Blázquez, C., Martínez-Palacio, J., Villanueva, C., Fernández-Aceñero, M. J., Huffman, J. W., & Guzmán, M. (2003). Inhibition of skin tumor growth and angiogenesis in vivo by activation of cannabinoid receptors. *The Journal of Clinical Investigation*, *111*(1), 43–50. https://doi.org/10.1172/JCI16116.

Chalmers, S., Garcia, S., Draganski, A., Doerner, J., Friedman, A., Friedman, J., & Putterman, C. (2018). Topical endocannabinoid administration protects MRL-Lpr/Lpr mice from cutaneous lupus erythematosus [abstract]. In *2018 ACR/ARHP annual meeting*. https://acrabstracts.org/abstract/topical-endocannabinoid-administration-protects-mrl-lpr-lpr-mice-from-cutaneous-lupus-erythematosus/.

Changoer, L., & Anastassov, G. (2018). *Method to treat psoriasis*. USA Patent No. F. P. Online.

Chelliah, M. P., Zinn, Z., Khuu, P., & Teng, J. M. C. (2018). Self-initiated use of topical cannabidiol oil for epidermolysis bullosa. *Pediatric Dermatology*, *35*(4), e224–e227. https://doi.org/10.1111/pde.13545.

ClinicalTrials.gov. (2017). *JBT-101 in systemic lupus erythematosus (SLE)*.

Costa, B., Comelli, F., Bettoni, I., Colleoni, M., & Giagnoni, G. (2008). The endogenous fatty acid amide, palmitoylethanolamide, has anti-allodynic and anti-hyperalgesic effects in a murine model of neuropathic pain: Involvement of CB(1), TRPV1 and PPARgamma receptors and neurotrophic factors. *Pain*, *139*(3), 541–550. https://doi.org/10.1016/j.pain.2008.06.003.

Costa, B., Trovato, A. E., Colleoni, M., Giagnoni, G., Zarini, E., & Croci, T. (2005). Effect of the cannabinoid CB1 receptor antagonist, SR141716, on nociceptive response and nerve demyelination in rodents with chronic constriction injury of the sciatic nerve. *Pain*, *116*(1-2), 52–61. https://doi.org/10.1016/j.pain.2005.03.043.

Darling, M. R. (2003). Cannabis abuse and oral health care: Review and suggestions for management. *SADJ*, *58*(5), 189–190.

Darling, M. R., & Arendorf, T. M. (1993). Effects of cannabis smoking on oral soft tissues. *Community Dentistry and Oral Epidemiology*, *21*(2), 78–81. https://doi.org/10.1111/j.1600-0528.1993.tb00725.x.

Dhadwal, G., & Kirchhof, M. G. (2018). The risks and benefits of cannabis in the dermatology clinic. *Journal of Cutaneous Medicine and Surgery*, *22*(2), 194–199. https://doi.org/10.1177/1203475417738971.

Dobrosi, N., Tóth, B. I., Nagy, G., Dózsa, A., Géczy, T., Nagy, L., & Bíró, T. (2008). Endocannabinoids enhance lipid synthesis and apoptosis of human sebocytes via cannabinoid receptor-2-mediated signaling. *The FASEB Journal*, *22*(10), 3685–3695. https://doi.org/10.1096/fj.07-104877.

Dogrul, A., Gul, H., Akar, A., Yildiz, O., Bilgin, F., & Guzeldemir, E. (2003). Topical cannabinoid antinociception: Synergy with spinal sites. *Pain*, *105*(1-2), 11–16. https://doi.org/10.1016/s0304-3959(03)00068-x.

Du, Y., Ren, P., Wang, Q., Jiang, S. K., Zhang, M., Li, J. Y., & Guan, D. W. (2018). Cannabinoid 2 receptor attenuates inflammation during skin wound healing by inhibiting M1 macrophages rather than activating M2 macrophages. *Journal of Inflammation (London)*, *15*, 25. https://doi.org/10.1186/s12950-018-0201-z.

Dvorak, M., Watkinson, A., McGlone, F., & Rukwied, R. (2003). Histamine induced responses are attenuated by a cannabinoid receptor agonist in human skin. *Inflammation Research*, *52*(6), 238–245. https://doi.org/10.1007/s00011-003-1162-z.

Eagleston, L. R. M., Kalani, N. K., Patel, R. R., Flaten, H. K., Dunnick, C. A., & Dellavalle, R. P. (2018). Cannabinoids in dermatology: A scoping review. *Dermatology Online Journal*, *24*(6).

Eberlein, B., Eicke, C., Reinhardt, H. W., & Ring, J. (2008). Adjuvant treatment of atopic eczema: Assessment of an emollient containing N-palmitoylethanolamine (ATOPA study). *Journal of the European Academy of Dermatology and Venereology*, *22*(1), 73–82. https://doi.org/10.1111/j.1468-3083.2007.02351.x.

Elphick, M. R., & Egertová, M. (2001). The neurobiology and evolution of cannabinoid signalling. *Philosophical Transactions of the Royal Society of London. Series B, Biological Sciences*, *356*(1407), 381–408. https://doi.org/10.1098/rstb.2000.0787.

Firth, N. A. (1997). Marijuana use and oral cancer: A review. *Oral Oncology*, *33*(6), 398–401. https://doi.org/10.1016/s0964-1955(97)00022-5.

Foster, E., Nguyen, C., & Norris, P. (2018). Contact buzz: Allergic contact dermatitis to cannabis. *Dermatitis*, *29*(4), 223–224. https://doi.org/10.1097/DER.0000000000000378.

Gaffal, E., Cron, M., Glodde, N., & Tüting, T. (2013). Anti-inflammatory activity of topical THC in DNFB-mediated mouse allergic contact dermatitis independent of CB1 and CB2 receptors. *Allergy*, *68*(8), 994–1000. https://doi.org/10.1111/all.12183.

Gaffal, E., Glodde, N., Jakobs, M., Bald, T., & Tüting, T. (2014). Cannabinoid 1 receptors in keratinocytes attenuate fluorescein isothiocyanate-induced mouse atopic-like dermatitis. *Experimental Dermatology*, *23*(6), 401–406. https://doi.org/10.1111/exd.12414.

GlobalNewswire. (2021). *Corbus pharmaceuticals announces topline results from DETERMINE Phase 3 Study of Lenabasum for Treatment of Dermatomyositis*. Global Newswire. Retrieved October 17th from: https://www.globenewswire.com/en/news-release/2021/06/24/2252567/0/en/Corbus-Pharmaceuticals-Announces-Topline-Results-from-DETERMINE-Phase-3-Study-of-Lenabasum-for-Treatment-of-Dermatomyositis.html.

Gonzalez, E. G., Selvi, E., Balistreri, E., Akhmetshina, A., Palumbo, K., Lorenzini, S., & Distler, J. H. (2012). Synthetic cannabinoid ajulemic acid exerts potent antifibrotic effects in experimental models of systemic sclerosis. *Annals of the Rheumatic Diseases*, *71*(9), 1545–1551. https://doi.org/10.1136/annrheumdis-2011-200314.

Gupta, A. K., & Talukder, M. (2021). Cannabinoids for skin diseases and hair regrowth. *Journal of Cosmetic Dermatology*, *20*(9), 2703–2711. https://doi.org/10.1111/jocd.14352.

Hillard, C. J. (2018). Circulating endocannabinoids: From whence do they come and where are they going? *Neuropsychopharmacology*, *43*(1), 155–172. https://doi.org/10.1038/npp.2017.130.

Inci, R., Kelekci, K. H., Oguz, N., Karaca, S., Karadas, B., & Bayrakci, A. (2017). Dermatological aspects of synthetic cannabinoid addiction. *Cutaneous and Ocular Toxicology*, *36*(2), 125–131. https://doi.org/10.3109/15569527.2016.1169541.

Jin, S., & Lee, M. Y. (2018). The ameliorative effect of hemp seed hexane extracts on the Propionibacterium acnes-induced inflammation and lipogenesis in sebocytes. *PLoS One*, *13*(8), e0202933. https://doi.org/10.1371/journal.pone.0202933.

Johanek, L. M., & Simone, D. A. (2004). Activation of peripheral cannabinoid receptors attenuates cutaneous hyperalgesia produced by a heat injury. *Pain*, *109*(3), 432–442. https://doi.org/10.1016/j.pain.2004.02.020.

Karsak, M., Gaffal, E., Date, R., Wang-Eckhardt, L., Rehnelt, J., Petrosino, S., & Zimmer, A. (2007). Attenuation of allergic contact dermatitis through the endocannabinoid system. *Science*, *316*(5830), 1494–1497. https://doi.org/10.1126/science.1142265.

Katsuyama, S., Mizoguchi, H., Kuwahata, H., Komatsu, T., Nagaoka, K., Nakamura, H., & Sakurada, S. (2013). Involvement of peripheral cannabinoid and opioid receptors in β-caryophyllene-induced antinociception. *European Journal of Pain*, *17*(5), 664–675. https://doi.org/10.1002/j.1532-2149.2012.00242.x.

Kim, H. J., Kim, B., Park, B. M., Jeon, J. E., Lee, S. H., Mann, S., & Jeong, S. K. (2015). Topical cannabinoid receptor 1 agonist attenuates the cutaneous inflammatory responses in oxazolone-induced atopic dermatitis model. *International Journal of Dermatology*, *54*(10), e401–e408. https://doi.org/10.1111/ijd.12841.

Klein, T. W. (2005). Cannabinoid-based drugs as anti-inflammatory therapeutics. *Nature Reviews. Immunology*, *5*(5), 400–411. https://doi.org/10.1038/nri1602.

Klein, T. W., Newton, C., Larsen, K., Lu, L., Perkins, I., Nong, L., & Friedman, H. (2003). The cannabinoid system and immune modulation. *Journal of Leukocyte Biology*, *74*(4), 486–496. https://doi.org/10.1189/jlb.0303101.

Kong, H. E., Pollack, B. P., & Blalock, T. W. (2021). Cannabinoids in dermatologic surgery. *Journal of the American Academy of Dermatology*, *85*(6), 1565–1570. https://doi.org/10.1016/j.jaad.2021.01.016.

Kupczyk, P., Reich, A., & Szepietowski, J. C. (2009). Cannabinoid system in the skin—A possible target for future therapies in dermatology. *Experimental Dermatology*, *18*(8), 669–679. https://doi.org/10.1111/j.1600-0625.2009.00923.x.

LaBuda, C. J., Koblish, M., & Little, P. J. (2005). Cannabinoid CB2 receptor agonist activity in the hindpaw incision model of postoperative pain. *European Journal of Pharmacology*, *527*(1-3), 172–174. https://doi.org/10.1016/j.ejphar.2005.10.020.

Leonti, M., Casu, L., Raduner, S., Cottiglia, F., Floris, C., Altmann, K. H., & Gertsch, J. (2010). Falcarinol is a covalent cannabinoid CB1 receptor antagonist and induces pro-allergic effects in skin. *Biochemical Pharmacology*, *79*(12), 1815–1826. https://doi.org/10.1016/j.bcp.2010.02.015.

Maccarrone, M., Di Rienzo, M., Battista, N., Gasperi, V., Guerrieri, P., Rossi, A., & Finazzi-Agrò, A. (2003). The endocannabinoid system in human keratinocytes. Evidence that anandamide inhibits epidermal differentiation through CB1 receptor-dependent inhibition of protein kinase C, activation protein-1, and transglutaminase. *The Journal of Biological Chemistry*, *278*(36), 33896–33903. https://doi.org/10.1074/jbc.M303994200.

Maekawa, T., Nojima, H., Kuraishi, Y., & Aisaka, K. (2006). The cannabinoid CB2 receptor inverse agonist JTE-907 suppresses spontaneous itch-associated responses of NC mice, a model of atopic dermatitis. *European Journal of Pharmacology*, *542*(1-3), 179–183. https://doi.org/10.1016/j.ejphar.2006.05.040.

Maida, V., & Corban, J. (2017). Topical medical cannabis: A new treatment for wound pain-three cases of pyoderma gangrenosum. *Journal of Pain and Symptom Management*, *54*(5), 732–736. https://doi.org/10.1016/j.jpainsymman.2017.06.005.

Majmudar, V., Azam, N. A., & Finch, T. (2006). Contact urticaria to Cannabis sativa. *Contact Dermatitis*, *54*(2), 127. https://doi.org/10.1111/j.0105-1873.2006.0560h.x.

Marks, M. A., Chaturvedi, A. K., Kelsey, K., Straif, K., Berthiller, J., Schwartz, S. M., & D'Souza, G. (2014). Association of marijuana smoking with oropharyngeal and oral tongue cancers: Pooled analysis from the INHANCE consortium. *Cancer Epidemiology, Biomarkers & Prevention*, *23*(1), 160–171. https://doi.org/10.1158/1055-9965.EPI-13-0181.

Mounessa, J. S., Siegel, J. A., Dunnick, C. A., & Dellavalle, R. P. (2017). The role of cannabinoids in dermatology. *Journal of the American Academy of Dermatology*, *77*(1), 188–190. https://doi.org/10.1016/j.jaad.2017.02.056.

Movahed, P., Evilevitch, V., Andersson, T. L., Jönsson, B. A., Wollmer, P., Zygmunt, P. M., & Högestätt, E. D. (2005). Vascular effects of anandamide and N-acylvanillylamines in the human forearm and skin microcirculation. *British Journal of Pharmacology*, *146*(2), 171–179. https://doi.org/10.1038/sj.bjp.0706313.

Muller, C., Morales, P., & Reggio, P. H. (2018). Cannabinoid ligands targeting TRP channels. *Frontiers in Molecular Neuroscience*, *11*, 487. https://doi.org/10.3389/fnmol.2018.00487.

Nam, G., Jeong, S. K., Park, B. M., Lee, S. H., Kim, H. J., Hong, S. P., & Kim, B. W. (2016). Selective cannabinoid receptor-1 agonists regulate mast cell activation in an oxazolone-induced atopic dermatitis model. *Annals of Dermatology*, *28*(1), 22–29. https://doi.org/10.5021/ad.2016.28.1.22.

Navarini, L., Bisogno, T., Mozetic, P., Piscitelli, F., Margiotta, D. P. E., Basta, F., & Maccarrone, M. (2018). Endocannabinoid system in systemic lupus erythematosus: First evidence for a deranged 2-arachidonoylglycerol metabolism. *The International Journal of Biochemistry & Cell Biology*, *99*, 161–168. https://doi.org/10.1016/j.biocel.2018.04.010.

Neff, G. W., O'Brien, C. B., Reddy, K. R., Bergasa, N. V., Regev, A., Molina, E., & Schiff, E. (2002). Preliminary observation with dronabinol in patients with intractable pruritus secondary to cholestatic liver disease. *The American Journal of Gastroenterology*, *97*(8), 2117–2119. https://doi.org/10.1111/j.1572-0241.2002.05852.x.

Odan, M., Ishizuka, N., Hiramatsu, Y., Inagaki, M., Hashizume, H., Fujii, Y., & Arimura, A. (2012). Discovery of S-777469: An orally available CB2 agonist as an antipruritic agent. *Bioorganic & Medicinal Chemistry Letters*, *22*(8), 2803–2806. https://doi.org/10.1016/j.bmcl.2012.02.072.

Oláh, A., Markovics, A., Szabó-Papp, J., Szabó, P. T., Stott, C., Zouboulis, C. C., & Bíró, T. (2016). Differential effectiveness of selected non-psychotropic phytocannabinoids on human sebocyte functions implicates their introduction in dry/seborrhoeic skin and acne treatment. *Experimental Dermatology*, *25*(9), 701–707. https://doi.org/10.1111/exd.13042.

Oláh, A., Tóth, B. I., Borbíró, I., Sugawara, K., Szöllõsi, A. G., Czifra, G., & Bíró, T. (2014). Cannabidiol exerts sebostatic and antiinflammatory effects on human sebocytes. *The Journal of Clinical Investigation*, *124*(9), 3713–3724. https://doi.org/10.1172/JCI64628.

O'Sullivan, S. E., & Kendall, D. A. (2010). Cannabinoid activation of peroxisome proliferator-activated receptors: Potential for modulation of inflammatory disease. *Immunobiology*, *215*(8), 611–616. https://doi.org/10.1016/j.imbio.2009.09.007.

Paradisi, A., Pasquariello, N., Barcaroli, D., & Maccarrone, M. (2008). Anandamide regulates keratinocyte differentiation by inducing DNA methylation in a CB1 receptor-dependent manner. *The Journal of Biological Chemistry*, *283*(10), 6005–6012. https://doi.org/10.1074/jbc.M707964200.

Petrosino, S., Verde, R., Vaia, M., Allarà, M., Iuvone, T., & Di Marzo, V. (2018). Anti-inflammatory properties of cannabidiol, a nonpsychotropic cannabinoid, in experimental allergic contact dermatitis. *The Journal of Pharmacology and Experimental Therapeutics*, *365*(3), 652–663. https://doi.org/10.1124/jpet.117.244368.

Phan, N. Q., Siepmann, D., Gralow, I., & Ständer, S. (2010). Adjuvant topical therapy with a cannabinoid receptor agonist in facial postherpetic neuralgia. *Journal der Deutschen Dermatologischen Gesellschaft*, *8*(2), 88–91. https://doi.org/10.1111/j.1610-0387.2009.07213.x.

Pucci, M., Pasquariello, N., Battista, N., Di Tommaso, M., Rapino, C., Fezza, F., & Maccarrone, M. (2012). Endocannabinoids stimulate human melanogenesis via type-1 cannabinoid receptor. *The Journal of Biological Chemistry*, *287*(19), 15466–15478. https://doi.org/10.1074/jbc.M111.314880.

Ramot, Y., Sugawara, K., Zákány, N., Tóth, B. I., Bíró, T., & Paus, R. (2013). A novel control of human keratin expression: Cannabinoid receptor 1-mediated signaling down-regulates the expression of keratins K6 and K16 in human keratinocytes in vitro and in situ. *PeerJ*, *1*, e40. https://doi.org/10.7717/peerj.40.

Robinson, E., Murphy, E., & Friedman, A. (2018). Knowledge, attitudes, and perceptions of cannabinoids in the dermatology community. *Journal of Drugs in Dermatology*, *17*(12), 1273–1278.

Roelandt, T., Heughebaert, C., Bredif, S., Giddelo, C., Baudouin, C., Msika, P., & Hachem, J. P. (2012). Cannabinoid receptors 1 and 2 oppositely regulate epidermal permeability barrier status and differentiation. *Experimental Dermatology*, *21*(9), 688–693. https://doi.org/10.1111/j.1600-0625.2012.01561.x.

Salinas-Santander, M., Sánchez-Domínguez, C., Cantú-Salinas, C., Gonzalez-Cárdenas, H., Cepeda-Nieto, A. C., Cerda-Flores, R. M., & Ocampo-Candiani, J. (2015). Association between PTPN22 C1858T polymorphism and alopecia areata risk. *Experimental and Therapeutic Medicine*, *10*(5), 1953–1958. https://doi.org/10.3892/etm.2015.2728.

Salzet, M., Breton, C., Bisogno, T., & Di Marzo, V. (2000). Comparative biology of the endocannabinoid system possible role in the immune response. *European Journal of Biochemistry*, *267*(16), 4917–4927. https://doi.org/10.1046/j.1432-1327.2000.01550.x.

Sánchez, M. G., Ruiz-Llorente, L., Sánchez, A. M., & Díaz-Laviada, I. (2003). Activation of phosphoinositide 3-kinase/PKB pathway by CB(1) and CB(2) cannabinoid receptors expressed in prostate PC-3 cells. Involvement in Raf-1 stimulation and NGF induction. *Cellular Signalling*, *15*(9), 851–859. https://doi.org/10.1016/s0898-6568(03)00036-6.

Scheau, C., Badarau, I. A., Mihai, L. G., Scheau, A. E., Costache, D. O., Constantin, C., & Caruntu, A. (2020). Cannabinoids in the pathophysiology of skin inflammation. *Molecules*, *25*(3). https://doi.org/10.3390/molecules25030652.

Schlosburg, J. E., O'Neal, S. T., Conrad, D. H., & Lichtman, A. H. (2011). CB 1 receptors mediate rimonabant-induced pruritic responses in mice: Investigation of locus of action [OriginalPaper]. *Psychopharmacology*, *216*(3), 323–331. https://doi.org/10.1007/s00213-011-2224-5.

Servettaz, A., Kavian, N., Nicco, C., Deveaux, V., Chéreau, C., Wang, A., & Batteux, F. (2010). Targeting the cannabinoid pathway limits the development of fibrosis and autoimmunity in a mouse model of systemic sclerosis. *The American Journal of Pathology*, *177*(1), 187–196. https://doi.org/10.2353/ajpath.2010.090763.

Sheriff, T., Lin, M. J., Dubin, D., & Khorasani, H. (2019). The potential role of cannabinoids in dermatology. *The Journal of Dermatological Treatment*, *1-7*. https://doi.org/10.1080/09546634.2019.1675854.

Soliman, E., & Van Dross, R. (2016). Anandamide-induced endoplasmic reticulum stress and apoptosis are mediated by oxidative stress in non-melanoma skin cancer: Receptor-independent endocannabinoid signaling. *Molecular Carcinogenesis*, *55*(11), 1807–1821. https://doi.org/10.1002/mc.22429.

Spiera, R., Hummers, L., Chung, L., Frech, T., Domsic, R., Hsu, V., … White, B. (2017). A phase 2 study of safety and efficacy of Anabasum (JBT-101), a cannabinoid receptor type 2 agonist. In *Diffuse Cutaneous Systemic Sclerosis—ACR Meeting Abstracts*. https://acrabstracts.org/abstract/a-phase-2-study-of-safety-and-efficacy-of-anabasum-jbt-101-a-cannabinoid-receptor-type-2-agonist-in-diffuse-cutaneous-systemic-sclerosis/.

Spiera, R., Kuwana, M., Khanna, D., Hummers, L., Frech, T., Stevens, W., & Denton, C. (2021). OP0171 Phase 3 trial of lenabasum, a CB2 agonist, for the treatment of diffuse cutaneous systemic sclerOSIS (DCSSC). *Annals of the Rheumatic Diseases*, *80*.

Spleman, L., Sinclair, R., Freeman, M., Davis, M., & Gebauer, K. (2018). 1061 The safety of topical cannabidiol (CBD) for the treatment of acne. *Journal of Investigative Dermatology*, *138*(5), S180.

Srivastava, B. K., Soni, R., Patel, J. Z., Joharapurkar, A., Sadhwani, N., Kshirsagar, S., & Patel, P. R. (2009). Hair growth stimulator property of thienyl substituted pyrazole carboxamide derivatives as a CB1 receptor antagonist with in vivo antiobesity effect. *Bioorganic & Medicinal Chemistry Letters*, *19*(9), 2546–2550. https://doi.org/10.1016/j.bmcl.2009.03.046.

Ständer, S., Reinhardt, H. W., & Luger, T. A. (2006). Topical cannabinoid agonists. An effective new possibility for treating chronic pruritus. *Hautarzt*, *57* (9), 801–807. https://doi.org/10.1007/s00105-006-1180-1.

Ständer, S., Schmelz, M., Metze, D., Luger, T., & Rukwied, R. (2005). Distribution of cannabinoid receptor 1 (CB1) and 2 (CB2) on sensory nerve fibers and adnexal structures in human skin. *Journal of Dermatological Science*, *38*(3), 177–188. https://doi.org/10.1016/j.jdermsci.2005.01.007.

Ständer, S., Weisshaar, E., & Luger, T. A. (2008). Neurophysiological and neurochemical basis of modern pruritus treatment. *Experimental Dermatology*, *17*(3), 161–169. https://doi.org/10.1111/j.1600-0625.2007.00664.x.

Stöckli, S. S., & Bircher, A. J. (2007). Generalized pruritus in a patient sensitized to tobacco and cannabis. *Journal der Deutschen Dermatologischen Gesellschaft*, *5*(4), 303–304. https://doi.org/10.1111/j.1610-0387.2007.06270.x.

Sugawara, K., Bíró, T., Tsuruta, D., Tóth, B. I., Kromminga, A., Zákány, N., & Paus, R. (2012). Endocannabinoids limit excessive mast cell maturation and activation in human skin. *The Journal of Allergy and Clinical Immunology*, *129*(3), 726–738.e728. https://doi.org/10.1016/j.jaci.2011.11.009.

Sun, Y., & Bennett, A. (2007). Cannabinoids: A new group of agonists of PPARs. *PPAR Research*, *2007*, 23513. https://doi.org/10.1155/2007/23513.

Szabó, I. L., Herczeg-Lisztes, E., Szegedi, A., Nemes, B., Paus, R., Bíró, T., & Szöllősi, A. G. (2019). TRPV4 is expressed in human hair follicles and inhibits hair growth in vitro. *The Journal of Investigative Dermatology*, *139*(6), 1385–1388. https://doi.org/10.1016/j.jid.2018.11.020.

Szepietowski, J. C., Szepietowski, T., & Reich, A. (2005). Efficacy and tolerance of the cream containing structured physiological lipids with endocannabinoids in the treatment of uremic pruritus: A preliminary study. *Acta Dermatovenerologica Croatica*, *13*(2), 97–103.

Szöllősi, A. G., Oláh, A., Bíró, T., & Tóth, B. I. (2017). Recent advances in the endocrinology of the sebaceous gland. *Dermato-Endocrinology*, *9*(1), e1361576. https://doi.org/10.1080/19381980.2017.1361576.

Taha, T., Meiri, D., Talhamy, S., Wollner, M., Peer, A., & Bar-Sela, G. (2019). Cannabis impacts tumor response rate to nivolumab in patients with advanced malignancies. *The Oncologist*, *24*(4), 549–554. https://doi.org/10.1634/theoncologist.2018-0383.

Telek, A., Bíró, T., Bodó, E., Tóth, B. I., Borbíró, I., Kunos, G., & Paus, R. (2007). Inhibition of human hair follicle growth by endo- and exocannabinoids. *The FASEB Journal*, *21*(13), 3534–3541. https://doi.org/10.1096/fj.06-7689com.

Tijani, A. O., Thakur, D., Mishra, D., Frempong, D., Chukwunyere, U. I., & Puri, A. (2021). Delivering therapeutic cannabinoids via skin: Current state and future perspectives. *Journal of Controlled Release*, *334*, 427–451. https://doi.org/10.1016/j.jconrel.2021.05.005.

Tóth, K. F., Ádám, D., Bíró, T., & Oláh, A. (2019). Cannabinoid signaling in the skin: Therapeutic potential of the "C(ut)annabinoid" system. *Molecules*, *24*(5). https://doi.org/10.3390/molecules24050918.

Tóth, B. I., Oláh, A., Szöllősi, A. G., & Bíró, T. (2014). TRP channels in the skin. *British Journal of Pharmacology*, *171*(10), 2568–2581. https://doi.org/10.1111/bph.12569.

Turkmenoglu, F. P., Kasirga, U. B., & Celik, H. H. (2015). Ultra-structural hair alterations of drug abusers: A scanning electron microscopic investigation. *International Journal of Clinical and Experimental Medicine*, *8*(6), 8803–8811.

Ueda, Y., Miyagawa, N., Matsui, T., Kaya, T., & Iwamura, H. (2005). Involvement of cannabinoid CB(2) receptor-mediated response and efficacy of cannabinoid CB(2) receptor inverse agonist, JTE-907, in cutaneous inflammation in mice. *European Journal of Pharmacology*, *520*(1-3), 164–171. https://doi.org/10.1016/j.ejphar.2005.08.013.

Van Dross, R., Soliman, E., Jha, S., Johnson, T., & Mukhopadhyay, S. (2013). Receptor-dependent and receptor-independent endocannabinoid signaling: A therapeutic target for regulation of cancer growth. *Life Sciences*, *92*(8-9), 463–466. https://doi.org/10.1016/j.lfs.2012.09.025.

Veronesi, B., & Oortgiesen, M. (2006). The TRPV1 receptor: Target of toxicants and therapeutics. *Toxicological Sciences*, *89*(1), 1–3. https://doi.org/10.1093/toxsci/kfj034.

Wang, L. L., Zhao, R., Li, J. Y., Li, S. S., Liu, M., Wang, M., & Guan, D. W. (2016). Pharmacological activation of cannabinoid 2 receptor attenuates inflammation, fibrogenesis, and promotes re-epithelialization during skin wound healing. *European Journal of Pharmacology*, *786*, 128–136. https://doi.org/10.1016/j.ejphar.2016.06.006.

Ward, S. J., Lefever, T. W., Rawls, S. M., Whiteside, G. T., & Walker, E. A. (2009). Age-dependent effects of the cannabinoid CB1 antagonist SR141716A on food intake, body weight change, and pruritus in rats. *Psychopharmacology*, *206*(1), 155–165. https://doi.org/10.1007/s00213-009-1592-6.

Werth, V., Hejazi, E., Pena, S., Haber, J., Okawa, J., Feng, R., … White, B. (2018). FRI0470 A phase 2 study of safety and efficacy of lenabasum (JBT-101), a cannabinoid receptor type 2 agonist, in refractory skin-predominant dermatomyositis. *Annals of the Rheumatic Diseases*. https://doi.org/10.1136/annrheumdis-2018-eular.3531.

Wilkinson, J. D., & Williamson, E. M. (2007). Cannabinoids inhibit human keratinocyte proliferation through a non-CB1/CB2 mechanism and have a potential therapeutic value in the treatment of psoriasis. *Journal of Dermatological Science*, *45*(2), 87–92. https://doi.org/10.1016/j.jdermsci.2006.10.009.

Williams, C., Thompstone, J., & Wilkinson, M. (2008). Work-related contact urticaria to Cannabis sativa. *Contact Dermatitis*, *58*(1), 62–63. https://doi.org/10.1111/j.1600-0536.2007.01169.x.

Wolkenstein, P., Misery, L., Amici, J. M., Maghia, R., Branchoux, S., Cazeau, C., & Taïeb, C. (2015). Smoking and dietary factors associated with moderate-to-severe acne in French adolescents and young adults: Results of a survey using a representative sample. *Dermatology*, *230*(1), 34–39. https://doi.org/10.1159/000366195.

Wollenberg, A., Seba, A., & Antal, A. S. (2014). Immunological and molecular targets of atopic dermatitis treatment. *The British Journal of Dermatology*, *170*(Suppl 1), 7–11. https://doi.org/10.1111/bjd.12975.

Yesilyurt, O., Dogrul, A., Gul, H., Seyrek, M., Kusmez, O., Ozkan, Y., & Yildiz, O. (2003). Topical cannabinoid enhances topical morphine antinociception. *Pain*, *105*(1-2), 303–308. https://doi.org/10.1016/s0304-3959(03)00245-8.

Yuan, C., Wang, X. M., Guichard, A., Tan, Y. M., Qian, C. Y., Yang, L. J., & Humbert, P. (2014). N-palmitoylethanolamine and N-acetylethanolamine are effective in asteatotic eczema: Results of a randomized, double-blind, controlled study in 60 patients. *Clinical Interventions in Aging*, *9*, 1163–1169. https://doi.org/10.2147/CIA.S65448.

Zhao, Z., Yang, J., Zhao, H., Fang, X., & Li, H. (2012). Cannabinoid receptor 2 is upregulated in melanoma. *Journal of Cancer Research and Therapeutics*, *8*(4), 549–554. https://doi.org/10.4103/0973-1482.106534.

Zheng, D., Bode, A. M., Zhao, Q., Cho, Y. Y., Zhu, F., Ma, W. Y., & Dong, Z. (2008). The cannabinoid receptors are required for ultraviolet-induced inflammation and skin cancer development. *Cancer Research*, *68*(10), 3992–3998. https://doi.org/10.1158/0008-5472.CAN-07-6594.

Zheng, J. L., Yu, T. S., Li, X. N., Fan, Y. Y., Ma, W. X., Du, Y., & Guan, D. W. (2012). Cannabinoid receptor type 2 is time-dependently expressed during skin wound healing in mice. *International Journal of Legal Medicine*, *126*(5), 807–814. https://doi.org/10.1007/s00414-012-0741-3.

Zhou, Y., Oudin, M. J., Gajendra, S., Sonego, M., Falenta, K., Williams, G., & Doherty, P. (2015). Regional effects of endocannabinoid, BDNF and FGF receptor signalling on neuroblast motility and guidance along the rostral migratory stream. *Molecular and Cellular Neurosciences*, *64*, 32–43. https://doi.org/10.1016/j.mcn.2014.12.001.

Chapter 40

Cannabis consumption in reproductive function and teratogenicity

Susana Sá[a,b] and Bruno M. Fonseca[c,d]

[a]*Department of Biomedicine, Unit of Anatomy, Faculty of Medicine, University of Porto, Porto, Portugal,* [b]*Center for Health Technology and Services Research (CINTESIS), Faculty of Medicine, University of Porto, Porto, Portugal,* [c]*UCIBIO – Applied Molecular Biosciences Unit, Laboratory of Biochemistry, Faculty of Pharmacy, University of Porto, Porto, Portugal,* [d]*Associate Laboratory i4HB – Institute for Health and Bioeconomy, Faculty of Pharmacy, University of Porto, Porto, Portugal*

Abbreviations

2-AG	2-arachidonoylglycerol
AEA	anandamide
BCRP	breast cancer resistance protein
cAMP	cyclic adenosine monophosphate
CB	cannabinoid receptor
CBD	cannabidiol
eCBs	endocannabinoids
ECS	endocannabinoid system
FAAH	fatty acid amide hydrolase
FSH	follicle-stimulating hormone
GnRH	gonadotropin-releasing hormone
HPG	hypothalamic-pituitary-gonadal
LH	luteinizing hormone
NO	nitric oxide
THC	Δ9-tetra-hidrocanabinol
WIN	R(+)-WIN 55,212-2 mesylate

Cannabis in reproduction and fertility

Cannabinoid agonism in the hypothalamic-pituitary-gonadal axis

Reproductive success is dependent on many neuropeptide and hormonal systems working to modulate gonadal functions. The hypothalamic-pituitary-gonadal (HPG) axis, which has a crucial role in endocrine reproductive function, governs the production of the primary sex hormones through the release of gonadotropin-releasing hormone (GnRH). The hypothalamus typically secretes GnRH in pulses that stimulate the anterior pituitary gland to secrete the gonadotropins, luteinizing hormone (LH), and follicle-stimulating hormone (FSH). Both LH and FSH are released into the systemic circulation and act in the gonads—the testes and ovaries—to promote gonadal hormone production and gamete development. Besides guiding reproductive function in the peripheral tissues, these sex hormones modulate upstream components of the HPG axis to control the secretion of GnRH, luteinizing hormone (LH), and follicle-stimulating hormone (FSH).

The HPG axis activity is regulated by a complex series of outside molecules as well. Cannabinoids, both endogenous and exogenous, modulate gonadal function primarily by activating hypothalamic cannabinoid receptor (CB) 1. This activation leads to reduced release or disruption of the normal pulsatility of GnRH secretion. This is, however, a subsidiary effect, as the reduction of GnRH secretion requires the inhibition of glutamate-releasing (glutamatergic) neurons (Liu & Herbison, 2013). As a consequence, cannabis also reduces the expression of receptors for GnRH in the pituitary, reducing the release of plasma LH levels in women and, in minor extent, of FSH from the pituitary gland and of testosterone or estradiol from the gonads. Cannabinoid-induced variations in LH concentrations depend on the phase of the menstrual

Cannabis Use, Neurobiology, Psychology, and Treatment. https://doi.org/10.1016/B978-0-323-89862-1.00035-0

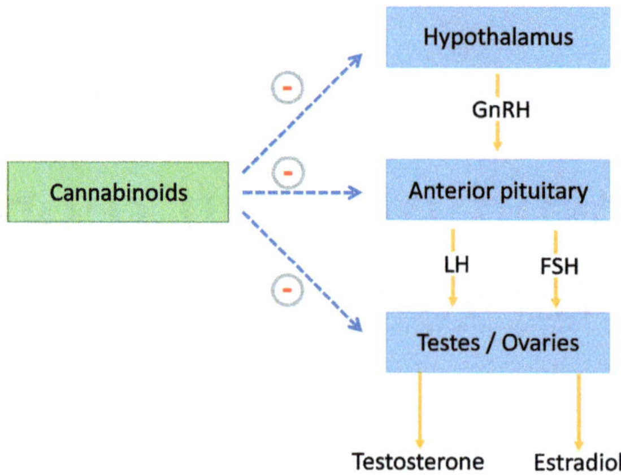

FIG. 1 Interaction between cannabinoids and the reproductive endocrine system. It is currently accepted that cannabinoids modify the secretion of LH and, possibly FSH, by inhibiting the release of GnRH in the hypothalamus. This, in turn, decreases the levels of sex hormones and results in diminished fertility for both males and females.

cycle. In fact, the reduction of plasma LH levels is more pronounced during the secretory/luteal phase of the menstrual cycle (Mendelson et al., 1986). Moreover, occasional marijuana users (up to three times in the last 3 months) had a longer follicular phase than nonusers (Jukic et al., 2007). On the contrary, chronic cannabis users show normal LH levels, whereas FSH levels are not different from the control subjects, suggesting that tolerance is an important mechanism underlying the effects of cannabinoids upon HPG axis functioning (Vescovi et al., 1992). The combined action of LH and FSH regulate follicular growth, ovulation, and maintenance of the corpus luteum. As a consequence, cannabis use may compromise the reproductive functions (Fig. 1).

Ovulation is a process dependent on the accumulation of cyclic adenosine monophosphate (cAMP). The specific inhibition of FSH-stimulated cAMP accumulation by Δ9-tetra-hidrocanabinol (THC) was demonstrated in rat granulosa cells (Treinen et al., 1993). In addition, the reduction in the levels of ovarian hormones in response to THC is reversed by human chorionic gonadotropin which indicates that cannabinoid actions in the central nervous system (CNS) are, at least partially, responsible for the suppression of GnRH release from the anterior pituitary gland (Almirez et al., 1983). Nevertheless, the central cannabinoid control over LH release also influences progesterone signaling. The latter mechanism explains the cannabis effects on sexual behavior by facilitating sexual receptivity and motivation and by delaying sexual maturation in female rats. However, some authors argue that cannabinoids may have a dual effect on sexual behavior: at low doses, THC increases both sexual receptivity and proceptivity in female hamsters (Turley & Floody, 1981), whereas at high doses, it decreases sexual receptivity (Gordon et al., 1978).

In males, both chronic and acute administration of THC reduce plasma LH and testosterone concentrations. In addition, as already mentioned for women, during continued cannabis use, the reduction on testosterone levels is also inverted by stimulation with human chorionic gonadotropin, indicating a central mechanism behind LH reduction. The combined action of LH and FSH promotes the development of the seminiferous tubules and stimulate the initiation of spermatogenesis and testosterone production. Thus, the interference of cannabinoids in HPG axis disturbs many reproductive functions. However, these effects are still controversial, as some authors report no effects on circulating testosterone, prolactin, LH, or FSH levels in response to THC or other cannabinoids. Accordingly, in a recent study, involving 758 male participants, neither cannabis use nor frequency of cannabis use was associated with reduced fertility (Kasman et al., 2018), highlighting the contrasting results in reproductive function in males exposed to cannabinoids. Furthermore, cannabis exposure modulates sexual behavior, with some authors reporting the ability of cannabis preparations to increase libido at short-term. Nevertheless, at long-term, cannabis use may be associated with pathological outcomes, namely gynecomastia, erectile disorders, and reduced erectile function (Pizzol et al., 2019).

The harmful effects of cannabinoids on reproductive function are not exclusively due to the action of cannabinoids on the central component of the HPG axis (Fig. 2). In effect, in the testis, spermatozoa, as well as Sertoli and Leydig cells express both cannabinoid receptors (Amoako et al., 2010; Bari et al., 2011; Rossato et al., 2005). As a result of cannabinoid's high lipophilic properties, the blood-testicular barrier is permeable to cannabinoids. The magnitude of cannabinoid-direct effects on the gonads differs but includes decreased production of sperm, reduced sperm motility (Kolodny et al., 1974), and inhibition of the processes needed to facilitate sperms' ability to achieve conception, as CB1 activation inhibits

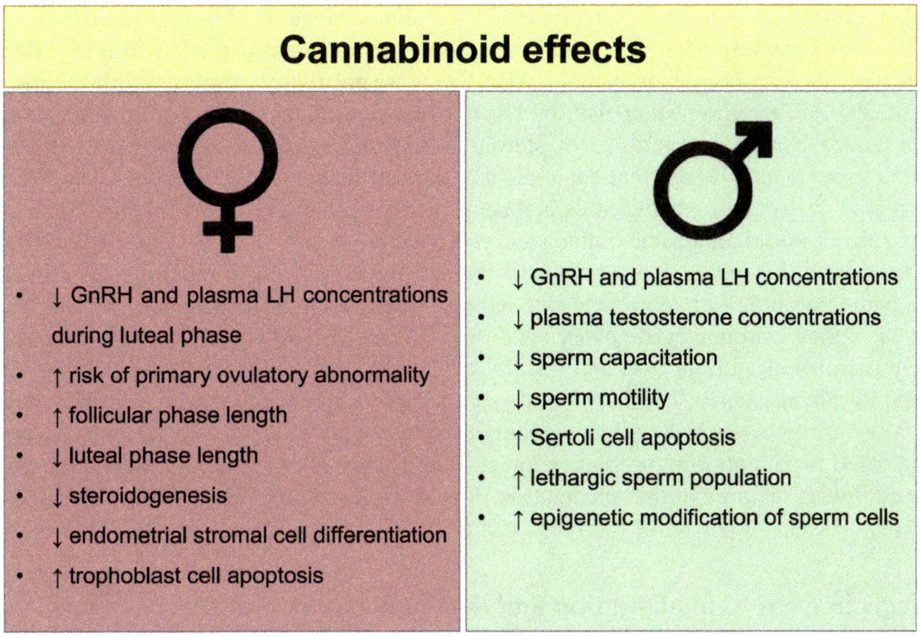

FIG. 2 Reproductive effects of cannabinoid exposure in both male and female fertility. Besides the reduction in female and male fertility, the use of cannabinoids during pregnancy was associated with increased odds of threatened preterm labor, early onset delivery, poor fetal growth, and stillbirth. The long-term effects of cannabis exposure in uterus also include deficits in cognitive functions, sleep and mood disorders, and increased risk of autism spectrum disorders.

sperm capacitation and reduces sperm mitochondrial activity (Rossato et al., 2005). Blocking CB1 with the specific antagonist SR141716 confirmed the specific effects of cannabinoids upon human sperm. In turn, activation of CB2 increases the lethargic sperm population (Agirregoitia et al., 2010). Furthermore, chronic administration of THC depressed testosterone synthesis of the testis (Banerjee et al., 2011; Jakubovic et al., 1979) and inhibited dihydrotestosterone binding to androgen receptors on in vitro rat prostate cells (Purohit et al., 1980). The mechanisms are unclear, but the study suggests that cannabinoid effects on male reproductive tract include other pathways of action besides the changes in testosterone levels. Interestingly, recent evidences suggest epigenetic modifications of sperm cells as a result of paternal activation of CB2 (Innocenzi et al., 2019). The exact significance of such evidence is unknown, but it highlights the potential hazards from an unidentified side: the father's premating exposure to cannabinoids.

As described in male fertility, the effects of cannabis exposure in woman also extend behind brain regulation. First, both ovary and endometrium express cannabinoid receptors and, thus, are potential targets for cannabinoid action. In addition, cannabis preparations may prevent the conversion of pregnenolone to progesterone (Walker et al., 2019) and inhibit steroidogenesis (Kostellow et al., 1980), explaining why cannabis users present reduced progesterone concentrations. Therefore, besides a great risk for women in the ovulatory period, cannabis consumption impacts menstrual cycle regularity. Second, accumulating evidence demonstrates that cannabis action in some reproductive functions result, at least partially, from disturbances in the crosstalk between sex steroid hormones and endocannabinoids (eCBs). This evidence is supported by the observation that anandamide (AEA), the major eCB, affect estrogen signaling (Neradugomma et al., 2019), whereas ovarian steroid hormones modulate the activity of eCB-metabolic enzymes and, consequently, eCB levels (Maia et al., 2017). Estradiol administration to ovariectomized female rats increases AEA levels in the medial basal hypothalamus (Scorticati et al., 2004). In turn, chronic administration of AEA decreases LH concentrations in pregnant rats (Habayeb et al., 2002).

Moreover, like estrogens, progesterone can also regulate eCB signaling. Besides the already mentioned central eCB control of progesterone release, cannabinoids may also control progesterone signaling by direct binding onto receptor sites on the corpus luteum. In addition, progesterone increases the expression and activity of fatty acid amide hydrolase (FAAH) in both T cells and immortalized human lymphoma U937 cells by interacting with a transcription factor in the promoter region of the *faah* gene (Maccarrone et al., 2001). On the contrary, progesterone downregulates FAAH activity in the mouse uterus (Maccarrone et al., 2000), an effect not seen in immortalized human neuroblastoma CPH100 cells (Maccarrone et al., 2004), suggesting the existence of different FAAH variants in human cells and highlighting the intricacies of eCB regulation.

It is now clear that, during the menstrual cycle, there is a dynamic change of AEA plasmatic levels. In women, serum AEA levels are higher during the follicular phase and lower during the luteal phase (Habayeb et al., 2004). The circulating AEA levels fluctuate with pituitary hormones, namely, AEA levels are positively correlated with serum estradiol, FSH, and LH (El-Talatini et al., 2010). Therefore, when disturbed by exogenous cannabinoids, aberrant cannabinoid signaling may lead to reproductive failure. However, regarding sex steroid hormones and eCB crosstalk, the evidence is still limited in humans. Moreover, despite large evidence that cannabis use can disrupt ovulation and menstrual regularity, large-scale cohort studies report a poor association between cannabis use and fecundability or sexual function.

Another relevant subject associating cannabinoids and reproductive health is the involvement of the endocannabinoid system (ECS) in reproductive physiology. It is now clear that either activation or overexpression of cannabinoid receptors is associated with the pathophysiology of some reproductive disorders. The CB1 expression and the levels of eCBs were significantly higher in women with polycystic ovary syndrome (Juan et al., 2015). In those patients, it was also observed a significant positive correlation with glucose and insulin levels, suggesting that eCB elements may be associated with energy regulation and insulin resistance. The ECS has also been implicated in endometrial disorders, since it was seen that patients diagnosed with endometriotic lesions and adenomyosis present lower expression of both cannabinoid receptors in endometrium as compared to patients with normal endometrium (Tanaka et al., 2020). Although the involvement of ECS elements in these pathologies requires further elucidation, it is now clear that eCBs involvement in reproductive system extends far beyond the central component of the HPG axis (Fig. 2).

Cannabinoid effects in embryo implantation and decidualization

Besides interfering with ovulation and menstrual cycle, cannabis preparations can seriously compromise the woman's ability to become pregnant by interfering with the ability of embryos to implant. After fertilization in the oviduct, the egg suffers mitotic divisions to form a morula that further develops into blastocyst. The later consists of an inner cellular mass that will originate the embryo, and the trophectoderm, that will form the placenta and extra-embryonic membranes. Studies in rodents have shown that the blastocyst contains CB1, making the embryo a target for natural and endogenous cannabinoids, increasing the importance of cannabinoid signaling in female fertility. Moreover, in culture, both THC and AEA interfere with blastocyst development and differentiation. Furthermore, blastocysts cultured in the presence of low levels of AEA showed rapid trophoblast differentiation, while higher AEA levels induced the opposite effect (Liu et al., 2002). This adverse outcome can be prevented by the addition of CB1 antagonists. The negative effect of cannabinoids on embryo development and implantation involves the inhibition of Ca^{2+} influx (Paria et al., 1998). During the early stage of pregnancy, in addition to embryo development, the transport of the blastocyst along the oviduct occurs in parallel and is another critical process for pregnancy establishment. Normal oviductal transport is one of the prerequisites for implantation and it has been shown, in rodents, that this transport is negatively influenced by the action of cannabinoids and that it is dependent on CB1 action (Wang et al., 2004). Furthermore, AEA-metabolic enzymes are differentially expressed in order to create the appropriate AEA levels during the implantation window. In fact, low concentrations of AEA at the implantation site have been found to be conducive to implantation: just prior to embryo implantation, AEA levels decrease, and this change contributes to the receptive state (Schmid et al., 1997).

The uterus is the target of many modifications before and after implantation of the embryo and decidualization takes part as an important process, both in humans and rodents. In humans, decidualization of endometrial stromal cells occurs in each menstrual cycle at late luteal phase to prepare the endometrium for a future pregnancy. The components of the ECS are also expressed in maternal tissues during pregnancy. Decidual cells are also involved in the development of surrounding trophoblastic, mesenchymal, and hematopoietic cells. In addition, they also supply growth factors and cytokines that enable the embryo's development and are implicated in supporting the formation of maternal blood vessels, so that nutrition and perfusion of the embryo may be possible throughout pregnancy. Therefore, decidualization of the endometrium is crucial for the existence and maintenance of a healthy pregnancy. In addition to the existence of a complete ECS in decidual cells, a negative effect on the function of these cells has been observed upon the activation of CB1 (Moghadam et al., 2005). Furthermore, the synthetic cannabinoid R(+)-WIN 55,212-2 mesylate (WIN) that mimics many of the actions of THC, was able to inhibit the cAMP-dependent induction of decidual cell differentiation by decreasing the levels of specific decidualization markers, such as prolactin, laminin, and insulin-like growth factor-binding protein 1 (Moghadam et al., 2005). All these effects were reversed by the addition of a CB1 antagonist, indicating that the activation of this receptor inhibits decidualization and stimulates apoptosis through a cAMP-dependent mechanism. Moreover, CB1 activation by AEA or 2-arachidonoylglycerol (2-AG) induces decidual cell death, which involves eCBs in uterine remodeling processes occurring during pregnancy establishment (Fonseca et al., 2009). In another study using rat decidual cells, activation of CB1 led to the accumulation of ceramide through de novo synthesis which resulted in phosphorylation of the p38-MAPK protein, followed by an induction of mitochondrial stress and production of reactive oxygen species, leading to programmed cell death (Fonseca et al., 2013).

Finally, there is also evidence that cannabinoids may affect nitric oxide (NO) synthesis. NO can function as a paracrine mediator and an important regulatory agent in several female reproductive processes. However, both cannabinoid receptors demonstrated opposite effects on NO synthesis activity. While CB1 has a stimulatory effect, CB2 has an inhibitory effect (Molina-Holgado et al., 2002). This opposite effect on NO synthesis may be relevant for the control of reproduction, as NO has several effects on fertility, namely female fertility. As the human endometrium expresses both receptors (Maia et al., 2020) it is conceivable that the two subtypes of receptors are activated at different times, oppositely modulating NO levels.

Cannabis and placentation

In the case of an insufficient trophoblast invasion of uterine spiral arteries, due to limitations provoked by decidual cells, low levels of oxygen are supplied to the fetus, which can lead to fetal prematurity, fetal growth restriction, and pre-eclampsia. In the later, trophoblast invasion is only superficial, the uterine spiral arteries remodeling is not total, and the apoptosis of decidual cells is reduced, which increases the resistance of the blood vessels and results in high blood pressure. On the other hand, an excessive trophoblast invasion resulting of deficient or absent decidua is equally dangerous, since it results in the formation of *placenta acreta* that can cause postpartum hemorrhage and usually yields to emergency hysterectomy or even maternal death (Fonseca et al., 2012).

Cannabis exposure impairs placental development and leads to changes in placenta structure and morphology. These changes are triggered by an inflammatory response and/or by poor supply of nutrients and oxygen. Cannabinoid receptors and the enzymatic machinery of AEA are expressed by cytotrophoblast and syncytiotrophoblast (Costa et al., 2015). While 2-AG interferes with signaling pathways that are important for placental endocrine function (Costa et al., 2016), AEA impairs the production of pregnancy-related proteins by human syncytiotrophoblast (Costa et al., 2015). In addition, both AEA and 2-AG can induce cytotrophoblast apoptosis and interfere with the expression and activity of syncytiotrophoblast-related proteins. Therefore, cannabis-associated miscarriage can be provoked by deviations in the expression of some components of the ECS. One example of these effects is the increased levels of AEA in plasma, induced by the reduction on FAAH activity, which has been associated with increased risk of miscarriage (Brents, 2016). Furthermore, THC impairs syncytialization, enhances syncytiotrophoblast mitochondrial activity, and inhibits human trophoblast cells turnover (Costa, 2016). Other authors tested the acute and chronic effects of THC upon the apical uptake of folic acid in primary trophoblast cells and in the trophoblast cell line BeWo cells and found that uptake of folic acid was reduced by chronic treatment of the cells with cannabinoids (Araujo et al., 2009).

Cannabis action on ontogeny

Midgestational pregnancy and term

The HPG axis is active in the midgestational fetus but silenced toward term because of the negative feedback effects mediated by placental hormones. At birth, the HPG axis is reactivated and, approximately between the ages of 1 year until puberty, it remains quiescent as gonadotropins and sex steroid hormone levels decrease. Because the fetal HPG axis plays an important role in the development of several vital organs, the modulation of the fetal HPG axis by cannabinoids may have endocrine consequences throughout gestation.

Although the use of cannabis during pregnancy is associated with increased risk of adverse birth outcomes, no association was observed between preconception cannabis use and pregnancy loss (Table 1). Cannabis users had higher rates of spontaneous preterm births (Saurel-Cubizolles et al., 2014). In addition, continued maternal marijuana use at 20 weeks' gestation was associated with spontaneous preterm births independent of cigarette smoking status and socioeconomic index (Leemaqz et al., 2016). Higher risks of pregnancy complications among pregnant women using cannabis have previously been reported. Cannabis use has been associated with reduced fetal head circumference for gestational age patterns compared with nonexposed fetuses (El Marroun et al., 2009). Heavy users presented even greater differences (Grzeskowiale et al., 2020), suggesting a dose-dependent effect of cannabis preparations on neonatal outcomes. However, some studies demonstrate that the length and intensity of cannabis use is relevant to the risk of adverse birth outcomes. Frequent cannabis users during pregnancy present even higher risk of fetal growth restriction in mid-and late pregnancy and reduced birth weight (Gabrhelik et al., 2021). Furthermore, the high lipophilicity of cannabinoids is responsible for their long half-life in body tissues, increasing fetal exposure to cannabinoids, as THC readily cross the placenta (Blackard & Tennes, 1984).

TABLE 1 Effects of cannabis consumption in prenatal, perinatal, and adolescent periods of development.

Study design	Number of participants	Exposure window	Cannabis exposure	Consequence of cannabis exposure	Reference
Obstetrical outcomes					
Case-control study	1932	During pregnancy	Self-reported	↑ stillbirth (OR 2.34, 95% CI 1.13–4.81)	Varner et al. (2014)
Retrospective cohort study	167	During pregnancy	Urine toxicology	↑ placenta abruption (OR 21.3, 95% CI 7.5–49.4)	Vestermark and Laier (2019)
Prospective cohort study	1228	Before pregnancy	Urine toxicology and self-reported	= pregnancy loss (OR 0.8, 95% CI 0.5–1.4)	Mumford et al. (2021)
Neonatal outcomes					
Prospective cohort study	3857	During pregnancy	Self-reported	↓ birth weight (OR 2.6, 95% CI 1.1–6.2) ↓ gestational age (OR 2.3, 95% CI 1.3–4.1) ↑ preterm delivery (OR 1.9, 95% CI 1.0–3.9)	Hatch and Bracken (1986)
Case-control study	1985	During pregnancy	Self-reported	↑ sudden infant death syndrome (OR 2.23, 95% CI 1.4–3.6)	Scragg et al. (2001)
Case-control study	478	During pregnancy	Self-reported	= sudden infant death syndrome (OR 2.0, 95% CI 0.6–6.5)	Klonoff-Cohen and Lam-Kruglick (2001)
Retrospective cohort study	24,874	Before and during pregnancy	Self-reported	↓ birth weight (OR 1.7; 95% CI 1.3–2.2) ↑ preterm labor (OR 1.5; 95% CI 1.1–1.9) ↓ gestational age (OR 2.2; 95% CI 1.8–2.7) ↑ neonatal intensive care unit admission (OR 2.0; 95% CI 1.7–2.4)	Hayatbakhsh et al. (2012)
Cross-sectional study	344	During pregnancy	Self-reported	↓ low birth weight (OR 6.5 95% CI 3.0–14.3) ↓ gestational age (OR 3.8, 95% CI 1.9–7.6) ↑ negative birth outcomes (OR 3.9, 95% CI 1.4–11.2)	Brown et al. (2016)
Case-control study	465	Following 1st trimester in opioid dependent pregnant women	Self-reported	↑ LR of treatment for NAS (27.6% in cannabis exposed infants vs. 15.7% in nonexposed) ↑ duration of infant hospital stays (7.7 days in cannabis exposed infants vs. 6.6 days in nonexposed infants)	O'Connor et al. (2017)
Prospective cohort study	34	During pregnancy	Self-reported	= adverse pregnancy outcome (aOR1.29; 95% CI 0.56–2.96)	Metz et al. (2017)
Retrospective cohort study	1206	Before and during pregnancy	Urine toxicology and self-reported	↑ composite adverse outcome (e.g., hypertensive disorders of pregnancy, spontaneous preterm delivery, small for gestational age, stillbirth) (aOR 1.69, 95% CI 1.22–2.34)	Rodriguez et al. (2019)

TABLE 1 Effects of cannabis consumption in prenatal, perinatal, and adolescent periods of development—cont'd

Study design	Number of participants	Exposure window	Cannabis exposure	Consequence of cannabis exposure	Reference
Retrospective cohort study	9427	Before pregnancy	Self-reported	↑ preterm birth <37 weeks gestation (RR 1.41, 95% CI 1.36–1.47) ↑ preterm birth <28 weeks gestation (RR 1.97, 95% CI 1.70–2.28) ↓ preeclampsia (RR 0.90, 95% CI, 0.86–0.95) ↓ gestational diabetes (RR 0.91, 95% CI, 0.86–0.96)	Corsi et al. (2019)
Prospective cohort study	344	Before and during pregnancy	Self-reported	= low birth weight (OR 6.5, 95% CI 3.0–14.3) = preterm birth (OR 2.3, 95% CI 0.6–2.6) ↓ gestational age (OR 2.0, 95% CI 2.3–3.3)	Michalski et al. (2020)
Prospective cohort study	2032	Before pregnancy	Self-reported	↑ preterm birth (aOR 6.65, 95% CI 1.92–23.09) ↓ birth weight (aOR 5.84, 95% CI 1.70–20.08)	Hines et al. (2021)
Prospective cohort study	363	During pregnancy	Urine toxicology and self-reported	= spontaneous preterm birth (RR 1.06, 95% CI 0.76–1.47) = time to delivery (RR 1.26, CI 0.84–2.00)	Klebanoff et al. (2021)
Long-term effects					
Case-control study	122	During pregnancy	Self-re ported	↑ ventricular septal defect (cOR 2.35, 95% CI 1.43–3.86)	Williams et al. (2004)
Prospective cohort study	563	During pregnancy	Self-reported	↓ onset of cannabis use	Day et al. (2006)
Retrospective cohort study	18,745	Before pregnancy	Self-reported	↑ anencephaly (OR 1.9, 95% CI 1.1–3.2) ↑ esophageal atresia (OR 1.7, 95% CI 1.0–2.9) ↑ diaphragmatic hernia (OR 1.8, 95% CI 1.1–3.0) ↑ gastroschisis (OR 1.7, 95% CI 1.2–2.3)	van Gelder et al. (2014)

aOR, adjusted odds ratio; *CI*, confidence interval; *cOR*, crude odds ratios; *LR*, likelihood ratio; *NAS*, neonatal abstinence syndrome; *OR*, odds ratio; *RR*, relative risk.

Although some findings associate maternal cannabis use and adverse outcomes, the whole range of short- and long-term health effects requires further studies.

The impact of cannabis in fetal development may also be manifested in other vital systems (Table 1). Recent studies links cannabis exposure to cardiovascular defects in neonates. During fetal development, cannabinoids exposure increases the risk of congenital heart defects, such as ventricular septal anomalies and malformation of the tricuspid valve, suggestive of the activation of CB1, highly expressed on the derivatives of the endocardial cushions. Other authors have found an association in gastrochisis and prenatal cannabis exposure which, similarly with cardiovascular observations, is consistent with high CB1 expression on umbilical vessels observed in gastrochisis patients (Corsi, 2020).

THC potentially influences the development of the fetal brain, especially in the first trimester. Because during fetal development the blood brain barrier is still maturing, cannabis consumption can impact fetal growth and brain development with adverse pregnancy outcomes. In addition, cannabinoids can cross the blood-brain barrier and, thus, reach the fetal brain that, in the second trimester of gestation already expresses CB1 (Wang et al., 2003). It is now clear that fetal

CB1 plays a crucial role in brain development, including neural stem cell function, glial support cells, white matter and central nervous system (CNS) tract formation, myelination, dendrite formation, synapse formation and neural migration. In addition, CB1 is highly expressed in endoplasmic reticulum and mitochondria of neuronal cells (Hebert-Chatelain et al., 2016), which indicates their potential involvement in the control of cell function, such as DNA repair pathways, by supplying energy, by metabolite shuttle, and by RNA signaling. Hence, it is not surprising that gestational cannabis impairs cognitive functions on several levels. Repeated CB1 activation during critical periods of brain development can affect the expression and functionality of dopaminergic receptors that are involved in higher cognitive functions. In effect, it was shown that infants exposed to cannabis in utero exhibit decreased dopamine receptor 2 mRNA in the *nucleus accumbens*, through mechanisms that regulate histone lysine methylation (DiNieri et al., 2011). The epigenetic modifications in individuals prenatally exposed to cannabis may contribute to addiction vulnerability later in life. Moreover, as far as we know, there are no studies about the potential effects on infants exposed to secondary cannabis smoke.

Childhood growth

The long-term actions of cannabis reflect its potential for chronic and cumulative dose effects. The developmental effects of cannabinoids exposure are poorly studied but can affect every body system. Prenatal exposure in rodents induces changes in the dopaminergic activity of the *corpus striatum,* which leads to attention deficit and hyperactivity disorders and alterations in locomotion, and on the prefrontal cortex, which cause cognitive impairment and emotional dysregulation (Trezza et al., 2012). There are also evidences that perinatal cannabis exposure affects the migration of glutamatergic neurons and GABAergic interneurons and induces alterations in the electrophysiological properties of the Purkinje neurons of the cerebellum (Shabani et al., 2011).

Although even few studies addressed the in utero exposure to cannabinoids in humans, it was reported an association between cannabis exposure and autism spectrum disorder (Corsi et al., 2020). Besides an increase in the incidence of intellectual disability and learning disorders on children exposed to cannabis in utero (Corsi et al., 2020), it was also reported alterations on sleep and affective symptoms in those children (Lisdahl et al., 2016). In addition to the deficits in cognitive function, children prenatally exposed to cannabis present thicker prefrontal cortex when compared with nonexposed children (El Marroun et al., 2016). In addition, adolescents at age 14 that were prenatally exposed to cannabis have an earlier onset of cannabis use and use cannabis more frequently compared to adolescents who were not exposed (Day et al., 2006).

Although a critical analysis of results from longitudinal studies examining the impact of prenatal cannabis exposure on multiple domains suggest that prenatal cannabis exposure alone is associated with clinically significant cognitive functioning impairments, the specific effects of prenatal cannabis exposure remain unclear and warrant further research (Torres et al., 2020).

The effects of cannabis use in lactation are largely unknown. Cannabinoids can be found in breast milk. THC concentrations of 27.5 ng/mL (0.8–190.5) and CBD of 1.2 ng/mL (0.5–17.0) were found in breast milk (Moss et al., 2021); however, no studies have evaluated the effects of cannabis on infants during lactation. Nevertheless, current data suggest an association of cannabis exposure with development and behavior.

Conclusion

Cannabinoid-induced hypogonadism is a well-established process. This reduction has multiple endocrine and metabolic consequences including sexual dysfunction and infertility. Cannabinoids interfere with the HPG axis, causing changes in the secretion pattern of endocrine glands. In men, hypogonadism results from failure of the testes to produce adequate levels of testosterone and a normal number of sperms, whereas in women it results from inadequate function of the ovaries, with impaired production of sex hormones. In women, the combined reduction in LH and FSH hormonal levels is also associated with ovulation disruption and abnormalities in menstruation and endometrial decidualization. Yet, large-scale prospective cohort studies have failed to demonstrate reduced fertility in women who frequently use cannabis.

During pregnancy, cannabis use is associated with a higher risk of spontaneous pre-term birth including lower birth weight, length, and head circumference. In addition, cannabis use during pregnancy has also been associated with adverse neonatal outcomes such as deficits in cognitive functions. Nevertheless, data on the potential health effects of cannabis use in women during pregnancy are inconsistently described.

There is paucity of randomized control trials assessing the effects of cannabinoids in women fertility and there are even fewer studies in men. THC, as its major metabolites, are present in the semen and demonstrated to reduce sperm quality in heavy or long-term cannabis users. However, the studies evaluating fertility in men are mostly observational, with conflicting results.

Current data indicate that increasing use of marijuana among young women of reproductive age is a major public health concern. We need more studies to better inform our treatment decisions and assess the impact of cannabinoids in this population. Despite our lack of understanding of the effects of cannabinoids on fertility, it might be reasonable to consider cannabis use in a comprehensive infertility evaluation and possibly sexual dysfunction symptoms. Thus, promoting cannabis cessation among youngsters and during pregnancy could restore human fertility and lead to improved perinatal and later childhood outcomes.

Applications to other areas

In this chapter, we have reviewed potentially effects cannabinoid's can have on reproductive health. So far, endogenous cannabinoids have been confirmed to participate in the regulation of various physiological processes of the body, and have a significant impact on the endocrine system, including the activity of the pituitary gland and gonads. In women, studies have shown that acute and chronic use of cannabinoids can affect sex hormones (i.e., follicle-stimulating hormone, luteinizing hormone), disrupt ovulation, and result in menstrual irregularities including amenorrhea. Similarly, a reduction in testosterone production, sperm motility, and fertility potential have been observed in man exposed to high levels of THC. In addition, both cannabinoid receptors are expressed in reproductive tissues and human gametes, which expands the potential detrimental actions of these compounds. These topics are introduced and most prominent clinical evidence about cannabis consumption in human reproductive potential and teratogenicity are reviewed.

Mini-dictionary of terms

Decidualization—Decidualization designates the morphological and functional differentiation of human endometrial stromal cells into specialized secretory decidual cells. It occurs in response to elevated levels of the ovarian steroid hormones and results from a complex interplay of transcription factors, morphogens, cytokines, cell cycle regulators, and signaling pathways. By providing a nutritive and immune-privileged matrix, decidual cells are essential for embryo implantation and placental development.

Hypothalamus—The hypothalamus is a small brain structure (less than 1% of the human brain volume) located just below the thalamus and above the pituitary gland. It consolidates signals derived from upper cortical inputs, autonomic function, environmental cues, and peripheral endocrine feedback. It is central to homeostasis, stress responses, energy metabolism, and reproductive and neuropsychiatric functions.

Hypothalamic-pituitary axis—The hypothalamic pituitary axis is a complex set of direct influences and feedback inter actions among the hypothalamus and the pituitary gland. The hypothalamus delivers precise signals to the pituitary gland, which then releases hormones that directly affects the functions of the thyroid, the adrenal gland, and the gonads, as well as influencing growth, milk production, and water balance.

Implantation—Embryo implantation is a coordinated biological process that requires the synchronization between a healthy embryo and a receptive endometrium through the production of growth factors, cytokines, and hormones. Implantation can be affected by many factors including sperm and oocyte quality, early development of the embryo, the endometrium, and the reciprocal communication between blastocysts and endometrium.

Placentation—Placentation is a reproductive adaptation that establishes the physical contact between the fetus and the mother for the transfer of nutrients and hormones, as well as the removal of waste products from the fetus. The placenta is comprised of specialized epithelial cell types, collectively referred as trophoblast cells. Trophoblast differentiation is crucial for proper placentation. Placenta also determines postnatal health and susceptibility to disease.

Key facts of cannabis in reproductive function

Key facts of cannabis incidence and prevalence among young adults

According to the EMCDDA (European Monitoring Centre for Drugs and Drug Addiction), cannabis is the most widely used illicit drug in Europe. Its use is increasing among the general population with a prevalence of 14.1% among individuals at reproductive age. In addition, the prevalence of cannabis use in pregnancy is also increasing, notwithstanding the awareness of fertility and pregnancy complications.

Key facts of cannabis in hypothalamic-pituitary-gonadal (HPG) axis

Cannabinoids have a direct action on the HPG axis, by inhibiting the normal pulsatility of GnRH secretion and by reducing the LH hormone secretion from the anterior pituitary. Cannabinoids inhibit the function of the HPG axis by activating cannabinoid receptor 1 in the hypothalamus. Additionally, endogenous cannabinoids normally bind to these receptors in hypothalamus and modulate the activity of the HPG axis.

Summary points

- Cannabinoids disrupt the normal pulsatility of GnRH secretion and reduce the release of LH hormone.
- At low doses, THC increases both sexual receptivity and proceptivity, whereas at high doses, it decreases sexual receptivity.
- The blood-testicular barrier is permeable to cannabinoids as result of its high lipophilic properties, which exposes male gonads to direct cannabinoid effects.
- Cannabinoids decrease the production of sperm cells as well as affect sperm motility and capacitation.
- Disturbances in the crosstalk between ovarian steroid hormones and endocannabinoids underlie some detrimental reproductive events associated with cannabinoids.
- Cannabinoids inhibit decidualization and interfere with blastocyst development and differentiation.
- Cannabinoids can cross the placenta and interfere with fetal endocannabinoid signaling during neurodevelopment, causing long-lasting deficits.

References

Agirregoitia, E., Carracedo, A., Subiran, N., Valdivia, A., Agirregoitia, N., Peralta, L., Velasco, G., & Irazusta, J. (2010). The CB(2) cannabinoid receptor regulates human sperm cell motility. *Fertility and Sterility*, *93*(5), 1378–1387. https://doi.org/10.1016/j.fertnstert.2009.01.153.

Almirez, R. G., Smith, C. G., & Asch, R. H. (1983). The effects of marijuana extract and delta 9-tetrahydrocannabinol on luteal function in the rhesus monkey. *Fertility and Sterility*, *39*(2), 212–217. https://www.ncbi.nlm.nih.gov/pubmed/6295828.

Amoako, A. A., Marczylo, T. H., Lam, P. M., Willets, J. M., Derry, A., Elson, J., & Konje, J. C. (2010). Quantitative analysis of anandamide and related acylethanolamides in human seminal plasma by ultra performance liquid chromatography tandem mass spectrometry. *Journal of Chromatography. B, Analytical Technologies in the Biomedical and Life Sciences*, *878*(31), 3231–3237. https://doi.org/10.1016/j.jchromb.2010.09.024.

Araujo, J. R., Goncalves, P., & Martel, F. (2009). Effect of cannabinoids upon the uptake of folic acid by BeWo cells. *Pharmacology*, *83*(3), 170–176. https://doi.org/10.1159/000192587.

Banerjee, A., Singh, A., Srivastava, P., Turner, H., & Krishna, A. (2011). Effects of chronic bhang (cannabis) administration on the reproductive system of male mice. *Birth Defects Research Part B-Developmental and Reproductive Toxicology*, *92*(3), 195–205. https://doi.org/10.1002/bdrb.20295.

Bari, M., Battista, N., Pirazzi, V., & Maccarrone, M. (2011). The manifold actions of endocannabinoids on female and male reproductive events. *Frontiers in Bioscience-Landmark*, *16*, 498–516. https://doi.org/10.2741/3701.

Blackard, C., & Tennes, K. (1984). Human placental-transfer of cannabinoids. *New England Journal of Medicine*, *311*(12), 797. <Go to ISI>://WOS:A1984TJ35800016.

Brents, L. K. (2016). Marijuana, the endocannabinoid system and the female reproductive system. *The Yale Journal of Biology and Medicine*, *89*(2), 175–191. https://www.ncbi.nlm.nih.gov/pubmed/27354844.

Brown, S. J., Mensah, F. K., Kit, J. A., Stuart-Butler, D., Glover, K., Leane, C., Weetra, D., Gartland, D., Newbury, J., & Yelland, J. (2016). Use of cannabis during pregnancy and birth outcomes in an aboriginal birth cohort: A cross-sectional, population-based study. *BMJ Open*, *6*(2).

Corsi, D. J. (2020). The potential association between prenatal cannabis use and congenital anomalies. *Journal of Addiction Medicine*, *14*(6), 451–453. https://doi.org/10.1097/Adm.0000000000000639.

Corsi, D. J., Donelle, J., Sucha, E., Hawken, S., Hsu, H., El-Chaar, D., Bisnaire, L., Fell, D., Wen, S. W., & Walker, M. (2020). Maternal cannabis use in pregnancy and child neurodevelopmental outcomes. *Nature Medicine*. https://doi.org/10.1038/s41591-020-1002-5.

Corsi, D. J., Walsh, L., Weiss, D., Hsu, H., El-Chaar, D., Hawken, S., Fell, D. B., & Walker, M. (2019). Association between self-reported prenatal cannabis use and maternal, perinatal, and neonatal outcomes. *Journal of the American Medical Association*, *322*(2), 145–152.

Costa, M. A. (2016). The endocannabinoid system: A novel player in human placentation. *Reproductive Toxicology*, *61*, 58–67. https://doi.org/10.1016/j.reprotox.2016.03.002.

Costa, M. A., Fonseca, B. M., Mendes, A., Braga, J., Teixeira, N. A., & Correia da Silva, G. (2015). The endocannabinoid anandamide affects the synthesis of human syncytiotrophoblast-related proteins. *Cell and Tissue Research*, *362*(2), 441–446. https://doi.org/10.1007/s00441-015-2236-2.

Costa, M. A., Fonseca, B. M., Mendes, A., Braga, J., Teixeira, N. A., & Correia-da-Silva, G. (2016). The endocannabinoid 2-arachidonoylglycerol dysregulates the synthesis of proteins by the human syncytiotrophoblast. *Biochimica et Biophysica Acta*, *1861*(3), 205–212. https://doi.org/10.1016/j.bbalip.2015.12.008.

Day, N. L., Goldschmidt, L., & Thomas, C. A. (2006). Prenatal marijuana exposure contributes to the prediction of marijuana use at age 14. *Addiction*, *101* (9), 1313–1322. https://doi.org/10.1111/j.1360-0443.2006.01523.x.

DiNieri, J. A., Wang, X., Szutorisz, H., Spano, S. M., Kaur, J., Casaccia, P., Dow-Edwards, D., & Hurd, Y. L. (2011). Maternal cannabis use alters ventral striatal dopamine D2 gene regulation in the offspring. *Biological Psychiatry*, *70*(8), 763–769. https://doi.org/10.1016/j.biopsych.2011.06.027.

El Marroun, H., Tiemeier, H., Franken, I. H. A., Jaddoe, V. W. V., van der Lugt, A., Verhulst, F. C., Lahey, B. B., & White, T. (2016). Prenatal cannabis and tobacco exposure in relation to brain morphology: A prospective neuroimaging study in young children. *Biological Psychiatry*, *79*(12), 971–979. https://doi.org/10.1016/j.biopsych.2015.08.024.

El Marroun, H., Tiemeier, H., Steegers, E. A., Jaddoe, V. W., Hofman, A., Verhulst, F. C., van den Brink, W., & Huizink, A. C. (2009). Intrauterine cannabis exposure affects fetal growth trajectories: The generation R study. *Journal of the American Academy of Child and Adolescent Psychiatry*, *48*(12), 1173–1181. https://doi.org/10.1097/CHI.0b013e3181bfa8ee.

El-Talatini, M. R., Taylor, A. H., & Konje, J. C. (2010). The relationship between plasma levels of the endocannabinoid, anandamide, sex steroids, and gonadotrophins during the menstrual cycle. *Fertility and Sterility*, *93*(6), 1989–1996. https://doi.org/10.1016/j.fertnstert.2008.12.033.

Fonseca, B. M., Correia-da-Silva, G., & Teixeira, N. A. (2009). Anandamide-induced cell death: Dual effects in primary rat decidual cell cultures. *Placenta*, *30*(8), 686–692. https://doi.org/10.1016/j.placenta.2009.05.012.

Fonseca, B. M., Correia-da-Silva, G., & Teixeira, N. A. (2012). The rat as an animal model for fetoplacental development: A reappraisal of the post-implantation period. *Reproductive Biology*, *12*(2), 97–118.

Fonseca, B. M., Correia-da-Silva, G., & Teixeira, N. A. (2013). The endocannabinoid anandamide induces apoptosis of rat decidual cells through a mechanism involving ceramide synthesis and p38 MAPK activation. *Apoptosis*, *18*(12), 1526–1535. https://doi.org/10.1007/s10495-013-0892-9.

Gabrhelik, R., Mahic, M., Lund, I. O., Bramness, J., Selmer, R., Skovlund, E., Handal, M., & Skurtveit, S. (2021). Cannabis use during pregnancy and risk of adverse birth outcomes: A longitudinal cohort study. *European Addiction Research*, *27*(2), 131–141. https://doi.org/10.1159/000510821.

Gordon, J. H., Bromley, B. L., Gorski, R. A., & Zimmermann, E. (1978). Delta9-tetrahydrocannabinol enhancement of lordosis behavior in estrogen treated female rats. *Pharmacology, Biochemistry, and Behavior*, *8*(5), 603–608. https://www.ncbi.nlm.nih.gov/pubmed/674265.

Grzeskowiale, L. E., Grieger, J. A., Andraweera, P., Knight, E., Leemaqz, S., Poston, L., McCowan, L., Kenny, L., Myers, J., Walker, J. J., Dekker, G. A., & Roberts, C. T. (2020). The deleterious effects of cannabis during pregnancy on neonatal outcomes. *Medical Journal of Australia*. https://doi.org/10.5694/mja2.50624.

Habayeb, O. M. H., Bell, S. C., & Konje, J. C. (2002). Endogenous cannabinoids: Metabolism and their role in reproduction. *Life Sciences*, *70*(17), 1963–1977. https://doi.org/10.1016/S0024-3205(01)01539-9.

Habayeb, O. M., Taylor, A. H., Evans, M. D., Cooke, M. S., Taylor, D. J., Bell, S. C., & Konje, J. C. (2004). Plasma levels of the endocannabinoid anandamide in women- -a potential role in pregnancy maintenance and labor? *The Journal of Clinical Endocrinology and Metabolism*, *89*(11), 5482–5487. https://doi.org/10.1210/jc.2004-0681.

Hatch, E. E., & Bracken, M. B. (1986). Effect of marijuana use in pregnancy on fetal growth. *American Journal of Epidemiology*, *124*(6), 986–993.

Hayatbakhsh, M. R., Flenady, V. J., Gibbons, K. S., Kingsbury, A. M., Hurrion, E., Mamun, A. A., & Najman, J. M. (2012). Birth outcomes associated with cannabis use before and during pregnancy. *Pediatric Research*, *71*(2), 215–219.

Hebert-Chatelain, E., Desprez, T., Serrat, R., Bellocchio, L., Soria-Gomez, E., Busquets-Garcia, A., Zottola, A. C. P., Delamarre, A., Cannich, A., Vincent, P., Varilh, M., Robin, L. M., Terral, G., Garcia-Fernandez, M. D., Colavita, M., Mazier, W., Drago, F., Puente, N., Reguero, L., … Marsicano, G. (2016). A cannabinoid link between mitochondria and memory. *Nature*, *539*(7630), 555-+. https://doi.org/10.1038/nature20127.

Hines, L. A., Spry, E. A., Moreno-Betancur, M., Husin, H. M., Becker, D., Middleton, M., Craig, J. M., Doyle, L. W., Olsson, C. A., & Patton, G. (2021). Cannabis and tobacco use prior to pregnancy and subsequent offspring birth outcomes: A 20-year intergenerational prospective cohort study. *Scientific Reports*, *11*(1).

Innocenzi, E., De Domenico, E., Ciccarone, F., Zampieri, M., Rossi, G., Cicconi, R., Bernardini, R., Mattei, M., & Grimaldi, P. (2019). Paternal activation of CB2 cannabinoid receptor impairs placental and embryonic growth via an epigenetic mechanism. *Scientific Reports*, *9*. https://doi.org/10.1038/s41598-019-53579-3.

Jakubovic, A., McGeer, E. G., & McGeer, P. L. (1979). Effects of cannabinoids on testosterone and protein synthesis in rat testis Leydig cells in vitro. *Molecular and Cellular Endocrinology*, *15*(1), 41–50. https://doi.org/10.1016/0303-7207(79)90069-8.

Juan, C. C., Chen, K. H., Wang, P. H., Hwang, J. L., & Seow, K. M. (2015). Endocannabinoid system activation may be associated with insulin resistance in women with polycystic ovary syndrome. *Fertility and Sterility*, *104*(1), 200–206. https://doi.org/10.1016/j.fertnstert.2015.03.027.

Jukic, A. M. Z., Weinberg, C. R., Baird, D. D., & Wilcox, A. J. (2007). Lifestyle and reproductive factors associated with follicular phase length. *Journal of Womens Health*, *16*(9), 1340–1347. https://doi.org/10.1089/jwh.2007.0354.

Kasman, A. M., Thoma, M. E., McLain, A. C., & Eisenberg, M. L. (2018). Association between use of marijuana and time to pregnancy in men and women: Findings from the National Survey of Family Growth. *Fertility and Sterility*, *109*(5), 866–871. https://doi.org/10.1016/j.fertnstert.2018.01.015.

Klebanoff, M. A., Wilkins, D. G., & Keim, S. A. (2021). Marijuana use during pregnancy and preterm birth: A prospective cohort study. *American Journal of Perinatology*, *38*, E146–E154.

Klonoff-Cohen, H., & Lam-Kruglick, P. (2001). Maternal and paternal recreational drug use and sudden infant death syndrome. *Archives of Pediatrics & Adolescent Medicine*, *155*(7), 765–770.

Kolodny, R. C., Masters, W. H., Kolodner, R. M., & Toro, G. (1974). Depression of plasma testosterone levels after chronic intensive marihuana use. *New England Journal of Medicine*, *290*(16), 872–874. https://doi.org/10.1056/Nejm197404182901602.

Kostellow, A. B., Ziegler, D., Kunar, J., Fujimoto, G. I., & Morrill, G. A. (1980). Effect of cannabinoids on estrous cycle, ovulation and reproductive capacity of female A/J mice. *Pharmacology*, *21*(1), 68–75. https://doi.org/10.1159/000137418.

Leemaqz, S. Y., Dekker, G. A., McCowan, L. M., Kenny, L. C., Myers, J. E., Simpson, N. A. B., Poston, L., Roberts, C. T., & Consortium, S. (2016). Maternal marijuana use has independent effects on risk for spontaneous preterm birth but not other common late pregnancy complications. *Reproductive Toxicology*, *62*, 77–86. https://doi.org/10.1016/j.reprotox.2016.04.021.

Lisdahl, K. M., Tamm, L., Epstein, J. N., Jernigan, T., Molina, B. S. G., Hinshaw, S. P., Swanson, J. M., Newman, E., Kelly, C., Bjork, J. M., & Grp, M. N. (2016). The impact of ADHD persistence, recent cannabis use, and age of regular cannabis use onset on subcortical volume and cortical thickness in young adults. *Drug and Alcohol Dependence*, *161*, 135–146. https://doi.org/10.1016/j.drugalcdep.2016.01.032.

Liu, W. M., Duan, E. K., & Cao, Y. J. (2002). Effects of anandamide on embryo implantation in the mouse. *Life Sciences*, *71*(14), 1623–1632. https://doi.org/10.1016/S0024-3205(02)01928-8.

Liu, X., & Herbison, A. E. (2013). Dopamine regulation of gonadotropin-releasing hormone neuron excitability in male and female mice. *Endocrinology*, *154*(1), 340–350. https://doi.org/10.1210/en.2012-1602.

Maccarrone, M., De Felici, M., Bari, M., Klinger, F., Siracusa, G., & Finazzi-Agro, A. (2000, May). Down-regulation of anandamide hydrolase in mouse uterus by sex hormones. *European Journal of Biochemistry*, *267*(10), 2991–2997. <Go to ISI>://WOS:000087209500019.

Maccarrone, M., Gasperi, V., Fezza, F., Finazzi-Agro, A., & Rossi, A. (2004). Differential regulation of fatty acid amide hydrolase promoter in human immune cells and neuronal cells by leptin and progesterone. *European Journal of Biochemistry*, *271*(23–24), 4666–4676. https://doi.org/10.1111/j.1432-1033.2004.04427.x.

Maccarrone, M., Valensise, H., Bari, M., Lazzarin, N., Romanini, C., & Finazzi-Agro, A. (2001). Progesterone up-regulates anandamide hydrolase in human lymphocytes: Role of cytokines and implications for fertility. *Journal of Immunology*, *166*(12), 7183–7189. https://doi.org/10.4049/jimmunol.166.12.7183.

Maia, J., Almada, M., Silva, A., Correia-Da-Silva, G., Teixeira, N., Sa, S. I., & Fonesca, B. M. (2017). The endocannabinoid system expression in the female reproductive tract is modulated by estrogen. *Journal of Steroid Biochemistry and Molecular Biology*, *174*, 40–47. https://doi.org/10.1016/j.jsbmb.2017.07.023.

Maia, J., Fonseca, B. M., Teixeira, N., & Correia-da-Silva, G. (2020). The fundamental role of the endocannabinoid system in endometrium and placenta: Implications in pathophysiological aspects of uterine and pregnancy disorders. *Human Reproduction Update*, *26*(4), 586–602. https://doi.org/10.1093/humupd/dmaa005.

Mendelson, J. H., Mello, N. K., Ellingboe, J., Skupny, A. S., Lex, B. W., & Griffin, M. (1986). Marihuana smoking suppresses luteinizing hormone in women. *The Journal of Pharmacology and Experimental Therapeutics*, *237*(3), 862–866. https://www.ncbi.nlm.nih.gov/pubmed/3012072.

Metz, T. D., Allshouse, A. A., Hogue, C. J., Goldenberg, R. L., Dudley, D. J., Varner, M. W., Conway, D. L., Saade, G. R., & Silver, R. M. (2017). Maternal marijuana use, adverse pregnancy outcomes, and neonatal morbidity. *American Journal of Obstetrics and Gynecology*, *217*(4).

Michalski, C. A., Hung, R. J., Seeto, R. A., Dennis, C. L., Brooks, J. D., Henderson, J., Levitan, R., Lye, S. J., Matthews, S. G., & Knight, J. A. (2020). Association between maternal cannabis use and birth outcomes: An observational study. *BMC Pregnancy and Childbirth*, *20*(1).

Moghadam, K. K., Kessler, C. A., Schroeder, J. K., Buckley, A. R., Brar, A. K., & Handwerger, S. (2005). Cannabinoid receptor I activation markedly inhibits human decidualization. *Molecular and Cellular Endocrinology*, *229*(1–2), 65–74. https://doi.org/10.1016/j.mce.2004.09.007.

Molina-Holgado, F., Molina-Holgado, E., Guaza, C., & Rothwell, N. J. (2002). Role of CB1 and CB2 receptors in the inhibitory effects of cannabinoids on lipopolysaccharide-induced nitric oxide release in astrocyte cultures. *Journal of Neuroscience Research*, *67*(6), 829–836. https://doi.org/10.1002/jnr.10165.

Moss, M. J., Bushlin, I., Kazmierczak, S., Koop, D., Hendrickson, R. G., Zuckerman, K. E., & Grigsby, T. M. (2021). Cannabis use and measurement of cannabinoids in plasma and breast milk of breastfeeding mothers. *Pediatric Research*. https://doi.org/10.1038/s41390-020-01332-2.

Mumford, S. L., Flannagan, K. S., Radoc, J. G., Sjaarda, L. A., Zolton, J. R., Metz, T. D., Plowden, T. C., Perkins, N. J., De Vilbiss, E. A., Andriessen, V. C., Purdue-Smithe, A. C., Kim, K., Yisahak, S. F., Freeman, J. R., Alkhalaf, Z., Silver, R. M., & Schisterman, E. F. (2021). Cannabis use while trying to conceive: A prospective cohort study evaluating associations with fecundability, live birth and pregnancy loss. *Human Reproduction*, *36*(5), 1405–1415.

Neradugomma, N. K., Drafton, K., Mor, G. G., & Mao, Q. C. (2019). Marijuana-derived cannabinoids inhibit uterine endometrial stromal cell decidualization and compromise trophoblast-endometrium cross-talk. *Reproductive Toxicology*, *87*, 100–107. https://doi.org/10.1016/j.reprotox.2019.05.064.

O'Connor, A. B., Kelly, B. K., & O'Brien, L. M. (2017). Maternal and infant outcomes following third trimester exposure to marijuana in opioid dependent pregnant women maintained on buprenorphine. *Drug and Alcohol Dependence*, *180*, 200–203.

Paria, B. C., Ma, W., Andrenyak, D. M., Schmid, P. C., Schmid, H. H., Moody, D. E., Deng, H., Makriyannis, A., & Dey, S. K. (1998, Jun). Effects of cannabinoids on preimplantation mouse embryo development and implantation are mediated by brain-type cannabinoid receptors. *Biology of Reproduction*, *58*(6), 1490–1495. https://doi.org/10.1095/biolreprod58.6.1490.

Pizzol, D., Demurtas, J., Stubbs, B., Soysal, P., Mason, C., Isik, A. T., Solmi, M., Smith, L., & Veronese, N. (2019). Relationship between cannabis use and erectile dysfunction: A systematic review and meta-analysis. *American Journal of Mens Health*, *13*(6). https://doi.org/10.1177/1557988319892464.

Purohit, V., Ahluwahlia, B. S., & Vigersky, R. A. (1980). Marihuana inhibits dihydrotestosterone binding to the androgen receptor. *Endocrinology*, *107*(3), 848–850. https://doi.org/10.1210/endo-107-3-848.

Rodriguez, C. E., Sheeder, J., Allshouse, A. A., Scott, S., Wymore, E., Hopfer, C., Hermesch, A., & Metz, T. D. (2019). Marijuana use in young mothers and adverse pregnancy outcomes: A retrospective cohort study. *BJOG : An International Journal of Obstetrics and Gynaecology*, *126*(12), 1491–1497.

Rossato, M., Ion Popa, F., Ferigo, M., Clari, G., & Foresta, C. (2005). Human sperm express cannabinoid receptor Cb1, the activation of which inhibits motility, acrosome reaction, and mitochondrial function. *The Journal of Clinical Endocrinology and Metabolism*, *90*(2), 984–991. https://doi.org/10.1210/jc.2004-1287.

Saurel-Cubizolles, M. J., Prunet, C., & Blondel, B. (2014). Cannabis use during pregnancy in France in 2010. *BJOG : An International Journal of Obstetrics and Gynaecology*, *121*(8), 971–977. https://doi.org/10.1111/1471-0528.12626.

Schmid, P. C., Paria, B. C., Krebsbach, R. J., Schmid, H. H., & Dey, S. K. (1997). Changes in anandamide levels in mouse uterus are associated with uterine receptivity for embryo implantation. *Proceedings of the National Academy of Sciences of the United States of America*, *94*(8), 4188–4192. http://www.ncbi.nlm.nih.gov/entrez/query.fcgi?cmd=Retrieve&db=PubMed&dopt=Citation&list_uids=9108127.

Scorticati, C., Fernandez-Solari, J., De Laurentiis, A., Mohn, C., Prestifilippo, J. P., Lasaga, M., Seilicovich, A., Billi, S., Franchi, A., McCann, S. M., & Rettori, V. (2004, Aug 10). The inhibitory effect of anandamide on luteinizing hormone-releasing hormone secretion is reversed by estrogen. *Proceedings of the National Academy of Sciences of the United States of America*, *101*(32), 11891–11896. https://doi.org/10.1073/pnas.0404366101.

Scragg, R. K. R., Mitchell, E. A., Ford, R. P. K., Thompson, J. M. D., Taylor, B. J., & Stewart, A. W. (2001). Maternal cannabis use in the sudden death syndrome. *Acta Paediatrica*, *90*(1), 57–60.

Shabani, M., Hosseinmardi, N., Haghani, M., Shaibani, V., & Janahmadi, M. (2011, Jan 13). Maternal exposure to the Cb1 cannabinoid agonist win 55212-2 produces robust changes in motor function and intrinsic electrophysiological properties of cerebellar Purkinje neurons in rat offspring. *Neuroscience*, *172*, 139–152. https://doi.org/10.1016/j.neuroscience.2010.10.031.

Tanaka, K., Mayne, L., Khalil, A., Baartz, D., Eriksson, L., Mortlock, S. A., Montgomery, G., McKinnon, B., & Amoako, A. A. (2020, Jan). The role of the endocannabinoid system in aetiopathogenesis of endometriosis: A potential therapeutic target. *European Journal of Obstetrics, Gynecology, and Reproductive Biology*, *244*, 87–94. https://doi.org/10.1016/j.ejogrb.2019.11.012.

Torres, C. A., Medina-Kirchner, C., O'Malley, K. Y., & Hart, C. L. (2020). Totality of the evidence suggests prenatal cannabis exposure does not Lead to cognitive impairments: A systematic and critical review [systematic review]. *Frontiers in Psychology*, *11*(816). https://doi.org/10.3389/fpsyg.2020.00816.

Treinen, K. A., Sneeden, J. L., & Heindel, J. J. (1993). Specific inhibition of FSH-stimulated cAMP accumulation by delta 9-tetrahydrocannabinol in cultured rat granulosa cells. *Toxicology and Applied Pharmacology*, *118*(1), 53–57. https://doi.org/10.1006/taap.1993.1008.

Trezza, V., Campolongo, P., Manduca, A., Morena, M., Palmery, M., Vanderschuren, L. J. M. J., & Cuomo, V. (2012). Altering endocannabinoid neurotransmission at critical developmental ages: Impact on rodent emotionality and cognitive performance. *Frontiers in Behavioral Neuroscience*, *6*. https://doi.org/10.3389/fnbeh.2012.00002.

Turley, W. A., & Floody, O. R. (1981). Delta-9-tetrahydrocannabinol stimulates receptive and proceptive sexual behaviors in female hamsters. *Pharmacology Biochemistry and Behavior*, *14*(5), 745–747. https://doi.org/10.1016/0091-3057(81)90142-8.

van Gelder, M. M. H. J., Donders, A. R. T., Devine, O., Roeleveld, N., Reefhuis, J., & National Birth Defects Prevention Study. (2014). Using Bayesian models to assess the effects of under-reporting of cannabis use on the association with birth defects, National Birth Defects Prevention Study, 1997-2005. *Paediatric and Perinatal Epidemiology*, *28*(5), 424–433.

Varner, M. W., Silver, R. M., Hogue, C. J. R., Willinger, M., Parker, C. B., Thorsten, V. R., Goldenberg, R. L., Saade, G. R., Dudley, D. J., Coustan, D., Stoll, B., Bukowski, R., Koch, M. A., Conway, D., Pinar, H., Reddy, U. M., & Eunice Kennedy Shriver National Institute of Child Health and Human Development Stillbirth Collaborative Research Network. (2014). Association between stillbirth and illicit drug use and smoking during pregnancy. *Obstetrics and Gynecology*, *123*(1), 113–125.

Vescovi, P. P., Pedrazzoni, M., Michelini, M., Maninetti, L., Bernardelli, F., & Passeri, M. (1992). Chronic effects of marihuana smoking on luteinizing hormone, follicle-stimulating hormone and prolactin levels in human males. *Drug and Alcohol Dependence*, *30*(1), 59–63. https://doi.org/10.1016/0376-8716(92)90036-c.

Vestermark, V., & Laier, G. H. (2019). Cannabis use in pregnancy and risk for placental abruption. *Medical Journal of Obstetrics and Gynecology*, *7*(1), 1125.

Walker, O. S., Holloway, A. C., & Raha, S. (2019). The role of the endocannabinoid system in female reproductive tissues. *Journal of Ovarian Research*, *12*(1), 3. https://doi.org/10.1186/s13048-018-0478-9.

Wang, X., Dow-Edwards, D., Keller, E., & Hurd, Y. L. (2003). Preferential limbic expression of the cannabinoid receptor mRNA in the human fetal brain. *Neuroscience*, *118*(3), 681–694. https://www.ncbi.nlm.nih.gov/pubmed/12710976.

Wang, H., Guo, Y., Wang, D., Kingsley, P. J., Marnett, L. J., Das, S. K., DuBois, R. N., & Dey, S. K. (2004). Aberrant cannabinoid signaling impairs oviductal transport of embryos. *Nature Medicine*, *10*(10), 1074–1080. https://doi.org/10.1038/nm1104.

Williams, L. J., Correa, A., & Rasmussen, S. (2004). Maternal lifestyle factors and risk for ventricular septal defects. *Birth Defects Research Part A: Clinical and Molecular Teratology*, *70*(2), 59–64.

Chapter 41

Phytocannabinoids have cytotoxic, antiproliferative, and antimigratory activities on cancer cells and cancer stem cells

Hadar Peeri and Hinanit Koltai
Department of Ornamental Horticulture and Biotechnology, Institute of Plant Sciences, ARO, Volcani Institute, Rishon LeZion, Israel

Abbreviations

ATF4	activating transcription factor 4
BCL-2	B cell lymphoma 2
BH3	BCL-2 homology 3
CB1/2	cannabinoid receptor type 1/type 2
CBC	cannabichromene
CBD	cannabidiol
CBG	cannabigerol
CBGA	cannabigerolic acid
CBN	cannabinol
CHOP	C/EBP homologous protein
CNS	central nervous system
CSCs	cancer stem cells
CTCL	cutaneous T-cell lymphoma
ECS	endocannabinoid system
ER	endoplasmic reticulum
GBM	glioblastoma multiforme
GPCR	G-protein-coupled receptor
GSCs	glioma stem cells
IC50	half maximal inhibitory concentration
Id1	inhibitor of DNA binding 1
MAPK	mitogen-activated protein kinase
MM	multiple myeloma
MMP	matrix metalloproteinases
MTORC1	mammalian target of rapamycin complex 1
PPARs	peroxisome proliferator-activated receptors
ROS	reactive oxygen species
THC	Δ^9-tetrahydrocannabinol
THCA	Δ^9-tetrahydrocannabinolic acid
THCV	Δ^9-tetrahydrocannabivarin
TIMP	tissue inhibitor of metalloproteinases
TMZ	temozolomide
TRIB3	tribbles homolog 3
TRPA	transient receptor potential ankyrin
TRPV	transient receptor potential vanilloid
UC	urothelial carcinoma
WHO	World Health Organization

Cannabis Use, Neurobiology, Psychology, and Treatment. https://doi.org/10.1016/B978-0-323-89862-1.00030-1

Introduction

Cannabis sativa L. (*C. sativa*) is being used worldwide to treat different ailments and medical conditions. Yet, in most cases, it is used as palliative or supportive care (Bridgeman & Abazia, 2017; Corroon et al., 2019). Historically, the medical use of cannabis was prevalent in China and India for pain relief, as anticonstipation, antibiotic and antiinflammatory, tranquilizer, and anesthetic (Mikuriya, 1969; Touw, 1981; Zuardi, 2006).

C. sativa produces more than 600 different secondary metabolites that are synthesized and stored mainly in epidermal glandular trichomes of the female inflorescence and bracts. The secondary metabolites include more than 160 phytocannabinoids (terpenophenolic compounds produced in the plant as prenylated aromatic carboxylic acids), multiple flavonoids, and hundreds of terpenes (Aizpurua-Olaizola et al., 2016; ElSohly et al., 2017; Hanuš et al., 2016). Decarboxylation of the carboxylic moiety and conversion of the phytocannabinoids into neutral homologs occur to some extent within the living plant inflorescence, but mostly following harvesting, when the inflorescence or extract thereof is catalyzed by heat (Aizpurua-Olaizola et al., 2016; Flores-Sanchez & Verpoorte, 2008). Among the most abundant (decarboxylated) phytocannabinoids are Δ9-tetrahydrocannabinol (THC), cannabidiol (CBD), cannabigerol (CBG), and cannabichromene (CBC). Yet, CBC is present in cannabis extract usually in smaller amounts, and CBG is abundant mostly in extracts of immature cannabis female inflorescence (Hanuš et al., 2016).

Cancer stem cells

Cancer, a complex family of diseases, is a results of abnormal growth of cells associated with genomic instability and imbalance in cell death (Ruddon, 2007). Several hallmarks of cancer were identified, including the sustaining of cell proliferative while avoiding growth suppressors, reduction of cell death, limitless potential for replication, induction of tissue invasion, and angiogenesis and metastasis (Hanahan & Weinberg, 2011).

In many cancer types, cancer stem cells (CSCs) can be identified in malignant solid tumors as a subpopulation of cells. These cells have clonogenic and tumorigenic potential, a capacity for self-renewal, and an ability to increases genetic heterogeneity within the tumor mass by multilineage differentiation. In fact, transitions in daughter cells from stem-to-nonstem and from nonstem-to-stem are taking place in various cancers, suggesting that CSC may not be considered as a discrete entity (Heft Neal et al., 2022). CSCs show resistance to conventional cancer therapies such as radio- or chemotherapy (Phi et al., 2018). Moreover, CSCs often stimulate tumor initiation and infiltration, angiogenesis, invasion, and contribute to metastasis and tumor recurrence (Zhou et al., 2021). Therefore, there is great importance in identifying novel treatments that target CSCs in particular.

Glioma and glioblastoma

Gliomas are primary brain tumors that originate from glial cells and account for most malignant brain tumors (Fernandes et al., 2017; Louis et al., 2016; McCutcheon & Preul, 2021). According to classification of the World Health Organization (WHO), gliomas are classified to four grades, with grade I being benign and grade IV being malignant (Louis et al., 2007). A newer classification includes various molecular parameters that define tumor entities (Louis et al., 2016). The WHO grade IV glioma, glioblastoma multiforme (GBM), is the most frequent, invasive, complicated, treatment-resistant, and lethal subtype (Louis et al., 2007, 2016).

The median survival rate for GBM patients is less than 2 years from diagnosis, despite aggressive multidisciplinary treatments, and not or only little significant improvements were achieved in decades (McCutcheon & Preul, 2021). Maximal surgical resectioning followed by chemo- and radiotherapy are used for GBM therapies (Fernandes et al., 2017; McCutcheon & Preul, 2021). Nevertheless, in recurrent or progressive GBM, treatments include, in addition to supportive care, surgery, combined modality therapy, re-irradiation and systemic therapies (Fernandes et al., 2017; McCutcheon & Preul, 2021).

One of the most significant pathologies associated with glioblastoma derived from the genetic heterogeneity within the tumor mass. Various subpopulations of cells develop in the glioma tumor, including glioma stem cells (GSCs) (Holland, 2000). GSCs are thought to promote gliomagenesis and invasiveness and to induce angiogenesis (Biserova et al., 2021). Moreover, GSCs are resistant to conventional cancer therapies such as chemo- or radiotherapy, thereby contributing to the high resistance of GBM tumors to current standard therapies (Biserova et al., 2021). GBM tumors and cell lines were demonstrated to contain a subpopulation of cells that can form tumor-neurospheres (Lathia, Mack, Mulkearns-Hubert, Valentim, & Rich, 2015; Yuan et al., 2004). These neurospheres are enriched with cells biased toward progenitor features and stem cell characteristics including multipotency, self-renewal, and tumor generation (Lathia et al., 2015; Yuan et al.,

2004). Also, it was suggested that the existence rates of GSC increase proportionally with the grade of gliomas (Kong et al., 2013). Subsequently, new effective therapies are urgently needed for the treatment of GBM patients. Agents that will target both GBM tumors and GSCs are preferred.

Cancer and the endocannabinoid system

The endocannabinoid system (ECS) consists of endogenous ligands termed endocannabinoids, cannabinoid receptors, and metabolic enzymes, and act as a signaling network. The ECS has an important role in homeostatic-balance maintenance and is widely distributed in the body. The ECS regulates various physiological processes including those of the immune system and synaptic transmission. Dysregulation of the ECS is involved, among others, with inflammation, cancer, depression, anxiety and posttraumatic stress disorder (Di Marzo & Piscitelli, 2015; Hillard, 2018). In accordance, pharmaceutical modulation of ECS activity has therapeutic effects on various medical conditions, including antitumor activity in various cancer models (Fraguas-Sánchez et al., 2018; Laezza et al., 2020).

The cannabinoid receptors (CB) type 1 and type 2 (CB1 and CB2) are activated by binding to *endo-*/phyto- or synthetic cannabinoids and are members of the superfamily of seven transmembrane G-protein coupled receptors (GPCR); GPCR are among the most abundant receptors in the body (Mackie, 2008; Pertwee et al., 2010). Other types of receptors that interact with cannabinoids, in addition to GPCR, include ion channels such as the TRP ankyrin (TRPA) family, the transient receptor potential vanilloid (TRPV) family, and peroxisome proliferator-activated receptors (PPARs) (Biringer, 2021; De Petrocellis et al., 2011).

Cannabinoids receptors were found to produce antitumor activity in different types of cancer. For example, selective CB2 antagonists prevented tumor regression in vivo (Sánchez et al., 2001). Also, CB1 expression was increased following THC treatment in T-cell lymphoma, whereas in prostate cancer, CB1 receptor expressed at a higher level than that seen in normal human prostate epithelial cells (Hermanson & Marnett, 2011). In GBM cell lines, CB2 receptor expression was positively correlate with grade of tumor malignancy, and it was highly expressed in tissue biopsies compared to normal tissues; normal tissues expressed mostly CB1 receptors (Ellert-Miklaszewska et al., 2020). Also, GSC derived from GBM biopsies highly expressed CB1 and CB2 receptors and other components of the ECS (Aguado et al., 2007).

Anticancer activities of cannabis compounds

In numerous studies, phytocannabinoids were shown to possess anticancer properties. They inhibit cell proliferation, migration, and angiogenesis, and they induce apoptosis in various cancer cells including skin, prostate, breast, lung, and glioma (Ramer & Hinz, 2017; Velasco et al., 2012). Various signaling pathways associated with cell death are triggered by phytocannabinoids treatments, including oxidative stress, cell cycle arrest, endoplasmic reticulum (ER) stress, autophagy, and apoptosis (Dumitru et al., 2018; Ellert-Miklaszewska et al., 2020).

THC inhibited the growth of some tumors, including glioma, lung carcinoma, lymphoma, leukemia, and skin carcinoma. THC-induced apoptotic cell death in diverse cancers cells in vitro and in vivo and inhibited angiogenesis (Blázquez, Carracedo, et al., 2008; Blázquez, Salazar, et al., 2008; Carracedo, Lorente, et al., 2006; Guzmán et al., 2006). Further, CBD inhibited the invasiveness of GBM and breast cancer cells at sublethal concentrations (Elbaz et al., 2015; Solinas et al., 2013). In vivo, CBD and/or THC reduced GBM tumor growth (Ellert-Miklaszewska et al., 2020). CBG too has anticancer activity, including in human oral epithelioid carcinoma cells, mouse melanomas, human breast carcinomas, and colorectal cancer cells (Tomko et al., 2020).

As for activity against GSCs, it was demonstrated that specific combinations of phytocannabinoids that are rich with THC or CBG have the potential to target key signaling pathways affecting these cells viability and motility. The combinations of compounds had, in addition, significant cytotoxic activity against GSCs from a GBM primary tumor (Peeri et al., 2021; Peeri & Koltai, 2022). Moreover, in sublethal concentrations, these combinations of compounds inhibited neurosphere formation in 2- and 3-dimensional models (Peeri et al., 2021; Peeri & Koltai, 2022).

Phytocannabinoids had significant cytotoxic activity also against CSC of other cancer types, including human breast and lung cancer cells. It was demonstrated that the activation of cannabinoid receptors alters in CSCs regulatory genes expression, stem cell proliferation and differentiation, and inhibits the invasiveness and tumorigenesis of these cells (Hamad & Olsen, 2021; Mohammadpour et al., 2017).

Moreover, evidences from clinical trial suggest phytocannabinoid may serve as a base for effective treatments against GBM (Dumitru et al., 2018). THC was shown to have a good safety profile in a pilot phase I clinical trial. Moreover, administration of THC in two of nine GBM patients led to a decreased tumor-cell proliferation (based on Ki67 immunostaining; Guzmán et al., 2006).

The entourage effect and synergies between cannabis compounds

In cannabis, as in other plant species, higher effectivity was characterized for treatments with natural combinations found in the plant over those with a single molecule. This phenomena was termed the "entourage effect" (Koltai & Namdar, 2020; Russo, 2019).

Several studies demonstrated intra-entourage, i.e., enhancement of the biological activity by the interaction of different phytocannabinoids (Koltai & Namdar, 2020). For example, in GBM cell lines, THC and CBD showed synergistic inhibition of cell proliferation (Marcu et al., 2010). Also, synergistic interaction between THC and CBC was identified for cell cycle arrest, apoptosis and inhibition of cell migration in human bladder urothelial carcinoma (UC) cells (Anis et al., 2021). In another example, a combination of phytocannabinoids that contained CBD, CBG, THC, and CBC was more cytotoxic to human cutaneous T-cell lymphoma (CTCL) cells than CBD, the primary molecule, only. This combined treatment led to apoptosis and upregulation of ER-stress-related gene expression (Mazuz et al., 2020). In addition, two different combinations of phytocannabinoids, with CBG or THC as the lead molecule, had significant cytotoxic activity against GBM cell lines and GSCs derived from tumor specimens (Peeri et al., 2021). In addition, in this case, the activity of the specific combinations was higher than that of the isolated lead molecule in each fraction. The active fractions induced the expression of ER-stress-associated genes and apoptosis. These activities were mediated via the CB2 receptor; when a CB2 inverse agonist was added to the treatments, their effects were significantly reduced. The active fractions also altered cell cytoskeleton, inhibited cell migration, reduced cell invasion, and colony formation (Peeri et al., 2021; Peeri & Koltai, 2022). Notably, also on GSC, the therapeutic ability of the combination of cannabis compounds to exert cytotoxic, antiproliferative, and antimigratory effects was demonstrated (Peeri et al., 2021).

In addition, treatment of colorectal cancer cell lines and colon polyps with synergistic mixture of a cannabigerolic acid (CBGA)- and a tetrahydrocannabinolic acid (THCA)-rich fractions resulted in higher activity compared to each fraction alone. The synergistic combination increased G0/G1 cell cycle arrest, induced apoptotic cell death, and led to identification of genes that were differentially expressed because of the treatment. Some of these genes are involved in Wnt and p53 signaling pathways (Nallathambi et al., 2018).

Additional studies on multiple myeloma (MM) and leukemia cells showed that the combination of THC and CBD for treatment was more effective than treatment with each compound on its own. When THC and CBD were combined at a 1:1 ratio in treatments of leukemic cells, activity was 2–3 fold higher compared to that of CBD or THC separately (Scott et al., 2017). Also in MM cells, THC and CBD combination showed higher activity in inducing autophagic cell death and cell cycle arrest (Nabissi et al., 2016).

A possible explanation to the mechanism behind the entourage or synergistic effect could be the simultaneous activation of multiple receptors by phytocannabinoids used for treatment (e.g., CB1, CB2), leading to intensification of the antitumor activity. An alternative explanation for synergy might be the parallel activation of several signaling pathways by phytocannabinoids used for treatments (Koltai & Namdar, 2020).

Mechanism of anticancer activity of phytocannabinoids

As mentioned above, phytocannabinoids were shown to affect a multitude of cancer-related signaling pathways. This extensive effect might be a result of phytocannabinoids ability to affect components of the ECS, which is widely involved with various physiological processes (Di Marzo & Piscitelli, 2015; Hillard, 2018; Laezza et al., 2020).

The ER-stress pathway is one of the common pathways activated by phytocannabinoids in different cancer types (Fig. 1). This is the main phytocannabinoid-induced mechanism of apoptosis in glioma, melanoma, astrocytoma, and pancreatic tumor cells (Carracedo, Gironella, et al., 2006; Hermanson & Marnett, 2011; Salazar et al., 2009; Velasco et al., 2005). Both CB1 and CB2 receptor agonists were shown to stimulate the synthesis and accumulation of ceramide in glioma models (Fig. 1; Herrera et al., 2006; Velasco et al., 2005). Ceramides are a proapoptotic lipid second messenger that leads to the induction of stress protein p8 (Fig. 1). Downstream, ER-stress-related genes are induced (Fig. 1). As a result, the intrinsic mitochondrial pathway is activated (Carracedo, Gironella, et al., 2006; Herrera et al., 2006).

In glioblastoma cells, in a CB2 activation-dependent manner, CBG-rich, and THC-rich combinations of phytocannabinoids induced transcription of transcription factor 4 (*ATF4*), C/EBP homologous protein (*CHOP*)*-10* (*GADD153/DDIT-3*) and Tribbles homolog 3 (*TRIB3*) (Fig. 1; Peeri et al., 2021; Peeri & Koltai, 2022). In accordance, *ATF4* is transiently induced after treatment with ER stressors (Iurlaro & Muñoz-Pinedo, 2016). In turn, ATF4 induce expression of *CHOP*, a transcription factor that regulates the expression of pro- and antiapoptotic genes (Fig. 1; Hu et al., 2019; Iurlaro & Muñoz-Pinedo, 2016). CHOP activates proapoptotic B cell lymphoma-2 (BCL-2) family proteins and represses antiapoptotic BCL-2 family proteins under ER-stress (Hu et al., 2019). TRIB3 is a pseudokinase and was found to facilitate

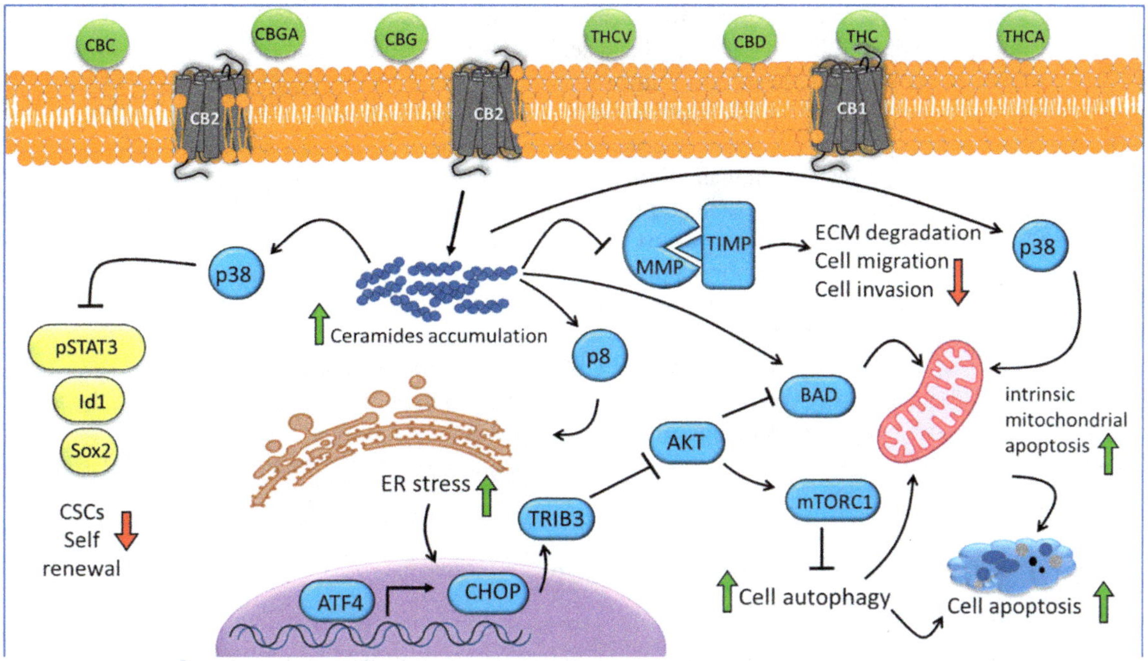

FIG. 1 Some of the molecular mechanisms underlying *Cannabis sativa* compounds antitumor effects on cancer cells and cancer stem cells. Phytocannabinoids (e.g., Cannabichromene [CBC], Cannabigerolic acid [CBGA], Cannabigerol [CBG], Δ^9-tetrahydrocannabivarin [THCV], Cannabidiol [CBD], Δ^9-tetrahydrocannabinol [THC], Δ^9-tetrahydrocannabinolic acid [THCA]) may activate CB receptors (CB1 or CB2) and stimulate the synthesis and accumulation of ceramides. This accumulation triggers the induction of p8. Induction of p8 leads to the inhibition of cell migration and invasion via downregulation of MMPs. Moreover, p8 leads to upregulation of ER-stress-related genes *TRIB-3, ATF-4,* and *CHOP* (blue shapes), following by inhibition of the Akt-mTORC1 axis. These activities initiate autophagy, which is upstream of apoptosis and intrinsic mitochondrial apoptosis. In addition, inhibition of Akt leads to the overexpression of BAD. Overexpression of BAD induces apoptosis via the intrinsic mitochondrial pathway. Another signaling pathway activated by ceramide accumulation is p38-MAPK, which activates apoptosis and downregulate stemness regulators such as p-STAT3, Id1 and Sox2 in CSCs (*yellow shapes*). This downregulation inhibits CSC self-renewal. *Green arrows* represent upregulation, and *red arrows* represent downregulation of biological processes. *Blue* or *yellow shapes* represent genes or proteins.

ER-stress-dependent apoptosis via the NF-κB pathway (Fang et al., 2014). TRIB3 inhibits the Akt-mTORC1 axis, and thus leads to the initiation of autophagy, whereas autophagy is upstream of intrinsic mitochondrial apoptosis (Fig. 1; Salazar et al., 2009).

Furthermore, the expression of BCL-2 homology 3 (BH3)-only family member BAD, a proapoptotic protein, was upregulated upon treatment with the cannabinoid-receptor synthetic agonist WIN-55,212-2, and this is in response to the serine/threonine kinase Akt downregulation and ceramide accumulation (Fig. 1; Ellert-Miklaszewska et al., 2020). Ceramide accumulation also activates the p38 mitogen-activated protein kinase (MAPK), and THC treatment on human leukemia cells activated this pathway while inducing apoptosis via CB2 receptor (Fig. 1; Herrera et al., 2005; McKallip et al., 2002).

Cannabis compounds have also, potentially, antimigrative and antiinvasive activities on GBM cells, demonstrated both in vitro and in vivo. Treatment with THC or CBD downregulated the expression of proteins associated with migration of tumor cells, in particular matrix metalloproteinases (MMP)-2, MMP-9, and MMP inhibitors (TIMP)-4, and TIMP-1 (Blázquez, Carracedo, et al., 2008; Solinas et al., 2013). Upregulation of TIMP-1 and several MMPs expression is associated with tumor malignancy and negative prognosis in different cancers (Quintero-Fabián et al., 2019). Notably, in the case of glioma tumor cells, THC or CBD was effective even at sublethal, low concentrations, which were insufficient to induce cell apoptosis (Gialeli et al., 2011; Kessenbrock et al., 2010). In glioma cell lines as well as in cultured human GBM primary cells, THC treatment reduced TIMP-1 and MMP-2 expression. In addition, downregulation of TIMP-1 and MMP-2 expression was obtained in glioma-bearing mice, subcutaneously, following local administration of THC. Similar results were observed in biopsies from tumors of two patients with recurrent GBM (Blázquez, Carracedo, et al., 2008; Blázquez, Salazar, et al., 2008).

Moreover, The CB2 receptor agonist JWH-133 inhibits MMP-2 expression in mouse gliomas suggesting that THC acts on glioma cells mostly via CB2 receptor activation (Blázquez, Salazar, et al., 2008). It was shown that in cultured glioma, THC inhibited MMP-2 expression and invasion cells via ceramide accumulation and p8 stress protein (Blázquez, Salazar, et al., 2008; Fig. 1). CBD treatment of breast-cancer cell lines and the relevant mouse model led to a significant reduction in metastatic nodules. This reduction was associated with inhibition of MMP2 and MMP9 secretion by the tumor cells (Elbaz et al., 2015).

In GSC, CBD was shown to increase the levels of reactive oxygen species (ROS). CBD also activated the p38-MAPK signaling pathway, leading to the inhibition of survival and self-renewal of cultured primary GSC. P38-MAPK activation was also associated with the downregulation of key regulators of stem cell such as Sox2, inhibitor of DNA binding 1 (Id1) and p-STAT3 (Fig. 1; Singer et al., 2015). CBD treatment stimulated in GBM the activation of caspase-3 in vivo and prolonged survival of mice bearing GSCs-derived, intracranial GBM xenografts (Singer et al., 2015).

GSCs isolated from GBM biopsies and human glioma cell lines express CB2 and other ECS elements such as MAGL and FAAH, the enzymes responsible for endocannabinoids degradation (Aguado et al., 2007). In this study, activation of CB receptors by synthetic cannabinoid agonists led to downregulation of genes involved in cell cycle progression and cell proliferation. The expression of, e.g., the tumor suppressor RBL1 was increased by treatment with cannabinoid agonist (HU-210), promoting GSC differentiation. As a result, the cells' ability to initiate generation and growth of glioma tumor in vivo was reduced (Aguado et al., 2007).

In addition, in both human breast cancer cell line and CSCs, the CB1 receptor agonist ACEA was shown to possess antiinvasion potential without affecting cell viability. In the case of a CB1 receptor antagonist, AM251, the invasiveness of CSCs was enhanced rather than repressed, and this is in a dose-dependent manner (Mohammadpour et al., 2017). In contrast, another CB1 receptor antagonist/inverse agonist, rimonabant, reduced the proliferation of CSCs derived from colorectal patients (Fiore et al., 2018).

Effect of phytocannabinoids on normal tissue

Normal brain cells, such as primary neurons and astrocytes, in contrast to malignant cells, do not present ceramide accumulation or undergo apoptosis in response to phytocannabinoid treatments (Carracedo, Lorente, et al., 2006; Guzmán et al., 2001). In addition, even at high doses, there is no sign of any neurotoxicity or normal brain damage following treatments with phytocannabinoids in vivo (Galve-Roperh et al., 2000). Hence, cannabinoid receptors may regulate differently cell survival and cell death in cancer cells and nontransformed cell (Guzmán, 2003). In addition, as detailed above, there are differences between normal tissue and cancer cells in the expression of cannabinoid receptors (Ellert-Miklaszewska et al., 2020; Hermanson & Marnett, 2011). Evidently, the CB1 receptor antagonist/inverse agonist rimonabant had no cytotoxic effect against normal human colon organoid cultures ex vivo, demonstrating the potential of the ECS in selectively targeting cancer cells (Fiore et al., 2018).

Summary

Due to lack of clinical trials, there is no solid basis for cannabis compounds therapeutic potential. This is despite numerous evidences on the anticancer activity of phytocannabinoids in various preclinical cancer models.

The studies presented in this chapter and additional ones suggest that cannabis may target malignant cells and CSCs, at least partially, via activation of cannabinoid receptor-dependent mechanisms. Indeed, the antitumor activity of the ECS is well established in various cancer models; however, its anti-CSC potential needs to be further examined.

Yet, improved and more targeted usage of cannabis should be considered. This improvement demands transformation of the beneficial effects of full-spectrum cannabis extracts into active plant ingredients (API) formulation. This transformation can be done via careful selection of composition of phytomolecules with optimal activity (Koltai & Namdar, 2020). First, since modern medicine requires standardization and dosing of drugs, the API-based products might be more compatible to be used therapeutically due to their expected compatibility with these requirements. Second, the mode of action of API might be characterized in large parts, and hence, API might be targeted to specific mechanisms involved with cancer. Lastly, API formulation might be combined with other pharmaceutical drugs to provide adjuvant therapy to complement and improve the current standard of care.

Application to other areas

In this chapter, we have reviewed the anticancer activities of different phytocannabinoids and their underlying mechanisms. Cannabis contains hundreds of chemically active compounds (Aizpurua-Olaizola et al., 2016; ElSohly et al., 2017; Hanuš et al., 2016), and the crude plant extract is often effective in treating various medical conditions (Bridgeman & Abazia, 2017; Corroon et al., 2019). However, there is a growing list of studies that show that the isolation of active fractions from the crude extract allows for better identification of specific active components and/or their synergistic combinations from cannabis (e.g., Anis et al., 2021; Peeri et al., 2021). Importantly, these compounds or combinations thereof are often more active than the crude extract. A similar approach for isolating active compounds and their combinations might be relevant for treating multiple medical conditions. Potential examples include inflammation, neurodegenerative disorders, chronic pain, anxiety and depression, insomnia, and other therapeutic targets (Corroon et al., 2019). In each case, isolation of active compounds may lay the basis for formulating active plant ingredients (API) for drug development, and deciphering modes of action.

Mini-dictionary of terms (alphabetical)

- **Apoptosis** is the mechanism of programmed cell death by which a cell actively undergoes controlled death in response to certain stimuli. This mechanism is necessary for both physiological and pathological conditions.
- **Glial cells** are the nonneuronal cells of the nervous system, surrounding neurons and providing both structural and physiological support, regulating key functions, and are involved in neuro-immune responses.
- **Phytocannabinoids** are cannabinoids produced by *Cannabis sativa* L.
- **Stem cells** are undifferentiated and unspecialized progenitor cells that have the potential to differentiate to specific types of cells.
- **Synergy** is the concept in which interactions between components lead to a greater effect than the sum of their individual effects.

Key facts about secondary metabolites

Plants produce a wide range of organic compounds, including primary and secondary metabolites. Although secondary metabolites are phytomolecules not directly essential for photosynthesis, respiration, and plant growth, they may play a key role in plant adaption to the environment and protection against pathogens. Based on their biosynthetic pathways, secondary metabolites are usually divided into three groups: flavonoids and phenolic compounds, terpenes and steroids, and alkaloids. Plant secondary metabolites may be sources for perfumes, dyes, flavor agents, insecticides, and drugs. In recent years, there has been increasing evidence that many of these compounds possess therapeutic potential against many medical conditions, including cancer, cardiovascular diseases, diabetes, and more.

Summary points

- Phytocannabinoids possess anticancer properties, including the inhibition of proliferation and migration and the induction of apoptosis in different types of cancer cells.
- Accumulating evidence suggests that mixtures of cannabis compounds are more active than single molecules.
- The antitumor effects of phytocannabinoids are mediated by targeting the phytocannabinoid receptors and activating various molecular mechanisms.
- Phytocannabinoids may activate the endoplasmic reticulum (ER) stress pathway in different cancer types.
- Recently, phytocannabinoids were demonstrated to target cancer stem cells in vitro.
- Phytocannabinoids may have therapeutic potential and should be further examined in clinical trials.

References

Aguado, T., Carracedo, A., Julien, B., Velasco, G., Milman, G., Mechoulam, R., Alvarez, L., Guzmán, M., & Galve-Roperh, I. (2007). Cannabinoids induce glioma stem-like cell differentiation and inhibit gliomagenesis. *Journal of Biological Chemistry*, *282*(9), 6854–6862.

Aizpurua-Olaizola, O., Soydaner, U., Öztürk, E., Schibano, D., Simsir, Y., Navarro, P., Etxebarria, N., & Usobiaga, A. (2016). Evolution of the cannabinoid and terpene content during the growth of *Cannabis sativa* plants from different chemotypes. *Journal of Natural Products*, *79*(2), 324–331.

Anis, O., Vinayaka, A. C., Shalev, N., Namdar, D., Nadarajan, S., Anil, S. M., Cohen, O., Belausov, E., Ramon, J., Mayzlish Gati, E., & Koltai, H. (2021). Cannabis-derived compounds cannabichromene and Δ9-tetrahydrocannabinol interact and exhibit cytotoxic activity against urothelial cell carcinoma correlated with inhibition of cell migration and cytoskeleton organization. *Molecules*, *26*(2), 465.

Biringer, R. G. (2021). Endocannabinoid signaling pathways: Beyond CB1R and CB2R. *Journal of Cell Communication and Signaling*, *15*(3), 335–360.

Biserova, K., Jakovlevs, A., Uljanovs, R., & Strumfa, I. (2021). Cancer stem cells: Significance in origin, pathogenesis and treatment of glioblastoma. *Cells*, *10*, 621.

Blázquez, C., Carracedo, A., Salazar, M., Lorente, M., Egia, A., González-Feria, L., Haro, A., Velasco, G., & Guzmán, M. (2008). Down-regulation of tissue inhibitor of metalloproteinases-1 in gliomas: A new marker of cannabinoid antitumoral activity? *Neuropharmacology*, *54*(1), 235–243.

Blázquez, C., Salazar, M., Carracedo, A., Lorente, M., Egia, A., González-Feria, L., Haro, A., Velasco, G., & Guzmán, M. (2008). Cannabinoids inhibit glioma cell invasion by down-regulating matrix metalloproteinase-2 expression. *Cancer Research*, *68*(6), 1945–1952.

Bridgeman, M. B., & Abazia, D. T. (2017). Medicinal cannabis: History, pharmacology, and implications for the acute care setting. *Pharmacy and Therapeutics*, *42*(3), 180.

Carracedo, A., Gironella, M., Lorente, M., Garcia, S., Guzmán, M., Velasco, G., & Iovanna, J. L. (2006). Cannabinoids induce apoptosis of pancreatic tumor cells via endoplasmic reticulum stress-related genes. *Cancer Research*, *66*(13), 6748–6755.

Carracedo, A., Lorente, M., Egia, A., Blázquez, C., García, S., Giroux, V., Malicet, C., Villuendas, R., Gironella, M., González-Feria, L., Piris, M.Á., Iovanna, J. L., Guzmán, M., & Velasco, G. (2006). The stress-regulated protein p8 mediates cannabinoid-induced apoptosis of tumor cells. *Cancer Cell*, *9*(4), 301–312.

Corroon, J., Sexton, M., & Bradley, R. (2019). Indications and administration practices amongst medical cannabis healthcare providers: A cross-sectional survey. *BMC Family Practice*, *20*(1), 1–12.

De Petrocellis, L., Ligresti, A., Moriello, A. S., Allarà, M., Bisogno, T., Petrosino, S., Stott, C. G., & Di Marzo, V. (2011). Effects of cannabinoids and cannabinoid-enriched Cannabis extracts on TRP channels and endocannabinoid metabolic enzymes. *British Journal of Pharmacology*, *163*(7), 1479–1494.

Di Marzo, V., & Piscitelli, F. (2015). The endocannabinoid system and its modulation by phytocannabinoids. *Neurotherapeutics*, *12*(4), 692–698.

Dumitru, C. A., Sandalcioglu, I. E., & Karsak, M. (2018). Cannabinoids in glioblastoma therapy: New applications for old drugs. *Frontiers in Molecular Neuroscience*, *11*, 159.

Elbaz, M., Nasser, M. W., Ravi, J., Wani, N. A., Ahirwar, D. K., Zhao, H., Oghumu, S., Satoskar, A. R., Shilo, K., Carson, W. E., & Ganju, R. K. (2015). Modulation of the tumor microenvironment and inhibition of EGF/EGFR pathway: Novel anti-tumor mechanisms of cannabidiol in breast cancer. *Molecular Oncology*, *9*(4), 906–919.

Ellert-Miklaszewska, A., Ciechomska, I. A., & Kaminska, B. (2020). Cannabinoid signaling in glioma cells. *Glioma Signaling*, *986*, 223–241.

ElSohly, M. A., Radwan, M. M., Gul, W., Chandra, S., & Galal, A. (2017). Phytochemistry of *Cannabis sativa* L. *Phytocannabinoids*, *103*, 1–36.

Fang, N., Zhang, W., Xu, S., Lin, H., Wang, Z., Liu, H., Fang, Q., Li, C., Peng, L., & Lou, J. (2014). TRIB3 alters endoplasmic reticulum stress-induced β-cell apoptosis via the NF-κB pathway. *Metabolism*, *63*(6), 822–830.

Fernandes, C., Costa, A., Osório, L., Lago, R. C., Linhares, P., Carvalho, B., & Caeiro, C. (2017). *Current standards of care in glioblastoma therapy* (pp. 197–241). Exon Publications.

Fiore, D., Ramesh, P., Proto, M. C., Piscopo, C., Franceschelli, S., Anzelmo, S., Medema, J. P., Bifulco, M., & Gazzerro, P. (2018). Rimonabant kills colon cancer stem cells without inducing toxicity in normal colon organoids. *Frontiers in Pharmacology*, *8*, 949.

Flores-Sanchez, I. J., & Verpoorte, R. (2008). Secondary metabolism in cannabis. *Phytochemistry Reviews*, *7*(3), 615–639.

Fraguas-Sánchez, A. I., Martín-Sabroso, C., & Torres-Suárez, A. I. (2018). Insights into the effects of the endocannabinoid system in cancer: A review. *British Journal of Pharmacology*, *175*, 2566–2580.

Galve-Roperh, I., Sánchez, C., Cortés, M. L., del Pulgar, T. G., Izquierdo, M., & Guzmán, M. (2000). Anti-tumoral action of cannabinoids: Involvement of sustained ceramide accumulation and extracellular signal-regulated kinase activation. *Nature Medicine*, *6*(3), 313–319.

Gialeli, C., Theocharis, A. D., & Karamanos, N. K. (2011). Roles of matrix metalloproteinases in cancer progression and their pharmacological targeting. *The FEBS Journal*, *278*(1), 16–27.

Guzmán, M. (2003). Cannabinoids: potential anticancer agents. *Nature Reviews Cancer*, *3*(10), 745–755.

Guzmán, M., Duarte, M. J., Blázquez, C., Ravina, J., Rosa, M. C., Galve-Roperh, I., Sánchez, C., Velasco, G., & González-Feria, L. (2006). A pilot clinical study of Δ9-tetrahydrocannabinol in patients with recurrent glioblastoma multiforme. *British Journal of Cancer*, *95*(2), 197–203.

Guzmán, M., Galve-Roperh, I., & Sánchez, C. (2001). Ceramide: A new second messenger of cannabinoid action. *Trends in Pharmacological Sciences*, *22*(1), 19–22.

Hamad, H., & Olsen, B. B. (2021). Cannabidiol induces cell death in human lung cancer cells and cancer stem cells. *Pharmaceuticals*, *14*(11), 1169.

Hanahan, D., & Weinberg, R. A. (2011). Hallmarks of cancer: The next generation. *Cell*, *144*(5), 646–674.

Hanuš, L. O., Meyer, S. M., Muñoz, E., Taglialatela-Scafati, O., & Appendino, G. (2016). Phytocannabinoids: A unified critical inventory. *Natural Product Reports*, *33*(12), 1357–1392.

Heft Neal, M. E., Brenner, J. C., Prince, M. E. P., & Chinn, S. B. (2022). Advancement in cancer stem cell biology and precision medicine-review article, head and neck cancer stem cell plasticity and the tumor microenvironment. *Frontiers in Cell and Development Biology*, *9*, 660210.

Hermanson, D. J., & Marnett, L. J. (2011). Cannabinoids, endocannabinoids, and cancer. *Cancer and Metastasis Reviews*, *30*(3), 599–612.

Herrera, B., Carracedo, A., Diez-Zaera, M., del Pulgar, T. G., Guzmán, M., & Velasco, G. (2006). The CB2 cannabinoid receptor signals apoptosis via ceramide-dependent activation of the mitochondrial intrinsic pathway. *Experimental Cell Research*, *312*, 2121–2131.

Herrera, B., Carracedo, A., Diez-Zaera, M., Guzmán, M., & Velasco, G. (2005). p38 MAPK is involved in CB2 receptor-induced apoptosis of human leukaemia cells. *FEBS Letters*, *579*(22), 5084–5088.

Hillard, C. J. (2018). Circulating endocannabinoids: From whence do they come and where are they going? *Neuropsychopharmacology*, *43*, 155–172.

Holland, E. C. (2000). Glioblastoma multiforme: The terminator. *Proceedings of the National Academy of Sciences*, *97*(12), 6242–6244.

Hu, H., Tian, M., Ding, C., & Yu, S. (2019). The C/EBP homologous protein (CHOP) transcription factor functions in endoplasmic reticulum stress-induced apoptosis and microbial infection. *Frontiers in Immunology*, *9*, 3083.

Iurlaro, R., & Muñoz-Pinedo, C. (2016). Cell death induced by endoplasmic reticulum stress. *The FEBS Journal*, *283*(14), 2640–2652.

Kessenbrock, K., Plaks, V., & Werb, Z. (2010). Matrix metalloproteinases: Regulators of the tumor microenvironment. *Cell*, *141*(1), 52–67.

Koltai, H., & Namdar, D. (2020). Cannabis phytomolecule "entourage": From domestication to medical use. *Trends in Plant Science*, *25*(10), 976–984.

Kong, B. H., Park, N. R., Shim, J. K., Kim, B. K., Shin, H. J., Lee, J. H., Huh, Y. M., Lee, S. J., Kim, S. H., Kim, E. H., Park, E. K., Chang, J. H., Kim, D. S., Kim, S. H., Hong, Y. K., Kang, S. G., & Lang, F. F. (2013). Isolation of glioma cancer stem cells in relation to histological grades in glioma specimens. *Child's Nervous System*, *29*(2), 217–229.

Laezza, C., Pagano, C., Navarra, G., Pastorino, O., Proto, M. C., Fiore, D., Piscopo, C., Gazzerro, P., & Bifulco, M. (2020). The endocannabinoid system: A target for cancer treatment. *International Journal of Molecular Sciences*, *21*, 747.

Lathia, J. D., Mack, S. C., Mulkearns-Hubert, E. E., Valentim, C. L., & Rich, J. N. (2015). Cancer stem cells in glioblastoma. *Genes & Development*, *29*, 1203–1217.

Louis, D. N., Ohgaki, H., Wiestler, O. D., Cavenee, W. K., Burger, P. C., Jouvet, A., Scheithauer, B. W., & Kleihues, P. (2007). The 2007 WHO classification of tumours of the central nervous system. *Acta Neuropathologica*, *114*(2), 97–109.

Louis, D. N., Perry, A., Reifenberger, G., Von Deimling, A., Figarella-Branger, D., Cavenee, W. K., Ohgaki, H., Wiestler, O. D., Kleihues, P., & Ellison, D. W. (2016). The 2016 World Health Organization classification of tumors of the central nervous system: A summary. *Acta Neuropathologica*, *131* (6), 803–820.

Mackie, K. (2008). Cannabinoid receptors: Where they are and what they do. *Journal of Neuroendocrinology*, *20*, 10–14.

Marcu, J. P., Christian, R. T., Lau, D., Zielinski, A. J., Horowitz, M. P., Lee, J., Pakdel, A., Allison, J., Limbad, C., Moore, D. H., Yount, G. L., Desprez, P. Y., & McAllister, S. D. (2010). Cannabidiol enhances the inhibitory effects of Δ9-tetrahydrocannabinol on human glioblastoma cell proliferation and survival. *Molecular Cancer Therapeutics*, *9*(1), 180–189.

Mazuz, M., Tiroler, A., Moyal, L., Hodak, E., Nadarajan, S., Vinayaka, A. C., Gorovitz-Haris, B., Lubin, I., Drori, A., Drori, G., Cauwenberghe, O. V., Faigenboim, A., Namdar, D., Amitay-Laish, I., & Koltai, H. (2020). Synergistic cytotoxic activity of cannabinoids from Cannabis sativa against cutaneous T-cell lymphoma (CTCL) in-vitro and ex-vivo. *Oncotarget*, *11*(13), 1141.

McCutcheon, I. E., & Preul, M. C. (2021). Historical perspective on surgery and survival with glioblastoma: How far have we come? *World Neurosurgery*, *149*, 148–168.

McKallip, R. J., Lombard, C., Fisher, M., Martin, B. R., Ryu, S., Grant, S., Nagarkatti, P. S., & Nagarkatti, M. (2002). Targeting CB2 cannabinoid receptors as a novel therapy to treat malignant lymphoblastic disease. *Blood, The Journal of the American Society of Hematology*, *100*(2), 627–634.

Mikuriya, T. H. (1969). Marijuana in medicine: Past, present and future. *California Medicine*, *110*(1), 34.

Mohammadpour, F., Ostad, S. N., Aliebrahimi, S., & Daman, Z. (2017). Anti-invasion effects of cannabinoids agonist and antagonist on human breast cancer stem cells. *Iranian Journal of Pharmaceutical Research*, *16*(4), 1479.

Nabissi, M., Morelli, M. B., Offidani, M., Amantini, C., Gentili, S., Soriani, A., Cardinali, C., Leoni, P., & Santoni, G. (2016). Cannabinoids synergize with carfilzomib, reducing multiple myeloma cells viability and migration. *Oncotarget*, *7*(47), 77543.

Nallathambi, R., Mazuz, M., Namdar, D., Shik, M., Namintzer, D., Vinayaka, A. C., Ion, A., Faigenboim, A., Nasser, A., Laish, I., Konikoff, F. M., & Koltai, H. (2018). Identification of synergistic interaction between cannabis-derived compounds for cytotoxic activity in colorectal cancer cell lines and colon polyps that induces apoptosis-related cell death and distinct gene expression. *Cannabis and Cannabinoid Research*, *3*(1), 120–135.

Peeri, H., & Koltai, H. (2022). Cannabis biomolecule effects on cancer cells and cancer stem cells: Cytotoxic, anti-proliferative, and anti-migratory activities. *Biomolecules*, *12*(4), 491.

Peeri, H., Shalev, N., Vinayaka, A. C., Nizar, R., Kazimirsky, G., Namdar, D., Anil, S. M., Belausov, E., Brodie, C., & Koltai, H. (2021). Specific compositions of *Cannabis sativa* compounds have cytotoxic activity and inhibit motility and colony formation of human glioblastoma cells in vitro. *Cancers*, *13*(7), 1720.

Pertwee, R. G., Howlett, A. C., Abood, M. E., Alexander, S. P. H., Di Marzo, V., Elphick, M. R., Greasley, P. J., Hansen, H. S., Kunos, G., Mackie, K., Mechoulam, R., & Ross, R. A. (2010). International Union of Basic and Clinical Pharmacology. LXXIX. Cannabinoid receptors and their ligands: Beyond CB1 and CB2. *Pharmacological Reviews*, *62*(4), 588–631.

Phi, L. T. H., Sari, I. N., Yang, Y. G., Lee, S. H., Jun, N., Kim, K. S., Lee, Y. K., & Kwon, H. Y. (2018). Cancer stem cells (CSCs) in drug resistance and their therapeutic implications in cancer treatment. *Stem Cells International*, *2018*.

Quintero-Fabián, S., Arreola, R., Becerril-Villanueva, E., Torres-Romero, J. C., Arana-Argáez, V., Lara-Riegos, J., Ramírez-Camacho, M. A., & Alvarez-Sánchez, M. E. (2019). Role of matrix metalloproteinases in angiogenesis and cancer. *Frontiers in Oncology*, *9*, 1370.

Ramer, R., & Hinz, B. (2017). Cannabinoids as anticancer drugs. *Advances in Pharmacology*, *80*, 397–436.

Ruddon, R. W. (2007). *Cancer biology*. Oxford University Press.

Russo, E. B. (2019). The case for the entourage effect and conventional breeding of clinical cannabis: no "strain," no gain. *Frontiers in Plant Science*, *9*, 1969.

Salazar, M., Carracedo, A., Salanueva, Í. J., Hernández-Tiedra, S., Egia, A., Lorente, M., Vázquez, P., Torres, S., Iovanna, J. L., Guzmán, M., Boya, P., & Velasco, G. (2009). TRB3 links ER stress to autophagy in cannabinoid antitumoral action. *Autophagy*, *5*(7), 1048–1049.

Sánchez, C., de Ceballos, M.L., del Pulgar, T.G., Rueda, D., Corbacho, C., Velasco, G., Galve-Roperh, I., Huffman, J.W., y Cajal, S.R., & Guzmán, M. (2001). Inhibition of glioma growth in vivo by selective activation of the CB2 cannabinoid receptor. Cancer Research, 61 (15), 5784–5789.

Scott, K. A., Dalgleish, A. G., & Liu, W. M. (2017). Anticancer effects of phytocannabinoids used with chemotherapy in leukaemia cells can be improved by altering the sequence of their administration. *International Journal of Oncology*, *51*(1), 369–377.

Singer, E., Judkins, J., Salomonis, N., Matlaf, L., Soteropoulos, P., McAllister, S., & Soroceanu, L. (2015). Reactive oxygen species-mediated therapeutic response and resistance in glioblastoma. *Cell Death & Disease*, *6*(1), e1601.

Solinas, M., Massi, P., Cinquina, V., Valenti, M., Bolognini, D., Gariboldi, M., Monti, E., Rubino, T., & Parolaro, D. (2013). Cannabidiol, a non-psychoactive cannabinoid compound, inhibits proliferation and invasion in U87-MG and T98G glioma cells through a multitarget effect. *PLoS One*, *8*(10), e76918.

Tomko, A. M., Whynot, E. G., Ellis, L. D., & Dupré, D. J. (2020). Anti-cancer potential of cannabinoids, terpenes, and flavonoids present in cannabis. *Cancers*, *12*(7), 1985.

Touw, M. (1981). The religious and medicinal uses of Cannabis in China, India and Tibet. *Journal of Psychoactive Drugs*, *13*(1), 23–34.

Velasco, G., Galve-Roperh, I., Sánchez, C., Blázquez, C., Haro, A., & Guzmán, M. (2005). Cannabinoids and ceramide: Two lipids acting hand-by-hand. *Life Sciences*, *77*(14), 1723–1731.

Velasco, G., Sánchez, C., & Guzmán, M. (2012). Towards the use of cannabinoids as antitumour agents. *Nature Reviews Cancer*, *12*(6), 436–444.

Yuan, X., Curtin, J., Xiong, Y., Liu, G., Waschsmann-Hogiu, S., Farkas, D. L., Black, K. L., & Yu, J. S. (2004). Isolation of cancer stem cells from adult glioblastoma multiforme. *Oncogene*, *23*(58), 9392–9400.

Zhou, H.-M., Zhang, J.-G., Zhang, X., & Li, Q. (2021). Targeting cancer stem cells for reversing therapy resistance: Mechanism, signaling, and prospective agents. *Signal Transduction and Targeted Therapy*, *6*(1), 1–17.

Zuardi, A. W. (2006). History of cannabis as a medicine: A review. *Brazilian Journal of Psychiatry*, *28*, 153–157.

Part IX

Treatments and strategies

Chapter 42

Motivational interviewing for cannabis use disorders: A narrative review from research to practice

Maria Teresa Pons Cabrera, Clara Oliveras Salvà, Pablo Rodrigo Guzmán Cortez, and Hugo López-Pelayo
Addictions Unit, Psychiatry and Psychology Service, Clinic Neurosciences Institute, Hospital Clínic de Barcelona, Grup de Recerca en Addiccions Clínic (GRAC), Institut d'Investigacions Biomèdiques August Pi i Sunyer (IDIBAPS), Barcelona, Spain

Abbreviations

ASSIST	Alcohol, Smoking, and Substance Involvement Screening Test
BAC	bood alcohol concentration
BI	brief Intervention
BZD	benzodiazepines
CBT	cognitive-behavioral therapy
CI	confidence interval
CM	contingency management
CUD	cannabis use disorder
DDMI	two-session motivational interview adapted for dually diagnosed psychotic and drug-related disordered patients
EPE	elicit-provide-elicit
GP	general practitioner
MAT	mutual aids therapy
MBCT	mindfulness-based cognitive therapy
MI	motivational interviewing
OR	odds ratio
RCT	randomized controlled trial(s)
RR	relative risk
SI	two-session standard psychiatric interview
SMD	standardized mean difference
TAU	treatment as usual
USA	United States of America

Introduction: Motivational interviewing: Definition, concepts, and key components

The concept of motivational interviewing (MI) was first introduced in 1983 by William R. Miller, describing his experience treating alcoholism. Before this work, according to the previous traditional model of motivation, all failures during the treatment process would be attributed to motivational properties of the patient's personality. On the contrary, therapeutic successes were commonly considered a result of the treatment programs and counselors. Miller remarked that it was not demonstrated that the patients denied or were unaware of the problems for which they consulted. He also observed in his practice that when a difficult choice was to be made, direct argument was ineffective in changing attitudes, and it generated counterstatements. Moreover, he stated that people grasp what they believe as they hear themselves talk, defend a position, and then they become more committed to it. MI proposes ideas to help someone with a motivational struggle, providing an alternative to the model of denial as a client personality trait (Miller, 1983):

Cannabis Use, Neurobiology, Psychology, and Treatment. https://doi.org/10.1016/B978-0-323-89862-1.00032-5

- De-emphasize on labeling, focusing on the problems the person is having, and what needs to be done.
- Return responsibility and capacity of making responsible decisions to the person, believing in his or her own wisdom, and ability to choose a healthful path if given sufficient support. This scope moves away from the moralistic point of view and is placed with humanistic psychology.
- Place the responsibility for the condition on the individual and give him or her credit for a change (a change for which the individual is responsible for is more long lasting).
- Recognize cognitive dissonances or inconsistencies within the individual's beliefs, feelings, and attitudes. When they emerge, a motivational condition arises to restore consistency. The therapist task is to add weight to this dissonance to drive it to a behavioral change.

In 1995, a full definition of what we now know as MI was published. Rollnick and Miller (1995) defined it as: "a directive, person-centred counselling style for eliciting behaviour change by helping clients to explore and resolve ambivalence". The counseling style is eliciting, never confronting, and does not push changes for which individuals are not ready. The directiveness is only focused on the client examining and resolving his or her own ambivalence. Readiness to change is seen as a fluctuating matter, to which the counselor is attentive in the sessions. Resistance and "denial" are also product of interpersonal interaction, and as such it can respond to the therapist behavior. The therapeutic exchange is more a partnership than expert/recipient roles. MI is related with methods as the drinker check-up, motivational enhancement therapy, brief intervention, and brief MI. It differs from more confrontational methods, being an alternative to them. Although it started for patients with problematic drinking, the help for "unmotivated patients" proved useful in many other areas (Table 1).

TABLE 1 Grade of recommendations of motivational interviewing in addictive disorders according to population and specific formats.

Type of intervention	Grade of recommendation	Literature in which is based the recommendation
MI in alcohol use disorders: aim reducing alcohol intake and preventing risky behavior in adults	A	One meta-analysis with positive results
BI in alcohol use disorders: aim reducing risky alcohol use	A	Two meta-analyses with positive results
MI and alcohol use disorders: prevention of alcohol use disorders and alcohol-related problems in young people	A	One meta-analysis with positive results
MI in tobacco use disorders: aim reducing use	A	One meta-analysis with positive results One meta-analysis with positive results but low-certainty estimates RCTs with positive results
MI in benzodiazepine use disorders: aim reducing use	C	One meta-analysis failed to find positive results
MI in cocaine use disorders	B	RCTs with positive results
MI in opioid use disorders: aim reducing overdose episodes	B	One RCT with positive results
BI in opioid use disorders: aim reducing opioid use	C	One meta-analysis failed to find positive results
MI in amphetamine use disorders: improving attendance to Specialist Addiction Services	C	One RCT failed to find positive results
MI in amphetamine use disorders: aim reducing amphetamine use	B	One RCT with positive results
MI in gambling: aim reducing gambling behavior	A	Two meta-analyses with positive results
MI in gambling: aim reducing expenditure in the short term	A	One meta-analysis with positive results
BI in gambling: aim reducing gambling behavior	A	One meta-analysis with positive results

A=High, B=Moderate; C=Low; D=Very low.
Based on "Grading of Recommendations Assessment, Development and Evaluation (GRADE) (Guyatt, G. H., Oxman, A. D., Kunz, R., Falck-Ytter, Y., Vist, G. E., Liberati, A., & Schünemann, H. J. (2008). GRADE: Going from evidence to recommendations. *BMJ, 336*(7652), 1049–1051. https://doi.org/10.1136/BMJ.39493.646875.AE)" (https://www.essentialevidenceplus.com/product/ebm_loe.cfm?show=grade).

The spirit

Later on the authors wrote about the importance of the spirit, without which MI becomes a series of techniques to drive the individual toward the "right choice". The importance to place the motivation for change from the individual and not imposed toward him or her was remarked, being the client's task and not the counselor's to articulate and resolve his or her ambivalence. Four key elements must be mentioned to understand the spirit of MI (Miller & Rollnick, 2013):

- Partnership: MI is done "for" and "with" a person. MI is an active collaboration between experts, without falling into the "expert trap," which refers to providing direction regardless of the patient's collaboration or expertise. Thinking to have the solution to the person's dilemma, the righting reflex of the therapist will arise.
- Acceptance, which involves 4 aspects:
  - Prizing the inherent absolute worth of every human being.
  - An active interest and effort to understand the other's internal perspective (accurate empathy).
  - Supporting each person's autonomy, their right and capacity of self-direction.
  - Appreciating and communicating the person's strengths and efforts with affirmations.
- Compassion: giving priority to the other's welfare and needs.
- Evocation: the scope of MI is based on what individuals already have much of what they need, and the task of the MI practitioner is to "call for it" (or evoke it). People with ambivalences toward a change already have the arguments supporting change and the ones that support status quo. Their own personal motivations for change are likely to be more persuasive than other's. The focus is not on detecting deficits to correct, but to evoke one's motivation and resources.

The techniques

In the last edition of the MI manual, four sequential processes have been formulated to comprise MI (Miller & Rollnick, 2013). See them summarized in Fig. 1.

Some core communication skills help proceeding through these processes. They are shared with other person-centered approaches:

- Asking open questions to help reflect and elaborate. They help the practitioner understand the person's frame of reference. Open questions have an important role all along.
- Affirming—recognizing and commenting—person's particular strengths, abilities, good intentions, and efforts. Accentuating the positive as opposite of "making people bad enough" to change (which proved pointless).
- Reflective listening with statements that make a guess about the clients' meaning, select where to put the attention, and allow them to hear their thoughts and feelings.
- Summarizing. MI practitioners reflect a collection of what people have been saying and propose links between the ideas. Summaries promote understanding and an opportunity to fill in, and they collect the clients' change talk.
- Informing and advising only if asked for it and always with permission. The person's perspectives and needs must be kept in mind, and he/she is free to agree, to consider it, and to implement it or not.

MI techniques, as every other ability, need to be practiced, and some authors point out that one main challenge in implementing MI is the one of letting go the righting reflex (the identification of a problem by a therapist and the willing of

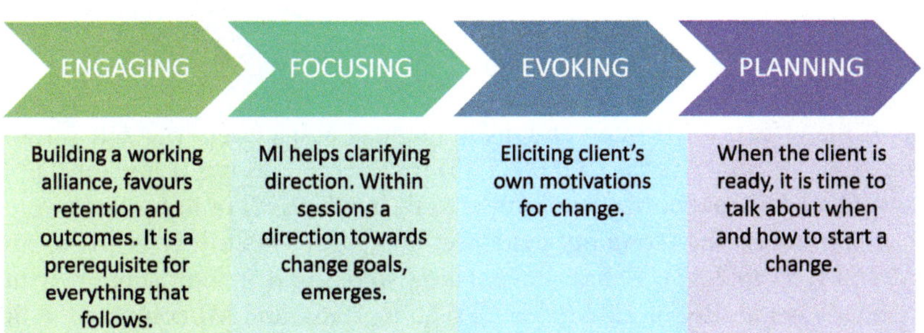

FIG. 1 Motivational Interviewing sequential processes.

solving it for the patient). MI narrative's is not something to apply "to" or "on" patients, but "with" or "for" them (Rollnick et al., 2010).

Miller put this system in a model of change (Prochaska & DiClemente, 1986) in which some stages are described (contemplation, preparation and action, maintenance, and relapse) and occur cyclically in those who pursue a change. But it is not a theory of change itself. MI was described as a "bottom-up" model from practical experience and clinical intuition in the field of alcohol and submitted afterward to the verification of the scientific method (Miller & Rollnick, 2012).

Research suggested that the psycholinguistic aspect that predicted behavioral change was the strength of committing language and that, at the end of a session of MI, change talk and commitment language was reinforced (Miller & Rollnick, 2004).

Years later, hundreds of RCT show positive effects across many health problems.

To conclude, MI is a person-centered approach of counseling that began in the field of addiction but extended to other health behaviors. It facilitates an ambient supporting competence, autonomy, and relatedness. This can promote the internalization and integration of the regulation of a new behavior so that it is engaged with the person's goals, values, and sense of self (Markland et al., 2005). MI focuses largely on the amount of change talk as a paramount element toward change (Deci & Ryan, 2012).

Section 1: Evidence from addictions field beyond cannabis use disorders

There is already a strong body of evidence supporting the efficacy of MI to improve the prognosis of a broad range of health-related conditions and health habits. In fact, a systematic review and meta-analysis of randomized controlled trials evaluating efficacy of MI in medical aid settings concluded that overall, patients receiving MI had one and half times the probability of improving on a wide variety of health measures (such as physical activity, weight loss, HIV viral load, blood pressure. or cholesterol) compared to control groups (Lundahl et al., 2013). But MI was born as a therapeutic approach in the field of addictive disorders (Miller, 1983).

MI and alcohol use disorders

A recent meta-analysis estimated the magnitude of the effect of MI interventions in health care settings was OR = 2.31 (95% CI 1.75–3.06) for reducing alcohol intake and OR = 1.83 (95% CI 1.33–2.53) for preventing risky behavior related to alcohol use (Lundahl et al., 2013). The efficacy of motivational interviewing for risky drinking has also been established as a brief intervention. A meta-analytic review of studies comparing brief MI with no treatment found an aggregate effect size of 0.18 (95% CI 0.07, 0.29). In this same meta-analysis, when comparing brief MI with another treatment (one of a diverse set of interventions), the aggregate effect size was of 0.43(95% CI 0.17, 0.70) (Vasilaki et al., 2006). A more recent meta-analysis investigated the effect of MI delivered in a brief intervention during an emergency care contact. In this study, the drinking frequency decreased significantly more after MI than after control interventions in young people with history of risky alcohol consumption ($SMD \leq -0.17$, $P \leq 0.03$) (Kohler & Hofmann, 2015).

Besides, a 2016 Cochrane Systematic Review included 84 randomized controlled trials in young adults comparing MIs with no intervention, assessment only, or alternative interventions for preventing alcohol misuse and alcohol-related problems. The risk of bias assessment showed that these studies provided moderate or low-quality evidence. At 4 or more month's follow-up, effects in favor of MI were found for the quantity of alcohol consumed (a reduction from 13.7 to 12.5 drinks/week); frequency of alcohol consumption (a reduction in the consumption days/week from 2.74 to 2.52 days); and peak blood alcohol concentration, or BAC (a reduction from 0.144% to 0.131%). Also, a marginal effect in favor of MI for alcohol problems was found (a reduction in an alcohol problems scale score from 8.91 to 8.18). (Foxcroft et al., 2016).

MI and tobacco use disorders

A recent meta-analysis estimated the magnitude of the effect of MI interventions in medical aid settings was OR = 1.34 (95% CI 1.05–1.70) for improving abstinence rates and OR = 1.18 (95% CI 0.96–1.45) for reducing its use (Lundahl et al., 2013). Also, a 2019 Cochrane Systematic Review included 37 randomized controlled trials in which MI or its variants were offered to smokers to assist smoking cessation. It was found low-certainty evidence, comparing the effect of MI to no treatment for smoking cessation (RR = 0.84, 95% CI 0.63 to 1.12). Enhancing existing smoking cessation support with additional MI, compared with existing support alone, gave an RR of 1.07 (95% CI 0.85 to 1.36), and MI compared with other forms of smoking cessation support gave an RR of 1.24 (95% CI 0.91 to 1.69), although both of these estimates were of low certainty (Lindson et al., 2019).

MI as a therapeutic approach for tobacco use disorders has even showed promising results in acute medical settings. For instance, a systematic review (including 17 studies) of smoking cessation interventions in emergency departments found that the 2 studies reporting significant differences in cessation both used motivational interviewing-based interventions (Pelletier et al., 2014).

MI and benzodiazepine use disorders

Regarding benzodiazepine use disorders, evidence is mixed and scarce. In 2015, a Cochrane Systematic Review included 25 randomized controlled trials examining the use of a psychosocial intervention versus pharmacological interventions, no intervention, placebo, or a different psychosocial intervention on reducing the use of BZD. Meta-analytic summary of the available studies (very low quality of evidence) showed that the effect of MI versus treatment as usual (TAU) was unclear post treatment, at 3-month follow-up, at 6-month follow-up, and at 12-month follow-up (Darker et al., 2015).

MI and cocaine use disorders

Regarding cocaine use disorders, evidence is also mixed and scarce.

A randomized controlled pilot trial of MI for patients with psychotic and drug use disorders in USA evaluated the efficacy of a two-session motivational interview adapted for dually diagnosed psychotic and drug-related disordered patients (DDMI) in comparison to a two-session standard psychiatric interview (SI). Among the diverse primary substances among the sample, DDMI resulted in significantly better primary drug treatment outcomes for the cocaine-using group. The between-treatment effect size was 0.50 (Martino et al., 2006).

Another example is a randomized trial with 6-month follow-up for 198 persons who used cocaine at least weekly in USA published in 2008. Participants were randomly assigned to a four-session motivational intervention or an assessment control group. Although in the full cohort, there were no significant intervention effects, among those using cocaine on 15 or more of the 30 previous days, motivational interviewing participants had a significantly larger mean reduction in cocaine use days (Stein et al., 2009).

MI and opioid use disorders

There are a few examples of studies showing promising results of MI in opiate use disorders. For instance, a single-blinded randomized-controlled trial of a repeated dose motivational interviewing intervention to reduce overdose versus treatment as usual was performed in the USA among adults 18–65 years of age with opioid use, active opioid use, opioid overdose within 5 years, and prior receipt of naloxone kits. Participants in the intervention group were less likely to experience any overdose and experienced fewer overdose events (Coffin et al., 2017). But a recent systematic review of randomized controlled trials assessing the effects of brief motivational interventions in adult patients currently using opioids concluded that there was limited, low quality evidence to conclude that it reduces opioid use in opioid-dependent patients (McNerney et al., 2020).

MI and amphetamine use disorders

There is a paucity of well-conducted studies among primary amphetamine users, and the results are inconclusive. In a randomized controlled trial of acute psychiatric inpatients with dual disorders, a single MI intervention failed to improve attendance to a Specialist Substance Misuse Service (Baker et al., 2002). In a more recent RCT, both single-session and multiple-session MI interventions showed significant decreases in methamphetamine use in dependent samples (Polcin et al., 2014).

MI and gambling

MI showed promising results for the treatment of behavioral addictions such as gambling (Di Nicola et al., 2020). A 2012 Cochrane Systematic Review included 4 randomized controlled trials evaluating MI therapy for pathological or problem gambling compared to "no treatment" controls, referral to Gamblers Anonymous and nonspecific treatment component controls. It concluded that there was preliminary evidence for some benefits from motivational interviewing therapy in terms of reduced gambling behavior, although not necessarily other symptoms of pathological and problem gambling (Cowlishaw et al., 2012). A recent meta-analysis found that motivational interviewing was associated with significant reduction in gambling frequency up to a year after treatment delivery. For gambling expenditure, MI yielded significant reductions in dollars spent gambling compared to nonmotivational controls at 1–3 months) (Yakovenko et al., 2015). Also, another meta-analysis results supported the efficacy of brief interventions for problem gambling compared to inactive control in the reduction of gambling behavior (Quilty et al., 2019).

Overall, evidence of MI's efficacy across different addictive behaviors does not always paint a clear picture (Diclemente et al., 2017). As a rule of thumb, in most addictive disorders, MI seems to be more effective than no treatment. Stronger evidence supports the use of MI for the treatment of alcohol and tobacco use disorders and gambling.

Section 2: Reviewing evidence of motivational interviewing for CUD

The effectiveness of MI for reducing alcohol and other drug use or improving healthy lifestyle is well demonstrated. As cannabis use and cannabis use disorder (CUD) are increasing worldwide, especially in Europe (Manthey et al., 2021), and cannabis-related consequences on health are concerning due to their impact on quality of life, functionality, and mortality (e.g., psychosis and schizophrenia, lung cancer) (Campeny et al., 2020), and availability of effective interventions for CUD is mandatory.

A recent meta-analysis of 40 studies confirmed effectiveness of motivational interviewing for CUD both in adults and adolescents. Likelihood of cannabis abstinence was 3 times higher in adults and 2 times in adolescents after 3 months of intervention (OR 3.8, 95% CI 2.4–6.2; OR 2.0 95% CI 1.4–2.9, respectively) compared to the control group. Even when the aim was reduction in frequency or quantity, adults reduced 0.7 joints per day and 4 days per month of use in the month 3 after intervention compared to the control group. However, this revision also highlighted some gaps in the field: (1) Is MI effective for adolescents in a risk reduction paradigm? (2) Does MI have an impact on quality of life, functionality, or reduction in symptoms of CUD (e.g., craving)? (3) Is there enough evidence for recommending MI in vulnerable groups as patients with psychosis? (4) How many sessions and how long each session are enough? (5) Who benefits the most of MI? (6) Are other outcomes such as readiness to change, adherence to treatment, or reduction in risky behaviors valid for research purpose in people with severe lack of insight of CUD? (Calomarde-Gómez et al., 2021).

As cannabis use is especially deleterious for brain development (Meruelo et al., 2017), adolescents and young adults are in the focus of interest for seeking evidence-based treatment aiming abstinence or at least a clinically significant reduction. Despite MI is promising in achieving abstinence according to one meta-analysis (see above), there is lacking evidence in favor of MI for reduction of use in adolescents (Steele et al., 2020).

Mental health disorders are very prevalent in people with CUD, and cannabis use may worsen the prognosis of psychiatric comorbidity (e.g., depression, bipolar disorder, schizophrenia, and anxiety (Campeny et al., 2020). MI has been proposed as a valid approach for people who use cannabis and suffers from depression, psychosis, bipolar disorder, or other severe mental illness. However, the evidence in this population is still inconclusive (Baker et al., 2010; Hunt et al., 2019).

MI plus other psychological approaches in CUD

MI is potentially compatible with other psychological approaches, as it is a style of communication and relationship with client/patient and is not a manualized treatment. Cognitive-Behavioral Therapy (CBT), Contingency Management (CM), and Mutual Aids Therapy (MAT) showed efficacy in treating CUD. CBT "involves efforts to change thinking and behavioural patterns." CBT has strong body of evidence for several disorders including substance use disorders (American Psychological Association, 2021). CM is based on "giving tangible rewards to reinforce positive behaviours" (National Institute on Drug Abuse, 2018). MAT "refers to the social, emotional and informational support provided by, and to, members of a group at every stage of their recovery from active alcohol and/or drug use and addiction. It is not a peer support network. It relies upon a structured program that is focused on recovery (Public Health England, 2015)."

Their combination with MI are promising strategies to improve results of each intervention, but further research is warranted (Cooper et al., 2015; Davis et al., 2015; Gates et al., 2017; Winters et al., 2021). Especially, research must focus on cost-effectiveness (e.g., number of sessions of each type of intervention) and personalized treatments (e.g., who will respond to each combination).

Mindfulness-based cognitive therapy (MBCT) has been proposed as a valid approach for substance use disorders including CUD (Dakwar & Levin, 2013; De Dios et al., 2012; Fortuna et al., 2018; Schneegans et al., 2021; Sipe & Eisendrath, 2012). MBCT is a combination of CBT elements with stress reduction through mindfulness techniques (e.g., present moment awareness, observe the thoughts). However, evidence is still low, and there is one pilot study (De Dios et al., 2012), up to our knowledge, about its combination with MI.

There are several psychological approaches tested for CUD and potentially complementary to MI, but due to its limited research and implementation, they are not included in this section.

Brief interventions (BI), MI and CUD

A brief intervention (BI) is a type of short therapeutic or preventive communication carried out by a healthcare professional aiming at reduction of alcohol or other drug use and the related problems (Vasilaki et al., 2006). Motivational brief interventions are opportunistic or specialized interventions based on acronym "FRAMES": (1) give **F**eedback of personal risk; (2) **R**esponsibility of the patient in the own change; (3) give **A**dvice to change; (4) propose a **M**enu of alternative goals and strategies to reduce drug use; (5) use **E**mpathic counseling and avoid confrontation; (6) promote **S**elf-efficacy (encourage patients' optimism that the chosen goals can be achieved) (Bien et al., 1993).

One meta-analysis of 26 studies (>6300 participants) showed excellent results of BIs for nonseeking treatment emerging adults with CUD. BI reduced symptoms of CUD and increased likelihood of abstinence (Halladay et al., 2019). Other meta-analysis of BIs in healthcare settings did not find a positive impact of opportunistic BIs in abstinence rates or other outcomes (Imtiaz et al., 2020).

Other meta-analysis of six brief school-based intervention trials showed a short-term reduction in cannabis abuse and long-term reduction in cannabis use and dependence (Carney et al., 2016).

Digital health and MI

Digital health is the use of any of the wide range of available technologies—usually linked to Internet—for improving physical, mental, and social well-being (López-Pelayo, Aubin, & Drummond, 2020). Digital mental health, and specifically the use of digital health for drug use disorder, is one of the most promising strategies in addictions area (Schaub et al., 2020) due to its potential to facilitate access to treatment in a very stigmatized population who claim for privacy, the capacity to provide treatment in rural areas and other populations with limited access to treatment, possibility of personalized treatment and complementarities with other treatments. Regarding CUD, digital health is especially interesting as implementation in young people is more feasible and easier (European Commission, 2018).

A meta-analysis of nine studies (>2900 participants) about computerized interventions informed of reduction in cannabis use (Olmos et al., 2018). Other meta-analysis from nine studies (> 4100 participants) found a small effect of Internet interventions for CUD (Tait et al., 2013). And still another meta-analysis found that a combination of online brief motivational interventions and cognitive behavioral CBT was short term effective in nonclinical settings (>1900 participants) (Hoch et al., 2016).

A recent systematic review found that combination of motivational, psycho educational, and cognitive enhancement therapy through technology-based tools was modestly effective in decreasing cannabis use or achieving abstinence in people with psychosis (Tatar et al., 2020).

Motivational interviewing is a useful strategy for reducing cannabis use and achieving abstinence in adults. Its potential is even higher if it is combined with other evidence-based approaches such as contingency management, cognitive-behavioral treatment, or mutual support therapy. Even low-dose motivational interviewing, as brief interventions, has an impact on well-being. Digital health approach based on motivational interviewing as an alternative to face-to-face interventions, especially if barriers exist is promising. Evidence of its utility in adolescents and people with mental illness is insufficient (e.g., summary of evidence in CUD is provided in Table 2).

Section 3: MI for CUD in the real world: Characteristics and case example

Long-term abstinence in people with CUD, which is the most desired outcome from a neurophysiologic addiction standpoint, seems to be a difficult target for some clinical researchers, calling for more focus on harm reduction and quality of life and symptom improvement (Loflin et al., 2020).

We will use an example of MI Therapy with a fictitious client ("Mohammed, 26 years old"), illustrating the processes of MI and skills that are involved. For illustrative purposes, our client will not have a comorbid mental health disorder or another addictive diagnosis. The therapy sessions are structured following the MI processes, plus a diagnostic session at the beginning where a questionnaire is employed to gauge CUD severity and rule out other important mental health issues. As a general rule, to reach the point of quitting or reducing cannabis use, a strong and successful engagement process with the client is required, while advancing through the processes of focusing and evoking in a continuous or overlapped way.

TABLE 2 Grade of recommendations of motivational interviewing in CUD according to population and specific formats.

Type of intervention	Grade of recommendation	Literature in which is based the recommendation
MI in adults: aim abstinence	A	One meta-analysis with positive results
MI in adolescents: aim abstinence	A	One meta-analysis with positive results
MI in adults: reduction in cannabis use	A	One meta-analysis with positive results
MI in adolescents: reduction in cannabis use	C	RCTs with positive results Meta-analyses failed to find positive results
MI in people affected by mental health problems	C	RCTs with positive results Meta-analyses failed to find positive results
Brief motivational interventions in emerging adults	A	One meta-analysis with positive results
Brief motivational interventions in healthcare settings (adults)	C	RCTs with positive results Meta-analyses failed to find positive results
Digital motivational interventions	A	Two meta-analyses with positive results and one meta-analysis with promising results
Combination with CBT, contingency management or mutual aid groups	B	RCTs with positive results. No meta-analyses.

A=High, B=Moderate; C=Low; D=Very low.
Based on "Grading of Recommendations Assessment, Development and Evaluation (GRADE) (Guyatt, G. H., Oxman, A. D., Kunz, R., Falck-Ytter, Y., Vist, G. E., Liberati, A., & Schünemann, H. J. (2008). GRADE: Going from evidence to recommendations. *BMJ, 336*(7652), 1049–1051. https://doi.org/10.1136/BMJ.39493.646875.AE)" (https://www.essentialevidenceplus.com/product/ebm_loe.cfm?show=grade).

Session 1

In the first session, the therapist needs to active listen to the client while creating a therapeutic environment for the client to feel comfortable to talk openly, which is a practical way to establish empathy and engagement. Mohammed tells the therapist that he has been in the country for 10 years, leaving his nuclear family behind to live with an aunt to study and work. This is reflected back by the therapist with "It has got to have been a difficult decision to come to this country, but you are looking to improve your own life and give your family a better future" (note that this reflection was ended on a positive note, making an implicit connection between the immigration process and the treatment process as both are part of the patient wanting to improve his life, adding an evocative quality to the reflection).

After that initial introduction, preliminary information about addiction and cannabis specificities are discussed, using Elicit-Provide-Elicit (EPE). The client is asked (Elicit) about what he or she knows about cannabis brain chemistry and the addictive process in general. Then, after asking permission, information is given (Provide) to contrast or complement the point of view of the client. Later, feedback is asked from the client to check on how the information was received (Elicit). Mohammed seems to be very informed of cannabis chemicals but not about the physiology of the addiction process. At this point, besides informing and providing psychoeducation, we are simultaneously establishing the stage of change of the transtheoretical model (Velasquez et al., 2005) that the client is located at the moment. Using open-ended questions like "What do you think has happened to you with cannabis?" "Why do you think this problem happened?" or "What would you like to do?" information is obtained about the disposition to change in a short term, which helps us not to rush beyond what the client is willing to do. Mohammed answers these questions expressing certainty that cannabis is harming him and the will to quit, although He does not feel totally prepared right now, so he will continue smoking, until he is prepared to completely stop in the near future. This sets him in the contemplative stage. The objective is to get him to cross to the preparation phase and then helping him to move to the action phase.

Session 2

At this point, the MI process of engagement is at a strong enough state for the therapist to move into the MI process of focusing. Focusing is a short stage for Mohammed because he has already verbalized his intentions to stop smoking cannabis, which is aligned to the interests of the therapist. In the case of patients that are ambivalent and have not fully decided to stop smoking cannabis, the therapy must help "developing discrepancy" (using the objectives and values of the patient to generate questions that put in display how the decision of smoking is an obstacle or contradiction to those objectives or values). For example, "You say you want to concentrate more and improve your grades. How does that fit in with smoking"? If the patient is too adamant on maintaining smoking, an intermediate focus could be needed like a reduction in usage, ideally framing it as a temporal stepping stone, leaving the possibility of abstinence for the future if the values or objectives are not met.

Going back to our example, after Focusing, the therapist can advance to the core process of MI, which is Evoking. Mohammed at this point feels comfortable and secure with the therapist and is willing to talk in depth about his cannabis use.

To help the patient visualize his or hers own reasons for smoking cessation while checking on his ambivalence, which is usually present, a decisional balance instrument can be used. It includes two columns, one with frequent, predefined, reasons to keep using cannabis and the other with frequent reasons to quit. Both columns have blank spaces to add other personal motives if needed. The patient is then asked to score by personal importance from 1 to 10 the reasons in both columns and a final score for each side is compared. Mohammed scores 20 points in the reasons for keep using column and 40 points on the reasons for change.

He got a higher score in the reasons to change column (40) but his reasons to change still got a high score (20), showing ambivalence but still giving more weight to the change motives. Mohammed says that he smokes because "It helps me sleep and relax after a day of stress." This is reflected to the client with "You have been struggling with sleeping and anxiety problems." In the other hand, he says that he is willing to change his use "to feel clearer minded and to have more energy." This is reflected with "You want to feel refreshed and active during the day" He responds with "Yes, because I am having problems with my university assignments right now." The therapist then makes a summary "You are saying that you are having sleeping problems and not feeling well with anxiety and want to feel better with yourself by quitting cannabis. One of the reasons is that you want to have a good academic performance." Mohammed agrees. The therapist ends with an affirmation "you are looking to improve your quality of life and trying to tackle a difficult habit that is worrying you. This is something that requires character and strength which you are showing already by coming to these sessions."

Session 3

Mohammed gets to know his own patterns of consumption and his personal triggers. He concludes that he mostly smokes alone in his room after work. At this point, the stage of change of the transtheoretical model is changing to preparation, detected by when the client starts making questions like "What should I do to not use cannabis when I get to my room," "What do you recommend to do in my situation?" The therapist dodges the trap of the "correction reflex" by not responding to these questions directly but reflecting back the client's questions "What do you think could help you in this situation?," "What are the alternatives for you at that time of the day?," providing Mohammed with a safe space to start proposing ideas that could help with the plan. This interaction involves the evoking process of MI because the solutions have to come from the client himself. The therapist then uses a MI self-efficacy instrument to gauge readiness to change by asking "From 1 to 10, what is your confidence level in succeeding?" He says 8. This is followed by "Why did you say 7 and not 5?

(Eliciting change talk about what is already there to make him say 7 and not on what is missing for him not to say a 9 or 10).

Session 4

In the fourth session, the plan to quit is finished in its practical details (he will throw the cannabis paraphernalia with the help of a friend), and the sequence of events that need to happen is verbalized by the patient as a commitment. This involves the planning process of MI. Possible abstinence symptoms are discussed and normalized which makes Mohammed feel empowered. The therapist affirms Mohammed for his commitment and persistence to keep coming to the sessions.

After the day of quitting

During this phase is common that the client relapses, the objective is to prevent relapses. In this critical period, the spirit of MI (including compassion and acceptance) is fundamental. In our example, Mohammed relapsed one time after he was invited by some friends to smoke and was too tempted to say no. Instead of directly giving nonasked advice on avoiding social situations, the therapist makes open questions: "What would you have done in a different way?" or "Is there anything that can help you prevent this happening again"? While also giving affirmation on how, despite the setback, the client has returned to the therapist office and been persistent with his efforts and commitment. This sets the mood for the client to elaborate a new plan that includes the risk of social situations which he has not considered before the relapse. During this period,. Mohammed said he felt more confident to talk with his family about his feelings and more secure in avoiding risky social situations. He maintained abstinence in the follow-up period.

Applications to other areas

In this chapter, we reviewed motivational interviewing and how it is applied to cannabis use disorders. But MI, a person-centered counseling style has been proven effective improving the prognosis of different addictive behaviors. Regarding alcohol use disorders, there is a long body of evidence concluding efficacy (Lundahl et al., 2013). It can be effective delivered in form of brief interventions (Vasilaki et al., 2006), even in an emergency context (Kohler & Hofmann, 2015), and specially for younger drinkers (Vasilaki et al., 2006). For tobacco addiction, MI can improve abstinence rates and amount of tobacco used in medical care settings (Lundahl et al., 2013), and it is useful in enhancing existing smoking cessation support (Lindson et al., 2019). For opiate use disorders, less research is found. However, results are promising to reduce overdose (Coffin et al., 2017). Regarding to amphetamines data are scarce and inconclusive in general, but a recent RCT must be mentioned, as it suggests that MI can reduce the use in dependent samples (Polcin et al., 2014). In behavioral addictions, some benefits have been found in reducing gambling (Cowlishaw et al., 2012). For cocaine and benzodiazepine use disorders, data are also mixed and scarce. To conclude, in most addictive disorders, MI seems to be more effective than no treatment. The strongest evidence for its use on addictions has been found on the treatment of alcohol and tobacco use disorders, gambling, as well as for cannabis use disorders (Calomarde-Gómez et al., 2021).

Regarding different health-related conditions and behaviors, MI's potential is noteworthy, since it as a collaborative style to reinforce the person's own motivation for change. Cardiovascular risk factors can be addressed with MI. Significant and clinically relevant decreases for the combined effect estimate of body mass index or systolic blood pressure have been found. Moreover, MI has been successfully applied by different counselor's profile such as psychologists, psychiatrists, physicians, GPs, nurses or dieticians (Rubak et al., 2005).

Key facts

Key facts of motivational interviewing

- MI was introduced by William R. Miller in 1983 describing his experience with people with alcohol use disorders.
- MI was first defined as "a directive, person-centred counselling style for eliciting behaviour change by helping clients to explore and resolve ambivalence" by Rollnick and Miller (1995).
- Some MI techniques used are the open questions, the affirmations, the reflective listening, and the summaries.
- The spirit of MI is based on partnership, compassion, acceptance, and evoking.
- The techniques and spirit of MI are applied though 4 steps in the process toward change: engaging, focusing, evoking, and planning.
- If analyzed from a psycholinguistic view, the change talk is the key element predicting behavioral change.

Key facts of motivational interviewing for cannabis use disorder

- The application of MI is effective for CUD. Data are inconclusive on people with CUD and severe mental disorders.
- Cognitive-Behavioral Therapy, Contingency Management, and Mutual Aids Therapy have been successfully combined with MI for treating CUD.
- BIs for nonseeking treatment emerging adults with CUD have shown useful.
- Brief motivational interventions and cognitive behavioral CBT delivered online have proven promising results for CUD.
- Digital health could be a feasible tool to get to young people with CUD.

Mini-dictionary of terms

MI Spirit: An embodied way of understanding the therapeutic relationship and human interaction in general, in which MI practice is based. It has four components: Partnership, Acceptance, Compassion, and Evocation.
Partnership: An interpersonal style in which the therapist collaborates with the client during the session while avoiding being directive or paternalist.
Acceptance: It has four aspects: affirmation (a conscious effort to pinpoint the values and efforts that the patient has or is striving for), accurate empathy (the ability to make someone feel safe to talk freely about anything), autonomy (to constantly remark the capacity for freedom that the other person has), and absolute worth (a deep respect for the patient just for being a human being).
Compassion: The genuine desire of the therapist to improve the patient's health and quality of life. It sustains the ethical usage of MI techniques.
Evocation: A deep trust in the process of obtaining useful therapeutic material from the patient's own biography, values, or ideas.
MI Processes: A conceptual framework to help therapists understand the four different stages that the patient usually goes through to change behavior (Engaging, Focusing, Evoking,
Planning). They are sequential, but there can be overlap, and detecting the necessity to reinforce a previous process is usual.
Engaging: First process of MI and the foundation of the therapeutic process. The goal is making the patient feel heard, understood, and validated in their efforts which makes them want to continue further ahead with the therapeutic process.
Focusing: Second process of MI, the patient is guided to the desired behavior change in a collaborative fashion. Ambivalence is decreased or resolved.
Evoking: Third and core process of MI, where specific techniques are applied to generate dialogue about the patient's DARNs (Desires, Abilities, Reasons, Needs), increasing the probability of the behavior change happening.
Planning: Fourth and final process of MI, where a plan is designed and the patient commits to it.
OARS: Acronym for the core skills of MI (Open-ended questions, Affirmations, Reflective listening, Summarizing listening). These skills are used in every process of MI and are the basis of most of the therapists statements during the course of the sessions.
Open-ended questions: Ability to generate open-ended questions for the patient to think more deeply on issues or communicate discrepancy between the patient values and behaviors.
Affirmations: A genuine statement of admiration made by the therapist, in which the strengths and efforts that the patient is implicitly or explicitly displaying are highlighted with the intention to evoke self-efficacy.
Reflective listening: A core skill of MI in which the therapist responds to the statement of the patient by repeating or rephrasing the same idea or a part of it, increasing the probability of behavior change happening later. This is known as a reflection, and it can be simple if the therapist repeats literally or paraphrases what the patient said, or complex when the reflection is about a specific aspect or an underlying emotion. The accuracy of the reflection is then conveyed to the therapist by the patient's verbal and nonverbal feedback, allowing the therapist to adjust the next reflection to be closer to what the patient is expressing, making the patient feel heard and understood in the process.
Summarizing listening: A skill that consists on generating a logical, cohesive statement from the most relevant parts of the previous three or four reflections and affirmations, with the objective of closing a segment of the conversation or allowing the patient to reflect more on what is being talked about. The patient then gives feedback on the accuracy of the summary to the therapist, who then knows if more clarification is needed or if it is appropriate to continue with the next step.

Summary points

- MI is a person-centered counseling style to facilitate behavior change by helping the person to explore and resolve ambivalence.
- Autonomy, relatedness, and competence link SBT and MI, and confrontational methods differ from them.
- Stronger evidence of MI supports its use for the treatment of gambling, and alcohol, tobacco, and cannabis use disorders.
- MI is also effective to promote other health behaviors or changes.
- MI in the field of CUD is effective delivered in different formats: as a preventive brief intervention, with online format, and as part of a structured program.

References

American Psychological Association. (2021). *What is cognitive behavioral therapy?*. https://www.apa.org/ptsd-guideline/patients-and-families/cognitive-behavioral#:~:text=CBT%20places%20an%20emphasis%20on,%2C%20problematic%20emotions%2C%20and%20behavior.

Baker, A. L., Hides, L., & Lubman, D. I. (2010). Treatment of cannabis use among people with psychotic or depressive disorders: A systematic review. *The Journal of Clinical Psychiatry*, *71*(3), 247–254. https://doi.org/10.4088/JCP.09R05119GRY.

Baker, A., Lewin, T., Reichler, H., Clancy, R., Carr, V., Garrett, R., Sly, K., Devir, H., & Terry, M. (2002). Motivational interviewing among psychiatric in-patients with substance use disorders. *Acta Psychiatrica Scandinavica*, *106*(3), 233–240. https://doi.org/10.1034/j.1600-0447.2002.01118.x.

Bien, T. H., Miller, W. R., & Tonigan, J. S. (1993). Brief interventions for alcohol problems: A review. *Addiction (Abingdon, England)*, *88*(3), 315–335.

Calomarde-Gómez, C., Jiménez-Fernández, B., Balcells-Oliveró, M., Gual, A., & López-Pelayo, H. (2021). Motivational interviewing for Cannabis use disorders: A systematic review and meta-analysis. *European Addiction Research*. https://doi.org/10.1159/000515667.

Campeny, E., López-Pelayo, H., Nutt, D., Blithikioti, C., Oliveras, C., Nuño, L., Maldonado, R., Florez, G., Arias, F., Fernández-Artamendi, S., Villalbí, J. R., Sellarès, J., Ballbè, M., Rehm, J., Balcells-Olivero, M. M., & Gual, A. (2020). The blind men and the elephant: Systematic review of systematic reviews of cannabis use related health harms. *European Neuropsychopharmacology*, *33*, 1–35. https://doi.org/10.1016/j.euroneuro.2020.02.003.

Carney, T., Myers, B. J., Louw, J., & Okwundu, C. I. (2016). Brief school-based interventions and behavioural outcomes for substance-using adolescents. In T. Carney (Ed.), *Vol. 1. Cochrane database of systematic reviews* (p. CD008969). John Wiley & Sons, Ltd. https://doi.org/10.1002/14651858.CD008969.pub3.

Coffin, P. O., Santos, G. M., Matheson, T., Behar, E., Rowe, C., Rubin, T., Silvis, J., & Vittinghoff, E. (2017). Behavioral intervention to reduce opioid overdose among high-risk persons with opioid use disorder: A pilot randomized controlled trial. *PLoS One*, *12*(10), 1–15. https://doi.org/10.1371/journal.pone.0183354.

Cooper, K., Chatters, R., Kaltenthaler, E., & Wong, R. (2015). Psychological and psychosocial interventions for cannabis cessation in adults: A systematic review short report. *Health Technology Assessment (Winchester)*, *19*(56), 1–129. https://doi.org/10.3310/HTA19560.

Cowlishaw, S., Merkouris, S., Dowling, N., Anderson, C., Jackson, A., & Thomas, S. (2012). Psychological therapies for pathological and problem gambling. *Cochrane Database of Systematic Reviews*. https://doi.org/10.1002/14651858.cd008937.pub2.

Dakwar, E., & Levin, F. R. (2013). Individual mindfulness-based psychotherapy for cannabis or cocaine dependence: A pilot feasibility trial. *The American Journal on Addictions*, *22*(6), 521–526. https://doi.org/10.1111/J.1521-0391.2013.12036.X.

Darker, C., Sweeney, B., Barry, J., Farrell, M., & Donelly-Swift, E. (2015). Psychosocial interventions for benzodiazepine harmful use, abuse or dependence. *Cochrane Database of Systematic Reviews*, *5*. https://doi.org/10.1002/14651858.CD009652.pub2.www.cochranelibrary.com.

Davis, M. L., Powers, M. B., Handelsman, P., Medina, J. L., Zvolensky, M., & Smits, J. A. J. (2015). Behavioral therapies for treatment-seeking cannabis users: A meta-analysis of randomized controlled trials. *Evaluation & the Health Professions*, *38*(1), 94–114. https://doi.org/10.1177/0163278714529970.

De Dios, M. A., Herman, D. S., Britton, W. B., Hagerty, C. E., Anderson, B. J., & Stein, M. D. (2012). Motivational and mindfulness intervention for young adult female marijuana users. *Journal of Substance Abuse Treatment*, *42*(1), 56–64. https://doi.org/10.1016/j.jsat.2011.08.001.

Deci, E. L., & Ryan, R. M. (2012). Self-determination theory in health care and its relations to motivational interviewing: A few comments. *International Journal of Behavioral Nutrition and Physical Activity*, *9*(1), 24. https://doi.org/10.1186/1479-5868-9-24.

Di Nicola, M., De Crescenzo, F., D'Alò, G. L., Remondi, C., Panaccione, I., Moccia, L., Molinaro, M., Dattoli, L., Lauriola, A., Martinelli, S., Giuseppin, G., Maisto, F., Crosta, M. L., Di Pietro, S., Amato, L., & Janiri, L. (2020). Pharmacological and psychosocial treatment of adults with gambling disorder: A meta-review. *Journal of Addiction Medicine*, *14*(4), e15–e23. https://doi.org/10.1097/ADM.0000000000000574.

Diclemente, C. C., Corno, C. M., Graydon, M. M., Wiprovnick, A. E., & Knoblach, D. J. (2017). Motivational interviewing, enhancement, and brief interventions over the last decade: A review of reviews of efficacy and effectiveness. *Psychology of Addictive Behaviors: Journal of the Society of Psychologists in Addictive Behaviors*, *31*(8), 862–887. https://doi.org/10.1037/ADB0000318.

European Commission. (2018). *eHealth: Digital health and care*. https://health.ec.europa.eu/ehealth-digital-health-and-care_en.

Fortuna, L. R., Porche, M. V., & Padilla, A. (2018). A treatment development study of a cognitive and mindfulness-based therapy for adolescents with co-occurring post-traumatic stress and substance use disorder. *Psychology and Psychotherapy*, *91*(1), 42–62. https://doi.org/10.1111/PAPT.12143.

Foxcroft, D. R., Coombes, L., Wood, S., Allen, D., Almeida Santimano, N. M. L., & Moreira, M. T. (2016). Motivational interviewing for the prevention of alcohol misuse in young adults. *Cochrane Database of Systematic Reviews*, *2016*(7). https://doi.org/10.1002/14651858.CD007025.pub4.

Gates, P. J., Sabioni, P., Copeland, J., Le Foll, B., & Gowing, L. (2017). Psychosocial interventions for cannabis use disorder (review) summary of findings for the main comparison. *Cochrane Database of Systematic Reviews*, *5*. https://doi.org/10.1002/14651858.CD005336.pub4. www.cochranelibrary.com.

Halladay, J., Scherer, J., MacKillop, J., Woock, R., Petker, T., Linton, V., & Munn, C. (2019). Brief interventions for cannabis use in emerging adults: A systematic review, meta-analysis, and evidence map. *Drug and Alcohol Dependence*, *204*(April), 107565. https://doi.org/10.1016/j.drugalcdep.2019.107565.

Hoch, E., Preuss, U. W., Ferri, M., & Simon, R. (2016). Digital interventions for problematic Cannabis users in non-clinical settings: Findings from a systematic review and meta-analysis. *European Addiction Research*, *22*(5), 233–242. https://doi.org/10.1159/000445716.

Hunt, G. E., Siegfried, N., Morley, K., Brooke-Sumner, C., & Cleary, M. (2019). Psychosocial interventions for people with both severe mental illness and substance misuse. *The Cochrane Database of Systematic Reviews*, *12*(12). https://doi.org/10.1002/14651858.CD001088.PUB4.

Imtiaz, S., Roerecke, M., Kurdyak, P., Samokhvalov, A. V., Hasan, O. S. M., & Rehm, J. (2020). Brief interventions for cannabis use in healthcare settings: Systematic review and meta-analyses of randomized trials. *Journal of Addiction Medicine*, *14*(1), 78–88. Lippincott Williams and Wilkins https://doi.org/10.1097/ADM.0000000000000527.

Kohler, S., & Hofmann, A. (2015). Can motivational interviewing in emergency care reduce alcohol consumption in young people? A systematic review and meta-analysis. *Alcohol and Alcoholism*, *50*(2), 107–117. https://doi.org/10.1093/alcalc/agu098.

Lindson, N., Thompson, T. P., Ferrey, A., Lambert, J. D., & Aveyard, P. (2019). Motivational interviewing for smoking cessation. *Cochrane Database of Systematic Reviews*, *2019*(7). https://doi.org/10.1002/14651858.CD006936.pub4.

Loflin, M. J. E., Kiluk, B. D., Huestis, M. A., Aklin, W. M., Budney, A. J., Carroll, K. M., D'Souza, D. C., Dworkin, R. H., Gray, K. M., Hasin, D. S., Lee, D. C., Le Foll, B., Levin, F. R., Lile, J. A., Mason, B. J., McRae-Clark, A. L., Montoya, I., Peters, E. N., Ramey, T., & Strain, E. C. (2020). The state of clinical outcome assessments for cannabis use disorder clinical trials: A review and research agenda. In *Vol. 212. Drug and alcohol dependence* Elsevier Ireland Ltd. https://doi.org/10.1016/j.drugalcdep.2020.107993.

López-Pelayo, H., Aubin, H. J., Drummond, C., et al. (2020). "The post-COVID era": Challenges in the treatment of substance use disorder (SUD) after the pandemic. *BMC Medicine*, *18*, 214. https://doi.org/10.1186/s12916-020-01693-9.

Lundahl, B., Moleni, T., Burke, B. L., Butters, R., Tollefson, D., Butler, C., & Rollnick, S. (2013). Motivational interviewing in medical care settings: A systematic review and meta-analysis of randomized controlled trials. *Patient Education and Counseling*, *93*(2), 157–168. https://doi.org/10.1016/j.pec.2013.07.012.

Manthey, J., Freeman, T. P., Kilian, C., López-Pelayo, H., & Rehm, J. (2021). Public health monitoring of cannabis use in Europe: Prevalence of use, cannabis potency, and treatment rates. *The Lancet Regional Health - Europe*, *10*, 100227. https://doi.org/10.1016/J.LANEPE.2021.100227/ATTACHMENT/92A316F9-6946-45B4-B4E1-4EB58C208B96/MMC2.ZIP.

Markland, D., Ryan, R. M., Tobin, V. J., & Rollnick, S. (2005). Motivational interviewing and self-determination theory. *Journal of Social and Clinical Psychology*, *24*(6), 811–831. https://doi.org/10.1521/jscp.2005.24.6.811.

Martino, S., Carroll, K. M., Nich, C., & Rounsaville, B. J. (2006). A randomized controlled pilot study of motivational interviewing for patients with psychotic and drug use disorders. *Addiction*, *101*(10), 1479–1492. https://doi.org/10.1111/j.1360-0443.2006.01554.x.

McNerney, N. A., Losensky, M. J., Lash, M. M., Rozaieski, K. R., Ortiz, D., Garcia, A. N., & Rethorn, Z. D. (2020). Low theoretical fidelity hinders the research on health coaching for opioid reduction: A systematic review of randomized controlled trials. *PLoS One*, *15*(10 October), 1–15. https://doi.org/10.1371/journal.pone.0241434.

Meruelo, A. D., Castro, N., Cota, C. I., & Tapert, S. F. (2017). Cannabis and alcohol use, and the developing brain. *Behavioural Brain Research*, *325*, 44–50. https://doi.org/10.1016/j.bbr.2017.02.025.

Miller, W. R. (1983). Motivational Interviewing with Problem Drinkers. *Behavioural Psychotherapy*, *11*, 147–172.

Miller, W. R., & Rollnick, S. (2004). Talking oneself into change: Motivational interviewing, stages of change, and therapeutic process. *Journal of Cognitive Psychotherapy*, *18*(4), 299–308. https://doi.org/10.1891/jcop.18.4.299.64003.

Miller, W. R., & Rollnick, S. (2012). Meeting in the middle: Motivational interviewing and self-determination theory. *International Journal of Behavioral Nutrition and Physical Activity*, *9*. https://doi.org/10.1186/1479-5868-9-25.

Miller, R., & Rollnick, S. (2013). *Motivational interviewing: Helping people change*. Guilford Press.

National Institute on Drug Abuse. (2018). *Principles of drug addiction treatment: A research-based guide* (3rd). Report NIDA. https://nida.nih.gov/sites/default/files/podat-3rdEd-508.pdf.

Olmos, A., Tirado-Muñoz, J., Farré, M., & Torrens, M. (2018). The efficacy of computerized interventions to reduce cannabis use: A systematic review and meta-analysis. *Addictive Behaviors*, *79*, 52–60. https://doi.org/10.1016/J.ADDBEH.2017.11.045.

Pelletier, J. H., Strout, T. D., & Baumann, M. R. (2014). A systematic review of smoking cessation interventions in the emergency setting. *American Journal of Emergency Medicine*, *32*(7), 713–724. https://doi.org/10.1016/j.ajem.2014.03.042.

Polcin, D. L., Bond, J., Korcha, R., Madhabika, B. N., Galloway, G. P., & Evans, K. (2014). Randomized trial of intensive motivational interviewing for methamphetamine dependence. *Journal of Addictive Disorders*, *33*(3), 253–265. https://doi.org/10.1080/10550887.2014.950029.Randomized.

Prochaska, J. O., & DiClemente, C. C. (1986). Toward a comprehensive model of change. In *Treating addictive behaviors: Processes of change* (pp. 3–27). Plenum Press. https://doi.org/10.1007/978-1-4613-2191-0_1.

Public Health England. (2015). *Improving mutual aid engagement: A professional development resource* (p. 2). https://www.drugsandalcohol.ie/24293/.

Quilty, L. C., Wardell, J. D., Thiruchselvam, T., Keough, M. T., & Hendershot, C. S. (2019). Brief interventions for problem gambling: A meta-analysis. *PLoS One*, *14*(4), 1–17. https://doi.org/10.1371/journal.pone.0214502.

Rollnick, S., Butler, C. C., Kinnersley, P., Gregory, J., & Mash, B. (2010). Motivational interviewing. *BMJ (Online)*, *340*(7758), 1242–1244. https://doi.org/10.1136/bmj.c1900.

Rollnick, S., & Miller, W. R. (1995). What is motivational interviewing? *Behavioral and Cognitive Psychotherapy*, *23*(1), 325–334.

Rubak, S., Sandbæk, A., Lauritzen, T., & Christensen, B. (2005). Motivational interviewing: A systematic review and meta-analysis. *British Journal of General Practice*, *55*(513), 305–312.

Schaub, M. P., Berman, A. H., Pelayo, H. L., Boumparis, N., Khadjesari, Z., Blankers, M., Gual, A., Riper, H., & Pas, L. (2020). E-INEBRIA special interest group roadmap for best practices for research on brief digital interventions for problematic alcohol and illicit drug use. *Journal of Medical Internet Research*, *22*(8), 1–4. https://doi.org/10.2196/20368.

Schneegans, A., Bourgognon, F., Albuisson, E., Schwan, R., Arfa, M., Polli, L., Moulard, M., Laprévote, V., & Schwitzer, T. (2021). Mindfulness-based relapse prevention for cannabis regular users: Preliminary outcomes of a randomized clinical trial. *L'Encéphale*. https://doi.org/10.1016/J.ENCEP.2021.02.015.

Sipe, W. E. B., & Eisendrath, S. J. (2012). Mindfulness-based cognitive therapy: Theory and practice. *Canadian Journal of Psychiatry*, *57*(2), 63–69. https://doi.org/10.1177/070674371205700202.

Steele, D. W., Becker, S. J., Danko, K. J., Balk, E. M., Saldanha, I. J., Adam, G. P., Bagley, S. M., Friedman, C., Spirito, A., Scott, K., Ntzani, E. E., Saeed, I., Smith, B., Popp, J., & Trikalinos, T. A. (2020). *Interventions for substance use disorders in adolescents: A systematic review*. https://doi.org/10.23970/AHRQEPCCER225.

Stein, M. D., Herman, D. S., & Anderson, B. J. (2009). A motivational intervention trial to reduce cocaine use. *Journal of Substance Abuse Treatment, 36* (1), 118–125. https://doi.org/10.1016/j.jsat.2008.05.003.

Tait, R. J., Spijkerman, R., & Riper, H. (2013). Internet and computer based interventions for cannabis use: A meta-analysis. *Drug and Alcohol Dependence, 133*(2), 295–304. https://doi.org/10.1016/J.DRUGALCDEP.2013.05.012.

Tatar, O., Bastien, G., Abdel-Baki, A., Huỳnh, C., & Jutras-Aswad, D. (2020). A systematic review of technology-based psychotherapeutic interventions for decreasing cannabis use in patients with psychosis. *Psychiatry Research, 288*, 112940. https://doi.org/10.1016/J.PSYCHRES.2020.112940.

Vasilaki, E. I., Hosier, S. G., & Cox, W. M. (2006). The efficacy of motivational interviewing as a brief intervention for excessive drinking: A meta-analytic review. *Alcohol and Alcoholism (Oxford, Oxfordshire), 41*(3), 328–335. https://doi.org/10.1093/alcalc/agl016.

Velasquez, M. M., Von Sternberg, K., Dodrill, C. L., Kan, L. Y., & Parsons, J. T. (2005). The transtheoretical model as a framework for developing substance abuse interventions. *Journal of Addictions Nursing, 16*(1–2), 31–40. https://doi.org/10.1080/10884600590917174.

Winters, K. C., Mader, J., Budney, A. J., Stanger, C., Knapp, A. A., & Walker, D. D. (2021). Interventions for cannabis use disorder. *Current Opinion in Psychology, 38*, 67–74. https://doi.org/10.1016/J.COPSYC.2020.11.002.

Yakovenko, I., Quigley, L., Hemmelgarn, B. R., Hodgins, D. C., & Ronksley, P. (2015). The efficacy of motivational interviewing for disordered gambling: Systematic review and meta-analysis. *Addictive Behaviors, 43*, 72–82. https://doi.org/10.1016/j.addbeh.2014.12.011.

Chapter 43

An overview of current pharmacologic interventions for management of cannabis withdrawal syndrome and the evidence for their use

David Wolinsky and Ryan Vandrey
Department of Psychiatry and Behavioral Sciences, Johns Hopkins Bayview Medical Center, Behavioral Biology Research Center, Baltimore, MD, United States

Abbreviations

ADHD	attention deficit and hyperactivity disorder
AUD	alcohol use disorder
CB1	cannabinoid type 1
CBD	cannabidiol
CUD	cannabis use disorder
CWS	cannabis withdrawal syndrome
DSM-V	Diagnostic and Statistical Manual of Mental Disorders, Fifth Edition
ECS	endocannabinoid system
FAAH	fatty acid amide hydrolase
GABA	gamma-aminobutyric acid
MAGL	monoacylglycerol lipase
RCT	randomized control trial
SUDs	substance use disorders
THC	delta-9-tetrahydrocannabinol

Introduction

Cannabis is a popular drug of abuse globally. Though cannabis and some of its chemical constituents have been developed as medications to help treat a wide array of health conditions, including seizure disorders (Gaston & Szaflarski, 2018) and nausea/emesis (Machado Rocha et al., 2008), its rate of use for nonmedicinal purposes ranks behind only alcohol and nicotine/tobacco in most developed countries (UNODC, 2021). Potential health risks associated with frequent cannabis use include liability for misuse (SAMHSA, 2019), risk of developing psychosis (Marconi et al., 2016), and a withdrawal syndrome seen in well over 50% of individuals with cannabis use disorder (CUD) who try to quit use (Budney et al., 2004). The symptoms that comprise cannabis withdrawal syndrome (CWS) yield a level of discomfort that rises to such a degree that there is evidence to suggest that CWS can drive CUD. (Chung et al., 2008; Copersino et al., 2006; Crowley et al., 1998; Levin et al., 2010). Management of CWS may therefore not only reduce the distress and impairment arising when one attempts to quit cannabis but may also attenuate the incidence of CUD.

Numerous pharmacologic interventions for CWS treatment have been investigated and tested in research studies involving human subjects; these treatments will be the focus of this chapter. Prior to this discussion, we will provide an overview of the endocannabinoid system and the proposed neurobiological adaptations underlying cannabis withdrawal followed by the symptoms constituting CWS and pertinent epidemiology. The chapter will conclude with a look at the medications and supplements that hold promise for management of CWS but have not yet been studied for this purpose.

Cannabis Use, Neurobiology, Psychology, and Treatment. https://doi.org/10.1016/B978-0-323-89862-1.00037-4

CWS: An overview

Neurobiological mechanism

Cannabis' effects are mediated by the endocannabinoid system (ECS), which consists of endogenous cannabinoid ligands that affect cannabinoid receptors to influence a variety of biological functions. Cannabinoid receptors can be found throughout the body, with the cannabinoid type 1 (CB1) receptor representing the most prominent cannabinoid receptor in the central nervous system (CNS). CB1 receptors regulate pain perception, memory, emotional regulation, appetite, and other processes mediated by CNS structures such as the hippocampus and amygdala predominantly through inhibition of the release of glutamate or gamma-Aminobutyric acid (GABA). Anandamide and 2-arachidonoyl glycerol (2AG)—the two most well-studied endocannabinoid ligands—can both act as CB1 agonists. Anandamide is degraded by the endogenous enzyme fatty acid amide hydrolase (FAAH) and 2AG is degraded by the endogenous enzyme monoacylglycerol lipase (MAGL) (Di Marzo et al., 2004; Lu & Mackie, 2016; Mackie, 2006). Both enzymes will be discussed later as possible targets for CWS.

Delta-9-tetrahydrocannabinol (THC), the most prevalent cannabinoid found in most cannabis plants, acts as a partial agonist at the CB1 receptor in a manner similar to anandamide (Lu & Mackie, 2016; Pertwee, 2008). Long-term daily use of cannabis containing THC can result in tolerance to the acute effects of the drug. The neurobiological mechanism that most likely mediates cannabis tolerance is CB1 receptor downregulation, which is reversed with abstinence (D'Souza et al., 2016; Hirvonen et al., 2012; Spindle et al., 2021).

It has been hypothesized that this downregulation of available CB1 receptors facilitates CWS once cannabis use is halted. Neuroimaging studies of individuals with CUD have consistently shown a link between heavy cannabis use and lower density of CB1 receptors, yet findings of a relation between reduced CB1 receptor presence and more severe cannabis withdrawal symptoms have been mixed. A 2012 fMRI study of the brains of male daily cannabis users confirmed that active users did have fewer CB1 receptors in cortical brain structures including the hippocampus and cerebellum compared to control subjects, but there was no association found between the number of CB1 receptors and the severity of withdrawal (Hirvonen et al., 2012). However, a similar imaging study of male adults with CUD from 2016 found a negative correlation between CB1 receptor availability and withdrawal symptom severity (D'Souza et al., 2016). A 2021 brain imaging study of females with CUD again identified decreased density of CB1 receptors in neurologic structures compared to female controls; CB1 receptor density was negatively correlated with increased Anger/Hostility during abstinence, but no other CWS symptoms (Spindle et al., 2021).

One limitation of all three studies was that withdrawal was measured in nontreatment-seeking individuals who volunteered to temporarily stop cannabis use while on a residential research unit. Past research has shown withdrawal severity in these settings—when cannabis cessation is time limited—tends to be less severe compared to those in an outpatient environment, when individuals are seeking to stop use indefinitely (Budney et al., 2004). This effect may relate to exposure to drug-related environmental cues as well as participant bias because those likely to experience severe withdrawal are less likely to volunteer and/or are less likely to complete the study and provide evaluable data.

Symptoms

CWS consists of symptoms including irritability, anger/aggression, insomnia, intense/vivid dreams, appetite decline, restlessness along with physical symptoms including abdominal distress, fever, chills, headache, sweating and tremulousness (Bonnet & Preuss, 2017; Schlienz & Vandrey, 2019). Symptoms typically peak within 2–6 days after last use of cannabis and mostly resolve within 2 weeks, though some symptoms persist for several weeks to months (Budney et al., 2003).

In individuals with CUD who attempt to abstain from cannabis, these symptoms can produce a level of distress that is intense enough to impair functioning (Allsop et al., 2012) and be observable to others (Budney et al., 2001). A retrospective study of hundreds of individuals with CUD attempting to quit use found that approximately 70% of those who did experience withdrawal eventually used cannabis to resolve their symptoms, illustrating a strong link between CWS and relapse onto cannabis use (Levin et al., 2010). CWS was included as a distinct disorder in the fifth edition of the Diagnostic and Statistical Manual of Mental Disorders (DSM-5); per this volume, patients must have at least three of the aforementioned symptoms (including one physical symptom) occur within a week of stopping marijuana for criteria to be met (American Psychiatric Association, 2013) (see Table 1 for further details).

Risk factors

Longstanding daily or near daily cannabis use naturally acts as the risk factor with the greatest predictive value for the presence and severity of CWS, yet additional characteristics have also been found to influence withdrawal severity.

TABLE 1 DSM-5 criteria for cannabis withdrawal syndrome (American Psychiatric Association, 2013).

A	**Cessation of cannabis use that has been heavy and prolonged (i.e., usually daily or almost daily use over a period of at least a few months)**
B	Three (or more) of the following signs and symptoms develop within approximately 1 week after criterion A: (1) Irritability, anger, or aggression (2) Nervousness or anxiety (3) Sleep difficulty (e.g., insomnia, disturbing dreams) (4) Decreased appetite or weight loss (5) Restlessness (6) Depressed mood (7) At least one of the following physical symptoms causing significant discomfort: abdominal pain, shakiness/tremors, sweating, fever, chills, or headache
C	The signs or symptoms in Criterion B cause clinically significant distress or impairment in social, occupational, or other important areas of functioning
D	The signs of symptoms are not attributable to another medical condition and are not better explained by another mental disorder, including intoxication or withdrawal from another substance.

Adolescents with cannabis dependence experienced milder withdrawal compared to adult individuals with cannabis dependence (Vandrey et al., 2005), while female sex (Herrmann et al., 2015) and family history of alcohol use disorder (AUD) or substance use disorders (SUDs) predicted a more severe withdrawal syndrome (Agrawal et al., 2008). Psychiatric symptoms were also linked to more severe withdrawal symptoms in a few studies (Budney et al., 1999; Cousijn & van Duijvenvoorde, 2018), although a 2017 study found that differences in CWS between those with psychiatric illness and those without psychiatric illness become insignificant after the first week of withdrawal (Schuster et al., 2017). A retrospective cohort study of adolescents in a juvenile detention facility found that Black adolescents reported significantly milder withdrawal symptoms than adolescents of different races (Soenksen et al., 2015), though no such difference was seen in a separate study (Agrawal et al., 2008).

Concurrent use of other substances may also impact CWS severity. Heavy tobacco smokers were significantly more irritable during cannabis withdrawal than non-tobacco smokers (Soenksen et al., 2015). Opioid use worsened the sleep disturbances of CWS in one retrospective study (Vorspan et al., 2010) while Opioid Dependence was not associated with significant differences in CWS in a different study (Chauchard et al., 2014). A 2016 review article exploring the role of the ECS in the development of AUD and alcohol withdrawal indicated that chronic alcohol use can increase endocannabinoid levels and subsequently downregulate CB1 receptors (Henderson-Redmond et al., 2016). Though this process resembles the proposed mechanism underlying CWS mentioned earlier, alcohol's role on CWS severity remains unclear.

CWS treatments

A variety of medications have been evaluated as potential CWS treatments. Research approaches have included outpatient randomized control trials (RCTs) with treatment-seeking volunteers, outpatient laboratory studies with nontreatment-seeking volunteers, and residential laboratory studies with nontreatment-seeking volunteers. The inpatient and outpatient laboratory models have been carefully developed to rapidly screen medications prior to evaluation in larger, more expensive RCTs. The predictive validity of the laboratory models for reducing CWS appears strong. However, in many trials, medications that reduce CWS do not significantly increase abstinence rates compared with placebo.

Cannabinoid agonists

As might be expected considering CWS is defined by an acute deficiency in cannabinoid activity, cannabinoid agonists have some of the strongest evidence for ameliorating CWS. The agonist approach to withdrawal suppression is similar to the use of nicotine replacement products (e.g., nicotine patch or nicotine gum) for smoking cessation and opioid maintenance therapy (e.g., methadone or buprenorphine) for Opioid Use Disorder. Dronabinol—a gelcap containing synthetic THC suspended in sesame oil and indicated for appetite stimulation and management of chemotherapy-induced nausea—dose-dependently reduced withdrawal symptoms (Budney et al., 2007; Haney et al., 2004; Levin et al., 2011; Schlienz et al., 2018), without producing intoxication in doses ranging from 20mg to 120mg per day (Vandrey et al., 2013).

Nabilone, an oral THC analog with indications for nausea treatment (Ware et al., 2008), also exhibits evidence for dose-dependent suppression of measured CWS symptoms—including irritability, poor sleep, and low appetite—when dosed at 8 mg a day (Haney et al., 2013). Nabilone's capacity for reversing the appetite loss and mood disruptions seen in CWS was enhanced when it was co-administered with the hypnotic agent zolpidem (Herrmann et al., 2016).

Unlike the synthetic cannabinoids dronabinol and nabilone, nabiximols is a botanically derived pharmaceutical that contains both THC and another phytocannabinoid, cannabidiol (CBD). Nabiximols is indicated in many countries for the treatment of multiple sclerosis-related spasticity. Nabiximols (dosed at 86.4 mg THC and 80 mg CBD) significantly reduced overall CWS in one inpatient RCT (Allsop et al., 2014), but no significant effects were seen in a later RCT investigating nabiximols as a treatment for CUD (Trigo et al., 2018).

Isolated CBD has not undergone the same degree of investigation as a treatment for CWS compared with THC. It could be hypothesized that CBD will have limited therapeutic effects for CWS because it has poor affinity for the CB1 receptor and acts primarily as a negative allosteric modulator of CB1 (Chung et al., 2008). That said, CBD can inhibit FAAH in high concentrations, which could increase anandamide levels and perhaps attenuate CWS. One published case report described an individual with cannabis dependence experiencing good control of CWS with CBD treatment (Crippa et al., 2013), but more rigorous studies are needed before any conclusions could be made regarding the efficacy of CBD to treat CWS (Table 2).

Inhibition of endocannabinoid catabolism

A second pharmacologic approach to the treatment of CWS that targets the ECS is increasing endogenous cannabinoid tone by inhibiting the catabolism of anandamide or 2AG. Catabolic enzyme inhibition could functionally act in a similar manner as administration of exogenous agonist or partial agonist medications such as dronabinol. Preclinical studies have shown that administration of the FAAH inhibitor URB597 (which increases anandamide levels) or the MAGL inhibitor JZL184 (which increases 2AG levels) attenuated precipitated withdrawal symptoms in mice chronically maintained on THC (Schlosburg et al., 2009).

Following this preclinical study, a recent RCT found that individuals with CUD who were chronically administered the FAAH inhibitor PF-04457845 (4 mg daily) during a 1-week period of supervised abstinence in a residential setting had reduced ratings of depression and anxiety, as well as improvements in subjective sleep quality and less alteration of sleep architecture compared to placebo (D'Souza et al., 2019). This study represents the only clinical study involving a FAAH inhibitor and no studies have been completed with a MAGL inhibitor, so the capability of inhibitors of endocannabinoid catabolism to treatment CWS remains an area for future research (Table 2).

SUDs treatments

Though currently no medications have FDA approval for the treatment of CUD, there do exist such treatments for other SUDs, including Tobacco Use Disorder and AUD. Nicotine patches have evidence for treating nicotine dependence, and daily administration of 7 mg patches reduced withdrawal-related negative affect in cannabis-dependent individuals who either did not use nicotine or only used nicotine occasionally (heavy smokers were not studied). No other measured withdrawal symptoms were impacted (Gilbert et al., 2020).

Alpha-agonists

Alpha-2 adrenergic agonists represent a class of medications used for managing an array of conditions, including hypertension and ADHD, via inhibition of CNS activation. Lofexidine, an alpha-agonist that received FDA approval for the management of opioid withdrawal symptoms in 2018, has been studied for its capacity to treat cannabis withdrawal. One study investigated lofexidine dosed at 2.4 mg daily and a combination of lofexidine 2.4 mg daily and THC 60 mg daily for CWS treatment. Lofexidine monotherapy significantly reduced chills and restlessness while improving subjective reports of sleep quality, but it also significantly worsened withdrawal-related anorexia and did not differ from placebo in other measured CWS symptoms, such as irritability. The combination of THC and lofexidine demonstrated similar results but unlike lofexidine alone, this combination was able to improve psychomotor task performance and marijuana craving compared to placebo. It also did not significantly worsen the loss of appetite seen in CWS compared to placebo (Haney et al., 2008).

The combination of lofexidine and THC was investigated again in a larger double-blind RCT in 2016, when subjects received dronabinol 20 mg three times daily and either placebo or lofexidine three times daily over 8 weeks. CWS was

TABLE 2 CWS interventions: Cannabinoid agonists and FAAH inhibitors.

Medication	Class	Results	Other comments
Dronabinol	Cannabinoid agonist	• Budney et al. (2007): Human laboratory study in which subjects received placebo, dronabinol 10mg three times a day, or dronabinol 30mg three times a day→dronabinol 10mg three times a day improved total withdrawal discomfort, irritability, aggression, and sleep difficulty), dronabinol 30mg three times a day improves these symptoms as well as depressed mood and cravings. • Haney et al. (2004): Residential lab study in which subjects received placebo or oral THC 50mg daily→oral THC improved withdrawal-related anxiety, chills, misery, loss of appetite, and cannabis cravings. • Levin et al. (2011): RCT in which subjects received oral dronabinol 20mg twice a day or placebo→significant reduction in withdrawal symptoms. • Schlienz et al. (2018): Residential lab study with placebo, oral dronabinol at 120mg a day, or oral dronabinol at 180–240mg a day→low dose and high dose dronabinol significantly reduced withdrawal symptoms, also reduced cannabis use. • Vandrey et al. (2013): Residential lab study with placebo or oral dronabinol at 30mg, 60mg, or 120mg daily→significant reduction of total withdrawal at 60 and 120mg doses.	• Intoxication not observed even at higher doses (Vandrey et al., 2013)
Nabilone	Cannabinoid agonist	• Haney et al. (2013): Residential lab study with placebo and nabilone at 0, 6, and 8mg a day→significant reduction in withdrawal-related irritability and sleep effects.	
Nabiximols	Cannabinoid agonist	• Allsop et al., 2014: RCT involving placebo and nabiximols buccal spray with max doses of 86.4mg THC and 80mg CBD→significant reduction in overall withdrawal; no effect on cessation. • Trigo et al. (2018): RCT with placebo or nabiximols up to 113.4mg THC/105mg CBD→no effects on withdrawal.	
PF-04457845	FAAH inhibitor	• D'Souza et al. (2019): RCT with placebo or 4mg capsule of PF-0447845→ significant reduction in withdrawal symptoms (including depression and irritability), improved subjective report of sleep quality, and optimization of slow-wave sleep during cannabis abstinence	Use of PF-04457845 also associated with reduced cannabis use after 4weeks and no serious adverse effects compared to placebo (D'Souza et al., 2019)

measured via weekly completion of a cannabis withdrawal checklist. The combination of lofexidine and dronabinol was not found to yield a significant benefit compared to placebo with respect to treatment of CWS (Levin et al., 2016).

Guanfacine, an alpha-agonist used for managing hypertension and ADHD, was administered to cannabis-dependent subjects at a dose of 2mg nightly to test its effects on CWS symptoms including irritability, poor sleep, and loss of appetite. Guanfacine improved CWS-related irritability, with study authors pointing out that this is the first non-cannabinoid agonist to do so. Sleep latency was also reduced with guanfacine compared to placebo, though subjective sleep effects in CWS did not differ between placebo and guanfacine. Guanfacine had no effects on CWS-related loss of appetite (Haney et al., 2019) (Table 3).

TABLE 3 CWS interventions: $Alpha_2$ agonists.

Medication	Class	Results	Other comments
Guanfacine	$Alpha_2$ agonist	• Haney et al. (2010): Residential lab study with placebo or guanfacine 2 mg capsule given at night → significant improvements in withdrawal-related irritability and sleep disruptions.	
Lofexidine	$Alpha_2$ agonist	• Haney et al. (2008): Residential lab study with placebo, oral THC capsules (60 mg daily), oral lofexidine capsules (2.4 mg daily), or a combination of THC and lofexidine → lofexidine improved restlessness, sleep, and chills but no other withdrawal symptoms and worsened withdrawal-related anorexia. • Levin et al. (2016): RCT with placebo or combination of dronabinol 20 mg and lofexidine 0.6 mg three times a day—no effects on global withdrawal symptoms or cannabis cessation.	

Antidepressants and antipsychotics

Several antidepressants have been tested as possible treatments for CWS, yet they have largely failed to demonstrate a significant effect on symptoms. Venlafaxine titrated up to 225 mg daily not only failed to attenuate CWS but in fact appeared to worsen withdrawal symptoms (Kelly et al., 2014). Bupropion 300 mg daily worsened withdrawal-related depression and amotivation compared to placebo in one study (Haney et al., 2001). A later randomized study focusing on bupropion 150 mg twice a day did not find that bupropion worsened withdrawal symptoms but also did not identify significant treatment benefits. Authors of the latter study hypothesized that their focus on treatment-seeking subjects akin to those who use bupropion to treat Tobacco Use Disorder mediated the differences from the former study, which involved nontreatment-seeking individuals in an inpatient study (Penetar et al., 2012).

Despite a preclinical trial demonstrating lithium's capacity to control CWS (Cui et al., 2001), lithium dosed at 500 mg twice a day did not significantly treat overall CWS compared to placebo in an RCT despite reducing certain symptoms, including anorexia and nightmares (Johnston et al., 2014). Mirtazapine dosed at 30 mg daily likewise improved poor appetite and insomnia in subjects with CWS without yielding a significant improvement in mood symptoms or cannabis cravings (Haney et al., 2010). The antipsychotic quetiapine, titrated up to 150 mg daily, improved insomnia, loss of appetite and abdominal pain seen in CWS but did not improve CWS-related mood symptoms and was actually linked to increased cannabis cravings (Cooper et al., 2013). Nefazodone 450 mg daily decreased anxiety and muscle pain but did not yield a difference in the other measured CWS symptoms, including irritability and sleep quality, compared to placebo (Haney et al., 2003) (Table 4).

Anticonvulsants

Divalproex—an anticonvulsant frequently used as a mood stabilizer—reduced cannabis cravings at a dose of 1500 mg daily. However, it also worsened several measures related to CWS, including low mood and insomnia (Haney et al., 2004).

More promising results were observed with gabapentin, a GABA-ergic agent, that has previously been shown to have benefits for treating alcohol withdrawal (Myrick et al., 2009). One double-blind RCT found that a daily dose of 1200 mg of gabapentin was associated with significantly decreased global CWS severity compared to placebo (Mason et al., 2012). (For more information on GABA, please see the "Key Facts of GABA" section included in this chapter) (Table 5).

Muscle relaxants and sedatives

Another medication with GABA-ergic properties that has been studied as a CWS treatment option is baclofen, a muscle relaxant with evidence for treating opioid withdrawal symptoms (Akhondzadeh et al., 2000). Baclofen use was not associated with any significant reduction in withdrawal symptoms, and it worsened cognitive performance and early morning waking. The authors did point out that CWS occurred so infrequently in both the active and control groups that it was difficult to make a firm assessment of baclofen's capacity to treat withdrawal (Haney et al., 2010).

As mentioned earlier, zolpidem helped improve withdrawal-related mood and sleep symptoms when combined with nabilone in one study over zolpidem alone in a 2016 placebo-controlled study (Herrmann et al., 2016). Zolpidem also

TABLE 4 CWS interventions: Antidepressants and antipsychotics.

Medication	Class	Results	Other comments
Bupropion	Antidepressant	• Haney et al. (2001): Residential laboratory study with placebo or oral bupropion 300mg daily → Increased irritability, depression, restlessness, and insomnia during withdrawal. • Penetar et al. (2012): Double blind placebo-controlled trial with placebo or bupropion 150mg twice a day → less withdrawal discomfort with bupropion.	
Lithium	Antidepressant/ mood stabilizer	• Johnston et al. (2014): RCT in which participants received either placebo or oral lithium 500mg twice a day → significant reduction in strange dreams, anorexia, and stomach pain during abstinence but otherwise no significant effects on CWS (Johnston et al., 2014)	
Mirtazapine	Antidepressant	• Haney et al. (2010): Residential lab study with placebo or mirtazapine 30mg daily → mirtazapine improved appetite and sleep during abstinence but otherwise no significant effects on withdrawal symptoms (Haney et al., 2010)	
Nefazodone	Antidepressant	Haney et al. (2003): Residential lab study with placebo or oral nefazodone 450mg daily → significantly decreased anxiety and muscle pain during abstinence but no other effects on withdrawal symptoms	
Quetiapine	Antipsychotic	Cooper et al. (2013): Residential lab study with placebo vs. placebo 200mg/day of quetiapine → significant improvement in low appetite and sleep disruptions seen in withdrawal.	Same study found a significant association between quetiapine use and marijuana cravings, as well as active use and money spent on marijuana
Venlafaxine	Antidepressant	Kelly et al. (2014): Secondary analysis of an RCT focused on venlafaxine's effects on cannabis use with participants receiving placebo vs oral venlafaxine up to 375mg daily → significantly worse withdrawal symptoms with venlafaxine, including shakiness, sleep difficulty, nervousness and nausea.	

TABLE 5 CWS interventions: Anticonvulsants.

Medication	Class	Results	Other comments
Divalproex	Anticonvulsant	• Haney et al. (2003): Residential lab study of placebo vs. oral divalproex titrated to 1500mg daily → worsened mood, anxiety, and subjective sleep quality related to withdrawal.	
Gabapentin	Anticonvulsant	• Mason et al. (2012): RCT with placebo vs oral gabapentin 1200mg daily → significant improvement in overall acute withdrawal symptoms as well as sleep disruption and depression during abstinence.	• Gabapentin also associated with less marijuana use (Mason et al., 2012)

yielded a significant improvement in sleep efficiency and reduced sleep latency in this study. A 2011 study placebo-controlled crossover study likewise found that zolpidem attenuated withdrawal-mediated effects on REM sleep and improved subjective sleep quality but saw no impact on sleep latency (Vandrey et al., 2011). Notably, zolpidem alone did not produce subjective effects associated with abuse but when combined with nabilone there was evidence of abuse liability. Thus, care must be taken to underscore the increased abuse potential of this medication when used in combination with cannabis or other cannabinoids.

The benzodiazepine nitrazepam has also been found to improve withdrawal-related sleep problems. A 2015 inpatient RCT focused on lithium's role in managing sleep symptoms among patients experiencing CWS, though nitrazepam could be requested on any three of the seven nights in which the study took place. Though lithium did not reduce CWS symptoms in this study, nitrazepam significantly improved sleep fragmentation and sleep efficiency on nights it was taken compared with nights it was not taken by a subset of study participants (Allsop et al., 2015). It should be noted that nitrazepam was not studied independently of lithium use in this study. Moreover, it is not clear if any such benefits would be maintained over time as individuals frequently develop tolerance to the sedative effects of benzodiazepines (File, 1985) (Table 6).

Future directions

In addition to the pharmaceuticals mentioned above with potential benefit in the treatment of CWS—benefits that would need to be confirmed in larger controlled studies for many of these interventions—there are medications that have demonstrated some capacity to treat CUD but whose effects on CWS have yet to be explored. Topiramate is one such example. A carbonic anhydrase inhibitor with evidence for the treatment of AUD (Kranzler et al., 2014) and Cocaine Use Disorder (Johnson et al., 2013), topiramate at 200mg daily reduced the amount of cannabis smoked among adolescents with CUD in one placebo-controlled RCT (Miranda Jr et al., 2017). *N*-acetylcysteine, a supplement typically used to treat acetaminophen overdose, was likewise associated with significantly reduced cannabis use (based on negative urine drug screens) in adolescents with CUD when it was dosed at 1200mg daily (Gray et al., 2012). Impacts on CWS were not explored in either of these trials and would need to be included in subsequent studies.

Progesterone—a hormone critical for ovulation—was evaluated as a potential treatment for CWS in a small placebo-controlled RCT in which either progesterone 200mg daily or placebo was administered to women with CUD during the follicular phase of their menstrual cycles. Subjects were then instructed to abstain from cannabis for 5 days. Despite the small sample size ($n=8$), progesterone use was associated with a nonsignificant reduction in cannabis craving. Based on this result, study authors suggested that progesterone could potentially manage CWS and called for larger clinical trials to verify this hypothesis (Sherman et al., 2019).

TABLE 6 CWS interventions: Miscellaneous.

Medication	Class	Results	Other comments
Baclofen	Muscle relaxant	• Haney et al. (2010): Residential lab study with placebo or baclofen either 60 or 90mg daily doses—baclofen worsened cognitive performance and one measure of sleep during withdrawal.	
Nicotine patch	Nicotine replacement therapy	• Gilbert et al. (2020): RCT involving placebo patch or 7mg nicotine patch administered daily→nicotine patch associated with significant decrease in withdrawal related affect; increased nausea in treatment group.	• Heavy smokers not included in study
Zolpidem	Sedative	• Vandrey et al. (2011): Residential laboratory study with placebo vs oral zolpidem 12.5mg at night→zolpidem significantly improved subjective sleep quality and attenuated withdrawal's effects on sleep.	Combination of zolpidem and nabilone attenuated withdrawal-related negative mood and low appetite (Herrmann et al., 2016)

Conclusions

CWS is a cluster of symptoms including irritability, anger/aggression, insomnia, intense/vivid dreams, appetite decline, restlessness along with physical symptoms including abdominal distress, fever, chills, headache, sweating, and tremulousness arising when individuals with CUD abruptly stop their use. It is suspected that CWS is mediated by the down-regulation of CB1 receptors in the CNS that occurs after chronic, daily, or near daily exposure to THC.

With this proposed mechanism in mind, it is not surprising that cannabinoid agonists such as dronabinol and nabilone possess a particularly robust evidence base for the management of CWS. Gabapentin and FAAH inhibitor designated as PF-04457845 likewise demonstrated efficacy in the management of global CWS symptoms. Several medications—including guanfacine, quetiapine, mirtazapine, zolpidem, and nicotine patches—improved specific CWS symptoms without impacting others, and their effects on global CWS remain unknown.

It bears noting that none of these medications have such a robust evidence base that they should be used as standard treatment of cannabis withdrawal. Additionally, even though there is strong evidence to show that cannabis agonists effectively manage CWS, its clinical utility compared to other forms of substitution therapy (i.e., nicotine replace therapy or methadone and buprenorphine for OUD) remains questionable. Future studies need to be performed both for verifying the treatment effects seen in these trials and for exploring the CWS-treatment capacity of medications and supplements that have already shown evidence for managing CUD. Until then, these are options that could be kept in mind for patients who are struggling with debilitating and intractable withdrawal symptoms that are facilitating cannabis use and CUD.

Applications to other areas

In this chapter, we have discussed CWS and several types of medications that could serve as treatments. Many of these medications already have FDA approval as treatments of other symptoms and illnesses, but FAAH inhibitors represent a novel class of agents that so far lack such a distinction. Yet FAAH inhibitors have been—and continue to be—explored for their capacity to treat conditions aside from CWS. Results from human trials indicate the potential of FAAH inhibitors to treat psychiatric conditions Social Anxiety Disorder (Schmidt et al., 2021) and Post-traumatic Stress Disorder (PTSD) (Mayo et al., 2020). Regarding its role as an analgesic, FAAH inhibitors did not significantly reduce osteoarthritis pain in one randomized clinical trial involving humans (Huggins et al., 2012), though there are other FAAH inhibitors in development that have shown promise in animal studies (Kiso et al., 2020). Between the possible clinical value of FAAH inhibitors and the medical roles for cannabinoid agonists such as dronabinol and Epidiolex, there could be a wealth of conceivable therapeutics arising from manipulation of the ECS.

Key facts of GABA

- Gamma-aminobutyric acid (GABA) represents the primary inhibitory neurotransmitter in adult humans, though there is evidence to suggest that it is primarily excitatory in developing human brains (Ben-Ari, 2002)
- Though it has some presence in different tissues (including the pancreas, where GABA is co-secreted with insulin to prevent the release of glucagon and subsequent increase in glucose (Rorsman et al., 1989)), GABA primarily exists within neurons.
- GABA receptors are composed of a $GABA_A$ unit and a $GABA_B$ unit. Benzodiazepines, barbiturates, and alcohol act on the $GABA_a$ receptor, while the muscle relaxants baclofen and sodium oxybate (GHB) act on the $GABA_B$ unit (Allen et al., 2021).
- Deficiencies in GABA production and/or signaling are associated with numerous neuropsychiatric illnesses, including epilepsy and anxiety disorders (Ting Wong et al., 2003). Benzodiazepines and barbiturates treat these conditions via their agonism of GABA receptors.
- Alcohol likewise acts on GABA receptors, and longstanding overuse of alcohol will yield a reduction of GABA receptors. This response also occurs with overuse of barbiturates and benzodiazepines. The withdrawal syndromes from these substances therefore arise from a deficiency of GABA and unopposed action of glutamate, the primary excitatory transmitter in mammals (Hoffman et al., 2021).
- Gabapentin is a GABA analog that has FDA approval for the treatment of seizures yet is frequently used off-label for the treatment of anxiety, neuropathic pain, and alcohol withdrawal. Despite its name, it does not bind at the GABA receptor or affect GABA transmission. Instead, gabapentin (as well as pregabalin, another gabapentinoid) binds at a subunit of voltage-dependent calcium channels, yielding inhibitory effects (Allen et al., 2021).

- The effects of cannabinoids are suspected to be mediated at least in part by interactions with the GABA system. CBD specifically has been shown to potentiate GABA activity via binding to $GABA_A$ receptors, which may explain potential anxiolytic and antiepileptic effects (Brunt & Bossong, 2020). THC, meanwhile, has been shown to decrease GABA levels in animal studies (Pistis et al., 2002)

Mini-dictionary of terms

1. Anandamide—endogenous cannabinoid found in animal tissues and some foods that affects pain control, anxiety, and appetite
2. Cannabidiol—phytocannabinoid found in the cannabis plant that has been explored for its clinical implications in treatment of symptoms including seizures, pain, and depression
3. Delta-9-tetrahydrocannabinol (THC)—phytocannabinoid highly prevalent in the cannabis plant that is a partial agonist at CB1 receptors and is thought to, among other functions, mediate most of cannabis' psychoactive effects
4. Dependence—physiologic state in which the body adapts to the presence of a substance and subsequently becomes reliant on it.
5. Fatty acid amide hydrolase (FAAH)—enzyme involved in the breakdown of anandamide
6. Withdrawal—symptoms that emerge after abrupt cessation of a substance on which one has developed dependence

Summary points

- Cannabis withdrawal syndrome (CWS) is a collection of symptoms (including irritability, abdominal distress, insomnia, and tremulousness) experienced by regular cannabis users within a week of abruptly stopping daily or near daily cannabis use.
- The proposed mechanism underlying CWS involves the downregulation of CB1 cannabinoid receptors that occurs with chronic exposure to THC.
- Although cannabis use inherently represents the most predictive risk factor for CWS, other elements that have been observed to correlate with more severe CWS included female sex, history of other substance use disorders, psychiatric symptoms, and tobacco use.
- Many medications have been studied as potential treatment agents for CWS and while there is currently no agent with FDA approval for treating CWS, many have demonstrated treatment effects.
- The cannabinoid agonists dronabinol and nabilone, anticonvulsant gabapentin, and fatty acid amide hydrolase (FAAH) inhibitor PF-04457845 were associated with significant treatment effects for overall CWS.
- The antidepressants mirtazapine and nefazodone, the antipsychotic quetiapine, alpha-agonist guanfacine, and sedative zolpidem all yielded significant benefits for at least one symptom seen in CWS.
- The benefits of using dronabinol or nabilone as agonist pharmacotherapies for reducing CWS or treating CUD is not as apparent as the benefits of nicotine replacement for smoking cessation or opioid substitution therapy for treating Opioid Use Disorder.
- Future studies need to be performed both for verifying the treatment effects seen in preliminary trials and for exploring the CWS-treatment capacity of medications and supplements that have already shown evidence for managing cannabis use disorder (CUD).

References

Agrawal, A., Pergadia, M. L., & Lynskey, M. T. (2008). Is there evidence for symptoms of cannabis withdrawal in the national epidemiologic survey of alcohol and related conditions? *American Journal on Addictions*, *17*(3), 199–208.

Akhondzadeh, S., Ahmadi-Abhari, S. A., Assadi, S. M., Shabestari, O. L., Kashani, A. R., & Farzanehgan, Z. M. (2000). Double-blind randomized controlled trial of baclofen vs. clonidine in the treatment of opiates withdrawal. *Journal of Clinical Pharmacy and Therapeutics*, *25*(5), 347–353.

Allen, M. J., Sabir, S., & Sharma, S. (2021). GABA receptor [Updated 2021 Feb 17]. In *StatPearls [Internet]*. Treasure Island (FL): StatPearls Publishing. Available from https://www.ncbi.nlm.nih.gov/books/NBK526124/.

Allsop, D. J., Bartlett, D. J., Johnston, J., Helliwell, D., Winstock, A., McGregor, I. S., & Lintzeris, N. (2015). The effects of lithium carbonate supplemented with nitrazepam on sleep disturbance during cannabis abstinence. *Journal of Clinical Sleep Medicine*, *11*(10), 1153–1162.

Allsop, D. J., Copeland, J., Lintzeris, N., Dunlop, A. J., Montebello, M., Sadler, C., & McGregor, I. S. (2014). Nabiximols as an agonist replacement therapy during cannabis withdrawal: A randomized clinical trial. *JAMA Psychiatry*, *71*(3), 281–291.

Allsop, D. J., Copeland, J., Norberg, M. M., Fu, S., Molnar, A., Lewis, J., & Budney, A. J. (2012). Quantifying the clinical significance of cannabis withdrawal. *PLoS ONE*, *7*(9), e44864. https://doi.org/10.1371/journal.pone.0044864.

American Psychiatric Association. (2013). *Diagnostic and statistical manual of mental disorders* (5th ed.). American Psychiatric Association. https://doi.org/10.1176/appi.books.9780890425596.

Ben-Ari, Y. (2002). Excitatory actions of gaba during development: The nature of the nurture. *Nature Reviews Neuroscience*, *3*(9), 728–739.

Bonnet, U., & Preuss, U. W. (2017). The cannabis withdrawal syndrome: Current insights. *Substance Abuse and Rehabilitation*, *8*, 9.

Brunt, T. M., & Bossong, M. G. (2020). The neuropharmacology of cannabinoid receptor ligands in central signaling pathways. *European Journal of Neuroscience*, *55*(4), 909–921.

Budney, A. J., Hughes, J. R., Moore, B. A., & Novy, P. L. (2001). Marijuana abstinence effects in marijuana smokers maintained in their home environment. *Archives of General Psychiatry*, *58*(10), 917–924.

Budney, A. J., Hughes, J. R., Moore, B. A., & Vandrey, R. (2004). Review of the validity and significance of cannabis withdrawal syndrome. *American Journal of Psychiatry*, *161*(11), 1967–1977.

Budney, A. J., Moore, B. A., Vandrey, R. G., & Hughes, J. R. (2003). The time course and significance of cannabis withdrawal. *Journal of Abnormal Psychology*, *112*(3), 393.

Budney, A. J., Novy, P. L., & Hughes, J. R. (1999). Marijuana withdrawal among adults seeking treatment for marijuana dependence. *Addiction*, *94*(9), 1311–1322.

Budney, A. J., Vandrey, R. G., Hughes, J. R., Moore, B. A., & Bahrenburg, B. (2007). Oral delta-9-tetrahydrocannabinol suppresses cannabis withdrawal symptoms. *Drug and Alcohol Dependence*, *86*(1), 22–29.

Chauchard, E., Goncharov, O., Krupitsky, E., & Gorelick, D. A. (2014). Cannabis withdrawal in patients with and without opioid dependence. *Substance Abuse*, *35*(3), 230–234.

Chung, T., Martin, C. S., Cornelius, J. R., & Clark, D. B. (2008). Cannabis withdrawal predicts severity of cannabis involvement at 1-year follow-up among treated adolescents. *Addiction*, *103*(5), 787–799.

Cooper, Z. D., Foltin, R. W., Hart, C. L., Vosburg, S. K., Comer, S. D., & Haney, M. (2013). A human laboratory study investigating the effects of quetiapine on marijuana withdrawal and relapse in daily marijuana smokers. *Addiction Biology*, *18*(6), 993–1002.

Copersino, M. L., Boyd, S. J., Tashkin, D. P., Huestis, M. A., Heishman, S. J., Dermand, J. C., & Gorelick, D. A. (2006). Cannabis withdrawal among non-treatment-seeking adult cannabis users. *The American Journal on Addictions*, *15*(1), 8–14.

Cousijn, J., & van Duijvenvoorde, A. C. K. (2018). Cognitive and mental health predictors of withdrawal severity during an active attempt to cut down cannabis use. *Frontiers in Psychiatry*, *9*, 301.

Crippa, J. A. D. S., Hallak, J. E. C., Machado-de-Sousa, J. P., Queiroz, R. H. C., Bergamaschi, M., Chagas, M. H. N., & Zuardi, A. W. (2013). Cannabidiol for the treatment of cannabis withdrawal syndrome: A case report. *Journal of Clinical Pharmacy and Therapeutics*, *38*(2), 162–164.

Crowley, T. J., Macdonald, M. J., Whitmore, E. A., & Mikulich, S. K. (1998). Cannabis dependence, withdrawal, and reinforcing effects among adolescents with conduct symptoms and substance use disorders. *Drug and Alcohol Dependence*, *50*(1), 27–37.

Cui, S.-S., et al. (2001). Prevention of cannabinoid withdrawal syndrome by lithium: Involvement of oxytocinergic neuronal activation. *Journal of Neuroscience*, *21*(24), 9867–9876.

D'Souza, D. C., Cortes-Briones, J. A., Ranganathan, M., Thurnauer, H., Creatura, G., Surti, T., & Skosnik, P. D. (2016). Rapid changes in cannabinoid 1 receptor availability in cannabis-dependent male subjects after abstinence from cannabis. *Biological Psychiatry: Cognitive Neuroscience and Neuroimaging*, *1*(1), 60–67.

Di Marzo, V., Bifulco, M., & De Petrocellis, L. (2004). The endocannabinoid system and its therapeutic exploitation. *Nature Reviews Drug Discovery*, *3* (9), 771–784.

D'Souza, D. C., Cortes-Briones, J., Creatura, G., Bluez, G., Thurnauer, H., Deaso, E., & Skosnik, P. D. (2019). Efficacy and safety of a fatty acid amide hydrolase inhibitor (PF-04457845) in the treatment of cannabis withdrawal and dependence in men: A double-blind, placebo-controlled, parallel group, phase 2a single-site randomised controlled trial. *The Lancet Psychiatry*, *6*(1), 35–45.

File, S. E. (1985). Tolerance to the behavioral actions of benzodiazepines. *Neuroscience & Biobehavioral Reviews*, *9*(1), 113–121.

Gaston, T. E., & Szaflarski, J. P. (2018). Cannabis for the treatment of epilepsy: An update. *Current Neurology and Neuroscience Reports*, *18*(11), 1–9.

Gilbert, D. G., Rabinovich, N. E., & McDaniel, J. T. (2020). Nicotine patch for cannabis withdrawal symptom relief: A randomized controlled trial. *Psychopharmacology*, *237*(5), 1507–1519.

Gray, K. M., Carpenter, M. J., Baker, N. L., DeSantis, S. M., Kryway, E., Hartwell, K. J., & Brady, K. T. (2012). A double-blind randomized controlled trial of N-acetylcysteine in cannabis-dependent adolescents. *American Journal of Psychiatry*, *169*(8), 805–812.

Haney, M., Cooper, Z. D., Bedi, G., Herrmann, E., Comer, S. D., Reed, S. C., & Levin, F. R. (2019). Guanfacine decreases symptoms of cannabis withdrawal in daily cannabis smokers. *Addiction Biology*, *24*(4), 707–716.

Haney, M., Cooper, Z. D., Bedi, G., Vosburg, S. K., Comer, S. D., & Foltin, R. W. (2013). Nabilone decreases marijuana withdrawal and a laboratory measure of marijuana relapse. *Neuropsychopharmacology*, *38*(8), 1557–1565.

Haney, M., Hart, C. L., Vosburg, S. K., Comer, S. D., Reed, S. C., Cooper, Z. D., & Foltin, R. W. (2010). Effects of baclofen and mirtazapine on a laboratory model of marijuana withdrawal and relapse. *Psychopharmacology*, *211*(2), 233–244.

Haney, M., Hart, C. L., Vosburg, S. K., Comer, S. D., Reed, S. C., & Foltin, R. W. (2008). Effects of THC and lofexidine in a human laboratory model of marijuana withdrawal and relapse. *Psychopharmacology*, *197*(1), 157–168.

Haney, M., Hart, C. L., Vosburg, S. K., Nasser, J., Bennett, A., Zubaran, C., & Foltin, R. W. (2004). Marijuana withdrawal in humans: Effects of oral THC or divalproex. *Neuropsychopharmacology*, *29*(1), 158–170.

Haney, M., Hart, C. L., Ward, A. S., & Foltin, R. W. (2003). Nefazodone decreases anxiety during marijuana withdrawal in humans. *Psychopharmacology*, *165*(2), 157–165.

Haney, M., Ward, A. S., Comer, S. D., Hart, C. L., Foltin, R. W., & Fischman, M. W. (2001). Bupropion SR worsens mood during marijuana withdrawal in humans. *Psychopharmacology*, *155*(2), 171–179.

Henderson-Redmond, A. N., Guindon, J., & Morgan, D. J. (2016). Roles for the endocannabinoid system in ethanol-motivated behavior. *Progress in Neuro-Psychopharmacology and Biological Psychiatry*, *65*, 330–339.

Herrmann, E. S., Cooper, Z. D., Bedi, G., Ramesh, D., Reed, S. C., Comer, S. D., & Haney, M. (2016). Effects of zolpidem alone and in combination with nabilone on cannabis withdrawal and a laboratory model of relapse in cannabis users. *Psychopharmacology*, *233*(13), 2469–2478.

Herrmann, E. S., Weerts, E. M., & Vandrey, R. (2015). Sex differences in cannabis withdrawal symptoms among treatment-seeking cannabis users. *Experimental and Clinical Psychopharmacology*, *23*(6), 415.

Hirvonen, J., et al. (2012). Reversible and regionally selective downregulation of brain cannabinoid CB 1 receptors in chronic daily cannabis smokers. *Molecular Psychiatry*, *17*(6), 642–649.

Hoffman, R. S., Weinhouse, G. L., Traub, S. J., & Grayzel, J. (2021). *Management of moderate and severe alcohol withdrawal syndromes. UpToDate.* https://www-uptodate-com.proxy1.library.jhu.edu/contents/management-of-moderate-and-severe-alcohol-withdrawal-syndromes?search=management%20of%20alcohol%20withdrawal&source=search_result&selectedTitle=1~150&usage_type=default&display_rank=1 (Accessed January 28, 2022).

Huggins, J. P., Smart, T. S., Langman, S., Taylor, L., & Young, T. (2012). An efficient randomised, placebo-controlled clinical trial with the irreversible fatty acid amide hydrolase-1 inhibitor PF-04457845, which modulates endocannabinoids but fails to induce effective analgesia in patients with pain due to osteoarthritis of the knee. *Pain*, *153*(9), 1837–1846.

Johnson, B. A., Ait-Daoud, N., Wang, X. Q., Penberthy, J. K., Javors, M. A., Seneviratne, C., & Liu, L. (2013). Topiramate for the treatment of cocaine addiction: A randomized clinical trial. *JAMA Psychiatry*, *70*(12), 1338–1346.

Johnston, J., Lintzeris, N., Allsop, D. J., Suraev, A., Booth, J., Carson, D. S., & McGregor, I. S. (2014). Lithium carbonate in the management of cannabis withdrawal: A randomized placebo-controlled trial in an inpatient setting. *Psychopharmacology*, *231*(24), 4623–4636.

Kelly, M. A., Pavlicova, M., Glass, A., Mariani, J. J., Bisaga, A., Sullivan, M. A., & Levin, F. R. (2014). Do withdrawal-like symptoms mediate increased marijuana smoking in individuals treated with venlafaxine-XR? *Drug and Alcohol Dependence*, *144*, 42–46.

Kiso, T., Watabiki, T., & Sekizawa, T. (2020). ASP8477, a fatty acid amide hydrolase inhibitor, exerts analgesic effects in rat models of neuropathic and dysfunctional pain. *European Journal of Pharmacology*, *881*, 173194.

Kranzler, H. R., Covault, J., Feinn, R., Armeli, S., Tennen, H., Arias, A. J., & Kampman, K. M. (2014). Topiramate treatment for heavy drinkers: Moderation by a GRIK1 polymorphism. *American Journal of Psychiatry*, *171*(4), 445–452.

Levin, K. H., Copersino, M. L., Heishman, S. J., Liu, F., Kelly, D. L., Boggs, D. L., & Gorelick, D. A. (2010). Cannabis withdrawal symptoms in non-treatment-seeking adult cannabis smokers. *Drug and Alcohol Dependence*, *111*(1–2), 120–127.

Levin, F. R., Mariani, J. J., Brooks, D. J., Pavlicova, M., Cheng, W., & Nunes, E. V. (2011). Dronabinol for the treatment of cannabis dependence: A randomized, double-blind, placebo-controlled trial. *Drug and Alcohol Dependence*, *116*(1–3), 142–150.

Levin, F. R., Mariani, J. J., Pavlicova, M., Brooks, D., Glass, A., Mahony, A., & Choi, J. C. (2016). Dronabinol and lofexidine for cannabis use disorder: A randomized, double-blind, placebo-controlled trial. *Drug and Alcohol Dependence*, *159*, 53–60.

Lu, H. C., & Mackie, K. (2016). An introduction to the endogenous cannabinoid system. *Biological Psychiatry*, *79*(7), 516–525.

Machado Rocha, F. C., Stefano, S. C., De Cassia Haiek, R., Rosa Oliveira, L. M. Q., & Da Silveira, D. X. (2008). Therapeutic use of *Cannabis sativa* on chemotherapy-induced nausea and vomiting among cancer patients: Systematic review and meta-analysis. *European Journal of Cancer Care*, *17*(5), 431–443.

Mackie, K. (2006). Mechanisms of CB1 receptor signaling: Endocannabinoid modulation of synaptic strength. *International Journal of Obesity*, *30*(1), S19–S23.

Marconi, A., Di Forti, M., Lewis, C. M., Murray, R. M., & Vassos, E. (2016). Meta-analysis of the association between the level of cannabis use and risk of psychosis. *Schizophrenia Bulletin*, *42*(5), 1262–1269.

Mason, B. J., Crean, R., Goodell, V., Light, J. M., Quello, S., Shadan, F., & Rao, S. (2012). A proof-of-concept randomized controlled study of gabapentin: Effects on cannabis use, withdrawal and executive function deficits in cannabis-dependent adults. *Neuropsychopharmacology*, *37*(7), 1689–1698.

Mayo, L. M., Asratian, A., Lindé, J., Morena, M., Haataja, R., Hammar, V., & Heilig, M. (2020). Elevated anandamide, enhanced recall of fear extinction, and attenuated stress responses following inhibition of fatty acid amide hydrolase: A randomized, controlled experimental medicine trial. *Biological Psychiatry*, *87*(6), 538–547.

Miranda, R., Jr., Treloar, H., Blanchard, A., Justus, A., Monti, P. M., Chun, T., & Gwaltney, C. J. (2017). Topiramate and motivational enhancement therapy for cannabis use among youth: A randomized placebo-controlled pilot study. *Addiction Biology*, *22*(3), 779–790.

Myrick, H., Malcolm, R., Randall, P. K., Boyle, E., Anton, R. F., Becker, H. C., & Randall, C. L. (2009). A double-blind trial of gabapentin versus lorazepam in the treatment of alcohol withdrawal. *Alcoholism: Clinical and Experimental Research*, *33*(9), 1582–1588.

Penetar, D. M., Looby, A. R., Ryan, E. T., Maywalt, M. A., & Lukas, S. E. (2012). Bupropion reduces some of the symptoms of marihuana withdrawal in chronic marihuana users: A pilot study. *Substance Abuse: Research and Treatment*, *6*. SART-S9706.

Pertwee, R. (2008). The diverse CB1 and CB2 receptor pharmacology of three plant cannabinoids: Δ9-tetrahydrocannabinol, cannabidiol and Δ9-tetrahydrocannabivarin. *British Journal of Pharmacology*, *153*(2), 199–215.

Pistis, M., Ferraro, L., Pira, L., Flore, G., Tanganelli, S., Gessa, G. L., & Devoto, P. (2002). Δ9-tetrahydrocannabinol decreases extracellular GABA and increases extracellular glutamate and dopamine levels in the rat prefrontal cortex: An in vivo microdialysis study. *Brain Research*, *948*(1–2), 155–158.

Rorsman, P., Berggren, P. O., Bokvist, K., Ericson, H., Möhler, H., Östenson, C. G., & Smith, P. A. (1989). Glucose-inhibition of glucagon secretion involves activation of GABA A-receptor chloride channels. *Nature*, *341*(6239), 233–236.

SAMHSA. (2019). *Key substance use and mental health indicators in the United States: Results from the 2018 National Survey on Drug Use and Health*. Rockville, MD: Center for Behavioral Health Statistics and Quality. Substance Abuse and Mental Health Services Administration. HHS Publication No. PEP19–5068, NSDUH Series H-54. Retrieved from https://www.samhsa.gov/data.

Schlienz, N. J., Lee, D. C., Stitzer, M. L., & Vandrey, R. (2018). The effect of high-dose dronabinol (oral THC) maintenance on cannabis self-administration. *Drug and Alcohol Dependence*, *187*, 254–260.

Schlienz, N. J., & Vandrey, R. (2019). Cannabis withdrawal. In *Cannabis use disorders* (pp. 93–102). Cham: Springer.

Schlosburg, J. E., Carlson, B. L., Ramesh, D., Abdullah, R. A., Long, J. Z., Cravatt, B. F., & Lichtman, A. H. (2009). Inhibitors of endocannabinoid-metabolizing enzymes reduce precipitated withdrawal responses in THC-dependent mice. *The AAPS Journal*, *11*(2), 342–352.

Schmidt, M. E., Liebowitz, M. R., Stein, M. B., Grunfeld, J., Van Hove, I., Simmons, W. K., & Drevets, W. C. (2021). The effects of inhibition of fatty acid amide hydrolase (FAAH) by JNJ-42165279 in social anxiety disorder: A double-blind, randomized, placebo-controlled proof-of-concept study. *Neuropsychopharmacology*, *46*(5), 1004–1010.

Schuster, R. M., Fontaine, M., Nip, E., Zhang, H., Hanly, A., & Evins, A. E. (2017). Prolonged cannabis withdrawal in young adults with lifetime psychiatric illness. *Preventive Medicine*, *104*, 40–45.

Sherman, B. J., Caruso, M. A., & McRae-Clark, A. L. (2019). Exogenous progesterone for cannabis withdrawal in women: Feasibility trial of a novel multimodal methodology. *Pharmacology Biochemistry and Behavior*, *179*, 22–26.

Soenksen, S., Stein, L. A. R., Brown, J. D., Stengel, J. R., Rossi, J. S., & Lebeau, R. (2015). Cannabis withdrawal among detained adolescents: Exploring the impact of nicotine and race. *Journal of Child & Adolescent Substance Abuse*, *24*(2), 119–124.

Spindle, T. R., Kuwabara, H., Eversole, A., Nandi, A., Vandrey, R., Antoine, D. G., & Weerts, E. M. (2021). Brain imaging of cannabinoid type I (CB1) receptors in women with cannabis use disorder and male and female healthy controls. *Addiction Biology*, *26*(6), e13061.

Ting Wong, C., Guin, T. B., & Carter Snead, O., III. (2003). Gaba, γ-hydroxybutyric acid, and neurological disease. *Annals of Neurology: Official Journal of the American Neurological Association and the Child Neurology Society*, *54*(S6), S3–S12.

Trigo, J. M., Soliman, A., Quilty, L. C., Fischer, B., Rehm, J., Selby, P., & Le Foll, B. (2018). Nabiximols combined with motivational enhancement/cognitive behavioral therapy for the treatment of cannabis dependence: A pilot randomized clinical trial. *PLoS One*, *13*(1), e0190768.

UNODC. (2021). *World Drug Report*. United Nations Publication. Sales No. E.21.XI.8). Accessed January 2, 2022.

Vandrey, R., Budney, A. J., Kamon, J. L., & Stanger, C. (2005). Cannabis withdrawal in adolescent treatment seekers. *Drug and Alcohol Dependence*, *78*(2), 205–210.

Vandrey, R., Smith, M. T., McCann, U. D., Budney, A. J., & Curran, E. M. (2011). Sleep disturbance and the effects of extended-release zolpidem during cannabis withdrawal. *Drug and Alcohol Dependence*, *117*(1), 38–44.

Vandrey, R., Stitzer, M. L., Mintzer, M. Z., Huestis, M. A., Murray, J. A., & Lee, D. (2013). The dose effects of short-term dronabinol (oral THC) maintenance in daily cannabis users. *Drug and Alcohol Dependence*, *128*(1–2), 64–70.

Vorspan, F., Guillem, E., Bloch, V., Bellais, L., Sicot, R., Noble, F., & Gorelick, D. A. (2010). Self-reported sleep disturbances during Cannabis withdrawal in Cannabis-dependent outpatients with and without opioid-dependence. *Sleep Medicine*, *11*(5), 499.

Ware, M. A., Daeninck, P., & Maida, V. (2008). A review of nabilone in the treatment of chemotherapy-induced nausea and vomiting. *Therapeutics and Clinical Risk Management*, *4*(1), 99.

Chapter 44

Managing cannabinoid hyperemesis syndrome in pediatrics: A new narrative review

Justin P. Reinert[a,b], Abagail Barazi[c], Norman E. Fenn, III[d,e], Rachel Leis[c], and Alison Paplaskas[c]

[a]*The University of Toledo College of Pharmacy and Pharmaceutical Sciences, Toledo, OH, United States,* [b]*The University of Toledo Medical Center, Toledo, OH, United States,* [c]*Mercy Health St. Vincent Medical Center, Toledo, OH, United States,* [d]*Manchester University College of Pharmacy & Health Sciences, Fort Wayne, IN, United States,* [e]*Parkview Women's & Children's Hospital, Fort Wayne, IN, United States*

Abbreviations

5-HT	5-hydroxytryptophan
CB1	cannabidiol receptor 1
CB2	cannabidiol receptor 2
CHS	cannabinoid hyperemesis syndrome
CUD	cannabis-use disorder
CVS	cyclic vomiting syndrome
GABA	gamma-aminobutyric acid
HPA	hypothalamic-pituitary axis
TRVP-1	transient receptor potential vanilloid-1

Introduction

With its rapidly fluctuating legal status and established medicinal benefit in select patient populations, it has been postulated that the use of cannabis and cannabinoid related products, both recreationally and medicinally, is the highest in decades and perhaps, ever. Though many societal and health benefits may be derived from cannabis products, it is important for clinicians and the general public to recognize that its use is not without risk, regardless of the pedigree of the purveyor. One such risk is the development of cannabinoid hyperemesis syndrome (CHS), which may present as a type of cannabis use disorder (CUD) and is frequently associated with cyclic vomiting syndrome (CVS). While these syndromes share similar pathologies, it is the use of cannabinoid products that exclusively supports a diagnosis of CHS. This chapter will seek to define relevant pathophysiology and pharmacotherapy options for pediatric patients afflicted with CHS, with special emphasis on the safety and efficacy of medications in this population.

Background and epidemiology

The Center for Disease Control and Prevention (CDC) estimates that over 48 million Americans have used marijuana in the year 2019, being the most commonly used illicit substance under federal law in the United States (Centers for Disease Control and Prevention, 2021). Approximately 3 in 10 users have cannabis use disorder (CUD), defined in Table 1 (Hasin, 2017). The National Epidemiologic Survey on Alcohol and Related Conditions found CUD was strongly associated with other substance use disorders including alcohol, nicotine, and drugs, as well as psychiatric disorders of mood, anxiety, and personality disorders (Hasin, 2017). Although not all cannabis users will develop CUD, there have been genetic and environmental components hypothesized to contribute to its development as well (Bogdan et al., 2016; Sherva et al., 2016; Sorensen et al., 2017). Cannabis exposure is seen in all age groups, with nearly half of high school students stating they have tried marijuana and close to 5% of women reporting to have used marijuana during their pregnancies (Centers for Disease

Cannabis Use, Neurobiology, Psychology, and Treatment. https://doi.org/10.1016/B978-0-323-89862-1.00010-6

TABLE 1 DSM-5 substance-use disorder (Hasin, 2017).

Two or more of the following within the past 12 months	Hazardous use
	Social/Interpersonal problems related to use
	Neglected major roles to use
	Withdrawal
	Tolerance
	Used larger amounts/longer
	Repeated attempts to quit/control use
	Much time spent using
	Physical/psychological problems related to use
	Activities given up due to substance use
	Cravings

Control and Prevention, 2017). Cannabis use is not just seen in the young, with an increase of use seen in those 55 years and older since 2014 (Azofeifa et al., 2016).

Cannabis use has been seen in the United States since the 1960s; however, the first case of CHS was described in literature in 2004 by Allen and colleagues (Allen et al., 2004). The prevalence of CHS is difficult to ascertain, as the symptoms do not always necessitate medical care or are not reported as CHS. It is also unknown how many patients with CUD are diagnosed with CHS. Given this, Habboushe and colleagues estimated in their urban emergency department population with self-reported daily marijuana use, 33% of patients surveyed met criteria for experiencing CHS (Habboushe et al., 2018). They further extrapolated using the 2014 CDC data that close to 3 million individuals may experience CHS symptoms in the United States. It is likely that, with an increased cannabis use in the United states since 2014, including the pediatric population, these figures are an underestimation. The National Poison Data System assessed the calls they received from 2017 through 2019 regarding cannabis use. Total exposures increased during this time period, with more exposures being from manufactured products than plant alone. Twenty seven percent of cases involved children under the age of 10 years; and in the cases of isolated cannabis ingestion, 38.9% of plant-based exposures had serious medical outcomes (Dilley, 2021). They also documented an overall increase in exposures via edibles, concentrates, and vaporized liquids as state legislature changed during this time period in the United States. These increases can also contribute to the hypothesis that CHS is being underestimated.

Diagnosis

CHS can be difficult to diagnose due to several factors including: the stigma associated with drug use, and a similar presentation of symptoms to several other medical disorders not limited to CVS, migraines, and hyperemesis of pregnancy (Zhu, Gonsalves, Issenman, & Kam, 2021). Previous literature has suggested CHS is exhibited in three phases: a prodrome, hyperemetic, and recovery phase (Ruffle et al., 2015). Patients may present for medical attention in any phase but most often are seen in the hyperemetic phase.

The prodrome or pre-emetic phase of CHS can last several months during which time patients may experience some abdominal discomfort and nausea (Ruffle et al., 2015). Some may experience early morning nausea and a fear of vomiting leading them to increase their cannabis use. The hyperemetic phase lasts 24 to 48 h, and the patients will have frequent nausea and vomiting, often severe in nature that can result in weight loss, acute dehydration, electrolyte abnormalities, and possibly even acute renal failure. Hydrophilia, a behavior in which the patient takes frequent hot showers, can be used to help alleviate some of the symptoms during this phase. The recovery phase begins once the severe nausea and vomiting subsides, but relapse can occur once the patient reinitiates cannabis use.

Several diagnostic criteria were hypothesized when CHS was first identified, including: a history of regular cannabis use for greater than 1 year, severe nausea and vomiting, vomiting that occurs in a cyclical pattern, resolution of symptoms when

TABLE 2 Rome-IV criteria for diagnosis (Rome Foundation, 2020).

1. Stereotypical episodic vomiting resembling cyclic vomiting syndrome in terms of onset, duration, and frequency
2. Presentation after prolonged use of cannabis
3. Relief of vomiting episodes by sustained cessation of cannabis use
May be associated with pathologic bathing behavior

the offending agent is removed, and symptom relief with hot baths (Rome Foundation, 2020). Male patients may be more likely to develop CHS than female patients. The Rome-IV developed diagnostic criteria seen in Table 2 support the notion that CHS is a variant of CVS (Perisetti, 2020).

Pathophysiology

Although CHS is now considered to be a subset of CVS, the pathophysiology is not fully understood. Nausea and emesis are regulated through the cortex and limbic system, medulla oblongata, gastrointestinal tract, and vestibular system. Within many of these areas, the G-protein-coupled cannabinoid receptors CB1 and CB2 can be found. Specifically, CB1 is found in the central nervous system and gastrointestinal tract and influences gastric secretion, motility, and inflammation. The CB2 receptor is found in lymphoid and peripheral tissues, is involved in immune regulation, and thought to play a smaller role in the pathophysiology of CHS (Richards, 2018). Transient receptor potential vanilloid 1 (TRPV-1) is another G-protein coupled receptor involved in the cannabinoid system found on nociceptive neurons that are connected to nausea and vomiting pathways (DeVuono & Parker, 2020; Lapoint et al., 2018).

The major active component of marijuana is 9-tetrahydrocannabinol (THC), which is used as an antiemetic. THC reaches its peak effect approximately 2 h after inhalation and only achieves 25%–30% of those concentrations when ingested orally. Due to its lipophilic properties, THC accumulates in cerebral fat with long-term use, and peak accumulation typically occurs in 5 days. When THC binds to CB1, it antagonizes 5-hydroxytryptamine-3 (5-HT_3) receptors to act as an antiemetic and works to prevent dopamine's emetogenic mechanisms. In addition, TRPV-1 receptors are upregulated with activation of the cannabinoid system, leading to disturbances in thermoregulation that may explain the relief some patients experience with bathing in hot water (Wallace, Martin, & Park, 2007). Generally, CHS is considered a paradoxical response to THC consumption based on its physiologic antiemetic properties (DeVuono & Parker, 2020; Heise, 2015; Richards, 2018; Sawni et al., 2016).

There are many proposed mechanisms that lead to CHS, including dose-dependent toxicity, toxicity from cannabis-derived metabolites, CB1 receptor desensitization or downregulation, disruption of the hypothalamic-pituitary-adrenal (HPA) axis, thermoregulation disturbances, and reduced gut motility (Fig. 1). Activation of the CB1 receptor leads to HPA axis inhibition and sympathetic nervous system activation. These functions then lead to inhibition of gastric motor function, which can cause nausea and vomiting. Disruption of the HPA axis may also lead to impaired thermoregulation via CB1 receptor stimulation in the hypothalamus (Sawni et al., 2016; Solis-Garcia et al., 2019). Additionally, the sympathetic and HPA-axis mechanisms are further reinforced by evidence that has shown symptoms of CHS are more prominent in the morning, when HPA-axis activity is at its highest (DeVuono & Parker, 2020).

Desensitization, in combination with downregulation of the CB1 receptors, is another mechanism that is supported, which may explain the paradoxical development of nausea and vomiting with CHS. Downregulation of CB1 receptors was evaluated in chronic daily cannabis smokers using positron emission tomography scan imaging. Scans were completed on chronic marijuana users during use as well as 4 weeks after continuous abstinence. It was found that CB1 receptors were approximately 20% lower in daily, chronic cannabis smokers than control patients in cortical, but not subcortical brain regions. Unfortunately, it is unclear whether this was due to a decrease in CB1 receptor density or affinity but was negatively correlated with patients who smoked for longer periods of time (Hirvonen et al., 2012). While there are many proposed mechanisms leading to CHS, no single mechanism has been proven as the cause. The pathophysiology of CHS is likely due to a combination of CB1 receptor desensitization and/or downregulation and decreased intestinal transit time, leading to nausea and vomiting.

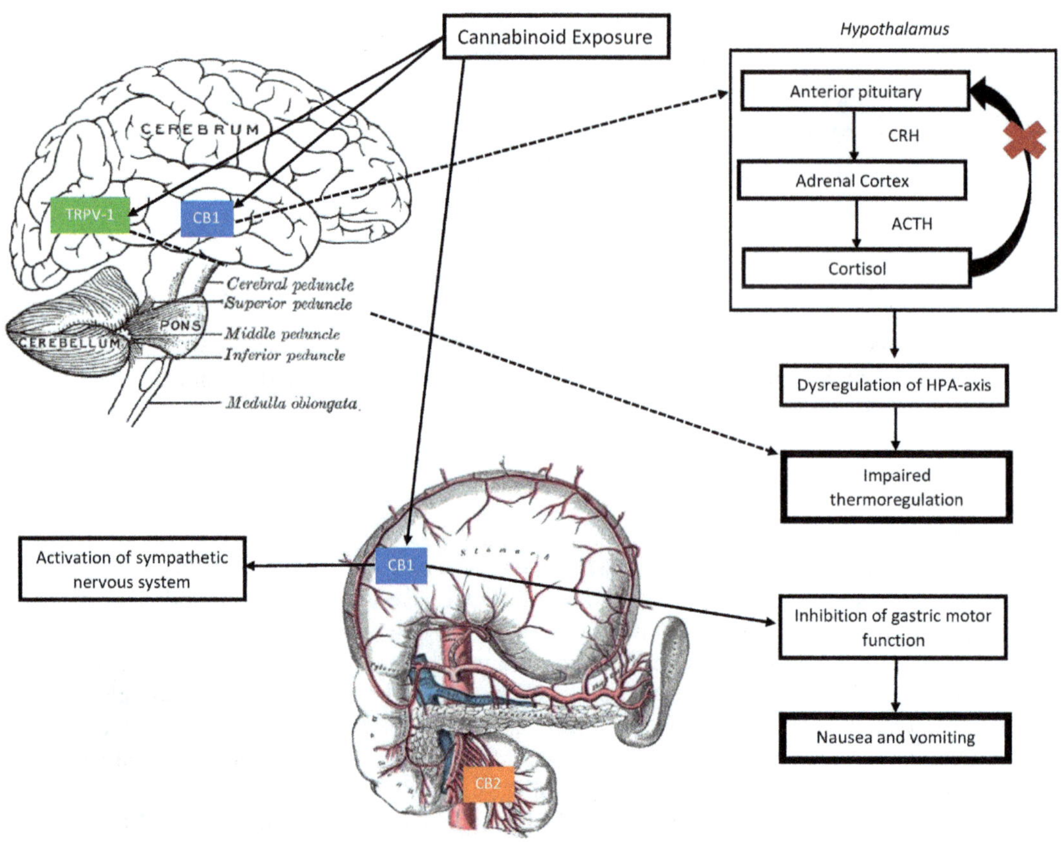

FIG. 1 Proposed pathophysiology of cannabinoid hyperemesis syndrome. *CB1*, cannabinoid receptor 1; *HPA*, hypothalamic-pituitary axis; *CRH*, corticotropin-releasing hormone; *ACTH*, adrenocorticotropic hormone; *TRPV-1*, transient receptor potential vanilloid 1; *CB2*, cannabinoid receptor 2 (Gray, 1918). *Anatomic illustrations obtained from open access publication Gray, H. (1918). Anatomy of the human body. Lewis, W.H. (Ed.) Philadelphia: Lea and Febiger.*

Antiemetic pharmacology review

Limited evidence exists on the types of supportive care measures and pharmacologic agents to successfully abort hyperemesis in this situation, with even fewer resources being focused on the pediatric population. Abstinence is the only definitive treatment for recurrent cannabinoid hyperemesis identified in the literature (Sorensen et al., 2017). Considered a home remedy, hot showers or baths are often used for symptom relief with higher water temperatures providing the greatest relief. While the precise pathway is unknown, it has been found that transient receptor potential vanilloid-1 receptor activity is augmented by heat, typically greater than 43°C (Stumpf & Williams, 2021). Several theories including compensation for the cannabinoid dose-dependent hypothermic effect on the hypothalamus and cutaneous vasodilation to alter core temperature as well as splanchnic circulation lessening abdominal pain have been postulated (Richards, 2018).

Benzodiazepines are frequently considered first-line agents for CVS, as cannabinoid use has been shown to reduce gamma-aminobutyric acid (GABA) neurotransmission, decrease extracellular glutamate, while increasing dopamine release and extracellular GABA (Richards, 2018). Although the exact mechanism of action is unknown, benzodiazepines appear to bind to GABA receptors, specifically $GABA_A$ between the alpha and gamma subunits, throughout the central nervous system and gastrointestinal tract (Bashashati & McCallum, 2014; Drug Facts and Comparison [Lexi]; Vinkers & Olivier, 2012). $GABA_A$ receptors exert antiemetic effects by inducing sedation, reducing anxiety, and potentially depressing the brainstem emetic nuclei (Bashashati & McCallum, 2014). As such, benzodiazepines, which agonist GABA receptors, have been found to be efficacious in CHS.

Dopamine antagonists act both centrally and peripherally. Centrally, these agents act by blocking the receptors within the chemoreceptor trigger zone of the medulla that are associated with emesis and peripherally have prokinetic effects on the gastrointestinal tract (Richards, 2018). Haloperidol, a first-generation antipsychotic, also functions as dopamine antagonist by nonselective blockade of D_2 receptors within the brain and is often used off-label as an antiemetic. Cannabis-induced increases of dopamine synthesis, turnover, efflux, and firing may explain the clinical improvement noted after the use of haloperidol and the lack of efficacy of other traditional dopaminergic medications, including metoclopramide, prochlorperazine, and promethazine. Second-generation antipsychotics, which commonly possess a mixed mechanism of action including dopamine and serotonin receptors, have not been found to be effective for CHS (Gajendran, Sifuentes, Bashashati, & McCallum, 2020; Sorensen et al., 2017).

Capsaicin, also known as 8-methyl-*N*-vanillyl-6-nonenamide, is a compound found in several types of chili peppers (Richards, 2018). Although the exact mechanism for use in cannabinoid hyperemesis is unknown, it is thought to be associated with the compound binding to the TVRP-1 cation channels, which in turn produce potent antiemetic effects (Stumpf & Williams, 2021). These receptors can be activated by low pH and high temperatures and may regulate substance P release. Substance P is an important mediator of nausea and vomiting (Richards, 2018). Prolonged cannabinoid exposure can desensitize TRPV-1, reduce receptor signaling and inhibit substance P release (Izzo & Sharkey, 2010; Stumpf & Williams, 2021). Capsaicin is available as an over-the-counter topical analgesic.

Serotonin antagonists such as ondansetron, dolasetron, and granisetron are identified in numerous case series as treatment for cannabinoid hyperemesis. The antiemetic effects are exerted by $5HT_3$ blockade in both the chemoreceptor trigger zone of the CNS and peripherally in the gastrointestinal tract (Hendren, Aponte-Feliciano, & Kovac, 2015).

Corticosteroids, specifically methylprednisolone and dexamethasone, are effective antiemetics for chemotherapy-induced nausea and vomiting as well as postoperative nausea and vomiting (Hendren et al., 2015). The exact mechanism has yet to be elucidated, however; theories include decreasing serotonin release, activating glucocorticoid receptors, antagonizing $5HT_3$ receptors, inhibiting prostaglandin synthesis, and preventing the release of endogenous opioids (Bashashati & McCallum, 2014).

Opioids are commonly given in combination with other therapies; however, since nausea and vomiting are common side effects of opioid therapy via stimulation of the delta and mu-receptors, these agents may worsen CHS and should be given with caution (Lapoint et al., 2018; Richards, 2018). Peripherally acting mu-receptor antagonists, such as alvimopan and methylnaltrexone, may reduce opioid-related nausea and vomiting by bowel stimulation postoperatively, though their use as antiemetics requires further investigation (Bashashati & McCallum, 2014).

The neurokinin-1 receptor antagonists, aprepitant and fosaprepitant, are indicated for the prevention of nausea and vomiting, although the role of these agents in CHS is unclear. Further study based on receptor interactions may be beneficial (Bashashati & McCallum, 2014).

Effective, ineffective, and supportive medication therapies in CHS

Very few effective therapies exist in the management of CHS, a deficit which is further exacerbated by the paucity of data in pediatric cohorts. It is important to note that the cessation of cannabis is the only remedy with proven efficacy for the prolonged avoidance of CHS, and that all other therapies described here are targeted interventions to assist with acute symptom management and to limit sequalae of continuous vomiting.

Effective therapies

A review of published literature has identified three pharmacologic interventions that have been found to be both efficacious and safe in pediatrics for the management of CHS. These data have been extrapolated and collated from case reports and case series reports, as no large observational cohort studies or randomized control trials have been conducted in this population to date. A summary of dosing considerations for effective therapies can be found in Table 3.

Benzodiazepines: Benzodiazepines have been prescribed more than most other therapies for the management of pediatric CHS. While lorazepam has been utilized as a quicker-acting benzodiazepine and successfully abated incidences of vomiting and nausea in pediatric patients, longer acting benzodiazepines have been more successful in treating patients beyond their initial presentation (Allen et al., 2004; Besli et al., 2015; Gammeter et al., 2016; Sawni et al., 2016). Limited data support the use of nightly alprazolam or clonazepam for a duration not to exceed 2 weeks (Gammeter et al., 2016; Sawni et al., 2016).

Haloperidol: A single case report documents the success of utilizing oral haloperidol for the management of acute nausea and vomiting associated with cannabis (Jones & Abernathy, 2016). Though only a single case report, the information from this clinical experience has many potential ramifications. The patient was successfully started on haloperidol in the emergency department and then transitioned to home on the same dose for 2 weeks. Second, the patient did not stop using cannabis but found that her vomiting had stopped, thereby indicating that haloperidol may have utility in patients for whom cessation is not a realistic goal.

TABLE 3 Dosing considerations for effective therapies.[a]

Medication	Dose	Duration	Additional Considerations
Benzodiazepines	0.5–1 mg lorazepam IV every 30 min until nausea/vomiting subsides 0.25–0.5 mg clonazepam by mouth nightly 0.25–0.5 mg alprazolam by mouth nightly	May repeat up to four times or a maximum of four milligrams for acute management Up to 2 weeks. Up to 2 weeks	Dosing considerations based on limited available data. Dosing will likely need adjustment for younger children or those who fall in the extremes of body habitus. Encourage cannabis cessation
Haloperidol	2.5–5 mg by mouth daily	Up to 2 weeks	Efficacy demonstrated without cannabis cessation. Encourage cannabis cessation
Topical Capsaicin	1-mm thick coating of 0.0.25%–0.075% applied to abdomen daily	Up to 2 weeks	Patient or caregiver should take contact precautions to avoid innocuous spread beyond desired treatment area. Encourage cannabis cessation

[a]*Dosing considerations based on limited available data. Dosing will likely need adjustment for younger children or those who fall in the extremes of body habitus.*

Topical capsaicin: The rapid cessation of nausea and vomiting has been reported following the application of topical capsaicin products to the abdomen (Graham et al., 2017). Variability exists as to the concentration of capsaicin in commercial products; however; no difference in efficacy has been unequivocally determined based on the active ingredient concentration.

Ineffective therapies

Though more data are needed to define treatment modalities for pediatric CHS, the following therapies have not been found to be efficacious.

Antiemetics: Traditional antiemetics have not been found to be efficacious in the cessation of nausea and vomiting associated with CHS. Available data have documented treatment failures with 5-HT_3 antagonists and D_2 receptor antagonists (Desjardins et al., 2015; Jones & Abernathy, 2016; Rajaram Manoharan et al., 2018; Pélissier et al., 2016; Wilson, 2016).

Acid suppressive therapies: Both H_2 receptor antagonists and proton pump inhibitors were not found to have any effect on nausea and vomiting episodes (Jones & Abernathy, 2016; Kelly & Van Opstal, 2019; Miller et al., 2010; Pélissier et al., 2016).

Atypical Antipsychotics: No evidence exists supporting the use of these agents. It is possible that their mixed mechanisms of action explain their lack of efficacy when compared to the mechanism of action of haloperidol, which has been found to be efficacious (Jones & Abernathy, 2016).

Supportive therapies

External thermoregulation: Though care is necessary to ensure that patients do not suffer from hot water burns, external thermoregulation by the way of bathing or showering has been found to be marginally effective to decrease nausea and vomiting episodes. Reported data indicate that patients found relief after multiple exposures to warm water over the course of a day (Desjardins et al., 2015; Solis-Garcia et al., 2019).

Fluid resuscitation: It is likely that patients who present with a significant history of vomiting will be dehydrated and may require intravenous fluid resuscitation. A crystalloid fluid is the most appropriate for resuscitation, though the amount of fluid is dependent upon physical assessment, review of laboratory data, and prescriber preference (Graham et al., 2017; Solis-Garcia et al., 2019). Conversely, if nausea and vomiting has been controlled, it may be reasonable to employ the Holliday-Seager method to determine an oral rehydration strategy in pediatric patients (Holliday et al., 2007).

Safety considerations

Very little information on adverse events when treating pediatric CHS are described in the literature. All medications utilized for this condition are done so off-label. While there is some level of risk in their general use, dosing of these agents for the management of pediatric CHS does not exceed current recommended maximum doses, which suggests an incidence comparable to what has been reported in alternative indications.

Benzodiazepines: The most common benzodiazepine adverse events including sedation, somnolence, fatigue, dizziness, and cognitive impairment. Patients who are on concomitant CNS depressants are at a higher risk of experiencing these adverse events. Pediatric patients, especially those with known hyperactive behavior, may be at higher risk of experiencing a paradoxical reaction such as hyperactive or aggressive behavior that requires pharmacologic intervention. Concomitant opioid use is not recommended unless the benefits outweigh the risk. Finally, anterograde amnesia has been reported with benzodiazepine use. Withdrawal symptoms are not anticipated due to the short duration of therapy.

Lorazepam—The liquid and injectable formulations of lorazepam contain propylene glycol, a known nephrotoxic agent. Patients with acute kidney injury or a sensitivity to this compound should be prescribed the tablet formulation if lorazepam is indicated.

Clonazepam—Clonazepam may also lower the seizure threshold in patients with a known seizure disorder and should be avoided.

Alprazolam—Alprazolam use should be avoided with concomitant strong CYP3A4 inhibitors excluding ritonavir. The short half-life of alprazolam may require more frequent dosing than lorazepam or clonazepam; tolerance may also develop quickly.

Butyrophenones: Class effects of butyrophenones, including haloperidol, include QTc prolongation, hypotension associated with intravenous administration, and extrapyramidal symptoms. These may also increase the risk of seizures in patients with a history of seizure disorder. Patients who are experiencing moderate-to-severe dehydration may be predisposed to cardiac arrhythmias; caution is recommended with the use of butyrophenones until hemodynamic stability is achieved. Tardive dyskinesia is a rare but serious adverse effect reported with even low doses of short duration.

Capsaicin: The most commonly reported adverse effect of capsaicin is application site reaction. Erythema, edema, pain, and a burning sensation at the site of application have all been described. While these should subside with repeated application in as little as 3 days, some patients have reported intolerable experiences leading to product discontinuation. Capsaicin should not be applied to areas involving the face, neck, or mucous membranes, nor should it be used immediately before or after showering.

Disease states

Patients with myasthenia gravis are cautioned against the use of alprazolam or butyrophenones. Patients with a seizure disorder may see an increase in seizure frequency with benzodiazepines or butyrophenones.

Mini-dictionary

- **9-Tetrahydrocannabidiol**. The active ingredient in cannabis and cannabinoid products.
- **Cannabinoid**. General term describing a product derived from cannabis.
- **Cannabis.** The physical plant, and more specifically leaves, from which cannabinoid products are manufactured and which may be ingested, smoked, or applied directly to the body for the purpose of experiencing the effects of 9-tetrahydrocannabidiol.
- **Cyclic vomiting syndrome**. Broad diagnostic category of intractable nausea and vomiting, in which CHS is frequently categorized.
- **Transient receptor potential vanilloid-1**. G-coupled protein receptor responsive to external thermoregulation, such as with hot showers or topical capsaicin products.

Key facts

- *Key facts of the pathophysiology*: Agonism of the CB1 and CB2 receptors results in disruption of the HPA, dysregulation of serotonergic and dopaminergic neurotransmission, and direct manipulation of the emetic center within the CNS.

- *Key facts of the treatment options*: Traditional anti-emetic medications, used as monotherapy or in combination with each other, are typically ineffective in the management of CHS. Benzodiazepines, haloperidol, and topical capsaicin have the most efficacy data in pediatric patients.

Summary points

- Use of cannabinoid products, both medically and recreationally, is increasing across the globe.
- The incidence of CHS is low but increasing, likely due to the increased use of cannabinoid products and due to knowledge of the separate pathophysiology by clinicians.
- Diagnostic criteria with the Rome-IV criteria are easily applied to patients with known cannabinoid use.
- Though mechanisms may exist, the pathophysiology of CHS relates directly to the agonism of the CB1 and CB2 receptors. As such, abstinence from cannabinoid products remains the only definitive cure.
- Traditional antiemetic medications, including 5-HT_3 antagonists such as ondansetron and dopaminergic antagonists such as promethazine, have not demonstrated efficacy in the management of CHS symptoms.
- Select benzodiazepines, including lorazepam, alprazolam, and clonazepam, have data supporting their use in pediatric CHS. These, along with haloperidol and topical capsaicin, represent the treatment modalities with the most supporting evidence for use.
- Care must be given as to the route of administration, dosing, and adverse effect profile for each pediatric patient.

References

Allen, J. H., de Moore, G. M., Heddle, R., & Twartz, J. C. (2004). Cannabinoid hyperemesis: Cyclical hyperemesis in association with chronic cannabis abuse. *Gut*, *53*(11), 1566–1570. https://doi.org/10.1136/gut.2003.036350.

Azofeifa, A., Mattson, M. E., Schauer, G., McAfee, T., Grant, A., & Lyerla, R. (2016). National estimates of marijuana use and related indicators—National survey on drug use and health, United States, 2002–2014. *MMWR Surveillance Summaries*, *65*(11), 1–28. https://doi.org/10.15585/mmwr.ss6511a1.

Bashashati, M., & McCallum, R. W. (2014). Neurochemical mechanisms and pharmacologic strategies in managing nausea and vomiting related to cyclic vomiting syndrome and other gastrointestinal disorders. *European Journal of Pharmacology*, *722*, 79–94.

Besli, G. E., Ikiz, M. A., Yildirim, S., & Saltik, S. (2015). Synthetic cannabinoid abuse in adolescents: A case series. *The Journal of Emergency Medicine*, *49*(5), 644–650. https://doi.org/10.1016/j.jemermed.2015.06.053.

Bogdan, R., Winstone, J. M., & Agrawal, A. (2016). Genetic and environmental factors associated with cannabis involvement. *Current Addiction Reports*, *3*(2), 199–213. https://doi.org/10.1007/s40429-016-0103-1.

Centers for Disease Control and Prevention. (2017). *National estimates of marijuana use and related indicators—National survey on drug use and health, United States, 2002–2014*. Centers for Disease Control and Prevention. Retrieved November 8, 2021, from: https://www.cdc.gov/mmwr/volumes/65/ss/ss6511a1.htm?s_cid=ss6511a1_e.

Centers for Disease Control and Prevention. (2021). *Marijuana data and statistics*. Centers for Disease Control and Prevention. Retrieved November 8, 2021, from: https://www.cdc.gov/marijuana/data-statistics.htm.

Desjardins, N., Jamoulle, O., Taddeo, D., & Stheneur, C. (2015). Cannabinoid hyperemesis syndrome in a 17-year-old adolescent. *The Journal of Adolescent Health*, *57*(5), 565–567. https://doi.org/10.1016/j.jadohealth.2015.07.019.

DeVuono, M. V., & Parker, L. A. (2020). Cannabinoid hyperemesis syndrome: A review of potential mechanisms. *Cannabis and Cannabinoid Research*, *5*(2), 132–144. https://doi.org/10.1089/can.2019.0059.

Dilley, J. A. (2021). *Trends in cannabis product exposures reported to US Poison Control Centers*. JAMA Network Open. Retrieved November 8, 2021, from: https://jamanetwork.com/journals/jamanetworkopen/fullarticle/2780068.

Gajendran, M., Sifuentes, J., Bashashati, M., & McCallum, R. (2020). Cannabinoid hyperemesis syndrome: Definition, pathophysiology, clinical spectrum, insights into acute and long-term management. *Journal of Investigative Medicine*, *68*(8), 1309–1316.

Gammeter, W. B., Duke, K. A., & Soundy, T. J. (2016). Case report of intractable vomiting and abdominal pain related to heavy daily cannabis use. *South Dakota Medicine: The Journal of the South Dakota State Medical Association*, *60–63*.

Graham, J., Barberio, M., & Wang, G. S. (2017). Capsaicin cream for treatment of cannabinoid hyperemesis syndrome in adolescents: A case series. *Pediatrics*, *140*(6). https://doi.org/10.1542/peds.2016-3795.

Gray, H. (1918). In W. H. Lewis (Ed.), *Anatomy of the human body*. Philadelphia: Lea and Febiger.

Habboushe, J., Rubin, A., Liu, H., & Hoffman, R. S. (2018). The prevalence of cannabinoid hyperemesis syndrome among regular marijuana smokers in an urban public hospital. *Basic & Clinical Pharmacology & Toxicology*, *122*(6), 660–662. https://doi.org/10.1111/bcpt.12962.

Hasin, D. S. (2017). US epidemiology of cannabis use and associated problems. *Nature News*. Retrieved November 8, 2021, from: https://www.nature.com/articles/npp2017198.

Heise, L. (2015). Cannabinoid hyperemesis syndrome. *Advanced Emergency Nursing Journal*, *37*(2), 95–101. https://doi.org/10.1097/tme.0000000000000062.

Hendren, G., Aponte-Feliciano, A., & Kovac, A. (2015). Safety and efficacy of commonly used antiemetics. *Expert Opinion on Drug Metabolism & Toxicology*, *11*(11), 1753–1767.

Hirvonen, J., Goodwin, R. S., Li, C.-T., Terry, G. E., Zoghbi, S. S., Morse, C., Pike, V. W., Volkow, N. D., Huestis, M. A., & Innis, R. B. (2012). Reversible and regionally selective downregulation of brain cannabinoid CB1 receptors in chronic daily cannabis smokers. *Molecular Psychiatry*, *17*(6), 342–349. https://doi.org/10.1038/mp.2011.82.

Holliday, M. A., Ray, P. E., & Friedman, A. L. (2007). Fluid therapy for children: Facts, fashions and questions. *Archives of Disease in Childhood*, *92*(6), 546–550. https://doi.org/10.1136/adc.2006.106377.

Izzo, A. A., & Sharkey, K. A. (2010). Cannabinoids and the gut: New developments and emerging concepts. *Pharmacology & Therapeutics*, *126*(1), 21–38.

Jones, J. L., & Abernathy, K. E. (2016). Successful treatment of suspected cannabinoid hyperemesis syndrome using haloperidol in the outpatient setting. *Case Reports in Psychiatry*, *2016*, 3614053. https://doi.org/10.1155/2016/3614053.

Kelly, E., & Van Opstal, E. (2019). Case 2: 16-year-old boy with intractable nausea and vomiting. *Pediatrics in Review*, *40*(6), 305–307. https://doi.org/10.1542/pir.2018-0114.

Lapoint, J., Meyer, S., Yu, C. K., Koenig, K. C., Lev, R., Thihalolipavan, S., Staats, K., & Kahn, C. A. (2018). Cannabinoid hyperemesis syndrome: Public health implications and a novel model treatment guideline. *The Western Journal of Emergency Medicine*, *19*(2), 380–386. https://doi.org/10.5811/westjem.2017.11.36368.

Miller, J. B., Walsh, M., Patel, P. A., Rogan, M., Arnold, C., Maloney, M., & Donnino, M. (2010). Pediatric cannabinoid hyperemesis: Two cases. *Pediatric Emergency Care*, *26*(12), 919–920. https://doi.org/10.1097/PEC.0b013e3181fe9189.

Pélissier, F., Claudet, I., Gandia-Mailly, P., Benyamina, A., & Franchitto, N. (2016). Cannabis hyperemesis syndrome in the emergency department: How can a specialized addiction team be useful? A pilot study. *The Journal of Emergency Medicine*, *51*(5), 544–551. https://doi.org/10.1016/j.jemermed.2016.06.009.

Perisetti, A. (2020). Cannabis hyperemesis syndrome: An update on the pathophysiology and management. *Annals of Gastroenterology*. https://doi.org/10.20524/aog.2020.0528.

Rajaram Manoharan, S. V. R., Aggarwal, R., & Taneli, T. (2018). Cannabinoid hyperemesis syndrome: A case report. *Asian Journal of Psychiatry*, *34*, 64. https://doi.org/10.1016/j.ajp.2018.04.013.

Richards, J. R. (2018). Cannabinoid hyperemesis syndrome: Pathophysiology and treatment in the emergency department. *The Journal of Emergency Medicine*, *54*(3), 354–363. https://doi.org/10.1016/j.jemermed.2017.12.010.

Rome Foundation. (2020). *Rome IV criteria*. Retrieved November 8, 2021, from: https://theromefoundation.org/rome-iv/rome-iv-criteria/.

Ruffle, J. K., Bajgoric, S., Samra, K., Chandrapalan, S., Aziz, Q., & Farmer, A. D. (2015). Cannabinoid hyperemesis syndrome. *European Journal of Gastroenterology & Hepatology*, *27*(12), 1403–1408. https://doi.org/10.1097/meg.0000000000000489.

Sawni, A., Vaniawala, V. P., Good, M., Lim, W. Y., & Golec, A. S. (2016). Recurrent cyclic vomiting in adolescents: Can it be cannabinoid hyperemesis syndrome? *Clinical Pediatrics*, *55*(6), 560–563. https://doi.org/10.1177/0009922815594233.

Sherva, R., Wang, Q., Kranzler, H., Zhao, H., Koesterer, R., Herman, A., … Gelernter, J. (2016). Genome-wide association study of Cannabis Dependence severity, novel risk variants, and shared genetic risks. *JAMA Psychiatry*, *73*(5), 472. https://doi.org/10.1001/jamapsychiatry.2016.0036.

Solis-Garcia, G., Gonzalez-Martinez, F., Urbiola, E., Carballo, J. J., Perez-Moreno, J., Rodriguez-Fernandez, R., Moreno, D., & Pina-Camacho, L. (2019). Cannabinoid hyperemesis in adolescents: A poorly diagnosed syndrome. *Journal of Paediatrics and Child Health*, *55*(10), 1264–1266. https://doi.org/10.1111/jpc.14479.

Sorensen, C. J., DeSanto, K., Borgelt, L., Phillips, K. T., & Monte, A. A. (2017). Cannabinoid hyperemesis syndrome: Diagnosis, pathophysiology, and treatment—A systematic review. *Journal of Medical Toxicology*. Retrieved November 8, 2021, from: https://www.ncbi.nlm.nih.gov/pmc/articles/PMC5330965/.

Stumpf, J. L., & Williams, L. D. (2021). Management of cannabinoid hyperemesis syndrome: focus on capsaicin. *Journal of Pharmacy Practice*, *34*(5), 786–793.

Vinkers, C. H., & Olivier, B. (2012). Mechanisms underlying tolerance after long-term benzodiazepine use: a future for subtype-selective receptor modulators? *Advances in Pharmacological Sciences*, *2012*, 1–19.

Wallace, D., Martin, A.-L., & Park, B. (2007). Cannabinoid hyperemesis: Marijuana puts patients in hot water. *Australasian Psychiatry*, *15*(2), 156–158. https://doi.org/10.1080/10398560701196778.

Wilson, L. (2016). Case 3: Recurrent vomiting and 60-lb weight loss in a 17-year-old girl. *Pediatrics in Review*, *37*(6), 264–266.

Zhu, J. W., Gonsalves, C. L., Issenman, R. M., & Kam, A. J. (2021). Diagnosis and acute management of adolescent cannabinoid hyperemesis syndrome: A systematic review. *The Journal of Adolescent Health*, *68*(2), 246–254. https://doi.org/10.1016/j.jadohealth.2020.07.035.

Chapter 45

Gabapentinoids for the treatment of cannabis use disorder: A focus on human laboratory research

Joshua A. Lile, William W. Stoops, and Michael J. Wesley

Department of Behavioral Science, University of Kentucky College of Medicine, Lexington, KY, United States

Abbreviations

CB	cannabinoid
CUD	cannabis use disorder
DSST	digit symbol substitution task
EXP	experiment
RAT	repeated acquisition task
THC	delta^9-tetrahydrocannabinol
THC-COOH	11-nor-9-carboxy-delta^9-tetrahydrocannabinol
SUD	substance use disorder
VAS	visual analog scale
VDCC	voltage-dependent calcium channel

Introduction

There is an established yet unmet need for highly effective treatments for cannabis use disorder (CUD). The most recent data from the United States National Survey on Drug Use and Health indicate that 11.8% of Americans aged 12 and older used cannabis in the past month, with 1.8% endorsing criteria for CUD (Center for Behavioral Health Statistics and Quality, 2020). Cannabis was reported as the primary substance for 13% of treatment admissions included in the most recent state agency data systems (Center for Behavioral Health Statistics and Quality, 2017), which is the third most common drug after opioids (34%) and alcohol (29%). Unlike opioids and alcohol, no FDA-approved medications are available to manage problematic use of cannabis, so efforts to identify medications are ongoing.

To date, over 20 placebo-controlled, randomized clinical trials have been conducted in adults to evaluate medications in treatment-seeking cannabis users for their ability to reduce cannabis use. The first placebo-controlled trial to show a positive signal was with gabapentin (Mason et al., 2012). In that pilot study, 1200 mg per day of gabapentin (Neurontin) decreased urinary levels of the 11-nor-9-carboxy metabolite of delta^9-tetrahydrocannabinol (THC), self-reported cannabis use, craving and depression, and improved performance on tests of executive function, in 25 nontreatment-seeking cannabis users relative to the 25 subjects who received placebo. The present chapter will begin with an overview of the pharmacology of gabapentin and its analog, pregabalin, relevant to their potential use as CUD medications. This chapter will then cover the available animal and human data supporting a CUD indication, with an emphasis on the human laboratory studies that were prompted by these promising pilot results.

Gabapentinoid pharmacology

Gabapentin is indicated for the treatment of neuropathic pain and seizures, and its therapeutic effects have been linked primarily to voltage-dependent calcium channels (VDCCs), particularly those containing $\alpha_2\delta$-1 subunits (Sills, 2006).

Cannabis Use, Neurobiology, Psychology, and Treatment. https://doi.org/10.1016/B978-0-323-89862-1.00042-8

Although the consequences of ligands binding to the $\alpha_2\delta$-1 subunit are not completely understood, there is evidence for a direct operational impact on calcium conductance, as well as VDCC trafficking to control the number of functional channels, with the overall result being a dampening of neuronal activity that impacts the release of multiple neurotransmitters (Gale & Houghton, 2011). Pregabalin (Lyrica) was developed as a "next-generation" gabapentinoid and is approved for the same indications, but is considered a product improvement. Compared to gabapentin, pregabalin has a better pharmacokinetic profile, with a more rapid (1 h versus 3–4 h) and complete ($\geq$90% versus 60%–33%) absorption that follows linear (i.e., first-order) kinetics, yielding a proportional dose-concentration relationship, rather than saturable (i.e., zero-order) kinetics like gabapentin (Bockbrader et al., 2010). The pharmacokinetic advantages of pregabalin over gabapentin appear to translate into an improved pharmacodynamic response. Meta-analyses indicate that pregabalin is more effective for refractory partial epilepsy and chronic post-surgical pain (Clarke et al., 2012; Delahoy et al., 2010), a therapeutic response to pregabalin is observed in some patients for whom gabapentin is not effective (e.g., Saldaña et al., 2012; Stacey et al., 2010; Toth, 2010), and United States-based cost-effectiveness analyses found that pregabalin offered cost savings for fibromyalgia patients due to its greater relative effectiveness and was associated with reduced medical expenses and increased workplace productivity in epilepsy patients (Kleinman et al., 2012; Lloyd et al., 2012).

Preclinical evidence supporting gabapentinoids as CUD medications

Examining gabapentinoids as potential CUD medications was warranted given the functional link in VDCC and cannabinoid (CB) systems and overlap in the behavioral and pharmacotherapeutic effects of VDCC and CB drugs. For example, one of the main consequences of CB receptor-mediated G-protein activation is the inhibition of VDCCs (Howlett et al., 2010). In addition, there is evidence for CB-receptor-independent modulation of VDCCs by CB ligands, through interaction with the plasma membrane lipid bilayer and by direct interaction with a binding site on the ion channel (Lozovaya et al., 2009). Preclinical studies comparing gabapentinoids and CB ligands have consisted mainly of animal models of neuropathic pain, which have generally found that gabapentin and pregabalin are comparable to CB agonists at reducing pain responses (e.g., Caprioli et al., 2012; Gunduz et al., 2011; Hasnie et al., 2007; Luszczki & Florek-Łuszczki, 2012; Palmer et al., 2008; Wallace et al., 2007) and produce synergistic effects when combined (Atwal et al., 2019). There only appears to be a single preclinical study that tested a gabapentinoid in a model directly relevant to CUD (Aracil-Fernández et al., 2013). In that study, mice were treated repeatedly with CP-55,940, a high-efficacy CB agonist, and then changes in brain gene transcription and motor and anxiety-like behavior were recorded during spontaneous withdrawal and treatment with pregabalin. Pregabalin attenuated anatomically specific changes in the expression of certain genes thought to be involved in CUD (i.e., tyrosine hydroxylase in the ventral tegmental area and CB_1 receptors in the nucleus accumbens) as well as the motor and anxiety responses induced by CB agonist cessation.

Clinical evaluation of gabapentinoids as CUD medications

Clinical research that tested both CBs and gabapentinoids has also been limited. One clinical study (Bestard & Toth, 2011) directly compared nabilone, a nonselective CB agonist, with gabapentin in an open-label study in neuropathic pain patients. Relative to baseline and across drugs at 3- and 6-month intervals, pain, sleep parameters, and anxiety were similarly improved by gabapentin and nabilone. Cross-study comparisons have also demonstrated the ability of CB agonists and gabapentinoids to improve outcomes related to neuropathic pain, sleep, and anxiety (e.g., de Paris et al., 2003; Hindmarch et al., 2005; Karst et al., 2010; Nakano et al., 1978; Roth et al., 2012; Ware et al., 2010). Worth noting is that sleep difficulties, anxiety, and physical discomfort are commonly observed during cannabis abstinence and often reported as reasons for continued use (Elkashef et al., 2008). Aside from the initial trial with gabapentin by Mason and colleagues, we are aware of only two other published clinical studies that evaluated the effects of gabapentinoids in cannabis users, which are reviewed here in detail. The first of those studies used human laboratory procedures to determine the discriminative-stimulus effects of acutely administered gabapentin, alone and in combination with THC, in cannabis users who learned to discriminate THC from placebo (Lile et al., 2016). The second study used human laboratory procedures to evaluate the ability of maintenance on pregabalin to reduce smoked cannabis self-administration (Lile et al., 2021).

Separate and combined effects of gabapentin and THC in cannabis users discriminating THC: Drug discrimination rationale and design

The purpose of this study (Lile et al., 2016) was to examine a potential mechanism of action of gabapentin to treat cannabis use disorder by determining the interoceptive effects of gabapentin alone and in combination with THC using a pharmacologically selective drug discrimination procedure in cannabis users. Drug discrimination was chosen as the primary outcome because the data from this procedure are concordant with the actions of a drug at the receptor level (Holtzman & Locke, 1988). Moreover, an extensive review of clinical studies that used drug discrimination and subjective drug effects (hereafter, subjective effects) questionnaires to examine the neuropharmacology of a range of drug classes (Kelly et al., 2003) indicated that drug discrimination data conformed more closely to the results from in vitro and in vivo preclinical studies.

The drug discrimination design used in this study was typical of clinical research (see Bolin et al., 2016 for an in-depth review) and had three phases (Table 1). First, subjects completed a sampling phase in which they received capsules containing the training drug (30 mg THC), identified by a letter code (e.g., Drug X). Oral administration of THC was chosen to eliminate the external cues and/or expectations associated with smoked cannabis. Subjects were not informed of the drug contents but were instructed to associate the drug effects with the letter code. Next, subjects completed a control phase,

TABLE 1 Separate and combined effects of gabapentin and THC in cannabis users discriminating THC: Design and events.

Session #	THC dose (mg)	GBP dose (mg)	Feedback
Sampling phase			
1	30	0	Drug X[a]
2	30	0	Drug X
Control phase[b]			
3	0	0	Not drug
4	30	0	Drug X
5	30	0	Drug X
6	0	0	Not drug
Test phase[c]			
7	5	1200	Test
8	15	0	Test
9	30	600	Test
10	0	0	Not drug X
11	5	0	Test
12	0	600	Test
13	15	600	Test
14	30	0	Drug X
15	0	0	Test
16	30	0	Test
17	15	1200	Test
18	5	600	Test
19	0	0	Not Drug X
20	0	1200	Test
21	30	0	Drug X
22	30	1200	Test

The order of dose conditions is representative. *GBP*, gabapentin.
[a]*The letter code was unique to each subject.*
[b]*Up to 12 total sessions.*
[c]*Additional control sessions added as needed.*

lasting between 4 and 12 sessions, in which they received the two control conditions (i.e., 30 mg THC and placebo), in randomized order. Sessions were identical to the sampling phase, except subjects were not informed which drug condition (i.e., Drug X or Not Drug X) was administered until the end of the session, and they received additional study earnings when they responded correctly. When subjects met the criterion for having acquired the discrimination ($\geq$80% correct responding for four consecutive sessions), they moved into a final test phase in which capsules containing 0, 5, 15, or 30 mg THC and 0, 600, or 1200 mg gabapentin were administered; all subjects received all possible dose combinations. For these 12 sessions, because there was no right or wrong response on the task, subjects received additional study earnings based on their average control session performance. The test phase also included four control sessions to monitor drug discrimination performance and provide feedback to subjects; if a subject responded incorrectly on a control session (i.e., <80% correct responding), additional control sessions were scheduled until the subject accurately identified both training conditions across consecutive sessions.

Separate and combined effects of gabapentin and THC in cannabis users discriminating THC: Methods

Eight adults (aged 18–31 years; mean=23) reporting current cannabis use (2–5 days per week; mean=5) completed this placebo-controlled, double-blind, randomized, within-subjects study. Prior to enrollment, subjects completed a screening procedure that assessed their medical, psychiatric, and drug use histories and included standard laboratory tests of physiological function. Subjects were required to attend two drug-free practice sessions prior to 21–26 (mean=23) experimental sessions over 8–18 weeks (mean=11). Experimental sessions lasting 6.5 h were conducted Monday through Friday, with 1–5 sessions occurring per week. In addition to the drug discrimination task, subjective effects questionnaires and measures of safety and tolerability (psychomotor performance tasks and cardiovascular measures) were also administered but will not be comprehensively covered here. Data were collected prior to capsule administration and then 1, 2, 3, 4, and 5 h after THC administration, with the exception that the drug discrimination task was completed at only the 3–5 h time points because of the slow onset of the effects of oral THC (e.g., Lile et al., 2013). THC doses were selected based on prior studies in frequent cannabis users (e.g., Lile et al., 2013), and the gabapentin doses were based on the pilot clinical trial by Mason et al. (2012). For reference, the acute recommended THC dosing range in adults for appetite stimulation, and the prevention of nausea and vomiting is 2.5 to 20 mg. The recommended starting therapeutic oral dose of gabapentin is 300–400 mg, administered three times a day, and titrated up to a clinically effective maintenance dose of 1800–3600 mg/day. Gabapentin and THC were administered concurrently, consistent with the onset of their peak concentrations following oral dosing (Gidal et al., 1998; Lile et al., 2013).

Separate and combined effects of gabapentin and THC in cannabis users discriminating THC: Results and discussion

Drug discrimination

The discriminative-stimulus effects of THC, alone and in combination with gabapentin, are shown in Fig. 1. During the test phase, THC alone dose dependently increased drug appropriate responding on the drug discrimination task. Both doses of gabapentin alone also significantly increased drug-appropriate responding and significantly shifted the discriminative-stimulus effects of THC leftward/upward. These results support the involvement of VDCCs in the interoceptive cue produced by THC, in agreement with the ability of CB agonists to impact VDCC function via CB_1 receptor-mediated G-protein activation as well as CB-receptor-independent mechanisms (Howlett et al., 2010; Lozovaya et al., 2009). These findings, together with the initial promising results from the pilot treatment trial with gabapentin, are also consistent with an agonist replacement approach to managing substance use disorders (SUDs).

Subjective effects

Several positive items from a subjective effects questionnaire using a 100-unit Visual Analog Scale (VAS) showed effects of gabapentin on the response to THC similar to those from the drug discrimination task; the items good effects, high and like drug are presented in Fig. 2. In general, at least one dose of both drugs alone increased ratings on these items, and the effects of the combination of THC and gabapentin were significantly greater compared to THC alone. In addition, the subject rated effects of the combinations were frequently a function of dose (i.e., the combination of 1200 mg gabapentin and THC produced larger effects than 600 mg gabapentin combined with the same THC dose). These results suggest that

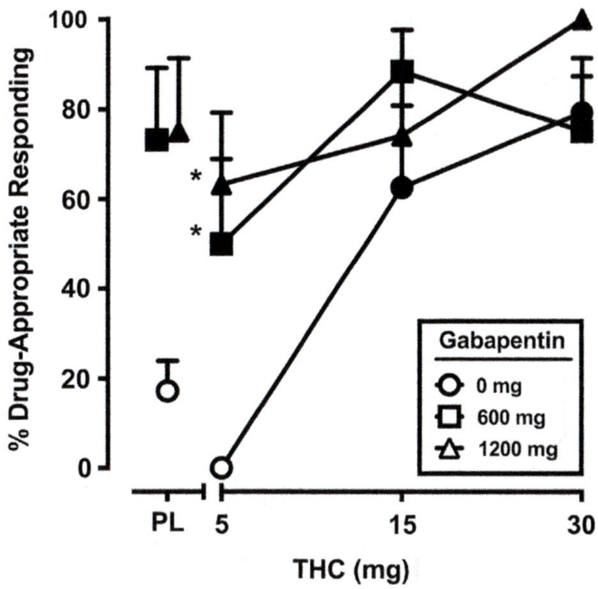

FIG. 1 Separate and combined effects of THC and gabapentin on THC-appropriate responding on the drug discrimination task. The *x*-axis represents the THC dose in mg; PL denotes placebo. Filled symbols indicate values that are significantly different from placebo. Asterisks indicate combinations of THC and gabapentin that are significantly different from that dose of THC alone. Unidirectional brackets indicate 1 SEM.

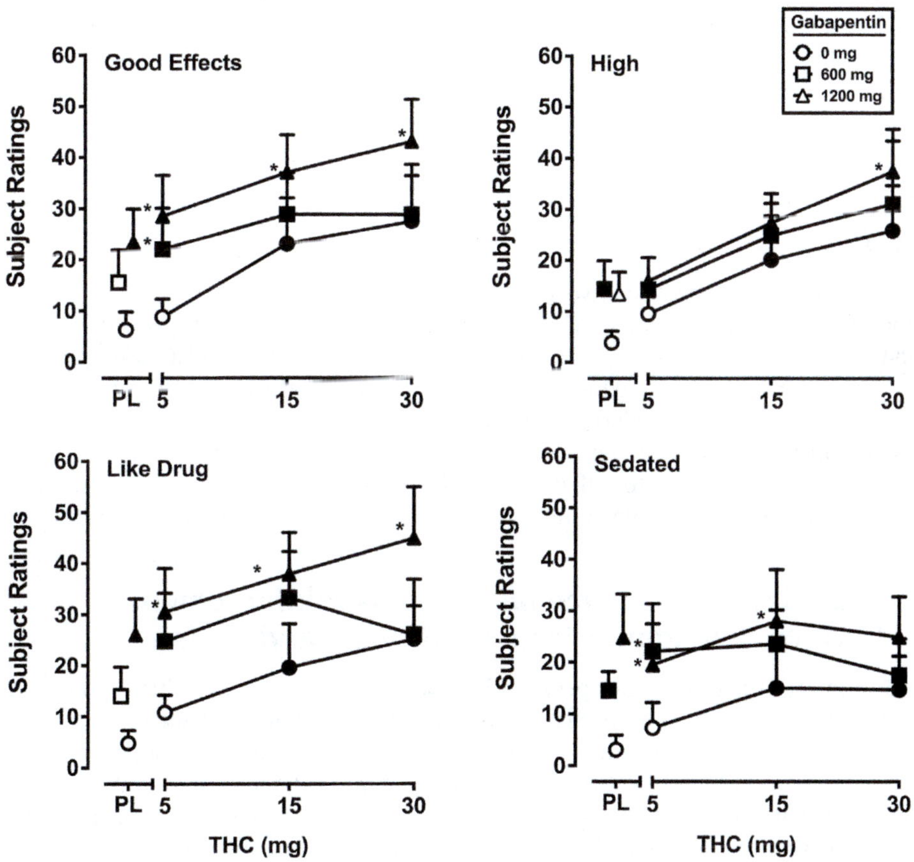

FIG. 2 Peak (maximum value) Visual Analog Scale ratings for THC and gabapentin, alone and in combination, on the subjective effects questionnaire items Good Effects, High, Like Drug, and Sedated. The *x*-axis represents the THC dose in mg; PL denotes placebo. Filled symbols indicate values that are significantly different from placebo. Asterisks indicate combinations of THC and gabapentin that are significantly different from that dose of THC alone. Unidirectional brackets indicate 1 SEM.

gabapentin has the potential to be abused. Consistent with these findings, a recent systematic review identified signals of gabapentin abuse from case studies and epidemiological reports, particularly in individuals with a history of substance use problems (Smith et al., 2016). However, an important counterpoint to consider is that any positive subjective effects of gabapentin and pregabalin might be advantageous, as they would likely enhance medication adherence.

Safety and tolerability

Gabapentin, alone and in combination with THC, produced impairment on two measures of psychomotor performance, the Repeated Acquisition Task (RAT) and the Digit Symbol Substitution Task (DSST). Relative to placebo, rate and accuracy on the RAT were impaired by the 600 and 1200 mg doses of gabapentin alone, the 5 and 15 mg doses of THC alone, and by all active THC doses when combined with gabapentin. In addition, 1200 mg gabapentin impaired DSST accuracy alone and when combined with every active dose of THC, compared to placebo, and DSST accuracy was impaired following administration of 600 mg gabapentin combined with 15 or 30 mg THC. With respect to the cardiovascular effects of THC and gabapentin combinations, 30 mg THC increased heart rate regardless of gabapentin dose, and the 15 mg dose of THC +1200 mg gabapentin dose condition increased heart rate, relative to placebo. In addition, the 1200 mg dose of gabapentin increased systolic blood pressure alone and when combined with the 30 mg dose of THC, relative to placebo. Lastly, the 1200 mg gabapentin+30 mg THC dose combination significantly elevated systolic blood pressure relative to that dose of THC alone. Diastolic blood pressure was not influenced by THC or gabapentin.

That gabapentin impaired psychomotor performance and elevated certain cardiovascular indices, and also exacerbated these effects of THC, might be of concern because this could translate to side effects in the context of therapeutic use. However, these results could be due to the administration of gabapentin doses above what is recommended at the start of treatment and without titration. Indeed, gabapentin therapy with recommended doses appears to have an acceptable safety profile. Moreover, several studies have demonstrated that maintenance on even large daily gabapentin doses ($\geq$2400 mg/day) had relatively minor, if any, impact on cognitive task performance across various domains (e.g., Leach et al., 1997; Salinsky et al., 2002). Lastly, executive function improved in cannabis users during treatment on gabapentin (Mason et al., 2012), although whether this was a direct consequence of gabapentin administration, or a result of reduced cannabis use is not clear.

Separate and combined effects of gabapentin and THC in cannabis users discriminating THC: Summary and conclusions

As predicted, oral THC served as a discriminative stimulus, produced positive subjective effects, elevated heart rate, and impaired psychomotor performance. Both doses of gabapentin substituted for the THC discriminative stimulus and engendered subjective and performance-impairing effects that overlapped with those of THC when administered alone. When administered concurrently, gabapentin shifted the discriminative-stimulus effects of THC leftward/upward, and combinations of THC and gabapentin generally produced larger effects on CB-sensitive outcomes relative to THC alone. These results suggested that gabapentin might facilitate cannabis abstinence by producing effects that overlap with CBs in keeping with effective agonist treatment strategies for other SUDs, which supported the advancement of gabapentinoids as CUD medications. A logical next step was to evaluate the ability of a gabapentinoid to reduce cannabis self-administration.

Influence of pregabalin maintenance on cannabis reinforcement and related behaviors: Self-administration rationale and design

The positive findings from the studies with gabapentin (Lile et al., 2016; Mason et al., 2012) prompted the conduct of two experiments to continue the investigation of the pharmacotherapeutic potential of gabapentinoids (Lile et al., 2021). Those studies used a human laboratory model that included outpatient medication maintenance (Wesley et al., 2018) to test the ability of two doses of pregabalin (300 and 450 mg/day) to impact experimentally controlled and naturalistic cannabis use behaviors, as well as the response to cannabis. Cannabis self-administration was chosen as the primary outcome because the ability of an intervention to reduce drug-taking behavior has been predictive of effective treatments for SUD (Comer et al., 2008; Haney & Spealman, 2008). Pregabalin was chosen as the maintenance drug because of its improved pharmacokinetic profile compared to gabapentin.

Various types of self-administration procedures and schedules of reinforcement have been used in human laboratory studies; see Jones and Comer (2013) for review. The present experiments used a procedure in which subjects completed sampling and self-administration sessions on separate days for each pregabalin and cannabis dose combination (Stoops et al., 2007). During sampling sessions and after pregabalin administration, subjects smoked four puffs from two cannabis cigarettes (i.e., 8 total puffs; 0% or 5.9% THC) to familiarize them with the effects of the cannabis they could work for in a subsequent session. Subjects were instructed to pay attention to the effects of the cannabis, as they would be given the opportunity to work to receive puffs from the same cannabis condition in the next session. During subsequent self-administration sessions, immediately after maintenance dosing, subjects completed a self-administration task to earn puffs of cannabis sampled the day prior and/or an alternative monetary reinforcer ($0.50), which were concurrently available on independent progressive-ratio schedules. A maximum of 8 reinforcers could be earned (e.g., 8 puffs of cannabis, $4 or some combination of cannabis puffs and money), and subjects were required to make a total of 8 choices. The initial response requirement for each reinforcer was 400 responses (i.e., mouse clicks). The completion of a response requirement for a given reinforcer (i.e., cannabis puffs or money) increased the response requirement for that reinforcer by 200. Once the task was complete, subjects self-administered the number of puffs earned using a paced smoking procedure. Table 2 illustrates the design and events for both experiments.

Influence of pregabalin maintenance on cannabis reinforcement and related behaviors: Methods

Thirteen adults (aged 19–32 years; mean=26) were enrolled in EXP 1 and completed at least one of the two maintenance phases; 11 subjects completed both phases. Fourteen subjects (aged 18–31 years; mean=25) were enrolled in EXP 2 and completed at least one of the two maintenance phases; 13 subjects completed both phases. Prior to enrollment, subjects completed a screening procedure that assessed their medical, psychiatric, and drug use histories and included standard laboratory tests of physiological function. The protocol consisted of two practice sessions followed by two outpatient pregabalin maintenance phases (11 days on 300 mg/day in EXP 1 and 15 days on 450 mg/day in EXP 2; doses titrated up to the target per product insert guidance) that concluded with four experimental sessions within each phase. Doses were based on those recommended for neuropathic pain and seizures (i.e., 150–600 mg/day). During experimental sessions, maintenance continued, and subjects completed two 2-day blocks of sampling and self-administration sessions to determine the reinforcing effects of smoked cannabis (0% and 5.9% THC).

On maintenance-only days, subjects completed online questionnaires to assess recent activities, medication side effects, sleep quality, cannabis craving, and other self-reported substance use (not comprehensively covered here). Subjects were administered the first of the two daily pregabalin doses on-site and were given take-home doses for nights and weekends. Maintenance day attendance was reinforced by escalating compensation that reset if an appointment was missed. Pregabalin dosing adherence outside of the laboratory was monitored via Wisepill technology, which is a capsule container that detects when it is opened (i.e., presumptive dosing) and makes use of mobile phone and Internet technologies to provide real-time medication management. Naturalistic cannabis use was also monitored using semi-quantitative urinalysis. To simulate the motivation to quit using cannabis, and in an effort to enhance the ability to detect pregabalin effects on naturalistic cannabis use, subjects were asked to attempt abstinence from cannabis use and were incentivized to abstain on maintenance days prior to experimental cannabis administration. Abstinence was determined by assaying urine samples for the THC metabolite 11-nor-9-carboxy-THC (THC-COOH) and creatinine. Creatinine-normalized THC-COOH ratios less than 100 ng/mL were considered negative for the purpose of determining whether a participant qualified for an incentive, consistent with methods used in clinical trial designs (e.g., Levin et al., 2013).

Practice and experimental (i.e., sampling and self-administration) sessions were conducted on weekdays and lasted 6.5 h. The maintenance day questionnaires and dosing procedures described above were also completed during sessions. Data were collected in fixed order, prior to, and following, cannabis administration for 4 h, at regular intervals, which varied by outcome measure. In addition to the self-administration task, a subjective effects questionnaire, the DSST, a measure of attentional bias (dot probe task coupled with eye-tracking technology) and cardiovascular indices were included in experimental sessions to evaluate the effects of pregabalin and cannabis combinations (not comprehensively covered here).

Influence of pregabalin maintenance on cannabis reinforcement and related behaviors: Results and discussion

Self-administration

Subjects in EXP 1 self-administered a significantly greater number of puffs of smoked cannabis containing active THC compared to placebo, but the number of puffs did not vary by pregabalin condition. In EXP 2, the difference in the number

TABLE 2 Influence of pregabalin maintenance on cannabis reinforcement and related behaviors: Design and events.

EXP 1. Design and events															
First maintenance phase (0mg/day pregabalin)								**0% THC**		**5.9% THC**					
Study Day	1	2	3	4	5	6	7	8	9	10	11				
Week Day	M	T	W	Th	F	S	Su	M	T	W	Th				
Activity	Maint	Maint	Maint	Maint	Maint	Maint	Maint	Sampling	Self-Admin	Sampling	Self-Admin				
Urinalysis	✓				✓			✓							
10-Day tapering dose and washout period															
Second maintenance phase (300mg/day pregabalin)								5.9% THC		0% THC					
Study Day	12	13	14	15	16	17	18	19	20	21	22				
Week Day	M	T	W	Th	F	S	Su	M	T	W	Th				
Activity	Maint	Maint	Maint	Maint	Maint	Maint	Maint	Sampling	Self-Admin	Sampling	Self-Admin				
Urinalysis	✓				✓			✓							
EXP 2. Design and events															
First maintenance phase (0mg/day pregabalin)												0% THC		5.9% THC	
Study Day	1	2	3	4	5	6	7	8	9	10	11	12	13	14	15
Week Day	Th	F	S	Su	M	T	W	Th	F	S	Su	M	T	W	Th
Activity	Maint	Maint	Maint	Maint	Maint	Maint	Maint	Maint	Maint	Maint	Maint	Sampling	Self-Admin	Sampling	Self-Admin
Urinalysis	✓				✓							✓			

TABLE 2 Influence of pregabalin maintenance on cannabis reinforcement and related behaviors: Design and events—cont'd

6-Day tapering dose and washout period															
Second maintenance phase (450 mg/day pregabalin)												5.9% THC		0% THC	
Study Day	16	17	18	19	20	21	22	23	24	25	26	27	28	29	30
Week Day	Th	F	S	Su	M	T	W	Th	F	S	Su	M	T	W	Th
Activity	Maint	Maint	Maint	Maint	Maint	Maint	Maint	Maint	Maint	Maint	Maint	Sampling	Self-Admin	Sampling	Self-Admin
Urinalysis	✓				✓							✓			

Maint, Maintenance; *Self-Admin,* Self-administration.
The order of pregabalin maintenance dose and %THC concentration is representative.
✓ indicates days on which urine samples were collected for semi-quantitative urinalysis.

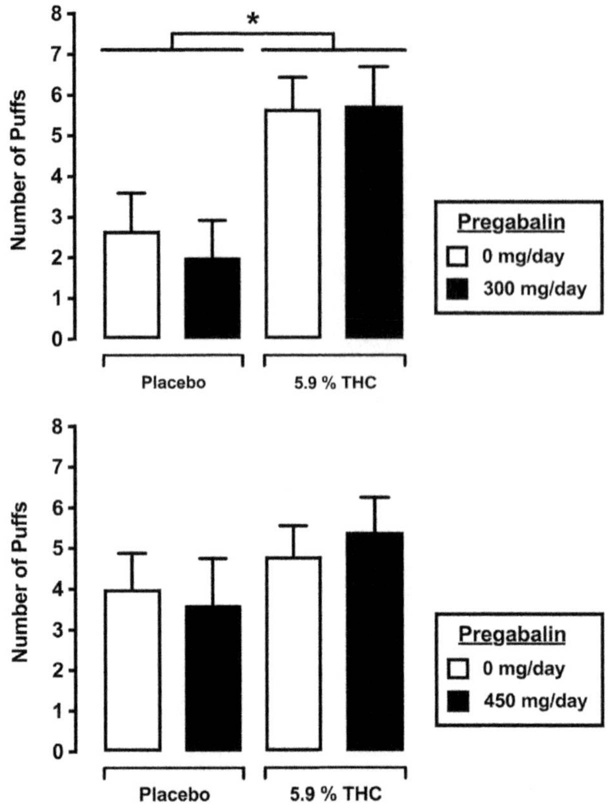

FIG. 3 Self-administration (number of puffs; *y*-axis) of smoked cannabis containing 0% THC (*x*-axis, left side) and 5.9% THC (*x*-axis, right side) during 0 mg/day (open bars) and 300 mg/day (filled bars, top panel) or 450 mg/day pregabalin (filled bars, bottom panel). Unidirectional brackets indicate 1 SEM. The asterisk denotes a significant effect of cannabis concentration.

of self-administered puffs of active versus placebo smoked cannabis approached significance ($p=0.06$) and did not appear to vary by pregabalin condition. Self-administration data from both experiments are shown in Fig. 3. Consistent with these results, naturalistic cannabis use, as measured by semi-quantitative urinalysis and self-report, was also not significantly changed by either dose of pregabalin. As noted, the ability of a medication to attenuate the reinforcing effects of abused drugs in laboratory studies has been predictive of therapeutic efficacy (Comer et al., 2008; Haney & Spealman, 2008). For example, maintenance on oral extended release d-amphetamine reduced intravenous and intranasal cocaine self-administration in human laboratory studies that enrolled nontreatment-seeking subjects (Lile et al., 2020), consistent with some clinical trials in patients seeking treatment for cocaine use disorder (reviewed in Tardelli et al., 2020). Similarly, medications that are clinically effective for the treatment of opioid, alcohol, and tobacco use disorders have been found to reduce self-administration of those drugs in laboratory studies (e.g., Comer et al., 2005; Davidson et al., 1999; McKee et al., 2012). That pregabalin failed to impact cannabis self-administration suggests that pregabalin is unlikely to be effective as a treatment for CUD by reducing the reinforcing effects of cannabis.

Important to note is that the predictive validity of human laboratory cannabis self-administration procedures for the treatment of CUD has yet to be demonstrated. To date, it appears that only the combination of THC and the α_{2A} adrenergic agonist lofexidine has been tested in both a human laboratory study and a treatment-oriented placebo-controlled clinical trial. The combination of these drugs improved sleep and decreased cannabis withdrawal, craving and relapse in nontreatment-seeking subjects in the laboratory (Haney et al., 2008), but did not promote the treatment success target of 3 weeks of abstinence, or reductions in use, compared to placebo in a clinical trial (Levin et al., 2016). Worth mentioning is that a fully powered, controlled clinical trial to evaluate 1200 mg/day gabapentin as a CUD medication in treatment-seeking patients has been completed but not published (NCT00974376). In that larger trial, gabapentin did not increase the percentage of negative urinary drug screens for recent cannabis use at 12 weeks relative to placebo, which is consistent with the results with pregabalin from the present study. Future clinical trials in treatment-seeking patients should be conducted to evaluate medications that have reduced cannabis self-administration in human laboratory studies (e.g., naltrexone, nabilone, nabilone+zolpidem, high-dose THC; Haney et al., 2013; Haney et al., 2015; Herrmann et al., 2016;

Schlienz et al., 2018), which would provide further information about the predictive validity of these procedures and hopefully advance one of these treatments to clinical use.

Subject ratings

In both experiments, cannabis significantly increased various positive VAS items, but these ratings were not impacted by either dose of pregabalin. Results for the questionnaire items any effect and stoned are presented in Fig. 4. These findings are inconsistent with the pregabalin findings described above, in which acutely administered gabapentin enhanced the discriminative-stimulus effects and peak ratings on most subjective effects questionnaire items increased by THC (Lile et al., 2016). We predicted, based on prior human laboratory studies that tested agonist replacement medications (i.e., compounds that shared a mechanism of action and had overlapping subjective and discriminative-stimulus effects with the abused drug) with demonstrated efficacy in clinical trials (Donny et al., 2005; Lile et al., 2020; Sobel et al., 2004), that pregabalin maintenance would attenuate the subjective response to cannabis, rather than enhance. Prior studies found that the subjective effects of the abused drug were reduced during maintenance, possibly via tolerance. In our studies, however, pregabalin did not produce an analogous attenuation of the subjective response to cannabis that would have been characteristic of an agonist replacement medication. One possibility is that treatment with a compound having a common direct mechanism of action (i.e., another CB receptor ligand) would be needed to alter the subjective effects of cannabis, and that maintenance on a drug having shared interoceptive effects, which might be the result of downstream receptor actions, is not sufficient.

Attentional bias

In both experiments, subjects exhibited an attentional bias to cannabis cues relative to neutral cues, as measured by a visual probe task and eye-tracking technology, which is consistent with prior studies using similar methods in cannabis users (Alcorn 3rd et al., 2019; Field et al., 2006). However, cannabis and/or pregabalin administration had minimal impact

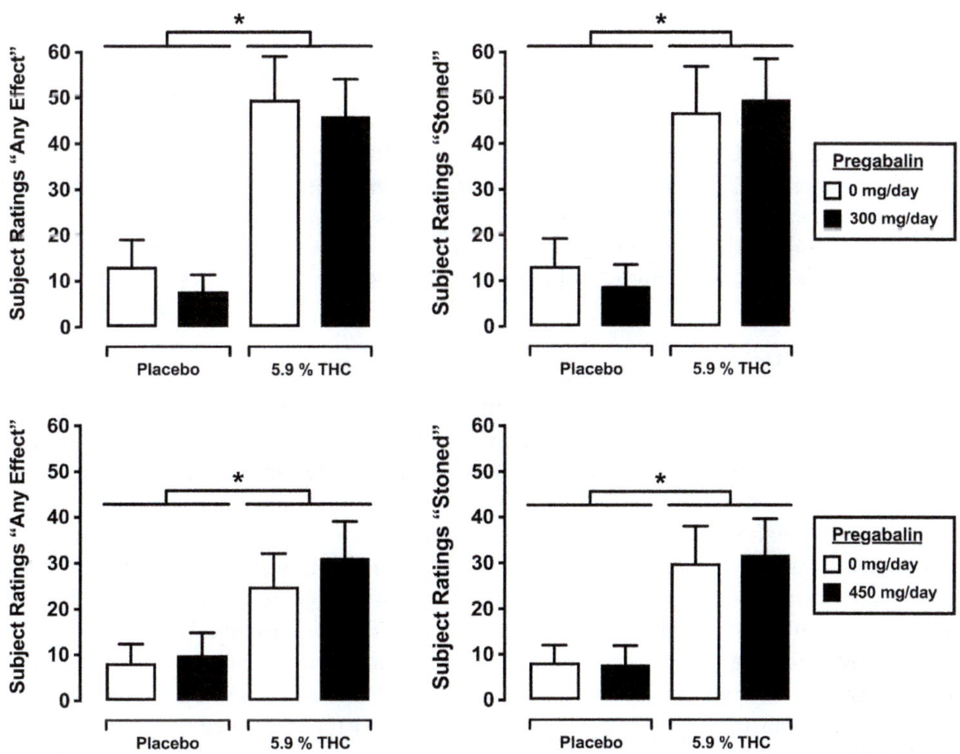

FIG. 4 Peak Visual Analog Scale ratings (out of 100) of any effect (*y*-axis; left panels) and stoned (*y*-axis; right panels) following administration of smoked cannabis containing 0% THC (*x*-axis, left side) and 5.9% THC (*x*-axis, right side) during 0mg/day (open bars) and 300mg/day (filled bars, top panels) or 450mg/day pregabalin (filled bars, bottom panels). Unidirectional brackets indicate 1 SEM. The asterisks denote a significant effect of cannabis concentration.

on gaze time or number of fixations on this task, suggesting that pregabalin would not be effective as a CUD medication by reducing attention to cannabis cues.

Safety and tolerability

In EXP 1, no significant effects of cannabis or pregabalin were detected on the DSST. In EXP 2, the number and percentage of correct responses on the DSST were significantly reduced by 450 mg/day pregabalin (approximately 2% and 4%, respectively). As for the cardiovascular outcomes, active cannabis significantly elevated heart rate in both experiments, and heart rate was increased by 450 mg/day pregabalin in EXP 2. In EXP 1, there was an interaction between cannabis and pregabalin on diastolic blood pressure. Active cannabis+300 mg/day pregabalin decreased diastolic blood pressure relative to active cannabis alone; however, active cannabis+placebo pregabalin increased diastolic blood pressure relative to placebo pregabalin alone. No effects of cannabis or pregabalin were observed on blood pressure in EXP 2. In general, pregabalin was safe and well tolerated. It produced only small magnitude changes on these outcomes alone and when combined with cannabis and typically only at the larger pregabalin dose.

Influence of pregabalin maintenance on cannabis reinforcement and related behaviors: Summary and conclusion

Prototypical cannabis effects were observed during experimental sessions; cannabis was self-administered and increased ratings on positive subjective effects questionnaire items and heart rate. However, pregabalin generally did not modulate these effects and only had small magnitude effects on a few experimental session outcomes. In addition, pregabalin maintenance did not impact naturalistic cannabis use and had minimal effects on outcomes thought to be related to cannabis use (e.g., sleep, cannabis craving; data not shown). These human laboratory results do not support the use of pregabalin as a stand-alone pharmacotherapy for CUD, although a controlled clinical trial in treatment-seeking patients who meet criteria for CUD would be needed to fully evaluate the efficacy of pregabalin for this indication.

Final conclusions

Gabapentinoids and CBs have overlapping pharmacology, therapeutic effects, and interoceptive/subjective effects and are safe and well tolerated alone and in combination in cannabis users. However, gabapentin and pregabalin do not appear to be effective at consistently reducing cannabis intake on their own. Nonetheless, given the interactions between these systems, hints of a positive signal (i.e., the pilot trial and the agonist-like profile of gabapentin) and lack of available alternatives, future research should consider combinations of gabapentinoids with other promising pharmacotherapies (e.g., CB ligands; D'Souza et al., 2019; Freeman et al., 2020; Lintzeris et al., 2019) as potential treatments for CUD. The use of complementary drug combinations has been explored as a treatment option for various SUDs and is common practice for the treatment of tobacco use disorder (e.g., Gomez-Coronado et al., 2018).

Applications to other areas

The literature reviewed here illustrates the challenges of medications development research. Ideally, a coordinated, bidirectional translational approach to medications development (i.e., animal laboratory⇆human laboratory⇆clinical trial) for SUDs should be used. For a given drug of abuse, compelling clinical findings should guide the development of predictive laboratory models (i.e., reverse translation), which can then be used to screen new compounds or combinations (i.e., forward translation). To date, laboratory drug self-administration has been the most predictive model with which to evaluate SUD treatment effectiveness. In the case of cannabis use disorder, however, the predictive validity of human laboratory cannabis self-administration procedures has yet to be demonstrated, and reliable THC self-administration procedures have not been successfully established in nonhuman animals. These goals should be a top priority for our field.

The data reviewed here regarding the abuse potential and safety/tolerability of gabapentinoids alone and combined with cannabis are also generalizable to nontreatment-seeking users of medical or nonmedical cannabis products who might be prescribed a gabapentinoid for one of their approved indications. The existing evidence suggests that using cannabis during treatment with a gabapentinoid will likely be safe and well tolerated, though caution is warranted to monitor for potentially larger effects than anticipated following combined administration. Moreover, given the overlap in the therapeutic profile of CBs and gabapentinoids for managing pain, insomnia, and anxiety, combinations of these two drug classes might be useful for one or more of these conditions without producing side effects above and beyond the constituents.

Mini-dictionary of terms

- **Drug discrimination.** An operant procedure in which responding is reinforced based on the presence or absence of receptor-selective interoceptive cues produced by a drug.
- **Interoception.** Perception of the internal state of the body, which cannot be directly observed.
- **Reinforcer.** A stimulus, such as drug, that increases the likelihood of the behavior that preceded its delivery.
- **Self-administration.** An operant procedure in which responding is reinforced by the delivery of a stimulus.
- **Subjective drug effects.** The myriad drug effects experienced internally by the user, which can only be measured by self-report. Contrast to objective drug effects that can be directly observed.

Key facts of Gabapentinoids for the treatment of cannabis use disorder: A focus on human laboratory research

- Cannabis use disorder is largely characterized by maladaptive drug use decisions at the expense of other choices, as well as tolerance, craving, and withdrawal upon cessation of use.
- Operant procedures (i.e., drug discrimination and drug self-administration) are valuable as primary outcomes because they provide an objective means of measuring the interoceptive and reinforcing effects of drugs.
- In the translational approach for medications development for SUDs, human laboratory research serves as a valuable link between preclinical nonhuman animal research and clinical trials of medication effectiveness in the target population.
- Human laboratory studies have the advantage of using within-subjects designs and smaller sample sizes to evaluate initial safety, tolerability, and efficacy of putative SUD pharmacotherapies.
- Laboratory models used to screen potential medications for SUD should have demonstrated predictive validity based on clinical effectiveness.

Summary points

- No medications are approved for the treatment of cannabis use disorder.
- Gabapentin and pregabalin are voltage-dependent calcium channel (VDCC) ligands.
- VDCCs are modulated directly and indirectly by CB ligands.
- CB agonists and VDCC ligands share therapeutic effects such as improvements in neuropathic pain, sleep, and anxiety.
- A pilot clinical trial in treatment-seeking cannabis users found that gabapentin treatment decreased urinary THC metabolite levels, self-reported cannabis use, craving, and depression and improved performance on tests of executive function, which prompted further research in this area.
- A human laboratory study demonstrated that gabapentin and THC have overlapping discriminative stimulus and subjective effects in cannabis users.
- In contrast, another human laboratory study that maintained nontreatment-seeking cannabis users on pregabalin found no effect on cannabis intake or effects.
- A larger follow-up trial of gabapentin for CUD yielded negative results.
- These data do not support the use of gabapentinoids as stand-alone treatments for CUD.

References

Alcorn, J. L., 3rd, Marks, K. R., Stoops, W. W., Rush, C. R., & Lile, J. A. (2019). Attentional bias to cannabis cues in cannabis users but not cocaine users. *Addictive Behaviors*, *88*, 129–136.

Aracil-Fernández, A., Almela, P., & Manzanares, J. (2013). Pregabalin and topiramate regulate behavioural and brain gene transcription changes induced by spontaneous cannabinoid withdrawal in mice. *Addiction Biology*, *18*, 252–262.

Atwal, N., Casey, S. L., Mitchell, V. A., & Vaughan, C. W. (2019). THC and gabapentin interactions in a mouse neuropathic pain model. *Neuropharmacology*, *144*, 115–121.

Bestard, J. A., & Toth, C. C. (2011). An open-label comparison of nabilone and gabapentin as adjuvant therapy or monotherapy in the management of neuropathic pain in patients with peripheral neuropathy. *Pain Practice*, *11*, 353–368.

Bockbrader, H. N., Wesche, D., Miller, R., Chapel, S., Janiczek, N., & Burger, P. (2010). A comparison of the pharmacokinetics and pharmacodynamics of pregabalin and gabapentin. *Clinical Pharmacokinetics*, *49*, 661–669.

Bolin, B. L., Alcorn, J. L., Reynolds, A. R., Lile, J. A., & Rush, C. R. (2016). Human drug discrimination: A primer and methodological review. *Experimental and Clinical Psychopharmacology*, *24*, 214–228.

Caprioli, A., Coccurello, R., Rapino, C., Di Serio, S., Di Tommaso, M., Vertechy, M., Vacca, V., Battista, N., Pavone, F., Maccarrone, M., & Borsini, F. (2012). The novel reversible fatty acid amide hydrolase inhibitor ST4070 increases endocannabinoid brain levels and counteracts neuropathic pain in different animal models. *Journal of Pharmacology and Experimental Therapeutics*, *342*, 188–195.

Center for Behavioral Health Statistics and Quality. (2017). *Treatment episode data set (TEDS): 2005-2015*. Rockville, MD: National Admissions to Substance Abuse Treatment Services. Substance Abuse and Mental Health Services Administration.

Center for Behavioral Health Statistics and Quality. (2020). *2019 National survey on drug use and health: Detailed tables*. Rockville, MD: Substance Abuse and Mental Health Services Administration.

Clarke, H., Bonin, R. P., Orser, B. A., Englesakis, M., Wijeysundera, D. N., & Katz, J. (2012). The prevention of chronic postsurgical pain using gabapentin and pregabalin: A combined systematic review and meta-analysis. *Anesthesia & Analgesia*, *115*, 428–442.

Comer, S. D., Ashworth, J. B., Foltin, R. W., Johanson, C. E., Zacny, J. P., & Walsh, S. L. (2008). The role of human drug self-administration procedures in the development of medications. *Drug and Alcohol Dependence*, *96*, 1–15.

Comer, S. D., Walker, E. A., & Collins, E. D. (2005). Buprenorphine/naloxone reduces the reinforcing and subjective effects of heroin in heroin-dependent volunteers. *Psychopharmacology*, *181*, 664–675.

D'Souza, D., Cortes-Briones, J., Creatura, G., Bluez, G., Thurnauer, H., Deaso, E., Bielen, K., Surti, T., Radhakrishnan, R., Gupta, A., Gupta, S., Cahill, J., Sherif, M. A., Makriyannis, A., Morgan, P. T., Ranganathan, M., & Skosnik, P. D. (2019). Efficacy and safety of a fatty acid amide hydrolase inhibitor (PF-04457845) in the treatment of cannabis withdrawal and dependence in men: A double-blind, placebo-controlled, parallel group, phase 2a single-site randomized controlled trial. *Lancet Psychiatry*, *6*, 35–45.

Davidson, D., Palfai, T., Bird, C., & Swift, R. (1999). Effects of naltrexone on alcohol self-administration in heavy drinkers. *Alcoholism: Clinical and Experimental Research*, *23*, 195–203.

de Paris, F., Sant'Anna, M. K., Vianna, M. R., Barichello, T., Busnello, J. V., Kapczinski, F., Quevedo, J., & Izquierdo, I. (2003). Effects of gabapentin on anxiety induced by simulated public speaking. *Journal of Psychopharmacology*, *17*, 184–188.

Delahoy, P., Thompson, S., & Marschner, I. C. (2010). Pregabalin versus gabapentin in partial epilepsy: A meta-analysis of dose-response relationships. *BioMed Central Neurology*, *10*, 104.

Donny, E. C., Brasser, S. M., Bigelow, G. E., Stitzer, M. L., & Walsh, S. L. (2005). Methadone doses of 100 mg or greater are more effective than lower doses at suppressing heroin self-administration in opioid-dependent volunteers. *Addiction*, *100*, 1496–1509.

Elkashef, A., Vocci, F., Huestis, M., Haney, M., Budney, A., Gruber, A., & el-Guebaly, N. (2008). Marijuana neurobiology and treatment. *Substance Abuse*, *29*, 17–29.

Field, M., Eastwood, B., Bradley, B. P., & Mogg, K. (2006). Selective processing of cannabis cues in regular cannabis users. *Drug and Alcohol Dependence*, *85*, 75–82.

Freeman, T. P., Hindocha, C., Baio, G., Shaban, N. D. C., Thomas, E. M., Astbury, D., Freeman, A. M., Lees, R., Craft, S., Morrison, P. D., Bloomfield, M. A. P., O'Ryan, D., Kinghorn, J., Morgan, C. J. A., Mofeez, A., & Curran, H. V. (2020). Cannabidiol for the treatment of cannabis use disorder: A phase 2a, double-blind, placebo-controlled, randomized, adaptive Bayesian trial. *Lancet Psychiatry*, *7*, 865–874.

Gale, J. D., & Houghton, L. A. (2011). Alpha 2 Delta (α(2)δ) ligands, gabapentin and pregabalin: What is the evidence for potential use of these ligands in irritable bowel syndrome. *Frontiers in Pharmacology*, *2*, 28.

Gidal, B. E., Maly, M. M., Kowalski, J. W., Rutecki, P. A., Pitterle, M. E., & Cook, D. E. (1998). Gabapentin absorption: Effect of mixing with foods of varying macronutrient composition. *Annals of Pharmacotherapy*, *32*, 405–409.

Gomez-Coronado, N., Walker, A. J., Berk, M., & Dodd, S. (2018). Current and emerging pharmacotherapies for cessation of tobacco smoking. *Pharmacotherapy*, *38*, 235–258.

Gunduz, O., Karadag, H. C., & Ulugol, A. (2011). Synergistic anti-allodynic effects of nociception/orphanin FQ and cannabinoid systems in neuropathic mice. *Pharmacology, Biochemistry and Behavior*, *99*, 540–544.

Haney, M., Cooper, Z. D., Bedi, G., Vosburg, S. K., Comer, S. D., & Foltin, R. W. (2013). Nabilone decreases marijuana withdrawal and a laboratory measure of marijuana relapse. *Neuropsychopharmacology*, *38*, 1557–1565.

Haney, M., Hart, C. L., Vosburg, S. K., Comer, S. D., Reed, S. C., & Foltin, R. W. (2008). Effects of THC and lofexidine in a human laboratory model of marijuana withdrawal and relapse. *Psychopharmacology*, *197*, 157–168.

Haney, M., Ramesh, D., Glass, A., Pavlicova, M., Bedi, G., & Cooper, Z. D. (2015). Naltrexone maintenance decreases cannabis self-administration and subjective effects in daily cannabis smokers. *Neuropsychopharmacology*, *40*, 2489–2498.

Haney, M., & Spealman, R. (2008). Controversies in translational research: Drug self-administration. *Psychopharmacology*, *3*, 403–419.

Hasnie, F. S., Breuer, J., Parker, S., Wallace, V., Blackbeard, J., Lever, I., Kinchington, P. R., Dickenson, A. H., Pheby, T., & Rice, A. S. (2007). Further characterization of a rat model of varicella zoster virus-associated pain: Relationship between mechanical hypersensitivity and anxiety-related behavior, and the influence of analgesic drugs. *Neuroscience*, *144*, 1495–1508.

Herrmann, E. S., Cooper, Z. D., Bedi, G., Ramesh, D., Reed, S. C., Comer, S. D., Foltin, R. W., & Haney, M. (2016). Effects of zolpidem alone and in combination with nabilone on cannabis withdrawal and a laboratory model of relapse in cannabis users. *Psychopharmacology*, *233*, 2469–2478.

Hindmarch, I., Dawson, J., & Stanley, N. (2005). A double-blind study in healthy volunteers to assess the effects on sleep of pregabalin compared with alprazolam and placebo. *Sleep*, *28*, 187–193.

Holtzman, S. G., & Locke, K. W. (1988). Neural mechanisms of drug stimuli: Experimental approaches. *Psychopharmacology Series*, *4*, 138–153.

Howlett, A. C., Blume, L. C., & Dalton, G. D. (2010). CB(1) cannabinoid receptors and their associated proteins. *Current Medicinal Chemistry*, *17*, 1382–1393.

Jones, J. D., & Comer, S. D. (2013). A review of human drug self-administration procedures. *Behavioural Pharmacology*, *24*, 384–395.

Karst, M., Wippermann, S., & Ahrens, J. (2010). Role of cannabinoids in the treatment of pain and (painful) spasticity. *Drugs*, *70*, 2409–2438.

Kelly, T. H., Stoops, W. W., Perry, A. S., Prendergast, M. A., & Rush, C. R. (2003). Clinical neuropharmacology of drugs of abuse: A comparison of drug discrimination and subject-report measures. *Behavioral and Cognitive Neuroscience Reviews*, *2*, 227–260.

Kleinman, N. L., Sadosky, A., Seid, J., Martin, R. C., & Labiner, D. M. (2012). Costs, work absence, and adherence in patients with partial onset seizures prescribed gabapentin or pregabalin. *Epilepsy Research*, *102*, 13–22.

Leach, J. P., Girvan, J., Paul, A., & Brodie, M. J. (1997). Gabapentin and cognition: A double blind, dose ranging, placebo controlled study in refractory epilepsy. *Journal of Neurology, Neurosurgery, and Psychiatry*, *62*, 372–376.

Levin, F. R., Mariani, J., Brooks, D. J., Pavlicova, M., Nunes, E. V., Agosti, V., Bisaga, A., Sullivan, M. A., & Carpenter, K. M. (2013). A randomized double-blind, placebo controlled trial of venlafaxine-extended release for co-occuring dependence and depressive disorders. *Addiction*, *108*, 1084–1094.

Levin, F. R., Mariani, J. J., Pavlicova, M., Brooks, D., Glass, A., Mahony, A., Nunes, E. V., Bisaga, A., Dakwar, E., Carpenter, K. M., Sullivan, M. A., & Choi, J. C. (2016). Dronabinol and lofexidine for cannabis use disorder: A randomized, double-blind, placebo-controlled trial. *Drug and Alcohol Dependence*, *159*, 53–60.

Lile, J. A., Alcorn, J. L., Hays, L. R., Kelly, T. H., Stoops, W. W., Wesley, M. J., & Westgate, P. M. (2021). Influence of pregabalin maintenance and on cannabis effects and related behaviors in daily cannabis users. *Experimental and Clinical Psychopharmacology*. https://doi.org/10.1037/pha0000464. ePub ahead of print.

Lile, J. A., Johnson, A. R., Banks, M. L., Hatton, K. W., Hays, L. R., Nicholson, K. L., Poklis, J. L., Rayapati, A. O., Rush, C. R., Stoops, W. W., & Negus, S. S. (2020). Pharmacological validation of a translational model of cocaine use disorder: Effects of d-amphetamine maintenance on choice between intravenous cocaine and a nondrug alternative in humans and rhesus monkeys. *Experimental and Clinical Psychopharmacology*, *28*, 169–180.

Lile, J. A., Kelly, T. H., Charnigo, R. J., Stinchcomb, A. L., & Hays, L. R. (2013). Pharmacokinetic and pharmacodynamics profile of supratherapeutic oral doses of Δ^9-THC in cannabis users. *Journal of Clinical Pharmacology*, *53*, 680–690.

Lile, J. A., Wesley, M. J., Kelly, T. H., & Hays, L. R. (2016). Separate and combined effects of gabapentin and Δ^9-tetrahydrocannabinol in humans discriminating Δ^9-tetrahydrocannabinol. *Behavioural Pharmacology*, *27*, 215–224.

Lintzeris, N., Bhardwaj, A., Mills, L., Dunlop, A., Copeland, J., McGregor, I., Bruno, R., Gugusheff, J., Phung, N., Montebello, M., Chan, T., Kirby, A., Hall, M., Jefferies, M., Luksza, J., Shanahan, M., Kevin, R., Allsop, D., & Agonist Replacement for Cannabis Dependence Study Group. (2019). Nabiximols for the treatment of cannabis dependence: A randomized clinical trial. *Journal of the American Medical Association Internal Medicine*, *179*, 1242–1253.

Lloyd, A., Boomershine, C. S., Choy, E. H., Chandran, A., & Zlateva, G. (2012). The cost-effectiveness of pregabalin in the treatment of fibromyalgia: US perspective. *Journal of Medical Economics*, *15*, 481–492.

Lozovaya, N., Min, R., Tsintsadze, V., & Burnashev, N. (2009). Dual modulation of CNS voltage-gated calcium channels by cannabinoids: Focus on CB_1 receptor-independent effects. *Cell Calcium*, *46*, 154–162.

Luszczki, J. J., & Florek-Łuszczki, M. (2012). Synergistic interaction of pregabalin with the synthetic cannabinoid WIN 55,212-2 mesylate in the hot-plate test in mice: An isobolographic analysis. *Pharmacological Reports*, *64*, 723–732.

Mason, B. J., Crean, R., Goodell, V., Light, J. M., Quello, S., Shadan, F., Buffkins, K., Kyle, M., Adusumali, M., Begovic, A., & Rao, S. (2012). A proof-of-concept randomized controlled study of gabapentin: Effects on cannabis use, withdrawal and executive function deficits in cannabis-dependent adults. *Neuropsychopharmacology*, *37*, 1689–1698.

McKee, S. A., Weinberger, A. H., Shi, J., Tetrault, J., & Coppola, S. (2012). Developing and validating a human laboratory model to screen medications for smoking cessation. *Nicotine & Tobacco Research*, *14*, 1362–1371.

Nakano, S., Gillespie, H. K., & Hollister, L. E. (1978). A model for evaluation of antianxiety drugs with the use of experimentally induced stress: Comparison of nabilone and diazepam. *Clinical Pharmacology and Therapeutics*, *23*, 54–62.

Palmer, J. A., Higuera, E. S., Chang, L., & Chaplan, S. R. (2008). Fatty acid amide hydrolase inhibition enhances the anti-allodynic actions of endocannabinoids in a model of acute pain adapted for the mouse. *Neuroscience*, *154*, 1554–1561.

Roth, T., Lankford, D. A., Bhadra, P., Whalen, E., & Resnick, E. M. (2012). Effect of pregabalin on sleep in patients with fibromyalgia and sleep maintenance disturbance: A randomized, placebo-controlled, 2-way crossover polysomnography study. *Arthritis Care and Research*, *64*, 597–606.

Saldaña, M. T., Pérez, C., Navarro, A., Masramón, X., & Rejas, J. (2012). Pain alleviation and patient-reported health outcomes following switching to pregabalin in individuals with gabapentin-refractory neuropathic pain in routine medical practice. *Clinical Drug Investigation*, *32*, 401–412.

Salinsky, M. C., Binder, L. M., Oken, B. S., Storzbach, D., Aron, C. R., & Dodrill, C. B. (2002). Effects of gabapentin and carbamazepine on the EEG and cognition in healthy volunteers. *Epilepsia*, *43*, 482–490.

Schlienz, N. J., Lee, D. C., Stitzer, M. L., & Vandrey, R. (2018). The effect of high-dose dronabinol (oral THC) maintenance on cannabis self-administration. *Drug and Alcohol Dependence*, *187*, 254–260.

Sills, G. J. (2006). The mechanisms of action of gabapentin and pregabalin. *Current Opinion in Pharmacology*, *6*, 108–113.

Smith, R. V., Havens, J. R., & Walsh, S. L. (2016). Gabapentin misuse, abuse and diversion: A systematic review. *Addiction*, *111*, 1160–1174.

Sobel, B. F., Sigmon, S. C., & Griffiths, R. R. (2004). Transdermal nicotine maintenance attenuates the subjective and reinforcing effects of intravenous nicotine, but not cocaine or caffeine, in cigarette-smoking stimulant abusers. *Neuropsychopharmacology*, *29*, 991–1003.

Stacey, B. R., Emir, B., Petersel, D., & Murphy, K. (2010). Pregabalin in treatment-refractory fibromyalgia. *Open Rheumatology Journal*, *4*, 35–38.

Stoops, W. W., Lile, J. A., Robbins, C. G., Martin, C. A., Rush, C. R., & Kelly, T. H. (2007). The reinforcing, subject-rated, performance, and cardiovascular effects of d-amphetamine: Influence of sensation-seeking status. *Addictive Behaviors*, *6*, 1177–1188.

Tardelli, V. S., Bisaga, A., Arcadepani, F. B., Gerra, G., Levin, F. R., & Fidalgo, T. M. (2020). Prescription psychostimulants for the treatment of stimulant use disorder: A systematic review and meta-analysis. *Psychopharmacology*, *237*, 2233–2255.

Toth, C. (2010). Substitution of gabapentin therapy with pregabalin therapy in neuropathic pain due to peripheral neuropathy. *Pain Medicine*, *11*, 456–465.

Wallace, V. C. J., Segerdahl, A. R., Lambert, D. M., Vandevoorde, S., Blackbeard, J., Pheby, T., Hasnie, F., & Rice, A. S. C. (2007). The effect of the palmitoylethanolamide analogue, palmitoylallylamide (L-29) on pain behaviour in rodent models of neuropathy. *British Journal of Pharmacology*, *151*, 1117–1128.

Ware, M. A., Fitzcharles, M. A., Joseph, L., & Shir, Y. (2010). The effects of nabilone on sleep in fibromyalgia: Results of a randomized controlled trial. *Anesthesia & Analgesia*, *110*, 604–610.

Wesley, M. J., Westgate, P. M., Stoops, W. W., Kelly, T. H., Hays, L. R., & Lile, J. A. (2018). Influence of tiagabine maintenance on cannabis effects and related behaviors in daily cannabis users. *Experimental and Clinical Psychopharmacology*, *26*, 310–319.

Chapter 46

Treating psychosis in users of synthetic cannabinoids

Tatiana Klimenko[a,b] and Svetlana Shakhova[c]

[a]*V. Serbsky Federal Medical Research Center of Psychiatry and Addiction, Ministry of Health, Russian Federation—National Scientific Center of Narcology (Branch), Moscow, Russia,* [b]*The All-Russian State University of Justice, Ministry of Justice of the Russian Federation, Moscow, Russia,* [c]*Department of Drug Addiction Clinical Research, V. Serbsky Federal Medical Research Center of Psychiatry and Addiction, Ministry of Health, Russian Federation—National Scientific Center of Narcology (Branch), Moscow, Russia*

Abbreviations

mg/day	mg per day
NPS	new psychoactive substances
PS	psychoactive substances
SC	synthetic cannabinoids

Introduction

In recent years, the structure of addictive disorders in the Russian Federation demonstrated tendency to change due to an increase in the proportion of people using smoking mixtures with synthetic cannabinoids (SC) for nonmedical purposes (National Institute on Drug Abuse, 2021). Most often, SC users are hospitalized in medical narcological and toxicological institutions with an acute psychoses (Bokhan & Selivanov, 2015; Fattore, 2016). The routine practice suggests that the relief of an intoxication psychosis, regardless of its psychopathological structure and the patient's condition, takes a short time, but many patients then insist on being discharged from the hospital and almost immediately resume SC use. After the relief of acute psychotic disorders, the patients are still addicted SC, demonstrate no criticism of SC use, and no intent to undergo treatment and refrain from SC use (Bondar et al., 2015).

Psychic disorders resulting from synthetic cannabinoid use

Regardless of the disease stage, the psychopathological structure of psychotic conditions resulting from SC use usually involved delirious-oneiroid disorders (Shakhova, 2018).

The clinical dynamics of the psychotic condition following SC intake included several successive stages: (a) the first or affective stage was characterized by a variable affect from emotional depression to euphoria or hypomanic state with unreasonable fun, excessive talkativeness, uncontrollable associations, excessive distraction of attention, physical agitation, fast and incoherent speech, and mental hypersthesia; (b) at the second illusory-hallucinatory stage, multiple, brightly colored, and extremely variable pareidolias appeared, which were easily transformed into colorful and dynamic visual hallucinations with single auditory, tactile, and general feeling deceptions of perception coherent with the hallucination content, with sensual delirium, with mobile and changeable facial expressions reflecting the content of psychopathological experiences, and with the affect lability; (c) at the third stage of deep mental activity disorganization, the patients present with complete detachment from the outside world, with severe disintegration of thinking, with fragmentary hallucinatory-delusional disorders, with fragmentary, incoherent, and not related to external factors verbal activity and the motor sphere deterioration.

In contrast to classic delirium, the delirious disorders resulting from SC use were characterized by rapid onset rather than by a gradual psychopathological symptoms manifestation. There was a fulminant development of anxiety and irritability, increased affect lability, followed by true visual hallucinations, usually scene-like, combined with verbal

Cannabis Use, Neurobiology, Psychology, and Treatment. https://doi.org/10.1016/B978-0-323-89862-1.00013-1

hallucinations and fragmented sensory delusions, accompanied by fear. Psycho-motor agitation, present at the initial stages of the development of the psychotic state, was observed for a short time and was replaced by a state of pronounced inhibition. This condition gradually, over several hours, deteriorated until the patient was completely detached from the surrounding reality. Experiences began to take on a fantastic character, patients were completely immersed in the "cinema" scenes "visible" to them, perceived themselves as living on other planets, felt themselves to be "superheroes," "chosen ones," "communicated with God," "participated," in atomic wars," were present "at the death of the universe. The sense of reality was completely lost. Self-consciousness was changed and deeply disintegrated. Quite often, the patients developed catatonic disorders in the form of a stupor, with individual short-term episodes of agitation within the bed. The patients were overwhelmed, and their environment did not attract their attention. The expression of surprise, horror, anxiety, or fear changed on the face in direct correlation with the content of the experiences.

The delayed onset psychoses development in some cases involves a prodromal period, which manifested itself with sleep disorders (superficial sleep with frequent awakenings, nightmares, frightening dreams), a variable affect with a predominant anxiety affect, asthenic disorders, in individual cases - large convulsive or abortive epileptic seizures.

A comparative clinical-dynamic analysis, taking into account the depth of the clouding of consciousness and its psychopathological structure suggests three clinical variants of a psychosis: (1) delirious; (2) delirious-oneiroid; and (3) amentive-type.

The delirious psychosis was manifested by a false orientation in the environment, multiple, colorful and dynamic pareidolias, scene-like visual hallucinations, single tactile and auditory perception deceptions. The patients observed the imaginary events, but did not get involved in them. There were fragmentary figurative delusional ideas, variable affect with a predominance of anxiety affect and fear accompanied by speech-motor agitation.

The clinical picture of the **delirious-oneiroid psychosis** was characterized by a dreamy consciousness clouding with an influx of dreamlike fantastic hallucinatory delusional experiences, disorientation in place and time and double orientation in one's own personality, and impaired perception of time. Self-consciousness was changed and deeply disintegrated. Patients perceived themselves as participants in fantastic events developing in their imagination, and were completely immersed in the images they experienced. The patients completely lost the feeling of reality, paid no attention to the environment, on their faces there was an expression of surprise, horror, anxiety, or fear, which was in direct connection with the content of psychopathological experiences. The patients developed catatonic disorders in the form of a substupor, catalepsy, and individual short-term episodes of agitation in the bed. The patients looked absent-minded, inhibited, sometimes wandered thoughtfully with an "enchanted smile," periodically froze, movements were scanty and slow, facial expressions were frozen, sometimes they made "flying or swimming" movements with their hands.

The amentive-type psychotic conditions were manifested with a deep consciousness impairment accompanied by the deterioration of all types of mental activity. The patients were deeply disoriented in place, time and self. They were confused, did not reflex over the events that were taking place, neither did they understand the events that were taking place; they perceived the environment in fragments, could not recall their names, did not recognize themselves in the mirror. The patients demonstrated individual incoherent visual and auditory hallucinations and elements of incoherent and fragmentary figurative delirium. There was a pronounced disintegration of all components of thinking and self-consciousness. Accelerated speech consisted of separate and incoherent words. Thinking and speaking were incoherent. Emotions were inadequate, inconsistent, and often change their polarity. The mood was unstable, the state of enthusiasm was alternated with tearfulness and/or emotional depression. Motor agitation in the bed manifested itself with monotonous stereotypical twitches, jerks, and purposeless, inconsistent, and often sweeping movements. No lucid episodes were observed; in the evening and night time, short-term episodes of delirious structure were often revealed.

The cessation of psychotic conditions in all patients was lytic, through prolonged deep sleep. The adequate therapy stopped the psychotic symptoms within several hours (6 ± 0.3 h) after drug-induced sleep. In some cases, the psychosis recovery was followed by 5 ± 1.7 days of post-psychotic asthenia in the form of drowsiness, lethargy, absent mindedness, and decreased cognitive functions. At the end of the oneiroid-type psychosis, residual delirium persisted for several days.

The end of the delirious and delirious-oneiroid type psychotic condition was not followed by amnesia, and the patients reproduced in sufficient detail the content of psychotic experiences, but did not remember the real situation. The amentive psychoses were followed only by isolated and fragmentary recollections of past psychopathological experiences.

The somatic status included: pronounced skin pallor, sclera injection, sweating, tachycardia and arterial hypertension rapidly alternating with persistent hypotension, hyperthermia, and urinary disorders. The neurological status included mydriasis, ataxia, impaired coordination, muscle hypotonia, and tremor.

SC dependent patients develop acute psychoses all disease phase conditions.

Psychosis duration

The duration of psychosis was different: during SC intoxication—from 12 to 24 h, in a withdrawal syndrome—up to 36 h, in delayed onset psychoses—up to 72 h. The average duration of psychosis was 24±10 h.

For a comparative clinical analysis of the selected SC-associated psychosis clinical variants and for the subsequent development of a psychosis prevention strategy, from all the studied signs (social-psychological, clinical-biological, clinical-psychopathological), the parameters were identified that conditionally characterize the SC associated psychotic condition development in one of the selected clinical variants

The delirious psychosis often followed SC intoxication or SC withdrawal in SC or other psychoactive substances dependence with a high level of basic tolerance and a relatively high actual single SC dose, in a combined personal and endogenous pathology without organic mental disorder, organic damage to the central nervous system and somatic problem in the morbid period.

The delirious-oneiroid psychosis correlated with the delayed onset development period, the combined personal and endogenous pathology, without drug abuse history or with single and nonsystemic SC and other PS use in history, with low basic tolerance to PS and without a combined clinically significant somatic and neurological pathology.

The amentive-type psychosis more often developed in withdrawal syndrome, in persons with SC dependence, combined somatic and neurological pathology, organic mental disorder, and low basic tolerance.

Patients with acute psychotic conditions following cannabinoid and especially SC intake should be transferred to the narcological emergency departments.

The treatment strategy selection in psychotic disorders in the structure of complicated cannabinoid withdrawal syndrome is determined by the structure of psychosis and the severity of somatic and neurological symptoms (Table 1).

Regardless of the cannabinoid addiction syndrome phase followed by the acute psychotic disorder, the treatment strategy should be determined by the psychosis psychopathological structure and the characteristics of somatic and neurological disorders.

The neurosis-like symptoms: restlessness, obsession, somatic and vegetative and moderate anxiety disorders require tranquilizers: 1 mg of Phenazepam 2–3 times a day, with an increase to 4–6 mg/day. The average daily dose is up to 4–6 mg/day.

An acute psychotic conditions with symptoms of fear, anxiety, pronounced psycho-motor agitation, aggressiveness, and hostility require drugs with a powerful general incisive effect, for example butyrophenones: Haloperidol 1.5–10 mg/day.

The hallucinatory-paranoid disorders in the structure of the psychotic condition: pseudo- and true hallucinations, and delusions of influence or persecution require antipsychotic drugs with a pronounced antihallucinatory and antidelusional effect, for example, Benzisoxazole derivatives: Risperidone 2–6 mg/day.

Regardless of the phase of the disease in which it developed, a delirious episode turning into a oneiroid requires detoxification therapy of 20–40 mL/kg of body weight at a rate of below 500 mL/h, the average volume of 3–4 L/day with monitoring of the central venous pressure, water-electrolyte balance, acid-base distance and diuresis as well as butyrophenones: Haloperidol 1.5–20 mg/day.

Patients with a high probability of acute psychotic conditions require intensive detoxification infusion therapy with potassium, magnesium, and B vitamins in high doses.

TABLE 1 Treatment strategy selection in psychotic disorders in the structure of complicated cannabinoid withdrawal syndrome.

Delirium	Psychosis of delirious structure: detoxification and strengthening therapy, traditional for intoxication psychosis, relieves psychosis within 24 h
Oneiroid	The deepening of the psychosis structure due to oneiroid and amentive-type disorders +antipsychotics
Amentive-type	1. The use of neuroleptics in this therapeutic situation is undesirable, since it significantly delays the relief of somatic and neurological disorders associated with psychosis 2. The use of Haloperidol in the therapy of psychoses slowed down the therapeutic dynamics of somatic and neurological disorders; when clinically apparent, Tiapride is a neuroleptic agent of choice

This table demonstrates the various treatment strategies depending on the type of psychosis.

Antipsychotics are used to treat the hallucinatory-delusional condition and psycho-motor agitation, psychopathic behavioral disorders and at the same time to reverse the pathological addiction to cannabinoids, especially synthetic ones. In the strong pathological addiction, preference should be given to derivatives of the butyrophenones: Haloperidol 1.5–20 mg/day, with a global incisive effect, gradually reducing various psychotic manifestations.

The treatment of withdrawal syndrome in patients addicted to several PS, including cannabinoids (synthetic), involves atypical antipsychotics from the group of substituted Benzamines: 100–800 mg/day of Tiapride which producessedative, antipsychotic and antialgic effects. The presence of antialgic effect of this drug is relevant in the treatment of withdrawal syndrome in patients with dependence on several PS, including opioids.

The duration of acute psychotic symptoms relief is from 1 to 3 days, depending on the phase of the condition when the symptoms developed.

The efficacy of psychotic condition treatment is determined by the productive psychotic symptoms relief, the recovery of a critical attitude towards the environment and one's own personality, the normalization of the patient's behavior, and the cessation of psycho-motor agitation.

The dynamics of psychopathological disorders throughout the course of treatment is determined by a Brief Psychiatric Rating Scale (Overall & Gorham, 1988).

The efficacy of traditional therapeutic approaches to the SC use associated psychosis treatment can be significantly increased by including antipsychotic drugs in the therapy regimen. At the same time, when the treatment strategy selection requires taking into account the ratio of mental and somatic and neurological disorders in the psychosis structure, since the inclusion of Haloperidol in the therapeutic regimen significantly improves the relief of mental disorders, and the inclusion of Tiapridal accelerates the compensation of somatic and neurological symptoms.

Neuroleptics in the complex therapy of patients with psychosis resulting the SC use and in antirelapse therapy improves the adherence of patients to therapy and increases the likelihood of patients to abstain from the SC use. The antirelapse potential of Haloperidol is significantly higher than that of Tiapridal.

Applications to other areas

In this chapter, we examined acute psychoses associated with SC use. We described the variants and treatment strategies of psychotic disorders. We dare to assume that this observation will be useful for both practitioners and researchers.

Mini-dictionary of terms

- **Acute psychoactive substance intoxication (F1x.0)** is a transient condition that occurs after the intake of alcohol or other psychoactive substances, leading to a disorder of consciousness, cognitive functions of perception, emotions, behavior or other psychophysiological functions or reactions (Dmitrieva, Krasnova, Neznanova, & Semke, 2009).
- **Addiction (dependence) syndrome** is a combination of somatic, behavioral and cognitive phenomena in which the use of a substance or a class of substances begins to occupy the first place in the human value system.
- **Delirium** is a syndrome characterized by a combined disorder of consciousness and attention, perception, thinking, memory, psycho-motor behavior, emotions and the rhythm of sleep-wakefulness. It is an acute psychotic (encephalopathic) disorder, where the main clinical psychopathological characteristic is confusion. It is accompanied by a disorientation in place and time, while maintaining orientation in one's own personality. Illusions, true hallucinations (auditory, visual, tactile) are important manifestations. Delirium clinical severity ranges from mild ("incomplete," abortive) to severe ("professional," exaggerated) with amentive and oneiroid psychopathological inclusions.
- **Mydriasis**. pupil dilation to over 2.5 mm in diameter.
- **New psychoactive substance (NPS)** is a group of synthetic drugs that appeared on the illegal drug market after 2000.
- **Withdrawal state (withdrawal syndrome)** is a group of symptoms of variable clustering and severity occurring on absolute or relative withdrawal of a psychoactive substance after persistent use of that substance. The onset and course of the withdrawal state are time limited and are related to the type of psychoactive substance and dose being used immediately before cessation or reduction of use (ICD-10).

Key facts of acute psychosis

- Psychoses can follow mental illness or develop independently. This is an acute condition, which is characterized by inappropriate behavior and reactions of a person to the surrounding reality. Psychosis often result from previous traumatic brain damage, brain diseases or alcohol and drugs use. A thorough diagnosis is always necessary to establish the causes of an acute condition.

- Acute psychotic states can manifest themselves both in individual hallucinatory, delusional and emotional disorders and in their combination.
- Data from epidemiological studies demonstrate show that the use of cannabinoids, and especially SC, increases the risk of psychotic disorders, and also indicates a dose-dependent effect, dependence on the duration of use and an increased risk of psychosis in the future.
- The psychopathological structure of psychotic conditions resulting from the SC use as a rule involved delirious-oneiroid disorders regardless of the disease stage.
- The causes of the psychotic conditions development in SC dependent patients without the dependence on other psychoactive substances remain unclear and is currently under active research. One of the working hypotheses is that SC, as full agonists of CB1 receptors, produce a greater effect on humans without dependence on other PS. At the same time, the increased tolerance in people who are addicted to several psychoactive substances serves as an "adaptive" mechanism and prevents the development of psychotic conditions in patients with significant history of the PS substance use. In this case, the dependence on several PS and the existing "fixed" disorders in the mesolimbic system decrease the SC effect while in patients with a shorter drug addiction history and with possible "initial stage" mesolimbic system disorders, the SC use leads to the full development of SC use effects and increases the probability of adverse side effects, including psychoses.

Summary points

- Psychoses resulting from the SC use develop both at the prenosological stage and at all clinical stages of the SC dependence syndrome, in SC intoxication, in the structure of the SC withdrawal syndrome and in a delayed onset psychosis.
- Psychoses is characterized by a transient consecutive psychopathological disorders development including illusory-delirious, delirious-oneiroid and amentive-type clinical and dynamic psychoses.
- Psychotic disorders resulting from the SC use are characterized by psychopathological polymorphism which should be considered in the treatment.
- The effective treatment strategy selection should always consider not only the psychopathological structure of psychosis, but also the severity of the psychosis-associated somatic and neurological symptoms.
- The therapeutic dynamics of psychosis is positive, psychotic symptoms are completely stopped through medication-induced sleep, in individual cases followed by a short-term post-psychotic asthenia.

References

Bokhan, N. A., & Selivanov, G. Y. (2015). Klinicheskaya tipologiya psikhopatologicheskikh rasstroistv u potrebitelei sinteticheskikh kannabinoidov (spaisov). *Sibirskii Vestnik Psikhiatrii i Narkologii*, *4*(89), 18–23 (In Russ.).

Bondar, I. V., Nadezhdin, A. V., Vyazovichenko, Y. E., Simonov, D. V., & Vishnyakov, D. A. (2015). *Ob aktual'nosti problemy kuritel'nykh smesei (spaisov) na sovremennom etape. Sbornik materialov Vserossiiskoi nauchno-prakticheskoi konferentsii «Protivodeistvie nezakonnomu oborotu narkoticheskikh sredstv, psikhotropnykh veshchestv, ikh analogov i prekursorov v sovremennoi rossii: ugolovno-pravovoi i kriminologicheskii aspekty»*. Pod obshch. red. Batyrshina I.I. M.: FKU NITs FSKN Rossii (pp. 27–32). (In Russ.).

Dmitrieva, T. B., Krasnova, V. N., Neznanova, N. G., & Semke, V. Y. (2009). Psychiatry: National guidelines. *GEOTAR-Media*, 1000 p.

Fattore, L. (2016). Synthetic cannabinoids—Further evidence supporting the relationship between cannabinoids and psychosis. *Biological Psychiatry*, *79*, 539–548.

National Institute on Drug Abuse. (2021). *Drug facts: K2/Spice («Synthetic Marijuana»)*. Available at: https://www.drugabuse.gov/drugs-abuse/synthetic-cannabinoids-k2spice. Accessed 27 Dec 2021. Accessed 10 May 2017.

Overall, J. E., & Gorham, D. R. (1988). The brief psychiatric rating scale (BPRS): Recent developments in ascertainment and scaling. *Psychopharmacology Bulletin*, *24*, 97–99.

Shakhova, S. M. (2018). Dependence on smoking mixtures containing synthetic cannabinoids: Clinic and therapy: Dissertation of the candidate of medical sciences. *V. Serbsky Federal Medical Research Center of Psychiatry and Addiction, Ministry of Health, Russian Federation.*, 239 p.

Further reading

Ivanets, N. N., Anokhina, I. P., & Vinnikova, M. A. (Eds.). (2016). *Narcology: National guidelines* (2nd ed.). GEOTAR-Media. revised and enlarged. M.: GEOTAR-Media. 944s.

Part X

Resources

Chapter 47

Recommended resources for studying and investigating cannabis use, neurobiology, psychology, and treatments

Rajkumar Rajendram[a,b], Daniel Gyamfi[c], Vinood B. Patel[d], and Victor R. Preedy[e]

[a]*College of Medicine, King Saud bin Abdulaziz University for Health Sciences, Riyadh, Saudi Arabia,* [b]*Department of Medicine, King Abdulaziz Medical City, King Abdullah International Medical Research Center, Ministry of National Guard Health Affairs, Riyadh, Saudi Arabia,* [c]*The Doctors Laboratory Ltd, London, United Kingdom,* [d]*School of Life Sciences, University of Westminster, London, United Kingdom,* [e]*School of Life Course and Population Sciences, Faculty of Life Science and Medicine, King's College London, London, United Kingdom*

Introduction

Cannabis contains many different cannabinoids including Δ9-tetrahydrocannabinol (THC) (Chandra et al., 2019). Synthetic cannabinoids have been available on the black market since the start of the 21st century. These molecules stimulate the endogenous cannabinoid system (Russo, 2016). The cannabinoid system regulates many neurotransmitter systems (e.g., dopamine), which is often involved in substance misuse (Covey et al., 2017). Thus, by stimulating the endogenous cannabinoid system, cannabinoids may affect cognition and behavior (Russo, 2016).

Anxiety is cannabis' most common adverse psychiatric effect (De Aquino et al., 2018). Cannabis use is also associated with a risk acute toxic psychosis (De Aquino et al., 2018; Konings et al., 2008).

The connections between cannabis use, neurobiology, and psychology is actively being investigated (Kroon et al., 2021). Thus, the literature on these areas has increased greatly in recent years. Information on treatments and strategies to curtail cannabis misuse has expanded significantly. We have previously highlighted resources to assist researchers and clinicians to stay up to date. Those embarking on research into cannabis use can also be guided by this chapter's resources which were recommended by active researchers and practitioners. The list of acknowledgements below includes all of the experts who provided input into the compilation of these valuable resources.

Resources

Tables 1–5 list the most up-to-date information on the regulatory bodies (Table 1), professional societies (Table 2), books (Table 3), emerging technologies and platforms (Table 4), and other resources of interest (Table 5) that are relevant to an evidence-based approach to cannabis use, neurobiology, psychology, and treatment. Some organizations are listed in more than one table as they occasional fulfill multiple roles.

Other resources

The Wellcome Collection (https://wellcomecollection.org/collections) and The British Library (https://www.bl.uk/) also hold material on topics related to biological medicine, including cannabis use, neurobiology, psychology, and treatment. There are also specialist journals devoted to medicinal usage of cannabis and cannabinoids.

Chapters on resources relevant to chemicals derived from plants (recommended by authors and practitioners) may also be relevant to cannabis use, neurobiology, psychology, and treatment (Rajendram et al., 2017, 2022, 2023a, 2023b, 2023c, 2023d; Rajendram & Preedy, 2019).

Disclaimer: This list of material in these tables is included to provide general information only. It does not constitute any recommendation or endorsement of the activities of these sites, facilities, or other resources listed in this chapter, by the authors or editors of this book.

Cannabis Use, Neurobiology, Psychology, and Treatment. https://doi.org/10.1016/B978-0-323-89862-1.00051-9

TABLE 1 Websites, regulatory bodies, or organizations dealing with cannabis use, neurobiology, psychology, and treatment or related fields and areas.

Regulatory body or organization	Web address
Agency for Healthcare Research and Quality (AHRQ)	https://www.ahrq.gov/
Barshop Institute for Longevity and Aging Studies	https://barshopinstitute.uthscsa.edu/
Canal Drogues: Generalitat de Catalunya	https://drogues.gencat.cat/ca/inici
Cannabis & Mental Health Projects in Canada: Mental Health Commission of Canada	https://mentalhealthcommission.ca/cannabis-mental-health-research-projects/
Cannabis in Canada: Government of Canada	https://www.canada.ca/en/services/health/campaigns/cannabis.html
Center for Research and Education Addressing Cannabis and Health: University of Colorado Boulder (CU REACH)	https://www.colorado.edu/center/reach/
Centers of Disease Control and Prevention	https://www.cdc.gov/
College on Problems of Drug Dependence	https://cpdd.org/
European Food Safety Authority	https://www.efsa.europa.eu/en
European Medicines Agency (EMA)	https://www.ema.europa.eu/en
European Monitoring Centre for Drugs and Drug Addiction (EMCDDA)	https://www.emcdda.europa.eu/
Evidencia Cannabis Tabaco (EVICT)	https://evictproject.org/
French National Agency for the Safety of Medicines and Health Products (ANSM)	https://ansm.sante.fr/
Institute of Cannabis Research: Colorado State University Pueblo	https://www.csupueblo.edu/institute-of-cannabis-research/
International Academy of Law and Mental Health	https://ialmh.org/
International Narcotics Control Board (INCB)	https://www.incb.org/
International Union of Basic and Clinical Pharmacology (IUPHAR)	https://iuphar.org/
Ministry of Health National Plan on Drugs, Spain	https://pnsd.sanidad.gob.es/
National Health Surveillance Agency (ANVISA): Ministry of Health, Brazil	https://www.gov.br/anvisa/pt-br
National Institute of Health	https://www.nih.gov/
National Institute on Aging	https://www.nia.nih.gov/
National Institute on Drug Abuse	https://nida.nih.gov/
National Institute on Drug Abuse: Synthetic Cannabinoids	https://nida.nih.gov/publications/drugfacts/synthetic-cannabinoids-k2spice
New Psychoactive Substances in Canada: Government of Canada	https://www.canada.ca/en/health-canada/services/publications/healthy-living/psychoactive-substances-canada-2020-2021.html
Psychotic Disorder and Cannabis Use: Government of Canada	https://www.canada.ca/en/public-health/services/reports-publications/health-promotion-chronic-disease-prevention-canada-research-policy-practice/vol-40-no-5-6-2020/psychotic-disorder-cannabis-use-canadian-hospitalization-trends-2006-2015.html
Schizophrenia Society of Canada: Cannabis and psychosis	https://cannabisandpsychosis.ca/
Spanish Observatory of Medicinal Cannabis	https://www.oedcm.com/
Substance Abuse and Mental Health Service Administration (SAMHSA)	https://www.samhsa.gov/

TABLE 1 Websites, regulatory bodies, or organizations dealing with cannabis use, neurobiology, psychology, and treatment or related fields and areas—cont'd

Regulatory body or organization	Web address
United States Drug Enforcement	https://www.dea.gov/
United States Food and Drug Administration	https://www.fda.gov/
US Centers for Disease Control and Prevention (CDC)	https://www.cdc.gov/
US Centers for Disease Control and Prevention (CDC): Synthetic Cannabinoids	https://www.cdc.gov/nceh/hsb/chemicals/sc/About.html/#symptoms
World Health Organization (WHO)	https://www.who.int/

This table lists the regulatory bodies and organizations involved with cannabis use, neurobiology, psychology, and treatment. Some of the links have indirect references to this topic. The links were accurate at the time of going to press but may move or alter. In these cases, the use of the "Search" tabs should be explored at the parent address or site. In some cases, links direct the reader to pages related to cannabis use, neurobiology, psychology, and treatment within parent sites. Some societies and organizations have a preference for shortened terms, such as acronyms and abbreviations, see also Table 2.

TABLE 2 Professional societies relevant to cannabis use, neurobiology, psychology, and treatment or related fields and areas.

Society name	Web address
American Association of Poison Control Centers	https://www.aapcc.org/
American Cannabis Nurses Association	https://www.cannabisnurses.org/
Brazilian Association of Support for Cannabis Hope (ABRACE)	https://abraceesperanca.org.br/
Brazilian Society for the Study of *Cannabis Sativa* (SBEC)	https://sbec.med.br/
British Pharmacology Society	https://www.bps.ac.uk/
Canadian Cannabis Association	https://www.associationcannabis.ca/en/
Canadian Consortium for the Investigation of Cannabinoids (CCIC)	https://ccic.net/
International Association for Cannabinoid Medicines	https://www.cannabis-med.org/
International Cannabinoid Research Society	https://icrs.co/
International Early Psychosis Association (IEPA): Early Intervention in Mental Health	https://iepa.org.au/
International Society for the Study of Emerging Drugs	https://issed.net/
Research Society on Marijuana (RSMJ)	https://www.researchmj.org/
Schizophrenia Society of Canada	https://schizophrenia.ca/
Society of Cannabis Clinicians	https://www.cannabisclinicians.org/
Society of Critical Care Medicine	https://www.sccm.org/
Spanish Society for Research on Cannabinoids (SEIC)	https://www.seic.es/
Spanish Society of Addictions (Socidrogalcohol)	https://socidrogalcohol.org/

This table lists the professional societies involved with cannabis use, neurobiology, psychology, and treatment. Some of the links have indirect references to this topic. The links were accurate at the time of going to press but may move or alter. In these cases, the use of the "Search" tabs should be explored at the parent address or site. In some cases, links direct the reader to pages related to cannabis use, neurobiology, psychology, and treatment within parent sites. Some societies and organizations have a preference for shortened terms, such as acronyms and abbreviations, see also Table 1.

TABLE 3 Books on cannabis use, neurobiology, psychology, and treatment or related fields and areas.

Book title	Authors or editors	Publisher	Year of publication
Analytical Chemistry of Cannabis	Thomas BF, Elsohly M	Elsevier	2015
Cannabinoids and Neuropsychiatric Disorders	Murillo-Rodriguez E, Pandi-Perumal SR, Monti JM	Springer	2021
Cannabis and Psychosis: An Early Psychosis Treatment Manual	Hinton MF, Early Psychosis Prevention and Intervention Centre—ORYGEN Youth Health Research Centre	EPPIC	2002
Cannabis in Medicine: An Evidence-Based Approach	Finn K	Springer	2020
Complex Connection between Cannabis and Schizophrenia	Compton M, Manseau M	Elsevier	2017
Handbook of Cannabis	Pertwee RG	Oxford	2015
Handbook of Cannabis and Related Pathologies	Preedy VR	Elsevier	2016
Handbook of Experimental Pharmacology	Banister SD, Connor M	Springer	2018
Health Effects of Cannabis and Cannabinoids: The Current State of Evidence and Recommendations for Research	National Academies of Sciences, Engineering, and Medicine	The National Academies Press	2017
New Tools to Interrogate Endocannabinoid Signaling: From Natural Compounds to Synthetic Drugs	Maccarrone M	Royal Society of Chemistry	2021
Psychosis with Coexisting Substance Misuse NICE Guideline on Assessment and Management in Adults and Young People	National Collaborating Centre for Mental Health	British Psychological Society	2011

This table lists books relevant to cannabis use, neurobiology, psychology, and treatment.

TABLE 4 Emerging techniques, instruments and analytical platforms or devices for investigating cannabis use, neurobiology, psychology, and treatment. or related fields and areas.

Organization or company name	Web address
App To Track Cannabis Use and Symptoms: Releaf App	https://releafapp.com/
App To Track Cannabis Use and Symptoms: Strainprint	https://strainprint.ca/app/
Cayman Chemical	https://www.caymanchem.com/
Clinical Toxicology and Environmental Biomonitoring Lab: University of California San Francisco (UCSF)	https://geronalab.ucsf.edu/
Illuminating the druggable genome: National Institute of Health	https://druggablegenome.net/DRGC_GPCR
Metabolomics Innovation Centre	https://www.metabolomicscentre.ca/
Redwood Toxicology Laboratory: Synthetic Cannabinoids Urine Testing	https://www.redwoodtoxicology.com/services/synthetic_cannabinoid_urine

This table lists technologies or platforms relevant to cannabis use, neurobiology, psychology, and treatment. Please note, occasionally the location of the websites or web address changes.

TABLE 5 Other resources of interest or relevance for health care professionals or patients related to cannabis use, neurobiology, psychology, and treatment. or related fields and areas.

Name of resource or organization	Web address
Agency for Healthcare Research and Quality (AHRQ): Cannabis-Based Products	https://www.ahrq.gov/data/infographics/cannabis.html
Canadian Centre on Substance Use and Addiction	https://www.ccsa.ca/cannabis
Cannabis and Cannabinoid Research	https://home.liebertpub.com/publications/cannabis-and-cannabinoid-research/633/overview
Cannabis Compound Database	https://cannabisdatabase.ca/
Cannabis Resources for Family Physicians: College of Family Physicians of Canada	https://www.cfpc.ca/en/education-professional-development/practice-tools-guidelines/cannabis-resources-for-family-physicians
Cannabis: Canadian Medical Association	https://www.cma.ca/cannabis
Cannabis: Government of Canada	https://www.canada.ca/en/health-canada/services/drugs-medication/cannabis.html
Centers of Disease Control and Prevention: Synthetic Cannabinoids	https://www.cdc.gov/nceh/hsb/chemicals/sc/default.html
Drugs@FDA: FDA-Approved Drugs (Cannabinoids)	https://www.accessdata.fda.gov/scripts/cder/daf/
European Monitoring Centre for Drugs and Drug Addiction (EMCDDA): New Psychoactive Substances	https://www.emcdda.europa.eu/publications/rapid-communication/update-eu-early-warning-system-2022_en
French Addictovigilance Network (FAN)	https://addictovigilance.fr/
Fundación CANNA: Scientific Research and Cannabis Testing	https://www.fundacion-canna.es/en
International Medical Cannabis Conference	https://cannx.org/
United States Drug Enforcement Administration: Cannabis	https://www.dea.gov/taxonomy/term/336

This table lists other resources of interest or relevance for health care professionals or patients relevant to cannabis use, neurobiology, psychology, and treatment. Please note, occasionally the location of the websites or web address changes.

Summary points

- Cannabis contains molecules that interact with the endogenous cannabinoid system.
- The cannabinoid system modulates several other neurotransmitters including dopamine.
- The most common psychiatric effect of cannabis is anxiety.
- Cannabis use has been associated with acute psychosis.
- This chapter lists resources relevant to cannabis and cannabinoids.

Acknowledgments

We thank the following authors for their contributions to the development of this resource. We apologize if some of the suggested material was not included in this chapter or has been moved to different sections.

AlTfaili, Hanana
Araújo Rolim, Larissa
Breivogel, Chris
Di Marco Pisciottano, Ilaria
Ginsburg, Brett C
Guerrero-Alba, Raquel
Howlett, Allyn
Jalali, Sara
López Pelayo, Hugo
Marichal-Cancino, Bruno A

Matheson, Justin
Morales, Paula
Reinert, Justin
Rudroff, Thorsten
Sachdev, Shivani
Santesteban-Echarri, Olga
Schoedel, Kerri Alexandra
Wishart, David Scott
Wolinsky, David

References

Chandra, S., Radwan, M. M., Majumdar, C. G., Church, J. C., Freeman, T. P., & ElSohly, M. A. (2019). New trends in cannabis potency in USA and Europe during the last decade (2008–2017). *European Archives of Psychiatry and Clinical Neuroscience*, *269*, 5–15.

Covey, D. P., Mateo, Y., Sulzer, D., Cheer, J. F., & Lovinger, D. M. (2017). Endocannabinoid modulation of dopamine neurotransmission. *Neuropharmacology*, *124*, 52–61.

De Aquino, J. P., Sherif, M., Radhakrishnan, R., Cahill, J. D., Ranganathan, M., & D'Souza, D. C. (2018). The psychiatric consequences of cannabinoids. *Clinical Therapeutics*, *40*, 1448–1456. https://doi.org/10.1016/j.clinthera.2018.03.013.

Konings, M., Henquet, C., Maharajh, H., Hutchinson, G., & Van Os, J. (2008). Early exposure to cannabis and risk for psychosis in young adolescents in Trinidad. *Acta Psychiatrica Scandinavica*, *118*, 209–213.

Kroon, E., Kuhns, L., & Cousijn, J. (2021). The short-term and long-term effects of cannabis on cognition: Recent advances in the field. *Current Opinion in Psychology*, *38*, 49–55. https://doi.org/10.1016/j.copsyc.2020.07.005.

Rajendram, R., Gyamfi, D., Patel, V. B., & Preedy, V. R. (2023a). Recommended resources on in relation to foods, plants, herbs and spices in cardiovascular disease. In R. Rajendram, V. R. Preedy, & V. B. Patel (Eds.), *Foods, plants, herbs and spices in health and disease: Cardiovascular disease* CRC Press/Taylor and Francis Group LLC, USA (in press).

Rajendram, R., Gyamfi, D., Patel, V. B., & Preedy, V. R. (2023b). Recommended resources on in relation to foods, plants, herbs and spices in cancer. In R. Rajendram, V. R. Preedy, & V. B. Patel (Eds.), *Foods, plants, herbs and spices in health and disease: Cancer* CRC Press/Taylor and Francis Group LLC, USA (in press).

Rajendram, R., Gyamfi, D., Patel, V. B., & Preedy, V. R. (2023c). Recommended resources on in relation to foods, plants, herbs and spices in diabetes. In R. Rajendram, V. R. Preedy, & V. B. Patel (Eds.), *Foods, plants, herbs and spices in health and disease: Diabetes* CRC Press/Taylor and Francis Group LLC, USA (in press).

Rajendram, R., Gyamfi, D., Patel, V. B., & Preedy, V. R. (2023d). Recommended resources on in relation to middle eastern foods, plants, herbs and spices. In R. Rajendram, V. R. Preedy, & V. B. Patel (Eds.), *Foods, plants, herbs and spices in health and disease: Middle East* CRC Press/Taylor and Francis Group LLC, USA (in press).

Rajendram, R., Patel, V. B., & Preedy, V. R. (2017). Cannabis neuropathology resources and recommended reading. In V. R. Preedy (Ed.), *Handbook of cannabis and related pathologies: Biology, pharmacology, diagnosis, and treatment* (pp. 1111–1113). Elsevier Inc.

Rajendram, R., Patel, V. B., & Preedy, V. R. (2022). Recommended resources, sites, and research groups for the neuroscience of anesthetics and analgesics. In R. Rajendram, V. R. Preedy, V. B. Patel, & C. Martin (Eds.), *The neurobiology, physiology, and psychology of pain* (pp. 419–426). Elsevier Inc.

Rajendram, R., & Preedy, V. R. (2019). Resources for the neuroscience of nicotine. In V. R. Preedy (Ed.), *Neuroscience of nicotine mechanisms and treatment* (pp. 529–535). Elsevier Inc.

Russo, E. B. (2016). Beyond cannabis: Plants and the endocannabinoid system. *Trends in Pharmacological Sciences*, *37*(2016), 594–605.

Index

Note: Page numbers followed by *f* indicate figures and *t* indicate tables.

D

W

X